PLANT NUTRITION – PHYSIOLOGY AND APPLICATIONS

Developments in Plant and Soil Sciences

VOLUME 41

The titles published in this series are listed at the end of this volume.

Plant Nutrition – Physiology and Applications

Proceedings of the Eleventh International Plant Nutrition Colloquium, 30 July – 4 August 1989, Wageningen, The Netherlands

Edited by
M.L. VAN BEUSICHEM
Department of Soil Science and Plant Nutrition
Wageningen Agricultural University
Wageningen, The Netherlands

Springer-Science+Business Media, B.V.

Library of Congress Cataloging in Publication Data

ISBN 978-94-010-6751-5 ISBN 978-94-009-0585-6 (eBook)
DOI 10.1007/978-94-009-0585-6

Softcover reprint of the hardcover 1st edition 1990

Contents

* Contributions indicated with an asterisk were first published in *Plant and Soil*, Volume 124 (1990).

Eleventh International Plant Nutrition Colloquium

Organizing and Editorial Committee

A. van Diest – chairman
M.L. van Beusichem – programme manager
W.M.F. Raijmakers – assistant manager
M. van Noordwijk
C. Sonneveld
B.W. Veen

Colloquium Assistants

A.B. Brader
R.H. de Bruin
A. de Groot
E. Hoffland
P. Jansen
E. van Loenen
J.A. Nelemans
H.L.M. Sijlmans
J.G. Slofstra

Colloquium Chairmen and Moderators

Y.P. Abrol – India
S.A. Barber – United States of America
M. Chino – Japan
A. van Diest – The Netherlands
G.R. Findenegg – The Netherlands
R.D. Graham – Australia
D.J. Greenwood – United Kingdom
Ch. Hecht-Buchholz – Federal Republic of Germany
U. Kafkafi – Israel
E.A. Kirkby – United Kingdom
H. Marschner – Federal Republic of Germany
M.R. Sarić – Yugoslavia
J. van de Vooren – The Netherlands

Eleventh International Plant Nutrition Colloquium

Assistance and financial support
Generous support and advice from the following Institutions is gratefully acknowledged.

Department of Soil Science and Plant Nutrition	– Wageningen
Wageningen Agricultural University	– Wageningen
Institute for Soil Fertility Research	– Haren
Centre for Agrobiological Research	– Wageningen
Glasshouse Crops Research Station	– Naaldwijk
Ministry of Agriculture and Fisheries	– The Hague
Ministry of Economic Affairs	– The Hague
Netherlands Fertilizer Institute	– The Hague
The Royal Society	– London
Kluwer Academic Publishers	– Dordrecht
Commonwealth Agricultural Bureaux	– Slough
Isover Glacerius de Sant-Roch	– Delft
Rockwool-Grodan BV	– Roermond
Brinkman BV	– 's-Gravenzande
Priva BV	– De Lier
Sierra Europe BV	– De Meern
AKZO Chemie BV	– Amersfoort
Skalar Analytical BV	– Breda
Windmill Holland BV	– Vlaardingen
Rhine Vacation Line BV	– Rotterdam
Burgers' Zoo	– Arnhem
International Agricultural Centre	– Wageningen
Arnhem Region Tourist Office	– Arnhem
Netherlands Convention Bureau	– Amsterdam

International Council on Plant Nutrition
as at 1 January 1990

M.L. van Beusichem	– The Netherlands
M. Chino	– Japan
M.M. El-Fouly	– Egypt
A.R. Ferguson	– New Zealand
R.D. Graham	– Australia
M.A. Gorshkova	– Union of Soviet Socialist Republics
D.J. Greenwood	– United Kingdom
V. Hernando F.	– Spain
A. Jungk	– Federal Republic of Germany
U. Kafkafi	– Israel
T. Kállay	– Hungary
E.A. Kirkby	– United Kingdom
R.F. Korcak	– United States of America
J.F. Loneragan	– Australia
E. Malavolta	– Brazil
H. Marschner	– Federal Republic of Germany
P. Martin-Prével	– France
W. Podlesak	– German Democratic Republic
K.L. Sahrawat	– India
M.R. Sarić	– Yugoslavia

Preface

Exactly 35 years after the first Colloquium was held, the Eleventh International Plant Nutrition Colloquium took place from 30 July to 4 August 1989 in Wageningen, The Netherlands. Although impressive progress has been made during the past decades in our understanding of the mechanisms of uptake, distribution and assimilation of nutrients in relation to crop yield and quality, there are still significant gaps in our insight into many fundamental aspects of plant mineral nutrition and related metabolic processes. In spite of improved knowledge of nutrient requirements of crops and improved fertilizer application strategies, the world population remains to be burdened with an enormous shortage of plant products for food, timber, fuel, shelter, and other purposes. The main challenge facing the plant nutrition research community is to at least alleviate the increasing world-wide need for applying scientific knowledge to practical problems in agriculture, horticulture, and forestry. It is therefore felt by many scientists that the Plant Nutrition Colloquia, which are intended to bring together scientists and to integrate knowledge and approaches acquired in plant physiology, biochemistry, soil science, agronomy and related disciplines, have indeed made a significant contribution to the advancement of our knowledge and understanding in this vital and interdisciplinary field of agrobiology.

About 260 scientists from 40 nations attended the Colloquium in Wageningen. A wide spectrum of topics was presented, but almost all contributions were directly or indirectly focussed on the following themes, which at the same time clearly reflect the major areas of contemporary concern in plant nutrition research:

– nutrient acquisition by plant roots (phosphate, micronutrients, *etc.*); – nutrient uptake, assimilation, and distribution processes in plants; – plant responses to stress factors in the root environment (aluminium, acidity, salinity, *etc.*); – fertilizer application in relation to yield and quality; – nutrient management decisions (diagnostic systems, modelling, *etc.*).

In the Colloquium the above themes were covered by 195 abstracts, 162 of which were actually presented either as an oral or a poster contribution. The present Proceedings contain 136 refereed full papers, providing an excellent overview of the latest developments and actual problems in pure and applied plant nutrition research.

I wish to thank the contributors for the generally high standard of their presentations and for conscientious preparation of their manuscripts. The serious attempts to meet the editorial standards and to satisfy the wishes of the referees are highly appreciated. I am very grateful to the other members of the Organizing Committee for their generous participation and cooperation in all possible ways, from the planning of the scientific programme and the arrangements for non-conventional accomodation, to their indispensable help in editing the Proceedings.

The success of a scientific meeting heavily depends on maintaining a balance between scientific and social aspects. An elaborate programme can not be realized without a rather large group of more or less back-stage assistants with special tasks and responsibilities. It is no exaggeration to state that the excellent support of the Colloquium Assistants has resulted in a very smooth progress of the meeting and in a relaxed atmosphere throughout the week; they may thus be highly commended for their contribution to the overall success of the Colloquium.

The Twelfth International Plant Nutrition Colloquium will be held in Perth, Western Australia, and is tentatively scheduled for September 1993. The local organizer will be Prof. J.F. Loneragan, Murdoch University, School of Biological and Environmental Sciences, Murdoch WA 6150, Australia.

Wageningen, April 1990 *M.L. van Beusichem*

A

Uptake, assimilation, and distribution of macronutrients: Mechanisms and concepts

M. L. van Beusichem (Ed.), *Plant nutrition – physiology and applications*, 3–8.

PLSO IPNC329

The soil to plant transfer of nutrients: Combining plant and soil characteristics

R. MERCKX, E. SMOLDERS and K. VLASSAK
Laboratory of Soil Fertility and Soil Biology, Faculty of Agricultural Sciences, K.U. Leuven, Kardinaal Mercierlaan 92, B-3030 Leuven, Belgium

Key words: assimilation, nitrogen, soil-solution, soil-plant transfer, *Spinacia oleracea* L.

Abstract

A mathematical expression is postulated describing the uptake and assimilation of nitrogen for soil-grown plants. The formalism uses soil solution data, plant growth characteristics and nitrogen assimilation rates. In order to establish the necessary kinetic parameters to describe the behaviour of nitrogen on its path from soil solution to reduced forms in the plant, plants were hydroponically grown in a controlled environment. The kinetic parameters obtained were used to predict the nitrogen content of soil-grown plants, thereby replacing the nutrient solution concentration by a weighted average soil solution concentration. The formalism was successful in predicting the total N content of spinach plants (*Spinacia oleracea* L.) in their exponential growth phase. One of the essentials of the model is that nitrogen assimilation is assumed to obey first order kinetics during this period in a concentration range relevant to agricultural practice.

Introduction

The traditional view on nutrient uptake mechanisms by plants from soils is that the process is controlled to a large if not exclusive extent by the concentrations or activities of that nutrient in the soil solution at the root surface (Barber, 1984; Nye and Tinker, 1977). Apart from more innovative approaches as forwarded more recently by De Willigen and Van Noordwijk (1987), classical mechanistic models reduce the role of the plant root to a growing surface thereby increasing the contact area with the soil solution with time. Within this approach uptake is frequently described by Michaelis-Menten kinetics (Nissen, 1980), implying a pure concentration dependency. Yet, evidence exists since a long time that plants are able to adapt uptake rates in answer to the prevailing demand for nutrients (Van Burg, 1968). As such it can be expected that nutrient uptake is growth phase dependent and can be tuned to meet the requirements.

Although it seems obvious that soil-plant relations can only be understood through a complete understanding of soil solution chemistry (Adams, 1974), it is only recently that soil-solution data are given full attention through various methods of soil-solution sampling (Amacher, 1984; Bingham *et al.*, 1983; Pavan *et al.*, 1982). In this paper, as an example, nitrogen contents were predicted on the basis of both soil solution concentrations and growth parameters. Due to the very nature of the nutrient uptake phenomena in soils we are confronted with a situation of decreasing soil solution concentrations along the growth period. Meanwhile it is during this later growth stages that the bulk of nutrient uptake occurs, as more dry matter is produced per unit of time. In the present approach a relatively simple solution to this problem is suggested. In order to quantify the importance of concentrations in the soil-solution at times of higher growth rate, we have adopted a time dependent average concentration concept in which the growth rate is used as a weight factor, very similar to the approach of Van Loon (1986) to quantify the soil to plant transfer of Technetium.

Methods

Plant growth

Spinach (*Spinacia oleracea* L. cv. Subito) was used as a test plant both in hydroponical and soil experiments. Environmental conditions were the same in both experiments: plants were grown in a growth chamber (Weiss 8'Sp/ + 5DU-Pi) with a 11 h light cycle and a photon flux density in the P.A.R waveband of $380 \pm 20\ \mu Es.s^{-1}.m^{-2}$ measured at canopy height. The day/night temperature regime was 21°C/15°C and the relative humidity regime was 70–85%. The hydroponical experiments were using a flowing nutrient solution system with a composition according to Steiner (1961). Seeds were germinated for 6 days in moistened paper and then transferred to the flowing nutrient solution. After 4 days a selection was made from this seedlings and transferred to a fresh solution. Soil-grown plants were germinated in the same way but directly transferred to the pots filled with a loam soil at 1000 g moist soil per pot. Before filling into pots the soil was amended with nutrients and remoistened. To reduce surface evaporation a 1 cm cover of gravel was layered on top of the soil surface. Soil moisture content was adjusted daily to the initial value of 23% (pF = 2.3).

Analyses

At regular time intervals (18, 23, 28, 32 and 37 days from germination) plants were harvested and samples withdrawn from the nutrient solution for analysis. Fresh and dry weight measurements were taken and the dry (70°C) material kept for nitrate and total nitrogen analyses. Both components were measured following standard colorimetric or destillation methods respectively. Analyses on the soil-grown plants were similar but harvests took place at 13, 17, 20, 24 and 28 days from germination. Subsamples of 30 g root-free soil were used for soil solution analysis. Soil solution was sampled by immiscible displacement with chloroform by centrifuging at 25000 g (Mubarak and Olsen, 1976). The supernatant was membrane-filtered (0.45 μm) and kept at −18°C pending nitrate analysis.

Parameters and equations

In short the following formalism is proposed to take into account uptake and assimilation processes. Firstly a constant relation between internal $(NO_3^-)_{in}$ and external $(NO_3^-)_o$ nitrate concentrations is postulated and described by an equilibrium constant $K = (NO_3^-)_{in}/(NO_3^-)_o$. In the plant nitrate converts into reduced forms of nitrogen by nitrate reductase. The concentration of reduced nitrogen forms results from two processes: the assimilation of nitrate by nitrate reductase and the dilution of nitrogen by plant growth. This can be formalized as:

$$\frac{d(N)}{dt} = \left[\frac{\delta(N)}{\delta N_{org}}\right]_W \frac{dN_{org}}{dt} + \left[\frac{\delta(N)}{\delta W}\right]_{N_{org}} \frac{dW}{dt} \qquad (1)$$

with

(N) = concentration of reduced N in the plant (mol/g fresh weight)

N_{org} = total amount of reduced N in the plant (mol/plant)

W = fresh weight of the plant (g)

The first term relates to the change of the concentration of nitrogen due to nitrate assimilation at a constant plant volume, the second term represents the dilution of reduced nitrogen due to plant growth. The term dN_{org}/dt in equation (1) can be written as $(k_0 + b \times t)N_{NO_3^-}$ with $N_{NO_3^-}$ the amount of nitrate in the plant (mol/plant). Therefore the assumption was made that nitrate reduction occurs by psuedo-first order kinetics with a constant k, itself being time dependent by $k = k_0 + b \times t$. The increase in k with time is postulated as a consequence of the increase in photosynthetic activity as plants develop (Kennedy and Johnson, 1981). As (N) in equation (1) is equal to N_{org}/W, $[\delta(N)/\delta N_{org}]_W$ is equal to $1/W$ and $[\delta(N)/\delta W]_{Norg} = -N_{org}/W^2$. Further is dW/dt the growth rate in an exponential growth curve $W = W_0 \times e^{RGR \times t}$ with RGR the relative growth rate and W_0 the fresh weight at time 0. Solving the differential equation for the nitrogen concentration at time t therefore gives (Van Loon, 1986):

$$(N) = k_0 \times K \times (NO_3^-)_o \times (1 - e^{-RGR \times t})/RGR + b \times K \times (NO_3^-)_o \times e^{-RGR \times t} \times (e^{RGR \times t}(RGR \times t - 1) + 1)/RGR^2 \quad (2)$$

In situations where nitrate concentrations are variable, as in soils, $(NO_3^-)_o$ is to be replaced by a weighted mean concentration, $(NO_3^-)_w$. This should enable to take into account the differential impact of a concentration at a given moment according to the actual growth phase. A possibility of doing this is by the following formula:

$$(NO_3^-)_w = \frac{\int_0^\tau (NO_3^-)_t \times e^{RGR \times t} \times dt}{\int_0^\tau e^{RGR \times t} \times dt} \quad (3)$$

with τ the time at which the weighted average concentration is to be determined.

Results

Growth curves from both types of experiments are given in Figure 1. The nutrient solution grown plants apparently have a longer "lag-phase" than the soil-grown plants, presumably due to the transplanting procedure which takes place at a later and thus more vulnerable stage than for the soil-grown plants. Relative growth rates as estimated by non-linear regression using the NLIN procedure of SAS statistical analysis software (SAS Institute Inc., 1987) were 0.24 d^{-1} and 0.26 d^{-1} for the solution and soil experiments respectively. An apparent growth delay of 8 days was maintained during the entire growth experiment.

The results of the nitrate and nitrogen analyses from the nutrient solution grown plants are given in Table 1.

The rate of organic nitrogen accumulation (dN_{org}/dt) is estimated for each harvest interval by calculating the difference between two successive N_{org} quantities and given in m*M*/plant/day. As a first approximation the psuedo-first-order rate constant can be obtained by dividing (dN_{org}/dt) by $N_{NO_3^-}$. It can be noticed that k increases from a value of 1.08 d^{-1} after 20 days to a value of 1.53 d^{-1} after 29 days. Between day 29 and day 34 the k constant appears to decrease again to a value of 1.37 d^{-1}. Confining ourselves to the first 29 days a linear increase in k of about

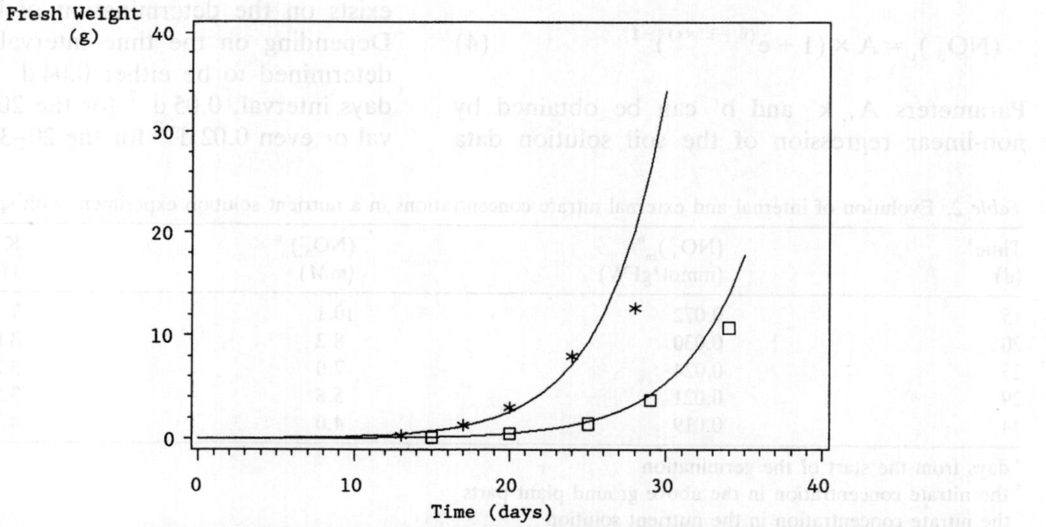

Fig. 1. Growth curves (fresh weights vs time) of the spinach plants grown in nutrient solution (□) and soils (*).

Table 1. Fresh weights, nitrate, total nitrogen contents and estimates of first order rate constants (k) for nitrate reduction in spinach plants grown on a nutrient solution

Time[a] (days)	FW/plant[b] (g)	$N_{NO_3^-}$[c] (mmol/pl)	N_{org}[d] (mmol/pl)	dN_{org}/dt[e] (mmol/d/pl)	k (d^{-1})
15	0.11	0.008	0.023		
20	0.50	0.015	0.105	0.016	1.08
25	1.40	0.033	0.317	0.042	1.28
29	3.70	0.077	0.788	0.118	1.53
34	10.69	0.203	2.177	0.278	1.37

[a] days from the start of the germination
[b] FW = fresh weight
[c] $N_{NO_3^-}$ = the amount of nitrate per plant
[d] N_{org} = the amount of organic N per plant
[e] dN_{org}/dt calculated for the previous time interval between two harvests.

$0.05\,d^{-1}$ can be obtained. Accordingly k_0 and b constants in the equation are estimated as $0.08\,d^{-1}$ and $0.05\,d^{-2}$ respectively.

Although the internal and external nitrate concentrations during the course of the experiment (Table 2) seem to oppose the existence of a constant relation between them, the average K can be calculated to be 4.51×10^{-3} 1/g fresh weight. The K values obtained vary by a factor of 2.12 over the entire growth period. If one omits the extremes at both first and last harvest, however, one obtains a K factor equal to 3.54×10^{-3} 1/g fresh weight which varies between minimum and maximum only by a factor 1.09.

To calculate the weighted average concentrations of nitrate in the soil solution with equation (3), the concentration data were described by a decreasing sigmoidal function:

$$(NO_3^-)_t = A \times (1 + e^{(b' - k' \times t)})^{-1} \qquad (4)$$

Parameters A, k' and b' can be obtained by non-linear regression of the soil solution data (SAS Institute Inc., 1987). Results of this are given in Figure 2. To calculate the weighted average concentration $(NO_3^-)_w$ the integrals in equation (3) were solved numerically. Results of this calculation are also given in Figure 2. Due to the weighing procedure more importance is given to the nitrate concentrations occurring at later growth stages. Therefore these concentrations are lower than calculated time-average concentrations. The weighted concentrations are then combined with the estimated K, k_0 and b constants in equation (2) to yield the results listed in Table 3. With the constants $k_0 = 0.08\,d^{-1}$ and $b = 0.05\,d^{-2}$ an overestimation of the (N) concentration is obtained. The ratio of predicted over measured concentrations at day 24 for instance was 1.67. The results in Table 1 however illustrate that considerable uncertainty exists on the determination of both k_0 and b. Depending on the time interval this b can be determined to be either $0.04\,d^{-2}$ for the 20–25 days interval, $0.05\,d^{-2}$ for the 20–29 days interval or even $0.02\,d^{-2}$ for the 20–34 days interval.

Table 2. Evolution of internal and external nitrate concentrations in a nutrient solution experiment with spinach plants

Time[a] (d)	$(NO_3^-)_{in}$[b] (mmol/gFW)	$(NO_3^-)_o$[c] (m*M*)	$K = (NO_3^-)_{in}/(NO_3^-)_o$[d] 1/gFW
15	0.072	10.1	7.15×10^{-3}
20	0.030	8.2	3.68×10^{-3}
25	0.024	7.0	3.37×10^{-3}
29	0.021	5.8	3.58×10^{-3}
34	0.019	4.0	4.75×10^{-3}

[a] days from the start of the germination
[b] the nitrate concentration in the above-ground plant parts
[c] the nitrate concentration in the nutrient solution
[d] K = the postulated equilibrium constant.

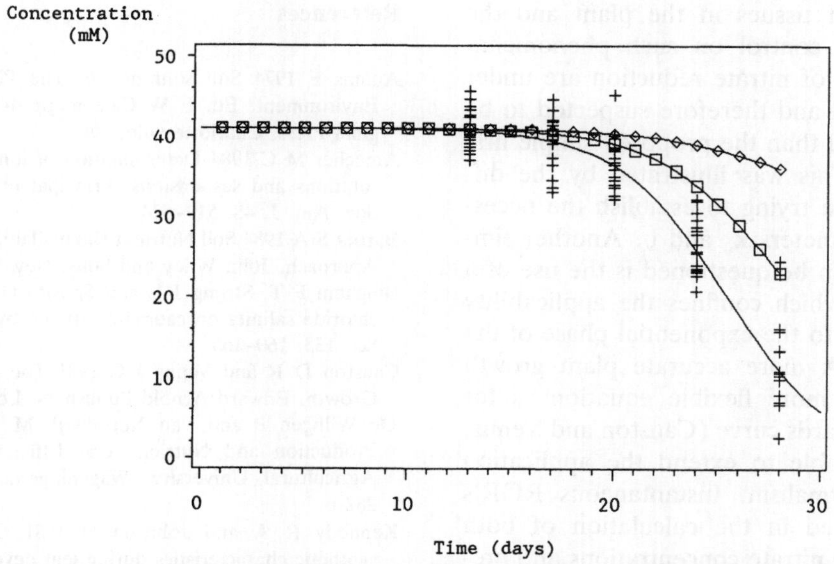

Fig. 2. Measured (+), predicted (−), weighted average (□) and time-average (◇) concentrations of nitrate in samples obtained by immiscible displacement from a loam soil during the entire growth period.

For this reason predictions are included with an average b factor of $0.03\ d^{-2}$ which clearly is in the range of the estimated parameters and gives a better agreement between measured and predicted data. For day 24 the ratio of predicted over measured concentration was only 1.05. To illustrate the importance of using weighted average concentrations rather than time-average concentrations, the predictions are listed using time-average concentrations as well. For this part $b = 0.03\ d^{-2}$ is used. Again the predicted (N) concentrations are overestimated because the small concentrations, typical for the final growth stages, are under-represented in a calculation of a time average concentration. In this case the ratio of predicted over measured concentration rose again to 1.21 at day 24.

Discussion

In this first attempt to predict plant nutrient concentrations by combining soil and plant characteristics in a relative simple formalism, analogous to the one used for the radioisotope Technetium (Van Loon, 1986), it was shown that the formalism can be adapted for nutrients as well. However, in order to improve the performance of the model in terms of both accuracy and application range, the following modifications will be necessary. Especially in the case of nitrate, several assumptions made may indeed be criticized in view of the experimental data. The existence of constant proportions between internal and external concentrations can be argued in view of the existing concentration differences

Table 3. Comparison between measured ($N_{org}m$) and predicted ($N_{org}p$) values of organic nitrogen concentrations in soil-grown spinach plants

Time (d)	$N_{org}m$ (mol/gFW)	(mol/gFW)	$N_{org}p$ (mol/gFW)	(mol/gFW)
13	3.28×10^{-4}	3.01×10^{-4}	1.98×10^{-4}	1.98×10^{-4}
17	2.78×10^{-4}	4.06×10^{-4}	2.61×10^{-4}	2.64×10^{-4}
20	2.70×10^{-4}	4.72×10^{-4}	3.00×10^{-4}	3.10×10^{-4}
24	3.00×10^{-4}	5.01×10^{-4}	3.15×10^{-4}	3.63×10^{-4}
28	2.84×10^{-4}	4.03×10^{-4}	2.52×10^{-4}	3.94×10^{-4}

[a] calculated with weighted average nitrate concentrations, $k_0 = 0.08\ d^{-1}$, $b = 0.05\ d^{-2}$, $K = 0.00354$ 1/gFW and $RGR = 0.26\ d^{-1}$.
[b] idem as a but with $b = 0.03\ d^{-2}$.
[c] idem as b but with time-average concentrations.

between different tissues in the plant and the strong metabolic control on such phenomena. Also the kinetics of nitrate reduction are under metabolic control and therefore suspected to be more complicated than the proposed simple first order kinetics. This was illustrated by the difficulties met when trying to establish the necessary kinetic parameters k_0 and b. Another simplification that can be questioned is the use of a constant RGR, which confines the applicability of the formalism to the exponential phase of the growth period. A more accurate plant growth analysis using a more flexible equation as for instance the Richards curve (Causton and Venus, 1981) should enable to extend the application range of the formalism. Instantaneous RGR's may then be used in the calculation of both weighted average nitrate concentrations and predicted (N) concentrations in the plant. Finally the formalism will contain a variable K, a variable RGR and a variable k. In this paper, we have deliberately chosen to use the formalism in its most simple appearance as a first test and in order to keep analytical solutions of the differential equation feasible.

Acknowledgements

This work was supported by a research grant from the "K.U. Leuven Onderzoeksfonds". E Smolders acknowledges a scholarship as "aspirant" from the "Nationaal Fonds voor Wetenschappelijk Onderzoek", Belgium. We also thank Mr F Schoovaerts and Mrs R Vinck for their technical assistance.

References

Adams F 1974 Soil solution. *In* The Plant Root and Its Environment. Ed. E W Carson. pp 441–471. Univ. Virginia Press, Charlottesville, VA.

Amacher M C 1984 Determination of ionic activities in soil solutions and suspensions: Principal limitations. Soil Sci. Soc. Am. J. 48, 519–524.

Barber S A 1984 Soil Nutrient Bioavailability: A Mechanistic Approach. John Wiley and Sons, New York, 398 p.

Bingham F T, Strong J E and Sposito G 1983 Influence of chloride salinity on cadmium uptake by Swiss chard. Soil Sci. 135, 160–165.

Causton D R and Venus J C 1981 The Biometry of Plant Growth. Edward Arnold Publishers, London.

De Willigen P and Van Noordwijk M 1987 Roots, Plant Production and Nutrient Use Efficiency. Doct. thesis, Agricultural University, Wageningen, The Netherlands. 282 p.

Kennedy R A and Johnson D 1981 Changes in photosynthetic characteristics during leaf development in apple. Photosynthesis Research 2, 213–223.

Nissen P, Fageria N K, Rayar A J, Hassan M M and Tang van Hai 1980 Multiphasic accumulation of nutrients by plants. Physiol. Plant. 49, 222–240.

Nye P H and Tinker P H 1977 Solute Movement in the Soil-Root System. Blackwell Scientific Publications, Oxford. 342 p.

Pavan M, Bingham F T and Pratt P F 1982 Toxicity of aluminium to coffee in Ultisols and Oxisols amended with $CaCO_3$, $MgCO_3$ and $CaSO_4.2H_2O$. Soil Sci. Soc. Am. J. 48, 1201–1207.

SAS Institute Inc. 1987 The NLIN procedure. In SAS/STAT Guide for Personal Computers, Version 6 Edition. pp 675–713. Cary, NC.

Steiner A A 1961 A universal method for preparing nutrient solutions of a certain desired composition. Plant and Soil 15, 134–154.

Van Burg P F J 1968 Nitrogen fertilizing of grassland in spring. Neth. Nitrogen Techn. Bull. 6, The Hague, 45 p.

Van Loon L 1986 Kinetic Aspects of the Soil-to-Plant Transfer of Technetium. Ph.D. thesis 150, Catholic University, Leuven, Belgium.

M. L. van Beusichem (Ed.), *Plant nutrition – physiology and applications*, 9–15.

PLSO IPNC412B

Simulation of nutrient uptake by a growing root system considering increasing root density and inter-root competition

E. HOFFLAND[1], H.S. BLOEMHOF[1], P.A. LEFFELAAR[2], G.R. FINDENEGG[1] and J.A. NELEMANS[1]
[1]*Department of Soil Science and Plant Nutrition, Wageningen Agricultural University, P.O. Box 8005, 6700 EC Wageningen, The Netherlands, and* [2]*Department of Theoretical Production Ecology, Wageningen Agricultural University, Bornsesteeg 65, 6708 PD Wageningen, The Netherlands*

Key words: *Brassica napus* L., CSMP, diffusion, inter-root competition, nitrate, mass flow, nutrient uptake, quartz sand, simulation model, zero-sink

Abstract

A simulation model is presented which describes uptake of a growth limiting nutrient from soil by a growing root system. The root surface is supposed to behave like a zero-sink. Uptake of the nutrient is therefore determined by the rate of nutrient supply to the root surface by mass flow and diffusion. Inter-root competition and time dependent root density are accounted for by assigning to each root a finite cylindrical soil volume that delivers nutrients. The radius of these cylinders declines with increasing root density. Experiments with rape plants grown on quartz sand were used to evaluate the model. Simulated nitrogen uptake agreed well with observed uptake under nitrogen limiting conditions. In case no nitrogen limitation occurred nitrogen uptake was overestimated by the model, probably because the roots did not behave like a zero-sink any more.

Introduction

Simulation models for nutrient uptake have frequently been used in the evaluation of the effect of soil and root characteristics on nutrient uptake. Uptake models allowing for growing roots were developed by Nye *et al.* (1975), Claassen and Barber (1976) and Cushman (1979). No effects of time dependent root density on inter-root competition are included in these models. This hampers their use in the simulation of experiments with rapidly increasing root densities, as usually found in pot experiments.

Claassen and Barber did not include inter-root competition because they assumed that roots exploit a cylindrical soil volume with a constant solute concentration at the outer boundary. Their model overestimates nutrient uptake if nutrients are mobile (Silberbush and Barber, 1983). In Cushman's model the cylinder wall is impermeable to nutrients and the nutrient concentration at the cylinder wall declines in consequence of uptake. Though root growth is considered in this model, root density is kept constant in time *i.e.* the soil volume available per unit root length does not change with increasing root length. Therefore, in a situation with high root growth rates, this model also overestimates nutrient uptake (Silberbush and Barber, 1983). Baldwin *et al.* (1973) proposed an equation to extend Nye's model in which the effect of increasing root density on inter-root competition is described: the radius of each finite soil volume surrounding a root is a function of root density, assuming that each newly formed root samples the mean nutrient concentration. The validity of this assumption is questionable, because it is well established that roots branch mostly in zones of highest nutrient concentration (Russell, 1977). Yet, Nye's model was only tested under conditions where inter-root competition was not expected so that the equation of Baldwin *et al.* was not included in the model. Overestimation of nutrient uptake by Nye's model was again attributed to inter-root competition (Brewster *et al.*, 1975).

The objective of this paper is to present a new simulation methodology that considers inter-root competition in a soil with increasing root density. Simulation results and experimental data will be compared.

Methods

Theoretical

The simulation model developed describes nutrient uptake by a root system that grows in a restricted soil volume. Each root is assigned a finite cylindrical soil volume delivering nutrients, and the soil volume per unit root length declines with increasing root density. Uptake of nutrients by the root, transport of nutrients to the root surface and the effect of increasing root density are considered as the three main components of this model and will be described subsequently.

Each timestep, the equation of continuity for cylindrical co-ordinates is solved:

$$\frac{\delta C}{\delta t} = -\frac{1}{r} \times \frac{\delta}{\delta r}(r \times F) + S \quad \text{(Eq. 1.9 in Nye and Tinker, 1977)}$$

The initial boundary condition is described by:

$$t = 0 \quad r > r_0 \quad C = C_i$$

The sink term S represents nutrient uptake by the root that is situated in the centre of a soil cylinder. All nutrients arriving at the root surface are supposed to be absorbed, *i.e.* the root surface is supposed to behave like a zero-sink. Root hairs are assumed to be so abundantly present that they are regarded to enlarge the root surfacial area to one located near their tips (Nye, 1966). The boundary condition at the tips of the root hairs is therefore:

$$t > 0 \quad r = r_0 \quad C_0 = 0$$

The rate of nutrient supply to the root surface by mass flow and diffusion is described according

List of symbols

symbol	definition	units
A	total surface area of the pot	cm^2
C	ion concentration in soil solution	$\mu mol \times cm^{-3}$
C_0	ion concentration in the soil solution at the root surface	$\mu mol \times cm^{-3}$
C_i	initial ion concentration in the soil solution	$\mu mol \times cm^{-3}$
D_0	diffusion coefficient of ion in free solution	$cm^2 \times day^{-1}$
D_e	effective diffusion coefficient	$cm^2 \times day^{-1}$
F	flux of nutrients to root surface	$\mu mol \times cm^{-2} \times day^{-1}$
F_1	total flux of the nutrient across the outer boundary of the soil cylinder	$\mu mol \times cm^{-2} \times day^{-1}$
f	tortuosity factor	–
L_c	critical root length	cm
M_t	amount of P on infinite sink	$\mu mol \times cm^{-2}$
n	number of plants per pot	–
r	radial distance from root axis	cm
r_0	root radius + root hair length	cm
r_1	radius of the soil cylinder surrounding each root	cm
S	sink term	$\mu mol \times cm^{-3} \times day^{-1}$
t	time	day
V_1	total volume of the pot or soil layer considered	cm^3
v	inward water flux	$cm^3 \times cm^{-2} \times day^{-1}$
v_0	water flux across the root surface	$cm^3 \times cm^{-2} \times day^{-1}$
v_1	water flux across the outer boundary of the soil cylinder	$cm^3 \times cm^{-2} \times day^{-1}$
θ	volumetric moisture content	$cm^3 \times cm^{-3}$

to

$$F = -D_e \times \frac{dC}{dr} + v \times C$$ (Eq. 1.5 in Nye and Tinker, 1977)

with

$$D_e = \theta \times f \times D_0$$ (Nye, 1968)

Buffering of the solute by the soil is not included.

Inter-root competition for nutrients is accounted for by assigning finite cylindrical volumes with radius r_1 to each root. Water but no nutrients can pass the cylinder wall, similarly to Cushman's model. The boundary condition at r_1 is therefore:

$$t > 0 \quad r = r_1 \quad F_1 = 0 \quad v_1 = r_0 \times v_0 / r_1$$

The initial radius of each soil cylinder is calculated by

$$t = 0 \quad r_1 = \sqrt{(A/(\pi \times n))}$$

assuming that each plant starts with one root growing in vertical direction. Each of these parallel soil cylinders is divided into a number of concentric compartments (shells). The time course of the concentration of nutrients in the soil solution in each of these shells is described according to the above mentioned equations.

Each time the actual root length exceeds a certain critical root length L_c, the outer shell of each soil cylinder is stripped off and their material is used to form new soil cylinders that are assigned to the newly developing roots. These new cylinders have the same radius as the older ones after stripping and they are divided into the same number of shells as left over in the stripped cylinders.

The solute concentration in the newly formed soil cylinders is initially the same as that in the outer stripped shells, where the nutrient concentrations are highest. This can be interpreted as new roots penetrating between older ones, in zones of highest nutrient concentration. After formation, a new time dependent nutrient concentration gradient is calculated in each soil cylinder.

The critical root length L_c is a function of the current radius of the soil cylinder surrounding each root, according to the following equation:

$$L_c = V_1/(\pi \times r_1^2)$$

Up to the first moment that a critical root length is reached and the stripping procedure is executed, the uptake calculation takes place with a root length that is half of the first critical root length. After reaching the first critical root length, the calculations are performed with the arithmetic average of the previous and newly calculated critical root length. Therefore, during the first half of each period nutrient uptake is overestimated, whereas during the last one it is underestimated. On the average this should yield reasonable results. This method spares CPU-time in comparison with a method by which uptake is calculated with the actual root length.

Additional conditions for which the model was developed are: neither temporal nor spatial gradients in volumetric moisture content occur, there is no nutrient production, no spatial gradient in root density and nutrient uptake is homogeneous along the root.

Plant parameters needed to run the model are: root length as a function of time, radius of the root, root hair length and the water uptake per unit root length as a function of time. Soil parameters needed are: the volumetric water content θ, the tortuosity factor f, D_0 and the initial concentration C_i of the nutrient.

The CSMP-III simulation model was executed on a VAX computer using the variable time step integration method of Runge-Kutta Simpson. A copy of the model is available at request from the first authoress (E.H.).

Experimental

Experiments were done to provide the above mentioned soil and plant parameters for the model and to evaluate the model by comparing predicted nitrate uptake with observed nitrate uptake by rape plants growing in cylindrical pots on quartz sand.

Measurement of f as function of θ The relation between the tortuosity factor f and θ was determined by a method similar to that described by Vaidyanathan and Nye (1966). Iron oxide paper (2 × 4 cm; see Van der Zee *et al.*, 1987) was used as an "infinite sink" for phosphate ions. Quartz sand was washed with 1.5 *M* HCl and demineralized water respectively and subsequently heated at 900°C to make it inert with respect to phosphate so that the behaviour of phosphate did not differ essentially from that of nitrate. The pretreated sand was mixed with a nutrient solution containing 5.0 m*M* KH_2PO_4 and packed into Petri dishes (dry bulk density 1.28 g × cm^{-3}) with a piece of iron oxide paper on the bottom. The moisture content of the iron oxide paper had previously been equilibrated with quartz sand that was moistened with demineralized water up to the desired θ. To prevent sagging of the nutrient solution, the dishes were placed in an end-over-end shaker. After about 3 hours contact time the iron oxide paper was removed, washed in demineralized water, air dried, and extracted in 5 ml 0.2 *M* H_2SO_4. The extract was analyzed for phosphate by the molybdenum-blue method. D_e was then calculated from the amount of phosphate on the paper by

$$D_e = \frac{\pi \times M_t^2}{4 \times C^2 \times t}$$ (derived from Eq. 3.15 in Crank, 1975)

and f by

$$f = \frac{D_e}{\theta \times D_0}$$ (derived from Nye, 1968) .

Plant growth Rape plants (*Brassica napus* L. cv. Jetneuf) were grown for 16 days on quartz sand in cylindrical pots at two nitrate levels. Ten plants of rape were grown in 3-l pots (ϕ 12 cm, height 27 cm). Each pot contained a mixture of 3.2 kg quartz sand (dry bulk density 1.28 g × cm^{-3}) and 575 ml (high moisture level) or 385 ml (low moisture level) nutrient solution. In the 1.5 m*M*-treatment 1.5 m*M* KNO_3 and 3.5 m*M* KCl were added to the nutrient solution which consisted of 5.0 m*M* $CaCl_2$, 2.0 m*M* $MgSO_4$, 2.0 m*M* KH_2PO_4 and trace elements (in mg L^{-1}): Fe (as FeEDTA) 4.6; B 0.5; Mn 0.5; Zn 0.05; Cu 0.02; Mo 0.01. In the 5.0 m*M*-treatment 5.0 m*M* KNO_3 was added. The plants were grown in a growth chamber at 20°C, a 16 h light (70 W × m^{-2})-8 h dark cycle and a relative humidity of about 80%. Evaporation (from blanc pots) and evapotranspiration were measured daily and the moisture level was readjusted daily. The beginning of the experiment, t = 0 was defined as the moment at which half of the plants was germinated. At each harvest, a number of pots was deep frozen and divided into five layers of 4.5 cm height by sawing. The layers are referred to as layer I through V, from top to bottom. Root length, θ and nitrate concentration in the soil solution were measured in each layer. Nitrate was measured after extraction of dried sand by automatic spectrophotometry after reduction to nitrite. From other pots, plant material was dried and analyzed for N after wet digestion in a H_2SO_4-Se-salicylic acid mixture with addition of H_2O_2. Total N was determined by the indophenol blue method.

Results

Experimental

Measurement of f as function of θ The value of f for different volumetric moisture contents is shown in Figure 1.

Plant growth Substantial differences in volumetric moisture content θ (Table 1) and in the nitrate concentration in the soil solution (Table 2) were found among the five layers of one pot. In pots without plants neither of these two gradients changed significantly during the experiment. Differences of θ with depth resulted from the poor water holding capacity of the quartz sand used. The difference in N concentration among layers is probably caused by the fact that evaporation from layer I was compensated by supplying demineralized water via a tube positioned with its bottom in layer II.

Also considerable difference in root length among the layers (Table 3) was found within one treatment. There was no difference among the five layers with respect to root radius (0.02 cm) and root hair length (±0.05 cm).

Water uptake per cm root declined from

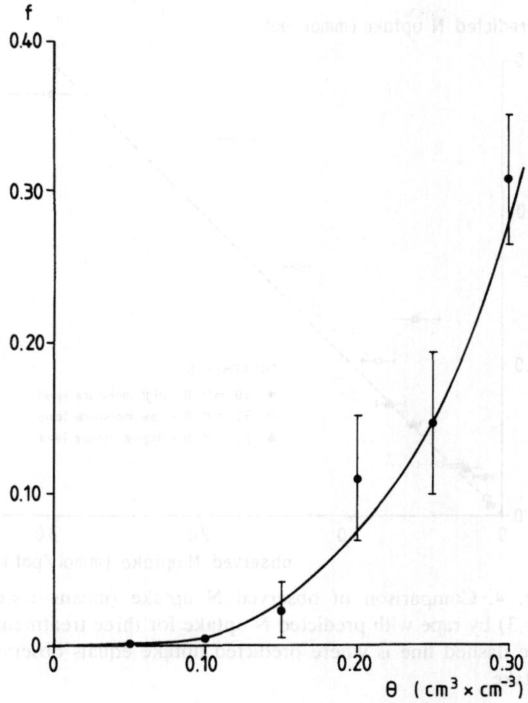

Fig. 1. Experimentally determined relation between f and θ (means ±s.d., $n = 5$) and relation used in the simulation model (——).

$0.028\ cm^3 \times day^{-1}$ at the beginning of the experiment to $0.002\ cm^3 \times day^{-1}$ at the end. No significant differences were found among the three treatments.

Total N uptake per pot and tissue N concentration as a function of time for each treatment are depicted in Figures 2 and 3, respectively. In the high moisture level treatments no nitrate was left in the soil solution at the end of the experiment, while in the low moisture level treatment about 30% of the added nitrogen was left in the nutrient solution (mainly in layers IV and V).

Table 1. Volumetric water content per layer for each treatment. Values are means of two replicates

N level (m*M*):	1.5	5.0	5.0
Moisture level:	high	high	low
Layer	Volumetric water content ($cm^3 \times cm^{-3}$)		
I	0.13	0.13	0.10
II	0.19	0.19	0.08
III	0.23	0.23	0.15
IV	0.28	0.28	0.17
V	0.30	0.30	0.21

Table 2. Concentration N in the soil solution per layer at $t = 0$ for each treatment. Values are means of two replicates

N level (m*M*):	1.5	5.0	5.0
Moisture level:	high	high	low
Layer	N concentration in soil solution (m*M*)		
I	2.6	9.6	7.7
II	0.5	2.7	1.9
III	1.2	3.9	4.9
IV	1.3	5.6	6.4
V	1.6	5.1	4.8

Simulation

To approximate the model conditions of absence of spatial gradients in moisture level and root density, the simulation model was run for each of the five soil layers considered in the experiment. Total N uptake per pot was calculated by summation.

Soil parameters presented in Table 1 and Figure 1 were used to run the model. The simulation was initialized with respect to the N concentration in the soil solution with data given in

Table 3. Root length per layer as a function of time for each treatment. Values are means of two replicates.

Layer	1.5 m*M* N high moisture level					5.0 m*M* N high moisture level					5.0 m*M* N low moisture level				
	Root length (m) after 0, 3, 7, 10 or 16 days														
	0	3	7	10	16	0	3	7	10	16	0	3	7	10	16
I	0.3	1.6	5.0	6.4	12.5	0.3	1.2	6.4	8.4	24.8	0.3	0.3	4.3	7.2	10.0
II	0.1	0.5	3.7	4.1	6.6	0.1	0.3	3.5	7.2	10.9	0.1	0.1	2.5	7.1	5.5
III	0.0	0.2	2.8	4.5	8.3	0.0	0.2	3.3	7.2	17.0	0.0	0.0	1.9	9.3	11.8
IV	0.0	0.1	1.4	4.2	11.7	0.0	0.1	2.3	5.7	19.1	0.0	0.0	1.4	8.0	17.7
V	0.0	0.0	0.8	2.0	19.7	0.0	0.1	0.8	4.0	41.7	0.0	0.0	0.6	3.9	25.1

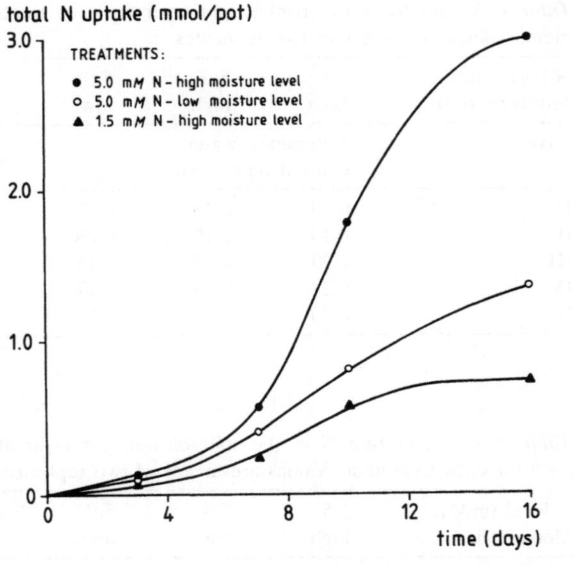

Fig. 2. N uptake by rape plants grown on quartz sand as a function of time. Values are means of three replicates.

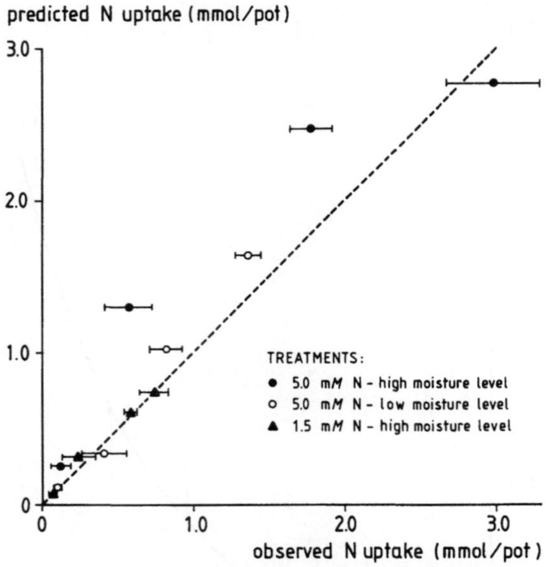

Fig. 4. Comparison of observed N uptake (means ±s.d., $n = 3$) by rape with predicted N uptake for three treatments. The dashed line is where predicted uptake equals observed uptake.

Table 2. Root length (Table 3) and water uptake per cm root as a function of time were used as forcing functions. The initial number of shells surrounding a root was set to 27. During the simulation period this number declined to 1 at t = 16 days in consequence of root growth.

Predicted and observed N uptake are shown in Figure 4.

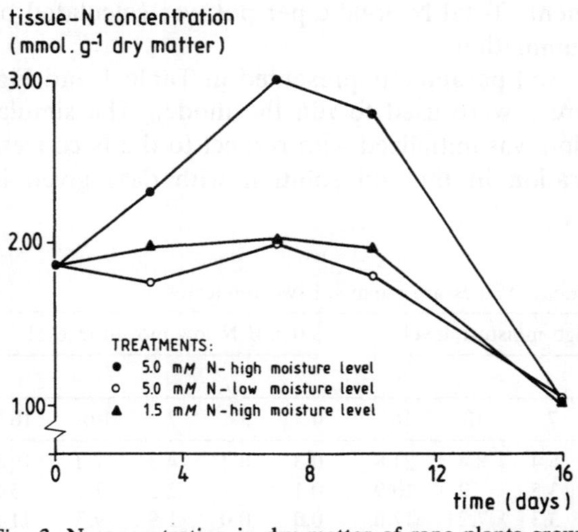

Fig. 3. N concentration in dry matter of rape plants grown on quartz sand as a function of time. Values are means of three replicates.

Discussion

Comparison of predicted N uptake with observed uptake (Fig. 4) shows good agreement for the 1.5 m*M*-high moisture level and 5.0 m*M*-low moisture level treatment. When predicted and observed N uptake are compared per layer (data not shown) for the 1.5 m*M*-high moisture level treatment, there is a close agreement in each layer. This means that both nutrient transport to the root surface and the effect of increasing root density on nutrient uptake are simulated well. In the 5.0 m*M*-low moisture level treatment a slight overestimation of N uptake occurs after 10 days of growth. This overestimation originates from layer IV and V, with relatively high amounts of N available in the soil solution.

The model overestimates N uptake for the 5.0 m*M*-high moisture level treatment (Fig. 4). Figure 3 shows that tissue N concentration in this treatment at t = 7 and t = 10 days is about 3 mmol $\times$ g^{-1} dry matter, which is the concentration of N sufficient rape plants (Hoffland *et al.*, 1989). Nitrate was apparently not growth limiting and therefore, it is very likely that the model assumption that roots act like a zero-sink is not met under these conditions. This will cause an overestimation of N uptake. The model should be extended with a description of biologically controlled nutrient uptake to simulate uptake under these conditions. No effort has been made to describe nutrient uptake by first order or Michaelis-Menten kinetics, because too little is known about the required parameters and their dependence on root age, state of development of the plant and nutrient status of the plant.

The satisfactory prediction of N uptake in cases where N is growth limiting throughout the experiment indicates that the presented equations that describe the effects of inter-root competition and increasing root density on nutrient uptake are powerful.

Acknowledgements

The authors are very grateful to Dr ir J Goudriaan for his valuable suggestions in the development of the simulation model. Thanks are also due to Mr E Heij for analytical assistance and to Dr ir B H Janssen for his useful comments on the manuscript.

References

Baldwin J P, Nye P H and Tinker P B 1973 Uptake of solutes by multiple root systems from soil. III. A model for calculating the solute uptake by a randomly dispersed root system developing in a finite volume of soil. Plant and Soil 38, 621–635.

Brewster J L, Bhat K K S and Nye P H 1975 The possibility of predicting solute uptake and plant growth response from independently measured soil and plant characteristics. III. The growth and uptake of onions in a soil fertilized to different initial levels of phosphate and a comparison of the results with model predictions. Plant and Soil 42, 197–226.

Claassen N and Barber S A 1976 Simulation model for nutrient uptake from soil by a growing plant root system. Agron. J. 68, 961–964.

Crank J 1975 The Mathematics of Diffusion, 2nd edition. Clarendon Press, Oxford, 414 p.

Cushman J H 1979 An analytical solution to solute transport near root surfaces for low initial concentration. I. Equations development. Soil Sci. Soc. Am. J. 43, 1087–1090.

Hoffland E, Findenegg G R and Nelemans J A 1989 Solubilization of rock phosphate by rape. I. Evaluation of the role of the nutrient uptake pattern. Plant and Soil 113, 155–160.

Nye P H 1966 The effect of the nutrient intensity and buffering power of a soil, and the absorbing power, size and root hairs of a root, on nutrient absorption by diffusion. Plant and Soil 25, 81–105.

Nye P H 1968 The use of exchange isotherms to determine diffusion coefficients in soil. 9th Int. Cong. Soil Sci. Trans. (Adelaide) 1, 117–126.

Nye P H and Tinker P B 1977 Solute Movement in the Soil-Root System. Blackwell Scientific Publications, Oxford, 342 p.

Nye P H, Brewster J L and Bhat K K S 1975 The possibility of predicting solute uptake and plant growth response from independently measured soil and plant characteristics. I. The theoretical basis of the experiments. Plant and Soil 42, 161–170.

Russell R S 1977 Plant Root Systems: Their Function and Interaction with the Soil. McGraw-Hill Book Company, London, 298 p.

Silberbush M and Barber S A 1983 Prediction of potassium uptake by soybeans with a mechanistic mathematical model. Soil Sci. Soc. Am. J. 47, 262–265.

Vaidyanathan L V and Nye P H 1966 The measurement and mechanism of ion diffusion in soils. II. An exchange resin paper method for measurement of the diffusive flux and diffusion coefficient of nutrient ions in soils. J. Soil Sci. 17, 175–183.

Van der Zee S E A T M, Fokkink L G J and Van Riemsdijk W H 1987 A new technique for assessment of reversibly adsorbed phosphate. Soil Sci. Soc. Am. J. 51, 599–604.

M. L. van Beusichem (Ed.), *Plant nutrition – physiology and applications*, 17–20.

PLSO IPNC734

Nitrate accumulation by wheat (*Triticum aestivum*) in relation to growth and tissue N concentrations

R.G. ZHEN and R.A. LEIGH
AFRC Institute of Arable Crops Research, Rothamsted Experimental Station, Harpenden, Herts. AL5 2JQ, UK

Key words: nitrate accumulation, nitrate mobilisation, plant growth, solution culture, *Triticum aestivum* L., wheat

Abstract

The accumulation of nitrate in relation to total N concentrations ($[N]_i$) in tissues of wheat (*Triticum aestivum* L., cv Sicco) grown in solution culture was investigated. Root, shoot and leaf tissues showed qualitatively similar relationships between internal nitrate concentrations and $[N]_i$, both expressed on a tissue water basis. At low $[N]_i$, no nitrate was detectable but once a particular $[N]_i$ was exceeded, nitrate accumulated as a linear function of $[N]_i$. The threshold $[N]_i$ values for nitrate accumulation were 110, 450, and 550 m*M* for roots, total shoot and leaf 4, respectively. The slope of the relationship between nitrate and $[N]_i$ indicated that in all tissues nitrate accounted for 50–55% of the extra N accumulated above the threshold $[N]_i$. All growth requirements for N were satisfied before nitrate accumulated.

Introduction

Nitrogen supply is often the most important factor limiting plant growth and crop yield. Enhanced N supply increases both growth and the concentration of N in dry matter (Leigh and Johnston, 1985). However, N will continue to accumulate beyond the level needed to achieve maximum growth and the extra N is accumulated in storage forms, particularly as nitrate. This nitrate is located in the vacuole (Granstedt and Huffaker, 1982; Martinoia *et al.*, 1981) and may be made available for metabolism if exogenous N becomes limiting.

Understanding the physiological basis for nitrate accumulation in agricultural crops is important for a number of reasons. Firstly, tissue nitrate might be useful as an indicator of sufficiency of N supply. Thus, understanding its accumulation in relation to supply and growth might provide a diagnostic tool that could be used to assist in the design of fertilizer regimes that maximise growth and yield but minimise environmental impact. Secondly, maximising nitrate storage in plant tissues offers a route for removing nitrate from the soil and so decreasing the opportunity for leaching.

In this paper, we describe the results of some experiments designed to investigate the relationships between nitrate and total N ($[N]_i$) concentrations in wheat. A particular aim was to determine whether nitrate only began to accumulate when particular $[N]_i$ concentrations were achieved. Concentrations of nitrate and total N were both expressed on the basis of tissue water, rather than as a % in dry matter, because the former has been shown to provide a more physiologically-relevant basis for expressing crop N concentrations (Leigh and Johnston, 1985).

Material and methods

Seeds of wheat (*Triticum aestivum* L. cv. Sicco) were germinated for 5 d on moist tissue paper at 25°C in continuous light and then were transplanted to 1-litre pots containing nutrient solution. In the first type of experiment, eleven levels of nutrient solution containing 0.05 to 20 m*M* nitrate were used and 9 seedlings were transplanted to each pot. In the second type of experiment, four treatments were imposed, HH: 10 m*M* nitrate provided throughout the experiment; HO: 10 m*M* nitrate provided for the first 2 weeks, with no N thereafter; LL: 2 m*M* nitrate provided throughout the experiment; LO: 2 m*M* nitrate provided for the first 2 weeks, with no N thereafter. Solutions were changed every 2 days. All experiments were conducted in a controlled environment at 20°C with a 16 h photoperiod, 70% and 90% day/night relative humidity, and a photon flux density of 450–480 $\mu E\ m^{-2}\ s^{-1}$ at plant level.

All sampling was begun 8 h after the start of the light period. In the first experiment, plants were harvested when leaf 4 was fully expanded. Harvested plants were divided into roots and shoots, and leaf 4 was removed for separate analysis. Fresh and dry weights were measured on all samples and the length and area of leaf 4 were determined. In the second experiment, plants were harvested after 10 and 14 days, and then the experimental solutions were changed to impose new treatments (see above) and further samples were taken on days 15, 17, 19 and 21. Harvested plants were divided into roots and shoots, and fresh and dry weights were measured.

Total-N content of dried plant material was determined using a Carlo Erba automatic nitrogen analyzer. Nitrate was extracted from dried plant material by boiling in water for 1 min, and was analyzed colorimetrically (Lichfield, 1967).

Results

Growth increased with external nitrate concentrations between 0.05 and 4 m*M* and then no further increase was observed (data not shown). This was accompanied by an increase in $[N]_i$. For whole shoots, $[N]_i$ expressed on a dry weight basis increased from 1.5 to 6.0%; on a tissue water basis (Leigh and Johnston, 1985) the increase was from 200 to 600 m*M*. For leaf and root tissue, the corresponding increases were from 300 to 800 m*M* and from 50 to 250 m*M*, respectively.

Figure 1 shows the relationship between internal nitrate concentrations and $[N]_i$ for shoots from the first experiment. At low $[N]_i$, no nitrate was detectable but above a particular, tissue-specific value of $[N]_i$, nitrate accumulated proportionately with changes in $[N]_i$. Similar relationships were observed for roots and leaf 4 but the threshold $[N]_i$ for nitrate accumulation were different. For roots, shoots and leaf 4, the threshold values of $[N]_i$ above which nitrate began to accumulate were 110, 450, and 550 m*M*, respectively. In all tissues, the slope of the relationship above the threshold indicated that nitrate represented 50–55% of the extra N accumulated after the threshold was reached. The threshold values of $[N]_i$ at which nitrate began to accumulate and the percentage of the extra N that was stored as nitrate were very similar in a number of experiments (Table 1).

Figure 2a shows the relationship between fresh weight of shoots and $[N]_i$. The vertical dashed line marks the threshold value at which nitrate begins to accumulate in shoots. The largest growth response to N occured before $[N]_i$ reached the threshold value; there was relatively little change in fresh weight at values of $[N]_i$ above the threshold. This indicates that nitrate

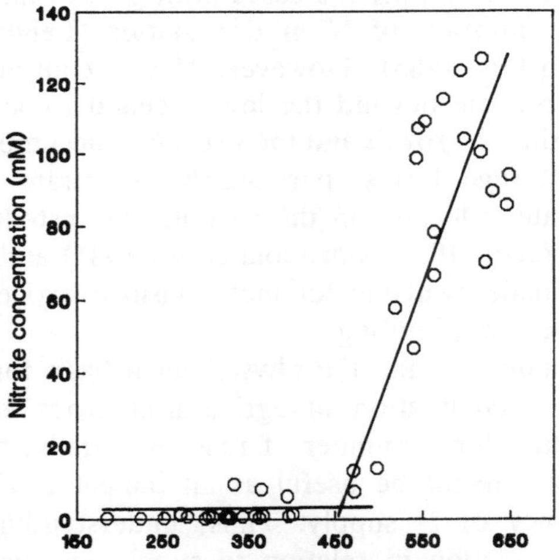

Fig. 1. Relationships between $[N]_i$ and internal nitrate concentrations in shoots of wheat plants grown continuously in nutrient solution containing different nitrate concentrations.

Table 1. Mean (±s.e.) values for the threshold $[N]_i$ at which nitrate begins to accumulate in different tissues of wheat, and the percentage of extra N accumulated above the threshold that was stored as nitrate

Tissue	Threshold $[N]_i$ (m*M*)	Percentage of extra N stored as nitrate	Number of experiments
Root	118 ± 5	55 ± 4	3
Shoots	443 ± 15	52 ± 5	3
Leaf 4	538	53	2

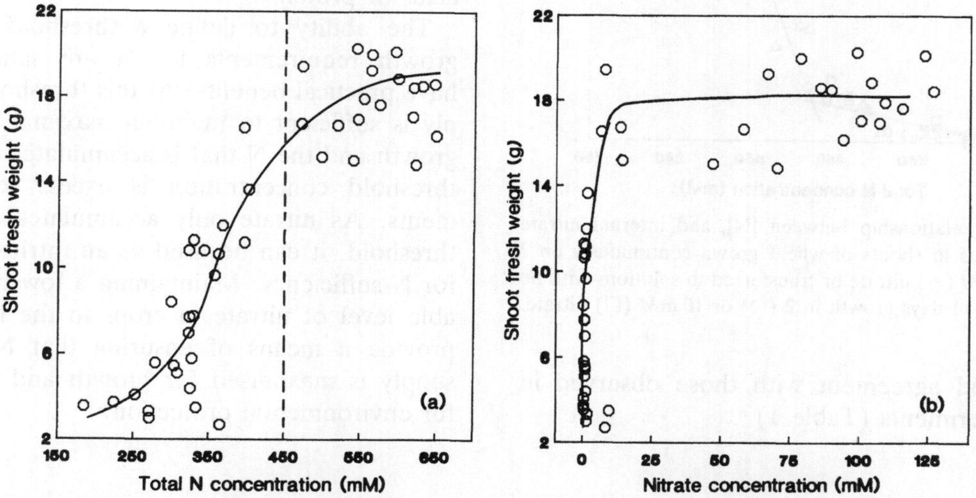

Fig. 2. The relationships between shoot fresh weight and (a) $[N]_i$; and (b) nitrate concentrations in shoots of wheat. The vertical dashed line in (a) indicates the threshold value of $[N]_i$ for nitrate accumulation in shoots.

only begins to accumulate once the N requirements of growth are satisfied. This is confirmed by the data in Figure 2b which show that there is little difference in shoot fresh weight between plants containing 10 or 120 m*M* nitrate.

The second type of experiment was designed to determine whether the relationship between $[N]_i$ and nitrate concentrations changed as plants mobilised nitrate. Plants were grown at either 2 or 10 m*M* external nitrate for 14 d and then some were transferred to solutions containing no nitrate. Transfer to nitrate-free solution caused a large decrease in internal nitrate concentrations over the next 3 days (Fig. 3). However, the relationships between $[N]_i$ and nitrate were not changed significantly compared to those in the first type of experiment in which different concentrations of nitrate were provided continuously (compare Figs. 1 and 4). The threshold values of $[N]_i$ measured in this depletion experiment were 425 and 120 m*M* for shoots and roots, respective-

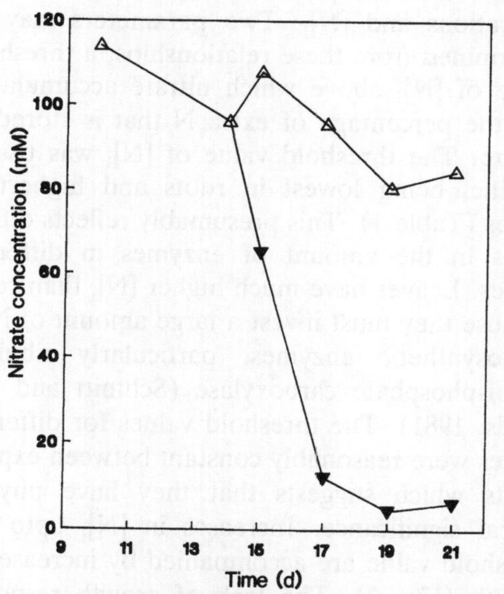

Fig. 3. Changes in the concentrations of nitrate in shoots of wheat grown continuously on nutrient solution containing 10 m*M* nitrate (△) or transferred after 14 days to solutions containing no nitrate (▼).

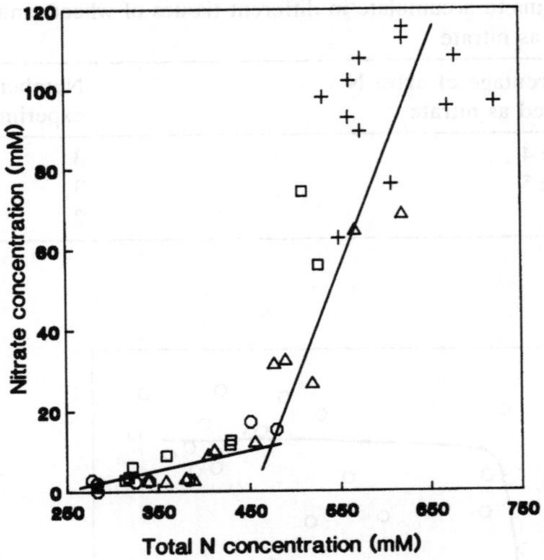

Fig. 4. The relationship between $[N]_i$ and internal nitrate concentrations in shoots of wheat grown continuously on 2 (△) or 10 m*M* (+) nitrate or transferred to solutions with no nitrate after 14 days growth in 2 (○) or 10 m*M* (□) nitrate.

ly, in good agreement with those observed in other experiments (Table 1).

Discussion

Roots, shoots and leaves all show qualitatively similar relationships between tissue nitrate concentrations and $[N]_i$. Two parameters may be determined from these relationships; a threshold value of $[N]_i$ above which nitrate accumulates, and the percentage of extra N that is stored as nitrate. The threshold value of $[N]_i$ was tissue-specific, being lowest in roots and highest in leaves (Table 1). This presumably reflects differences in the amount of enzymes in different tissues. Leaves have much higher $[N]_i$ than roots because they must invest a large amount of N in photosynthetic enzymes, particularly ribulose 1,5-bisphosphate carboxylase (Schmitt and Edwards, 1981). The threshold values for different tissues were reasonably constant between experiments which suggests that they have physiological significance. Increases in $[N]_i$ upto the threshold value are accompanied by increases in growth (Fig. 2). The lack of growth response once the threshold for nitrate accumulation is exceeded indicates that storage of N only occurs once growth requirements are satisfied, as expected if stored N is excess to metabolic requirements. Surprisingly, the proportion of the extra N that is stored as nitrate is similar in all tissues, despite the large differences in $[N]_i$ between tissues. The form of the N that is not stored as nitrate was not determined in these experiments but it was presumably accumulated as free amino acids or proteins.

The ability to define a threshold at which growth requirements for N are satisfied may have practical benefits. At this threshold, N supply is sufficient to maintain maximal vegetative growth and the N that is accumulated above the threshold concentration is excess to requirements. As nitrate only accumulates above the threshold, it can be used as an intrinsic marker for N-sufficiency. Maintaining a low but detectable level of nitrate in crops in the field could provide a means of ensuring that N fertilizer supply is maximised for growth and minimised for environmental protection.

Acknowledgements

We thank Ms. Ruth Skilton and Mr. Simon Driscoll for assistance with the chemical analyses.

References

Granstedt R C and Huffaker R C 1982 Identification of the leaf vacuole as a major nitrate storage pool. Plant Physiol. 70, 410–413.

Leigh R A and Johnston A E 1985 Nitrogen concentration in field-grown spring barley: An examination of the usefulness of expressing concentrations on the basis of tissue water. J. Agric. Sci. Camb. 105, 397–406.

Lichfield M H 1967 The automated analysis of nitrite and nitrate in blood. The Analyst 92, 312–316.

Martinoia E, Heck U and Wiemken A 1981 Vacuoles as storage compartments for nitrate in barley leaves. Nature 289, 292–293.

Schmitt M R and Edwards G E 1981 Photosynthetic capacity and nitrogen use efficiency of maize, wheat and rice: a comparison between C_3 and C_4 photosynthesis. J. Exp. Bot. 32, 459–466.

M. L. van Beusichem (Ed.), *Plant nutrition – physiology and applications*, 21–27.

PLSO IPNC491B

Effect of varied shoot/root ratio on growth of maize (*Zea mays*) under nitrogen-limited conditions: Growth experiment and model calculations

G.R. FINDENEGG

Department of Soil Science and Plant Nutrition, Wageningen Agricultural University, P.O. Box 8005, 6700 EC Wageningen, The Netherlands

Key words: functional equilibrium, growth model, net assimilation rate, nitrogen nutrition, shoot/root ratio, *Zea mays* L.

Abstract

Young maize plants have been grown for two weeks on a perlite/sand mixture under controlled high light conditions at two suboptimal nitrogen levels. The relationships between [1] root dry weight (RDW) and dry weight of the total plants, [2] RDW and total root length (TRL), and [3] TRL and daily N-uptake during the two weeks were different for the two N-levels. In contrast, the relationships between [4] the N-concentrations in shoots (NCS) and in total plants, [5] shoot dry weight and leaf area, and [6] NCS and net assimilation rate (NAR) were similar at both N-levels. Mathematical descriptions of the six experimentally derived relationships were combined in a growth model. Varying relation [1] in the model allowed predictions about the effect of different shoot/root ratio (SRR) on growth. Both experimentally observed SRR-relations were nearly optimal for the respective N-levels. Therefore,the SRR established by the plants was close to the 'functional equilibrium' proposed for root and shoot growth.

Introduction

Deficiencies of water or nutrients can strongly decrease the shoot/root ratio (SRR) of plants (Brouwer, 1962). In this way growth depression resulting from the limiting factor can be minimized. For example, when nitrogen is deficient, increased root development will lead to an increase in nitrogen uptake, because larger soil volumes can be exploited.

The mechanism of SRR adaptation has been disputed (Van Andel *et al.*, 1983; Wilson, 1988). The classical theory of a 'functional equilibrium' between root and shoot states that under all growth conditions the roots have a priority in the use of the nitrogen taken up whereas the shoot has a similar priority for the products of photosynthesis. Consequently, when nitrogen supply falls short, both the relative shortage of nitrogen and the relative excess of carbohydrates in the plant will affect root development less seriously than shoot development. The envisaged "equilibrium" is stable, *i.e.* any deviation from optimum SRR will be corrected immediately and the growth rate of the plant will be maintained at its maximum.

Davidson (1969) has formulated that according to this principle

$$\frac{\text{specific root activity} \times \text{root mass}}{\text{specific shoot activity} \times \text{shoot mass}} = \text{constant} \qquad (1)$$

Since the development of this theory our knowledge about priorities in nitrogen allocation within the plant has been extended considerably (Pate, 1980). Recent investigations do not support the simple equilibrium theory (Lambers, 1983). Rather, the nutrient status of plants may influence the production of plant hormones (Kuiper *et al.*, 1989) that in turn determine the distribution of assimilates (Marschner, 1986) and nitrogen (Simpson *et al.*, 1982). Therefore, the self-optimalisation of the SRR cannot be taken as guaranteed any more.

In relation with crop production the question arises therefore, whether the growth of plants under nutrient limited conditions may be improved by a manipulation of SRR (*e.g.* by application of hormones).

This question is difficult to answer experimentally. Application of hormones may cause other physiological responses besides a shift in SRR. The effect of root pruning or defoliation is only transient, because plants quickly return to their original SRR (Brouwer, 1962). Evaluating formula (1) with experimental data is embarrassed by the fact that usually SRR and the activities of roots and shoots are continuously changing during growth (Cooper and Thornley, 1982; Hunt and Burnett, 1973).

The aim of the present study was to evaluate the effect of different SRR's on growth rate by means of dynamic simulation. Nitrogen uptake rate was chosen as the factor limiting root activity and photosynthetic dry matter increase was considered as the principal shoot activity. Maize plants were grown at two N-levels, both growth limiting, so that different SRR's were established. The net assimilation rate as a function of leaf nitrogen and the nitrogen uptake rate as a function of root length were followed during two weeks. Subsequently the effect of various SRR's could be estimated for both N-levels by means of a dynamic simulation model.

Methods

Maize seedlings (*Zea mays* L. LG 11) have been grown in 17 L-containers on a sand/perlite mixture (1:2 v/v). Once a day the pots were rinsed with an excess of nutrient solution which replaced the total moisture of the pot. The composition of the basic nutrient solution was: 5 m*M* $CaCl_2$, 2 m*M* $MgSO_4$, 2 m*M* KH_2PO_4, 2 m*M* K_2SO_4, Fe-EDTA and micronutrients.

There were two nitrogen treatments: 15 pots were treated with the basic solution containing 0.2 m*M* $Ca(NO_3)_2$ ('N-level 0.4'), and 15 pots with solution containing 1.0 m*M* Ca $(NO_3)_2$ ('N-level 2.0').

Growth conditions were 25°C, a light-dark cycle of 16:8 hours (light intensity 90 Watt/m^2), and a relative air humidity of 85%. After one week three replicate pots (with 4 plants each) were harvested from both N-levels (=day 0). The following harvests were on day 3, 7, 10, and 14, when three pots with 2, 1, 1 and 1 plants were harvested, respectively.

Leaf area was determined by a leaf area meter, root length by the line intersect counting method, dry weight after drying at 70°C, and total-N concentrations in destructed root and shoot material with the indophenol blue method (Novozamsky *et al.*, 1974).

Results and discussion

Basic growth data

In Figure 1 dry weight and total N content of roots and shoots, leaf area and root length are plotted for both N-levels against time. The solid lines shown in the figure are polynomial regression lines, calculated using the formula

$$\ln x(t) = a + b.t + c.t^2 + d.t^3 \qquad (2)$$

where lnx(t) is the logarithm of the parameter x at time t (in days), and a, b, c, and d are the polynomial coefficients. The correlation coefficients (r) are indicated in the figure.

Growth relationhips

Relationships between growth parameters, as deduced from basic growth data were used for the growth model. In this section it is described how these relationships were derived. Regression data of the basic parameters as presented in Figure 1, with daily intervals were used for the calculations. Root dry weight was plotted against total dry weight (Fig. 2). As expected, the SRR at the N-level 2.0 was higher than at the N-level 0.4. However, SRR was not constant but increased with age at both N-levels. This resulted in positive intercepts of the regression lines on the DMR-axis (Fig. 2). The line linking root dry weight and root length (Fig. 3) is steeper at the

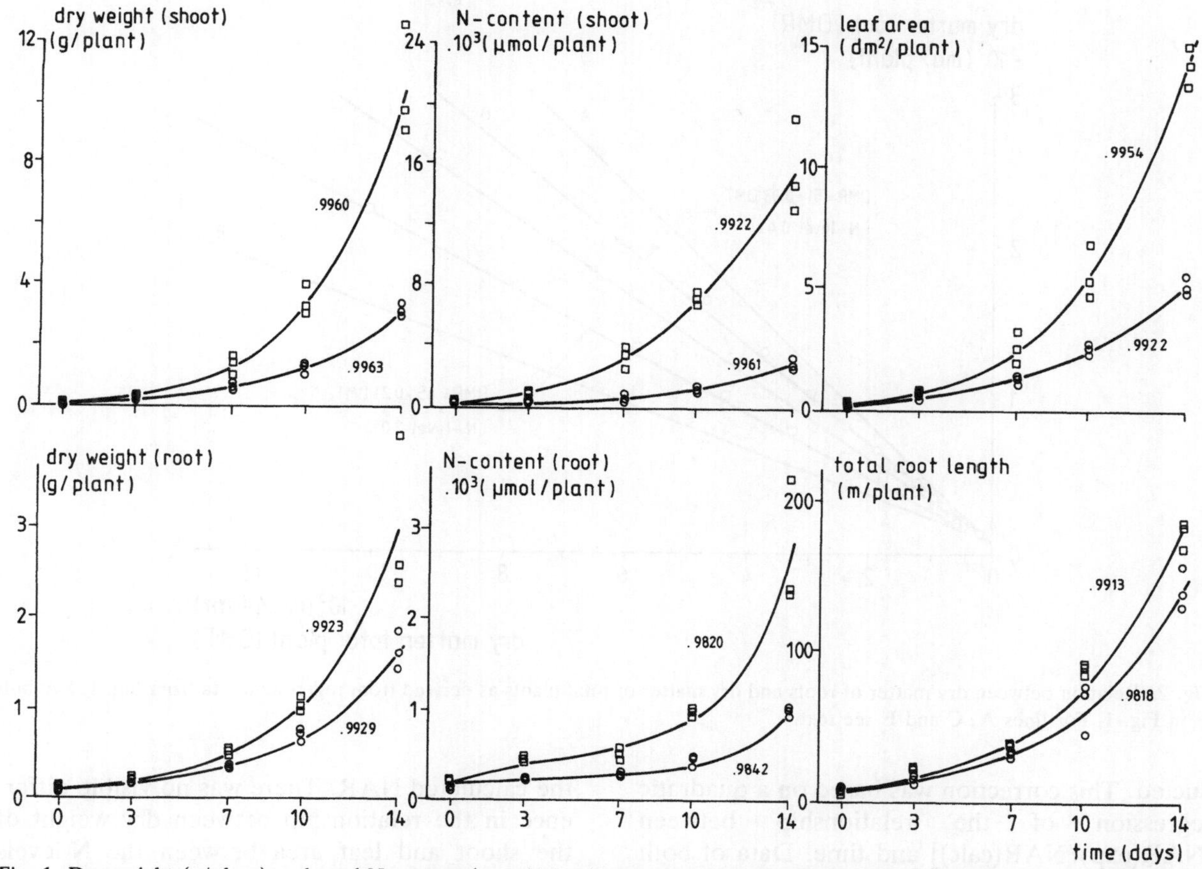

Fig. 1. Dry weight (g/plant) and total N-content (μmol/plant) of roots and shoots, leaf area (dm^2/plant) and total root length (m/plant) of young maize plants during two weeks of growth under controlled conditions. ○ N-level 0.4; □ N-level 2.0. Polynomial regression lines; the numbers indicated are correlation coefficients.

0.4 compared to the 2.0 N-level, indicating that thinner roots were formed at the lower N-level.

The daily rinsing of the root medium with nutrient solution prevented depletion of the rooting medium and led to a rather constant daily N-uptake rate per cm root length (ΔNT) as seen from the nearly proportional increasing lines in Figure 4. The uptake rate (slope of the line) was roughly 5 times higher for the N-level 2.0 compared to the level 0.4. This corresponds with the fivefold NO_3-concentration in the nutrient solution of this treatment. In Fig. 5 the nitrogen concentration in the shoot (NCS) is plotted as a function of the nitrogen concentration in the total plant (NCT). NCS can fairly well be described as $1.16 \times$ NCT for both N-levels.

The net assimilation rate (NAR) has been calculated by dividing the daily dry weight increase ΔDMT by the leaf area. The relation between NCS and the NAR is presented in Figure 6. For the description a modified Michaelis-Menten curve has been used that allows for a minimum concentration in the tissue.

Apparently, the dependence of NAR from NCS is influenced by age. For both N-levels the NAR was overestimated at the beginning of the growth period whereas it was underestimated at the end (solid arrows). There are several explanations for this. For example, the mean light intensity experienced may have increased with plant size, because of the decreasing distance from the light source. Further, the ratio of leaf to stem weight changed during the experiment and the conversion of assimilates into leaf and stem material may occur with different efficiency.

A correction for this age effect has been con-

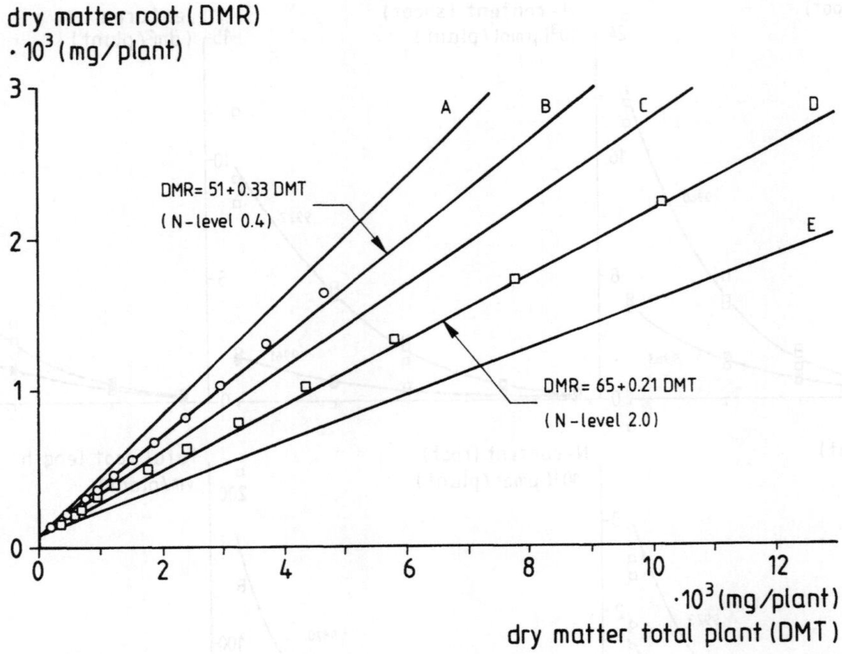

Fig. 2. Relation between dry matter of roots and dry matter of total plants as derived from regression data from Fig. 1. Symbols as in Fig. 1. For lines A, C and E see text.

ducted. This correction was based on a quadratic regression of the relationship between [NAR(exp)/NAR(calc)] and time. Data of both N-levels were pooled. This correction removed 64% of the deviations of the experimental from the calculated NAR. There was no visible difference in the relationship between dry weight of the shoot and leaf area between the N-levels (Fig. 7), suggesting that there was little effect of nitrogen supply on specific leaf area.

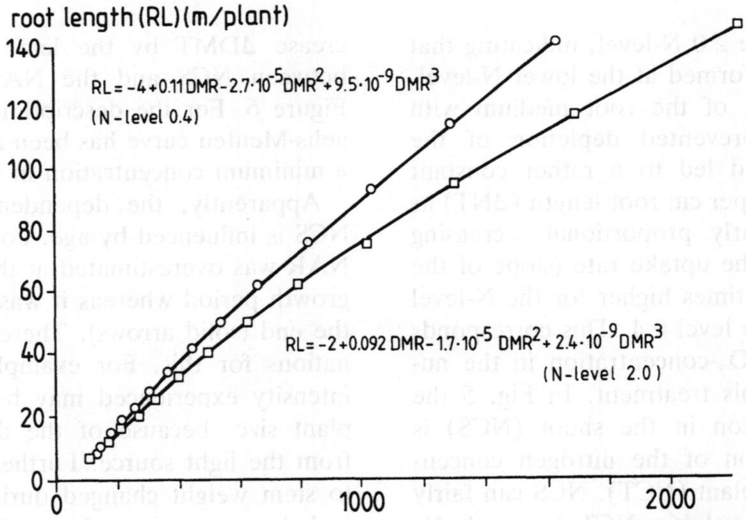

Fig. 3. Relation between the dry matter of roots and total root length as derived from regression data from Fig. 1. Symbols as in Fig. 1.

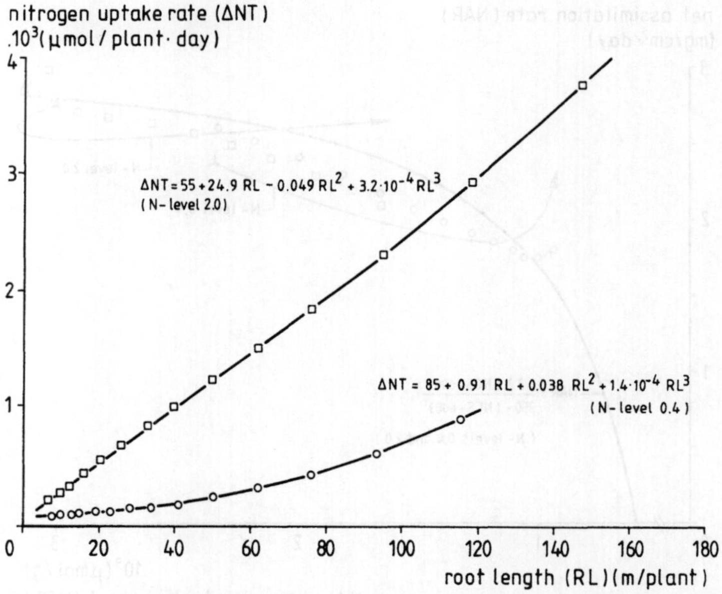

Fig. 4. Relation between daily nitrogen uptake and root length as derived from regression data from Fig. 1. Symbols as in Fig. 1.

Growth model construction

The model starts with the total dry matter (DMT) and the total amount of nitrogen of one plant (NT) at day 0. The variation of SRR is introduced into the model by dividing DMT into root and shoot weight (DMR and DMS, resp.) according to the relationships A to E depicted in Figure 2. Besides the relations B (derived from N-level 0.4) and D (N-level 2.0) an intermediate relationship (C) as well as two extreme ones (A and E) have been used. Subsequently root length RL is calculated from DMR according to Figure 3 and daily nitrogen uptake ΔNT from RL according to Figure 4.

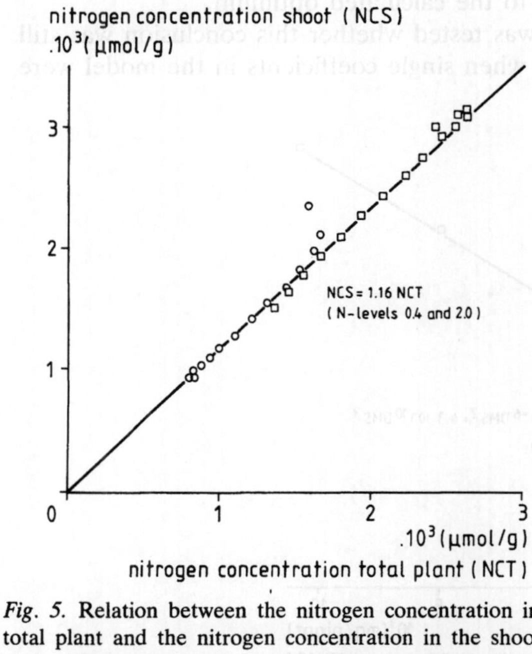

Fig. 5. Relation between the nitrogen concentration in the total plant and the nitrogen concentration in the shoots as derived from regression data from Fig. 1. Symbols as in Fig. 1.

Next the nitrogen concentration of the total plant is calculated by dividing NT by DMT, and the nitrogen concentration in the shoot (NCS) is calculated as a function of NCT (Fig. 5). The net assimilation rate (NAR) is subsequently calculated from NCS (Fig. 6).

Total leaf area (LA) is obtained from DMS according to Figure 7. Finally the daily dry matter increase ΔDMT is calculated as the product of LA and NAR, and DMT and NT are updated by adding ΔDMT and ΔNT, respectively. The cycle is repeated for each day of the experiment.

Results of model calculations

The model was run with the parameters calculated from plants from both N-levels, using theDMT/DMR relationships A to E in either case. When the "real" DMT/DMR relations have been used the calculated growth data did

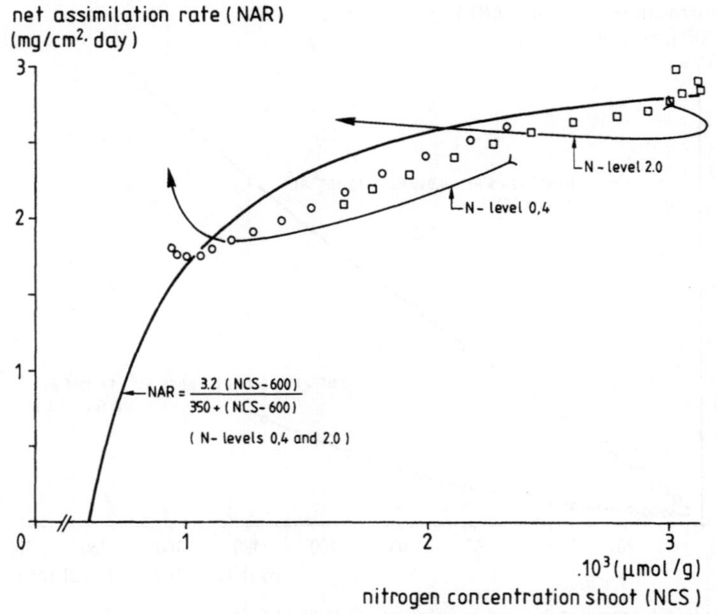

Fig. 6. Relation between the nitrogen concentration in the shoot and the net assimilation rate as derived from regression data from Fig. 1. The relation has been corrected for the time effect (symbols as in Fig. 1) or not corrected (solid arrows).

not deviate too much from the experimental. Typical deviations were 10%; corresponding to the variation between the experimental replicates (Fig. 1). This was true for both nitrogen levels and throughout the experimental period, including final values (Fig. 8). When SRR was varied at the N-level 0.4, DMT was not affected very much. There was a broad range of SRR where nearly optimal growth was realized. The actual SRR of the N-level 0.4 (relation B) was close to the optimum. At the N-level 2.0 the effect of SRR on final weight was much more prounced but again the actual relation (D) was close to the calculated optimum.

It was tested whether this conclusion was still valid when single coefficients in the model were

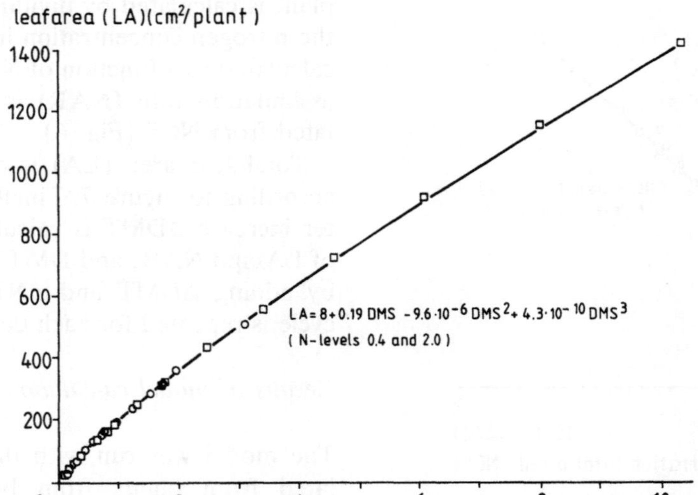

Fig. 7. Relation between the leaf area and the dry matter of shoots as derived from regression data from Fig. 1. Symbols as in Fig. 1.

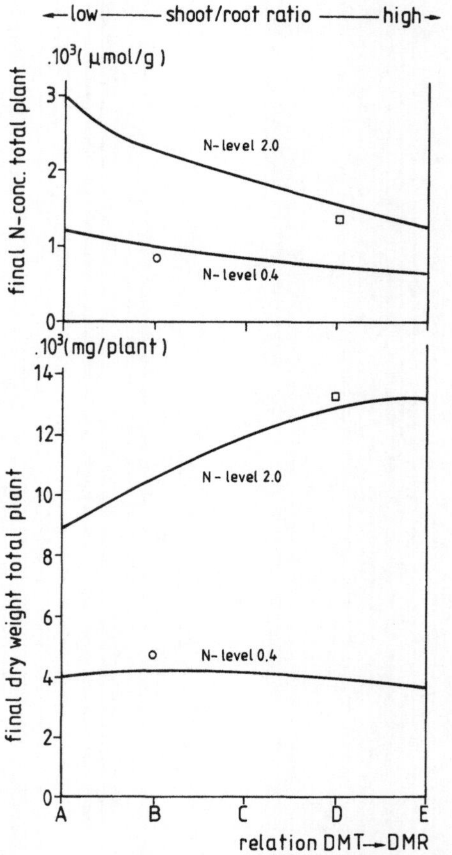

Fig. 8. Solid lines: calculated effect of shoot/root ratio on final dry weight and final N-concentration of plants. Shoot/root ratio according to relations A to E from Fig. 2. Calculations based on a growth model using the growth relations depicted in Figs. 3 to 7. Symbols (see Fig. 1): experimental regression data.

changed. Generally, the model was not very sensitive to altered coefficients, except for the constants used in the NCS-NAR relationship (Fig. 6). Especially the value of the minimum concentration affected the position of the optimum SRR at the N-level 0.4. The value used (600 μmol/g), has been measured directly in another experiment where maize plants have ceased to grow following nitrogen exhaustion in the growth medium under similar growth conditions. Therefore this value is reliable and the conclusion safe. It has also been ascertained that the position of the optimum SRR at both N-levels did not change significantly whether the time correction of the NCS-NAR relationship (Fig. 6) has been conducted or not.

Concluding, the model calculations indicate that the SRR realized by the plants was close to the optimum value. Thus, although the SRR established in plants is probably not the result of a 'functional equilibrium' but rather of a hormonal regulation (see Introduction), the realized SRR value does not significantly deviate from the value that would have been established by a 'functional equilibrium' mechanism. One would not expect therefore that a manipulation of the SRR of crops under nitrogen limited conditions will improve their nitrogen efficiency significantly.

References

Andel O M van, Soekarjo R and Verkaar H J P A (Eds.) 1983 Functional equilibrium between shoots and roots. Neth. J. Agric. Sci. 31, 283–356.

Brouwer R 1962 Nutritive influences on the distribution of dry matter in the plant. Neth. J. Agric. Sci. 10, 399–408.

Cooper A J and Thornley J H M 1976 Response of dry matter partitioning, growth and carbon and nitrogen levels in the tomato plant to changes in root temperature: Experiment and theory. Ann. Bot. 40, 1139–1152.

Davidson R L 1969 Effect of root/leaf temperature differentials on root/shoot ratios in some pasture grasses and clover. Ann. Bot. 33, 561–569.

Hunt R and Burnett J A 1973 The effects of light intensity and external potassium level on root/shoot ratio and rates of potassium uptake in perennial ryegrass (*Lolium perenne* L.). Ann. Bot. 37, 519–537.

Lambers H 1983 The functional equilibrium, nibbling on the edges of a paradigm. Neth. J. Agric. Sci. 31, 305–311.

Kuiper D, Schuit J and Kuiper P J C 1989 Effects of internal and external cytokinin concentrations on root growth and shoot to root ratio of *Plantago major* ssp *pleiosperma* at different nutrient conditions. *In* Structural and Functional Aspects of Transport in Roots. Eds. B C Loughman, O Gašpaříková and J. Kolek. pp 183–188. Kluwer Academic Publishers, Dordrecht, The Netherlands

Marschner H 1986 Mineral Nutrition of Higher Plants. Academic Press, London. 674 p.

Novozamsky I, Van Eck R, Van Schouwenburg J Ch and Walinga I 1974 Total nitrogen determination in plant material by means of the indophenol blue method. Neth. J. Agric. Sci. 22, 3–5.

Pate J S 1980 Transport and partitioning of nitrogenous solutes. Annu. Rev. Plant Physiol. 31, 313–340.

Simpson R J, Lambers H and Dalling M J 1982 Kinetin application to roots and its effect on uptake, translocation and distribution of nitrogen in wheat (*Triticum aestivum*) grown with a split root system. Physiol. Plant. 56, 430–435.

Wilson J B 1988 A review of evidence on the control of shoot: root ratio, in relation to models. Ann. Bot. 61, 433–449.

M. L. van Beusichem (Ed.), *Plant nutrition – physiology and applications*, 29–32.

PLSO IPNC638

Influence of nitrate placement on morphology and physiology of maize (*Zea mays*) root systems

K. THOMS and B. SATTELMACHER
Institute for Plant Nutrition and Soil Science, Christian-Albrechts-University, Olshausen Str. 40, D-2300 Kiel 1, FRG

Key words: cell division, ^{14}C-assimilate translocation, ^{3}H-Methylthymidine, maize, ^{15}N-NO_3-uptake, nitrate placement, *Zea mays* L.

Abstract

The effect of a local NO_3^--supply on morphology and physiology of maize root systems were investigated in water culture.

If NO_3^- supply is restricted to a part of the root system only (zone of supply), a stimulation of extension growth of first order laterals is observed 4 days after onset of the treatment. However, number of first order laterals remains relatively unaffected.

From the very beginning of the treatment $^{15}NO_3^-$-absorption rate in the zone of supply is considerably higher than in control plants, *i.e.* NO_3^--supply to the entire root system.

Accumulation of ^{14}C-labelled assimilates in the zone of supply at the third day of the treatment indicates a higher assimilate demand in this root region. It is assumed that phloem mobile phytohormones, possibly auxins, may lead to an endogenous shift in the hormone balance, responsible for the described morphogenetic reaction. This idea is supported by the fact that accelerated cell division rate in the zone of supply (measured as incorporation of ^{3}H-Methylthymidine) does not occur before the fifth day after onset of the treatment, *i.e.* two days after accumulation of ^{14}C-assimilates in the supply zone.

Introduction

Localizing the supply of mineral nutrients to only part of the root system causes morphogenetic effects. These effects have been studied for NO_3^-, NH_4^+, and P by several authors (Drew *et al.*, 1973; Drew and Saker, 1978; Hackett, 1968; Sattelmacher and Thoms, 1989a,b; Wiersum, 1958). In contrast, a localized K-supply has no or hardly any effect on root morphology (Drew, 1975).

Because of the restricted root surface area in contact with the nutrient considered, uptake rate is enhanced considerably in case of a local nutrient supply (Drew and Saker, 1975; De Jager, 1984).

A higher nutrient uptake leads to an increased respiration rate (De Jager, 1985) and thus to a higher assimilate demand in the zone of supply (Barta, 1976; Sattelmacher and Thoms, 1989a). The fact that by raising root zone temperature a comparable effect may be induced (Sattelmacher and Thoms, 1989b), rises the question whether root morphogenesis is controlled by mineral nutrients directly, or more unspecific by mediation of metabolic activity. The involvement of phytohormones in these processes is suggested.

The presented experiments were carried out to gain further information of the physiological mechanisms responsible for the morphogenetic reaction.

Methods

Preculture of plants, experimental set-up and composition of nutrient solution used for our investigations were identical to those described earlier (Sattelmacher and Thoms, 1989a;b).

Control plants received 4 m*M* NO_3^- exposed to the entire root system; locally supplied plants recieved 4 m*M* NO_3^- applied only to a 3 cm wide zone of their root system (zone of supply).

Morphological studies: For determination of root length and number a digitizer (Summagraphics) connected to a Tandon computer was used. Data presented are the mean of 15 plants.

NO_3^--uptake experiments: For uptake studies, ^{15}N-labelled $Ca(NO_3)_2$ (^{15}N-enrichment = 5%) were used.

N-content was determined by Kjeldahl-method (Nelson and Sommers, 1973) using a distilling unit (Tecator) and a titrator (Schott) connected to an Epson computer.

^{15}N-content of samples was measured with the aid of a ^{15}N-spectrometer (NOI-6E). Natural ^{15}N-abundance was checked with reference samples. Data of ^{15}N-measurements and total nitrogen are the mean of 15 replications.

Double labelling experiments: Experimental units with 15 plants were enclosed into polyethylene bags, containing $NaH^{14}CO_3$. $^{14}CO_2$ ($5\mu Ci = 1{,}85 \times 10^5$ Bq) was liberated by addition of 1 mL HCl (1*M*) with a syringe. ^3H-Methyl-thymidine ($250\mu Ci/1 = 9{,}25 \times 10^6$ Bq) was added to the nutrient solution and was circulated continuously by a peristaltic pump (Ismatec). Labelling period was 5 h.

At the end of the labelling period, plants were washed and separated into shoot and three root segments. After combustion in an oxidizer (Packard Tri-Carb), ^3H- and ^{14}C-activity was determined by liquid scintillation counting. Data of labelling experiments are the mean of 3 replications, 5 plants, each.

Results

Beginning from the fourth day dry weight of the zone of supply rises considerably if compared to the same root zone of the control plants (Fig. 1a). Since number of first order laterals in the zone of supply remains relatively unaffected by a local NO_3^--supply (Fig. 1c), this increase in root dry weight is basically a result of enhanced extension growth of first order laterals starting also from the fourth day after onset of the treatment (Fig. 1b).

^{15}N-NO_3^--uptake rate of roots having received a localized NO_3^--supply is notably higher than it is the case in control plants. Differences diminished however with time (Fig. 2a).

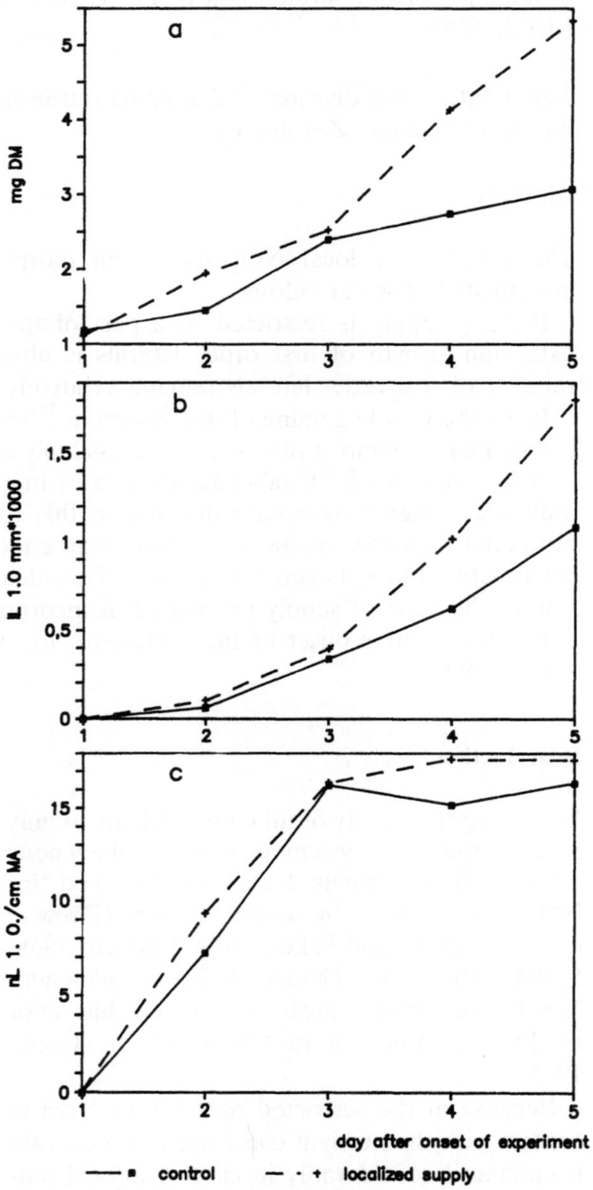

Fig. 1. Influence of a localized supply of nitrate on root dry matter (mg DM) (**a**), on length of 1. order laterals (1L 1.0. mm) (**b**) and on number of first order laterals per cm of main axis (nL 1.0./cm MA) (**c**) Control: 4 m*M* NO_3^- was applied to the whole root system. Localized supply: 4 m*M* NO_3 was supplied to the middle zone only.

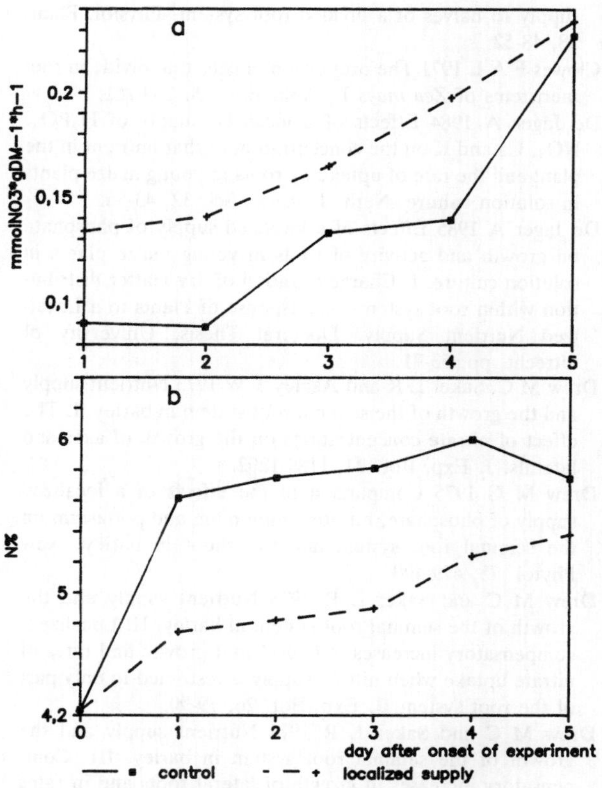

Fig. 2. Influence of a localized supply of nitrate on NO_3^--uptake rate of roots in the middle zone (mmol $NO_3^- \times gDM^{-1} \times h^{-1}$) (**a**) and on total nitrogen content of the whole plants (N%) (**b**). Experimental treatments were as described above.

Until the first day after onset of the treatment total nitrogen content of control plants is remarkably higher than that of locally supplied plants. As already observed for $^{15}N\text{-}NO_3^-$-absorption, differences between treatments were smaller later on (Fig. 2b).

^{14}C-labelled assimilates are translocated predominantly into the zone of local nitrate supply. On day 3, locally supplied plants show significant accumulation of ^{14}C-labelled assimilates in the supply zone. The subsequent decrease of ^{14}C-activity in both treatments is a reflection of growth. (Fig. 3a).

Considerably higher ^{3}H-Methylthymidine incorporation into first order lateral roots of locally supplied plants, compared with controls, occurs at day 5 after start of the experiment (Fig. 3b). Since DNA is the sole Thymidine-pool, 3H-Methylthymidine is considered as a marker for cell division rate (Clowes, 1971).

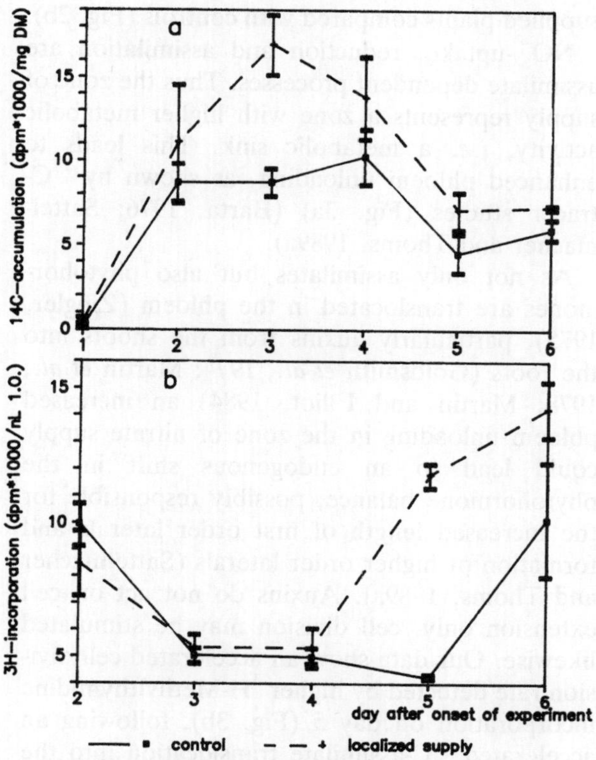

Fig. 3. ^{14}C-accumulation (dpm/mg DM) (**a**) and ^{3}H-Methylthymidine-incorporation into first order lateral roots in the middle root zone (dpm/nL 1.0.) (**b**) as influenced by a local nitrate supply. Experimental treatments were as described above (bars = standard deviation).

Discussion

The observed effects of a local nitrate supply on root morphology is in general agreement with data in the literature (Drew *et al.*, 1973; Drew, 1975; Drew and Saker, 1978; Hackett, 1968; Sattelmacher and Thoms 1989a,b).

In case of a local nutrient supply, plants have to meet their nutrient demand from the restricted zone of supply. Our data demonstrate accelerated nitrate uptake rate in the zone of supply from the very beginning of the experiment (Fig. 2a). Similar effects of locally supplied nitrate on uptake rate were recorded for barley (Drew and Saker, 1975), ryegrass (Barta, 1976), and maize (De Jager 1984).

Although uptake rate of roots of locally supplied plants is risen considerably, demand of plants is not met sufficiently. This is especially apparent immediately after onset of the experiment, resulting in lower total N-content of locally supplied plants compared with controls (Fig. 2b).

NO_3^--uptake, reduction and assimilation are assimilate dependent processes. Thus the zone of supply represents a zone with higher metabolic activity, *i.e.* a metabolic sink. This leads to enhanced phloem unloading, as shown by ^{14}C-tracer studies (Fig. 3a) (Barta, 1976; Sattelmacher and Thoms, 1989a).

As not only assimilates but also phytohormones are translocated in the phloem (Ziegler, 1975), particularly auxins from the shoots into the roots (Goldsmith *et al.*, 1974; Martin *et al.*, 1978; Martin and Elliot, 1984) an increased phloem unloading in the zone of nitrate supply could lead to an endogenous shift in the phytohormone balance, possibly responsible for the increased length of first order laterals and formation of higher order laterals (Sattelmacher and Thoms, 1989a). Auxins do not act on cell extension only, cell division may be stimulated likewise. Our data show an accelerated cell division rate detected by higher ^{3}H-Methylthymidine incorporation on day 5 (Fig. 3b), following an accelerated ^{14}C-assimilate translocation into the supply zone on day 3 (Fig. 3a). These results are in general agreement with our hypothesis expounded above.

According to this hypothesis, all factors stimulating metabolic activity of roots should have similar morphogenetic effects. This has been demonstrated for temperature (Sattelmacher and Thoms, 1989b), but could not be detected for a K-placement (Drew, 1975). To check our hypothesis more detailed experiments involving K and examination of phytohormones are therefore required.

Acknowledgement

The authors wish to thank Uwe Rabsch for his friendly assistance in determination of isotopes and Dr H Künnemann who kindly supplied us with an isotope growth cabinet. This work was supported by the DFG.

References

Barta A L 1976 Transport and distribution of $^{14}CO_2$-assimilate in *Lolium perenne* in response to varying nitrogen supply to halves of a divided root system. Physiol. Plant. 38, 48-52.

Clowes F A L 1971 The proportion of cells that divide in root meristems of *Zea mays* L. Ann. Bot. 35, 249-261.

De Jager A 1984 Effects of a localized supply of H_2PO_4, NO_3, Ca and K on the concentration of that nutrient in the plant and the rate of uptake by roots in young maize plants in solution culture. Neth. J. Agric. Sci. 32, 43-56.

De Jager A 1985 Effects of a localized supply of phosphate on growth and activity of roots in young maize plants in solution culture. I. Characterization of dry matter distribution within root system. *In* Response of Plants to a Localized Nutrient Supply. Doctoral Thesis, University of Utrecht, pp. 53-71.

Drew M C, Saker L R and Ashley T W 1973 Nutrient supply and the growth of the seminal root system in barley. I. The effect of nitrate concentrations on the growth of axes and laterals. J. Exp. Bot. 24, 1189-1202.

Drew M C 1975 Comparison of the effects of a localized supply of phosphate, nitrate, ammonium and pottasium on the seminal root system and the shoot in barley. New Phytol. 75, 479-490.

Drew M C and Saker L R 1975 Nutrient supply and the growth of the seminal root system in barley. II. Localized, compensatory increases in lateral root growth and rates of nitrate uptake when nitrate supply is restricted to only part of the root system. J. Exp. Bot. 26, 79-90.

Drew M C and Saker L R 1978 Nutrient supply and the growth of the seminal root system in barley. III. Compensatory increases in growth of lateral roots and in rates of phosphate uptake in response to localized supply of phosphate. J. Exp. Bot. 29, 435-451.

Goldsmith M H, Cataldo D A, Karn J, Brennemann T and Trip P 1974 The rapid non-polar transport of auxin in the phloem of intact coleus plants. Planta 116, 301-317.

Hackett C 1968 A study of the root system of barley. I. Effects of nutrition on two varieties. New Phytol. 67, 287-299.

Martin H V and Elliot M C 1984 Ontogenetic changes in the transport of indol-3yl-acetic acid into maize roots from the shoots and caryopsis. Plant Physiol. 74, 971-974.

Martin H V, Elliot M C, Wangermann E and Pilet P E 1978 Auxin gradient along the root of the maize seedling. Planta 141, 179-181.

Nelson D W and Sommers L E 1973 Determination of total nitrogen in plant material. Agr. J. 65, 109-112.

Sattelmacher B and Thoms K 1989a Root growth and ^{14}C-translocation into the roots of maize (*Zea mays* L.) as influenced by local nitrate supply. Z. Pflanzenernähr. Bodenkd. 152, 7-10.

Sattelmacher B and Thoms K 1989b Morphology of maize root systems as influenced by a local supply of nitrate or ammonia. Proceedings of the ISSR-meeting 1988; *In press*

Wiersum L K 1958 Density of root branching as affected by substrate and separate ions. Acta Bot. Neerl. 7, 174-190.

Ziegler H 1975 Nature of substances in phloem. *In* Encyclopedia of Plant Physiology. Vol. 1, pp 59-138. Springer-Verlag, Berlin, Heidelberg, New York.

M. L. van Beusichem (Ed.), *Plant nutrition – physiology and applications*, 33–37.

PLSO IPNC636

Influence of nitrogen form and concentration on growth and ionic balance of tomato (*Lycopersicon esculentum*) and potato (*Solanum tuberosum*)

J. GERENDÁS and B. SATTELMACHER
Institute for Plant Nutrition and Soil Science, Christian-Albrechts-University, Olshausen Str. 40, D-2300 Kiel 1, FRG

Key words: ionic balance, *Lycopersicon esculentum* L., polyamines, root morphology, *Solanum tuberosum* L.

Abstract

The influence of N form (NO_3^- vs NH_4^+) and concentration (5 m*M* vs 0.05 m*M*) on growth of potato (*Solanum tuberosum* L.) and tomato plants (*Lycopersicon esculentum* L.) were studied in water culture. While a high NH_4^+ concentration strongly inhibits total plant growth, a low concentration may stimulate plant development.

In further experiments the influence of N form on the ionic balance and on the polyamine content of tomato plants was investigated. In ammonium-grown plants the difference between cations and anions (C-A), which correlates with the content of organic acids, is lower than in nitrate-grown plants, especially in leaves.

High ammonium concentrations cause a 10-fold increase of the putrescine content in the leaves. However, putrescine did not contribute significantly to the ionic balance. There are indications that low ammonium concentrations induce a small increase in free polyamine content. This possibly is involved in the growth stimulation described above.

Introduction

Nitrogen form and concentration have pronounced effects on plant growth. While at high concentrations ammonium inhibits plant growth (Wilcox *et al.*, 1985) at low concentrations (20 to 60 μM) ammonium may stimulate plant development (Cox and Reisenauer, 1973).

Several hypothesis were developed to explain the effects of N form on plant development. The growth inhibition was related to ammonium-induced mineral deficiency (Barker *et al.*, 1967; Wilcox *et al.*, 1973), the decoupling of photosynthesis (Puritch and Barker, 1967), and the depletion of soluble sugars due to detoxification of ammonium (Breteler, 1973).

The stimulating effect of low NH_4^+ concentrations however has been attributed either to shifts in the phytohormone balance (Buban *et al.*, 1978), or to the lower energy demand of ammonium-grown plants (Cox and Reisenauer, 1973).

The nitrogen form also reveals a predominant influence on the ionic balance of plants. Thus, some authors assumed, based on the pH of leaf homogenates, that the N form influences cytoplasmic pH (Allen and Smith, 1986; Kirkby and Mengel, 1967). In this connexion the high polyamine (PA) content observed in ammonium-grown plants appears of interest. According to Murty *et al.* (1971) these organic polycations may contribute significantly to the ionic balance. However, high PA concentrations are phytotoxic (Massé *et al.*, 1985) and have been related to symptoms of potassium deficiency (Coleman and Richards, 1956).

The experiments reported here were carried out to contribute to the question in how far polyamines and the ionic balance may be responsible for the morphogenetic effect described above.

Materials and methods

Experiments with potato cuttings

Potato cuttings (var. LT-1) were grown for three weeks at two N concentrations (5 m*M*, 0.05 m*M*) and two N forms (NO_3^-, NH_4^+). Standard nutrient solution (5 m*M* NO_3^-) contained 2 m*M* $Ca(NO_3)_2$, 0.5 m*M* $MgSO_4$, 0.5 m*M* KH_2PO_4, 1 m*M* KNO_3. Micronutrients in mgL^{-1}: Fe 5, Mn 0.2, Zn 0.2, Cu 0.1, B 0.3, Mo 0.1. Ammonium source was 2.5 m*M* $(NH_4)_2SO_4$, and Ca was balanced as $CaSO_4$. Medium pH was buffered with $CaCO_3$, and solution was changed three times a week.

Experiments with tomato seedlings

Tomato seedlings (var. Hellfrucht × Stamm 1280) were cultivated in a complete nutrient solution (5 m*M* NO_3^-) for four weeks. Thereafter, they were transferred to a flowing solution system (6 plants per 100 L) and exposed for three weeks to the nutrient solutions described above. Medium pH of the nutrient solution was adjusted daily to pH 6, and solution was changed once a week.

Plants were harvested, divided into roots, stems and leaves, chilled in liquid nitrogen and lyophylized. Ca, K, Mg and P were determined by standard procedures. Nitrate was analyzed following the method of Cataldo (1975). Equivalents of cations (K^+, Ca^{2+}, Mg^{2+}) and anions ($H_2PO_4^-$, NO_3^-) were calculated according to De Wit *et al.* (1963). Polyamines were analyzed as dansyl derivatives by HPLC according to Börner (1987).

For morphological studies preparation of plants was carried out as described by Sattelmacher (1987).

Results and discussion

Morphological effects

As can be seen in Figure 1A, a concentration of 5 m*M* NH_4^+ inhibits growth of potato cuttings strongly, while low NH_4^+ concentrations stimulate growth (Fig. 1B). Since root growth is stimulated stronger than shoot growth, a decrease in the shoot-root ratio was observed. This is mainly due to an increase in number and length of first order laterals (Table 1). Growth stimulation induced by NH_4^+ has also been demonstrated for wheat and has been attributed to lower energy requirements by ammonium-grown plants (Cox and Reisenauer, 1973). Evaluating the role of root tips in cytokinine synthesis (Torrey, 1976), the increase in number of root tips shown in our experiment (Table 1) appears of interest. While the inhibiting effect of high NH_4^+ concentration may be observed in potato and tomato (compare Fig. 1A and 2), growth stimulations were not achieved in the latter plant species. Since tomato has been used by several workers as a model plant to study the effect of N

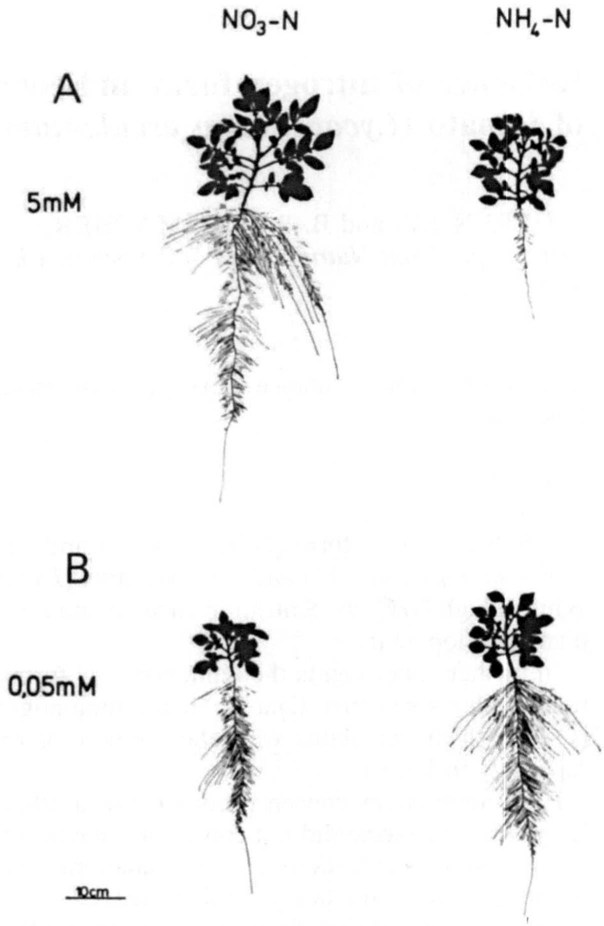

Fig. 1. Morphogenetic effect of N form and concentration on potato cuttings. One representative adventitious root per cutting was chosen for preparation.

Table 1. Influence of N form and concentration on the partitioning of dry weight and parameters of root growth of potato cuttings. Data are the mean of five replications and expressed on the basis of adventitious root

	Treatment			
	5 m*M*		0.05 m*M*	
	NO_3^-	NH_4^+	NO_3^-	NH_4^+
Shoot DW (g)	2.7	1.5	0.8	1.1
Root DW (g)	1.3	0.1	0.6	1.0
Shoot/root ratio	2.1	13.0	1.3	1.1
Length of main axis (cm)	48.3	22.6	34.7	42.4
Length of 1st order laterals	962	37	409	807
Length of 2nd order laterals	120	0	0	47
Number of root tips	320	52	108	187
Root surface area (dm^3)	87	8	36	65

forms (Barker *et al.*, 1967; Jungk, 1970; Kirkby and Mengel, 1967), this plant species was chosen for subsequent experiments.

Changes in the ionic balance

In roots NH_4^+ reduces the cation-anion difference (C-A) considerably at high N concentration, while at 0.05 m*M* N only small differences are apparent (Table 2). In stems of plants grown at 0.05 m*M* N the C-A values show the same reaction (lower C-A values in nitrate-grown plants), while at 5 m*M* N the opposite effect can be observed. The lower C-A values in stems of nitrate-grown plants are due to the accumulation of nitrate (data not shown) in xylem parenchyma cells. In leaves ammonium N reduces the C-A value significantly, irrespective of concentration.

Changes in the ionic balance should be reflected in different cytoplasmic pH values, but data reported in the literature are contradictory. While Kirkby and Mengel (1967) as well as Allen and Smith (1986) found higher pH values in homogenates of nitrate-grown plants, Andrade and Anderson (1986) could not prove a significant influence of N form on cytoplasmic pH of corn root tips (^{31}P NMR). In our own experiments an influence of N form on this parameter could only be demonstrated with ammonium-based nutrient solutions at pH 8. This relatively minor influence of N form on cytoplasmic pH is a reflection of the formation of

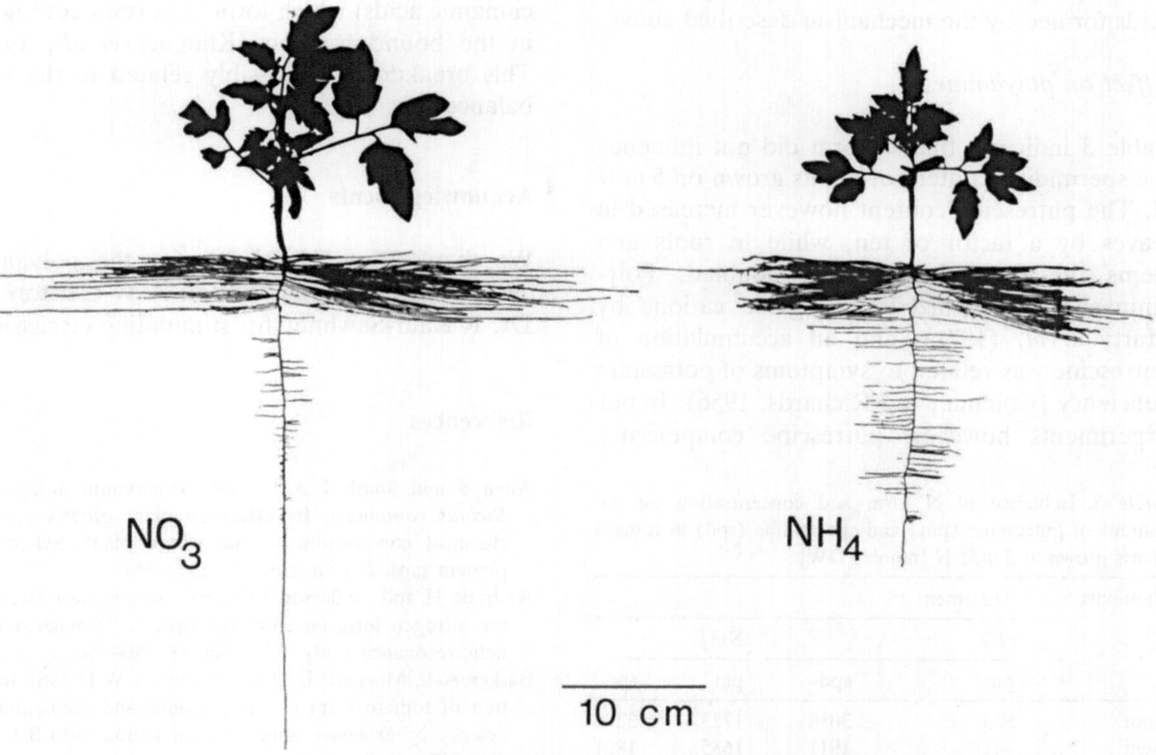

Fig. 2. Influence of N form on the morphology of tomato seedlings, treated with 10 m*M* N for one week.

Table 2. Influence of N form and concentration on the content of cations (K^+, Ca^{2+}, Mg^{2+}), anions (NO_3^-, $H_2PO_4^-$) and their difference (C-A) in tomato plants [mval/100 g DW]

Plant part	Ion equivalents	Treatment 5 m*M* NO_3^-	5 m*M* NH_4^+	0.05 m*M* NO_3^-	0.05 m*M* NH_4^+
Root	cations	244	128*	220	208
	anions	118	53*	58	60
	(C-A)	126	76*	162	148
Stem	cations	404	266*	290	194*
	anions	213	36*	51	36
	(C-A)	191	231*	239	158*
Leaf	cations	338	158*	278	188*
	anions	63	43*	38	35*
	(C-A)	274	114*	240	153*

* Indicates significant differences ($P = 0.05$).

organic acids by the biochemical pH-stat (Raven and Smith, 1976). The close correlation between C-A values and the organic acid content has frequently been verified (Jungk, 1970; Kirkby and Mengel, 1967). It is therefore suggested that the higher C-A values observed with nitrate-grown plants are compensated for by organic acids formed by the mechanism described above.

Effect on polyamines

Table 3 indicates that N form did not influence the spermidine content of plants grown on 5 m*M* N. The putrescine content however increased in leaves by a factor of ten, while in roots and stems the putrescine content doubled. Polyamines are recognized as organic cations by Murty *et al.* (1971), and an accumulation of putrescine was related to symptoms of potassium deficiency (Coleman and Richards, 1956). In our experiments however, putrescine compensates

Table 3. Influence of N form and concentration on the content of putrescine (put) and spermidine (spd) in tomato plants grown at 5 m*M* N [nmol/g DW]

Plant part	Treatment NO_3^- put	NO_3^- spd	NH_4^+ put	NH_4^+ spd
Root	823	3088	1723	3271
Stem	941	1911	1685	1821
Leaf	467	2912	4058	2768

just 0.45% of the difference in the C-A values. Thus, a significant contribution of putrescine to the ionic balance of tomato plants could not be demonstrated.

Comparable polyamine levels were detected in K deficient tobacco plants (Smith, 1970). Although in our experiments no toxicity symptoms appeared it can be assumed that putrescine concentrations of 4000 nmol/g dry weight may have a negative effect on plant growth. A large increase of the putrescine content, with only moderate changes for higher PA, is a typical reaction for plants under stress conditions (Flores *et al.*, 1985).

Polyamines are considered also as secondary messengers and were related to growth phenomena (Galston, 1983) and flower induction (Klinguer *et al.*, 1986). Thus, the involvement of PA in the growth stimulation at low NH_4^+ concentrations (Fig. 1, Table 1) may be assumed. Preliminary experiments with pea indicate a small increase of free PA at the expense of bound forms. Similar results were described by Priebe *et al.* (1978). A possible explanation for this shift may be the breakdown of organic acids (hydroxy cinnamic acids) which form important conjugates in the bound fraction (Klinguer *et al.*, 1986). This breakdown is possibly related to the ionic balance.

Acknowlegements

We thank Prof. H Börner for the polyamine determination as well as Prof. A W Galston and Dr. R Kaur-Sawhney for stimulating discussions.

References

Allen S and Smith J A C 1986 Ammonium nutrition in *Ricinus communis*: Its effect on plant growth and the chemical composition of the whole plant, xylem and phloem saps. J. Exp. Bot. 37, 1599–1610.

Andrade H and Anderson I C 1986 Physiological effects of the nitrogen form on corn root tips: A ^{31}P nuclear magnetic resonance study. Crop Sci. 26, 293–296.

Barker A V, Maynard D N and Lachman W H 1967 Induction of tomato stem and leaf lesions and potassium deficiency by excessive ammonium nutrition. Soil Sci. 103, 319–327.

Börner H 1987 Veränderungen im Polyamingehalt während

der Keimung der Sklerotien und der Apothezienentwicklung bei *Sclerotinia sclerotiorum*, dem Erreger der Weißstengeligkeit des Rapses. Z. Pflanzenkrankheiten Pflanzenschutz 94, 190–196.

Breteler H 1973 A comparison between ammonium and nitrate nutrition of young sugar-beet plants grown in nutrient solutions at constant acidity. 1. Production of dry matter, ionic balance and chemical composition. Neth. J. Agric. Sci. 21, 227–244.

Cataldo D A, Haroon M, Schrader L E and Youngs V L 1975 Rapid colorimetric determination of nitrate in plant tissue by nitration of salicylic acid. Commun. Soil Sci. Plant Anal. 6, 71–80.

Coleman R G and Richards F J 1956 Physiological studies in plant nutrition. XVIII. Some aspects of nitrogen metabolism in barley and other plants in relation to potassium deficiency. Ann. Bot. 20, 393–409.

De Wit C T, Dijkshoorn W and Noggle J C 1963 Ionic balance and growth of plants. Agric. Res. Rep. 69, 15.

Cox W J and Reisenauer H M 1973 Growth and ion uptake by wheat supplied nitrogen as nitrate, or ammonium, or both. Plant and Soil 38, 363–380.

Flores H E, Young N D and Galston A W 1985 Polyamine metabolism and plant stress. Cell. Molec. Biol. of Plant Stress 22, 93–114.

Galston A W 1983 Polyamines as modulators of plant development. Bioscience 33, 382–388.

Jungk A 1970 Wechselwirkungen zwischen Stickstoffkonzentrationen (NH_4, NH_4NO_3 und NO_3) und pH der Nährlösung auf Wuchs und Ionenhaushalt von Tomatenpflanzen. Gartenbau-Wissenschaften 35, 13-28.

Kirkby E A and Mengel K 1967 Ionic balance in different tissues of the tomato plant in relation to nitrate, urea, or ammonium nutrition. Plant Physiol. 42, 6–14.

Klinguer S, Martin-Tanguy J and Martin C 1986 K-Nutrition, growth bud formation, and amine and hydroxycinnamic acid contents in leaf explants of *Nicotiana tabacum* variety Xanthi n.c. cultivated *in vitro*. Plant Physiol. 82, 561–565.

Massé J, Laberche J-C and Jeanty G 1985 Influence de polyamines exogènes sur la croissance de plantules de maïs (*Zea mays* L.). C.R. Acad. Sc. 301, Serie III, 27–32.

Maynard D N and Barker A V 1969 Studies on the ammonium tolerance of plants. J. Am. Soc. Hort. Sci. 94, 235–239.

Murty K S, Smith T A and Bould C 1971 The relation between the putrescine content and potassium status of black currant leaves. Ann. Bot. 35, 687–695.

Priebe A, Klein H and Jäger H-J 1978 Role of polyamines in SO_2-polluted pea plants. J. Exp. Bot. 29, 1045–1050.

Puritch G S and Barker A V 1967 Structure and function of tomato leaf chloroplasts during ammonium toxicity. Plant. Physiol. 42, 1229–1238.

Raven J A and Smith F A 1976 Nitrogen assimilation and transport in vascular land plants in relation to intracellular pH regulation. New Phytol. 76, 415–431.

Sattelmacher B 1987 Methods for measuring root volume and for studying root morphology. Z. Pflanzenernachr. Bodenkd. 150, 54–55.

Smith T A 1970 Putrescine, spermidine and spermine in higher plants. Phytochem. 9, 1479–1486.

Torrey J G 1976 Root hormones and plant growth. Annu Rev. Plant Physiol. 27, 435–459.

Wilcox G E, Hoff J E and Jones C E 1973 Ammonium reduction of calcium and magnesium content of tomato and sweet corn leaf tissue and influence on the incidence of blossom end rot of tomato fruit. J. Am. Soc. Hort. Sci. 98, 86–89.

Wilcox G E, Magalhaes J R and Da Silva F L I M 1985 Ammonium and nitrate concentration as factors in tomato growth and nutrient uptake. J. Plant Nutr. 8, 989–998.

M. L. van Beusichem (Ed.), *Plant nutrition – physiology and applications*, 39–43.

PLSO IPNC564

Growth responses of Plantago to ammonium nutrition with and without pH control: Comparison of plants precultivated on nitrate or ammonium

S.R. TROELSTRA, R. WAGENAAR and W. SMANT
Institute for Ecological Research, P.O. Box 317, 3233 ZG Oostvoorne, The Netherlands

Key words: acidity control, ammonium nutrition, ionic balance, low-pH stress, nitrate, *Plantago lanceolata* L., *Plantago media* L., proton efflux

Abstract

Plantago species precultivated on either NO_3^- ('nitrate plants') or NH_4^+ ('ammonium plants') were grown on 100% NH_4^+, with and without pH adjustment. Growth, proton excretion, and changes in chemical composition were studied in detail. Under the same pH regime total net H^+ effluxes were approximately equal for nitrate and ammonium plants. Nitrate plants showed higher increases in dry-matter yield and in content of organic N than ammonium plants and the adverse effects upon plant growth induced by the ammonium treatment without pH adjustment were delayed in nitrate plants. Nitrate plants appeared to benefit from ther internal nitrate pool in more than one way. In addition to the increase in organic N via nitrate reduction, NH_4^+-derived organic N was also higher in nitrate plants which corresponded to an 'extra' uptake and utilization of ammonium. This 'extra' utilization of NH_4^+ in nitrate plants could be related directly to the observed changes in nitrate and organic-anion contents.

Introduction

At external pH values <9–10, the H^+ efflux is an active process (Smith and Raven, 1979) and this electrogenic H^+ extrusion may be considered as the driving force for ion uptake and a primary step in the regulation of pH and electroneutrality in plant cells (Israel and Jackson, 1982; Van Beusichem, 1984). Plant species may respond differently to high acidities (Islam *et al.*, 1980) or develop different acidities when grown on NH_4^+ (Barker and Mills, 1980; Maynard and Barker, 1969; Troelstra *et al.*, 1985; 1987). The ability of the proton pump to cope with increasing external acidity determines the lower pH limit tolerated by plants (Smith and Raven, 1979), *i.e.* that pH value where the plant can still just generate a positive net outward H^+ gradient. When put on 100 per cent NH_4^+ without pH adjustment, plants will decrease the pH of their immediate root environment just beyond this lower pH limit, thereby blocking ion uptake completely and ultimately impairing further plant growth (*e.g.* Troelstra *et al.*, 1985; 1987). However, growth may still continue for some time by internal reallocation and utilization of nutrients.

The primary aim of the present study was to establish in greater detail the development and chemical composition of plants under a condition of low-pH stress that was induced by the omission of pH adjustment in the NH_4^+ nutrition. Two species of Plantago were chosen: *P. lanceolata*, which occurs over a wide range of soil habitats (from acidic to calcareous), and *P. media*, which is more restricted to calcareous sites. As the process of nitrate reduction is an important factor in the internal pH regulation of a plant, responses of two types of precultivated plants were studied, *viz*. NO_3^--fed and NH_4^+-fed plants.

Materials and methods

Plantago lanceolata L. and *Plantago media* L. were grown in the greenhouse and 24- or 36-day-old seedlings were precultivated for 18 or 24 days (*P. lanceolata* and *P. media*, respectively) on either NO_3^- or NH_4^+ at pH 6.5, after which the plants were transferred to the final NH_4^+ treatment without further pH control ('−pH'). The culture solution contained the following macronutrients (m*M*): NH_4^+ 4; K^+ 2.5; Ca^{2+} 1.25; Mg^{2+} 0.5; $H_2PO_4^-$ 0.5; Cl^- 2.5; SO_4^{2-} 3.5. Micronutrients were present in the following concentrations (mg L^{-1}): Fe (as FeEDTA) 5; B and Mn 0.5; Zn and Mo 0.05; Cu 0.02. Dicyandiamide was added (7.5 mg L^{-1}) as nitrification inhibitor. Initial pH was adjusted to 7.0 with NaOH. There were 8 or 12 plants per pot (*P. lanceolata* and *P. media*, respectively) and 12 pots of both 'nitrate plants' and 'ammonium plants', each pot containing 10.6 liter of nutrient solution. Nutrient solutions were aerated continuously. Within each set, plants were transferred in such a way that each pot supported a similar range of plant sizes with a similar total plant dry weight per pot. Nitrate plants had slightly lower initial total dry weights than ammonium plants: 0.89 and 1.08; and 0.30 and 0.37 g.plant^{-1} for *P. lanceolata* and *P. media*, respectively. *P. lanceolata* was in the generative stage, almost from the start of the experiment. Plants (one pot per set, selected randomly) were harvested at varying intervals between 0 and 25 days after the start of the final treatment. Per set of plants, three additional pots served to follow the growth of plants on NH_4^+ with daily pH adjustment at about 6.5 with NaOH ('+pH'), and these plants were harvested at days 7, 14, and 21 (or 25); +pH solutions were renewed weekly (*P. media*) or twice weekly (*P. lanceolata*).

At each harvest, plants were divided into leaves, spikes (or shoots) and roots, and dried for 48 h at 70°C. Samples were analyzed for total N and the quantitatively important constituents of the ionic balance (NH_4^+, K^+, Na^+, Ca^{2+}, Mg^{2+}, NO_3^-, $H_2PO_4^-$, Cl^-, SO_4^{2-}), as described previously (Troelstra, 1983). Organic nitrogen (N_{org}) was estimated by subtraction of NO_3^- plus NH_4^+ from total N. Total organic anions (C − A) were calculated as the difference between sum totals of cations (C) and inorganic anions (A). Net H^+ efflux of 'non-adjusted' plants was measured by the base addition needed to bring the solution pH to its original value (as indicated by 'control' pots without plants); the net H^+ efflux of 'adjusted' plants was estimated from data on biomass and chemical composition.

Time courses of biomass and constituents of −pH plants were fitted exponentially, linearly, by a 2nd-degree polynomial, or combinations.

Results

Growth and nitrogen utilization

The final pH without acidity control was reached after 4 (*P. lanceolata*) or 7 days (*P. media*), being 3.4 for each of the two species and of the two types of precultivated plants. Final H^+ efflux was 1.0 and 0.6 meq.plant^{-1} for *P. lanceolata* (Fig. 1) and *P. media*, respectively.

Nitrate plants (−pH) showed stronger yield increases than ammonium plants (−pH; Table 1), particularly during the period of pH decrease ($P < 0.05$ in the case of *P. lanceolata*; Fig. 1). The slopes of the linear increases after the final pH was reached were also steeper for nitrate plants, the difference being significant ($P < 0.05$) in the case of *P. media* (data not shown). Yield increases of +pH plants were still higher for nitrate plants of *P. lanceolata*, but not for those of *P. media* (Table 1).

Increases in N_{org} were higher in nitrate plants, especially in the case of *P. lanceolata* ($P < 0.01$). This was in part due to the presence of a larger NO_3^- pool in *P. lanceolata* (caused by the plants being larger as well as having higher tissue NO_3^- concentrations compared to *P. media*). This NO_3^- became largely reduced: in *P. lanceolata* (−pH) NO_3^- decreased from 1.43 to 0.16 meq.plant^{-1} (Fig. 1); in *P. media* (−pH) from 0.21 to 0.03 meq.plant^{-1}. Together with the disappearance of NO_3^-, the other inorganic anions (Cl^-, and to a lesser extent $H_2PO_4^-$ and SO_4^{2-}) showed an increase in the tissues of nitrate plants, especially during the period of pH decrease (data not shown). However, the observed increase in N_{org} in nitrate plants (−pH) was higher than the sum of NO_3^- reduction in nitrate plants and N_{org} increase in ammonium plants (−pH). Analyses of the nutrient solutions also indicated an 'extra' NH_4^+ uptake by nitrate plants: 0.62 and

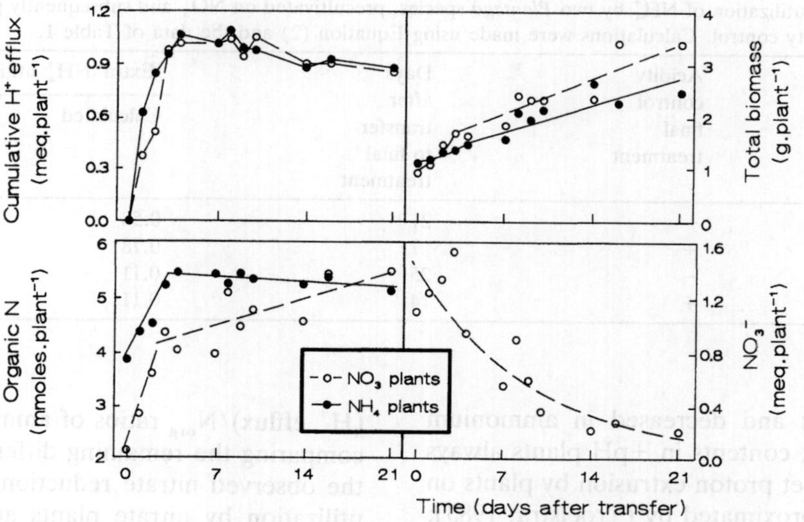

Fig. 1. Time courses of cumulative proton efflux, dry weight of total biomass, and contents of organic nitrogen and nitrate in *Plantago lanceolata* following transfer to 100 per cent NH_4^+ nutrition without acidity control. Plants were precultivated on either NO_3^- or NH_4^+.

0.10 meq.plant^{-1} for *P. lanceolata* and *P. media*, respectively. After the pH decrease, N_{org} contents of ammonium plants did not further increase, whereas nitrate plants still demonstrated a gradual linear increase (Fig. 1).

Organic-anion contents, proton effluxes and 'extra' NH_4^+ utilization by nitrate plants

Without acidity control, organic-anion contents remained at the same level or increased slightly

Table 1. Increases in total dry-matter yield, and changes in contents of organic N, NO_3^- and organic anions (C − A) in two *Plantago* species grown on NH_4^+, either without or with acidity control. Plants were precultivated on either NO_3^- or NH_4^+. Proton effluxes were calculated according to Equation (1)

Acidity control final treatment	Days after transfer to final treatment	Precultivation	Increase or decrease in Total dry-matter yield (g.plant^{-1})	N_{org} (mmoles.plant^{-1})	NO_3^- (meq.plant^{-1})	C − A	Calculated H^+ efflux
Plantago lanceolata							
−	21	NO_3^-	2.59	3.34	−1.27	0.38	1.00
		NH_4^+	1.64	1.39	−	−0.41	0.90
+	7	NO_3^-	1.79	4.65	−0.74	0.07	2.99
		NH_4^+	1.26	3.07	−	0.03	2.93
	21	NO_3^-	5.70	11.7	−1.34	1.49	9.88
		NH_4^+	4.79	9.69	−	1.44	10.6
Plantago media							
−	25	NO_3^-	0.65	1.09	−0.18	−0.002	0.67
		NH_4^+	0.51	0.76	−	−0.08	0.64
+	14	NO_3^-	0.62	2.35	−0.08	0.36	2.42
		NH_4^+	0.69	2.04	−	0.37	2.30
	25	NO_3^-	1.55	4.71	−0.14	1.05	5.23
		NH_4^+	1.66	3.94	−	0.86	4.59

Table 2. The 'extra' utilization of NH_4^+ by two *Plantago* species, precultivated on NO_3^- and subsequently grown on NH_4^+, either without or with acidity control. Calculations were made using Equation (2) and the data of Table 1.

Species	Acidity control final treatment	Days after transfer to final treatment	'Extra' NH_4^+ utilization ($meq.plant^{-1}$)	
			Calculated	Measured
P. lanceolata	–	21	0.58	0.62
	+	7	0.78	–
P. media	–	25	0.11	0.10
	+	14	0.11	–

in nitrate plants and decreased in ammonium plants (Table 1); contents in +pH plants always increased. The net proton extrusion by plants on NH_4^+ can be approximated by (Troelstra, 1983):

$$H^+ \text{ efflux} = 0.946\,\Delta N_{org} + \Delta(\text{C-A}) + 2\,\Delta NO_3^- \tag{1}$$

where the Δ signs refer to changes (meq or $mmol.plant^{-1}$) in N_{org}, C – A, and previously absorbed NO_3^-, a decrease being negative. Proton effluxes thus calculated (Table 1) showed to be in good agreement with the measured values of the –pH plants (see above).

By applying Eq. (1) to both nitrate and ammonium plants, the 'extra' utilization of NH_4^+ by nitrate plants *at equal net H^+ efflux* is given by:

$$(\Delta(\text{C-A})_{NH_4} - \Delta(\text{C-A})_{NO_3} - 1.054\,\Delta NO_3^-)/0.946 \tag{2}$$

where the subscripts refer to the type of precultivated plants. Some calculations for the cases of Table 1 with similar H^+ effluxes (difference $\leq 0.1\ meq.plant^{-1}$) for nitrate and ammonium plants are shown in Table 2. Since time courses of C – A in +pH plants appeared to be less influenced by the presence of an internal nitrate pool, nitrate reduction becomes solely responsible for and should approximately equal the 'extra' NH_4^+ utilization under these conditions (compare Tables 1 and 2). Taking into account the differences in H^+ efflux between nitrate and ammonium plants (+pH) at t = 21 and t = 25 (Table 1), N_{org} increases of these plants can be 'transformed' to an equal H^+ efflux using the (H^+ efflux)/N_{org} ratios of ammonium plants. By comparing the remaining difference in N_{org} with the observed nitrate reduction, the 'extra' NH_4^+ utilization by nitrate plants at equal H^+ efflux can be simply estimated: 1.33 (*P. lanceolata*, t = 21) and 0.08 $meq.plant^{-1}$ (*P. media*, t = 25), confirming the above reached conclusion.

Discussion

A simultaneous utilization of NO_3^- and NH_4^+ may be of advantage in terms of internal pH regulation (*e.g* Raven and Smith, 1976; Runge, 1983). The present results indicated that a supply of NO_3^- prior to the NH_4^+ treatment without acidity control delayed the eventual adverse effects of this treatment. Nitrate supply has been reported to alleviate conditions of ammonium injury in tomato plants (Ikeda and Yamada, 1984). Nitrate plants most likely maintain their internal pH at proper levels during a longer period than ammonium plants due to the OH^-/HCO_3^- production accompanying the nitrate reduction.

Referring to the conceptual model on ion uptake of Israel and Jackson (1982), increases in N_{org} can be divided into the following categories:

(1) Utilization of NH_4^+ taken up with a concurrent net H^+ efflux; cytoplasmatic OH^-/HCO_3^- (generated by the operation of the ATPase-mediated proton extrusion) is used for the disposal of H^+ originating from the utilization of NH_4^+.

(2) Utilization of NH_4^+ taken up in combination with uptake of inorganic anions; cytoplasmatic OH^-/HCO_3^- (originating either directly

from the NO_3^- reduction or indirectly via decarboxylation of previously formed carboxylates) is used for H^+ disposal. The inorganic anions (Cl^-, $H_2PO_4^-$, SO_4^{2-}) replace the reduced NO_3^- or decarboxylated organic anions.

(3) Utilization of reduced NO_3^-.

Because of the occurrence of category (3) in nitrate plants, category (2) becomes quantitatively more important in these plants than in ammonium plants, thus giving rise to an 'extra' NH_4^+ utilization in nitrate plants under these low-pH-stress conditions. However, the effect can be also operative under non-stress conditions (+pH). Apart from possible advantages of compartmental separation of NO_3^- and NH_4^+ assimilation within the plant (in the shoot and root, respectively), this phenomenon may well contribute to the often reported relatively high yields of plants and relatively high rates of N acquisition (Troelstra *et al.*, 1987) when there is a mixed supply of NH_4^+ and NO_3^-.

References

Barker A V and Mills H A 1980 Ammonium and nitrate nutrion of horticultural crops. Hortic. Rev. 2, 395–423.

Ikeda M and Yamada Y 1984 Palliative effect of nitrate supply on ammonium injury of tomato plants: Growth and chemical composition. Soil Sci. Plant Nutr. 30, 485–493.

Islam A K M, Edwards D G and Asher C J 1980 pH optima for crop growth results of a flowing solution culture experiment with six species. Plant and Soil 54, 339–357.

Israel D W and Jackson W A 1982 Ion balance, uptake and transport processes in N_2-fixing and nitrate- and urea-dependent soybean plants. Plant Physiol. 69, 171–178.

Maynard D N and Barker A V 1969 Studies on the tolerance of plants to ammonium nutrition. J. Am. Soc. Hort. Sci. 94, 235–239.

Raven J A and Smith F A 1976 Nitrogen assimilation and transport in vascular land plants in relation to intracellular pH regulation. New Phytol. 76, 415–431.

Runge M 1983 Physiology and ecology of nitrogen nutrition. *In* Encyclopedia of Plant Physiology. New Series 12C, Physiological Plant Ecology III. Eds. O.L. Lange, P S Nobel, C B Osmond and H Ziegler. pp 163–200. Springer-Verlag, Berlin, Heidelberg, New York.

Smith F A and Raven J A 1979 Intracellular pH and its regulation. Annu. Rev. Plant Physiol. 30, 289–311.

Troelstra S R 1983 Growth of *Plantago lanceolata* and *Plantago major* on a NO_3/NH_4 medium and the estimation of the utilization of nitrate and ammonium from ionic balance aspects. Plant and Soil 70, 183–197.

Troelstra S R, Van Dijk C and Blacquière T 1985 Effects of N source on proton excretion, ionic balance and growth of *Alnus glutinosa* (L.) Gaertner: Comparison of N_2 fixation with single and mixed sources of NO_3 and NH_4. Plant and Soil 84, 361–385.

Troelstra S R, Blacquière T, Wagenaar R and Van Dijk C 1987 Ionic balance, proton efflux, nitrate reductase activity and growth of *Hippophaë rhamnoides* L. ssp. *rhamnoides* as influenced by combined-N nutrition or N_2 fixation. Plant and Soil 103, 169–183.

Van Beusichem M L 1984 Non-ionic Nitrogen Nutrition of Plants. Doct. Thesis, Wageningen Agricultural University, 141 pp.

M. L. van Beusichem (Ed.), *Plant nutrition – physiology and applications*, 45–51.

PLSO IPNC113

Rhizosphere pH along different root zones of Douglas-fir (*Pseudotsuga menziesii*), as affected by source of nitrogen

A.J. GIJSMAN
Institute for Soil Fertility Research, P.O. Box 30003, 9750 RA Haren, The Netherlands

Key words: Douglas-fir, H^+/OH^- excretion, microelectrodes, nitrogen source, *Pseudotsuga menziesii* [Mirb] Franco, rhizosphere pH, root zones

Abstract

Douglas-fir, grown on strongly acid soil (pH-H_2O 3.87), was fertilized with ammonium, nitrate or ammonium nitrate as N-source. Rhizosphere pH along the root axis was measured with microelectrodes. Pure ammonium supply resulted in acidification of the rhizosphere, almost along the entire root axis; only the extreme root tip was slightly alkaline compared with the bulk soil. With nitrate supply, the alkalization of the rhizosphere at the root tip was considerably stronger than with ammonium supply and the length of the alkalization zone greater, extending over the entire growth zone of the root. Acidification of the rhizosphere along the older parts of the root was less pronounced in the case of nitrate. It is concluded that nitrate nutrition enables the plant to protect its most essential root zone from the adverse effects of strong acidity by locally raising the rhizosphere pH.

Introduction

One of the mechanisms which enables a root to function in strongly acid soil is to increase the pH of its immediate root environment. This ability largely depends on the source of nitrogen taken up; ammonium nutrition always leads to excretion of H^+ ions by roots, and thus to acidification of the rhizosphere, while with nitrate acidification as well as alkalization can occur, depending on the level of supply. As gradients in H^+ concentration in the rhizosphere can be very steep (Nye, 1981), exact measurement of the pH at the soil/root interface is difficult. Riley and Barber (1969) introduced a method for measuring the rhizosphere pH by separate collection of rhizosphere and bulk soil. By this method rhizosphere pH is averaged over the whole root system, and possible differences between root zones are lost. Recently, new techniques have been developed which make pH measurements in the rhizosphere possible on a smaller scale (Häussling *et al.*, 1985; Marschner *et al.*, 1982; Pilet *et al.*, 1983; Schaller and Fischer, 1981; Weisenseel *et al.*, 1979). These techniques showed that large differences in rhizosphere pH can exist between distinct zones along a root. Schaller and Fishcer (1985), working with maize on a soil of pH 5.5, found that the rhizosphere pH varied from 3 to 8, depending on the type of N-nutrition. Even within one single treatment, rhizosphere pH values of root tip and root hair zone differed by 4 units. It is obvious, therefore, that determination of mean values for the rhizosphere pH of the whole root system can only give very limited information, since it obscures differences on a smaller scale.

In earlier experiments (Gysman, 1990a, b) it was shown that Douglas-fir on a very acid soil (pH 3.87) grew considerably better with ample nitrate than without nitrate, even when the ammonium supply was low in the last situation. Ionic balance calculations showed that ammonium nutrition substantially reduced net carboxylate production in the plant. This probably led to a shortage of organic substrate in the plant root to detoxify ammonium ions, and to a reduced capacity to eliminate H^+ ions produced during ammonium assimilation. It was also shown that with ammonium supply the roots excreted considerable amounts of H^+ ions, while with nitrate or ammonium nitrate much less H^+ or mostly OH^- was excreted. For all N-sources, the mean rhizosphere pH of the whole root system, however, was only slightly different from the bulk pH and variation among treatments was very small. The experiment presented here was set up to determine whether rhizosphere pH indeed varied little with N-source, or that differences did exist on a smaller scale, but were obscured in the earlier experiment.

Materials and methods

The experiment was carried out with three-year old Douglas-fir (*Pseudotsuga menziesii* [Mirb.] Franco, provenance Arlington 202), planted in rectangular boxes (size 50 × 38 × 8.5 cm) with removable perspex front and back walls; these walls were normally covered by an aluminium plate to shield soil and roots from light. A strongly acid sandy soil was used (pH-H_2O 3.87; pH-KCl 3.28). The particle-size distribution was as follows: <2μm, 3.7%; 2–50 μm, 5.6%; 50–210 μm, 58.9% and >210 μm, 31.8%. The organic matter content as determined by loss on ignition was 3.2 g per 100 g air-dry soil. The soil was fertilized with one of three sources of nitrogen at an N level of 50 mg.kg^{-1} (on an oven dry soil weight basis): ammonium as $(NH_4)_2SO_4$, nitrate as $Ca(NO_3)_2$ or the mixed source as NH_4NO_3. Several other nutrients and trace elements were added as basal fertilizer, and – together with the nitrogen – mixed with the soil at the start of the experiment. The basal fertilization consisted of the following nutrients (per kg oven-dry soil) 30 mg P as $Ca(H_2PO_4)_2$, 50 mg K as K_2SO_4, 70 mg Ca as $CaSO_4$ (not on the pots receiving nitrate as sole N-source), 25 mg Mg as $MgSO_4$, 5 mg Fe as FeEDTA, 0.8 mg of each of the following trace elements: B as $Na_2B_4O_7{\cdot}10\,H_2O$, Mo as $Na_2MoO_4{\cdot}2\,H_2O$, Cu as $CuSO_4{\cdot}7\,H_2O$, Mn as $MnSO_4{\cdot}5\,H_2O$, Zn as $ZnSO_4{\cdot}7\,H_2O$. To prevent conversion of ammonium into nitrate, 10 mg N-Serve 24 E (2-chloro-6-trichloromethylpyridine) was added per kg moist soil. This was repeated every 10 weeks by injecting 100 ml N-Serve solution with a 20 cm long needle into the soil at different depths.

A 10 cm layer of small alkathene pellets was placed on top of the pots to reduce evaporation of water from the soil surface. Soil moisture content was kept at about 0.20 $cm^3\,cm^{-3}$, measured halfway up the growth box, with an estimated variation of 0.05 $cm^3\,cm^{-3}$. Water was added from below through an opening in one of the sides of the box.

The buffer capacity of the soil was determined by suspending 10 g of soil in 50 ml 1 *M* KCl solution, to which different amounts of HCl or $Ca(OH)_2$ were added. During successive days the suspension was shaken regularly and pH was measured several times during periods of up to 96 hours. The relation between H^+/OH^- addition and resulting pH, was found to be almost lineair, and was expressed as a buffer value according to Hartikainen (1986). The short- and long-term buffer values, determined after 15 minutes and 96 hours, respectively, were 24.10^{-3} mmol g^{-1} and 36.10^{-3} mmol g^{-1} (on an air-dry soil weight basis).

The trees were planted in the autumn of 1987 and measurements were made between June and August of the following year. Two boxes each were used for both the ammonium and nitrate treatments; the ammonium nitrate treatment was carried out in triplicate.

The method used to measure the rhizosphere pH was similar to the one described by Häussling *et al.* (1985). It consisted of the combined use of an agar sheet, containing a pH indicator to visualize zones with different pH along the root, and a microelectrode to measure the pH. Sheets of agar were made by pouring boiled agar solution (7.5 g L^{-1}), mixed with bromophenol blue (0.06 g L^{-1}) and adjusted to pH 3.87, onto a plate for cooling. The thickness of the sheets was about 2 mm.

One of the walls of a slightly tilted growth container was removed, so that roots growing at the soil surface along this wall were exposed for measurements. The root and its rhizosphere to be measured were then covered by a small (about 6 mm) strip of agar (Fig. 1), while the rest of the soil surface was covered with moist tissues to prevent drying out of soil and roots. After

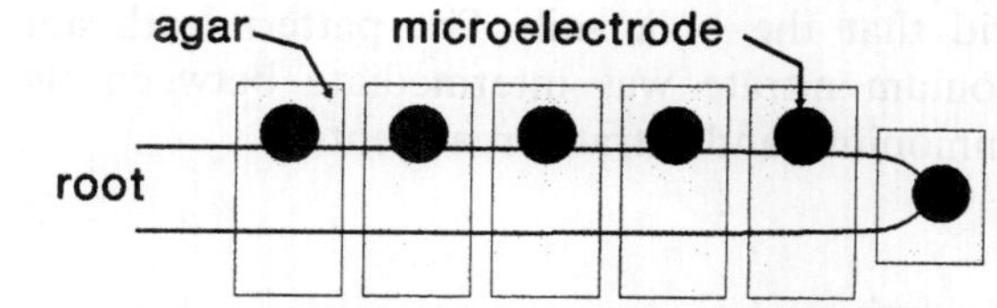

Fig. 1. Positioning of the agar strip and the microelectrode on root and rhizosphere.

about 15 minutes, pH measurements were made by mechanically pushing a microelectrode through the agar with a micromanipulator (Märzhäuser DC3) almost to the soil/agar interface. Starting at the root tip, measurements were made every 5 mm along the root axis up to a maximum distance of 55 mm from the tip. Using this method rather than covering the whole soil surface with a large agar sheet prevented local pH differences from being evened out in the agar. As H^+/OH^- diffusion in agar occurs at a much higher rate than in soil, pH gradients – which can be very steep around a root (Nye, 1981) – would then be less pronounced. For the same reason the strip of agar was interrupted transversely at several points (Fig. 1).

The electrode used was a 1.2 mm diameter glass microelectrode (Microelectrodes MI 405), with a separate standard reference electrode (Orion). To improve contact of the reference electrode with the soil, it was placed in a glass tube with a much larger sintered glass contact area (about 65 mm^2). The tube was filled with a KCl solution of low molarity (0.01 *M*) to prevent liquid junction potentials with the soil solution, which would disturb the stability of the electrode potential (Linnet, 1970). About one quarter of the microelectrode's diameter was placed onto the root and three quarters onto the rhizosphere soil (Fig. 1), so the width of the rhizosphere measured was less than 1 mm.

Several authors described the use of antimony microelectrodes for pH measurements in the rhizosphere (Häussling *et al.*, 1985; Hauter and Mengel, 1988; Schaller and Fischer, 1981), which seemed to be promising, since these electrodes can be made with a very small diameter (up to about 0.5 mm). Despite repeated trials, however, we could not obtain a steady mV-reading when these electrodes were used either in soil or in agar. Antimony electrodes are subject to many interferences. According to Ives (1961) and Kolthoff and Elving (1978) the solution to be measured must be free of interfering solutes which include oxidizing (oxygen!) or reducing agents, salts of acids which form complexes with antimony such as citric, oxalic, tartaric and phenylacetic acids, certain amino acids, and salts of metals more noble than antimony, notably copper. It is therefore not surprising that, in a soil containing a variety of organic molecules and other substances, these electrodes did not produce a stable potential. The instability during measurements in agar probably resulted from the change in partial oxygen pressure in the agar, as oxygen entered through the hole the electrode created when it was pushed through the agar.

Results

Soon after growth of the plants started in early spring, the first roots became visible behind the transparent wall. The appearance of the roots in the ammonium treatment strongly differed from that in the other two treatments. With nitrate or ammonium nitrate supply the root tips were white over a distance of at least 1 cm, but with ammonium the roots were discoloured much closer to the tip and often no white tips could be found at all. Also, older parts of the roots were darker – often completely black – and generally looked rather scabby. The most striking effect of ammonium on root morphology was the claviform appearance of the root tips, while growth of these roots had ceased.

Determination of the soil nitrate content of the ammonium-fertilized boxes showed that no nitrate had formed during the experiment, so it can be concluded that the N-Serve completely blocked nitrification. This was supported by the observation that the pH of the bulk soil hardly varied, neither in time, nor among treatments.

For measurement of the rhizosphere pH, roots were needed which grew along the perspex window for at least a few centimeters. Since root growth was very poor and roots, after having grown at the soil/perspex interface for some distance, often turned away from the perspex, no more than 4 or 5 roots per box could be used for pH measurements. Figure 2 shows the pattern of

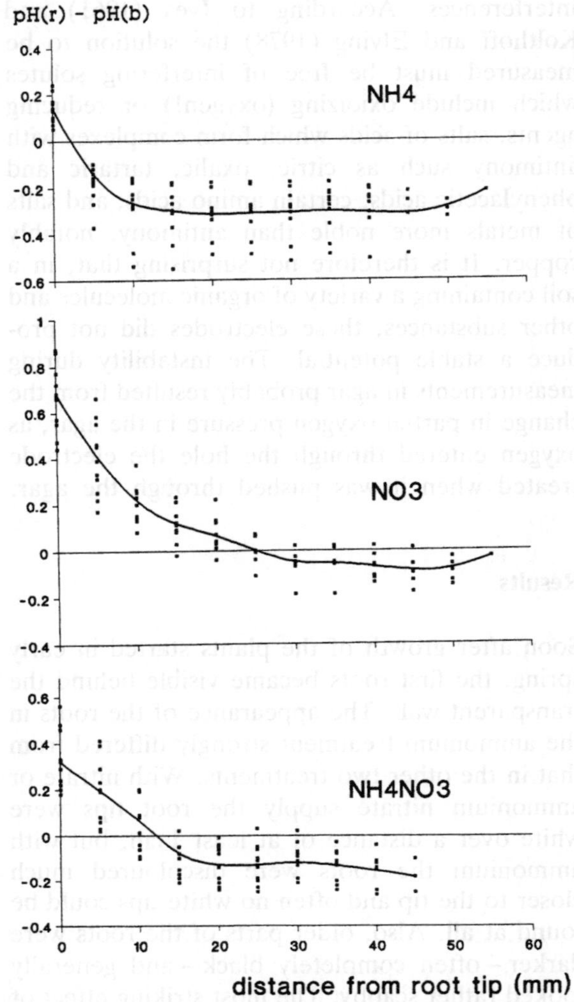

Fig. 2. Difference between rhizosphere pH (pH_r) and bulk soil pH (pH_b) along the axis of the root with ammonium, nitrate or ammonium nitrate fertilization.

rhizosphere pH values along the root axis. In the ammonium treatment, the rhizosphere pH of all root zones measured (excluding the extreme tip) was lower than the bulk soil pH. Differences varied from 0.2 to 0.55 pH unit, with a mean of about 0.3 unit below bulk soil pH. At the extreme root tip the rhizosphere was a little more alkaline than the bulk soil, although differences were very small. With nitrate, the alkalization at the root tip was much stronger and the length of the alkalization zone greater. Maximum pH increase of the rhizosphere compared with the bulk soil was about 0.9 unit. From about 25 mm behind the tip the rhizosphere was slightly more acid that the bulk soil. The pattern with ammonium nitrate was intermediate between the ammonium and nitrate treatments.

Discussion

The experiment showed that, depending on the form of nitrogen absorbed, distinct zones along the root with different rhizosphere pH could be distinguished. Largest differences between treatments were observed at the root tip and at the first 20–25 mm behind it. As this is the zone of root growth, the ability to reduce the acidity stress in the immediate root environment can be of considerable importance for the roots to function in very acid soils. The root morphology of the ammonium fed plants indicated that the external conditions were detrimental to the root and ammonium supply probably affected root functioning too. It can be expected that at these low pH values the aluminium concentration in the soil solution was increased sharply, and that membrane stability was impaired by aluminium ions. With nitrate and – to a lesser extent – ammonium nitrate, the plant was apparently able to increase the pH of the rhizosphere near its most sensitive and important root zone, and so to protect it from the adverse soil conditions. With an increase of 0.7 pH unit at the root apex of nitrate fed plants, the rhizosphere pH reached a value at which the aluminium concentration was no longer harmful (critical pH = 4.5). Along other root zones, pH differences among treatments were small, but, judging from observations of root morphology, these differences could have been important.

In an experiment carried out at the same time under identical conditions, the effect of source and level of N-supply on mean rhizosphere pH of the whole plant was determined (Gijsman, 1990b). Following Riley and Barber (1969), the rhizosphere pH was measured using soil particles adhering onto the roots. No distinction could be made between different root zones. With ammonium, the rhizosphere pH averaged over the whole root system, was always lower than the bulk soil pH, irrespective of N-level. With low levels of nitrate or ammonium nitrate, rhizosphere acidification still occurred, but with increasing supply a shift to a more alkaline rhizosphere was observed. Mean differences between rhizosphere and bulk pH, however, were very small and did not exceed 0.15 unit. The present experiment shows that much larger differences in rhizosphere pH do exist along different root zones.

Most of the work on rhizosphere pH pattern along a root has been carried out with annual species, where, irrespective of N-source, the root tip was always acid compared with the bulk soil (Hauter and Mengel, 1988; Römheld, 1983, 1986; Römheld and Marschner, 1986; Schaller and Fischer, 1985; Schaller, 1986). By contrast, the scarce reports on conifers showed the root tip to be more alkaline than the bulk soil. For 60–80 year old *Picea abies* trees on strongly acid soil (pH 3.5, 4.0 and 4.1) Marschner *et al.* (1985) found an alkalization of 0.25–0.36 pH unit at the root tip. The adjacent white and brown root zones were also more alkaline than the bulk soil. Analysis of the soil solution in these plots showed that less than 5% of the mineral nitrogen was in the ammonium form. In an experiment with much younger trees on a slightly less acid soil (pH 4.5), a small acidification zone between the alkalizing root tip and also alkalizing brown root zone was found. In these plots the ammonium levels in the soil solution were much higher, but nitrate levels also were, so still the contribution of ammonium in total N-supply was negligible. When the ammonium fraction rose to about 11%, the brown root zone also acidified its environment, resulting in a rhizosphere which was more alkaline than the bulk soil around the root tip, and more acid around the rest of the root. Likewise, Häussling *et al.* (1985) found for Norway Spruce that nitrate fertilization led to a more alkaline rhizosphere around the root tip, followed by a zone where pH was slightly lower than the bulk pH, followed in turn by an alkaline zone. With ammonium fertilization the lowering of the pH was stronger and the length of the zone of pH lowering greater; it was not clear whether there also was a second alkalization zone.

The pattern of an alkalizing root tip, followed by a zone of acidification, corresponds with the results reported here. The degree of alkalization at the root tip and acidification along the other root zones, as well as the length of the alkalizing zone, depended on the N-source. Whether there was also an alkalizing zone at the more basal part of the root system could not be determined, since roots growing at the soil/wall interface over such a distance were seldom found. The scarce measurements made on more basal root zones (data not shown), however, indicate that the pH occurring about 50 mm from the tip is representative of older root zones also.

The pH patterns observed along the root axis in the three treatments differ in: a) degree of alkalization at the root tip, b) distance behind the root tip where rhizosphere pH equals bulk soil pH, and c) degree of acidification along the older part of the root. Interpretation of this pattern is complicated by the fact that root growth in the ammonium treatment was much poorer than that in the nitrate treatment, so root segments at the same distance behind the root tip were of a different age in each treatment. We cannot directly conclude at what distance behind the tip or age of the root tissue acidification started and to what extent the alkalization just behind the root tip results from a buffering action of the soil, once the alkalization pulse from the root tip is over. We hope to analyse this further with model calculations on H^+ transport in and buffering of the soil, with a slightly modified version of Nye's (1981) rhizosphere pH model (Gijsman, 1989).

There are several hypotheses to explain differences in rhizosphere pH along the root axis. 1. The ratio of cation to anion uptake may vary between different parts of the rooting system (Blanchar and Lipton, 1986), which is supported by the observation that for some nutrients rate of uptake varies with root zone (Brouwer, 1954; Häussling *et al.*, 1988). Also differences in rate of depletion and replenishment of nutrients in the soil can influence cation/anion uptake balance of different root zones, as *e.g.* nutrient availability for old root zones will differ from that for a young root just invading a new soil volume. 2. Nitrate reduction leads to production of OH^- ions in the plant, which, when reduction occurs in the roots, are largely excreted. A relation between the localization of the nitrate reductase enzyme in the roots and the rhizosphere pH therefore is obvious. 3. Exudates at the root apex (Horst *et al.*, 1982; Mulkey and Evans, 1981) or root exudates related to Fe or P deficiencies (Brown, 1976; Gardner *et al.*, 1983; Jayman and Sivasubramaniam, 1975) often are (anions of) organic acids or other organic molecules, which can affect the pH of the rhizosphere. 4. The region of cell division and cell elongation is often associated with H^+ excretion (Mulkey and Evans, 1981; O'Neill and Scott, 1983; Pilet *et al.*, 1983; Weisenseel *et al.*, 1979).

For Douglas-fir, the localization of nitrate reductase enzym along the root axis probably is important in increasing the rhizosphere pH at the root apex. Li *et al.* (1972) were unable to detect any nitrate reductase in older root segments of Douglas-fir. Bigg and Daniel (1978) found that nitrate reductase activity (NRA) was highest in the apical 60 mm of the roots, but they did not test smaller root segments, so further specification might be possible. For *Picea abies* NRA in the apical 1 cm of the root was twice as high as in the next 1 cm (Peuke, 1987); in older root segments NRA was less than 20% of that in the apex. The distorted appearance of the roots in the ammonium treatment, probably connected with impaired nutrient uptake by the roots, and the larger differences in cation/anion uptake balance between N-sources (Gijsman, 1990a), support the hypothesis that differences in nutrient uptake pattern also contribute to the variation in rhizosphere pH along different root zones.

It can be concluded that uptake of nitrate as the main nitrogen source enables the plant to protect its root growth zone by locally raising the rhizosphere pH. When ammonium is the predominant source of nitrogen – *e.g.* due to atmospheric ammonia deposition – the degree of nitrification therefore can be a determining factor in the functioning of the root, even though the nitrification process itself leads to a further acidification of the soil.

Acknowledgements

The author wants to thank Dr A de Jager for the work he did in starting up this research and raising funds for it. Prof. P J C Kuiper, University of Groningen, gave some helpful suggestions for improvement of the text.

References

Bigg L W and Daniel T W 1978 Effects of nitrate, ammonium and pH on the growth of conifer seedlings and their production of nitrate reductase. Plant and Soil 50, 371–385.

Blanchar R W and Lipton D S 1986 The pe and pH in alfalfa seedling rhizospheres. Agron. J. 78, 216–218.

Brouwer R 1954 The regulating influence of transpiration and suction tension on the water and salt uptake by roots of intact *Vicia faba* plants. Acta Bot. Neerl. 3, 264–312.

Brown J C 1976 Mechanism of iron uptake by plants. Plant Cell Environ. 1, 249–257.

Gardner W K, Barber D A and Parbery D G 1983 The acquisition of phosphorus by *Lupinus albus* L. III. The probable mechanism by which phosphorus movement in the soil/root interface is enhanced. Plant and Soil 70, 107–124.

Gijsman A J 1989 PHGRAD, a simulation model for the description of pH gradients in the rhizosphere. (*In Dutch with English summary*). Dutch Priority Programme on Acidification, Report 83–02, RIVM, Bilthoven.

Gijsman A J 1990a Nitrogen nutrition of Douglas-fir (*Pseudotsuga menziesii*) on strongly acid sandy soil. I. Growth, nutrient uptake and ionic balance. Plant and Soil 126 (*In press*).

Gijsman A J 1990b Nitrogen nutrition of Douglas-fir (*Pseudotsuga menziesii*) on strongly acid sandy soil. II. Proton excretion and rhizosphere pH. Plant and Soil 126 (*In press*).

Hartikainen H 1986 Acid and base titration behaviour of Finnish mineral soils. Z. Pflanzenernaehr. Bodenkd. 149, 522–532.

Häussling M, Jorns C A, Lehmbecker G, Hecht-Buchholz Ch and Marschner H 1988 Ion and water uptake in relation to root development in Norway spruce (*Picea abies* (L.) Karst.). J. Plant Physiol. 133, 486–491.

Häussling M, Leisen E, Marschner H and Römheld V 1985 An improved method for non-destructive measurements of the pH at the root-soil interface (rhizosphere). J. Plant Physiol. 117, 371–375.

Hauter R and Mengel K 1988 Measurement of pH at the root surface of red clover (*Trifolium pratense*) grown in soils differing in proton buffer capacity. Biol. Fertil. Soils 5, 295–298.

Horst W J, Wagner A and Marschner H 1982 Mucilage protects root meristems from aluminium injury. Z. Pflanzenphysiol. 105, 435–444.

Ives D J G 1961 Oxide, oxygen and sulfide electrodes. *In* Reference Electrodes, Theory and Practice. Eds. D J G Ives and G J Janz. pp 336–351. Academic Press, New York, London.

Jayman T C Z and Sivasubramaniam S 1975 Release of bound iron and aluminium from soils by the root exudates of tea (*Camellia sinensis*) plants. J. Sci. Food Agric. 26, 1895–1898.

Kolthoff I M and Elving P J 1978 Treatise on Analytical Chemistry. Part 1, Volume 1. John Wiley and Sons, New York/Chichester/Brisbane/Toronto.

Li C Y, Lu K C, Trappe J M and Bollen W B 1972 Nitrate

reducing capacity of roots and nodules of *Alnus rubra* and roots of *Pseudotsuga menziesii*. Plant and Soil 37, 409–414.

Linnet N 1970 pH Measurements in Theory and Practice, 1st edition. Radiometer. Copenhagen.

Marschner H, Häussling M and Leisen E 1985 Rhizosphere pH of Norway Spruce trees grown under both controlled and field conditions. Proc. Workshop EC and Kernforschungsanlage Jülich GmbH: 'Effects of air pollution on terrestrial aquatic ecosystems'; working party 1, 113–118.

Marschner H, Römheld V and Ossenberg-Neuhaus H 1982 Rapid method for measuring changes in pH and reducing processes along roots of intact plants. Z. Pflanzenernaehr. Bodenkd. 105, 407–416.

Mulkey T J and Evans M L 1981 Geotropism in corn roots: Evidence for its mediation by differential acid efflux. Science 212, 70–71.

Nye P H 1981 Changes of pH across the rhizosphere induced by roots. Plant and Soil 61, 7–26.

O'Neill R A and Scott T K 1983 Proton flux and elongation in primary roots of barley. Plant Physiol. 73, 199–201.

Peuke A D 1987 Der Effekt von Schwefeldioxid-, Ozon- und Stickstoffdioxid-Begasung auf den Stickstoffmetabolismus steril kultivierter Fichtenkeimlinge (*Picea abies* (L.) Karst.). Ph.D. Thesis, University of Göttingen, FRG.

Pilet P E, Versel J M and Mayor G 1983 Growth distribution and surface pH patterns along maize roots. Planta 158, 398–402.

Riley D and Barber S A 1969 Bicarbonate accumulation and pH changes at the soybean (*Glycine max* (L.) Merr.) soil root interface. Soil Sci. Soc. Am. Proc. 33, 905–908.

Römheld V 1983 pH-Veränderungen in der Rhizosphäre in Abhängigkeit vom Nährstoffangebot. Landwirtschaftliche Forsch. Sonderheft 40, 226–230.

Römheld V 1986 pH-Veränderungen in der Rhizosphäre verschiedener Kulturpflanzenarten in Abhängigkeit vom Nährstoffangebot. Kali-Briefe (Büntehof) 18, 13–30.

Römheld V and Marschner H 1986 A simple method for non-destructive measurements of pH and root exudates at the root-soil interface (rhizosphere). Transactions of the XIII Congress Int. Soc. Soil Sci., Hamburg, pp 937–938.

Schaller G 1986 Einige Ursachen und Wirkungen der pH-Änderungen in der Rhizosphäre. Kali-Briefe (Büntehof) 18, 1–12.

Schaller G and Fischer W R 1981 Die Verwendung von Antimon-Elektroden zur pH-Messung in Boden. Z. Pflanzenernaehr. Bodenkd. 144, 197–204.

Schaller G and Fischer W R 1985 pH-Änderungen in der Rhizosphäre von Mais- und Erdnusswurzeln. Z. Pflanzenernaehr. Bodenkd. 148, 306–320.

Weisenseel M H, Dorn A and Jaffe L F 1979 Natural H^+ currents traverse growing roots and root hairs of barley (*Hordeum vulgare* L.). Plant Physiol. 64, 512–518.

M. L. van Beusichem (Ed.), *Plant nutrition – physiology and applications*, 53–59.

PLSO IPNC704

Effects of variation in nitrogen nutrition on growth of poplar (*Populus trichocarpa*) clones

A.D. PEUKE and R. TISCHNER
Institute of Plant Physiology, Georg-August University, Untere Karspüle 2, D-3400 Göttingen, FRG

Key words: balsam poplar, hydroponic culture, nitrogen nutrition, *Populus trichocarpa*

Abstract

Green cuttings of six balsam poplar clones were cultivated in a hydroponic medium in a growth chamber under controlled conditions. The nitrogen nutrition was varied with regard to concentration, nature of N-source and nitrate/ammonium ratio. Production of biomass, pH changes in the rhizosphere and the consumption of nitrate and ammonium were investigated. Balsam poplar is sensitive to NH_4^+. The plants grew best without or at low NH_4^+ concentrations. In NH_4^+-only nutrient solution (1.8 m*M*) the plants died within 3–6 weeks, dependent on the clone. In the nutrient solution, pH shifts were found to be correlated with variation in the use of the two N sources. We exclude acidification of the rhizosphere as sole reason for plant death.

Introduction

The last few decades balsam poplar, especially clones of *Populus trichocarpa*, has become more important for forestry in the Federal Republic of Germany. This is due to a greater ability of balsam poplar to adapt to site differences and low light intensity, in contrast to black poplar. For the six clones used in our experiments data of cultivation for several years are available. Thus clones "Columbia River" and "Brühl 6" showed average growth, whereas "Muhle Larsen" and "Scott Pauley" gave enhanced yields and "9/60" and "Heimburger" produced 30% less biomass.

In the present investigation we examined the effect of differences in N-nutrition on growth of poplar clones in hydroponic cultures. We measured biomass production (fresh weight), consumption of nitrogen and changes in pH of the nutrient solution. The aim of this and further investigations will be to detect differences in efficiency of nitrogen use.

Methods

Cuttings of balsam poplar clones (Muhle Larsen ML, Scott Pauley SP, Columbia River CR, Brühl 6 B6, 9/60 96 and Heimburger HB) were collected in the Hessische Forstliche Versuchsanstalt Hannoversch-Münden, West Germany. From these woody cuttings green cuttings were prepared. Shoot tips were cut off (8 cm) and all leaves, except the 2 youngest, were removed. The cuttings were placed in water containing 0.01 mg L^{-1} α-naphthaleneacetic acid. After 2 weeks they had rooted and were transferred to a hydroponic system.

On each black plastic pot containing 3.6 L continuously aerated nutrient solution, four cuttings were placed. The plants were cultivated for 6 weeks in a growth chamber at 22°C (18°C at night), 75% humidity (80% at night) and with a 16 h photoperiod (350–550 μE m^{-2} s^{-1} from basis to top of the plants). The standard nutrient solution contained 0.3 m*M* NH_4^+ and 1.5 m*M* NO_3^- (1.8 m*M* total nitrogen) and had the following composition (μM), based on Long Ashton solution, Hewitt, 1966: KNO_3 400, $Ca(NO_3)_2$ 400, NH_4NO_3 300, NaH_2PO_4 150, $MgSO_4$ 150, $FeCl_3$ 10, NaCl 17 and 1/5 Long Ashton trace elements pH 5.5. This standard solution was varied with regard to N concentration, N source and NO_3^-/NH_4^+ ratio, as indicated in the figures. Ammonium was added as a mixture of NH_4Cl and $(NH_4)_2SO_4$ (4:1 mol ratio). In nutrient solutions with reduced NO_3^- concentrations, cation concentrations were compensated for by adding KCl and $CaCl_2$.

The nutrient solutions were changed twice weekly, and pH changes and daily ion uptake values were calculated from measurements made on the spent solution. Nitrate was determined by HPLC and NH_4^+ and potassium by ion chromatography (Peuke, 1987).

Results

Plant growth

With variations in N concentration (5 NO_3^-:1 NH_4^+ ratio) between 0.36 and 9 mM no influence on the growth of the balsam poplar clones was observed (Fig. 1). Only at 1.8 mM N B6 and CR produced significantly higher yields (Fig. 1b).

When the N-source was varied (N-concentration constant at 1.8 mM) the best growth was obtained with sole NO_3^- nutrition, compared to a combination of NO_3^- and NH_4^+ (Fig. 2). If only NH_4^+ was supplied the plants died. After 10 days the first symptoms of chlorosis and root damage occurred, and after 3 weeks the first plants of HB and 96 died (Fig. 2c).

In the experiment with varying NO_3^-: NH_4^+ ratios the plants grew best at the lowest NH_4^+ concentration (0.15 mM, Fig. 3a). This effect became pronounced after 4 weeks. The clones CR and HB were damaged most severely by higher NH_4^+ concentration. HB died within the cultivation time at the highest NH_4^+ concentration (0.9 mM). In contrast, ML grew as well in this nutrient solution as at higher NO_3^- concentrations.

pH changes

The changes in pH of the nutrient solutions were as observed previously with other plants: when NH_4^+ was used the pH decreased, and pH increases occurred with NO_3^- (Fig. 4–6). Acidification was more pronounced when more NH_4^+ was

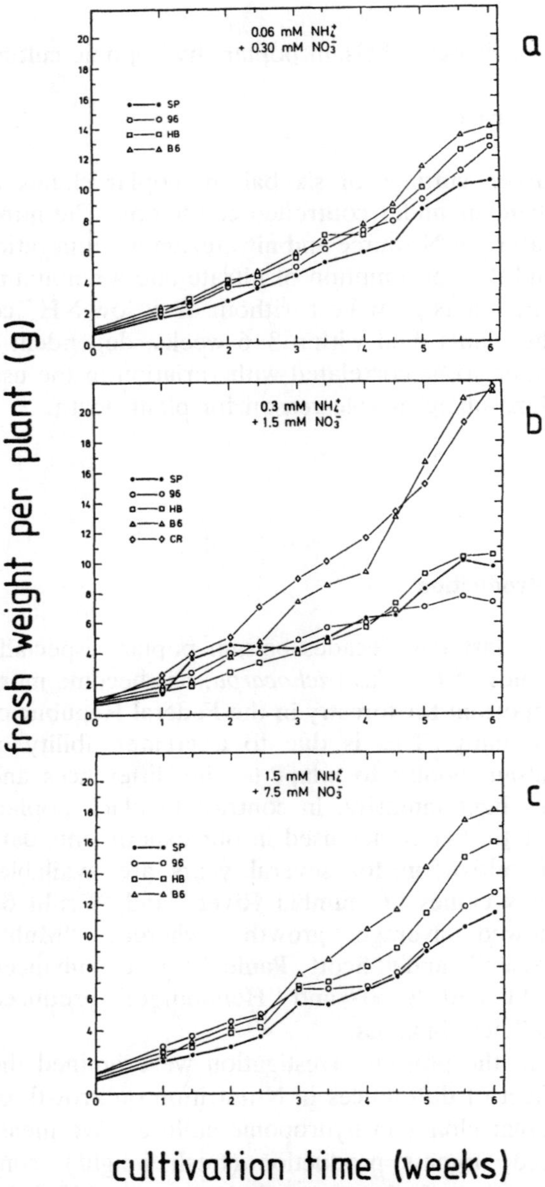

Fig. 1. Average fresh weight production per plant, as dependent on variation in total N-concentration ($NO_3^- + NH_4^+ =$ 5 + 1).

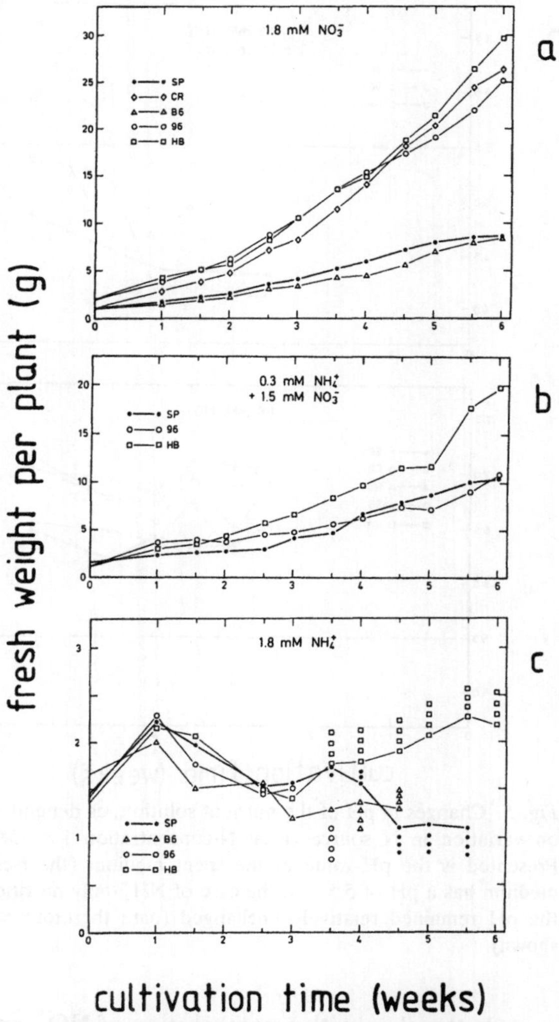

Fig. 2 Average fresh weight production per plant, as dependent on variation in N source (total N-concentration 1.8 m*M*). Curves end before week 6, when plants have died.

supplied (Fig. 6), but with NH_4^+ as sole N source the pH remained relatively unchanged.

Consumption of nitrogen

pH changes were correlated with rates of N consumption. When both N sources were supplied, NH_4^+ was used first (in some cases NH_4^+ was taken up totally) and the consumption of NO_3^- was inhibited (Fig. 8). Later when the plants became larger, NO_3^- was used more strongly (Fig. 9). Exceptions were the experiments with high NH_4^+ concentrations (0.9 and 1.5 m*M*, Fig. 7, 9 and 10). Here N consumption (both NO_3^- and NH_4^+) and changes in pH rested relatively on the same level within the cultivation period. In the experiments with 1.5 m*M* + 0.3 m*M* and 1.65 m*M* + 0.15 m*M* NO_3^- and NH_4^+ ML and CR grew similarly, but CR needed twice as much nitrogen to reach the same biomass production (1.6 mmol N per g fresh weight).

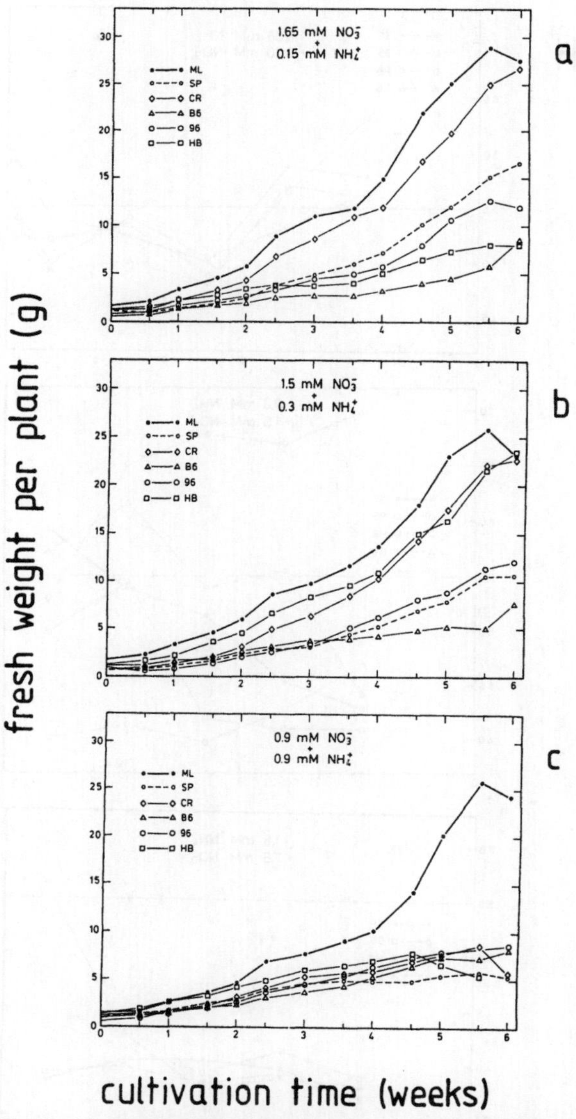

Fig. 3. Average fresh weight production per plant, as dependent on variation in NO_3^- : NH_4^+ ratio (total N-concentration 1.8 m*M*). In one case, "Heimburger" died after 5.5 weeks.

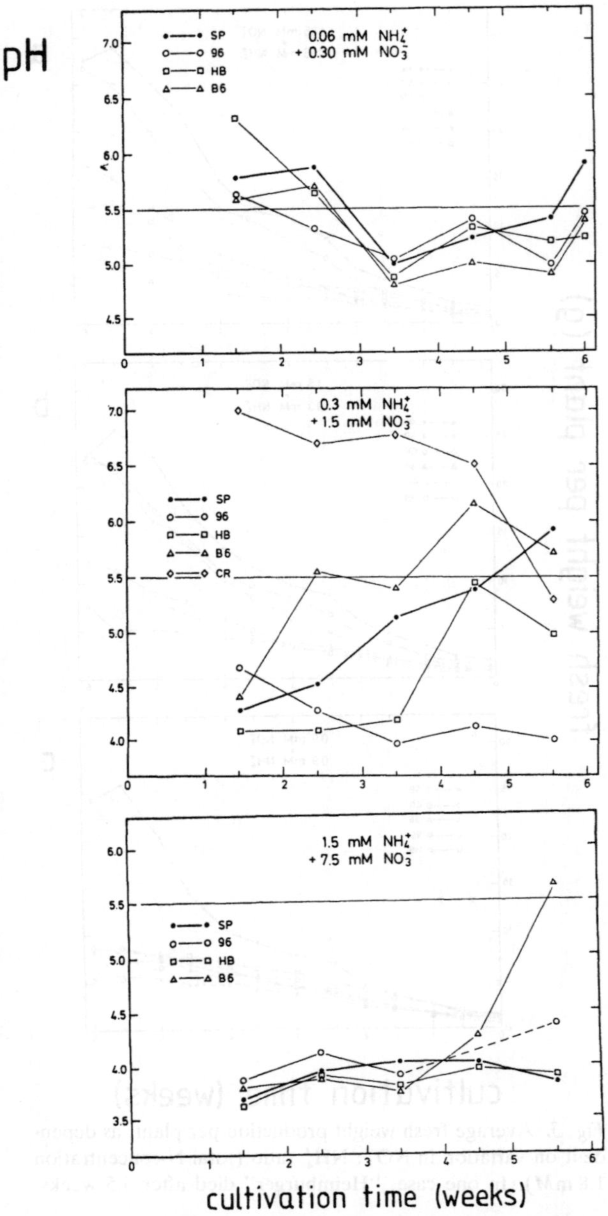

Fig. 4. Changes in pH of the nutrient solution, as dependent on variation in total N-concentration ($NO_3^- + NH_4^+ = 5 + 1$). Presented is the pH-value of the spent medium (the fresh medium has a pH of 5.5).

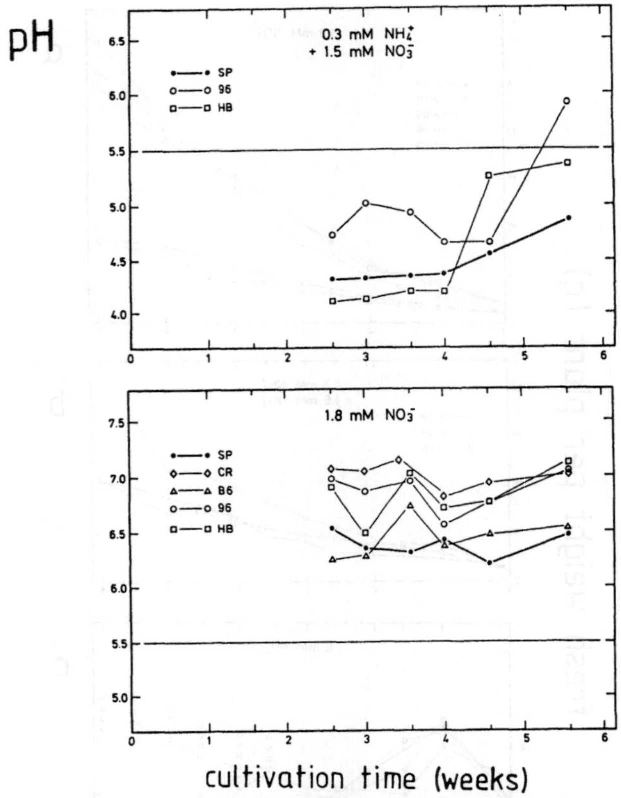

Fig. 5. Changes in pH of the nutrient solution, as dependent on variation in N source (total N-concentration 1.8 m*M*). Presented is the pH-value of the spent medium (the fresh medium has a pH of 5.5). In the case of NH_4^+-only nutrition the pH remained relatively unchanged (data therefore not shown).

Discussion

Balsam poplar is sensitive to NH_4^+. In experiments with combined NO_3^- and NH_4^+ nutrition growth was best with low NH_4^+ concentration in the nutrient solution. With NO_3^- only the clones grew better than with a combination of NO_3^- and NH_4^+. In a NH_4^+-only medium the plants died within six weeks. The clones responded differently to NH_4^+, with 96 and HB being the most sensitive clones.

Evers (1963) found similar effects on growth of *Populus euramericana* cv. Missouriensis when varying the N-nutrition in sand culture. However, in these experiments the negative effect of NH_4^+-only nutrition was not present at pH 8.0. The two N sources are known to have contrasting effects on pH in the rhizosphere and intracellularly, on cation – anion balance, on mineral cation uptake and on energy – and carbohydrate metabolism (Marschner, 1986; Runge, 1983). Based on previous results, rhizosphere acidification cannot be seen as sole cause of NH_4^+ toxicity in balsam poplar. The pH could decline

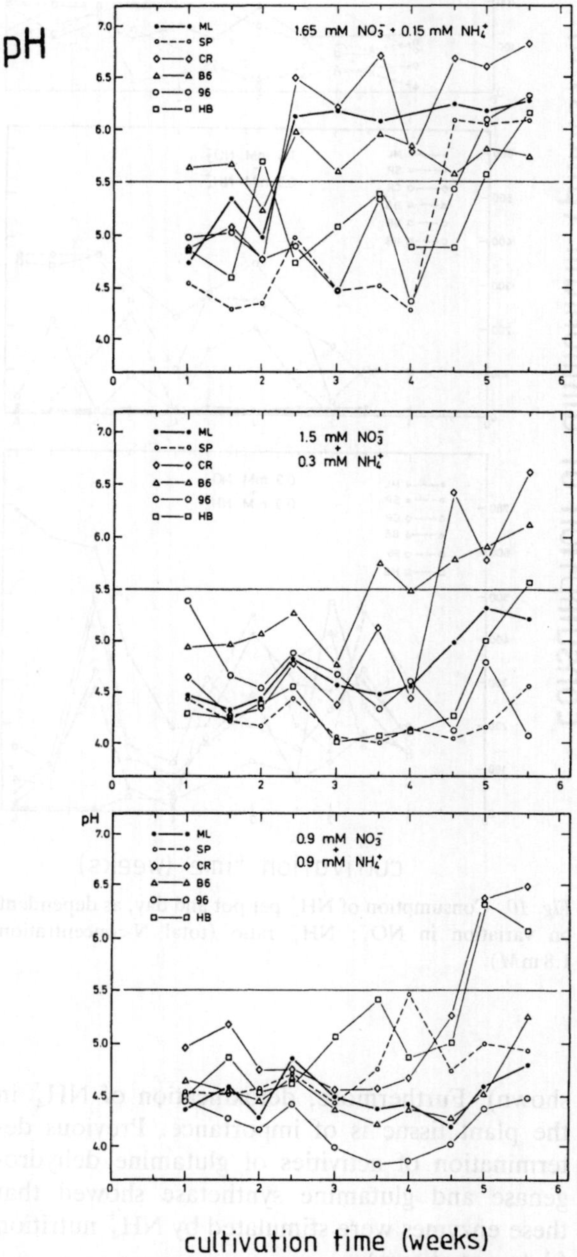

Fig. 6. **Changes in pH of the nutrient solution, as dependent on variation in NO_3^-: NH_4^+ ratio (total N-concentration 1.8 m*M*). Presented is the pH-value of the spent medium (the fresh medium has a pH of 5.5).**

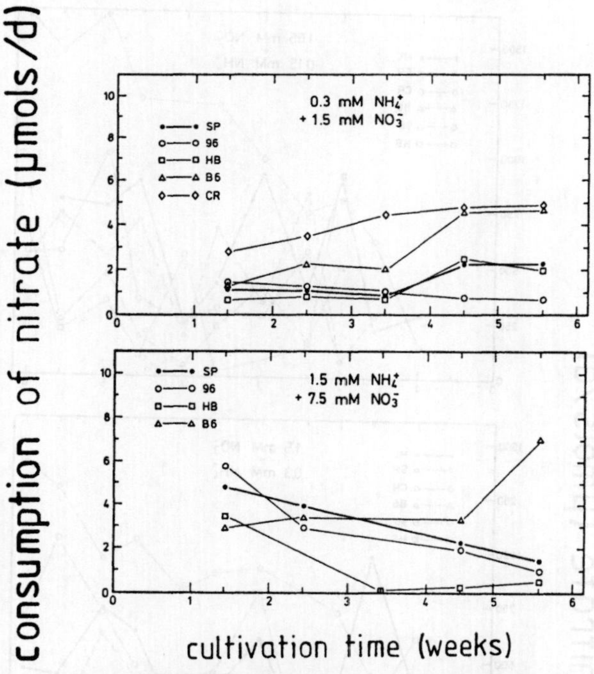

Fig. 7. Consumption of NO_3^- per pot and day, as dependent on variation in total N-concentration ($NO_3^- + NH_4^{+-} = 5 + 1$). In the case of 0.36 m*M* N NO_3^- was taken up totally from the medium (data therefore not shown).

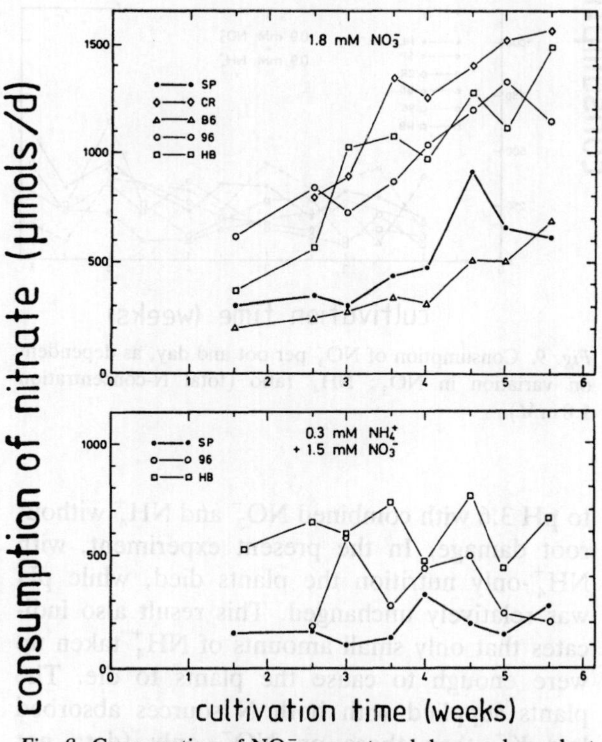

Fig. 8. Consumption of NO_3^- per pot and day, as dependent on variation in N source (total N-concentration 1.8 m*M*).

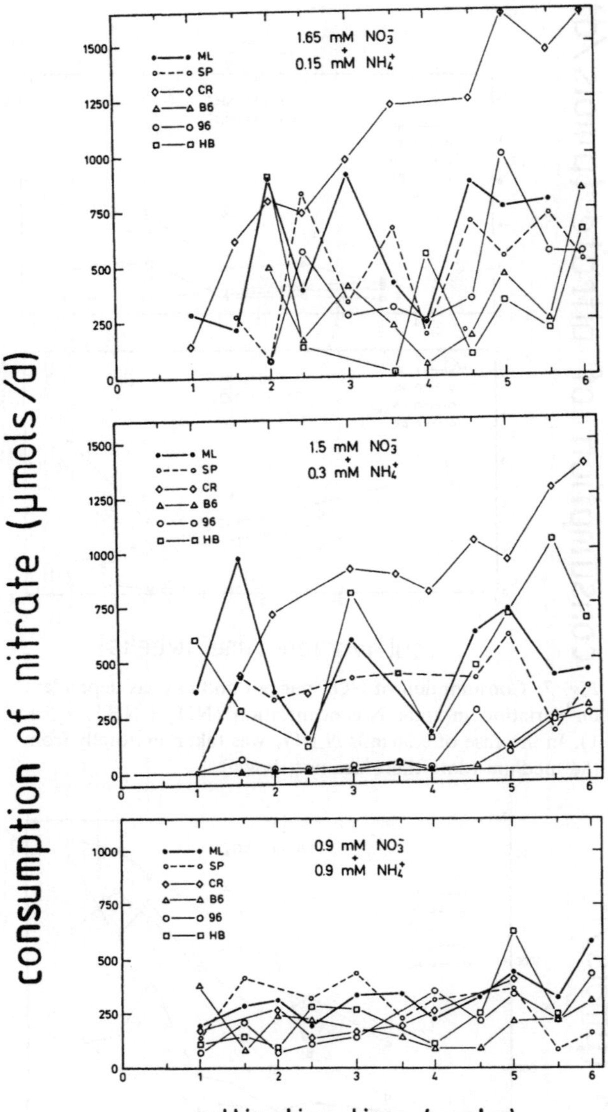

Fig. 9. Consumption of NO_3^- per pot and day, as dependent on variation in NO_3^-: NH_4^+ ratio (total N-concentration 1.8 m*M*).

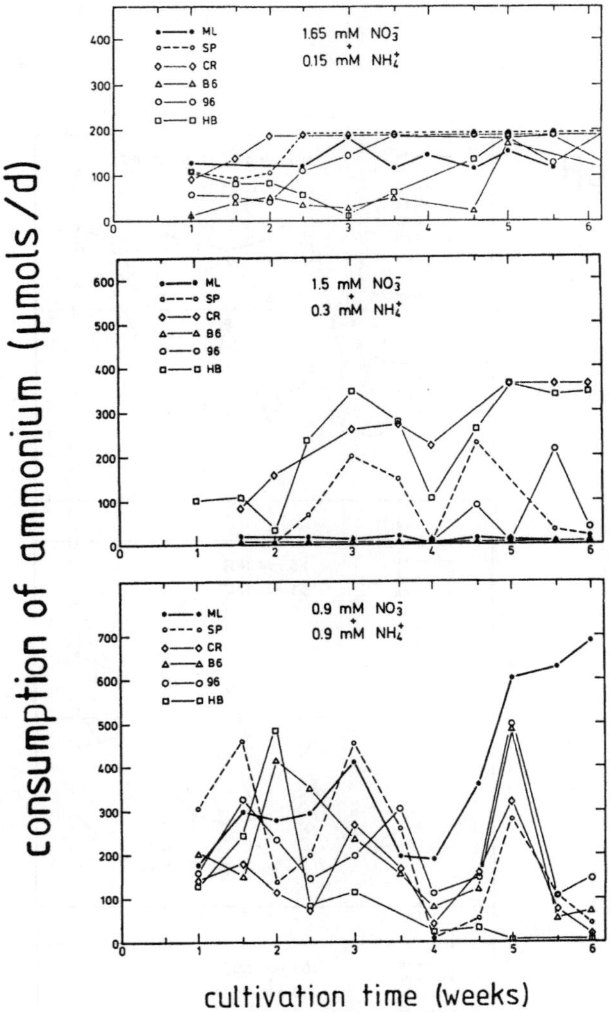

Fig. 10. Consumption of NH_4^+ per pot and day, as dependent on variation in NO_3^-: NH_4^+ ratio (total N-concentration 1.8 m*M*).

to pH 3.6 with combined NO_3^- and NH_4^+ without root damage. In the present experiment, with NH_4^+-only nutrition the plants died, while pH was relatively unchanged. This result also indicates that only small amounts of NH_4^+ taken up were enough to cause the plants to die. The plants supplied with both N sources absorbed less K^+ than those on NO_3^- only (data not shown). Furthermore, detoxification of NH_4^+ in the plant tissue is of importance. Previous determination of activities of glutamine dehydrogenase and glutamine synthetase showed that these enzymes were stimulated by NH_4^+ nutrition (data not shown).

The cause of the NH_4^+ toxicity in balsam poplar is yet not clear, multicausal effects cannot be excluded. In further experiments, metabolic changes caused by NH_4^+ and detoxification mechanisms will be investigated.

Acknowledgements

This work was supported by a grant from the Bundesministerium für Forschung und Technologie, Federal Republic of Germany. We thank the Hessische Forstliche Versuchsanstalt, Hann.-Münden for delivery of balsam poplar cuttings.

References

Evers F H 1963 Die Wirkung von Ammonium- und Nitratstickstoff auf Wachstum und Mineralstoffhaushalt von Picea und Populus: Das Wachstum bei abgestufter Azidität und verschiedener Ca-Konzentration im Nährmedium. Z. Bot. 51, 61–79.

Hewitt E J 1966 Sand and Water Culture Methods Used in the Study of Plant Nutrition. Commonw. Bur. Hortic. Plant Crops G. B. Tech. Commun. 22.

Marschner H 1986 Mineral Nutrition in Higher Plants. Academic Press, London, 674 p.

Peuke A D 1987 Der Effekt von Schwefeldioxid-, Ozon- und Stickstoffdioxid-Begasung auf den Stickstoffmetabolismus steril kultivierter Fichtenkeimlinge (*Picea abies* (L.) Karst.). Dissertation, Göttingen, 105 p.

Runge M 1983 Physiology and ecology of nitrogen nutrition. *In* Encyclopedia of Plant Physiology 12 C Physiological Plant Ecology. III. Eds. O L Lange, P S Nobel, C B Osmond and H Ziegler. pp 163–200. Springer-Verlag, Berlin, Heidelberg. New York.

M. L. van Beusichem (Ed.), *Plant nutrition – physiology and applications*, 61–67. PLSO IPNC153

Adaptation of growth rate of *Populus euramericana* to light and nitrate proceeds via the vascular system

G.A. PIETERS and M.E. VAN DEN NOORT
Department of Plant Physiological Research, Wageningen Agricultural University, Gen. Foulkesweg 72, 6703 BW Wageningen, The Netherlands

Key words: absolute growth rate, adaptation, growing shoot, irradiance, nitrate, *Populus euramericana* 'Robusta', vascular system

Abstract

Shoots of *Populus euramericana* (Dode) Guinier cv. 'Robusta' were cultivated in subirrigated gravel culture at irradiances of 7.5 and 30.0 W m^{-2} and at 3 nitrate dosages in proportionality to irradiance. It is shown that (i) a poplar shoot adapts to a linear nitrate dosage in a similar way as to irradiance, viz. via adaptation of the volume of the growing shoot (GS), (ii) nitrate concentration of the organs, chlorophyll content, nitrate reductase activity of the leaves and shoot-root ratio are mainly a function of the ratio irradiance/nitrate-dosage (I/N-ratio) and (iii) the proposed morphogenic model provides a better basis for comprehension of these effects than the exponential model. It is suggested that the volume of GS is a reflection of the development of the vascular system. The constancy of the optimal I/N-ratio may point to a morphogenetic significance of protein synthesis for the enlargement of the vascular system.

Introduction

Ingestad (1982) gave a new impulse to the research on the relation between mineral nutrition and plant growth by designing the method of the relative addition rate coupled to the relative growth rate (RGR) of the whole plant to maintain a constant nitrate concentration. He also showed that in a young plant RGR and nitrate content are linearly correlated up to a certain optimum. The method of Ingestad is based on a model, assuming exponential growth. It appeared, however, that the period of exponential growth is short. Ingestad and Lund (1986) explain this by selfshading, although it has to be ascribed also to a declining ratio of growing and mature tissue. This model of growth has therefore a limited significance.

Pieters (1986) and Pieters and Van den Noort (1985) developed a morphogenic model of growth, in which each individual shoot, after a period of increasing absolute growth rate during adaptation to a given irradiance, continues to grow linearly with time. This model has a more general significance.

Preliminary results (Pieters and Van den Noort, 1985) indicated that sunflower and poplar adapt to a given, limiting and linear nitrate dosage in a similar way as to irradiance. Consequently, it must be possible to reach a steady state by growing shoots with a constant absolute growth rate, adapted to the limiting and linear nitrate dosage. While Ingestad adapts the nitrate dosage to the growth of the plants, in our experiments plants should adapt to the nitrate dosage. Pieters and Van den Noort (1985) found also that the nitrate requirement of a shoot is linearly related to irradiance (cf. Ingestad and Lund, 1986).

The morphogenic model of growth is based on an intensive study of the growth of individual leaves and internodes (Pieters and Van den Noort, 1988). It was established that these organs grow according to organ-specific patterns.

A growth pattern can be defined as a 'fixed' relation between relative growth rate and age of the organ. If growth patterns are constant, differences in final size of *e.g.* leaves can only be explained by differences in the size of the initiated leaf primordia: primordia with equal size do not necessarily have an equal age.

Growth analysis also revealed that the initiation rate (1/P, leaves/day) of leaves with larger mature size (L_m, cm) was proportionally faster. At a temperature of 20°C this relation was: $1/P = a \times L_m$, in which a is 0.042 leaves/(day × cm). Consequently the plastochron (P, =time unit between repetitive events) is inversely proportional to mature leaf length (L_m). Leaf area production rate (A) equals the production of one mature leaf per plastochron, thus:

A = c × leaf length × leaf width/plastochron, in which: c = 0.65 (leaf area coefficient, Pieters and Van den Noort, 1985). Because also leaf width is proportional to L_m, A is proportional to L_m^3 or to a volume.

Since primordia are produced by the apex, we concluded that this volume might reflect the volume of the apex and measurement of the growing part of the primary stem (the growing shoot, GS) corroborated this (Pieters and Van den Noort, 1988). Mean RGR, calculated over GS, is a measure of the absolute growth rate of the average cell and remained constant throughout the whole growth period of the shoot. Differences in absolute growth rate of the shoot are due to the increase of the volume of GS. A larger apex produces proportionally larger primordia at a proportionally larger rate. In other experiments we found that diameter and length of GS increased linearly with time with a rate proportional to irradiance: the volume of GS is a power function of its linear dimensions and consequently total leaf and stem weight per shoot do not increase exponentially. When the increase of the size of GS stops, a shoot grows linearly with time. The increasing volume of GS was correlated with an increase of phyllotactic order.

There is a striking resemblance between the development of GS and that of the vascular system as described by Larson (1975, 1977, 1980). In a seed the vascular system contains two vascular bundles. During the development of the seedling the number of vascular bundles increases by systematic branching of the bundles to 3, 5, 8 and 13, while phyllotactic order increases to 1/3, 2/5, 3/8 and 5/13. At the same time the length and presumably also the diameter of each individual bundle increases proportionally. All this can fully explain the increase of the size of GS. The relation between structure and function is that bundles with a larger diameter produce larger primordia, while the correlated increase in the number of bundles increases leaf initiation rate. On the average leaf production rate per bundle appeared to be constant. It is generally accepted that growth vigour and phyllotactic order are correlated.

Larson (1975) also observed that long before a new primordium is visible at the apex, its primordial vascular bundle (procambial trace) branches off from the preceding mother trace deeper down under the apical dome and develops acropetally through the apical tissue. We may define the moment of initiation of a new leaf primordium as the moment of branching off of a new procambial trace. There are more observations, indicating that the development of the vascular system precedes the development of the shoot. Therefore we suggest that growth is a reflection of the development of the vascular system.

The aim of this report is to show that (i) the morphogenic model of growth is a better basis for understanding the effects of linear dosaging on plant properties than the exponential model; (ii) the adaptation of plant growth to a linear nitrate dosage proceeds via GS and (iii) the irradiance/nitrate-dosage is a determinant of deficiency symptoms.

Methods

Fresh cuttings (with low levels of reserves) of *Populus euramericana* (Dode) Guinier c.v. Robusta, on which one shoot was allowed to grow, were cultivated in growth rooms at 22°C, 60% R.H. and a day length of 16 h on a nitrate-free subirrigated gravel culture solution. Nitrate was added daily as a solution of KNO_3, $Ca(NO_3)_2$ and $Mg(NO_3)_2$ (molar ratio: 35:45:20), according to the scheme in Table 1.

Table 1. Scheme of experimental treatments

Irradiance		Code for I/N-ratio[a]		
		1	2	3
		Nitrate-dosage		
$W\,m^{-2}$	$\mu mol\,m^{-2}\,s^{-1}$	mmol	$shoot^{-1}$	day^{-1}
7.5	32.5	1.0	0.25	0.0
30.0	130.0	4.0	1.0	0.0

[a] Note that the ratio between irradiance and nitrate dosage is equal for both irradiances at each code.

The shoots were irradiated from above and from two sides. At the 2 highest nitrate dosages 6 shoots and at the lowest nitrate dosage 4 shoots per treatment were used. Three times per week leaf length and diameter and length of internodes were measured. The other measurements were performed on harvested shoots: two times at 0.0 mmol nitrate and three times at the higher nitrate dosages.

Chlorophyll determination was performed after Bruinsma (1963), the chemical analyses after methods described by the Department of Soil Science and Plant Nutrition, Agricultural University, Wageningen.

Results

Although there were large differences in plant dry weight between the different treatments, relative growth rates at half mature length (RGR_{50}) of leaves and internodes (Table 2) were similar, except at a nitrate dosage of 0.0 mmol nitrate $shoot^{-1}$ day^{-1} (code-3).

The amount of nitrate remaining in the nutrient solution just before the weekly change, appeared to diminish with the increasing growth rate of the plants. Adding up the weekly nitrate contents of the nutrient solution and total nitrate content of the harvested plants per treatment, recovery percentages between 95 and 106% were found at the two highest nitrate dosages (code 1 and 2). For plants grown in nitrate-free solution the recovery percentages were 97 and 57%. This accentuates the importance of uniformity in the plant material at the start. Small differences in the thickness of a cutting lead to large differences in the ratio between the nitrate poor wood and the nitrate rich cortex tissue. Because we calculated an average nitrate concentration per g dry weight of cutting, large errors resulted in the recovery percentage of plants which did not get any additional nitrate.

The development of the dimensions of the stems at different treatments in the course of time is shown in Figure 1. Each curve represents the form of the stem at a given measurement date; normally the measurements were done three times per week, as discernible from the figures. Each curve can be divided into a part with primary (GS) and a part with secondary growth, as indicated by the fitted line through measurement points, which are not shown. The higher irradiance and nitrate dosage, the higher was the rate of enlargement of GS. At the nitrate dosage of 0.0 mmol nitrate $shoot^{-1}$ day^{-1} GS became smaller during the short period of growth before dormancy was induced.

The distribution of N_{total} concentration over the various organ groups (Fig. 2), chlorophyll content (Fig. 3) and nitrate reductase activity (NRA) of the leaves (Fig. 4) and shoot-root ratio (Fig. 5) are related to the I/N-ratio, as can be seen by comparing the right and the left side of the figures. A complication is that some plants at 7.5 W m^{-2} became dormant at limiting nitrate dosage.

Table 2. Mean plant dry weight in g and relative growth rates at half mature length RGR_{50} of leaves and internodes in % per day

	Irradiance ($W\,m^{-2}$)					
	7.5	30.0	7.5	30.0	7.5	30.0
			RGR_{50} of			
Code	Dry weight		Leaves		Internodes	
1	22.2	75.1	14.9	15.5	22.4	22.8
2	17.0	54.9	14.9	15.9	22.1	21.9
3	4.5	4.9	13.1	12.9	16.5	18.0

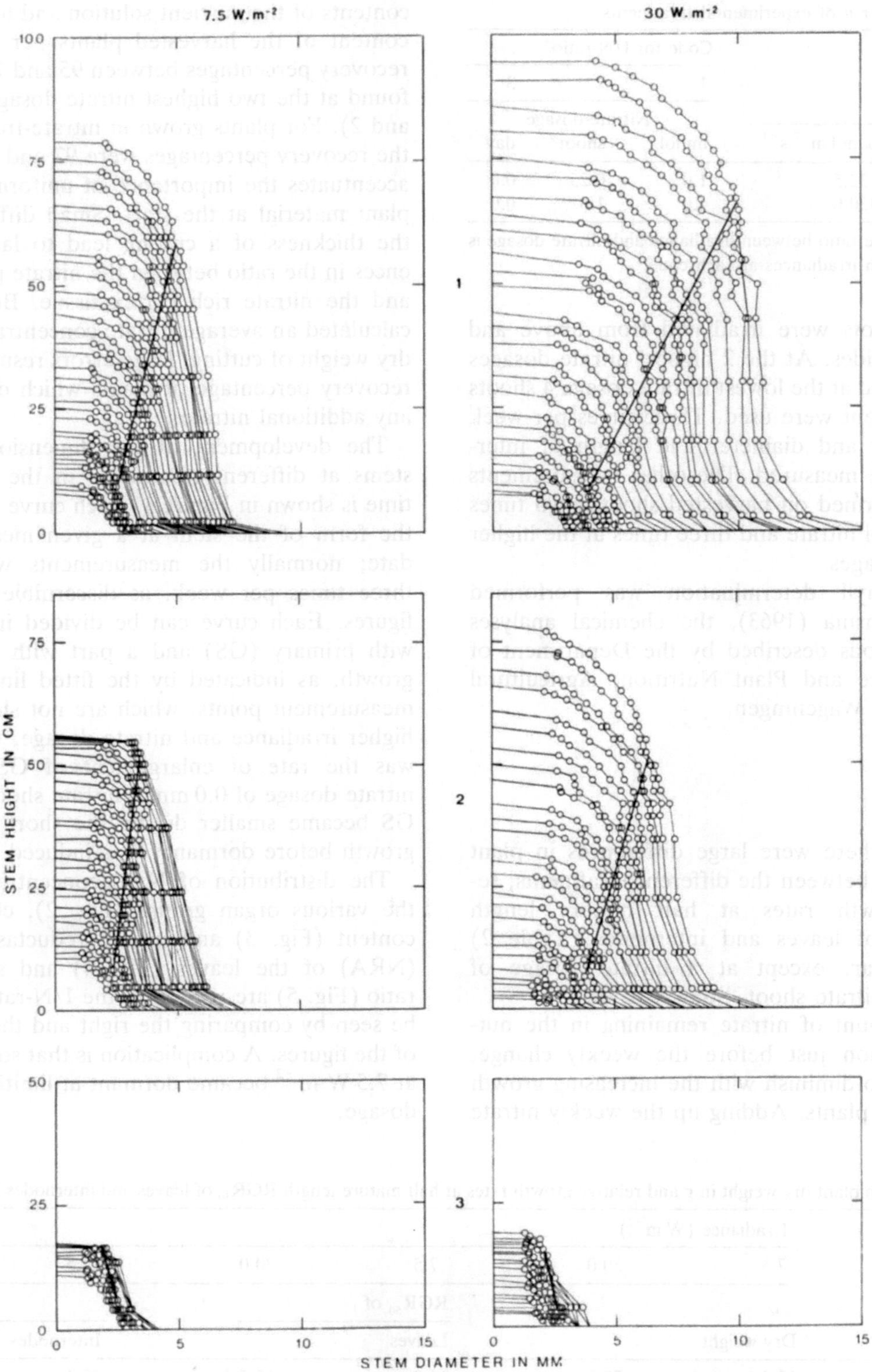

Fig. 1. The development of stem form in shoots from various treatments in the course of time. Each curve represents the mid-internode diameter versus cumulative internode length at a certain measurement date. The straight lines represent the base of the growing shoot (GS) and thus also the border between primary and secondary growth. Numbers between Figures indicate code for I/N ratio.

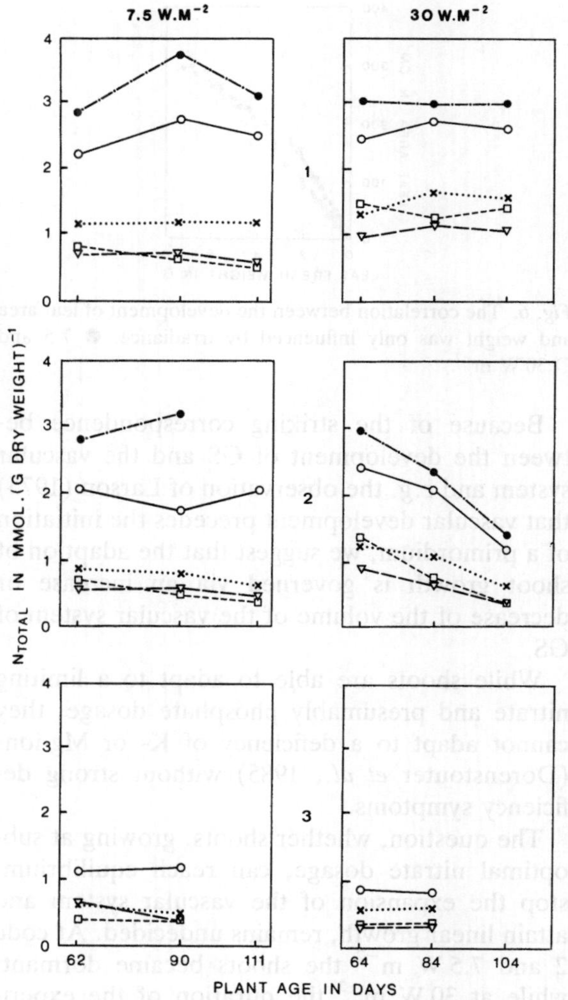

Fig. 2. Distribution of N_{total}-concentration over organ groups of plants of the various treatments against shoot age. Legends: ○ mature leaves; ● growing leaves; ▽ petioles; □ stem; × roots. For numbers between Figures: See Figure 1.

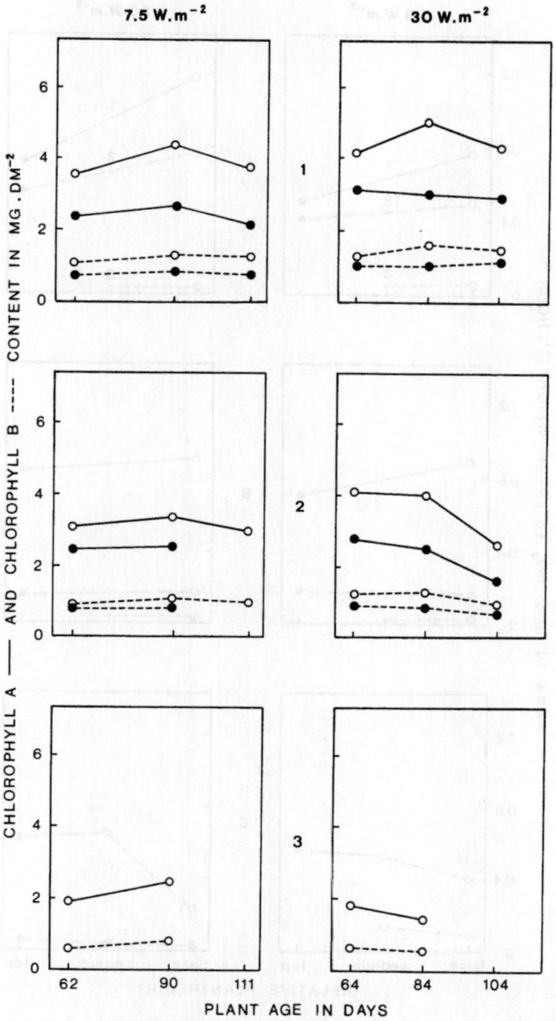

Fig. 3. Chlorophyll a and b content of growing (●) and mature (○) leaves of plants of the various treatments against shoot age. For numbers between Figures: See Figure 1.

Under conditions of strong nitrate deficiency the growing leaves tended to get a higher NRA than the mature leaves. The NRA of the roots is negligible.

The development of leaf weight is strongly related to the development of leaf area (Fig. 6). The slope of the correlation is a measure of leaf thickness and depended only on irradiance.

Discussion

The growth patterns of leaf length and width and internode length and diameter were independent of irradiance and nitrate dosage (Table 2). In many experiments we observed that shoots, growing in constant conditions, maintained their growth patterns or became dormant. Shoots, cultivated at 0.0 mmol nitrate $shoot^{-1}$ day^{-1}, however, were not able to maintain their growth patterns, presumably because the nitrate reserve in the cutting was used up before maturation and redistribution of nitrate could not sustain a high RGR_{50}. The adaption of absolute growth rate to irradiance, nitrate- and presumably (Ingestad, 1987) phosphate-dosage thus proceeds mainly via the volume of the growing shoot (GS, Fig. 1), as expected in the morphogenic model.

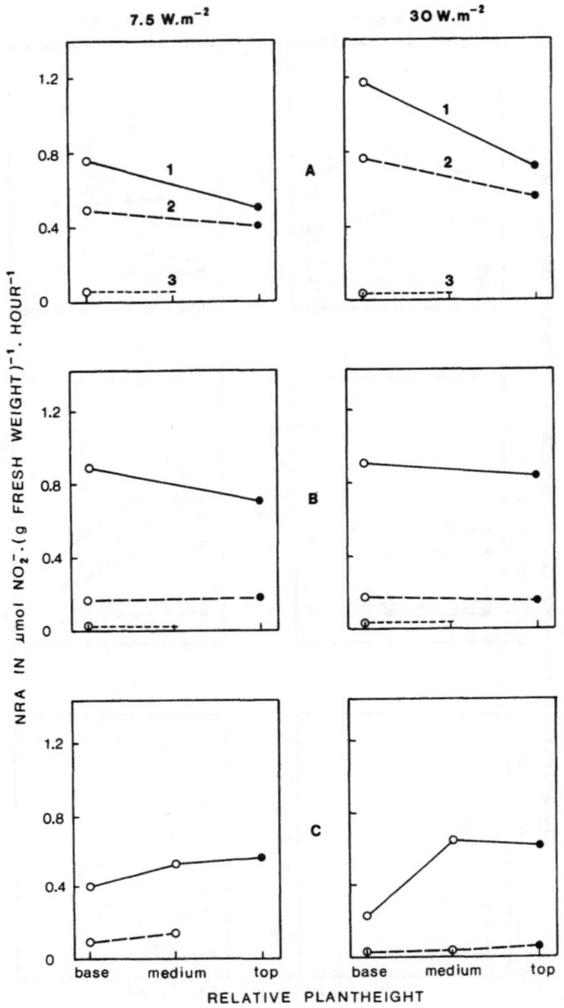

Fig. 4. Nitrate reductase activity (NRA) in growing (top, ●), respectively young (medium, ○) and old (base, ○) mature leaves of plants of the various treatments. **A**, **B** and **C** indicate the first, second and third harvest.

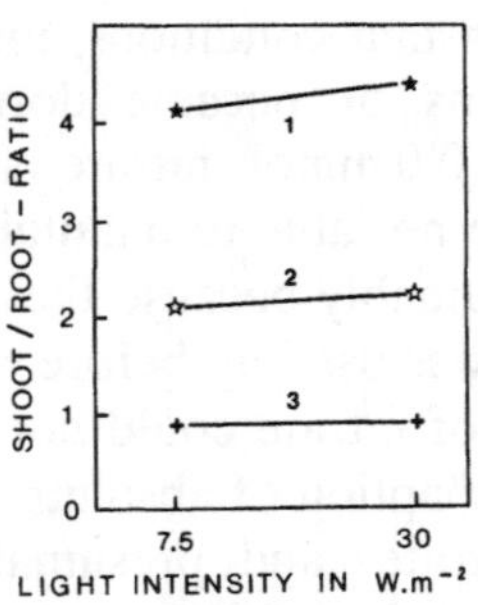

Fig. 5. The shoot-root ratio of plants of the various treatments was determined mainly by irradiance/nitrate ratio (I/N-ratio). Last harvest.

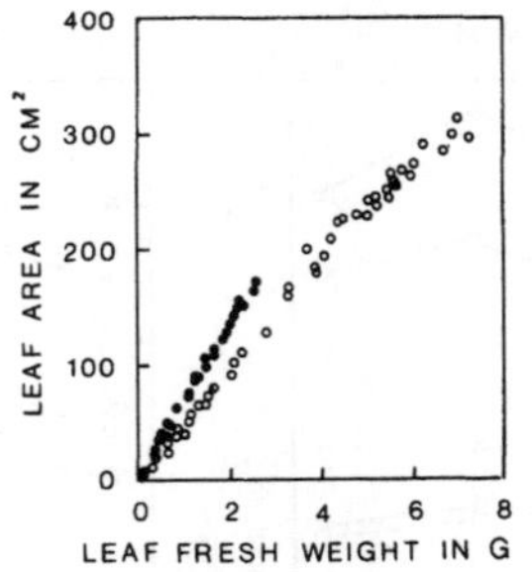

Fig. 6. The correlation between the development of leaf area and weight was only influenced by irradiance. ● 7.5 and ○ 30 W m^{-2}.

Because of the striking correspondence between the development of GS and the vascular system and *e.g.* the observation of Larson (1975) that vascular development precedes the initiation of a primordium, we suggest that the adaption of shoot growth is governed via an increase or decrease of the volume of the vascular system of GS.

While shoots are able to adapt to a limiting nitrate and presumably phosphate dosage, they cannot adapt to a deficiency of K- or Mg-ions (Dorenstouter *et al.*, 1985) without strong deficiency symptoms.

The question, whether shoots, growing at suboptimal nitrate dosage, can reach equilibrium, stop the expansion of the vascular system and attain linear growth, remains undecided. At code 2 and 7.5 W m^{-2} the shoots became dormant, while at 30 W m^{-2} the duration of the experiment was too short. As expected, also the shoots, grown at the highest nitrate dosage (code 1), did not reach linear growth, because an ever increasing illuminated leaf area (see Methods) leads to a continuous increase of GS within the duration of the experiment. Dry matter distribution and chemical composition and activity of the plants do not depend on the absolute irradiance or nitrate- (presumably also phosphate-) dosage, but on the ratio of these factors, as indicated by the code. In the present experiments the lowest I/N-ratio was the best and, in comparison to plants from other experiments growing without any nitrate limitation, the plants were growing nearly optimally. This means that there exists an optimal I/N-ratio. Although we did not yet analyze this ratio in respect to the resulting relative amounts of C-, N- an P-compounds in the plant, we suggest that the constancy of the optimal ratio points to a morphogenetic significance of protein synthesis for the enlargement of the vascular system.

Linear dosaging of nitrate in relation to irradiance is an interesting alternative for studying the effects of ion nutrition on the growth of plants.

Acknowledgement

The members of the Department of Soil Science and Plant Nutrition of Agricultural University, Wageningen are duly acknowledged for their help and advice.

References

Bruinsma J 1963 The quantitative analysis of chlorophylls a and b in plant extracts. Photochem. Photobiol. 2, 241–249.

Dorenstouter H, Pieters G A and Findenegg G R 1985 Distribution of magnesium between chlorophyll and other photosynthetic functions in magnesium deficient 'sun-' and 'shade-' leaves of poplar. J. Plant Nutr. 8, 1089–1101.

Ingestad T 1982 Relative addition rate and external concentration: Driving variables used in plant nutrition research. Plant Cell Environ. 5, 443–453.

Ingestad T and Lund A 1986 Theory and techniques for steady state mineral nutrition and growth of plants. Scand. J. For. Res. 1, 439–453.

Ingestad T 1987 New concepts on soil fertility and plant nutrition as illustrated by research on forest trees and stands. Geoderma 40, 237–252.

Larson P R 1975 Development and organization of the primary vascular system in *Populus deltoides* according to phyllotaxy. Am. J. Bot. 62, 1084–1099.

Larson P R 1977 Phyllotactic transitions in the vascular system of *Populus deltoides* Bartr. as determined by ^{14}C-labeling. Planta 134, 241–249.

Larson P R 1980 Interrelations between phyllotaxis, leaf development and primary-secondary vascular transition in *Populus deltoides*. Ann. Bot. 46, 757–769.

Pieters G A 1986 Dimensions of the growing shoot and the absolute growth rate of a poplar shoot. Tree Physiol. 2, 283–288.

Pieters G A and Van den Noort M E 1985 Leaf area coefficient of some *Populus euramericana* strains. Photosynthetica 19, 189–193.

Pieters G A and Van den Noort M E 1988 Effect of irradiance and plant age on the dimensions of the growing shoot of poplar. Physiol. Plant. 74, 467–472.

M. L. van Beusichem (Ed.), *Plant nutrition – physiology and applications*, 69–72. PLSO IPNC621

Optimal nutrition in two forest stands exposed to acid atmospheric deposition

P.H.B. DE VISSER and N. VAN BREEMEN
Department of Soil Science and Geology, Wageningen Agricultural University, P.O. Box 37, 6700 AA Wageningen, The Netherlands

Key words: acid atmospheric deposition, irrigation, nitrogen supply, optimal nutrition, *Pinus sylvestris* L., *Pseudotsuga menziesii* (Mirb.) Franco

Abstract

A Douglas fir (*Pseudotsuga menziesii* (Mirb.) Franco) and a Scots pine (*Pinus sylvestris* L.) stand on acidified soils are manipulated by means of irrigation and optimal nutrition. The aim is to assess effects of different soil conditions on tree growth in order to quantify indirect effects of acid deposition on trees. High atmospheric inputs of nitrogen in the soil are observed, resulting in soil acidification, high N availability, high aluminium concentrations and unbalanced nutrient supply. The soil solution composition is changed gradually in the manipulated soils. After one year treatment no significant growth differences have been observed yet.

Introduction

The effects of adverse soil conditions on forest ecosystems, resulting from acid atmospheric deposition are difficult to quantify with present-day knowledge. A major problem is the lack of variation in deposition rates at one place, including a reference situation without air pollution. Optimizing the soil conditions for tree growth and preventing the acid deposition from reaching the soil are ways to simulate a soil reference situation. In such a plot only direct effects of air pollution on foliage will affect tree growth. These direct effects and the interactions between direct effects and indirect, soil related effects of air pollution can then be studied.

This paper describes a method that aims at elimination of adverse soil conditions in forest ecosystems affected by acid atmospheric deposition. In forests in the Netherlands high atmospheric inputs of nitrogen and sulphur result in soil acidification, unbalanced nutrient supply and high concentrations of aluminium in the soil solution. When deficiencies of water and nutrients are eliminated as growth limiting factors, with or without acid deposition inputs, the effects of controlled, optimal soil conditions on tree growth can be quantified. In existing mature stands a combination of irrigation and addition of repeated, small amounts of complete fertilizers have resulted in dramatic growth responses (Aronsson *et al.*, 1977; Linder *et al.*, 1987). In the present experiment the same kind of fertilization is used, based on the nutrient flux density and nutrient productivity concepts (Ingestad, 1988). The aim is to achieve a steady-state optimal nutrition with a stable, internal nutrient status. The nutrients must therefore be supplied according to the uptake rate in the plants. Optimum nutrient status is assumed when all nutrients in the plant are present in certain proportions to nitrogen (Ingestad, 1979; 1988). The supply of all nutrients will be proportional to the nitrogen demand of the trees. Some preliminary calculations on the nitrogen availability and storage in the trees will be presented here.

Materials and methods

Site

The exeriments are done in an even-aged forest stand of Douglas-fir (*Pseudotsuga menziesii* (Mirb.) Franco) and a stand of Scots pine (*Pinus sylvestris* L.). Both stands are located in the central, forested area of the Netherlands, the 'Veluwe'. The age of the trees is 37 years and 38 years for Scots pine and Douglas-fir respectively. In the Douglas-fir stand no understorey vegetation is present. The soil is a well-drained Plaggic Dystrochrept (Soil Taxonomy, 1975) on loamy, fine sand from eolian origin. In the pine stand an abundant vegetation of mainly Deschampsia flexuosa and Prunus serotina is present; the soil is a well-drained Typic Udipsamment on coarse sand from fluvioglacial origin.

Experimental set-up

In both stands the effect of optimal water supply is studied separately and in combination with optimal nutrition (Table 1). The 30 × 30 m^2 plots are divided in 4 replicates of 15 × 15 m^2 each. On smaller plots (10 × 10 m^2) the amount of acid loads reaching the soil, are manipulated. In all the treatments a mini sprinkler system is used to achieve a good spatial distribution of irrigation water and nutrients. The irrigation and fertilization system is completely computer-controlled and all measurement and control data are stored on diskette daily (Fig. 1). The computer software is developed by Ingestad and Lund (1989).

Optimal water and nutrient supply

The water supply is optimized by creating a daily infiltration rate of 3 mm, *i.e.* throughfall water and additional irrigation with demineralized water. The daily amount of throughfall water is measured by a tipping-bucket device.

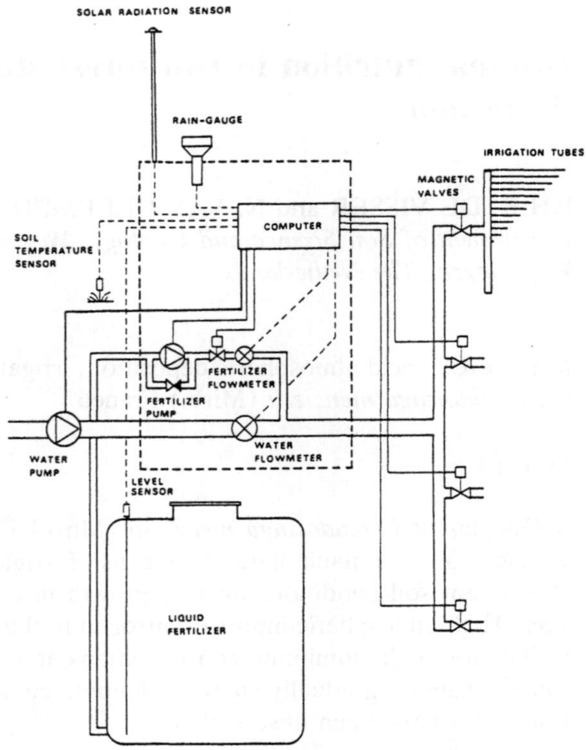

Fig. 1. A diagram of the fertilization and irrigation system including a weather station for measurement of solar radiation, soil temperature and precipitation. Figure from Ingestad (1988).

The optimal nutrition is carried out with daily additions of liquid fertilizer, injected in the irrigation water. The fertilizer is divided into two solutions to avoid precipitation of salts.

The nutient proportions in the fertilizer are derived from literature on optimal nutrition of tree seedlings (Ingestad, 1979; Van den Burg, 1971) and are shown in Table 2.

The nitrogen addition must equal the uptake capacity of the tree stand for nitrogen. The

Table 1. Manipulation experiments in two forest stands. Stand, treatment, addition frequency and plot area are given

Stand	Douglas-fir	Scots pine
Treatment:	Control >900 m^2	Control 900 m^2
	Irrigation daily 900 m^2	Irrigation daily 900 m^2
	Irrigation and optimal nutrition daily 900 m^2	Irrigation and optimal nutrition daily 900 m^2
	Irrigation and optimal nutrition. Withdrawal of throughfall water 100 m^2	extra $(NH_4)_2SO_4$ weekly 100 m^2
	Demineralization and irrigation of intercepted throughfall water weekly 100 m^2	

Table 2. Optimum weight proportions as a quotient of nitrogen. Values are from Ingestad (1979) and Van den Burg (1971)

Element	N	P	K	Ca	Mg	S	Fe	Mn	B	Cu	Zn	Cl	Mo	Na
Scots pine	100	13	65	7	8.5	9	0.7	0.4	0.6	0.06	0.03	0.03	0.007	0.003
Douglas-fir	100	30	50	4	5	9	0.7	0.4	0.6	0.03	0.03	0.03	0.007	0.003

demand for N was estimated from growth measurements on above-ground biomass (see below).

The supply of nitrogen is calculated as the flux of available nitrogen in the soil from N deposition rates at the site and the estimated mineralization rate. Added fertilizer-N will increase the nitrogen flux density and balance supply and demand. The nutrients are added daily in proportion to the calculated mineralization rate (M_d) (Ingestad and Agren, 1984) on that specific day (d):

$$M_d = M_0 \cdot d^z(d_v - d) ,$$

where d_v is the number of days in the vegetation period. z is a curve form factor and Mo is the initial mineralization rate.

Manipulation of acid load inputs

In the canopy throughfall water at the sites high amounts of N and S are present from acid atmospheric deposition. On some $10 \times 10\ m^2$ plots in the Douglas-fir stand throughfall water is removed by a roof construction at 3 m above the land surface.

Under the roof two treatments are carried out: (1) the throughfall water is deionized and sprinkled on the plot in the same amounts weekly; only small amounts of seasalt and nutrients leached from the canopy are added; (2) daily irrigation and optimal nutrition with nutrients added to deionized water. On a third roof plot the original throughfall water is returned to the plot, to evaluate effects due to the roof.

In one plot in the Scots pine stand an excess supply of ammonia is realized by weekly additions of dissolved $(NH_4)_2SO_4$ amounting to an annual dose of 120 kg/ha N.

Measurements

Biomass dynamics are studied intensively in all treatments. During the growing season shoot samples are analysed frequently for shoot length, needle retention, nutrient contents (N, P, K, Ca, Mg; Houba *et al.*, 1985), amino acids and water potential. Litterfall is collected monthly in litter traps ($1\ m^2$) and analysed for nutrient contents. Fine root growth is measured in ingrowth cores. Diameter at breast height (dbh) and tree height are recorded annualy.

The soil solution is sampled with ceramic cups fortnightly at four depths for chemical analysis.

Photosynthesis is measured in the optimal nutrition plot of Douglas-fir using small assimilation chambers. The response of shoot photosynthesis to air with different levels of SO_2, NO_x and O_3 is being studied.

Results

In the first year no significant growth response of trees to the treatments has been observed yet. Increased growth of the understorey vegetation was observed visually in the irrigated plots in the Scots pine stand for 1989. 1-year-old needles show high N content in pine and insufficient P and K supply in Douglas-fir (Van den Burg, 1985). In Scots pine P and K contents are low relative to N (Table 3).

Gross needle increment is estimated 6 t/ha and 7.6 t/ha for the Scots pine and the Douglas-fir stand respectively. Including net increment of stem and bark the uptake of N is estimated 118 kg/ha · yr for Scots pine and 135 kg for Douglas-fir. Uptake in fine-root biomass is not considered, assuming a turn-over time of 1 year or less.

In 1988 litterfall of 5.2 t/ha in pine and 2.2 t/ha in Douglas-fir was measured, resulting in a nitrogen flux of 64 respectively 33 kg/ha, according to chemical analysis.

Atmospheric deposition of all N-forms, as measured in throughfall water, amounts to 30 kg/ha in the Scots pine stand and 39 kg/ha in Douglas-fir in 1988.

Table 3. Nutrient status of 1-year old needles in Scots pine and Douglas-fir. Absolute values in % of dry weight

	N	P	K	Ca	Mg
Scots pine	2.21	0.14	0.62	0.32	0.09
low:	1.1–1.4	0.1–0.13	0.3–0.5	0.1–0.2	0.04–0.07
Relative to N	100	6	28	14	4
Douglas-fir	1.83	0.13	0.58	0.30	0.15
low:	1.1–1.4	0.1–0.14	0.4–0.6	0.15–0.2	0.04–0.07
Relative to N	100	7	32	16	8

Discussion

At present, the treatment period has been too short to expect any significant changes in tree growth. Linder (1986) found growth response of Scots pine on irrigation and liquid fertilization from the second year onwards. Nevertheless water and nutrient additions already stimulated understorey growth after one year treatment. As water retention capacity is very low at both sites, irrigation is expected to improve water availability considerably. Supply of the complete fertilizer may result in higher needle nutrient contents, with respect to P and K.

According to the concepts of nutrient flux density (Ingestad *et al.*, 1981), total annual fertilizer additions have to decrease with time. Only in the treatment where a roof intercepts the atmospheric inputs of N and S, complete control on nutrient supply can be expected in the long run.

Acknowledgements

This research is financially supported by the Dutch Ministry of Environment and the Commission of European Communities. T Befenati is thanked for her work on litterfall and needle analyses. I thank Dr T Ingestad and Dr A Lund for their help in installing the computer program. R van Woerkom helped a great deal in building up the field system. I also acknowledge Drs H van Dijk and Dr A W Boxman of the Catholic University of Nijmegen for their important contribution in the roof experiments.

References

Aronsson A, Elowson S and Ingestad T 1977 Elimination of water and mineral nutrition as limiting factors in a young Scots pine stand. 1. Experimental design and some preliminary results. Swed. Conif. For. Techn. Rep. 10, 1–38.

Houba V J G Novozamsky I, Van der Lee J J, Van Vark W and Nab E 1985 Chemische analyse van gewassen. Internal Rep. Dept. Soil Science and Plant Nutrition, Wageningen Agric. Univ., The Netherlands.

Ingestad T 1979 Mineral nutrient requirements of *Pinus sylvestris* and *Picea abies* seedlings. Physiol. Plant. 45, 373–380.

Ingestad T 1988 A fertilization model based on the concepts of nutrient flux density and nutrient productivity. Scand. J. For. Res. 3, 115–131.

Ingestad T and Agren G I 1984 Fertilization for long-term maximum production. *In* Ecology and Management of Forest Biomass Production Systems. Ed. K Perttu. Dept. Ecol. and Environ. Res., Swed. Univ. Agric. Sci. Rep. 15, pp 155–165.

Ingestad T and Lund A 1989 Handbook of computerized fertilization in field experiments. AB Biokonsult, Uppsala, Sweden, 30 p.

Linder S 1986 Responses to water and nutrients in coniferous ecosystems. *In* Potentials and Limitations of Ecosystem Analysis. Eds. E D Schulze and H Zwölfer. Ecol. Stud. 68, pp 180–202.

Soil Taxonomy 1975 U.S. Dept. Agric., Soil Conservation Service. U.S. Government Printing Office, Washington, DC, 754 p.

Van den Burg J 1971 Some experiments on the mineral nutrition of forest tree seedlings. Inst. Res. on Forestry and Landscape 'De Dorschkamp', Wageningen, Internal Rep. 8, 67 p.

Van den Burg J 1985 Foliar analysis for determination of tree nutrient status – compilation of literature data. Inst. Res. on Forestry and Landscape 'De Dorschkamp', Wageningen, Internal Rep. 414.

M. L. van Beusichem (Ed.), *Plant nutrition – physiology and applications*, 73–79.

PLSO IPNC485

Steady state nutrition by transpiration controlled nutrient supply

W. G. BRAAKHEKKE and D. A. LABE[1]
Department of Soil Science and Plant Nutrition, Wageningen Agricultural University, P.O. Box 8005, 6700 EC Wageningen, The Netherlands. [1]*Present address: Institute for Agricultural Research, Ahmadu Bello University, P.M.B. 1044, Zaria, Nigeria*

Key words: nitrogen nutrition, nitrogen productivity, *Pennisetum americanum* (L.) Leeke, programmed nutrient addition, relative growth rate, steady state nutrition, transpiration coefficient, transpiration controlled nutrient supply

Abstract

Programmed nutrient addition with a constant relative addition rate has been advocated as a suitable research technique for inducing steady state nutrition in exponentially growing plants. Transpiration controlled nutrient supply is proposed as an alternative technique for plants with a short or no exponential growth phase. A two-weeks experiment with transpiration controlled nitrogen supply to *Pennisetum americanum* was carried out to evaluate this method.

After an adaptation phase a constant plant N-concentration was maintained, while the relative growth rate decreased rapidly. The transpiration coefficient was almost constant in time and insensitive to moderate N-stress, but increased sharply when plant N-concentration dropped below 1760 mmol/kg DW. Relative growth rate and nitrogen productivity showed a steep decline at the lowest N-concentrations (about 1000 mmol/kg DW). Nitrogen productivity was optimal at about 1760 mmol/kg DW.

The results show that transpiration controlled nutrient supply is applicable in research and gives accurate results in growth analysis. When the transpiration coefficient is known, the nutrient solution can be adjusted to give any desired plant N-concentration, except for the lowest concentrations.

Introduction

In studying physiological and morphological responses of plants to nutrient supply, it is essential to be able to grow plants with a well-defined, constant, suboptimal nutrient status. This condition, referred to as steady state nutrition (SSN), is usually characterized by a constant internal concentration of the growth limiting nutrient. It can be achieved by supplying nutrients at a rate proportional to the actual growth rate of the plant. For plants that grow exponentially, Programmed Nutrient Addition (PNA) with a constant relative addition rate (RAR; see the appendix) has proven to be a suitable technique (Ingestad and Lund, 1986).

In experiments that outlast the exponential growth phase PNA with constant RAR fails to maintain SSN. When, for some internal or external cause (like a change in biomass allocation or self shading), the potential relative growth rate of a plant decreases, a constant RAR will result in a gradually improving nutrient status, eventually leading to luxurious consumption and elimination of the nutritional control of growth.

To maintain SSN, the nutrient supply should be adjusted to the changing potential growth rate. One can think of several ways to achieve this. RAR could be made a function of time, plant size or development stage. The rate of nutrient addition could also be controlled by the actual growth rate of the plant as calculated by simultaneous computer simulation (Van Koninckxloo, 1986; Schapendonk *et al.*, 1990). Both approaches require detailed information on the growth and nutrient demand of the plant.

In this paper we present an alternative approach, which requires little information on plant growth. We use the transpiration rate of the plant to control the supply of the limiting nutrient. The transpiration coefficient (TC) has been found to be rather constant in time and independent of nutrient status in many situations, provided the nutrient level is not 'too low' (Van Keulen and Seligman, 1987; De Wit, 1958). If this holds, transpiration controlled nutrient supply results in a situation where the limiting nutrient is supplied at a rate that is proportional to the actual growth rate:

$$d(ANPL)/dt = 0.001 \cdot d(DWPL)/dt \cdot CN \cdot TC \quad (1)$$

(abbreviations are listed in the appendix).

The plants are grown in an initial solution that contains all essential nutrients except the limiting nutrient to be studied. This nutrient is supplied by daily compensating for the transpiration with a complete nutrient solution. Different levels of nutrient stress are induced by varying the concentration of this nutrient solution (CN). The overall plant N-concentration (NCPL) approaches $CN \cdot TC$ after a short adaptation phase:

$$NCPL \rightarrow 1000 \cdot d(ANPL)/d(DWPL) = CN \cdot TC \quad (2)$$

Here we present the results of an experiment with transpiration controlled nitrogen supply to Pearl millet (*Pennisetum americanum* (L.) Leeke), using a communicating vessel system designed to allow frequent measurement of transpiration and non-destructive observation of several plant characteristics. The aim was to evaluate the proposed method of maintaining steady state nutrition and to test its applicability in a study on the influence of internal nitrogen status and age of the plant on its transpiration coefficient, relative growth rate and nitrogen productivity.

Materials and methods

Plant material

Seeds of pearl millet *Pennisetum americanum* (L.)Leeke (= *P. typhoides* S. & H.) were germinated in moist quartz sand without additional nutrients. Five days after sowing (DAS) the seedlings were transplanted to a container with a complete nutrient solution with 5.2 m*M* nitrogen. At the start of the experiment (12 DAS) 12 plants were selected for equal size and used for the experiment.

The communicating vessel system

The plants were mounted in glass cylinders (4 cm diameter, 20 cm length). Each cylinder was connected to a burette (50 mL by a flexible tube (Fig. 1). Cylinder and burette were clamped next to each other on a board and protected from light. Aeration rate was low and the relative humidity of the pressurized air was increased by leading it through a washing bottle to reduce evaporation from the solution to about 1 mL/d.

Daily, the level in the cylinders was adjusted to a mark by raising the burette (accuracy of adjustment was 0.1 to 0.5 mL). Evapotranspiration was read from the burette. Evaporation (as estimated from 3 blanc cylinders) was compensated for by adding water. After lowering the burette, a volume of nutrient solution was added into the cylinder, equal to the transpired volume.

The experiment was carried out in a fytotron. Growth conditions were: light period 13 h; light intensity 120 W/m^2 at plant height; day/night temperature 27/20°C; 80/80% relative humidity.

Nutrient solutions

At the start the cylinders were filled with a nitrogen free initial solution containing (in m*M*): 5.2 K; 0.4 Ca; 0.4 Mg; 2.6 H_2PO_4; 1.3 SO_4; 1.6 Cl. Trace element concentrations were (in $mg\,L^{-1}$): 4.6 Fe; 0.5 B; 0.5 Mn; 0.05 Zn; 0.02 Cu; 0.01 Mo. Fe was added as a mixture of Fe-sulphate and citric acid.

Four treatments (replicated 3 times) were used, receiving 10, 20, 30 and 40 mmol N per litre transpiration (CN) respectively. Nitrogen

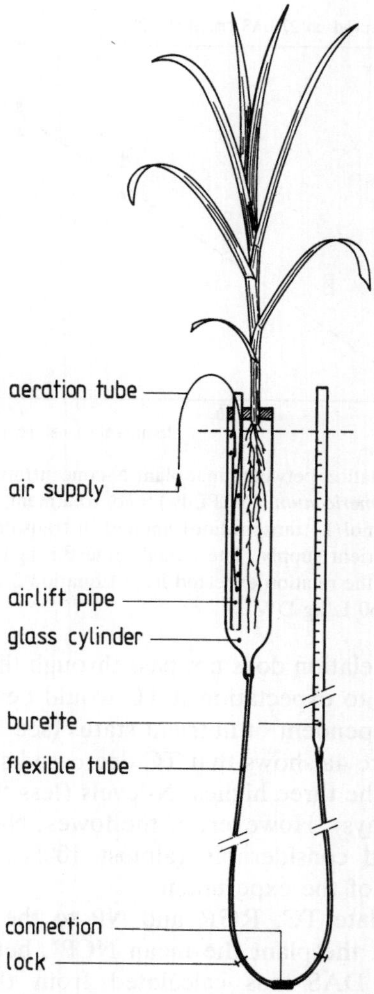

Fig. 1. Schematic representation of the communicating vessel system used for short term experiments with transpiration controlled nutrient supply. (Further explanation in text.)

was added with complete nutrient solutions. Macro-nutrient concentrations in the first treatment solution (CN = 10) were (in m*M*): 5.0 K; 0.3 Ca; 0.3 Mg; 2.8 NH_4; 7.2 NO_3; 0.6 H_2PO_4; 0.3 SO_4; 0.6 Cl. Trace element concentrations in this solution were one tenth of those in the initial solution. Concentrations in the other treatment solutions were respectively 2, 3 and 4 times as high. The macro-nutrient composition of these solutions was in accordance with earlier determined uptake ratios of Pennisetum plants. The NH_4/NO_3 ratio was adjusted to balance the cation/anion uptake ratio and minimize plant-induced pH changes in the initial solution.

Observations

Transpiration of each plant was measured daily. On 12, 14, 17, 21, 24 and 27 DAS the following observations were made: fresh weight of the plant including water adhering to the roots after a standard dripping time (FWPLwet), root volume, number of young, mature and dead leaves, length and width of each leaf, tiller number, length of the main stem. Solution pH and NO_3 and NH_4 concentration in the cylinder were measured also. These non-destructive observations allow estimation of RGR, ANPL and NCPL (based on wet fresh weight). RGR and nitrogen productivity (NP) were calculated using the following equation:

$$RGR, NP = GR/[(X2 - X1)/\ln(X2/X1)] \quad (3)$$

where GR is the absolute growth rate (g/plant/d) and X1 and X2 are the values of FWPLwet resp. ANPL at the start (X1) and the end (X2) of the time interval.

On 27 DAS the plants were harvested. Leaf area, fresh and dry weight of the roots, stems and leaves were determined. Whole plants were analyzed for NO_3, total N, P and K (methods according to Novozamsky *et al.*, 1983a; b).

Results

Only results relevant for the evaluation of transpiration controlled nutrient supply are shown in this paper. Most graphs and calculations are based on FWPLwet. When relevant, values based on estimates of fresh weights without adhering water (FWPL) and dry weights (DWPL) are given in the text.

Analysis of the nitrogen concentration in the cylinders indicated complete uptake of the nitrogen added within 24 hours in all four treatments. Therefore, the course of the NCPL (based on fresh weights) can be estimated from initial ANPL, the cumulative N-additions and FWLPwet. Final NCPL estimated in this way on 27 DAS ($NCPL_{est}$) correlated well with the results of the chemical analysis ($NCPL_{meas}$), but there was a small systematic deviation

($NCPL_{est} = NCPL_{meas} \times 1.15 - 128$; $r^2 = 0.974$). Final plant weights differed a factor five (0.89, 1.81, 2.74 and 4.42 gDW/plant at CN = 10, 20, 30 and 40 m*M* respectively).

Adaptation of NCPL and RGR to the treatment took about 5 days at the highest N-level and 9 days at the lowest N-level (Fig. 2). Thereafter, changes in NCPL were relatively small (mean coefficient of variance (CV) between harvests was 14%; mean CV between replicates was 6%), whereas RGR showed a substantial and steady decrease after the adaptation phase (decrease during the last week with 50% at the highest N-level and 75% at the lowest N-level; mean CV between replicates was 8%).

Final NCPL was highly correlated with the N-concentration in the nutrient solution (Fig. 3),

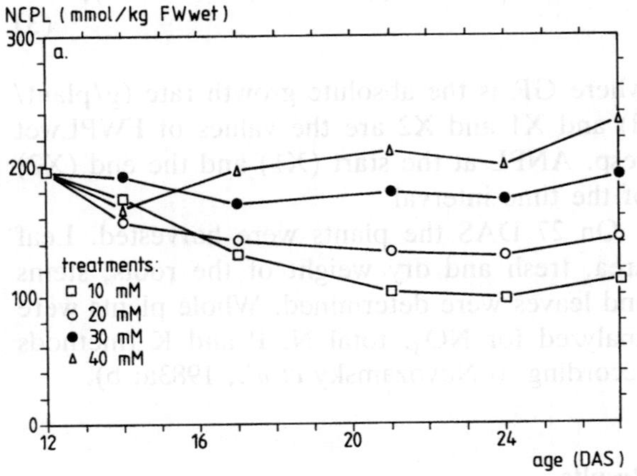

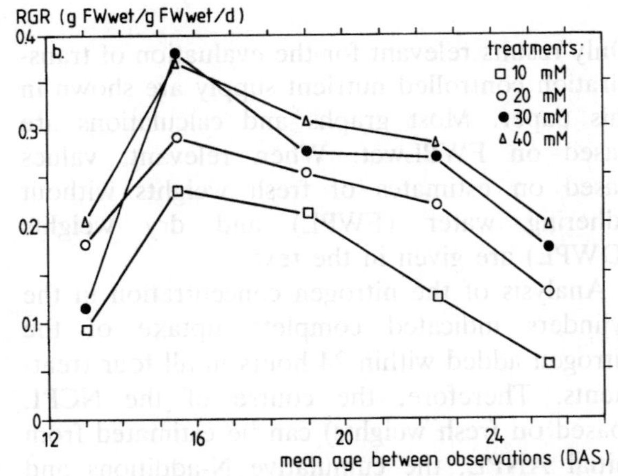

Fig. 2. Course of the plant N-concentration (NCPL; **a**) and relative growth rate (RGR; **b**) of *Pennisetum americanum* with age. Means of three replicates are shown. Four N-levels were induced using transpiration controlled nutrient supply with 10, 20, 30 and 40 mmol N per litre transpiration.

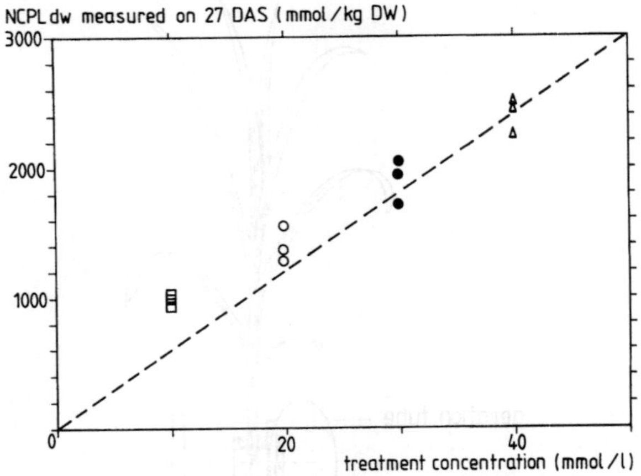

Fig. 3. Relation between final plant N-concentration of *Pennisetum Americanum* (NCPLdw) and treatment N-concentration (mmol/L transpiration) applied in transpiration controlled nutrient supply. The straight line through the origin represents the relation expected from Equation 2 when TC is constant (60 L/kg DW).

but the relation does not pass through the origin, contrary to expectation if TC would be constant and independent of nutrient status (see Equation 2). Figure 4a shows that TC decreased slightly in time in the three highest N-levels (less than 10% in ten days). However, at the lowest N-level TC increased considerably (almost 100%) towards the end of the experiment.

To relate TC, RGR and NP to the nutrient status of the plant the mean NCPL between 24 and 27 DAS was calculated from the mean FWPLwet and the mean ANPL during the interval. The latter means were calculated using the last term in Equation 3.

Figure 4b shows that TC was hardly influenced by the substantial N-stress at CN = 20 m*M*, where the final yield was only 40% of the final yield at CN = 40 m*M* and NCPL was about 140 mmol/kg FWwet (170 mmol/kg FW or 1760 mmol/kg DW). This affirms the conclusion of De Wit (1958) and Tanner and Sinclair (1983) that transpiration efficiency changes little, if at all, until nutrient deficiency reduces yield to about half that on well-fertilised soil. TC increased sharply at CN = 10 m*M*, where the final yield was reduced to 20% of the highest N-level and NCPL dropped to 100 mmol/kg FWwet (about 120 mmol/kg FW or 1000 mmol/kg DW).

TC is about 6 L/kg FWwet (6.5 L/kg FW; 65 L/kg DW) at the two highest N-levels. This is

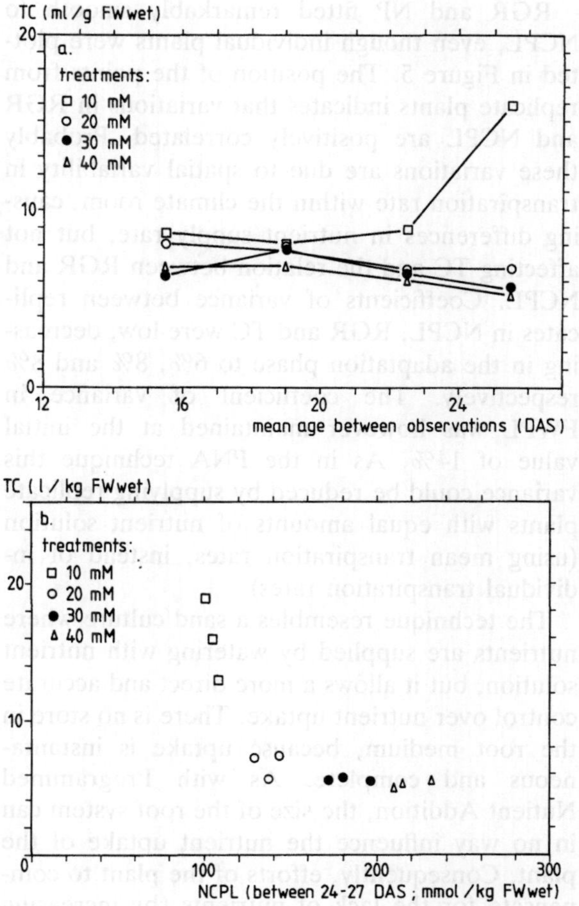

Fig. 4. **Transpiration coefficient (TC) of *Pennisetum americanum* plotted against age (means of three replicates are plotted between the observations dates (DAS); a) and against plant N-concentration (NCPL; TC of individual plants is plotted against mean NCPL between 24 and 27 DAS; b).**

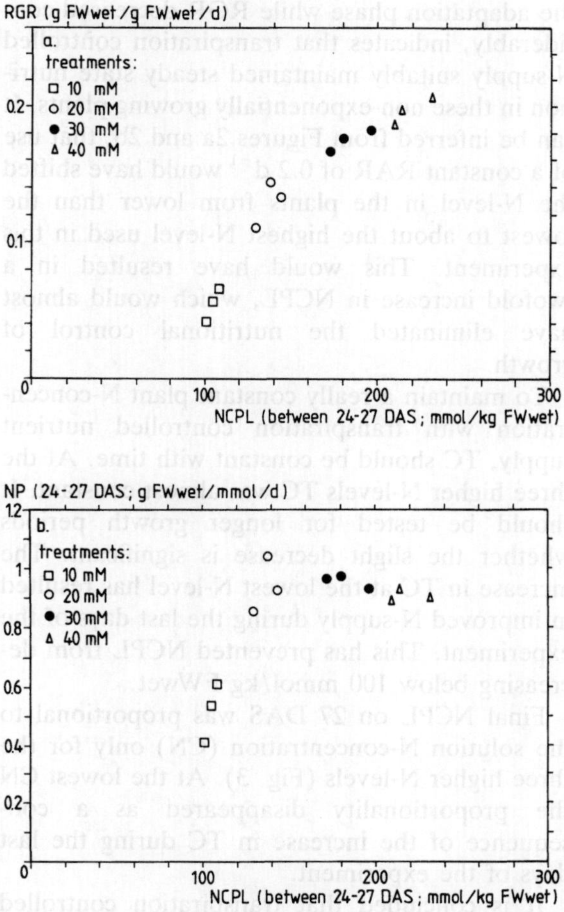

Fig. 5. **Relation between relative growth rate (RGR; a) and nitrogen productivity (NP; b) with plant N-concentration (NCPL; RGR of individual plants is plotted against the mean NCPL between 24 and 27 DAS).**

low compared to values commonly found in literature, because TC is expressed on a whole plant basis, millet is a C4 species, and the evaporative demand in the climate room was low.

RGR depends strongly on NCPL (Fig. 5a). The relation is not linear, as is often found (Ingestad and Lund, 1986), nor does it resemble the relation found by Hirose (1988) and Hirose *et al*. 1988). However, when NCPL is calculated for organic N instead of total N (not shown), a curvilinear relation is found, resembling the curve presented by Hirose (1988). Below 140 mmol/kg FWwet RGR bends down. By extrapolation the minimum NCPL is found to be about 90 mmol/kg FWwet. The N-productivity (NP) reaches its optimum at CN = 20 m*M* (NCPL = about 140 mmol/kg FWwet; Fig. 5b). A similar curve was found by Hirose (1988) and Hirose *et al.* (1988).

Discussion

The fact that NCPL was almost constant after

the adaptation phase while RGR decreased considerably, indicates that transpiration controlled N-supply suitably maintained steady state nutrition in these non-exponentially growing plants. It can be inferred from Figures 2a and 2b, that use of a constant RAR of 0.2 d^{-1} would have shifted the N-level in the plants from lower than the lowest to about the highest N-level used in this experiment. This would have resulted in a twofold increase in NCPL, which would almost have eliminated the nutritional control of growth.

To maintain a really constant plant N-concentration with transpiration controlled nutrient supply, TC should be constant with time. At the three higher N-levels TC was almost constant. It should be tested for longer growth periods whether the slight decrease is significant. The increase in TC at the lowest N-level has resulted in improved N-supply during the last days of the experiment. This has prevented NCPL from decreasing below 100 mmol/kg FWwet.

Final NCPL on 27 DAS was proportional to the solution N-concentration (CN) only for the three higher N-levels (Fig. 3). At the lowest CN the proportionality disappeared as a consequence of the increase in TC during the last days of the experiment.

It is concluded that transpiration controlled nutrient supply allows control of plant N-concentrations, except at the lowest N-levels. When the actual TC is known CN can be adjusted to give any desired plant N-concentration. This possibility may be interesting for research purposes, as well as for commercial greenhouse crop growing.

The communicating vessel technique used for transpiration controlled nutrient supply is basically simple, though it requires more time per plant and accuracy of handling than other possible techniques, using transpiration measurements by weighing the pots. This was compensated by using non-destructive observations instead of sequential harvesting, which allowed a considerable reduction in the number of plants. Correction for the increase in root volume is not considered necessary, because it causes only a small systematic error of about 3% of the volume transpired.

RGR and NP fitted remarkably smooth to NCPL, even though individual plants were plotted in Figure 5. The position of the points from replicate plants indicates that variations in RGR and NCPL are positively correlated. Probably these variations are due to spatial variability in transpiration rate within the climate room, causing differences in nutrient supply rate, but not affecting TC and the relation between RGR and NCPL. Coefficients of variance between replicates in NCPL, RGR and TC were low, decreasing in the adaptation phase to 6%, 8% and 8% respectively. The coefficient of variance in FWPL was however maintained at the initial value of 14%. As in the PNA technique this variance could be reduced by supplying replicate plants with equal amounts of nutrient solution (using mean transpiration rates, instead of individual transpiration rates).

The technique resembles a sand culture where nutrients are supplied by watering with nutrient solution, but it allows a more direct and accurate control over nutrient uptake. There is no store in the root medium, because uptake is instantaneous and complete. As with Programmed Nutient Addition, the size of the root system can in no way influence the nutrient uptake of the plant. Consequently, efforts of the plant to compensate for the lack of nutrients (by increasing biomass investments in roots) cannot reduce the differences in nutrient supply between experimental treatments, as may happen with other culture techniques. This is an advantage in studying the effect of nutrient status on allocation and internal use efficiency of nutrients. On the other hand, the technique cannot be used in studying the 'reward' for the increased biomass investment in roots in terms of uptake and growth.

Acknowledgements

The authors thank Friederike de Mol and Vincent Deenen for carrying out the experiments with great diligence and accuracy. We thank G R Findenegg, E Hoffland, H van Keulen and B W Veen for helpful discussions during experimentation and preparation of the manuscript.

References

De Wit C T 1958 Transpiration and crop yields. Agric. Res. Rep. 64.6. Pudoc, Wageningen, 88 p.

Hirose T 1988 Modelling the relative growth rate as a function of plant nitrogen concentration. Physiol. Plant. 72, 185–189.

Hirose T, Freijsen A H J and Lambers H 1988 Modelling of the responses to nitrogen availability of two *Plantago* species grown at a range of exponential nutrient addition rates. Plant Cell Environ. 11, 827–834.

Ingestad T and Lund A-B 1986 Theory and techniques for steady state mineral nutrition and growth of plants. Scand. J. For. Res. 1, 439–453.

Novozamsky I, Houba V J G, Van Eck R and Van Vark W 1983a A novel digestion technique for multi-element plant analysis. Commun. Soil Sci. Plant Anal. 14, 239–249.

Novozamsky I, Houba V J G, Van der Eyk D and Van Eck R 1983b Notes on determination of nitrate in plant material. Neth. J. Agric. Sci. 31, 239–248.

Schapendonk A H C M, Spitters C J T and De Vos A L F 1990 Comparison of nitrogen utilization of diploid and tetraploid perennial ryegrass genotypes using a hydroponic system. *In* Genetic Aspects of Plant Mineral Nutrition. Eds N El Bassam, M Dambroth and B C Loughman. pp 299–306. Kluwer Academic Publishers, Dordrecht, The Netherlands.

Tanner C B and Sinclair T R 1983 Efficient water use in crop production: research or re-search? *In* Limitations to Efficient Water Use in Crop Production. Eds. H M Taylor, W R Jordan and T R Sinclair. pp 1–27. ASA Monographs Inc., Madison, WI.

Van Keulen H and Seligman N G 1987 Simulation of Water Use, Nitrogen Nutrition and Growth of a Spring Wheat Crop. Simulation Monographs. Pudoc, Wageningen, 310 p.

Van Koninckxloo M 1986 Application of the nitrogen productivity concept to the fertilisation of maize. Agro 5, 13–28.

Appendix

Abbreviations:

ANPL:	total amount of N in plant + pot (mmol/plant);
CN:	concentration of N in the nutrient solution (mmol/L);
CV:	coefficient of variance (%);
DAS:	days after sowing;
DWPL:	plant dry weight (g DW/plant);
FWPL:	plant fresh weight (g FW/plant);
FWPLwet:	FWPL including water adhering to the roots (g FWwet/plant);
GR:	growth rate; $GR = (FWPLwet_2 - FWPLwet_1)/(t_2 - t_1)$
N:	the growth limiting nutrient being nitrogen in the present experiment;
NCPL:	internal plant N-concentration (mmol/kg);
NP:	N productivity (g/mmol/d);
PNA:	technique of programmed nutrient addition;
RAR:	relative addition rate (1/d); RAR = d(ANPL)/(dt.ANPL)
RGR:	relative growth rate (1/d);
SSN:	the condition of steady state nutrition;
t:	time in DAS;
TC:	transpiration coefficient (L/kg increment).

M. L. van Beusichem (Ed.), *Plant nutrition – physiology and applications*, 81–85.

PLSO IPNC302

A culture system for steady state nutrition in optimum and suboptimum treatments

A.H.J. FREIJSEN, A.J. DE ZWART and H. OTTEN
Institute for Ecological Research, P.O. Box 40, 6666 ZG Heteren, The Netherlands

Key words: culture system, hydroponic culture, *Plantago lanceolata* L., programmed nutrient addition, steady state nutrition

Abstract

The paper describes a culture system for programmed nutrient addition, consisting of growth units and computer equipment. Roots hang in a mist of nutrient solution. The composition of this solution is controlled by a computer program. Two regimes of nutrient (nitrate) availability can be applied sequentially: exponential addition and maintenance of the ambient concentration. Exponential addition rates are used in the exponential growth phase; in the linear growth phase the external concentration established during exponential growth is maintained. This sequence of nutrient regimes is assumed to be particularly apt to accomplish steady state nutrition in optimum as well as suboptimum treatments.

Introduction

Steady state nutrition is attained, if the ratio between nutrient demand by the plant and nutrient availability in the medium is constant throughout the experimental treatment (Asher and Blamey, 1987; Ingestad, 1977). Stable optimum nutrition can be realized by periodically supplying quantities of nutrients which match the demand of the plant. In practice an excess of nutrients is often supplied. Suboptimum treatments can be established by supplying only fractions of the maximum dose per unit of time. Plant parameters are stable under conditions of steady state nutrition, which makes it easier compared to unstable conditions to assess values for plant parameters relevant to a certain phase of growth.

Exact control of nutrient availability – a prerequisite for establishing stable conditions – is possible in hydroponic culture systems. Ingestad and Lund (1979, 1986) developed a 'growth unit' for laboratory purposes. Plant roots were sprayed with nutrient solution and the availability of nutrients was controlled by a computer program. The method was applied to bring about constant exponential growth rates, maximum as well as reduced rates (*e.g.* Ericsson, 1981; Ingestad, 1981; Jia and Ingestad, 1984). The parameters of the experimental plants (*e.g.* the relative growth rate and the internal N concentration) were constant; the existence of steady state could also be concluded from the absence of deficiency symptoms. Freijsen and Otten (1987) used a mechanical device for the exponential addition of nutrient to culture vessels. The main results were similar. Unlike the system of Ingestad the nutrient addition was continuous, resulting in constant (very low) ambient concentrations of the added nutrient (Freijsen *et al.*, 1989).

Exponential addition can only be applied, when the potential growth rate of the plant is exponential *i.e.* in the juvenile developmental stage. For later stages other addition schemes are necessary. Asher and Blamey (1987) presented a computer program to calculate daily nutrient additions to be applied in cases of exponential as well as linear or sigmoidal growth. The optimum regime ('non-stressed control plants') was based on the known or expected (intrinsic) growth curve and the corresponding internal nutrient concentrations; suboptimum regimes could be effected by adding fixed fractions of the optimum daily doses. After the exponential growth phase the relative growth rate (RGR) of plants decreases and a constant RGR is no longer applicable as a criterion for steady state growth as is the case in the juvenile phase. In some culture experiments described by Freijsen and Veen (1990) a constant plant N concentration was handled instead. By using methods similar to those of Asher and Blamey (1987) plant N concentrations were obtained that were equal in the exponential and subsequent linear phase of growth, at optimum as well as suboptimum supply. Above-mentioned methods to establish steady state growth have been applied under optimum or at least constant climatic conditions. In greenhouse experiments nutrient additions could be based on the photosynthetic activity attained by the experimental plants at the prevailing irradiance (Schapendonk and De Vos, 1988). In the optimum treatment nitrate additions kept pace with potential growth, in the suboptimum treatment 30% of the optimum addition rate was applied. Nutrient treatments were applied using a computer program for control.

In this report a culture system for programmed addition of nutrients under laboratory conditions is described. The description deals with the technical equipment and the computer program to control its functioning. Addition rates can be adapted to various growth patterns. Contrary to above-mentioned systems the ambient nutrient concentration can be applied as a variable. The system was developed to grow plants with exponential nutrient (nitrate) addition in the juvenile (exponential) phase of growth and at constant ambient (nitrate) concentrations afterwards. The ambient concentrations are those which are established in preceding treatments with exponential addition.

Results of some preliminary culture experiments are reported with emphasis on technical aspects.

Material and methods

Plant material

The culture experiments were carried out using *Plantago lanceolata* L. Seeds were collected from a population occurring in a dune grassland near Oostvoorne (The Netherlands). After germination on wetted glass beads for 8 days the seedlings were transferred to the culture system.

Growth unit

The culture system is composed of a number of 'growth units' and the computer equipment. Growth units are trolleys (62 × 71 cm) each carrying a container to hold the experimental plants and provided with accessories. The containers (poly-ethylene) are 87 cm deep and covered with a PVC lid (43 × 43 cm). The lid has 144 holes to hold plants (unused holes can be closed with marbles). Roots of experimental plants hang in the interior. A nozzle assembly (Spraying Systems Co.) is mounted in the centre of the container filling it during operation with a mist of nutrient solution. After passing a filter solution dropped from the roots and walls is conveyed through an outlet near the bottom and recirculated. On the trolley several accessories are mounted for various tasks during experiments. A gear pump (Micropump) circulates the nutrient solution from the outlet to the nozzle assembly and (via a bypass) to a measuring chamber with electrodes and a pressure switch. The electrodes are a nitrate-ionselective electrode (Radiometer F2412 NO_3), a glass electrode (Yokogawa SM21/AS4) and a reference electrode (non-flow type, Yokogawa SR20/AP24). Two diaphragm pumps (ProMinent) dose stock solutions (KNO_3 and H_2SO_4) into the container. A level sensor checks the volume of nutrient solution and activates a magnetic valve for the supply of water to compensate for losses. When the pressure in the solution circuit drops below a minimum the switch deactivates dosing pumps and activates an alarm. Tubes connecting the container with the accessory equipment can be disconnected without loss of solution and consequently containers can be easily removed from the trolleys.

Nutrient solution

Each growth unit contained 5 litres of nutrient solution. The basic solution contained non-restrictive quantities of the mineral nutrients other than N (see Freijsen and Otten, 1984). During the experimental period N as NO_3 was added by a dosing pump controlled by a computer program. A second pump added H_2SO_4 to maintain a pH of 5.

Electronic equipment and software

The electrodes and the dosing pumps are connected to a Data Acquisition/Control Unit (Hewlett Packard, model 3412A) to measure electrode potentials and to control the operation of the pumps. This unit, in turn, is connected to a PC (HP-86), which executes the program during experiments.

The computer program is in BASIC. Every hour a procedure is carried out, which includes the measurement of the NO_3 ion concentration and the pH of the nutrient solution followed by calculation and execution of possible additions of stock solutions. Two modes of NO_3 supply can be implemented sequentially: exponential supply and concentration controlled supply. Exponential supply is programmed according to the formula:

$$N = N_0 \times (\exp(RAR/24)-1)$$

where N = the hourly dose, N_0 = the quantity of N in the plants, RAR = the relative addition rate. RAR, having a value equal to or smaller than the maximum relative growth rate, is in fact the treatment variable (Ingestad, 1982, 1987). When the measured NO_3 concentration exceeds a set value which is based on the prevailing concentration during the exponential phase of growth, the program switches over to the second regime. Accumulation of NO_3 can be expected at the end of the exponential growth phase, when the RAR exceeds the decreasing RGR. In the second regime the set value of the NO_3 concentration is maintained. Hourly quantities of added NO_3 are automatically changed to adjust the concentration measured during the preceding procedure. Maintenance of pH is programmed in the same way. The program provides for periodic print-outs of values measured and volumes of solutions added.

Climatic conditions

Experiments were carried out in a growth cabinet. The conditions were: 23°C, 70% RH, 80 W m^{-2} (14 h light period) + 6 Philinea lamps (24 h). Mutual shading was avoided by periodic harvesting.

Results

P. lanceolata was grown in the growth units till flowering. When the dry weight of plants was about 600 mg, exponential growth (relative growth rate = 0.29 and 0.19/d) ended and was followed by linear growth (Fig. 1). Root branches of each root system clung together and did not intermingle with other plants so that harvesting of individual root systems was easy. Ambient nitrate concentrations were rather high *e.g.* 260 $\mu M\ L^{-1}$ when the RGR was 0.19/d. Possibly, higher concentrations were needed than in water culture, because the inner parts of root systems were not directly wetted by the solution mist (*cf.* Freijsen *et al.* 1989).

From figures on the ionic balance of harvested plants it could be calculated that about 40% of the absorbed NO_3 was counterbalanced by efflux

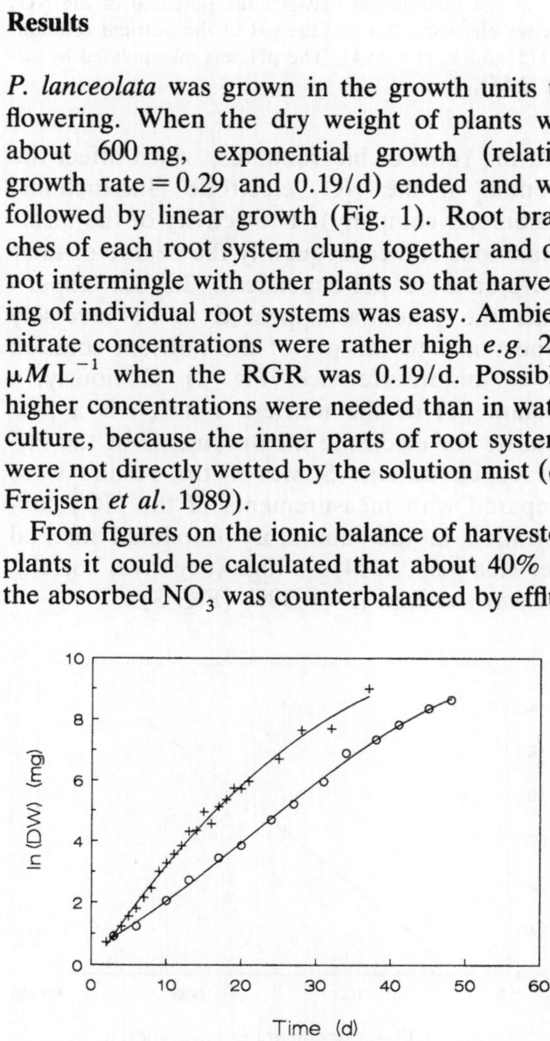

Fig. 1. Growth curves of *Plantago lanceolata* grown with programmed nitrate addition. Exponential growth of 0.29 (+) or 0.19/d (0) was followed by linear growth. The transition from exponential to linear growth occurred when the dry weight of plants was about 600 mg (ln (DW) = 6.4).

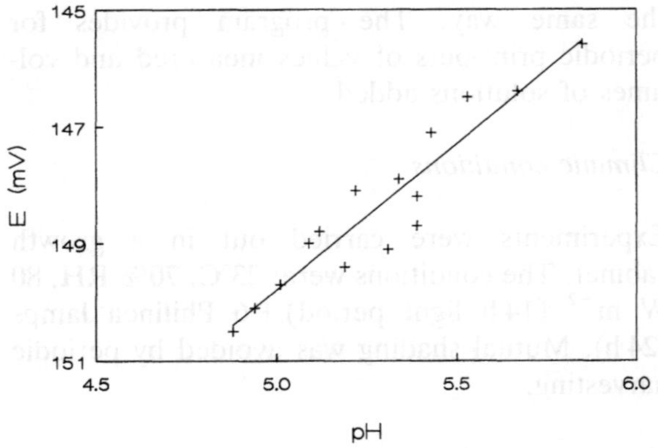

Fig. 2. The relationship between the potential of the NO_3 selective electrode (E) and the pH of the nutrient solution; $y = 175 - 5.0\,x$, $r^2 = 0.941$. The pH was manipulated by adding H_2SO_4.

from the roots of bicarbonates, which affect the potential of the NO_3 electrode (Instructions, Radiometer company). The activity of the bicarbonate ions and consequently the degree of their interference is pH-dependent. In our experiments the potential appeared to be inversely proportional to the pH of the nutrient solution due to this interference (Fig. 2). Obviously, a constant pH of the nutrient solution is a prerequisite for electrode measurements of the nitrate concentration. Some of the results were compared with measurements of the NO_3 concentration in the laboratory (modified method after Kempers, 1974). A high (negative) correlation was assessed (r = 0.999) (Fig. 3).

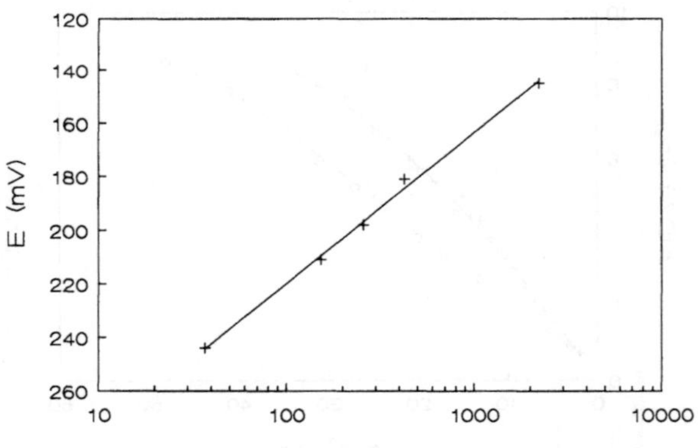

Fig. 3. The relationship between the potential of the NO_3-electrode (E) and the ambient nitrate concentration as assessed after Kempers (1974); $y = 333 - 56.4\,x$, $r^2 = 0.998$.

Discussion

The culture system appeared to be well-suited to grow plants from the seedling stage till flowering under optimum and suboptimum conditions. In our program continuation of steady state nutrition in the linear phase of growth is based on the maintenance of the ambient nitrate concentration established in the exponential phase. This means that steady state is defined in terms of an external factor. Constant internal N concentrations are a (theoretically) possible response of the plant to this treatment. In other approaches (see Introduction) the internal concentration is manipulated; constancy of the internal concentration during the development of the plant is *a priori* taken as a criterion for steady state. However, it is not well-known that internal concentrations of N tend to be constant during the life-cycle of plants. Further experiments with the described culture equipment can be conclusive with respect to the choice of treatments that are relevant for steady state nutrition.

References

Asher C J and Blamey F P 1987 Experimental control of plant nutrient status using programmed nutrient addition. J. Plant Nutr. 10, 1371–1380.

Ericsson T 1981 Effects of varied nitrogen stress on growth and nutrition in three *Salix* clones. Physiol. Plant. 51, 423–429.

Freijsen A H J and Otten H 1984 The effect of nitrate concentration in a flowing solution system on growth and nitrate uptake of two *Plantago* species. Plant and Soil 77, 159–169.

Freijsen A H J and Otten H 1987 A comparison of the responses of two *Plantago* species to nitrate availability in culture experiments with exponential nutrient addition. Oecologia (Berlin) 74, 389–395.

Freijsen A H J, Troelstra S R, Otten H and Van der Meulen M A 1989 The relationship between the ambient nitrate concentration and plant parameters of *Plantago lanceolata* in culture experiments with exponential nutrient addition. Plant and Soil 117, 121–127.

Freijsen A H J and Veen B W 1990 Phenotypic variation in growth as affected by N-supply: Nitrogen productivity. *In* Causes and Consequences of Variation in Growth Rate and Productivity of Higher Plants. Eds. H Lambers, M L

Cambridge, H Konings and T L Pons. pp, 19–33. SPD Academic Publishing bv., The Hague.
Ingestad T 1977 Nitrogen and plant growth: Maximum efficiency of nitrogen fertilizers. Ambio 6, 146–151.
Ingestad T 1981 Nutrition and growth of birch and grey alder seedlings in low conductivity solutions and at varied relative rates of nutrient addition. Physiol. Plant. 52, 454–466.
Ingestad T 1982 Relative addition rate and external concentration: Driving variables used in plant nutrition research. Plant Cell Environ. 5, 443–453.
Ingestad T 1987 New concepts on soil fertility and plant nutrition as illustrated by research on forest trees and stands. Geoderma 40, 237–252.
Ingestad T and Lund A B 1979 Nitrogen stress in birch seedlings. I. Growth technique and growth. Physiol. Plant. 45, 137–148.
Ingestad T and Lund A B 1986 Theory and techniques for steady-state mineral nutrition and growth of plants. Scand. J. For. Res. 1, 439–453.
Jia H and Ingestad T 1984 Nutrient requirements and stress response of *Populus simonii* and *Paulownia tomentosa*. Physiol. Plant. 62, 117–124.
Kempers A I 1974 Determination of sub-microquantities of ammonium and nitrates in soils with phenol, sodium-nitroprusside and hypochlorite. Geoderma 12, 201–206.
Schapendonk A H C M and De Vos A L F 1988 Implications of selecting for persistency on hydroponics in timothy (*Phleum pratense* L.). Euphytica S, 131–139.

M. L. van Beusichem (Ed.), *Plant nutrition – physiology and applications*, 87–92.

PLSO IPNC707

Interaction between nitrate assimilation in shoots and nitrate uptake by roots of soybean (*Glycine max*) plants: Role of carboxylate

B. TOURAINE, N. GRIGNON and C. GRIGNON
Laboratoire de Biochimie et Physiologie Végétales, Institut National de la Recherche Agronomique, (CNRS-URA 573), Ecole Nationale Supérieure Agronomique, Place Viala, F-34060 Montpellier Cedex 1, France

Key words: carboxylate, *Glycine max* L., nitrate uptake, OH^- excretion, phloem transport, shoot-root interaction, xylem transport

Abstract

The rates of accumulation of K^+, NO_3^- and reduced N (N_r) in shoots, relative to that of Ca^{2+}, were compared to the ratios K^+/Ca^{2+}, NO_3^-/Ca^{2+} and N_r/Ca^{2+} in the xylem sap of NO_3^--fed soybean plants in order to estimate the xylem transport of K^+, NO_3^- and N_r. Then, phloem transport of K^+ and N_r could be calculated. The quantification of NO_3^- flux in the xylem and NO_3^- accumulation in shoots allowed us to estimate the contribution of roots and shoots in the reduction of NO_3^-. Subtraction of the inorganic anion content from the main cation content estimated the accumulation of carboxylate anions (R^-). The export of R^- to roots was obtained by subtracting R^- accumulation in shoots from the theoretical production associated to NO_3^- reduction in these organs. The OH^- net excretion in the medium was monitored. Alkalinization of the medium was stopped by interruption of phloem conduction to the roots, and was not restored by glucose addition to the medium. Excretion of ^{14}C by roots after incorporation of $^{14}CO_2$ in leaves was greater than that from urea-fed plants. These results suggest that the OH^- equivalents excreted in the medium originated from R^- produced in shoots. The supply of K-malate to the shoots via the transpiration flow stimulated NO_3^- uptake. This result indicates that the circulation of R^- to the roots could be a mechanism by which NO_3^- assimilation in shoots controls NO_3^- uptake by roots.

Introduction

During NO_3^- nutrition, plants take up anions in excess of cations, and the balance of ionic absorption is maintained by a net influx of H^+ equivalents (true H^+ transport, or excretion of OH^- or of HCO_3^-) (Pitman, 1970). The assimilation of NO_3^- releases OH^- equivalents, which cannot be excreted in the medium when produced in shoots, because of the limited ability of phloem to transport OH^- or HCO_3^- (Raven, 1985). This results in a production of organic carboxylic anions (R^-) in shoots during NO_3^- assimilation. Since only a part of the produced R^- is retrieved in leaves, it has been hypothesized that R^- are transported to the roots, where they are decarboxylated, and that the HCO_3^- released are excreted in the medium (Ben Zioni *et al.*, 1971). This model presents a mechanism by which NO_3^- assimilation in shoots may stimulate NO_3^- uptake by roots (HCO_3^- ions released by decarboxylation of R^- are exchanged with the NO_3^- absorbed). In this model, K^+ is the accompanying cation of NO_3^- in xylem and of R^- in phloem.

The validity of this hypothesis has been questioned for barley (Cram, 1976), corn (Keltjens, 1981), tomato (Kirkby and Knight, 1977), and inoculated soybean (Israel and Jackson, 1982), which were reported to accumulate R^- in amounts approximately equivalent to NO_3^- reduction in shoots. In castor oil, the predominant ions transported are K^+ and NO_3^- in xylem sap, and K^+ and R^- in phloem sap (Kirkby and Armstrong, 1980), which is in accordance with the above hypothesis. These plants excreted more than 50% of the HCO_3^- they produced by reducing NO_3^-. On the basis of *in vitro* nitrate reductase assays, Van Beusichem *et al.* (1985, 1988) estimated that, in these plants, the roots were the site of about 44% of total NO_3^- reduction. Then, the decarboxylation of shoot-borne R^- accounted only for a little part of excreted HCO_3^-. Thus, these authors concluded that the cycling model of K^+/NO_3^- and K^+/R^- between roots and shoots was of minor importance (Van Beusichem *et al.*, 1985). Evidence for this model, and for its effect on NO_3^- uptake, must be looked for in plants which reduce NO_3^- mainly in shoots. Furthermore, it is important to measure accurately NO_3^- reduction in both roots and shoots. Unfortunately, nitrate reductase assays give poor estimates of actual NO_3^- reduction.

This report presents first a quantification of roots to shoots and shoots to roots transports, which allows to calculate the NO_3^- reduction in each organ, and then a study of the relationship between R^- phloemic translocation and NO_3^- uptake.

Materials and methods

All the experiments were conducted with *Glycine max* L., cv Kingsoy, grown for 20 days on a complete solution, where nitrogen was provided as 4 m*M* NO_3^-. The whole culture took place in a growth chamber with 25/20°C during the 14/10 h light/dark cycle. The ionic contents of shoots and roots were determined after drying of tissues and HCl extraction. K^+, Ca^{2+} and Mg^{2+} were assayed by flame spectrometry. NO_3^- was colorimetrically determined after reduction on a cadmium column on an autoanalyser. Total N_r contents were estimated by NH_4^+ colorimetric assay after volatilization of NO_3^- with H_2O_2 and Kjeldahl digestion of N_r. To obtain xylem sap, plants were decapitated 1 cm below the cotyledonary node, the root systems were introduced in tight pots containing the culture solution, and a 150 kPa pressure was applied with a 20%/80% O_2/N_2 mixture. The xylem exudates were collected in tubes connected to the hypocotyl sections with tygon tubing and assayed as tissues extracts. Net H^+ transport rates between the medium and the roots were measured in fresh medium bubbled with decarbonated air. The rates of net H^+ exchanges were estimated from the recorded automatic delivery of 10 m*N* KOH or H_2SO_4 necessary to maintain the pH of the medium at 5.50. In some experiments, hypocotyls were steam girdled to interrupt phloem conduction; we verified that this treatment did not affect xylem transport.

Results and discussion

Quantification of ionic and N transports between shoots and roots

From day 6 to day 20, the accumulations of K^+, Ca^{2+}, NO_3^- and N_r in roots and shoots were periodically assayed. When expressed as logarithm of contents and plotted against time, the results gave linear graphs. These regressions were used to calculate the accumulation rates of K^+ Ca^{2+}, NO_3^- and N_r in roots and shoots.

To collect the xylem sap exuding from excised root systems, we applied positive pressure on the medium bathing the roots to simulate the transpiration flux. We obtained an exudation rate equal to the transpiration rate with a pressure of 150 kPa. The K^+, Ca^{2+}, NO_3^- and N_r concentrations in the xylem exudates collected each 3 h during the day 16 were constant between 5 min and 50 min after shoot excision and root pressurization. We calculated the ratios K^+/Ca^{2+}, NO_3^-/Ca^{2+} and N_r/Ca^{2+} in the xylem exudates of 16-day-old plants. The values obtained did not depend upon the pressure applied (between 50 and 300 kPa). Moreover the ratios K^+/Ca^{2+} and NO_3^-/Ca^{2+} in the sap extracted from stem segments with gentle pressurization were equal to the ones calculated in the root exudates.

The accumulation rates and concentrations ratios in the xylem allowed us to calculate both xylem and phloem transports of the chemical species studied. Although Ca^{2+} is probably present in phloem tissue, it is certainly not transported in the sieve sap (Raven, 1977); for example, in castor oil, the K^+/Ca^{2+} ratio in phloem exudates was 2500 (Allen and Raven, 1987). Thus, we used the accumulation of this cation in shoots as an estimate of its transport by the xylem. The transport of K^+, NO_3^- and N_r in the xylem were calculated as the products of the Ca^{2+} flux in the xylem and the K^+/Ca^{2+}, NO_3^-/Ca^{2+} and N_r/Ca^{2+} ratios in the xylem exudate, respectively. The transport of K^+ in the phloem was estimated from the difference between its transport in the xylem and its accumulation in the shoots. As NO_3^- had not been found in appreciable amounts in phloem exudates of castor oil (Allen and Raven, 1987) and soybean (Fellows *et al.*, 1978; our assays), we assume that NO_3^- was not transported in the phloem. Then, the difference between xylem transport and shoot accumulation of $(NO_3^- + N_r)$ gave the rate of N_r export in the phloem. The results are summarized in Figure 1. NO_3^- reduction in shoots, estimated as the difference between NO_3^- import rate by the xylem and NO_3^- accumulation rate, accounted for 93% of the reduction in the whole plant. The proportion of N_r in xylem N in these plants was 32%. The main part of xylemic N_r originated from shoots. This result agrees with the studies made on wheat (Cooper *et al.*, 1986) and soybean (Rufty *et al.*, 1982) with ^{15}N. In the latter case, where shoot contribution in NO_3^- reduction was more than 80–90%, 40–60% of xylemic N was in reduced form.

On day 16, 203 μmol NO_3^- per plant were reduced, corresponding theoretically to an equal production of R^-. The difference between the total cationic content $(K^+ + Ca^{2+} + Mg^{2+})$ of the plants and their total inorganic anions content (total $P + SO_4^{2-} + NO_3^-$) estimated the accumulation of R^-. It accounted for only 38% of the negative charges released by NO_3^- reduction. Thus, the plants must have excreted the difference (127 μeq per plant) during the day 16. The roots accumulated virtually no R^-, so the 14 μeq negative charges released by NO_3^- reduction in these organs were excreted. The other negative charges excreted in the medium (113 μeq per plant) originated from the shoots. They accounted for 89% of total base excretion.

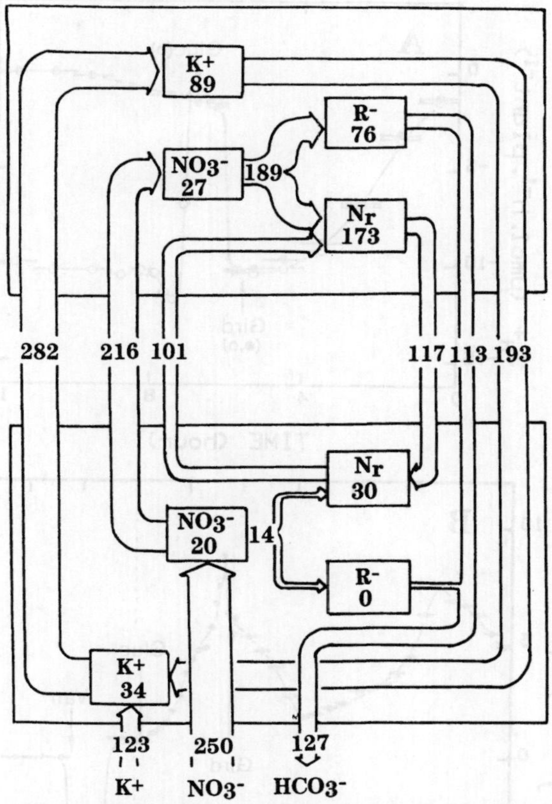

Fig. 1. Balance sheet of net transports in 16-day-old soybean plants grown on NO_3^- medium. Upper compartment represents the shoots, and lower one the roots. The data are expressed as μmol per day and per plant.

NO_3^- uptake by roots and efflux of OH^- equivalents to the medium

On the NO_3^- medium, the plants alkalinized the solution. On the contrary, both urea-fed plants during the whole culture, and plants grown on the NO_3^- medium but transferred on a N free medium for 5 d produced a H^+ net efflux (Fig. 2*A*). The OH^- net efflux observed in NO_3^- medium increased during the light period (with a maximum at the end of the period), and decreased during the dark period, after a 2 h lag (Fig. 2*B*). The OH^- net efflux, integrated over day 16, amounted for 110 μeq per plant, a value close to the prediction of the circulation model (127 μeq: see Fig. 1). This effect of light on the OH^- net excretion rate is in good accordance with the known kinetics of NO_3^- reduction in soybean (Rufty *et al.*, 1987). The study of the transport between roots and shoots led to the

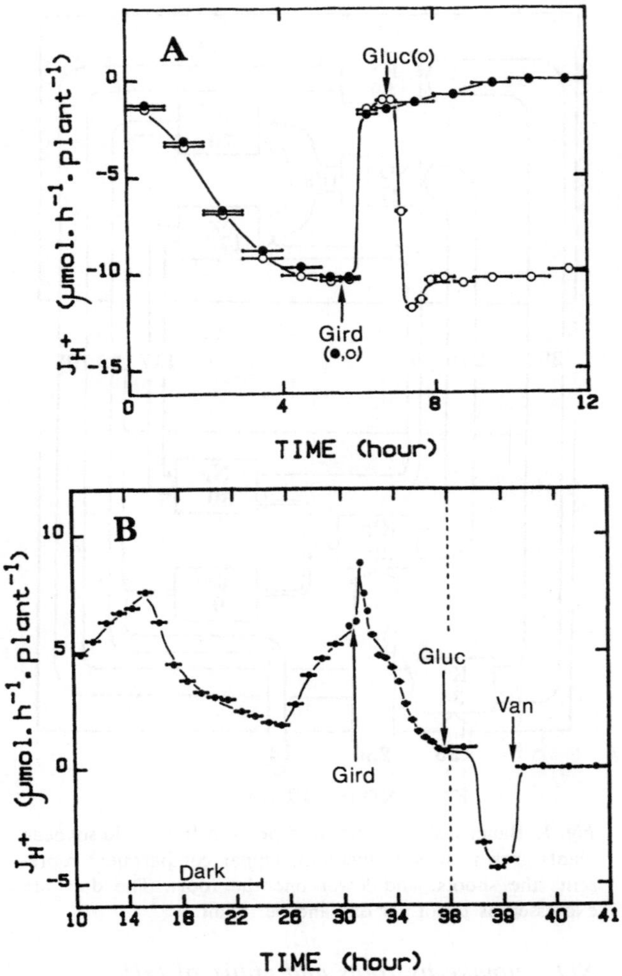

Fig. 2. Net H^+ transport rate into roots of intact soybeans. **A.** The plants were grown on NO_3^- medium for 11 d and on Cl^- (N free) medium for 5 d. **B.** The plants were grown on NO_3^- medium for 16 d. Gird: girdling of the base of the stem with water vapor stream; Gluc: addition of 100 m*M* glucose to the medium; Van: addition of 0.5 m*M* vanadate to the medium. Positive values of J_{H^+} correspond to an alkalinization of the medium.

conclusion that 89% of the excreted base equivalents originated in the leaf. To verify this point, we girdled the hypocotyls in order to interrupt the transport of R^- in the phloem. This treatment inhibited 85% of the OH^- net efflux within 6 h (Fig. 2*B*), which is very close to the 89% predicted. But this inhibition could have been caused by energy shortage in roots, due to the interruption of carbohydrate supply to the roots via the phloem sap. Indeed, girdling of NO_3^--deprived plants, which displayed a H^+ net efflux, stopped the acidification of the solution (Fig. 2*A*), presumably by inhibiting the proton pump activity. This was confirmed by restoration of the H^+ net efflux upon addition of 0.1 *M* glucose to the medium (Fig. 2*A*). However, when glucose was added to the NO_3^- solution after girdling of the plants, the alkalinization was not restored, but a H^+ net efflux appeared (Fig. 2*B*). It was inhibited by 0.5 m*M* vanadate, which indicates that the proton pump was reactivated by glucose. These results show that the inhibition of the OH^- net efflux on NO_3^- medium by girdling was not due to energy shortage, but to interruption of R^- supply. This conclusion implies that the OH^- net efflux corresponds to HCO_3^- excretion. This was verified by enclosing the roots of an intact plant in a sealed pot aerated with air passing through an ethanolamine CO_2 trap. A 30 min pulse of $^{14}CO_2$ was applied on a mature leaf. Excretion of volatile ^{14}C by the roots was greater in NO_3^- medium than in urea medium, even if we took into account a maximum isotopic dilution due to the respiration of all the C from urea absorbed (data not shown). Moreover, the ^{14}C excretion rate increased with NO_3^- concentration in the medium.

Transport of carboxylate in the phloem and NO_3^- uptake

The circulation model and the results on the OH^- net efflux in NO_3^--fed soybean plants support the hypothesis of a cycling of R^- and K^+ in plants which reduce NO_3^- in their leaves. The question then is the physiological role of such a circulation: is it involved in the regulation of NO_3^- uptake by roots? We calculated the net NO_3^- uptake rate by plants after interrupting R^- transport by steam girdling and restoring energy supply by addition of glucose to the medium. Girdling inhibited NO_3^- uptake by 90%, but glucose restored 50% to 80% of it (cf. figure 8 in Touraine *et al.*, 1988). This result suggests that the carboxylate translocation from shoots to roots is not essential for NO_3^- uptake. The HPLC assay of the acid fraction of phloem exudates collected in 20 m*M* EDTA showed that the main R^- transported from shoots to roots in our NO_3^--fed soybean plants were malate. Thus, we

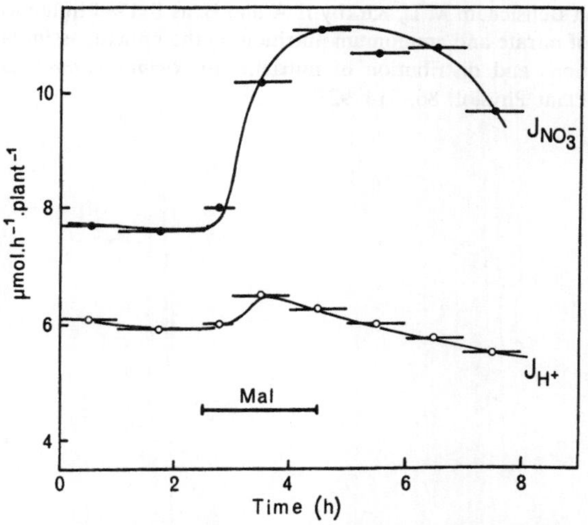

Fig. 3. Effect of introduction of malate in the transpiration flow on NO_3^- absorption. The curves are NO_3^- net uptake ($J_{NO_3^-}$) and H^+ net influx (J_{H^+}) by a part of the root system of a soybean plant installed on a split root device. Mal, addition of 5 m*M* K-malate in the medium bathing the other part of the root system.

provided malate to shoots via the xylem sap. Plants were installed in a split root device, and supplied with 0.5 m*M* KNO_3 plus 0.1 m*M* $CaSO_4$ on one side and with only 0.1 m*M* $CaSO_4$ on the other side. The NO_3^--deprived roots were excised 10 cm from the apices to ensure free access of the solution to the xylem, and 5 m*M* K-malate was then added in the medium. The NO_3^- uptake by the other part of the root system was calculated from the decrease of NO_3^- concentration in the medium. The treatment stimulated the net uptake of NO_3^-, but not the net efflux of OH^- (Fig. 3). When the experiment was done with Ca-malate, there was no significant effect, either on NO_3^- uptake or on OH^- net efflux. Perhaps, this was due to the sequestering of malate in leaves, which were unable to recirculate Ca^{2+} as an accompanying cation in the phloem.

Conclusion

Our results suggest that an increase in the malate pool in leaves, resulting from reduction of NO_3^-, may stimulate NO_3^- net uptake by roots. This effect is dependent on transport of both K^+ and NO_3^- to the leaves. This stimulation did not, however, alter OH^- net efflux, which indicates that an increased absorption of K^+ must have occurred so as to maintain charge balance. This suggests that the observed effect is more complex than a stimulation of NO_3^-/HCO_3^- antiport by malate decarboxylation in roots.

References

Allen S and Raven J A 1987 Intracellular pH regulation in *Ricinus communis* grown with ammonium or nitrate as N source: The role of long distance transport. J. Exp. Bot. 38, 580–596.

Ben Zioni A, Vaadia Y and Lips S H 1971 Nitrate uptake by roots as regulated by nitrate reduction products of the shoot. Physiol. Plant. 24, 288–290.

Cooper H D, Clarkson D T, Johnston M G, Whiteway J N and Loughman B C 1986 Cycling of amino-nitrogen between shoots and roots in wheat seedlings. Plant and Soil 91, 319–322.

Cram W J 1976 The regulation of nutrient uptake by cells and roots. *In* Transport and Transfer Processes in Plants. Eds. I F Wardlaw and J B Passioura. pp 113–124. Academic Press, New York.

Fellows R J, Egli D B and Leggett J E 1978 A pod leakage technique for phloem translocation studies in soybean (*Glycine max* L.] Merr.). Plant Physiol. 62, 812–814.

Israel D W and Jackson W A 1982 Ion balance, uptake, and transport processes in N_2-fixing and nitrate- and urea-dependent soybean plants. Plant Physiol. 69, 171–178.

Keltjens W G 1981 Absorption and transport of nutrient cations and anions in maize roots. Plant and Soil 63, 39–46.

Kirkby E A and Armstrong M J 1980 Nutrient uptake by roots as regulated by nitrate assimilation in the shoot of castor oil plants. Plant Physiol. 65, 286–290.

Kirkby E A and Knight A H 1977 Influence of the level of nitrate nutrition on ion uptake and assimilation, organic acid accumulation, and cation-anion balance in whole tomato plants. Plant Physiol. 60, 349–353.

Pitman M G 1970 Active efflux from cells of low-salt barley roots during salt accumulation. Plant Physiol. 45, 787–790.

Raven J A 1977 H^+ and Ca^{2+} in phloem and symplast: Relation of relative immobility of the ions to the cytoplasmic nature of the transport paths. New Phytol. 79, 465–480.

Raven J A 1985 pH regulation in plants. Sci. Prog. 69, 495–509.

Rufty T W, Volk R J, MacClure P R, Israel D W and Raper C D 1982 Relative content of NO_3^- and reduced N in xylem exudate as an indicator of root reduction of concurrently absorbed $^{15}NO_3^-$. Plant Physiol. 69, 166–170.

Rufty T W, Volk R J and MacKown C T 1987 Engodenous NO_3^- in the root as a source of substrate for reduction in the light. Plant Physiol. 84, 1421–1426.

Touraine B, Grignon N and Grignon C 1988 Charge balance in NO_3^--fed soybean: Estimation of K^+ and carboxylate recirculation. Plant Physiol. 88, 605–612.

Van Beusichem M L, Baas R, Kirkby E A and Nelemans J A 1985 Interacellular pH regulation during NO_3^- assimilation in shoot and roots of *Ricinus communis*. Plant Physiol. 78, 768–773.

Van Beusichem M L, Kirkby E A and Baas R 1988 Influence of nitrate and ammonium nutrition on the uptake, assimilation, and distribution of nutrients in *Ricinus communis*. Plant Physiol. 86, 914–921.

M. L. van Beusichem (Ed.), *Plant nutrition – physiology and applications*, 93–99. PLSO IPNC598A

Diurnal changes of nitrate content and nitrate reductase activity in different organs of *Atriplex hortensis* (C_3 plant) and *Amaranthus retroflexus* (C_4 plant)

G. GEBAUER
Institut für Botanik und Mikrobiologie, Technische Universität München, Arcisstr. 21, D-8000 München 2, FRG. Present address: Lehrstuhl für Pflanzenökologie, Universität Bayreuth, Universitätsstr. 30, D-8580 Bayreuth, FRG

Key words: *Amaranthus retroflexus* L., *Atriplex hortensis* L., C_3 and C_4 plant, diurnal changes, nitrate, nitrate reductase, transpiration

Abstract

Diurnal fluctuations of nitrate content and of nitrate reductase activity (NRA) in leaves, petioles + shoot axes and roots of a C_3 plant (*Atriplex hortensis* L.) and a C_4 plant (*Amaranthus retroflexus* L.) were investigated under natural light conditions. In addition, transpiration of the leaves and photosynthetically active radiation were monitored. Irrespective of the plant organ and species, nitrate content and NRA showed inverse diurnal changes, with highest nitrate concentrations per g dry weight being evident in the morning and highest NRA per g dry weight in the early afternoon. The fluctuations, however, were more pronounced in the shoot organs than in the roots. Irrespective of a much higher transpiration of the leaves of the C_3 plant, the absolute fluctuations of the nitrate concentrations and of the NRA were quite similar in the two species, indicating that the transpiration rate was of minor importance for the rate of nitrate supply to the leaves.

Introduction

Nitrate is the most important mineral nitrogen source in the majority of naturally occurring plant communities (Ellenberg, 1977) and for most agricultural plants. The reduction of nitrate catalyzed by the enzyme nitrate reductase is the first and rate limiting step of nitrate assimilation by plants (cf. Lee and Stewart, 1978). Nitrate reductase activity (NRA) is heavily influenced by the amount of nitrate present and by light (Beevers and Hageman, 1983). In most plant species nitrate is preferentially reduced in leaves (Gebauer *et al.*, 1988; Pate, 1980), but some NRA is also found in roots and other plant organs (cf. Gebauer *et al.*, 1988).

Up to now most investigations on diurnal changes of NRA have been focused on leaves only, and measurements of concommitant diurnal changes of nitrate concentrations in plant organs have been omitted. Furthermore, most experiments have been carried out under controlled, but artificial environmental conditions at constant radiation intensities during the light period.

The aim of the present investigation was to compare the diurnal fluctuations of NRA and nitrate concentrations under natural light conditions not only in leaves, but in all organs of two plant species having different transpiration rates and thus presumably exhibiting different velocities of passive nitrate transport from the roots to the leaves in the xylem. To this end a C_3 plant with a high transpiration rate and a C_4 plant with a typically lower transpiration rate were chosen for investigation.

Materials and methods

Plant material

The experiments were carried out with *Atriplex hortensis* L. (C_3 plant) and *Amaranthus retroflexus* L. (C_4 plant). These two species are annual weeds found in nitrogen-rich habitats and are of similar size and exhibit similar ecological response. The seeds were germinated in the greenhouse and the seedlings transplanted into asbestos cement boxes (150 × 80 × 80 cm) in the open air when 4–6 cm tall. Only equally tall seedlings were employed. 98 seedlings per species and box were planted at regular spacings. The plants were cultured for 27 days in soil containing 70% "Fruhstorfer Erde T" (Industrieerdenwerk E. Archut, Lauterbach/Fulda, FRG) and 30% sand. "Fruhstorfer Erde T" is a peaty soil with a 40% clay content. At the beginning of the experiment the inorganic nitrogen content (nitrate + ammonium) was 63.3 $\mu g\,N \times g$ dry $soil^{-1}$. 83% of this inorganic nitrogen was nitrate.

Experimental procedures

Four plants of each species were selected randomly for harvest every three hours from 6^{00} h a.m. until 3^{00} h a.m. (Central European Summer Time) of the following day. Each plant was separated into three samples: laminae, petioles + shoot axis and roots. Each sample was analyzed for NRA, dry weight and nitrate content. During the light period transpiration of leaves of non-harvested plants was measured every hour in five replications.

Methods

NRA was measured according to a modified version of the *in vivo*-test described by Jaworski (1971). In addition to the test procedure described by Gebauer *et al.* (1984), the flasks containing buffer + plant material were vacuum infiltrated and incubated under a N_2 atmosphere for 2 h in the dark at 30°C in incubation chambers. With this procedure a significantly higher nitrite release was found (see Gebauer *et al.* 1988). An incubation buffer containing 0.08 *M* KNO_3, 0.25 *M* KH_2PO_4, 1.5% (v/v) propanol, pH 7.5 was found to be optimal for the determination of NRA in both *Atriplex hortensis* and *Amaranthus retroflexus*. Nitrate in plant samples was determined by anion exchange HPLC and UV detection as described by Gebauer *et al.* (1984). Transpiration was measured with the Steady State Porometer LI-1600 and photosynthetically active radiation (PAR) above the plants was monitored continuously with the self-registrating solar monitor LI-1776 from LI-COR.

Results

Transpiration

During the light period transpiration of the leaves in a ventilated cuvette ranged between mean minimum values of 3.11 (Atriplex at 7^{00} h) or 2.17 $\mu mol\ H_2O \times cm^{-2} \times s^{-1}$ (Amaranthus at 7^{00} h) and mean maximum values of 14.14 (Atriplex at 13^{00} h) or 8.59 $\mu mol\ H_2O \times cm^{-2} \times s^{-1}$ (Amaranthus at 13^{00} h) (Fig. 1). From 7^{00} h until 20^{00} h the mean transpiration of Atriplex leaves was always higher than that of Amaranthus leaves at similar points of time. These differences in transpiration between the two species were statistically significant from 10^{00} h until 19^{00} h ($P < 0.05$ according to Student's t-test). Mean transpiration of Amaranthus leaves during the whole light period was only 60% of that found for Atriplex leaves.

Diurnal changes of nitrate content

The highest nitrate concentrations per g of total plant material (calculated from the nitrate concentrations and the dry weights of the individual plant organs) were found in both species at 9^{00} h (for Atriplex 386 and for Amaranthus 700 $\mu mol\ NO_3^- \times g$ dry wt^{-1}). At that time radiation and transpiration were comparatively low (Fig. 1). During the period of highest radiation and most rapid transpiration (9^{00} h–15^{00} h) the nitrate concentrations in both species decreased considerably. Between 15^{00} h and 21^{00} h nitrate concentrations stagnated at minimum levels of about 200 (Atriplex) or 400 $\mu mol\ NO_3^- \times g$ dry wt^{-1} (Amaranthus). During the night nitrate concen-

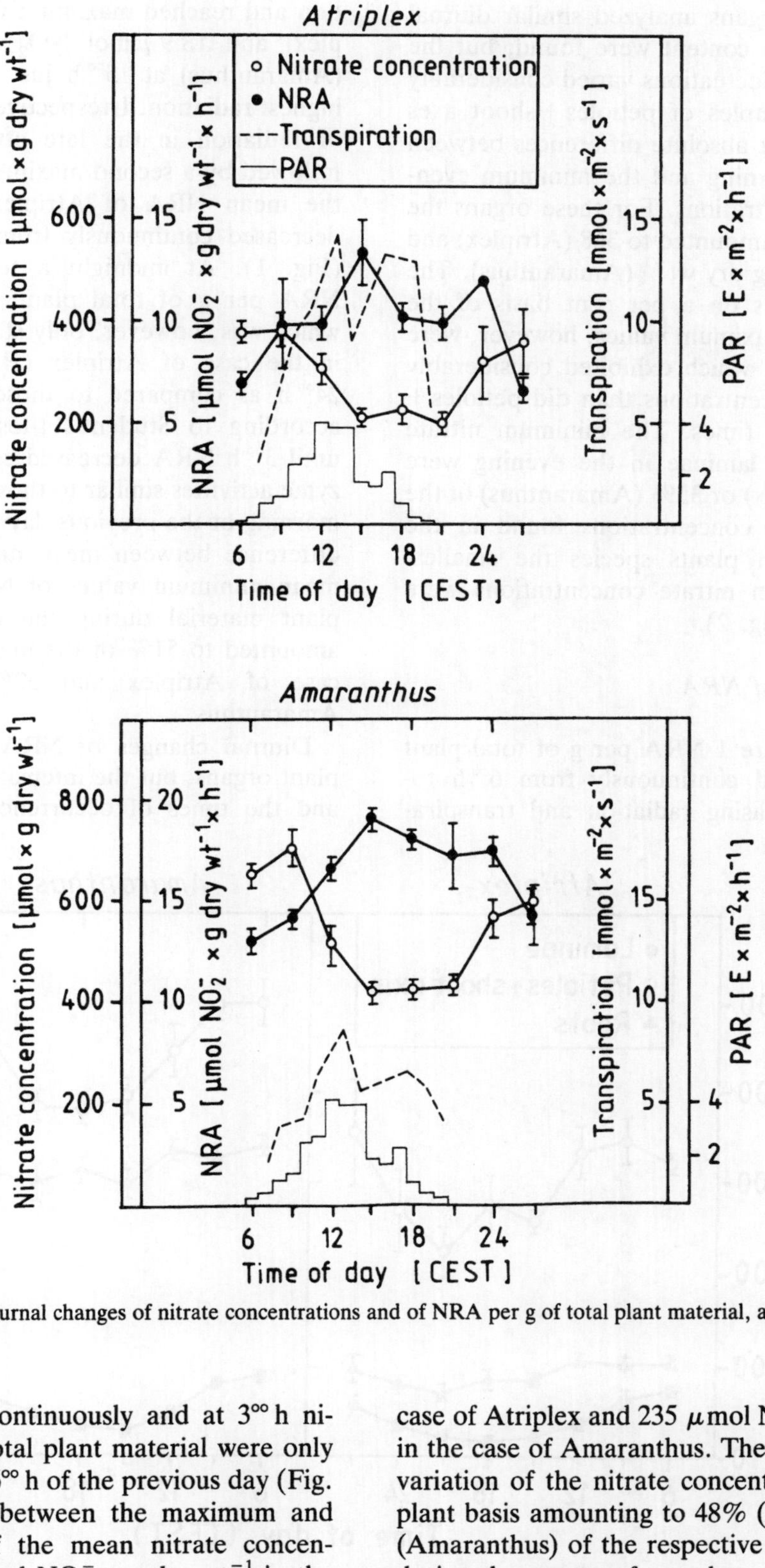

Fig. 1. Mean (±1 s.e.) diurnal changes of nitrate concentrations and of NRA per g of total plant material, as well as transpiration of leaves and PAR.

trations increased continuously and at 3^{00} h nitrate levels of the total plant material were only slightly less than at 6^{00} h of the previous day (Fig. 1). The difference between the maximum and minimum values of the mean nitrate concentrations was 185 μmol $NO_3^- \times$ g dry wt^{-1} in the case of Atriplex and 235 μmol $NO_3^- \times$ g dry wt^{-1} in the case of Amaranthus. These data indicate a variation of the nitrate concentration on a total plant basis amounting to 48% (Atriplex) or 34% (Amaranthus) of the respective maximum values during the course of one day.

For all plant organs analyzed similar diurnal rhythms of nitrate content were found, but the intensities of the fluctuations varied considerably (Fig. 2). The samples of petioles + shoot axes showed the largest absolute differences between the maximum morning and the minimum evening nitrate concentrations. For these organs the mean differences amounted to 368 (Atriplex) and 357 μmol NO_3^- $\times$ g dry wt^{-1} (Amaranthus). The largest fluctuations on a per cent basis of the individual daily maximum values, however, were found in laminae, which exhibited considerably lower nitrate concentrations than did petioles + shoot axes at all times. The minimum nitrate concentrations in laminae in the evening were only 25% (Atriplex) or 35% (Amaranthus) of the maximum nitrate concentrations found in the morning. In both plants species the smallest diurnal changes in nitrate concentrations were found in roots (Fig. 2).

Diurnal changes of NRA

As shown in Figure 1 NRA per g of total plant material increased continuously from 6^{00} h together with increasing radiation and transpiration and reached maximum values of 13.4 (Atriplex) and 18.9 μmol NO_2^- $\times$ g dry wt^{-1} $\times$ h^{-1} (Amaranthus) at 15^{00} h just after the period of highest radiation. Irrespective of a slight increase of radiation in the late afternoon which was followed by a second maximum of transpiration, the mean NRA of Atriplex and Amaranthus decreased continuously from 15^{00} h until 21^{00} h (Fig. 1). At midnight a second maximum of NRA per g of total plant material was found which was, however, only statistically significant in the case of Atriplex ($P < 0.05$ for data of 24^{00} h as compared to those of 21^{00} h or 3^{00} h according to Student's t-test). From midnight until 3^{00} h NRA decreased considerably and enzyme activities similar to those found in the early morning of the previous day were observed. The difference between mean maximum values and mean minimum values of NRA per g of total plant material during the course of one day amounted to 51% of the maximum value in the case of Atriplex and 32% in the case of Amaranthus.

Diurnal changes of NRA were found in all plant organs, but the intensity of the fluctuations and the times of occurrence of the maximum

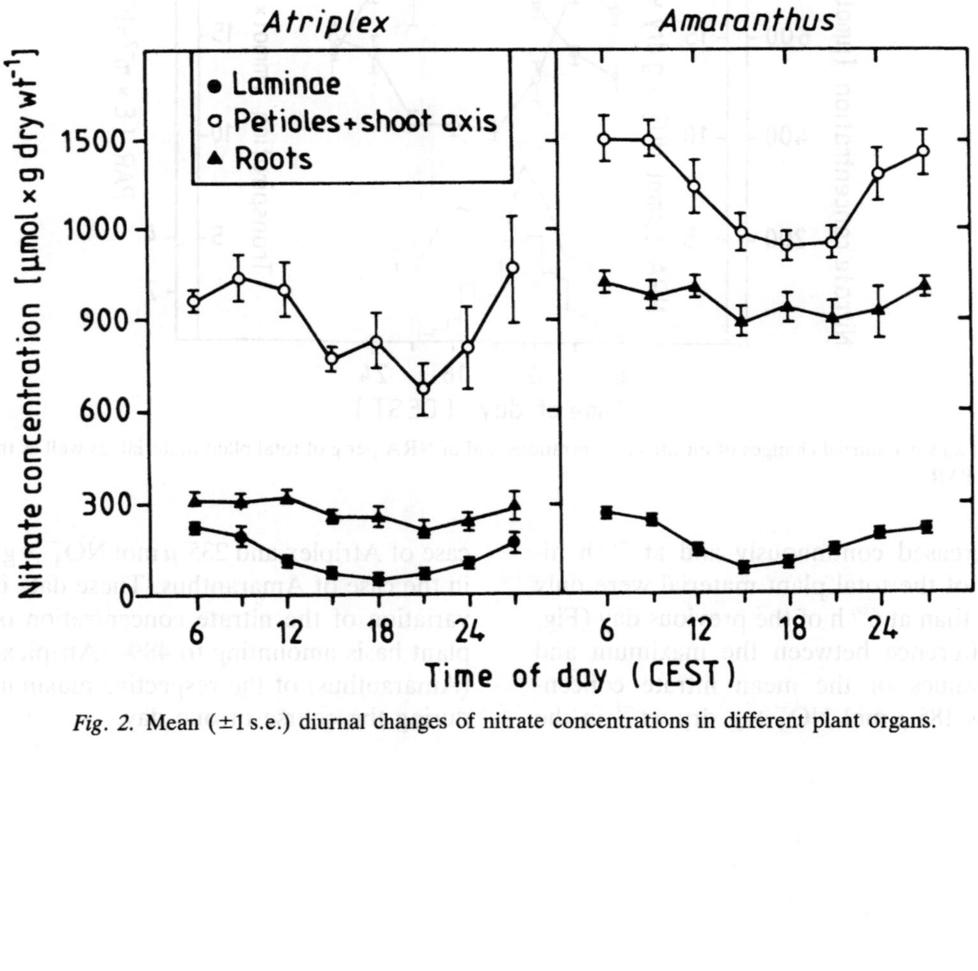

Fig. 2. Mean (±1 s.e.) diurnal changes of nitrate concentrations in different plant organs.

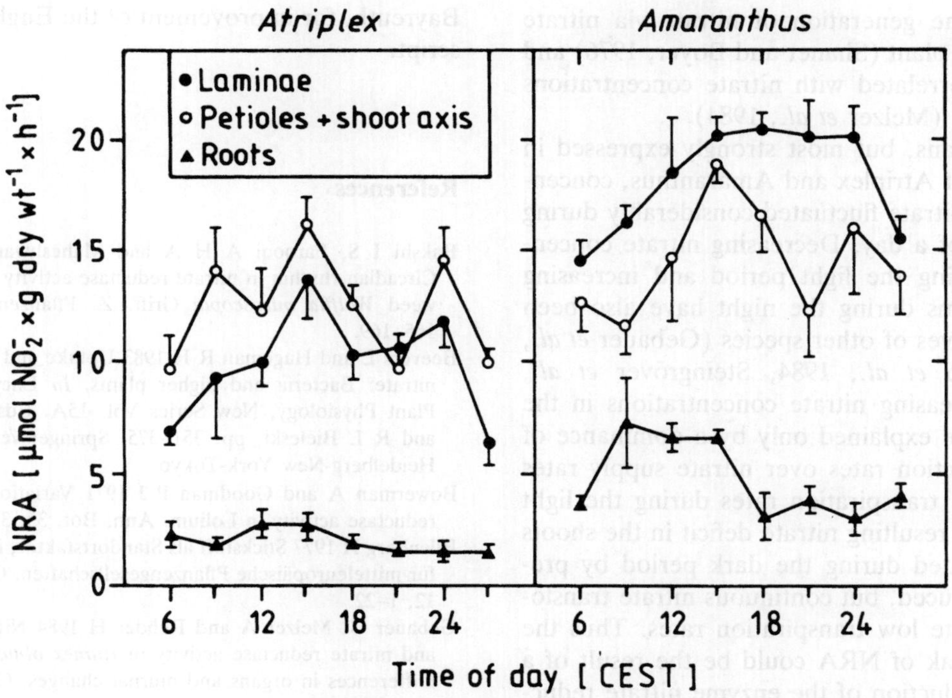

Fig. 3. Mean (±1 s.e.) diurnal changes of NRA in different plant organs.

values were different for the individual organs analyzed (Fig. 3). In roots of both species maximum NRA occurred earlier than in laminae or in petioles and shoot axes. In general, NRA in roots was considerably lower than in shoot organs of both species. In petioles + shoot axes of Atriplex NRA was even higher than in laminae. A similar result had been found in prior experiments with young individuals of this species (see Gebauer *et al.*, 1987). The midnight peak of NRA was most clearly seen in the samples of petioles + shoot axes of both species, but it was not found in roots.

Discussion

Previous investigations on diurnal changes of NRA yielded entirely contradictory results depending on experimental conditions: Under natural light conditions NRA in leaves was always highest at noon or in the early afternoon (*e.g.* Bowerman and Goodman 1971, Gebauer *et al.*, 1984; Hipkin *et al.*, 1984). Under controlled laboratory conditions at constant light intensities, however, no diurnal changes of NRA in leaves (Steingröver *et al.*, 1986) or maximum values at the beginning (Bakshi *et al.*, 1978), in the middle (Janiesch, 1973) or at the end of the light period (Lillo and Henriksen, 1984) have been observed. In cell suspension cultures the appearance of diurnal changes of NRA additionally depended on the age of the cell cultures (Renner and Beck, 1988).

The results found with laminae and petioles + shoot axes of Atriplex and Amaranthus are in accordance with results of similar experiments under natural conditions and point to the conclusion that a maximum of NRA in the shoot organs of plants immediately follows the period of highest radiation under natural light conditions (Gebauer *et al.*, 1984; Hipkin *et al.*, 1984). As also shown by Keltjens and Nijenstein (1987), the maximum of NRA in roots, however, seems to occur somewhat earlier. A second maximum of NRA at midnight, which was most clearly seen in petioles + shoot axes of Atriplex and Amaranthus, was also found by Steer (1974) in leaves of *Capsicum annuum*. This maximum can not be explained by changes in light intensity, but in addition to the light-dependent generation of reducing equivalents, NRA is also regulated by the generation of nitrate via nitrate fluxes in the plant (Shaner and Boyer, 1976) and is closely correlated with nitrate concentrations accumulated (Melzer *et al.*, 1984).

In all organs, but most strongly expressed in the shoots of Atriplex and Amaranthus, concentrations of nitrate fluctuated considerably during the course of a day. Decreasing nitrate concentrations during the light period and increasing concentrations during the night have also been found in leaves of other species (Gebauer *et al.*, 1984; Kallio *et al.*, 1984, Steingröver *et al.*, 1986). Decreasing nitrate concentrations in the shoot can be explained only by a dominance of nitrate reduction rates over nitrate supply rates despite high transpiration rates during the light period. The resulting nitrate deficit in the shoots is compensated during the dark period by presumably reduced, but continuous nitrate translocation despite low transpiration rates. Thus the midnight peak of NRA could be the result of a substrate induction of the enzyme nitrate reductase during the refilling period. This substrate induction, however, might be terminated by a limited generation of reducing equivalents during the night. The coincidence of the midnight maximum of NRA and the replenishment of nitrate concentrations was most obvious in petioles + shoot axes. The large diurnal fluctuations of nitrate concentrations that we found also in laminae of Atriplex and Amaranthus contradict the opinion of Steer (1974), who presumed that most nitrate in leaves is fixed in vacuoles and therefore is hardly accessible for nitrate reduction.

Irrespective of considerably different transpiration rates in leaves of the C_3 plant *Atriplex hortensis* and of the C_4 plant *Amaranthus retroflexus*, absolute diurnal fluctuations of NRA and of nitrate concentrations were quite similar in all organs of both species. This observation indicates that lower transpiration rates in the leaves of the C_4 plant were presumably compensated for by higher nitrate concentrations per mol of water transported in the xylem.

Acknowledgements

The author thanks Miss Gisela Schmidt for skillful assistance and Dr P Ziegler, Universität Bayreuth, for improvement of the English manuscript.

References

Bakshi I S, Farooqi A H A and Maheshwari S C 1978 Circadian rhythm in nitrate reductase activity in the duckweed *Wolffia miroscopia* Griff. Z. Pflanzenphysiol. 90, 165–169.

Beevers L and Hageman R H 1983 Uptake and reduction of nitrate: Bacteria and higher plants. *In* Encyclopedia of Plant Physiology, New Series Vol. 15A. Eds. A Läuchli and R L Bieleski, pp. 351–375. Springer-Verlag, Berlin-Heidelberg-New York-Tokyo.

Bowerman A and Goodman P J 1971 Variation in nitrate-reductase activity in Lolium. Ann. Bot. 35, 353–366.

Ellenberg H 1977 Stickstoff als Standortsfaktor, insbesondere für mitteleuropäische Pflanzengesellschaften. Oecol. Plant. 12, 1–22.

Gebauer G, Melzer A and Rehder H 1984 Nitrate content and nitrate reductase activity in *Rumex obtusifolius* L. I. Differences in organs and diurnal changes. Oecologia 63, 136–142.

Gebauer G, Rehder H and Wollenweber B 1988 Nitrate, nitrate reduction and organic nitrogen in plants from different ecological and taxonomic groups of Central Europe. Oecologia 75, 371–385.

Gebauer G, Schuhmacher M I, Krstic B, Rehder H and Ziegler H 1987 Biomass production and nitrate metabolism of *Atriplex hortensis* L. (C_3 plant) and *Amaranthus retroflexus* L. (C_4 plant) in cultures at different levels of nitrogen supply. Oecologia 72, 303–314.

Hipkin C R, Al Gharbi A and Robertson K P 1984 Studies on nitrate reductase in British angiosperms. II. Variations in nitrate reductase activity in natural populations. New Phytol. 97, 641–651.

Janiesch P 1973 Beitrag zur Physiologie der Nitrophyten, Nitratspeicherung und Nitratassimilation bei *Anthriscus sylvestris* Hoffm. Flora 162, 479–491.

Jaworski E G 1971 Nitrate reductase assay in intact plant tissue. Biochem. Biophys. Res. Commun. 43, 1274–1279.

Kallio H, Rousku R, Salminen A and Tikanmäki E 1984 Diurnal variation in nitrate content of red beets. J. Agric. Sci. Finland 56, 239–243.

Keltjens W G and Nijenstein J H 1987 Diurnal variations in uptake, transport and assimilation of NO_3^- and efflux of OH^- in maize plants. J. Plant Nutr. 10, 887–900.

Lee J A and Steward R G 1978 Ecological aspects of nitrogen assimilation. Adv. Bot. Res. 6, 2–43.

Lillo C and Henriksen A 1984 Comparative studies of diurnal variations of nitrate reductase activity in wheat, oat and barley. Physiol. Plant. 62, 89–94.

Melzer A, Gebauer G and Rehder H 1984 Nitrate content and nitrate reductase activity in *Rumex obtusifolius* L. II. Responses to nitrogen starvation and nitrogen fertilization. Oecologia 63, 380–385.

Pate J S 1980 Transport and partitioning of nitrogenous solutes. Annu. Rev. Plant Physiol. 31, 313–340.

Renner U and Beck E 1988 Nitrate reductase activity of photoautotrophic suspension culture cells of *Chenopodium rubrum* is under the hierarchial regime of NO_3^-, NH_4^+ and light. Plant Cell Physiol. 29, 1123–1131.

Shaner D L and Boyer J S 1976 Nitrate reductase activity in maize (*Zea mays* L.) leaves. I. Regulation by nitrate flux. Plant Physiol. 58, 499–504.

Steer B T 1974 Control of diurnal variations in photosynthetic products. II. Nitrate reductase activity. Plant Physiol. 54, 762–765.

Steingröver E, Ratering P and Siesling J 1986 Daily changes in uptake, reduction and storage of nitrate in spinach grown at low-light intensity. Physiol. Plant. 66, 550–556.

M.L. van Beusichem (Ed.), *Plant nutrition – physiology and applications*, 101–106 PLSO IPNC598B

Nitrate assimilation and nitrate content in different organs of ash trees (*Fraxinus excelsior*)

G. GEBAUER and J. STADLER
Lehrstuhl für Pflanzenökologie, Universität Bayreuth, Universitätsstr. 30, D-8580 Bayreuth, FRG

Key words: biomass, *Fraxinus excelsior* L., nitrate, nitrate reductase, plant organs

Abstract

Nitrate reductase activity (NRA), nitrate content and biomass proportions of leaflets, leaf stalks, shoot axis segments and fine and large roots of 10–13 years old ash trees (*Fraxinus excelsior* L.) growing in their natural habitats were investigated. Nitrate distribution between different organs of *Fraxinus excelsior* was quite similar to that found in herbaceous plants. The highest nitrate concentrations per g dry weight were found in the leaf stalks. In contrast to general textbook knowledge, the most pronounced nitrate assimilation was not found in roots, but in the shoot of this tree species. The highest NRA per g dry weight and also per total biomass was found in the leaflets, although root biomass was much higher than total leaflet biomass. The implications for nitrate use in tree species are discussed.

Introduction

Nitrate is usually the most important form of inorganic nitrogen available to higher plants. Before being assimilated into amino acids, nitrate must be reduced to ammonium. The first and rate limiting step of this reduction process is catalyzed by the enzyme nitrate reductase (cf. Lee and Steward, 1978). In principle, nitrate reduction is possible in roots as well as in aboveground organs of higher plants. Most herbaceous plants reduce nitrate preferently in leaves (Gebauer *et al.*, 1988; Pate, 1980). Woody plants (trees and shrubs), however, have commonly been considered to reduce nitrate nearly exclusively in roots (see *e.g.* Andrews, 1986; Larcher, 1980; Marschner, 1986; Mengel, 1984; Pate, 1980; Runge, 1983). These findings can be traced back to xylem sap analyses by Bollard (1956; 1960), who found that most nitrogen is transported to the shoots of apple trees as amino acids. Smirnoff *et al.* (1984) and Al Gharbi and Hipkin (1984) were the first to show that trees of various taxonomic groups (including gymnosperms and angiosperms) have also nitrate reductase activity in leaves under natural growing conditions. But they did not analyze the relative proportions of shoot and root nitrate reduction which are not only a function of nitrate reductase activity, but also of total biomass and of nitrate concentrations.

The aim of our investigation was thus to evaluate the proportions of nitrate reductase activity in different organs of a common tree species together with results of analyses of biomass proportions and of nitrate concentrations in the respective organs.

It must be pointed out that the *in vivo*-test of nitrate reductase activity will not guarantee the reflection of the nitrate reduction occurring in intact plants with absolute accuracy, since manipulations which are necessary to perform the test create artificial conditions (see discussions by Gebauer *et al.*, 1984; and Van Beusichem *et al.*, 1987). These artificial conditions, however, are the same for all samples analyzed. Thus data obtained with different organs in one experiment should at least be comparable among each other.

Materials and methods

Four 10–13 years old trees of *Fraxinus excelsior* L. were used for investigation. In July 1988 the trees were excavated at their natural habitat in the "Studentenwald" near Bayreuth, which is a

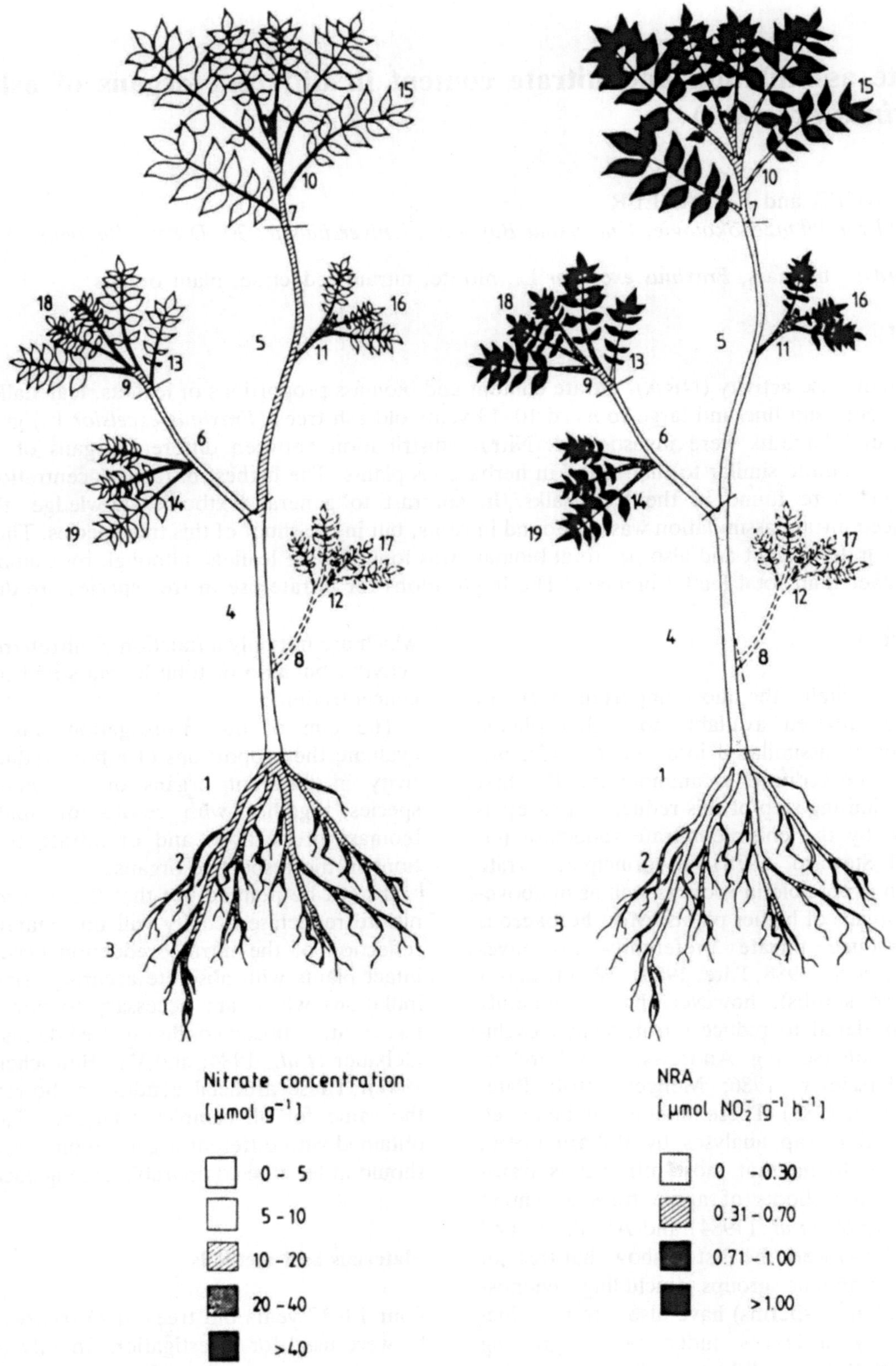

Fig. 1. Distribution of nitrate and NRA within 10–13 years old ash trees. The numbers indicated refer to the organs listed in the Table 1 and 2. Dotted compartments (numbers 8, 12 and 17) were present in two of the four trees only and thus are not considered.

nutrient-rich decidous forest, and were separated into 19 compartments as shown in Figure 1. Each compartment was analyzed for total dry weight, and subsamples of each compartment were analyzed for nitrate reductase activity (NRA) and for nitrate concentration per g dry weight. From these data nitrate content and NRA per compartment were calculated.

NRA was measured according to a modified version of the *in vivo*-test described by Jaworski (1971). For details of the test and the optimization procedures of the incubation buffer see Gebauer *et al.* (1984) and Gebauer (1990). An incubation buffer of the following composition was found to be optimal for the determination of NRA in *Fraxinus excelsior*: 0.06 *M* KNO_3, 0.15 *M* KH_2PO_4, 1.0% (v/v) propanol, pH 7.0. Nitrate in the plant samples was determined by anion exchange HPLC and UV detection as described by Gebauer *et al.* (1984). For dry weight determinations the plant material was dried at 105°C until constant weight.

Results

Nitrate content in different plant organs

Table 1 shows the mean nitrate concentrations (per g dry weight), biomass proportions and absolute nitrate contents of all plant compartments analyzed. In addition it also shows grouped data of nitrate contents of different compartments. Figure 1 illustrates the distribution of compartments with high or low nitrate concentrations within the trees. The highest nitrate concentrations were found in the leaf stalks (39–53 $\mu mol \times g^{-1}$), irrespective of their posi-

Table 1. Mean nitrate concentrations (on dry weight basis) and dry weights, including their standard errors, of 19 compartments of 10–13 years old ash trees (n = 4). Nitrate contents per compartment are calculated from these data. The numbers of the compartments refer to their position within the trees (see Fig. 1)

	Nitrate concentration [$\mu mol\ g^{-1}$]		Dry weight [g]		Nitrate content per compartment [μmol]			
No.	x̄	s.e.	x̄	s.e.	x̄			
Leaflets								
19	3.36	1.20	7.85	4.82	26.38			
18	2.91	1.11	6.03	2.19	17.55			
17	4.25	---	1.80	---	7.65	165.31		
16	3.20	1.67	24.20	6.24	77.44			
15	5.60	2.73	6.48	2.94	36.29		676.01	
Leaf stalks								
14	43.49	21.86	1.62	0.79	70.45			
13	49.63	28.40	1.32	0.27	65.51			
12	53.32	---	0.64	---	34.12	510.70		3199.39
11	39.37	12.71	6.01	1.59	236.61			
10	42.11	25.32	2.47	0.50	104.01			
Current year stem								
9	14.86	0.91	1.09	0.38	16.20			
8	3.08	---	2.94	---	9.06	97.72		
7	17.01	6.70	4.26	1.37	72.46		2523.38	
Old stem								
6	8.55	0.89	35.35	11.71	302.24			
5	11.71	3.93	88.84	21.76	1040.32	2425.66		
4	5.82	2.53	186.10	79.00	1083.10			
Roots								
3	37.09	24.19	1.02	0.23	37.83			
2	38.48	17.29	13.98	2.58	537.95	2777.85	2777.85	2777.85
1	18.13	5.83	121.46	27.46	2202.07			

tion within the canopy. Only root tips and fine roots exhibited nitrate concentrations nearly as high as those found for leaf stalks. The lowest nitrate concentrations were found in the leaflets. The mean nitrate concentration in leaflets was always less than 15% of that found for corresponding leaf stalks. This great difference in nitrate concentrations between leaf stalks and leaflets reveals the leaflets to represent a strong nitrate sink. Nitrate concentrations in large roots and in different segments of the stem ranged between those found for fine roots and for leaflets. Absolute nitrate contents of roots (2778 μmol NO_3^- in 137 g dry substance) and of stem material (2523 μmol NO_3^- in 319 g dry substance) were quite similar, while the total nitrate content of leaf material and also leaf biomass were much smaller (676 μmol NO_3^- in 58 g dry substance). Leaflets had a total nitrate content of only 165 μmol NO_3^- in 46 g dry substance.

It must be pointed out that the nitrate concentrations varied greatly among the four trees analyzed. This is indicated by the wide range of standard errors and reflects the different nutritional status of the trees. The gradients between the different organs shown in Figure 1, however, were similar for all trees analyzed.

NRA in different plant organs

The highest NRA per g dry weight was found in leaflets (Table 2). Mean NRA of leaflets was quite similar for all positions within the canopy (Fig. 1). It ranged only between 1.34 and 1.53 μmol $NO_2^- \times g^{-1} \times h^{-1}$. Very low NRA per g dry substance was detected in large roots and in old stem segments (0.04 – 0.17 μmol NO_2^- ×

Table 2. Mean NRA (NO_2^- production rate per unit dry weight under specific conditions) and dry weights, as well as their standard errors, of 19 compartments of 10-13 years old ash trees (n = 4). NRA per compartment is calculated from these data. The numbers of the compartments refer to their position within the trees (see Fig. 1)

	NRA [μmol NO_2^- $g^{-1}h^{-1}$]		Dry weight [g]		NRA per compartment [μmol NO_2^- h^{-1}]			
No.	x̄	s.e.	x̄	s.e.		x̄		
Leaflets								
19	1.34	0.186	7.85	4.82	10.52			
18	1.53	0.115	6.03	2.19	9.23			
17	1.42	---	1.80	---	2.56	67.83		
16	1.49	0.115	24.20	6.24	36.06			
15	1.46	0.232	6.48	2.94	9.46		77.38	
Leaf stalks								
14	0.79	0.081	1.62	0.79	1.28			
13	0.72	0.097	1.32	0.27	0.95			
12	1.18	---	0.64	---	0.76	9.55		109.13
11	0.82	0.225	6.01	1.59	4.93			
10	0.66	0.110	2.47	0.50	1.63			
Current year stem								
9	0.51	0.104	1.09	0.38	0.56			
8	1.16	---	2.94	---	3.41	5.67		
7	0.40	0.070	4.26	1.37	1.70		31.75	
Old stem								
6	0.10	0.030	35.35	11.71	3.54			
5	0.17	0.074	88.84	21.76	15.10	26.08		
4	0.04	0.008	186.10	79.00	7.44			
Roots								
3	0.70	0.071	1.02	0.23	0.71			
2	0.51	0.053	13.98	2.58	7.13	17.56	17.56	17.56
1	0.08	0.008	121.46	27.46	9.72			

$g^{-1} \times h^{-1}$). NRA per g of fine roots and root tips was at most only half of that found for leaflets. Even the mean NRA per g of leaf stalks was in most cases higher than the NRA in roots.

When calculated on a compartment basis, NRA of leaves (leaflets + leaf stalks) amounted to 61% of the total tree NRA, although leaves accounted for only 11% of the total tree dry weight (Table 2). Roots, on the other hand, represented 27% of the total tree dry weight, but the root proportion of total tree NRA was only 14%. Despite the low NRA per g of stem material, total stem NRA was considerable (25% of total tree NRA), since stem material accounted for a high proportion of the biomass (62% of total tree dry weight).

Discussion

The results of the *in vivo*-test of NRA indicate that nitrate is preferently reduced in leaves of *Fraxinus excelsior*, whereas reduction in the root of this woody species is of only minor importance. Furthermore, the pattern of nitrate distribution within *Fraxinus excelsior* is similar to that found in herbaceous species (see *e.g.* Austenfeld, 1972; Gebauer *et al.*, 1984; Janiesch, 1973), which have been shown to reduce nitrate preferently in leaves. In particular, far lower nitrate concentrations were found in leaf laminae than in leaf stalks for herbaceous plants, as well as for *Fraxinus excelsior*. This observation reveals the leaflets of ash trees to be the major sink for nitrate.

At this point the question arises for the reason of different results found with analyses of the nitrate to organic nitrogen ratio in the xylem sap of trees (Bollard, 1956; 1960) and with the *in vivo*-test of NRA. The following considerations could help to answer this question: Organic nitrogen compounds in the xylem sap can originate from the metabolism of inorganic nitrogen in the roots, but, in principle, they can also originate from a shoot-to-root translocation and a phloem-to-xylem flux of organic nitrogen in the roots (see Pate, 1980; Van Beusichem *et al.*, 1987). If the latter possibility also holds true for trees, xylem sap analyses are not suitable for measuring the contribution of roots to total nitrate reduction.

Since only a few investigations concerning nitrate assimilation in trees have been performed, further experiments will have to show whether the results presented above are general with respect to other tree species.

Acknowledgements

The authors wish to thank Miss Michaela Albrecht and Miss Petra Dietrich for skilful technical assistance and Dr. P. Ziegler, Universität Bayreuth, for improvement of the English. The critical reading of the manuscript by Prof Dr E.-D Schulze, Universität Bayreuth, is gratefully acknowledged. The study is part of a research programme supported by the Deutsche Forschungsgemeinschaft (to G.G.).

References

Al Gharbi A and Hipkin C R 1984 Studies on nitrate reductase in British angiosperms. I. A comparison of nitrate reductase activity in ruderal, woodland-edge and woody species. New Phytol. 97, 629–639.

Andrews M 1986 The partitioning of nitrate assimilation between root and shoot of higher plants. Plant Cell Environ. 9, 511–519.

Austenfeld F A 1972 Untersuchungen zur Physiologie der Nitratspeicherung und Nitratassimilation von *Chenopodium album* L. Z. Pflanzenphysiol. 67, 225–270.

Bollard E G 1956 Nitrogenous compounds in plant xylem sap. Nature 178, 1189–1190.

Bollard E G 1960 Transport in the xylem. Annu. Rev. Plant Physiol. 11, 141–166.

Gebauer G 1990 Diurnal changes of nitrate content and nitrate reductase activity in different organs of *Atriplex hortensis* L. (C_3 plant) and *Amaranthus retroflexus* L. (C_4 plant). *In* Plant Nutrition – Physiology and Applications. Ed. ML van Beusichem. pp. 93–99. Kluwer Academic Publishers, Dordrecht, The Netherlands.

Gebauer G, Melzer A and Rehder H 1984 Nitrate content and nitrate reductase activity in *Rumex obtusifolius* L. I. Differences in organs and diurnal changes. Oecologia 63, 136–142.

Gebauer G, Rehder H and Wollenweber B 1988 Nitrate, nitrate reduction and organic nitrogen in plants from different ecological and taxonomic groups of Central Europe. Oecologia 75, 371–385.

Janiesch P 1973 Beitrag zur Physiologie der Nitrophyten: Nitratspeicherung und Nitratassimilation bei *Anthriscus sylvestris* Hoffm. Flora 162, 479–491.

Jaworski E G 1971 Nitrate reductase assay in intact plant tissue. Biochem. Biophys. Res. Commun. 43, 1274–1279.

Larcher W 1980 Ökologie der Pflanzen. Ulmer Verlag, Stuttgart, 399 p.

Lee J A and Stewart G R 1978 Ecological aspects of nitrogen assimilation. Adv. Bot. Res. 6, 2–43.

Marschner H 1986 Mineral Nutrition of Higher Plants. Academic Press, London. 674 p.

Mengel K 1984 Ernährung und Stoffwechsel der Pflanze. Gustav Fischer Verlag, Stuttgart, 431 p.

Pate J S 1980 Transport and partitioning of nitrogenous solutes. Annu. Rev. Plant Physiol. 31, 313–340.

Runge M 1983 Physiology and ecology of nitrogen nutrition. *In* Encyclopedia of Plant Physiology, New Series Vol. 12C. Eds. O L Lange, P S Nobel, C B Osmond and H Ziegler. pp 163–200. Springer-Verlag, Berlin, Heidelberg, New York.

Smirnoff N, Todd P and Stewart G R 1984 The occurrence of nitrate reduction in the leaves of woody plants. Ann. Bot. 54, 363–374.

Van Beusichem M L, Nelemans J A and Hinnen M G J 1987 Nitrogen cycling in plant species differing in shoot/root reduction of nitrate. J. Plant Nutr. 10, 1723–1731.

M. L. van Beusichem (Ed.), *Plant nutrition – physiology and applications*, 107–109.

PLSO IPNC486

The effect of cadmium on nitrate reductase activity in sugar beet (*Beta vulgaris*)

N. PETROVIĆ, R. KASTORI[1] and I. RAJČAN
Institute of Field and Vegetable Crops, Faculty of Agriculture, University of Novi Sad, YU-21000 Novi Sad, Yugoslavia. [1]Corresponding author.

Key words: *Beta vulgaris* L., cadmium, nitrate reductase

Abstract

The effects of variable levels of Cd supply on the content of total N, nitrate, and actual and potential nitrate reductase activity (NRA) in roots and leaves of young sugar beet plants were investigated. It was found that a low concentration of Cd (10 μmol L^{-1}) stimulated slightly both NRA and nitrate accumulation, whereas high Cd concentrations acted inhibitorily. The same effect of Cd on NRA was confirmed by experiments with excised segments of sugar beet roots and leaves. The results indicated that Cd may affect the assimilation of nitrate and that the action of Cd depends on its concentration.

Introduction

The increasing interest in environmental pollution has led to investigations of Cd uptake, its distribution in different plant species (Jarvis *et al.*, 1976), and its effect on plant metabolism. Plant metabolism may be affected by Cd ions in different ways. Cd ions were found to be effective inhibitors of chlorophyll biosynthesis (Stobart *et al.*, 1985), photosynthesis (Weigel, 1985), respiration and the activities of several enzymes (Lee *et al.*, 1976), *i.e.*, processes important for the NRA. This paper deals with an investigation on the effects of Cd ions on NRA, in order to gain more information about *in vivo* effects of Cd ions on nitrate assimilation.

Materials and methods

Experiments were conducted on young plants of the sugar beet hybrid NS-Hy 11. The effect of Cd on NRA was examined in leaf and root segments which had been treated for two hours with 10, 100 and 1000 μmol Cd $\cdot$ dm^{-3}, as well as in intact plants grown for 7 days on 1/2 Hoagland's nutrient medium which contained the above Cd concentrations. Actual NRA was determined in a phosphate buffer pH 7.4, and potential NRA in the same buffer which contained 0.10 μmol $KNO_3 \cdot dm^{-3}$. Enzymic activity was examined *in vivo* using sulphanilamide and N-(1-naphtyl)-ethylendiamindihydrochloride. The plants were dried and weighed to obtain dry matter content. The NO_3 concentration in plant material was determined spectrophotometrically, total N by the Kjeldahl's method and Cd concentration by AAS.

Results

With increasing Cd supply in the nutrient solution, Cd concentrations increased significantly in leaves and especially in roots, while their dry matter content decreased (Table 1). Concentrations of nitrate and total N were slightly increased by the low Cd applies. Conversely, high

Table 1. Effects of cadmium levels in nutrient solution on cadmium concentration and dry matter content in leaves and roots of young sugar beet plants

Treatment μmol Cd · dm^{-3}	Cd in plant dry matter (μg · g^{-1})		Dry matter content (g · plant^{-1})	
	Leaves	Roots	Leaves	Roots
0	0,34	1,20	1,70	0,66
10	10,46	35,20	1,69	0,59
100	16,13	374,22	1,55	0,54
1000	59,15	3.542,37	1,16	0,53
LSD 5%	1,33	84,93	0,025	0,020

Cd supplies reduced the concentrations of nitrate and total N, except for the concentration of total N in roots (Table 2). Low Cd supplies resulted in slight increases NRA in both young plants and leaf and root segments. The highest Cd supply was toxic to the plants and resulted in a large reduction in NRA (Tables 3 and 4).

Discussion

Jarvis *et al.* (1976) found that Cd accumulates more readily in the fibrous roots than in the leaves of fodder beet and other plant species. The deposition of Cd in roots may be considered as a form of detoxification. Different compounds and ions are involved in the precipitation of Cd within plants, thus rendering it metabolically inactive. Changes in the concentrations of nitrate and total N at increased Cd concentrations in the nutrient solution may be brought about by changes in the intensity of N uptake, dry matter production and the intensity of the NRA. Nitrate reductase of higher plants requires $NADH_2$ as an electron donor. The source of the reductant is either from photosynthetic electron transport or mitochondrial acticity. Cd is known to inhibit mitochondrial swelling and the light reactions of

Table 2. Effect of cadmium supply on nitrate and total nitrogen concentration in dry matter of leaves and roots of young sugar beet plants

Treatment μmol Cd · dm^{-3}	NO_3 (μg · g^{-1})		N(%)	
	Leaves	Roots	Leaves	Roots
0	599	1047	2.31	2.02
10	806	1080	2.41	2.15
100	1062	1857	1.43	2.21
1000	541	1231	1.88	2.23
LSD 5%	148.1	217.7	0.19	0.06

Table 3. Effect of cadmium supply on nitrate reductase activity in leaves and roots of young sugar beet plants

Treatment μmol Cd · dm^{-3}	μmol NO_2 · g^{-1} fresh weight · h^{-1}			
	Leaves		Roots	
	Actual	Potential	Actual	Potential
0	2.44	14.79	0.20	2.40
10	3.09	13.41	0.31	2.77
100	2.37	11.35	0.21	2.20
1000	0.74	2.36	0.12	1.52
LSD 5%	0.46	0.72	0.06	0.21

Table 4. Effect of cadmium supply on nitrate reductase acticity in excised root and leaf segments of young sugar beet plants

Treatment μmol Cd · dm^{-3}	μmol NO_2 · g^{-1} fresh weight · h^{-1}			
	Leaves		Roots	
	Actual	Potential	Actual	Potential
0	1.00	8.69	0.25	2.80
10	1.13	9.87	0.29	3.03
100	0.64	7.06	0.21	2.14
1000	0.45	3.75	0.15	1.85
LSD 5%	0.33	0.63	0.05	0.16

photosynthesis (Bazzaz and Govindjee, 1974), and to accelerate the oxidation of $NADH_2$ (Miller *et al.*, 1973). Thus, the effect of Cd on these metabolic processes reduces the production of $NADH_2$. Inadequacy of the reducing power of $NADH_2$ would lead to inhibition of NRA by Cd. The stimulative effects of low Cd concentrations on the NRA may be considered as non-specific, since it has been observed in other enzymes (Bai Baozhang *et al.*, 1988), as well as in the synthesis of NR, proteins and DNA (Melnishuk *et al.*, 1982).

References

Bai Baozhang, Zhang Guixia and Li Wei 1988 Study on the effect of water cadmium pollution on the growth and development of rice. I. The response of germinating rice seeds to water cadmium pollution. Acta Agricul. Univers. Jilinensis. 10, 6–10.

Bazzaz M B and Govindjee 1974 Effects of cadmium nitrate on spectral characteristics and light reactions of chloroplasts. Environ. Lett. 6, 1–12.

Jarvis S C, Jones L H and Hopper M J 1976 Cadmium uptake from solution by plants and its transport from roots to shoots. Plant and Soil 44, 179–191.

Lee K C, Cunningham B A, Paulsen G M, Liang G H and Moore R B 1976 Effects of cadmium on respiration rate and activities of several enzymes in soybean seedlings. Physiol. Plant. 36, 4–6.

Melnishuk Yu P, Lishko A K and Kalinin F I 1982 Effect of cadmium on early stages of seed germination, RNA, protein and DNA synthesis in the radicale meristem. Plant Physiol. 29, 655–660.

Miller R J, Bittel J E and Koeppe D E 1973 The effect of cadmium on electron and energy transfer reactions in corn mitochondria. Physiol. Plant 28, 166–171.

Stobart A K, Griffiths W T, Ameen-Bukhari I and Sherwood R P 1985 The effect of Cd^{2+} on the biosyntheis of chlorophyll in leaves of barley. Physiol. Plant. 63, 293–298.

Weigel H J 1985 The effect of Cd^{2+} on photosynthetic reactions of mesophyll protoplasts. Physiol. Plant. 63, 192–200.

M. L. van Beusichem (Ed.), *Plant nutrition – physiology and applications*, 111–115.

PLSO IPNC491A

Transport of glutamine into the xylem of sunflower (*Helianthus annuus*)

G. R. FINDENEGG[1], W. PLAISIER[1], M. A. POSTHUMUS[2] and W. CH. MELGER[2]
[1]*Department of Soil Science and Plant Nutrition, Wageningen Agricultural University, P.O. Box 8005, 6700 EC Wageningen, The Netherlands, and* [2]*Department of Organic Chemistry, Wageningen Agricultural University, P.O. Box 8026, 6700 EG Wageningen, The Netherlands*

Key words: ammonium nutrition, glutamine, *Helianthus annuus* L., ^{15}N, phloem-to-xylem transfer, xylem exudate

Abstract

Sunflower (*Helianthus annuus* L.) plants were grown on nutrient solution with ammonium nitrogen. After 12 days of growth the ammonium in the nutrient solution was labeled with ^{15}N (99%). Three hours later glutamine-N in the xylem exudate was labeled for 56% as shown by GC-MS; this percentage increased to 63% after 8, and to 69% after 24 hours of incubation. When the xylem exudate had been collected from the epicotyl instead of the hypocotyl, 15-N abundances were 52%, 56% and 63% respectively. Results are consistent with an import of glutamine into the transpiration stream during its ascension in the xylem. On basis of the differences in abundance of double-labeled, single-labeled and unlabeled glutamine between the two sampling sites it was estimated that at least 20% of the xylem glutamine was imported into xylem along this distance (~4 cm).

Introduction

The occurrence of a xylem-to-phloem transfer of nitrogen compounds in plants is well established: Nitrogen taken up by plant roots is primarily transported to the shoot via the xylem transpiration stream. Growing points have a high nitrogen demand, but generally low rates of transpiration. Therefore, on the way from roots to young leaves, part of the nitrogen must interchange from xylem to phloem (Simpson, 1986). Export of nitrogen from the xylem can be studied by measuring the removing of (labeled) nitrogen compounds from the transpiration stream (Van Bel *et al.*, 1979; Dickson *et al.*, 1985). It has been found that amides like glutamine and asparagine are exported specifically and efficiently (Sharkey and Pate, 1975).

Much less is known about a possible import of nitrogen compounds into the xylem. However, such a transfer may affect determinations of root nitrate reduction (Rufty and Volk, 1986) and recirculation of nitrogen (Simpson *et al.*, 1982) when such determinations are based on the composition of the xylem sap.

Rufty and Volk (1986) measured considerable percentages of endogeneous organic nitrogen (*i.e.* organic nitrogen derived from nitrate absorbed some time before the experiment) in the xylem sap, but argued that this could also be explained by assuming that nitrogen assimilation in the root causes a delay of its transport into the xylem. Rowland (1986) could detect organic ^{15}N in the xylem sap after exposure of barley plants to atmospheric $^{15}NO_2$.

In this paper a substantial import of glutamine into the xylem of sunflower along the stem is indicated. ^{15}N-abundance in glutamine has been shown to decrease in the ascending xylem sap.

Methods

Seeds of sunflower (*Helianthus annuus* L. var. Relax) were germinated in moist quartz sand. After one week seedlings were transferred to nutrient solution (3 m*M* $(NH_4)_2SO_4$, 1 m*M* K_2SO_4, 1 m*M* $CaCl_2$, 1 m*M* $MgSO_4$, 0.25 m*M* KH_2PO_4, Fe-EDTA and micronutrients).

Throughout the period of germination, growth, and ^{15}N-incubation the nutrient solution was aerated, mixed and kept at pH 6.0 by automatic titration with diluted NH_4OH. The growth conditions were kept constant at 20° C and 80% air humidity with a photoperiod of 16 hours (60 W m^{-2}).

After 12 days of growth on nutrient solution 18 plants of equal size were selected and transferred to a smaller container with 1.6 L nutrient solution labeled with $^{15}NH_4$ (99%). Xylem exudate was collected during 30 minutes from plants cut freshly ~2 cm above or ~2 cm below the cotyledons and immediately deep frozen.

For GC-MS analysis of the xylem exudates the procedure of MacKenzie and Tenaschuk (1985) was used. Three samples of 40 μL from different plants were freeze-dried. Subsequently 4 μL tri-methylamine and 60 μL N-methyl-N-(tert. butyl-dimethylsilyl)-trifluoroacetamide (MTBSTFA; Pierce Chemical Company) was added. The closed reaction vessel was heated to 75°C for 30 min. 2 μL were injected into the Pye Unicam 204-series gaschromatograph equipped with a CP-Sil 19CB capillary column of 0.25 mm i.d. × 25 m. Temperature program was 2 min at 180°C followed by a rise to 260C at a rate of 4°C min^{-1}. Mass spectra were obtained from a connected VG7070F mass spectrometer using 70 eV electron impact ionization. During the chromatographic analysis of the derivatized xylem exudate the intensities of the mass spectral peaks corresponding to the $(GLN(TBDMSi)_3-COOSiC_6H_{15})^+$-fragment (masses 329.24, 330.24 and 331.23; see also Rhodes *et al.*, 1989) were recorded. The obtained integrated peak intensities were corrected for the natural abundances of ^{2}H, ^{13}C, ^{15}N, ^{17}O, ^{18}O, ^{29}Si and ^{30}Si.

Amino acids in the xylem sap were determined with an amino acid analyzer.

Results and discussion

Abundance of glutamine and ^{15}N in the xylem exudate

The amino acid analysis of the xylem exudate is given in Table 1. Glutamine is by far the most abundant amino acid.

The ^{15}N-abundance in glutamine from the xylem exudate (NX) during the incubation with ^{15}N-ammonium is shown in Figure 1. When

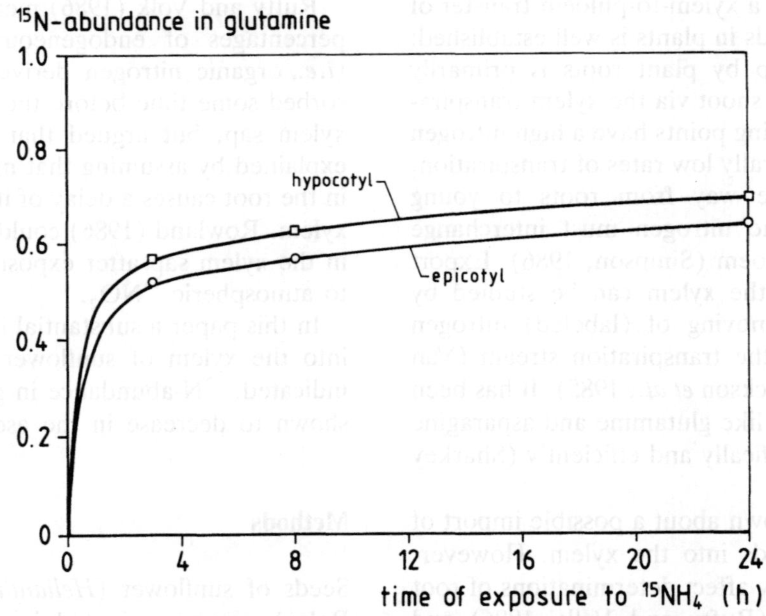

Fig. 1. ^{15}N abundance of glutamine in the xylem exudate of sunflower plants. Xylem exudate has been collected from hypocotyl (□) or epicotyl (○). Means of 3 replicates.

Table 1. Concentration of amino acids in the xylem exudate of sunflower. Means of two exudate samples collected from the hypocotyl

Amino acid	Concentration (m*M*)
alanine	1.9
asparagine	23.1
aspartic acid	0.6
glutamic acid	3.8
glutamine	149.3
glycine	0.5
histidine	1.0
isoleucine	0.8
leucine	0.9
lysine	0.5
phenylalanine	0.4
serine	6.0
threonine	2.2
valine	2.1

Table 2. Abundance of unlabeled (X0), single labeled (X1) and double labeled (X2) glutamine in the xylem exudate of sunflower. Means of 3 replicates

Collection site	Time of exposure to $^{15}NH_4$ (hours)	X0	X1	X2
Hypocotyl	3	0.208	0.456	0.336
	8	0.170	0.404	0.427
	24	0.120	0.371	0.509
Epicotyl	3	0.252	0.454	0.294
	8	0.218	0.441	0.344
	24	0.168	0.390	0.442

xylem sap has been collected from the hypocotyl, 56% of the glutamine-N was labeled 3 hours after application of ^{15}N, *i.e.* nearly half of the glutamine N present in the xylem has been taken up by the plant more than 3 hours before. Later this percentage increased only slightly (to 63% after 8, and to 69% after 24 hours). The initial rapid and the later slow increase in ^{15}N abundance is in line with the results of Rufty and Volk (1986) and means that the nitrogen-pool equilibrating with xylem-glutamine is great, so that its ^{15}N abundance does not increase quickly.

When the xylem sap has been collected from the epicotyl, consistently lower ^{15}N-abundances were recorded (Fig. 1). This indicates an import of unlabeled glutamine along the stem.

Abundance of unlabeled, single labeled and double labeled glutamine

By comparing the size of the 329, 330, and 331 peak of the mass spectrometer, abundances of unlabeled, single and double labeled glutamine molecules (X0, X1 and X2) could be calculated (Table 2). It was found that at any given ^{15}N-abundance in the xylem (NX) the percentage of single labeled glutamine was less than expected on basis of a stochastic distribution of label over the amino and amido group of glutamine. With a stochastic distribution the abundance of double labeled glutamine X2 should be NX^2; however the experimentally derived value as $NX^{1.87\pm0.04}$ (Fig. 2). This is consistent with the results of Rhodes *et al.*, (1989). Apparently glutamine molecules have been synthesized at sites of different ^{15}N-abundance. Glutamine synthesized at sites with high ^{15}N-abundance has an increased chance to get labeled at both nitrogen atoms whereas glutamine synthesized at sites with low abundances had an increased chance to stay totally unlabeled. Long-distance transport of intermediate products (*e.g.* glutamate) during glutamine synthesis thus seems to be of minor importance.

Analysis of variance

An analysis of variance has been conducted for the ^{15}N-abundance data. Significance levels of the effects of the factor 'xylem sap collection site' and of the covariate 'time of incubation' are shown in Table 3. NX, X0, X1 and X2 were significantly influenced by the time of incubation whereas the decline of ^{15}N-abundance between hypo- and epicotyl was significant only with respect to the increase of unlabeled glutamine.

The fact that the exponent EX, linking NX to X2, did not increase significantly during the experiment or between the collection sites means that there was no indication of a later metabolization of glutamine after its initial production. Such a metabolization should randomize the preferential formation of double labeled glutamine and consequently increase EX towards a value of 2. Sharkey and Pate (1975) reported that

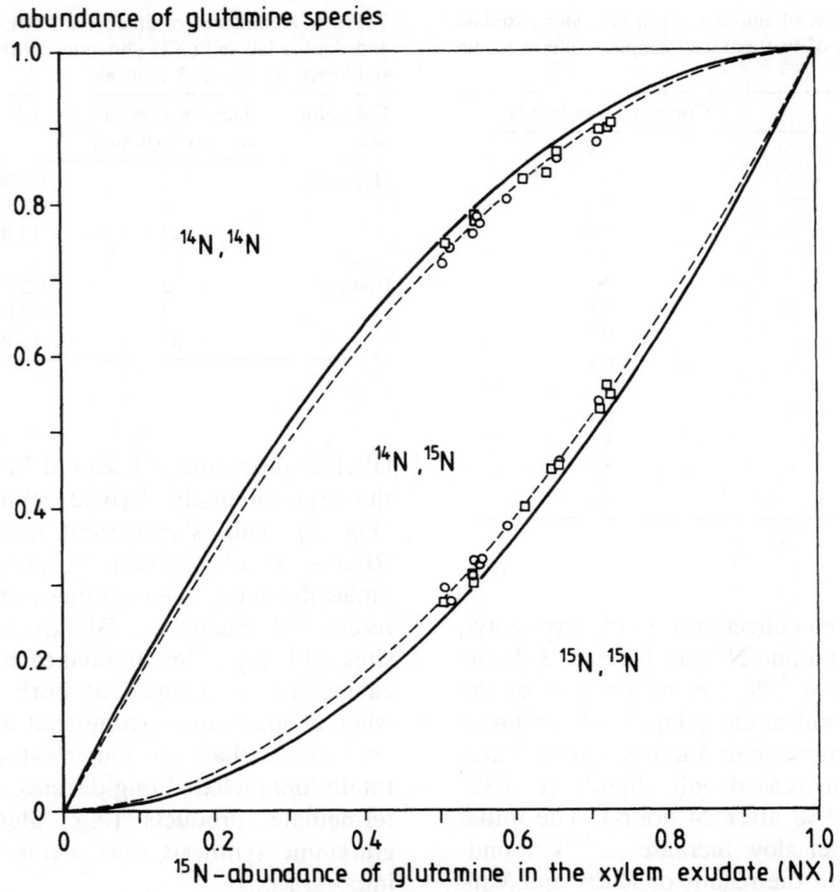

Fig. 2. Abundance of unlabeled (X0), single labeled (X1) and double labeled (X2) glutamine in the xylem exudate of sunflower plants as a function of the ^{15}N-abundance of glutamine (NX). Xylem exudate has been collected from the hypocotyl (□) or epicotyl (○). Solid lines represent stochastic distribution of ^{15}N between amino and amido group, broken lines are drawn according to the formula $X2 = NX^{1.87}$.

Table 3. Analysis of variance of data on the abundance of unlabeled (X0), single labeled (X1) and double labeled (X2) glutamine in the xylem exudate, the ^{15}N-abundance of glutamine in the xylem exudate (NX) and the exponent linking X2 to NX, (EX). Significance levels: ** <0.01; * <0.05; NS >0.05

	Effect of	
Analysis for	Covariate: time	Factor: site of collection
X0	**	*
X1	**	NS
X2	**	NS
NX	**	NS
EX	NS	NS

glutamine exported from the xylem was largely unmetabolized when recovered in the phloem sap.

Quantitative estimation of glutamine import into the xylem

From the labeling pattern of the glutamine molecules at the two collection sites the percentage of glutamine imported between hypocotyl and epicotyl can be estimated. From Table 2 it can be calculated that after 3 hours of incubation X0 at the hypocotyl was 17% lower than at the epicotyl. Because a possible export of glutamine from the xylem should not affect the ratios between X0, X1 and X2, it can be concluded that 17% of the glutamine present in the epicotyl must have been replaced by unlabeled glutamine between the collection sites. Data after 8 and 24 hours incubation gave a differences in X0 of 22 and 29%, respectively. Glutamine in the exchange pool may have got slightly labeled during the experiment, so that the import rate obtained in this way may be an underestimate. Therefore it is reasonable to assume that at least 20% of the glutamine found in the xylem sap of the epicotyl has been imported along the ~4 cm-distance between the collection sites.

It should be noted that the import rate obtained in this way is about double the rate which would have been calculated by simply comparing NX of the two collection sites (7.7 to 10.1%). This discrepancy is due to the fact that differences of X0 between the sampling sites are more pronounced than those of NX.

It is unlikely that the import of glutamine into the xylem described here can be explained by a xylem-to-xylem transfer as postulated by Layzell *et al.*, (1981). In our experiment external ^{15}N crossed the root into the xylem within 3 hours. Regarding this high rate of transport one would not expect that glutamine transfer from one xylem vessel to another is so slow that the imported glutamine is still largely unlabeled after 24 hours. Export of glutamine from the cotyledons via phloem to the xylem of the stem is much more likely. According to our measurements such a transfer may be much more intense than indicated by nitrogen balance studies (Layzell *et al.*, 1981).

References

Dickson R E, Vogelman Th C and Larson P R 1985 Glutamine transfer from xylem to phloem and translocation to developing leaves of *Populus deltoides*. Plant Physiol. 77, 412–417.

Layzell D B, Pate J S, Atkins C A and Canvin D T 1981 Partitioning of carbon and nitrogen and the nutrition of root and shoot apex in a nodulated legume. Plant Physiol. 67, 30–36.

MacKenzie S L and Tenaschuk D 1985 Gas liquid chromatographic assay for asparagine and glutamine. J. Chromatogr. 322, 228–235.

Rhodes D, Rich P J and Brunk D G 1989 Amino acid metabolism of *Lemna minor* L. IV. ^{15}N-labeling kinetics of the amide and the amino groups of glutamine and asparagine. Plant Physiol. 89, 1161–1171.

Rowland A J 1986 Nitrogen uptake, assimilation and transport in barley in the presence of atmospheric nitrogen dioxide. *In* Fundamental, Ecological and Agricultural Aspects of Nitrogen Metabolism in Higher Plants. Eds. H Lambers, J J Neeteson and I Stulen. pp 211–214. Martinus Nijhoff Publishers, Dordrecht, The Netherlands.

Rufty T W and Volk R J 1986 Alterations in enrichment of NO_3^- and reduced-N in xylem exudate during and after extended plant exposure to $^{15}NO_3$. Plant and Soil 91, 329–332.

Sharkey P J and Pate J S 1975 Selectivity in xylem to phloem transfer of amino acids in fruiting shoots of white lupine (*Lupinus albus* L.). Planta 127, 251–262.

Simpson R J, Lambers H and Dalling M J 1982 Translocation of nitrogen in a vegetative wheat plant (*Triticum aestivum*). Physiol. Plant. 56, 11–17.

Simpson R J 1986 Translocation and metabolism of nitrogen: Whole plant aspects. *In* Fundamental, Ecological and Agricultural Aspects of Nitrogen Metabolism in Higher Plants. Eds. H Lambers, J J Neeteson and I Stulen. pp 71–96. Martinus Nijhoff Publishers, Dordrecht, The Netherlands.

Van Bel A J E, Mostert E and Borstlap A C 1979 Kinetics of L-alanine escape from the xylem vessels. Plant Physiol. 63, 244–247.

M. L. van Beusichem (Ed.), *Plant nutrition – physiology and applications*, 117–120.

PLSO IPNC013A

Influence of fusicoccin and xylem wall adsorption on cation transport into maturing wheat (*Triticum aestivum*) ears

H. KUPPELWIESER and U. FELLER[1]
Institute of Plant Physiology, University of Bern, Altenbergrain 21, CH-3013 Bern, Switzerland.
[1]*Corresponding author*

Key words: fusicoccin, internode, phloem, rubidium, solute transfer, *Triticum aestivum* L., wheat, xylem

Abstract

Detached wheat ears with 18 cm stem were incubated for 3 days standing in 2 m*M* $RbCl + 2$ m*M* $SrCl_2$. Strontium was mainly translocated with the transpiration stream to the glumes while rubidium, similar to K, accumulated in the lower part of the stem. The selective elimination of Rb from the xylem was stimulated by fusicoccin. Our result suggest that the rapid transfer from the xylem to surrounding cells was driven by a membrane potential difference.

Introduction

Potassium and rubidium (frequently used as a tracer for potassium) are highly mobile in the phloem and are easily redistributed in maturing wheat (Haeder and Beringer, 1984a; Marschner and Schimansky, 1971; Martin, 1982). Strontium, which behaves in higher plants similar to the macronutrient calcium, is not mobile in the phloem (Marschner, 1986). Potassium (but not calcium) can be remobilized in considerable quantities from the leaves and translocated to the maturing grains. Rubidium fed into the xylem through the cut internode of detached wheat shoots is removed from the transpiration stream and accumulates in the stem. A rapid transfer of rubidium from the xylem to the phloem was proposed previously (Feller, 1989; Haeder and Beringer, 1984a).

Negative charges in the xylem walls could allow a reversible ion exchange and may contribute to the retention of cations in the stem. Such interactions with charged surfaces require no transport through a membrane. On the other hand, the rapid transfer of rubidium or potassium may be based on the carrier-mediated transport across a membrane (Klotz and Erdei, 1988). The selectivity depends on the properties of carriers present. Membrane potential differences are important in the accumulation of ions on one side of a membrane and can be affected by the fungitoxin fusicoccin, which causes hyperpolarization (Marrè, 1979). The objectives of the work reported here were to identify the relative importance of the different mechanisms mentioned above for the retention of cations in wheat internodes.

Methods

Winter wheat (*Triticum aestivum* L., cv. 'Arina') was grown in a field near Bern. Plants were detached above the soil surface 14 days after anthesis, recut submerged in distilled water and transported to the laboratory standing in distilled water. The submerged stem was cut 18 cm below the ear with a razor blade prior to starting the experiments.

Each plant was transferred to a small flask containing 15 mL feeding solution (2 m*M*

RbCl + 2 m*M* $SrCl_2$). In some cases fusicoccin was added in a final concentration of 10 μM from a stock solution (367 μM in 0.6% v/v ethanol) prepared according to De Boer *et al.* (1985). The shoots were incubated in a culture room in a light/dark cycle with 14 h light (120 $\mu E\ m^{-2}\ sec^{-1}$ from 4 Philips TL 40 W/33 and 2 Osram Fluora fluorescent tubes; 24–26°C) and 10 h dark (21–23°C) per day. Five separately incubated shoots were analyzed for each treatment. The different parts were dried at 105°C and analyzed for Rb and Sr contents by atomic absorption spectrophotometry as described previously (Feller, 1989).

Results and discussion

Strontium fed into the cut stem of detached wheat shoots entered the ear with the transpiration stream (Fig. 1). Within the ear strontium accumulated mainly in the glumes, while the contents in the maturing grains remained low. Only minor quantities of strontium were detected in stem segments. In contrast to strontium, rubidium was retained in the lower stem segments and only trace amounts reached the ear. Within the ear rubidium was found mostly in the rachis and the grains, but remained very low in the glumes. These results are consistent with the hypothesis that rubidium is rapidly eliminated from the xylem sap and that strontium is mainly distributed by the transpiration stream. Fusicoccin stimulated the accumulation of rubidium in the lowest stem segments (0–4 cm) and further decreased the contents in the ear. These findings suggest that the rubidium transfer from the xylem vessels to surrounding cells is driven by a membrane potential difference. The strontium contents in the lower stem segments were not affected by fusicoccin. The differences between rubidium and strontium transport can be explained by the specificities of carriers which transport rubidium but not strontium. The enhanced strontium contents in the uppermost stem parts of fusicoccin-treated shoots were most likely caused by a stimulated transpiration of the photosynthetically active

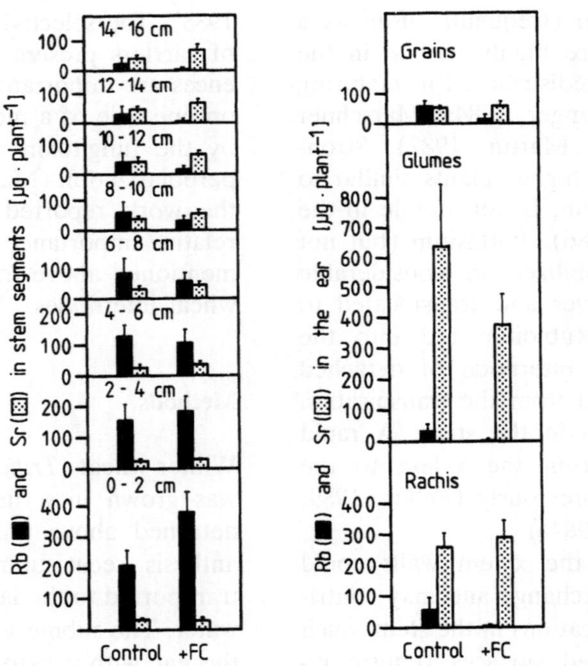

Fig. 1. Effects of fusicoccin on Rb and Sr fluxes in detached wheat. Ears with 18 cm stem were incubated for 72 h standing in 2 m*M* RbCl + 2 m*M* $SrCl_2$. Fusicoccin (10 μM) was added to this solution when indicated (+FC). The shoots were then divided into grains, glumes (containing glumes, lemmas and paleas), rachis (including the uppermost part of the stem) and stem segments (beginning with 0–2 cm at the base). Means and standard deviations of 5 plants are shown.

part of the stem below the ear. In the presence of fusicoccin only trace amounts of rubidium reached the upper stem segments and the ear due to the accelerated elimination from the xylem sap in the lower stem parts.

Essentially no rubidium was remobilized from the lowest stem segments after transferring the shoots to $CaCl_2$ solution (Fig. 2). The decrease in the uppermost segments might be due to the export of rubidium through the phloem into the ear. The high rubidium content present in the basal part of the stem was not accessible for ion exchange with the xylem solution. Rubidium was most likely transferred to the phloem (symplast). Since the major phloem sources (the leaves) were removed it must be assumed that no regular flow in the phloem was possible in this artificial system (Feller, 1989). The strontium content in the lowest segments (0–4 cm) decreased by about 50% after the transfer to the $CaCl_2$ solution. These results suggest that some strontium was reversibly bound to charged surfaces. The higher strontium contents in the upper stem segments remained relatively constant. The transpiration of these photosynthetically active parts of the stem may be responsible for the irreversible elimination of strontium from the xylem. It remains open to question whether this strontium remained in the apoplast or passed through a membrane.

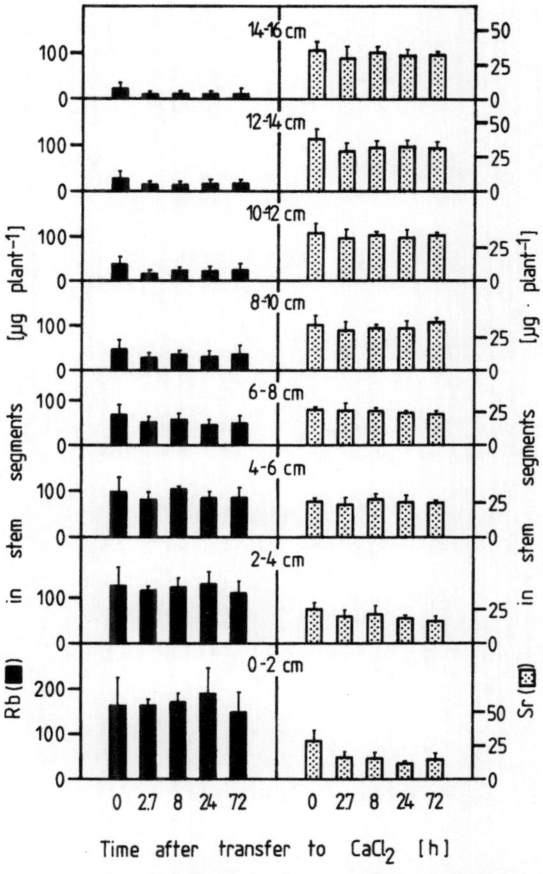

Fig. 2. Retention of Rb and Sr in pre-fed wheat stems. Ears with 18 cm stem were fed for 72 h with 2 m*M* RbCl + 2 m*M* $SrCl_2$ and then transferred to 10 m*M* $CaCl_2$ (without Rb and Sr). Stem segments (beginning with 0–2 cm at the base) were analyzed for Rb and Sr contents 0, 2.7, 8, 24 and 72 h after the transfer. Means and standard deviations of 5 replications are shown. Different scales are used for Rb and Sr.

The rapid elimination of rubidium from the xylem in wheat internodes is most likely driven by a membrane potential difference. A previous report from Haeder and Beringer (1984b) indicates that these transfer processes for rubidium and potassium are also relevant in intact wheat plants. The control of solute transfer between the two long-distance transport systems is not yet satisfactorily understood. Our results lead to the conclusion that membrane potentials and concentration gradients may play a major role in the regulation of transfer processes between xylem and phloem which allow a selective solute channelling in maturing wheat.

Acknowledgements

We thank the Landwirtschaftliche Schule Rütti in Zollikofen for growing plants and Dr A Fleming for improving the English of the manuscript. This work was supported by Swiss National Science Foundation (Project 3100–009407).

References

De Boer A H, Katou K, Mizuno A, Kojima H and Okamoto H, 1985. The role of electrogenic xylem pumps in K^+-absorption from the xylem of *Vigna unguiculata*: The effects of auxin and fusicoccin. Plant Cell Environ. 8, 579–586.

Feller U 1989 Transfer of rubidium from the xylem to the phloem in wheat internodes. J. Plant Physiol. 133, 764–767.

Haeder H E and Beringer H 1984a Long-distance transport

of potassium in cereals during grain filling in detached ears. Physiol. Plant. 62, 433–438.
Haeder H E and Beringer H 1984b Long-distance transport of potassium in cereals during grain filling in intact plants. Physiol. Plant. 62, 439–444.
Klotz M G and Erdei L 1988 Effects of tentoxin on K^+ transport in winter seedlings of different K^+-status. Physiol. Plant. 72, 298–304.
Marrè E 1979 Fusicoccin: A tool in plant physiology. Annu. Rev. Plant Physiol. 30, 273–288.
Marschner H 1986 Mineral Nutrition of Higher Plants. Academic Press, London, 674 p.
Marschner H and Schimansky CH 1971 Suitability of using rubidium-86 as a tracer for potassium in studying potassium uptake by barley plants. Z. Pflanzenernaehr. Bodenkd. 128, 129–143.
Martin P 1982 Stem xylem as a possible pathway for mineral retranslocation from senescing leaves to the ear in wheat. Aust. J. Plant Physiol. 9, 197–207.

M. L. van Beusichem (Ed.), *Plant nutrition – physiology and applications*, 121–125.

PLSO IPNC013B

Effect of phloem interruption on leaf senescence and nutrient redistribution in wheat (*Triticum aestivum*)

D. SCHENK and U. FELLER[1]
Institute of Plant Physiology, University of Bern, Altenbergrain 21, CH-3013 Bern, Switzerland.
[1]*Corresponding author*

Key words: grain filling, nutrient redistribution, phloem interruption, *Triticum aestivum* L., senescence, steam-girdling, wheat

Abstract

The phloem of field-grown wheat was interrupted shortly after anthesis by steam-girdling below the ear or below one of the 2 uppermost leaf nodes. Senescence and nutrient remobilisation was delayed in leaves without phloem connection to the ear. Dry matter, phosphorus and potassium accumulation in the ear strongly depended on the number of leaves with an intact phloem to the head. Mobile nutrients were further redistributed within the ear from the glumes to the grains, while calcium increased in all parts. The capacities of sources, sinks and transport systems may contribute to the overall regulation of nutrient fluxes in intact plants.

Introduction

Potassium, nitrogen and phosphorus can be translocated through the xylem and the phloem, while the long-distance transport of calcium is essentially restricted to the xylem (Jeschke *et al.*, 1987; Martin, 1982; Marschner, 1986; Simpson and Dalling, 1981; Wolf and Jeschke, 1987). Furthermore, mobile nutrients can be transferred between xylem and phloem during translocation (Feller, 1989; Haeder and Beringer, 1984; McNeil *et al.*, 1979; Pate, 1975; Sharkey and Pate, 1975). The phloem transport depends on loading (sources), unloading (sinks) and on solute transfer between the sieve tubes and surrounding cells (Delrot and Bonnemain, 1985; Pate, 1975; Vreugdenhil, 1985). The long-distance transport in the symplast is important for the redistribution of nutrients from senescing leaves to maturing grains of cereals (Feller and Keist, 1986; Martin, 1982). Leaf senescence can be affected by ear removal, but the mechanisms involved are not yet satisfactorily understood (Crafts-Brandner *et al.*, 1984). The phloem represents a link between the leaves and the maturing wheat grains. The redistribution of solutes through the symplast can be terminated by steam-girdling the stem (Martin, 1982). The objectives of this work were to identify regulatory effects of phloem interruption at different positions on leaf senescence and grain filling in field-grown wheat.

Methods

Plants of field-grown winter wheat (*Triticum aestivum* L., 'Arina') were steam-girdled as described by Martin (1982). The phloem was interrupted by such treatments 2 weeks after ear emergence either below the ear, below the flag leaf node or below the second node from the top. For the treatments as well as for untreated controls samples were taken in 4 replicates at the day of steam-girdling and at intervals of one week throughout the maturation period. The leaf spreads were stored at −20°C prior to extraction, while the other parts of the shoots were separated and dried at 105°C.

For the extraction one leaf lamina was cut into small pieces and homogenized in 5 ml buffer

(50 m*M* Tris/HCl pH 7.5 containing 1% polyvinylpolypyrrolidone and 0.1% mercaptoethanol) with a Polytron mixer as reported previously (Feller, 1987). The homogenate was filtered through Miracloth (Calbiochem, San Diego). Total chlorophyll was detected as described by Strain *et al.* (1971), total proteins in the crude extract with Coomassie Brilliant Blue (Bradford, 1976) and free amino groups with a ninhydrin reagent (Cramer, 1958).

For P, K and Ca contents dry samples were heated for several hours at 550°C and after cooling 0.2 mL 10 *N* HCl and 7.8 mL H_2O were added. Leaf extracts were mineralized by mixing 1 mL extract with 0.1 mL 30% H_2O_2 and heating the tubes to 105°C until the samples were completely dry. The addition of 0.1 mL 30% H_2O_2 and the heating were repeated once. After cooling 0.1 mL 10 *N* HCl and 3.9 mL H_2O were added to the samples. Phosphate was detected in

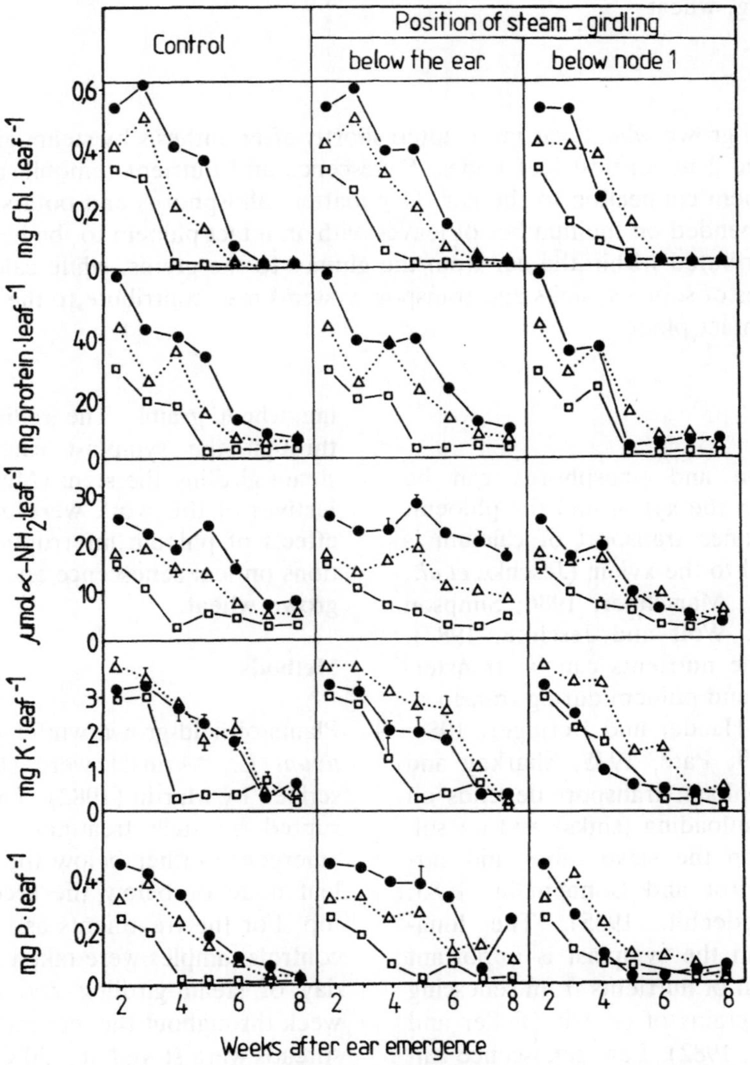

Fig. 1. Effect of phloem interruption on senescence and nutrient remobilization in leaves of field grown wheat. The phloem was interrupted 2 weeks after ear emergence by steam-girdling the stem below the ear or below the flag leaf node. Chlorophyll, total proteins, free amino groups, potassium andphosphorus were analysed in weekly intervals in the lamina of the flag leaf ((●—●), of the second (△···△) and of the third (□—□) leaf from the top. The symbols represent the mean of 4 replicates. Standard errors are shown when exceeding the size of the symbol (on one side only for clarity).

these solutions with a vanadate-molybdate reagent. K and Ca were measured by atomic absorption spectrometry after appropriate dilution with 1.267 gL^{-1} CsCl suprapur in 0.1 *N* HCl (for K) or with 13.37 gL^{-1} $LaCl_3$ 7 H_2O in 0.1 *N* HCl (for Ca).

Results

The net degradation of proteins and of chlorophyll in the uppermost 3 leaves was slightly delayed by steam-girdling below the ear (Fig. 1). On the other hand interruption of the phloem below the flag leaf node accelerated senescence and nutrient remobilisation in the flag leaf lamina and simultaneously delayed senescence in the leaves below the girdling position. As a consequence the flag leaf senesced in such plants more rapidly than the second leaf from the top. No major accumulation of amino acids was detected in the leaves of steam-girdled plants. These results indicate that the transport capacity was sufficient for the amino acids deriving from catabolic processes. In general phosphorus and potassium contents decreased slower in leaves below the girdling position.

The accumulation of dry matter and of nutrients in the ear strongly depended on an intact phloem (Fig. 2). The dry matter increased with the number of leaves connected through the phloem with the head.

The redistribution of the mobile elements phosphorus and potassium from the vegetative to the reproductive plant parts was considerably affected by steam-girdling (Fig. 2). The phloem interruption below the ear was most effective. Within the ear these elements normally accumulated in the grains (Fig. 3). The increase of the P content in the grains was paralleled by a decrease in the glumes. At the time of harvest the P content in the glumes remained highest in plants with a poor grain-filling (steam-girdled below the ear). These results suggest a control of the phosphorus redistribution form the glumes to the grains by the sink demand.

The potassium content in the ear of plants girdled below the ear decreased during maturation (Fig. 2). Within these ears the K content of the grains slightly increased, but the content in

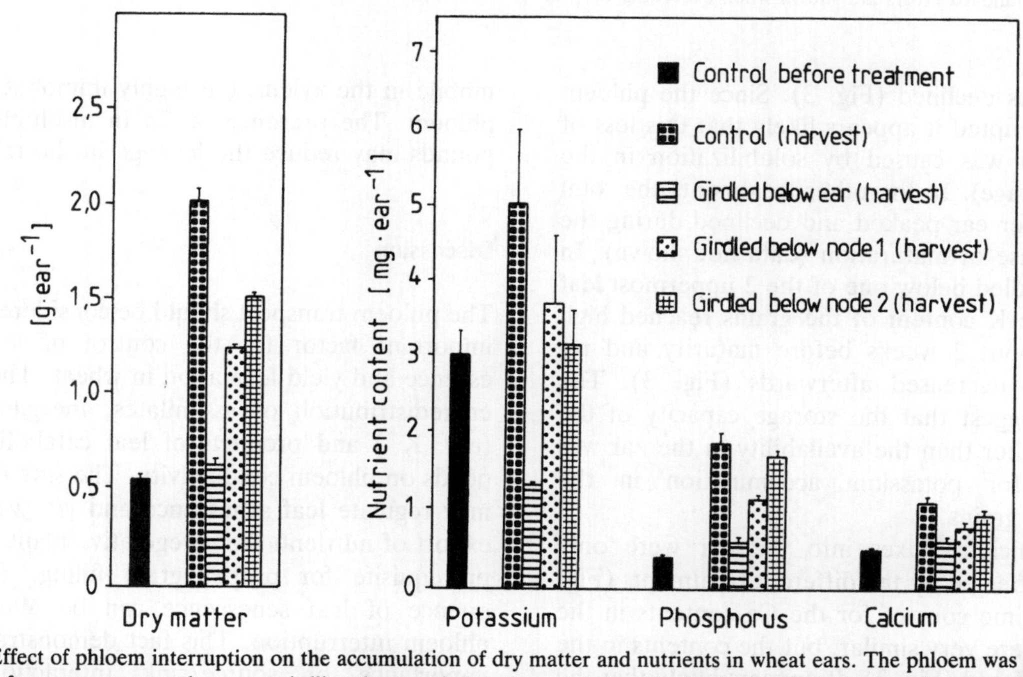

Fig. 2. Effect of phloem interruption on the accumulation of dry matter and nutrients in wheat ears. The phloem was interrupted 2 weeks after ear emergence by steam-girdling the stem below the ear, below the flag leaf node (node 1) or below the node of the second leaf from the top. Ears were analysed immediately before girdling and at maturity. Means and standard errors of 4 replicates are shown.

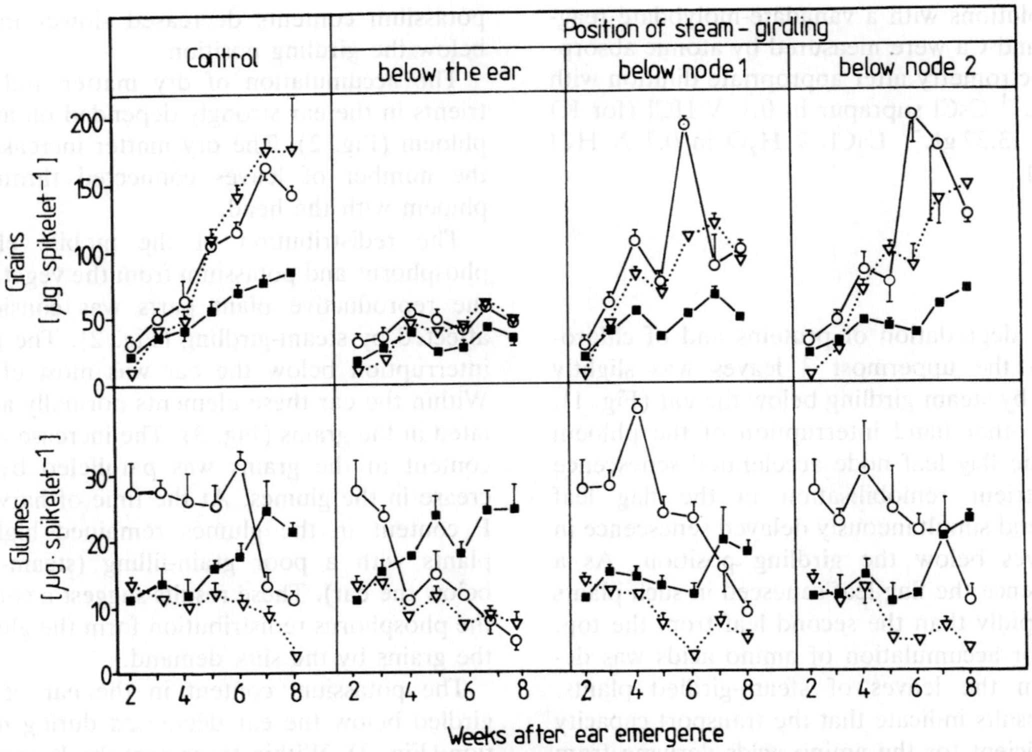

Fig. 3. Effect of phloem interruption on nutrient contents in glumes and grains of maturing wheat ears. The phloem was interrupted 2 weeks after ear emergence by steam-girdling the stem below the ear, below the flag leaf node (node 1) or below the node of the second leaf from the top. K (○—○), P (▽···▽) and Ca (■—■) contents per spikelet represent the mean of 4 replicates. Standard errors are shown when exceeding the size of the symbol (on one side only for clarity).

the glumes declined (Fig. 3). Since the phloem was interrupted it appears likely that this loss of potassium was caused by solubilization in the rain (leakage). In the other treatments the total content per ear peaked and declined during the latest phase of maturation (data not shown). In plants girdled below one of the 2 uppermost leaf nodes the K content of the grains reached high values about 2 weeks before maturity and remarkably decreased afterwards (Fig. 3). The results suggest that the storage capacity of the grains rather than the availability in the ear was limiting for potassium accumulation in the maturing grains.

The calcium fluxes into the ear were only slightly affected by the different treatments (Fig. 2). The time courses for the Ca contents in the glumes were very similar, but the contents in the grains differed (Fig. 3). It appears likely that the calcium flux to the ear was lowered indirectly by a reduced transpiration, because this element is mobile in the xylem, but highly immobile in the phloem. The presence of Ca in insoluble compounds may reduce the leakage in the rain.

Discussion

The phloem transport should be considered as an important factor for the control of leaf senescence and yield formation in wheat. The proper redistribution of assimilates, inorganic ions (*e.g.* K^+) and products of leaf catabolism depends on phloem conductivity. The sink demand may regulate leaf senescence and *vice versa* the export of nutrients from vegetative plant parts is prerequisite for proper grain filling. The sequence of leaf senescence can be altered by phloem interruption. This fact demonstrates the importance of source/sink interactions for senescence control. Source pressure, sink demand and translocation itself, which differ in their regulatory properties, may influence nutrient redistribution and senescence on a whole plant level.

Acknowledgements

We thank the Landwirtschaftliche Schule Rütti in Zollikofen for supplying the plant material, Dr. A. Fleming for assistance in preparing the English manuscript and Swiss National Science Foundation for the support of this work (Project 3100–009407).

References

Bradford M M 1976 A rapid and sensitive method for the quantitation of microgram quantities of protein using the principle of protein dye binding. Anal. Biochem. 72, 248–254.

Crafts-Brandner S J, Below F E, Harper J E and Hageman R H 1984 Differential senescence of maize hybrids following ear removal. I. Whole plant. Plant Physiol. 68, 360–367.

Cramer F 1958 Papierchromatographie, Verlag Chemie, Weinheim, 102 p.

Delrot S and Bonnemain J L 1985 Mechanism and control of phloem transport. Physiol. Vég. 23, 199–220.

Feller U and Keist M 1986 Senescence and nitrogen metabolism in annual plants. *In* Fundamental, Ecological and Agricultural Aspects of Nitrogen Metabolism in Higher Plants. Eds. H Lambers, J J Neeteson and I Stulen. pp 219–234. Martinus Nijhoff Publishers, Dordrecht, The Netherlands.

Feller U 1987 Nitrogen remobilisation and proteolytic enzymes in the flag leaves of detached wheat shoots: Effect of vascular connections and environmental conditions. Agricoltura Mediterranea 117, 331–337.

Feller U 1989 Transfer of rubidium from the xylem to the phloem in wheat internodes. J. Plant Physiol. 133, 764–767.

Haeder H E and Beringer H 1984 Long-distance transport of potassium in cereals during grain filling in detached ears. Physiol. Plant. 62, 433–438.

Jeschke W D, Pate J S and Atkins C A 1987 Partioning of K^+, Na^+, Mg^{2+} through xylem and phloem to component organs of nodulated white lupin under mild salinity. J. Plant Physiol. 128, 77–93.

Marschner H 1986 Mineral Nutrition of Higher Plants. Academic Press, London, pp 71–102.

Martin P 1982 Stem xylem as a possible pathway for mineral retranslocation from senescing leaves to the ear in wheat. Aust. J. Plant Physiol 9, 197–207.

McNeil D L, Atkins C A and Pate J S 1979 Uptake and utilisation of xylem-borne amino compounds by shoot organs of a legume. Plant Physiol. 63, 1076–1081.

Pate S 1975 Exchange of solutes between phloem and xylem and circulation in the whole plant. *In* Transport in Plants. I. Phloem Transport. Eds. M H Zimmermann and J A Milburn. Encyclopedia of Plant Physiology, New Series, Vol. 1, pp 451–473. Springer-Verlag, Berlin.

Sharkey P J and Pate J S 1975 Selectivity in xylem to phloem transfer of amino acids in fruiting shoots of white lupin (*Lupinus albus* L.). Planta 127, 251–262.

Simpson R J and Dalling M J 1981 Nitrogen redistribution during grain growth in wheat (*Triticum aestivum* L.). Planta 151, 447–456.

Strain, H H, Cope B T and Svec W A 1971 Analytical procedures for isolation, identification, estimation and investigation of the chlorophylls. *In* Methods in Enzymology, Vol 23. Eds. S P Colowick and N O Kaplan. pp 452–476. Academic Press, New York.

Vreugdenhil D 1985 Source-sink gradient of potassium in the phloem. Planta 163, 238–241.

Wolf O and Jeschke W D 1987 Modeling of sodium and potassium flows via phloem and xylem in the shoot of salt-stressed barley. J. Plant Physiol. 128, 371–386.

M. L. van Beusichem (Ed.), *Plant nutrition – physiology and applications*, 127–131.

PLSO IPNC182

Efficiency of Azospirillum inocula in peat formulations: Maize (*Zea mays*) root colonization and effects on initial plant growth

P. MORANDINI, R. VIGNOLA and P. VITTORIOSO
Agrobiotechnological Research Area, Enichem Agricoltura, Via E. Ramarini 32, I-00015 Monterotondo (RM), Italy

Key words: *Azospirillum brasilense*, peat formulation, root colonization, *Zea mays* L.

Abstract

Pot trials were carried out to find the optimal inoculum conditions for an *Azospirillum brasilense* strain isolated from Italian soil. We examined different maize genotypes and increasing concentrations of Azospirillum (from 10^6 to 10^9 cells/plant); peat was used as a carrier, the formulation obtained being pre-tested for assaying microbial survival. The influence of two levels of nitrogen fertilizer was also evaluated. Efficiency of the peat formulation in root colonization was estimated by determining the number of viable cells inside the root at three different phases of plant growth. Two months after sowing both fresh and dry shoot weight were measured to evaluate the effects of inoculum on initial plant growth. Our results prove that peat formulation is a useful method for inoculating *Azospirillum brasilense* on maize. It is also shown that different maize genotypes have different growth responses although the number of microorganisms found inside the root can be comparable.

Introduction

Since their isolation from roots of tropical grasses by Döbereiner and Day (1976), several species of Azospirillum have been the subject of particular attention by many researchers working in scientific institutes and industries.

The scientific interest in Azospirillum comes from the possibility of studying a new nitrogen-fixing system in which the microorganisms associate with roots without getting the plants to build up external structures, such as nodules (like in the case of Rhizobium-legume symbiosis).

Industrial interest is based on hope that the success obtained with Rhizobium inocula can be repeated with Azospirillum. However, earlier attempts (Albrecht *et al.*, 1981; Barber *et al.*, 1979; Okon, 1984) to increase yields of several cereal crops in field experiments with Azospirillum inocula gave variable results. Only about sixty per cent (Mertens and Hess, 1984; Smith *et al.*, 1984) of the field trials showed an effect of inoculum, consisting of crop yield increases ranging from 20% to 30%. There is reason to suppose, according to other research groups (Favilli *et al.*, 1984; Okon and Kapulnik, 1986), that the variability in effects may be due to association specificity and to variations in inocula supply techniques.

The aim of the present work is to determine the optimal concentration of the inoculum, to verify the efficiency of peat as a carrier for *Azospirillum brasilense*, and to analyse the effects of the peat formulation on two different maize genotypes.

Materials and methods

Bacterial strains and growth conditions

The present study was carried out with an *A. brasilense* strain isolated from an alluvial, fresh soil in our laboratory (strain 11). Bacteria were grown on the medium described by Okon (Okon *et al.*, 1977) for 24 hours at 32°C in a CF 3000 Chemap pilot fermenter with pH constantly maintained at 6.8 by adding 30% DL-malic acid. Cells were then harvested by centrifugation (4500 *g* per 20 minutes), washed with 60 m*M* phosphate saline buffer, pH 6.7, and resuspended with the same buffer at a predetermined volume.

Peat formulation

The bacterial suspension was mixed with pretreated German peat moss at a final concentration of 10^8 Colony Forming Units (CFU)/g and 10^{10} CFU/g. Before inoculum preparation, peat was finely sieved and neutralized with a 1% (w/w) $CaCO_3$ solution at pH 6.7, then sterilized by autoclaving at 121°C for 2 hours.

Inoculation with dead bacterial cells, sterilized at 110°C for 30 minutes, and equivalent before sterilization to 10^8 CFU/g, was practiced as a control, as described for live bacteria.

Plant growth conditions

Two different genotypes of *Zea mays* (lines A3/b73 and D3/b37, supplied by Istituto di Genetica Vegetale, Università del Sacro Cuore, Piacenza) were used for inoculum with *A. brasilense*. These genotypes had been obtained by backcrossing of two commercial hybrids with a naturally occurring population of maize and selected for their associative capacity towards Azospirillum.

Plants were grown in a greenhouse (min. 15°C, max. 25°C) in 7-kg plastic pots filled with a 10-mm sieved mixture (2.8:1) of two soils ('Alberone', an alluvial clay soil and 'Maccarese', a very light sandy soil). Four pre-germinated (25°C for 72 h) maize seeds were sown in each pot (3 cm depth) and then inoculated, by covering the seeds with the *A. brasilense*-peat formulation. To obtain four different inoculum concentrations, 10 or 100 mg/seed of both peat formulations (10^8 and 10^{10} CFU/g) were used. After sowing, the soil in the pots was carefully hand-irrigated and gradually brought up to 80% of field capacity. On full emergence of the second leaf the plants were thinned to three per pot and grown for 60 days. At the end of the experiment, foliage fresh and dry weights (after drying in a forced-draft oven at 60°C for 72 h) were measured.

Experimental design

The variables examined in the trial were:

a) Inoculum concentration: four different rates, ranging from 10^6 to 10^9 CFU/plant, besides a control inoculated with steam sterilized peat formulation;

b) Nitrogen fertilization: two levels of nitrogen, 200 and 400 mg N/pot.

Overall ten different treatments, 5-fold replicated in a randomized block design, were considered. A bifactorial analysis of variance allowed levels of significance of differences ($p < 0.05$) to be calculated.

Determination of viable cells in maize roots

Six to ten plants of each treatment were sampled 14, 30, and 60 days after inoculation, for enumeration of *A. brasilense* inside the roots.

Soil particles were removed from the roots by slight rinsing in sterile water; the washed roots (1 g, fresh weight) were shaken in 10 mL of sterile water for 10 minutes to release bacteria from the root surface. The roots were surface-sterilized for 1 minute in ethanol, washed three times with sterile water, and aseptically crushed in 10 mL of water with mortar and pestle. The suspensions were immediately diluted in 10-fold steps. After each dilution, three 0.1-mL replicate samples were transferred to cotton-plugged bottles containing 5 mL of a selective nitrogen free medium (NFB) (Okon *et al.*, 1977) with 0.1% (w/v) yeast extract and 0.05% (w/v) agar added.

The Most Probable Number (MPN) of microorganisms was calculated using the probability tables of McGrady (Postgate, 1969). Additionally, a liquid sample (0.1 mL) was serially diluted, plated on NFB solid broth and incubated for 24–48 h at 32°C. These plates were inspected for morphological characteristics of *A. brasilense* colonies.

Microbial survival in peat formulations

Peat formulations were stored at 4°C and at room temperature (20°C), and survival of our strain was regularly determined at both temperatures over a six-month period. Numbers of viable cells were evaluated with the MPN method and according to the McGrady's tables (Postgate, 1969).

Results

Cell viability levels inside the roots

In Figure 1 the results of root colonization by *A. brasilense* at different times after inoculation are shown. These values show that during the earlier phase of colonization, the two genotypes respond differently to Azospirillum invasion. Later on the bacterial concentrations inside the roots seem to be similar for the two genotypes. The results do not show any significant differences among levels of bacterial inoculum, *i.e.* different initial bacterial concentrations do not lead to different bacterial levels inside the roots, at least not for the employed inoculum concentrations.

Effect of inoculation on plant growth

The effects of Azospirillum on initial growth of maize are shown in Figure 2. The responses of the two genotypes are quite similar. At the low N level (200 mg/pot) a slight, but not significant, increase in dry weight was obtained with inoculations of 10^6 and 10^7 CFU/seed; with 10^8 CFU/seed the increase was higher in line A3/b73 (+7.1%) but disappeared in line D3/b37; at 10^9 CFU/seed no differences between inoculated and uninoculated treatments were observed.

At the high N level (400 mg/pot) the effects seem to decrease: in most of the treatments very slight differences between inoculated and control plants were observed. Only line A3/b73, when inoculated at 10^6 CFU/seed, showed an increase of 6.3% over the control.

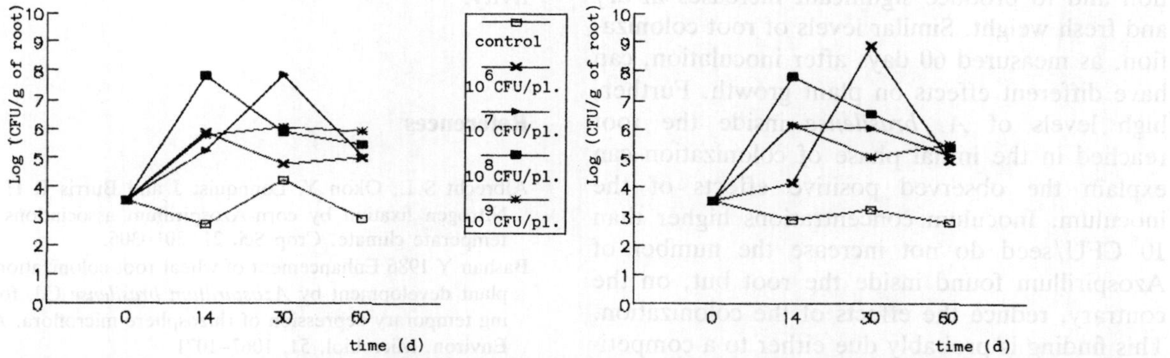

Fig. 1. Root colonization of *Azospirillum brasilense* (strain 11), at different times after inoculation, for D3/b37 (*left*) and A3/b73 (*right*) maize genotypes.

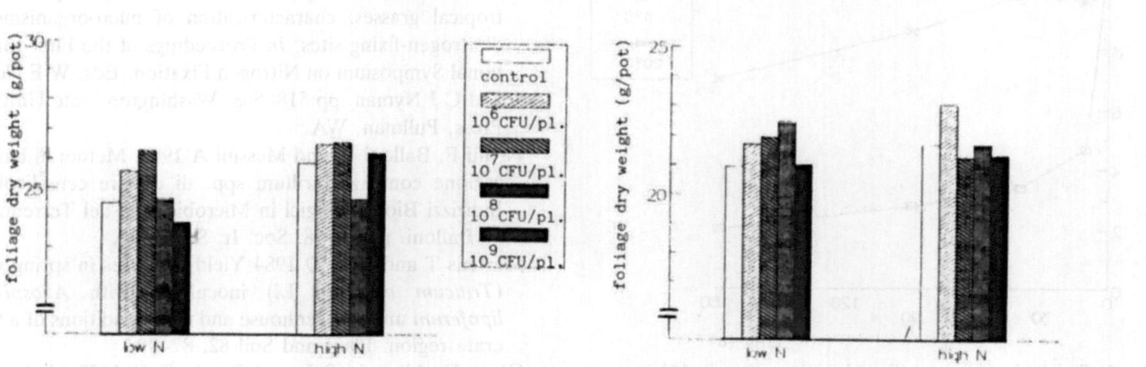

Fig. 2. Effects of treatments with *Azospirillum brasilense* (strain 11) for D3/b37 (*left*) and A3/b73 (*right*) maize genotypes.

Azospirillum brasilense *survival in peat formulations*

The efficiency of peat as a carrier of *A. brasilense* at different temperatures (4°C and 20°C) was also tested. The results are shown in Figure 3.

The survival curves show that *A. brasilense* can be satisfactorily conserved in the peat formulation at room temperature up to six months. On the contrary, when the formulation is stored at 4°C, a rapid decrease in the number of viable bacterial cells, even after a very short period, is observed.

Discussion

The use of peat as a carrier for *A. brasilense* inocula proved to be a useful, practical method for conservation and field inoculum of this beneficial rhizosphere bacterium.

For maize, inoculum rates of 10^6–10^7 CFU/seed are sufficient to assure proper root colonization and to produce significant increases in dry and fresh weight. Similar levels of root colonization, as measured 60 days after inoculation, can have different effects on plant growth. Further, high levels of *A. brasilense* inside the root reached in the initial phase of colonization can explain the observed positive effects of the inoculum. Inoculum concentrations higher than 10^8 CFU/seed do not increase the number of Azospirillum found inside the root but, on the contrary, reduce the effects of the colonization. This finding is probably due either to a competitive action exerted by the unassociated Azospirillum upon any other plant-promoting rhizosphere microorganism, or to a specific inhibiting action of the surplus Azospirillum upon the roots, as found in our hydroponic trials (data not shown). As for the two nitrogen levels employed, it seems that the higher level reduces inoculum effectiveness. In fact, there are reasons to state that, at the lower N level, it can be obtained with Azospirillum a yield similar to the one obtained with the higher N level without Azospirillum. It should also be considered that the yields of the two control treatments (respectively 200 and 400 mg N/pot) do not differ largely. It could mean that, at the fertilization levels employed, the plant response to N is already decreasing.

All results of the present study were obtained with our strain 11 of *A. brasilense*, but experiments with some control strains (ATCC 29710 and 29145) and some other ones from our laboratory, are in progress. Moreover, field trials are in progress with the aim to confirm the positive effect of Azospirillum on maize productivity.

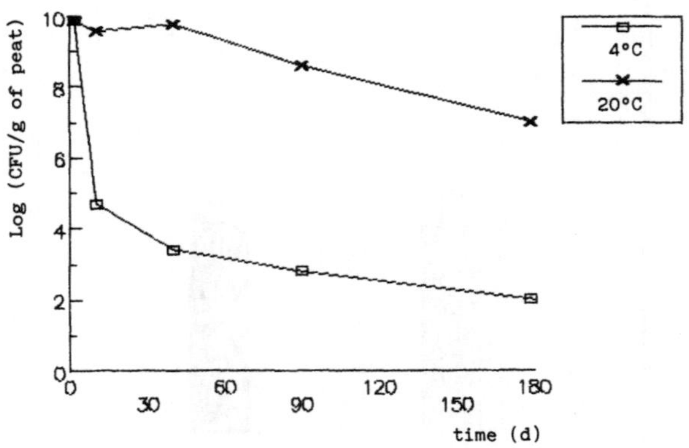

Fig. 3. Survival of *Azospirillum brasilense* (strain 11) in peat formulation at different temperatures.

References

Albrecht S L, Okon Y, Lonnquist J and Burris R H 1981 Nitrogen fixation by corn-Azospirillum associations in a temperate climate. Crop Sci. 21, 301–306.

Bashan Y 1986 Enhancement of wheat root colonization and plant development by *Azospirillum brasilense* Cd. following temporary depression of rhizosphere microflora. Appl. Environ. Microbiol. 51, 1067–1071.

Barber L F, Russell S A and Evans H J 1979 Inoculation of millet with Azospirillum. Plant and Soil 52, 49–57.

Döbereiner J and Day J M 1976 Association symbioses in tropical grasses: characterization of microorganisms and dinitrogen-fixing sites. *In* Proceedings of the First International Symposium on Nitrogen Fixation. Eds. W E Newton and C J Nyman. pp 518–538. Washington State University Press, Pullman, WA.

Favilli F, Balloni W and Messini A 1984. Metodi di batterizzazione con Azospirillum spp. di colture cerealicole. *In* Indirizzi Biotecnologici in Microbiologia del Terreno. Ed. W Balloni. pp 83–98. Soc. It. Sc. Suolo.

Mertens T and Hess D 1984 Yield increases in spring wheat (*Triticum aestivum* L.) inoculated with *Azospirillum lipoferum* under greenhouse and field conditions of a temperate region. Plant and Soil 82, 87–99.

Okon Y, Albrecht S L and Burris R H 1977 Methods for growing *Spirillum lipoferum* and for counting it in pure

culture and in association with plants. Appl. Environ. Microbiol. 33, 85–88.
Okon Y 1984 Response of cereal and forage grasses to inoculation with N_2-fixing bacteria. *In* Advances in Nitrogen Fixation Research. Eds. C Veeger and W E Newton. pp. 303–309. Nijhoff/Junk, Dordrecht, The Netherlands.
Okon Y and Kapulnik Y 1986 Development and function of Azospirillum-inoculated roots. Plant and Soil 90, 3–16.
Postgate J R 1969 Viable counts and viability. *In* Methods in Microbiology. Eds. J R Norris and D W Ribbons. pp 611–628. Academic Press, London.
Smith R L, Schank S C, Milan J R and Baltensperger A A 1984 Response of *Sorghum* and *Pennisetum* species to the N_2-fixing bacteria *Azospirillum brasilense*. Appl. Environ. Microbiol. 47, 1331–1336.

M. L. van Beusichem (Ed.), *Plant nutrition – physiology and applications*, 133–134.

PLSO IPNC539

Cellular organization of symbiotic nitrogen fixation and nitrogen assimilation in leguminous plants

I. N. ANDREEVA and S. F. ISMAILOV
Timiriazev Institute of Plant Physiology, Academy of Sciences, Botanicheskaya 35, SU-127276, Moscow, USSR

Key words: bacteroid, legumes, lupin, *Lupinus luteus* L., nitrogen fixation, peribacteroid membrane, peribacteroid space, root nodule, ultrastructure

Abstract

Information on the ultrastructure and enzyme composition of the space between bacteroids and the peribacteroid membrane, the so-called peribacteroid space (PBS), in leguminous root nodules are presented. In the period of high nitrogenase activity nitrogen- and carbon-metabolizing enzymes were found in the PBS, whereas during senescence the activity of acid phosphatase was localized at peribacteroid membrane and in PBS. Thus, in the process of development of a nitrogen-fixing symbiotic system the enzyme pattern and the composition of PBS and its role were found to be changing.

Introduction

Due to its ecological safety, biological nitrogen fixation is of great importance in agriculture. To increase the effectiveness of the symbiotic nitrogen fixing systems in leguminous plants, a thorough knowledge of the interrelations of macro- and microsymbiont is necessary.

The interaction of Rhizobium bacteria and host plants in the legume family results in the development of root nodules. The Rhizobium bacteroids in the host cell cytoplasm remain surrounded by a membrane envelope known as the peribacteroid membrane (PBM) derived from the plant's membrane material. The peribacteroid space (PBS) is located between the bacteroid cell wall and the PBM. The role of PBS as an intermediary zone between the partners of symbiosis remains still unclear (Werner *et al.*, 1988).

Materials and methods

To study the PBS-PBM system we used both the electron-microscope approach (Andreeva *et al.*, 1989) and a biochemical procedure. A special technique was developed to isolate the soluble constituents of PBS for the investigation of its substrate and enzyme profile. The first step in this procedure is the isolation of membrane-enclosed bacteroids from yellow lupin root nodules. The enclosing PBM was ruptured by a mild osmotic shock treatment. In this manner cytosol, bacteroids, PBM and the PBS constituents were obtained. To characterize the degree of purity of these fractions and to examine the native state of the PBM and of the bacteroids, an electron-microscope technique and special biochemical tests were developed. These tests included an examination of the influence of the osmotic properties of the media on the PBM and of the permeability of the PBM to silver ions (Ismailov *et al.*, 1989). The specific activity of asparagine synthetase was expressed in nmol ^{14}C-asparagine, glutamine synthetase in nmol γ-glutamyl-hydroxamate, glutamate dehydrogenase and alanine dehydrogenase in nmol NAD, all per mg of protein per minute (Ismailov, 1986).

Table 1. Activities of enzymes of ammonium assimilation in subcellular fractions of lupin root nodules

Enzyme	Plant cell cytosol (nmol per mg of protein per min)	Bacteroids, % of sp. act. in cytosol	PBS, % of sp. act. in cytosol
Asparagine synthetase	0.11	21.2	0
Glutamine synthetase	115.80	1.6	96.0
Glutamate dehydrogenase	11.58	39.0	0
Alanine dehydrogenase	3.50	19.7	56.3

Results and discussion

In electron-microscope investigations it was shown that in the root nodules of various leguminous plants the volume of the PBS, the number of bacteroids per unit PBS, the character and quantity of membrane-fibrillar, vesicular and other inclusions are different. These parameters change significantly in leguminous species in the course of ontogenesis, thus characterizing PBS as a functionally active zone. The endomembrane system of plant cells is involved in the increase of PBM surface after the endocytosis of Rhizobium bacteroids from the infection thread and subsequent proliferation in the course of the development of nitrogen fixation. It was noticed that a fusion of the vesicles of various electron density derived from Golgi dictyosomes and from the cisternae of rough endoplasmic reticulum with PBM takes place. In the period of high nitrogen fixation and in the course of root nodule senescence the rough endoplasmic reticulum is of primary importance in PBM formation.

In the PBS of lupin root nodules we found small quantities of free amino acids and of protein, but a glucan-type polysaccharide appeared to be the main structural component. In our experiments we could detect the activities of α-mannosidase, α-glucosidase and β-glucosidase in the PBS of lupin nodules. These enzymes are known to possess not only a hydrolyzing activity, but a transglycosylation activity as well.

The specific activities of enzymes of nitrogen metabolism are much lower in PBS than in plant cytosol (Table 1), but in PBS a high specific activity of glutamine synthetase was found. This finding suggests that ammonia formed in bacteroids through N_2-fixation can be incorporated not only in plant cytosol but in PBS as well. In bacteroids we determined reasonable activities of asparagine synthetase, alanine dehydrogenase (aminating), and glutamate dehydrogenase (aminating). Thus it can be hypothesized that the assimilation of a part of the fixed nitrogen takes place in bacteroids. The main portion of fixed nitrogen is assimilated in plant cell cytosol. These investigations were carried out in a period of active nitrogen fixation. In the period of declining symbiotic activity, acid phosphatase activity was localized cytochemically on PBM and in PBS, which can be seen as evidence that lytic processes are going on inside the PBS and afterwards in bacteroids.

References

Andreeva I N, Kozlova G I, Livanova G I, Zhiznevskaya G Ya and Ismailov S F 1989 Peribacteroid space in legume root nodules: An electron-microscopic study. Fiziol. Rast. 36, 551–560.

Ismailov S F 1986 Nitrogen Metabolism in Plants. Nauka, Moscow, 320 p.

Ismailov S F, Radyukina N L, Agibetov K A, Kadyrov R M, Andreeva I N, Fedorova E E and Zhiznevskaya G Ya 1989 Peribacteroid space in yellow lupin root nodules: Isolation of its contents and intactness of the peribacteroid membrane. Fiziol. Rast. 36, 309–317.

Werner D, Mörschel E, Garbers C, Bessarab S and Mellor R B 1988 Particle density and composition of the peribacteroid membrane from soybean root nodules as affected by mutation in the microsymbiont *Bradyrhizobium japonicum*. Planta 174, 263–270.

M. L. van Beusichem (Ed.), *Plant nutrition – physiology and applications*, 135–142.

PLSO IPNC076

Influence of phosphate status on phosphate uptake kinetics of maize (*Zea mays*) and soybean (*Glycine max*)

A. JUNGK[1], C.J. ASHER[2], D.G. EDWARDS[2] and D. MEYER[1]
[1]*Institute of Agricultural Chemistry, Georg-August University, Von-Siebold-Str. 6, D-3400 Göttingen, FRG, and* [2]*Department of Agriculture, University of Queensland, St. Lucia, Queensland 4067, Australia.*

Key words: C_{min}, depletion curve, flowing solution culture, *Glycine max* L., I_{max}, K_m, maize, phosphate uptake kinetics, soybean, *Zea mays* L.

Abstract

To obtain plants of different P status, maize and soybean seedlings were grown for several weeks in flowing nutrient solution culture with P concentrations ranging from 0.03–100 μmol P L^{-1} kept constant within treatments. P uptake kinetics of the roots were then determined with intact plants in short-term experiments by monitoring P depletion of a 3.5 L volume of nutrient solution in contact with the roots. Results show maximum influx, I_{max}, 5-fold higher in plants which had been raised in solution of low compared with high P concentration. Because P concentrations in the plants were increased with increase in external P concentration, I_{max} was negatively related to % P in shoots. Michaelis constants, K_m, were also increased with increased pretreatment P concentration, only slightly with soybean, but by a factor of 3 with maize. The minimum P concentration, C_{min}, where net influx equals zero, was found between 0.06 and 0.3 μmol L^{-1} with a tendency to increase with pretreatment P concentration. Filtration of solutions at the end of the depletion experiment showed that part of the external P was associated with solid particles.

It was concluded that plants markedly adapt P uptake kinetics to their P status, essentially by the increase of I_{max}, when internal P concentration decreases. Changes of K_m and C_{min} were of minor importance.

Introduction

Quantitative aspects of nutrient uptake kinetics have often been described in terms of the Michaelis-Menten equation. This equation as modified by Nielsen and Barber (1978) describes net influx, I_n, of a nutrient into plant roots as a function of external concentration of this nutrient:

$$I_n = \frac{I_{max}(C - C_{min})}{K_m + C - C_{min}} \tag{1}$$

where I_{max} = maximum influx at high concentration
K_m = the Michaelis constant; concentration where influx is $\frac{1}{2}$ of I_{max}
C_{min} = the concentration where net influx is zero

It was found earlier that tomato plants, resupplied with P after a period of P starvation, increased in I_{max}, whereas K_m and C_{min} tended to decrease (Jungk, 1974). The change of I_{max} was later confirmed with several plant species and other experimental conditions (Clarkson *et al.*, 1978; Drew *et al.*, 1984; Fist, 1987; Lefebvre and Glass, 1982). Since Michaelis-Menten kinetics is used as part of a mechanistic model (Barber, 1984; Claassen *et al.*, 1986) to simulate nutrient transfer from soil into plants, information is needed on uptake kinetics of plants under conditions of varying nutrient supply.

The objective of the present work was to study the influence of external P supply on the parameters of P uptake kinetics of the roots of intact plants.

Materials and methods

The experiments consisted of a combination of two procedures

1. To obtain plants of different P status, seedlings were grown for 3 weeks in flowing nutrient solutions at a range of constant P concentrations.
2. P uptake kinetics of roots were determined by placement of these plants in small volumes of nutrient solution with a common initial P concentration and by monitoring P depletion of the solutions until C_{min} was attained.

Flowing solution culture

To keep P concentrations at the root surface constant even at very low P concentrations, the flowing nutrient solution culture system developed by Asher and Edwards (1978), installed in a greenhouse near Brisbane, Qld, Australia, was used. Each unit had a total volume of 2400 L nutrient solution which was circulated from a storage tank through 60 pots each of 1.5 L capacity in which plants were grown. The pots were arranged in parallel with respect to solution flow. Nutrient solution was pumped at a rate of 1.6 L/min through these pots, a rate sufficiently high to prevent substantial P depletion while roots absorbed P (Edwards and Asher, 1974).

Nutrient solutions

The concentrations of basal nutrients employed in all treatments were (μmol L^{-1}): NO_3^- 1000, SO_4^{2-} 900 – 1000, Ca^{2+} 1000, K^+ 500, Mg^{2+} 200, Fe 20 (Chel – 138), B 3, Mn 0.5, Zn 0.5, Cu 0.1, Mo 0.02. pH of solutions was kept between 5.0 and 4.5 by feeding in automatically 0.1 *M* KOH or HNO_3; temperature was regulated at 24 ± 1°C. Nutrient solutions were analysed every second day and nutrients resupplied if necessary.

Treatments

P concentrations, μmol L^{-1}: 0.03, 0.1, 0.3, 1, 3, 10, 30, 100. Daily determinations of P concentration in nutrient solutions were made using the malachite-green method (Motomizu *et al.*, 1983). P concentrations were maintained at the planned level by continuous injection at a rate aimed to balance the rate of uptake by the plants.

Plants

Maize cv. "Pioneer 3906" and soybean cv. "Fitzroy" were germinated in moist filter paper at 28°C and transferred into the flowing culture system when roots were 2 cm long. Four plants were grown in each pot. They were kept in polyethylene baskets of 10 cm diameter which were filled with a 2 cm depth of black polyethylene beads to support the plants and to exclude light from the nutrient solutions. Two harvests were made, each consisting of 4 replications; root length and plant P content were measured. The mean rate of P absorption under these conditions was calculated using a formula similar to that of Williams (1948) with root length being used instead of root weight:

$$I_n = \frac{M_2 - M_1}{RL_2 - RL_1} \cdot \frac{\ln(RL_2/RL_1)}{t_2 - t_1} \quad (2)$$

where M is P content of the plant, RL is root length and t is time. The subscripts 1 and 2 refer to the first and the second harvest of plants respectively. Harvests were made on the following days after seedlings had been placed into flow culture:

	Harvest 1	*Harvest 2*
Maize	13	20
Soybean	15	27

P depletion experiments

P uptake kinetics were determined by the method of Claassen and Barber (1974) after maize plants were grown in the flowing nutrient solutions for 18 days and after soybean plants were grown for 26 days. One or two plant baskets, each containing 4 plants, were transferred from each flowing culture unit to a pot of 3.5 L capacity containing the same basal nutrient solution. The initial phosphate concentration was 15

μmol L^{-1} with maize and 12.5 μmol L^{-1} with soybean. ^{32}P phosphoric acid was added as a tracer to give an activity of about 2000 Bq/mL to allow P concentration to be monitored radiometrically. Solutions were aerated and samples were pumped out continuously at a rate of 1 mL/min. Using a fraction collector, samples were taken every 20 minutes and analysed for ^{31}P and ^{32}P.

^{31}P concentration was determined photometrically. When the P concentration was >2 μmol L^{-1}, the molybdenum blue method (Murphy and Riley, 1962) with 1 cm cells was used, while at concentrations <2 μmol L^{-1}, the more sensitive malachite green method (Motomizu *et al.*, 1983) with 4 cm cells was used. In this way, P concentrations as low as about 0.02 μmol L^{-1} could be measured.

^{32}P concentration of nutrient solutions was determined with a Packard Tricarb Scintillation counter (Model 1900 CA). 10 mL samples of nutrient solution (or 1 mL samples plus 9 mL water where ^{32}P concentration was >100 Bq/mL) were transferred into scintillation vials, mixed with 0.2 mL of 2.5% ANDA as a wave length shifter, and the Cerenkov radiation measured.

Rates of P uptake were determined from P solution depletion. To obtain the rate of P uptake per unit root length (=influx), differences in P concentration of consecutive samples were used to calculate the loss of P per pot (=uptake of plants) and divided by root length (cm) and the time interval (s). At the end of the experiment, root length was determined with a modified line-incercept method of Newman (1966) in order to calculate net influx of P. In addition, total P content of the roots and shoots was determined.

Results and discussion

Plant growth was affected by P concentration in the very low range only. A P concentration of 1 μmol L^{-1} nutrient solution was adequate for maximum growth of both maize and soybean. This range is in agreement with results of Jintakanon *et al.* (1982) and Fist *et al.* (1987). Mean rates of P absorption as determined with plants grown continually in flowing nutrient solution

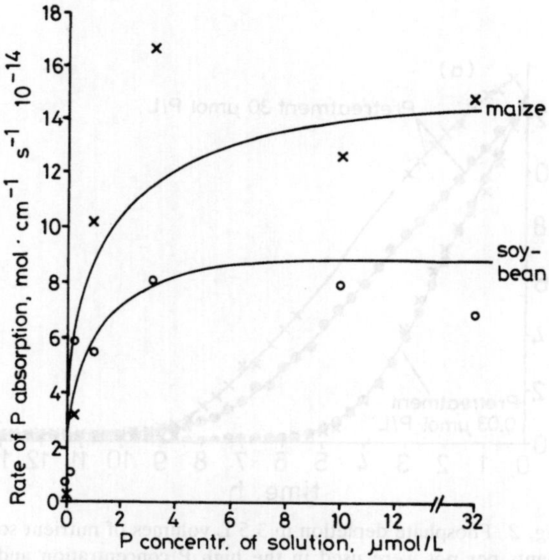

Fig. 1. Mean rates of phosphate absorption between the first and the second harvest into roots of maize and soybean grown continually in flowing nutrient solutions of constant P concentration.

using the Williams equation (Eq. 2) were strongly influenced by external P concentration only below 3 μmol P L^{-1} (Fig. 1). According to these data, mean rates of P absorption (in mol $cm^{-1}s^{-1}$ 10^{-14}) of about 6 for soybean and 10 for maize were sufficient for maximum growth. For the 'short-term' P uptake experiments, plants from 4 treatments out of the flowing solution cultures were selected: 0.1, 1, 10, 100 μmol P L^{-1} with maize, and 0.03, 0.3, 3 and 30 μmol P L^{-1} with soybean.

An example of the results obtained by the depletion technique is shown in Figure 2a for soybean plants grown at pretreatment P concentrations of 0.03 and 30 μmol L^{-1}. In the case of ^{32}P, an almost linear decrease of concentration with time occurred down to a level of about 4 μmol L^{-1}. The high P concentration pretreatment showed a smaller slope than the other, reflecting a lower rate of P uptake. Below about 4 μmol P L^{-1}, the slope gradually levelled off, indicating a decrease of P influx with the further decrease of solution P concentration. Similar effects were observed with maize (Fig. 2b).

Comparison of the ^{32}P and ^{31}P data shows that for both soybean (Fig. 2a) and maize (Fig. 2b) almost identical curves can be drawn in the case

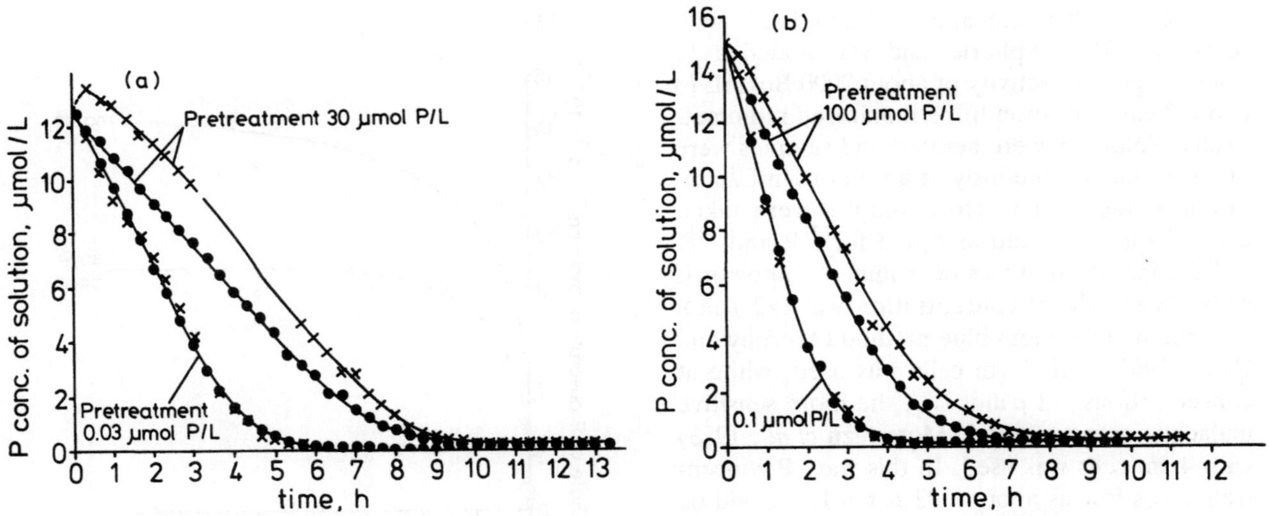

Fig. 2. Phosphate depletion in 3.5 L volumes of nutrient solution by **(a)** soybean and **(b)** maize of differing P status. Two sets of plants per pot were used in the high P concentration and only one set in the low P concentration pretreatment (Method of measurement shown thus: x ^{31}P, •^{32}P).

of the low P concentration pretreatments. However, the chemically measured ^{31}P values seem to have somewhat higher variability. In contrast, marked differences between P concentrations monitored with ^{32}P and ^{31}P were obtained with the high P pretreatments, the ^{31}P values being consistently higher than the ^{32}P values. These higher ^{31}P concentrations at any given point of time can be attributed to the release, in the first 15–30 min only, of P from the free space of roots or efflux of P from the inner space of the plants. As shown in Table 1, shoot P concentrations increased with increase in pretreatment solution P concentration. Particularly in plants grown in the high pretreatment P concentrations, it would appear that more inorganic P was present, which may be more subject to efflux. The ^{32}P data can thus be regarded as a measure of influx, whereas the ^{31}P data are a measure of net influx (influx-efflux).

The depletion data were transformed into rates of P uptake (influx) per unit root length. From these influxes, the values of I_{max} and K_m were obtained by means of the Hanes plot (Fig.

Table 1. Influence of P concentration of flowing nutrient solutions on shoot and root P concentration and P uptake parameters of soybean and maize roots

Solution P conc. (μmol L^{-1})	P (%) Shoot	Root	I_{max} mol cm^{-1} s^{-1} 10^{-14}	K_m (μmol L^{-1})	C_{min} [a]	C_{min} [b]
Soybean						
0.03	0.22	0.23	17.6	1.6	0.06	0.01
0.3	0.34	0.30	16.9	1.7	0.10	0.03
3	0.59	0.56	6.5	1.2	0.16	0.08
30	0.66	0.90	3.7	1.0	0.16	0.06
Maize						
0.1	0.22	0.20	37.0	6.1	0.17	0.01
1	0.68	0.45	21.4	3.9	0.24	0.02
10	1.08	1.00	6.6	1.9	0.28	0.04
100	1.16	1.35	7.1	3.4	0.24	0.02

[a] Solutions unfiltered.
[b] Solutions filtered through 0.45 μm membrane filter.

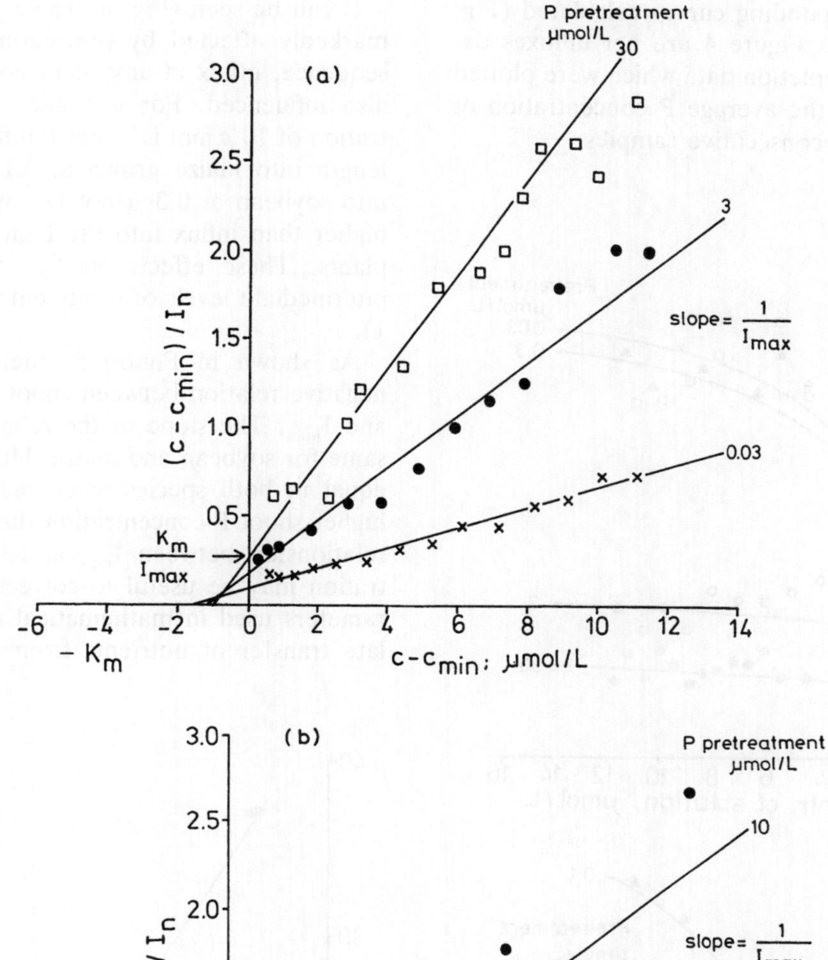

Fig. 3. Hanes plot of P influx into (a) soybean and (b) maize of differing P status. Data obtained from P depletion experiments.

3). The Hanes plot was chosen because the results are least biased by the variability of the original values as suggested by Cornish-Bowden (1979, p. 26). As can be seen from Figure 3, the fitted regression lines give a reasonable approximation to the data. It can thus be concluded that influx follows Michaelis-Menten kinetics. From this plot, I_{max} can be obtained from the slope of the line and K_m as the intercept on the negative X-axis. C_{min} was directly read from depletion curves. The values for I_{max}, K_m and C_{min} (summarized in Table 1) were inserted into equation

1 and the corresponding curves calculated (Fig. 4). The points in Figure 4 are net influxes derived from the depletion data which were plotted as a function of the average P concentration of the relevant two consecutive samples.

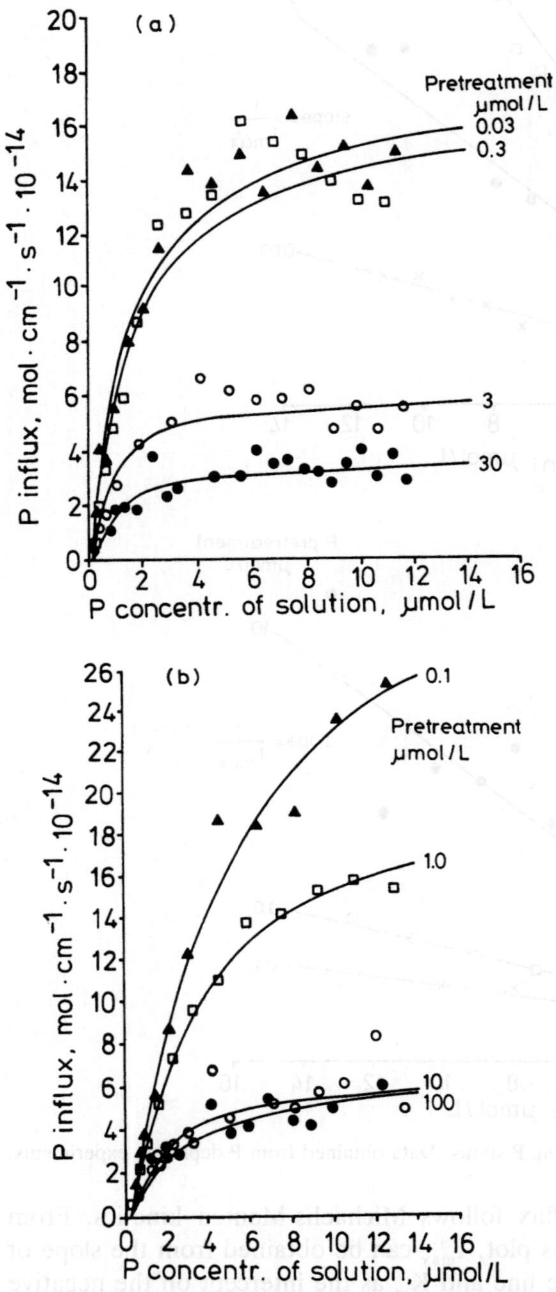

Fig. 4. Phosphate uptake kinetics of **(a)** soybean and **(b)** maize of differing P status. (Curves were calculated with Equation 1 using I_{max} and K_m values obtained from Hanes plots).

It can be seen (Fig. 4, Table 1) that I_{max} was markedly affected by pretreatment. As a consequence, influx at any other concentration was also influenced. For example, at a P concentration of 10 μmol L^{-1}, net P influx per unit root length into maize grown at 0.1 μmol L^{-1} and into soybean at 0.3 μmol L^{-1} was about 4-fold higher than influx into the high P pretreatment plants. These effects on I_{max} were largest at intermediate levels of treatment P (Fig. 4, Table 1).

As shown in Figure 5, there was a linear negative relation between shoot P concentration and I_{max}. The slope of the relationship was the same for soybean and maize. However, I_{max} was equal in both species when maize has a 0.6% higher shoot P concentration than soybean. This relationship between I_{max} and shoot P concentration may be useful to correct the kinetic parameters used in mathematical models to simulate transfer of nutrients from soil into plants.

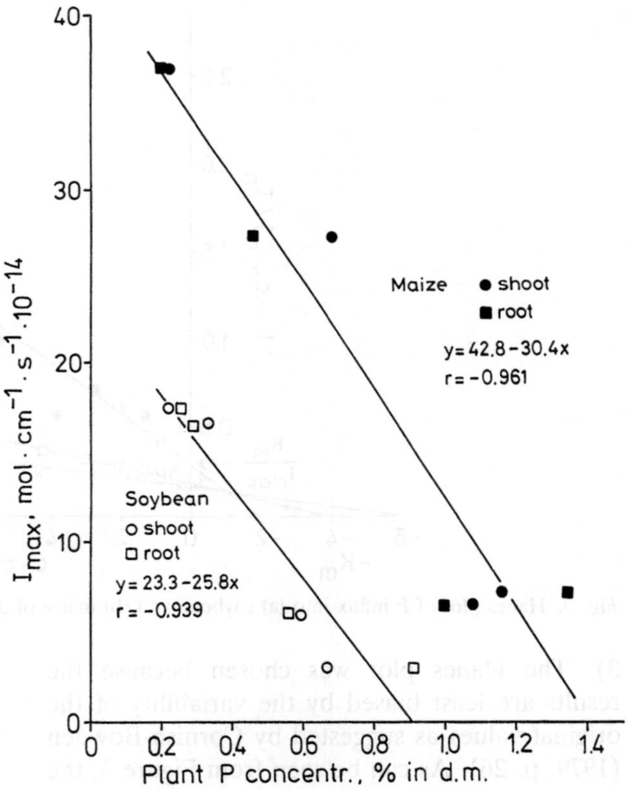

Fig. 5. Maximum P influx, I_{max}, in relation to P concentration of soybean and maize.

The results show that the uptake capacity of the roots is regulated over a wide range by the P status of the plants. The increase of uptake efficiency of the roots at low P status was, however, not due principally to changes in K_m. The Michaelis constant, K_m, generally interpreted as the affinity parameter of the uptake system, did not decrease with the decrease of P concentration as might be expected. On the contrary, K_m increased slightly in P deficient soybean but by a factor of 3 in P deficient maize, thus indicating a decrease of affinity of the P uptake system. Drew *et al.* (1984) using barley and Fist (1987) using tropical grain legumes have not found substantial changes of K_m between plants of different P status.

C_{min} was always below 0.3 μmol P L^{-1} and showed a tendency to decrease with decreased pretreatment P concentration (Table 1). Filtration of nutrient solutions through 0.45 μm membrane filter at the end of the depletion experiments showed that part of the P which remained was associated with solid particles. The revised (and lower) C_{min} values following removal of particulate P are presented in Table 1. They show that plants are able to deplete phosphate from solution around their roots to very low concentrations. This is one of the mechanisms to mobilize P from the solid soil phase because it disturbs the equilibrium between dissolved and adsorbed P, leading to P desorption.

The main conclusion of this study is that none of the three parameters used to describe P uptake kinetics of soybean and maize roots can be regarded as constant. They all depend on the P concentrations the plants were exposed to and hence on their internal P concentration. The change of I_{max} has the greatest influence. It increased markedly with decrease of the pretreatment P concentration, thus indicating a feed-back between P status and P uptake kinetics of roots. The decline in K_m with increased shoot P concentration is at variance with the observation of Lee (1982), Cogliatti and Clarkson (1983) and Fist (1987) and does not provide any support for allosteric regulation of P influx as suggested by Lefebvre and Glass (1982). However, efflux may also be a substantial component of the regulation of net P accumulation in plants as concluded by Elliott *et al.* (1984).

Finally, the results of the long-term experiment may be compared with those of the short-term uptake experiments. It shows that mean rate of P absorption between the two harvests of soybean obtained by the modified Williams equation (Fig. 1) was fairly close to the curve obtained with the depletion technique for the soybean plants with the low P concentration pretreatment (Fig. 4) in the range of P concentrations below 3 μmol P L^{-1}. This is, according to Figure 1, the range where P concentration limited P influx. In contrast, P influx of soybean grown continually in solutions of higher P concentrations was lower than the influx measured with the depletion technique. This is the range where net influx seems to have been limited by the growth rate of the plants rather than by P concentration. Comparison of the maize data leads to a similar picture, particularly for the range beyond 1 μmol P L^{-1}. In the lower P concentrations, the data of the long-term experiments even exceeded those of the low P pretreatment in the short-term experiment. Therefore, net P absorption rates by plants grown at concentrations below 3 μmol P L^{-1} in flowing nutrient solutions were on average between the two harvests equal to the influx into soybean and greater than the influx into maize in the short-term depletion experiment, even when the most efficient plants which had been grown under P deficient conditions were considered. Also K_m values of plants grown continually in flowing solution were found considerably lower than those measured by the depletion technique. These K_m values may be estimated from Figure 1 to be about 0.5 μmol L^{-1} for maize and 0.2 μmol L^{-1} for soybean.

These discrepancies may be explained as follows:

— Adaption of P uptake kinetics at the start or during the course of the depletion experiment as suggested by Shock and Williams (1984) for sulfate uptake.

— Efflux of P. Using ^{32}P in depletion experiments, the influx measured is gross influx. This may markedly differ from net influx which was measured in the long-term experiment. Comparison with ^{31}P data showed that considerable efflux occurred with the plants grown in the high P pretreatments. This is in agreement with Elliott *et al.* (1984) who measured a considerable P efflux particularly with plants grown in high P concentration.

Acknowledgement

Thanks are expressed to the Deutsche Forschungsgemeinschaft for financial support.

References

Asher C J and Edwards D G 1978 Relevance of dilute solution culture studies to problems in low fertility tropical soils. *In* Mineral Nutrition of Legumes in Tropical and Subtropical Soils. Eds. C.S. Andrew and E.J. Kamprath. pp 131–152. CSIRO, Melbourne.

Barber S A 1984 Soil Nutrient Bioavailability: A Mechanistic Approach. John Wiley and Sons, New York, 398 p.

Claassen N and Barber S A 1974 A method for characterizing the relation between nutrient concentration and flux into roots of intact plants. Plant Physiol. 54, 564–568.

Claassen N, Syring K M and Jungk A 1986 Verification of a mathematical model by simulating potassium uptake from soil. Plant and Soil 95, 209–220.

Clarkson D T and Scattergood C B 1982 Growth and phosphate transport in barley and tomato plants during the development of, and recovery from, phosphate stress. J. Exp. Bot. 33, 865–875.

Cogliatti D H and Clarkson D T 1983 Physiological changes in, and phosphate uptake by potato plants during development of, and recovery from phosphate deficiency. Physiol. Plant. 58, 287–294.

Cornish-Bowden A 1979 Fundamentals of Enzyme Kinetics. Butterworths, London-Boston.

Drew M C, Saker L R, Barber S A and Jenkins W 1984 Changes in the kinetics of phosphate and potassium absorption in nutrient-deficient barley roots measured by a solution depletion technique. Planta 160, 490–499.

Edwards D G and Asher C J 1974 The significance of solution flow rate in flowing culture experiments. Plant and Soil 41, 161–175.

Elliott G C, Lynch J and Läuchli A 1984 Influx and efflux of P in roots of intact maize plants. Plant Physiol. 76, 336–341.

Fist A J 1987 Regulation of phosphorus transport in tropical grain legumes. PhD Thesis, St. Lucia, Queensland, Australia.

Fist A J, Smith F W and Edwards D G 1987 External phosphorus requirements of five tropical grain legumes growing in flowing-solution culture. Plant and Soil 99, 75–84.

Jintakanon S, Edwards D G and Asher C J 1982 An anomalous, high external phosphorus requirement for young cassava plants in solution culture. *In* Proc. 5th Int. Symp. Trop. Root and Tuber Crops. (Manila, 1979). Eds. E H Belen and M Villaneuva. pp 507–518. PCARRD, Los Baños, Philippines.

Jungk A 1974 Phosphate uptake characteristics of intact root systems in nutrient solution as affected by plant species, age and P supply. *In* Plant Analysis and Fertilizer Problems. Ed J Wehrmann. pp 185–196. Proc. 7th Internat. Colloq. Hannover. German Society of Plant Nutrition, Hannover, FRG.

Lee R B 1982 Selective and kinetics of ion uptake by barley plants following nutrient deficiency. Ann. Bot. 50, 429–449.

Lefebvre D D and Glass A D M 1982 Regulation of phosphate influx in barley roots: effects of phosphate deprivation and reduction of influx with provision of orthophosphate. Physiol. Plant. 54, 199–206.

Motomizu S, Wakimoto T and Toei K 1983 Spectrometric determination of phosphate in river waters with molybdate and malachite green. Analyst 108, 361–367.

Murphy J and Riley J P 1962 A modified single solution method for the determination of phosphate in natural waters. Anal. Chim. Acta 27, 31–36.

Newman E J 1966 A method of estimating the total length of root in a sample. J. Appl. Ecol. 3, 133–145.

Nielsen N E and Barber S A 1978 Differences among genotypes of corn in the kinetics of P uptake. Agron. J. 70, 695–698.

Shock C C and Williams W A 1984 Sulfate uptake kinetics of three annual range species. Agron. J. 76, 35–40.

Williams R F 1948 The effects of phosphorus supply on the rates of intake of phosphorus and nitrogen and upon certain aspects of phosphorus metabolism in gramineous plants. Aust. J. Sci. Res. B 1, 333–361.

M. L. van Beusichem (Ed.), *Plant nutrition – physiology and applications*, 143–146.

PLSO IPNC079

Using a mechanistic model to evaluate the effect of soil pH on phosphorus uptake

S.A. BARBER and JEN-HSHUAN CHEN
Agronomy Department, Purdue University, West Lafayette, IN 47907, USA

Key words: buffer power, effective diffusion, exchangeable phosphorus, mechanistic model, soil solution phosphorus, *Zea mays* L.

Abstract

The objective of this research was to show how a mechanistic uptake model that accurately predicts phosphorus (P) uptake by maize (*Zea mays* L.) in a pot experiment may be used to evaluate the reasons for the differences in P availability observed when soil pH is varied. The model predicts P uptake by integrating soil P supply by mass flow and diffusion; size, shape and growth rate of roots; and P uptake kinetics of the root. The P supply parameters of the model that may be affected by soil pH are P_{li}, initial P concentration in the soil solution; b, the buffer power of P in the soil, P_{si}, for P_{li}, and D_e, effective diffusion coefficient. The effect of these changes on P uptake was predicted with the model by using measured values of the three soil supply parameters and of size, shape, and growth rate of roots and keeping the other parameters at values characteristic of maize. Values for three soil supply parameters can be calculated from measurements of P_{li}, P_{si}, and θ, volumetric water content. The predictions of the model closely agreed with observed uptake when form of P present at the higher pH's was accounted for. There was a significant positive correlation ($r = 0.94$) between P_{li} and observed P uptake and a significant negative correlation (-0.93) between P_{si} and observed P uptake. The use of the model demonstrated the significance of P form and the importance of P_{li} in P uptake. It also showed importance of root growth rate.

Introduction

Soil pH has been shown to affect P uptake by plants. The effect may vary with soil and the reason for the differences are not well understood. The effect of soil pH on P availability has been measured with plant growth experiments in which increases in either plant weight or uptake of P have been the measure of the effect of pH. Development of a mechanistic uptake model (Barber and Cushman, 1982; Barber, 1984) allows investigation of the effect of soil pH on the mechanisms involved in the effect of soil pH on P supply to the plant. This mechanistic model describes nutrient uptake by integrating the effects of soil P supply to the root surface by mass flow and diffusion, changes in root geometry and size due to growth, and the relation of P uptake rate to P concentration in solution at the root surface. Root supply of P to the root by mass flow and diffusion is described using P concentration in the soil solution P_{li}, buffer power, b, of the adsorbed P, P_{si}, for P_{li}, effective diffusion coefficient, D_e and v_0, rate of water uptake. The model has been verified with both pot and field experiments for P uptake by corn and soybeans from P fertilized soils (Barber, 1984) with correlation coefficients ranging from 0.85 to 0.98 and most regression coefficients between 0.95 and 1.08. Additional parameters used in the model include the plant root growth parameters of L_0, initial root length; r_0, mean root radius; k, rate of root growth; and r_1, half distance between adjacent root axes. The uptake of P by the root as related to P concentration in solution at the root surface is described by: I_{max}, maximum influx at high ion concentrations in solution; Km, the ion concentration in solution where net influx-C_{min} is one-half of I_{max}; and C_{min}, the ion concentration in solution where net influx is zero. The mean rate of water uptake, v_0, is required for measurement of mass flow. The nutrient influx at any time will be related to P concentration at the root surface which results from the balance between supply to the root by mass flow and diffusion and the rate of uptake as related to ion concentration in solution at the root surface.

The supply of P to the root surface depends on C_{li}, b, D_e, and v_0. A sensitivity analysis of the model (Silberbush and Barber, 1983) for P uptake by soybean (*Glycine max* L.) shows that changes in values for the soil supply have more influence on the rate of uptake than changes in the values for the rate of uptake as related to P concentration at the root surface since rate of P supply to the root surface is more limiting than rate of uptake by the root.

The objective of the research reported in this paper is to show how a mechanistic uptake model can be used to evaluate the effect of changing soil pH on the mechanisms governing P uptake by plant roots.

Methods

A pot experiment using Chalmers silt loam (Typic Haplaquolls) limed to five pH values was used to verify the uptake model. The soil, initially at pH 4.7, was limed to pH's of 5.7, 6.5, 7.6, and 8.2. The experiment was completely randomized with two replicates. Six-day-old maize (*Zea mays* L.) plants were transplanted into pots with 3 kg of soil. The plants were grown for 10 days in a controlled climate facility with a 16 h day of 6000 $\mu w\ cm^{-2}$ irradiance and a temperature of 25°C. At harvest, shoot weight, root weight, shoot P content, root P content, root length were measured and mean root diameter, and mean half-distance between root axes calculated. Soil measurements made were pH; C_{si}, anion-exchangeable P, C_{li}, concentration of P in the soil solution, θ, volumetric water at −33 kPa tension. The pH values were measured in 1:1 soil-water. The C_{li} values were measured by placing 500 g of soil in a 7.5-cm diameter plexiglas column with a perforated bottom. The soil was brought to −33 kPa tension, covered with filter paper, allowed to stand for 24 h, then water was added at a rate of 4 to 8 ml per h to the surface and 50 ml of water displaced. The displaced solution was filtered through a 0.45 μm filter and analyzed for P.

Anion-exchange P used for the value of C_{si} was determined by shaking one g of Dowex 2 × 8 Cl-saturated resin for 24 h with one g of soil in 100 ml of deionized water. The soil was separated from the resin by sieving and then the P was displaced from the resin with warm 1.0 *M* NaCl by shaking for 6 to 8 h. A subsample of the solution was filtered through a 0.45 μm filter and then analyzed for P. The buffer power b was obtained from $\Delta C_s/\Delta C_l$. The value for D_e was determined from the relation $D_e = D_e\theta f_1/b$ where D_e is the diffusion of P in water, θ is volumetric water, f_1 is the impedence factor that is related to θ and b is the buffer power.

The plant material was wet digested with H_2SO_4 and H_2O_2 and the digestate made up to 100 ml and analyzed for P. All measurements for P were made with the Murphy and Riley (1962) procedure. Roots were washed from the soil and length determined by the root intersect method (Tennant, 1975). Fresh roots was assumed to have a density of 1.0 so that g of roots equaled cc of root volume and this value was used together with root length to obtain mean root radius. Calculation of the values used in the model is shown in Schenk and Barber (1979).

Results

Phosphorus uptake calculated by the model assuming no effect of pH on P uptake is compared with P uptake as measured in the plants in Table 1. There was close agreement between observed and calculated uptake except that calculated uptake at pH's 7.5 and 8.3 was much greater than observed. At these pH levels, much of the P in solution is as the HPO_4 ion rather than the H_2PO_4 ion (Table 2). Hendrix (1967) found that the uptake rate for the HPO_4 ion was about one-tenth that of the H_2PO_4 ion. When this assumption was made in calculating predicted P

Table 1. The effect of soil pH on the relation between predicted P uptake and observed P uptake by maize growing in a pot experiment

Soil pH	Observed P uptake	Predicted P uptake[a]	Predicted P uptake[b]
		μmol pot^{-1}	
4.7	274	250	250
5.7	219	205	205
6.5	232	257	221
7.6	36	105	39
8.3	26	117	19

[a] Predicted P uptake calculated assuming the rate of P absorption was not affected by pH.
[b] Predicted P uptake assuming HPO_4 ions were absorbed at one tenth the rate of H_2PO_4 ions.

Table 2. The effect of soil pH on solution concentration of H_2PO_4 ion and HPO_4 ion, exchangeable P and length of maize roots after 16 days

Soil pH	Solution P H_2PO_4	HPO_4	Exchangeable P	Root length
	μmol L^{-1}		mmol/kg	m pot^{-1}
4.7	25.2	0	5.16	138
5.7	13.2	0	8.03	124
6.5	14.4	2.7	6.55	137
7.6	3.9	9.4	9.55	82
8.3	0.97	12.6	11.65	79

uptake, there was close agreement between observed and predicted uptake (Table 1). Hence when maize was grown in soil, P uptake was affected by the valence of the P ion present in solution.

Since the model accurately predicted P uptake, we can now look at the effect of soil pH on the parameters of the model to see what factors were influenced by changes in soil pH and determine their relation to the effect of soil pH change on P uptake and plant growth.

Discussion

The main parameters of the model that affect P uptake are P_{li}, P_{si}, root length, and root surface area which includes the effect of root radius plus root length. When observed P uptake was correlated with these parameters, the results shown in Table 3 were obtained. There was an expected correlation between P uptake and root length or

Table 3. The correlation of observed P uptake by maize with soil and plant parameters used in the model

	r
P uptake *vs* root length	0.97
P uptake *vs* root surface	0.99
P uptake *vs* C_{li}	0.94
P uptake *vs* C_{si}	−0.93

surface area since as root surface increases there is increased area for absorption of P from the soil. Hence when root growth was greater at certain soil pH levels, P uptake was increased.

To obtain the soil supply parameters of the model, it is necessary to measure P_{li}, the initial P concentration in soil solution, and P_{si}, the initial P concentration adsorbed on the solid phase that equilibrates with P_{li}. Soil tests for available P usually measure values closely related with P_{si}. Measurements of P_{li} are seldom made. The correlation between P uptake and P_{li} and P_{si} were interesting. They both gave correlation values above 0.9 but the correlation between P_{si} and P uptake was negative, which indicates that although there was a correlation, it was without meaning since if P_{si} was affecting P supply to the root we would expect a positive correlation. The positive correlation between P_{li} and P uptake indicates that P_{li} is a major soil parameter influencing the supply of P to the root. This result has considerable interest since P_{si} is a value similar to that obtained by many soil tests. Increasing pH increased P_{si} which would indicate that soil P availability should be increased, however, the opposite effect occurred, hence for this study, measurement of P_{si} alone would not be a good measure of P availability.

The results of this experiment show that the Barber-Cushman (1981) uptake model is a useful tool for investigating the influence of soil pH on P uptake. Use of the model demonstrated that form of P ion should be considered at pH levels above 6.7 and that the level of P in the soil solution and the amount of root growth were the dominant factors determining P uptake.

References

Barber S A and Cushman J H 1981 Nitrogen uptake model for agronomic crops. *In* Modeling Wastewater Renovation-land Treatment. Ed. I K Iskander. pp 382-409. John Wiley and Sons, New York.

Barber S A 1984 Soil Nutrient Bioavailability: A Mechanistic Approach. John Wiley and Sons, New York, 398 p.
Hendrix J E 1967 The effect of pH on the uptake and accumulation of phosphate and sulfate ions by bean plants. Am. J. Bot. 54, 560–564.
Murphy J and Riley J D 1962 A modified single solution method for the determination of phosphate in natural waters. Anal. Chem. Acta 27, 31–36.
Schenk M K and Barber S A 1979 Phosphate uptake by corn as affected by soil characteristics and root morphology. Soil Sci. Soc. Am. J. 43, 880–883.
Silberbush M and Barber S A 1983 Sensitivity of simulated phosphorus uptake by parameters used by a mechanistic-mathematical model. Plant and Soil 74, 93–100.
Tennant D 1975 A test of a modified line intersect method of estimating root length. J. Ecol. 63, 995–1001.

M. L. van Beusichem (Ed.), *Plant nutrition – physiology and applications*, 147–151.

PLSO IPNC167B

Phosphorus diffusion to barley (*Hordeum vulgare*) roots as influenced by moisture and phosphorus content of soils

K.R. VÉGH[1], G.Y. FÜLEKY[2,4] and T. VARRÓ[3]
[1]*Institute for Soil Science and Agricultural Chemistry, Hungarian Academy of Sciences, Budapest, Hungary,* [2]*Soil Science Department, University of Agricultural Sciences, Gödöllö, Hungary and* [3]*Colloid Chemistry Section, Chemistry Institute, Kossuth Lajos University, Debrecen, Hungary.*
[4]*Corresponding author*

Key words: barley, crop phosphate-uptake, effect of soil texture, *Hordeum vulgare* L., moisture content, phosphate-diffusion coefficient, soil depletion volume

Abstract

Rapid diffusion in the soil macropores and a much slower one in the micropores have been shown. Effects of soil moisture- and P-content on the effective diffusion coefficients are also obtained. The soil volumes in which P moves towards the roots and the amount of P moving by rapid and slow diffusion were estimated by the effective diffusion coefficients, root growth dynamics and soil P-fractions, obtained in different moisture and nutrient states.

In the soil with low P-buffering power the rapid diffusion process determines P-transport, whereas in more buffered soil the depletion volume is much smaller and the slow diffusion process is significant in P-supply.

Introduction

The P amount actually taken up by the plants depends on the properties of the soil P-transport and root density (De Willigen and Van Noordwijk, 1988). Both are influenced by watering and fertilization. Depletion volume can be an important factor determining P-recovery of plants. P-transport by the diffusion towards the roots can be characterized by the effective diffusion coefficient (D_e) (Barber, 1984). D_e and the path length of P-transport are influenced by the moisture content, structure, porosity and P-buffering capacity of the soil (Nye, 1979).

This paper is concerned with the soil depletion volume and the available P-content change, while the root system is growing in different soil textures and under different water and fertilization conditions.

Methods

Four fertilization levels were set up on a sandy and a loamy soil (Tables 1 and 2). Soils were then incubated at field water capacity for two weeks. The water-holding capacity was determined as a function of suction for the soils, and soil moisture contents were adjusted to pF 3.7, 2.3 and 2.0 suctions and maintained during the experiments.

Effective diffusion coefficients (D_e) were determined by ^{32}P radioabsorption method at each moisture and nutrient level. By the two-detector technique D_e can be determined in the range between $10^{-9} - 10^{-19}$ m^2s^{-1} (Varró *et al.*, 1985).

The treated soils were used in a pot experiment. One spring barley (cv. SK 1775) plant was grown in one pot containing 1700 g soil, in controlled environment. Roots and shoots were

Table 1. Characterization of the soils used in the study

Soil	pH_{H_2O}	$CaCo_3$ %	b mg/cm³	Organic matter %	Sand %	Silt %	Clay %	Bulk density g/cm³	Total porosity %
Sandy soil	7.2	0.5	43.8	1.10	73.0	22.0	5.0	1.53	42.6
Loam	7.8	8.8	85.7	2.16	33.9	56.4	9.7	1.20	52.1

Table 2. Soil phosphorus contents after application of increasing amounts of phosphate fertilizer

P-levels	P-content mg/kg			
	Sandy soil		Loam	
	H_2O-P	AL-P	H_2O-P	AL-P
P0	11.43	196.2	4.29	111.5
P1	31.61	269.2	17.14	184.6
P2	70.96	369.2	21.29	319.2
P3	111.26	492.3	23.23	453.8

sampled on ten occasions during the first 30 days. Root length (L), fresh and dry weight, shoot dry weight and P-content were measured.

P-concentration in the soil solution and the amount of available-P were characterized by water- and ammonium-lactate-acetate extraction respectively. Phosphorus buffering capacity (b) was determined using Freundlich isotherms.

Results and discussion

The effective diffusion coefficients of ^{32}P show a fast and a slower process of P-transport. In the first 5400 sec P-transport is fairly rapid and is considered a fast diffusion with coefficient values of the order of 10^{-11} m^2s^{-1}, but after 1.5 hours the rate of transport is reduced and is controlled by a slower diffusion process with coefficient values of 10^{-14} m^2s^{-1} and a lifetime of 8 days (Table 3). The fast diffusion is characteristic for the transport in soil macropores and the slower one in the micropores (Varró *et al.*, 1985). The rate of diffusion in the macropores of the sandy soil is found higher than that of the loam, whereas transport rate in the micropores of the finer textured soil is much higher, as is shown in Table 3. The diffusion coefficients are consistent in showing that the rate of P-transport is raised by increasing soil moisture and P-content both in the micro- and macropores. P-mobility in the dry P-fertilized soil is similar to that in moist or wet unfertilized soils. At the highest fertilization level De values are markedly reduced presumably due to the harmful effect of the high salt concentration on the soil structure.

The maximum distance of P-transport (r) was calculated by the effective diffusion coefficients, and the lifetime of the process as $r = (2Dt)^{1/2}$. On the basis of the growth characteristics (r_0, L) we estimated the depletion soil volumes in which P was transported to the roots (V_d). V_d was calculated as a volume of a soil cylinder having a radius of $R = r + r_0$ and a length equal to the root length (L) in a given time, and reduced by the volume of the root. Root elongation was considered exponential in the experimental period. Root growth rate (k) was calculated as $(\ln L_2 - \ln L_1)/(t_2 - t_1)$ and used in the estimation of V_d.

The volume of the soil depleted by one barley plant growing in sandy soil is shown in Figure 1. P-fertilization of the dry soil has a similar effect as watering on unfertilized soil, leading to similar depletion volumes. The lower the D_e value the longer the root needed to deplete the same soil volume, as shown in Figure 2. Consequently competition could arise because of higher root density but, if D_e is small, on account of the small transport distance this does not come about.

The amount of P available for a single barley plant within the study period, *i.e.* the amount of P transported to the roots by diffusion, was calculated. Soil solution P content (water-extractable P) for rapid transport in macropores and the amount of available-P (AL-P) for the slow transport in micropores were considered. We assumed that the rate of transport is controlled by diffusion. The amount of P diffusible

Table 3. Effective diffusion coefficients of ^{32}P at two different periods after P supply

	Sandy soil			Loam		
	pF = 3.7	pF = 2.3	pF = 2.0	pF = 3.7	pF = 2.3	pF = 2.0
P-levels	Macropores t = 5400 s $D_e \times 10^{-11}$ m^2s^{-1}					
P0	5.4	11	12	3.3	6.8	7.2
P1	8.3	9.9	24	4.2	5.3	14
P2	12.2	13	25	4.6	12	13
P3	5.8	9.3	17	4.1	7.0	7.5
	Micropores t = 8 days $D_e \times 10^{-14}$ m^2s^{-1}					
P0	0.3	3.8	3.7	3.8	4.9	4.9
P1	3.6	4.5	9.1	4.5	8.9	14
P2	4.1	4.5	9.1	4.5	13	18
P3	1.5	3.9	5.4	4.5	4.1	13

in the micropores, in the whole V_d, and P taken up by the plants are shown in Figure 3 and 4. The significance of the rapid and slow diffusion is different in the two soils. While the 70–80% of the total-P is transported by the slow diffusion process in the loam, it is only 20–30% in sandy soil, irrespective of the soil moisture content. In loam, due to the higher P-buffering capacity (b), P-concentration in the soil solution and D_e values for rapid diffusion in macropores are relatively low. The slow process mainly determines the P-supply. In contrast, the high P-concen-

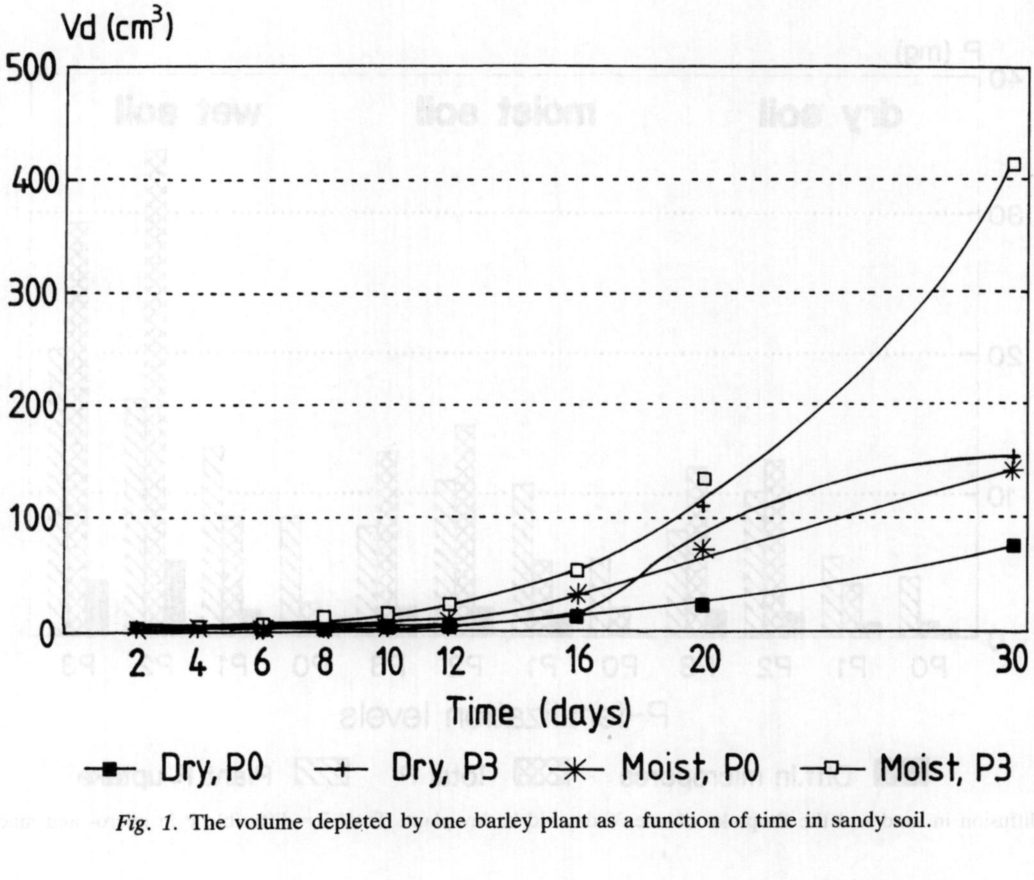

Fig. 1. The volume depleted by one barley plant as a function of time in sandy soil.

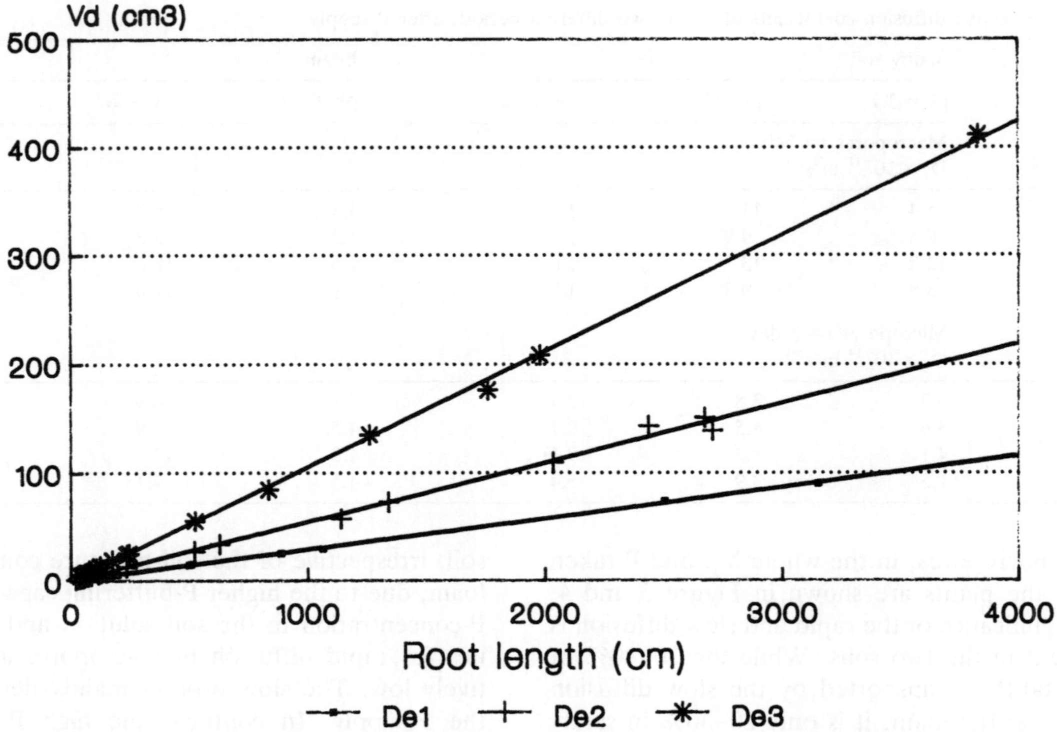

Fig. 2. Depletion volume as a function of root length. $D_1 = 5.4 \times 10^{-11}$; $D_2 = 12 \times 10^{-11}$; $D_3 = 25 \times 10^{-11}$.

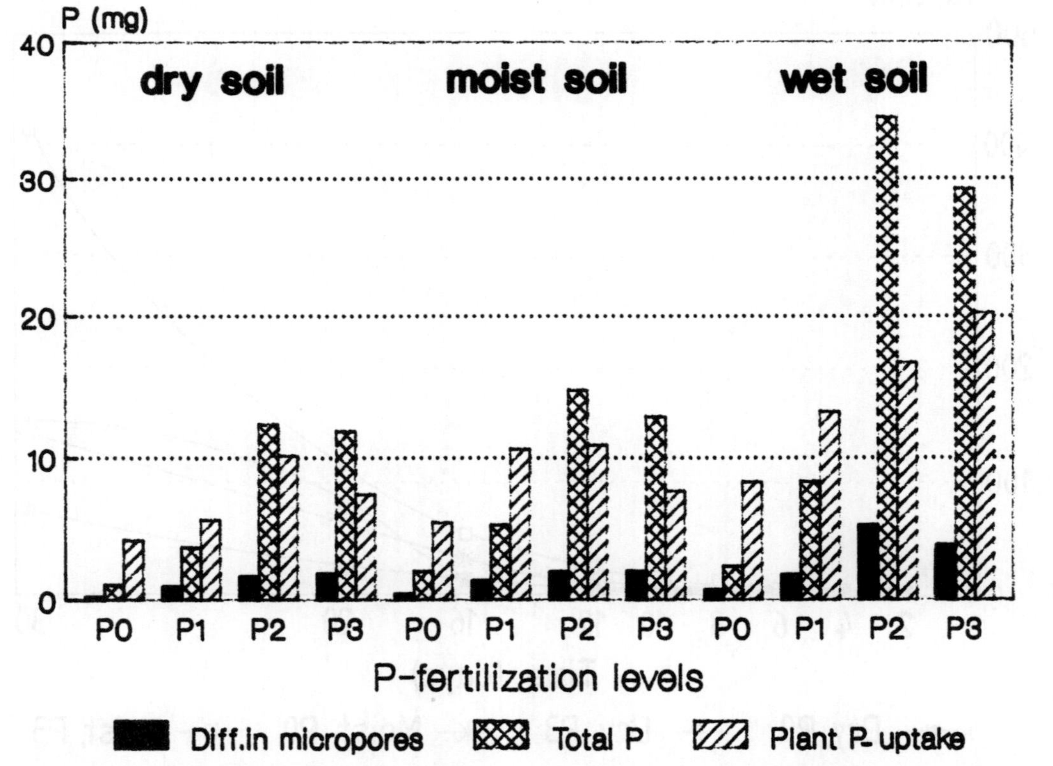

Fig. 3. P-diffusion in sand and the P-uptake of one 30-day-old barley plant. Total P = diffusible P in micro- and macropores.

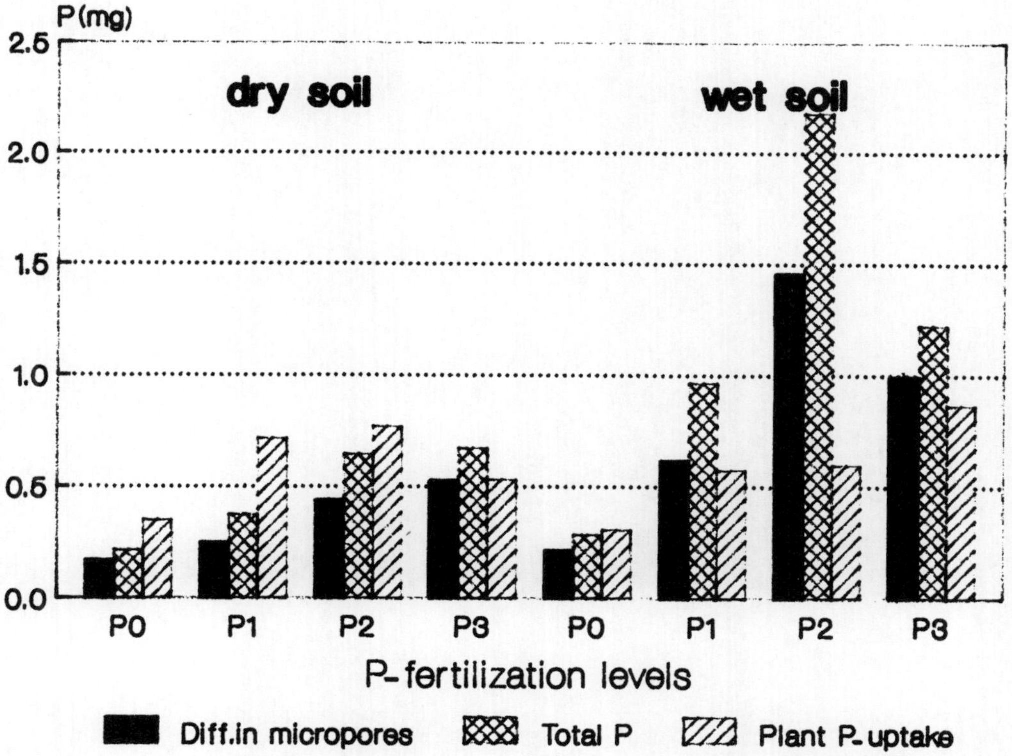

Fig. 4. P-diffusion in loam and the P-uptake of one 20-day-old barley plant. Total P = diffusible P in micro- and macropores.

tration in soil solution and the high D_e of rapid diffusion process maintains a high-rate P-transport in the macropores of the sandy soil. In the unfertilized soils the amount of P taken up by the plant was higher, than the available-P estimated at even the higher moisture levels.

The results indicate that this approach is quite suitable for studying P-supply at high P-fertility; at low P-supply additional factors are apparently involved that were not considered in our estimations. The plant can increase V_d through the root hairs and by chemically affecting the soil roots can raise the amount of diffusible phosphorus in V_d. These estimates of diffusible-P based on experimentally determined D_e values are subject to numerous uncertainties (*e.g.* root hair growth and chemical effects of roots were not considered, complete soil-root contact was assumed along the roots). Some special features of the soil-plant system significant for rapid and slow diffusion in P-transport, the changes of V_d and its diffusible P-content during root growth were discussed. P recovery by plants from the soil is found closely related to the size of V_d, which depends on soil properties, moisture- and P-content, and any condition affecting root growth.

References

Barber S A 1984 Soil Nutrient Bioavailability: A Mechanistic Approach. John Wiley and Sons, New York, 398 p.

De Willigen P and Van Noordwijk M 1988 Roots, Plant Production and Nutrient Use Efficiency. Doct. Thesis, Agricultural University, Wageningen, 282 p.

Nye P H 1979 Diffusion of ions and uncharged solutes in soils and soil clays. Adv. Agron. 31, 225–271.

Varró T, Filep G and Rédly M 1985 Ion transport in soils of different moisture content and porosity. (*In Hungarian*). Agrokémia és Talajtan 34, 343–354.

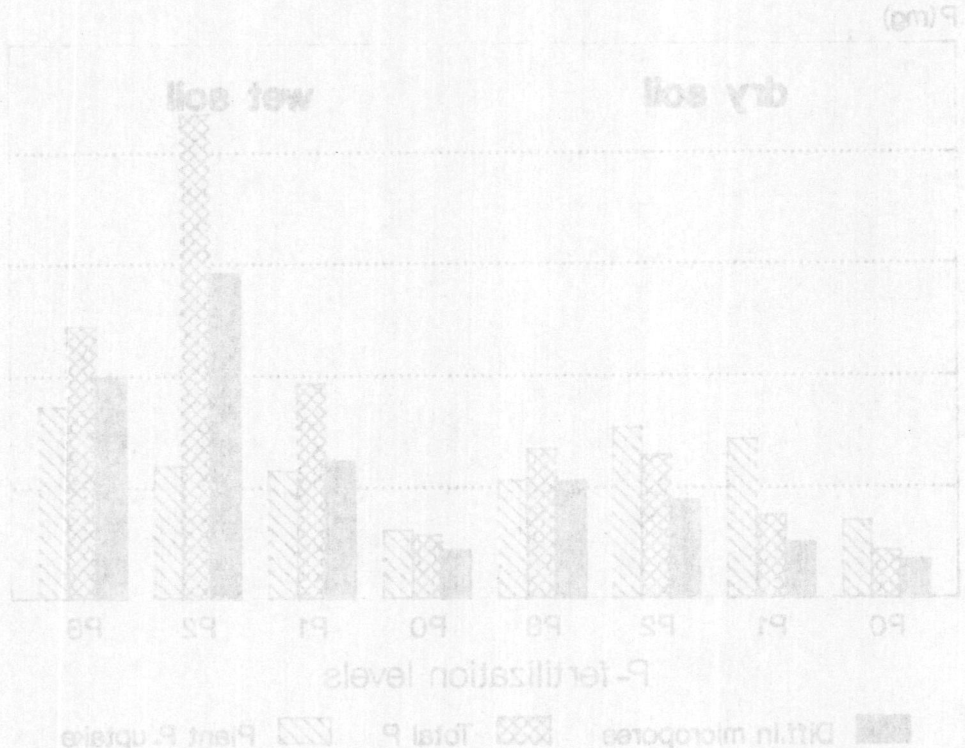

M. L. van Beusichem (Ed.), *Plant nutrition – physiology and applications*, 153–159.

PLSO IPNC124

Effects of *Glomus fasciculatum* and isolated rhizosphere microorganisms on growth and phosphate uptake of *Plantago major* ssp. *pleiosperma*

R. BAAS
Institute for Ecological Research, Department of Dune Research, P.O. Box 317, 3233 2G Oostvoorne, The Netherlands. Present address: Research Station for Floriculture, Linneauslaan 2A, 1431 JV Aalsmeer, The Netherlands

Key words: *Glomus fasciculatum*, phosphate uptake, *Plantago major* ssp. *pleiosperma*, rhizosphere microorganisms, VA mycorrhizas

Abstract

An experiment was set up in order to study 1) the relationship between net P uptake and dry matter production in mycorrhizal and non-mycorrhizal plants and 2) the effects of isolated rhizosphere bacteria and fungi on net P uptake and growth of *P. major* ssp. *pleiosperma*. A similar relationship between net P uptake and dry matter production was found for both mycorrhizal and non-mycorrhizal plants, although the regression lines differed in intercept.

Compared to non-inoculated treatments, inoculation with bacteria slightly decreased dry matter production and P uptake of *P. major*, whereas inoculation with fungi or bacteria + fungi showed no effect. The results are discussed in terms of competition for available P and host photosynthates between host plant and rhizosphere microorganisms.

Introduction

The impact of vesicular-arbuscular mycorrhizal (VAM) infection on plant dry weight can be affected by several factors. For instance, high P levels (Bethlenfalvay *et al.*, 1983; Koide, 1985) and low light intensities (Son and Smith, 1988) have been shown to change the symbiosis into a parasitic relationship, probably as a result of the carbon demand of the symbiosis.

Under natural conditions, the contribution of mycorrhiza to plant growth may also be decreased as a result of interactions with other soil microorganisms (Hetrick *et al.*, 1986; Mosse *et al.*, 1969; Howeler *et al.*, 1987). Several explanations have been suggested to explain this phenomenon, such as the competition with ineffective strains of VAM fungi for infection sites (Howeler *et al.*, 1987), reduced infection levels due to substrate competition with other rhizosphere microorganisms (Hetrick *et al.*, 1986), grazing of external hyphae by collembola and nematodes (Fitter, 1985), and microbial suppression of VAM fungal sporulation (Ross, 1970). In *Plantago major* ssp. *pleiosperma* the suppression of VAM-stimulated growth was believed to be caused by reduction in P uptake due to biotic factors in rhizosphere soil (Baas *et al.*, 1989a). However, in that publication no identification of the organisms that were responsible for the observed effects could be given.

The purpose of the present study was to determine 1) whether P uptake can indeed be related to growth of VAM infected and non-infected *P. major* ssp. *pleiosperma* plants and 2) whether rhizosphere microorganisms can be isolated and subsequently reinoculated in order to simulate the effects of rhizosphere soil on growth and P uptake of *P. major* ssp. *pleiosperma*.

Materials and methods

The experiment had a 3-factorial design with mycorrhizal treatments, phosphate treatments, and inoculum treatments as independent variables.

The bulk soil used in the experiments originated from the 5–25 cm layer of a former beach plain, which was embanked in 1966 (location Oostvoornse Meer). The soil was sieved (5 mm), and subsequently sterilized by means of gamma-irradiation (2.5 Mrad). The soil was amended with 1.67 gram of $Ca_{10}(PO_4)_6(OH)_2$ per kg soil (dry weight basis).

Rhizosphere soil was collected from the root inhabiting soil layer at the same location. *P. major* plants were grown in the greenhouse in a 10% rhizosphere soil-90% sterilized soil mixture. Roots from this pot culture were harvested after 4 weeks and washed in demineralized water, chopped and subsamples were sonicated in a physiological salt solution during 30 seconds. Root samples and sonication solution were subsequently plated either on 1.5% soy-agar (containing 0.01% cycloheximide as a fungicide) or on potato-dextrose agar (containing 0.01% chloroamphenicol as a bactericide). Plates were incubated at 25°C. Four different fungi were isolated from the rhizosphere soil and identified as *Penicillium corylophilum* Dierckx, *Fusarium avenaceum* (Fr) Sacc., *Cladosporium tenuissimum* Cooke, and *Plectosphaerella cucummerina* Kleb. Fungal spores and mycelium were collected from the plates using sterilized water and pipettes and were subsequently suspended in solutions, which were counted using a counting chamber (Bürker Türk). These solutions were combined to give a final suspension containing (spores per ml): *P. corylophilum*, 1.4×10^6; *F. avenaceum*, 3.1×10^5; *C. tenuissimum*, 4.1×10^4; and *P. cucumerina*, 5.4×10^5. The bacteria that were isolated were not identified, and are referred to as SA-bacteria. The SA-bacteria were also collected and counted in a similar manner as the fungi. The final suspension contained 1.7×10^9 bacteria per mL.

Spores of *Glomus fasciculatum* (Thaxt. sensu Gerdemann) Gerdemann and Trappe were collected from a white clover culture on a 45 μm sieve using a fluidizing column (Trudgill *et al.*, 1972), and were suspended in 1 liter demineralized water. Spores were counted in 1 mL aliquots of this suspension under a binocular at a magnification of 16 ×. From this suspension 10 mL aliquots were pipetted and mixed with the basic soil to give a final concentration of 9200 spores per kg soil in the +VAM treatments.

Plastic pots were filled with 600 g of the prepared bulk soil (dry weight). The pots were covered with tinfoil to reduce contamination and evaporation during the experiment. There were six replicate pots per treatment.

Phosphate treatments were applied by adding 150 (P3 treatment) and 1500 μmol KH_2PO_4 (P4 treatment) per pot.

The inoculum treatments consisted of a control treatment (C), an isolated rhizosphere bacteria treatment (B), an isolated rhizosphere fungi treatment (F), and a mixture treatment of bacteria + fungi (BF). The B and F treatments were applied by pipetting 2 × 1mL of the respective suspensions in 1 cm deep planting holes. In the BF treatment both 2 × 1mL bacteria and 2 × 1mL fungi suspensions were applied.

Seeds of an inbred line of *Plantago major* L. ssp. *pleiosperma* Pilger were germinated on moistened glass beads. Two seedlings were planted per pot through holes in the tinfoil. After 14 days the plants were thinned to one plant per pot. The pots received weekly 10 mL of a phosphate-free nutrient solution as described previously (Baas *et al.*, 1989a). The water content during the experiment was kept between 10 and 20% (w/w) using demineralized water. Pots were placed in a greenhouse under natural daylight conditions (March to April 1988) and a controlled temperature of 23°C.

Plants were harvested 26 and 46 days after planting. Fresh weights and dry weights (48 h at 70°C) of shoots and roots were determined. Subsamples of the roots were stained with Chlorazol Black E (Brundrett *et al.*, 1984), and mycorrhizal infection percentages were obtained with the line-intersect method (Giovanetti and Mosse, 1980). In the shoots and roots concentrations of total P and total N were determined. Samples from day 26 were pooled per treatment for analytical reasons.

Calculations and statistical anaysis

Total net uptake of P per plant at day 46 was calculated by multiplying individual dry weights of shoots and roots with their respective P concentrations. Least squares regression analysis was used to relate total net uptake of P with total dry weight. Data per harvest were statistically analyzed using analysis of variance. If variances were not homogeneous according to Cochran's Q-test, data were either ln- or square root- transformed. Comparisons of means were performed using Tukey's HSD test at the $P = 0.05$ level. Specific net P uptake rates between 26 and 46 days were calculated using the formula of Williams (1948). Analysis of covariance was used for comparison of regression lines (Sokal and Rohlf, 1981).

Results

Day 26

Dry weights of both shoot and roots in the P3 treatment (150 μmol KH_2PO_4 per pot) were lower than in the P4 treatment (1500 μmol KH_2PO_4 per pot) (Table 1; $P < 0.001$). The +VAM treatments showed lower root dry weight ($P < 0.01$). As a result, the shoot-root ratio was

Table 1. Effects of *Glomus fasciculatum*, phosphate, and isolated rhizosphere microorganisms (C, control; B, bacteria; F, fungi; BF, bacteria + fungi) on dry weight, shoot-root ratio, P concentrations in shoot and roots, and total net P uptake at day 26

Treatment	Dry weight (g)		Shoot-root ratio $g\,(g\,DW)^{-1}$	P $\mu mol (g\,DW)^{-1}$		P-uptake $\mu mol(plant)^{-1}$
	Shoot	Roots		Shoot	Roots	
CP3 − VAM	0.03 abc	0.03 ab	1.23 a	52	49	3
CP3 + VAM	0.02 ab	0.01 a	1.95 abcde	89	116	4
CP4 − VAM	0.09 e	0.05 d	1.82 abcde	74	60	9
CP4 + VAM	0.06 cde	0.03 abcd	2.33 cde	78	90	7
BP3 − VAM	0.02 a	0.02 a	1.26 ab	50	50	2
BP3 + VAM	0.03 abc	0.02 a	2.00 abcde	96	116	5
BP4 − VAM	0.07 de	0.04 bcd	1.65 abcde	77	66	8
BP4 + VAM	0.05 bcde	0.03 abcd	1.92 abcde	81	95	7
FP3 − VAM	0.03 abc	0.04 abc	1.16 a	54	58	4
FP3 + VAM	0.02 a	0.02 a	1.62 abcd	95	95	3
FP4 − VAM	0.07 e	0.05 d	1.50 abc	75	61	8
FP4 + VAM	0.07 e	0.03 bcd	2.13 bcde	97	108	10
BFP3 − VAM	0.03 abc	0.02 a	1.36 a	78	60	3
BFP3 + VAM	0.04 abcd	0.01 a	2.52 d	104	114	5
BFP4 − VAM	0.07 e	0.04 bcd	1.68 abcde	81	68	9
BFP4 + VAM	0.08 e	0.03 cd	2.47 e	83	94	10
ANOVA Table						
VAM	ns	**	***			
P	***	***	**			
CBF	ns	ns	*			
VAMXP	ns	ns	ns			
VAMXCBF	*	ns	ns			
PXCBF	ns	ns	ns			
VAMXPXCBF	ns	ns	ns			

Means in each column followed by the same letter are not significantly different according to Tukey's HSD test ($P < 0.05$). Significance symbols in table of analysis of variance: ns, not significant; *, $P < 0.05$; **, $P < 0.01$; ***, $P < 0.001$

increased by both P-addition ($P<0.01$) and VAM infection ($P<0.001$). There were no significant effects of inoculation treatments on dry weights of shoots and roots. However, a slightly significant ($P<0.05$) effect on the shoot-root ratio was apparent.

P concentrations in the +VAM treatments were higher than in the −VAM treatments; however, due to reduced root weights, total net uptake of P was hardly affected.

Day 46

Mycorrhizal infection was well established in all +VAM treatments and no significant differences among different +VAM treatments in infection levels were observed (Table 2). Dry weights of both shoots and roots were higher in the P4 treatments than in the P3 treatments ($P<0.001$). Also, the +VAM treatment showed an overall increase in dry weight of shoot and roots. However, the significant ($P<0.001$) VAM × P interaction indicates that the +VAM effect was not similar in both P treatments. The +VAM treatment had a greater effect in the P3 treatment than in the P4 treatment, which was also demonstrated by the relatively large increase in shoot-root ratio in the P3 treatment.

The inoculation treatments differed in their effect on shoot dry weight ($P<0.01$). The BP3-VAM treatment showed a significantly lower yield than the CP3-VAM and FP3-VAM treat-

Table 2. Effects of *Glomus fasciculatum*, phosphate, and isolated rhizosphere microorganisms (C, control; B, bacteria; F, fungi; BF, bacteria + fungi) on dry weight, shoot-root ratio, P concentrations in shoot and roots, and total net P uptake, and mycorrhizal infection of the roots at day 46

Treatment	Dry weight (g)		Shoot-root ratio $g\,(g\,DW)^{-1}$	P ($\mu mol(gDW)^{-1}$)		P-uptake $\mu mol(plant)^{-1}$	%VAM
	Shoot	Roots		Shoot	Roots		
CP3 − VAM	0.15 b	0.16 b	0.93 a	32 a	27 a	9 a	0 a
CP3 + VAM	0.50 cd	0.28 cd	1.80 de	54 b	55 b	42 d	56 b
CP4 − VAM	0.63 def	0.41 fg	1.54 bc	35 a	28 a	33 bcd	0 a
CP4 + VAM	0.70 ef	0.32 cde	2.18 g	63 bc	60 b	63 e	68 b
BP3 − VAM	0.05 a	0.06 a	0.88 a	32 a	32 a	4 a	0 a
BP3 + VAM	0.44 c	0.26 cd	1.66 cd	56 bc	56 b	38 cd	60 b
BP4 − VAM	0.58 cde	0.41 fg	1.43 bc	32 a	26 a	29 b	0 a
BP4 + VAM	0.69 ef	0.33 def	2.06 fg	60 bc	58 b	60 e	70 b
FP3 − VAM	0.12 b	0.13 ab	0.94 a	32 a	28 a	8 a	0 a
FP3 + VAM	0.47 cd	0.26 cd	1.81 ed	53 b	57 b	40 d	61 b
FP4 − VAM	0.55 cde	0.39 efg	1.41 b	36 a	27 a	30 bc	0 a
FP4 + VAM	0.76 f	0.37 ef	2.06 fg	60 bc	58 b	67 e	60 b
BFP3 − VAM	0.11 ab	0.10 ab	1.04 a	33 a	31 a	7 a	0 a
BFP3 + VAM	0.45 c	0.24 c	1.85 def	60 bc	58 b	41 d	58 b
BFP4 − VAM	0.67 ef	0.45 g	1.49 bc	34 a	25 a	34 bcd	0 a
BFp4 + VAM	0.63 def	0.32 de	1.94 efg	64 c	60 b	59 e	59 b
ANOVA Table							
VAM	* * *	* * *	* * *	* * *	* * *	* * *	* * *
P	* * *	* * *	* * *	* * *	ns	* * *	ns
CBF	* *	ns	*	ns	ns	* *	ns
VAMXP	* * *	* * *	* * *	*	* *	ns	ns
VAMXCBF	*	*	ns	ns	ns	ns	ns
PXCBF	*	*	* *	ns	ns	ns	ns
VAMXPXCBF	*	*	ns	ns	ns	ns	ns

Means in each column followed by the same letter are not significantly different according to Tukey's HSD test ($P<0.05$). Significance symbols in table of analysis of variance: ns, not significant; *, $P<0.05$; * *, $P<0.01$; * * *, $P<0.001$

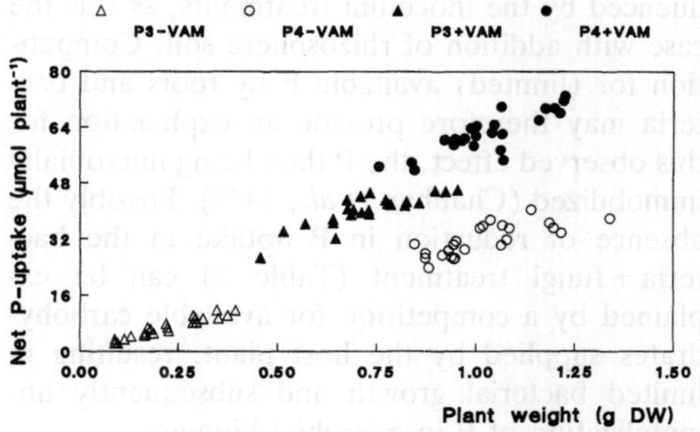

Fig. 1. Relation between total dry weight and total net P uptake in mycorrhizal (filled symbols) and non-mycorrhizal (open symbols) *P. major* ssp. *pleiosperma* at 150 μmol KH_2PO_4 per pot (P3) and 1500 μmol KH_2PO_4 per pot (P4).

ments, and tended to have a lower shoot-root ratio.

P concentrations in both shoots and roots were almost twice as high in the +VAM treatments as in the −VAM treatments. In contrast, only the P concentration in the shoot was enhanced by the P treatment. There were no effects of the inoculation treatments on P concentrations. Total uptake of P was affected by the inoculation treatments ($P < 0.01$). This could be ascribed largely to an overall lower uptake of P in the B treatments compared to the C and F treatments.

The average relative growth rate (RGR) and specific P uptake rate (SPUR) of pooled inoculum treatments between days 26 and 46 (Table 3) was higher in the +VAM treatments than in the −VAM treatments, especially in the P3 treatment. Relations between P uptake and total dry weight were significant in all pooled treatments (Fig. 1). However, analysis of covariance showed that in contrast to the slopes, intercepts in the +VAM treatments were different from those in the −VAM treatments.

Discussion

Availability of P was limiting growth of plants in the P3 treatments, since dry weights in these treatments were lower than those in the P4 treatments (Tables 1, 2). However, considering the higher RGR of the +VAM treatments between the two harvests compared to the P4 − VAM treatments (Table 3), and the relatively low P concentrations in the P4 treatments at day 46 (Table 2), P availability was probably also limiting growth in the P4 treatment between days 26 and 46.

The decrease in root dry weights in the +VAM treatments at day 26 indicate a parasitic effect of the VAM fungus in the early stages of infection, as has been observed before (Cooper, 1975). Despite this parasitism, P concentrations were higher in the +VAM treatments at day 26, suggesting that the VAM symbiosis was already effectively translocating P into the host plant. Therefore, total net uptake of P was not decreased by VAM infection, and dry matter production in the +VAM treatments was not correlated to P uptake in a similar manner as in the −VAM treatments. The same applied to day 46 (Table 3). It is apparent that within each treatment significant relations are found (Fig. 1), again indicating that P indeed was limiting growth. However, the relationships for the −VAM treatments differed from those of the +VAM treatments in intercept. Possible causes for this difference in intercept are storage of P in the arbuscules of the fungal tissue (preventing the plant to benefit from the increased P uptake) and the high rate of P inflow in VAM infected plants (Table 3) causing (temporary) luxury consumption.

The increase in P concentrations was associated with an increase in shoot-root ratio

Table 3. Mean relative growth rate and specific net P uptake rate between 26 and 46 days of the experiment, and relations between total net P uptake and total dry weight at day 46 in the pooled VAM/P combinations

	RGR mg (g DW)$^{-1}$ d^{-1}	SPUR μmol (g DW)$^{-1}$ d^{-1}	P-uptake μmol plant^{-1}	r^2
P3 − VAM	71	4	0.3 + 29 × dry wt.	(0.92)
P3 + VAM	144	22	16.7 + 32 × dry wt.	(0.71)
P4 − VAM	108	7	3.9 + 27 × dry wt.	(0.59)
P4 + VAM	121	21	21.5 + 40 × dry wt.	(0.71)

(Table 1, 2), as was observed before in *P. major* ssp. *pleiosperma* (Baas and Lambers, 1988). This increase in shoot-root ratio was found to be an important factor in explaining the increase in RGR due to VAM infection under P limiting conditions (Baas *et al.*, 1989b). Therefore, both the higher P uptake and higher shoot-root ratio suggest that the P4 + VAM plants will eventually reach a higher dry weight than the P4-VAM plants. This time-dependency is an important characteristic of the VAM symbiosis (Harris *et al.*, 1985), and indicates the need for frequent harvesting when conclusions on the effect of VAM infection on plant growth are to be drawn.

From previous experiments (Baas *et al.*, 1989a) it was concluded that biotic factors, other than VAM fungi, in rhizosphere soil had a negative effect on growth and P uptake of *P. major* ssp. *pleiosperma*. However, the ubiquitous nature of VAM fungi caused that the treatments inoculated with rhizosphere soil also to become inoculated with VAM fungi, so that effects of the harmful biotic factors and those of the VAM fungi could not be separated. Therefore, in the present experiment bacteria and fungi were isolated from the rhizosphere soil and reinoculated. Soil microorganisms were also believed to suppress mycorrhizal growth responses in big bluestem grass, since growth in non-sterilized soil was less improved than in sterilized soil (Hetrick *et al.*, 1986). In a later study (Hetrick *et al.*, 1988), addition of non-sterile soil sievings as well as certain groups of isolated bacteria from non-sterile soil were shown to suppress the mycorrhizal infection and growth response. The uptake of labeled P was decreased 10-fold due to addition of non-sterile soil. In the present experiment only relatively small effects were found compared to effects of rhizosphere soil in previous experiments (Baas *et al.*, 1989a). Some causes for these relatively small effects of the inoculation treatments may be the inability to isolate harmful rhizosphere microorganisms, inadequate inoculation methods and (minor) differences in growth conditions, compared to the previous experiments with rhizosphere soil. However, the bacteria treatment showed an overall decrease in P uptake and growth compared to the control treatment, especially in the P3-VAM treatment (Table 2). VAM infection levels were not influenced by the inoculum treatments, as was the case with addition of rhizosphere soil. Competition for (limited) available P by roots and bacteria may therefore provide an explanation for this observed effect, the P then being microbially immobilized (Chauhan *et al.*, 1979). Possibly the absence of reduction in P uptake in the bacteria + fungi treatment (Table 2) can be explained by a competition for available carbohydrates supplied by the host plant, resulting in limited bacterial growth and subsequently immobilization of P in microbial biomass.

Acknowledgements

The help of Jacques Mangelaars during the experiment is greatly acknowledged.

References

Baas R and Lambers H 1988 Effects of vesicular-arbuscular mycorhizal infection and phosphate on *Plantago major* ssp. *pleiosperma* in relation to the internal phosphate concentration. Physiol. Plant. 74, 701–707.

Baas R, Van Dijk C and Troelstra S R 1989a Effects of rhizosphere soil, vesicular-arbuscular mycorrhizal fungi and phosphate on *Plantago major* L. ssp. *pleiosperma* Pilger. Plant and Soil 113, 59–67.

Baas R, Van der Werf A and Lambers H 1989b Root respiration and growth in *Plantago major* as affected by VA mycorrhizal infection. Plant Physiol. 91, 227–232.

Bethlenfalvay G J, Bayne H G and Pacovsky R S 1983 Parasitic and mutualistic associations between a mycorrhizal fungus and soybean: The effect of phosphorus on host plant-endophyte interactions. Physiol. Plant. 57, 543–548.

Brundrett M C, Piché Y and Peterson R C 1984 A new method for observing the morphology of vesicular-arbuscular mycorrhizae. Can. J. Bot. 62, 2128–2134.

Chauhan B S, Stewart J W B and Paul E A 1979 Effects of carbon additions on soil labile inorganic, organic and microbially held phosphate. Can J. Soil Sci. 59, 387–396.

Cooper K M 1975 Growth responses to the formation of endotrophic mycorrhizas in *Solanum leptospermum* and New Zealand ferns. *In* Endomycorrhizas. Eds. F E Sanders, B Mosse and P B Tinker. pp 391–407. Academic Press, London.

Fitter A H 1985 Functioning of vesicular-arbuscular mycorrhiza's under field conditions. New Phytol. 99, 257–265.

Giovanetti M and Mosse B 1980 An evaluation of techniques for measuring vesicular-arbuscular mycorrhizal infection in roots. New Phytol. 84, 489–500.

Harris D, Pacovsky R S and Paul E A 1985 Carbon economy

of soybean-*Rhizobium-Glomus* associations. New Phytol. 101, 427–440.

Hetrick B A, Kitt D G and Wilson G T 1986 The influence of phosphorus fertilization, drought, fungal species, and non-sterile soil on mycorrhizal growth response in tall grass prairie plants. Can. J. Bot. 64, 1199–1203.

Hetrick B A, Thompson G and Kitt D G 1988 Effects of soil microorganisms on mycorrhizal contribution to growth of big bluestem grass in non-sterile soil. Soil Biol. Biochem. 20, 501–507.

Howeler R H, Sieverding E and Saif S 1987 Practical aspects of mycorrhizal technology in some tropical crops and pastures. Plant and Soil 100, 249–283.

Koide R 1985 The nature of growth depressions in sunflower caused by vesicular-arbuscular mycorrhizal infection. New Phytol. 99, 449–462.

Mosse B, Hayman D S and Ide G J 1969 Growth responses of plants in unsterilized soil to inoculation with vesicular-arbuscular mycorrhiza. Nature 224, 1031–1032.

Ross J P and Harper J A 1970 Effect of *Endogone* mycorrhiza on soybean yields. Phytopathology 60, 1552–1556.

Sokal R R and Rohlf F J 1981 Biometry. 2nd Edition. The Principles and Practice of Statistics in Biological Research. Freeman and Co., New York.

Son C L and Smith S E 1988 Mycorrhizal growth responses: Interactions between photon irradiance and phosphorus nutrition. New Phytol. 108, 305–314.

Trudgill D L, Evans K and Faulkner G 1972 A fluidising column for extracting nematodes from soil. Nematologica 18, 469–475.

Williams R F 1948 The effect of phosphorus supply on the rates of intake of phosphorus and nitrogen and upon certain aspects of phosphorus metabolism in gramineous plants. Aust. J. Sci. Res. 1, 353–375.

M. L. van Beusichem (Ed.), *Plant nutrition – physiology and applications*, 161–164.

PLSO IPNC587

Soluble carbohydrates in roots of leek (*Allium porrum*) plants in relation to phosphorus supply and VA mycorrhizas

F. AMIJEE[1], D.P. STRIBLEY[2] and P.B. TINKER[3]
[1]*School of Agriculture, University of Aberdeen, Aberdeen, AB9 1UD, Scotland, UK*, [2]*AFRC Institute of Arable Crops Research, Rothamsted Experimental Station, Harpenden, Herts., AL5 2JQ, UK, and* [3]*Natural Environment Research Council, Polaris House, Swindon, Wilts., SN2 1EU, UK*

Key words: *Allium porrum* L., *Glomus mosseae* L., leek, phosphorus supply, soluble carbohydrates, VA mycorrhizas

Abstract

Leek plants (*Allium porrum* L.) inoculated with *Glomus mosseae* were raised on sterilized soil/sand medium amended with $Ca(H_2PO_4)_2.H_2O$ to test the hypothesis that high concentration of soil P inhibits formation of vesicular-arbuscular (VA) mycorrhizas by reducing concentration of soluble carbohydrate in the root. When P supply was increased, from either P addition or VA mycorrhizal infection, there was initially also an increase in concentration of soluble carbohydrate in the root. At the concentration of soil P at which infection was reduced, concentration of soluble carbohydrate was at its maximum. Therefore the above hypothesis is discounted. An increased delay in infection establishment and a greater number of abortive entry points would suggest that high concentration of soil P reduces VA mycorrhizal infection by changing the anatomy of the root to make it resistant to fungal penetration.

Introduction

The study of processes that control colonization of root systems by vesicular-arbuscular (VA) mycorrhizal fungi is fundamental to the understanding of effects of mycorrhizas upon the physiology of the host plant. The development of VA mycorrhizal root systems is a highly dynamic process, with the fungus colonizing a host organism which itself is growing. Although the infection process is well understood, many experiments have shown that formation of VA mycorrhizas can be influenced by various physical and environmental factors (Harley and Smith, 1983). In particular, high concentration of P in soil markedly inhibits infection: a clear example is presented by Stribley *et al.* (1980). A study by Sanders (1975) in which leaves of *Allium cepa* L. were foliar fed with P clearly showed that it was the internal concentration of P in the plant which controls infection, but the mechanism of this effect is not understood.

In a recent series of papers on studies of P supply and VA mycorrhizal infection, Robson and colleagues have suggested that there is a correlation between the concentration of soluble carbohydrates in roots and the extent of VA mycorrhizal infection (Jasper *et al.*, 1979; Same *et al.*, 1983; Thompson, *et al.*, 1986). They found the concentration of soluble carbohydrates in roots changed inversely with increased P supply. Presumably, although not explicitly stated by these authors, the rate of fungal growth can be controlled by the concentration of carbohydrates in the root.

In the present study, we have identified the components of the infection process most affected by P addition, and measured concentrations of root soluble carbohydrate to test whether they correlate with changes in VA mycorrhizal colonization.

Methods

Non-mycorrhizal (NM) and mycorrhizal (M) leek plants (*Allium porrum* L.) infected with *Glomus mosseae* were raised on a sterilized soil/sand medium amended with $Ca(H_2PO_4)_2.H_2O$ to give 22 (P_0), 75 (P_1), 140 (P_2), 208 (P_3), 276 (P_4) and 344 (P_5) mg P kg^{-1} soil (bicarbonate-soluble P), respectively. Plants were raised in a controlled environment cabinet (20°C/16°C, 14 h photoperiod) and sequentially harvested over a period of 52 d from transplanting to measure growth of shoots, growth of roots, development and spread of infection, total P content of shoots, and concentration of ethanol-soluble carbohydrates of roots (fructose, glucose and sucrose: by gas chromatography). All the methods used here are fully described by Amijee (1986).

Results

As expected, addition of P to soil initially increased the shoot mass of NM and M plants (Table 1). Mycorrhizal infection also increased mass of shoots on soil deficient in P, but the effect declined with increasing concentration of soil P (Table 1). This pattern of response to VA mycorrhizal infection is consistent with other studies (*e.g.* Stribley *et al.*, 1980): the decline in response to infection is due to a combination of a reduction in infection as soil P increases, and the decline in the response of the plant to added P.

Addition of P increased the total length of root per plant not because the rate of extension of individual root members was greater, but because it stimulated the initiation of first-order laterals (Amijee *et al.*, 1989a). An increased rate of branching is a general response of plant root systems to increased supply of major nutrients (Fitter, 1982).

There was little effect of P on any aspect of fungal development or spread up to P_2, hence the initial increase with P in VA mycorrhizal root (Table 1) resulted from an increase in the length of root available for the fungus to colonize. However, when the bicarbonate-soluble P in soil exceeded 140 mg P kg^{-1} soil (P_2), the length of infected root declined abruptly (Table 1). It was accompanied by an increase in the delay of establishment of infection, a decrease in the rate of lateral extension of infection, and a decrease in the intensity of internal colonization (Amijee *et al.*, 1989b). It is important to emphasize that at these levels of soil P (P_3 and P_4) root growth continued to increase with added P, so that the sharp decline in the length of infected root was not a result of a reduction in the availability of the host root (Amijee *et al.*, 1989b).

The concentration of soluble carbohydrates in roots of NM and M plants *increased* with soil P

Table 1. Length of VA mycorrhizal roots, dry mass and P concentration of shoots, and soluble carbohydrate concentration in roots of nonmycorrhizal (NM) and mycorrhizal (M) plants at 52 days as affected by P additions. SE in parentheses

Soil soluble P (mg kg^{-1})	Shoot dry mass (g)		Shoot P (%)		Root carbohydrate (mg kg^{-1})		VA mycorrhizal root (cm)
	NM	M	NM	M	NM	M	M
22(P_0)	0.17	0.40	0.19	0.26	5.36	8.29	212
	(0.030)	(0.023)	(0.008)	(0.027)	(0.022)	(0.881)	(18.9)
75(P_1)	0.35	0.48	0.21	0.35	9.23	14.20	302
	(0.042)	(0.048)	(0.029)	(0.010)	(1.807)	(1.459)	(27.5)
140(P_2)	0.54	0.56	0.35	0.42	12.16	26.82	287
	(0.049)	(0.021)	(0.006)	(0.029)	(2.189)	(5.513)	(5.6)
208(P_3)	0.88	0.76	0.43	0.47	14.08	29.72	144
	(0.092)	(0.116)	(0.020)	(0.041)	(1.141)	(1.368)	(11.3)
276(P_4)	0.68	0.36	0.49	0.49	22.63	20.05	73
	(0.068)	(0.016)	(0.018)	(0.026)	(0.768)	(0.912)	(10.1)
344(P_5)	0.27	0.29	0.58	0.55	8.80	12.41	43
	(0.031)	(0.059)	(0.003)	(0.012)	(1.165)	(3.636)	(12.9)

to a maximum at P_3 and then declined concomitantly with a decrease in mass of shoots (Table 1). This decline in the carbohydrate concentration did not occur at the P concentration which decreased the length of root colonized by VA mycorrhizas (Table 1).

Discussion

Our results contrast sharply with Jasper *et al.* (1979), Same *et al.* (1983) and Thompson *et al.* (1986), but agree with studies of Siqueira *et al.* (1984) and Ocampo and Azcon (1985) who also found that soluble carbohydrates in roots of NM and M plants are increased by improved P nutrition. We therefore discount the idea that in leek, the reduction in colonization of VA mycorrhizal fungi by high P supply results from lack of soluble carbohydrate in the host tissue. Indeed, modern work on the relationship between photosynthesis and the concentration of P in the cytoplasm of leaf cells (Walker and Sivak, 1986) strongly implies that improved P nutrition should increase the supply of soluble carbohydrates within the plant. The results of Robson and his colleagues are therefore difficult to interpret. The increased delay in infection establishment and the observation of a greater number of abortive entry points would suggest that the mechanism by which VA mycorrhizal colonization is reduced could be related to anatomical changes of the root caused by high P concentration.

Lewis (1975) stated that it is important to distinguish cause and effect when considering the role of carbohydrates in the control of mycorrhizal infection. Our data exemplify this *caveat.* We reject a causal mechanism, but instead suggest that infection increases the supply of carbohydrates to the root (Table 1). This might be expected because of the improved P nutrition of the VA mycorrhizal host. However in addition, there appears to be a unique effect of VA mycorrhizal infection, for over a range of P supply. Roots of M plants contained a higher concentration of soluble carbohydrates than did NM plants of similar shoot P concentration (Table 1). It is not unexpected that diversion of host assimilates to the site of infection should occur in M plants, for such altered patterns of translocation are observed in different types of infection of vascular plants by other biotrophic fungi (Farrar, 1984).

Acknowledgements

F Amijee was in receipt of an Agricultural and Food Research Council grant during the course of this work, and received financial support from the Natural Environment Research Council, Tansley Fund, and BP Venture Research to attend this colloquium.

References

Amijee F 1986 Colonization of root systems by vesicular-arbuscular mycorrhizal fungi. Ph.D. Thesis, University of Leeds. 175p.

Amijee F, Tinker P B and Stribley D P 1989a Effects of phosphorus on the morphology of VA mycorrhizal root system of leek (*Allium porrum* L.). Plant and Soil 119, 334–336.

Amijee F, Tinker P B and Stribley D P 1989b The development of endomycorrhizal root systems. VII. A detailed study of effects of soil phosphorus on colonization. New Phytol. 111, 435–446.

Farrar J F 1984 Effects of pathogens on plant transport systems. *In* Plant Diseases: Infection, Damage and Loss. Eds. R K S Wood and G J Jellis. pp. 87–104. Blackwell Scientific Publications, Oxford.

Fitter A H 1982 Morphometric analysis of root systems: Application of the technique and influence of soil fertility on root system development in two herbaceous species. Plant Cell Environ. 5, 313–322.

Harley J L and Smith S E 1983 Mycorrhizal Symbiosis. Academic Press, London, 483 p.

Jasper D A, Robson A D and Abbott L K 1979 Phosphorus and the formation of vesicular-arbuscular mycorrhizas. Soil Biol. Biochem. 11, 501–505.

Lewis D H 1975 Comparative aspects of the carbon nutrition of mycorrhizas. *In* Endomycorrhizas. Eds. F E Sanders, B Mosse and P B Tinker pp. 119–148. Academic Press, London.

Ocampo J A and Azcon R 1985 Relationship between the concentration of sugars in the roots and VA mycorrhizal infection. Plant and Soil 86, 95–100.

Same B I, Robson A D and Abbott L K 1983 Phosphorus, soluble carbohydrates and endomycorrhizal infection. Soil Biol. Biochem. 15, 593–597.

Sanders F E 1975 The effect of foliar-applied phosphate on mycorrhizal infections of onion roots. *In* Endomycorrhizas. Eds. F E Sanders, B Mosse and P B Tinker, pp. 261–276. Academic Press, London.

Siqueira J O, Hubbell D H and Valle R R 1984 Effects of phosphorus on formation of the vesicular-arbuscular mycorrhizal symbiosis. Pesq. Agrop. Brasil. 19, 1465–1474.

Stribley D P, Tinker P B and Snellgrove R C 1980 Effect of vesicular-arbuscular mycorrhizal fungi on the relations of plant growth, internal phosphorus concentration and soil phosphate analyses. J. Soil Sci. 32, 655–672.

Thompson B D, Robson A D and Abbott L K 1986 Effects of phosphorus on the formation of mycorrhizas by *Gigaspora calospora* and *Glomus fasciculatum* in relation to root carbohydrates New Phytol. 103, 751–765.

Walker D A and Sivak M N 1986 Photosynthesis and phosphate: A cellular affair? Trends Biochem. Sci. 11, 176–179.

M. L. van Beusichem (Ed.), *Plant nutrition – physiology and applications*, 165–170.

PLSO IPNC296B

Benefit and cost analysis and phosphorus efficiency of VA mycorrhizal fungi colonizations with sorghum (*Sorghum bicolor*) genotypes grown at varied phosphorus levels

P.S. RAJU[1] R.B. CLARK[1,3], J.R. ELLIS[1], R.R. DUNCAN[2] and J.W. MARANVILLE[1]
[1] *Department of Agronomy and U.S. Department of Agriculture, Agricultural Research Service, University of Nebraska, Lincoln, NE 68583, USA, and* [2] *Department of Agronomy, University of Georgia, Georgia Agricultural Experiment Station, Griffin, GA 30223-1797, USA.* [3] *Corresponding author*

Key words: *Glomus fasciculatum*, phosphorus nutrition, *Sorghum bicolor* (L.) Moench, VA mycorrhizas

Abstract

Sorghum [*Sorghum bicolor* (L.) Moench] was grown in a greenhouse in a low P (3.6 mg kg^{-1}) soil (Typic Argiudolls) inoculated with the vesicular-arbuscular mycorrhizal fungi (VMAF) *Glomus fasciculatum* and P added at 0, 12.5, 25.0, and 37.5 mg kg^{-1} soil to determine the effects of VAMF-root associations on plant growth, benefit and cost analysis, and P efficiency (dry matter produced/unit P absorbed). Root colonization with VAMF and shoot growth enhancements decreased with increased soil P applications. Mycorrhizal plants were less P efficient than nonmycorrhizal plants. Shoot dry matter differences between mycorrhizal and nonmycorrhizal plants were considered the benefit derived by plants from VAMF-root associations. Shoot dry matter differences between mycorrhizal and nonmycorrhizal plants with similar P concentrations were considered the costs paid by plants for VAMF-root associations. Values of benefit and cost analysis for VAMF-root associations were highest when soil P was lowest and decreased with increasing P applications. Genotypic differences for calculated costs were pronounced, but not benefits. Benefit and cost analysis may be helpful to evaluate host plant genotypes and VAMF species to optimize efficiencies of VAMF symbiosis in different soil environments.

Introduction

Root associations with vesicular-arbuscular mycorrhizal fungi (VAMF) normally benefit plant growth, particularly through enhanced P uptake (Gerdemann, 1964; Janos, 1987; Stribely, 1987). However, host plants must provide carbohydrates to VAMF for development and growth. The relationships of roots with VAMF are important and their interactions to improve plant growth and at the same time nourish the VAMF have been described in benefit and cost analysis concepts (Koide and Elliott, 1989).

Benefit and cost analysis concepts based on carbon accumulation, P uptake, and the loss of carbon due to respiration, exudation, leaching, death of plant parts, herbivory, and symbiotic organisms have been described (Koide and Elliott, 1989). The hypothesis was that VAMF may enhance P uptake, but that colonization by VAMF with roots may not be the most efficient way to bring P into the plant. These concepts may be theoretical approaches, but they try to address some of the important issues of root-VAMF associations.

VAMF associations with plant roots not only

benefit growth and mineral element uptake, but VAMF infected plants can give greater tolerance to root pathogens, drought, low soil temperatures, adverse soil pH, and transplant shock. VAMF-root associations have great potential in land reclamation and agriculture practices on arid and acid lands, where drought, low soil fertility (especially P deficiency), and high soil salinity and/or toxicity elements can be major constraints to crop production.

Several crop plants like sorghum [*Sorghum bicolor* (L.) Moench] absorbed more P from soils when colonized with VAMF than nonmycorrhizal plants (Krishna and Bagyaraj, 1981). Soil P applications promoted growth of nonmycorrhizal plants more than growth of mycorrhizal plants and inhibited VAMF-root colonizations (Powell, 1980).

The symbiotic interactions between VAMF and host plants in a variety of soil conditions need to be studied to understand them and to optimize beneficial effects of VAMF. Potential decreases in biomass yield (costs paid) by crop plants which support VAMF associations also need to be understood. This study was conducted to determine benefit and cost analysis and P efficiency (dry matter produced/unit P absorbed) of *Glomus fasciculatum* colonization with sorghum roots when genotypes were grown at different soil P levels.

Materials and methods

The experiment was conducted in a greenhouse (November/December) with 14 h light at 28 ± 3°C. Metal halide lamps provided supplemental light to extend the light period beyond the normal day length.

Seeds of two sorghum genotypes (SC6 and SC97) were planted in 11 kg Burchard clay loam (fine-loamy, mixed, mesic Typic Argiudolls) contained in polyvinyl chlorine (PVC) tubes (60 cm tall and 15 cm inner diameter). The genotypes were chosen because of their differences to P uptake and efficiency (Raju, 1985). The soil contained 3.6 mg P kg^{-1} (Bray-I extracted P), was calcareous (2.3% $CaCO_3$), and had a pH of 8.3 (1 soil:1 water). The soil was steam sterilized (autoclaved 2 h) to kill indigenous organisms including any VAMF. To ensure that elements were not limiting to plant growth, supplements of 26.9 NO_3–N, 23.1 NH_4–N, 55.0 K, 3.2 S, and 3.1 Mg in mg kg^{-1}, and 251.0 Fe, 48.6 B, 88.6 Mn, 27.3 Zn, 14.1 Mo, and 6.9 Cu in $\mu g\ kg^{-1}$ soil were added.

Four soil P levels, two VAMF inocula, and two sorghum genotypes formed a $4 \times 2 \times 2$ factorial in a randomized complete block design with five replications. The P treatment levels consisted of 0, 12.5, 25.0 and 37.5 mg P kg^{-1} soil (equivalent to 0, 25, 50, and 75 kg P ha^{-1}, respectively) as KH_2PO_4. Intact VAMF inoculum cores (25 g) were placed 2 cm below the sorghum seeds. The inoculum cores consisted of *Glomus fasciculatum* (Thaxter sensu Gerd.) Gerd. and Trappe colonized sudangrass [*Sorghum bicolor* (L.) Moench var. *sudanense*] root fragments and spores (42 g^{-1} inoculum) mixed in sand (1 roots:40 sand). A control VAMF treatment consisted of the same amount of autoclaved inoculum.

Seedlings were thinned to two per tube seven days after planting. Distilled water was added as needed to maintain soil moisture near field capacity. Plants were harvested 48 days after planting when plants were in the 10-leaf stage of growth. Shoots were severed 1 cm above the soil, dried at 70°C, weighed, ground to pass a 0.5 mm sieve, and analyzed for P by energy dispersive x-ray fluorescence (Knudsen *et al.*, 1981). Roots were washed free from soil and representative fresh samples were stained with trypan blue (Phillips and Hayman, 1970) and percent root colonized with VAMF was estimated under a light microscope using a grid intersect method (Giovannetti and Mosse, 1980).

Potential shoot dry matter (DM) yields of nonmycorrhizal plants (nm) that had the same shoot P concentration as mycorrhizal plants (m), and benefit and cost analysis for plants with VAMF-root associations were calculated according to formulas described (Yocum, 1981; D H Yocum, personal communications) which were as follows:

$$\text{Potential DMnm} = \frac{\text{PUEnm}}{\text{PUEm}} \times \text{DMnm where},$$

$$\text{PUE} = \text{Phosphorus use efficiency (g DM g}^{-1}\text{ shoot P)} \quad (1)$$

$$\text{Benefit} = \text{DMm} - \text{DMnm; and \% benefit} = \frac{\text{benefit}}{\text{DMnm}} \times 100 \quad (2)$$

Cost = Potential DMnm − DMm; and % cost

$$= \frac{\text{cost}}{\text{DMnm}} \times 100 \quad (3)$$

Results and discussion

Nonmycorrhizal sorghum plants grown on the low P soil without added P or VAMF showed severe P deficiency symptoms. Mycorrhizal plants showed no visible P deficiency symptoms and had higher shoot dry matter yields and P concentrations and contents than nonmycorrhizal plants grown without supplemental P (Table 1). Percent root colonization with VAMF was good (52 to 60%) in the soil without added P. The increased absorption surface area offered by the fungal hyphae external to roots (Gerdemann, 1968) might have increased P supply and alleviated P deficiency of sorghum plants and promoted shoot growth.

Phosphorus applications to soil enhanced shoot growth and P content for nonmycorrhizal plants more than for mycorrhizal plants (Table 1). Optimal shoot dry matter yields were obtained in mycorrhizal plants at 12.5 but not in nonmycorrhizal plants until 25.0 mg P kg^{-1} soil was applied. Percent root colonization with VAMF decreased markedly with the application of 12.5 mg P kg^{-1} soil, and decreased even more but at a slower rate with additional P applications. Promotion of shoot growth attributed to the VAMF-root association decreased with P applications. This might have occurred because of the reduced VAMF-root colonization at the higher soil P levels with subsequent lower effect of VAMF on plant growth.

Mycorrhizal plants had higher shoot P concentrations than nonmycorrhizal plants at each soil P level (Table 1). Mycorrhizal plants had higher amounts of P inside the plant to produce a unit of shoot dry matter [a physiological definition of P efficiency (Clark *et al.*, 1978; Gerloff, 1976)] than nonmycorrhizal plants (Table 2). Using this definition of P efficiency, mycorrhizal sorghum plants were less P efficient than nonmycorrhizal plants. This might have been because VAMF used C for their development and growth rather than for enhancing host plant growth. VAMF absorb P from soil and supply much of it to the host plant. In turn, the host plant supplies carbohydrates to the fungi. VAMF can act as respiratory and growth sinks and may cause some drain on host plant C resources. Because of such a C drain, P/C ratios in plants would likely increase

Table 1. Root colonization, shoot dry matter yield, and shoot concentrations and contents, of nonmycorrhizal (nm) and mycorrhizal (m) sorghum genotypes grown with different levels of applied soil P

Applied soil P (mg/kg)	Sorghum genotype	Root colonization (%)	Dry matter yield (g/plant) nm	Dry matter yield (g/plant) m	P concn (mg/g DM) nm	P concn (mg/g DM) m	P content (g/plant) nm	P content (g/plant) m
0	SC6	60.0[a]	0.46	2.90	0.97	1.45	0.45	4.22
	SC97	52.2	0.43	2.59	0.68*	1.80	0.29	4.66
12.5	SC6	33.2	2.43	4.07	1.93	3.12	4.77	12.57
	SC97	26.6	2.77	4.33	1.78	2.38	5.09	10.28
25.0	SC6	19.0	3.41	3.88	2.68	3.50	9.19	13.51
	SC97	20.6	3.76	4.50*	2.29	2.52*	8.44	11.33
37.5	SC6	12.0	3.66	4.51	3.10	3.63	11.01	16.32
	SC97	9.0	3.21	4.17	2.52	2.93*	7.61*	12.39*
P significance	df							
Mycorrhiza (M)	1	–		<0.01		<0.01		<0.01
P level (P)	3	<0.01		<0.01		<0.01		<0.01
M × P	3	–		<0.01		>0.20		0.03
Genotype (G)	1	0.04		>0.20		<0.01		<0.01
M × G	1	–		>0.20		>0.20		0.18
P × G	3	>0.20		0.04		0.05		<0.01
M × P × G	3	–		>0.20		0.10		>0.20

[a] No colonization occurred on plants grown with sterile VAMF.
* Differences between genotypes are significant at $P < 0.05$.

Table 2. Physiological phosphorus use efficiency (PUE) of nonmycorrhizal (nm) and mycorrhizal (m) plants, potential shoot dry matter (DM) yield of nonmycorrhizal plants, and VAMF benefit and cost calculations of sorghum genotypes with different levels of applied soil P

Applied soil P ($mg\ kg^{-1}$)	Sorghum genotype	PUE ($g\ DM\ g^{-1}$) nm	PUE ($g\ DM\ g^{-1}$) m	Potential DM(nm) ($g\ plant^{-1}$)	Calculated Benefit ($g\ plant^{-1}$)	Calculated Benefit (%)	Calculated Cost ($g\ plant^{-1}$)	Calculated Cost (%)
0	SC6	1031	690	4.33	2.44	85	1.43	49
	SC97	1471*	556*	6.85	2.16	83	4.26*	164*
12.5	SC6	518	321	6.57	1.64	40	2.50	61
	SC97	562	420*	5.79	1.56	36	1.46*	34*
25.0	SC6	373	286	5.06	0.47	12	1.18	30
	SC97	437	397*	4.95	0.74*	16*	0.45*	10*
37.5	SC6	323	275	5.30	0.85	19	0.79	18
	SC97	397	338	4.90	0.96	23	0.73	18
P significance	df							
Mycorrhiza (M)	1	<0.01		–	–		–	
P level (P)	3	<0.01		<0.01	<0.01		<0.01	
M × P	3	<0.01		–	–		–	
Genotype (G)	1	0.01		0.16	>0.20		<0.01	
M × P	1	0.01		–	–		–	
P × G	3	<0.01		0.03	<0.01		<0.01	
M × P × G	3	<0.01		–	–		–	

* Differences between genotypes are significant at $P < 0.05$.

(Cooper, 1984). Mycorrhizal leek (*Allium porrum* L.) plants transported 7% more total fixed C from shoots to roots than similar sized nonmycorrhizal plants (Snellgrove *et al.*, 1982). The extra photosynthate translocated was accounted for by VAMF growth, increased root respiration, and C losses to the soil.

Although beneficial effects of VAMF on host plant growth and P uptake are important, losses of C by plants (costs paid) for VAMF colonization with roots are important for an overall understanding of VAMF symbiosis. The calculated benefit and cost values for sorghum plants were highest when soil P was low and decreased with P application (Table 2). Cost values were lowest at 25.0 and 37.5 $mg\ P\ kg^{-1}$ soil. The benefit and cost values might have decreased because of the reduced root colonization with VAMF and the reduced dependence of the sorghum plants on VAMF for P uptake when soil P availabilities were higher. Using similar concepts, onion (*Allium cepa* L.) plants were found to derive a net dry weight benefit of 14% from VAMF, and paid a cost of 25% for a VAMF association (Yocum, 1981).

Nonmycorrhizal SC97 sorghum had higher P efficiency than SC6 when P was applied (Table 2). However, mycorrhizal SC6 was more P efficient than SC97 when P was not applied. The calculated benefits conferred by *G. fasciculatum* on the two sorghum genotypes appeared to be similar at each soil P level. However, the calculated costs for SC6 were considerably lower than those for SC97 when P was not applied, but higher when P was applied at 12.5 and 25.0 $mg\ kg^{-1}$ soil.

Using our definition of P efficiency and the benefit and cost analysis calculations, SC6 might be considered a better genotype choice for growth with VAMF in a low P soil with no added P, and SC97 might be more appropriate when 12.5 mg or higher amounts of $P\ kg^{-1}$ soil are applied. The host plant, the VAMF, and the soil environment can influence potential benefit and cost values and P efficiency in VAMF symbiosis. Benefit and cost analyses might be useful in selecting appropriate host plant genotypes and VAMF species to enhance VAMF symbiosis efficiencies when plants are grown under different soil conditions.

It should be emphasized, however, that the cost analysis for plant dry matter depressions due to VAMF-root associations may only be temporary since C losses by the host plant may be compensated by enhanced photosynthesis and other metabolic processes (Allen *et al.*, 1981; Brown and Bethlenfalvay, 1988; Losel and Cooper, 1979; Snellgrove *et al.*, 1986). Nevertheless, C losses as exudates, respiration, or other purposes need to be considered (Koide and Elliott, 1989). Root losses of C may also benefit soil properties since these C sources have been shown to enhance soil aggregation (Thomas *et al.*, 1986). In the absence of such a compensation, plant dry matter losses may occur. Host plant growth depressions resulting from VAMF-root associations have been reported (Abbott and Robson, 1984; Buwalda and Goh, 1982; Cooper, 1975; Hall *et al.*, 1977; Mosse 1973; Raju *et al.*, 1988), and might be explained using cost and benefit analysis.

Acknowledgement

This research was supported in part by a grant from the International Sorghum/Millet Collaborative Research Support Program (INTSORMIL) through the U.S. Agency for International Development (USAID grant AID/DAN-1254-G-55-5065-00 and published as Paper No. 8992, Journal Series, Nebraska Agricultural Research Division, Lincoln. Current address of P S Raju is Department of Agricultural Engineering, Federal University of Paraiba, 58 100 Campina Grande, PB, Brazil.

References

Abbott L K and Robson A D 1984 The effect of VA mycorrhizae on plant growth. *In* VA Mycorrhiza. Eds. C Ll Powell and D J Bagyaraj. pp 113–130. CRC Press, Boca Raton, FL.

Allen M F, Smith W K, Moore T S and Christensen M 1981 Comparative water relations and photosynthesis of mycorrhizal and non-mycorrhizal *Bouteloua gracilis* H.B.K. Lag Ex Steud. New Phytol. 88, 683–693.

Brown M S and Bethlenfalvay G J 1988 The *Glycine-Glomus-Rhizobium* symbiosis. VII. Photosynthetic nutrient-use efficiency in nodulated, mycorrhizal soybeans. Plant Physiol. 86, 1292–1297.

Buwalda G J and Goh K M 1982 Host-fungus competition for carbon as a cause of growth depression in vesicular-arbuscular mycorrhizal ryegrass. Soil Biol. Biochem. 14, 103–106.

Clark R B, Maranville J W and Gorz H J 1978 Phosphorus efficiency of sorghum grown with limited phosphorus. *In* Plant Nutrition. Eds. A R Ferguson, R L Bieleski and I B Ferguson. pp 93–99. New Zealand Dept. Sci. Indus. Res., Inform. Ser. No. 134, Wellington, New Zealand.

Cooper K M 1975 Growth responses to the formation of endotrophic mycorrhizas in *Solanum*, *Leptospermum*, and New Zealand ferns. *In* Endomycorrhizas. Eds. F E Sanders, B Mosse and P B Tinker. pp 391–407. Academic Press, New York.

Cooper K M 1984 Physiology of VA mycorrhizal associations. *In* VA Mycorrhiza. Eds. C Ll Powell and K J Bagyaraj. pp 155–186. CRC Press, Boca Raton, FL.

Gerdemann J W 1964 The effects of mycorrhizae on the growth of maize. Mycologia 56, 342–349.

Gerdemann J W 1968 Vesicular-arbuscular mycorrhiza and plant growth. Annu. Rev. Phytopathol. 6, 397–418.

Gerloff G C 1976 Plant efficiencies in the use of nitrogen, phosphorus, and potassium. *In* Plant Adaptation to Mineral Stress in Problems Soils. Ed. M J Wright. pp. 161–173. Cornell Univ. Agric. Exp. Stn., Ithaca, NY.

Giovannetti M and Mosse B 1980 An evaluation of techniques for measuring vesicular arbuscular mycorrhizal infection in roots. New Phytol. 84, 489–500.

Hall I R, Scott R S and Johnstone P D 1977 Effect of vesicular-arbuscular mycorrhizas on response to 'Grasslands Huia' and 'Tamar' white clovers to phosphorus. N. Z. J. Agric. Res. 20. 349–355.

Janos D P 1987 VA mycorrhizas in humid tropical ecosystems. *In* Ecophysiology of VA Mycorrhizal Plants. Ed. G R Safir. pp 107–134. CRC Press, Boca Raton, FL.

Koide R and Elliott G 1989 Cost, benefit and efficiency of the vesicular-arbuscular mycorrhizal symbiosis. Funct. Ecol. 3, 252–255.

Krishna K P and Bagyaraj D J 1981 Note on the effect of VA mycorrhiza and soluble phosphate fertilizer on sorghum. Indian J. Agric. Sci. 51, 688–690.

Knudsen D, Clark R B, Denning J L and Pier P A 1981 Plant analysis of trace elements by x-ray. J. Plant Nutr. 3, 61–75.

Losel D M and Cooper K M 1979 Incorporation of ^{14}C-labelled substrates by uninfected and VA mycorrhizal roots of onion. New Phytol. 83, 415–426.

Mosse B 1973 Plant growth responses to vesicular-arbuscular mycorrhiza. IV. In soil given additional phosphate. New Phytol. 72, 127–136.

Phillips J M and Hayman D S 1970 Improved procedures for clearing roots and staining parasitic and vesicular-arbuscular mycorrhizal fungi for rapid assessment of infection. Trans. Brit. Mycol. Soc. 55, 158–161.

Powell C Ll 1980 Phosphate response curves of mycorrhizal and nonmycorrhizal plants. I. Responses to superphosphate. N. Z. J. Agric. Res. 23, 225–231.

Raju P S 1985 Differential phosphorus nutrition in sorghum genotypes. MSc Thesis, Dept. of Agronomy, Univ. of Nebraska, Lincoln, NE.

Raju P S, Clark R B, Ellis J R and Maranville J W 1988 Effects of VA mycorrhizae on growth and mineral uptake of sorghum grown at varied levels of soil acidity. Commun. Soil Sci. Plant Anal. 19, 919–931.

Snellgrove R C, Splittstoesser W E, Stribley DP and Tinker

P B 1982 The distribution of carbon and the demand of the fungal symbiont in leek plants with vesicular-arbuscular mycorrhizas. New Phytol. 92, 75–87.

Snellgrove R C, Stibley D P, Tinker P B and Lawlor D W 1986 The effect of vesicular-arbuscular mycorrhizal infection on photosynthesis and carbon distribution in leek plants. *In* Physiological and Genetical Aspects of Mycorrhizae. Eds. V Gianinazzi-Pearson and S Gianinazzi. pp 421–424. Nat. Inst. Agron. Res., Paris.

Stribley D P 1987 Mineral nutrition. *In* Ecophysiology of VA Mycorrhizal Plants. Ed. G R Safir. pp 59–70. CRC Press, Boca Raton, FL.

Thomas R S, Dakessian S, Ames R N, Brown M S and G J Bethlenfalvay 1986 Aggregation of the silty clay loam soil by mycorrhizal onion roots. Soil Sci. Soc. Am. J. 50, 1494–1499.

Yocum D H 1981 Quantifications of costs and benefits accrued by onion plants from vesicular-arbuscular mycorrhizal associations. *In* Proc. Fifth North Am. Conf. Mycorrhizae. p 21. University Laval, Quebec, Canada.

M. L. van Beusichem (Ed.), *Plant nutrition – physiology and applications*, 171–177.

PLSO IPNC293

Phosphorus absorption by chickpea (*Cicer arietinum*) as affected by VA mycorrhiza and carboxylic acids in root exudates

Y. OHWAKI and H. HIRATA[1]
Faculty of Agriculture, Tokyo University of Agriculture and Technology, Fuchu, Tokyo 183, Japan.
[1]*Corresponding author*

Key words: andosol, carboxylic acids, chickpea, *Cicer arietinum* L., legumes, phosphate absorption, root exudation, VA mycorrhizas

Abstract

Phosphorus absorption pattern in chickpea grown on andosol that has high P-fixing capacity was studied in a pot experiment and compared with soybean, cowpea and kidneybean.

Among the four legumes, chickpea showed the greatest ability to extract P from soil low in available P (2.33 mg/kg by Bray II), in spite of its relatively poor root development and nearly the same degree of VA mycorrhizal infection at later stage. Accumulated P (up to 1.7%) in shoots of chickpea was severely released from roots towards pod maturity, probably due to P toxicity resulting from supraoptimal P concentration.

Carboxylic acids in root exudates of P-depleted legumes grown in water culture were compared. Malonic, succinic, fumalic, malic, citric and t-aconitic acids were detected in the exudates of chickpea. Large amounts of these carboxylic acids, especially citric and malic acid, were exuded from the roots of chickpea as compared with those of soybean, cowpea, and kidneybean.

The possible role of the extremely high amounts of citric and malic acid exuded from chickpea roots is discussed in relation to its unique absorption pattern of P.

Introduction

It is widely known that phosphorus absorption by plants is fundamentally controlled by its diffusion rate in soil solution and by the morphological characteristics of roots and the root system including VA mycorrhizae (Cradus, 1980; Nye and Tinker, 1977; Tinker, 1975). On the other hand, it has been postulated that P absorption by lupins, having neither large effective absorbing root surface nor VA mycorrhizal infection, may correlate with the exudation of citrate from roots to solubilize P in the soil through forming ferric hydroxy P polymers in soil solution (Gardner *et al.*, 1983).

This paper describes the absorption and distribution of P in chickpea (*Cicer arietinum* L., cv.CPS-1) grown on an andosol as compared with soybean (*Glycine max* Merr.,cv.Enrei), cowpea (*Vigna sinensis* Endl.,cv.Kuromame) and kidneybean (*Phaseolus vulgaris* L.,cv.Kurodane-Kinugasa) in relation to VA mycorrhizal infection. Moreover, carboxylic acid exudation from the roots of chickpea as compared with the other legumes grown in a P-depleted water culture was studied.

Materials and methods

Soil culture

A greenhouse experiment was conducted in undrained plastic pots filled with 2200 g of dry sub-soil of andosol to which two different levels of P(200 and 2000 mg) were applied as $Ca(H_2PO_4)_2$. Besides P, $N((NH_4)_2SO_4)$ 50,

K(KCl) 250, Ca($CaSO_4$) 1290, Mg($MgSO_4$) 90, Fe($FeCl_3$) 15, B(H_3BO_4) 3, Cu($CuSO_4$) 3, Mn($MnCl_2$) 9, Zn($ZnSO_4$) 9, Mo($(NH_4)_6Mo_7O_{24}$) 3 were mixed uniformly in each pot (mg/pot). Fresh top soil from the same farm field was also mixed uniformly (100 g/pot) as an inoculum for VA mycorrhizal fungi (mainly Glomus spp., spore number 200–600).

Four pre-germinated seeds were transplanted into each pot and inoculated with a suitable Rhizobium strain. After several days the number was reduced to two plants and cultivated to reach different growth stages; vegetative, flowering, pod-filling, and maturation. Soil moisture was maintained at 60% of the water-holding capacity using deionized water. When sampling at each growth stage, root parts were carefully taken out from the pot with soil soon after cuttng off the shoots, and isolated softly from soil aggregates by hand, being soaked into deionized water. Each part sampled at each growth stage was oven-dried, weighed and digested with H_2SO_4 and H_2O_2 for colorimetric determination of P. Prior to oven-drying of the roots, a part of the roots was subjected to estimation of VA mycorrhizal infection, using the grid-line intersect method (Giovanetti and Mosse 1980), after staining with 0.05% trypanblue in lactophenol.

Extractable P (Bray II; 0.1 *N*-HCl and 0.03 *N*-NF_4F) in the soil applied with 200 and 2000 mg P after 10 days incubation at 30°C were 2.33 and 9.27 mg/kg, respectively. The pH(H_2O) of the soil was 6.0 in both P treatments.

Water culture

Plant material. Surface sterilized seeds of the four legumes were germinated and grown on a vermiculite bed, providing adequate moisture for 10 days. Carefully isolated seedlings were transferred to a 35 L container filled with continuously aerated culture solution. Composition of nutrient solution (mg/L); NH_4-N 5.0, NO_3-N 15.6 (NH_4NO_3, $NaNO_3$), P 1.0(KH_2PO_4), K 38.6 (K_2SO_4), Ca 71.4($CaCl_2$), Mg 24.2($MgSO_4$), Fe 2.4(EDTA-Fe), Mn 0.16($MnCl_2$), Zn 0.1 ($ZnSO_4$), Cu 0.032 ($CuSO_4$), Mo 0.013(Na_2MoO_4), and B 0.017(H_3BO_3). The solution was renewed every two days, and its pH was adjusted at 6.0 every day.

After growing the plants for 21-26 days under glasshouse conditions, they were supplied with a phosphorus-free solution for 3 days before collecting root exudates. Fresh weights (g/plant) of shoots and roots in four legumes at the time of root exudate collection were; 2.86, 1.37(soybean), 4.13, 1.82(cowpea), 4.82, 2.08(kidneybean), 1.03, 1.40(chickpea), respectively.

Collection and analysis of root exudates. One plant from each of the four legumes with three replicates was submerged in 100 mL of an aerated aqueous solution of 0.5 m*M* $CaCl_2$ containing 0.05 mg/mL rifampcin and 0.025 mg/mL tetracycline to prevent bacterial contamination of the root exudates. After 2 h in the antibiotic solution, the roots were rinsed in sterile water, then reimmersed in aerated sterilized water containing 0.5 m*M* $CaCl_2$ for 6 h, followed by a transfer to a second flask of the same water for another 6 h. They were then placed in a growth chamber at a temperature of 25°C and light intensity of 500 $\mu E/m^2/sec$. The contents of the two flasks were pooled for each plant and evaporated to about 1-2 mL under reduced pressure below 40°C on a rotary evaporator.

Concentrated samples were passed through a 5 mL bed volume column with 1 cm diameter of anion exchange Sephadex QAE-25. Acidic compounds were eluated with 30 mL of 4% formic acid after washing the neutral ones with 20 mL of distilled water. The eluate was evaporated and dried completely over silica-gel under vacuum. It was resolved with 100 μL of anhydrous-pyridine and converted to trimethylsilyl derivatives using 50 μL of BSA(N-0-bistrimethylsilylacetoamide). A few μL of the clear supernatant solution was injected into the gas chromatograph (Shimazu GC-9A). A stainless column, 200 cm long with 2 mm internal diameter, was equipped and packed with 5% SE-30(Chromosorb WHP, 80 mesh) as a carrier material. The temperature of the chromatographic oven was programmed linearly from 90 to 220°C at the rate of 4°C/min with 40 mL/min of N_2 gas. Tartartic acid was used as an internal standard.

Results

Soil culture

Both growth and P accumulation patterns varied markedly with time among the four legumes as indicated in Figures 1 and 2.

In the 200 mg P treatment, dry matter production in each part was generally higher in soybean and cowpea than in kidneybean and chickpea (Fig. 1), in which the majority of the leaves dropped at maturity. Only chickpea showed a continuous increase in vegetative parts after the pod-filling stage.

While, the P accumulation pattern was quite different, namely maximum P absorption level was the highest in chickpea as compared to the other three legumes. P absorption in soybean and cowpea increased steadily as growth progressed, whereas chickpea and kidneybean scarcely absorbed P before flowering. This was followed by a sharp increase and decrease during the pod-filling stage and maturation, respectively, indicating that a severe loss of accumulated P from stems and leaves through decaying roots occurred towards maturation.

In the 2000 mg P treatment, similar patterns in dry matter production were observed, although at somewhat higher levels (Fig. 2). Maturation in cowpea and chickpea was accelerated as compared with the 200 mg P treatment. P accumulation in each part of soybean and cowpea was higher than the corresponding value in the 200 mg P application, whereas that of chickpea and kidneybean did not increase, except before flowering.

A sharp increase in P accumulation between flowering and pod-filling stage in chickpea applied with 200 mg P resulted in an extremely high P content in shoots – 1.8% in leaves and 1.7% in stems – at the pod-filling stage, followed by a sharp decrease (0.2% in shoots) towards maturity.

VA mycorrhizal infection was well established in the four legumes applied with 200 mg P, reaching a maximum of between 70-94% after flowering, although the infection level at flowering stage in kidneybean and chickpea was fairly lower than that in soybean and cowpea. The 2000 mg P application generally suppressed VA mycorrhizal infection (Table 1).

Water culture

As shown in Table 2, huge amounts of carboxylic acids were exuded from the roots of chickpea as compared with the other legumes. All types of carboxylic acids examined here were found to be present in the exudates of chickpea roots. The respective exudation rate of citric and malic acid from chickpea roots was nearly 35 and 16 times higher than that from soybean roots, and amounted to 54% and 19% of total carboxylic acids exudated, respectively. Succinic and t-aconotic acids were only detected in chickpea exudates.

Besides these carboxylic acids, quinic acid was detected in the exudates of chickpea roots, though not estimated quantitatively.

Table 1. Development of VA mycorrhizal infection (%) along with growth stage in four legumes. Stage I: Vegetative, II: Flowering, III: Pod-filling, IV: Maturation, Figures in parentheses indicate number of days after sowing. ND: Not determined due to the degradation of root tissue during KOH treatment

Legume	P applied (mg/pot)	VA mycorrhizal infection in roots (%)			
		I	II	III	IV
Soybean	200	3 b (21)	70 ab (44)	56 cd (65)	ND (99)
	2000	1 b (21)	33 d (44)	30 f (44)	ND (104)
Cowpea	200	7 a (19)	74 a (44)	67 bc (65)	94 a (87)
	2000	3 b (19)	53 bc (44)	49 de (65)	60 c (77)
Kidneybean	200	1 b (19)	24 d (36)	84 a (52)	82 b (67)
	2000	1 b (19)	20 d (36)	40 ef (52)	32 d (67)
Chickpea	200	3 b (22)	39 cd (39)	74 ab (57)	ND (111)
	2000	1 b (22)	30 d (39)	56 cd (57)	ND (77)

Statistical separation in columns by Duncan's Multiple Range Test ($P < 0.05$)

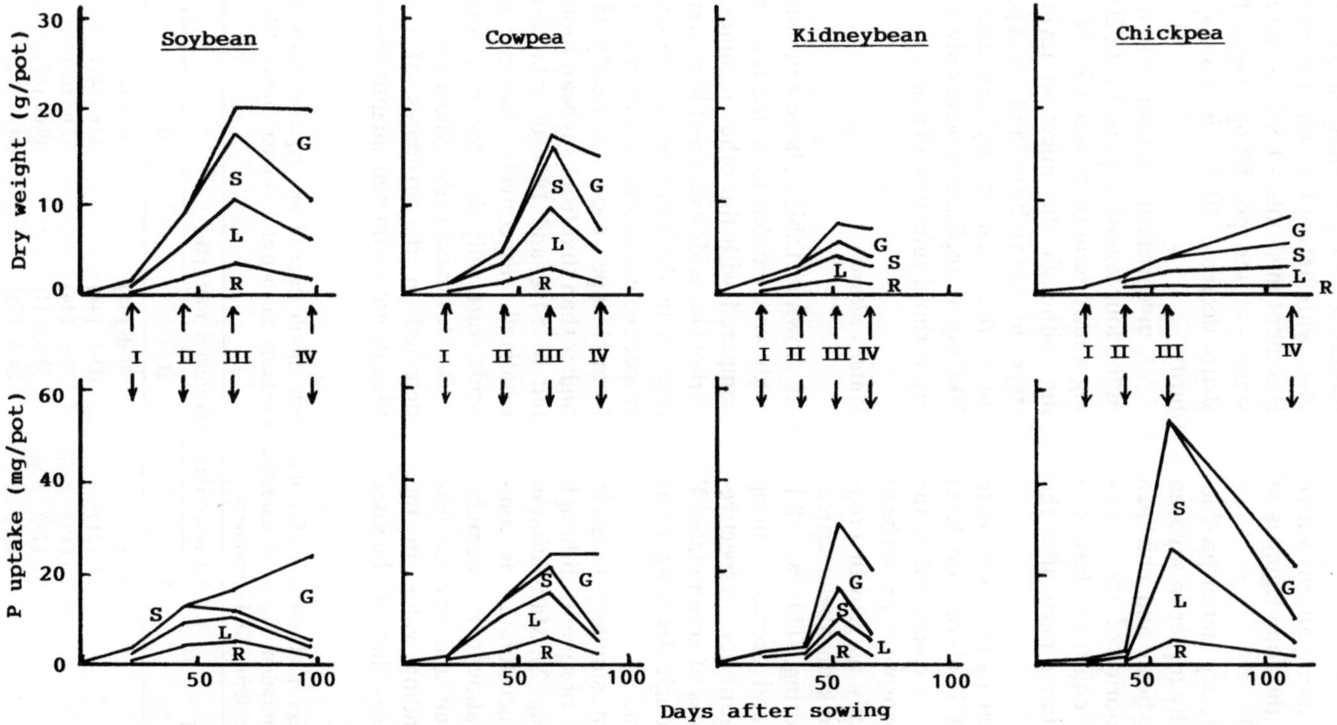

Fig. 1. Changes with time in dry weight and phosphorus accumulation of each part of four legumes applied with 200 mg P/pot.

Stage (I: Vegetative, II: Flowering, III: Pod-filling, IV: Maturation,

Symbol (G: Grain + husks, S: Stems + petioles, L: Attached and fallen leaves, R: Roots + nodules,

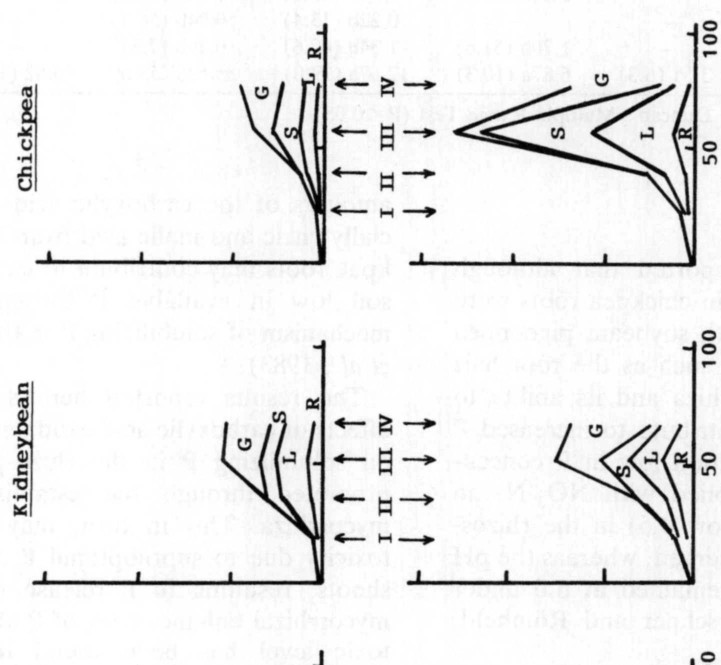

Fig. 2. Changes with time in dry weight and phosphorus accumulation of each part of four legumes applied with 2,000 mg P/pot. Stages and symbols are the same as in Figure 1.

Table 2. Exudation rates of carboxylic acids from roots of four legumes. Nanomole/g fresh weight of roots/12 h, Values in parentheses represent % individual carboxylic acid of the total detected

Legume	Malonic	Succinic	Fumaric	Malic	Citric	t-Aconitic	Total
Soybean	–	–	1.03b (36.4)	0.78b (27.6)	1.02b (36.0)	–	2.83 (100)
Cowpea	0.51b (30.4)	–	–	0.22b (13.4)	0.94b (56.2)	–	1.67 (100)
Kidneybean	–	–	1.70b (51.6)	1.34b (40.6)	0.26b (7.8)	–	3.30 (100)
Chickpea	7.04a (10.6)	3.51 (5.3)	6.87a (10.3)	12.67a (19.0)	35.63a (53.6)	0.82 (1.2)	66.54 (100)

Statistical analysis in columns by Duncan's Multiple Range Test ($P < 0.05$)

Discussion

Recently, Itoh (1987) reported that although relatively low Km values in chickpea roots were observed as compared with soybean, pigeonpea and maize, other factors, such as the root hair development, VA mycorrhiza and its ability to solubilize soil P, may contribute to increased P uptake from the soil solution low in P concentration. Even when supplied with NO_3-N, an extreme drop in pH (below 4.5) in the rhizosphere of chickpea was observed, whereas the pH of soybean rhizosphere remained at 6.0 under the same condition (Marschner and Römheld, 1983).

Figure 1 shows a marked increase in P accumulation by chickpea applied with 200 mg P between flowering and pod-filling stage, corresponding well to extensive progress of VA mycorrhizal infection during the same stages (Table 1). This results may indicate that the development of VA mycorrhizal infection definitely controlled P absorption in this condition. There was a sharp decline in P content in chickpea shoots between pod-filling and maturity, probably due to release of P from the roots into growth medium. These tendencies of P accumulation pattern and VA mycorrhizal infection were also shown in kidneybean at somewhat mild level. P release from chickpea roots during pod-filling to maturation was also observed in the 2000 mg P treatment.

A significant increase in citrate exudation from the roots of P-stressed alfalfa was confirmed using an apparatus specially designed for aseptic collection of root exudates (Lipton *et al.*, 1987). Hoffland *et al.* (1989) also reported that distinctly higher exudation of malic and citric acid in P-starved rape roots might coincide with the local acidification of the rhizosphere. Huge amounts of the carboxylic acid exudates, especially citric and malic acid from P-depleted chickpea roots may contribute to extract P from the soil low in available P through the possible mechanism of solubilizing P in the soil (Gardner *et al.*, 1983).

The results reported here suggest that the effects of carboxylic acid exudates from the roots on solubilizing P in the rhizosphere could be promoted through the establishment of VA mycorrhiza. This in turn, may bring about P toxicity due to supraoptimal P concentration in shoots, resulting in P release from roots. VA mycorrhizal enhancement of P absorption to the toxic level has been found in onion plants (Mosse, 1973), and P toxicity was described with P concentration of around 1% in shoots of clover which was strongly enhanced by zinc deficiency (Lonergan *et al.*, 1979). Although it was not possible to identify P toxicity in this experiment, severe loss of P between pod-filling stage and maturation in chickpea could correlate with P toxicity, probably due to nutrient imbalance and/or other physiological reasons.

Acknowledgements

We thank International Crops Research Institute for the Semi-Arid Tropics (ICRISAT) for providing the chickpea seeds and the rhizobium.

References

Cradus J R 1980 Distinguishing between grass and legume species for efficiency of phosphorus use. N. Z. J. Agric. Res. 23, 75–81.

Gardner W K, Barber D A and Parbery D G 1983 The acquisition of phosphorus by *Lupinus albus* L. III. The probable mechanism by which phosphorus movement in the soil/root interface is enhanced. Plant and Soil 70, 107–124.

Giovannetti M and Mosse B 1980 An evaluation of techniques for measuring VA mycorrhizal infection in roots. New Phytol. 84, 489–500.

Hoffland E, Findenegg G R and Nelemans J A 1989 Solubilization of rock phosphate by rape. II. Local root exudation of organic acids as a response to phosphorus starvation. Plant and Soil 113, 161–166.

Itoh S 1987 Characteristics of phosphorus uptake of chickpea in comparison with pigeonpea, soybean and maize. Soil Sci. Plant Nutr. 33, 417–422.

Lipton D S, Blancher R W and Blevins D G 1987 Citrate, malate and succinate concentration in exudates from P-sufficient and P-stressed *Medicago sativa* L. seedlings. Plant Physiol. 85, 315–317.

Lonergan J F, Grove T S, Robson A D and Snowball K 1979 Phosphorus toxicity as a factor in zinc – phosphorus interactions in plants. Soil Sci. Soc. Am. J. 43, 966–972.

Marschner H and Römheld V 1983 *In vivo* measurement of root-induced pH changes at the soil-root interfaces: Effect of plant species and nitrogen source. Z. Pflanzenphysiol. 111, 241–251.

Mosse B 1973 Plant growth repsonses to vesicular-arbuscular mycorrhiza. IV. In soil given additional phosphate. New Phytol. 72, 127–136.

Nye P H and Tinker P B 1977 Solute Movement in the Soil-Root System. Blackwell Osney Mead, Oxford, 342 p.

Tinker P B 1975 Effect of vesicular-arbuscular mycorrhizas on higher plants. Symp. Soc. Exp. Biol. 29, 325–350.

M. L. van Beusichem (Ed.), *Plant nutrition – physiology and applications*, 179–183.

PLSO IPNC412A

Origin of organic acids exuded by roots of phosphorus stressed rape (*Brassica napus*) plants

E. HOFFLAND, J.A. NELEMANS and G.R. FINDENEGG
Department of Soil Science and Plant Nutrition, Wageningen Agricultural University, P.O. Box 8005, 6700 EC Wageningen, The Netherlands

Key words: *Brassica napus* L., citrate, $^{14}CO_2$, malate, organic acids, phosphorus deficiency, phloem transport, phosphoenolpyruvate carboxylase, root exudation

Abstract

To determine the origin of organic acids exuded by the roots of P deficient rape plants, phosphoenolpyruvate carboxylase (PEPC) activity was measured in plants after deprivation of P in the nutrient solution. PEPC activity in the shoot increased as a reaction to P stress. This increase coincided with accumulation of citrate in the shoot and with a higher citrate/sugar ratio in the phloem. Application of $^{14}CO_2$ to the shoots resulted in a ninefold increase in specific activity of organic acids exuded by −P roots as compared with +P roots. These results indicate that exuded organic acids resulted from increased PEPC activity in the shoot of P stressed rape plants.

Introduction

Roots of phosphorus deficient rape plants acidify their rhizosphere by exudation of malic and citric acid. The efficient use of rock phosphates by rape has been attributed to this phenomenon. Exudation is restricted to a root segment of 1.5 cm behind the root tip and coincides with higher tissue concentrations of malic and citric acid in this segment (Hoffland *et al.*, 1989).

The purpose of this study was to investigate whether root exudation originates from an increased rate of organic acid synthesis induced by P deficiency. Phosphoenolpyruvate (PEP) carboxylation and subsequent reduction of oxaloacetate to malate by malate dehydrogenase was regarded as the most obvious anaplerotic pathway for organic acid synthesis in plant tissue (Latzko and Kelly, 1983).

We determined the origin of organic acids exuded by P deficient rape plants by measuring PEP carboxylase (PEPC) activity in root tissue. PEPC activity in shoot tissue and phloem levels of organic acids were measured to establish the role of shoot-borne organic acids in root exudation.

Materials and methods

Plant cultivation

Seeds of rape (*Brassica napus* L. var. Jetneuf) were germinated in quartz sand. After 6 days the seedlings were placed on 50-L containers filled with nutrient solution. The plants were grown in a growth chamber at 20°C, a light intensity of 70 W m^{-2}, a photoperiod of 16 h and a relative humidity of ±80%.

Nutrient media

The −P nutrient solution consisted of 1.25 m*M* $Ca(NO_3)_2$, 1.25 m*M* KNO_3, 0.5 m*M* $MgSO_4$, and trace elements, in mg L^{-1}: Fe (as FeEDTA) 4.6; B 0.5; Mn 0.5; Zn 0.05; Cu 0.02; Mo 0.01. In the +P nutrient solution 0.25 m*M* KH_2PO_4 was added.

P concentration

Plant material was analyzed for P after wet digestion in a H_2SO_4-Se-salicylic acid mixture with addition of H_2O_2 (Novozamsky *et al.*, 1983). P was determined by the molybdenum-blue method (Murphy and Riley, 1962).

PEPC activity

The determination of PEPC activity by coupling the carboxylation reaction to NADH oxidation was described elsewhere (Arnozis *et al.*, 1988). MDH and LDH were added to the standard assay medium according to Meyer *et al.* (1988).

Collection of phloem sap

Phloem sap was collected by the method of King and Zeevaart (1974). The stems of three rape plants were cut just above the roots while submerged under 20 m*M* K-EDTA, pH 7.5. The shoots were attached in an Eppendorf reaction vessel filled with 500 μL of the K-EDTA solution. After 1 h incubation (24°C, 100% humidity, light intensity 65 W m^{-2}) the solution was immediately analyzed for malate, citrate and sugar.

Malate, citrate and sugar concentration

Sample preparation
To determine malate and citrate in plant material, the extracts for the determination of PEPC activity were used (see above).

Nutrient solutions of the labeling experiment were freeze dried, resolved in 0.5 mL 80% methanol and centrifuged. From the supernatants 25 μL was used for chromatography (see below). The other part was dried under N_2, resolved in 500 μL water and analyzed for malate and citrate.

Determinations
Enzymatic procedures provided by Boehringer Mannheim GmbH were used for the determination of malate, citrate (Anonymous, 1989) and glucose + sucrose (Anonymous, 1986) concentrations. The glucose + sucrose concentration will further be referred to as sugar concentration.

Labeling of the shoots

Two shoots of intact rape plants grown for 7 days on nutrient solution were enclosed with gum in a 100-mL glass tube. From three or two sets of two plants the roots were put in 15 mL nutrient solution. $^{14}CO_2$ was liberated inside the tubes by adding HCl to $Na_2\,^{14}CO_3$ (specific activity 49.3 MBq/mmol). To two shoots 100 kBq was added.

The shoots were exposed to $^{14}CO_2$ for 6 h during the light period. All analyses were carried out with material collected 27 h after the start of the labeling.

Radioactivity

To count radioactivity in plant material, the samples were ground in a mortar in 96% ethanol. To part of the samples of nutrient solution 0.1 mL 1 *M* HCl was added per mL to remove CO_2.

Radioactivity of plant extracts and nutrient solutions was counted in a Packard Liquid Scintillation Counter.

Chromatography

From the above mentioned supernatants of nutrient solutions (see 'Malate, citrate and sugar concentration') 25 μL was chromatographed on a cellulose TLC plate developed with 2-pentanol/formic acid/water (48.8/48.8/2.4). Formic acid removed by heating the plate at 120°C for 1 h. The plate was stained with bromocresolegreen (0.04% (w/v) in 96% ethanol, pH 13). The acid and radioactive spots with r_f values equal to those of standards of radioactive malic acid and citric acid were collected and radioanalyzed.

Results

PEPC activity and organic acids in tissue and phloem

Rape plants were precultured for 7 days on +P nutrient solution. Then, at the moment defined as t = 0, the roots were washed and the plants were transferred to a −P nutrient solution. Analyses in shoots and roots were done 0, 6, 11 and 14 days after transfer.

Tissue P concentration declined steadily after t = 0 in both shoots and roots (Fig. 1), but more rapidly in the shoots. In the shoots this decline coincided with a substantial increase in PEPC activity and an accumulation of citrate while the malate concentration declined slightly. No such effects could be detected in the roots.

Citrate accumulation in the shoot might cause, with an overflow mechanism, increased transport of organic acids towards the roots via the phloem. Therefore phloem sap was analyzed for malate, citrate and sugars 0, 4, 6 and 8 days after transfer from +P to −P nutrient solution.

After t = 0 an increase in the citrate/sugar ratio was determined (Table 1). This was caused by both a decrease in the amount of sugar and an increase in the amount of citrate detected. No significant change in the malate/sugar ratio was found.

In other experiments plants were precultured for 7 days on −P nutrient solution and then transferred to +P nutrient solution. In such plants PEPC activity and citrate concentration in the shoot and phloem transport of citrate decreased after transfer (results not shown). The above described phenomena can therefore be considered as reactions to P stress, and not as being a consequence of aging.

Labeling experiment

Rape plants were grown for 7 days with or without P. It has been shown that PEPC activity in −P shoots was 214% of that of +P shoots. Shoots of those plants were exposed to $^{14}CO_2$. Results of this experiment are given in Table 2.

The radioactivity fixed per gram fresh weight was about 2.5 times higher in −P shoots than in +P shoots. This proportion was even higher in the roots. This indicates an enhanced transport of labeled compounds from the shoot towards the root in −P plants. Because no significant difference in radioactivity of samples of the nutrient solution with and without addition of HCl was found, no $^{14}CO_2$ had been available to the roots. The radioactivity in the roots therefore had to originate from the shoots.

More organic acids, especially malate, were

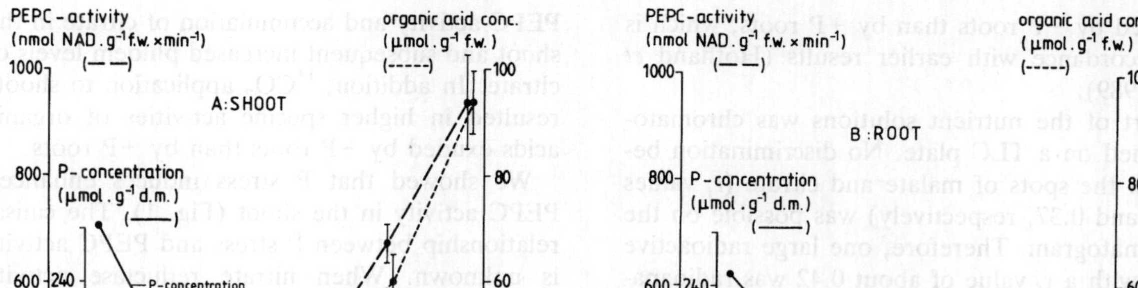

Fig. 1. P concentration, PEPC-activity and organic acid concentrations in shoots (A) and roots (B) of rape plants as a function of time. The plants were precultured on +P nutrient solution and at t = 0 transferred to −P nutrient solution. Values of PEPC-activity and organic acids are means ± s.d. (n = 3).

Table 1. Citrate, malate and sugar in phloem exudates of three rape plants, collected in 500 μL 20 m*M* K-EDTA during 1 h. At t = 0 rape plants were transferred from +P to −P nutrient solution. Values are given ±s.d. (n = 6)

Time (days)	Sugar (nmol)	Citrate (nmol)	Citrate / Sugar	Malate (nmol)	Malate / Sugar
0	732 ± 78	32 ± 4	0.04 ± 0.00	104 ± 29	0.14 ± 0.04
4	601 ± 254	42 ± 14	0.07 ± 0.03	106 ± 37	0.18 ± 0.07
6	468 ± 107	72 ± 21	0.16 ± 0.06	97 ± 26	0.21 ± 0.11
8	296 ± 126	101 ± 7	0.34 ± 0.13	53 ± 12	0.18 ± 0.05

Table 2. Results of the labeling experiment. Shoots of rape plants grown for 7 days on nutrient solution were exposed to $^{14}CO_2$. If two values are given, the first one results from an experiment with 6 plants, the second one from 4 plants per 15 mL nutrient solution. If one value is given, it results from the first experiment

Material	Quantity	Unit	Sample	Treatment			
				+P		−P	
Plant material	Fresh weight	mg × plant^{-1}	shoot	525	486	237	213
			root	75	76	109	76
	Radioactivity	kBq × g^{-1} fw	shoot	47	35	103	119
			root	14	12	40	41
		kBq × plant^{-1}	shoot	24.7	17.0	24.4	25.3
			root	1.0	0.9	2.9	3.1
Nutrient solution	Radioactivity	kBq × plant^{-1}		0.6	0.7	1.9	2.1
		Bq × plant^{-1}	Malate + citrate	5.9	4.4	90.8	72.5
	Amounts exuded	nmol × plant^{-1}	malate	4.7		7.5	
			citrate	2.7		5.0	
	Specific activity	kBq × mmol^{-1}	malate + citrate	0.80		7.27	

exuded by −P roots than by +P roots, which is in accordance with earlier results (Hoffland *et al.*, 1989).

Part of the nutrient solutions was chromatographed on a TLC plate. No discrimination between the spots of malate and citrate (r_f values 0.47 and 0.37, respectively) was possible on the chromatogram. Therefore, one large radioactive spot with a r_f value of about 0.42 was radioanalyzed. In combination with the results of the concentration determinations the specific activities of malate + citrate could be calculated. The −P plants exuded malate and citrate with a much higher specific activity than the +P plants (Table 2).

Discussion

The results presented indicate that organic acids exuded by roots of P stressed rape plants originate from enhanced PEPC activity in the shoot. Increase of P stress coincided with increase of PEPC activity and accumulation of citrate in the shoot and subsequent increased phloem levels of citrate. In addition, $^{14}CO_2$ application to shoots resulted in higher specific activities of organic acids exuded by −P roots than by +P roots.

We showed that P stress induces enhanced PEPC activity in the shoot (Fig. 1). The causal relationship between P stress and PEPC activity is unknown. When nitrate reductase activity (NRA) increases, PEP carboxylation can be increased, resulting in the production of malate, which plays a key-role in the intracellular pH-stat (Smith and Raven, 1979) or in the supply of reductants for NRA (Naik and Nicholas, 1986). However, Moorby *et al.*, (1988) demonstrated that the P concentrations found in our P stressed rape plants cause a reduction in NRA. A direct influence of P_i on PEPC is speculative, but cannot be excluded. Data on the effect of P_i on PEPC from C_4 plants *in vitro* are conflicting (Walker *et al.*, 1988). So far, effects of P deficiency on PEPC activity in C_3 plants have not been reported.

Malate must be regarded as the major product

of PEP carboxylation (Lance and Rustin, 1984). Therefore, citrate accumulation in P stressed rape shoots and subsequent root exudation of malate seems remarkable. However, a change from malate to citrate as the predominant organic acid in combination with higher tissue concentrations was demonstrated before by Landsberg (1981) in roots of several C_3 plants as a response to Fe deficiency. The same appeared to occur in P stressed rape shoots. In the exuding root zones, however, where increased malate concentrations were determined (Hoffland *et al.*, 1989), the opposite should occur.

Accumulation of organic acids in P stressed rape shoots could be caused by decreased oxidation rates due to decreased transport into the mitochondria. Wiskich (1975) reported that influx of malate and especially citrate into isolated mitochondria of *Brassica oleracea* L. is stimulated by P_i, and that the rate of entry of these acids can limit the rate of mitochondrial oxidation. It remains to be investigated whether this phenomenon plays a role in accumulation of organic acids in P stressed rape plants.

The suggestion that exudation is not caused by leakage but by increased synthesis of organic acids is confirmed by the fact that the ratio of labeled compounds in the root to that in the nutrient solution did not differ between −P and +P plants. The increased transport of labeled compounds from shoot to root in −P plants and the increased specific activity of organic acids exuded by −P roots are strong indications that these acids originiate from increased PEP carboxylation in the shoot.

Further research should establish which of the above reactions to P stress are specific for rock phosphate mobilizing species like rape in order to understand fully the exudation of organic acids.

Acknowledgements

The authors are very grateful to Mr Hans Overbeek for his hospitality and his essential suggestions during the performance of the labeling experiment. This was carried out entirely at his laboratory at the Centre for Agrobiological Research, Wageningen, which is thanked for the supply of chemicals.

Mr Erik Heij and Mr Karel van Gaalen are thanked for analytical and technical assistance, and Dr Rien van Beusichem for his comments on the manuscript.

References

Anonymous 1986 Methods of biochemical Analysis and Food Analysis using Test-combinations. Boehringer Mannheim GmbH, Mannheim, FRG.

Anonymous 1989 Methods of biochemical Analysis and Food Analysis using single Reagents. Boehringer Mannheim GmbH, Mannheim, FRG.

Arnozis P A, Nelemans J A and Findenegg G R 1988 Phosphoenolpyruvate carboxylase activity in plants grown with either NO_3^- or NH_4^+ as inorganic nitrogen source. J. Plant Physiol. 132, 23–27.

Hoffland E, Findenegg G R and Nelemans J A 1989 Solubilization of rock phosphate by rape. II. Local root exudation of organic acids as a response to P-starvation. Plant and Soil 113, 161–165.

King R W and Zeevaart J A D 1974 Enhancement of phloem exudation from cut petioles by chelating agents. Plant Physiol. 53, 96–103.

Lance C and Rustin P 1984 The central role of malate in plant metabolism. Physiol. Vég. 22, 625–641.

Landsberg E-Ch 1981 Organic acid synthesis and release of hydrogen ions in response to Fe deficiency stress of mono- and dicotyledonous plant species. J. Plant Nutr. 3, 579–591.

Latzko E and Kelly G J 1979 The many-faceted function of phosphoenolpyruvate carboxylase in C_3 plants. Physiol. Vég. 21, 805–815.

Meyer C R, Rustin P and Wedding R T 1988 A simple and accurate spectrophotometric assay for phosphoenolpyruvate carboxylase activity. Plant Physiol. 86, 325–328.

Moorby H, Nye P H and White R E 1988 The effect of phosphate nutrition of young rape plants on nitrate reductase activity and xylem exudation, and their relation to H ion efflux from the roots. Plant and Soil 105, 257–263.

Murphy J and Riley J P 1962 A modified single-solution method for the determination of phosphate in natural waters. Anal. Chim. Acta 27, 31–36.

Naik M S and Nicholas D J D 1986 Malate metabolism and its relation to nitrate assimilation in plants. Phytochemistry 25, 571–576.

Novozamsky I, Houba V J G, Van Eck R and Van Vark W 1983 A novel digestion technique for multi-element plant analysis. Commun. Soil Sci. Plant Anal. 14, 239–248.

Smith F A and Raven J A 1979 Intracellular pH and its regulation. Annu. Rev. Plant Physiol. 30, 289–311.

Walker G H, Ku M S B and Edwards G E 1988 The effect of phosphorylated metabolites and divalent cations on the phosphate and carboxylase activity of maize leaf phosphoenolpyruvate carboxylase. J. Plant Physiol. 133, 144–151.

Wiskich J T 1975 Phosphate-dependent substrate transport into mitochondria. Plant Physiol. 56, 121–125.

M. L. van Beusichem (Ed.), *Plant nutrition – physiology and applications*, 185–187.

PLSO IPNC429A

Grain development in wheat (*Triticum aestivum*) ears cultured in media with different concentrations of phosphorus and sucrose

G.D. BATTEN and K. SLACK
NSW Agriculture and Fisheries, Yanco, Agricultural Institute, Yanco, N.S.W. 2703, Australia

Key words: ear culture, phosphorus, sucrose, *Triticum aestivum* L., wheat

Abstract

Ears were detached from wheat plants below the peduncular node and grains allowed to develop in a liquid culture system. Grains accumulated more dry matter when cultured in media which contained phosphorus if the sucrose concentration was between 10 and 40 g L^{-1}. Sucrose at 80 g L^{-1} caused early senescence of the ears, less response to phosphorus and smaller grains.

Introduction

It is not possible to manipulate grain nutrients without affecting other plant parts, so therefore it is not possible to describe the effects of the nutrient *per se*. Grain development and nutrient content will be confounded with the effects of plant size or environmental constraints. For example, phosphorus fertilizer can increase grain yield by increasing the number of tillers per plant, the number of grains per ear, and the grain weight. However in a dryland environment soil water may be exhausted before the potential grain weight is achieved. In dry seasons the grains of low-P plants can be heavier than the grains of well fertilized plants (Batten *et al.*, 1984).

We have investigated head culturing as a system which allows the supply of phosphorus to the grain to be manipulated while other variables are kept constant. This paper reports the effects of phosphorus supply on grain development.

Methods

Wheat (*Triticum aestivum* L. cv. Dollarbird) was grown in the glasshouse in sand with a low-P nutrient regime as used by Batten and Wardlaw (1987). Ears were detached below the peduncular node, just before or just after anthesis, and the flag leaf removed. The stem was sterilized using 1% hypochlorite solution and placed in 50 mL of culture medium at 1°C in a room set at a temperature of 25°C. Fluorescent lights gave a photon flux density of 300 μmoles m^{-1} s^{-1} at the ear level for 12 hours per day.

Table 1. Composition of media used to culture wheat ears

Major elements	*Salt* (*g* L^{-1})
$CaCl_2.2H_2O$	0.20
KNO_3 nil-P only	0.46
K_2HPO_4 control-P	0.40
KH_2PO_4 high-P	0.62
$MgSO_4.7H_2O$	0.37
Minor elements	*Salt* (*mg* L^{-1})
H_3BO_3	6.20
$MnSO_4.4H_2O$	22.3
$ZnSO_4.4H_2O$	8.60
KI	0.83
$Na_2MoO_4.2H_2O$	0.25
$CuSO_4.5H_2O$	0.25
$CoCl_2.6H_2O$	0.25
Fe-EDTA	15.32
myo-Inositol	100
Thiamin HCl	0.80

Nitrogen was supplied as glutamine at 8 g L^{-1}. The media were adjusted to pH 5.8 using HCl or KOH before being filtered.

The culture media were based on that used by Donovan and Lee 1977 (Table 1). Nil-P medium was prepared by using KNO_3; high-P medium was prepared using KH_2PO_4 in place of K_2HPO_4. Sucrose was added to the nutrient solution prior to filter sterilization using a 0.45 μm membrane filter.

Results and discussion

Effects of phosphorus in the culture media and time in culture media on grain development

Grain weight was increased by phosphorus in three of the four studies conducted to date. In the first experiment the differences in grain weight between nil-P, control-P and twice control-P were not significant. Figure 1 illustrates the effect of phosphorus supply on grain development for plants cultured from 10 days after anthesis, while Figure 2 shows the effect of phosphorus on grains of ears cultured from before, at and post-anthesis.

Effects of sucrose supply in the culture media on grain development

Grain development was most rapid when detached ears were cultured in a solution with 20 or 40 g sucrose L^{-1} (Fig. 3a). This is consistent with grain development in head culture studies by Singh' and Jenner (1983). The mature grains were smaller in size than those on attached ears.

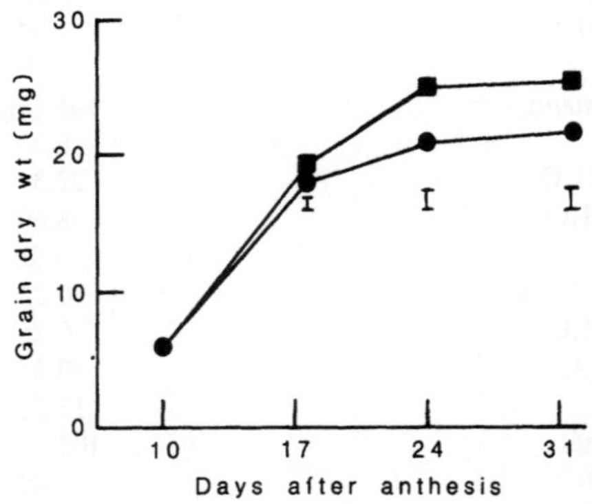

Fig. 1. Development of grains in ears of wheat cultured in media with nil (●) and control (■) phosphorus. Bars indicate twice the standard error.

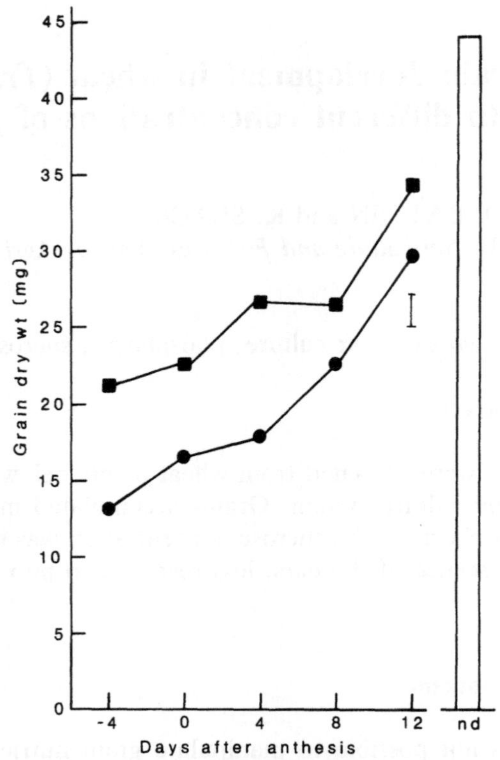

Fig. 2. Effects of time commencing culturing and phosphorus concentration in the media on the dry weight of wheat grains at physiological maturity. Symbols as in Fig. 1; nd indicates ears not detached.

Phosphorus promoted the growth of grains cultured with 10 to 40 g sucrose L^{-1} but had little effect on grains cultured with nil or 80 g sucrose L^{-1}.

The water content of the grains decreased as the sucrose concentration in the media increased (Fig. 3a). This was consistent with the visual observation that sucrose enhanced the onset of senescence. Barlow *et al.* (1983) reported high accumulation of sucrose in non-grain ear tissues and reduced transpiration of water when high concentrations of sucrose were fed to detached ears. The effects of higher concentrations of sucrose on grain dry weight and grain water content appear to be greater in the present study.

Grain water content changed significantly due to increased phosphorus supply on only one occasion (from 69 to 71% at 7 days into the culturing phase in the first experiment).

The only significant interaction effects between sucrose and phosphorus on grain water

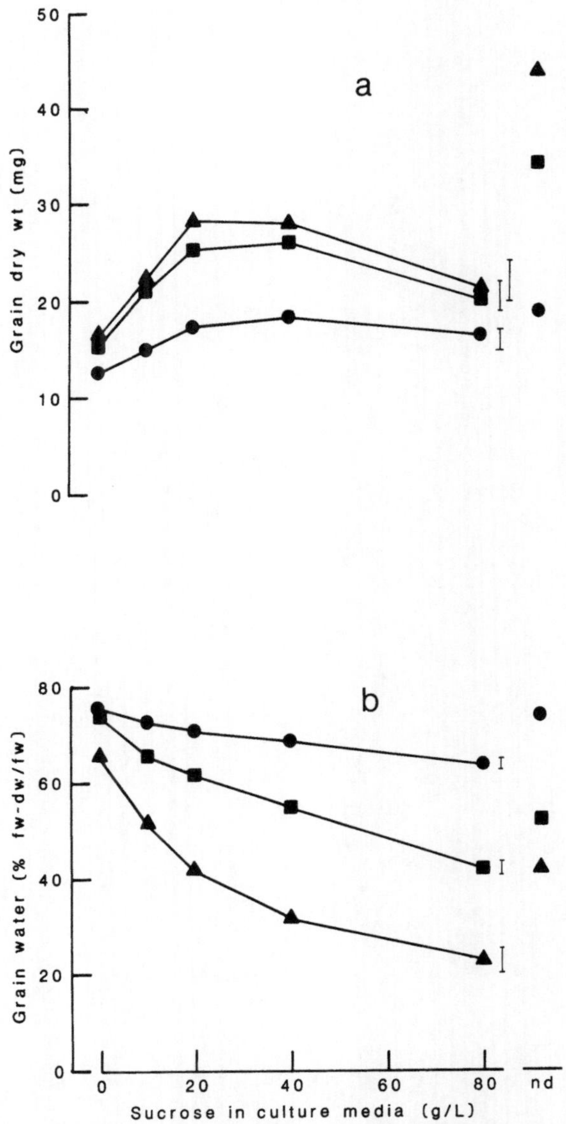

Fig. 3. Effects of sucrose concentrations in the media on dry weight (**a**) and water content (**b**) in grains of wheat ears after 7 (●), 14 (■) and 21 (▲) days of culturing, or not detached (nd) from the plant. Error bars indicate twice the standard error.

content occurred 21 days after culturing commenced. In the period of 14 to 21 days of culturing there was more rapid loss of water from the grains developed with the nil-P-nil-sucrose treatment.

The culturing of detached ears of cereal plants allows comparisons between treatments imposed during the grain filling period to be made in the absence of confounding effects such as uptake of nutrients from the soil, translocation of nutrients from lower stem and root tissues, and environmental factors.

This study has demonstrated that phosphorus supplied during the grain development phase promotes grain dry weight if sucrose is supplied at 10 to 40 g L^{-1}.

Acknowledgements

We thank Ms Victoria Jones for assistance with the experimental work and the Wheat Research Council (Project DAN102) for financial support.

References

Barlow E W R, Donovan G R and Lee J W 1983 Water relations and composition of wheat ears grown in liquid culture: Effect of carbon and nitrogen. Aust. J. Plant. Physiol. 10, 99–108.

Batten G D, Khan M A and Cullis B R 1984 Yield responses by modern wheat genotypes to phosphate fertilizer and their implications for breeding. Euphytica 33, 81–89.

Batten G D and Wardlaw I F 1987 Senescence and grain development in wheat plants grown with contrasting phosphorus regimes. Aust. J. Plant Physiol. 14, 253–65.

Donovan G R and Lee J W 1977 The growth of detached wheat ears in liquid culture. Plant Sci. Lett. 9, 107–113.

Singh B K and Jenner C F 1983 Culture of detached ears of wheat in liquid culture: Modification and extension of the method. Aust. J. Plant. Physiol. 10, 227–236.

M. L. van Beusichem (Ed.), *Plant nutrition – physiology and applications*, 189–192.

PLSO IPNC464A

Nodule activity and nodule phosphorus in white clover (*Trifolium repens*) in the presence of mineral nitrogen

A.L. HART
Grasslands Division, Department of Scientific and Industrial Research, Private Bag, Palmerston North, New Zealand

Key words: nitrogen, nitrogen fixation, nodules, phosphorus fractions, phosphorus, *Trifolium repens* L., white clover

Abstract

In white clover growing in nutrient solution and dependent on nitrogen fixation, nodules contained high concentrations of phosphorus. Acetylene reducing activity was closely related to nodule phosphorus. When ammonium nitrate was added, nodule activity was strongly inhibited and the concentration of nodule phosphorus fractions fell. The concentration of leaf phosphorus fractions was unaffected. There was a decrease in the ratio of root phosphorus and dry matter to that in the shoot, and the shoots of plants given nitrogen were larger than those left to rely on nitrogen fixation. Addition of nitrogen relieved the plants of the necessity to maintain high concentrations of phosphorus in the nodules and there was a change in the relative strength of sinks for phosphorus, although where the external supply of phosphorus was stopped, redistribution of phosphorus was insufficient to compensate for the shortfall in supply.

Introduction

Nodules are important sinks for carbon, to the extent that the respiration associated with nitrogen (N) fixation is considered to reduce the growth of N-fixers relative to plants with access to mineral N (*e.g.* Brugge and Thornley, 1984). Nodules are also important sinks for phosphorus (P). P is directed to them in feeding experiments (Hoshino, 1974), concentrations of it in them may be higher than in other parts of the plant, and they may contain a significant proportion of root P (Hart *et al.*, 1981).

On the application of mineral N at levels sufficient to inhibit N fixation, the importance of nodules as sinks for carbon is reduced. Carbon is directed to other sinks, and the added N may compensate for any previous shortfall in N availability (Allos and Bartholomew, 1959; Ryle *et al.*, 1981). Less attention has been paid to the effects of N application on the nodules as sinks for P. In this paper a description is given of the consequences of mineral N application on P concentrations in nodules and leaves and on the distribution of P and dry matter between root and shoot in white clover.

Methods

Stolon tip cuttings of a genotype of *Trifolium repens* cv. 'Grasslands Huia' were placed in 1-litre pots of nutrient solution. The initial pH of the solution was 6.0, the P and N concentrations being 0.1 mol/m^3 and 1.43 mol/m^3 respectively; the solution was changed weekly. Twenty-two days later, the rooted cuttings were inoculated with a thick suspension of *Rhizobium trifolii* (NZP 560) (N was removed from the solution). On day 29, each plant was transferred to a 6-litre bucket of aerated nutrient solution. The solution was changed weekly, the pH also being adjusted mid-weekly. There were 4 levels of P supply: 0.96, 1.86, 3.60 and 7.20 mg P/plant/week. The other components of the solution were 0.1 mol/m^3 Morpholinoethanesulfonic acid and Ca 1000, Mg 1000, K 1456, SO_4 1940, Fe 54, B 46, Mn 9, Zn 1.5, Cu 1.6, Mo 0.15, Co 0.50 mmol/m^3.

Beginning on day 57, half the plants were given 10 mol/m^3 as NH_4NO_3 at the solution changes. At the same time, P was removed from the solution surrounding half the plants given N and half those dependent on N fixation. This gave a factorial combination of 4 initial P levels, 2 final P treatments and 2 N treatments, which was replicated 3 times.

Harvests began on day 71. The concentration of P-fractions in leaves and nodules, the acetylene reducing activity of the roots and the total P content in tissue were estimated as in Hart (1989).

All differences mentioned are significant at the 5% level or less.

Results

In plants reliant on N-fixation throughout the experiment, acetylene reduction activity increased with P supply to the plants, the increase being about 2.7 fold. There appeared to be a close relationship between root activity and total nodule P and inorganic-P (P_i) (Fig. 1). The relationships between activity and nodule lipid- and residual-P were weaker.

The addition of NH_4NO_3 had a profound effect on acetylene reducing activity. The mean activity in the absence of NH_4NO_3 was 369.3 nmol/min/g (root dw), in its presence 10.3. Nodules of plants exposed to NH_4NO_3 were small and especially at the lower levels of P supply had a withered, shrunken appearance. The concentration of all the P fractions in these nodules decreased compared to those in the N-fixing plants (Table 1) (despite this effect, the concentration of P in these nodules still showed a positive relationship with P supply). In contrast, the concentration of P in the youngest mature leaves was unaffected by the addition of NH_4NO_3 (Table 1).

Over the experiment as a whole, the final mean weight of the shoots grown at the highest P level was about 1.7 times that at the lowest; the roots increased to a lesser extent, about 1.2 fold.

Table 1. The mean concentrations of P (μmol/g(dw)) in various fractions in white clover leaves and nodules. The plants were either reliant on nitrogen fixation (−N) or given mineral nitrogen (+N)

	Leaves			Nodules		
	+N	−N	SEM (n = 24)	+N	−N	SEM (n = 24)
P_t^a	99.9	99.1	3.9	98.2	194.6	5.5
P_i	10.7	12.5	0.6	14.4	19.5	0.8
P_l	17.0	18.3	0.7	11.7	40.1	1.2
P_{res}	49.7	47.1	1.5	54.0	81.3	7.9

[a] P_t: total-P, P_i: inorganic-P, P_l: lipid-P, P_{res}: residual-P.

The +N plants provided with P throughout were the largest (Table 2). Root weight was unaffected by the form of N supply, but +N plants had larger shoots than their N-fixing counterparts (Table 2), so there was a difference in root: shoot ratio (+N, 0.28 *vs* − N, 0.34; SEM = 0.01, n = 12). Where the P supply was discontinued, the difference (Table 2) between the shoots of +N and −N plants was insufficient to change the root: shoot ratio. The root: shoot ratio decreased in all plants with increasing P supply, to a greater extent in +N than in −N plants (data not shown).

In +N plants, the ratio of root to shoot P decreased as the P supply increased, regardless of whether or not the supply was discontinued at the time of N application (Table 3). The ratios for plants always fixing N also decreased with

Table 2. Mean root, shoot and total dry weights (g) of white clover plants. −N: reliant on N fixation; +N: given mineral N; P: P provided throughout; P*: P removed in last two weeks

	+N, P	−N, P	+N, P*	−N, P*	SEM (n = 12)
Root	1.22	1.23	1.23	1.22	0.06
Shoot	4.73	3.74	4.26	3.98	0.20
Total	5.95	4.97	5.49	5.20	0.25

Table 3. The mean ratios of root P to shoot P in white clover plants given mineral N (+N) or dependent on N fixation (−N) at various levels of P supply. P: P provided throughout; P*: P withheld during final two weeks. SEM = 0.03; n = 3

P supply[a]	+N, P	+N, P*	−N, P	−N, P*
0.96	0.45	0.50	0.47	0.48
1.86	0.29	0.34	0.36	0.38
3.60	0.26	0.27	0.28	0.39
7.20	0.18	0.26	0.44	0.27

[a] mg P/plant/week.

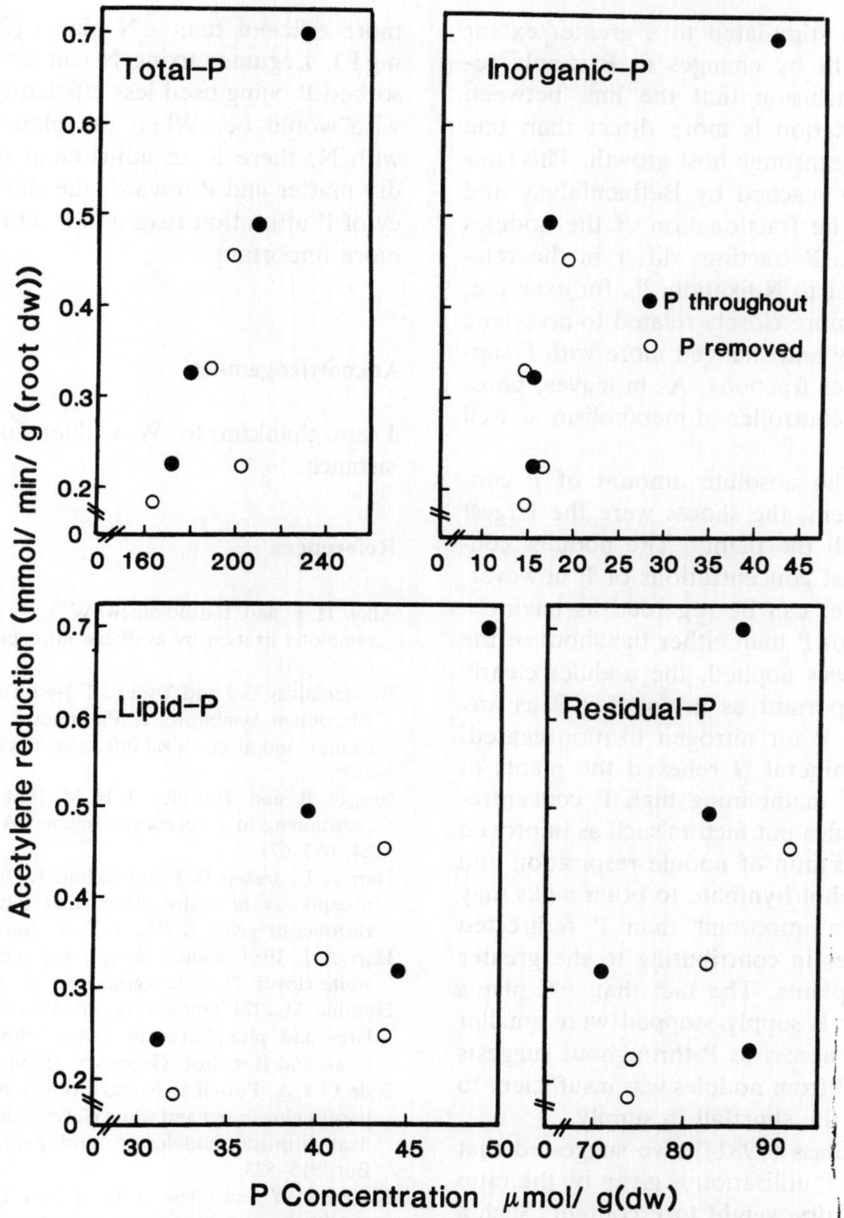

Fig. 1. Plots of the acetylene reducing activity of white clover root systems *vs* the concentration of P fractions in the nodules. Plants dependent on N fixation throughout.

increases in P supply, except in the case of those plants given P throughout at the highest level, where the ratio remained relatively high (Table 3).

The application of N had no significant effect on the concentration of shoot P. Shoot and root P concentration rose with P supply. Concentrations in the nodules were higher than in the roots: mean root %P for all N-fixing plants = 0.26 vs 0.60 for nodules.

Discussion

The relationship between nodule P and activity (Fig. 1) is essentially only correlative but the fact that activity was stimulated to a greater extent than plant growth by changes in P supply reinforces the conclusion that the link between supply and N-fixation is more direct than one mediated entirely through host growth. This conclusion was also reached by Bethlenfalvay and Yoder (1981). The fractionation of the nodules suggests that the P fractions differ in the relationship they bear to N-fixation. P_i, for example, appeared to be more closely related to acetylene reduction activity and changed more with P supply than the other fractions. As in leaves, phosphate may be a controller of metabolism as well as a substrate.

In terms of the absolute amount of P contained within them, the shoots were the largest sinks for P in all the plants. The nodules contained the highest concentrations of P however, and in this sense, can be regarded as having a higher priority for P than either the shoot or the roots. Once N was applied, the nodules clearly became less important as a sink for P as any requirements of P for nitrogen fixation ceased.

Addition of mineral N relieved the plants of the necessity of maintaining high P concentrations in the nodules but factors such as improved N nutrition, cessation of nodule respiration and re-direction of photosynthate to other sinks may have been more important than P redirected from the nodules in contributing to the greater growth of +N plants. The fact that +N plants which had their P supply stopped were smaller than those with access to P throughout suggests that P redirected from nodules was insufficient to compensate for the shortfall in supply.

Siddiqi and Glass (1981) have suggested that the efficiency of P utilisation is given by the ratio of the square of dry weight to P content. Such a calculation for plants given P throughout at the highest level, shows that the +N plants were more efficient than −N plants (23.1 *vs* 17.7 g^2/mg P). Legumes fixing N can be N limited, absorbed P being used less efficiently than it otherwise would be. When the plants are provided with N, there is an adjustment of allocation of dry matter and P towards the shoot, the efficiency of P utilisation rises and P limitation becomes more important.

Acknowledgement

I am thankful to W Collier for technical assistance.

References

Allos H F and Bartholomew W V 1959 Replacement of symbiotic fixation by available nitrogen. Soil Sci. 87, 61–66.

Bethlenfalvay G J and Yoder J F 1981 The Glycine-Glomus-Rhizobium symbiosis. I. Phosphorus effect on nitrogen fixation and mycorrhizal infection. Physiol. Plant. 52, 141–145.

Brugge R and Thornley J H M 1984 Shoot-root-nodule partitioning in a vegetative legume: A model. Ann. Bot. 54, 653–671.

Hart A L, Jessop D J and Galpin J 1981 The response to phosphorus of white clover and lotus inoculated with rhizobia or given KNO_3. N.Z. J. Agric. Res. 24, 27–32.

Hart A L 1989 Nodule phosphorus and nodule activity in white clover. N.Z. J. Agric. Res. 32, 145–149.

Hoshino M 1974 Translocation and accumulation of assimilates and phosphorus in Ladino clover. Bull. National Grassland Res. Inst. (Japan) 5, 35–84.

Ryle G J A, Powell C E and Gordon A J 1981 Assimilate partitioning in red and white clover either dependent on N_2 fixation in root nodules or utilising nitrate nitrogen. Ann. Bot. 515–523.

Siddiqi M Y and Glass A D M 1981 Utilisation index: A modified approach to the estimation and comparison of nutrient utilisation efficiency in plants. J. Plant Nutr. 4, 289–302.

M. L. van Beusichem (Ed.), *Plant nutrition – physiology and applications*, 193–194.

PLSO IPNC244

Properties of ecto Ca-ATPase isolated from barley (*Hordeum vulgare*) roots

N. I. TIKHAYA, T. N. STEKHANOVA, M. D. FEDOROVSKAYA and D. B. VAKHMISTROV
Timiriazev Institute of Plant Physiology, Academy of Sciences, Botanicheskaya 35, SU-127273, Moscow, USSR

Key words: ATPase, barley roots, *Hordeum vulgare* L., microsomal fraction

Abstract

Ca-dependent ATPase was found to be extracted when the microsomal fraction treated with 0.5 *M* KCl. Properties of membrane-bound and extracted enzymes were found to be similar. The possible participation of Ca-ATPase in contractile systems as myosin ATPase is suggested.

Introduction

In the microsomal fraction enriched with plasma membrane fragments and on the surface of barley root cells (Tikhaya *et al.*, 1985) ATPase was found with main properties similar to those of the plasmamembrane-bound ecto Ca-ATPase of animal cells (Tuana and Dhalla, 1982). The microsomal Ca-ATPase from barley root cells did not require Mg-ions for its activity as is the case with Ca-stimulated Mg-dependent ATPase from maize (Deiter and Marme, 1981).

The present study was aimed at extracting ATPase from barley roots which is activated by millimolar concentrations of divalent cations, but exhibits a relatively high affinity to Ca-ions.

Methods

Extraction of ATPase

The microsomal fraction isolated from 5-day old barley roots (*Hordeum vulgare* L.) and purified on the sucrose gradient was treated with 0.5 *M* KCl in 1 m*M* Tris-MES pH 7.2 for 1 hr in a cold room. After centrifugation at 150000 g for 1 hr the pellet was discarded and the supernatant was used as enzyme preparation.

ATPase assay

Ca-ATPase (Mg-ATPase) activity was assayed in a total volume of 1 m*M* containing 1–2 μg protein, 33 m*M* Tris-MES pH 6, 1 m*M* EDTA, 100 m*M* KCl, 3 m*M* $CaCl_2$ (3 m*M* $MgCl_2$), 1 m*M* Tris-ATP. The reaction was started by the addition of ATP, measured at 20°C for 15 min and stopped by 30% cold trichloroacetic acid treatment. The ATP hydrolysis that occurred in the absence of Ca-ions (Mg-ions) was subtracted for a calculation of the activity due to Ca-ATPase (Mg-ATPase).

Results and discussion

Ca-ATPase was found to be extractable when the microsomal fraction was treated with 0.5 *M* KCl. It was observed that Mg-ATPase was also released under this condition. For both Ca-ATPase (2.88 μmoles $P_i \cdot mg^{-1}$ protein $\cdot$ min^{-1}) and Mg-ATPase (2.14 μmoles $P_i \cdot mg^{-1}$ protein $\cdot$ min^{-1}) the values of enzyme recovery were similar (60% and 56%, respectively). These enzyme activities in the extracted preparations were 5 times higher than those present in the microsomal fraction. The question arose whether the hydrolysis of ATP in barley root cells in the pres-

Table 1. Effects of inhibitors on the Ca- and Mg-ATPase activities of microsomal and extracted preparations.

Inhibitor	ATPase activity, % of control			
	Microsomal preparation		Extracted preparation	
	Ca	Mg	Ca	Mg
Blank	100	100	100	100
DES, $5 \cdot 10^{-5}$ *M*	85	81	84	88
DCCD, $5 \cdot 10^{-5}$ *M*	92	86	99	96
Vanadate, $5 \cdot 10^{-5}$ *M*	99	95	94	95
Sodium fluoride, $5 \cdot 10^{-3}$ *M*	71	77	77	75
p-CMB, $1 \cdot 10^{-4}$ *M*	103	95	92	95
Ammonium molybdate, $1 \cdot 10^{-4}$ *M*	91	97	90	83

ence of Ca- or Mg-ions is due to a single enzyme or to two different enzymes. It is thereby interesting to compare not only properties of microsomal and extracted Ca-ATPase, but also those of Ca- and Mg-ATPases of both preparations. It was shown that these enzyme activities were observed in a wide range of temperatures (10–70°C with optimum at 40°C) and pH values (4–9 with optimum at pH 6). K_m values calculated according to Lineweaver-Burk for Ca_{free} (7 μM) and Mg_{free} (17 μM) were the same for microsomal and extracted ATPases. Ca- or Mg-ATPase activity was not stimulated by the monovalent cations Na and K.

Microsomal and extracted Ca-ATPase (Mg-ATPase) hydrolysed nucleotidetriphosphates and ADP, whereas AMP, β-glycerophosphate or pyrophosphate were not utilized by these enzymes.

Ca- and Mg-ATPase of microsomal and extracted preparations were insensitive to inhibitors such as p-CMB, DCCD, vanadate and ammonium molybdate while DES and sodium fluoride decreased ATPase activity (Table 1).

Thus Ca-ATPase was found to be extractable from barley roots and a similarity between the properties of Ca-ATPases of microsomal and enzyme preparations was found. Since Mg-ATPases was also released under these conditions and the extracted enzyme was identical in its properties to membrane-bound Mg-ATPase and also to membrane-bound and extracted Ca-ATPase, it appears that ATPase activity of barley roots indused by Ca- or Mg-ions may be due to a single enzyme. The function of the divalent cation-dependent ATPase is not known, but its extraction in concentrated salt solutions suggests its possible participation in contractile systems as myosin ATPase.

References

Deiter P and Marme D 1981 A calmodulin-dependent microsomal ATPase from corn (*Zea mays* L.). FEBS Letters 125, 245–248.

Tikhaya N I and Vakhmistrov D B 1989 Ecto-Ca-ATPase of plasma membrane fraction from barley roots. *In* Structural and Functional Aspects of Transport in Roots. Eds. B C Loughman, O Gasparikova and J Kolek. pp 85–87. Kluwer Academic Publishers, Dordrecht, The Netherlands.

Tuana B S and Dhalla N S 1982 Purification and characterization of a Ca-dependent ATPase from rat heart sarcolemma. J. Biol. Chem. 257, 14440–14445.

B

Micronutrients and toxic elements in plant and crop nutrition

M. L. van Beusichem (Ed.), *Plant nutrition – physiology and applications*, 197–206.

PLSO IPNC672

Responses of chickpea (*Cicer arietinum*) to iron stress measured using a computer-controlled continuous-flow hydroponic system

G. A. ALLOUSH and F. E. SANDERS
Department of Pure and Applied Biology, University of Leeds, Leeds LS2 9JT, UK

Key words: cation-anion balance, *Cicer arietinum* L., chickpea, iron stress

Abstract

Non-nodulated chickpea plants of a Fe-efficient variety were grown in a computer-controlled hydroponic system capable of measuring uptake rates of NO_3^-, K^+, $H_2PO_4^-$, Ca^{++}, Mg^{++} and Na^+ from a nutrient solution and replenishing those ions taken up. Plants were subjected to two regimes of Fe supply (A: 0–20.5 days, 5.6 ppm Fe; 20.5–35.5 days, 0 ppm Fe. B: the reverse of A). Plants were harvested at intervals over the experimental period and their growth and nutrient contents were measured.

Iron stress significantly reduced plant growth and uptake of nutrients, having a greater effect on anions than cations, and gave rise to efflux of H^+. Unstressed plants took up more anions than cations and there was a net efflux of OH^-. Effluxes of H^+ and OH^- differed from those predicted by the balance of cation and anion uptake, this difference was probably due to exudation of organic acids. Stressed plants showed oscillatory behaviour, with alternating high and low rates of K^+ and NO_3^- uptake.

The response to the relief of Fe stress was immediate, indicating that changes in nutrient uptake in response to iron supply may be regulated in the roots.

Introduction

A number of different strategies are employed by plants to overcome iron stress (Marschner, 1986). In the case of dicots, reactions to iron stress usually include acidification of the rhizosphere and excretion of organic acids which lead to enhanced mobilisation and transport of Fe^{3+} from the soil to sites of reduction and absorption in the roots.

The effects of iron stress on the uptake of other nutrients are less clear, but include decreased uptake of anions and decreased assimilation of nitrogen (Van Egmond and Aktas, 1977; Venkat Raju *et al.*, 1972).

The purpose of this investigation was to define more clearly the changes in cation and anion uptake induced by iron stress in chickpea and to relate these to pH changes in the external medium. Of particular interest were the dynamics of responses to iron stress. The computer-controlled hydroponic system was therefore used to follow changes in rates of nutrient uptake resulting from changes in the supply of iron.

Materials and methods

Plant cultivation

Seeds of chickpea (cv. ILC-195) were surface sterilised and pregerminated. Young seedlings were transferred to troughs along which nutrient solution was circulated from reservoir tanks. After six days of preculture, during which no iron was supplied, one of two treatment regimes (A or B) was imposed. In A, plants were supplied with iron at 5.6 mg L^{-1} until 20.5 days when the supply of iron was discontinued, thus creating iron-stress. In B, severe iron stress was allowed to develop by continuing to withold iron until 20.5 days when iron was introduced to the solution at the same concentration as in A.

Treatment	Phase I (0–20.5 d)	Phase II (20.5–35.5 d)
A	5.6 mg Fe L^{-1}	0 mg Fe L^{-1}
B	0 mg Fe L^{-1}	5.6 mg Fe L^{-1}

Plants were grown in a controlled environment chamber (day 25°C, night 20°C; photoperiod 16h at 620 μEinsteins m^{-2} s^{-1}).

Every 12 hours, concentrations of NO_3, H_2PO_4, K, Ca and Na were measured automatically under computer control. The methods of analysis were UV absorption (200 nm) for nitrate (Cawse, 1967), the ammonium molybdate/malachite green method for phosphate (Itaya and Ui, 1966) and flame photometry for the cations. Iron concentrations were measured manually by atomic absorption.

On the basis of the concentration measurements, further stock nutrient solutions were added under computer control to restore ion concentrations to set levels (40 mg L^{-1} NO_3-N, 40 mg L^{-1} K, 2 mg L^{-1}P). When necessary, Fe was added manually as FeEDTA.

Trace elements were supplied in the water added under computer control to compensate for evaporation and transpiration at the start of each cycle of analysis. Concentrations of trace elements were as in the Long Ashton formula (Hewitt, 1966).

At the end of each cycle of analysis pH was adjusted to 6.0 by addition of 0.02 *M* H_2SO_4 or a saturated solution of $Ca(OH)_2$. Acid was added automatically under computer control. Addition of $Ca(OH)_2$ was by hand.

Procedure at harvest

Plants (3 or more) were harvested from each treatment at selected times. Fresh and dry weights of roots and shoots were recorded. Shoot and root samples from each harvest were bulked according to treatment prior to grinding for subsequent analysis.

Plant analysis

Total N was measured using a CHN analyser (Carlo Erba 1106). Total K, P, Ca, Mg, Na, S and Fe were determined using an ICPS (ARL 3580) after dry-ashing of subsamples, digestion in 6M HCl and final preparation in 2% HCl. Nitrate and chloride in the dry plant tissue were determined by ion chromatography (Dionex 4000i) of hot water extracts of subsamples.

Calculations

Uptake of each nutrient element over each 12h period was calculated from the difference between the final concentration after nutrient addition and the concentration measured after the subsequent period of depletion × the volume of solution in each trough system and reservoir tank. This was approximately 13 litres but could not be controlled more accurately than +/−1 litre. Both this and analytical error contributed to error in calculated uptakes.

In cases where measurements made under computer control were thought to be inaccurate, values obtained by manual analysis of samples of the nutrient solution were substituted.

Specific absorption rates (SAR, meq/g root dry weight/day) were calculated for each major nutrient ion. Log-transformed data for root dry weights and nutrient contents at each harvest time were plotted against time and third degree polynomials were fitted by the method of least squares. The equations of the fitted lines were used to calculate instantaneous values of SAR as follows:

$$\log_e y = a + bt + ct^2 + dt^3$$
$$dy/dT = (b + 2ct + 3dt^2).y$$

where y is nutrient content (X, mg) or root dry weight (W, g), a, b, c and d are the polynomial coefficients and t is time in days.
Then:

$$SAR_t = (dX/dT)_t/W_t \, .$$

Values of SAR were also estimated using rates of

nutrient uptake calculated from data on depletion of the nutrient solution:

$$SAR_t = 2.\ (\text{Uptake of nutrient over 12h period}/W_t)$$

Results

Plant growth

Plant dry weights plotted against time for the two treatments are shown in Figure 1a. Plants given iron (treatment A) were indistinguishable from stressed plants (treatment B) up to 13.5 days (harvest 3). Thereafter, plants given treatment A were always larger. There was no rapid response in either treatment to changing the iron supply at 20.5 days.

Symptoms of iron deficiency (general yellowing of leaves and ultimately necrosis of the shoot tips) became apparent just before the second harvest (10.5 days) in B plants. Symptoms in the roots were a general stunting, production of short, thick laterals with swollen tips and a brown colour. Recovery from stress in B plants was in the form of new growth of normal lateral shoots and roots which started approximately 8 days after stress was relieved. A plants showed only yellowing of leaves after 28 days, and there were no severe effects on root growth.

In phase I, dry matter content was significantly greater in iron deficient plants (B) but declined sharply during phase II (Fig. 1b). In A plants, there was no sharp change on withdrawal of iron.

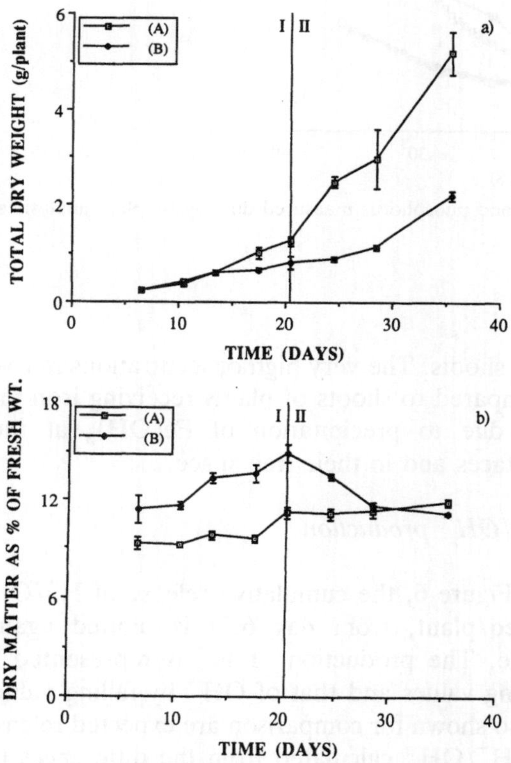

Fig. 1a, b. Total plant dry weight and percentage dry matter. Bars are standard errors.

Plant nutrient contents and concentrations

The uptakes of N, P, and K measured by the computer were in good agreement with those measured by analysing harvested plants (Fig. 2). Uptakes of Mg, Ca and S are shown in Figure 3.

Uptakes of N and S by B plants were considerably depressed by iron stress and did not recover quickly when this was relieved. There was no obvious effect of removal of iron supply in A plants. In phase I, concentrations of N in the dry matter, while tending to decline with time, were higher in A than B plants. In B plants, supplying iron at 20.5 days had no immediate effect, although the concentration of N eventually increased before the last harvest (Fig. 4a). In A plants, the concentration of N declined throughout the period of measurement. In the case of S (Fig. 4c), concentration rose during phase I in A plants while that in B plants remained steady. Reverse trends became apparent early in phase II. Concentrations of P (Fig. 4b) did not differ greatly between the two sets of plants and generally declined with time.

There was no large difference between A and B plants in their contents of Ca and Mg during phase I (Fig. 3a, b). After this time, A plants took up increasing quantities compared to B plants. During phase I, concentrations in the dry matter tended to be higher in plants given treatment B and then to fall during phase II. Conversely, concentrations in plants given treatment A seemed eventually to rise in response to removal of the iron supply at 20.5 days (Fig. 4e, f).

K content (Fig. 2b) showed a similar overall pattern of increase to that of Ca. Concentrations of K rose slightly over the period of measure-

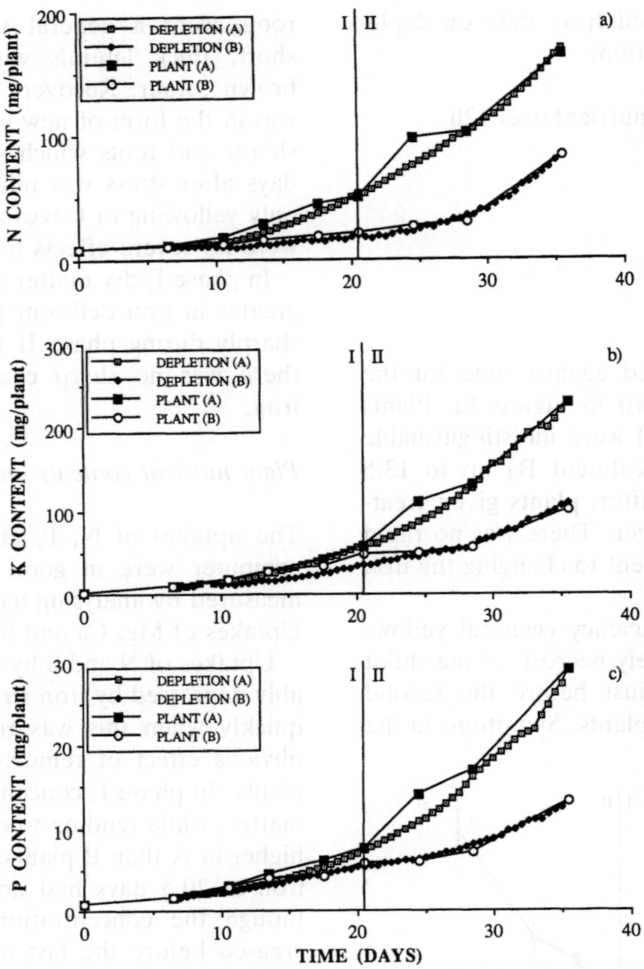

Fig. 2a–c. **Comparison between plant uptakes of nitrogen, potassium and phosphorus measured directly by plant analysis and calculated from measurements of depletion of the nutrient solution.**

ment but here there was no major difference between the two treatments (Fig. 4d). Uptakes of Na were small.

Plant Fe contents and concentrations in shoots and roots are shown in Fig. 5. During phase I, Fe content of plants in treatment A increased rapidly. Slight uptake by plants in treatment B was probably due to trace contamination of the deionised water supply. When the iron supply was discontinued, uptake continued until day 24.5, presumably due to mobilisation of Fe precipitated in the troughs, pipework and tanks. In treatment B, uptake of Fe was rapid during phase II, concentration in the roots rose rapidly and there was a slower rise of concentration in the shoots. The very high concentrations in roots compared to shoots of plants receiving iron may be due to precipitation of $Fe(OH)_3$ at their surfaces and in their free space.

H^+/OH^- *production*

In Figure 6, the cumulative release of H^+/OH^- (meq/plant, from day 6.5) is plotted against time. The production of H^+ is represented by rising values and that of OH^- by falling values. Also shown for comparison are expected releases of H^+/OH^- calculated from the differences between cation and anion uptakes obtained from plant analysis (see discussion).

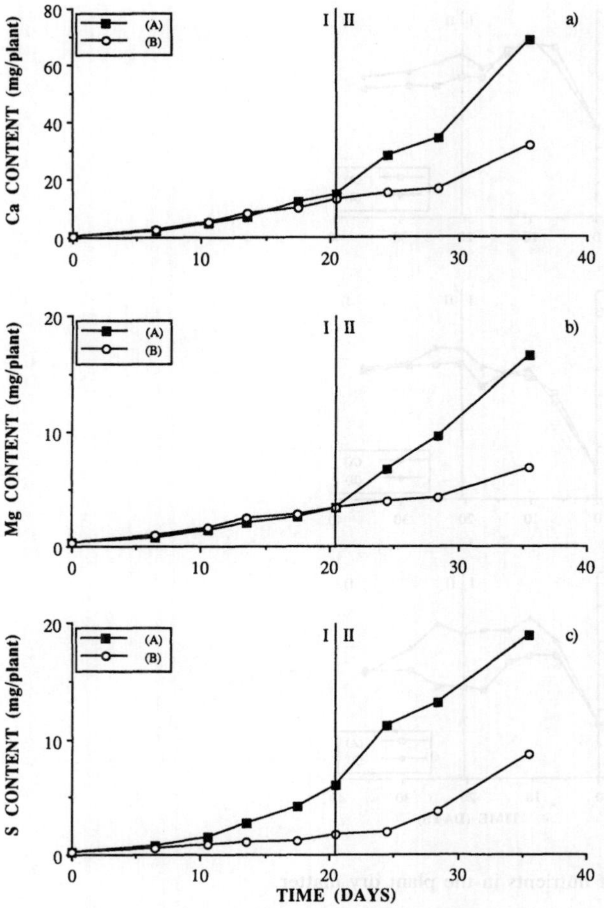

Fig. 3a–c. Uptakes of calcium, magnesium and sulphur.

Discussion

During preculture (day −6 to 0) the seedlings had developed few roots and probably took up little nutrient. After day 0 until the first harvest at day 6.5, concentration of all nutrients except N increased rapidly. The largest increase was observed in the case of iron in the A plants. In the B plants, uptake of Fe was negligible and concentrations in roots and shoots remained low until phase II. It is probable that both sets of plants had already experienced Fe stress during the preculture period because both acidified the solution in the period before 6.5 days.

This Fe stress was quickly relieved in the A plants after day 0 while B plants developed progressively more severe symptoms of stress during phase I. During phase II, A plants developed slight visible symptoms of iron stress only towards the end, presumably after exhaustion of the Fe remaining in the trough system and internal dilution had allowed tissue concentrations to drop below some critical level. During phase II, B plants started to take up iron rapidly after its resupply but recovery of growth was delayed for a further 8 days.

During phase I, B plants released H^+ at a steady net rate, reflected in the rising accumulation line (Fig. 6). Immediately Fe was resupplied, the net rate of release decreased, reaching zero two days later and thereafter there was a net release of OH^-, reflected in the falling accumulation line. The rapidity of this change is revealed by plotting the pH reached after each 12h uptake period against time (Fig. 7). In the case of the A plants which released OH^- at an increasing net rate during phase I, there was also an effect of changed Fe supply but this was slower to develop. Thus during phase II, there was a decreasing net rate of OH^- release which became zero at around 27 days. Rather than switching to a net release of H^+, a subsequent oscillation between release of H^+ and OH^- with a period of about 2 days was seen (Fig. 7), so that on average there was no long-term net release of either species and the accumulation curve became horizontal.

These differences between A and B plants were linked to differing cation-anion balance of the nutrients taken up. In Table 1, total uptakes after harvest 1 of cations and anions, and the differences between them (C–A), all expressed in milliequivalents (meq) are calculated. Also given for comparison are the accumulated total releases of H^+/OH^- after harvest 1.

The values of C–A and accumulated H^+/OH^- are in fair agreement, indicating that much of the H^+/OH^- released was accounted for by differences in the uptake of anions and cations between Fe-stressed and unstressed plants. However the agreement is not exact. On average, the measured net release of H^+ from stressed B plants is greater than can be accounted for by C–A. Conversely, in unstressed A plants, the

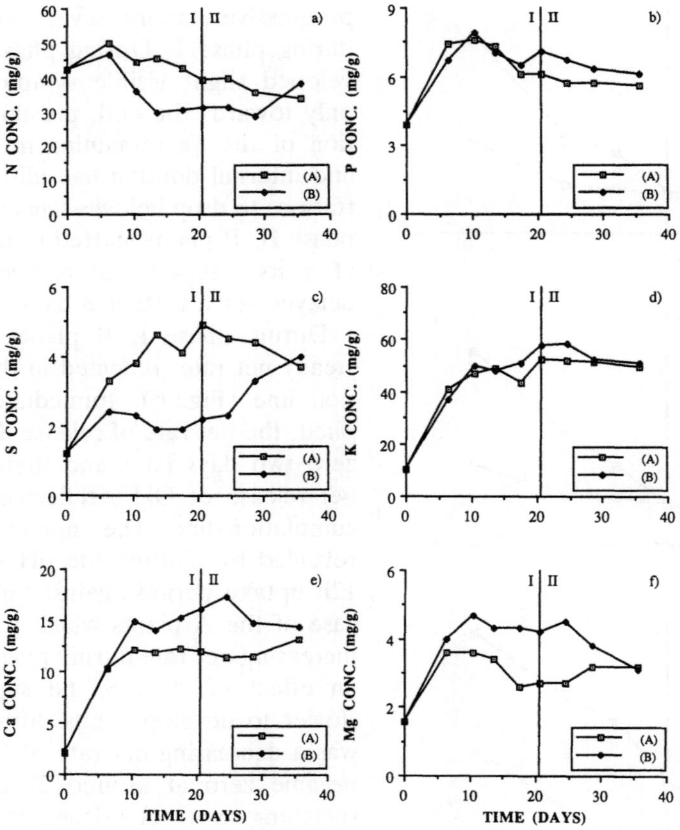

Fig. 4a–f. Average concentrations of nutrients in the plant dry matter.

Table 1. Calculated values (meq) of uptake of total cations, anions, cation-anion imbalance, H^+/OH^- release and organic acid accumulation from harvest 1, and calculated concentrations of organic acid in the dry matter. Treatments A and B

Time	Cations[a]		Anions[b]		C–A		Measured H^+/OH^-		Org. Acid[c]		Org. Acid Conc.[d] meq/g	
	A	B	A	B	A	B	A	B	A	B	A	B
6.5	0	0	0	0	0	0	0	0	0	0	1.1	1.2
10.5	0.42	0.43	0.56	0.19	−0.14	0.24	−0.09	0.24	0.36	0.43	1.5	2.0
13.5	0.84	0.90	1.5	0.50	−0.66	0.40	−0.34	0.44	0.73	0.90	1.6	2.0
17.5	1.5	1.2	2.8	0.75	−1.3	0.45	−0.86	0.78	1.3	1.1	1.5	2.1
20.5	2.1	2.0	3.3	0.98	−1.2	1.0	−1.4	1.1	1.7	2.0	1.6	2.8
24.5	4.4	2.0	7.3	1.3	−2.9	0.70	−2.1	1.1	3.8	1.9	1.6	2.5
28.5	5.5	2.2	7.8	1.7	−2.3	0.50	−2.3	0.59	5.0	2.0	1.8	2.1
35.5	10.3	4.3	13.2	6.0	−2.9	1.7	−2.2	1.1	9.3	3.8	2.0	1.9

[a] $\Sigma C = K + Ca + Mg + Na + Fe$.
[b] $\Sigma A = N + P + S + Cl$.
[c] Org. A. $= \Sigma C\text{-}Cl\text{-}NO_3$.
[d] Org. A. Conc. = Org. A. Content/plant dry wt.

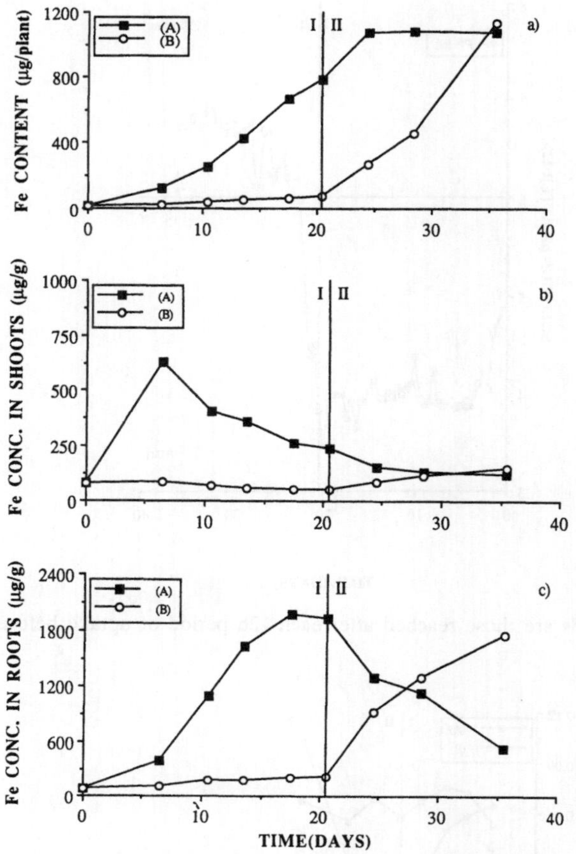

Fig. 5a–c. Uptakes of iron and its concentrations in dry shoot and root material.

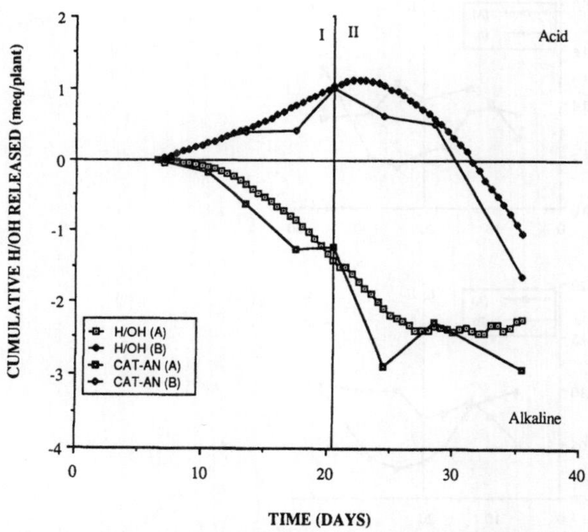

Fig. 6. Cumulative release of H^+/OH^- (measured from harvest 1) compared to that predicted from the calculated anion-cation imbalance.

measured net release of OH^- is less than predicted by C–A.

The most probable explanation for this discrepancy is that in both A and B plants, there was a release of organic acid anions. In B plants this release was balanced by H^+ so that the titratable acidity of the solution was increased. In A plants, the excess of anion over cation uptake was partially balanced by release of organic acid anions. From the slopes of lines fitted to plots of C–A against H^+/OH^- release for the A and B plants, it can be estimated that about 30% of the measured H^+ release was balanced by the excretion of $orgA^-$ in stressed B plants, while about 20% of the difference C–A in unstressed A plants was balanced by release of $orgA^-$ and 80% by production of OH^-.

Organic acid contents in the tissues were estimated and are presented as accumulated totals from harvest 1 (Table 1). Also shown are calculated tissue concentrations (meq $orgA^-$/g dry weight). Organic acid contents calculated in this way are likely to be overestimates since no account has been taken of other soluble inorganic anions (*e.g.* SO_4^{2-}) that may have been present in the vacuoles. In phase I, after an initial increase, organic acid concentration seemed to remain fairly constant in the A plants but increased with developing Fe stress during phase II. In the B plants, organic acid concentrations were greater during phase I, decreasing during phase II after stress was relieved. On the basis of this and other experiments in which organic exudates were directly measured, we may conclude that chickpea always releases organic acids from its roots but at a greater rate per unit weight of root during iron stress.

The differences in C–A between stressed and unstressed plants can be further analysed by calculation of specific absorption rates of the major nutrient cations and anions. The results of these calculations are shown in Figures 8, 9. SARs are averages for the whole root system, and wide local variation might be expected. However, experiments with plants grown under similar nutrient regimes using agar and indicators have shown both pH changes and the reduction of Fe^{3+} to occur more or less uniformly over the whole root system.

Third degree polynomials fitted the data well.

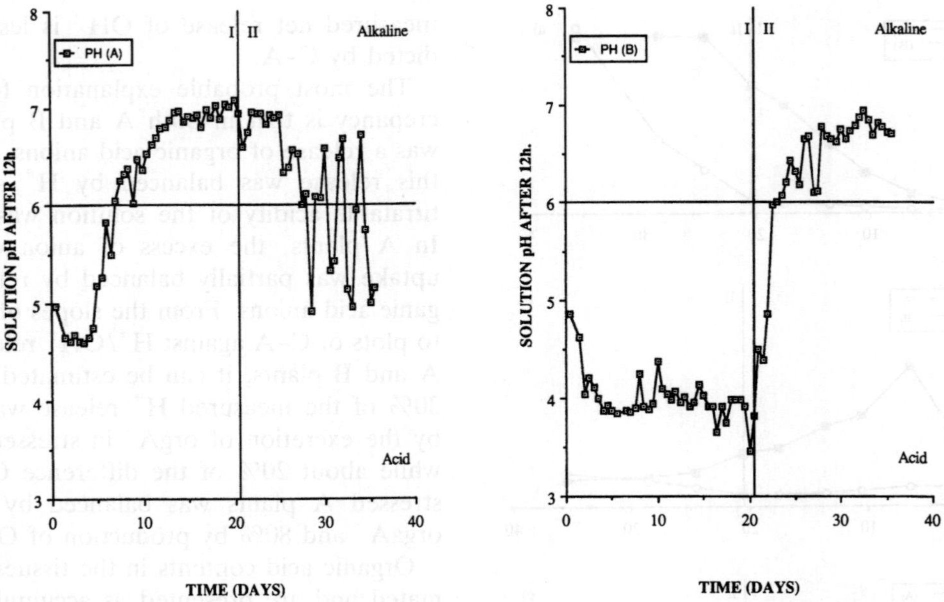

Fig. 7. **Changes in solution pH with time for A and B plants. pHs are those reached after each 12h period of uptake before adjustment back to pH 6.**

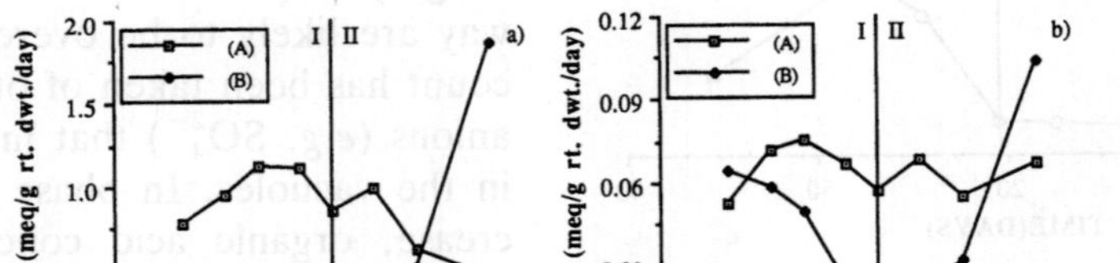

Fig. 8a–f. **Specific absorption rates of nitrogen, phosphorus, sulphur, potassium, calcium and magnesium calculated from plant nutrient contents.**

Fits were least good in the cases of Fe and Na content. As a consequence of the calculation procedure, SAR values may be expected to be least reliable at the beginning of Phase I and the end of Phase II.

For the A plants, SARs were fairly steady for K, Ca, Mg and P and rose and then declined for N and S. For B plants, SARs tended to drop below those of the A plants and then to rise above them before the last harvest. The change was proportionally largest in the case of Mg.

The difference C–A expressed as a specific rate was also plotted to enable comparison with directly measured rates of H^+/OH^- production (Fig. 9). SARs calculated in this way agreed closely with those calculated from depletion of the nutrient solution (Figs. 10, 11).

While much detail remains to be explained, it is clear that most of the variation in H^+/OH^- release resulting from the two treatments can be accounted for by greater variation in the rate of nitrate uptake compared to the cations.

The 12h estimates of uptake rate show smoother changes in A than in B plants. While individual points may not be highly accurate, and evidence of ion extrusion (negative SAR values) not therefore firm, there is a clear association of Fe stress with oscillatory behaviour. In B plants, oscillation in uptake of K^+ and NO_3^- is seen to be closely in phase; this may account for the greater smoothness of the curve for SAR of H^+/OH^-. Similar oscillation may be occurring in A plants at the beginning of phase I when they were probably recovering from Fe stress during

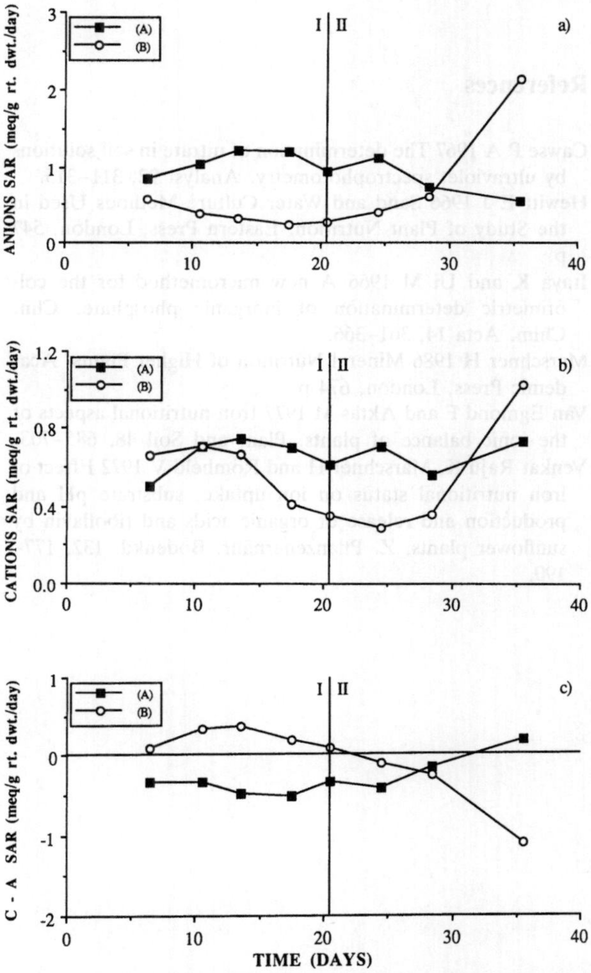

Fig. 9a–c. Specific absorption rates of total anions, total cations and their difference.

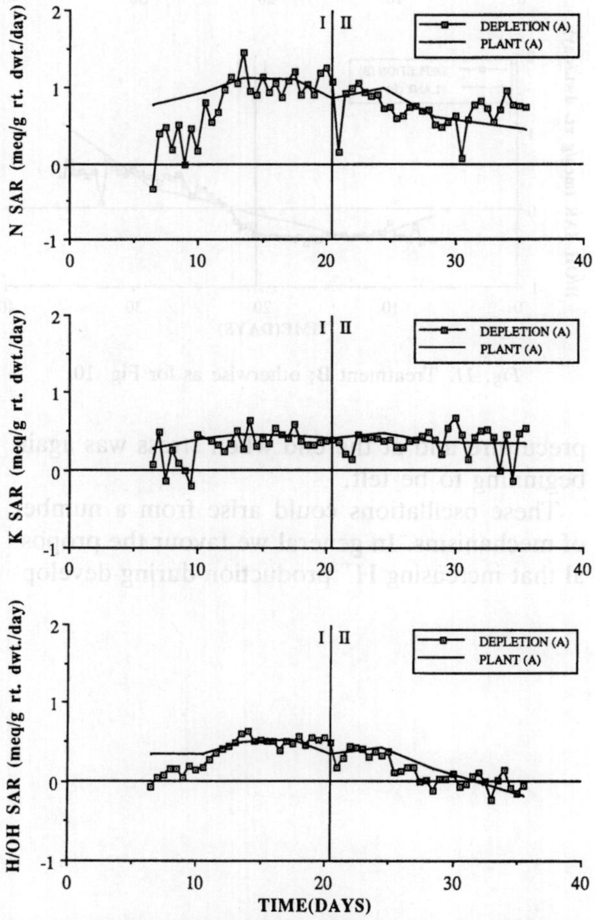

Fig. 10. Specific absorption rates of nitrogen, potassium and phosphorus for treatment A, calculated from depletion of the nutrient solution compared to the values calculated from plant nutrient contents.

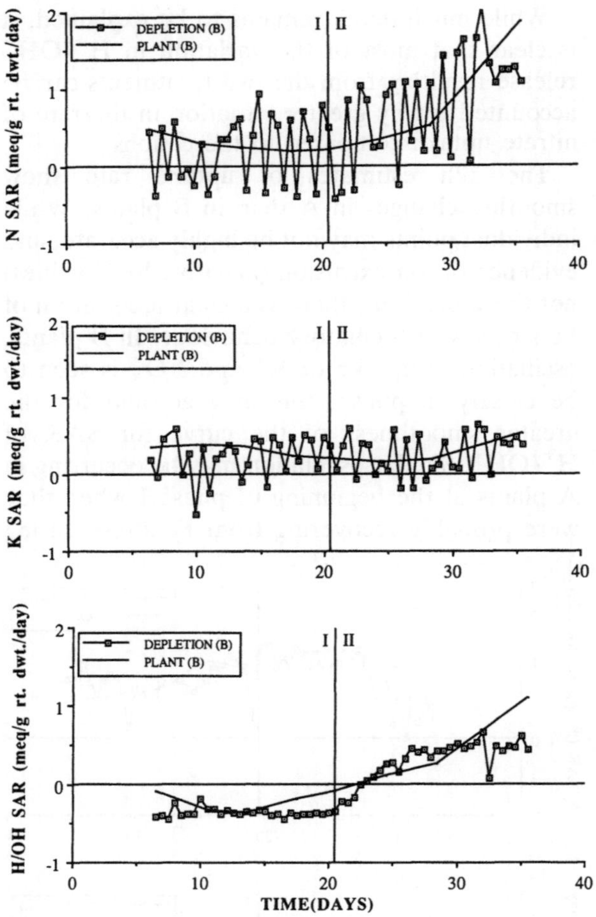

Fig. 11. Treatment B; otherwise as for Fig. 10.

preculture and at the end when stress was again beginning to be felt.

These oscillations could arise from a number of mechanisms. In general we favour the proposal that increasing H^+ production during developing Fe stress allows mobilisation and uptake of sufficient iron to provide temporary relief. This allows a relative increase in anion uptake and hence of solution pH which again causes Fe stress to be felt, followed by the stress reaction. Whether these responses are controlled locally in the roots or at a higher level remains an open question, but their rapidity tends to imply that long distance transport of signals from the shoots may not be immediately important.

Acknowledgements

We wish to thank F Buckley, A Kelly, A Lyon-Joyce and Dr J Rooke for their help in carrying out the plant analyses and J Le Bot and Dr D Pilbeam for helpful discussion and criticism.

References

Cawse P A 1967 The determination of nitrate in soil solutions by ultraviolet spectrophotometry. Analyst 92, 311–315.

Hewitt E J 1966 Sand and Water Culture Methods Used in the Study of Plant Nutrition. Eastern Press, London, 547 p.

Itaya K and Ui M 1966 A new micromethod for the colorimetric determination of inorganic phosphate. Clin. Chim. Acta 14, 361–366.

Marschner H 1986 Mineral Nutrition of Higher Plants. Academic Press, London, 674 p.

Van Egmond F and Aktas M 1977 Iron-nutritional aspects of the ionic balance of plants. Plant and Soil 48, 685–703.

Venkat Raju K, Marschner H and Römheld V 1972 Effect of iron nutritional status on ion uptake, substrate pH and production and release of organic acids and riboflavin by sunflower plants. Z. Pflanzenernähr. Bodenkd. 132, 177–190.

M. L. van Beusichem (Ed.), *Plant nutrition – physiology and applications*, 207–211.

PLSO IPNC737

Responses of pepper (*Capsium annum*) plants to iron deficiency: Solution pH and riboflavin

G.W. WELKIE, H. HEKMAT-SHOAR and G.W. MILLER
Department of Biology and Agricultural Experiment Station, Utah State University, Logan, UT 84322-5305, USA

Key words: *Capsium annum* L., chlorophyll, iron deficiency, iron stress, solution pH, pepper, riboflavin

Abstract

Pepper (*Capsium annum* L.) plants were grown in all nitrate N nutrient solutions with no Fe and four concentrations of Fe as Fe EDDHA. Riboflavin accumulation was maximal at the lowest concentration of added Fe, about half as much with no Fe and much lower at decreasing amounts for the remaining higher concentrations of Fe. The pH of nutrient solutions decreased the most when plants received no Fe, but the pH did not reach 4.0. All other Fe treatments resulted in pH increases of the nutrient solutions. Although riboflavin excretion was progressively less at increasing added Fe, the solution pH, mass of shoots and of roots, and apical leaf chlorophyll were about the same at the three highest Fe concentrations. Excess riboflavin may substitute for ferredoxin in the form of flavodoxin with Fe stress.

Introduction

Iron deficiency of dicotyledonous species is known to evoke a number of root morphological and physiological responses, some of which have been interpreted as mechanisms for enhancing the ability of the plant to acquire more Fe and recover from chlorosis (Bienfait, 1988; Brown and Jolley, 1988). Among the more frequently reported responses are plant induced acidification of a nitrate N nutrient medium, increased reducing capacity of the nutrient medium and roots, an increased root content of certain organic acids, phenolics and riboflavin, and the excretion of riboflavin and other products into the nutrient medium. Of some interest has been the time sequence of events relative to each other. For example, in the presence of no Fe the maximum decrease in solution pH and reductive activity and root reductive capacity occurred in 7 days and in 9 or 10 days respectively for Fe efficient tomato and soybean (Camp *et al.*, 1986). At 5 days the maximum pH decrease of Fe-stressed pepper plants coincided with maximum reduction of roots and citric and malic acid content of roots (Landsberg, 1986). Riboflavin reached a maximum concentration in Fe-stressed nutrient cultures of sunflower when the pH reached a minimum value of 18 days (Venkat Raju *et al.*, 1972).

We are interested in the temporal aspects of the root medium acidification and the excretion of riboflavin with respect to the question of whether the former is a prerequisite for the latter or whether they are merely coincident and independent events associated with Fe stress. For sunflower, evidence was presented that the pH decrease preceded the release of riboflavin but that the root content of riboflavin increased with iron stress and was released only when the solution was sufficiently acidified (Venkat Raju *et al.*, 1972). However, significant quantities of riboflavin were detected in Fe control solutions in which the pH increased. Our tobacco results suggested that the pH decrease was not a prerequisite for riboflavin excretion but was largely coincident with it (Welkie and Miller, 1988). Our results with sugar beet also reflect considerable riboflavin excretion from Fe stressed plants with only moderate pH decreases of the nutrient solution (to be published in J. Plant Nutr., 1989). During a screening of several species for riboflavin excretion in response to Fe stress, pepper plants appeared to excrete considerable amounts with little pH change. This species was studied in more detail at several Fe concentrations.

Materials and methods

Pepper (*Capsicum annum* L. cv. Italian) seeds were germinated in vermiculite containing half-strength Hoagland's No. 1 nutrient solution with Fe ethylenediamine di (o-hydroxy phenylacetate). After 18 days, seedlings were transferred to 8-L containers of full strength nutrient solution with 0.5 mg Fe L^{-1}. The plants were grown in a growth chamber with fluorescent and incandescent lamps on a 16 h photoperiod with 645 $\mu E\ m^{-2}$ at plant height. After 18 days, 30 plants with 4 leaves were selected and transferred to individual 2-L containers (black plastic-covered polyethylene) with full strength nutrient solution. Six replicate treatments of Fe^{3+} at 0.000, 0.031, 0.125, 0.500 and 2.000 mg Fe L^{-1} with an initial nutrient solution pH of 5.0 were used. The plants were grown in a greenhouse with supplemental fluorescent light on a 16 h photoperiod. The nutrient solutions were brought up to volume each day prior to the samples being taken. After pH and fluorescence measurements were taken, the samples were returned to their respective containers.

Fluorescence was measured with an Aminco Bowman Spectrophotofluormeter at optimum riboflavin activation and emission wave lengths of 460 and 520 nm respectively. The pH was measured with a combination electrode.

Total chlorophyl concentrations were determined on samples from the youngest leaf or leaves from which 4 discs, 9 mm in diameter, could be removed. Plants were sampled at 3 stages of growth and chlorophyll was determined by the method of Moran (1982).

Iron reduction of intact roots were determined at harvest 32 days after initiation of Fe treatments, using the method of Römheld and Marschner (1983).

Results

No significant pH changes of nutrient solutions by pepper plants growing at various iron treatments were noted until 10 days after treatment initiation (Fig. 1). The average pH of the 0.00 mg L^{-1} Fe treatment had increased from the initial pH of 5.0 to a high of 5.40 on the 7th day and decreased to 5.32 on the 10th day. On this latter date the pH of the nutrient solutions containing 0.031, 0.125, 0.50 and 2.0 mg Fe L^{-1} had respective pH values of 5.65, 5.66, 5.73 and 5.72; these were only significantly different from the value of solutions with 0.00 mg Fe L^{-1}. It was not until the 17th day that the pH of the solution with 0.031 Fe L^{-1} became significantly different from the solutions with higher Fe concentrations. It reached an average peak pH of 5.83 on day 13, decreased to 5.76 on day 17, and decreased to a low of 5.53 on day 18 before slowly increasing again. The final pH of solutions with no Fe and increasing Fe values after 32 days were 4.50,

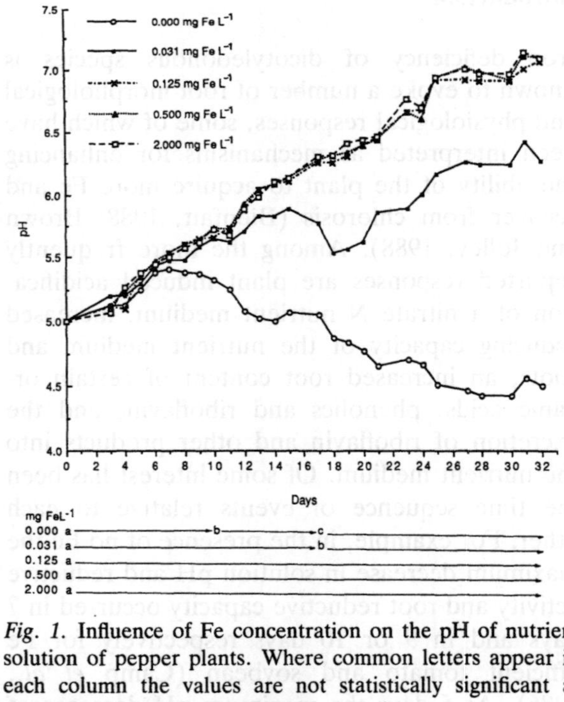

Fig. 1. Influence of Fe concentration on the pH of nutrient solution of pepper plants. Where common letters appear in each column the values are not statistically significant at $P = 0.05$ as determined by LSD.

6.26, 7.06, 7.08 and 7.08 with no further change in significance between treatments.

In contrast to the slow pH changes the same Fe treatments resulted in some riboflavin fluorescence values that were significantly different by the 3rd day of growth, although the values were extremely low (Fig. 2). During the first several days the solutions with 0.031 mg Fe L^{-1} that supported pepper growth had more rapid riboflavin fluorescence increases than those with either 0.00 or 0.125 mg Fe^{-1}. Although the amount of riboflavin fluorescence was less in solutions with no Fe than in those with 0.125 or 0.031 mg Fe^{-1}, it then exceeded the former by day 7 and the latter by day 9 but was later exceeded in those solutions with 0.031 mg Fe L^{-1}. By day 32 the respective riboflavin fluorescence values for the plants cultured with 0.00, 0.031, 0.125, 0.5 and 2.0 mg Fe L^{-1} were 0.695, 2.110, 0.128, 0.027 and 0.006.

No significant plant growth differentials were obtained in response to the three highest Fe concentrations as expressed in fresh and dry mass of roots and combined stems and leaves (Table 1). Significant decreases in tissue mass were obtained when the Fe was decreased from 0.125 mg L^{-1} to 0.031 and from the latter to no Fe. These results reflected the differences noted for pH, but did not reflect those of the riboflavin fluorescence of the nutrient solutions for the different Fe treatments.

The chlorophyll content of the apical leaves at 10 days were very low from plants receiving no Fe and were significantly different from plants grown at all higher Fe concentrations for the three dates tested (Table 2). The presence of 0.031 mg Fe L^{-1} in solution resulted in an intermediate amount of chlorophyll in apical leaves

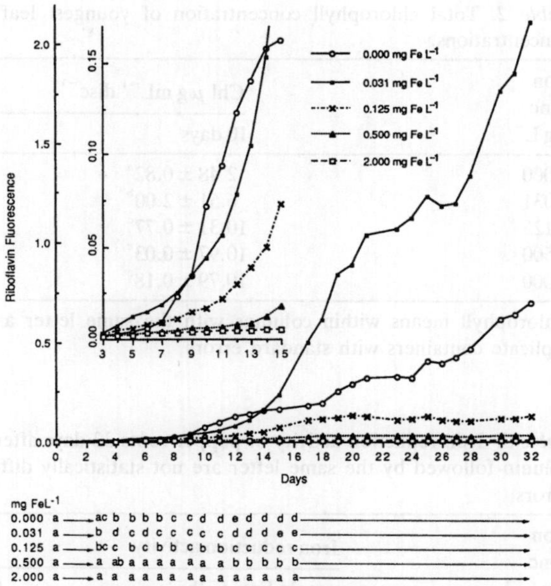

Fig. 2. Influence of Fe concentration of nutrient solution of pepper plants on riboflavin accumulation in nutrient solution from root excretion. Letters in columns that are not common are statistically significant at $P = 0.05$ calculated by LSD from ANOVA of log conversion of fluorescence values. A fluorescence value of 0.13 corresponds to 1 mg L^{-1} riboflavin.

that was between the amount of those for no Fe solutions and those with 0.125 mg Fe L^{-1}. By 18 days the apical leaf chlorophyll of plants in solution with 0.031 mg Fe L^{-1} increased to those of the higher iron concentrations, but then decreased to an intermediate value again at 32 days.

The reduction by the roots tested at the 32-day harvest suggested that reductive capacity increased with decreasing iron concentrations from 0.5 mg Fe L^{-1}, but variability within treatments resulted in only a limited number values between treatments that were significant (Table 3).

Table 1. Fresh and dry mass values of pepper plants 32 days after growth with different iron concentrations.

Iron conc.	Stems and leaves		Roots	
mg L^{-1}	Fresh mass	Dry mass	Fresh mass	Dry mass
0.000	3.31 ± 0.2^{a}**	0.37 ± 0.04^{a}**	1.30 ± 0.09^{a}**	0.10 ± 0.02^{a}**
0.031	19.37 ± 0.9^{b}**	1.70 ± 0.05^{b}*	3.84 ± 0.20^{b}**	0.26 ± 0.02^{b}**
0.125	27.98 ± 1.0^{c}	2.17 ± 0.08^{c}	6.10 ± 0.32^{c}	0.30 ± 0.04^{c}
0.500	28.15 ± 1.0^{c}	2.16 ± 0.12^{c}	6.10 ± 0.24^{c}	0.32 ± 0.04^{c}
2.000	29.91 ± 0.9^{c}	2.23 ± 0.18^{c}	6.30 ± 0.28^{c}	0.31 ± 0.02^{c}

Values are means of six replicates, with standard error. Treatment means within column followed by the same letter are not statistically different at $P = 0.05$. ** = significantly different at $P = 0.01$, * = significantly different at $P = 0.05$.

Table 2. Total chlorophyll concentration of youngest leaf of pepper plant after growth periods (days) with different iron concentrations.

Iron conc. mg L^{-1}	Chl $\mu g\ mL^{-1}\ disc^{-1}$		
	10 days	18 days	31 days
0.000	2.48 ± 0.82^a	0.04 ± 0.125^a	0.33 ± 0.1^a
0.031	6.51 ± 2.00^b	13.80 ± 02^c	8.90 ± 0.02^b
0.125	10.31 ± 0.77^c	12.70 ± 0.69^{bc}	11.72 ± 1.13^c
0.500	10.97 ± 0.03^c	11.04 ± 0.25^b	12.48 ± 0.43^c
2.000	10.79 ± 0.18^c	12.41 ± 0.19^{bc}	13.24 ± 0.18^c

Chlorophyll means within columns with the same letter are not statistically different at $P = 0.05$. Values are means of two replicate containers with standard errors

Table 3. Iron reduction activity of pepper roots 32 days after growth with different iron concentrations. Treatment means within column followed by the same letter are not statistically different at $P = 0.05$. Values are means of six replicates with standard errors

Iron conc. mg L^{-1}	Iron reduction activity			
	μmole $Fe^{2+}\ g^{-1}h^{-1}$ (Fresh mass)	% of Control	μmole $Fe^{2+}\ g^{-1}h^{-1}$ (Dry mass)	% of Control
0.000	225.1 ± 95.1^a	372.0	2920.9 ± 843.2^a	238.2
0.031	144.1 ± 44.5^{ab}	234.0	2098.9 ± 484.9^{ab}	171.2
0.125	95.1 ± 27.1^{bc}	154.6	1981.2 ± 633.3^{abc}	161.6
0.500	62.9 ± 11.6^d	102.4	1231.5 ± 261.3^c	100.4
2.000	61.5 ± 10.5^d	100.0	1226.2 ± 186.8^c	100.0

Discussion

In Landsberg's (1986) iron stress study with pepper the pH decreased to a greater extent (7 to 3.7) in a shorter time (ca. 4.66 days) than it did in these experiments. Also it increased again to pH 6.5 by about 5.32 days. Several methods varied from those used in these experiments: a different cultivar was used; the Fe source was EDTA rather than EDDHA; more iron was used in preculture; a much smaller volume of treatment solution was used and roots were aerated. It has been established that preculture with Fe as EDTA results in excessive root precipitation of iron as pH increases. This Fe becomes available as the pH is lowered in nutrient solution treatments devoid of Fe, leading to Fe uptake and an increase of pH (Landsberg, 1986). In the present experiments the Fe as EDDHA did not likely dissociate and precipitate. Increasing the roots per volume has been shown to result in much more pronounced decreases in pH with soybean (Jolley *et al.*, 1986). We noted more pronounced and earlier pH decreases when pepper seedlings were transferred to 0 Fe solution of 25 ml instead of 2 L. Although there were no significant differences in average pH values of nutrient solutions for plants grown at the three highest Fe concentrations, their riboflavin fluorescence values were markedly different.

The results from nutrient solution pH changes and fluorescence increases resulting from pepper plants being cultured under Fe-stress conditions do not support the contention that a pH decrease in the nutrient solution is a critical prerequisite to the excretion of riboflavin as once suggested (Venkat Raju, 1972), but do indicate that riboflavin is able to be excreted and accumulated in solutions when plants are exhibiting no pH decreases, and even when they exhibit continuous increases in pH. The maximum accumulation of riboflavin was not obtained in conjunction with the maximum decrease in nutrient solution pH obtained with no added Fe, but was with solutions containing 0.031 mg Fe L^{-1}, which never reached a pH value below the initial value of 5.0.

The ability for plants to excrete considerable amounts of riboflavin when supplied with 0.125 mg Fe L^{-1} was of interest in that there were no differences in values for pH, chlorophyll and mass for these plants versus those receiving the maximum Fe. If the excretion of riboflavin is an indication of the presence of root flavodoxin formation that is substituting for decreased ferredoxin in the root cells, pepper plants would appear to have a high capacity to maintain critical reduction processes in roots, and perhaps in leaves, essential for root growth and chlorophyll formation. When Fe becomes so limited that hemeproteins are also depleted, the presence of flavodoxin would no longer substitute in electron transfer for these molecules.

Iron stressed pepper plants seem to resemble sugar beet plants with regard to the maximum riboflavin being produced with a very low amount of added iron. They do not resemble sugar beet in that they produce the lowest solution pH with no Fe rather than with very low iron. More direct comparisons of these and other species for these and other parameters of iron stress response need to be made to establish how and to what degree the internal flavin may participate in plant resistance to Fe-deficiency chlorosis.

Landsberg (1986) noted that when pepper roots grown in nutrient solution devoid of added Fe were sectioned and examined under a microscope with epifluorescence optics and appropriate filters, they exhibited fluorescence in the rhizodermal layer, as did the expanded sections of whole root tips. The color of the fluorescence was not noted but the light sources and filters used would both excite riboflavin and permit some fluorescence over the broad fluorescence band having a peak about 520–530 nm. In our studies the riboflavin concentration in adventitous roots that emerged above the solution layer of Fe-stressed plants was sufficiently concentrated that roots were yellow with normal light and extremely fluorescent (yellow-green) with long wave ultraviolet light. When these roots came in contact with the solution most of the riboflavin was released from the roots, but they remained fluorescent. This suggests that the rhizodermal cells that are so modified in structure and reducing capacity are the likely sites of rapid riboflavin formation and perhaps the major site where flavodoxins could substitute for depleted ferredoxins.

Acknowledgement

This research was supported by the Agricultural Experiment Station, Utah State University, Logan, Utah *84322-4845.* Approved as journal paper No. 3869.

References

Bienfait H F 1988 Mechanisms in Fe efficiency reactions of higher plants. J. Plant Nutr. 11, 605–629.

Brown J C and Jolley V D 1988 Strategy I and strategy II mechanisms affecting iron availability to plants may be established too narrow or limited. J. Plant Nutr. 11, 1077–1098.

Camp S D, Jolley V D and Brown J C 1987 Comparative evaluation of factors involved in iron stress response in tomato and soybean. J. Plant Nutr. 10, 423–442.

Jolley V D, Brown J C, Davis T D and Walker R H 1986 Increased Fe-efficiency in soybeans through plant breeding related to increased response to Fe-deficiency stress. I. Iron stress response. J. Plant Nutr. 9, 375–386.

Landsberg E-C 1986 Function of rhizodermal transfer cells in the Fe stress response mechanism of *Capsicum annum* L. Plant Physiol. 82, 511–517.

Moran R 1982 Formulae for determination of chlorophylous pigments with N, N-dimethylformamide. Plant Physiol. 69, 1376–1381.

Römheld V and Marschner H 1983 Mechanism of iron uptake by peanut plants. Plant Physiol. 71, 949–954.

Venkat Raju K, Marschner H and Römheld V 1972 Effect of iron nutritional status on iron uptake substrate pH and release of organic acids and riboflavin by sunflower plants. Z. Pflanzenernaehr. Bodenkd. 132, 178–191.

Welkie G W and Miller G W 1988 Riboflavin excretion from roots of iron-stressed and reciprocally grafted tobacco and tomato plants. J. Plant Nutr. 11, 6190–200.

M. L. van Beusichem (Ed.), *Plant nutrition – physiology and applications*, 213–217.

PLSO IPNC610

Iron requirement of cereals and legumes in solution culture

N.K. FAGERIA
EMBRAPA/CNPAF, Cx.P 179, 74000 Goiania GO, Brazil

Key words: iron concentration, iron uptake, shoot/root weight

Abstract

A growth chamber experiment was conducted to study the response of five crop species to varying levels of iron in nutrient solution. The plant species tested were alfalfa (*Medicago sativa* L.), red clover (*Trifolium pratense* L.), common bean (*Phaseolus vulgaris* L.), rice (*Oryza sativa* L.) and wheat (*Triticum aestivum* L.). The iron levels applied were 20, 40, 80, 160, 320 and 640 μM. The crop species adapted showed various degrees of response to growth medium Fe levels. The Fe levels had no significant effects on shoot and root growth of alfalfa and common beans and root growth of red clover and rice. With a few exceptions, concentration and uptake increased with increasing Fe levels. With the exception of alfalfa, the rate of Fe uptake increased with increasing Fe levels, and in alfalfa the rate of Fe uptake increased up to the 160 μM Fe level then declined. In general, higher Fe concentrations in the growth medium decreased the uptake of most essential nutrients.

Introduction

Iron deficiency or toxicity in crop plants often represent a serious constraint to stabilizing and/or increasing crop yields. Iron deficiency is very common in calcareous soils around the world (Chen and Barak, 1982; Korcak, 1987; Vose, 1982). Calcareous soils cover over 30% of the earth's land surface and iron deficiency in extreme cases may lead to complete crop failure on these soils (Chen and Barak, 1982). Factors that can contribute to Fe deficiency in plants include low Fe supply from the soil; high lime and P applications; high levels of heavy metals such as Zn, Cu, and Mn; low and high temperatures; high levels of nitrate nitrogen; high organic matter content; poor aeration; calcium carbonate in the soil; bicarbonate in the soil or irrigation water; high light intensities; unbalanced cation ratios; and roots infected by nematodes (Korcak, 1987; Wallace and Lunt, 1960). Iron deficiency occurs in a variety of soils from sandy to fine-textured, mucks and peats (Brown, 1961).

Iron toxicity is not as common as Fe-deficiency. On acid soils, where Fe is more available, Fe^{2+} can become toxic to plants. Iron toxicity can be found in rice soils where unfavorable factors such as poor drainage, highly reducing conditions, and high sulphide content can occur (Foy *et al.*, 1978; Ponnamperuma, 1972). Rice is one of the most important food crops in the world and is widely distributed throughout the tropical, subtropical, and temperate zones of all continents. Iron toxicity is a serious problem for flooded rice (Fageria and Rabelo, 1987).

The objective of this investigation was to study the growth and nutrient uptake of alfalfa, red clover, common bean, rice and wheat over a wide range of Fe concentrations in solution culture.

Materials and methods

A solution culture experiment was conducted to study the response of alfalfa (*Medicago sativa* L. cv. Arc), red clover (*Trifolium pratense* L. cv. Kenstar), *common bean* (*Phaseolus vulgaris* L. cv. Carioca), rice (*Oryza sativa* L. cv. Rio Paranaiba), and wheat (*Triticum aestivum* L. cv.

Yecorra Rojo) to various levels of Fe under controlled conditions. The climatic conditions in the growth chamber during the experiment were 14 h of 530 μmoles S^{-1} m^{-2} light intensity, day temperature was 28°C and the relative humidity was 60%. At night, the temperature and relative humidity were 22°C and 80%, respectively.

The nutrient solution used in the experiment had the following composition in μM: KNO_3 5000; $Ca(NO_3)_2{\cdot}4H_2O$ 5000; $MgSO_4{\cdot}7H_2O$ 2000; KH_2PO_4 400; H_3BO_3 46; $MnCl_2{\cdot}4H_2O$ 9; $ZnSO_4{\cdot}7H_2O$ 0.80; $CuSO_4{\cdot}5H_2O$ 0.32; and $(NH_4)_6Mo_7O_{24}{\cdot}4H_2O$ 0.07. The nitrogen equivalent supplied by KNO_3 and $Ca(NO_3)_2{\cdot}4H_2O$ was 15 mM N as NO_3^-. The Fe concentrations used were 20, 40, 80, 160, 320 and 640 μM. These concentrations were supplied through FeEDDHA for legumes and through FeHEDTA for cereals as recommended by Chaney and Bell (1987). Seeds of the five crop species were germinated in germinating paper. One week old seedlings of bean and wheat and 10 day old seedlings of alfalfa, red clover and rice were transplanted in 1.7 liter plastic pots. In case of bean, there were 4 seedlings per pot, while 8 seedlings per pot were used for the other crop species. The pH of the solution was maintained at 5.5 and the solution was not changed during the experiment. Deionized water was used to balance water loss due to plant transpiration. The beans and wheat were harvested after 15 days of growth in the Fe treatments. Rice and alfalfa were harvested after 17 days growth and red clover after 28 days growth in the nutrient solution. A factorial design was used with 5 crop species and 6 Fe levels. The treatments were replicated 3 times.

At the time of harvesting, tops and roots were separated and washed with deionized water several times. Total root length was measured with a Comair root length scanner. Plant material (roots and tops) was dried to constant weight in a forced-draft oven at 65°C and then milled. Plant material was wet digested in a $HNO_3/HClO_4$ (4:1) mixture. Elemental determinations were made by inductively coupled plasma emission

Table 1. Influence of iron on growth parameters of five crop species

	Fe levels (μM)					
Crop species	20	40	80	160	320	640
Alfalfa						
Shoot dry wt. (g/pot)	2.70a	3.03a	2.89a	2.53a	2.66a	2.50a
Root dry wt. (g/pot)	0.68a	0.79a	0.77a	0.62a	0.69a	0.68a
Root length (m/pot)	23ab	22ab	18ab	18ab	22ab	27a
Red Clover						
Shoot dry wt. (g/pot)	6.48a–c	5.74bc	6.55ab	6.94a	6.12a–c	5.58c
Root dry wt. (g/pot)	0.89a	0.58a	0.75a	0.91a	0.90a	0.72a
Root length (m/pot)	55a	60a	65a	68a	68a	49a
Common Bean						
Shoot dry wt. (g/pot)	8.79a	10.32a	9.11a	9.43a	8.62a	8.87a
Root dry wt. (g/pot)	2.04a	2.31a	2.39a	2.27a	2.43a	2.28a
Root length (m/pot)	123b	213a	222a	169ab	100ab	212a
Rice						
Shoot dry wt. (g/pot)	4.09a	3.60ab	3.86ab	3.74ab	3.00b	2.01c
Root dry wt. (g/pot)	1.24a	1.04a	1.19a	1.20a	1.01a	0.69a
Root length (m/pot)	104bc	120ab	135a	105bc	84cd	68d
Wheat						
Shoot dry wt. (g/pot)	2.59a	2.35ab	2.57a	2.31ab	1.91b	1.24c
Root dry wt. (g/pot)	0.53ab	0.57a	0.61a	0.45b	0.32c	0.17d
Root length (m/pot)	35a	41a	37a	31ab	23b	22b

For a given growth parameter means along the Fe levels followed by the same letter are not significantly different at the 5% level of probability by Duncan's Multiple Range Test.

spectroscopy (ICP). Statistical Analysis System (SAS) programs were used to calculate regression equations and correlation coefficients relating plant growth and nutrient uptake.

Results and discussion

The Fe levels had no significant effects on shoot and root wt. of alfalfa and common beans and root wt. for red clover and rice (Table 1). Higher shoot wt. for red clover and rice were recorded at 160 μM Fe and 20 μM Fe levels, respectively. Increasing Fe levels beyond 160 μM reduced root length of rice and wheat, however, Fe levels had no significant effect on root length of alfalfa and red clover.

With a few exceptions, increasing Fe levels increased the concentration and uptake of Fe in roots and shoots of all the species (Table 2). At all Fe levels, shoot uptake of Fe was highest in common bean and lowest in alfalfa and wheat. The different species showed various degrees of concentration and uptake responses to the levels of Fe in the growth medium.

The rate of uptake of Fe (calculated as the differences between the Fe originally applied and that remaining in the solution at harvest) of each crop species is presented in Table 3. Rate of Fe uptake per unit of root dry wt. day^{-1} increased with increasing Fe levels of the growth medium for all crop species with the exception of alfalfa. In alfalfa the rate of Fe uptake increased up to 160 μM Fe level and then decreased. In various

Table 2. Influence of iron on root and shoot iron concentration and uptake in five crop species

Fe levels μM	Alfalfa		Red Clover		Common Bean		Rice		Wheat	
	Conc. $mg \cdot kg^{-1}$	Uptake $\mu g \cdot pot^{-1}$	Conc. $mg \cdot kg^{-1}$	Uptake $\mu g \cdot pot^{-1}$	Conc. $mg \cdot kg^{-1}$	Uptake $\mu g \cdot pot^{-1}$	Conc. $mg \cdot kg^{-1}$	Uptake $\mu g \cdot pot^{-1}$	Conc. $mg \cdot kg^{-1}$	Uptake $\mu g \cdot pot^{-1}$
Roots										
20	119f	81c	152c	134c	85c	174c	172c	217c	400cd	207ab
40	152e	120c	444bc	273c	169c	406c	214c	223c	264d	151b
80	282d	213b	281bc	214c	215c	532c	313bc	374c	299d	174b
160	413c	256b	527bc	482a–c	306c	694c	705ab	905c	597bc	268a
320	618b	428a	673b	592ab	663b	1658b	741a	707ab	655b	206ab
640	626a	428a	1065a	773a	1198a	2716a	830a	556a–c	1196a	202ab
Shoots										
20	49c	131b	46d	299d	61c	528c	56c	231c	48c	124b
40	45c	136b	48d	272d	66c	679c	74bc	262bc	58c	136b
80	78a–c	223ab	56cd	367cd	75c	689c	85bc	327bc	62c	159b
160	67bc	166b	92bc	638bc	84c	789c	169b	627a	83c	195b
320	121a	325a	130b	792b	163b	1403b	147bc	441ab	160b	306a
640	106ab	261ab	253a	1422a	445a	3937a	283a	542a	248a	311a

Concentration = Fe content per unit dry weight and uptake = conc. X dry weight. Means in the same column followed by same letter are not significantly different at 5% probability level by Duncan's Multiple Range Test.

Table 3. Rate of iron uptake by five crop species

Fe conc. μM	Alfalfa	Red Clover	Bean	Rice	Wheat
	$\mu M\ g^{-1}$ root dry wt. day^{-1}				
20	0.20c	0.27e	0.21c	0.32c	0.98b
40	0.23c	0.64de	0.37bc	0.52c	1.55b
80	0.43bc	0.97cd	0.61bc	1.48c	2.02b
160	4.73a	1.38c	1.23bc	2.26c	4.85b
320	2.92ab	3.02b	1.56b	5.49b	10.44b
640	2.58a–c	6.62a	4.39a	9.38a	40.99a

Means in the same column followed by the same letter are not significantly different at the 5% probability level by Duncan's Multiple Range Test.

Table 4. Nutrient uptake in the shoots of five crop species as influenced by iron levels

FE Level μM	P	K	Ca	Mg	S	Mn	Zn	Cu
	mg pot⁻¹					μg pot⁻¹		
Alfalfa								
20	7.75a	100a	29a	3.36ab	5.27a	211a	139a	40a
40	7.43a	86ab	25ab	3.16ab	4.88a	148b	109a	48a
80	8.07a	72ab	26ab	3.42ab	5.59a	137bc	90b	43a
160	7.00a	61b	22b	2.93b	4.67a	87c	82b	45a
320	7.53a	85ab	27ab	3.63a	5.47a	117bc	110ab	45a
640	7.32a	65ab	25ab	3.29ab	4.91a	100bc	108ab	44a
Red Clover								
20	10.02ab	147a	72a	13.82ab	13.14a	154a	179a	58ab
40	9.26a–c	132ab	62ab	12.58ab	10.12b	127ab	115b	53b
80	9.16a–c	140a	63ab	13.53ab	11.53ab	120b	154ab	58ab
160	10.20a	142a	75a	14.49a	13.76a	123ab	158ab	88a
320	8.81bc	94b	64ab	14.03ab	10.91ab	99bc	143ab	73ab
640	8.04c	93b	50b	10.85b	9.67b	78c	155ab	57ab
Common bean								
20	167a	230a	215a	273a	23a	401ab	310ab	98a
40	165a	240a	255a	288a	23a	489a	376a	112a
80	161a	142a	246a	292a	23a	396ab	232b	121a
160	150a	250a	227a	255a	24a	359b	208b	90a
320	161a	209a	245a	297a	24a	400ab	230b	117a
640	156a	164a	228a	320a	26a	459ab	337ab	110a
Rice								
20	15a	63a	24a	25a	18a	657a	151a	80a
40	15a	61a	23a	24a	16a	604a	180a	58a
80	14a	55a	25a	26a	16a	665a	161a	75a
160	13ab	63a	24a	26a	15a	595a	104b	64a
320	11b	63a	18b	18b	11b	602a	111b	62a
640	9c	49a	13c	11c	8b	903a	53c	58a
Wheat								
20	49a	14a	16a	4.41a	7.47a	190a	101a	36a
40	58a	12a	13a	3.80a	7.02ab	185a	83a	28ab
80	62a	14a	16a	4.35a	7.78a	177ab	78a	32a
160	71a	13a	15a	4.00a	6.43ab	126bc	81a	28ab
320	86a	13a	15a	3.74a	5.22b	99cd	98a	33a
640	45a	8b	7b	2.07b	3.58c	46d	56a	11b

For a given crop species means in the same column followed by the same letter are not significantly different at the 5% probability level by Duncan's Multiple Range Test.

Fe treatments, wheat showed the highest rate of Fe uptake and alfalfa showed the lowest rate of Fe uptake. Among legumes, red clover showed the highest rate of Fe uptake when compared to alfalfa and common beans.

The influence of Fe on the uptake of other nutrients in the shoot of the five crop species is shown in Table 4. In general, the uptake of P, K, Ca, Mg and S was decreased by increasing Fe levels in the growth medium. However, the uptake of Ca, Mg and S by common beans did not show any significant differences with varying levels of Fe. In alfalfa, red clover and wheat, uptake of Mn, Zn and Cu were decreased with higher levels of Fe.

Acknowledgements

The author wishes to thank Dr R L Chaney for

his help in selection of desirable Fe chelates. The author also thanks B Woolum and J Lilly for their expert technical assistance.

References

Brown J C 1961 Iron chlorosis in plants. Adv. Agron. 13, 329–369.

Chen Y and Barak P 1982 Iron nutrition of plants in calcareous soils. Adv. Agron. 35, 217–240.

Chaney R L and Bell P F 1987 Complexity of iron nutrition: Lessons for plant-soil interaction research. J. Plant Nutr. 10, 963–994.

Fageria N K and Rabelo N A 1987 Tolerance of rice cultivars to iron toxicity. J. Plant Nutr. 10, 653–661.

Foy C D, Chaney R L and White M C 1978 The physiology of metal toxicity in plants. Annu. Rev. Plant Physiol. 29, 511–566.

Korcak R F 1987 Iron deficiency chlorosis. Hortic. Rev. 9, 133–186.

Ponnamperuma F N 1972 The chemistry submerged soils. Adv. Agron. 24, 29–96.

Sposito G and Mattigod S V 1980 A computer program for the calculation of chemical equilibrium in soil solution and other natural water systems. Kearney Foundation of Soil Science, University of California, Riverside.

Vose P B 1982 Iron nutrition in plants: A world review. J. Plant Nutr. 5, 233–249.

Wallace A and Lunt O R 1960 Iron chlorosis in horticulture plants: A review. J. Am. Soc. Hortic. Sci. 75, 819–841.

M. L. van Beusichem (Ed.), *Plant nutrition – physiology and applications*, 219–222.

PLSO IPNC188

A transport mutant of pea (*Pisum sativum*) for the study of iron absorption in higher plant roots

M.A. GRUSAK, L.V. KOCHIAN and R.M. WELCH[1]
U.S. Plant, Soil and Nutrition Laboratory, USDA-ARS, Cornell University, Ithaca, NY 14853, USA.
[1]Corresponding author

Key words: iron absorption, iron reduction, membrane reductase, *Pisum sativum* L.

Abstract

Investigations have been conducted on a single-gene pea (*Pisum sativum* L.) mutant which accumulates toxic Fe levels in its older leaves (>1.0% dry weight Fe). Root-associated Fe reductase activity and Fe absorption were determined in +Fe- and −Fe-grown mutant (E107) and normal ('Sparkle') pea genotypes to elucidate the physiological differences which cause excess Fe accumulation in the mutant. Iron influx in roots of E107 (+/−Fe-grown) or −Fe-grown Sparkle was about four-fold higher than that measured in +Fe-Sparkle. Compared to +Fe-grown Sparkle, roots of the other three growth types also exhibited a greater root-associated reductase activity (4–6 fold higher). The physiological similarities of +Fe-grown E107 with the Fe-deficient E107 or Sparkle, suggest that the E107 mutant functions as a positively induced Fe-deficient dicotyledonous plant.

Introduction

The mechanism of Fe absorption by higher plant roots is not completely understood. For dicots and the non-graminaceous moncots, Fe is absorbed as the free Fe(II) ion by plant roots (Chaney *et al.*, 1972). Since Fe is found predominantly in the Fe(III) form in well aerated soils, there exists an obligatory need for the reduction of Fe(III) to Fe(II) prior to uptake by these types of plants. Iron(III) reduction is thought to occur primarily through the functioning of a root membrane-bound reductase system (Bienfait, 1985) which transfers electrons from NADH (Buckhout *et al.*, 1989) or NADPH (Sijmons *et al.*, 1984) in the root-cell cytoplasm, to Fe(III) in the root apoplasm.

Much of our understanding of the Fe(III) reducing system(s) comes from work with Fe-deficient plants, for which Fe deficiency (in dicots and non-graminaceous monocots) generally results in an enhancement of Fe-reducing capability and/or the induction of a secondary, high-activity, reductase system (Bienfait, 1985). Other physiological changes are also induced, and the specific regulators of these changes as well as the impact of these changes on the absorption of other ions are currently under investigation.

Recently, a single-gene mutant (E107) of pea (*Pisum sativum* L.) has been discovered (Kneen *et al.*, 1990), which has been shown to accumulate extremely high levels of Fe when compared to plants of its parent genotype – cv. 'Sparkle' (Welch and LaRue, 1990). We have been studying this mutant in order to characterize its root-associated Fe-reducing activity and ability to absorb Fe. In this paper, evidence is presented from studies with both +Fe- and −Fe-grown E107 and Sparkle, which suggest that the E107 mutant functions as though it is continuously Fe-deficient (*i.e.* various Fe deficiency-induced responses appear to be constantly activated). Our results are used to speculate on the mechanisms(s) of Fe uptake by root cells, and the usefulness of the E107 mutant to broaden our understanding of both the mechanism and regulation of Fe(III) reduction and Fe(II) absorption in higher plants.

Material and methods

Plant material and culture

The parental pea (cv. 'Sparkle') was obtained from a commercial source, while the mutant line E107 was the result of mutagenesis with ethyl methane sulfonic acid (Kneen *et al.*, 1990). Pea seeds were first imbibed overnight (day 0) in aerated deionized water, before placement on moistened filter paper (with deionized water) in the dark (day 1). On day 3, four seedlings were placed individually in specially designed plastic cups which supported them over a polyethylene container with 3.5 L of aerated nutrient solution (enclosed in back polyethylene), and were covered with black polyethylene beads. The nutrient solution contained the following macronutrients in mol m^{-3}: KNO_3, 1.2; $Ca(NO_3)_2$, 0.8; $NH_4H_2PO_4$, 0.1; $MgSO_4$, 0.2, and the following micronutrients in mmol m^{-3}: $CaCl_2$, 25; H_3BO_3, 25; $MnSO_4$, 2; $ZnSO_4$, 2; $CuSO_4$, 0.5; H_2MoO_4, 0.5; $NiSO_4$, 0.1. Nutrient solutions were changed on days 8, 10, 12, and 14. Iron-treated plants received Fe as 1 mmol m^{-3} Fe(III)-EDDHA (N,N'-ethylenebis[2-(2-hydroxyphenyl)glycine]). Plants were grown in a controlled environment chamber with a 16 h, 20°C/8 h, 15°C day–night regime and 580 μE m^{-2} s^{-1} photosynthetically active radiation.

Experimental techniques

Root-associated Fe(III) reduction assays were performed with intact root systems using the procedure of Chaney *et al.* (1972). The assay solution consisted of the macro- and micronutrients listed above, plus 5 mol m^{-3} MES (4-morpholineethanesulfonic acid) buffer (pH 5.0), 0.1 mol m^{-3} Fe(III)-EDTA (ethylenediaminetetraacetic acid), and 0.3 mol m^{-3} BPDS (bathophenanthrolinedisulfonic acid). The assay was conducted under low light conditions and was terminated after 15 min. In order to monitor daily changes in reductase activity, assays were performed on the same plants on consecutive days. Root weights were measured using the Archimedes' technique (Lang and Thorpe, 1989). Plants were grown to day 15 with either 1 mmol m^{-3} Fe(III)-EDDHA (Fe first added on day 3; +Fe-treated plants) or no added Fe (−Fe-treated plants). Begining on Day 8, 1 mol m^{-3} MES buffer (pH 5.0) was also added to all nutrient solutions to prevent extreme acidification of the growth media.

Short-term Fe flux measurements were carried out using a short segment (0.5 cm) of the seminal root from 15-day-old plants, having 3–4 primary lateral roots (13–16 cm in length) attached. The plants were grown as described for the Fe(III) reduction assays. The excised roots were placed in plexiglass tubes containing nutrient solution with 5 mol m^{-3} MES buffer (pH 5.0) for 10 min, followed by 10 min in a fresh solution of the same composition plus 0.1 mol m^{-3} Fe(III)-EDDHA (and +/−0.3 mol m^{-3} BPDS). At the end of the 10 min pretreatment, ^{59}Fe(III)-EDDHA was added to yield a specific activity of 18.5 kBq $mmol^{-1}$. All solutions were gently bubbled with air. After an uptake period of 20 min, the radioactive solution was evacuated from the uptake tube, and the root system was rinsed for 20 sec with 5 mol m^{-3} $MgCl_2$. This was followed by a desorption to remove Fe(III) precipitates from the root apoplasm using a modification of the Bienfait *et al.* (1985) technique. Roots were 'desorbed' in a solution, purged continuously with N_2, containing 5 mol m^{-3} $MgCl_2$, 5 mol m^{-3} $Na_2S_2O_4$ (sodium dithionite), and 0.1 mol m^{-3} $FeNH_4(SO_4)_2$ for 8 min. The roots were then gently blotted, weighed, placed in scintillation vials, and were radioassayed for τ-ray activity using an autogamma spectrophotometer.

Results

Root-associated reduction of Fe(III) (0.1 mol m^{-3} Fe(III)-EDTA) was about 6-fold higher in +Fe-grown E107 than in +Fe-grown Sparkle (Table 1). Both E107 (at all time periods) and Sparkle (after day 12) grown without Fe also exhibited higher Fe(III) reduction rates than +Fe-treated Sparkle (Table 1). Interestingly, both −Fe treatments exhibited slightly lower

Table 1. Iron(III) reduction rates[a] determined with intact root systems of either +Fe- or −Fe-grown mutant (E107) and normal ('Sparkle') pea genotypes

Plant treatment	Days after seed imbibition		
	10	12	15
	μmol Fe(II)-BPDS$_3$ g FW^{-1} h^{-1}		
E107 +Fe	1.51 ± 0.20	1.58 ± 0.08	1.37 ± 0.01
Sparkle +Fe	0.26 ± 0.14	0.26 ± 0.02	0.21 ± 0.01
E107 −Fe	1.21 ± 0.35	1.10 ± 0.10	0.90 ± 0.06
Sparkle −Fe	0.14 ± 0.14	1.12 ± 0.11	1.10 ± 0.15

[a] Values presented are mean ± standard error of the mean (n = 3).

Fe(III) reduction rates than that determined for the +Fe-treated E107. In separate experiments, where plants were initially supplied with Fe, the enhanced or inducible reductase system in Sparkle was found to be fully activated after 3 days without Fe, while no enhancement of reducing activity was observed in E107 not receiving Fe (data not shown).

Short-term flux measurements indicated enhanced (4–5 fold) Fe influx for +Fe-treated E107, −Fe-treated E107, and −Fe-treated Sparkle relative to +Fe-grown Sparkle (Fig. 1). Iron influx measurements in the presence of the Fe(II) chelator, BPDS, showed an apparent inhibition of 44% for +Fe-treated Sparkle, and ranged from 74–79% inhibition for the other three growth types (Fig. 1).

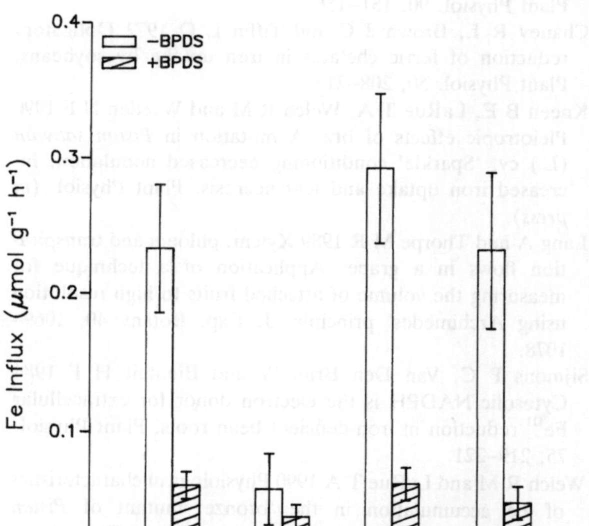

Fig. 1. Short-term Fe influx measurements with excised primary lateral roots of 15 day old pea plants, exposed to 100 μmol m^{-3} ^{59}Fe-labelled, Fe(III)-EDDHA. For the +BPDS treatments, the Fe(II) chelator was added at a concentration of 0.3 mol m^{-3}. The absorption and post-absorption protocol are described in the text under Material and Methods. Values are expressed on a g fresh weight of root basis. Error bars represent the mean ± standard error of the mean (n = 3).

Discussion

We have studied the physiology of a single-gene mutant of pea (E107) which can accumulate toxic levels of Fe in its older leaves (Welch and LaRue, 1990). Our working hypothesis has been that this mutant might have an enhanced ability to reduce Fe(III), which would provide increased levels of Fe(II) ions for absorption by root cells. Most dicots (and non-graminaceous monocots) can enhance their rates of Fe(III) reduction when subjected to Fe-deficiency stress (Bienfait, 1988). Possibly, this increased reducing capacity is the result of the induction of a membrane-bound Fe(III)-reducing system ('Turbo'-reductase), which functions in parallel to a normally occurring, non-Fe(III) reducing system (Bienfait, 1985). Our studies show that roots of E107 do exhibit an elevated reducing capacity (relative to +Fe-grown Sparkle; see Table 1), which occurs even when the plants are given an adequate supply of Fe for growth. Although our results might be explained by an alteration of E107's non-Fe(III) reducing system, such that it now reduces Fe(III), it is more likely the result of an activation of the so-called 'Turbo' reductase system, since Fe-deficient E107 does not show an enhanced reducing capacity over that of +Fe-treated E107 (Table 1). If E107 has not lost its ability to induce a 'Turbo'-reductase system (c.f. −Fe treated Sparkle, Table 1), then one would expect a greater capacity for Fe(III) reduction from −Fe-treated E107, if the reducing capacity in +Fe-treated E107 were the result of an altered non-Fe(III) reducing system. We suggest, therefore, that the E107 mutant performs as if it were continuously experiencing Fe deficiency stress.

Other results also support this view: rhizosphere acidification (data not shown) and Fe influx (Fig. 1) are both enhanced in +Fe-treated E107, relative to +Fe-treated Sparkle. The +Fe-treated E107 increases were similar to those for both −Fe-treated E107 and Sparkle, and indeed, both responses have previously been documented as Fe deficiency induced responses in other plants (Bienfait, 1988). With regard to the Fe flux data, these three growth types also had similar degrees of inhibition when Fe influxes were determined in the presence of BPDS (Fig. 1). Presently, we refer to the BPDS data as only apparent inhibitions, due to the fact that the roots retained a pink color (indicative of the Fe(II)-$BPDS_3$ complex) after removal from the post-absorption solution ('desorption' did not completely remove the coloration). Thus, the possible presence of apoplasmic ^{59}Fe may have led to an underestimation of the BPDS inhibition. Others (Chaney *et al.*, 1972), studying *Glycine max*, have reported a much higher degree of BPDS inhibition on Fe uptake. Nonetheless, it is still interesting to note that the extent of the BPDS inhibition was much lower in +Fe-treated Sparkle, suggesting that the actual kinetic/mechanistic activities of the Fe(II) uptake system are similar in +/−Fe-treated E107, and −Fe-treated Sparkle, and are different (or enhanced?) from the system in +Fe-treated Sparkle. Further experiments are in progress to clarify these differences between growth types.

In summary, we have found that the mutant E107 has many of the characteristics of a classic Fe-deficient dicot plant, because of physiological similarities to −Fe-grown Sparkle or E107 and its differences with +Fe-treated Sparkle. In the future, we wish to use this mutant to further our understanding of not only Fe(III) reduction and uptake systems in plants, but also the regulation of these processes as well. In this regard, we plan to isolate plasmalemma-bound proteins from roots, and to identify the reductase system responsible for the increased Fe(III) reduction in this single-gene mutant.

Acknowledgement

We wish to gratefully acknowledge the excellent technical expertise of Jon E. Shaff, who assisted with much of this project.

References

Bienfait H F 1985 Regulated redox processes at the plasmalemma of plant root cells and their function in iron uptake. J. Bioenerg. Biomembr. 17, 73–83.

Bienfait H F 1988 Mechanisms in Fe-efficiency reactions of higher plants. J. Plant Nutr. 11, 605–629.

Bienfait H F, Van Den Briel W and Mesland-Mul N T 1985 Free space iron pools in roots. Generation and mobilization. Plant Physiol. 78, 596–600.

Buckhout T J, Bell P F, Luster D G and Chaney R L 1989 Iron-stress induced redox activity in tomato (*Lycopersicum esculentum* Mill.) is localized on the plasma membrane. Plant Physiol. 90, 151–156.

Chaney R L, Brown J C and Tiffin L O 1972 Obligatory reduction of ferric chelates in iron uptake by soybeans. Plant Physiol. 50, 208–213.

Kneen B E, LaRue T A, Welch R M and Weeden N F 1990 Pleiotropic effects of brz. A mutation in *Pisum sativum* (L.) cv. 'Sparkle' conditioning decreased nodulation, increased iron uptake and leaf necrosis. Plant Physiol. (*in press*).

Lang A and Thorpe M R 1989 Xylem, phloem and transpiration flows in a grape: Application of a technique for measuring the volume of attached fruits to high resolution using Archimedes' principle. J. Exp. Botany 40, 1069–1078.

Sijmons P C, Van Den Briel W and Bienfait H F 1984 Cytosolic NADPH is the electron donor for extracellular Fe^{III} reduction in iron-deficient bean roots. Plant Physiol. 75, 219–221.

Welch R M and LaRue T A 1990 Physiological characteristics of Fe accumulation in the 'bronze' mutant of *Pisum sativum* L., cv 'Sparkle' E107 (brz brz). Plant Physiol. (*in press*).

M. L. van Beusichem (Ed.), *Plant nutrition – physiology and applications*, 223–228.

PLSO IPNC490

The significance of the magnesium to manganese ratio in plant tissues for growth and alleviation of manganese toxicity in tomato (*Lycopersicon esculentum*) and wheat (*Triticum aestivum*) plants

J. LE BOT[1], M.J. GOSS[2], M.J.G.P.R. CARVALHO[3], M.L. VAN BEUSICHEM[4] and E. A. KIRKBY[1,5]
[1]*Department of Pure and Applied Biology, University of Leeds, Leeds LS2 9JT, UK,* [2]*Soils and Agronomy Department, AFRC Institute of Arable Crops Research, Rothamsted Experimental Station, Harpenden, Herts. AL5 2JQ, UK,* [3]*University of Evora, P-7000, Portugal, and* [4]*Department of Soil Science and Plant Nutrition, Wageningen Agricultural University, P.O. Box 8005, 6700 EC Wageningen, The Netherlands.* [5]*Corresponding author*

Key words: *Lycopersicon esculentum* L., magnesium-manganese interaction, magnesium : manganese ratio, manganese toxicity, tomato, *Triticum aestivum* L., wheat

Abstract

Results are reported for tomato (*Lycopersicon esculentum* L. var. Ailsa craig) and wheat (*Triticum aestivum* L. cv. Mara) which demonstrate that increasing concentrations of Mg in the plant raises plant tolerance to Mn toxicity.

Water culture experiments with tomato show that under conditions of high Mn supply (200 μM, Mn), not only does increasing Mg application (0.75 mM to 15 mM) depress Mn uptake, but the higher Mg concentrations in the shoot counteract the onset of Mn toxicity when the concentrations of Mn in the shoot are also high. The ratio of Mg : Mn in the tissues is a better indicator of the appearance of toxicity symptoms than Mn concentration alone. Toxicity symptoms were observed when the Mg : Mn ratio in the shoot tissue was from 1.13 to a value between 3.53 and 6.54. The corresponding Mg : Mn ratio in the older leaves was from 0.82 to between 2.27 and 3.51.

For wheat grown in soil, analyses of leaves revealed that growth could be expressed by the following relationship: $Y = A + B\exp(-kX)$, where Y = growth, X = Mg : Mn ratio, A, B and k = constants. Growth was significantly reduced when the Mg : Mn ratio fell below 20 : 1. From a measurement of this ratio it is therefore possible to predict the appearance of Mn toxicity and its influence on growth.

Introduction

Manganese toxicity has long been recognized as an important factor limiting plant growth on acid and waterlogged soils. Different plant species and cultivars of the same species differ in tolerance to Mn. As well as these genetic differences, environmental factors including nutrition can also be important in conferring tolerance. In particular other nutrients such as Fe, Ca and Mg in the growth medium can modify the uptake of Mn from solution (Chinnery and Harding, 1980; Maas *et al.*, 1969).

In the older literature especially there are frequent references to the interaction between Mn and Mg as plant nutrients. The beneficial effects of Mg in the nutrient medium in depressing Mn toxicity was reported by Löhnis (1960) and more recently Hecht-Buchholtz *et al.* (1987) have observed that in Norway spruce seedlings excess Mn is taken up when Mg is in short supply. Interrelationships between Mn toxicity and Mg deficiency have also been studied in melon (Elamin and Wilcox, 1986a; 1986b; Simon *et al.*, 1986).

The beneficial influence of Mg appears to result in part from depressing Mn uptake, though there are also indications that a higher concentration of Mg in plant tissues confers tolerance to high concentrations of Mn. In this paper we compare these factors in relation to the effects of Mg in raising the tolerance of tomato and wheat plants to Mn.

Materials and methods

Two sets of data are reported, one for tomato in experiments at the University of Leeds UK, and the other for wheat grown in pots of soil obtained from Portugal and known to produce symptoms of manganese toxicity in wheat (Goss and Carvalho, 1989).

Two water culture experiments used tomato plants (*Lycopersicon esculentum* L. var. Ailsa craig). The first was to study concentrations of Mn of 10 μM, 50 μM, 100 μM and 300 μM in nutrient solutions each provided with 0.75 mM $MgSO_4$, 2 mM $Ca(NO_3)_2$, 0.65 mM K_2SO_4 and 0.15 mM KH_2PO_4. The micronutrient solution contained FeNa-EDTA 50 μM, $CuSO_4$ 0.95 μM, $ZnSO_4$ 0.65 μM, H_3BO_3 29.6 μM and Na_2MoO_4 0.52 μM. The pH of the nutrient solution was adjusted to 5.5 with a saturated solution of $Ca(OH)_2$. Ten plants were grown from the 4th leaf stage for each of the 4 Mn treatments in 50 litre containers and the aerated nutrient solutions were renewed every 4 days. Two harvests were taken one of 5 plants after 9 days and the other of the remaining 5 plants after 17 days. Each plant was separated into leaves, petioles, stems and roots and the shoot organs were divided into old and young tissues. The samples were oven dried (95°C) and weighed.

The second experiment investigated increasing concentrations of Mg in the nutrient medium (0.75 mM, 1.5 mM, 7.5 mM and 15 mM as $MgSO_4$) at a Mn concentration of 200 μM which the result of the previous experiment indicated would normally be toxic. The basic nutrient solution was the same as in the first experiment except that the K_2SO_4 and KH_2PO_4 concentrations were slightly increased to 0.75 mM K_2SO_4 and 0.25 μM KH_2PO_4 respectively. Two harvests were taken, one of 5 plants after 11 days and the remaining 5 plants after 20 days. The same harvesting procedure was adopted as in the first experiment. In both experiments results of the second harvest are reported.

The dried and ground plant material was analyzed for K by flame photometry and for Ca, Mg and Mn by atomic absorption spectrometry.

In the experiments with wheat (*Triticum aestivum* L. cv. Mara), a coarse-textured Portuguese soil derived from quartzdiorite was amended or not with $CaCO_3$ (limed or unlimed soil). Manganese and magnesium concentrations in the soil solution were modified by the addition of $CaCO_3$ and of increasing quantities of $MgSO_4$. These modified soils were then packed in columns (65 mm diameter and 250 mm height) and fertilized with NH_4NO_3 (31.6 mg N kg^{-1} of air-dry soil) and KH_2PO_4 (16.6 mg P and 20.8 mg K kg^{-1} of air-dry soil). The Mg:Mn ratio in the soil solution of these soils ranged between 1.6 (no $CaCO_3$ nor $MgSO_4$ added to the soil) and 347.4 (550 mg $CaCO_3$ kg^{-1} and 336 mg $MgSO_4$ kg^{-1} of air-dry soil). Four pre-germinated seeds were planted into each pot which was maintained at 17°C in a controlled environment chamber. After 3 weeks growth, shoots were cut off and weighed after oven drying at 60°C for 24 hours. Shoot material was digested in nitric acid and acidified to produce solutions with 5% v/v HCl which matched the calibration standards for the Inductively Coupled Plasma-Optical Emission Spectrometer (ARL 3400) used for measuring the Mg and Mn contents.

Results

Increasing Mg in the nutrient medium under conditions of high Mn supply (200 μM Mn) increased total dry matter yields of tomato plants and alleviated Mn toxicity symptoms in the two higher Mg treatments (Table 1). Both total Mn uptake per plant and the Mn concentration expressed on a whole plant dry matter basis were depressed by Mg. Magnesium uptake was markedly increased but that of Ca was depressed and

Table 1. The influence of increasing Mg supply (as $MgSO_4$) on the uptake of cations by tomato plants growing in a nutrient solution containing 200 μM Mn. (results expressed as $mg.g^{-1}$ dry weight and mg uptake per plant)

Treatment[a] (m*M* Mg)	Relative dry matter yields[b] (g)	K	Ca	Mg	Mn	Mn uptake[b] ($mg\ plant^{-1}$)
		($mg\ g^{-1}$)[b]				
0.75*	72 c	63.4 b	22.2 a	5.6 d	3.5 a	65.5 a
1.5*	70 c	68.9 a	20.0 b	7.6 c	3.2 b	57.9 ab
7.5	87 b	62.4 b	13.0 c	15.4 b	2.2 c	49.2 bc
15	100 a	54.2 c	7.8 d	19.2 a	1.8 d	45.3 c

[(a)] *denotes the presence of Mn toxicity symptoms.
[(b)] Values followed by different letters within a column are significantly different ($P = 0.05$).

K was little affected except in the highest Mg treatment in which it was somewhat decreased.

That the influence of Mg in preventing the onset of Mn toxicity symptoms and increasing dry matter yield was not simply the result of a decrease in Mn uptake, and hence also in Mn tissue concentration, can be seen from Table 2. Raising the Mn supply to 100 μM whilst retaining Mg concentration in the nutrient medium constant, depressed shoot yields by about 25% and markedly increased Mn concentration in the shoot, where toxicity symptoms were first observed in the older leaves. However, even when the Mn concentration in the nutrient medium was doubled to 200 μM Mn by also raising the Mg supply tenfold (to 7.5 m*M* Mg) the plants remained healthy despite having a somewhat higher concentration of Mn in the shoots than in the plants suffering from Mn toxicity.

From this evidence we conclude that a high concentration of Mg in plant tissues can alleviate Mn toxicity even though Mn tissue concentration is also high. If this is so, the Mg : Mn ratio should be a better indicator of plant Mn status in relation to Mn toxicity than Mn concentration alone. This is illustrated by comparing Figures 1 and 2. In Figure 1, Mn concentrations of 40 samples of old tomato leaves obtained from 5 replicate plants from 8 different treatments of Mg and Mn supply are evaluated in relation to the appearance of Mn toxicity symptoms. These older leaves which are the first plant tissues to show toxicity symptoms, have concentrations ranging from 658 $\mu g\ Mn\ g^{-1}$ to 8579 $\mu g\ Mn\ g^{-1}$ and the concentration is positively related to the extent to which the leaves show toxicity. However in the intermediate range there is no correlation, as leaves of 3758 $\mu g\ Mn\ g^{-1}$ show toxicity whereas those with 5270 $\mu g\ Mn\ g^{-1}$ are healthy. When these data are considered in terms of the Mg : Mn ratio (Fig. 2), the healthy leaves always have higher ratios than the leaves showing toxicity. The critical Mg : Mn ratio for the appearance of toxicity symptoms in old tomato leaves lies between 2.27 and 3.51. Corresponding values for the shoot are 3.53 and 6.54.

Measurements of Mn in the shoots of wheat plants grown on an unlimed Mn toxic soil showed that shoot growth declined as Mn concentration increased above 100$\mu g\ g^{-1}$ (Fig. 3).

Table 2. The influence of magnesium and manganese supply in the nutrient medium at given ratios Mg : Mn on the relative shoot yields, Mg and Mn concentrations in shoots and on the ratios of Mg : Mn in the shoot of tomato plants. (Ratios calculated in terms of $\mu g\ g^{-1}$ dry weight)

Treatment[a]	Relative dry matter yield of shoots (g)	Mg concentration in shoot[a] ($\mu g\ g^{-1}$)	Mn	Mg : Mn ratio in shoot (in terms of $\mu g\ g^{-1}$)	Mg : Mn ratio in solution (in terms of $\mu g\ ml^{-1}$)
10 μM Mn, 0.75 m*M* Mg	100	6746	286	23.6	33.2
100 μM Mn, 0.75 m*M* Mg*	74.7	6061	1800*	3.4	3.3
200 μM Mn, 7.5 m*M* Mg	—	16190	1905	8.5	16.6

[a]* denotes the presence of Mn toxicity symptoms.

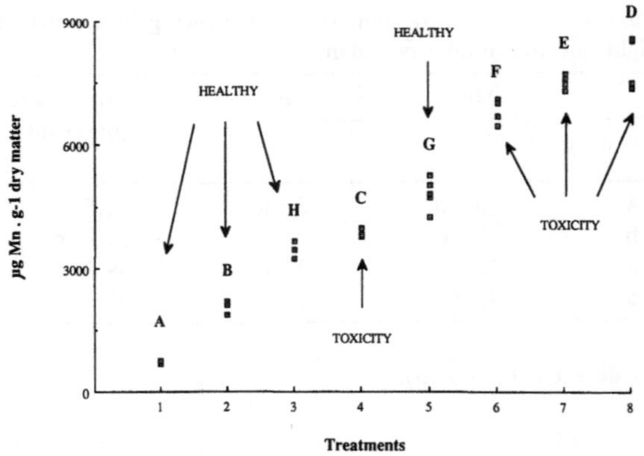

Fig. 1. Manganese concentrations in the old leaves obtained from 5 replicate tomato plants from 8 different nutritional regimes of magnesium and manganese. Treatments: A,0.75 m*M* Mg, 10 μ*M* Mn; B, 0.75 m*M* Mg, 50 μ*M* Mn; C, 0.75 m*M* Mg, 100 μ*M* Mn; D, 0.75 m*M* Mg, 300 μ*M* Mn; E, 0.75 m*M* Mg, 200 μ*M* Mn; F, 1.5 m*M* Mg, 200 μ*M* Mn; G, 7.5 m*M* Mg, 200 μ*M* Mn; H, 15 m*M* Mg, 200 μ*M* Mn

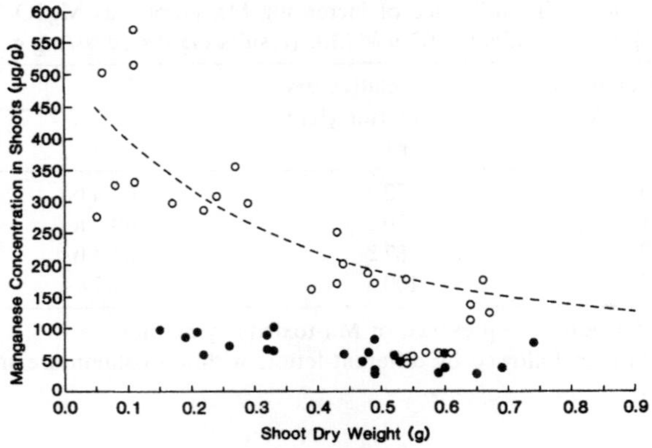

Fig. 3. Relationship between shoot dry weights and manganese concentration in the leaves of wheat. Closed symbols denote plants grown in soil with calcium lime added.

When the soil was limed, however, the Mn concentrations were decreased and fairly constant despite considerable differences in shoot growth. There was thus no satisfactory relationship between growth and the concentration of Mn (or Mg) in the shoots. By expressing the growth as a function of the ratio Mg:Mn concentration in shoot tissues obtained from plants grown on two different soils (limed and unlimed soil) (Fig. 4), an exponential equation of the form $Y = A + B \exp(-kX)$ was obtained in which: Y = shoot growth, X = Mg:Mn ratio in the shoot and A, B and k = constants. Using this relationship it was possible to predict shoot growth from Mg:Mn ratios in the leaves. Growth was significantly reduced when the ratio fell below 20:1.

A similar curve between Mg:Mn ratio and dry matter yield may also be derived from the data of Elamin and Wilcox (1986a; b) (Fig. 5), who investigated Mg/Mn interactions in melons. Again a shoot ratio of about 20:1 is required for optimum yield.

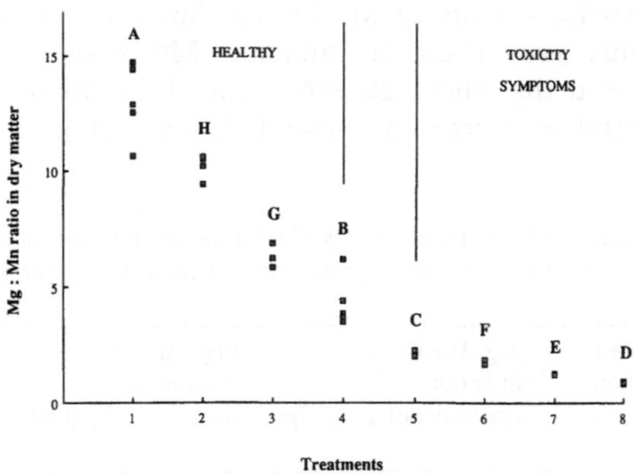

Fig. 2. Magnesium:manganese ratios [expressed in terms of $mg.g^{-1}$ dry weight] in the old leaves obtained from 5 replicate tomato plants from 8 different nutritional regimes of magnesium and manganese. (Treatments as for Fig. 1).

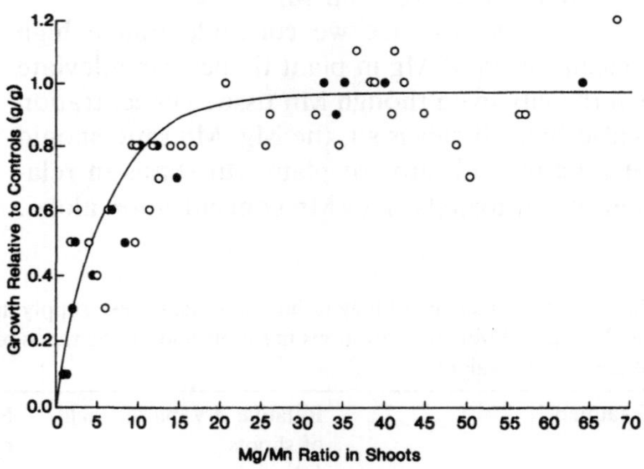

Fig. 4. Relationship between shoot growth and the magnesium : manganese ratio in wheat from limed (closed symbols) and unlimed soils (open symbols). The fitted curve is for the combined data obtained from both soils; values for constants: A = 0.97 B = −A (0.028) k = 0.167 (0.019) (numbers in brackets denote standard errors)

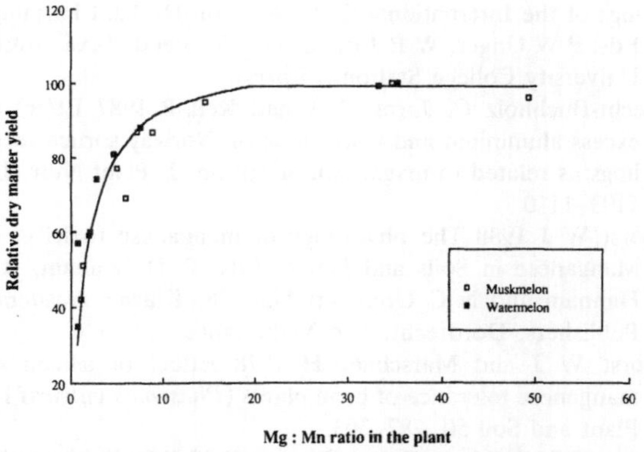

Fig. 5. Relationship between relative yields of muskmelon and watermelon and the magnesium : manganese ratio in the plants. (recalculated from the data of Elamin and Wilcox, 1986a, b).

Discussion

Our results with tomato show clearly that increasing Mg supply in the nutrient medium depressed Mn uptake, in accordance with earlier findings for other plant species (Elamin and Wilcox, 1986a; 1986b; Maas *et al.*, 1969). However this was not the only influence of Mg in alleviating Mn toxicity in tomato. Plants rich in Mg are better able to tolerate high Mn tissue concentrations and the Mg:Mn ratio in shoot tissue is a good indicator in predicting the presence or absence of Mn toxicity symptoms in older leaves (Table 2, Figs. 1 and 2). We therefore agree with the conclusion of Horst (1988) that for a given species or cultivar it is not possible to identify a Mn tissue concentration at which toxicity will occur.

In tomato we were not able to measure the effect of the Mg:Mn ratio on yield, as in each of the two experiments there were only 4 treatments. However for wheat when growth was expressed as a function of the ratio of Mg to Mn concentrations in the leaves, a relationship of the form $Y = A + B \exp(-kX)$ was obtained. This relationship indicates that growth was significantly reduced when the ratio of the shoot Mg:Mn fell below 20:1. From this evidence it is possible to predict that the limitation of growth depends on the ratio of the two ions in the shoot and not the absolute concentration of Mn.

The Mg:Mn leaf ratio of 20:1 for optimum growth of wheat is similar to values for melons derived from the results of Elamin and Wilcox (1986a, b) but is higher than the ratio which we found in healthy leaves of tomato. The lower ratio for tomato may indicate the much greater tolerance of this crop to Mn (Edwards and Asher, 1982; Le Bot *et al.*, 1990). Also the ratios for tomato relate to the appearance of toxicity symptoms rather than to optimum growth.

The effect of increasing Mg supply in the nutrient medium in depressing Mn uptake relates to the competition of the two divalent cations in membrane transport. It is of interest in this respect that competition for binding sites in the roots is more than a 1:1 competition (in μM terms) in favour of Mn. According to Marschner (1986) Mn^{2+} not only competes more effectively but also blocks binding sites for Mg^{2+} uptake. The competitive effect of Mg^{2+} in depressing Ca^{2+} and to a lesser extent K^+ uptake has been reported by other workers, and may be interpreted as general cation competition for cellular anion charge (Mengel and Kirkby, 1987).

It is still largely a matter of speculation why increasing concentrations of Mg in the plant should raise plant tolerance to Mn. A similar effect of Si on Mn tolerance of bean plants has been reported by Horst and Marschner (1978). These workers suggest that increased tolerance results from the altered and more homogeneous microdistribution of Mn brought about by Si. In the case of Mg and Mn it is well established that both ions are similar and to a certain extent interchangeable in biochemical behaviour. The ratio of Mg:Mn in plant cells may therefore also affect intracellular Mn distribution thereby increasing tissue tolerance. When the ratio is high, even in the presence of high concentrations of Mn^{2+}, Mg^{2+} may replace Mn^{2+} from physiologically active sites in the cytoplasm and Mn^{2+} may be sequestered in cell walls and vacuoles and thus rendered harmless.

Recent evidence of Houtz *et al.* (1988) from experiments with tobacco also indicates that the ratio of Mg:Mn in leaf tissue can influence growth by a direct effect on the rate of net photosynthesis, as ribulose bisphosphate (RuBP) carboxylase/oxygenase is activated by either Mg^{2+} or Mn^{2+}. Under normal circumstances when Mg^{2+} is dominant, carboxylation is favoured but when Mn^{2+} is high relative to Mg^{2+}

the oxidative rather than the reductive photosynthetic cycle becomes operative. One of the earliest physiological symptoms of Mn toxicity is thus a fall in the net rate of photosynthesis. The alleviating effect of high Mg tissue concentration on Mn toxicity may therefore depend on this relationship, a high Mg : Mn ratio being required for RuBP carboxylase activity, and hence also net photosynthesis and growth. The presence of Mn oxides as a symptom of Mn toxicity in the plant tissues may also result from the shift towards oxidative processes when the Mg : Mn ratio is low.

References

Chinnery L E and Harding C P 1980 The effect of ferrous ion on the uptake of manganese by *Juncus effusus* L. Ann. Bot. 46, 409–412.

Edwards D G and Asher C J 1982 Tolerance of crop and pasture species to manganese toxicity. *In* Plant Nutrition 1982 Ed. A Scaife. pp 145–150. Proc. 9th. Intern. Colloq., Coventry. Commonw. Agric. Bur., Slough, UK.

Elamin O M and Wilcox G E 1986a Effect of magnesium and manganese nutrition on muskmelon growth and manganese toxicity. J. Am. Soc. Hortic. Sci. 111, 582–587.

Elamin O M and Wilcox G E 1986b Effect of magnesium and manganese nutrition on watermelon growth and manganese toxicity. J. Am. Soc. Hortic. Sci. 111, 588–593.

Goss M J and Carvalho M J G P R 1989 Causes in variation in yields of wheat under dryland farming in the Alentejo region of Portugal and some future prospects. *In* Proceedings of the International Conference on Dryland Farming. Eds. P W Unger, W R Jordan and I V Sneed. Texas A&M University College Station. *In press*.

Hecht-Buchholz C, Jorns C A and Keil P 1987 Effect of excess aluminium and manganese on Norway spruce seedlings as related to magnesium nutrition. J. Plant Nutr. 10, 1103–1110.

Horst W J 1988 The physiology of manganese toxicity. *In* Manganese in Soils and Plants. Eds. R D Graham, R J Hannam and N C Uren. pp 175–188. Kluwer Academic Publishers, Dordrecht, The Netherlands.

Horst W J and Marschner H 1978 Effect of silicon on manganese tolerance of bean plants (*Phaseolus vulgaris* L.) Plant and Soil 50, 287–303.

Houtz R L, Nable R O and Cheniae G M 1988 Evidence for effects on the *in vivo* activity of ribulose bisphosphate carboxylase/oxygenase during development of Mn toxicity in tobacco. Plant Physiol. 86, 1143–1149.

Le Bot J, Kirkby E A and Van Beusichem M L 1990 Manganese toxicity in tomato plants: Effects on cation uptake and distribution. J. Plant Nutr. (*In Press*).

Löhnis M P 1960 Effects of magnesium and calcium supply on the uptake of manganese by various crop plants. Plant and Soil 12, 339–375.

Maas E V, Moore D P and Mason B J 1969 Influence of calcium and magnesium on manganese absorption. Plant Physiol. 44, 796–800.

Marschner H 1986 Mineral Nutrition of Higher Plants. Academic Press, London 674 p.

Mengel K and Kirkby E A 1987 Principles of Plant Nutrition, 4th Edition. International Potash Institute, Bern, Switzerland, 687 p.

Simon J E, Wilcox G E, Simini M, Elamin O M and Decoteau D R 1986 Identification of manganese toxicity and magnesium deficiency on melons grown in low pH soils. HortScience 21, 1383–1386.

M. L. van Beusichem (Ed.), *Plant nutrition – physiology and applications*, 229–233.

PLSO IPNC011

Effects of rhizosphere processes on the solubilization of manganese as revealed with radioisotope techniques

R.A. YOUSSEF[1] and M. CHINO
Agricultural Chemistry Department, University of Tokyo, Tokyo 113, Japan. [1]Present address: Soils and Water Use Department, National Research Centre, Dokki, Egypt

Key words: barley, *Glycine max* L., *Hordeum vulgare* L., heavy metals, rhizobox, rhizosphere pH, soybean genotypes

Abstract

The contributions of rhizosphere processes to Mn solubilization was evaluated in a rhizobox system. The results indicate that soil pH changed remarkably across the rhizosphere. Rhizosphere processes are responsible for the solubilization of Mn and Fe compounds in soil. Apparently, soybean has a greater ability than barley to solubilize soil Mn.

Introduction

Factors involved in solubilizing Mn in the rhizosphere are very incompletely understood (Geering *et al.*, 1969). In our rhizosphere research, it was observed that plant roots greatly affect soil pH in their vicinity, which in turn may affect heavy metal availability across the rhizosphere (Youssef and Chino, 1989). So far, no success has been achieved experimentally in determining whether the effect of root exudate in the rhizosphere soil may or may not be directly involved in the solubilization of heavy metals (Loneragan, 1975).

The experiments reported here were conducted to evaluate the contribution of processes in the rhizosphere of different plant species to the solubilization of Mn and Fe compounds in soil, with use made of ^{54}Mn and ^{59}Fe.

Materials and methods

Experiment 1: Preparation of rhizosphere soil of different plant species

The rhizoboxes were filled with a clay loam soil previously treated with fertilizer (0.5 g N as $(NH_4)_2SO_4$, 0.5 g P_2O_5 as NaH_2PO_4, and 0.5 g K_2O as KCl for 3 kg soil). Briefly, the rhizobox system consists of several soil compartments differing in thickness, and separated by 500-mesh nylon cloth. Seedlings (barley and soybean) were planted in the central compartment (C.C.), as shown in Figure 1 (Youssef and Chino, 1988).

Two cultivars of soybean (*Glycine max* L. Merr), Hawkeye and Bride B216 and one variety of barley (*Hordeum vulgare* L. var. Uzuakasinriki) were used. Hawkeye (HA) is known as an Fe-efficient and Bride B216 (BB) as an Fe-inefficient cultivar. Seeds were germinated and eight of them for each plant species were transplanted into the C.C. of a rhizobox. The plants grew in a growth chamber with natural light, a day/night temperature regime of 25/20°C, and a water content of the soil kept constant at 50% of WHC (Youssef and Chino, 1988).

One month after planting, all rhizoboxes were dismantled and the soil was peeled away from the nylon cloth. Soil samples were collected from the C.C. and from the compartments 1 mm and 2 mm away from the C.C. Soil from these compartments was designated as C.C.−, 1 mm−, and 2 mm soil and considered to be rhizosphere soil. Other soil samples were collected from the compartments >5 mm away from the C.C. These samples were designated as bulk soil. The pH of

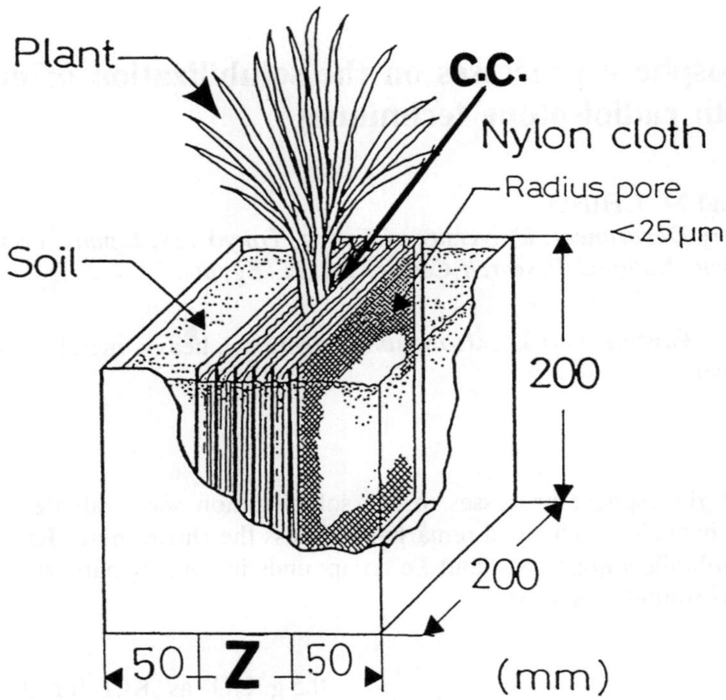

Fig. 1. Schematic diagram of rhizobox. C.C.: Central soil compartment, Z: Several soil compartment of 1 mm thickness.

each soil sample was determined. With use made of the rhizosphere soils, the following two experiments (Exps. 2 and 3) were carried out.

Experiment 2: Solubilization of Fe and Mn in rhizosphere soil

To examine processes in rhizosphere soil affecting the solubility of Mn and Fe, radioisotopes-labelled clay loam soil was prepared. Forty uCi of $^{54}Mn(Cl)_2$ or $^{59}Fe(Cl)_3$ in water was mixed thoroughly with 200 g dry soil. The labelled soil (10 g) was mixed with 10 g of soil from either the C.C., the 1 mm− or the 2 mm compartment or with bulk soil.

The solubilities of ^{54}Mn and ^{59}Fe in an ammonium acetate-acetic acid (AA-AA) solution were evaluated and their concentrations were estimated with use made of the Auto Well Gamma Scintillation Counter.

Experiment 3: Solubilization of MnO_2 and Fe_2O_3 in rhizosphere soil

To investigate processes in rhizosphere soil promoting the solubilization of MnO_2 and Fe_2O_3. $^{54}MnO_2$ and $^{59}Fe_2O_3$ were prepared from $^{54}Mn(Cl)_2$ and $^{59}Fe(Cl)_3$. Ten mg of the labelled oxides (either $^{54}MnO_2$ or $^{59}Fe_2O_3$) were mixed with 10 g of soil from each compartment. The increases in solubility of $^{54}MnO_2$ and $^{59}Fe_2O_3$ were evaluated and the concentrations of radioisotopes were estimated as described in Expt. 2.

Results

Experiment 1: pH of the rhizosphere

The pH values in the rhizosphere and bulk soil are recorded in Table 1. The pH in the rhizosphere was decreased by as much as 2 units compared with that of the bulk soil. The pH drop in the rhizosphere of soybean was greater than that of barley.

The rhizosphere pH of the Fe-efficient soybean genotype (HA) was always lower than that of the Fe-inefficient one (BB).

Table 1. Changes in pH values across the rhizosphere of different plant species

Plant species	pH values at different distance from the root			
	C.C.	1 mm	2 mm	bulk (>5 mm)
Barley	4.90	5.48	6.28	6.93
Soybean (HA)	4.15	4.78	5.40	6.92
Soybean (BB)	4.35	5.00	6.16	6.92

Experiment 2: Effect of rhizosphere soil on the solubility of the radio-isotopes in soil

Soluble ^{54}*Mn*

Solubilization of soil Mn in bulk soil and rhizosphere soil was compared. The amount of ^{54}Mn dissolved was considerably greater in rhizosphere soil than in bulk soil (Fig. 2). Processes promoting the solubilization of ^{54}Mn in rhizosphere soil were more intense for soybean than for barley (Fig. 2). Soluble ^{54}Mn in the rhizosphere of soybean was 3 to 9 time higher than in the bulk soil. In case of barley, the solubility of ^{54}Mn was also increased, but less so than for soybean.

Soluble ^{59}*Fe*

Marked differences in soluble Fe^{59} in soil were observed between rhizosphere soil and bulk soil, with concentrations being more than twice as high in the former than in the latter (Fig. 3). This rhizosphere effect was found to be restricted almost to the C.C., although pH changes also occurred beyond the C.C., as shown by the data of Table 1.

Experiment 3. Solubilization of labelled MnO_2 *and* Fe_2O_3 *in the rhizosphere soil*

Soluble $^{54}MnO_2$

The amount of $^{54}MnO_2$ dissolved in the rhizosphere soil of the C.C. was considerably greater than that dissolved in the bulk soil. Solubilization was more pronounced in the rhizosphere soil of soybean than in that of barley (Table 2), and more so for the BB than for the HA cultivar, in contrast to the relative abilities of these cultivars, to lower the pH of their rhizospheres (Table 1).

Soluble $^{59}Fe_2O_3$

The amount of $^{59}Fe_2O_3$ dissolved in rhizosphere soil (C.C.) was more than twice as high as in bulk soil (Table 2). In this case the Fe-efficient HA soybean cultivar exceeded the non-efficient BB cultivar in solubilizing capacity.

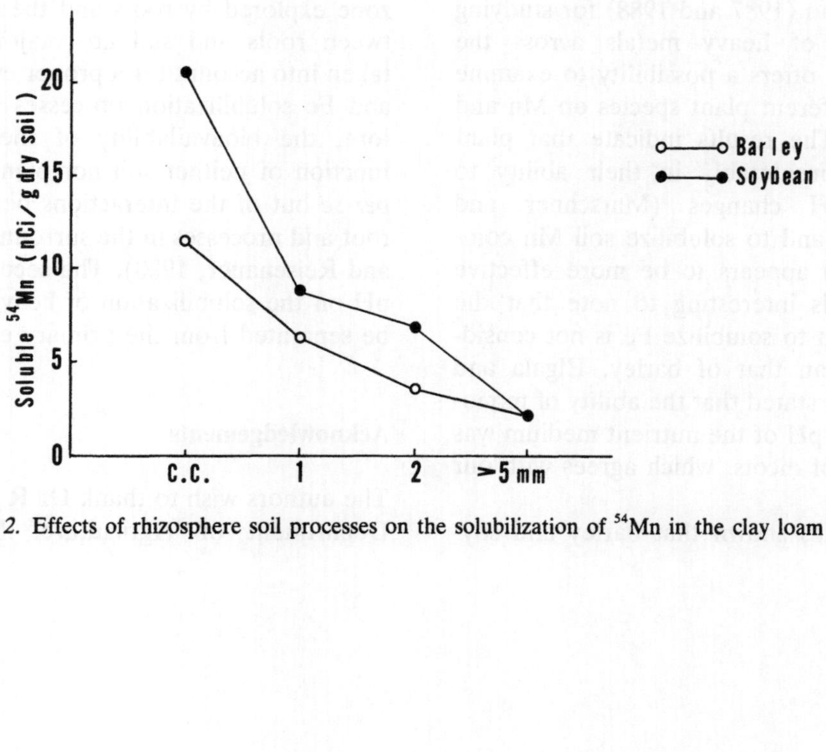

Fig. 2. Effects of rhizosphere soil processes on the solubilization of ^{54}Mn in the clay loam soil.

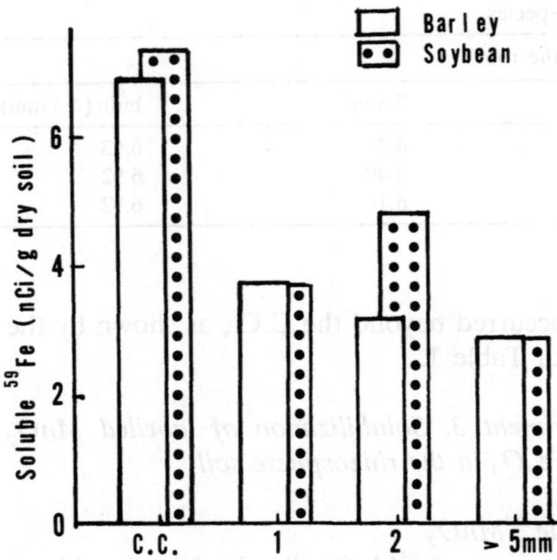

Fig. 3. Effects of rhizosphere soil processes on solubilization of ^{59}Fe in the clay loam soil.

Table 2. Effect of rhizospheric soil on the solubilization of labelled MnO_2 and Fe_2O_3

Soil sample		Soluble RI (nCi/g soil)	
		^{54}Mn	^{59}Fe
Bulk soil	>5 mm	31	0.18
Barley	(C.C)	43	0.38
Soybean (HA)	(C.C)	58	0.55
Soybean (BB)	(C.C)	81	0.38

Discussion

The rhizobox system which was developed by Youssef and Chino (1987 and 1988) for studying the distribution of heavy metals across the rhizosphere, also offers a possibility to examine the effects of different plant species on Mn and Fe solubilities. The results indicate that plant species differ considerably in their ability to bring about pH changes (Marschner and Römheld, 1983) and to solubilize soil Mn compounds. Soybean appears to be more effective than barley. It is interesting to note that the ability of soybean to solubilize Fe is not considerably higher than that of barley. Elgala and Amberger (1988) stated that the ability of monocots to lower the pH of the nutrient medium was lower than that of dicots, which agrees with our findings.

However, it was shown that barley and soybean were almost equally effective in solubilizing Fe in the soil. For barley it is known that mugineic acid, a phytosiderophore, is excreted to dissolve Fe in soil (Shi *et al.*, 1988). Thus, root excretion may be an important factor involved in increasing Fe solubility in the rhizosphere. The increase in Mn solubility is observed across the rhizosphere. This is in accordance with an earlier observation (Bromfield, 1958), that root washings contained substances capable of dissolving MnO_2. Rhizosphere flora has also been shown to be involved in the solubilization of sparsely soluble nutrients (Jones and Leeper, 1951).

Soybean cultivar HA was found to dissolve more Fe than cultivar BB. Brown and Jones (1962) reported that the reductive capacity of soybean (HA) roots was larger than that of cultivars susceptible to Fe deficiency, which was reconfirmed in our experiment. Surprisingly, soybean BB dissolved more Mn than HA even though the rhizosphere pH of the former was higher than that of the latter (Table 2). Thus in soybean rhizosphere the mechanism of Fe solubilization may be different from that of Mn.

The results indicate that the extents of pH changes as affected by root activities are not always closely correlated with Fe- and Mn-solubilizing capacities (Table 1 and Figs. 2 and 3). This suggests that other factors, such as root excretion and/or microbial activity are more involved in the solubilization of Mn and Fe than are pH changes in the rhizosphere.

The results show that the extent of the soil zone explored by roots and the interactions between roots and soil are major factors to be taken into account for a proper evaluation of Mn and Fe solubilization processes in soils. Therefore, the bioavailability of these metals is a function of neither soil nor plant characteristics *per se* but of the interactions between the plant root and processes in the surrounding soil (Godo and Reisenauer, 1980). The secondary effects of pH on the solubilization of heavy metals should be separated from the primary effects.

Acknowledgements

The authors wish to thank Dr R L Chaney, U.S. Department of Agriculture, Maryland, who kindly provided us soybean seeds, and for his interest in and fruitful discussions on this work.

References

Bromfield S M 1958 The properties of a biologically formed manganese oxide, its availability to oats and its solution by root washings. Plant and Soil 9, 325–337.

Brown J C and Jones W E 1962 Absorption of Fe, Mn, Zn, Ca, Rb and phosphate ions by soybean roots that differ in their reductive capacity. Soil Sci. 94, 173–179.

Elgala A M and Amberger A 1988 Root exudate and the ability of corn to utilize insoluble sources of iron. J. Plant Nutr. 11, 677–699.

Geering H R, Hodgson J F and Sdano C 1969 Micronutrients cation complexes in soil solution. IV. The chemical state of manganese in soil solution. Soil Sci. Soc. Am. Proc. 33, 81–85.

Godo G H and Reisenauer H M 1980 Plant effects on soil manganese availability. Soil Sci. Soc. Am. J. 44, 993–995.

Jones L H P and Leeper G W 1951 The availability of various manganese oxides to plants. Plant and Soil 3, 141–153.

Krauskopf K B 1972 Geochemistry of micronutrients. *In* Micronutrients in Agriculture. Eds J J Mortved, P M Giordano and W L Lindsay. pp 7–36. Soil Science Society of America, Inc. Madison, WI.

Leeper G W and Swaby R J 1940 The oxidation of manganous compounds by mircroorganisms in soil. Soil Sci. 49, 163–169.

Loneragan J F 1975. The availability of and adsorption of trace elements in soil-plant systems and their relation to the movement and concentration of trace elements in plants. *In* Trace Elements in Soil-Plant-Animal Systems. Eds. D J D Nicholas and A R Egan. pp 109–134. Acad. Press, Inc., New York.

Marschner H and Römheld V 1983 *In-vivo* measurement of root-induced pH changes at the soil-root interface: Effect of plant species and nitrogen source. Z. Pflanzenernaehr. Bodenkd. 149, 441–456.

Shi W, Chino M, Youssef R A, Mori S and Takagi S 1988 The occurrence of mugineic acid and in the rhizosphere soil of barley plant. Soil Sci. Plant Nutr. 34, 585–592.

Youssef R A and Chino M 1987 Studies on the behavior of nutrients in the rhizosphere. 1. Establishment of a new rhizobox system to study nutrient status in the rhizosphere. J. Plant Nutr. 10, 1185–1195.

Youssef R A and Chino M 1988 Development of a new rhizobox system to study the nutrient status in the rhizosphere. Soil Sci. Plant Nutr. 34, 461–465.

Youssef R A and Chino M 1989 Root-induced changes in the rhizosphere of plants. 1. pH changes in relation to the bulk soil. Soil Sci. Plant Nutr. 35, 461–468.

M. L. van Beusichem (Ed.), *Plant nutrition – physiology and applications*, 235–239.

PLSO IPNC086B

Effect of manganese supply on development of wheat (*Triticum aestivum*) roots

A.J. WEBB and B. DELL[1]
School of Biological and Environmental Sciences, Murdoch University, Murdoch, WA 6150, Australia.
[1]*Corresponding author*

Key words: anatomy, lignin, manganese, nutrient deficiency, root growth, *Triticum aestivum* L., wheat

Abstract

Wheat seedlings were grown in nutrient solution containing 1 μM Mn or nil Mn. Witholding Mn depressed root growth parameters including total root length, number of lateral roots, mean root length, mean extension rate, relative extension rate, and relative multiplication rate. Mn deficiency increased the number of lateral root primordia, many of which failed to emerge. Lignification of protoxylem vessels was impaired in Mn deficient plants. Withholding Mn affected lignification 3 to 5 days in advance of the appearance of symptoms of Mn deficiency in the shoot or depression in root and shoot dry matter production.

Introduction

Manganese deficiency is widely encountered in cereal growing areas of southern Australia (Donald and Prescott, 1975). In these areas grain yields may be reduced by the impairment of metabolic processes such as photosynthesis (Campbell and Nable, 1988) as well as by increased incidence of disease (Graham and Rovira, 1984; Wilhelm *et al.*, 1985). Furthermore, towards maturity, reduced root growth may limit the plants' access to water if roots fail to grow into deeper soil horizons low in available Mn. There is little detailed information, however, on the effect of Mn supply on root growth in cereals. Data for broadleaved species such as tomato (Abbott, 1967) and subterranean clover (Nable and Loneragan, 1984) suggest that the external Mn supply strongly affects root growth. The present experiment was undertaken to investigate effects of withholding Mn on both the structure and growth of roots in young wheat seedlings grown in water culture.

Materials and methods

A solution culture experiment was conducted in water baths at 20–22°C in the glasshouse with wheat (*Triticum aestivum* cv Gamenya). Procedures for the preparation of nutrient solutions and basal nutrient levels were as described previously (Webb and Loneragan, 1985), except for the following changes; 2 M $Ca(NO_3)_2$, 200 μM K_2HPO_4/KH_2PO_4, and 2 M NH_4NO_3. Mn was supplied as 1 μM $MnSO_4$.

The effect of Mn deficiency on root growth was studied by comparing plants grown in a complete nutrient solution (Mn+), with plants grown in a complete nutrient solution minus Mn (Mn−). Seeds were pregerminated in aerated DDI water in darkness and were transferred into treatments when the roots associated with the scutellar node had emerged (D 1). Plants were harvested on D 5, D 10, D 15, D 20 and D 25. Root branching configuration was classified according to Barley (1970). The root system was divided into seminal and nodal roots; root axes and lateral roots. Second order lateral roots were only observed on Mn+ roots at D 25. The following measurements were taken: total root length, root axis length, lateral root length, number of root axes, number of emerged first order lateral roots, number of lateral initials (or lateral root primordia, as seen by the unaided eye), root and shoot fresh and dry weights. Relative multiplication rates (RMR), relative extension rates (RER), mean extension rates (MER) and mean root lengths (MRL) were calculated using equations from the root growth model developed by May *et al.* (1965), Mn analysis was carried out on nitric-perchloric acid digests using atomic absorption spectrophotometry.

To assess anatomical effects, segments of roots were taken from a range of morphological positions in the root system (Fig. 2a). These were fixed overnight in 2% gluteraldehyde in nutrient solution at 21°C, postfixed in 1% OsO_4 for 2 h, dehydrated in ethanol and propylene oxide, and gradually infiltrated with araldite over a period of several days. Sections 0.7 μm thick were stained with toluidine blue O (Chroma-Gessellschaft, 1B481) in 1% borax. Unstained sections, 2.0 μm thick, with the resin removed (Imai *et al.* 1968), were examined for lignin using autofluorescence of the cell walls (Zeiss Catalogue No. 487718). Lignification of protoxylem vessels was assessed on a scale from 1 (purple with toluidine blue, no fluorescence) to 5 (blue-green, intense fluorescence).

Results and discussion

Plant growth

Plants supplied with 1 μM Mn grew well over the experimental period. Foliar symptoms of Mn deficiency, as described by Snowball and Robson (1983), were first evident in the Mn− treatment on D 13, and by D 15 there was a depression in dry matter accumulation in the Mn− seedlings. At D 25, growth of roots and shoots of plants without an external supply of Mn was depressed by approximately 80% and 70%, respectively (Fig. 1a). Mn concentrations were depressed at D 5 (Fig. 1b) and by D 15 the Mn concentrations of the shoot of the Mn− plants were below the critical concentration of the whole shoot (11–13 μg/g Mn dry wt, Graham *et al.*, 1985).

Effect of withholding Mn on the growth and development of the root system

Withholding Mn depressed the overall growth and development of the root system. For example, at D 25 total root length was reduced 5 fold from 20.4 m (Mn+) to 3.9 m (Mn−). Lateral roots contributed significantly to the depression in total root length: at D 25 lateral roots were 85% and 43% of the total root length in Mn+ and Mn− plants, respectively. Furthermore, nodal root initiation was delayed by 10 days (D 25) in Mn− plants, compared with Mn+ plants. The delay in production of nodal roots was due to delayed tillering in these plants.

The effects of Mn deficiency on root growth and development were most severe in the lateral roots: reduced root number (Fig. 1c), total root length (Fig. 1d), and mean root length (Fig. 1e). In Figure 1e, the MRL of the lateral roots in the seminal axes increased slowly for 10 days after initiation, prior to an almost exponential increase in MRL of the laterals on the Mn+ roots axes, whereas the MRL of the Mn− laterals decreased marginally. The MRL of the seminal and nodal root axes was also depressed in response to Mn deficiency (Fig. 1f). Clearly, Mn deficiency severely reduced both lateral root initiation and lateral root elongation.

Other measures of root growth (RMR, RER and MER) also indicated that Mn deficiency depressed the rate of root growth. The RMR, RER and MER values for lateral roots produced on seminal root axes are given in Table 1. The values for Mn+ plants are comparable with those given by Tennant (1976) for wheat seedlings supplied with standard levels of N, P and K. Another characteristic of Mn deficiency was an increase in the number of lateral primordia, which failed to penetrate through the cortex and epidermis of the main root axis (Fig. 1g). The density of branching, the number of lateral roots per cm of root axis, was also affected by the Mn supply. There were fewer first order lateral roots on the Mn− root axes than on the Mn+ root axes (Fig. 1h).

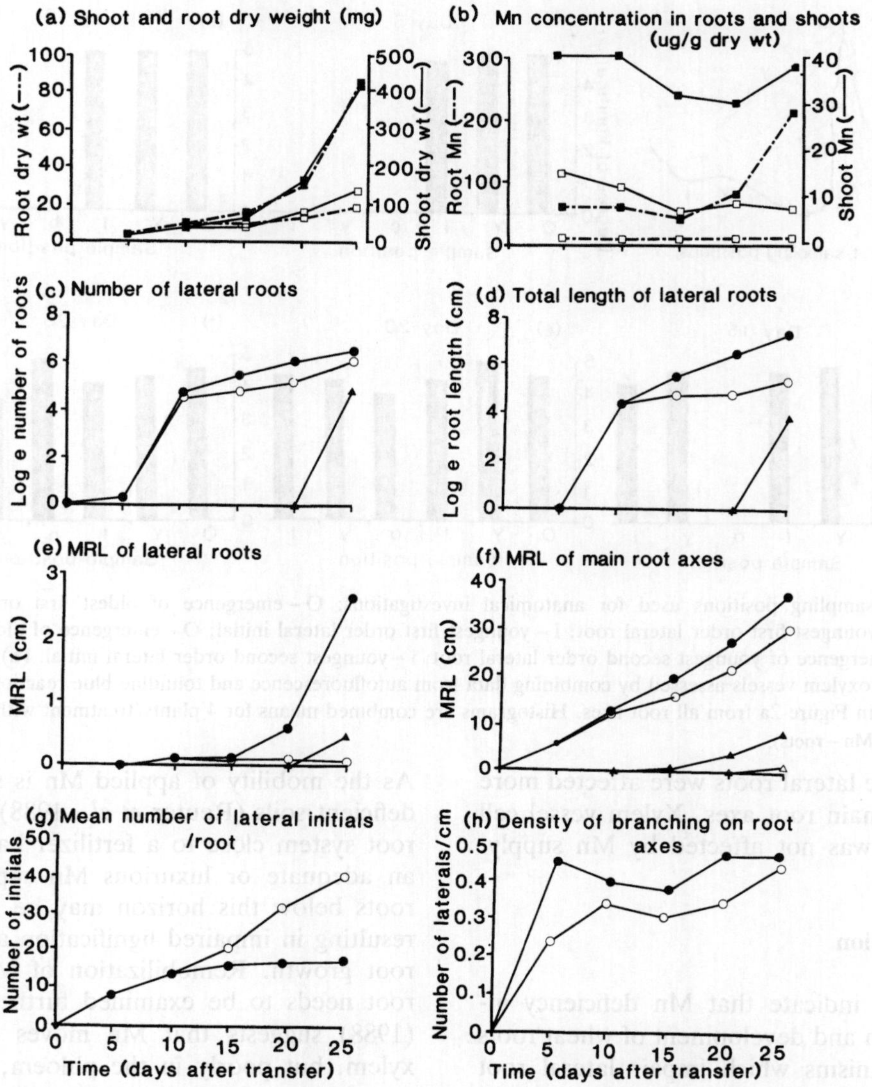

Fig. 1. Effect of Mn supply on: (**a**) dry weight of roots and shoots; (**b**) Mn concentration in roots and shoots; (**c**) number of first order lateral roots on seminal and nodal root axes; (**d**) total length of lateral roots on seminal and nodal roots; (**e**) mean root lengths of lateral roots and (**f**) main root axes; (**g**) mean number of lateral root initials (primordia) per root axis; and (**h**) the number of lateral roots per cm of main root axis. Data from 6 replicate plants per treatment. (Mn + roots ■- - -■ ; Mn – roots □- - -□; Mn + shoots ■——■; Mn – shoots □——□; Mn + seminal roots ●——● ; Mn – seminal roots ○——○ ; Mn + nodal roots ▲——▲ ; Mn – nodal roots △——△).

Table 1. Effect of Mn supply on constant rates of increase for relative multiplication rates, relative extension rates and mean extension rates for lateral roots produced on seminal root axes, from D 15 to D 25. Values in parentheses are standard errors

	Mn+	Mn−
RMR	0.13 (0.01)	0.10 (0.01)
RER	0.20 (0.02)	0.07 (0.02)
MER	0.28 (0.03)	0.06 (0.01)

Effect of withholding Mn on root anatomy

Withholding Mn prevented complete lignification in vascular tissues. By D 10 lignification was impaired in protoxylem cell walls (Fig. 2). As seed Mn reserves were depleted, the degree of lignification decreased, with the lignified cell walls of the younger parts of the root system being more susceptible to Mn deficiency than the

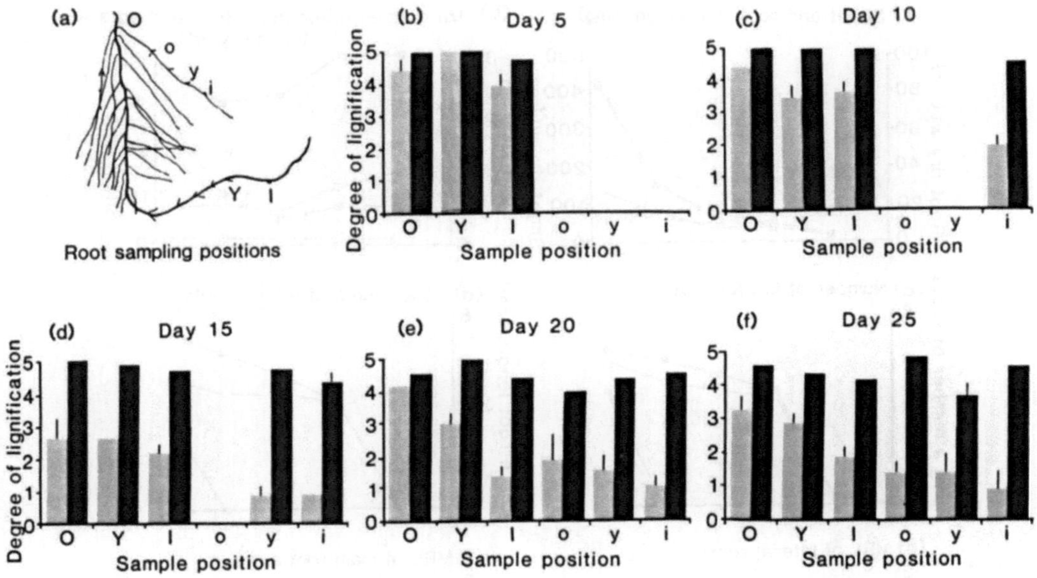

Fig. 2. (**a**) Root sampling positions used for anatomical investigations: O – emergence of oldest first order lateral root; Y – emergence of youngest first order lateral root; I – youngest first order lateral initial; O – emergence of oldest second order lateral root; y – emergence of youngest second order lateral root; i – youngest second order lateral initial. (**b**) to (**f**) Degree of lignification of protoxylem vessels assessed by combining data from autofluorescence and toluidine blue reactions. Samples were taken as indicated in Figure 2a from all root axes. Histograms are combined means for 4 plants/treatment with standard errors. (■, Mn + roots; ▨, Mn – roots).

older parts. The lateral roots were affected more severely than main root axes. Xylem vessel cell wall thickness was not affected by Mn supply.

General discussion

Present results indicate that Mn deficiency affects the growth and development of wheat roots through mechanisms which impair lateral root initiation, root elongation and lignification of cell walls. Abbott (1967) reported that in tomato, Mn deficiency inhibited lateral meristem production and main axis growth in excised roots. In wheat, lignification was depressed before the expression of foliar Mn deficiency symptoms. Mn concentrations in the roots of Mn− plants were below the critical concentrations reported by Graham *et al.* (1985) for the YEB. However it is not known whether the Mn in the roots was uniformly distributed in the current experiment.

Plants grown in the field may not experience such low soil solution levels of Mn as in the Mn− treatment. However, the root system is likely to be exposed to horizontal layering of nutrients as a result of application of Mn enriched fertilizers. As the mobility of applied Mn is slight in most deficient soils (Reuter *et al.*, 1988), parts of the root system close to a fertilizer band may have an adequate or luxurious Mn supply, whereas roots below this horizon may be Mn deficient, resulting in impaired lignification and depressed root growth. Remobilization of Mn within the root needs to be examined further. Loneragan (1988) suggests that Mn moves freely in the xylem, but poorly in the phloem, and rates of translocation of Mn from Mn adequate parts of the root to Mn deficient parts may not be sufficient to maintain vigorously growing roots.

Impaired lignification of protoxylem within the root caused by Mn deficiency, (as suggested by Campbell and Nable, 1988; Burnell, 1988), may render the plant susceptible to attack by pathogens (Brown *et al.*, 1984; Huber and Wilhelm, 1988).

Acknowledgements

This work was supported by the Australian Wheat Research Council. Thanks to Dr K E Finucane.

References

Abbott A J 1967 Physiological effects of micronutrient deficiencies in isolated roots of *Lycopersicum escultentum*. New Phytol. 66, 419–437.

Barley K P 1970 The configuration of the root system in relation to nutrient uptake. Adv. Agron. 22, 159–201.

Brown P H, Graham R D and Nicholas D J D 1984 The effects of manganese and nitrate supply on the levels of phenolics and lignin in young wheat plants. Plant and Soil 81, 437–440.

Burnell J N 1988 The biochemistry of manganese in plants. *In* Manganese in Soil and Plants. Eds. R D Graham, R J Hannam and N C Uren. pp 125–137. Kluwer Academic Publishers, Dordrecht, The Netherlands.

Campbell L C and Nable R O 1988 Physiological functions of manganese in plants. *In* Manganese in Soils and Plants. Eds. R D Graham, R J Hannam and N C Uren. pp. 139–154. Kluwer Academic Publishers, Dordrecht, The Netherlands.

Donald C M and Prescott J A 1975 Trace elements in Australian crop and pasture production 1924–1974. *In* Trace Elements in Soil-Plant-Animal Systems. Eds. D J D Nicholas and A R Egan. pp 7–37. Academic Press, New York.

Graham R D and Rovira A D 1984 A role for manganese in the resistance of wheat plants to take-all. Plant and Soil 78, 441–444.

Graham R D, Davies W J and Ascher J S 1985 The critical concentration of manganese in field-grown wheat. Aust. J. Agric. Res. 36, 145–155.

Huber D M and Wilhelm N S 1988 The role of manganese in resistance to plant diseases. *In* Manganese in Soils and Plants. Eds. R D Graham, R J Hannam and N C Uren. pp 155–173. Kluwer Academic Publishers, Dordrecht, The Netherlands.

Imai Y I, Sue A and Yamaguchi A 1968 A removing method of resin from epoxy-embedded sections for light microscopy. J. Electr. Microsc. 17, 84–85.

Loneragan J F 1988 Distribution and movement of manganese in plants. *In* Manganese in Soils and Plants. Eds. R D Graham, R J Hannam and N C Uren. pp 113–124. Kluwer Academic Publishers, Dordrecht, The Netherlands.

May L H, Chapman F H and Aspinall D 1965 Quantitative studies of root development. I. The influence of nutrient concentration. Aust. J. Biol. Sci 18, 25–35.

Nable R O and Loneragan J F 1984 Translocation of manganese in subterranean clover (*Trifolium subterranean* L. cv. Seaton Park). I. Redistribution during vegetative growth. Aust. J. Plant Physiol. 11, 101–111

Snowball K and Robson A D 1983 Symptoms of Nutrient Deficiencies: Subterranean Clover and Wheat. University of Western Australia Press, Nedlands, Western Australia.

Reuter D J, Alston A M and McFarlane J D 1988 Occurrence and correction of manganese deficiency in plants. *In* Manganese in Soils and Plants. Eds. R D Graham, R J Hannam and N C Uren. pp 205–224. Kluwer Academic Publishers, Dordrecht, The Netherlands.

Tennant D 1976 Root growth of wheat. I. Early patterns of multiplication and extension of wheat roots including effects of levels of nitrogen, phosphorus and potassium. Aust. J. Agric. Res. 183–196

Webb M J and Loneragan J F 1985 Importance of environmental pH during root development on phosphate absorption. Plant Physiol. 79, 143–148.

Wilhelm N S, Fisher J M and Graham R D 1985 The effect of managanese deficiency and cereal cyst nematode infection on the growth of barley. Plant and Soil 85, 23–32

M. L. van Beusichem (Ed.), *Plant nutrition – physiology and applications*, 241–249.

PLSO IPNC408

Growth enhancement by silicon in cucumber (*Cucumis sativus*) plants depends on imbalance in phosphorus and zinc supply

H. MARSCHNER[1], H. OBERLE[1], I. CAKMAK[2] and V. RÖMHELD[1]
[1]*Institute of Plant Nutrition, University of Hohenheim, P.O. Box 70 05 62, D-7000 Stuttgart 70, FRG, and* [2]*Institute of Soil Science and Plant Nutrition, University of Cukurova, Adana, Turkey*

Key words: *Cucumis sativus* L., chlorosis, phosphorus toxicity, zinc deficiency, silicon supply

Abstract

Based on results from water culture experiments with tomato and cucumber plants where severe leaf chlorosis and depression in flower and fruit formation occurred without silicon (Si) supply, Miyake and Takahashi (1978; 1983) concluded that Si is an essential mineral element for these two plant species. Using the same nutrient solution which is high in phosphorus (P) but low in zinc (Zn) we could confirm these results. Severe chlorosis occurred in cucumber when Si was omitted, and the addition of Si prevented these visual symptoms. Simultaneously the concentrations of P drastically decreased in the leaves and the proportions of water extractable Zn increased. Normal growth and absence of chlorosis were, however, also obtained without the addition of Si when either the external concentration of P was lowered or of Zn was increased. Short-term experiments revealed that Si has no direct effect on uptake or translocation of P to the shoot. According to these results, the experimental evidences so far are insufficient for the classification of Si as an essential mineral element for cucumber. Instead, Si may act as beneficial element under conditions of nutrient imbalances, for example, in P and Zn supply and corresponding P-induced Zn deficiency. The mechanism by which Si increases the physiological availability of Zn in leaf tissue is not yet clear.

Introduction

Improvement of plant growth by addition of silicon (Si) may occur for the following reasons: increased tissue tolerance to high manganese (Mn) concentrations (Horst and Marschner, 1978), increased resistance to fungal diseases (Leusch and Buchenauer, 1988a; Miyake and Takahashi, 1983; Volk *et al.*, 1958) or higher mechanical stability of stems and leaf blades and thus better light interception (Yoshida *et al.*, 1969). In a number of silicophile plant species (Si accumulators, Takahashi and Miyake, 1977), Si seems to meet the criteria of a plant nutrient (Marschner, 1986), for example in lowland rice (Okuda and Takahashi, 1965; Takahashi and Miyake, 1977) or in sugar cane (Elawad *et al.*, 1982a; b). Moreover, in greenhouse experiments without Si supply severe growth and fruit yield reduction and chlorosis were observed by Miyake and Takahashi also in non-accumulator species such as tomato (Miyake and Takahashi, 1978), cucumber (Miyake and Takahashi, 1983), soybean and strawberry plants (Miyake and Takahashi, 1985; 1986). Without Si supply, in these plant species the phosphorus (P) concentrations in the leaf dry matter were consistently high (~2% and more). In tendency, the growth stimulating effect of Si was inversely related to the P concentrations in the leaves of plants grown without Si supply.

Excessive accumulation of P in leaves is, however, also typical for zinc (Zn) deficient plants (Cakmak and Marschner, 1986; 1987; Lonergan *et al.*, 1982; Marschner and Cakmak, 1986). In the papers of Miyake and Takahashi (see above) no informations are given for the Zn concentrations in the plants. But in their nutrient solutions the concentrations of Zn were low (1×10^{-4} mM or less) and the concentrations of P were high ($0.23 - 2.3$ mM). We therefore suspected that Zn deficiency ('P-induced Zn deficiency') was at least involved in reported effects of Si on growth and on the observed 'Si deficiency symptoms' in these non-accumulator plant species. In order to test this we studied the effects of varied P and Zn concentrations in the presence and absence of Si supply on growth, visual disorder symptoms and on mineral element concentrations in cucumber plants.

Material and methods

Seeds of cucumber (*Cucumis sativus* L., cv. 'chinesische Schlange') were germinated on wet filter paper and then transferred into aerated nutrient solution containing the following composition (mM): $Ca(NO_3)_2$ 1.2; $(NH_4)_2SO_4$ 0.2; $MgSO_4$ 0.6; KH_2PO_4 0.05 to 1.2, depending on the treatments (see below); H_3BO_3 4.6×10^{-2}; Fe citrate 1.8×10^{-2}; $MnSO_4$ 7.3×10^{-3}; $CuSO_4$ 1.6×10^{-4}; $ZnSO_4$ 1.5×10^{-4}; or 1.0×10^{-3}, depending on the treatments (see below). Silicon was supplied at a concentration of 1.7 mM (~100 ppm SiO_2), as silicic acid after passing Na silicate through a H^+ loaded Dowex 50 W × 8 cation exchanger resin. Bidestilled water was used for the nutrient solution, and the pH of the nutrient solution was adjusted to 5.5.

The plants grew up to 38 days under controlled climatic conditions (photoperiod 16/8; 25/23°C; relative humidity 75%, light intensity 220 μE $m^{-2}s^{-1}$ Sylvania FR 96 T fluorescence tubes). The nutrient solutions were replaced every 3-4 days, depending on the size of the pots (volume of the nutrient solution was between 0.6 and 2.5 L per plant) and growth stage of the plants. At harvest roots, shoots and leaves of different age were separated, dried at either 80°C or freeze-dried, ground and ashed at 500°C for subsequent mineral element analyses: P by vandate-molybdate colorimetric method; K by flame photometry, Ca and Zn by atomic absorption spectrometry. For determination of water soluble Zn, 0.2-0.3 g freeze-dried ground samples were extracted according to Cakmak and Marschner (1987) with 10 mL MES buffer (10 mM, pH 6) for 5 h and then filtrated for Zn analysis. Silicon was determined in the plant dry matter by neutron activation analysis. Phosphorus uptake experiments were performed according to Cakmak and Marschner (1986). In order to study the effect of Si on the distribution of Zn in leaves, ^{65}Zn (specific activity 4.9 GBq ^{65}Zn/mmol Zn; 1.0×10^{-3} mM) was supplied in the nutrient solution for 6-24 h. The plants were harvested, freeze dried and exposed to Osray M3 Agfa-Gevaert X-ray films for preparing autoradiographs.

Data shown in tables and figures are means of at least three replications ± SD.

Results

After about 3 weeks growth in nutrient solutions with high P supply (1.2 mM), interveinal chlorosis developed on mature leaves of plants grown without Si supply and at the low Zn level (1.5×10^{-4} mM). As the chlorosis symptoms became more severe, growth was retarded in comparison to the healthy looking plants grown either with Si or without Si supply at the high Zn level (1.0×10^{-3} mM). In Figure 1 representative plants of three treatments are shown. Details on

Fig. 1. Effect of Si and Zn on the growth of 30-day-old cucumber plants grown in nutrient solutions with 1.2 mM P; $Zn_1 = 1.5 \times 10^{-4}$; $Zn_2 = 1.0 \times 10^{-3}$ mM. +Si = 1.7 mM Si.

Fig. 2. Details on the effects of Si and Zn supply on visual symptoms in 30-day-old cucumber plants (further informations see Fig. 1). **A**: Plants grown at low Zn level (Zn_1) without (left) or with (+Si) Si supply **B**: Plants grown without Si supply and low Zn level (Zn_1) and high Zn level (Zn_2).

Fig. 3. Leaf size and symptoms of interveinal chlorosis in fully expanded leaves of 38-day-old cucumber plants grown in nutrient solutions with different supply of P, Zn and Si. P concentrations: $P_1 = 0.05$ mM; $P_2 = 0.6$ mM; $P_3 = 1.2$ mM; Zn concentrations: $Zn_1 = 1.5 \times 10^{-4}$ mM; $Zn_2 = 1.0 \times 10^{-3}$ mM; Si concentrations: +Si = 1.7 mM **A**: effect of increasing P concentrations at Zn_1 **B**: Effect of Si and Zn at the high P concentration.

1
-Si
Zn 1
+Si
Zn1
-Si
Zn 2
2A
-Si
+Si
2B
Zn1
Zn 2
3A
P1 -Si
P2-Si
P3-Si
3B
P3-Si Zn1
P3+Si Zn1
P3-Si Zn2

the effects of Si on visual symptoms in plants with the low Zn level (Zn_1) are given in Figure 2A, and on the effects of high Zn supply ($Zn_2 = 1.0 \times 10^{-3}$ mM) in absence of Si supply are given in Figure 2B.

The effect of Si on shoot growth and interveinal chlorosis in mature leaves depended also on the P concentrations in the nutrient solution (Fig. 3). Chlorosis was absent in mature leaf blades from plants grown at the lowest P level (P_1) and low Zn level without Si supply, but became progressively severe with increased P levels (Fig. 3A). Chlorosis could be however, prevented also at the high P level (P_3) by either supply of Si or increasing the Zn level (Fig. 3B).

Shoot dry weight was affected by the levels of P and Zn, as well as by supply of Si (Fig. 4). At the lowest P level (P_1), the plants suffered from P deficiency, as indicated by the increase in dry weight from treatment P_1 to P_2 and also indicated by the shift in the shoot-root dry weight ratio. Silicon supply increased shoot dry weight at both P_1 and P_2 but not at P_3, whereas the highest shoot dry weight was obtained with the high level of Zn but without Si supply. Despite the severe chlorosis observed in the plants at P_2 and

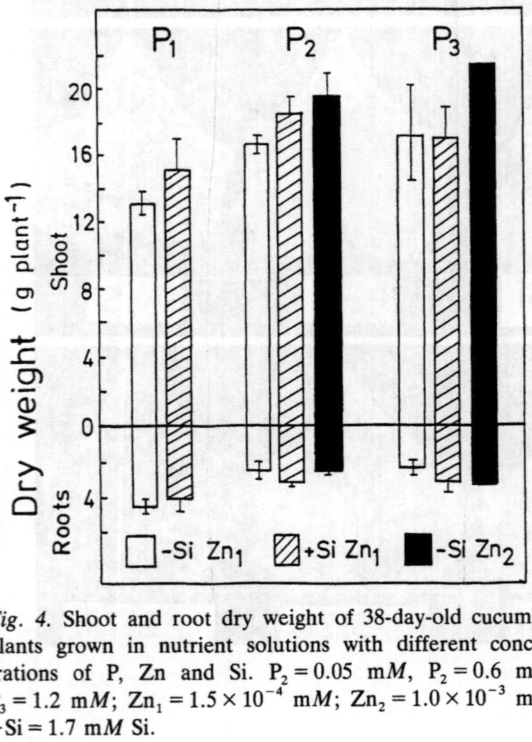

Fig. 4. Shoot and root dry weight of 38-day-old cucumber plants grown in nutrient solutions with different concentrations of P, Zn and Si. $P_2 = 0.05$ mM, $P_2 = 0.6$ mM; $P_3 = 1.2$ mM; $Zn_1 = 1.5 \times 10^{-4}$ mM; $Zn_2 = 1.0 \times 10^{-3}$ mM; +Si = 1.7 mM Si.

P_3, without supply of Si and at the low Zn level, differences in shoot dry weight between the various treatments were still relatively small at this stage of growth. Therefore the mineral element concentrations in the plants of the various treatments could be compared and interpreted more easily.

In Fig. 5, the concentrations of P, Zn and Ca in leaves of different age are shown. At P_1, the P concentrations in the leaves were in the range between suboptimal and deficient levels (Bergmann, 1988). With an increase in P supply, the P concentrations in the leaves increased and reached or exceeded critical toxicity levels (~1.5% P in leaf dry matter); particularly in mature and old leaves of plants without Si supply and at the low Zn level (Zn_1). Both, supply of Si and to a greater extent, increases in the Zn level, drastically decreased the P concentrations in the leaves. This decrease can only partially be explained by a 'dilution effect', *i.e.* increase in dry weight (see Fig. 4).

With exception of the P deficient plants (P_1), the Zn concentrations in the leaves of plants supplied with the low Zn level were found to be in the deficiency range (Fig. 5). The supply of Si had no effect on the concentration of total Zn in the leaves. With an increase in Zn supply from Zn_1 to Zn_2, the concentrations of Zn in the leaves increased by a factor of 2-5, particularly in the old leaves.

Generally, the concentration of Ca increased with leaf age and was not much affected by the various treatments (Fig. 5). The K concentration tended to be higher in both, plants grown without Si supply and at the low Zn level (data not shown).

With increasing P supply, the Zn concentrations decreased only slightly in the leaves (Fig. 5). However, in plants with very high P concentrations the concentrations of total Zn is not a good indicator for Zn bioavailability. Therefore, in Table 1 also the concentrations and proportions of water soluble Zn are presented. With increasing P concentrations water soluble Zn decreased, and at each P level in the plants with Si supply the concentrations and proportions of water soluble Zn were higher than in the plants without Si supply.

Despite of the Si effects on the binding form

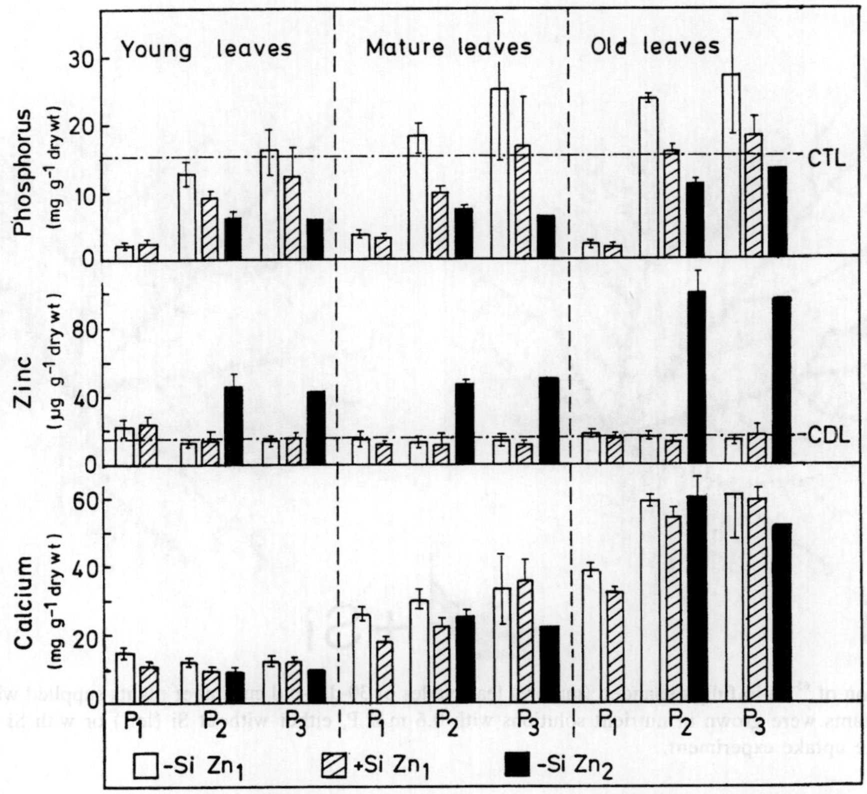

Fig. 5. Concentrations of P, Zn and Ca in leaves of 38-day-old cucumber plants grown in nutrient solutions with different levels of P, Zn and Si. $P_1 = 0.05$ m*M*; $P_2 = 0.6$ m*M*; $P_3 = 1.2$ m*M*; $Zn_1 = 1.5 \times 10^{-4}$ m*M*; $Zn_2 = 1.0 \times 10^{-3}$ m*M*; +Si = 1.7 m*M* Si CTL = critical level for P toxicity CDL = critical level for Zn deficiency.

of Zn in the leaves, Si did not influence the distribution of ^{65}Zn in the leaf blades (Fig. 6). In both absence and presence of Si supply, the ^{65}Zn was not evenly distributed within the leaf blades but mainly confined to the veines. This distribution pattern was independent of the duration of ^{65}Zn supply (6-24 h) to the roots. The plants used for the uptake experiment with ^{65}Zn were free of visual symptoms.

The Si concentrations in the plants supplied with Si are shown in Table 2. The concentrations are low in the stem and increase steeply with leaf age. The Si concentrations are higher in the P deficient plants (P_1).

Table 1. Effects of Si and P concentrations in the nutrient solution on the concentrations of total Zn and water soluble Zn in mature leaves of 38-days-old cucumber plants supplied with 1.5×10^{-4} m*M* Zn.

Treatment (m*M* P)		Zn concentration (µg Zn g⁻¹ dw) Total	H_2O soluble	Proportion (%) of H_2O soluble Zn
P_1 (0.05)	−Si	14.6 (±5.6)	6.8 (±0.6)	50
	+Si	11.6 (±1.4)	7.4 (±0.2)	64
P_2 (0.6)	−Si	12.6 (±1.7)	6.0 (±0.8)	46
	+Si	11.2 (±6.2)	6.6 (±3.7)	59
P_3 (1.2)	−Si	13.7 (±4.3)	4.8 (±0.9)	36
	+Si	10.9 (±2.5)	5.1 (±1.1)	47

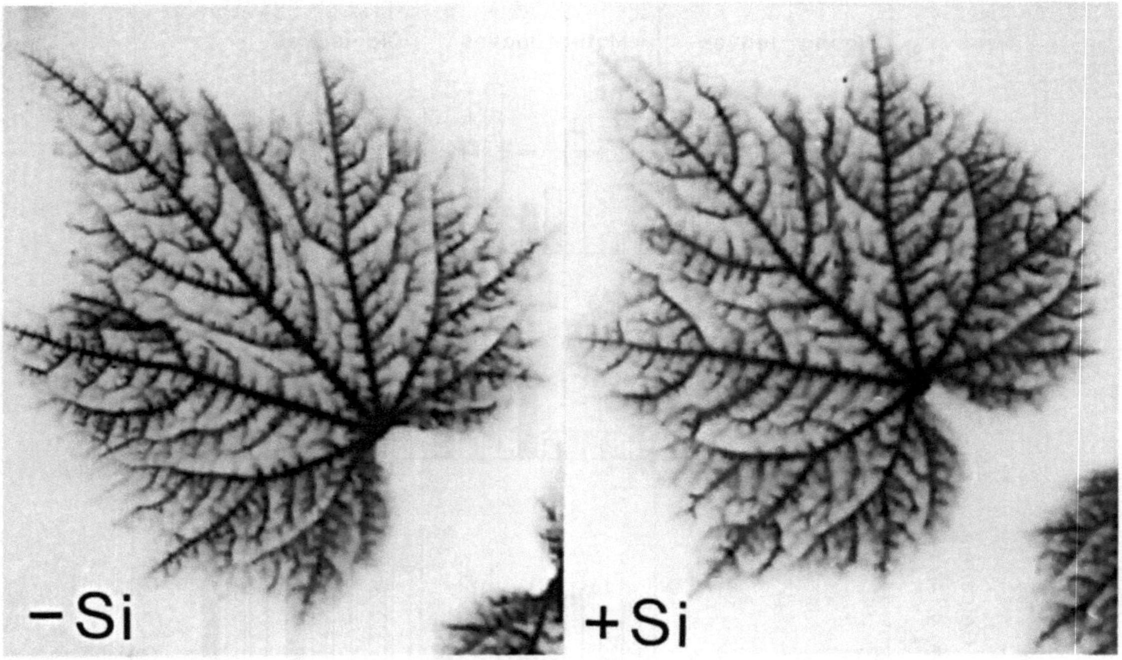

Fig. 6. Distribution of ^{65}Zn in fully expanded (mature) leaf blades of 30-day-old cucumber plants supplied with ^{65}Zn (1.0×10^{-3} m*M*) for 12 h. Plants were grown in nutrient solutions with 0.6 m*M* P, either without Si (left) or with Si (right) both during preculture and the uptake experiment.

Table 2. Concentrations of Si (mg Si g^{-1} dry wt) in leaves[a] and stem of 38-day-old cucumber plants supplied with different levels of P in the nutrient solution. Supply of Si 1.7 m*M* and of Zn 1.5×10^{-4} m*M*

Shoot organs	P supply (m*M*)		
	P_1(0.05)	P_2(0.6)	P_3(1.2)
Stem	4.8	4.0	4.3
Leaves			
Young	16.1	11.3	10.5
Mature	22.0	18.7	18.5
Old	36.4	25.9	23.0

[a]Si concentrations in leaves of plants without Si supply were below 0.9 mg g^{-1} dry weight.

The much lower P concentrations in the leaves of Zn_1 plants supplied with Si (Fig. 5) are not the result of a direct inhibitory effect of Si on the P uptake (Table 3). In short-term experiments the uptake rate of P in plants supplied with Zn_2 was not affected by the supply of Si in the external solution (Table 3; compare line ① and ③). In the Zn_1 plants without Si supply, the P uptake rate was much higher than with Si supply. Increases in the Zn level (Zn_2) was similarly effective as supply of Si in decreasing the P uptake rates. The results of this short-term experiment-decrease in P uptake rates by either Si supply or increase in the Zn level-are compar-

Table 3. Effect of Si and Zn supply on short-term uptake rate of P by cucumber plants grown in nutrient solutions with 0.6 m*M* P

	Si treatment (1.7 m*M* Si)		Uptake rate of P (mg P g^{-1} root dw $5h^{-1}$)	
	Pretreatment (growth medium)	Uptake period	Zn_1 (1.5×10^{-4} m*M*)	Zn_2 (1.0×10^{-3} m*M*)
1	−Si	−Si	6.4 ± 0.8	2.6 ± 0.2
2	+Si	+Si	2.2 ± 0.6	2.3 ± 0.2
3	−Si	+Si	n.d.	2.5 ± 0.4

able to those found for the P concentrations in the leaves in the long-term experiment (Fig. 5).

Discussion

The results from Miyake and Takahashi, demonstrating that supply of Si prevents chlorosis and excessive P accumulation in leaves of various dicotylodenous plant species has been confirmed in the present study with cucumber, if similar high P and low Zn concentrations in the nutrient solution were used as in the experiments of Miyake and Takahashi. However, chlorosis and excessive P accumulation in leaves could also be prevented in absence of Si supply when the concentration of Zn was increased in the plants from the critical deficiency level (15–20 μg Zn g^{-1} leaf dry wt.; Bergmann, 1988) to the sufficient level (Fig. 5). This effect of Zn on the P concentrations in leaves of cucumber is in full accordance with results in other plant species demonstrating that excessive P accumulation and P toxicity in leaves is a typical feature of Zn deficient plants supplied with high P concentrations in the nutrient solution (Cakmak and Marschner, 1986; Loneragan *et al.*, 1982; Marschner and Cakmak, 1986).

Chlorosis of mature leaves and leaf senescence are enhanced under Zn deficiency, especially at high light intensity (Marschner and Cakmak, 1989). This light effect is presumably causally related to the functions of Zn in protecting biomembranes from toxic oxygen radicals (Cakmak and Marschner, 1988b; Marschner and Cakmak, 1989). Leaf senescence in cucumber plants (Adatia and Besford, 1986) and in tomato plants can be delayed by Si, especially under high light intensity (Miyake and Takahashi, 1978). Furthermore, 'leaf freckling', described as symptom of Si deficiency in sugar cane, occurs predominantely under high light intensity and is difficult to reproduce under low light intensity (Gascho, 1977).

Malformation of young leaves has been described as another symptom of Si deficiency (Miyake and Takahashi, 1978; 1983; Takahashi and Miyake, 1982), but it is also a well known symptom in Zn deficient plants, especially under high light intensity (Marschner and Cakmak, 1989). In the nutrient solutions used by Miyake and Takahashi the P concentrations were very high but the Zn concentrations very low, for example, 1.5×10^{-4} m*M* (Miyake and Takahashi, 1978; 1983; 1985) or even as low as 2×10^{-5} m*M* (Miyake and Takahashi, 1986). Thus, classification of Si as a plant nutrient based on experiments in these nutrient solutions seems not justified unless the effects of Si on growth, chlorosis and senescence are also confirmed in Zn sufficient plants. The results of the present study emphasize this necessity, although in the experiments the reproductive stage of the plants was not examined where the effects of Si supply are more pronounced (Miyake and Takahashi, 1978; 1983; 1985).

Silicon did not affect the distribution of Zn (^{65}Zn) in the leaf blades as characterized by autoradiographic methods (Fig. 6). The distinct accumulation of ^{65}Zn in the veines suggest an impairment of the short distance transport of Zn in the apoplasm of the leaf tissue. The formation of sparingly soluble Zn phosphates in the leaf apoplasm of plants grown at high P levels (P_2) is likely to be a factor responsible for this accumulation of Zn in the veines. This assumption is supported by the high P concentrations in the xylem exudate of cucumber plants under these growing conditions (2-4 m*M*; data not shown).

For plants grown at high P levels, the fraction of water soluble Zn reflects the availability of Zn in leaves better than the total Zn (Cakmak and Marschner, 1987). For plants supplied with a suboptimal Zn concentration, Si increased the physiological availability of Zn in the leaves as indicated by the concentrations of water soluble Zn (Table 1). This influence of Si on the Zn availability is presumably responsible for the distinct decrease in the P concentrations in leaves of the plants with the low Zn supply (Fig. 5). In Zn deficient plants a shoot control signal is impaired which in Zn sufficient plants prevents excessive P uptake by the roots (Table 3) and P transport to the shoots (Fig. 5; Cakmak and Marschner, 1986; Marschner and Cakmak, 1986).

The mechanism responsible for this Si effect on the physiological availability of Zn is not clear. Similar effects of Si as described here for the Zn availability have been found for manganese (Mn) in soybean grown in nutrient solutions under Mn deficient conditions (Kluthcouski and Nelson, 1980). In shoot tissue most of the Si is deposited in the cell walls (apoplasm) as amorphous silica ($SiO_2 \times nH_2O$). In addition, Si may crosslink with cell wall constituents such as o-diphenols and thereby modify the cell wall structure (Weiss and Herzog, 1978) and increase the tissue elasticity (Emadian and Newton, 1989).

High Si concentrations in the cell walls of the conducting vessels and of leaf cells may affect transport and uptake of solutes into the leaf cells in various ways. High Si concentrations may favour the exchange adsorption of cations or act as a pH buffer in the apoplasm for the protons extruded by a light-stimulated pump at the plasma membrane of leaf cells (Petzold and Dahse, 1988). This H^+ efflux pump seems to play an important role in 'xylem unloading' and uptake of amino acids (Wilson *et al.*, 1988) or iron (Mengel and Geurtzen, 1988) into the leaf cells. Interestingly, senescence of leaves is correlated with decrease in acidification of the apoplasm (Canny, 1988). In view of the requirement of Zn for membrane integrity in general and of the plasma membrane in particular (Cakmak and Marschner, 1988a), conditions in the apoplasms and at the external surface of the plasma membrane of leaf cells require more attention in plant mineral nutrition in general and for understanding the role of Si in particular. It well might be that at least part of the observed beneficial Si effects described in the literature and in the present study are related to interactions between Si and Zn in the apoplasm of leaf cells. Such interactions with Si in the apoplasm are presumably also important for Mn deficiency (Kluthcouski and Nelson, 1980) and Mn toxicity (Horst and Marschner, 1978), as well as in the defense reactions of plants to fungal infections (Leusch and Buchenauer, 1988b).

Acknowledgement

The authors thank Dr V Cerkasov, Department of Physics, University Hohenheim, for advices and assistance in neutron activation analysis and Dr G Banuelos for correcting the English text.

References

Adatia M H and Besford R T 1986 The effect of silicon on cucumber plants grown in recirculating nutrient solution. Ann. Bot. 58, 343–351.

Bergman W 1988 Ernährungsstörungen bei Kulturpflanzen. Gustav Fischer Verlag, Jena.

Cakmak I and Marschner H 1986 Mechanism of phosphorous-induced zinc deficiency in cotton. I. Zinc deficiency-enhanced uptake rate of phosphorous. Physiol. Plant. 68, 483–490.

Cakmak I and Marschner H 1987 Mechanism of phosphorus-induced zinc deficiency in cotton. III. Changes in physiological availability of zinc in plants. Physiol. Plant. 70, 13–20.

Cakmak I and Marschner H 1988a Increase in membrane permeability and exudation in roots of zinc-deficient plants. J. Plant Physiol. 132, 356–361.

Cakmak I and Marschner H 1988b Enhanced superoxide radical production in roots of zinc-deficient plants. J. Exp. Bot. 39, 1449–1460.

Canny M J 1988 Bundle sheath tissues of legume leaves as a site of recovery of solutes from the transpiration stream. Physiol. Plant. 73, 457–464.

Elawad S H, Gascho G J and Street J J 1982a Response of sugarcane to silicate source and rate. I. Growth and Yield. Agron. J. 74, 481–484.

Elawad S H, Street J J and Gascho G J 1982b Response of sugarcane to silicate source and rate. II. Leaf freckling and nutrient content. Agron. J. 74, 484–487.

Emadian, S F and Newton R J 1989 Growth enhancement of lob-lolly pine (*Pinus taeda* L.) seedlings by silicon. J. Plant Physiol. 134, 89–103.

Gascho G J 1977 Response of sugarcane to calcium silicate slag. I. Mechanisms of response in Florida. Soil Crop Sci. Soc. Fla. Proc. 37, 55–58.

Horst W J and Marschner H 1978a Effect of silicon and manganese tolerance of bean plants (*Phaseolus vulgaris* L.). Plant and Soil 50, 287–303.

Kluthcouski J and Nelson L E 1980 The effect of silicon on the manganese nutrition of soybeans (*Glycine max* (L.) Merill). Plant and Soil 56, 157–160.

Leusch H J und Buchenauer H 1988a Einfluß von Bodenbehandlungen mit siliziumreichen Kalken und natriumsilikat auf den Mehltaubefall von Weizen. Kali-Briefe 19, 1–11.

Leusch H J und Buchenauer H 1988b Si-Gehalte und Si-Lokalisation im Weizenblatt und deren Bedeutung für die Abwehr einer Mehltauinfektion. Kali-Briefe 19, 13–24.

Loneragan J F, Grunes D L, Welch R M, Aduay E A, Teugah A, Lazar V A and Cary E E 1982b Phosphorus accumulation and toxicity in leaves in relation to zinc supply. Soil Sci. Soc. Am. J. 46, 345–352.

Marschner H 1986 Mineral Nutrition of Higher Plants. Academic Press, London, pp 351–359.

Marschner H and Cakmak I 1986 Mechanism of phosphorus-induced zinc deficiency in cotton. II. Evidence for impaired shoot control of phosphorus uptake and translocation under zinc deficiency. Physiol. Plant. 68, 491–496.

Marschner H and Cakmak I 1989 High light intensity enhances chlorosis and necrosis in leaves of zinc, potassium, and magnesium deficient bean (*Phaseolus vulgaris*) plants. J. Plant Physiol. 134, 308–315.

Mengel K and Geurtzen G 1988 Relationship between iron chlorosis and alkalinity in *Zea mays*. Physiol. Plant. 72, 460–465.

Miyake Y and Takahashi E 1987 Silicon deficiency of tomato plant. Soil Sci. Plant Nutr. 24, 175–189.

Miyake Y and Takahashi E 1983 Effect of silicon on the growth of solution-cultured cucumber plant. Soil Sci. Plant Nutr. 29, 71–83.

Miyake Y and Takahashi E 1985 Effect of silicon on the growth of soybean plants in a solution culture. Soil Sci. Plant Nutr. 31, 625–636.

Miyake Y and Takahashi E 1986 Effect of silicon on the growth and fruit production of strawberry plants in a solution culture. Soil Sci. Plant Nutr. 32, 321-326.

Okuda A and Takahashi E 1965 The role of silicon. *In* The Mineral Nutrition of the Rice Plant. pp 123–146. Proc. Symp. Internat. Rice Res. Inst. (IRRI). John Hopkins Press, Baltimore, MD.

Petzold U and Dahse I 1988 Proton extrusion by leaf discs of *Vicia faba* L.: Light- and ion-stimulated H^+ release. Biol. Plantarum (Praha) 30, 124–130.

Takahashi E and Miyake Y 1977 Silica and plant growth. Proc. Int. Seminar Soil Environment Fertil. Manag. Intensive Agric. Tokyo, pp 603–611.

Takahashi E and Miyake Y 1982 The effect of silicon on the growth of cucumber plants: Comparative studies on the silicon nutrition. *In* Plant Nutrition 1982. Ed. A. Scaife. pp 664–669. Proc. IXth Intern. Colloq. (Coventry) Commonwealth Agriculture Bureaux. Slough, U.K.

Volk R J, Kahn R P and Weintraub R L 1958 Silicon content of the rice plant as a factor influencing its resistance to infection by the blast fungis, *Piricularia aryzae*. Phytopathology 48, 179–184.

Weiss A and Herzog A 1978 Isolation and characterization of a silicon-organic complex from plants. *In* Biochemistry of Silicon and Related Problems. Eds. G Gendz and I Lindgrist, pp 109–127. Plenum Press, New York.

Wilson T P, Canny M J and McCully M E 1988 Proton pumps activity in bundle sheath tissues of broad-leaved trees in relation to leaf age. Physiol. Plant. 73, 465–470.

Yoshida S, Nasavero S A and Ramirez E A 1969 Effect of silica and nitrogen supply on some leaf characters of the rice plant. Plant and Soil 31, 48–56.

M. L. van Beusichem (Ed.), *Plant nutrition – physiology and applications*, 251–255.

PLSO IPNC556

The effect of zinc and magnesium application on ear density, grain development and grain yield of winter and spring barley (*Hordeum vulgare*) crops on some Irish soils

F.S. MacNAEIDHE and G.A. FLEMING
Agriculture and Food Development Authority, Johnstown Castle Research Centre, Wexford, Ireland

Key words: ear number per m^2, EDTA – extractable zinc, foliar application, grain number per ear, grain yield, *Hordeum vulgare* L., Hydromag, magnesium deficiency, magnesium EDTA, magnesium sulphate, 1000-grain weight, randomized block experiment, zinc deficiency, zinc EDTA, zinc sulphate, Zintrac

Abstract

Grain yield increases of 0–60% were obtained with foliar application of several zinc (Zn) compounds in six soils with EDTA – extractable Zn concentrations of 0.6-2.0 mg/kg. The largest increase (104–157%) was obtained at a Zn concentration of 0.6 mg/kg. No increase was obtained at 2.0 mg/kg. Zn treatment increased the ear numbers and this was the main factor responsible for the increased yield. There was no consistent effect of Zn application on grain numbers per ear or the 1000 grain weight. The three Zn compounds, Zn sulphate, Zn EDTA and Zintrac were equally effective in correcting Zn deficiency. In magnesium (Mg) deficient soils application of Zn sulphate at 2.0 kg/ha + Mg sulphate at 5.0 kg/ha (in spring and autumn) and Zintrac at 3.0 1/ha + Hydromag at 12.0 kg/ha increased the yield.

Introduction

Zn deficiency was first identified in spring barley cv. Fleet in the north-east of the Irish Republic in 1985 (MacNaeidhe *et al.*, 1986). The main symptoms were stunting of the crop, marginal chlorosis of the younger foliage and marginal necrosis of the older foliage. The symptoms occurred in patches throughout the field. These patches varied in size from a few m^2 to several hectares. In severely affected patches many of the deficient plants died. In less severe cases the crop made a recovery by mid-summer. Zn deficiency was not recorded previously in Ireland. Tunney (1969) has described the different trace element deficiencies in Irish soils but does not mention Zn deficiency as a problem in tillage crops. Fleming (1982) has noted that little research work has been carried out on Zn in Ireland. The results of preliminary investigations on Zn deficiency carried out in 1985 are reported elsewhere (MacNaeidhe, 1988; MacNaeidhe and Fleming, 1988). These papers describe the response of barley to Zn application on one soil type. An analysis of 700 soil samples taken on the corners and in the centre of a 10 km × 10 km grid in late 1985 has shown that other soil types throughout the country have low soil Zn concentrations. In 1986 and in 1987 investigations were extended to these areas. The experiments were conducted in soils with different concentrations of EDTA – extractable Zn to test the range of soil Zn values at which a response to Zn application is obtained. The results are described in this paper.

Materials and methods

The 1986 trials were carried out at two Zn deficient dites in the north-midlands. The soils at these sites were grey-brown podzolics of shale origin (Gardiner and Radford, 1980). Winter barley cv. Panda was grown at the first site.

Spring barley cv. Fleet was grown at the second site. The 1987 trials were carried out at two Zn deficient sites in the north-east, at a third slightly deficient site in the south-midlands and at a fourth non-deficient site in the south east. The soil at the north-east sites were acid brown earths which were formed from a shale drift. The soils in the south-midlands and south-east were grey-brown podzolics of limestone origin. Winter barley cv. Panda was sown at the two north-east sites and spring barley cv. Klaxton was sown at the south-midlands and south-east sites. The treatments at each site were replicated four times. All treatments were applied as foliar sprays is 450 L water per ha using a pressure retaining knapsack sprayer. At the two north-east sites the Zn and Mg treatments were applied twice, at growth stage 20 in the autumn and at growth stage 30-35 in the spring (Zadoks *et al.*, 1974). At all other sites the treatments were applied at growth stage 30–35. Plot size was 2m × 10m. The Zn compounds used were Zn sulphate (22.8% Zn), an ethylene diamine tetra acetate of Zn (EDTA) containing 7 per cent Zn and an inorganic salt formulation containing 70 per cent Zn (Zintrac). At the two north-east sites (Sites 1 and 2) and at the south-midlands site (Site 3) the exchangeable soil Mg was low and some Mg applications were included among the treatments. The Mg compounds used were Mg sulphate (16.4% Mg) and Hydromag, an inorganic Mg salt formulation containing 38.4% Mg. The soil pH and the Zn and Mg concentrations are given in Table 1. The soil pH was determined in distilled water using a digital pH meter. Exchangeable soil Mg was extracted with 5% Morgan's solution. Available soil Zn was extracted with 0.05 *M* EDTA. Mg and Zn were estimated by atomic absorption spectrophotometry (Byrne, 1979).

Table 1. Soil pH and zinc and magnesium contents of soils at the experimental sites

Year	Site	pH	Mg/kg	
			Magnesium	Zinc
1986	1	7.5	54	0.6
	2	6.4	49	1.4
1987	1	7.3	69	1.2
	2	7.1	53	1.4
	3	6.8	35	1.9
	4	7.2	458	2.0

Crop yield was measured by harvesting 12 × 0.2 m^2 quadrats in each plot. The ears were threshed in a Hege 16 – single ear thresher. The grain was dried to 15% moisture and weighed. The percentage marketable grain was assessed by passing five 200 g samples through a mechanically operated 2.5 mm Glasblaserii sieve. The yield of marketable grain was assessed using these data. The number of ears per m^2 was calculated in 8 × 0.2 m^2 quadrats placed at random in each plot. The grain number per ear was assessed by threshing 20 randomly selected ears and counting the threshed grains in a Simon Countmaster seedcounter. The 1000 grain weight was calculated by recording the weight of 5 × 200 grains which were randomly selected from the harvested grain lots.

Results

In the 1986 experiments Zn deficiency symptoms were severe at Site 1 and were moderate to severe at Site 2. The results are given in Table 2. At Site 1 Zn treatment increased the ear number per m^2, the grain number per ear and the grain yield. At Site 2 treatment with Zn gave a smaller increase in ear density, grain number per ear and grain yield. The 1000 grain weight was also slightly increased.

In the 1987 experiments deficiency symptoms occurred at Sites 1 and 2 in October at growth stage 15. The symptoms were most severe at Site 1. All treatments except Zn sulphate at 2.0 kg/ha increased the ear number per m^2 (Table 3) but this latter treatment gave the largest 1000 grain weight and grain yield. At Site 2 the largest grain yield was obtained with Zn sulphate at 2.0 kg + Mg sulphate at 5.0 kg/ha. At Site 3 there were no symptoms of Zn deficiency in the crop. Zintrac at 3.0 L + Hydromag at 12.0 kg/ha increased ear number per m^2, 1000 grain weight and grain yield. Hydromag at 12.0 kg/ha increased ear density and grain yield (Table 4). At Site 4 the application of Zintrac at 3.0 L/ha increased the grain number per ear. None of the

Table 2. The effect of zinc and magnesium application on ear density and grain yield in 1986

	Site 1. Winter barley cv. Panda				Site 2. Spring Barley cv. Fleet			
	Ear number per m^2	Grain number per ear	1000 grain weight (g)	Grain yield (tonnes/ha)	Ear number per m^2	Grain number per ear	1000 grain weight (g)	Grain yield (tonnes/ha)
Zn sulphate (2.5 kg/ha)	994	18.7	40.6	6.4	896	21.6	35.9	7.0
Zn sulphate (5.0 kg/ha)	1032	16.1	43.3	6.6	1011	22.2	35.9	8.0
Zn EDTA (3.0 L/ha)	866	16.1	45.1	5.7	1026	20.8	36.0	7.7
Zn EDTA (6.0 L/ha)	1002	18.6	45.1	7.2	1046	21.3	34.5	7.7
Control	494	9.4	46.4	2.8	738	20.3	36.1	5.4
S.E.[a] (df[b] = 24)	49.71	1.52	1.07	1.09	28.4	0.40	0.27	0.98

[a] standard error of the mean; [b]degrees of freedom.

Table 3. The effect of zinc and magnesium application on ear density and grain yield of winter barley cv. Panda, 1987

	Time of application[c]	Ear number per m^2		Grain number per ear		1000 grain weight (g)		Grain yield (tonnes/ha)	
		Site 1	Site 2	Site 1	Site 2	Site 1	Site 2	Site 1	Site 2
Zn sulphate (2.0 kg/ha)	A + B	600	579	17.8	18.9	50.2	46.3	7.1	6.1
Mg sulphate (5.0 mg/ha)	A + B	723	681	18.4	15.3	49.0	47.7	6.1	5.3
Zn sulphate (2.0 kg/ha) + Mg sulphate (5.0 mg/ha)	A + B	739	689	19.8	17.7	48.8	45.9	6.8	6.5
Zn EDTA (0.5 L/ha) + Zn EDTA (2.5 L/ha)	A B	613	634	19.8	17.8	48.8	48.2	6.8	5.9
Control		581	551	22.0	19.8	47.1	46.2	5.8	5.7
S.E.[a] (df[b] = 27)		44.39	21.11	0.425	1.55	0.513	0.248	0.248	0.145

[a,b] as in Table 1; [c]A, autumn application, B, spring application.

Table 4. The effect of zinc and magnesium application on ear density and grain yield of spring barley cv. Klaxton, 1987

Treatment	Ear number per m^2	Grain number per ear	1000 grain weight (g)	Grain yield (tonnes/ha)
Mg sulphate (10.0 kg/ha)	567	21.2	49.2	6.7
Mg sulphate (10.0 kg/ha) + Zn sulphate (5.0 kg/ha)	629	19.7	49.2	7.1
Hydromag (12.0 kg/ha)	682	18.2	48.6	7.4
Zintrac (3.0 L/ha) + Hydromag (12.0 kg/ha)	665	22.7	49.2	8.1
Control	591	20.2	48.2	7.0
S.E.[a] df[b] = 27	30.91	0.630	0.363	0.261

[a,b] as in Table 1.

other treatments increased ear density, grain number per ear, 1000 grain weight or grain yield (Table 5).

Discussion

The EDTA – extractable Zn concentration in the soils of the six experimental sites was 0.6–2.0 mg/kg. Zn deficiency symptoms have already been described in other publications (MacNaeidhe, 1988; MacNaeidhe and Fleming, 1988). In the experiments described here the symptoms were most severe at 0.6 mg/kg, were slight to severe at 1.2–1.4 mg/kg and were absent at 1.9–2.0 mg/kg. The symptoms occurred in patches and there was a variation in crop growth

Table 5. The effect of zinc application on ear density and grain yield in spring barley cv. Klaxton in 1987

Treatment	Ear number per m^2	Grain number per ear	1000 grain weight (g)	Grain yield (tonnes/ha)
Zn sulphate (5.0 kg/ha)	635	17.9	43.7	5.8
Zn EDTA (3.0 L/ha)	627	18.4	43.6	5.7
Zintrac (3.0 L/ha)	641	20.2	43.7	5.7
Control	588	17.6	44.8	5.7
S.E.[a] ($df^b = 27$)	25.61	0.495	0.291	0.223

[a,b] as in Table 1.

from one plot to another. Even in cases where the deficiency was severe it was difficult to measure responses to treatment accurately due to this variation.

The main effect of Zn treatment was an increase in ear density and grain yield in deficient crops. At an EDTA - extractable Zn concentration of 0.6 mg/kg treatment with different Zn compounds resulted in an increase in ear density and grain number per ear of 72–109% and an increase in grain yield of 104–157%. At a Zn concentration of 1.2–1.4 mg/kg Zn treatment increased ear density by 3–42% and increased grain yield by 3–48%. Zn sulphate gave a smaller increase in ear density than the other compounds. There was no increase in ear density or grain yield at a soil Zn concentration of 2.0 mg/kg. Zn treatment increased the grain number per ear only when the EDTA - extractable soil Zn was 0.6 mg/kg.

At soil Zn concentrations greater than 1.0 mg/kg the grain number per ear and the 1,000 grain weight was frequently decreased by Zn application. This decrease can reasonably be attributed to the influence of compensatory factors within the crop. Due to the higher ear density in the treated plots the overall grain number was very much larger leading to a corresponding reduction in the number of grains per ear and in the 1000 grain weight.

A characteristic of soils with low Zn in these experiments was the consistently low exchangeable Mg concentration. In 1987 soils with an exchangeable Mg concentration of less than 70 mg/kg were treated with Mg to ensure that adequate amounts of this element were available to the crop for healthy growth. The application of Mg sulphate did not increase grain yield. The application of Zn sulphate at 2.0 mg/kg + Mg sulphate at 5.0 mg/kg increased the grain yield at two deficient sites in the north-east but did not increase the yield at the slightly Zn deficient site in the south-midlands. Zintrac at 3.0 L/ha + Hydromag at 12.0 kg/ha increased the grain yield at this latter site. Hydromag at 12.0 mg/kg also gave a good increase in grain yield. These results indicate that this compound may be more effective as a control for Mg deficiency than Mg sulphate. Further investigations of Zn deficient soils are being conducted to evaluate any general relationship which may occur between low Zn and low Mg in soils.

The three Zn compounds, Zn sulphate, Zn EDTA and Zintrac were equally effective in controlling Zn deficiency in winter and spring barley. Zn EDTA at 6.0 L/ha gave a larger grain yield than Zn EDTA at 3.0 L/ha at a soil Zn concentration of 0.6 mg/kg but did not give a larger grain yield when the soil Zn concentration was 1.4 mg/kg. The excellent response to Zintrac when used in combination with Hydromag has already been discussed. The results reported here show that there is a yield response to Zn treatment at an EDTA - extractable soil Zn concentration of less than 1.5 mg/kg. On the basis of these results treatment of spring and winter barley crops with a foliar spray of Zn is recommended in soils with a Zn concentration below this value. When soil Zn is less than 1.0 mg/kg the crop deteriorates rapidly after emergence and immediate treatment with a Zn spray is recommended. Failure to apply Zn in such a situation can result in a yield reduction of more than 50%. There is no response to Zn treatment at a soil concentration of greater than 2.0 mg/kg and Zn treatment is not recommended in soils with a Zn concentration above this value. The most outstanding benefit of Zn treatment in a deficient soil is an increase in tiller production and a larger number of ears per unit area at maturity.

Acknowledgements

The authors wish to thank Messrs. F Codd and A N O'Sullivan for technical assistance. Thanks are also due to Barclay/ABM Chemicals and Topland/Phosyn Ltd. for providing the trace element carriers gratis.

References

Byrne E 1979 Chemical Analysis of Agricultural Materials. An Foras Taluntais, Dublin, 94 p.

Fleming G A 1982 Trace element investigations in Ireland: A half-century of research in soils, plant nutrition and soil-plant-animal interrelationships. J. Life. Sci. R. Dublin Soc. 4, 9–25.

Gardiner M J and Radford T 1980 Soil Associations of Ireland and their Land Use Potential. Soil Survey Bull. No. 36. An Foras Taluntais, Dublin, 142 p.

MacNaeidhe F S 1988 Zinc deficiency in spring barley crops in Ireland. Proc. 3rd Int. Symp. on the Role of Micronutrients in Agric. Brussels. Sept. 1988.

MacNaeidhe F S and Fleming G A 1988 A response in spring cereals to foliar sprays of zinc in Ireland. Irish J. Agric. Res. 27, 91–97.

MacNaeidhe F S, Fleming G A and Parle P J 1986 Zinc deficiency, first time in cereals in Ireland. Farm and Fd. Res. 17, 57–58.

Tunney H 1969 Micronutrients and magnesium fertilisation for agricultural crops. An Foras Taluntais, Tech. Bull. Agric. Ser. Index 2. An Foras Taluntais, Dublin, 8 p.

Zadoks J C, Chang T T and Konzac C F 1974 A decimal code for the growth stages of cereals. Weed Res. 14, 415–421.

M. L. van Beusichem (Ed.), *Plant nutrition – physiology and applications*, 257–260.

PLSO IPNC481

Control of zinc deficiency in apple orchards in southern Brazil

C. BASSO, A. SUZUKI, F.W.W. WILMS[1] and H. STUKER
Research Station of Caçador, P.O. Box D-1, 89500 Caçador, SC, Brazil. [1]*Present address: Deutsche Gesellschaft für Technische Zusammenarbeit (GTZ), P.O. Box 5180, D-6236 Eschborn, FRG*

Key words: apple, micronutrients, necrotic leaf spot, zinc deficiency, zinc sprays

Abstract

$ZnSO_4$ 5% plus phytosanitary pesticides at the dormancy stage, $ZnSO_4$ 1.5% plus break dormancy chemicals at the silver tip stage, and $ZnSO_4$ 0.2% with and without $Ca(OH)_2$ 0.2% or ZnO 0.1% or Zn chelate (7% Zn), 3L/ha, as 3 forthnightly sprays during the spring were sprayed in a Zn-deficient Golden Delicious apple orchard to control Zn deficiency and to evaluate possible phytotoxicity. All treatments significantly reduced Zn deficiency symtoms and none caused any visible phytotoxicity. Only spring sprays reduced necrotic leaf spot symptoms and increased Zn in the leaves at high levels suggesting leaf contamination. Leaf analysis was found not to be efficient to evaluate the Zn nutritional status of apple orchards. Sprays at the dormancy stage increased Zn in the top 20 cm soil layer suggesting rainfall leaching from the trees.

Introduction

Apple growing has become an important industry in Southern Brazil in the last 20 years. Acid soils with low fertility levels and a warm and wet climate are common in that region. This situation makes it easy to understand that the growing of this deciduous fruit under such conditions has been requiring special attention from research, extension service and growers. 'Fuji', 'Gala' and 'Golden Delicious' are the main cultivars being planted in this region.

Nutritional unbalance is present in many orchards. The main problems are related to high concentrations of N and K and low concentrations of Ca and Zn in leaves (Basso and Wilms, 1988). Worldwide, Zn deficiency is a major micronutrient problem in apple orchards. It reduces shoot growth, leaf area and affects yield and fruit quality. Attempts to supply Zn as soil amendment have been made with some success (Orphanos, 1982; Tomar *et al.*, 1970).

However, spraying is the most common way to supply Zn to the trees (Hoffman and Samish, 1966; Koch, 1981; Terblanche *et al.*, 1974). Zn-containing fungicides also can be a source of Zn (Young, 1983).

Necrotic leaf spot and premature leaf drop in susceptible cultivars are increased by Zn deficiency (Ducroquet, 1987). Zn is required in the synthesis of tryptophan which is a precursor of indole acetic acid (IAA). Zn-deficient leaves have a low level of IAA which predisposes them to early senescence and drop (Berton and Melzer, 1987). Simon (1986) found that necrotic leaf spot is a result of factors affecting cell membrane permeability like chemical unbalance and changes in tissue water content caused by sudden changes in the environmental conditions.

Zinc sulfate has been the main chemical used in the spray program in Southern Brazil. Zinc oxide and Zn chelate are also available. Since many questions arise on the use of Zn sulfate as a spray, an experiment was carried out to evaluate the efficiency of Zn sources and time of spraying on the control of Zn deficiency and necrotic leaf spot.

Materials and methods

The field trial was carried out in a severely Zn-deficient Golden Delicious / MM 106 orchard in Videira County during 3 growing seasons, starting in 1985 when the trees were 13 years old. The treatments comprise sprays in late July (dormancy), in September at the silver tip stage, and 3 fortnightly sprays in November-December (spring), as follows: 1. control; 2. $ZnSO_4$ 5% + mineral oil 1%; 3. $ZnSO_4$ 5% + surfactant; 4. $ZnSO_4$ 5% + CuO 0.3% + mineral oil 2%; 5. $ZnSO_4$ 5%; 6. $ZnSO_4$ 1.5% + mineral oil 4% + DNBP 0.12%; 7. $ZnSO_4$ 0.2%; 8. $ZnSO_4$ 0.2% + $Ca(OH)_2$ 0.2%; 9. ZnO 0.1%; 10. Zn chelate (7% Zn) 3 L ha^{-1}. The treatments 2 through 5 were applied at dormancy, 6 at silver tip and 7 through 10 as spring sprays. The experiment was set up as a completely randomized block design with 1 tree per plot and 6 replications. Leaf samples for analysis were collected on mid shoots before spring sprays in November and in the first week of February. The soil was sampled in May 1988. The occurrence of symptoms of Zn deficiency was evaluated in November before spring sprays and in February. Necrotic leaf spot was evaluated in December 1986 and in February 1988. Zn deficiency and necrotic leaf spot were visually evaluated, as follows: 0 = no symptoms; 1 = low; 2 = moderate; 3 = severe.

Results and discussion

Since no statistical differences were found among the dormancy-and among the spring treatments, the results are presented as the means of each group. The results of the treatment at silver tip stage were included in the dormany group because no differences were observed, except for Zn in the soil.

No treatment resulted in visually phytotoxic symptoms on the leaves. The same was found by Koch (1981) testing different compounds, but he stated that absence of phytotoxicity does not imply that damage will also be absent when doses other than the recommended ones are applied.

Soil analysis

Soil samples collected and analysed in May 1988 had the following results: pH = 6.5; P = 8.9 mg/kg; K = 277 mg/kg and O.M. = 3.6%. P and K were extracted from soil by 0.025 *N* H_2SO_4 + 0.05 *N* HCl. These values are representative of a good soil for apple culture, except for the very high K availability.

The results for Zn extracted with 0.1 *N* HCl are presented in Table 1. The availability of Zn may be considered high, but even so the trees showed severe Zn deficiency symptoms. Young (1983) condsidered Zn levels to be quite high when more than 4 mg/dm^3 was extracted from soil by 0.2 *N* NH_4Cl + 0.2 *N* HOAc + 0.015 *N* NH_4F + 0.012 *N* HCl at pH 2.5. He found that Zn in soil increased with the number of sprays with metiran, a Zn-containing fungicide. In the present experiment, only treatments $ZnSO_4$ 5% at dormancy increased the Zn level in soil. This may be explained as resulting from a possible leaching from the trees by rain, since much more Zn was applied at the dormancy stage than at the silver tip- or the spring stage.

Leaf analysis

Data in Table 2 show that spraying Zn at the dormancy- or at the silver tip stage did not increase Zn levels in leaves. Similar results were found by Hoffman and Samish (1966) and by

Table 1. Zinc levels in soil after 3 years of zinc spraying

Time of sprays	Zn concentrations (mg/kg) at two sampling depths	
	0–10 cm	10–20 cm
Control	4.7 b	2.7 b
Dormancy	12.1 a	6.5 a
Silver tip	7.1 b	3.3 b
Spring	6.7 b	3.0 b

Values followed by the same letter within the column do not significantly differ according to Duncan's multiple range test at $P = 0.05$.

Table 2. Effect of Zn sprays on Zn levels in leaves

Time of sprays	Zn concentrations(mg/kg) at five sampling periods				
	Feb 86	Nov 86	Feb 87	Nov 87	Feb 88
Control	27.8 b	13.8 a	18.8 b	10.7 a	19.5 b
Dormancy + silver tip	33.2 b	14.7 a	19.5 b	12.2 a	16.8 b
Spring	206.9 a	13.9 a	164.9 a	12.0 a	204.5 a

Values followed by the same letter within the column do not significantly differ according to Duncan's multiple range test at $P = 0.05$.

Orphanos (1982). Although the treatments reduced the symptoms of Zn deficiency, the Zn levels in leaves remained in the insufficient range according to current standards for leaf analysis interpretation (Basso and Wilms, 1988). Leaves sampled in February had higher Zn levels when sprayed in spring than at the dormancy- or at silver tip stages. Since leaves were not washed for analysis, it is supposed that a part of the sprayed Zn remained on the leaf surface as a contaminant. Young (1983) found that Zn in leaves increased linearly with the number of sprays of Zn-containing fungicides, and that the increment was lower in previously washed leaves, suggesting removal of surface contamination. The fact that the treatments failed to increase the Zn levels in leaves sampled in November may be due to the severity of the deficiency in that orchard, according to Hoffman and Samish (1966). Leaf analysis is a useful tool for detecting needs for Zn sprays, but its usefulness is questionable wherever Zn-containing compounds have been sprayed during the growing season.

Zinc deficiency symptoms

The results of Zn deficiency ratings (Table 3) reveal that all sprays were efficient in reducing the symptoms at the evaluations in February. This is in agreement with the results of other investigations which demonstrated the efficiency of different compounds and/or times of spraying (Hoffman and Samish, 1966; Koch, 1981; Terblanche *et al.*, 1974; Tomar *et al.*, 1970). For the evaluations in November, when the spring sprays were not yet applied, the best results were obtained with sprays at the dormancy- and at silver tip stages. This is explained by the fact that only trees treated at those stages had sufficient Zn to ensure normal growth for shoots and leaves. None of the sprays were capable of completely eliminating the deficiency symptoms because of their severity. In such cases it would be better to combine a spray at the dormancy- or at silver tip stage with those in spring. Depending on the severity of the deficiency, more than 3 sprays may be required in spring. Since there were no differences in efficiency of the compounds tested

Table 3. Effect of Zn sprays on the severity of Zn deficiency symptoms in leaves

Time of sprays	Ratings obtained at six stages					
	Nov 85	Feb 86	Nov 86	Feb 87	Nov 87	Feb 88
Control	2.17 a	2.33 a	2.83 a	2.33 a	1.78 a	2.57 a
Dormancy + silver tip	1.07 b	1.33 b	1.00 c	0.50 b	0.86 b	0.94 b
Spring	2.00 a	1.44 b	1.72 b	0.38 b	1.35 a	0.32 b

Values followed by the same letter within the column do not significantly differ according to Duncan's multiple range test at $P = 0.05$.

Table 4. Effect of Zn sprays on the degree of incidence of necrotic leaf spot

Time of sprays	Ratings obtained	
	Dec 86	Feb 88
Control	2.00 a	1.83 a
Dormancy + silver tip	2.33 a	1.70 a
Spring	1.04 b	0.93 b

Values followed by the same letter within the column do not significantly differ according to Duncan's multiple range test at $P = 0.05$.

anyone can be used. The choice may be based on cost and on compatibility with other chemicals to be sprayed.

Necrotic leaf spot

Necrotic leaf spot symptoms were reduced by 50% with spring sprays (Table 4). Sprays at the dormancy- or silver tip stage did not reduce the incidence of this physiological disorder. It seems that only leaves directly receiving Zn benefited from the sprays. This confirms the observations of Sutton and Clayton (1974) and Ducroquet (1987) that necrotic leaf spot can be controled by spraying compounds containing Zn. Thus, whenever necrotic leaf spot is expected to be a problem, it can be partially reduced by spraying Zn-containing compounds in the spring. Since this phenomenon seems to be related to factors such as moisture regime and nutritional imbalance (Berton and Melzer, 1987; Ducroquet, 1987; Simon, 1986), it is possible that the high availability of K in the soil and its high level in the leaves may have affected its occurrence.

Acknowledgements

The authors thank Dr A P Camilo for reviewing the manuscript, and Mr J M Hawerroth and Mr J C Werner for help in the field and in the laboratory.

References

Basso C and Wilms F W W 1988 Nutritional status of apple orchards in Southern Brazil. Acta Hortic. 232, 187–192.

Berton O and Melzer R 1987 Podridão amarga e queda antecipada de folhas de macieira. Inf. SBF 6, 11–12.

Ducroquet J P 1987 Control of necrotic leaf blotch and premature leaf drop in 'Golden Delicious' apples. HortScience 22, 574–575.

Hoffman M and Samish R M 1966 The control of zinc deficiency in apple. Israel J. Agric. Res. 16, 105–114.

Koch B 1981 Plot study seeks evaluation of foliar zinc sprays on Red Delicious apples. Goodfruit Grower 32, 28–29.

Orphanos P I 1982 Spray and soil application of zinc to apples. J. Hortic. Sci. 57, 359–366.

Simon P J 1986 Necrotic leaf spot on apple leaf tissue: *In vitro* measurements. Scientia Hortic. 29, 147–154.

Sutton T B and Clayton C N 1974 Necrotic leaf blotch of Golden Delicious apples. NC Agr. Expt. Sta., Tech. Bul. 224, 24p.

Terblanche J H, Piennar W J and Van Niekerk P E R 1974 Trace elements deficiencies in apple trees. Fruit and Fruit Tech. Res. Inst. Pretoria. 3 p.

Tomar N K, Dev. G and Randhawa N S 1970 Zinc deficiency, a major problem for apples in Kulu Valley. Indian Hortic. 15, 5.

Young E 1983 Assessment of the nutritional status by foliar analysis and soil testing. *In* Integrated Pest and Orchard Management System for Apples in North Carolina. Eds. G C Rock and J L Apple. pp 27–41. North Carolina State Univ., Raleigh.

M. L. van Beusichem (Ed.), *Plant nutrition – physiology and applications*, 261–265.

PLSO IPNC386

Zinc deficiency and pollen fertility in maize (*Zea mays*)

P. N. SHARMA, C. CHATTERJEE, S. C. AGARWALA and C. P. SHARMA
Botany Department, Lucknow University, Lucknow-226007, India

Key words: maize, pollen fertility, *Zea mays* L., zinc

Abstract

Zinc deficiency decreased pollen viability in maize (*Zea mays* L. cv. G2) grown in sand culture. On restoring normal zinc supply to zinc-deficient plants before the pollen mother cell stage of anther development, the vegetative yield of plants and pollen fertility could be recovered to a large extent, but the recovery treatment was not effective when given after the release of microspores from the tetrads. If zinc deficiency was induced prior to microsporogenesis it did not significantly affect vegetative yield and ovule fertility, but decreased the fertility of pollen grains, even of those which visibly appeared normal. If the deficiency was induced after the release of microspores from the tetrads, not only vegetative yield and ovule fertility but pollen fertility also remained unaffected.

Introduction

Deficiency of zinc has been reported to induce pollen sterility in wheat (Sharma *et al.*, 1979) and to repress male sexuality in maize (Sharma *et al.*, 1987). This paper reports the relative fertility of pollen and ovules as influenced by withholding zinc from plants given adequate zinc or supplying zinc to plants, deficient in zinc from the beginning of the experiment, at the pollen mother cell (PMC) stage and after completion of microsporogenesis in the anthers.

Materials and methods

Maize (*Zea mays* L. cv. G2) was grown in refined sand in a glasshouse with 0.13 mg Zn L^{-1} (adequate Zn) and 0.0065 mg Zn L^{-1} (low Zn) supply. The composition of the minus-zinc nutrient solution and the methods of purification of sand, nutrients and water were the same as described earlier (Sharma *et al.*, 1987). The contribution of zinc from purified water, nutrients, sand and containers was not more than 0.0001 mg L^{-1}. The pots with normal and low zinc levels were arranged in three sets each. Of the three sets of plants receiving adequate zinc, one continued to receive an adequate zinc supply (N); the zinc supply was withheld from the second set from day 36 on, three days after appearance of tassel initials (ND_1); and from the third set from day 45 on, one day after release of microspores from pollen tetrads (ND_2). To remove zinc from the ND_1 and ND_2 treatments the pots were flushed thoroughly with deionised water followed by glass-distilled water and finally with the minus-zinc nutrient solution. Of the three sets of zinc-deficient plants one set continued to receive a low zinc supply (D); the second set received adequate zinc from day 37 on, three days after the appearance of tassel initials (DN_1); and the third set received adequate zinc from day 53 on, one day after release of microspores from pollen tetrads (DN_2). Each treatment consisted of ten pots of 10-L capacity with a central drainage hole in the bottom of each pot through which the excess solution was drained. Eight seeds of maize were sown in each pot. Plants were harvested at different stages of growth, when they were needed for assessing the stage of anther development.

At the time of initiation of differential treatments there were two plants in each pot. These plants were grown to maturity.

Besides observing visible symptoms, determinations were made of pollen-producing capacity of anthers, pollen grain size, *in vitro* germination of pollen grains (Agarwala *et al.*, 1979) and pollen fertility. The latter was determined with the so-called seed-setting method by pollinating cobs of plants adequate in zinc with pollen produced in the N, ND_1, ND_2, D, DN_1, DN_2 treatments (Poehlman and Borthakur, 1969). Cobs of ND_1 and ND_2 plants were pollinated with pollen from plants adequate in zinc. Seed setting in the cobs of D, DN_1 and DN_2 plants could not be tested due to late silking in these plants. Cobs were air dried for one week and dry weight of cobs and grains was determined.

Concentration of zinc in pollen grains was determined in the nitric-perchloric acid (15:1) digest (Piper, 1942) by atomic absorption spectrophotometry. Activities of carbonic anhydrase (EC 4.2.1.1) and ribonuclease (EC 3.1.1.22) were assayed in the crude extracts of anthers by the methods of Rickli *et al.* (1964) and Tuve and Anfinsen (1960), respectively. For expressing enzyme activity on a protein basis, soluble proteins in the anther extracts were determined by the method of Lowry *et al.* (1951).

Results

Plant growth and yield

Plants deficient in zinc from the initiation of the experiment showed depression in growth from day 15 on and developed characteristic zinc deficiency symptoms visible from day 18 on. These plants developed tassel initials at day 34, a day later than in plants adequate in zinc. Zinc deficiency markedly retarded anther and pollen development and delayed anthesis by 13 days. Plants subjected to zinc deficiency after receiving normal zinc for 36 or 45 days showed a marginal decrease in dry matter yield (Table 1) and appeared visibly healthy, except for the ND_1 plants which showed mild purple pigmentation of the subterminal leaves and premature senescence after anthesis.

Initially, recovery from zinc deficiency in DN_1 plants was slow, but later these plants exhibited luxurious growth and appeared almost as healthy as plants receiving adequate zinc from the beginning. There was a significant delay in the onset of anthesis in these plants. Recovery from zinc deficiency effects was very poor in DN_2 plants (Table 1).

Development and viability of pollen grains

Plants deficient in zinc from the start of the experiment produced few and small pollen grains (Table 1), most of which had sparse cytoplasmic contents, appeared shrivelled and deflated. These pollen grains, and also those of plants in the ND_1 treatment, showed very poor *in vitro* germination and failed to set seeds in the cobs of normal plants. Induction of zinc deficiency after completion of microsporogenesis had little effect on pollen fertility. While the recovery treatment DN_2 was ineffective, pollen fertility was recovered substantially in DN_1 plants (Table 1). Pollen from plants adequate in zinc could set 65 and 57% seeds, respectively, in the cobs of ND_1 and ND_2 plants, as compared to the cobs of plants adequate in zinc (Table 2).

Zinc status of pollen grains and activities of carbonic anhydrase and RNase in mature anthers

Plants grown with low zinc from the initiation of the experiment on showed a drastic (79%) reduction in zinc concentration of pollen grains. Omission of zinc from the nutrient medium after the PMC stage or the release of microspores from the tetrads slightly decreased zinc concentration in the pollen grains, but the zinc concentration was much higher than in pollen grains from plants deficient in zinc from the beginning.

Restoration of normal zinc supply to the deficient plants resulted in significantly higher (77%) recovery when the recovery treatment was given at the PMC stage (DN_1), as compared to the recovery observed when the treatment was given after the release of microspores from the tetrads (Table 1).

Carbonic anhydrase activity in mature anthers from different treatments more or less paralleled

Table 1. Effect of zinc deficiency and recovery therefrom on plant yield, pollen-producing capacity (PPC) and activities of carbonic anhydrase and RNase in anthers, pollen grain size, *in vitro* germination and concentration of zinc in pollen grains of maize (*Zea mays* L. cv. G2) grown in refined sand

Treatment	Dry matter yield ($g\ plant^{-1}$)[a]	Cob yield ($g\ plant^{-1}$)[a]	PPC (Pollen grains $anther^{-1}$)[b]	Carbonic anhydrase (Units mg^{-1} protein)[a]	RNase (Units mg^{-1} protein)[a]	Pollen grain size (μm)[c]	*In vitro* germination (%)[b]	Zinc concentration ($\mu g\ g^{-1}$) air dry pollen)[a]
N	74.3	70.0	2839 ± 134	147.4	150.0	88.1 ± 1.86	84.5 ± 5.4	75.15
ND_1	66.9	17.6	2695 ± 168	28.3	343.0	81.9 ± 2.14	19.6 ± 4.3	27.05
ND_2	65.2	34.4	2881 ± 132	81.3	210.6	83.7 ± 1.64	70.5 ± 8.3	45.10
D	23.0	—	796 ± 92	18.4	355.7	72.1 ± 1.40	10.8 ± 4.6	16.52
DN_1	102.8	35.6	2149 ± 185	75.5	208.1	90.0 ± 2.35	68.1 ± 8.4	48.63
DN_2	36.2	—	1141 ± 144	40.9	189.2	81.8 ± 1.83	16.9 ± 4.8	23.46
$LSD_{0.05}$	7.7	10.6		7.4	26.3			6.45

[a] Mean values based on duplicate estimations.
[b] Mean values based on 10 assays.
[c] Mean values based on 50 measurements.

Table 2. Effect of zinc deficiency and recovery therefrom on the seed-setting capacity of pollen and development of grains in maize (*Zea mays* L. cv. G2) grown in refined sand

Cross ♂ × ♀	Cob weight (g)	Grain weight (g cob^{-1})	Grain formation (grains cob^{-1})	100 grain weight (g)
N N	154.2	116.7	385	30.4
ND_1 N	28.0	—	—	—
ND_2 N	71.1	42.5	138	30.8
D N	30.4	—	—	—
DN_1 N	88.5	54.2	171	31.7
DN_2 N	35.8	1.7	4	43.3
N ND_1	84.1	54.4	252	21.6
N ND_2	94.7	64.5	220	29.3
$LSD_{0.05}$	8.3	10.2	37	2.6

zinc concentration in pollen grains (Table 1). Zinc deficiency caused a marked increase in the activity of RNase in the anthers. The effect was particularly marked in D and ND_1 plants (Table 1).

Discussion

Continued zinc deficiency had little effect on the onset of the reproductive phase, but severely retarded the development of tassels, anthers and pollen grains. The observed decrease in pollen-producing capacity of anthers and pollen fertility in low-zinc plants indicate suppression of male sexuality due to zinc deficiency. The loss of fertility of pollen from zinc-deficient plants was underestimated by the *in vitro* germination of pollen grains, as compared to the seed-set method.

Subnormal zinc at any stage of anther development prior to microsporogenesis induced male sterility. In an earlier study, induction of zinc deficiency before the onset of the reproductive phase resulted in anther contabescence including a tendency for reversal of anthers to foliar appendages (Sharma *et al.*, 1987). Deficiency of zinc from the beginning of the experiment delayed microspore development and caused pollen sterility. When zinc was withheld at the PMC stage of another development, the development of pollen grains seemed normal, but those pollen grains were sterile. A significant recovery in the development, fertility and zinc status of pollen upon restoration of normal zinc supply to the deficient plants before the PMC stage of anther development further suggests that the requirement of zinc for developing anthers was crucial at the time of microsporogenesis. The requirement was not critical after microsporogenesis as a supply of zinc after this stage of anther development did not restore the fertility of pollen.

Pollen sterility in zinc-deficient plants may also be attributed to the derangement in RNA metabolism induced by zinc deficiency in the anthers. Aberrant RNA metabolism has been suggested to have an important role in pollen sterility (Heslop-Harrison, 1972). Involvement of zinc in RNA metabolism, as indicated by a marked increase in the activity of RNase, is well documented (Prask and Plocke, 1971; Sharma *et al.*, 1987).

While zinc deficiency, induced before the PMC stage, drastically reduced pollen fertility, it had little effect on ovule fertility. This is evident from the seeds set by normal pollen on the cobs of plants in the zinc-deficiency treatment ND_1. While low unit-grain weight due to poor grain filling in cobs of ND_1 plants could be attributed to decreased photosynthesis owing to low zinc (Seethambaram and Das, 1985), increased unit-grain weight in the cobs of normal plants pollinated with pollen from DN_2 plants, where seed setting was poor, could be attributed to the presence of a smaller sink with a relatively large source.

References

Agarwala S C, Chatterjee C, Sharma P N, Sharma C P and Nautiyal N 1979 Pollen development in maize plants subjected to molybdenum deficiency. Can. J. Bot. 57, 1946–1950.

Heslop-Harrison J 1972 Sexuality of angiosperms. *In* Plant Physiology: A treatise. Ed. F C Steward. pp. 133–289. Academic Press, New York.

Lowry O H, Rosebrough N J, Farr A L and Randall R J 1951 Protein measurement with the Folin phenol reagent. J. Biol. Chem. 193, 265–275.

Piper C S 1942 Soil and Plant Analysis. Waite Agric. Res. Inst., The University, Adelaide.

Poehlman J M and Borthakur D 1969 Breeding Asian field crops with special reference to crops of India. pp 151–181. Holt, Rinehart and Winston, Inc., New York.

Prask J A and Plocke D J 1971 A role of zinc in the structural integrity of the cytoplasmic ribosomes of *Euglena gracilis*. Plant Physiol. 48, 150–155.

Rickli E E, Ghazanfar B H, Gibbons B H and Edsall J T 1964 Carbonic anhydrase from human erythrocytes: Preparation and properties of two enzymes. J. Biol. Chem. 239, 1065–1078.

Seethambaram Y and Das V S R 1985 Photosynthesis and activities of C_3 and C_4 photosynthetic enzymes and zinc deficiency in *Oryza sativa* L. and *Pennisetum americanum* L. Photosynthetica 19, 72–79.

Sharma P N, Chatterjee C, Sharma C P, Nautiyal N and Agarwala S C 1979 Effect of zinc deficiency on the development and physiology of wheat pollen. J. Indian Bot. Soc. 58, 330–334.

Sharma P N, Chatterjee C, Sharma C P and Agarwala S C 1987 Zinc deficiency and anther development in maize. Plant Cell Physiol. 28, 11–18.

M. L. van Beusichem (Ed.), *Plant nutrition – physiology and applications*, 267–273.

PLSO IPNC397

Micronutrient foliar intake by different crop plants, as affected by accompanying urea

M.M. EL-FOULY[1], A.F.A. FAWZI[1], Z.M. MOBARAK[1], E.A. ALY[2] and F.E. ABDALLA[1]
[1] *Botany Department and* [2] *Air-Pollution Department, National Research Centre, Dokki-Cairo, Egypt*

Key words: EDTA-chelates, fababean, foliar application, *Glycine max* L., micronutrients, pea, *Pisum sativum* L., soybean, sulphate salts, *Triticum aestivum* L., urea, *Vicia faba* L., wheat

Abstract

EDTA-chelated or sulphate salts of Mn-Zn-Fe combination were sprayed at different concentrations with or without urea on plants 30 days after sowing in pot experiments. Results showed that mixing urea with chelated and non-chelated micronutrient compounds increased intake of micronutrients in comparison with either micronutrients without urea or control treatment.

It was concluded that adding urea to micronutrient spray solution makes it possible to reduce doses of applied micronutrients, without influencing their effect.

Introduction

Soil with high pH, low organic matter, irregular moisture and compactness have low availability of the micronutrients Zn, Mn, and Fe to crop plants (El-Fouly, 1983). It is impractical to supply plants with these elements through soil. Foliar application of micronutrients was successfully used for correcting their deficits in crops (Alexander, 1986). The micronutrient chemical carrier plays an important role in determining the efficiency of the foliar fertilizer (Stein and Storey, 1986). Urea is known to increase penetration of accompanying nutrients (Wittwer and Bukovac, 1969). Wallace and Wallace (1983) and Miller and Warnick (1986) indicated that urea increases the foliar intake of micronutrients. Volk and McAnliffe (1957) and Bowman and Paul (1989) reported that urea is one of the compounds most rapidly absorbed by leaves. The objective of the present work was to investigate effects of combining urea with EDTA-chelated micronutrients or their sulphate salts at foliar application on some legumes and cereals under semi-arid conditions.

Experimental

Pot experiments were twicely conducted in the vegetation house at the National Research Centre, Dokki-Cairo, Egypt on soybean, fababean, pea, wheat and maize. Seeds were sown on the proper date for every crop in Mitscherlich pots filled with 7 kg loamy soil (Nile-alluvium)/pot. Ten days after sowing, plants were thinned to 10/pot (except for wheat which was thinned to 20/pot) and fertilized with NPK. Treatments were replicated 4 times with 4 pots/replicate. Plants were sprayed 3 weeks after sowing with micronutrients.

The following compounds were used as foliar spray:
(1) EDTA-chelated powders: (a) Mn-stressed: (Wadi) containing 4.5% Mn + 3% Zn + 1.5% Fe (element); (b) Zn-stressed: (Sahara) containing 3.5% Mn + 5% Zn + 1.7% Fe (element). (2) Solutions of sulphate-salts: (a) Mn-stressed: (FIE-1) containing 5% Mn + 3.5% Zn + 2% Fe; (b) Zn-stressed: (FIE-2) containing 3.5% Mn + 5% Zn + 1.5% Fe (element). Zn-stressed compounds were used only on wheat. Compounds

used were formulated upon the previous soil and leaf analyses of the Project "Micronutrients and Other Plant Nutrition Problems in Egypt". Concentrations used were selected upon experience with every crop under pot experiment conditions. They ranged from 1 to 5 g L^{-1} for EDTA-chelated compounds and from 3 to 17 mL L^{-1} for the sulphate salts. Tap-water was used for preparing the spraying solutions; as well as for spraying control plants. EDTA-chelated or sulphate-salt compounds were sprayed without or with 1% urea (0.1 g urea/10 mL solution). Values of pH in water and spraying solutions were measured. The above-ground plant parts were harvested 10 days after spraying, washed with a sequence of tap water, 0.01 *M* HCl-acidified bidistilled water and bidistilled water. Plants were oven-dried at 70°C, weighed and ground in a stainless steel mill. Total Mn, Zn and Fe contents (μg/g dry matter) were determined using wet digestion with a mixture of nitric, perchloric and sulphuric acid (8:1:1, v/v), according to Chapman and Pratt (1961). The experimental error including sampling and analytical errors was below 5%.

Results

Maize did not show the same response in the 2 experiments, therefore both were excluded from results.

Dry weight and micronutrient content

In control plants, relatively low dry weight was almost accompanied by high contents of micronutrients (Table 1). However, Zn in fababean was an exception.

Table 1. Plant dry weight and nutrient content in control plants

Plant	Dry weight (mg/plant)	Nutrient content μg.g^{-1} dry weight		
		Mn	Zn	Fe
Soybean	870	43	26	97
Fababean	520	68	69	288
Pea	184	69	28	279
Wheat	90	90	78	358

Effect of micronutrient compounds

Contents of Mn, Zn and Fe in treated plants increased over those of control ones; in response to spraying either EDTA-chelated or sulphate-salt compounds (Table 2). In almost all treatments the micronutrient content in plant was increased with increasing concentration applied. No relation was observed between level of micronutrient content in control plants and the percentage of its increase due to the treatment. Comparison between effects of EDTA-chelated and sulphate-salt compounds at concentration of 0.3% from each (Table 2), shows that EDTA-chelate led to more Fe as well as less Zn and Mn concentration percent increases than those occurred due to sulphate-salt compounds. On the average, percent increments of micronutrient concentration due to both compounds reached maximum in Zn of soybean and pea, Mn of wheat and Fe of fababean (Table 2). Pattern of those increments in terms of micronutrient ratio varied with crop. Ratio of Mn/Zn/Fe relative increase in response to both compounds was Zn > Fe > Mn in soybean, Fe > Zn > Mn > in fababean, Zn > Mn > Fe in pea and Mn > Zn > Fe in wheat.

Effect of urea

Combining 1% urea with both micronutrient compounds raised considerably the induced increases of average micronutrient contents in soybean, fababean, pea and wheat (Figs. 1, 2, 3). On the average, urea led to increments of 16%, 26% and 33% with EDTA-chelated compounds and 13%, 48% and 21% with sulphate-salt ones in Mn, Zn and Fe contents; respectively. Results in Table 3 show that these average increases varied with concentrations of compound used. At the lowest concentrations, urea induced more Mn and Zn and the same Fe increment with EDTA compounds, as compared to those with sulphate-salt ones. At the highest concentrations urea effect was the same on Mn, more on Fe and less on Zn with EDTA than that with sulphate salts.

With both compounds and concentrations (Table 3) maximal percent increments by urea

Table 2. Effect of spraying micronutrient compounds on micronutrient content ($\mu g\ g^{-1}$-dry weight) in soybean, fababean, pea and wheat

Compound concentration	EDTA-chelated				Sulphate-salts			
(%)	Soybean	Fababean	Pea	Wheat	Soybean	Fababean	Pea	Wheat
Mn								
0.00	43	68	71	80	43	68	67	100
0.10	47	68	74	89				
0.20	47	70	75	91				
0.30	57	75	77	90	58	69	68	136
0.40	59		79	87				
0.50	61		78	103	58	69	69	148
0.70					61	74	71	166
0.90					77	75	79	193
1.30						76	80	224
1.70						83	83	265
Average increase (%)	26	2	8	16	45	8	10	81
Zn								
0.00	26	70	28	72	26	69	28	83
0.10	27	70	29	72				
0.20	33	71	30	74				
0.30	41	81	31	79	41	69	33	103
0.40	46		32	83				
0.50	50		37	89	43	73	34	115
0.70					66	73	42	115
0.90					67	74	45	129
1.30						82	46	167
1.70						88	49	169
Average increase (%)	51	6	14	10	109	11	48	60
Fe								
0.00	97	262	308	350	97	295	247	366
0.10	129	282	325	381				
0.20	123	308	343	384				
0.30	126	313	347	391	129	317	249	379
0.40	125		351	389				
0.50	139		323	403	174	332	247	379
0.70					191	337	263	383
0.90					231	336	266	386
1.30						348	267	394
1.70						382	281	444
Average increase (%)	32	11	7	10	87	16	6	5

occurred in Fe of soybean, Fe and Zn of fababean and Zn of both pea and wheat. Meanwhile, highest increments by urea occurred in Fe with EDTA compounds; especially on soybean and those in Zn and Mn with sulphate salts.

Effect of micronutrient compounds and urea on pH

Results in Table 4 indicate that both EDTA-chelated and sulphate-salt compounds decreased

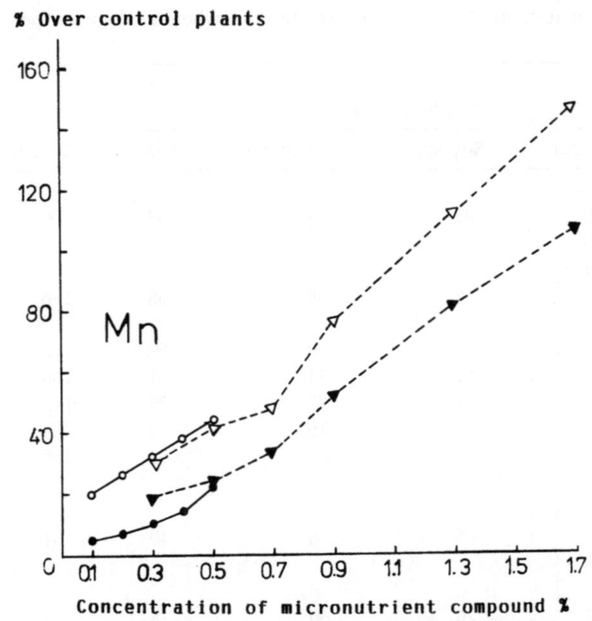

Fig. 1. Percent increase of average Mn concentration in soybean, fababean, pea and wheat, as affected by micronutrient – urea combination.
(mean of 2 experiments)

Compound	without urea	with urea(1 %)
EDTA-chelated	●——●	○——○
Sulphate salts	▲---▲	△---△

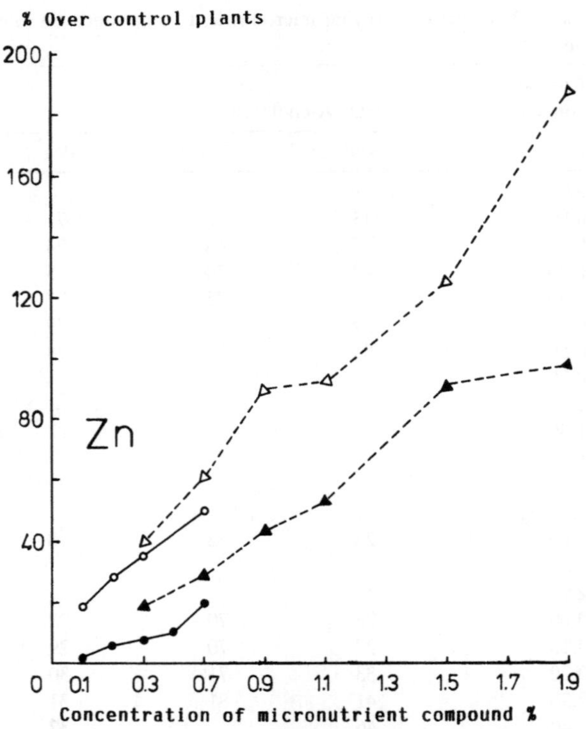

Fig. 2. Percent increase of average Zn concentration in soybean, fababean, pea and wheat, as affected by micronutrient – urea combination.
(mean of 2 experiments)

Compound	without urea	with urea(1 %)
EDTA-chelated	●——●	○——○
Sulphate salts	▲---▲	△---△

pH of spraying solutions. Urea addition increased pH of both water and solutions of both compounds.

Discussion

Results indicate that more dry matter in control plants led to a so-called "dilution effect" mentioned by Marschner (1986). The positive reponse of Mn, Zn and Fe contents in plants to foliar application of these micronutrients agrees with results reported in Egypt on fababean and wheat in field experiments (Fawzi *et al.*, 1983; Firgany *et al.*, 1983). Yield increments in response to micronutrients foliar application were mentioned by the same authors as well as by Abo-Khadra *et al.* (1983), El-Fouly (1983) and Abd El-Hadi *et al.* (1986) on legumes and cereals in field trials. Each EDTA-chelate and sulphate salt exerted its specific effect on micronutrient intake by treated plants. EDTA that induces more Fe penetration of cuticle and translocation in plant than sulphates (Basiony and Biggs, 1976; Wallace and Wallace, 1982, 1983) led to more Fe intake in spite of its lower amount than those of accompanying Mn and Zn in the compounds. Penetration of Mn and Zn was mentioned by Mederski and Hoff (1958) and Jooste (1965) to be dependent on concentration rather than on chelation. This would explain the results of relatively more Mn or Zn intake from sulphate compounds than from EDTA ones. In this respect, Haile-Mariam (1965) found that sulphates of Mn or Zn penetrates cuticle more efficiently than chelated ones; probably due to more cuticular sorption (Ferrandon and Chamel, 1986). Such effects of the compound formulation were modified by specific plant needs. Selectivity of every crop to the different elements applied (Mn, Zn and Fe) was indicated by the pattern of increases. Maximum increase of content occurred in Fe with Mn-stressed compounds and in

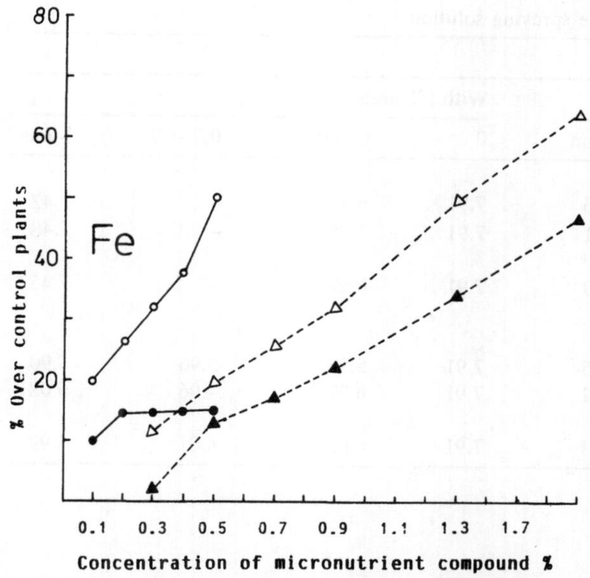

Fig. 3. Percent increase of average Fe concentration in soybean, fababean, pea and wheat, as affected by micronutrient – urea combination.
(mean of 2 experiments)

Compound	without urea	with urea(1 %)
EDTA-chelated	●——●	○——○
Sulphate salts	▲---▲	△---△

Mn with Zn-stressed ones. Further, pattern of relative content increments between micronutrients was Fe > Zn > Mn in soybean and fababean Zn > Fe > Mn in pea and Mn > Zn > Fe in wheat. High demands of soybean and fababean on Fe, wheat on Mn and pea on Zn were mentioned by several authors (*e.g.* El-Fouly *et al.*, 1984; Gattier *et al.*, 1986; Shorrocks, 1986).

Addition of urea to the micronutrient foliar spray increased plant intake of Mn, Zn and Fe from both EDTA-chelated and sulphate-salt compounds. The same effect was indicated by, Wallace and Wallace (1983) and Adams and Brooks (1986). Results of Hsu and Ashmead (1984) as well as Bowman and Paul (1989) showed increased leaf nutrient absorption with urea spraying. Further, Abo-Khadra *et al.* (1983) and Ashour *et al.* (1983) found that foliar micronutrients-induced-yield increments were raised by adding urea. It seems from the micronutrients and urea effect on pH of the spray-solution that pH values mediated the absorption of micronutrients; especially from sulphates; which is in harmony with results of Chamel (1986). Yet, whether urea-induced increases of Zn intake from sulphates was a function of relative pH increase (Ferrandon and Chamel, 1986) can neither be denied nor concluded under conditions of this work and needs further investigation.

Conclusion

It might be concluded that sulphate salts can induce similar effects as EDTA-chelates of mi-

Table 3. Increase of micronutrient content in plant in response to combining urea with micronutrient compounds (% of micronutrient compound without urea treatment)

Plant	Mn		Zn		Fe	
	EDTA	Sulphate	EDTA	Sulphate	EDTA	Sulphate
At the lowest concentration used						
Soybean	55	21	11	22	32	10
Fababean	4	2	17	13	10	7
Pea	5	16	48	3	5	9
Wheat	6	5	10	23	3	16
Mean	17.5	10.7	21.5	15.2	12.5	12
At the highest concentration used						
Soybean	38	14	28	9	79	9
Fababean	7	20	4	8	9	15
Pea	13	11	27	94	36	9
Wheat	8	23	22	51	32	16
Mean	16.5	17	20	40	39	12

Table 4. Effect of micronutrient compounds and urea on pH of the spraying solution

Compound	Micronutrient compound concentration (%)							
	Without urea				With 1% urea			
	0	0.1–0.5	0.7–1.7	Mean	0	0.1–0.5	0.7–1.7	Mean
1. EDTA-chelated								
Wadi	7.76	6.77	–	7.26	7.91	6.93	–	7.42
Sahara	7.76	6.87	–	7.31	7.91	7.05	–	7.48
Mean	7.76	6.82	–	7.29	7.91	6.99	–	7.45
2. Sulphate salts								
FIE-1	7.76	5.57	3.63	5.65	7.91	5.84	3.96	5.90
FIE-2	7.76	5.98	3.72	5.82	7.91	6.27	4.06	6.08
Mean	7.76	5.77	3.67	5.73	7.91	6.05	4.01	5.99

cronutrient foliar fertilizers to improve micronutrient status in some field crops at early stages of development. Combining urea with sulphates as well as EDTA-chelated elements can increase their utilization by plant and economize the use of foliar micronutrient fertilizer.

Acknowledgement

This paper was conducted as a part of the Egypto-German Project "Micronutrients and Other Plant Nutrition Problems in Egypt" excuted by the National Research Centre (Coordinator Prof. Dr. M M El-Fouly) and the Institute for Plant Nutrition, Technical University, Munich (Head Prof. Dr. A Amberger). The project was supported by the Egyptian Academy of Scientific Research and Technology and the German Federal Ministry of Technical Cooperation through the German Agency for Technical Cooperation (GTZ).

References

Abd El-Hadi A H, Asy A H, Döring H W, Khadr M S, Mohamed Y H, Mostafa A A and Taha M E 1986 Effect of foliar fertilization in different crops under Egyptian conditions. *In* Foliar Fertilization in Different Crops. Ed. A. Alexander pp. 126–141. Proc. 1st Internatl. Symp. Foliar Fertilization, Alexandria, Egypt.

Abo-Khadra S H, Hussein M M and El-Zeiny H A 1983 Effect of foliar nutrition on productivity of wheat fertilized with different rates of nitrogen. Proc. 1st Conf. of Agron, Egypt. pp 347–362.

Adams G P and Brooks J 1986 An integrated approach to trace element deficiency diagnosis and treatment. *In* Proceedings 2nd Intern. Symp. Role of Micronutrients in Agriculture. Ed. P. Morard. pp. 235–245. Toulouse, France.

Alexander A 1986 Crop needs specific foliar application of micronutrients. *In* Proceedings 2nd Intern. Symp. Role of Micronutrients in Agriculture. Ed. P. Morard. pp. 309–323. Toulouse, France.

Ashour N I, Saad A O M, Thalooth A T 1983 A preliminary study on the effect of foliar fertilization with urea combined with Fe, Zn or Mn on maize production under calcareous soil conditions Proc. 1st Conf Agron, Egypt. pp. 257–272.

Basiony F M and Biggs R M 1976 Penetration on ^{59}Fe through isolated cuticles of citrus leaves, Hort. Sci. 11, 417–419.

Bowman D C and Paul J L 1989 The foliar absorption of urea-N by Kentucky bluegrass turf. J Plant Nutr. 12, 659–673.

Chamel A 1986 Survey of different approaches to determine the behaviour of chemicals directly applied to aerial parts of plants. *In* Foliar Fertilization in Different Crops. Ed. A. Alexander. pp. 66–86. Proc. 1st Internatl. Symp. Foliar Fertilization. Alexandria, Egypt.

Chapman H D and Pratt P 1961 Methods of Analysis of Soil, Plant and Water, University of California, Div. Agric. Sci., CA. pp 1–309.

El-Fouly M M 1983 Micronutrients in arid and semiarid areas: Levels in soils and plants and the need for fertilizers with reference of Egypt. Proc. 15th Coll. Int. Potash Inst. (Bern). pp. 163–173.

El-Fouly M M, Fawzi A F A Firgani A H and El-Baz F K 1984 Micronutrient status of crops in selected areas in Egypt. Commun. Soil Sci. Plant Anal. 15, 1175–1189.

Fawzi A F A, Firgany A H, Rezk A I, Kishk M A and Shaaban M M 1983 Response of *Vicia faba* bean to K and micronutrient fertilizers. Egypt. J. Bot. 26, 113–121.

Ferrandon M and Chamel A 1986 Etude de comportment des oligoelements (Zn, Mn, Fe) fournis par voie foliaire. *In* Proc. 2nd Intern. Symp. Role of Micronutrients in Agriculture. Ed. P. Morard. pp. 299–308. Toulouse, France.

Firgany A H, Rezk A I, Fawzi A F A and El-Sayed A A 1983 Response of nutritive status and grain yield of wheat to nutrient supply in Egypt. Egypt. J. Bot. 26, 101–112.

Gattier S W, Martens D C and Brumback T E 1986 Timing of foliar manganese application for certain manganese deficiencies in soybean. Agron. J. 77, 627–630.

Haile-Mariam S M 1965 Mechanisms of Foliar Penetration and Translocation of Mineral Ions with Special Reference to Coffee (*Coffea arabica* L.) Ph.D Thesis, Michigan State University.

Hsu H H and Ashmead H D 1984. Effect of urea and ammonium nitrate on the uptake of iron through leaves. J. Plant Nutr. 7, 291–299.

Jooste J H 1965 Some factors affecting foliar uptake and distribution of zinc in apple and apricot seedlings. S. Afr. J. Agric. Sci. 3, 899–908.

Marschner H 1986 Diagnosis of deficiency and toxicity of mineral nutrients. *In* Mineral Nutrition of Higher Plants. pp. 393–404. Academic Press, London.

Mederski H J and Hoff D I, 1958 Factors affecting absorption of foliar applied manganese by soybean plants. Agron. J. 50, 175–178.

Miller G W and Warnick K 1986 Foliar applications of micronutrient, in correcting deficiencies in oats. *In* Foliar Feeding of Plants with Amino Acid Chelates. Ed. H. Dewayne Ashmead. pp. 300–320. Publ. Noyes Publications, Mill Road, Park Ridge, NJ, USA.

Shorrocks V M 1986 Boron, copper, iron, manganese molybdenum, selenium and zinc: A global appraisal of recent developments. *In* Proceedings 2nd Intern. Symp. Role of Micronutrients in Agriculture. Ed. P. Morard. pp. 397–416. Toulouse, France.

Stein L A and Storey J B and 1986 Influence of adjuvants on foliar absorption of nitrogen and phosphorus by soybeans. J. Am. Soc. Hort. Sci. 111, 824–832.

Volk R J and McAnliffe C 1957 Factors affecting the foliar absorption of ^{15}N labelled urea by tobacco. Soil Sci. Soc. Am. Proc. 18, 308–312.

Wallace A and Wallace G A 1982 Micronutrient uptake by leaves from foliar sprays of EDTA-chelated metals. J. Plant Nutr. 975–987.

Wallace S and Wallace G A 1983 Foliar fertilization with metalosates. J. Plant Nutr. 6, 551–559.

Wittwer S H and Bukovac M J 1969 The uptake of nutrients through leaf surface. *In* Handbuch der Pflanzenernährung und Düngung, Ed. Hans Linser, pp. 235–261. Springer-Verlag, Wien, New York.

M. L. van Beusichem (Ed.), *Plant nutrition – physiology and applications*, 275–280.

PLSO IPNC363A

Internal boron requirements of green gram (*Vigna radiata*)

R.W. BELL, L. McLAY, D. PLASKETT, B. DELL and J.F. LONERAGAN
School of Biological and Environmental Sciences, Murdoch University, Murdoch, WA 6150, Australia

Key words: boron deficiency, flowering, leaf blade elongation, pod development, pod abscission, seed development, *Vigna radiata* (L.) Wilczek

Abstract

Internal boron (B) requirements of green gram during reproductive development were studied in a solution culture experiment in the glasshouse by temporarily withdrawing the external B supply at each of two growth stages (at the appearance of first flower buds, and at early podding). B was re-supplied to −B plants when their leaf blade elongation rates declined below those in control plants. Subsequent effects of the −B treatment on flower and pod development were determined. Pod and seed development was more sensitive to −B than was development of the flowers. As −B plants grew into B deficiency, B concentrations in expanding pods and unfolding leaves declined rapidly. B concentrations less than 14 mg B/kg in leaf blades restricted their elongation and in pods caused withering and abscission and decreased the number of fully developed seeds.

Introduction

Green gram is an important crop in Asia but little is known about its nutrient requirements (Lawn and Ahn, 1985). Recently, B deficiency has been found to restrict pod set in green gram in northern Thailand and to depress shoot dry matter (DM) (Rerkasem, 1986), and seed yield (Pridisripipat and Rerkasem, unpublished data). Since there is increasing evidence of widespread B deficiency in soils of Thailand (Keerati-Kasikorn *et al.*, 1987; Netsangtip *et al.*, 1985; Rerkasem *et al.*, 1988), the present study was conducted to establish internal B requirements of green gram during reproductive growth to aid in the diagnosis of B deficiency. The study uses decreases in the elongation rate of young leaf blades (LBER) to indicate the onset of B deficiency in green gram shoots during reproductive growth: in soybean, this measure was a sensitive early indicator of B deficiency (Kirk and Loneragan, 1988).

Materials and methods

Details of experiments

The effects of interrupted B supply during two periods of reproductive growth on LBER, shoot DM, floral development, pod and seed set, and on B concentrations in plant parts were examined in a solution culture experiment conducted in the glasshouse with green gram (*Vigna radiata* (L.) Wilczek cv. Kampaengsaen 1).

B deficiency at early flower bud appearance
B treatments were imposed by interrupting B supply (−B) to one set of plants 24 days after germination when flower buds were first observed: control plants received a continuous B supply (control; 10 μM H_3BO_3) throughout the experiment. Two plants per pot were harvested at the commencement of the B treatment period (0 days after treatments commenced: 0 DAT). Plants were maintained in these treatments until

4 DAT when LBER in −B plants was depressed below the rate in control plants: a further two plants per pot were harvested, and B was re-supplied at 10 μM H_3BO_3 to −B pots. The remaining two plants per pot were harvested on 17 DAT (replicates II and IV) and 28 DAT (replicates I and III).

B deficiency at early podding

B treatments were imposed on a second set of plants by interrupting B supply (−B) 38 days after germination at early pod set: control plants received a continuous B supply (control: 10 μM H_3BO_3). Plants were maintained in these treatments until LBER in −B plants was depressed below the rate in control plants (9 DAT). Two plants were harvested at the commencement (0 DAT) and again at the end (9 DAT) of the −B treatment period when B was re-supplied at 10 μM H_3BO_3 to −B pots: the remaining two plants were grown until final harvest (27 DAT).

General procedures

Procedures for the preparation of nutrient solutions and the propagation of seedlings for the experiment are described by Kirk and Loneragan (1988). Each pot initially contained 10 plants which were thinned to 6 plants on day 14. Boron treatments were imposed by transferring B adequate plants into complete nutrient solutions (Kirk and Loneragan, 1988) containing 0 (−B) or 10 μM (control) H_3BO_3 after washing roots three times in 1 mM $Ca(NO_3)_2$ solutions.

At harvests at the commencement (0 DAT) and the end of the −B treatment period (4 or 9 DAT), plants were partitioned into the following parts for B analysis:

UL – unfolding leaf (see Kirk and Loneragan, 1988)
UL + 1 – the leaf immediately older than the UL
YFEL – youngest fully expanded leaf
Flower buds – calyx longer than corolla
Flowers – corolla longer than calyx
Pods < 2 cm
Pods ⩾ 2 cm

Between the end of the −B treatment period and the final harvest, and at approximately weekly intervals between, each inflorescence was examined and each flower bud, flower, and pod was categorized into one of the following groups:

Flower buds
Flowers
Pods < 2 cm length – newly set pods
Pods ⩾ 2 cm length – rapidly expanding pods
Unripened full size pods *i.e.* full length pods with indentations between seeds
Ripe pods with black colour developing on pod wall
Abscissed and withered pods

Pods present on plants at harvests were dissected and seeds were categorized as fully developed if they completely filled the cavity in the pod mesocarp, or small if they did not.

From 4 days after germination, pots were kept in temperature controlled water baths at 27°C in a glasshouse: mean maximum and minimum air temperatures during the experiment were 30.8 ± 0.6 and 17.9 ± 0.5°C, respectively. Nutrients were resupplied regularly during the experiment according to plant requirements for growth (Asher and Blamey, 1987). Solution pH was adjusted to 5.5 daily by additions of 0,2 M H_2SO_4.

Plant parts were dry ashed at 500°C for 8 h and their B concentrations were determined using the Azomethine-H method (Lohse, 1982).

Results

B deficiency at early flower bud appearance

Symptoms of B deficiency similar to those which appeared in B deficient soybean (Kirk and Loneragan, 1988) appeared in roots of −B plants 3 DAT; one day later, LBER was depressed in −B plants. Boron concentrations in the UL and flower buds of −B plants were depressed to 13 and 32 mg B/kg DM, respectively, at 4 DAT (Table 1): there were no effects of −B treatment on flower development during treatment or on flower development, pod set, pod development and seed set following restoration of B supply (data not shown).

Table 1. Effect of interrupted B supply for 4 days at early flower bud appearance and for 9 days at early podding on B concentration (mg kg^{-1}) in plant parts. Values are means of four replicates (SE)

Plant part[a]	B supply	
	control	−B
Flower bud appearance		
UL	51 (1)	13 (1)
Flower buds[b]	55	32
Early podding		
UL	34 (2)	11 (1)
YFEL	24 (2)	14 (1)
Pods (≥2 cm)	36 (2)	11 (1)
Pods <2 cm	50 (3)	19 (1)
Flowers	61 (4)	30 (7)
Flower buds	59 (1)	26 (2)

[a] See materials and methods for description of plant parts.
[b] Replicates bulked for analysis.

B deficiency at early podding

Boron deficiency symptoms appeared on roots of −B plants 4 DAT, but LBER was not depressed for a further 5 days. Boron concentrations were depressed in all plant parts of −B plants, but especially in young leaves and pods ≥ 2 cm which both dropped below 11 mg B/kg DM (Table 1). Boron concentrations in −B plants remained relatively high in flowers and flower buds.

During treatment, −B had no effects on flower development, pod development or shoot DM. Seven days after B was resupplied to −B plants (16 DAT), unripened full size pods began to wither and absciss on −B plants. AT 17 DAT, there was a four-fold increase in the number of abscissed pods on −B plants and a 50% decrease in the number of unripened full sized pods (Fig. 1). After a further six days (23 DAT), the number of ripe pods in −B plants decreased 50% reflecting the maturation of pods that were unripened full size pods at 17 DAT. At maturity (27 DAT), interrupting B supply depressed the number of ripe pods and increased numbers of withered/abscissed pods subtended at nodes 6–7 on the main stem (Fig. 2): at higher nodes on the main stem, it had the opposite effect. Furthermore, at maturity the unripened full size pods were unaffected by −B regardless of nodal position on the main stem.

Table 2. Effects of interrupted B supply for 9 days at early podding on the percentage of pods and flowers which subsequently developed into ripe pods (A) or withered and abscissed pods (B) at 27 days after treatments commenced. Values are means of four replicates (SE).

	Flowers	Pods (<2 cm)	Pods (>2 cm)
(A) Ripe pods			
Control	0 (0)	44 (15)	94 (3)
−B	25 (25)	78 (13)	77 (10)
(B) Withered and abscissed pods			
Control	0 (0)	56 (15)	6 (3)
−B	0 (0)	22 (13)	23 (10)

Interrupting B supply increased withering and abscission particularly in those pods which were ≥2 cm length and expanding rapidly at 9 DAT when B was re-supplied (Table 2). In −B plants, 23% of pods ≥2 cm on 9 DAT subsequently withered or abscissed, whereas in control plants only 6% withered. By contrast, interrupting B supply decreased the % of withered pods which developed from pods <2 cm on 9 DAT.

In addition to depressing early pod development, −B decreased the number of mature seeds from 8.0 ± 0.2 to 7.2 ± 0.1 per pod and inhibited seed development in ripe pods by decreasing from 64 ± 2 to 55 ± 2% the proportion of seeds that were full sized. Interrupting B supply had no effect on seed size in immature pods which developed subsequent to the resupply of B to −B plants. Interrupting B supply had no effect on germination % of seed in ripe pods (data not presented) or on numbers of seeds set per pod which averaged 12–13 per pod.

Discussion

Pod and seed development appeared to be the most sensitive phase of reproductive growth in green gram to low B in shoots. Its sensitivity to low B can be attributed to the fact that B concentrations declined to low values more rapidly in expanding pods (≥2 cm length) than did B concentrations in newly set pods (<2 cm length), flowers and flower buds.

By contrast with its effects on pod development, seed development, and on LBER, inter-

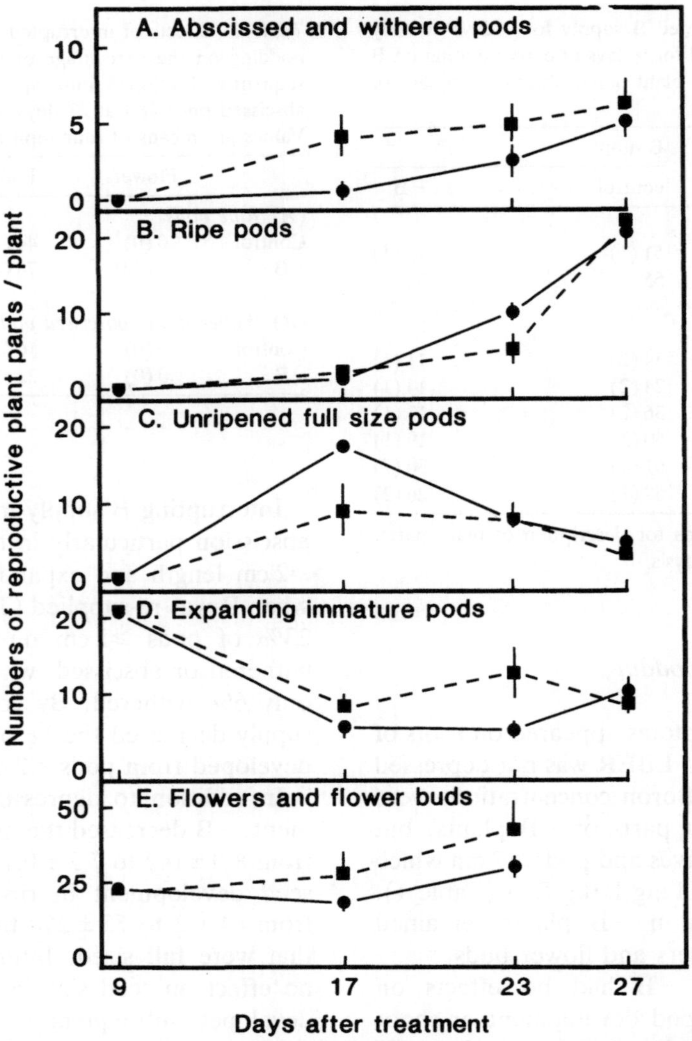

Fig. 1. Effects of interrupted B supply (●, control; ■, −B) at early podding on numbers of abscissed and withered pods (**A**), ripe pods (**B**), unripened full size pods (**C**), expanding immature pods (**D**), and flower and flower buds (**E**) per plant at 9, 17, 23 and 27 days after treatments commenced (DAT). Values are means of four replicates (SE). Note that B was re-supplied to −B plants at 9 DAT.

rupting B supply had no effect on flower development, on pod set, or on seed set in the present study. This may be attributed to the fact that whereas B concentrations in larger pods dropped below 11 mg B/kg DM, they remained higher than 18 mg B/kg DM in flowers, flower buds, and newly set pods. The results suggest that in greengram, low B in the shoots does not limit reproductive growth through pollen sterility unlike several other species (Garg *et al.*, 1979; Vaughan, 1977).

Pod development was as sensitive to low B in the shoot as was leaf blade elongation rate. Indeed in both plant parts, B concentrations in the range 10–14 mg B/kg appeared to inhibit their growth. The similarity of B requirements for growth of pods and the unfolding leaf suggests a functional B requirement for the processes of cell division and expansion which is common to all growing tissues in green gram. The apparent functional B requirement for pod and leaf expansion in green gram was also similar to the requirement for LBER in soybean (12 mg B/kg; Kirk and Loneragan, 1988).

The rapid decline in B concentrations in the unfolding leaf and the pods ≥2 cm may be a

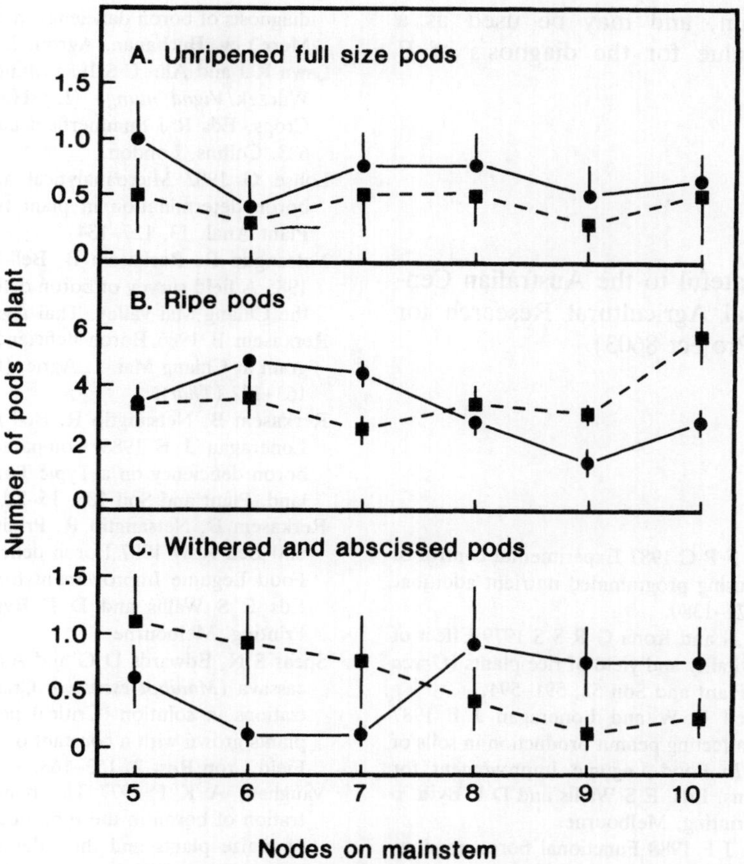

Fig. 2. Effects of interrupted B supply for (●, control; ■, −B) 9 days at early podding on distribution of unripened full size pods (**A**), ripe pods (**B**), and withered and abscissed pods (**C**) per pod on inflorescences subtended by main stem nodes at 27 days after treatments commenced. Values are means of four replicates (SE).

consequence of their high growth rates coupled with a rapidly declining external B supply. In the present experiment, the external B supply was suddenly terminated by transferring plants into −B solutions. Within 3 or 4 days of transfer, solutions were completely depleted of B since roots displayed characteristic symptoms of B deficiency. Subsequent B supply to the developing organs was dependent on translocation of B present in the xylem and retranslocation of B stored in roots and stems. Once all this B was depleted, B supply to developing organs would decline rapidly. Further increases in DM of plant parts would be limited by the amount of B present in their cells. Thus plant parts with the highest absolute DM increment would be the first to deplete their endogenous B supply and become B deficient. By contrast, distribution of B among developing organs may be more important in determining their relative sensitivity to B deficiency when plants have a continuous suboptimal B supply from the roots. The apparent sensitivity of the pods >2 cm to B deficiency may thus be a peculiar consequence of the sudden interruption to the external B supply experienced in the present study. The distribution of B among reproductive plant parts and the relative sensitivities to B deficiency of the different processes involved in reproductive growth should be re-examined in culture systems such as flowing solution culture (Spear *et al.*, 1978) or programmed nutrient addition (Asher and Blamey, 1987) which maintain constant B supply to the plant. Nevertheless, the present results suggest that B concentrations <14 mg/kg DM depress the expansion of developing leaves and pods in green gram, and may be used as a tentative critical value for the diagnosis of B deficiency.

Acknowledgements

The authors are grateful to the Australian Centre for International Agricultural Research for financial support (Project 8603).

References

Asher C J and Blamey F P C 1987 Experimental control of plant nutrient status using programmed nutrient addition. J. Plant Nutr. 10, 1371–1380.

Garg D K, Sharman A N and Kona G R S S 1979 Effect of boron on the pollen vitality and yield of rice plants (*Oryza sativa* L. var. Jaya). Plant and Soil 52, 591–594.

Keerati-Kasikorn P, Bell R W and Loneragan J F 1987 Nutrient deficiencies affecting peanut production in soils of northeast Thailand. *In* Food Legume Improvement for Asian Farming Systems. Eds. E S Wallis and D E Byth. p 261. Ramsay Ware Printing, Melbourne.

Kirk G and Loneragan J F 1988 Functional boron requirement for leaf expansion and its use as a critical value for diagnosis of boron deficiency in soybean (*Glycine max* (L.) Merr.) cv. Buchanan. Agron. J. 80, 758–762.

Lawn R J and Ahn C S 1985 Mung bean (*Vigna radiata* (L.) Wilczek/*Vigna mungo* (L.) Hepper). *In* Grain Legume Crops. Eds R J Summerfield and E H Roberts. pp 584–623. Collins, London.

Lohse G 1982 Microanalytical azomethine-H method for boron determination in plant tissue. Commun. Soil Sci. Plant Anal. 13, 127–134.

Netsangtip R, Rerkasem B, Bell R W and Loneragan J F 1985 A field survey of boron deficiency in peanut grown in the Chiang Mai valley. Thai Agric. Res. J. 3, 171–175.

Rerkasem B 1986 Boron deficiency in sunflower and green gram at Chiang Mai. J. Agric. (Chiang Mai University) 2, 163–172 (*Thai*).

Rerkasem B, Netsangtip R, Bell R W, Hiranburana N and Loneragan J F 1988 Comparative species responses to boron deficiency on a Typic Tropaqualf in northern Thailand. Plant and Soil 106, 15–21.

Rerkasem B, Netsangtip R, Pridisripipat S, Loneragan J F and Bell R W 1987 Boron deficiency in grain legumes. *In* Food Legume Improvement for Asian Farming Systems. Eds E S Wallis and D E Byth. p 267. Ramsay Ware Printing, Melbourne.

Spear S N, Edwards D G and Asher C J 1979 Response of cassava (*Manihot esculenta* Crantz) to potassium concentrations in solution: Critical potassium concentrations in plants grown with a constant or variable potassium supply. Field Crop Res. 2, 153–168.

Vaughan, A K F 1977 The relation between the concentration of boron in the reproductive and vegetative organs of maize plants and their development. Rhod. J. Agric. Res. 15, 163–170.

M. L. van Beusichem (Ed.), *Plant nutrition – physiology and applications*, 281–285.

PLSO IPNC363b

Effects of seed and soil boron on early seedling growth of black and green gram (*Vigna mungo* and *V. radiata*)

B. RERKASEM[1], R.W. BELL[2,3] and J.F. LONERAGAN[2]
[1]*Faculty of Agriculture, Chiang Mai University, Chiang Mai, Thailand and* [2]*School of Biological & Environmental Sciences, Murdoch University, Murdoch, WA 6150, Australia.* [3]*Corresponding author*

Key words: abnormal seedlings, sowing date, seed boron concentration, soil boron concentration, *Vigna mungo* (L.) Hepper, *Vigna radiata* (L.) Wilczek

Abstract

A field study was conducted to investigate the effect of seed B on early seedling growth in cool season plantings of black and green gram at Chiang Mai, Thailand. In December, low B seed of black gram [≤9 mg B/kg dry matter (DM)] produced over 75% abnormal seedlings when sown into low B soil (−B), and 9–22% in B treated soil (+B); with increasing seed B to 14 mg/kg DM, the % abnormal seedlings did not change at −B but at +B it decreased to <5%. In January, the same seed produced fewer abnormal seedlings in all treatments: low B seed in −B soil had >40% abnormal seedlings and 0–3% in +B soil: increasing B to 14 mg/kg seed DM in −B soil eliminated abnormal seedlings. In black gram, the lower levels of abnormal seedlings in each seed and soil treatment of the sowing in January compared with December were associated with higher temperatures during growth, suggesting that low temperatures may have inhibited B supply to the shoots.

Green gram seed sown in January behaved similarly to black gram seed sown at the same time: green gram seed with 5 to 9.5 mg B/kg DM produced 55 to 33% abnormal seedlings in −B soil: either increasing seed B to 15 mg B/kg DM or sowing seed in +B soil eliminated abnormal seedlings.

Introduction

Boron deficiency depresses seed dry matter (DM) of black gram (*Vigna mungo* (L.) Hepper) and green gram (*Vigna radiata* (L.) Wilczek) in northern Thailand (Rerkasem *et al.*, 1988). In addition, low B in shoots has been found to depress germination and vigour of black gram seeds in laboratory tests even when seed DM was not depressed (Bell *et al.*, 1989). In the present study, the effects of seed and soil B levels and planting date on early seedling growth of black and green gram were examined in two field experiments planted in the cool season at Chiang Mai, Thailand.

Materials and methods

Two experiments were carried out under irrigation on a Tropaqualf at Chiang Mai University. The first examined the effect of seed boron on the growth of black gram seedlings sown in soil at 2 levels of boron on December 14 during the cool season. In the second, the plots were divided into 2 subplots which were resown on January 22, one with the same black gram seed and the other with green gram seed.

In Experiment 1, a factorial combination of 2 soil and 8 seed treatments of black gram were imposed in triplicate with soil treatments as main plots and seed as sub-plots. In Experiment 2, the main plots were halved for sowing black and green gram as sub-plots in a factorial combination of 8 seed levels as sub-sub-plots.

Soil B treatments were imposed by application of nil (−B) or 32 kg borax/ha (+B) to a black gram experiment 18 months earlier: prior to the December sowing, hot water soluble boron levels (Dible and Berger, 1954) in the top 30 cm were 0.08 and 0.36 mg B/kg soil, respectively.

Seed B treatments were imposed by using eight lots of seed of green gram cv. Uthong and black gram cv. Regur obtained from a boron rates experiment (Pridisripipat, 1988). The seeds were dry ashed and B concentrations were determined in them using an Azomethine-H procedure (Lohse, 1982): they were 7.7, 8.0, 8.1, 8.3, 9.0, 10.0, 13.8 and 14.0 mg/kg DM for black gram and 5.0, 6.7, 7.8, 9.5, 10.0, 12.6, 15.3 and 16.3 mg/kg DM for green gram. For each seed boron level, a 20 g sample containing approximately 400 seeds was sown at 4 cm depth in two rows each of one metre.

At 20 days after sowing, all seedlings from each plot were classified as normal or abnormal. Normal seedlings had 2–4 sets of trifoliate leaves and an active growing point on the main stem. Seedlings were classed as abnormal if they had any of the following symptoms:

- entire epicotyl missing, *i.e.* only cotyledons present on main shoots;
- main shoot missing above unifoliate leaves;
- ragged trifoliate leaves
- arrested apical growth accompanied by premature lateral branching at cotyledonary or unifoliate node.

Data on the percentage abnormal green gram seedlings was transformed (arcsin) for statistic analysis to normalize treatment variances.

Results and discussion

Seed B concentration, soil B treatment, and time of sowing all affected the % abnormal seedlings produced from black gram seed (Fig. 1).

Abnormal seedlings from low B seed and their response to soil B and time of sowing

When sown into −B soil in December, low B black gram seeds (9 mg or less B/kg DM) produced more than 75% abnormal seedlings (Fig. 1A). When sown at the same time into +B soil, the same seed produced only 9–22% abnormal seedlings, indicating that soil B had corrected the deleterious effects of low B in most of the seed.

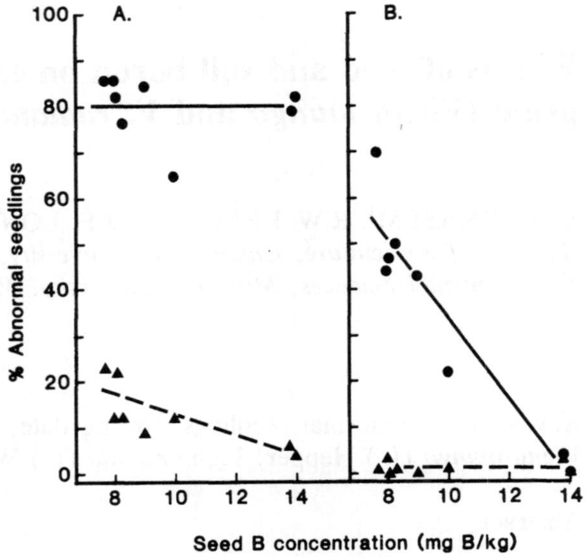

Fig. 1. Effect of sowing time (**A**, 14 December; **B**, 22 January), seed boron concentration and soil boron level (● −B; ▲ +B) on % of abnormal black gram seedlings 20 days after sowing on a Tropaqualf at Chiang Mai. Values are means of three replicates. Seed obtained from a boron rates experiment. Fitted regression equations are as follows:

December sowing	−B soil Y = 33.6 − 2.1x ;	$R^2 = 0.63^*$
	+B soil Y = 80 ;	$R^2 = 0.04^{NS}$
January sowing	−B soil Y = 0.05 ;	$R^2 = 0.01^{NS}$
	+B soil Y = 121 − 8.84x;	$R^2 = 0.88^{**}$

When sown in January, low B black and green gram seed produced about 50% abnormal seedlings in − B soil and almost none in +B soil (Fig. 1B; Table 1). Thus the harmful effects of low seed B were ameliorated in the January sowing and almost eliminated by higher soil B at this time.

Since increasing root temperature from 19 to 28°C is known to increase B absorption by some species (Forno *et al.*, 1979), it appears likely that the higher air and soil temperatures prevailing during the January compared with the December germination (Fig. 2) may have ameliorated the incidence of abnormal seedlings in low B seed by increasing their absorption of B.

Table 1. Effects of seed boron and soil boron on percent abnormal seedlings of green gram cv. Uthong 1 grown for 20 days on a Tropaqualf at Chiang Mai. Values are means of three replicates. Means followed by the same letter are not significantly different ($P = 0.05$) using a Duncan's Multiple Range Test on $\sqrt{}$arcsin transformed data

Seed boron (mg/kg)	Soil B treatments −B	+B
5.0	55a	0c
6.7	57a	2c
7.8	32b	2c
9.5	33b	0c
10.0	6c	0c
12.6	2c	0c
15.3	0c	0c
16.3	0c	0c

The effects of increasing seed B on abnormal seedlings and their response to environmental B and time of sowing

For the December 14 sowing, the effect of seed boron was minimal in −B soil with 70–80% abnormal seedlings at all levels of seed B. With a higher level of soil B, the % of abnormal seedlings was strongly depressed, and was further decreased to almost zero by increasing seed B concentrations from 7.7 to 14 mg B/kg DM.

For the January 22 sowing of black gram seed containing 7.7 mg B/kg DM, 70% of seedlings were abnormal in soil containing 0.08 mg/kg HWSB: the % abnormal seedlings decreased with increasing seed boron contents to none with 14 mg B/kg seed DM. Thus, increasing seed B alone eliminated abnormal seedlings in the January sowing and ameliorated their development on +B soil in the December sowing. In green gram also, increasing seed B alone eliminated abnormal seedlings in the January sowing (Table 1). A similar effect of B has been reported for peas in which apical bud abortion was eliminated by addition of B to the germination medium (Leggatt, 1948). However, in black gram seed with lower concentrations of B than those of the present study (<6 v 7–14 mg B/kg DM), the presence of B in the germinating medium did not prevent the development of as many as 50% abnormal seedlings (Bell *et al.*, 1989).

Bell *et al.* (1989) suggested that at very low concentrations of B in seed the embryo axis may be damaged irreversibly by B deficiency during its formation as appears to be the case for peanut seed affected by hollow heart (Harris and Brolmann, 1966). At somewhat higher concentrations, the embryo may have sufficient B for normal formation in the seed but insufficient for its normal development following germination;

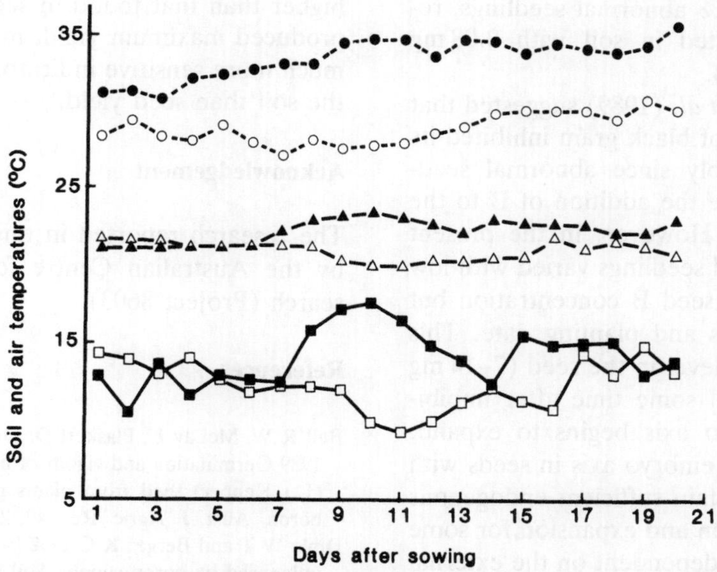

Fig. 2. Daily maximum (○ ●) and minimum (□ ■) air temperatures and soil temperatures taken at 0830 h at 10 cm depth (△ ▲) during growth of black gram seedlings planted at Chiang Mai on 14 December (open symbols) or 22 January (closed symbols).

such seed could be expected to produce abnormal seedlings unless germinated in an environment containing B.

General discussion

The results of the present study confirm and extend those reported by Bell *et al.* (1989) who showed that low B seeds when germinated produced a substantial proportion of seedlings which were abnormal. The present results demonstrate that low B concentrations in black gram seeds depress vigour of seedlings sown in the field as well as in laboratory tests and that low B in seeds can depress seedling growth in green gram as well as black gram. Furthermore, recent studies by Keerati-Kasikorn *et al.* (1988) showed that low B in peanut seeds depressed seedling emergence when seeds were sown in potted soil with or without B fertilizer although no specific symptoms of abnormality in the seedlings were reported. The levels of seed boron which have been found to cause abnormal seedling development in the present study are those found in seeds of high yielding crops showing no visual evidence of B deficiency. Black gram and green gram seed which contained 8 and 9 mg B/kg, respectively, was obtained from plots with maximum seed yield (Pridisripipat, 1988), yet they produced 22% and 33% abnormal seedlings, respectively, when planted in soil with 0.08 mg B/kg (Fig. 1, Table 1).

The results of Bell *et al.* (1989) suggested that <6 mg B/kg in seeds of black gram inhibited its development irreversibly since abnormal seedlings developed despite the addition of B to the germination medium. However, in the present study, the % abnormal seedlings varied with low seed B not only with seed B concentration but also with soil B levels and planting date. This suggests that a low B level in the seed (7–14 mg B/kg) becomes critical some time after imbibition when the embryo axis begins to expand. Growing points in the embryo axis in seeds with high B levels may contain sufficient endogenous B to sustain cell division and expansion for some time before becoming dependent on the external supply. In addition, the radicle probably absorbs too little B in the initial stages of germination to supply shoot meristems with B unless levels of B in the external medium are high. Thus any factor which limits absorption of B by the radicle such as low soil temperature (Forno *et al.*, 1979), low concentrations of B in the soil solution, and low levels of soil water and those factors which inhibit radicle elongation such as extremes of temperature, poor aeration, or soil acidity may cause critical shortages in the amounts of B delivered to shoot meristems for their cell division and expansion.

Regardless of the mechanism by which low seed B levels affect seedling growth, there are important practical consequences for crop production. Depressed early seedling growth may limit crop establishment, especially under adverse conditions such as water and temperature stresses or weed competition. In another study we have shown that low seed B levels in soybean not only retarded early growth and development but also eventually depressed seed yield even when high levels of B were applied to the soil at sowing (Rerkasem *et al.*, unpublished data). Thus, seed boron contents can influence the incidence of boron deficiency in crops, just as it had been shown in the case of molybdenum, copper and zinc (Hewitt Bolle-Jones and Miles, 1954). The level of seed boron required to prevent abnormalities varies with soil B levels and with planting date but in any case can be much higher than that found in seeds from crops that produced maximum yield, making seed vigour a much more sensitive indicator of low B supply in the soil than seed yield.

Acknowledgement

The research reported in this paper was funded by the Australian Centre for Agricultural Research (Project 8603).

References

Bell R W, McLay L, Plaskett D, Dell B and Loneragan J F 1989 Germination and vigour of black gram (*Vigna mungo* (L.) Hepper) seed from plants grown with and without boron. Aust. J. Agric. Res. 40, 273–279.

Dible W T and Berger K C 1952 Boron content of alfalfa as influenced by boron supply. Soil. Sci. Soc. Am. Proc. 16, 60–62.

Forno D, Asher C J and Edwards D G 1979 Boron nutrition

of cassava, and the boron × temperature interaction. Field Crops Res. 2, 265–279.

Hewitt E J, Bolle-Jones E W and Miles P 1954 The production of copper, zinc and molybdenum deficiencies in crop plants grown in sand culture with special references to some effects of water supply and seed reserves. Plant and Soil, 5, 205–222.

Keerati-Kasikorn P, Panya P, Plaskett D, Bell R W and Loneragan J F 1988 Effect of boron on seed quality of peanut Tainan 9. Proc. 26th National Conf. at Kasetsart University, Bangkok Thailand. pp 31–38. Kasetsart University, Bangkok.

Legatt C W 1948 Germination of boron-deficient peas. Scientific Agric. 28, 131–139.

Lohse G 1982 Microanalytical azomethine-H method for boron determination in plant tissue. Commun. Soil Sci. Plant Anal. 13, 127–134.

Pridisripipat S 1988 Responses to boron applications in Vigna. M.Sc. Thesis Faculty of Agriculture Chiang Mai University Thailand.

Rerkasem B, Netsangtip R, Bell R W, Loneragan J F and Hiranburana N 1988 Comparative species responses to boron deficiency on a Typic Tropaqualf in Northern Thailand. Plant and Soil 106, 15–21.

M. L. van Beusichem (Ed.), *Plant nutrition – physiology and applications*, 287–290.

PLSO IPNC725

The effects of short-term deficiency of boron on potassium, calcium, and magnesium distribution in leaves and roots of tomato (*Lycopersicon esculentum*) plants

A.M. RAMÓN, R.O. CARPENA-RUIZ[1] and A. GÁRATE
Departamento de Química Agrícola, Geología y Geoquímica, Universidad Autónoma de Madrid, E-28049 Madrid, Spain. [1]*Corresponding author*

Key words: boron deficiency, calcium, *Lycopersicon esculentum* Mill., magnesium, potassium, tomato

Abstract

Boron deficiency frequently affects long distance transport of nutrients in plants. In this investigation K, Ca and Mg distributions in shoots and roots of tomato plants under conditions of short-time B deficiency were studied. Plants were grown in water cultures with variation in B supply: 0.0 (−B) and 0.5 (+B) mg L^{-1}. Soluble and insoluble fractions of nutrients were obtained from leaves and roots sampled after 1, 2, 7 and 10 days of B treatments. Ca concentrations were found to be most affected by absence of B. The insoluble Ca fraction in roots increased by 188% over the control after 2 days without B. Simultaneously, the soluble Ca fraction in −B folioles was only 84% of that in control treatment. Similar trends were observed in both organs throughout the experimental period. K and Mg appeared to be only slightly affected by short-time B deficiency. The results suggest a specific effect of B on Ca translocation and on its incorporation into insoluble forms, *i.e.* as cell wall components in tomato plants.

Introduction

The interaction of B with other nutrients, such as K, Ca and Mg, has been studied by many researchers (Alvarez-Tinaut *et al.*, 1979; Brennan and Shive, 1948; Carpena and Carpena, 1987; Reeve and Shive, 1944). Furthermore, B deficiency has been frequently involved with the transport system of plants (Lewis, 1980). Yamauchi *et al.* (1986) reported that B deficiency inhibits the translocation of calcium to the upper leaves of tomato plant. The Ca-B relationship has been studied by many workers (Brown, 1979; Tang and De la Fuente, 1986).

The objective of the present investigation was to study K, Ca and Mg distributions in shoot and root of tomato plants under short-time B-deficiency conditions.

Materials and methods

Plant material and growth conditions

Seeds of *Lycopersicon esculentum* Mill. cv. Carmelo (Tm VFN-F1 Hybrido, Sluis and Groot) were germinated on moistened filter paper at 28°C. After 7 days, seedlings were placed in a perforated tray with 2 L of a modified Hoagland nutrient solution and under conditions as described previously by Carpena-Ruiz *et al.* (1989).

Seventeen day-old plants were transferred to nutrient solutions with two boron concentrations: 0.00 (−B) and 0.50 (+B) mg L^{-1}. Plants were sampled after 1, 2, 7 and 10 days. Each sample consisted of all the roots and folioles of three plants. The experiment was conducted in triplicate.

Plant material fractionation. Analytical determinations

Fresh plant material was homogenized and after an 80% ethanol extraction (40°C, 6 h) soluble fractions were obtained. The insoluble residue was digested in a Kjeldahl procedure. Soluble and isoluble K, Ca and Mg fractions were determined by atomic absorption spectrophotometry (Carpena-Ruiz *et al.*, 1990).

Results and discussion

Fresh weights of tomato roots and leaves were little affected by the two first days of B starvation (Table 1). Root weight of −B plants was reduced by 15–20% on the 7th day. Both roots and leaves were strongly affected after 10 days without B. Concentrations of chlorophyll a and b showed trends similar to those of fresh weights after 7 days of B absence (Table 2). Chlorophyll a showed the highest decrease after 10 days without B.

K and Mg soluble fractions in roots showed no differences among B treatments for the first 7 days of the experiment (Table 3), but on the 10th day accumulations of both fractions were observed (increases up to 30%). The Ca-soluble fraction (Table 4) in −B roots increased progressively from the 2nd day on, reaching its highest increment (63% compared to +B) on the last sampling date. The Ca-insoluble fraction in −B roots was even more increased (Table 4). For instance, the insoluble-Ca concentration in −B plants on the 10th day was almost double that in +B plants.

The concentrations of soluble K, Mg and also Ca in leaves did not show consistent trends in the two treatments (Tables 4 and 5). However, the insoluble Ca concentration in −B leaves decreased, although the rate of decrease was lower than in the +B plants (12% and 21% after 7 and 10 days, respectively).

The results show that short-time boron deficiency alters the Ca distribution, as evidenced by an accumulation of Ca in roots of young tomato plants from the 2nd day on. This was the first effect of −B after the start of the experiment. Similarly, Tang and De la Fuente (1986) detected an accumulation of water-extractable Ca in sunflower hypocotyls after 2 days of B

Table 1. The effects of variation in B supply on total fresh weight of roots and leaves of tomato plants

Treatment	Fresh weight (g/plant) at different sampling times (days)				
	0	1	2	7	10
Leaves					
+B	0.223	0.226	0.287	0.507	0.913
−B	–	0.236	0.236	0.505	0.583
Roots					
+B	0.194	0.205	0.193	0.355	0.477
−B	–	0.191	0.163	0.295	0.270

Table 2. The effects of variation in B supply on chlorophyll concentrations in tomato leaves

Treatment	Chlorophyll concentrations (mg/g F.W.) at different sampling times (days)				
	0	1	2	7	10
Chlorophyll a					
+B	1.25	1.28	1.39	1.23	1.50
−B	–	1.24	1.39	1.04	0.91
Chlorophyll b					
+B	0.35	0.44	0.50	0.48	0.53
−B	–	0.45	0.54	0.38	0.49

Table 3. The effects of variation in B supply on soluble K, Ca and Mg concentrations in tomato roots

Treatment	Concentrations ($\mu g/g$ F.W.) at different sampling times (days)				
	0	1	2	7	10
Potassium					
+B	2964	2786	2732	2447	3259
−B	–	2804	2746	2418	4284
Calcium					
+B	266	349	394	260	346
−B	–	308	467	349	565
Magnesium					
+B	1072	1183	1105	954	1212
−B	–	1011	1206	826	1641

Table 4. The effects of variation in B supply on insoluble Ca concentration in roots and leaves of tomato plants

Treatment	Concentrations ($\mu g/g$ F.W.) at different sampling times (days)				
	0	1	2	7	10
Leaves					
+B	3124	4108	2275	2834	2695
−B	–	3893	2348	2492	2150
Roots					
+B	447	382	416	603	1191
−B	–	195	783	763	2287

Table 5. The effects of variation in B supply on soluble K, Ca and Mg concentrations in tomato leaves

Treatment	Concentrations ($\mu g/g$ F.W.) at different sampling times (days)				
	0	1	2	7	10
Potassium					
+B	2789	3254	3159	2693	2896
−B	–	3429	2816	2802	2833
Calcium					
+B	258	360	360	716	270
−B	–	344	304	575	297
Magnesium					
+B	812	923	908	607	664
−B	–	794	1014	582	676

deficiency. Moreover, Yamauchi *et al.* (1986) suggested a possible B-Ca interaction leading to a larger incorporation of Ca into cell walls of −B plants. In agreement with these findings, we observed an increase in both soluble and insoluble Ca in −B roots. In contrast, K and Mg did not appear to be clearly affected in the early stages of B absence.

On the other hand, disturbances found in −B plants from the 7th day on in the form of reductions in fresh weight and in chlorophyll content and increases in soluble K and Mg in roots, could be described in consequence of the primary effects of B starvation on plant development.

The tendencies of increasing Ca concentrations in roots and decrease concentrations in leaves as observed in this investigation are in agreement with the results of other experiments (Carpena and Carpena, 1987; Carpena *et al.*, 1989). Changes in Ca distribution could be explained as resulting from greater incorporation of insoluble Ca into roots and from impeded Ca translocation caused by B deficiency (Carpena *et al.*, 1989; Yamauchi *et al.*, 1986).

References

Alvarez-Tinaut M C, Leal A, Agui I and Recalde L 1979 Efectos fisiológicos en la interacción Boro-Manganeso en plantas de tomate. II. Absorción y distribución de macroelementos. Anal. Edaf. Agrob. 5–6, 991–1012.

Brennan E G and Shive J W 1948 Effect of Ca and B nutrition of the tomato on the relation between these elements in the tissues. Soil Sci. 66, 65–75.

Brown J C 1979 Effects of boron stress on copper enzyme activity in tomato. J. Plant Nutr. 1, 39–53.

Carpena-Artés O and Carpena-Ruiz R O 1987 Effects of boron in tomato plant. I. Leaf evolutions. Agrochim. 31, 391–400.

Carpena-Ruiz R O, Ramón A M, Manzanares M and Gárate A 1990 Determination of N and P fractions in folioles and roots of tomato plants. Agrochim. 34 (*In press*).

Carpena-Ruiz R O, Ramón A M and Gárate A 1989 Podredumbre apical en frutos de *Lycopersicon esculentum* Mill. inducida por la ligera deficiencia de boro. Proceed. III. Congreso Soc. Española Ciencias Hortícolas. Tenerife. Actas de Horticultura 2 (*In press*).

Lewis D H 1980 Boron, lignification and the origin of vascular plants: A unified hypothesis. New Phytol. 84, 209–229.

Reeve E and Shive J W 1944 Potassium-boron and calcium-boron relationships in plant nutrition. Soil Sci. 57, 1–4.

Tang P M and De la Fuente R K 1986 The transport of indole-3-acetic acid in boron- and calcium-deficient sunflower hypocotyl segments. Plant Physiol. 81, 646–650.

Yamauchi T, Hara T and Sonoda Y 1986 Effects of boron deficiency and calcium supply on the calcium metabolism in tomato plants. Plant and Soil 93, 223–230.

M. L. van Beusichem (Ed.), *Plant nutrition – physiology and applications*, 291–295.

PLSO IPNC423

Effect of excess grain boron concentrations on early seedling development and growth of several wheat (*Triticum aestivum*) genotypes with different susceptibilities to boron toxicity

R.O. NABLE[1] and J.G. PAULL[2]
[1]*CSIRO Division of Soils, PMB* 2, *Glen Osmond, SA 5064, Australia, and* [2]*Waite Agricultural Research Institute, PMB* 1, *Glen Osmond, SA 5064, Australia*

Key words: boron, genotypic variation, germination, seeds, toxicity, *Triticum aestivum* L., wheat

Abstract

The effect of excess grain B concentrations on early seedling development and growth was examined in three wheat genotypes with markedly different susceptibilities to B toxicity. Despite grain B concentrations on a dry weight basis of up to 20 mg kg^{-1} (20 times normal), there was no detectable effect on seedling emergence. Nor was growth of seedlings, as measured by shoot dry weight and height, affected. Neither excess B during grain development nor excess B from the grain affected seedling development and growth in any genotype, irrespective of their relative susceptibilities to B toxicity.

The amount of B in the shoots of seedlings was similar irrespective of the amount of B in the grain. The data suggest that the excess B in the grain was not mobilized to the developing seedling.

Introduction

When cereal crops are grown on soils containing excess B, an obvious effect of B toxicity is substantially decreased plant growth and grain yield (Paull *et al.*, 1988). Less obvious, though potentially important effects could result from the B that accumulates in the grain, with B concentrations 5–10 times normal being reported (Cartwright *et al.*, 1987; Paull *et al.*, 1990). These concentrations could produce: a) abnormal grain development, which would affect both the value of the grain as a food source and its value as a seed source for growing future crops (Welch, 1983); or b) an excess B supply to developing seedlings, independent of B supply from the soil, which would also decrease the value of affected grain as a seed source.

These potential effects have implications for the breeding of cereals having increased B tolerance, especially in testing of the progeny of plants selected at high B concentrations. The present experiment was undertaken to examine the effect of high grain B concentrations on the early growth of three wheat genotypes having a wide range of susceptibility to B toxicity.

Materials and methods

Three wheat (*Triticum aestivum* L.) genotypes were used that had been previously shown to differ markedly in susceptibility to B toxicity: (Wl*MMC)/Wl/10, very susceptible; Warigal, moderately susceptible; and Halberd, moderately resistant (Cartwright *et al.*, 1987; Nable, 1988; Paull *et al.*, 1988).

Grain was obtained from plants cultured in a glasshouse in pots containing topsoil from a red-brown earth to which B, as H_3BO_3, was applied in solution at concentrations of 0, 20 and 60 mg B kg^{-1} soil. Full details of the culture procedures are provided elsewhere (Paull *et al.*, 1990). Weights and B concentrations and contents of grain used are shown in Table 1. The B concentrations in the grain of all genotypes used in the

Table 1. Weight and B composition of the grain used in the present experiment

B treatment of parent plants (mg B kg^{-1} soil)	Genotype Halberd	Warigal	(Wl*MMC)/W1/10)
	1000 grain weight (g)		
0	48.0	50.9	45.5
20	42.8	54.7	51.9
60	44.6	53.2	47.5
	B concentration (mg kg^{-1} dry matter)		
0	1.1	1.2	1.3
20	4.1	3.1	3.4
60	21.5	17.3	16.2
	B content (ng per seed)		
0	53	61	59
20	175	170	177
60	959	920	770

present experiment are comparable with concentrations in grain from field-grown plants from soils containing normal, high and extremely high levels of B. Only rarely does field-grown wheat grain have B concentrations in excess of 10 mg kg^{-1} dry matter.

The grain was surface sterilized for 5 min with Na-hypochlorite (5 g L^{-1} chlorine). Five grains per pot were sown into non-draining pots containing 330 g of potting mix (UC Mix C, Asher, 1978; 0.5 mg B kg^{-1} mix) that was maintained at 15°C and 10% saturation. The experiment was arranged as a completely randomized design with 6 replicates of each genotype * B combination.

Seedling emergence was recorded 3, 4, 6, 8 and 11 d after sowing. Development of plants was recorded non-destructively by measuring individual plant height to the tip of the youngest expanded blade 4, 6, 8, 11, 14, 21 and 28 d after sowing. Plant shoots were harvested 28 d after sowing, all plants from each pot combined, dried in a fan forced oven for 48 h at 70°C, then weighed and analysed for B (Zarcinas and Cartwright, 1983).

Results and discussion

There was no significant ($P > 0.05$) effect of grain B concentration on seedling emergence for any of the three wheat genotypes examined (Fig. 1). Nor was the growth of seedlings as measured by shoot height (Fig. 2) or shoot dry matter (Table 2) significantly ($P > 0.05$) affected by grain B concentrations.

Two conclusions can be drawn from these results. Firstly, high B supply to the parent plant did not cause physiological/developmental changes in the developing grain that adversely affected subsequent germination or seedling development and growth. Secondly, high levels of B deposited in grain had no detectable effect during seedling development and growth. Thus despite marked differences between genotypes in relative susceptibility to B toxicity during vegatative growth, there were not detectable differences in terms of germination and early seedling growth. Perhaps this reflects the fact that although B concentrations from 15 to 20 mg kg^{-1} dry matter are unusually high for wheat grain, in other tissues (*e.g.* leaves and roots) such B concentrations are normal and would not cause adverse effects.

These results are very important for the breeding of cereals having increased B tolerance, as it presents no impediment to progeny testing of plants selected at high B concentrations. On a much larger scale, the results indicate there is unlikely to be adverse effects on crop yield from sowing with relatively high B grain. Therefore, farmers in districts where B-rich soils occur need not undertake the expensive task of acquiring

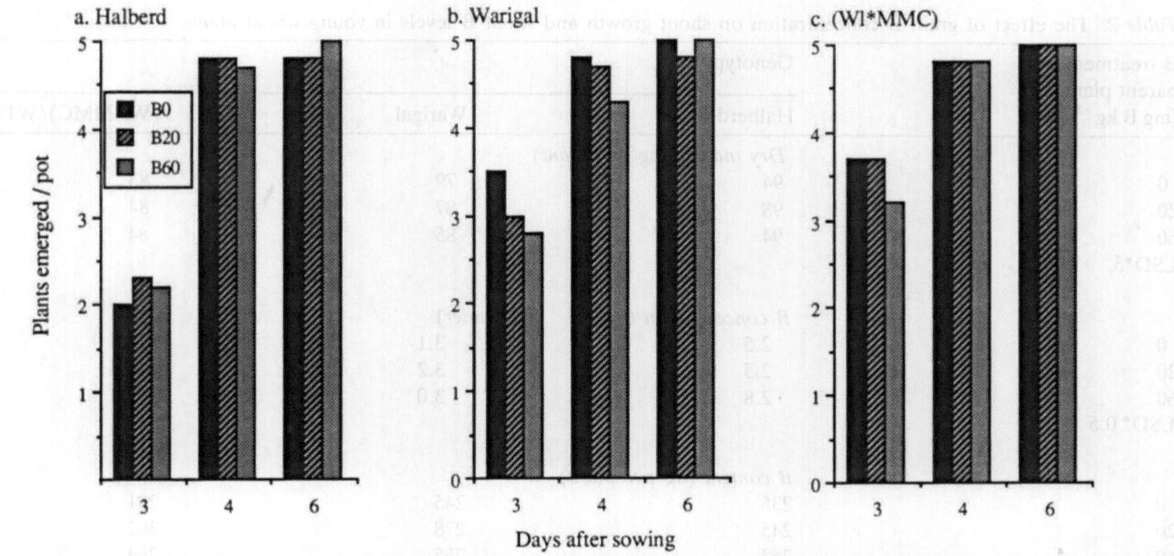

Fig. 1. Effect of grain B concentrations on seedling emergence of three wheat genotypes: (a) (WI*MMC)/W1/10, (b) Warigal, and (c) Halberd. Seed was produced from plants grown at three levels of B supply, B0, B20 and B60, as described in Materials and methods.

seed stocks from districts with soils containing lower levels of B to ensure optimal crop establishment.

The concentrations and contents of B in the shoots of developing seedlings were apparently unaffected by grain B concentrations (Table 2). Rather, the shoot B concentrations were independent of grain B and directly reflected the relative susceptibility of genotypes to B toxicity, as shown previously (Nable 1988; Paull *et al.* 1988). It appears therefore that either little B moves from the grain into the developing shoot or the amount which is mobilized is determined by the developing plants requirements, which

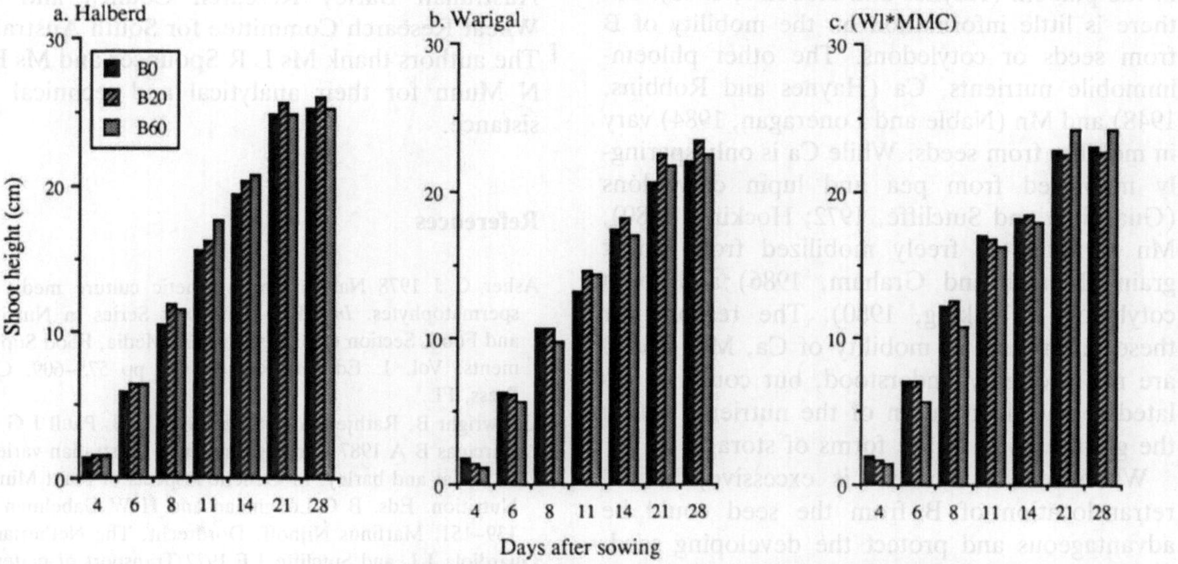

Fig. 2. Effect of grain B concentrations on seedling shoot height of three wheat genotypes: (a) (WI*MMC)/W1/10, (b) Warigal, and (c) Halberd. Seed was produced from plants grown at three levels of B supply, B0, B20 and B60, as described in Materials and methods.

Table 2. The effect of grain B concentration on shoot growth and shoot B levels in young wheat plants

B treatment of parent plants (mg B kg^{-1} soil)	Genotype		
	Halberd	Warigal	(Wl*MMC)/W1/10)
	Dry matter (mg per plant)		
0	94	79	83
20	98	87	84
60	94	85	84
LSD*5			
	B concentration (mg kg^{-1} dry matter)		
0	2.5	3.1	3.5
20	2.5	3.2	3.6
60	2.8	3.0	3.5
LSD* 0.5			
	B content (ng per plant)		
0	235	245	291
20	245	278	302
60	263	255	294
LSD* 49			

* LSD = least significant difference for comparisons between individual means ($p < 0.05$).

can be met from both the grain and the external medium. For example, while B60 grain contained from 770 to 1021 ng B (Table 1), the B60 shoots contained only 255 to 294 ng B, the same as B0 shoots that developed from grain with only 50 to 61 ng B (Table 2).

It is well known that B is not retranslocated in the shoots of plants and is regarded as immobile in the phloem (Haynes and Robbins, 1948), but there is little information on the mobility of B from seeds or cotyledons. The other phloem-immobile nutrients, Ca (Haynes and Robbins, 1948) and Mn (Nable and Loneragan, 1984) vary in mobility from seeds. While Ca is only sparingly mobilized from pea and lupin cotyledons (Guardiola and Sutcliffe, 1972; Hocking, 1980), Mn is relatively freely mobilized from wheat grain (Marcar and Graham, 1986) and lupin cotyledons (Hocking, 1980). The reasons for these differences in mobility of Ca, Mn, and B are not presently understood, but could be related to the distribution of the nutrients within the grain as well as the forms of storage.

When seed B content is excessive, limited retranslocation of B from the seed could be advantageous and protect the developing seedling. However, such immobility would imply that developing seedlings must acquire B from an external supply. If this external supply is low, then B deficiency could develop very rapidly, even though seed B content is normal. This possibility requires further examination.

Acknowledgements

The work was supported by grants from the Australian Barley Research Council and the Wheat Research Committee for South Australia. The authors thank Ms L R Spouncer and Ms F C N Munn for their analytical and technical assistance.

References

Asher C J 1978 Natural and synthetic culture media for spermatophytes. *In* CRC Handbook Series in Nutrition and Food. Section G: Diets, Culture Media, Food Supplements, Vol. 1. Ed. M. Rechcigl Jr. pp 575–609. CRC Press, FL.

Cartwright B, Rathjen A J, Sparrow D H B, Paull J G and Zarcinas B A 1987 Boron tolerance in Australian varieties of wheat and barley. *In* Genetic Aspects of Plant Mineral Nutrition. Eds. B C Loughman and H W Gabelman. pp 139–151. Martinus Nijhoff, Dordrecht, The Netherlands.

Guardiola J L and Sutcliffe J F 1972 Transport of materials from the cotyledons during germination of seeds of the garden pea (*Pisum sativum* L.). J. Exp. Bot. 23, 322–337

Haynes J L and Robbins W R 1948 Calcium and boron as

essential factors in the root environment. J. Am. Soc. Agron. 40, 795–803

Hocking P J 1980 Redistribution of nutrient elements from cotyledons of two species of annual legumes during germination and seedling growth. Ann. Bot. 45, 383–396

Marcar N E and Graham R D 1986 Effect of seed manganese content on the growth of wheat (*Triticum aestivum*) under manganese deficiency. Plant and Soil 96, 165–173

Nable R O 1988 Resistance to boron toxicity amongst several barley and wheat cultivars – a preliminary examination of the resistance mechanism. Plant and Soil 112, 45–52

Nable R O and Loneragan J F 1984 Translocation of manganese in subterranean clove (*Trifolium subterraneum* L. cv. Seaton Park). II Effects of leaf senescence and of restricting supply of manganese to part of a split root system. Aust. J. Plant Physiol. 11, 113–118

Paull J G, Cartwright B and Rathjen A J 1988 Responses of wheat and barley genotypes to toxic concentrations of soil boron. Euphytica 39, 137–144

Paull J G, Rathjen A J, Cartwright B and Nable R O 1990 Selection parameters for assessing the tolerance of wheat to high concentrations of boron. *In* Genetic Aspects of Plant Mineral Nutrition. Eds N El Bassam, M Dambroth and B C Loughman. pp 361–369. Kluwer Academic Publishers, Dordrecht, The Netherlands

Welch R M 1983 Effects of nutritional deficiencies on seed production and quality. *In* Advances in Plant Nutrition, Vol 2. Eds. B Tinker and A Läuchli. pp 205–247. Praeger, New York

Zarcinas B A and Cartwright B 1983 Analysis of soil and plant by inductively coupled optical emission spectrometry. CSIRO Div. Soils Tech. Paper No. 45

M. L. van Beusichem (Ed.), *Plant nutrition – physiology and applications*, 297–301.

PLSO IPNC499

Genetic differences in the copper efficiency of cereals

W. PODLESAK[1], T. WERNER[1], M. GRÜN[1], R. SCHLEGEL[2] and E. HÜLGENHOF[2]
[1]*Institut für Pflanzenernährung und Ökotoxikologie der Akademie der Landwirtschaftswissenschaften der DDR, Naumburger Str. 98, DDR-6909 Jena, DDR, and* [2]*Zentralinstitut für Genetik und Kulturpflanzenforschung der Akademie der Wissenschaften der DDR, Gatersleben, DDR*

Key words: chromosome, copper concentration, copper efficiency, copper uptake, copper utilization, genetic information, wheat-rye addition line, wheat-rye translocation line

Abstract

The aim of the present study was to investigate the chromosomal location of the genetic information responsible for the Cu efficiency of rye. Pot experiments were carried out with a set of wheat-rye addition lines derived from hybridization of winter wheat 'Holdfast' and winter rye 'King II' and a 4Bβ/5Rl translocation line selected from the winter wheat 'Viking'. The results of the addition lines indicate that the genetic location of Cu efficiency is on the long arm of rye chromosome 5(5Rl). Compared with 'Viking', the Cu efficiency of the 4Bβ/5Rl translocation line was significantly increased. The Cu concentration and Cu uptake in the translocation line suggest that the low internal Cu requirement may play an important role in the establishment of Cu efficiency.

Introduction

There are considerable differences between species and varieties of agricultural crops with regard to the requirements on the Cu supply of the soil. These differences are particularly distinct between wheat and rye. Whereas wheat reacts to Cu deficiency with visible symptoms and striking yield losses, yield depressions in rye caused by limited Cu supply are by far unknown.

The Cu efficiency of rye is genetically determined and can be passed on to the wheat-rye hybrid triticale. According to investigations into wheat-rye addition lines as well as wheat-rye translocation lines in the genetic background of the spring wheat 'Chinese Spring' and of Australian locally adapted wheat varieties, the genetic information responsible for the expression of Cu efficiency is situated on the long arm of the rye chromosome 5 (5Rl) (Graham, 1984; Graham *et al.*, 1987a;b).

The present paper aims at

1. confirming the localization of the genetic information for Cu efficiency in another genetic material, and at
2. analyzing the Cu concentrations in different plant organs as well as the Cu uptake by above ground plant matter in a Cu-efficient wheat-rye translocation line compared to its unchanged original variety.

Materials and methods

In order to determine the Cu efficiency of different genotypes of cereals, the grain yield obtained under Cu deficiency is related to the yield in the case of adequate Cu supply. When several genotypes are compared, the genotype which reacts to inadequate Cu supply with the lowest relative yield depression has the highest Cu efficiency. Two pot experiments with genetically manipulated wheat genotypes were carried out:

Experiment 1

Detection of the Cu efficiency of wheat-rye addition lines in order to determine the chromosome pair(s) which carry genetic information responsible for Cu efficiency.

Experiment 2

Comparison of the Cu efficiency of a wheat-rye translocation line and its genetically unchanged original variety.

The lines 1-7R of the wheat-rye addition line set 'Holdfast' (*Triticum aestivum* L.) × 'King II' (*Secale cereale* L.) were used for the determination of the chromosome(s) responsible for Cu efficiency.

Only the ditelosome line 3 Rs of the chromosome 3R was at our disposal. Furthermore, the ditelosome lines 5Rl and 5 Rs of the rye chromosome 5 were made use of. Parents and addition lines are winter forms. In all single plant descendants of the tested genotypes the chromosome numbers were checked by means of mitotic cells in squash preparations of root tips after staining with Feulgen's and Rossenbeck's (1924) staining procedure.

In the second experiment the winter wheat 'Viking' and a 4Bβ/5Rl translocation line, which occurred spontaneously in this variety and which was described by Riley *et al.* (1970), were made use of. The structural changes of the chromosome 4B caused by the translocation were detected by N-banding (Schlegel and Gill, 1984). In both experiments, 3 kg of a Cu-deficient peaty soil per pot (0.7 mg Cu L^{-1} air dried soil according to Westerhoff) with the following Cu supplies were used:

Experiment 1 (3 replications/variant)

mg Cu/pot (as $CuSO_4 \cdot 5H_2O$)

0 = severe Cu deficiency

5 = latent Cu deficiency

50 = adequate Cu

Experiment 2 (4 replications/variant)

0 and 50 mg Cu/pot (as $CuSO_4 \cdot 5H_2O$)

All pots received the same basic fertilization: 1 g N (NH_4NO_3); 0.6 g P (KH_2PO_4); 2 g K ($K_2SO_4 + KH_2PO_4$); 0.3 g Mg ($MgSO_4 \times 7 H_2O$); 3.0 mg B ($Na_2B_4O_7 \times 10 H_2O$); 100 mg Mn ($MnSO_4$). Before planting, all nutrients including Cu were added to the soil. A second N gift (1 g N, NH_4NO_3)was applied during stem elongation. 8 plants were cultivated in each pot. Soil humidity amounted to 60% of the water capacity. In both experiments, the plants were harvested at maturity; the grain and straw yields were determined; the plants were separated into grain, chaff, flag leaf, remaining leaves, stem and root.

After dry ashing (480 °C), Cu was determined by means of atomic absorption spectrometry (AAS 3 of the Enterprise VEB Carl Zeiss Jena).

Results

When the grain yields of main heads, side heads and the total grain yield of addition lines and their parent forms were compared, different reactions of the genotypes occurred depending on Cu supply. Since, the main heads are hardly subjected to cytological and physiological disorders and since the genotype-specific expression of the grain yields was most distinct in them, they are presented in the following (Table 1). The yields obtained with 'Holdfast' and 'King II' show wheat- or rye-specific behaviour.

The Cu-inefficient wheat 'Holdfast' showed distinct yield losses in the case of low Cu (0 and 5 mg Cu/pot). No yield depressions caused by low Cu were detectable in the rye 'King II'. Due to different yield potential of the tested genotypes, the comparison of Cu efficiency had to be carried out on the basis of relative yields.

The relative yields represented in Table 1 demonstrate that all addition lines reacted with a considerable depression of the grain yield, at least in the variant without Cu. The addition lines 1R, 2R, 3Rs, 4R, 5Rs, 6R and 7R proved to be as sensitive or even more sensitive to a low Cu supply as 'Holdfast'. On the other hand, the lines 5R and 5Rl tolerated the low Cu supply better than 'Holdfast'. In the variant with latent deficiency, the line 5R had a yield loss of only 29%, the line 5Rl achieved the same grain yield as in the case of adequate Cu supply. Since, however, the small number of plants/pot resulted in considerable standard deviation, only the last reaction is statistically significant. Only

Table 1. Grain yields of the main heads of the set of wheat-rye addition lines 'Holdfast' × 'King II' with different Cu supply

Genotype	Cu supply (mg Cu/pot)				LSD (Tukey, $P<0.05$)
	50	50	5	0	%
	Grain yields				
	g/pot	%	%	%	
Holdfast	6.7	100	54	4	genotype and Cu supply variable: 58
King II	6.4	100	115	106	
1R	3.8	100	25	1	
2R	2.0	100	8	0	
3Rs	5.2	100	51	7	
4R	3.8	100	27	0	
5R	3.8	100	71	14	
5Rl	3.3	100	108	37	
5Rs	5.6	100	31	1	
6R	3.7	100	34	2	
7R	5.7	100	37	5	

the addition of 5Rl and not that of 5Rs led to an improvement of Cu efficiency compared to 'Holdfast'. The grain yields of side heads as well as the total grain yield confirm this result.

Due to the genetic instability and the low yield potential of the addition lines, they are only partly suited for physiological investigations and cannot be used in practical breeding. These disadvantages can be circumvented by the incorporation of alien genes by means of translocation. Therefore, the results obtained in the first experiment were confirmed by a translocation line. The determination of the total grain yields of the wheat variety 'Viking' and of the 4Bβ/5Rl translocation line isolated therefrom demonstrated that there are distinct differences between both genotypes in their tolerance towards a low Cu supply (Table 2). The original variety 'Viking' proved to be very sensitive to a low Cu supply. There was an almost complete yield loss in the variant without Cu, only 3% of the yields were obtained compared to adequate Cu (50 mg Cu/pot). On the other hand, the yield depression in the translocation line is much less striking in the 0 variant. Even under the conditions of severe Cu deficiency, this line still produces 51% of the yield obtained with adequate Cu.

The Cu concentrations and Cu uptake of plants were determined in order to investigate the physiological reasons for Cu efficiency.

Independent of the Cu supply, the Cu concentrations in the organs did not show any dependence on the genotype (Table 3).

The good accordance of the contents in grain, chaff, flag leaf and remaining leaves in the 0 variant is particularly striking. Apart from chaff and flag leaf, all organs of 'Viking' and the translocation line reacted to the low Cu supply with a significant reduction of the Cu concentration compared to the variant with 50 mg Cu/pot.

When the Cu uptakes are calculated from the Cu concentrations and the dry matter yields (Table 4), a very good accordance of the values

Table 2. Grain yields of the winter wheat 'Viking' and the 4Bβ/5Rl translocation line

Genotype	Cu supply (mg Cu/pot)			LSD (Tukey, $P<0.05$)
	50	50	0	%
	Grain yield			
	g/pot	%	%	
'Viking'	35.3	100	3	genotype and Cu supply variable: 17
4Bβ/5Rl	30.8	100	51	

Table 3. Cu concentrations (μg/g d.m.) in the organs of 'Viking' and the 4Bβ/5Rl translocation line depending on Cu supply

Organ	Genotype	Cu supply (mg Cu/pot) 0	50	LSD Tukey, ($P<0.05$)
Grain	'Viking'	1.8	5.0	genotype and Cu supply variable: 0.69
	4Bβ/5Rl	1.3	5.7	
Chaff	'Viking'	2.6	2.1	genotype and Cu supply variable: 0.91
	4Bβ/5Rl	2.1	2.1	
Flag leaf	'Viking'	6.3	4.8	genotype and Cu supply variable: 2.2
	4Bβ/5Rl	4.7	4.3	
Remaining leaves	'Viking'	5.3	7.1	genotype and Cu supply variable: 1.9
	4Bβ/5Rl	4.0	6.7	
Stem	'Viking'	2.3	5.5	genotype and Cu supply variable: 2.3
	4Bβ/5Rl	2.7	4.5	
Root	'Viking'	4.4	12	genotype and Cu supply variable: 3.0
	4Bβ/5Rl	5.6	12	

Table 4. Cu uptake of the winter wheat 'Viking' and the 4Bβ/5Rl translocation line from 'Viking' due to different Cu supply

Genotype	Cu supply (mg Cu/pot) 50	50	0	LSD (Tukey, $P<0.05$) %
	Cu uptake μg	%	%	
'Viking'	469	100	27	genotype and Cu supply variable: 14
4Bβ/5Rl	332	100	32	

for both genotypes is detectable in the variant without Cu as well.

Due to the higher total dry matter production, however, 'Viking' achieves significantly higher Cu uptakes in the variant with adequate Cu (50 mg Cu/pot).

Discussion

The results demonstrate that the genetic information responsible for Cu efficiency lies on the long arm of the rye chromosome 5. When this information is transferred into the genome of wheat, it improves the Cu efficiency of this species. This finding is in good accordance with Graham's (1981, 1984) results found in spring forms of rye and genetically manipulated wheats. Since parents and addition lines of the combination 'Holdfast' × 'King II' as well as the 4Bβ/5Rl translocation and 'Viking' are winter forms, it can be assumed that 5Rl is the critical chromosome arm both in spring and winter rye.

Neither line 5R nor line 5Rl, however, showed as high a Cu efficiency as the rye parent 'King II' (cf. Table 1), *i.e.* there may be genetic information in the genome of wheat which counteract Cu efficiency (suppressor genes) or genes for Cu efficiency may be situated on other rye chromosomes and were not detected with this kind of experiment or which could not be detected since they are caused cytoplasmatically.

The fact that the yield potential of the addition lines with adequate Cu was lower than that of the parent varieties illustrates that the addition of single rye chromosome pairs is a decisive disruption of the genetic balance of wheat. Occurring interactions taking an adverse effect on vitality and performance may also be a cause for the lower Cu efficiency of lines 5R and 5Rl compared to 'King II'.

The wheat-rye addition lines are only of importance for basic research because of the low yield potential and the genetic instability.

On the other hand, a wheat genotype was found with the 4Bβ/5Rl translocation line from the winter wheat 'Viking' in which the genetic information from the long arm of the rye chromosome 5 is firmly incorporated into the wheat genome. This incorporation of the alien genes did not lead to a considerable reduction of the yield potential. This was demonstrated by the good accordance of the grain yields of the original variety and the translocation line in the variant with 50 mg Cu/pot (cf. Table 2). Thus, the 4Bβ/5Rl translocation line from 'Viking' seems suited as breeding material for the insertion of Cu efficiency into high-yielding wheat.

The comparison of the Cu concentrations in different plant organs of the 4Bβ/5Rl translocation line and the original 'Viking' as well as the calculation of the Cu uptakes led to investigation of the physiological backgrounds of Cu efficiency.

The good accordance of the Cu concentrations of both genotypes in the 0 variant points to the fact that the Cu utilization in the plant plays an important part for the expression of Cu efficiency.

In the case of equally low Cu concentrations in the tissues (Table 3) and of similar Cu uptake, the translocation line has a better capacity of producing grain yield than the original variety.

In accordance with Graham and Pearce (1979) as well as Graham *et al.* (1981) it was found in the case of wheat, rye and triticale that efficient forms have mostly increased Cu concentrations or at least distinctly higher Cu uptakes. Therefore Cu uptake and translocation as well as Cu utilization must be taken into consideration as causes for Cu efficiency.

Acknowledgements

We thank T. Miller (IPSR Cambridge) for placing the addition lines and the parent forms as well as the translocation line used in the experiments at our disposal.

References

Feulgen R and Rossenbeck N 1924 Mikroskopisch-chemischer Nachweis einer Nukleinsäure vom Typus einer Thymonukleinsäure und die darauf beruhende selektive Färbung von Zellkernen in mikroskopischen Präparaten. Z. Physiol. Chem. 135, 203–240.

Graham R D 1981 Genetics of the copper efficiency factor in rye. *In* Copper in Soils and Plants. Eds. J F Loneragan, A D Robson and R D Graham. p 358. Academic Press, Sydney, New York, London, Toronto, San Francisco.

Graham R D 1984 Breeding for nutritional characteristics in cereals. Adv. Plant Nutr. 1, 57–107.

Graham R D and Pearce D T 1979 The sensitivity of hexaploid and octoploid triticales and their parent species to copper deficiency. Aust. J. Agric. Res. 30, 791–799.

Graham R D, Anderson G D and Ascher J S 1981 Absorption of copper by wheat, rye and some hybrid genotypes. J. Plant Nutr. 3, 679–686.

Graham R D, Ascher J S, Ellis P A and Shepherd K W 1987a Transfer to wheat of the copper efficiency factor carried on rye chromosome arm 5Rl. Plant and Soil 99, 107–114.

Graham R D, Ascher J S, Ellis P A and Shepherd K W 1987b Transfer to wheat of the copper efficiency factor carried on rye chromosome arm 5Rl. *In* Genetic Aspects of Plant Mineral Nutrition. Eds. W H Gabelman and B C Loughman. pp 405–412. Martinus Nijhoff Publishers, Dordrecht, Boston, Lancaster.

Riley R, Chapman V and Miller T E 1970 Hairy necked 'Viking'. Ann. Rep. Plant Breeding Inst. Cambridge, 98 p.

Schlegel R and Gill B S 1984 N-banding analysis of rye chromosomes and the relationship between N-banded and C-banded hetero-chromatin. Can. J. Genet. Cytol. 26, 765–769.

M. L. van Beusichem (Ed.), *Plant nutrition – physiology and applications*, 303–306.

PLSO IPNC139

Effect of arsenic and molybdenum on plant response of cauliflower (*Brassica oleracea*) grown in sand culture

C.R. BLATT
Agriculture Canada, Research Station, Kentville, Nova Scotia, Canada B4N 1J5

Key words: arsenic, *Brassica* spp., cauliflower, glasshouse, heavy metals, molybdenum, sand culture

Abstract

Two cultivars of cauliflower were grown in two sand culture experiments in which arsenic (As) and molybdenum (Mo) were treatment variables. Generally, as As in solution increased, leaf As increased and leaf phosphorus (P), boron (B) and head weight decreased. The head weight response of each cultivar was different to increasing As and Mo in solution with c Idol Original reflecting increased tolerance to As at 0.2 mg L^{-1} Mo in solution and cv Fortuna not responding to increased Mo. There was some plant stunting and off color appearance of leaves at 50 mg L^{-1} As in solution, however there did not appear to be any P, B, or Mo deficiency symptoms in any cultivar. It did not appear that the typical whiptail Mo deficiency symptom of cauliflower could be induced by As toxicity.

Introduction

Tree fruit acreage, especially apples, has decreased appreciably during the last few decades in the Annapolis Valley of Nova Scotia and former apple orchard areas that had been treated with arsenicals for insect control are currently being used for a wide variety of crops. Bishop and Chisholm (1962) have documented total and water-soluble soil arsenic levels over a wide variety of soil series and locations in the Annapolis Valley and concluded that in some former orchard sites the arsenic accumulation could be detrimental to the growth of arsenic sensitive crops such as peas and beans. Chisholm (1972) reported lead, arsenic and copper levels in the edible portion of several vegetable crops and although the arsenic levels in crops grown on treated plots reflected lead arsenate applications and were consistently higher than those in crops grown on untreated plots, they were well below the Canadian tolerance of 1.0 mg/kg in fresh vegetables (Food and Drug Directorate, 1970). Recently, some local vegetable growers have experienced irregular growth of cauliflower that followed the row pattern in fields formerly planted to orchard. Although Woolson (1973) has characterized the *Brassica* spp. (cabbage) as being more tolerant to available arsenic than beans, spinach, radish, and tomato, it is possible that specific nutrient interactions and not arsenic toxicity *per se* could be responsible for the mis-shapen leaves, small heads and generally poor plant growth in these areas.

Phosphorus is similar to arsenic in chemical and physical characteristics and several studies have reported the interaction of these two elements with respect to phytotoxicity, availability and plant uptake (Rumberg *et al.*, 1960; Sadiq, 1986; Woolson *et al*, 1973). Aluminium, iron and calcium have also been reported to influence arsenic availability in soil (Woolson *et al.*, 1973).

Although soil arsenic levels were not determined in the problem areas for each field, soil phosphorus and pH levels were in the recommended ranges for cauliflower growing in these sandy loam to loam soil types. Plant symptoms, resembling molybdenum deficiency, became evident during early head development and applications of foliar molybdenum were reasonably effective in preventing further plant deterioration. As a result of these observations the objectives of this study were to determine: (1) the effect of an increasing rate of available arsenic on cauliflower plant growth, yield and tissue nutrient status; and (2) whether arsenic could influence plant molybdenum level and induce the characteristic whiptail molybdenum deficiency systems.

Methods

Uniform transplants of cauliflower (*Brassica oleracea* L. var. *botrytis* L. 'Fortuna' and 'Idol Original'), one per pail, were set in acid-washed 8-mesh silica grits in plastic 3.8 liter pails. Each experimental unit of 4 plants had a single nutrient solution (pH 5.50–6.00) supplied by a continuous drip automatic recycling technique with nutrient solutions renewed at weekly intervals. There were 6 experimental units (treatments) on each of two benches with one cultivar per bench and treatments randomly assigned. The nutrients were supplied by $Ca(NO_3)_2$, KH_2PO_4, KNO_3, $MgSO_4$, H_3BO_3, Fe-DTPA, $MnSO_4$, $CuSO_4$, $ZnSO_4$, at the following concentration in mg L^{-1}: N, 70; P, 50; K, 102; Ca, 80; Mg, 24; S, 32; Fe, 5; Mn, 0.5; B, 1.0; Cu, 0.1; and Zn, 0.05. Molybdenum (Mo) was supplied as Na_2MoO_4 at 0.02 mg L^{-1} in exp 1 and 0.02 and 0.2 mg L^{-1} in exp 2. Arsenic (As) was supplied as As_2O_5 at 0, 5, 10, 15, 20 and 25 mg L^{-1} in exp 1 and at 0, 25 and 50 mg L^{-1} in exp 2. During each 2–3 month experimental period a 16-h photoperiod was maintained and temperatures were approx 21°C (day) and 15°C (night). At the termination of each experiment, data recorded included: head weight, total top fresh weight, fresh weight of mature and young leaves of each plant. All leaves were dried at 75°C for dry weights and ground in a Wiley mill with sub-samples taken for N determination on a Leyco automated Dumas System and P, K, Ca, Mg, B, Fe, Mn, Cu and Zn determined with an inductively coupled argon plasma unit. Duplicate sub-samples were digested by refluxing with HNO_3, evaporated to near dryness, made up to final volume with distilled deionized water and As determined by an automated hydride-molybdenum blue procedure and Mo determined using a Perkin Elmer 560 atomic absorption spectrophotometer equipped with graphite furnace.

The data were subjected to analysis of variance for a factorial design with 4 samples per treatment, two cultivars in both years, with As(3) X Mo(2) in exp 2, but 6 levels of As at a single level of Mo in exp 1. Polynomial regressions were calculated on the levels of As and Mo, as well as for the interactions.

Results

Experiment 1

Fresh weights of total plant top and mature leaves and dry weight of mature leaves decreased as As in solution increased from 0 to 10 mg L^{-1}; however these values increased to levels equivalent to those at 0 mg L^{-1} As as solution As increased to 15 mg L^{-1} (Table 1). The cv. Idol Original continued to increase its total plant top fresh weight at 20 mg L^{-1} As in solution and this resulted in a concurrent head weight increase. In contrast, the plant top fresh weight, mature leaf fresh and dry weights and head weight of the cv. Fortuna decreased at As solution levels >15 mg L^{-1}. Arsenic in young and mature leaves increased, P and B in mature leaves decreased as solution As increased from 0 to 25 mg L^{-1}. There was no interaction between As and Mo and Mo levels were 5–6 mg L^{-1} in mature leaves (data not shown).

Experiment 2

Total plant top fresh weight of the cv Fortuna decreased as solution As increased from 25 to 50 mg L^{-1} (Table 2). Head weight of the cv. Idol Original decreased significantly as solution As increased from 0 to 50 mg L^{-1} with 0.02 mg L^{-1} Mo in solution; however, the head weight decrease was not significant with 0.2 mg L^{-1} Mo in solution. The cv. Fortuna exhibited the opposite response with a significant head weight decrease at 0.2 mg L^{-1} Mo as solution As increased from 0 to 50 mg L^{-1}, but no significant head weight reduction with 0.02 mg L^{-1} Mo in solution. Similar to experiment 1, As in mature leaves increased and leaf P decreased as solution As increased. Leaf B decreased with increasing As in solution only in the cv Idol Original. Leaf Mo

Table 1. Effect of 6 rates of arsenic (0, 5, 10, 15, 20, 25 mg L^{-1} As) combined with molybdenum at 0.02 mg L^{-1} Mo on the total leaf fresh weight, fresh and dry weight of mature leaves, head fresh weight, arsenic content in young and mature leaves, and phosphorus and boron content in mature leaves of the cauliflower cultivars Fortuna and Idol Original

Arsenic in Sol'n mg L^{-1}	Total leaf F.W. (g/pl)		F.W. (g/pl)	D.W. (g/pl)	Head weight (g/pl)		Arsenic (mg L^{-1})		Phosphorus	Boron
	Fortuna	Idol Original	mature leaves	mature leaves	Fortuna	Idol Original	Young leaves	Mature leaves	mature leaves (%)	mature leaves (mg L^{-1})
0	490	458	362	53.0	18.5	38.1	<0.3	<0.3	0.51	85.8
5	272	429	235	35.4	12.4	52.3	9.4	11.5	0.41	97.2
10	182	265	127	21.4	13.6	38.9	14.8	15.5	0.46	73.7
15	547	470	382	50.1	21.3	48.3	20.5	18.3	0.33	66.0
20	255	564	283	39.1	9.5	76.5	36.7	32.0	0.28	65.0
25	370	335	252	37.7	11.0	37.3	74.3	50.8	0.21	72.2
SEM	52.4		29.1	4.09	8.00		10.07	5.87	0.021	3.26
(n, df)	(4, 36)		(8, 36)	(8, 36)	(4, 36)		(8, 32)	(8, 36)	(8, 32)	(8, 36)

Table 2. Effect of 3 rates of arsenic (0, 25, 50 mg L^{-1} As) interacted with 2 rates of molybdenum (0.02, 0.2 mg L^{-1} Mo) on total leaf fresh weight, head fresh weight, arsenic, phosphorus and boron content in mature leaves of the cauliflower cultivars Fortuna and Idol Original

Arsenic in Sol'n mg L^{-1}	Total leaf F.W. (g/pl)		Head weight (g/pl)				Arsenic mature leaves (mg L^{-1})	Phosphorus mature leaves (%)		Boron mature leaves (mg L^{-1})	
	Fortuna	Idol Original	Fortuna Mo 0.02	(mg L^{-1}) 0.2	Idol Mo 0.02	Original (mg L^{-1}) 0.2		Fortuna	Idol Original	Fortuna	Idol Original
0	593	391	248.3	325.4	328.8	267.7	<5.0	0.36	0.36	103.0	114.8
25	505	370	212.6	194.5	206.6	245.0	36.5	0.33	0.28	124.3	108.2
50	380	340	185.0	135.1	149.0	192.7	112.9	0.28	0.19	115.9	88.9
SEM	34.1		30.34				14.33	0.020		5.33	
(n, df)	(8, 36)		(4, 32)				(4,5)	(8, 36)		(8, 36)	

values (1–10 mg L^{-1}) reflected solution Mo level and were not affected by solution or leaf As (data not shown).

Discussion

The P-to-As ratio has been studied with corn seedlings (Woolson *et al.*, 1973) and wheat (Hurd-Karrer, 1939) with the finding that a 10:1 treatment ratio (P/As) increased corn growth over the check on a given soil type and with wheat growing in solution culture a 5:1 ratio (P/As) was sufficient to render As harmless. In experiment 1, increasing solution As decreased P-to-As ratio and plant top growth until a ratio of 3.3:1 was reached with cv. Fortuna that significantly increased top growth in comparison to ratios of 10:1, 5:1, 2.5:1 and 2:1 (Table 1). The cv. Idol Original had plant top growth at a P-to-As ratio of 2.5:1 that was significantly greater than top growth at ratios of 5:1 and 2:1. For both cultivars, these increases in plant growth were reflected in head weight yield. The realtionship between leaf As and P reported in these 2 experiments is in agreement with previous studies, however as leaf P decreased with increasing solution and leaf As levels there was no indiction of P deficiency at the lowest leaf P levels.

Although there were no As–Mo interactions with regard to plant fresh and dry weights or leaf analysis in either experiment there was a cultivar response for head weight with increasing As and Mo in solution (Table 2). Even though the high (0.2 mg L^{-1}) level of Mo increased cv. Fortuna head weight at 0 mg L^{-1} solution As, head weight significantly decreased with added As. This As-Mo interaction was similar for the head weight values of cv. Idol Original at the low (0.02 mg L^{-1}) Mo level, however at the high Mo level decreases in head weight were not significant and head weight values were higher at both solution As levels with 0.2 mg L^{-1} Mo compared with 0.02 mg L^{-1} Mo. In this experiment, the general tendency for head weight to decrease was a response of plants being stunted and off color as solution As approached 50 mg L^{-1}. Apparently, there can be cultivar differences in response to As and Mo, however increasing level of solution As had no effect on leaf Mo or incidence of whiptail Mo deficiency symptoms in these cultivars.

Acknowledgements

The author gratefully acknowledges the technical assistance of A G Sponagle in conducting the experiments; Dr Ross McCurdy, Environmental Chemistry Lab, Victoria General Hospital, Halifax, N.S. for arsenic and molybdenum determinations; and Mr Ben Harnish, Chemistry Lab, Soils and Crops Branch, Nova Scotia Department of Agriculture and Marketing, Truro, N.S. for macro and micronutrient tissue analysis.

References

Bishop R F and Chisholm D 1962 Arsenic accumulation in Annapolis Valley orchard soils. Can. J. Soil Sci. 42, 77–80.

Chisholm D 1972 Lead, arsenic and copper content of crops grown on lead arsenate treated and untreated soils. Can. J. Plant Sci. 52, 583–588.

Hurd-Karrer A M 1939 Antagonism of certain elements essential to plants toward chemically related toxic elements. Plant Physiol. 14, 9–29.

Rumberg C B, Engel R E and Meggitt W F 1960 Effect of phosphorus concentration on the absorption of arsenate by oats from nutrient solution. Agron. J. 52, 452–453.

Sadiq M 1986 Solubility relationships of arsenic in calcareous soils and its uptake by corn. Plant and Soil 91, 241–248.

Woolson E A 1973 Arsenic phytotoxicity and uptake in six vegetable crops. Weed Sci. 21, 524–527.

Woolson E A, Axley J H and Kearney P C 1973 The chemistry and phytotoxicity of arsenic in soils. II. Effects of time and phosphorus. Soil Sci. Soc. Am. Proc. 37, 254–259.

M. L. van Beusichem (Ed.), *Plant nutrition – physiology and applications*, 307–312.

PLSO IPNC708

Correlation between extractable chromium, chromium uptake and productivity of beans (*Phaseolus vulgaris*) grown on tannery sludge-amended soil

B. GUNSÉ, CH. POSCHENRIEDER[1] and J. BARCELÓ
Lab. Fisiología Vegetal, Facultad de Ciencias, Universidad Autónoma de Barcelona, E-08193 Bellaterra, Spain. [1]*Corresponding author*

Key words: chromium, critical toxicity concentration, heavy metals, *Phaseolus vulgaris* L., phytotoxicity, sequential soil extraction, tannery sludge, Vertic Ustochrept

Abstract

The effects of different tannery sludge (0.27% Cr f.w.) doses (0, 8.7, 17.4, 34.8 g/kg soil) on growth and Cr content of *Phaseolus vulgaris* L. cv. Délinel grown on potted soil in a growth chamber were analyzed. The soil was a vertic Ustochrept (pH 8.05, organic matter 2.26%, CEC 8.42 meq/100 g). Total soil Cr concentrations after treatment were 60, 88, 112 and 160 mg/kg. All sludge treatments significantly decreased plant growth and yield. Chromium toxicity symptoms developed in plants grown on soil with 160 mg Cr/kg. Total soil Cr concentrations were significantly correlated with yield, but not with leaf Cr concentrations. Chromium VI was not detected either in sludge or in sludge-treated soil. Among the different extractants, viz. phosphate buffer, acetic acid, ammonium acetate, DTPA, acetic acid-extractable Cr showed the best correlation with both plant Cr concentration and yield. However, the critical toxicity concentration (CTC) for 10% growth reduction of acetic acid-extractable Cr was not significantly different from the Cr concentration associated with maximum yield. Sequential soil extraction showed that sludge Cr was incorporated mainly into the moderately reduceable fraction, extractable with acid oxalate. Acid oxalate-extractable Cr was significantly correlated with yield. Acid oxalate may be a useful extractant for distinguishing naturally occurring soil Cr from waste Cr, which can cause phytotoxicity under certain soil conditions.

Introduction

It was generally held that soil Cr is unimportant as a phytotoxic agent (Woolhouse, 1983). Trivalent chromium, the main oxidation form in the field, is only slightly soluble in soils at pH higher than 4 (Cary *et al.*, 1977). Most of the Cr absorbed by plants is retained in roots and generally very small amounts of Cr are translocated to upper plant parts (Barceló *et al.*, 1987; Vázquez *et al.*, 1987). Tannery sludge, a CrIII-rich waste product, was considered to be a useful nitrogen source for crop plants, that could be used without harm on farmland (Weber, 1981). These considerations were based on different field and pot studies which reported absence of toxicity at soil Cr concentrations up to 500 or 1000 mg Cr/kg soil (Kick and Braun, 1977; Mortvedt and Giordano, 1975). Recently, however, the EEC Council proposed to limit permissible Cr concentrations in soils to 100–200 mg/kg and in sludges for agricultural purpose to 1000–1750 mg/kg (EEC, 1988).

The critical toxicity concentration (CTC) for soil Cr may greatly differ for different plant species, soil types and environmental conditions. Detailed studies have shown that the behavior of Cr in soils is very intricate. Complex formation of CrIII with citric acid, DTPA or fulvic acids may lead to increased solubility of CrIII in soils, even at pH higher than 5.5 (James and Bartlett, 1983). Under certain conditions (*e.g.* high pH and presence of oxidized Mn), CrIII in soils can be oxidized to the hexavalent form (Bartlett and

James, 1979). Chromium toxicity seems possible in steam-sterilized soils or in alkaline soils with low organic matter content under high-humidity conditions (Gauglhofer, 1984).

The difficulties in evaluating the potential risks of phytotoxicity by waste Cr are due, at least partially, to problems in both assessing the plant-available fraction of soil Cr and correlating extractable soil Cr with Cr content and productivity of plants (Williams, 1988). In the present study we intended to establish such correlations, experimenting with different extractants as well as with sequential soil extraction.

Materials and methods

Soil and sludge properties

The experimental soil was a Vertic Ustochrept (pH in H_2O 8.05; contents in air-dry soil: organic matter 2.26%; organic C 1.31%; carbonates 35%; NO_3^- 138 mg/kg; N 0.27%; CEC 8.42 meq/100 g soil; Cr determined by HF/$HClO_4$ digestion 60 mg/kg). The undigested tannery sludge had the following composition in dry weight: organic matter 65.06%; organic C 32.53%; NO_3^- 25 mg/kg; N 0.45%; in fresh weight: total Cr 0.27%, CrVI not detectable; dry matter 17%.

Air-dried soil was sieved (2 mm) and thoroughly mixed with sludge at rates of 0 (control), 8.7, 17.4 or 34.8 g sludge/kg soil. All treatments were replicated three times. Sludge-amended and control soils were moistened with distilled water to field capacity and stored in the laboratory for seven days prior to Cr analysis and sowing.

Chromium analysis

To assess Cr associated with distinct soil fractions the following solvents were used: distilled water (1 g air-dried soil in 20 ml water, 2h agitation), 0.2 *M* KH_2PO_4 + 0.2 *M* K_2HPO_4 pH 7.4 (1:20, 2 h), CH_3COOH 2.5% (1:20, 2 h), 0.005 *M* DTPA + 0.1 *M* Triethyl amine (TEA) + 0.01 *M* $CaCl_2$ pH 7.3 (1:20, 2 h), digestion with $HNO_3:HClO_4 = 10:1$, and digestion with $HF:HClO_4 = 10:1$. Sequential fractionation was performed according to Förstner (1981), using 1 *M* ammonium acetate (1:20, 2 h) for exchangeable, 0.1 *M* hydroxylamine hydrochloride (HH) + 0.01 *M* HNO_3 (1:100, 24 h) for easily reduceable, 0.2 *M* ammonium oxalate + 0.2 *M* oxalic acid (1:100, 24 h in darkness) for moderately reduceable, 30% H_2O_2 (90°C) + 1 *M* ammonium acetate pH 2.5 (1:100, 24 h) for organic- and sulfide-associated, and HF/$HClO_4$ digestion for residual Cr fractions.

Chromium concentrations in the different extractants were determined by atomic absorption spectrophotometry (Perkin Elmer 703). Soil and sludge, as well as sludge-amended soils were analyzed for CrVI by the diphenylcarbazide method (Bartlett and James, 1979). Plant chromium concentrations were analyzed in the acid-digested plant material ($HNO_3:HClO_4:H_2SO_4 = 10:1:1$) by atomic absorption spectrophotometry.

Plant material and growth conditions

Studies on plant growth and Cr uptake were performed in a growth chamber (photon fluence rate 144 μM photon $m^{-2}s^{-1}$; day/night photoperiod 16/8 h; day/night temperature 27/23°C; day/night relative humidity 60/75%). Bush bean plants (*Phaseolus vulgaris* L. cv. Délinel) were grown in plastic pots (5 L capacity, 5 plants per pot) on control or sludge-amended soil. Soil properties and sludge treatments were as indicated above. Each treatment was replicated three times. Pots were watered daily to maintain soil moisture near field capacity.

At the preflowering stage (15 d from germination) samples were taken for growth and Cr analysis. Fruit weight and fruit Cr concentration were analyzed 45 d after germination. Given results are averages of three replicate determinations. Significance of differences between treatments were determined by ANOVA.

Results

Total soil Cr content (HF/$HClO_4$ digestion) increased linearly with sludge supply. Total soil Cr concentrations in all treatments were either

Table 1. Chromium extracted from control and sludge-amended soil with different solvents and by sequential soil extraction

Extractants	Cr concentrations (μg/g soil) after varying Cr doses (g/kg) applied							
	Control		8.7		17.4		34.8	
Total soil extraction								
Distilled water	<0.05		<0.05		<0.05		<0.05	
Phosphate buffer	<0.05		<0.05		<0.05		<0.05	
Acetic acid	1.2	(2.0)[a]	1.3	(1.5)	1.5	(1.3)	2.5	(1.6)
DTPA + TEA	1.8	(3.0)	1.2	(1.4)	0.1	(0.9)	0.8	(0.5)
$HF/HClO_4$	60.0	(100)	89.0	(100)	112.0	(100)	160.0	(100)
Sequential extraction								
Ammonium acetate	1.6	(2.5)	2.2	(2.3)	3.1	(2.8)	2.3	(1.4)
HH[b]	2.0	(3.1)	1.4	(1.4)	1.3	(1.2)	1.0	(0.6)
Acid oxalate	3.0	(4.6)	25.5	(26.4)	40.7	(37.3)	79.5	(50.2)
H_2O_2/NH_4Ac[c]	4.5	(6.9)	13.0	(13.5)	17.9	(16.4)	31.0	(19.6)
$HF/HClO_4$	53.9	(82.9)	54.3	(56.3)	46.1	(42.2)	44.6	(28.2)
Sum of fractions	65.0	(100%)	96.4	(100%)	109.1	(100%)	158.4	(100%)

[a] Values in parentheses represent % of total soil Cr as of sum of fractions.
[b] Hydroxylamine hydrochloride.
[c] Ammonium acetate.

below or within the ranges of maximally permissible Cr concentrations proposed by the EEC (1988) (Table 1). No hexavalent Cr was detected. Among the weak extractants, distilled water and phosphate buffer did not yield any detectable Cr. Only 2% or less of total Cr was extractable by acetic acid (Table 1), but the extracted Cr concentration linearly increased with sludge supply. DTPA + TEA was mostly effective in control soil, but extracted less Cr from sludge-amended soils (Table 1).

Among the strong acids, $HNO_3/HClO_4$ digestion was less effective than $HF/HClO_4$ digestion in control soils and in soils with low sludge amounts, but both combinations of acids gave similar results in the higher sludge treatments. Sequential soil extraction showed that the distribution of total soil Cr among the various soil fractions was quite different in control and sludge-amended soils. In control soil, Cr concentrations in the extractants increased along the extractant sequence used: ammonium acetate < HH < acid oxalate < H_2O_2 < $HF/HClO_4$. About 83% of total soil Cr was in the residual fraction. Chromium concentrations in acid oxalate and H_2O_2 (moderately reduceable and organic- and sulfide-associated fractions, respectively) proportionally increased with sludge supply. In sludge-treated soils the residual fraction only represented between 56 and 28% of total Cr. Even for the lowest sludge dose a more than 8-fold increase in acid oxalate-extractable Cr was found.

Plant growth linearly decreased with increasing sludge amounts (Fig. 1). Chromium-toxicity symptoms, in the form of severe chlorosis in trifoliolate leaves, were observed for the highest sludge dose. Plants grown on sludge-amended soil showed significantly increased Cr concentrations in roots, stems and fruits (Fig. 1). In leaves, significantly increased Cr concentrations were only found for the highest sludge dose (Fig. 1c). Table 2 shows the correlation coefficients for linear relations between soil Cr concentrations and plant growth or plant Cr content. Only acetic acid-extracted Cr was significantly correlated with Cr concentration in stems and leaves. Fruit Cr concentrations (direct) and plant dry weight (indirect) were significantly correlated with Cr in $HF/HClO_4$, acetic acid, acid oxalate and H_2O_2. Critical toxicity concentrations for a 10% yield reduction (CTC) are listed in Table 3. Values are given only for those extractants and organs presenting Cr concentrations which were significantly correlated with plant yield. Highest percent increases for CTC of Cr and Cr concentrations associated with maximum yield were found for acid oxalate-extracted soil Cr and for stem Cr concentration, respectively.

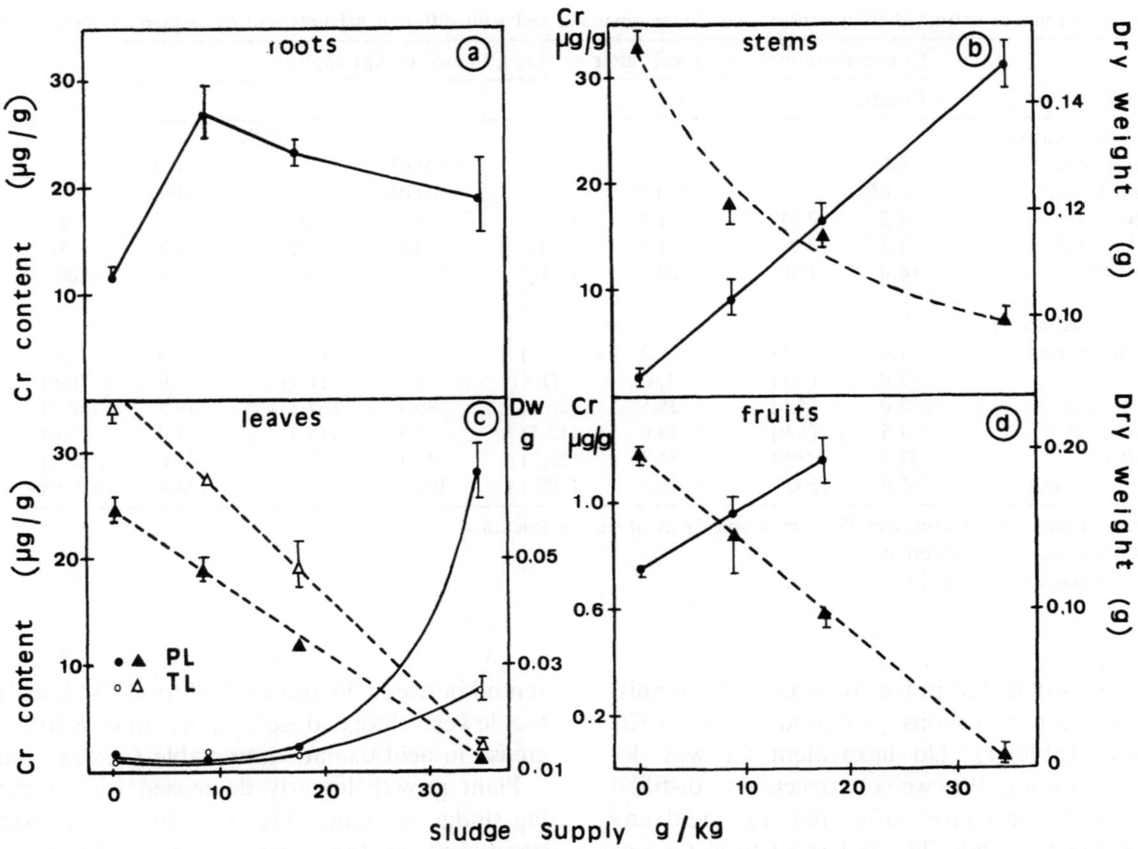

Fig. 1. Chromium concentration (µg/g) and dry weights (g) of organs of bean plants grown on soil amended with different amounts of tannery sludge. a, roots; b, stems; c, leaves (PL primary leaves, TL trifoliolate leaves); d, fruits. Solid lines: Chromium content. Dashed lines: Dry weight.

Discussion

Our results indicate that under certain experimental conditions, *viz*. pot-grown dicot species, alkaline soil, high humidity, Cr in waste materials can cause severe phytotoxicity at total soil Cr concentrations generally considered harmless. Therefore, total soil Cr may not always be a useful indicator of Cr toxicity. Analysis by the diphenylcarbazide method did not render detectable amounts of CrVI. However, the possibility of highly phytotoxic CrVI occurring in the rhizosphere soil cannot be excluded.

Significant growth reduction was found for sludge amounts which did not significantly increase leaf Cr levels. Similar results have been found by others (Williams, 1988) and cast doubt upon the relevance of determining CTC based on leaf Cr concentrations.

Leaf Cr concentrations, as well as leaf dry weight and fruit yield were highly correlated with acetic acid-extractable Cr, but CTC based on acetic-acid extracted Cr was not significantly different from Cr concentrations associated with maximum yield (Table 3). As for the other extractants, acid oxalate extractable soil Cr was significantly correlated with plant growth and with values for highest increase in Cr concentration associated with a 10% growth reduction in comparison with the concentration providing maximum yield.

Acid oxalate has been shown to extract metals associated with amorphous iron oxides or hydroxides, gel hydrous oxides and fulvic but not humic iron. Acid oxalate seems to have little effect on crystalline minerals (Beckett, 1989). Our results on sequential soil extraction clearly indicate that Cr of tannery sludge was principally

Table 2. Correlation coefficients for linear relationships between Cr concentration in soil or soil extracts and plant Cr concentration and growth

	Total soil extraction		Sequential soil extraction					
	$HF/HClO_4$	$HNO_3/HClO_4$	Acetic acid	DTPA	Ammonium acetate	Hydroxylamine hydrochloride	Acid oxalate	H_2O_2/NH_4Ac[a]
Plant Cr concentration								
Roots	0.2754	0.3069	0.0010	−0.5715	0.6524	−0.6046	0.2800	0.3079
Stems	0.8629	0.9421[b]	0.9461[b]	0.4431	0.0450	−0.7485	0.8807	0.8769
Primary leaves	0.8710	0.9056[b]	0.9811[b]	0.4389	0.0417	−0.6784	0.8815	0.8708
Trifoliolate leaves	0.8783	0.9354[b]	0.9730[b]	0.4381	0.0515	−0.7263	0.8925	0.8854
Fruits	0.8959[b]	0.9345[b]	0.8832	−0.2107	0.6567	−0.9577[b]	0.9804[b]	0.9825[b]
Plant dry weight								
Stems	−0.9433[b]	−0.9346[b]	−0.8101	0.2235	−0.6578	0.9984[b]	−0.9446[b]	−0.9538[b]
Primary leaves	−0.9985[b]	−0.9815[b]	−0.9626[b]	−0.0107	−0.4735	0.9208[b]	−0.9994[b]	−0.9981[b]
Trifoliolate leaves	−0.9992[b]	−0.9721[b]	−0.9589[b]	0.0150	−0.4946	0.9177[b]	−0.9986[b]	−0.9968[b]
Fruit (yield)	−0.9999[b]	−0.9749[b]	−0.9470[b]	0.0442	−0.5213	0.9346[b]	−0.9994[b]	−0.9999[b]

[a] Ammonium acetate.
[b] Significant at the 0.05 level.

Table 3. Critical toxicity concentrations (CTC) of Cr in different extractants and plant organs giving 10% growth reduction

Extractant/Organ	CTC ug Cr/g soil	Relative increase (%) over Cr conc. in control
$HF/HClO_4$ total soil	69.0	115.0
Acetic acid	1.1	92.0
Acid oxalate	9.9	330.0
H_2O_2/NH_4Ac[a]	7.2	160.0
Fruits	0.8	110.0
Stems	4.4	241.0

[a] Ammonium acetate.

incorporated into the acid oxalate extractable soil fraction (Table 1). As acid oxalate extracted less than 5% of total Cr in control soils, it may be a useful extractant for distinguishing between naturally occurring Cr in soil (more than 80% in the residual fraction) and waste Cr, which under certain conditions may cause phytotoxicity. Further studies are undertaken to determine to what extent our results obtained with potted soil and plants may also apply to plant species grown under field conditions.

Acknowledgements

We are grateful to the CIRIT of Catalonia for financial support and to the Spanish MOPU for a personal grant to Dr Gunsé.

References

Barceló J, Poschenrieder Ch and Gunsé B 1987 The impact of chromium in the environment. II. Chromium in living organisms. Circular Farmacéutica (Barcelona) 293, 31–48.

Bartlett R and James B 1979 Behavior of chromium in soils. III. Oxidation. J. Environ. Qual. 8, 31–35.

Beckett P H T 1989 The use of extractants in studies on trace metals in soils, sewage sludges and sludge-treated soils. *In* Advances in Soil Science, Vol. 9. Ed. B A Stewart, pp 143–176. Springer-Verlag, New York.

Cary E E, Allaway W H and Olson O E 1977 Control of chromium concentrations in food plants. II. Chemistry of chromium in soils and its availability to plants. J. Agric. Food Chem. 25, 305–309.

EEC 1988 Proposal for modification of the Council Directive 86/278/EEC with respect to chromium in sewage sludge used in agriculture. Comm. Eur. Communities COM (88) 624 final.

Förstner U W, Calmano K, Conrado H, Jaksch H, Schimkus C and Schoer J 1981 Chemical speciation of heavy metals in solid waste materials (sewage sludge, mining wastes, dredged materials, polluted sediments) by sequential extraction. Proc. Int. Conf. Heavy Metals Environ. pp. 698–704. CEP Consultants Ltd. Edinburgh.

Gauglhofer J 1984 Chrom. *In* metalle in der Umwelt. Ed. E Merian. pp. 409–421. Verlag Chemie, Weinheim.

James B and Bartlett R 1983 Behavior of chromium in soils. VI. Interactions between oxidation-reduction and organic complexation. J. Environ. Qual. 12, 173–176.

Kick H and Braun B 1977 Wirkung von chromhaltigen Gerbereischlämmen auf Wachstum und Chromaufnahme bei verschiedenen Nutzpflanzen. Landw. Forsch. 30, 160–173.

Mortvedt J J and Giordano P M 1975 Response of corn to zinc and chromium in municipal wastes applied to soil. J. Environ. Qual. 4, 170–174.

Vázquez M D, Poschenrieder Ch and Barceló J 1987 Chromium VI-induced structural and ultrastructural changes in bush bean plants (*Phaseolus vulgaris* L.) Ann. Bot. 59, 427–438.

Webber J 1981 Trace metals in agriculture. *In* Effect of Heavy Metal Pollution on Plants, Vol. 2, Ed. N W Lepp pp. 159–184. Applied Science Publishers, London.

Williams J H 1988 Chromium in sewage sludge applied to agricultural land. Final report. Office for Official Publications of The European Communities, Luxembourg, 63 p.

Woolhouse H W 1983 Toxicity and tolerance in the response of plants to metals. *In* Physiological Plant Ecology. III. Encyclopedia of Plant Physiology, New Series Vol. 12C. Eds. O L Lange, P S Nobel, C B Osmond H Ziegler. pp 245–300. Springer-Verlag, Berlin.

M. L. van Beusichem (Ed.), *Plant nutrition – physiology and applications*, 313–316.

PLSO IPNC475

Plant uptake of cadmium as affected by variation in sorption parameters

A.E. BOEKHOLD and S.E.A.T.M. VAN DER ZEE
Department of Soil Science and Plant Nutrition, Wageningen Agricultural University, P.O. Box 8005, 6700 EC Wageningen, The Netherlands

Key words: cadmium, crop quality, heavy metals, modelling, soil heterogeneity

Abstract

The effect of accumulation of cadmium in the topsoil on cadmium contents in crops is evaluated for field scale situations, using a model that links cadmium input, plant uptake, and leaching to cadmium accumulation in the rootzone. Measurements of pH and organic matter content, which regulate sorption behaviour to a large extent, show significant field-scale variability. Taking this heterogeneity into account, the probability that the cadmium concentration in part of the plants exceeds quality standards is compared with exceedance of the distribution average.

Introduction

Accumulation of heavy metals in an arable soil may cause acceptable heavy metal concentrations in plants to be exceeded. To assess the effect of elevated heavy metal contents in the topsoil, models can be used that describe the relevant processes involved. The process of heavy metal uptake is complex, because of the many interacting soil parameters that regulate availability of heavy metals to plant roots (Bingham *et al.*, 1983; Bjerre and Schierup, 1985; McBride *et al.*, 1981), next to the influence of the plant itself on its local root environment (Linehan *et al.*, 1985; Treeby *et al.*, 1989). For field scale predictions an additional complication arises, because natural soil systems are highly heterogeneous (Biggar and Nielsen, 1976; Jury *et al.*, 1987).

To investigate the impact of soil heterogeneity on model predictions of heavy metal concentration distributions over a field, a simplified physical description of the most important processes is used. The model includes a constant input rate of cadmium, equilibrium sorption of heavy metal onto the soil, and a relationship between heavy metal content of the soil and heavy uptake by plants. Stochastic theory is implemented in the model, to account for variability of parameters that are known to vary significantly throughout a field soil.

Taking cadmium uptake of barley as an example, the impact of variability of pH and organic matter content on crop quality is shown, comparing field-averaged cadmium concentration in plants with deterministic model predictions. The distribution of cadmium concentrations in all plants in the field is analyzed comparing percentiles of the frequency distribution of the plant uptake rate with official standards for crop quality.

Theory

The soil is considered to be homogeneous in vertical direction. The plough layer is assumed equivalent with rooting depth and the main processes in this soil compartment that regulate cadmium accumulation T [μmol.m^{-3}] in soil are cadmium input I [μmol.m^{-3}], leached amount J [μmol.m^{-3}] at the lower boundary of the system, and plant uptake of cadmium P [μmol.m^{-3}]:

$$\frac{dT}{dt} = \frac{dI}{dt} - \frac{dJ}{dt} - \frac{dP}{dt} = I_t - J_t - P_t \, . \quad (1)$$

Solute flux J_t equals soil water flux v [m.y^{-1}] times solute concentration c [μmol.m^{-3}], with correction for soil compartment thickness L [m] and water content θ [m^3.m^{-3}] of the soil. For a high distribution ratio and assuming equilibrium Freundlich sorption, J_t can be approximated with:

$$J_t = \frac{v\theta c}{L} = \frac{v\theta}{L}\left[\frac{T}{\rho k_1}\right]^{1/n}, \qquad (2)$$

in which ρ is the soil dry bulk density and k_1 and n are parameters that define the adsorption isotherm.

Soil chemical parameters as pH and organic matter content influence the shape of the adsorption isotherm. This is accounted for by Van der Zee and van Riemsdijk (1987) who included proton activity (H^+) mol.L^{-1}] and organic carbon content oc [g.g^{-1}.%] in the Freundlich equation:

$$k_1 = k_a oc(H^+)^{-1/2}, \qquad (3)$$

in which k_a is the adjusted adsorption constant, excluding effects of oc and (H^+).

Plant uptake P_t can also be expressed as a function of T, although total content of heavy metals generally does not reflect bioavailability. However, literature shows little quantitative information on the relationship between c and P_t. Mathematical relationships between T and P_t of the type:

$$P_t = k_2 T^m \qquad (4)$$

are proposed by Kuboi *et al.* (1986) and will be used instead.

We assume a constant cadmium input rate:

$$I_t = I_0 . \qquad (5)$$

Substitution of eq. (2) and (4) in eq. (1) yields a differential equation in terms of T:

$$\frac{dT}{dt} = I_0 - k_2 T^m - \frac{v\theta}{L}\left[\frac{1}{\rho k_1}\right]^{1/n} T^{1/n}. \qquad (6)$$

For some values of m and n this equation can be solved analytically. For our example a numerical solution is obtained.

Variability of soil parameters is included in this deterministic model using stochastic theory. Model predictions are then based on the assumption that an ensemble of measured parameter values on different locations in a field are representative for the underlying distribution at each location (ergodicity). The distribution of the variable soil parameters is represented by a probability density function (PDF), instead of single values. This implies that model predictions are also represented by PDF's. The parameters that describe the PDF are dependent on the sensitivity of the model to the variable soil parameter. Van der Zee and van Riemsdijk (1987) assumed a lognormal PDF for proton activity and organic carbon content:

$$f_y = [y s_x \sqrt{2\pi}]^{-1} \exp\left\{-0.5\left[\frac{x - m_x}{s_x}\right]^2\right\}, \qquad (7)$$

where y is (H^+) or oc, and x is ln y. m_x is the mean of the (normal) distribution of x and s_x the standard deviation of x. f_y is denoted $\Lambda(m_x, s_x^2)$.

The distribution of P_t can be characterized using percentiles of the distribution: The value where i% of the distribution is smaller or equal to, is the i-th percentile.

Results and discussion

Model calculations are done for a sandy soil in De Kempen region in the south of the Netherlands, that contains elevated average cadmium and zinc concentrations (4 mg/kg and 200 mg/kg respectively). A Cd/Zn ratio of 0.01 is assumed, for which parameter values for the adsorption isotherm are given by Chardon (1984). Soil samples were taken on a grid of size 17 × 8, with gridpoint distance 6 m. Measured values of (H^+) and oc showed a lognormal distribution (Fig. 1).

The plant uptake parameter m is chosen to be 1 in agreement with Kuboi *et al.* (1986), who found an average of m = 0.999 for 34 different plant species. Van Luit (1984) derived a value for k_2. Annual cadmium input I_0 consists of estimated Cd-input due to atmospheric deposition and phosphate fertilization of the soil (Ferdinandus, 1987). It is assumed that initially there is no cadmium present in the soil. Table 1 sum-

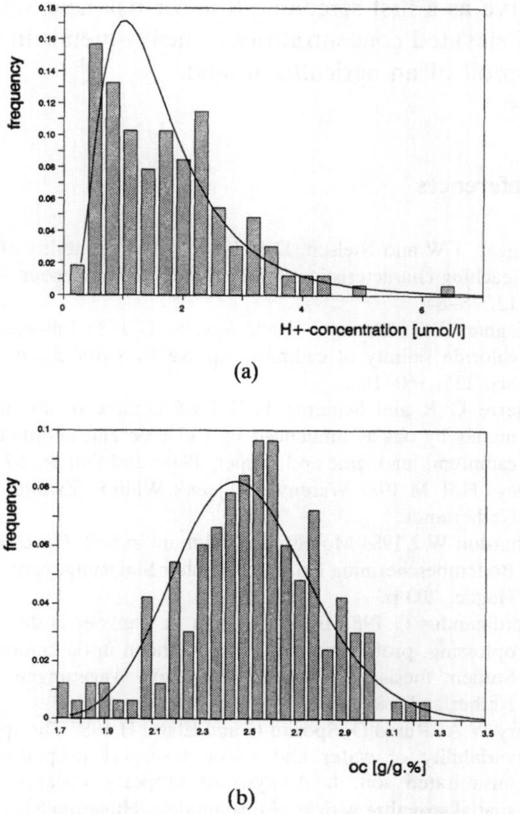

Fig. 1. Measured (bars) and fitted (line) frequency distribution for H^+-concentration (**1a**) and organic carbon content (**1b**).

marizes the parameter values used during simulation.

The stochastic model used the PDF's of Fig. 1 in Monte Carlo simulation. With (H^+) and oc negatively correlated, a lower buffer capacity for protons is assumed when the soil is lower in

Table 1. Parameter values used for calculations

I_0	$= 15.5$ [g.ha^{-1}.y^{-1}]
ka	$= 6.91 \times 10^{-3}$ [μmol$^{1-1/n}$.l$^{1/n}$.kg^{-1}]
k_2	$= 1.25 \times 10^{-4}$ [−]
L	$= 0.3$ [m]
m	$= 1$ [−]
n	$= 0.6847$ [−]
v	$= 0.83$ [m.y^{-1}]
θ	$= 0.3$ [m^3.m^{-3}]
ρ	$= 1400$ [kg.m^{-3}]
oc	: $\Lambda(0.915, 0.117^2)$, $m_{oc} = 2.51$, $s_{oc} = 0.29$ [g/g.%].
H^+	: $\Lambda(-13.45, 0.599^2)$, $m_{H^+} = 1.72 \times 10^{-6}$, $s_{H^+} = 1.13 \times 10^{-6}$ [mol.L^{-1}].

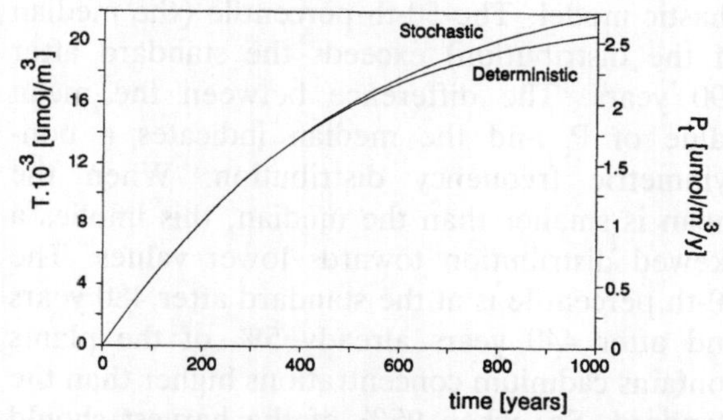

Fig. 2. Development of total content T and plant uptake P_t as a function of time for both the stochastic and the deterministic model.

organic matter. The mean values of these distributions are used as fixed numbers in the deterministic model.

Figure 2 gives T and P_t as a function of time, for both the deterministic (results as fixed numbers) and the stochastic (mean values of the PDF of T and P_t) model. The difference between both models is small, only after a long period of time the values diverge. Comparing P_t with the official Dutch standard for cadmium in grains (Bleys, 1987), which equals 0.15 mg Cd/kg fresh weight (equivalent with 2.22 μmol.m^{-3}.y^{-1}), the average plant uptake of cadmium is below the standard for approximately 610 years. Average behaviour of cadmium in the heterogeneous soil is very similar to cadmium behaviour in an equivalent homogeneous soil.

However, also extreme values for plant uptake are present in the PDF of P_t. Figure 3 shows percentiles of the distribution of P_t for the sto-

Fig. 3. Development of percentiles of P_t as a function of time when oc and (H^+) are negatively correlated.

chastic model. The 50-th percentile (the median of the distribution) exceeds the standard after 590 years. The difference between the mean value of P_t and the median indicates a non-symmetric frequency distribution. When the mean is smaller than the median, this implies a skewed distribution towards lower values. The 80-th percentile is at the standard after 480 years and after 440 years already 5% of the plants contains cadmium concentrations higher than the standard. So, when 95% of the harvest should remain below quality standards, exceedance of the standard occurs 170 years earlier than the deterministic model would predict.

For crop like barley, where all grains are mixed during the harvest, extreme high cadmium concentrations in some of the grains are averaged out. For crops that are consumed as separate entities, the extremes in the distribution of P_t are important. The deterministic approach does not give an adequate indication of crop quality in the latter cases.

Conclusions

The difference between deterministic modeling and stochastic modeling is small for the example used here, when the field averaged cadmium concentration in plants is considered. Extremes in the frequency distribution of cadmium concentration in plants cause the 95-th percentile of the distribution to reach the quality standard for cadmium in plants much earlier than the average cadmium concentration in plants. When pH and organic carbon content show profound variability, soil heterogeneity should be taken into account explicitly in models that predict crop quality. When accurate data are available that describe the variability of those parameters the model is sensitive for, stochastic modeling can serve as a first screening tool for risk assessment of elevated concentrations of heavy metals in the topsoil of an agricultural land.

References

Biggar, J W and Nielsen, D R 1976 Spatial variability of the leaching characteristics of a field soil. Water Resour. Res. 12, 78–84.

Bingham F T, Strong J E and Sposito G 1983 Influence of chloride salinity of cadmium uptake by swiss chard. Soil Sci. 135, 160–165.

Bjerre G K and Schierup H-H 1985 Uptake of six heavy metals by oat as influenced by soil type and additions of cadmium, lead, zinc and copper. Plant and Soil 88, 57–69.

Bleys H T M 1987 Warenwet. Tjeenk Willink, Zwolle, The Netherlands.

Chardon W J 1984 Mobility of Cadmium in Soil. (*In Dutch*). Bodembescherming No. 36. Publisher Staatsuitgeverij, The Hague, 200 p.

Ferdinandus G 1987 Beperking van de aanvoer is de enige oplossing, problemen met zware metalen in de landbouw. Student thesis, Agricultural University Wageningen, The Netherlands, 87 p.

Jury W A, Russo D, Sposito G and Elabd H 1987 The spatial variability of water and solute transport properties in unsaturated soil. 1. Analysis of property variation and spatial structure with statistical models. Hilgardia 55, 1–32.

Kuboi T, Noguchi A and Yazaki J 1986 Family-dependent cadmium accumulation characteristics in higher plants. Plant and Soil 92, 405–415.

Linehan D J, Sinclair A H and Mitchell M C 1985 Mobilisation of Cu, Mn and Zn in the soil solutions of barley rhizospheres. Plant and Soil 86, 147–149.

McBride M B, Tyler L D and Hovde D A 1981 Cadmium adsorption by soils and uptake by plants as affected by soil chemical properties. Soil Sci. Soc. Am. J. 45, 739–744.

Treeby M, Marschner H and Römheld V 1989 Mobilization of iron and other micronutrient cations from a calcareous soil by plant-borne, microbial, and synthetic metal chelators. Plant and Soil 114, 217–226.

Van der Zee, S E A T M and Van Riemsdijk W H 1987 Transport of reactive solute in spatially variable soil systems. Water Resour. Res. 23, 2059–2069.

Van Luit B 1984 Cadmium uptake by crops (*In Dutch*). Landbouwkundig Tijdschrift 96, 19–20.

M. L. van Beusichem (Ed.), *Plant nutrition – physiology and applications*, 317–322.

PLSO IPNC300

Interactive effects of cadmium, copper, manganese, nickel, and zinc on root growth of wheat (*Triticum aestivum*) in solution culture

G.J. TAYLOR and K.J. STADT
Department of Botany, University of Alberta, Edmonton, Alberta, Canada T6G 2E9

Key words: cadmium, copper, heavy metals, manganese, metal interactions, nickel, root growth, *Triticum aestivum* L., zinc

Abstract

The interactive effects of cadmium (Cd), copper (Cu), manganese (Mn), nickel (Ni), and zinc (Zn) on growth of *Triticum aestivum* L. in solution culture were evaluated using conventional analysis of variance (ANOVA) of both root weight data and a root weight index (RWI) designed to identify antagonistic, synergistic, and multiplicative interactions. In all trials, Ni (0 to 60 μM) served as the primary metal stress, producing near complete inhibition of root growth at 60 μM. A single concentration of either Cd, Cu, Mn, or Zn provided a secondary stress which further reduced root growth in all but the highest Ni treatments. Analysis of variance of root weight data indicated significant Ni × Cd, Ni × Mn, and Ni × Zn interactions. Such interactions, however, may not have been indicative of biological interactions. When relative root growth was expressed as root weight above the empirical growth minimum (the RWI), only the Ni × Mn interaction was significant, indicating an antagonistic interaction (growth was greater than predicted by the multiplicative model). Differences in interpretation of root weight data and the derived RWI suggest caution is needed when using primary growth data to detect possible interactions between phytotoxic metals.

Introduction

In recent years, a number of authors have documented responses of plants to combinations of metals in soils or growth solutions, but a clear understanding of potential interactions between phytotoxic metals has yet to appear. While differences in the chemistry of various metals and the biology of various plant species might be expected to lead to observed differences in interactive effects, Taylor (1989) suggested that our view of multiple metal stress may be clouded by continuing inconsistency in definitions of, and techniques used to differentiate between, additive, multiplicative, antagonistic, and synergistic effects. To illustrate this point, Taylor (1989) examined the phytotoxic effects of Ni and aluminium (Al) on *Triticum aestivum*. In this study, conventional ANOVA of root weight data indicated a significant Ni × Al interaction which could be interpreted as antagonistic, since growth was greater than predicted by the additive model. However, when data were interpreted in light of the full dose response relationship for Ni, the data were adequately represented by a multiplicative model. While it may be presumptuous to suggest that differences in analytical methods or terminology could account for the lack of a unified view of metal-metal interactions, such inconsistencies could nonetheless be hindering our understanding of multiple metal stress. The objectives of this study were (i) to document the effects of Cd, Cu, Mn, and Zn on the response of *Triticum aestivum* to varying concentrations of Ni in solution culture, and (ii) to examine the nature of potential metal-metal interactions using conventional ANOVA of root weight data and Taylor's (1989) root weight index.

Methods

A series of preliminary experiments were conducted to determine the dose response of *Triticum aestivum* L. cv. Neepawa to each of Cd, Cu, Mn, Ni, and Zn supplied as a single phytotoxic metal (data not reported). In the experiments reported here, plants were exposed to a range of Ni concentrations (selected to produce a growth response from no inhibition of growth to near complete inhibition), in the presence or absence of a second metal (a concentration selected to reduce yield by approximately 30%). For each of four experiments, a randomized block, factorial design with three replicates, 10 Ni treatments (0, 5, 10, 15, 20, 25, 30, 40, 50, 60 μM), and two competing metal treatments (basal nutrient solution concentration or 0.75 μM Cd, 6 μM Cu, 125 μM Mn, or 100 μM Zn supplied as sulfate or chloride (Mn) salts) was used.

To prepare plants for experimentation, seeds were surface sterilized in 1.2% sodium hypochlorite for 20 minutes, and germinated overnight immersed in a solution of 0.005 g L^{-1} Vitavax to prevent fungal growth. Seedlings were grown for 3 days in a solution containing (μM) Ca (1000), Mg (300), NO_3^- (2900), and NH_4^+ (300), and for 5 days in a complete nutrient solution containing (μM) Ca (1000), Mg (300), K (800), NO_3^--N (3300), NH_4^+-N (300), PO_4^{2-} (100), SO_4^{2-} (101), Cl (34), Na (20), Fe (10), B (6), Mn (2), Zn (0.5), Cu (0.15), and Mo (0.1). Iron was supplied as Fe-EDTA prepared from equimolar amounts of $FeCl_3$ and Na_2EDTA.

Metal treatments were superimposed over the full nutrient solution described above, except for the Ni × Cd and Ni × Mn experiments where an additional 300 μM NH_4NO_3 was added to maintain a low pH of solutions throughout the duration of the experimental period. The pH of aerated nutrient solutions were adjusted initially to either 4.8 (Ni × Cd, Ni × Mn) or 5.0 (Ni × Cu, Ni × Zn) with HCl or KOH. To begin experiments, eight uniform, nine-day-old seedlings were mounted on Plexiglas covers of each of 60 polyethylene containers of 10 L capacity (growth containers were covered to inhibit algal growth). Seedlings were grown in a pair of controlled-environment rooms with temperature maintained between 19 and 25°C during a 16 hr light period and between 17 and 19°C during darkness. Solution temperatures within each chamber were maintained between 16 and 19°C by standing all experimental pots in a common water bath. Relative humidity was maintained between 50 and 60% during the light period and 80 and 90% during darkness. Both growth rooms were illuminated by 12 HID mercury halide (400 W) and 4 HID high pressure sodium (400 W) lamps located 1.3 m above the plant bases, however, differences between growth rooms resulted in differences between experiments in photosynthetic photon flux density (PPFD) at plant base level (300 ± 4.1 $\mu mol\ m^{-2}\ s^{-1}$ for Ni × Cd and Ni × Cu; 350 ± 3.8 $\mu mol\ m^{-2}\ s^{-1}$ for Ni × Mn and Ni × Zn). Nutrient solutions were adjusted periodically to 10 L with distilled water to compensate for water loss by evaporation and transpiration.

After 14 days of treatment, plants were harvested, divided into roots and leaves, dried at 50°C, and weighed. Statistical analyses of the data were performed using ANOVA available on Statistical Graphics Corporation's Statgraphics Version 2.6. To achieve homogeneity of variance, ANOVA was performed on the log transformation of root weight data, and on the arcsin (square root) transformation of the RWI data. Significance was defined at the 95% confidence level.

Results

Analysis of variance of root weight data from all four experiments indicated significant main effects due to both the primary (Ni) and secondary (Cd, Cu, Mn, and Zn) metal treatments, as well as significant Ni × Cd, Ni × Mn, and Ni × Zn interactions. The Ni × Cu interaction was not significant (Table 1). In the absence of a secondary metal stress, root weight declined rapidly as the concentration of Ni in the culture solution increased from 0 to 40 μM. Above 40 μM, little

Table 1. Significance levels of main and interaction effects from ANOVA of the log transformation of root weight data and the arcsin (square root) transformation of RWI data

	Root weight data			RWI data		
	Main effects		Int.	Main effects		Int.
	Nickel	2° Metal	Effect	Nickel	2° Metal	Effect
Ni × Cd	0.000	0.000	0.000	0.000	0.351	0.524
Ni × Cu	0.000	0.000	0.644	0.000	0.007	0.298
Ni × Mn	0.000	0.001	0.016	0.000	0.010	0.040
Ni × Zn	0.000	0.000	0.000	0.000	0.001	0.289

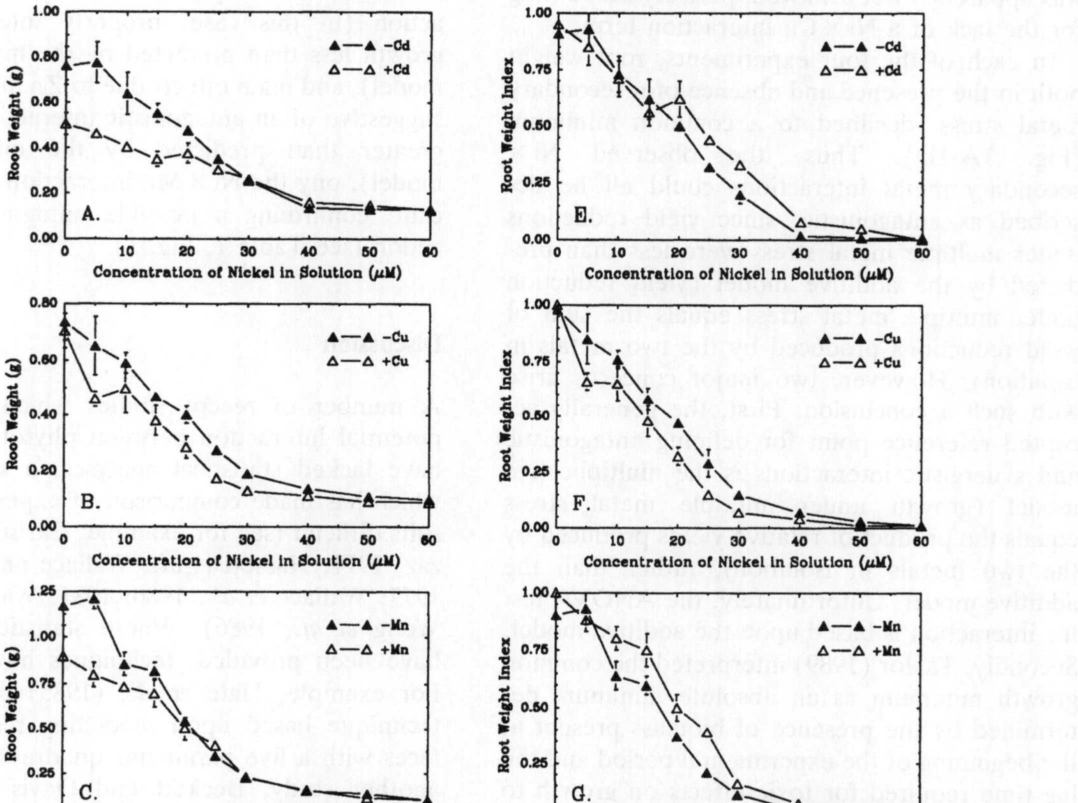

Fig. 1. Root growth of *Triticum aestivum* under conditions of Ni × Cd (A, E), Ni × Cu (B, F), Ni × Mn (C, G), and Ni × Zn (D, H) stress, as measured by root dry weight (g pot^{-1}; A–D) and the root weight index (RWI; E–H).

additional growth reduction was observed (Fig. 1A-1D). Supply of 0.75 μM Cd, 6 μM Cu, 125 μM Mn, or 100 μM Zn as a secondary metal stress further reduced root weight when Ni was supplied between 0 and 40 μM. Above 40 μM Ni, the secondary metal stress had little additional effect (Fig. 1A-1D). This pattern of growth accounted for significant Ni × Cd, Ni × Mn, and Ni × Zn interaction terms. While 6 μM Cu was selected to produce a 30% growth reduction compared to control, the desired toxic effect was apparently not achieved, possibly accounting for the lack of a Ni × Cu interaction term.

In each of the four experiments, root weight both in the presence and absence of a secondary metal stress, declined to a common minimum (Fig. 1A-1D). Thus, the observed Ni × secondary metal interactions could all be described as antagonistic, since yield reductions under multiple metal stress were less than predicted by the additive model (yield reduction under multiple metal stress equals the sum of yield reductions produced by the two metals in isolation). However, two major concerns arise with such a conclusion. First, the generally accepted reference point for defining antagonistic and synergistic interactions is the multiplicative model (growth under multiple metal stress equals the product of relative yields produced by the two metals in isolation), rather than the additive model. Unfortunately, the ANOVA test for interaction is based upon the additive model. Secondly, Taylor (1989) interpreted the common growth minimum as an absolute minimum determined by the presence of biomass present at the beginning of the experimental period and the lag time required for toxic effects on growth to be manifest. To account for both of these concerns, Taylor (1989) suggested the use of a root weight index (RWI) defined as relative root weight expressed as a fraction of potential root weight above the absolute minimum ($RWI = (RW - RW_{min})/(RW_{max} - RW_{min})$). Values of RWI are calculated independently for each replicate and secondary metal treatment group, thus for each replicated treatment group (+ or − the secondary metal) values of TWI will range from 0.0 to 1.0. By definition, the RWI for treatments with and without the secondary metal will be equal if the interaction between the primary and secondary metal is multiplicative.

In contrast to the results obtained with the root weight data, ANOVA of the RWI data indicated significant main effects of the primary metal Ni, the secondary metals Cu, Mn, and Zn, and the Ni × Mn interaction term, but main effects due to Cd, and the Ni × Cu, Ni × Cd, and Ni × Zn interactions were not significant (Table 1). Once again, the RWI data indicated that root weight declined rapidly with 0 to 40 μM Ni in solution, with little additional effect above 40 μM Ni (Fig. 1). While significant main effects due to Cu were suggestive of a synergistic interaction (in this case, properly interpreted as growth less than predicted by the multiplicative model), and main effects due to Zn and Mn were suggestive of an antagonistic interaction (growth greater than predicted by the multiplicative model), ony the Ni × Mn interaction was significant, confirming a possible antagonistic interaction (see Table 1, Fig. 1).

Discussion

A number of recent studies which tested for potential interaction between phytotoxic metals have lacked statistical analyses, a shortcoming which has made comparison of experimental results difficult (see for example, Carlson and Bazzaz, 1977; Hassett *et al.*, Wallace and Romney, 1977; Wallace *et al.*, 1980; 1981; Wallace, 1982; Wong *et al.*, 1986). Where statistical analyses have been provided, techniques have differed. For example, Hale *et al.*, (1985) developed a technique based upon modelling response surfaces with a five parameter quadratic model. In another study, Beckett and Davis (1978) developed a quantitative technique using critical tissue concentrations to differentiate between models of metal interaction. While the latter technique has been utilized by subsequent authors (Carlson and Rolfe, 1979), neither of these techniques have been used widely in the literature, perhaps because of their complexity.

Less complex tests for possible metal-metal interactions have also been devised, but these tests are based upon the additive model of interaction. For example, Lepp (1977) developed a numerical interaction index which, by definition, was equal to zero if an interaction was additive. Conventional statistical analysis was then used to determine if the interaction index was significantly greater (antagonistic), or less (synergistic) than zero. The most commonly used test to identify interactions, however, has been ANOVA of primary growth data. A sufficient number of studies which used this technique are available for comparative purposes. Unfortunately, these studies have failed to produce a consistent pattern of metal-metal interaction. For example, Wu and Antonovics (1975) and Davis and Carlton-Smith (1984) reported synergistic Zn × Cu interactions on root elongation in a Zn, Cu co-tolerant ecotype of *Agrostis stolonifera* and on above-ground yield of *Lolium perenne* respectively (growth less than predicted by the additive model). In contrast, Miller *et al.* (1977) reported an antagonistic Pb × Cd interaction on leaf weight in *Zea mays*, and McGrath *et al.*, (1980) reported an antagonistic Al × Cd interaction on root elongation in an Al-tolerant, Cd-sensitive ecotype of *Holcus lanatus* (growth greater than predicted by the additive model). Interestingly, in *Holcus lanatus* the Al × Cd term was not significant for an Al-sensitive, Cd-tolerant ecotype, indicating that effects of Al and Cd on root elongation in this ecotype were additive. Additive effects have also been observed in *Picea sitchensis* under conditions of Cd, Cu, and Ni stress (Burton *et al.*, 1986), and in *Lolium hybridum* under conditions of Cd, Ni, and Pb stress (Allison and Dzialo, 1981).

As suggested earlier, the lack of consistency between studies could reflect differences in the chemistry of various metals and the biology of various plant species, but the possibility that experimental error may be involved should not be overlooked. Our results confirm Taylor's (1989) suggestion that a significant ANOVA interaction term may not be indicative of a true biological interaction, hence, conventional ANOVA of root weight data must be used and interpreted with caution. In addition to the problem of basing interpretation of data on the additive model (Poole, 1974), the results reported here suggest that use of ANOVA without taking into account full dose response information can lead to erroneous interpretation of experimental data. Clearly, conclusions regarding the type of interaction between Ni and the secondary metals in these experiments varied with the methods of data analysis and, in several cases, apparent interactions were an artifact of the absolute growth minimum. Using the Ni × Cd experiment as an example, ANOVA of primary root weight data indicated significant main effects of Ni and Cd, as well as a signficant Ni × Cd interaction. This interaction could be interpreted as antagonistic, because observed growth reductions under conditions of Ni × Cd stress were less than predicted by the additive model (less than the multiplicative model as well). Indeed, the degree of antagonism appeared to increase with increasing concentrations of Ni in the growth solutions. In contrast, the results of ANOVA of the RWI provided little support for a Ni × Cd antagonism. Effects due to Cd and the Ni × Cd interaction term were not significant, indicating that the pattern of RWI response to Ni was independent of the presence of Cd in growth solutions. Thus, it must be concluded tha the interaction between Ni and Cd was multiplicative, a result previously reported for Ni × Al stress in *Triticum aestivum* (Taylor, 1989).

Results of the experiment with Zn were ambiguous, possibly highlighting a weakness of the RWI technique. In the Ni × Zn experiment, ANOVA of the RWI data indicated significant main effects due to Zn, but the Ni × Zn interaction was not significant. While significant main effects (Zn) were suggestive of an antagonistic interaction, this was not confirmed by the presence of a significant Ni × Zn interaction. This raises an important question. When interpreting the ANOVA of RWI data, which term is a more reliable indicator of potential interaction, the main effects due to the secondary metal, or the metal-metal interaction term?

The Ni × Mn experiment provided more concrete evidence of an antagonistic interaction. The ANOVA of both the root weight and RWI data indicated significant effects due to Ni and Mn, as well as a Ni × Mn interaction term. In isolation, these data would appear to provide substantive support for an antagonistic Ni × Mn interaction. When results are compared from each of the four experiments, however, the similar change in the RWI between treatment groups with and without the secondary metals (differences in the RWI response to Ni) raises some question about the importance of the significant Ni × Mn interaction term (or alternatively, the lack of a significant interaction term in other experiments). A question that arises is to what extent does the increased variance associated with dividing root weight values by an estimate of maximum potential root weight (RW_{max}) affect the ANOVA of the RWI data? Also, can inaccurate estimation of RW_{max} affect the absolute value of the RWI sufficiently to incorrectly identify antagonistic or synergistic interactions? These questions are important because variance increased with increasing mean root weight. While the RWI transformation may be essential for using ANOVA to interpret data in reference to a multiplicative model, the increased variance associated with the transformation may reduce the sensitivity and reliability of the technique.

In conclusion, it would appear that caution is required when using ANOVA of primary growth data to detect possible interactions between phytotoxic metals. Failure to take into account full dose response information can lead to erroneous interpretation of experimental data. While use of the RWI index as a tool to differentiate between additive, multiplicative, and synergistic interactions does provide additional insight into the phenomenon of metal-metal interactions, its use is not completely free of potential problems. To overcome these problems, we are now evaluating the possibility of using non-linear mathematical functions for modelling the effect of metals on growth and the potential for interactive effects between phytotoxic metals.

Acknowledgements

Financial support for this research was provided by the Natural Sciences and Engineering Research Council of Canada and the University of Alberta.

References

Allison D W and Dzialo C 1981 The influence of lead, cadmium, and nickel on the growth of ryegrass and oats. Plant and Soil 62, 81–89.

Beckett P H T and Davis R D 1978 The additivity of the toxic effects of Cu, Ni and Zn in young barley. New Phytol. 81, 155–173.

Burton K W, Morgan E and Roig A 1986 Interactive effects of cadmium, copper and nickel on the growth of Sitka spruce and studies of metal uptake from nutrient solutions. New Phytol. 103, 549–557.

Carlson R W and Bazzaz F A 1977 Growth reduction in American sycamore (*Plantanus occidentalis* L.) caused by Pb-Cd interaction. Environ. Pollut. 12, 243–253.

Carlson R W and Rolfe G L 1979 Growth of ryegrass and fescue as affected by lead-cadmium-fertilizer interaction. J. Environ. Qual. 8, 348–352.

Davis R D and Carlton-Smith C H 1984 An investigation into the phytotoxicity of zinc, copper and nickel using sewage sludge of controlled metal content. Environ. Pollut. Ser. B 8, 163–185.

Hale J C, Ormrod D P, Laffey P J and Allen O B 1985 Effects of nickel and copper mixtures on tomato in sand culture. Environ. Pollut. Ser. A 39, 53–69.

Hassett J J, Miler J E and Koeppe D E 1976 Interaction of lead and cadmium on maize root growth and uptake of lead and cadmium by roots. Environ. Pollut. 11, 297–302.

Lepp N W 1977 Interactions between cadmium and other heavy metals in affecting the growth of lettuce (*Lactuca sativa* L. c.v. Webbs Wonderful) seedlings. Z. Pflanzenphysiol. 84, 363–367.

McGrath S P, Baker A J M, Morgan A N, Salmon W J and Williams M 1980 The effects of interactions between cadmium and aluminium on the growth of two metal-tolerant races of *Holcus lanatus* L. Environ. Pollut. (Series A) 23, 267–277.

Miller J E, Hassett J J and Koeppe D E 1977 Interactions of lead and cadmium on metal uptake and growth of corn plants. J. Environ. Qual. 6, 18–20.

Poole R W 1974 An Introduction to Quantitative Ecology. McGraw-Hill, Inc., Toronto, 532 p.

Taylor G J 1989 Multiple metal stress in *Triticum aestivum* L. Differentiation between additive, multiplicative, antagonistic, and synergistic effects. Can J. Bot. 67, 2272–2276.

Wallace A 1982 Additive, protective, and synergistic effects on plants with excess trace metals. Soil Sci. 133, 319–323.

Wallace A and Romney E M 1977 Synergistic trace metal effects in plants. Commun. Soil Sci. Plant Anal. 8, 699–707.

Wallace A, Romney E M and Alexander G V 1981 Multiple trace element toxicities in plants. J. Plant Nutr. 3, 257–263.

Wallace A, Romney E M, Kinnear J and Alexander G V 1980 Single and multiple trace metal excess effects on three different plant species. J. Plant Nutr. 2, 11–23.

Wong M K, Chuah G K, Ang K P and Koh L L 1986 Interactive effects of lead, cadmium and copper combinations on the uptake of metals and growth of *Brassica chinensis*. Environ. Exp. Bot. 26, 331–339.

Wu L and Antonovics J 1975 Zinc and copper uptake by *Agrostis stolonifera*, tolerant to both zinc and copper. New Phytol. 75, 231–237.

M. L. van Beusichem (Ed.), *Plant nutrition – physiology and applications*, 323–326.

PLSO IPNC452

Long-term effects of fertilization and diffuse deposition of heavy metals on soil and crop quality

S.E.A.T.M. VAN DER ZEE, H.N.M. FERDINANDUS, A.E. BOEKHOLD and F.A.M. DE HAAN
Department of Soil Science and Plant Nutrition, Wageningen Agricultural University, P.O. Box 8005, 6700 EC Wageningen, The Netherlands

Key words: balance, deposition, fertilization, heavy metals, quality standards

Abstract

Atmospheric deposition and application of heavy metals in fertilizer result in heavy metal accumulation in soil, plants and increased leaching to ground water. Calculations of the balance between input and acceptable removal indicate that current immision rates of Cd and Pb are on average too high to preserve on the long term a good soil quality in The Netherlands, as standards for crop and ground water quality may be exceeded in the future.

Introduction

Agricultural application of fertilizer and pesticides have become a major threat for ground water quality in The Netherlands during the past decades. At the same time diffuse source soil contamination threatens agriculture itself. Monitoring networks have revealed that atmospheric deposition of *e.g.* heavy metals on soil is significant (Ferdinandus, 1989). Furthermore heavy metals are applied to soil due to their presence in manure and commercial fertilizer (De Boo, 1988). Due to this immission heavy metals accumulate in soils, which is accompanied by increasing concentrations in plant tissue (Busch, 1985) as well as increased leaching rates into ground water.

In this paper we give a methodology to assess whether current immission rates of heavy metals pose a problem with respect to soil, ground water and agricultural crop quality. This methodology is illustrated with the mean situation for The Netherlands. Information that is usually lacking for the approach suggested is identified.

Theory

Heavy metals deposited on the soil either accumulate in soil, in plants by uptake, or are leached down into groundwater. We can therefore give a balance equation by

$$\frac{dT}{dt} = \frac{dI}{dt} - \frac{dJ}{dt} - \frac{dP}{dt} = I_t - J_t - P_t \qquad (1)$$

(for symbols see Notation). To solve this balance we need to know how the accumulation (T), input (I), amount leached (J) and uptake by plants (T) are related. For most heavy metals the information available precludes the evaluation of eq. (1), however, for cadmium enough data are available for an approximation. For a humic soil plough layer with a density $\rho = 1380\ kg/m^3$ and a thickness of $\Delta z = 0.35$ m the accumulation rate can be given by ($\rho\ \Delta z\ dG/dt$ (mg/m^2y), where $\rho\ \Delta z = 480\ kg/m^2$. The removal rate, in the same units, in harvested product depends on the crop and using uptake data of Van Luit (1984) for continuous cropping of lettuce is equal to bGY where $b = 0.13$ and the yield is $Y = 8.4\ kg/m^3$

Table 1. Calculated input and removal rates (in mg/m^2 year). Removal rates are in agreement with current Dutch standards for crop quality, and the stand still principle for ground water quality. (After: Ferdinandus *et al.* 1989)

	Cd	Cu	Pb	Zn
Input				
Agricultural crops				
– commercial fertilizer (cf)	0.55	1.7	4.7	15.5
– sewage sludge + cf	0.84	91.0	55.0	250
– cattle manure + cf	0.38	6.0	6.2	21.0
– pigs manure + cf	0.21	30.0	5.0	59.5
– poultry manure + cf	0.15	15.5	4.6	60.5
Horticultural crops (cf)	1.35–1.6	3.6–4.3	5.5–7.5	27.5–330
Precipitation	0.2	3.2	13.0	20.0
Removal				
Agricultural crops for				
– human consumption	0.30–0.25	3.6–11.0	0.1–0.5	36.0–110
– animal fodder	0.15–0.75	3.6–13.5	7.0–31.0	36.0–135
Horticultural crops	0.43–0.85	3.7–7.5	1.7–4.5	37.0–75.0
Groundwater recharge	0.1	1.3	1.3	7.5

(De Jonge, 1981). In agreement with data of Van Luit (1984) and Busch (1985) a linear relation between uptake and concentration in soil was assumed. The removal by leaching can be approximated as $R(G/k)^{1/n}$ where the net percolation rate is $R = 0.25$ m/y. According to Chardon (1984) a Freundlich sorption isotherm for Cd (*i.e.*, $G = kc^n$) is realistic. For our case the Freundlich coefficient is $k = 1$ and in the presence of zinc the Freundlich power is $n = 0.5$ (Chardon, 1984). Technical details concerning e.g. conversion of units to those applied here can be inferred from Ferdinandus *et al.* (1989). For the input rate (dI/dt) a value of 0.79 mg Cd/m^{2}y) was taken, which is realistic for The Netherlands (Table 1). In summary, we obtain a balance equation in terms of G only, given by

$$\frac{dG}{dt} = w + vG + uG^2 \tag{2}$$

where $w = r/a$, $v = -bY/a$, and $u = -P/a$. The transient solution of (2) can be solved in closed form (in other cases numerical methods have to be used)

$$G(t) = \frac{(v + \alpha^{0.5})(v - \alpha^{0.5})(\exp(\tau) - 1)}{2u(v + \alpha^{0.5} - (v - \alpha^{0.5})\exp(\tau))} \tag{3a}$$

$$\alpha = v^2 - 4uw\,; \qquad \tau = \alpha^{0.5}t \tag{3b}$$

When a sound description of the relations describing sorption and plant uptake as a function of G is lacking, the balance (1) can not be calculated for small times. To evaluate current immission is then only possible by comparison with the acceptable removal of heavy metals in harvested crops and in leaching, in view of standards for crop and ground water quality. These standards differ for different countries and were reviewed for the Dutch situation by Ferdinandus *et al.* (1989). The current mean input depends on the atmospheric deposition and on fertilization scenarios, which differ with the crop and fertilizer used. In Table 1 the input is given assuming that manure or sewage sludge were used in combination with commercial fertilizer such that the N, P, and K requirements for the crop were just met. Also given is the removal by leaching (net precipitation is 0.25 m/y) when ground water concentrations of heavy metals do not further increase (the stand still principle). Removal in harvested crops is at maximum in view of yields (Y) and crop quality standards.

Results

In Figure 1 the transient balance for Cd is shown. The accumulation in soil increases slowly. In the steady state condition (large time) G = 0.63 mg/kg, which is larger than in agreement with a good soil quality (0.5 mg/kg). Cur-

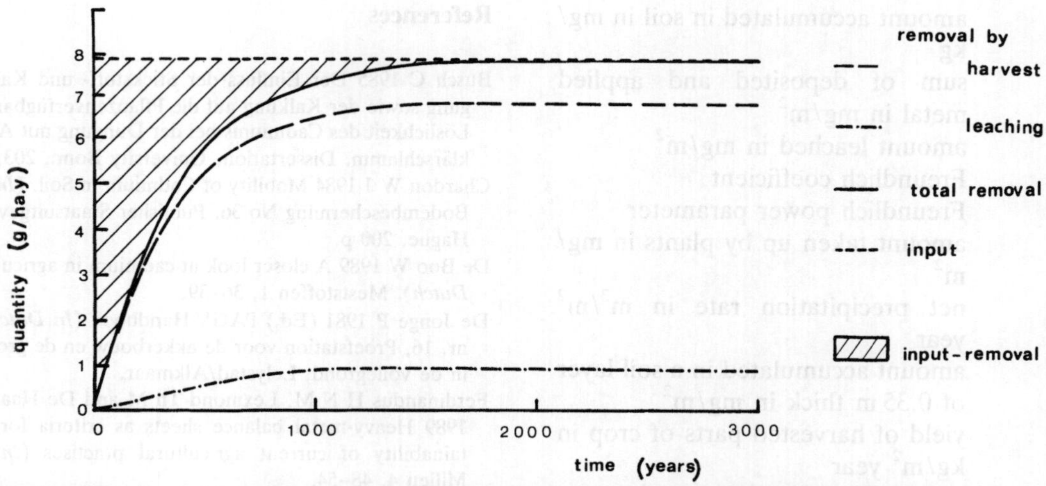

Fig. 1. Development of cadmium amounts as a function of time. Shown are the rates of input (atmospheric deposition and application in fertilizer), of the removal in harvested crop, and water leached to ground water, the total removal rate, and the difference of input and removal rates (accumulation rate in soil).

rently Dutch soils on average have accumulated approximately 0.3 mg Cd/kg already. In some parts, such as The Kempen area, G-values may be well above this mean already. For ground water recharge no further deterioration occurs for the case of Figure 1, while steady state crop quality (0.08 mg/kg) is acceptable in view of the standards (0.1 mg/kg). The transient balance depends much on the assumed sorption and uptake relations, which differ for different soils and crops. For different conditions a much faster, or slower approach of steady state may occur, while the equilibrium may be on a different level.

The results shown in Table 1 indicate currently the immission rates are larger than the removal rates, when crop and ground water recharge are in agreement with quality standards. Hence, such deposition rates of heavy metals may on the long term give rise to removal in crops and leaching, that are larger than according to standards. Notably the use of commercial fertilizer leads to large Cd immissions, and Cd in (P-) fertilizer should preferably be reduced (De Boo, 1988). The use of sewage sludge should be disencouraged as otherwise too much of all four heavy metals is applied. Pigs and chicken manure may give rise to problems with Cu. For lead the atmospheric deposition is significantly larger than is in agreement with e.g. ground water quality (according to the stand still principle), while most fertilization scenarios will on the long term make crops unacceptable for human consumption.

Conclusions

With balance considerations for agricultural soil, we established that diffuse atmospheric deposition and commonly accepted fertilization practise may give rise to deterioration of crop and/or ground water quality on the long term. For a sustainable agriculture it is therefore necessary to limit the input of heavy metals. To evaluate when approximately standards will be exceeded, and which of crop, soil or ground water standards will be exceeded first, detailed information of the sorption and uptake of heavy metals as a function of content in soil is needed. In most cases this information is yet lacking. Due to geographical differences in atmospheric deposition, soil, and crop rotation schemes and fertilization, the average situation depicted here may be pessimistic for some regions and overly optimistic for other regions. Spatial variability needs therefore be taken into account in future.

Notation

a	conversion parameter equal to $\rho\,\Delta z$ kg/m^2
b	parameter

G	amount accumulated in soil in mg/kg
I	sum of deposited and applied metal in mg/m^2
J	amount leached in mg/m^2
k	Freundlich coefficient
n	Freundlich power parameter
P	amount taken up by plants in mg/m^2
R	net precipitation rate in m^3/m^2 year
T	amount accumulated in a soil layer of 0.35 m thick in mg/m^2
Y	yield of harvested parts of crop in kg/m^2 year
Δz	ploughed layer thickness, in m
ρ	dry bulk density of soil in kg/m^3
u, v, m, α, τ	parameters

References

Busch C 1985 Der Einfluss der Stickstoff- und Kaliumdüngung sowie der Kalkung auf die Pflanzenverfügbarkeit und Löslichkeit des Cadmiums bei der Düngung mit Abwasserklärschlamm. Dissertation, University Bonn, 203 p.

Chardon W J 1984 Mobility of Cadmium in Soil. (*In Dutch*). Bodembescherming No.36. Publisher Staatsuitgeverij, The Hague, 200 p.

De Boo W 1989 A closer look at cadmium in agriculture. (*In Dutch*). Meststoffen 1, 36–39.

De Jonge P 1981 (Ed.) PAGV-Handbook (*In Dutch*). Publ. nr. 16, Proefstation voor de akkerbouw en de groenteteelt in de vollegrond, Lelystad/Alkmaar.

Ferdinandus H N M, Lexmond Th M and De Haan F A M 1989 Heavy-metal balance sheets as criteria for the sustainability of current agricultural practises (*In Dutch*). Milieu 4, 48–54.

Van Luit B 1984 Cadmium uptake by crops. (*In Dutch*). Landbouwkundig Tijdschrift 96, 19–20.

C

Plant responses to acidity, salinity and other stress factors in the root environment

M. L. van Beusichem (Ed.), *Plant nutrition – physiology and applications*, 329–334.
 PLSO IPNC306A

Effects of low activities of aluminium on soybean (*Glycine max*)

I. Early growth and nodulation

D.J. BRADY[1], CH. HECHT-BUCHHOLZ[2], C.J. ASHER[1] and D.G. EDWARDS[1]
[1] *Department of Agriculture, University of Queensland, St Lucia, Queensland 4067, Australia, and*
[2] *Institute of Plant Nutrition, Technical University of Berlin, Lentzeallee 55–57, D-1000 Berlin 33*

Key words: aluminium toxicity, *Glycine max* L., nodulation, soybean

Abstract

Soybean plants grown at 3 levels of aluminium (0, 8 and 12 μM summed activity of Al monomers (Σa_{Almono})) were transferred to Al-free solution at intervals after inoculation to identify the stage at which the nodulation process is most sensitive to Al stress. Dry weight of both shoots and roots 14 d after planting was reduced by transfer to 12 μM Al. Taproot nodule number and nodule fresh weight were greatly reduced by both 8 μM and 12 μM Al, relative to the control, in plants transferred as early as 12 *h* after inoculation.

In a second experiment, dry matter yields were reduced in plants grown in 3.5 or 5 μM Al solutions after 3 d cultivation in Al-free solution compared to plants grown only in 0 μM Al solution. Total nodule number was significantly lower in plants transferred into or grown continuously in 5 μM Al solution than in plants transferred into either 0 or 3.5 μM Al solution. Total nodule weight was reduced in plants transferred from 0 to 3.5 or 5 μM Al below that on plants grown continuously in either 0 or 5 μM Al solution.

Introduction

Several studies have shown that, in most cases, nodulation is more sensitive to Al toxicity than plant growth (Alva *et al.*, 1987; Carvalho *et al.*, 1982b; Kim *et al.*, 1985; Suthipradit, 1988). Carvalho *et al.* (1982b) showed that the early stages of the nodulation process in *Stylosanthes* were more sensitive to Al than subsequent development of the nodule. Nodule development and functioning appeared to be independent of Al even at high concentrations (Carvalho *et al.*, 1982a; Franco and Munns, 1982).

In soybean, the early events of the nodulation process can be separated into five stages identifiable with light microscopy: 1. Attachment of bacteria to the root hair (within 3 h after inoculation), 2. Root hair curling (12 h), 3. Formation of the infection thread (24 h), 4. Progression of the infection thread to the base of the root hair (48 h), 5. Development of the nodule meristem (3 d) (Turgeon and Bauer, 1981). It is unknown at which stage Al toxicity disrupts the process, leading to nodulation reduction or failure.

The objective of the present experiments, under solution culture conditions in which low activities of Al were strictly controlled, was to determine at which stage of the nodulation process Al is inhibitory.

Materials and methods

General

Both experiments were conducted in flowing solution culture in an evaporatively cooled glasshouse. The design, performance and method of operation of the flowing culture equipment have been described previously (Asher and Edwards, 1978; Asher, 1981).

Solid $CaSO_4.2H_2O$ to give a final concentration of 500 μM was added to flowing culture units filled with deionized water and allowed to equilibrate for 3 d. Other basal nutrients were then added to give final concentrations (μM) of 200 N (160 as KNO_3 and 40 as $(NH_4)_2SO_4$), 250 K 160 KNO_3 and 90 as K_2SO_4), 100 Mg as $MgSO_4.7H_2O$, 5 Si as $Na_2SiO_3.5H_2O$, 3 B as H_3BO_3, 2 P as NaH_2PO_4, 2 Fe as FeEDTA, 0.5 Zn as $ZnSO_4.7H_2O$, 0.25 Mn as $MnSO_4$.-$5H_2O$, 0.1 Cu as $CuSO_4.5H_2O$, 0.05 Mo as Na_2-$MoO_4.2H_2O$ and 0.04 Co as $CoSO_4.7H_2O$. pH was automatically maintained at 4.5 ± 0.2 using 0.05 M H_2SO_4 or KOH as the titrant.

The concentrations of phosphate (Motomizu *et al.*, 1983) and monomeric Al (Kerven *et al.*, 1989) were determined daily. From the first day continuous additions of dilute P solution were required to maintain the phosphate concentration.

The concentrations of Mg, Zn, Mn and Cu (ICP), Ca (AAS) and K (flame photometry) were measured regularly and adjusted when necessary during the experiments. Solution pH and electrical conductivity were recorded daily. Molar ionic strength was estimated as $0.013 \times$ EC ($mScm^{-1}$) (Griffin and Jurinak, 1973). The summed activity of monomeric Al species ($\Sigma\, a_{Almono}$) was calculated from solution parameters. (Blamey *et al.*, 1983). Additions of 0.05 M $Al_2(SO_4)_3.18H_2O$ solution were made necessary to maintain the nominal activities.

Soybean (cv. Fitzroy) seeds were germinated on paper towels moistened with 200 μM $CaSO_4.2H_2O$ solution. After 3 d incubation at 28°C, four seeds (radicle length approximately 3.5 cm) were planted and covered with a thin layer of black polyethylene beads. A dense suspension (10^8–10^9 cells ml^{-1}) of *Bradyrhizobium japonicum* CB1809 was used as inoculum at a rate of 5 ml per pot.

Experiment 1

Two solution culture units were treated with $Al_2(SO_4)_3.18H_2O$ solution to give mean activities of 8 ± 2 and 12 ± 1 μM $\Sigma\, a_{Almono}$ over the experimental period and a third unit was left untreated. Three days after planting, circulation of the nutrient solution in these units was halted for 1 h and each pot was inoculated. After 3 h, a set of four replicate cultures from each unit was rinsed twice in Al-free nutrient solution and transferred to uninoculated Al-free nutrient solution in a fourth flowing culture unit. Additional sets were transferred at 12, 24, 48 and 72 h, leaving a final set in each of the three original units.

Plants were harvested 14 d after planting and separated into roots, shoots and youngest fully expanded leaf blade (YFEB). Fresh roots were stored at 4°C until nodules could be removed. Nodules were separated into taproot and lateral root fractions and fresh weights recorded. Plant samples were oven-dried at 70°C for 1 week prior to determining dry weights.

Experiment 2

Plants were grown for 3 d in Al-free nutrient solution (<1 μM $\Sigma\, a_{Almono}$). Some plants were then transferred to units treated with $Al_2(SO_4)_3.18H_2O$ solution to give mean activities of 3.5 ± 0.5 or 5.0 ± 0.6 μM $\Sigma\, a_{Almono}$ ($Al_{0\rightarrow3.5}$ and $Al_{0\rightarrow5}$ treatments) and inoculated immediately. Plants were also grown continuously in 0 (<1 μM) or 5.0 ± 0.8 μM $\Sigma\, a_{Almono}$ solutions (Al_0 and Al_5 treatments) and inoculated after 3 d and in an uninoculated Al-free solution (Al_0(-inoc) treatment). Fifteen replicates of each treatment were harvested 21 d after planting and separated into roots, nodules, shoots and YFEB. Plant material was processed as for Experiment 1.

Results

Experiment 1

Growth of the host plant

In plants subjected to Al treatment for the full 14 d, both root and shoot yields were significantly reduced at 12 μM, whereas only shoot yield was significantly reduced at 8 μM (Table 1). Exposure of roots to Al for only a portion of the experimental period either had no significant effect on final yields or caused a small yield reduction in the 12 μM treatment. Increasing the time of exposure for up to 72 h after inoculation did not increase the detrimental effect on host plant growth.

Table 1. Effects on shoot and root dry weight (g/plant) of transfer of soybean plants to solutions containing no Al or Bradyrhizobium at 3, 12, 24, 48 or 72 h after inoculation. "Continuous" plants were not transferred

Transfer time (h)	Σa_{Almono} (μM) 0	8	12
	Shoot dry weight (g/plant)		
3	1.91	2.10	1.51
12	2.17	1.99	1.74
24	2.07	2.06	1.86
48	2.04	2.00	1.81
72	2.35	2.18	1.58
Cont.	1.93	1.34	0.98
		$LSD_{0.05}$	
		Al	0.10
		transfer time	0.15
		Al × transfer time	0.25
	Root dry weight (g/plant)		
3	0.59	0.60	0.50
12	0.67	0.60	0.50
24	0.61	0.68	0.56
48	0.62	0.56	0.52
72	0.66	0.66	0.47
Cont.	0.58	0.57	0.31
		$LSD_{0.5}$	
		Al	0.03
		transfer time	0.04
		Al × transfer time	0.08

Plants exposed to Al showed inhibition of root elongation, stunting of lateral roots, yellowing of root tips, microscopic changes in root cell structure and impairment of root hair development, which are described in detail elsewhere (Hecht-Buchholz *et al.*, 1990). In the continuous 12 μM treatment, symptoms of Al-induced Ca deficiency and Al-induced Mg deficiency were also observed on shoots.

Nodulation

The first nodules of Al-treated plants appeared on the lateral roots in the air space between the solution and the plant support basket. While these roots would have been splashed by solution due to vigorous aeration, it is unlikely that they were subject to the same Al stress as the taproot which was continuously bathed in solution from the time of transplanting. Nodulation on lateral roots was therefore variable and did not show significant effects of the Al treatments (data not shown).

Continuous exposure of the roots to 8 or 12 μM effectively prevented nodulation of the taproot (Table 2). At 12 μM, delaying transfer to Al-free solution until 12 h after inoculation significantly reduced nodule number and nodule weight. The data suggest that, 12 μM Al, increasing the time from inoculation to transfer to Al-free solution from 12 to 72 h further reduced nodulation. Data were more variable for the 8 μM treatment, but nodule weights were significantly less for plants transferred at 12, 24, 48 and 72 h after inoculation than for plants transferred at 3 h. In the absence of Al, although data were again variable, continuous exposure to Bradyrhizobium did not give a higher nodule number or weight than 3 h exposure.

Experiment

Growth of the host plant

Plant roots and shoots in all treatments were free of macroscopic symptoms. Root and shoot dry

Table 2. Effects on nodule number and fresh weight (mg/plant) of transfer of soybean plants to solutions containing no Al or Bradyrhizobium at 3, 12, 24, 48 or 72 h after inoculation. "Continuous" plants were not transferred

Transfer time (h)	Σa_{Almono} (μM) 0	8	12
	Nodule number per plant		
3	13	14	12
12	25	5	7
24	19	10	6
48	14	12	5
72	16	11	1
Cont.	16	1	0
		$LSD_{0.05}$	
		Al	3.2
		transfer time	4.6
		Al × transfer time	7.9
	Nodule fresh weight (mg/plant)		
3	29	27	17
12	25	12	11
24	20	11	9
48	19	18	9
72	22	9	4
Cont.	25	<1	0
		$LSD_{0.05}$	
		Al	5.3
		transfer time	7.6
		Al × transfer time	13.1

weights did not differ significantly between Al_0 and Al_0(-inoc) treatments (Table 3). Shoot dry weights of $Al_{0\rightarrow3.5}$ and $al_{0\rightarrow5}$ were reduced compared to Al_0, while root dry weights of the Al_0, $Al_{0\rightarrow3.5}$, $Al_{0\rightarrow5}$ and Al_0(-inoc) treatments were not significantly different. Both shoot and root dry weight of the Al_5 treatment were slightly greater ($P = 0.01$) than those of the Al_0 treatment.

Nodulation

No nodules were observed on plants grown in the Al_0(inoc) treatment, indicating that rhizobial contamination of the other treatments was unlikely to have occurred. In this experiment, as all lateral roots were completely immersed in the nutrient solution, total nodule number and nodule weight are reported (Table 3). In plants transferred from 0 to 5 μM Al, both nodule

Table 3. Mean dry weight of shoots, roots and nodules (g/plant) and nodule number of soybean plants grown continuously for 21 days in 0 or 5 μM Σa_{Almono} solution, or transferred from 0 to 3.5 or 5 μM Σa_{Almono} solution at inoculation

Treatment	Shoot dry wt	Root dry wt	Nodule dry wt	Nodule number
Al_0	2.45	1.19	0.11	99
$Al_{0\rightarrow3.5}$	2.09	1.08	0.08	91
$Al_{0\rightarrow5}$	2.07	1.21	0.07	68
Al_5	2.81	1.33	0.12	57
Al_0(-inoc)	2.34	1.18	0	0
$LSD_{0.05}$	0.21	0.10	0.03	24
$LSD_{0.01}$	0.28	0.14	0.04	32

number and nodule weight were reduced. However, with plants transferred from 0 to 3.5 μM Al, nodule weight alone was slightly reduced. Plants continuously maintained in 5 μM Al showed depressed nodule number, but a similar nodule weight to Al_0 plants.

Discussion

Comparison of nodule number and weight across transfer times in the 0 Al treatment in Experiment 1 (Table 2) shows that 3 h in inoculated solution was equivalent to continuous exposure to Bradyrhizobium, suggesting that 3 h was an adequate time for attachment of bacteria to root surfaces.

According to the scheme of Turgeon and Bauer (1981) the stages of the nodulation process which occur within 12 h after inoculation are attachment of rhizobia to root hairs and curling of root hairs. However, the adverse effect of increasing time of exposure, as was observed at 12 μM Al, tends to preclude disruption of a specific stage of the nodulation process within 12 hours.

A feature of roots exposed to both Al treatments in this experiment was a reduction in the number of root hairs (Hecht-Buchholz *et al.*, 1990). As all treated plants were exposed to Al prior to inoculation, and hence would have suffered Al effects on root hair formation, it is unclear why nodulation was not completely prevented in the transferred plants as it was in continuously exposed plants. It is suggested that after transfer to 0 Al solution, healthy root growth resumed and new root hairs emerged. The zone of emerging root hairs is known to be the infective region of the root (Calvert *et al.*, 1984). Bacterial cells attached to the damaged section of the root may have colonized the new section leading to successful infections. It is possible that plants transferred soon after inoculation may have been advantaged in terms of bacterial viability or proximity of attached bacteria to new healthy sections of root. In the continuously exposed plants no new root hairs were produced and nodulation was completely prevented.

Aluminium inhibition of root hair formation represents a host plant effect which directly affects nodulation. In Experiment 2, it was hoped that early growth of plants in Al-free solution prior to inoculation in the presence of Al would allow development of normal roots and root hairs and allow the subsequent effects of Al on nodulation to be assessed.

Nodule number was significantly reduced in both the Al_5 and $Al_{0\rightarrow5}$ treatments. Plants in the Al_5 treatment appeared to recover from Al stress, and while nodule number was greatly reduced, nodule weight was equal to that of Al_0 control plants. This observation is consistent with the findings of Carvalho *et al.* (1982b) and Franco and Munns (1982) that nodule development is independent of Al stress. The lowered nodule number in both the $Al_{0\rightarrow5}$ and Al_5 μM treatments compared to the control indicates that they probably had fewer root hairs, and therefore also fewer potential sites for infection, at the time of inoculation. That Al_5 plants grown continuously in Al may have become accustomed to the adverse root environment during the early stages of nodule initiation is suggested by the increased nodule weight of this treatment over the $Al_{0\rightarrow5}$. In general terms, these observations are in agreement with those of Suthipradit (1988) who found that the critical activity of monomeric Al (1.0 μM) for nodulation of soybean cv. Fitzroy plants grown in Al-free solution for 3 d and then exposed to Al for 2 d before inoculation was higher than that determined for the same cultivar exposed continuously to Al (0.4 μM, Alva *et al.*, 1987).

It is difficult to explain the overall improved growth of plants in the continuous 5 μM treatment in Experiment 2. Beneficial effects of Al have been observed in a number of species (Foy *et al.*, 1978); however, there has often been doubt over maintenance of toxic levels of Al during the experimental period. In the present experiments, Al activity and concentrations of basal nutrients were carefully controlled. Nutrient concentrations in the YFEB for all treatments in Experiment 2 (data not shown) were in the adequate range for growth of soybean (Reuter and Robinson, 1986). It was therefore concluded that the improved growth of plants in the Al_5 treatment was not due to alleviation of either a toxicity or an essential element deficiency

Results reported by Hecht-Buchholz *et al.* (1990), in which root hair growth and cell structure were examined, may help to elucidate the adverse effects of low activities of Al on infection and hence observed nodule number in soybean.

Acknowledgement

This work was conducted as part of an ACIAR-funded project "The Management of Soil Acidity for Sustained Crop Production".

References

Alva A K, Edwards D G, Asher C J and Suthipradit S 1987 Effects of acid soil infertility factors on growth and nodulation of soybean. Agron. J. 79, 302–306.

Asher C J 1981 Limiting external concentrations of trace elements for plant growth: use of flowing solution culture techniques. J. Plant Nutr. 3, 163–180.

Asher C J and Edwards D G 1978 Relevance of dilute solution culture studies to problems of low fertility tropical soils. *In* Mineral Nutrition of Legumes in Tropical and Subtropical Soils. Eds. C S Andrew and E J Kamprath. pp 131–152. CSIRO, Melbourne, Australia.

Blamey F P C, Edwards D G and Asher C J 1983 Effects of aluminum, OH:Al and P:Al molar ratios, and ionic strength on soybean root elongation in solution culture. Soil Sci. 136, 197–207.

Calvert H E, Pence M K, Pierce M, Malik N S A and Bauer W D 1984 Anatomical analysis of the development and distribution of Rhizobium infections in soybean roots. Can J. Bot. 62, 2375–2383.

Carvalho M M de, Asher C J, Edwards D G and Andrew C S 1982a Lack of effect of toxic aluminium concentrations on nitrogen fixation by nodulated *Stylosanthes* species. Plant and Soil 66, 225–231.

Carvalho M M de, Edwards D G, Asher C J and Andrew C S 1982b Effects of aluminium on nodulation of two *Stylosanthes* species grown in nutrient solution. Plant and Soil 64, 141–152.

Foy C D, Chaney R L and White M C 1978 The physiology of metal toxicity in plants. Annu. Rev. Plant Physiol. 29, 511–566.

Franco A A and Munns D N 1982 Acidity and aluminum restraints on nodulation, nitrogen fixation and growth of *Phaseolus vulgaris* in solution culture. Soil Sci. Soc. Am. J. 46, 296–301.

Gillman G P and L C Bell 1978 Soil solution studies on weathered soils from tropical north Queensland. Aust. J. Soil Res. 16, 67–77.

Griffin G P and Jurinak J J 1973 Estimation of activity coefficients from the electrical conductivity of natural aquatic systems and soil extracts. Soil Sci. 116, 26–30.

Hecht-Buchholz Ch, Brady D J, Asher C J and Edwards D G (1990) Effects of low activities of aluminium on soybean (*Glycine max*). II. Root cell structure and root hair development. *In* Plant Nutrition – Physiology and Applications. Ed. ML Van Beusichem. pp. 335–343. Kluwer Academic Publishers, Dordrecht, The Netherlands.

Kerven G L, Edwards D G, Asher C J, Hallman P S and Kokot S 1989 Aluminium determination in soil solution. II. Short-term colorimetric procedures for the measurement of inorganic monomeric aluminium in the presence of organic acid ligands. Aust. J. Soil Res. 27, *In press.*

Kim M-K, Asher C J, Edwards D G and Date R A 1985 Aluminium toxicity: Effects on growth and nodulation of subterranean clover. *In* Proc 15th Int Grassland Congr, Kyoto, Japan. pp 501–503. Sci Council Japan and Japanese Soc Grassland Sci. Tochigiken, Japan.

Motomizu S, Wakimoto T and Toei K 1983 Spectrophotometric determination of phosphate in river waters with molybdate and malachite green. Analyst 108, 361–367.

Suthipradit S 1988 Effects of aluminium on growth and nodulation of some tropical grain legumes. Ph.D Thesis, University of Queensland.

Turgeon B G and Bauer W D 1981 Early events in the infection of soybean by *Rhizobium japonicum*: Time course and cytology of the initial infection process. Can J. Bot. 60, 152–161.

Reuter D J and Robinson J B 1986 Plant Analysis: An Interpretation Manual. Inkata Press, Melbourne, Australia.

M. L. van Beusichem (Ed.), *Plant nutrition – physiology and applications*, 335–343.

PLSO IPNC306B

Effects of low activities of aluminium on soybean (*Glycine max*)

II. Root cell structure and root hair development

CH. HECHT-BUCHHOLZ[1], D.J. BRADY[2], C.J. ASHER[2] and D.G. EDWARDS[2]
[1]*Institute of Plant Nutrition, Technical University of Berlin, Lentzeallee 55–57, D-1000 Berlin 33, and*
[2]*Department of Agriculture, University of Queensland, St. Lucia, Queensland 4067, Australia*

Key words: aluminium toxicity, cell structure, *Glycine max* L., nodulation, root hairs, root growth

Abstract

Soybean plants exposed to low activities of monomeric aluminium (8 and 12 μM total activity of Al monomers ($\Sigma a_{Al\ mono}$)) showed symptoms of Al toxicity, including destruction of root epidermal and cortical cells resulting in disintegration of the outer root surface, premature vacuolization and accumulation of polyphenolic depositions in the meristematic cells of the root tip, and impairment of root hair formation. At 2.5 and 5 μM Al, root hair development was more inhibited than root elongation. The relationship between inhibition of root hair development and nodulation failure is discussed.

Introduction

Environmental factors which have been found to inhibit early steps in nodulation are low pH *per se*, low Ca (Munns, 1968; 1970) and toxic Al concentrations (Carvalho *et al.*, 1982).

In a previous experiment with soybean (Brady *et al.*, 1990), taproot nodule number on plants grown in flowing solution culture was greatly reduced by low activities of monomeric Al species. This appeared to be due to inhibition of early steps of the nodulation process occurring within 12 h after inoculation. Early steps of the nodulation process which occur within this period include attachment of bacteria to the root hairs, curling of root hairs, and formation of the infection thread in the root hair (Turgeon and Bauer, 1982). This and the fact that emerging root hairs are required for infection of soybean, led to the assumption that the Al-induced inhibition of nodulation might be related to impaired root hair formation.

The present experiments with soybean were conducted parallel to the previous experiment (Brady *et al.*, 1990) to find out whether the sensitivity of the host to Al contributes to nodulation failure. The effects of Al on root cell structure and root hair formation were investigated by means of light and electron microscopy. The relationship between inhibition of root hair formation and nodulation failure is discussed.

Material and methods

Plant culture

Experiment 1
Experiment 1 was conducted to identify the stage at which the nodulation process is most sensitive to Al (previous paper, Brady *et al.*, 1990) and to study the effects of Al on the host (present paper). This experiment was conducted in flowing solution culture according to the method of Asher and Edwards (1981) and Asher (1981).

Detailed information about nutrient solution composition was given in the previous paper (Brady *et al.*, 1990).

Solid $CaSO_4{\cdot}2H_2O$ to give a final concentration of 500 μM was added to flowing culture units filled with deionized water and allowed to equilibrate for 3 d. Other basal nutrients were then added to give final concentrations (μM) of 200 N (160 as KNO_3 and 40 as $(NH_4)_2SO_4$), 250 K (160 KNO_3 and 90 as K_2SO_4), 100 Mg as $MgSO_4 \cdot 7H_2O$, 5 Si as $Na_2SiO_3 \cdot 5H_2O$, 3 B as H_3BO_3, 2 P as NaH_2PO_4, 2 Fe as FeEDTA, 0.5 Zn as $ZnSO_4 \cdot 7H_2O$, 0.25 Mn as $MnSO_4 \cdot 5H_2O$, 0.1 Cu as $CuSO_4 \cdot 5H_2O$. 0.05 Mo as $Na_2MoO_4 \cdot 2H_2O$ and 0.04 Co as $CoSO_4 \cdot 7H_2O$. pH was automatically maintained at 4.5 ± 0.2 using 0.05 *M* H_2SO_4 or KOH as the titrant.

The solution culture units were treated with $Al_2(SO_4)_3 \cdot 18H_2O$ to give mean activities of 8 ± 2 and 12 ± 1 μM $\Sigma a_{Al\ mono}$, one unit was left untreated with Al. The pH was automatically maintained at 4.5 ± 0.2. Soybean (c.v. Fitzroy) seeds were germinated on paper towels, moistened with 200 μM $CaSO_4 \cdot 2H_2O$ solution. After 3 d incubation at 28°C seedlings with a radicle length of approximately 3.5 cm were chosen for planting. After 3 d of exposure to Al, when macroscopic symptoms of Al toxicity were observed, root segments were prepared for light and electron microscopy.

Experiment 2

Experiment 2 was conducted to study the effect of Al on root hair development at low levels of Al. This experiment was conducted in 24 L drums of aerated nutrient solution. The nutrient composition was the same as used in the flowing solution culture experiments 1 and 3: in μM: 500 Ca, 200 N, 750 S, 250 K, 100 Mg, 5 Si, 2 P, 3 B, 2 Fe, 0.5 Zn, 0.25 Mn, 0.1 Cu and 0.02 Mo. The pH was adjusted to 4.8 prior to Al treatments, Al was added by slow (dropwise) addition of $Al_2(SO_4)_3 \cdot 18H_2O$ solution. The pH was then adjusted to 4.5. The concentration of monomeric Al was measured after 24 h (Kerven *et al.*, 1989) and the summed activities of monomeric species ($\Sigma a_{Al\ mono}$) calculated (Blamey *et al.*, 1983). The Al concentrations immediately prior to planting were 0, 1.25, 2.5 and 5 μM $\Sigma a_{Al\ mono}$. Activities did not alter significantly over 3 days. Soybean (cv. Fitzroy) seeds were germinated as described above and seedlings were planted 3 d after incubation of the seeds.

Experiment 3

Experiment 3 was conducted to study the effect on both nodulation (Brady *et al.*, 1990, experiment 2) and root development at low levels of Al. This experiment was conducted, as it had been described for experiment 1, in units of the flowing culture system. Measurements of root length and length of the root hair zone had been undertaken in plants which had been continuously exposed to 0 and 5.0 ± 0.6 $\mu M_{\Sigma a\ mono}$Al.

Light and electron microscopy

Preparation

Root segments sectioned at different distances from the hypocotyl were fixed in 2.5% glutaraldehyd buffered with 0.05 *M* sodium cacodylate, washed with buffer and were used for the estimation of root hair number. Other root segments were postfixed in buffered 2% osmium tetroxide, dehydrated in a graded series of ethanol and embedded in Spurr's resin. Semi-thin and ultrathin sections were prepared using an ultramicrotome (Reichert). The ultra-thin sections were stained with uranyl acetate and lead citrate, the semi-thin sections with 1% toluidine blue in 1% borax solution. For scanning electron microscopy (SEM), samples were fixed in glutaraldehyde as above, washed in cacodylate buffer, postfixed in 1% osmium tetroxide, blockstained in 0.5% uranylacetate, dehydrated through ethanol, immersed in alcohol, critical point dried, and coated with gold (Bender *et al.*, 1987).

Root hairs

The number of root hairs per 2 mm root length were estimated on 2 sides of the root marginally in focus (Fig. 1) in a light microscope field of defined size (magnification 10 × 10). Numbers registered are means of at least 80 measurements in the root hair zone with mature root hairs on at least 4 plants per treatment.

Results

Symptoms of Al toxicity

Obvious macroscopic Al toxicity symptoms which occurred after 3 d of exposure to 12 μM Al included reduced taproot elongation (data not presented), and the development of stubby lateral roots with swollen, yellow root tips. The pres-

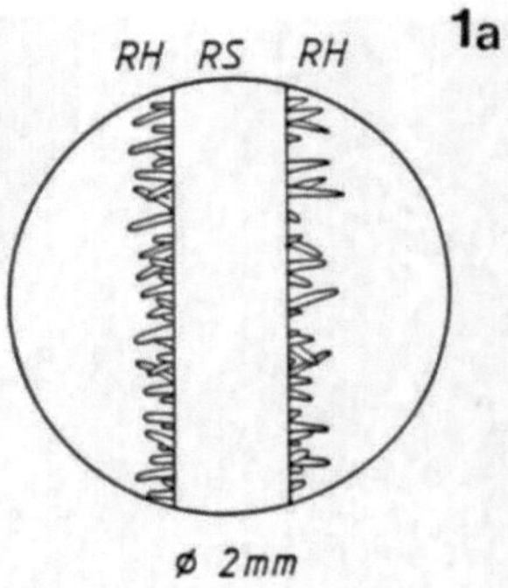

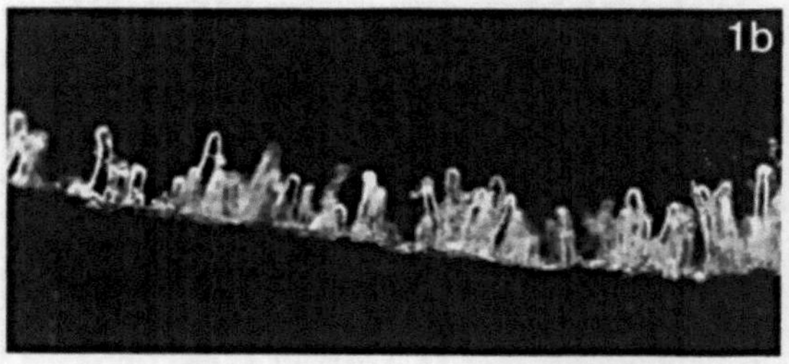

Fig. 1. **a.** Light microscopic field (10 × 10) with root segment (RS) of 2 m length and root hairs (RH) randomly in focus. **b.** Root hairs randomly in focus.

ence of lateral roots directly behind the taproot tip indicated a loss of apical dominance. Microscopic examination indicated that 12 μM Al deformed cortical and epidermal cells (Fig. 2), resulting in development of a disintegrated root-surface and impairment of root hair formation (Fig. 3). Other observed destructive effects of Al included premature vacuolisation of meristematic cells in the root tip (Fig. 4) and increased accumulation of electron dense inclusions, probably polyphenolic depositions were observed in both, meristematic cells at 5 μM Al (Fig. 4) and in mature cells of the cortex at 12 μM Al (Fig. 2).

Effects of Al on root hair development

In order to examine more closely the effect on root hair development at lower levels of Al (1.25, 2.5, 5.0 μM), an attempt was made to quantify root hair numbers and length of the root hair zone by light microscopic examination. Root hair zone length was estimated from the boundary between the hypocotyl and the root (5 cm from the cotyledons) in an apical direction

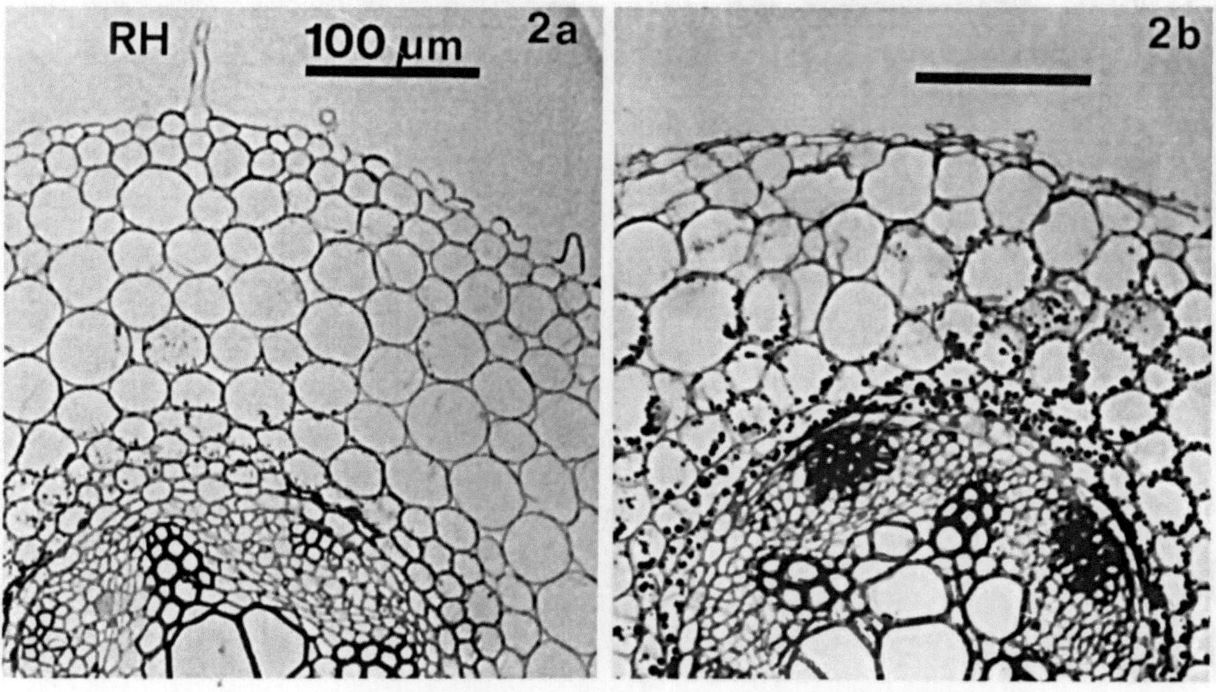

Fig. 2. Transverse sections of soybean roots 230x. **a.** 0 μM Al (RH = root hair). **b.** 12 μM Al. Epidermal and cortical cells are deformed and inner cortical cells contain dense inclusions probably polyphenolic depositions.

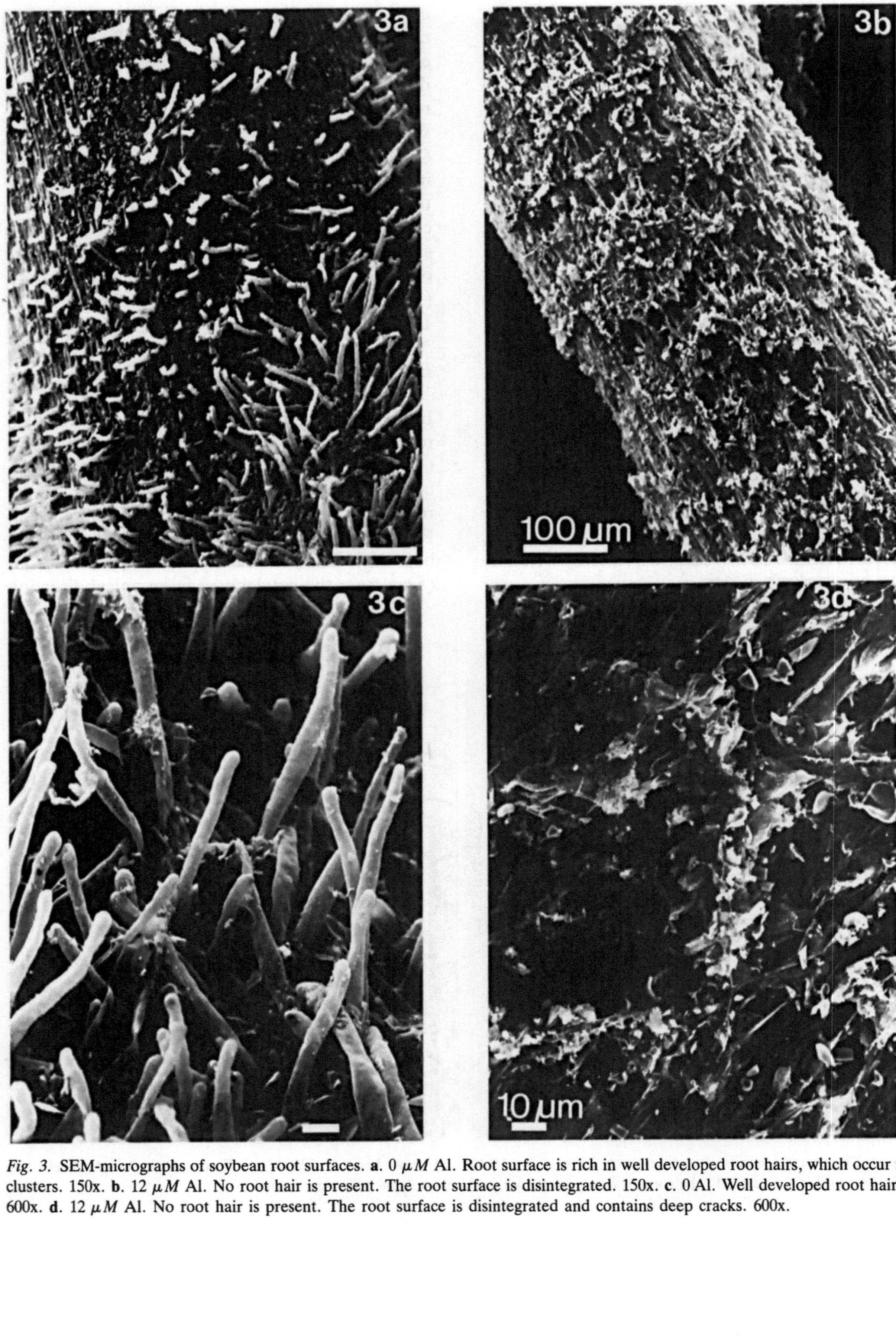

Fig. 3. SEM-micrographs of soybean root surfaces. **a**. 0 μM Al. Root surface is rich in well developed root hairs, which occur in clusters. 150x. **b**. 12 μM Al. No root hair is present. The root surface is disintegrated. 150x. **c**. 0 Al. Well developed root hairs. 600x. **d**. 12 μM Al. No root hair is present. The root surface is disintegrated and contains deep cracks. 600x.

Fig. 4. Meristematic cells of soybean root tips. 16 000x. **a.** 0 μM Al. **b.** 5 μM Al. In comparison with 0 Al there are larger vacuoles and an increased accumulation of polyphenolic inclusions.

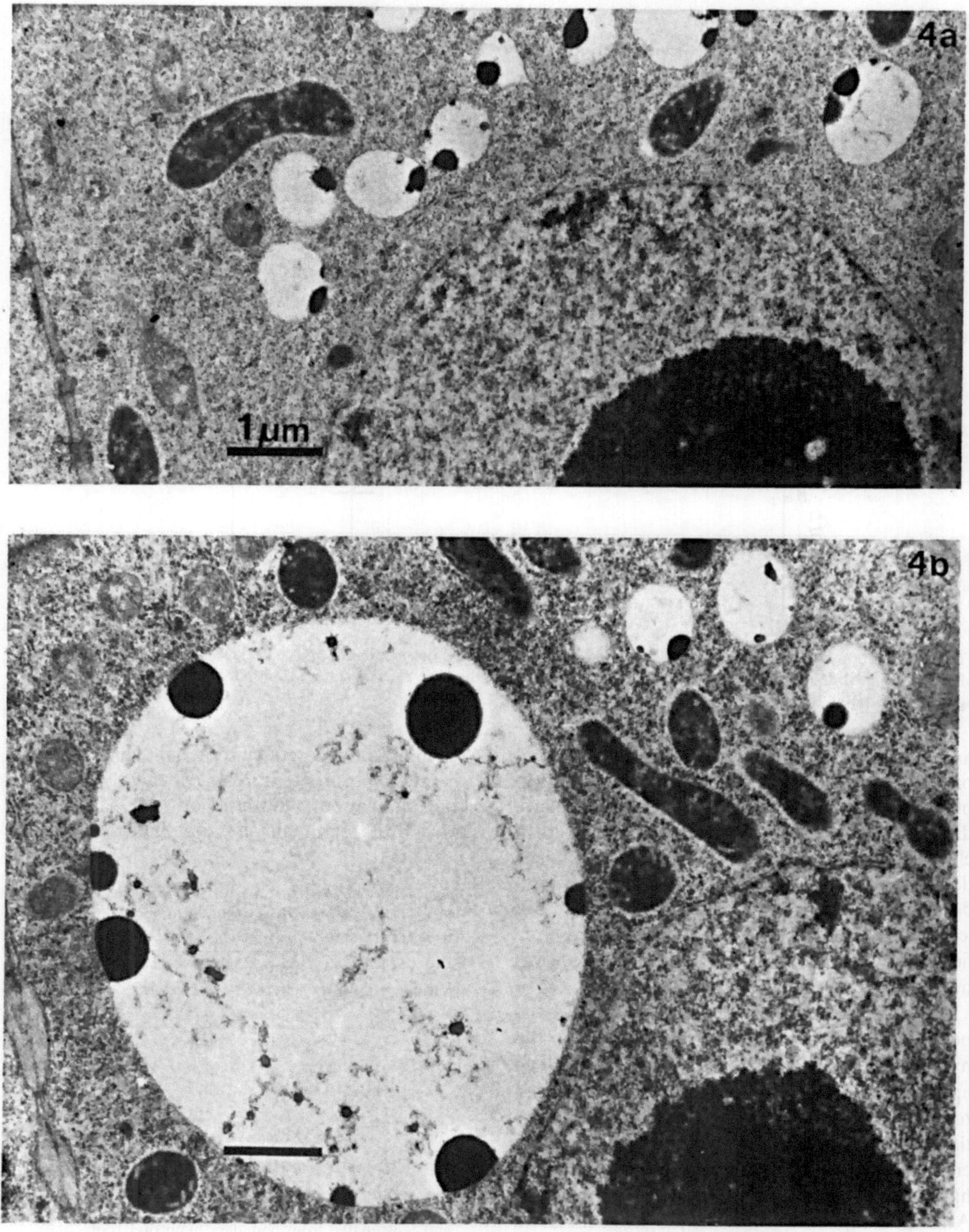

Fig. 4. Meristematic cells of soybean root tips. 16 000x. **a.** 0 μM Al. **b.** 5 μM Al. In comparison with 0 Al there are larger vacuoles and an increased accumulation of polyphenolic inclusions.

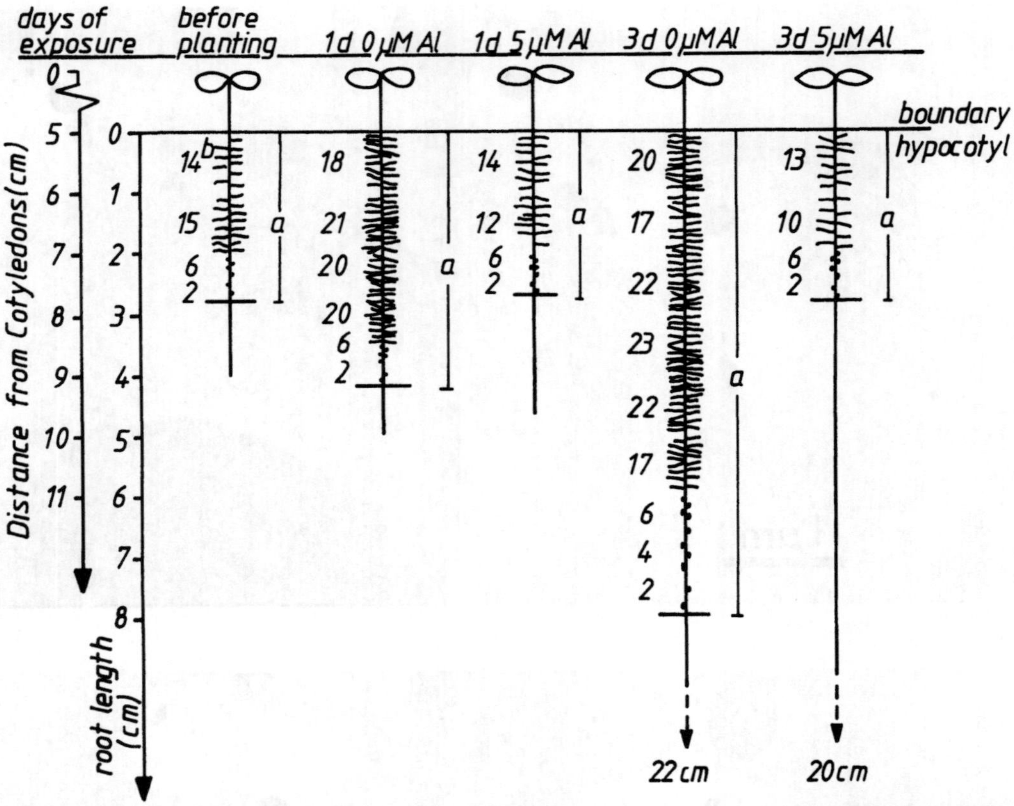

Fig. 5. Effect of exposure to 0 and 5 μM Al on (1 d and 3 d 0 μM on root length and root hair development in soybean. Length of the root hair zone (a), number of root hairs per 2 mm root length in the zone of mature root hairs randomly in focus under the light microscope at 10 × 10 (b).

and included the zone of emerging root hairs in proximity of the root tip (Fig. 5). It had been observed that the root hairs were not regularly distributed on the root surface but occurred in clusters (Fig. 3a). For comparison of root hair density between the treatments, means of a number of clustered root hairs per 2 mm of root length marginally focused in the light microscope field were chosen. The results of these observations are shown in Fig. 5 and Table 1.

The highest root hair density (up to 20 hairs per 1 mm root length) was observed near the hypocotyl. Root hair density decreased in an apical direction to 1 emerging root hair per 1 mm root length. After 24 h in Al-free nutrient solution, the root length of the seedlings had increased from 4.0 to 5.2 cm and the length of root hair zone had increased from 2.6 to 4.1 cm. After 24 h exposure to 1.25 or 2.5 μM Al (Table 1), root elongation appeared not to be inhibited compared with the control but there was a marked inhibitory effect on root hair zone development. Exposure to 5 μM Al slightly inhibited root elongation, but with both the 2.5 μM and 5 μM Al treatments the root hair zone length

Table 1. Root length, root hair zone length and root hair density of soybean seedlings exposed to aluminium for 1 or 3 days (Experiment 2)

Time of exposure (days)	Al (μM)	Taproot[a] length (cm)	Length of root hair zone[b] (cm)	Root hair density[c]
0	0	4.0	2.6	14
1	0	5.2	4.1	18
	1.25	5.5	3.8	17
	2.5	5.7	2.6	16
	5.0	4.7	2.7	14
3	0	22	7.9	20
	5.0	20	2.7	12

[a] and [b], estimated from boundary hypocotyl/root.
[c] means of root hair number per 2 mm root length randomly in focus in the light microscope at 10 × 10 (see Fig. 5).

Table 2. Root length and length of the root hair zone at time of inoculation 3 days after exposure to aluminium (Experiment 3)

Al (μM)	Taproot[a] length (cm)	Length of root hair zone[b] (cm)
0	20.1	7.3
5	19.0	4.7

[a] and [b] estimated from boundary hypocotyl/root.

had remained the same as it was at the time of planting into the Al treatments. After 3 d, root elongation had continued to about 20 cm in both the 0 μM and 5 μM Al treatments. However, the root hair zone was clearly affected by Al, and had remained the same as at the beginning of the experiment. After exposure to 5 μM Al root hair density had decreased, probably due to damage of root hairs.

In experiment 3 root hair development at the time of inoculation (after 3 d exposure to 5 μM A1) was examined using the above method. Again the length of the root hair zone and root hair density (data not presented) were inhibited by 5 μM Al whereas root elongation was not affected (Table 2). Root hairs present on plant roots exposed to Al in both experiments appeared to be shorter than those on control plant roots; they also seemed to be swollen. This is considered to be due to inhibited development rather than to direct physical damage.

Discussion

Al toxicity symptoms

Symptoms characteristic of Al toxicity such as premature vacuolisation of cells in the root tip, increased accumulation of phenolic compounds, swelling and damage of the root cells and disintegration of the epidermal and outer cortical cells have been observed in several plant species by numerous workers including, Fleming and Foy (1968) and Henning (1975) with wheat, Keser *et al.* (1977) with sugarbeet, Baier *et al.* (1976) and Bennett *et al.* (1985) with maize, Hecht-Buchholz and Foy (1981) and Hecht-Buchholz and Schuster (1987) with barley and Hecht-Buchholz *et al.* (1987) with Norway spruce. Recently Wagatsuma *et al.* (1987) have described microscopic Al toxicity symptoms in barley, oat, rice, maize and pea.

In the present experiments, characteristic Al toxicity symptoms occurred at rater low activities of monomeric Al. Even Al concentrations as low as 5 μM caused destructive effects of Al such as premature vacuolisation and increased accumulation of phenolic compounds in the root tip meristem. Whether those effects contribute to nodulation failure is not clear. However, it is likely that the inhibited root hair formation might play an important role in inhibition of early steps of nodulation.

Root hair formation

In the present experiments with soybean, root hair formation was limited at 2.5 μM Al and totally inhibited at 12 μM Al. At 2.5 μM Al, root hair formation was more sensitive than root elongation to Al. This result is in general agreement with those of Wood *et al.* (1984b) who reported for white clover that the critical Al concentration for inhibition of root elongation was 30 μM Al, while that for root hair formation was 10 μM Al.

Root hair formation and nodulation

The question whether nodulation failure is to be attributed to inhibition of root hair formation has been studied by several workers. In experiments with *Trifolium subterraneum* (Lowther and Loneragan, 1968) and *Trifolium repens* (Wood *et al.*, 1984a,b) reduced hair formation was found following low Ca, low pH *per se* and Al exposure. However, there was no clear relationship between inhibited root hair formation and reduced nodule number. As the fast-growing *R. trifolii* strains used for inoculation of Trifolium are known to be extremely sensitive to Al and low pH, Wood *et al.* (1984b) could not exclude the possibility that inhibition of nodulation was due to the limited survival and multiplication of rhizobia.

According to Munns (1968) the prevention of nodulation by acidity in Medicago sativa could be attributed to prevention of root hair curling;

and as Munns (1970) found out in further experiments with *M. sativa*, inhibition of nodulation by acidity or by low Ca were clearly not due to lack of root hairs.

A positive correlation between root hair length and Rhizobium infection has been observed by Franco and Munns (1982) in *Phaseolus vulgaris*. There was a drastic change in root morphology at low pH and after plants were transferred from vermiculite to solution culture.

In the present experiment, it was obvious that the formation of emerging root hairs was inhibited at Al concentrations as low as 2.5 μM Al, as indicated by a constant or diminished root hair zone length (Table 1). In addition, previous results (Brady *et al.*, 1990) indicated that exposure to 5 μM Al resulted in reduced nodule number parallel to cessation of root hair formation (Table 2). These results support the assumption that inhibition of emerging root hairs causes nodulation failure.

Regarding the question of whether inhibited survival and multiplication of rhizobia contributed to nodulation failure in the previous experiment, it should be mentioned that soybean-specific rhizobium strains such as *Bradyrhizobium japonicum* are slow growing and known to be less sensitive to low pH and Al than fast growing *R. trifolii* strains (Keyser and Munns (1979)). In the previous experiment (Brady *et al.*, 1990), nodulation was not prevented at low pH (pH 4.5). Keyser and Munns (1979) reported that growth of free living *Bradyrhizobium japonicum* CB 1809 was limited by a concentration of 25 μM Al. Whether the low activities of Al (2.5 and 5 μM Al) affect free living *B. japonicum* is unknown and needs further investigation.

Acknowledgement

This work was conducted as part of an ACIAR funded project 'The Management of Soil Acidity for Sustained Crop Production' and was supported by the Deutsche Forschungsgemeinschaft. We are grateful to Dr. Maret L. Vesk, University of Sydney, for the SEM micrographs.

References

Asher C J 1981 Limiting external concentrations of trace elements for plant growth: Use of flowing solution culture techniques. J. Plant Nutr. 3, 163–180.

Asher C J and Edwards D G 1978 Relevance of dilute solution culture studies to problems of low fertility tropical soils. *In* Mineral Nutrition of Legumes in Tropical and Subtropical Soils. Eds. C S Andrew and E J Kamprath. pp 131–152. CSIRO, Melbourne, Australia.

Baier R, Münnich H, Heinke F and Göring H 1976 Zytologische Untersuchungen zur Wirkung von Aluminiumionen auf Maiswurzein. Wissenschaftliche Zeitschrift der Humboldt-Universität zu Berlin. Math.-Nat. R. XXV 6, 840–844.

Bender G L, Nayudu M, Goydych W and Rolfe B G 1987 Early infection events in the nodulation of the non-legume *Parasponia andersonii* by Bradyrhizobium. Plant Science 51, 285–293.

Bennett R J, Breen C M, Foy M V 1985 The primary site of aluminum injury in the roots of *Zea mays* L. S. Afr. J. Plant Soil 2, 8–17.

Blamey F P C, Edwards D G and Asher C J 1983 Effects of aluminum, OH:Al and P:Al molar ratios, and ionic strength on soybean root elongation in solution culture. Soil Sci. 136, 197–207.

Brady D J, Hecht-Buchholz Ch, Asher C J and Edwards D G 1990 Effects of low activities of aluminium on soybean (*Glycine max*). I. Early growth and nodulation. *In* Plant Nutrition – Physiology and Applications. Ed. M L Van Beusichem. pp. 329–334. Kluwer Academic Publishers, Dordrecht, The Netherlands.

Carvalho M M de, Edwards D G, Asher C J and Andrew C S 1982 Effects of aluminium on nodulation of two *Stylosanthes* species grown in nutrient solution. Plant and Soil 64, 141–152.

Fleming A L and Foy C D 1968 Root structure reflects differential aluminium tolerance in wheat varieties. Agron. J. 60, 172–176.

Franco A A and Munns D N 1982 Nodulation and growth of *Phaseolus vulgaris* in solution culture. Plant and Soil 66, 149–160.

Hecht-Buchholz Ch and Foy C D 1981 Effect of aluminium toxicity on root morphology of barley. Plant and Soil 63, 93–95.

Hecht-Buchholz Ch 1983 Light and electron microscopic investigations of the reactions of various genotypes to nutritional disorders. Plant and Soil 72, 151–165.

Hecht-Buchholz Ch and Schuster J 1987 Responses of Al-tolerant Dayton and Al-sensitive Keney barley cultivars to calcium and/or magnesium during Al stress. Plant and Soil 99, 47–61.

Hecht-Buchholz Ch, Jorns C A and Keil P 1987 Effect of excess aluminium and manganese on Norway spruce seedlings as related to magnesium nutrition. J. Plant Nutr. 10, 1103–1110.

Henning S J 1975 Aluminum toxicity in the primary meristem of wheat roots. Ph.D. Thesis, Oregon State University, Corvallis.

Kerven G L, Edwards D G, Asher C J, Hallman P S and Kokot S 1989 Aluminium determination in soil solution. II. Short-term colorimetric procedures for the measurement of inorganic monomeric aluminium in the presence of organic acid ligands. Aust. J. Soil Res. 27, *In press*.

Keser M, Benedict F, Neubauer F, Hutchinson E and Verrill D B 1977 Differential aluminium tolerance of sugarbeet cultivars as evidenced by anatomical structure. Agron. J. 69, 347–350.

Keyser H H and Munns D N 1979 Effects of calcium, manganese and aluminium on growth of rhizobia in acid media. Soil Sci. Soc. Am. J. 43, 500–503.

Lowther W L and Loneragan J F 1968 Calcium and nodulation in subterranean clover (*Trifolium subterraneum* L.). Plant Physiol. 43, 1362–1366.

Munns D N 1968 Nodulation of *Medicago sativa* in solution culture. I. Acid-sensitive steps. Plant and Soil 28, 129–146.

Munns D N 1970 Nodulation of *Medicago sativa* in solution culture. V. Calcium and pH requirements during infection. Plant and Soil 32, 90–102.

Turgeon B G and Bauer W D 1982 Early events in the infection of soybeans by *Rhizobium japonicum*: Time course and cytology of the initial infection process. Can. J. Bot. 60, 152–161.

Wagatsuma T, Kaneko M and Hayasaka Y 1987 Destruction process of plant root cells by aluminium. Soil Sci. Plant Nutr. 33, 161–175.

Wood M, Cooper J E and Holding A J 1984a Soil acidity factors and nodulation of *Trifolium repens*. Plant and Soil 78, 367–379.

Wood M, Cooper J E and Holding A J 1984b Aluminium toxicity and nodulation of *Trifolium repens*. Plant and Soil 78, 381–391.

M. L. van Beusichem (Ed.), *Plant nutrition – physiology and applications*, 345–349.

PLSO IPNC425

Role of root cell walls in iron deficiency of soybean (*Glycine max*) and aluminium toxicity of wheat (*Triticum aestivum*)

D.L. ALLAN[1], J.R. SHANN[2] and P.M. BERTSCH[3]
[1]*Department of Soil Science, University of Minnesota, St. Paul, MN 55108, USA,* [2]*Department of Biological Sciences, University of Cincinnati, Cincinnati, OH 45221, USA, and* [3]*Department of Agronomy, University of Georgia, Athens, GA 30602, USA*

Key words: adsorption, aluminium-toxicity, cell walls, *Glycine max* L., iron-deficiency, proton titrations, soybean, *Triticum aestivum* L., wheat

Abstract

Cell walls of soybean and wheat roots were isolated by a non-destructive technique. Titration and adsorption experiments were performed to determine whether varietal differences in tolerance to Al toxicity or Fe stress were related to differences in surface chemical properties of the cell walls. Results of the soybean experiments indicate some differences in total binding capacity and Fe adsorption of the walls, but no differences in amounts of Ca adsorbed. The wheat cultivars do show a correlation of increased titratable acidity and Al adsorption with higher tolerance, but no differences in their relative preferences for Al or Ca.

Introduction

Apoplastic accumulation of ions has been implicated as a possible mechanism of tolerance to metal toxicity (Nishizono *et al.*, 1987; Taylor, 1987) or adaptation to nutrient deficiencies (Linehan, 1984; Longnecker and Welch, 1986). Ions which accumulate in the apoplast may be bound to the cell walls or may precipitate, in the case of Fe or Al, as hydroxide or phosphate salts. The objective of this research was to determine whether cultivars differing in tolerance to Fe deficiency or Al toxicity differed in their cell wall surface chemistry and consequent accumulation of these metals.

Evidence for the role of the cell wall in Al tolerance has been reviewed by Taylor (1988). Results have been contradictory, and it is not clear whether Al tolerant cultivars show greater or less accumulation of Al than sensitive cultivars. Several researchers have found positive correlations between root CEC and Al sensitivity, and suggest that the higher CEC causes more Al to bind to the wall, resulting in higher uptake or sensitivity (Foy *et al.*, 1967; Mugwira and Elgawhary, 1979; Vose and Randall, 1962). Higher CEC could also increase relative cation uptake, thus increasing rhizosphere acidity, and therefore Al uptake (Taylor, 1988). Other researchers (Memon, 1981; Kennedy, 1986) found the reverse relationship, and assumed that binding in the walls prevented Al from reaching critical metabolic sites in the cell. While Al has been shown to bind to the pectic substances in cell walls (Wagatsuma, 1983), other researchers have postulated surface hydrolysis of adsorbed hydroxy aluminium species, polymerization, and preciptitation as mechanisms of accumulation (Clarkson, 1967; Matsumoto *et al.*, 1977).

Plants also differ in their accumulation of iron in the root apoplast (Bienfait *et al.*, 1985; Longnecker, 1986; Römheld and Marschner, 1983) and in their ability to utilize this apoplastic iron (Bienfait *et al.*, 1985). Romheld and Marschner (1983) found a correlation between root cation exchange capacity and ^{59}Fe adsorption in different plant species, and there is microautoradiographic evidence for iron accumulation in cell walls (Clarkson and Sanderson, 1978). Longnecker (1986) demonstrated genotypic variation in amounts of Fe in the extracellular pool of soybean roots in early stages of plant development. Apoplastic accumulation of Fe correlated with resistance for the two varieties she tested, but the mechanism of accumulation was not investigated.

To investigate the hypothesis that differences in chemical behavior of the cell walls might affect metal accumulation and hence stress tolerance, cell walls were isolated from eight cultivars which differ in their tolerance to Fe deficiency induced chlorosis (soybean) or Al toxicity (wheat). Proton titrations and metal adsorption experiments were conducted on the isolated cell wall materials.

Methods

Cell wall isolation technique

Cell walls were isolated from 4 cultivars each of soybean (*Glycine max*) and wheat (*Triticum aestivum*). The soybean cultivars and their chlorosis ratings were: Maple Amber (5.0), Dawson (2.0), Corsoy (4.7), and A7 (1.7). Wheat cultivars in order of tolerance to Al toxicity were Yecorra Rojo, Titan, Caldwell and Wampum. Soybean varieties were grown in aeroponic culture in a growth chamber for 21-28 d after germinating on blotter paper for 3-4 d. Wheat was germinated and grown on a screen above solution-filled containers and roots were harvested after 2d. Harvested roots were frozen in liquid nitrogen, shattered to 1-2 mm fragments, and stored at −80°C until needed. Cell walls were isolated as described elsewhere (Allan and Jarrell, 1989). Briefly, the frozen material was defrosted in phosphate buffer, ruptured in a Parr cell disruption bomb, filtered and washed several times. Two modifications were made in the technique: after discharging the contents from the bomb, the filtered material was sonicated for 5 min with a 50% pulsed cycle; and after several water rinses, the cell walls were rinsed with 0.01 *N* HCl and then deionized water until Cl^- disappeared (usually 3-4 rinses).

Proton titrations (soybean)

Titrations were conducted on the soybean root cell walls and analyzed as described elsewhere (Allan and Jarrell, 1989). After washing and filtration, the filtered walls (approximately 1.3 g wet weight) were suspended in about 60 ml 0.02 *N* NaCl and titrated on a Metrohm titrator at 25°C under nitrogen gas. Initial pH of the cell wall suspensions ranged from 4.3 to 4.5. The suspensions and blank solution were brought to pH 9.8 with 0.02 *N* NaOH, and then back titrated with standardized 0.02 *N* HCl at an addition rate of 0.02 ml per minute. Formation functions were calculated by figuring concentrations of H and OH added to the titration vessel (OH_a and H_a) and remaining in solution (OH and H), then subtracting $((OH_a\text{-}OH)\text{-}(H_a\text{-}H))$ to get the net H neutralized for each point on the titration curve. The formation function for the blank titration was also calculated, then subtracted from each cell wall formation function and divided by the dry weight of cell walls to determine the mmols H neutralized per g cell wall during the titration experiment.

Fe and Ca adsorption experiments (soybean)

Rapid batch equilibrium experiments were used to investigate differences in Fe and Ca binding by the cell walls of the different soybean varieties. Walls were washed as above, and suspended in deionized water. The procedure of Allan and Jarrell (1989) was followed, with some minor modifications. For the Fe adsorption experiments, additions were made to 50 mL polyallomer centrifuge tubes as follows: 5 g cell wall suspension; 4 g 0.05 *N* $NaClO_4$; 2 g 500 ppm $Fe(ClO_4)_3$; $HClO_4$ or NaOH to adjust the pH from 1.5 to 3; and deionized water to bring the total weight to 20 g. To avoid precipitation of Fe, we used high concentrations, maintained the pH below 3, and made base additions as dilute as possible. For the Ca adsorption experiments, the sequence of additions was: 5 g cell wall suspension; 4 g 0.05 *N* $NaClO_4$; 2 g 200 ppm $Ca(ClO_4)_2$; $HClO_4$ or NaOH to adjust the pH from 2 to 8; and deionized water to bring the total weight to 20 g. Samples were centrifuged, filtered under vacuum and analyzed immediately for final pH and within 24 h for Fe or Ca on the atomic absorption spectrophotometer. Amounts sorbed by the cell wall were determined by difference between amount added and amount remaining in solution.

Al and Ca adsorption experiments (wheat)

An Al/Ca exchange experiment was conducted to investigate differences between Al tolerant (Yecorra Rojo) and susceptible (Caldwell) wheat varieties in their preferences for Ca and Al. Isolated walls were saturated wth unlabeled Ca (0.02 *M* $CaCl_2$) and placed in centrifuge tubes (100 mg dry weight per tube) with 30 ml unlabeled Ca/Al treatment solutions. Tubes were shaken for 30 min, centrifuged, and the solutions replaced. This process was repeated four times; then the solutions were replaced with ^{45}Ca/Al solutions. After 12 hours, tubes were centrifuged, supernatant decanted, and the walls filtered and weighed. Al and Ca were removed from walls with 1 *N* HCl. Al in solution and removed from the walls was measured by AA and ion chromatography, ^{45}Ca by scintillation counting.

Results and discussion

Soybean and Fe chlorosis

Formation functions calculated from the proton titration data (Fig. 1) demonstrate no significant differences in the shapes of the titration curves for the four soybean cultivars, although the magnitude varied. Total acidity values from pH 3 to 9.5 for the two Fe-deficiency chlorosis resistant cultivars, Dawson and A7, were 0.40 and 0.48 mol/kg, respectively. For the chlorosis susceptible cultivars, values were 0.51 (Maple Amber) and 0.34 (Corsoy). From the shapes of the titration curves it is evident that inflection points occur at about pH 6.5 and again at about pH 9. There was no difference between varieties in the proportion of their titratable groups falling into each of these regions of the curves.

These formation functions were normalized by selecting the endpoint for each region of the curve (using graphical methods), and setting it

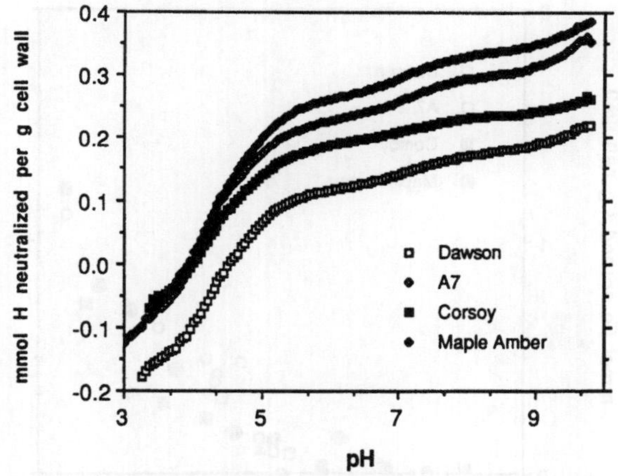

Fig. 1. Formation function describing mmols of H neutralized per g cell wall during titrations of soybean root cell walls. Samples suspended in 20 m*M* NaCl were brought to pH 9.8 and back titrated to pH 3.

equal to the point of maximum dissociation. In Figure 2, the carboxyl acidity region of the formation function curves has been normalized. Generally the curves are similar, and do not reflect any significant difference between the susceptible and resistant cultivars.

The results of the metal binding experiments are presented in Figures 3, 4 and 5. Only small differences in adsorption of Fe and no differences in adsorption of Ca were apparent in the constant [M^{2+}], variable pH experiments (Fig. 3 and 5). However, significantly more (30-60%) Fe was adsorbed by the resistant variety where

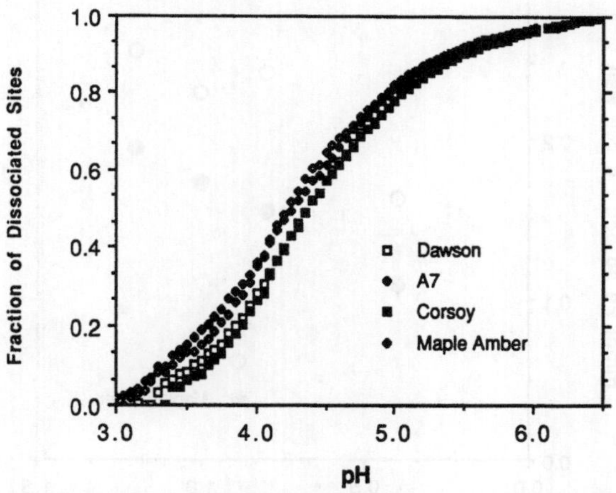

Fig. 2. Fraction of dissociated sites as a function of pH. Calculated from formation functions plotted in Figure 1 with inflection point at approximately pH 6.5 considered equivalent to 100% dissociation of carboxyl sites.

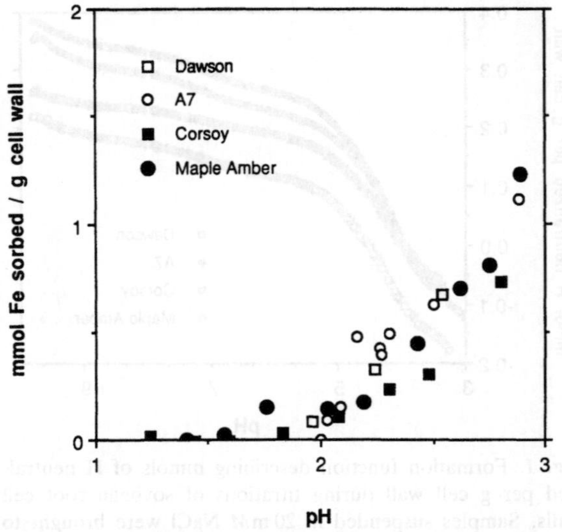

Fig. 3. pH dependence of the adsorption of Fe ions to soybean root cell walls. $Fe_T = 0.8$ m*M*, cell wall concentration = 0.4 g/L and initial pH of cell wall suspension = 2.0.

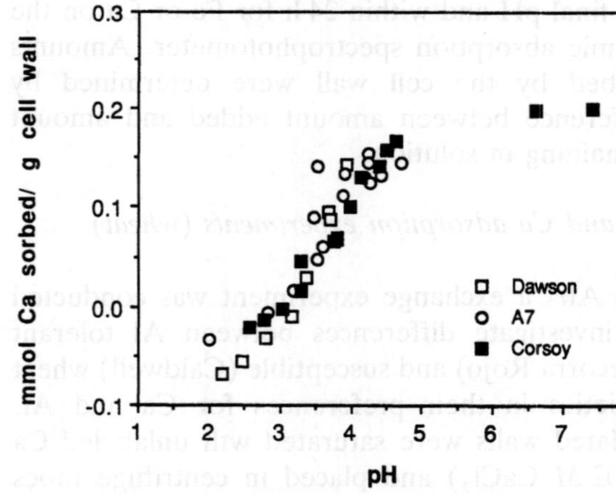

Fig. 5. pH dependence of the adsorption of Ca ions to soybean root cell walls. Suspension concentrations of cell walls and total Ca were: Dawson, 0.5 g cell wall/L and 0.47 m*M*; A7 and Corsoy, 0.8 g cell wall/L and 0.26 m*M*. Initial pH of the cell wall suspension was 4.6 for Dawson and 3.8 for A7 and Corsoy.

pH was held constant at 2.5 and Fe concentration ranged from 0.4-1.2 m*M* (Fig. 4).

Wheat and Al toxicity

Results of base titrations of the four wheat cultivars are presented in Figure 6. The curves demonstrate that total acidity increases with Al tolerance, such that the most tolerant variety (Yecorra Rojo) has nearly twice the titratable groups of the least tolerant variety (Wampum). These results contradict the findings of other researchers (Foy *et al.*, 1967; Mugwira and Elgawhary, 1979; Vose and Randall, 1962), who have correlated increased root CEC with increased sensitivity to Al. The increased capacity

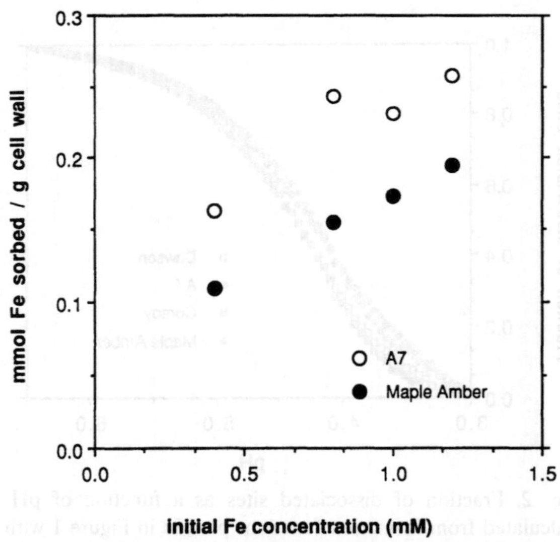

Fig. 4. Concentration dependence of Fe adsorption to soybean root cell walls. pH = 2.5, I = 0.01 *M*, cell wall concentration = 0.48 g/L (A7) and 0.44 g/L (Maple Amber).

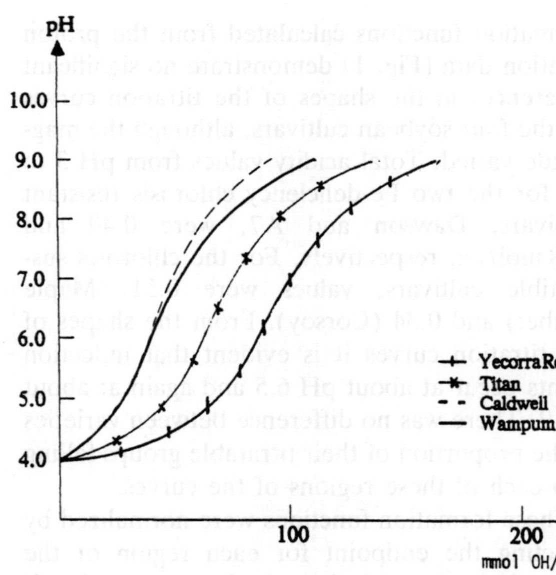

Fig. 6. Forward titration of wheat root cell walls in 20 m*M* NaCl.

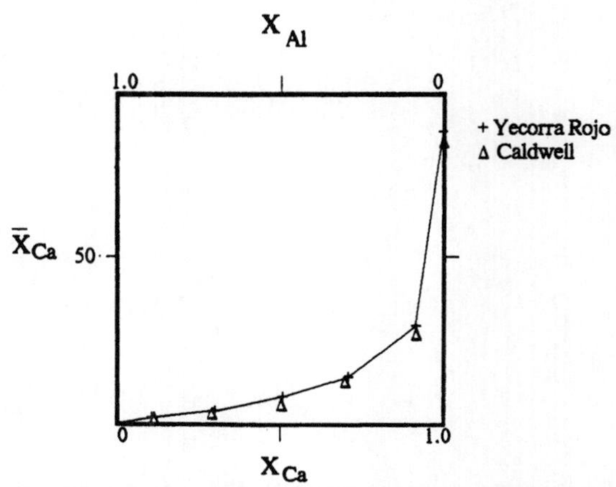

Fig. 7. Exchange isotherm for Ca-Al exchange on root cell walls of Al tolerant (Yecorra Rojo) and susceptible (Caldwell) wheat cultivars. Ca measured on walls and in solution was labeled. Equivalent fractions are expressed as molar ratios. Total amounts of Al adsorbed when $X_{Al} = 1.0$ were 157 mmol/kg for Yecorra Rojo and 111 mmol/kg for Caldwell.

did correlate with a 50% higher rate of Al adsorption by Yecorra Rojo root cell walls compared to Caldwell in the experiment depicted in Figure 7. Although there were differences in total amounts of Al bound, the figure indicates that both varieties show the same high preference for Al over Ca. Further investigations are required to determine whether the enhanced binding of Al by the tolerant variety is localized to certain regions of the root.

In summary, these results indicate that there may be a correlation between titratable acidity of wheat roots and tolerance to Al toxicity. This higher charge density may explain the higher binding of Al by these roots, as has been demonstrated previously by Mugwira and Elgawhary (1979) and Wagatsuma (1983). However, there is no difference between tolerant and susceptible varieties in their relative preference for Al over Ca. While differences in titratable acidity also were evident for the soybean cultivars, they did not correspond with resistance or susceptibility to Fe deficiency induced chlorosis. The results of Fe binding-experiments suggest that cultivar differences do occur, but they do not appear to correlate with differences in total acidity. Further experimental work is planned to confirm total and Fe exchange capacity and apoplastic Fe accumulation as a response to deficiency among these soybean cultivars.

References

Allan D L and Jarrell W M 1989 Proton and copper adsorption to maize and soybean root cell walls. Plant Physiol. 89, 823-832.

Bienfait H F, Van den Briel W, and Mesland-Mul N T 1985 Free space iron pools in roots: Generation and mobilization. Plant Physiol. 78, 596-600.

Clarkson D T and Sanderson J 1978 Sites of absorption and translocation of iron in barley roots: Tracer and microautoradiographic studies. Plant Physiol. 61, 731-736.

Clarkson D T 1967 Interactions between aluminum and phosphorus on root surfaces and cell wall material. Plant and Soil 27, 347-356.

Foy C D, Fleming A L, Burns G R, and Armiger W H 1967 Characterization of differential aluminum tolerance among varieties of wheat and barley. Soil Sci. Soc. Am. Proc. 31, 513-521.

Kennedy C W, Smith, Jr. W C, Ba M T 1986 Root cation exchange capacity of cotton cultivars in relation to aluminum toxicity. J. Plant Nutr. 9, 1123-1133.

Linehan D J 1984 Micronutrient cation sorption by roots and uptake by plants. J Exp. Bot. 35, 1571-1574.

Longnecker N and Welch R 1986 The relationships among iron-stress response, iron efficiency and iron-uptake of plants. J. Plant Nutr. 9, 715-727.

Longnecker N E 1986 A comparison of the resistance of soybean and sunflower to iron-deficiency induced chlorosis. Ph.D. thesis. Cornell University. Ithaca, NY, 56 p.

Matsumoto M, Morimura A, and Takahashi E 1977 Less involvement of pectin in the precipitation of Al in pea root. Plant Cell Physiol. 18, 325-335.

Memon A R, Ito S, and Yatazawa Y 1981 Microdistribution of aluminum and manganese in the tea leaf tissues as revealed by X-ray analyses. Commun. Soil Sci. Plant Anal. 12, 441-452.

Mugwira L M and Elgawhary S M 1979 Aluminum accumulation and tolerance of triticale and wheat in relation to root cation exchange capacity. Soil Sci. Soc. Am. J. 43, 736-740.

Nishizono H, Ichikawa H, Suziki S and Ishii F 1987 The role of the root cell wall in the heavy metal tolerance of *Athyrium yokoscense*. Plant and Soil 101, 15-20.

Ritchey K D, Baligar V C and Wright R J 1988 Wheat seedling responses to soil acidity and implications for subsoil rooting. Commun. Soil Sci. Plant Anal 19, 1285-1293.

Römheld V and Marschner H 1983 Mechanism of iron uptake by peanut plants. I. Fe-III reduction, chelate splitting, and release of phenolics. Plant Physiol. 71, 949-954.

Taylor G J 1988 The physiology of aluminum tolerance in higher plants. Commun. Soil Sci. Plant Anal. 19, 1179-1194.

Taylor G J 1987 Exclusion of metals from the symplasm: A possible mechanism of metal tolerance in higher plants. J. Plant Nutr. 10, 1213-1222.

Vose P B and Randall P J 1962 Resistance to aluminum and manganese toxicities in plants related to variety and cation-exchange capacity. Nature 196, 85-86.

Wagatsuma T 1983 Characterization of absorption sites for aluminum in the roots. Plant and Soil 47, 257-262.

M. L. van Beusichem (Ed.), *Plant nutrition – physiology and applications*, 351–355.

PLSO IPNC150

Mechanical impedance increases aluminium tolerance of soybean (*Glycine max*) roots

W.J. HORST, F. KLOTZ and P. SZULKIEWICZ
Institute for Plant Nutrition, University of Hannover, Herrenhäuser Str. 2, D-3000 Hannover 21, FRG

Key words: aluminium tolerance, *Glycine max* L., mechanical impedance, root growth, soybean

Abstract

The effect of aluminium (Al) on root elongation was studied in solution culture and sand culture. Compared to solution culture, in sand culture a ten times higher Al supply was necessary to inhibit root elongation to a comparable degree. This was due to a much lower Al uptake into the 5 mm root tips in sand culture. Fe concentrations in root tips were also lower in sand culture. Ca concentrations were higher and less depressed by Al, whereas Mg and K concentrations were not affected by the culture substrate. Regressions of Al concentrations in root tips versus inhibition of root elongation by Al revealed root damage at lower Al concentrations in sand culture. The effect of culture substrate on Al tolerance was independent of N source and could also be shown in flowing solution culture with and without sand. The results indicate that mechanical impedance in sand culture decreased Al uptake. This may be due to enhanced exudation of organic complexors thus reducing activites of monomeric Al species.

Introduction

Soybean may be classified as sensitive to 'soil acidity' compared to other plant species (Munns and Fox, 1977). In acid mineral soils, Al toxicity is among the most important soils factors limiting soybean growth. It is therfore expected that selection for genotypic Al tolerance will improve the adaption of soybean to acid soils (Devine *et al.*, 1979; Sapra *et al.*, 1982).

However, it has been reported that genotypic Al tolerance in solution culture was not related to 'soil acidity' tolerance under field conditions (Horst and Klotz, 1990). Among the reasons for the lack of correlation, of special importance could be the effect of the culture substrate on Al tolerance of the roots growing in a solid compared to a liquid substrate. The objective of the present study was to clarify the role of mechanical impedance on the inhibition of root growth by Al, by comparing growth in sand and solution cultures.

Material and methods

Soybean (*Glycine max* L.) seeds were kindly provided by the Institute of Crop Science, University of Hohenheim, D-7000 Stuttgart 70, and Kleinwanzlebener Saatzucht AG, D-3352 Einbeck. Due to shortage of seeds different cultivars as indicated were used which was thought to be justified owing to the much larger effect of the culture substrate on Al tolerance compared to the cultivar differences described elsewhere (Klotz and Horst, 1988a).

Growing conditions of plants

Solution and sand culture of the plants are described elsewhere (Klotz and Horst, 1988a; Klotz and Horst, 1988b). The set up of the solution and sand flowing-culture systems is shown in Figure 1. The following nutrient solutions with different N sources have been used (μM): NO_3^- solution: KNO_3 750; $Mg(NO_3)_2$ 325. NH_4NO_3

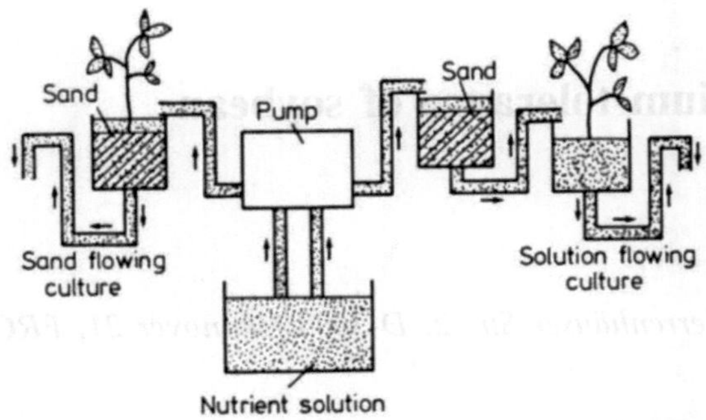

Fig. 1. Set up of sand and solution flowing-culture systems.

solution: KCl 750; $MgCl_2$ 325 NH_4NO_3 700. NH_4^+ soulution: KCl 750; $MgCl_2$ 325; NH_4Cl 1400, and $CaSO_4$ 250; FeEDDHA 20; KH_2PO_4 10; H_3BO_3 8; $CuSO_4$ 0.2; $ZnSO_4$ 0.2; $MnSO_4$ 0.2; $(NH_4)_6$ Mo_7 $(O_2)_4$ 0.2; Al was added as $AlCl_3$. The pH of the nutrient solution was adjusted to 4.2 in solution and flowing culture and to 4.0 in sand culture.

Measurement of total root length

At harvest, roots were spread out on a transparency film and photocopied against a black background. Root length was then measured with a graphic analyser (Digiplan AMO2).

Mineral element analysis

Total Al in solution was determined by atomic absorption spectroscopy, monomeric Al by the aluminon method of Kerven *et al.*, 1989 using membrane filtered solutions. Harvested roots were rinsed for 10 min in deionized water, pH 4.2, (room temperature). Secondary root tips of 5 mm length (approximately 30 root tips per sample) were then excised and dried on a filter-paper (celluloseacetate) at 80°C until constant weight, and wet ashed in boiling, concentrated HNO_3. Roots and shoots of the plants were dried (80°C), weighed and dry ashed at 450°C overnight. Al, Cu, Mg, K and Fe were measured in 1 : 30 (v/v) HNO_3 by atomic absorption spectroscopy.

Results

Plants grown in solution culture were very sensitive to Al supply as indicated by the reduction of root fresh weights (Fig. 2) which were highly significantly correlated to total root length. Plants grown in sand culture with the same nutrient solution replaced 2 times per day were much less affected by Al even at more than 10 times higher Al supply. Higher Al tolerance in sand culture was not due to precipitation or adsorption of Al, since concentrations of total Al and monomeric Al as well in the leachates were very similar to those of the original solutions (not shown). However, the inhibition of root elongation in solution culture at much lower Al supply (Table 1A) was related to higher Al uptake into the 5 mm root tips, the sites of cell division and elongation (Table 1B).

Much lower concentration in root tips in sand culture compared to solution culture were also found for Fe (Table 2). Ca, Mg and K concen-

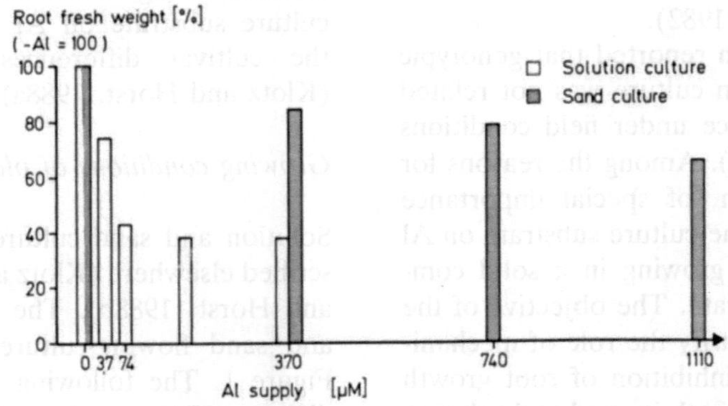

Fig. 2. Effect of Al supply on relative root fresh weights of soybean (cv. Sito) in solution and sand culture. Duration of Al treatment 4 days.

Table 1. Effect of Al supply on total root length (A) and Al concentrations (B) in different plant part of soybean (mean of cultivars Ronda and Maple Arrow) in solution and sand culture. Duration of Al treatment 4 days. Numbers followed by different letters indicate significant ($P_{0.05}$) differences between substrates and Al supplies.

A	Root length		
	−Al (cm)	+Al (cm)	+Al (% of −Al)
Solution culture (74 μM Al)	189a	39d	21
Sand culture (741 μM Al)	114b	50c	44
B	**Al concentration (mg g^{-1} dry wt)**		
	Shoot	Root	Root tips
Solution culture (74 μM Al)	0.03	1.95	3.86
Sand culture (741 μM Al)	0.02	2.63	0.85

Table 2. Effect of Al supply on mineral-element concentrations in root tips (5 mm) of soybean plants (Means of cultivars Ronda and Maple Arrow) in solution and sand culture. Duration of Al treatment 4 days. Numbers followed by different letters indicate significant ($P_{0.05}$) differences between substrates and Al supplies

Mineral element	Al supply	Concentration (mg g^{-1} dry wt.)	
		Solution culture (74 μM Al)	Sand culture (741 μM Al)
Ca	−Al	0.69 a	1.56 b
	+Al	0.36 a	1.22 b
Mg	−Al	1.37 a	1.39 a
	+Al	0.47 c	1.02 b
K	−Al	43.5 c	35.1 b
	+Al	28.2 a	27.4 a
Fe	−Al	0.91 b	0.39 c
	+Al	2.81 a	0.33 c

trations (Table 2) were depressed by Al much more in solution culture than in sand culture. The levels of K and Mg concentrations were comparable for both substrates. However, Ca concentrations were much higher in sand culture.

Using results of experiments with different cultivars and sources of N, regressions of Al concentrations in root tips versus inhibition of root elongation by Al were calculated. Figure 3A clearly indicate separate regressions for solution and sand culture. Independent of genotype and N source, Al concentrations were much lower in sand compared to solution culture. But root elongation was inhibited to a comparable degree in sand culture at much lower Al concentrations than in solution culture. Regressions of Fe concentrations versus Al-induced inhibition of root elongation (Fig. 3B) were also different for solution and sand cultures and resembled those shown for Al. The relationships between Ca concentrations and relative root elongation in solution and sand cultures could be described by the same highly significant regression (Fig. 3C).

In order to have better control of the Al supply to the root tips in sand culture and to exclude interaction between the sand and the nutrient solution as a factor modifying Al toxicity, sand and solution flowing-culture systems were set up (compare Fig. 1). Again root elongation of plants growing in sand flowing-culture was less inhibited by Al than in solution culture (Table 3).

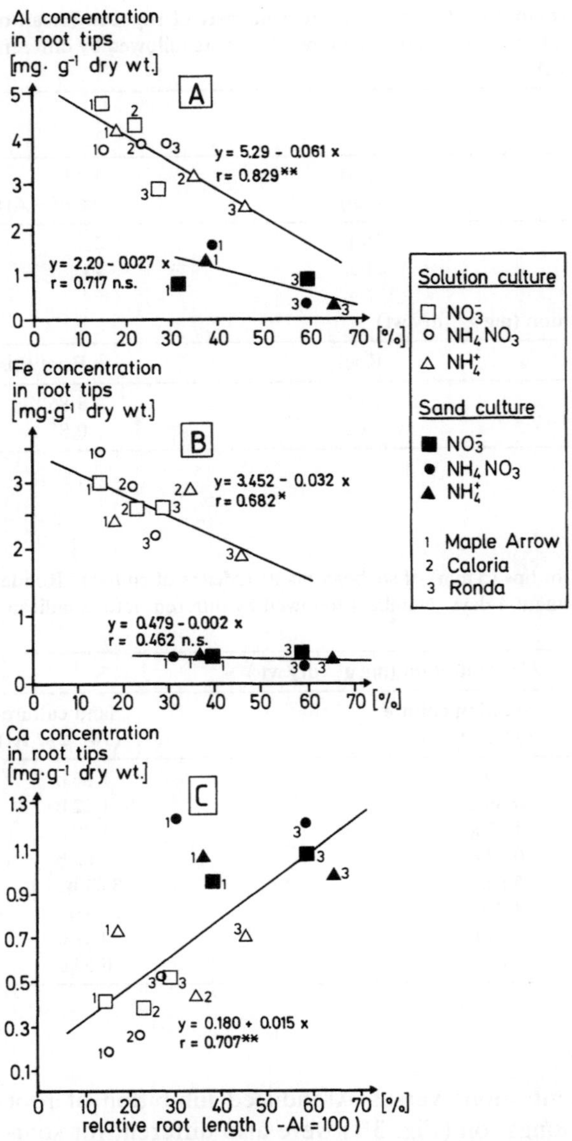

Fig. 3. Relationships between (A) Al, (B) Fe, (C) Ca concentrations in 5 mm root tips and relative root length with Al supply ($-Al = 100\%$) in solution and sand culture. N supply as NO_3^-, NH_4NO_3 or NH_4^+. Duration of Al treatment 4 days (74 μM Al in solution and 741 μM Al in sand culture).

Discussion

Higher Al tolerance of soybean plants grown in sand compared to solution culture (Fig. 2) was due to lower Al uptake into the root tips, the sites of cell division and cell elongation (Fig. 3). Reasons for this might be (1) slower Al movement to the roots by diffusion in sand compared to convection in aerated solution culture, (2) increase of pH at the root surface leading to precipitation of Al (Häußling *et al.*, 1985; Taylor and Foy, 1985), (3) enhanced release of root exudates (Barber and Gunn, 1974) which bind/inactivate Al (Horst *et al.*, 1982; Suhayda and Hung, 1986).

Lower Al uptake into the root tips (Table 1) is unlikely to be the consequence of lower Al supply to these sites in sand culture, because (1) the concentrations of Al in solution (total and monomeric Al) necessary to inhibit root growth comparably to solution culture were more than ten times higher, (2) concentrations of Ca, which is expected to behave similary to Al in sand culture, was even higher in the root tips of plants grown in sand culture (Table 2).

There is also good evidence that increase of pH in the rhizosphere owing to an acid uptake pattern is of minor importance because (1) higher Al tolerance and lower Al uptake into the root tips was independent on N source (NH_4^+ versus NO_3^-) supplied, (2) higher Al tolerance could also be found in sand compared to solution flowing-culture where built up of such pH gradients is less likely to occur.

It is therefore strongly suggested that an enhanced release of root exudates by root tips detoxifying Al was the main reason for higher Al tolerance in sand culture. Greater rates of exudation by roots growing at higher mechanical impedance (Barber and Gunn, 1974) might be

Table 3. Effect of Al supply on total root length of soybean (cv. Gieso) grown in solution and sand flowing-culture. Duration of Al treatment 4 days. Numbers following by different letters indicate significant ($P_{0.05}$) differences between substrates and Al supplies

Flowing-culture system	Root length (cm)		
	−Al	+Al (74 μM)	+Al (% of −Al)
Solution	213 a	95 d	45
Sand	193 b	133 c	69

explained by (1) higher allocation of assimilates to the roots as indicated by higher root/shoot ratio (Petterson and Barber, 1981 (2) thicker roots (Lindberg and Petterson, 1985) and thus more favorable conditions for root exudation, and (3) enhanced release of cell contents owing to damage of roots cells by the coarse sand. The comparable behaviour of Fe and Al (Fig. 3) suggest that both are complexed in a similar manner by root tip exudates.

Whereas uptake of Al into the root tips is clearly inhibited in sand culture, the toxicity of Al taken up into the root was considerably increased (Fig. 3). This is probably mainly due to lower binding of Al in insensitive sites of the roots especially in the apoplast.

The results show, that mechanical impedance as a factor modifying Al tolerance of soybean plants has to be taken into consideration when results with different culture substrates are compared.

Acknowledgements

We thank Dr D L Godbold, University of Göttingen, for critical reading of the manuscript and the Commission of the European Communities for financial support.

References

Barber D A and Gunn K B 1974 The effect of mechanical forces on the exudation of organic substances by the roots of cereal plants grown under sterile conditions. New Phytol. 73, 39–45.

Devine T E, Foy C D, Mason L and Fleming A L 1979 Aluminium tolerance in soybean germplasm. Soybean Genetic Newsletter 6.

Häußling M, Leisen E, Marschner H and Römheld V 1985 An improved method for non-destructive measurements of the pH at the root-soil interface (rhizosphere). J. Plant Physiol. 117, 371–375.

Horst W J, Wagner A and Marschner H 1982 Mucilage protects root meristems from aluminium injury. Z. Pflanzenphysiol. 105, 435–444.

Horst W J and Klotz F 1990 Screening soybean for aluminium tolerance and adaption to acid soils. *In* Genetic Aspects of Plant Mineral Nutrition. Eds. N El Bassan, M Dambroth and B C Loughman. pp 355–360. Kluwer Academic Publishers, Dordrecht, The Netherlands.

Kerven G L, Edwards D G, Asher C J, Hallman P S and Kokot S 1989 Aluminium determination in soil solution. II. Short-term colorimetric procedures for the measurement of inorganic monomeric aluminium in the presence of organic acid ligands. Aust. J. Soil Res. (*In press*)

Klotz F and Horst W J 1988a Genotypic differences in aluminium tolerance of soybean (*Glycine max* L.) as affected by aluminium and nitrate-nitrogen nutrition. J. Plant Physiol. 132, 702–707.

Klotz F and Horst W J 1988b Effect of ammonium- and nitrate-nitrogen nutrition on aluminium tolerance of soybean (*Glycine max* L.) Plant and Soil 111, 59–65.

Lindberg S and Petterson S 1985 Effects of mechanical stress on uptake and distribution of nutrients in barley. Plant and Soil 83, 295–309.

Munns D N and Fox R L 1977 Comparative lime requirements of tropical and temperate legumes. Plant and Soil 46, 533–548.

Petterson W R and Barber S A 1981 Soybean root morphology and K uptake. Agron. J. 73, 316–319.

Sapra V T, Mebrathu T and Mugwira L M 1982 Soybean germplasm and cultivar aluminium tolerance in nutrient solution and bladen clay loam soil. Agron. J. 74, 687–690.

Suhayda C G and Hung A 1986 Organic acids reduce aluminium toxicity in maize root membranes. Physiol. Plant. 68, 189–195.

Taylor G J and Foy C D 1985 Mechanisms of aluminium tolerance in Triticum aestivum (wheat). IV. The role of ammonium and nitrate nutrition. Can. J. Bot. 63, 2181–2186.

M. L. van Beusichem (Ed.), *Plant nutrition – physiology and applications*, 357–363.

PLSO IPNC482

Aluminium toxicity tolerance in rice (*Oryza sativa*) seedlings

V.P. CORONEL, S. AKITA and S. YOSHIDA
The International Rice Research Institute, P.O. Box 933, Manila, Philippines

Key words: aluminium tolerance, critical aluminium, hematoxylin, *Oryza sativa* L., root length, seedling age

Abstract

A phytotron study evaluated the response of selected rice cultivars in graded Al levels in nutrient solution. Root length reduction, the primary effect of Al toxicity, was interactive with nutrient concentration, cultivar, and seedling age. The critical solution concentration for Al toxicity was around 2 mg L^{-1} for susceptible rice cultivars, Cica4 and IR45. Hematoxylin staining of Al-affected roots showed Al to be localized in the epidermal and cortical cells. A rapid and nondestructive screening technique using 30 mg L^{-1} Al in full-strength nutrient solution was developed to identify Al-tolerant cultivars.

Introduction

Low productivity in acid upland and acid sulfate soils has been a major concern of rice scientists throughout the world. In these soils, rice growth and yield are often limited by the presence of toxic levels of Al.

Marked physiological changes occur when rice is exposed to toxic Al levels. Root growth and shoot growth are stunted, a result of the interaction among several plant and soil factors. The significant factors include cultivar characteristics, plant age, and nutrient concentration. The evaluation of these interactive factors would facilitate a better understanding of the nature of Al toxicity and provide a suitable parameter in assessing Al toxicity tolerance in rice seedlings.

Reports are few and conflicting regarding the critical solution concentration of Al. Critical level varied from 1 to 100 mg L^{-1} added Al and was apparently affected by the total salt concentration (Table 1). The association between Al concentration and total salt concentration has to be examined to establish the critical Al level under typical soil solution condition.

Several studies have traced the path of Al from the soil solution to the free spaces and meristematic cells of the root and its deposition on the cell wall and plasmalemma of roots (Martinez, 1976; Mohiddin, 1982). A simple histological test using hematoxylin would provide a direct evidence on the localization of Al in the affected root cells.

The nutrient solution technique has been used extensively in evaluating rice tolerance for Al. Three nutrient solution techniques have been examined – the absolute root length, root regrowth, and hematoxylin staining (Howeler and Cadavid, 1976; Martinez, 1976; Polle *et al.*, 1978). In the absolute root length technique, the indices for Al tolerance are the maximum and relative root lengths. In the root regrowth technique, the criterion is root recovery after Al stress. The hematoxylin technique involves visual detection of the stained root for Al tolerance. The method has not been tested on rice seedlings. A comparative evaluation of these techniques is necessary in the development of a suitable technique for screening for Al tolerance in rice seedlings.

This study sought to 1) examine the interaction among cultivar, nutrient concentration, seedling age, and Al concentration, and to identify a suitable parameter for assessing Al tolerance; 2) establish the critical solution concentration for Al toxicity; 3) delineate the localiza-

Table 1. Reported critical levels in nutrient solution for rice cultivars

Added Al conc. (mg L^{-1})	Salt conc. (m*M*)	Remarks	Author
1	Zero	Al added from $Al_2(SO_4)_3$, 14-day-old seedlings, 7-day treatment, British Guiana variety 6047	Cate and Sukhai (1964)
100	9.2	Al from $Al_2(SO_4)_3$, pH 4.0, 20% root reduction 25-day-old Peta, 14-day treatment	Tanaka and Navasero (1966)
0.05	1.4	At pH 3.5, 4.3 and 5.0, 4 mg P L^{-1}, Pwang ngeon variety, 18-day treatment, 33% root and 42% shoot reduction	Thawornwong and Van Diest (1974)
30	9.0	Al from $AlCl_3 \cdot 6H_2O$, pH 4.0, 830 rice varieties, 3-week treatment	Howeler and Cadavid (1976)
60	12.6	Al from $AlCl_3 \cdot 6H_2O$, pH 4.0, 110 varieties, 6-day treatment	Konzak *et al.* (1976)
20	1.1	Al from $Al_2(SO_4)_3 \cdot 18H_2O$, pH 4.0, 7 varieties, 4-day treatment	Martinez (1976)

tion of Al in the root cells using hematoxylin; and 4) develop a suitable screening technique for identifying Al-tolerant cultivars at the seedling stage.

Materials and methods

The study was conducted in the glasshouse room of the Phytotron with a 29/21°C day/night temperature and 70% relative humidity. The nutrient solution was renewed weekly and pH maintained at 4.0 unless specified.

Interaction among cultivar, nutrient concentration, seedling age, and Al concentration

Nine rice cultivars that represent the entire Al tolerance range were tested: E425, OS6, Monolaya, M1-48, and Columbia 1 as tolerant, 20A and Dawk Mali as moderately tolerant, and IR45 and Cica4 as susceptible cultivars. Seeds were soaked in water for 24 h and germinated for 48 h at 30°C. Ten uniform seeds per cultivar were sown on nylon nets attached to a styrofoam block. Each block has 20 2-x 12-cm holes that can hold 10 varieties. The block floats on 4 L of nutrient solution in a plastic tray. The treatments were combinations of 0, 8, and 30 mg L^{-1} Al and 1/10, 1/5, and full-strength nutrient solution. The composition of the full-strength nutrient solution followed the recommendation of Yoshida *et al.* (1976). The treatments were applied at 0–2 and at 2–4 wk after sowing. Al was supplied from 1 *M* $AlCl_3 \cdot 6H_2O$ solution. The four-factor experiment was in split-split-split-plot design with 4 replications.

Critical solution concentration for aluminum toxicity

Two tests that approximate actual acid soil condition – seminal root growth and root growth in 1/10 strength nutrient solution – were used.

In the seminal root growth method, OS4 and Cica4 seeds were soaked in water for 24 h and germinated for 24 h at 30°C. Seeds were sown on a nylon net in a 50-ml test tube filled with 0, 1, 2, and 3 mg L^{-1} Al in 10^{-4} *M* $CaCl_2$ solution at one seed per cultivar. Six replicates were used. Seminal root growth was measured after 66 h at 30°C.

Root growth in 1/10 strength nutrient solution with 0, 2.5, 5.0, 10.0 and 15.0 mg L^{-1} Al was determined at 2 wk after sowing for IAC3, IR45, and Cica4. This was done in two replicates.

Localization of Al in root cells

The staining solution consisted of 2 g hematoxylin and 0.2 g $NaIO_3$ dissolved in a liter of distilled water (Gill *et al.*, 1974).

Al-treated roots were thoroughly washed with distilled water and stained with 0.2% hematoxylin solution for 15 min. The roots were rinsed with water and dipped in 25% ethylene glycol for 1 h. Cross-sections of the stained roots were made by free hand. The sections were mounted with 70% glycerine and examined with an Olympus model BHB microscope at 100× magnification.

Developing a screening method for Al toxicity tolerance

Twenty test cultivars were used. The procedure of Polle *et al.* (1978) was followed for the hematoxylin staining technique. Five germinated seeds per cultivar were sown in 1/10 strength nutrient solution. After 32 h, Al from 1 *M* $AlCl_3 \cdot 6H_2O$ was added at 0, 30, 60, and 120 mg L^{-1}. The Al treatment was for 16 h. The roots were then stained with 0.2% hematoxylin solution. The stained roots were examined visually and with a microscope at 40× magnification.

In the regrowth technique, the procedure of Martinez (1976) was followed, with some modifications. Initial growth was at full-strength nutrient solution using ten germinated seeds per cultivar. After 48 h, the seedlings were transferred to 1/10 strength nutrient solution with and without 20 mg L^{-1} Al. After 5 d, the seedlings were sampled and maximum root length was measured. The seedlings were then regrown in full-strength nutrient solution for 5 d. Root length was again measured to determine root length recovery. Treatments were replicated twice.

In the absolute root length technique, ten germinated seeds per cultivar were grown in full-strength nutrient solution at 0 and 30 mg L^{-1}. After 14 days, maximum root length of five samples was measured. A two-factor experiment in split-plot design with two replications was used.

Results and discussion

Interaction among cultivar, nutrient concentration, seedling age, and Al concentration

Generally, shoot and root growth declined with increasing Al concentration and decreasing strength of nutrient solution (Table 2). Root growth was most affected by Al, and shoot growth by dilute nutrient concentration. Root length consistently declined with increasing Al concentration up to 30 mg L^{-1} and so did root weight but at a much less extent. The highest root length reduction, 87%, occurred in 2-wk-old seedlings at 1/10 strength nutrient concentration and 30 mg L^{-1} Al. This means the balance between Al and nutrient concentration can affect the degree of Al toxicity. The results support the observation in highly degraded acid soils where low levels of Al were found toxic to rice seedlings (Cate and Sukhai, 1964). On the other hand, detoxification of Al occurred when high levels of nutrients, especially Ca and P, were found (Clarkson, 1966; Clarkson and Sanderson, 1971). It is possible that Ca competes with Al for the exchange sites in the roots and P binds with Al by precipitation to form $AlPO_4$ or by adsorption on the phosphate molecule.

Young seedlings (2 wk old) were more sensitive to Al toxicity whereas older seedlings (4 wk old) were more sensitive to low nutrient concentration. Chlorosis occurred in dilute nutrient solution especially with 4-wk-old seedlings possibly due to nitrogen deficiency. The greater tolerance of older rice seedlings means we can minimize Al stress by using older seedlings in acid soils.

Varietal response was distinct in four treatments: 8 mg L^{-1} Al and 1/10 strength nutrient solution, 8 and 1/5, 30 and 1/5, and 30 and full. Distinct difference between tolerant and susceptible cultivars was best expressed in root length and to some degree in root weight. In full-strength nutrient solution, the more susceptible

Table 2. Relative response of 9 rice cultivars to varying Al and nutrient solution levels

Nutrient solution	Al ($mg\ L^{-1}$)	Relative response (%)[a] Root length	Shoot length	Root weight	Shoot weight
		2-week-old seedlings			
Full strength	0	100	100	100	100
	8	97	101	95	102
	30	70	95	79	98
1/5	0	79	71	79	64
	8	51	73	63	64
	30	29	58	42	49
1/10	0	79	57	68	49
	8	38	55	53	49
	30	13	40	23	36
		4-week-old seedlings			
Full strength	0	100	100	100	100
	8	102	103	91	98
	30	83	99	94	101
1/5	0	112	56	59	38
	8	74	59	54	42
	30	63	56	43	43
1/10	0	94	42	46	27
	8	60	44	38	31
	30	57	42	30	30

[a] Average of 9 rice cultivars, with full-strength nutrient solution and Al-free treatment as reference.

cultivars were IR45 and Cica4 and the tolerant cultivars were E425, OS6, Monolaya, and M1-48. Generally, the degree of Al tolerance was consistent with findings of Howeler and Cadavid (1976). The tolerant cultivars that were identified can be utilized in varietal improvement programs to enhance rice yield in problem soils.

Critical solution concentration for Al toxicity

Acid upland soils that are Al toxic are generally low in nutrients. To approximate such condition, we evaluated critical Al concentration in solution by two methods – seminal root growth and root growth in 1/10 strength nutrient solution.

Seminal root growth of both OS4 and Cica4 in 10^{-4} *M* $CaCl_2$ solution was severely impaired at 2 $mg\ L^{-1}$ Al (Fig. 1). Root length reduction was greater in Cica4. At 3 $mg\ L^{-1}$, root growth of both cultivars was extremely stunted. In 1/10 strength nutrient solution, root growth of IR45 and Cica4, Al-susceptible cultivars, was retarded at 2.5 $mg\ L^{-1}$ Al, whereas that of Al-tolerant IAC3 was impaired at 10 $mg\ L^{-1}$ Al (Fig. 2). Thus, the critical solution concentration for Al toxicity is about 2 $mg\ L^{-1}$ for susceptible cultivars like IR45 and Cica4.

Localization of Al in the root tissues

The reaction that occurs during the staining process using 0.2% hematoxylin involves $NaIO_3$

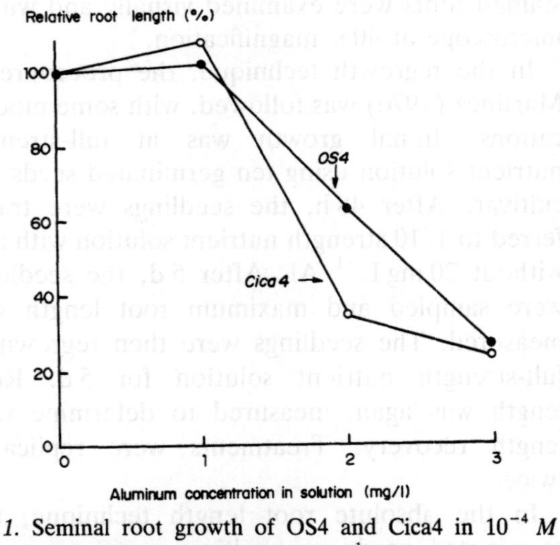

Fig. 1. Seminal root growth of OS4 and Cica4 in 10^{-4} *M* $CaCl_2$ solution with 0, 1, 2 and 3 $mg\ L^{-1}$ Al.

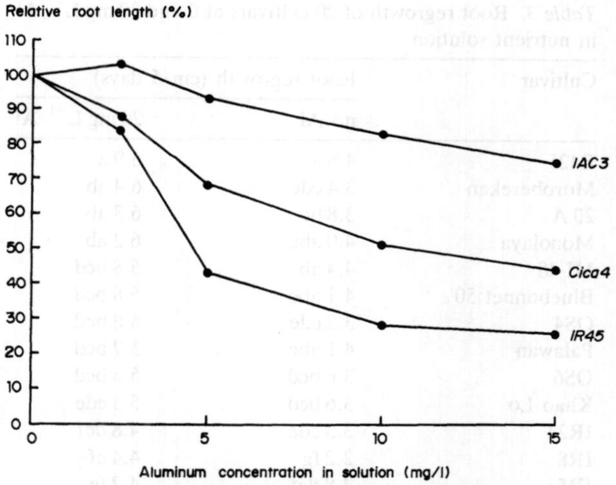

Fig. 2. Root growth of IR45, Cica4 and IAC3 in 1/10 strength nutrient solution with 0, 2.5, 5, 10 and 15 mg L^{-1} Al.

which half-oxidize hematoxylin to hematein. The hematein in turn binds with the Al ion found in the root cells, giving a purple stain.

The distribution of Al in the root cells is indicated by the appearance of purple stain (Plate 1). Al was mainly localized on the epidermal and cortical cells and hardly in the vascular system. Abnormal cortical cells were observed in Al-susceptible Cica4 and IR8. The appearance of the abnormal cells coincided with marked root length reduction.

Al deposition in the cell walls of the epidermal and cortical cells in rice is similar to that observed in barley (Wright and Donahue, 1953). Al was found to accumulate in the cell walls by binding with the carboxyl group of the pectin matrix. This rendered the matrix rigid and restricted cell wall expansion. Thus, the primary effect of Al is established – it inhibits root elongation by restricting cell division and elongation.

Developing a screening method for Al toxicity tolerance

In the hematoxylin method, the purple stain appeared around the root tip region. When Al concentration was increased to 120 mg L^{-1}, a darker, and bigger stain developed. However, varietal difference was detectable only by examining with a microscope (40×). Visual grouping of cultivars by its sensitivity to Al was difficult.

Although tedious and impractical for screening

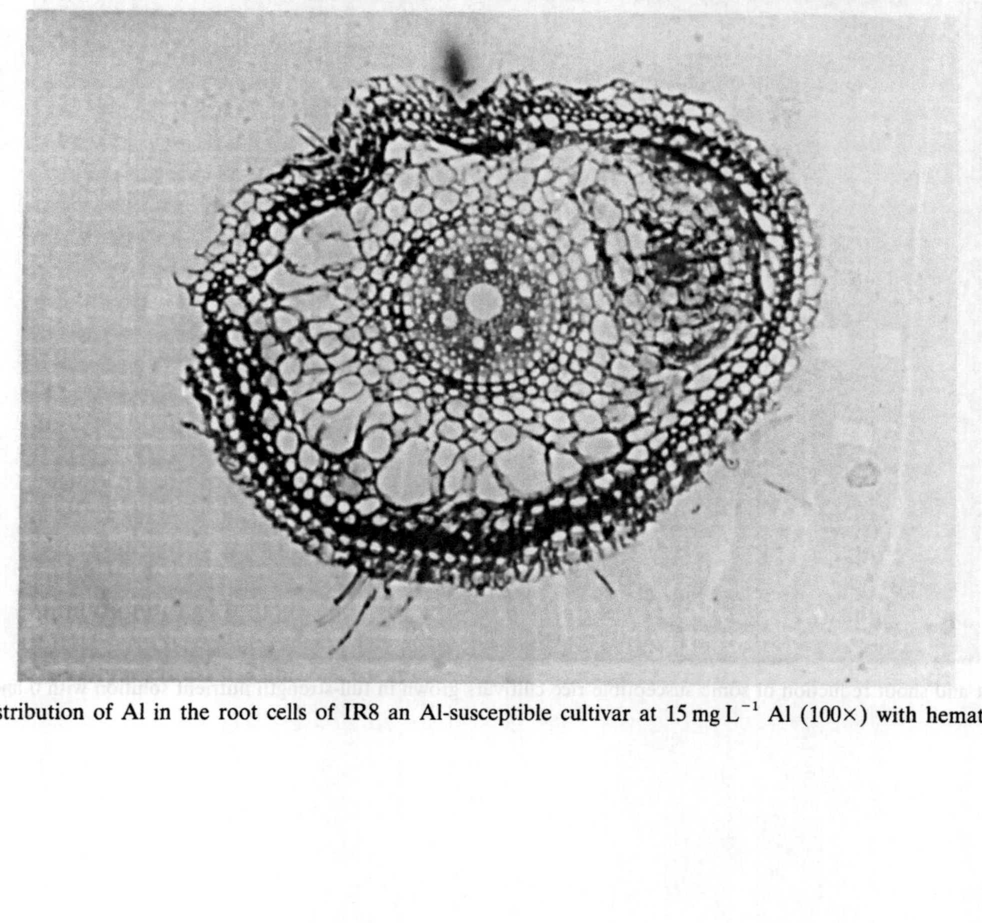

Plate 1. Distribution of Al in the root cells of IR8 an Al-susceptible cultivar at 15 mg L^{-1} Al (100×) with hematoxylen stain.

purposes, the hematoxylin method was useful in detecting Al deposition in the root cells.

In the regrowth method, root regrowth was expressed as the difference between root length after 5 days in Al treatment and root length after 5 days in full-strength nutrient solution.

Significant cultivar differences in root regrowth were observed in Al and Al-free treatments (Table 3). E425 gave the highest root regrowth and IR20, the lowest. Generally, cultivars with low root regrowth in Al-free treatments had lower root regrowth at 20 mg L^{-1} Al. The relative tolerance of a few test cultivars obtained by this method deviated from those obtained from the hematoxylin and absolute root length methods and from early reports (IRRI, 1978).

The ability of rice seedlings to recover from Al stress would not be a reliable indicator of a variety's tolerance for Al because Al stress is always present in Al-toxic soils. There is also the tedious work of taking two root measurements.

In the absolute root length method, significant cultivar differences were observed in both 0 and 30 mg L^{-1} Al treatments. Root length difference

Table 3. Root regrowth of 20 cultivars at 0 and 20 mg L^{-1} Al in nutrient solution

Cultivar	Root regrowth (cm/5 days)	
	no Al	20 mg L^{-1} Al
E425	4.8 a	6.9 a
Moroberekan	3.4 cde	6.4 ab
20 A	3.8 bc	6.3 ab
Monolaya	4.0 abc	6.2 ab
MI-48	4.4 ab	5.8 bcd
Bluebonnet 50	4.1 abc	5.8 bcd
OS4	3.2 cde	5.8 bcd
Palawan	4.1 abc	5.7 bcd
OS6	3.6 bcd	5.5 bcd
Khao Lo	3.6 bcd	5.3 cde
IR24	3.3 cde	4.8 def
IR8	2.2 fg	4.4 efg
IR5	2.8 def	4.2 fg
Cica4	3.8 bc	3.6 gh
Columbia 1	1.6 g	3.6 gh
Dawk Mali	2.5 efg	3.5 gh
IR22	3.3 cde	3.2 hi
IR36	2.6 ef	2.4 ij
IR442	3.2 cde	2.3 ij
IR20	2.1 fg	1.8 j

Within an Al level, means followed by the same letter are not significantly different at 5% level by Duncan's Multiple Range Test.

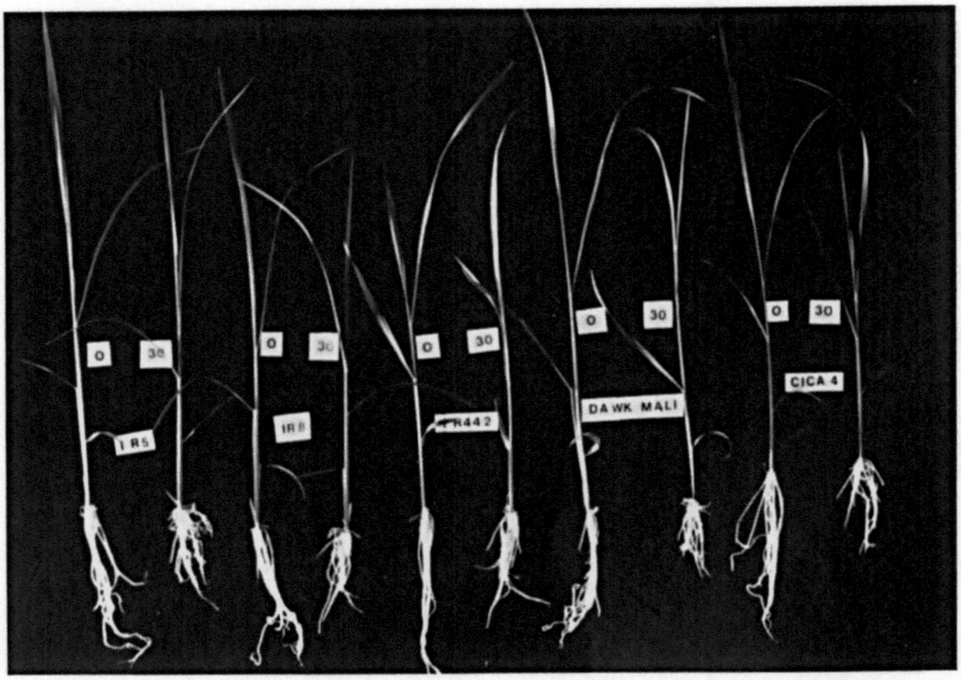

Plate 2. Root and shoot reduction of some susceptible rice cultivars grown in full-strength nutrient solution with 0 and 30 mg L^{-1} Al.

between 0 and 30 mg L^{-1} was significant only for the moderately tolerant and susceptible cultivars.

To eliminate genetic differences in root growth, the relative root length (RRL) was calculated by

$$\mathrm{RRL} = \frac{\text{root length in } 30\,\mathrm{mg\,L^{-1}\,Al}}{\text{root length in } 0\,\mathrm{mg\,L^{-1}\,Al}}$$

The classification of cultivars by relative tolerance to Al based on RRL agreed with that reported in IRRI (1978).

Compared with hematoxylin and regrowth methods, the absolute root length method has significant advantages. Response to Al was more distinct in 30 mg L^{-1} Al. Difference in root length was apparent (Plate 2). Therefore visual scoring can be used instead of the tedious root length measurement. Although root measurement is more precise, visual scoring can provide a convenient and rapid assessment of Al tolerance especially when large numbers of cultivars are tested.

Generally, the absolute root length method is relatively simple, nondestructive – the seedlings are kept intact and can be transplanted – and reproducible. It can be suitable for screening for Al-tolerant rice seedlings using 30 mg L^{-1} Al in full-strength nutrient solution.

References

Cate R B and Sukhai A P 1964 A study of Al in rice soils. Soil Sci. 98, 85–93.

Clarkson D T 1966 Effect of aluminum on the uptake and metabolism of phosphorous by barley seedlings. Plant Physiol. 41, 165–172.

Clarkson D T and Sanderson J 1971 Inhibition of the uptake and long-distance transport of calcium by aluminium and other polyvalent cations. J. Exp. Bot. 22, 837–851.

Gill G W, Frost J K and Miller K A 1974 A new formula for a half-oxidized hematoxylin solution that neither overstains nor requires differentiation. Acta Cytol. 18, 300–311.

Howeler R H and Cadavid L F 1976 Screening of rice cultivars for tolerance to Al toxicity in nutrient solutions as compared with a field screening method. Agron. J. 68, 551–555.

International Rice Research Institute 1978 Annual Report for 1977. Los Baños, Philippines, 548 p.

Konzak C F, Polle E and Kittrick J 1976 Screening several crops for aluminum tolerance. *In* Plant Adaptation to Mineral Stress in Problem Soils. Eds. M J Wright and S A Ferrari. pp 311–329. Cornell University Press, Ithaca, NY.

Martinez C P 1976 Aluminum Toxicity Studies in Rice (*Oryza sativa* L.) Ph.D. Thesis, Oregon State University, Corvalis, OR, 113 p.

Mohiddin M Y B H 1982 The Influence of Aluminum on the Growth of Rice. Ph.D. Thesis, University of Western Australia, Australia, 308 p.

Polle E, Konzak C F and Kittrick J A 1978 Visual detection of aluminum tolerance levels in wheat by hematoxylin staining of seedling roots. Crop Sci. 18, 823–827.

Tanaka A and Navasero S A 1966 Al toxicity of the rice plant under water culture conditions. Soil Sci. Plant Nutr. 12, 55–60.

Thawornwong N and Van Diest A 1974 Influence of high acidity and aluminum on the growth of lowland rice. Plant and Soil 41, 141–159.

Wright K E and Donahue B A 1953 Aluminum toxicity studies with radioactive phosphorus. Plant Physiol. 28, 674–680.

Yoshida S, Forno D A, Cock J A and Gomez K A 1976 Routine procedures for growing rice plants in culture solution. *In* Laboratory Manual for Physiological Studies of Rice. 3rd edition. pp 61–66. IRRI, Los Baños, Philippines.

M. L. van Beusichem (Ed.), *Plant nutrition – physiology and applications*, 365–374.

PLSO IPNC663

Aluminium tolerance of the velvet beans *Mucuna pruriens* var. *utilis* and *Mucuna deeringiana*

I. Effects of aluminium on growth and mineral composition

K. HAIRIAH[1], I. STULEN[2,3] and P.J.C. KUIPER[2]
[1]*Brawijaya University, Faculty of Agriculture, Malang 65145, Indonesia and* [2]*Department of Plant Physiology, University of Groningen, PO Box 14, 9750 AA Haren, The Netherlands.* [3]*Corresponding author*

Key words: aluminium tolerance, *Mucuna deeringiana* L., *Mucuna pruriens* L.

Abstract

Although Al is generally regarded a nonessential element, in nutrient solution studies at pH 4.2 with NO_3^- as nitrogen source an Al^{3+}-concentration of 0.33 meq L^{-1}, increased root fresh weight of *M. p. utilis* and *M. deeringiana* but not shoot fresh weight. The two species did not differ essentially in their response to aluminium.

Average root diameter of both *Mucuna* species increased continuously with Al^{3+} level (0, 0.33, 0.56, 1.12 meq L^{-1}), total root length (m/plant) and root surface area were reduced at 1.12 meq L^{-1} of Al^{3+}.

In roots total P increased with Al^{3+} levels of 0.56 meq L^{-1} and $H_2PO_4^-$ concentration decreased, but in the shoot total P and H_2PO_4 were reduced. Al^{3+} was found to accumulate in the roots. Al-PO_4 can account for 30–60% of the Al^{3+} in the roots in the 0.56 and 1.12 meq L^{-1} Al^{3+} treatment.

Hypotheses explaining the stimulated root growth by low Al^{3+}-levels are discussed, as well as possible Al-tolerance mechanisms.

Introduction

In tropical farming systems leguminous cover crops can be an important component. Hairiah and Van Noordwijk (1986, 1989) tested several leguminous cover crop species. The velvet bean (*Mucuna pruriens* (L) DC. var. *utilis* (Wall. ex Wight) Waker ex Burck.) gave the highest biomass production and it was the most suitable in control of the weed *Imperata cylindrica* (H.B.K.) Hitchc. under acid soil conditions in Nigeria and Indonesia, but it had a very superficial root system. Van Eijk-Bos (1986) reported that *M. deeringiana* (Bort) Small, is an effective cover crop for *Imperata* control in S. America. No information on its root development is available. A shallow root system under acid soil conditions as such might be due to Al-toxicity in the subsoil. Under field conditions at pH values below 5 the availability of Al^{3+} and Mn^{2+} increases and that of Ca^{2+}, Mg^{2+} and P decreases (Marschner, 1986; Sanchez, 1976).

Foy (1974) described that the first signs of Al-toxicity appear in the root system, which becomes stubby as a result of the inhibition of elongation of the main axis and lateral roots. Plant species and varieties may differ in tolerance to high Al^{3+} concentrations. In nutrient solutions, Al^{3+} concentrations of 0.33 meq L^{-1} (3 ppm) can lead to a strong reduction in root and shoot dry weight of *Sorghum bicolor* (L) (Keltjens and Van Ulden, 1986) under a NH_4NO_3 nutrition regime. Although Al^{3+} is generally regarded a nonessential element, in a certain concentration range Al^{3+} can stimulate dry matter production of certain plants. Al^{3+}

concentrations up to $0.55\ \text{meq}\ L^{-1}$ stimulate *Deschampsia flexuosa* and rice (Hacket, 1962; Van Hai *et al.*, 1989), and up to $0.15\ \text{meq}\ L^{-1}$ stimulate sugar maple (*Acer saccharum* Marsh.) (Thornton *et al.*, 1986).

Our aim was to analyze the performance of velvet bean species as tropical leguminous cover crops for acid soils. In this first study we started experiments to measure the response of two *Mucuna* species to Al^{3+} in order to analyze toxicity in the high concentration range and tolerance mechanisms in the lower range. The experiments were performed in nutrient solution culture.

Material and methods

All experiments were carried out in a glass house in Haren, the Netherlands, with automatic control of temperature and humidity. Day temperature was set at 27°C and night temperature at 20°C, with a relative humidity of the air of about 80%. Artificial light was supplied by Philips TLMF 140W/33RS double flux in addition to natural day light, resulting in a daily photoperiod of 16 h.

Seeds of *M. p. utilis* were obtained from Kediri, East Java, Indonesia, and seeds of *M. deeringiana* were collected from Uraba, Columbia. Seeds were soaked in warm water for a few hours before being sown in trays with sterilized vermiculite. During germination the trays were watered with rain water. After about 2 weeks, seedlings were transferred to 5 L pots containing well-aerated nutrient solution, a modified 1/4-strength Hoagland solution as described by Smakman and Hofstra (1982). The solution contained the following concentrations of macronutrients (m*M*): KNO_3 1.25, $Ca(NO_3)_2$ 1.25, $MgSO_4$ 0.50, KH_2PO_4 0.15 and of micro nutrients (μM) H_3BO_3 11.5, $MnCl_2$ 2.3, $ZnSO_4$ 0.20, $CuSO_4$ 0.08, $NaMoO_4$ 0.13; Fe was given as ferro rexenol (22.5 μM). Al was given as $AlCl_3$, according to the specific treatment.

Experiment 1 was carried out in January-February 1988; the effect of solution pH on root and shoot growth of *M. p. utilis* and *M. deeringiana* was tested. The pH values were daily adjusted to the initial pH, 3.5, 4.2 and 5.0 respectively, by adding either 0.1 *N* H_2SO_4 or 0.1 *N* KOH. Experiment 2 was performed in February-March 1988; the effect of Al^{3+} on both species was investigated at levels of 0, 0.11, 0.33, $0.56\ \text{meq}\ L^{-1}$. A third experiment was carried out in June-July 1988; Al^{3+} levels were 0, 0.33, 0.56, and $1.12\ \text{meq}\ L^{-1}$.

Based on experiment 1, in subsequent experiments the pH of the nutrient solution was daily adjusted to 4.2 (±0.05) for each individual pot. The nutrient solutions were renewed once a week. Water lost by evapotranspiration was replenished by addition of deionized water. The amount of water, added to every pot, was recorded.

The experiments were arranged in randomized blocks of 3 replicates for experiment 1, 5 for experiment 2, and 6 for experiment 3. The twining stems of the plants were led up and down a rope of about 90 cm long, under a slope of 30° with the horizontal plane, to avoid interference between plants.

Root and shoot fresh weight were weekly recorded in a non-destructive way by determining the plants fresh weight (root + shoot); roots were subsequently immersed in water and shoot weight was determined separately (Kuiper and Staal, 1987). At harvesting time root and shoot fresh weight were recorded destructively. There were no significant differences between our weight measurements, when determined nondestructively or destructively.

Six weeks after planting, plants were harvested. Roots were washed 3 times in deionized water, blotted between two layers of tissue paper, dried in an oven at 70°C for 48 hours, and weighed. Two repetitions of dry plant material from experiment 2, and three repetitions from experiment 3 were analysed for K^+, Ca^{2+}, Mg^{2+}, Fe^{3+}, Al^{3+}, total P and total N. Dry material was extracted with a mixture of H_2SO_4, HNO_3 and $HClO_4$; K^+, Ca^{2+}, Mg^{2+} and Fe^{3+} were measured by atomic absorption spectrophotometry; Al^{3+} was coloured with aluminon and measured by auto analyser; total P was measured by the molybdate blue method (modified from Murphy and Riley, 1962). Total N was determined with the Kjeldahl method (Bailey, 1962). Other replicate plants were immediately immersed in liquid nitrogen and subsequently stored at −20°C, before anion concentrations ($H_2PO_4^-$, Cl^-, NO_2^-, NO_3^-, SO_4^{2-}) were determined in a watery extract by means of high performance liquid chromatography (HPLC) analysis (Maas *et al.*, 1985). Concentrations of ions in plant tissue are expressed on the basis of 'plant solution', *i.e.* fresh weight minus dry weight. Total N content is also expressed on a dry weight basis.

Root diameter was measured directly on fresh root samples with the aid of a microscope fitted with an ocular micrometer. Subsequently root length was measured according to the line intersection technique of Tennant (1975). Specific and total root surface area was calculated with the following formulas (Van Noordwijk, 1987):

$$A_{rw} = \pi \times 0.001 \times D_r \times L_{rw} \quad m^2/g$$

and

$$A_{rp} = A_{rw} \times W_{rp} \quad m^2/plant$$

where,

A_{rw} = specific root surface area (surface area per unit dry weight), in m^2/g
D_r = average root diameter, in mm
L_{rw} = specific root length (root length per unit dry weight), in m/g
A_{rp} = root surface area per plant, in m^2/plant
W_{rp} = root dry weight per plant, in g/plant.

Localisation of aluminium in the roots was determined by making cross sections of roots of control and Al-treated plants. The sections were stained with aluminon (Haridasan *et al.*, 1986) and viewed under a light microscope.

Results were analysed with ANOVA (analysis of variance) by using the GENSTAT 5 computer program (Payne *et al.*, 1987), and student t-test ($P < 0.05$) when significant treatment effects were found. In case of uneven variances, a logarithmic transformation of data was used.

Results

Effects on Al on shoot and root growth

Figure 1 shows that at pH 4.2 of the nutrient solution the highest root and shoot fresh weight were obtained for both species. This pH value is also suitable for studying Al-toxicity, since at this pH Al^{3+} is the dominant aluminum form. The following experiments were performed at pH 4.2. In nutrient solutions with 0.11, 0.33 and 0.56 meq L^{-1} of Al^{3+} root fresh weight of both *Mucuna* species was significantly higher than in the control ($P < 0.05$) in experiment 2 (Fig. 2). In experiment 3, 0.33 meq L^{-1} of Al^{3+} again significantly ($P < 0.05$) increased root fresh weight. In experiment 3 an Al^{3+} concentration of 0.56 meq L^{-1} reduced shoot but not root fresh weight; 1.12 meq L^{-1} of Al^{3+} reduced both root and shoot fresh weight. Shoot: root ratio (on a fresh weight basis) continuously decreased with increasing Al-concentration.

Since the results of experiment 2 and experiment 3 were similar, we will only present further data for experiment 3, which covered a wider range of Al concentrations.

Statistical analyses of the data showed that for most of the parameters observed, no significant interaction existed between species-difference and Al-treatment. We will therefore only present data for the two main effects, Al-treatment and species-difference. The lack of significant interactions shows that the two species do not differ essentially in Al-response.

Table 1 showed that *M.p. utilis* produced significantly ($P < 0.05$) more dry weight of both shoot and root than *M. deeringiana*. In both species increasing Al^{3+} concentrations decreased shoot dry weight; a concentration of 0.56 meq L^{-1} already gave a significant reduction. Dry matter percentage of the shoots only increased at the highest Al^{3+} concentration of 1.12 meq L^{-1}.

Although at 0.33 meq L^{-1} Al^{3+} root fresh weight was increased, root dry weight and total root length were not significantly influenced for both *Mucuna* species. At Al^{3+} concentrations of 1.12 meq L^{-1} root dry weight and total root length were significantly reduced, while root dry matter percentage was significantly increased.

At Al^{3+} concentrations of more than 0.33 meq L^{-1} roots became stubby, and dark stripes (not shown here) were formed about 5 mm behind the root tips each time the solution was refreshed; at 6 weeks after planting Al-treated roots were lighter in colour than roots of control plants. Average root diameter of both

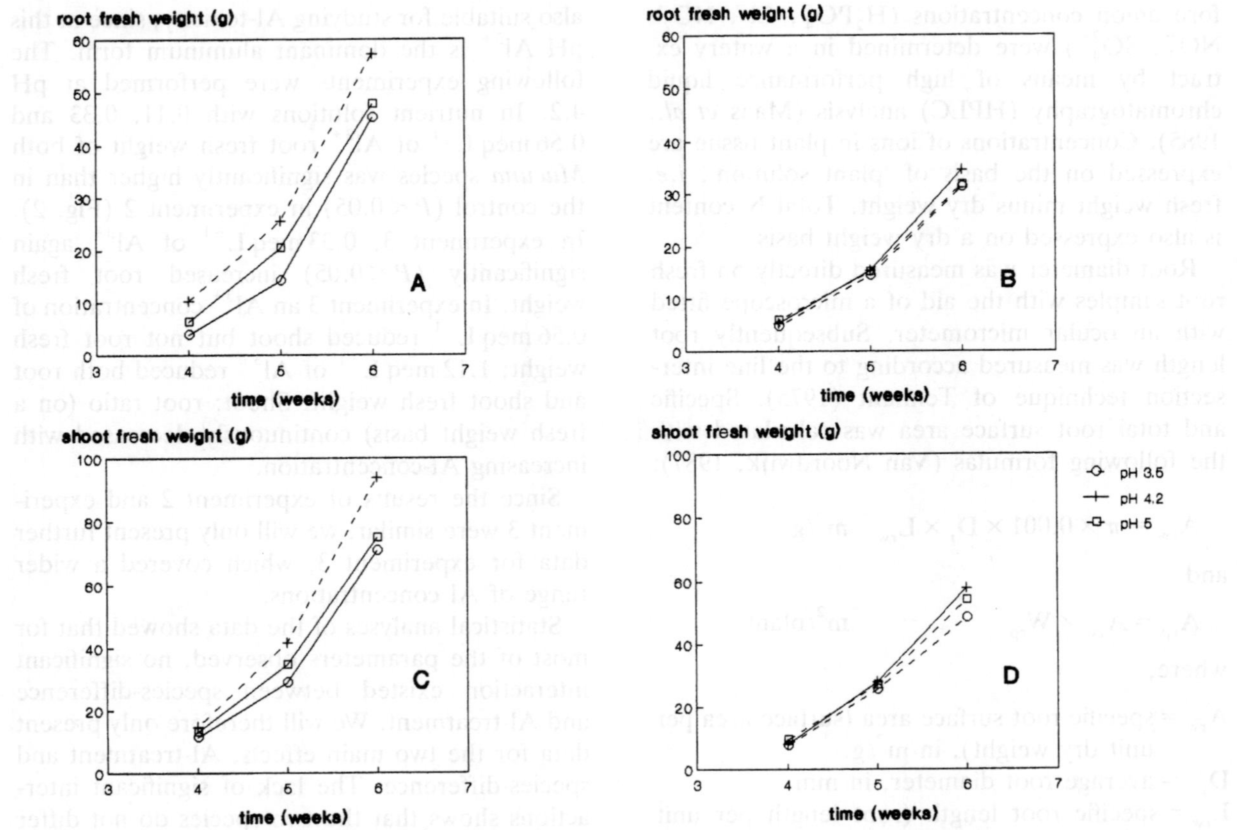

Fig. 1. Effect of pH on root and shoot fresh weight of *M. p. utilis* (**A**, **C**), and of *M. deeringiana* (**B**, **D**).

species increased with increasing Al concentration; specific root length (m/g) and specific root surface area (m^2/g) significantly decreased at 1.12 meq L^{-1} (Table 1).

Effects of Al on mineral composition

Figure 3 shows that in the shoot of both *Mucuna* species an increasing concentration of Al in the nutrient solution reduced the concentration of total P as well as ionic $H_2PO_4^-$; statistical analysis showed that Al^{3+} at 0.33 meq resulted in a significantly lower total P concentration in the shoot, but ionic $H_2PO_4^-$ was only affected at higher level 0.56 and 1.12 meq L^{-1}. 'Other forms of P' (total P minus ionic $H_2PO_4^-$) were not

Table 1. Effect of Al^{3+} on various root parameters (numbers followed by different letters indicate significant ($P < 0.05$) differences, NS = not significant different ($p < 0.05$), root fresh weight data are based on 6, all other data on 3 repetitions), D_r = root diameter (mm), L_{rw} = specific root length (m/g), A_{rw} = specific root surface area (m^2/g), L_{rp} = root length per plant (m/pl), A_{rp} = root surface area (m^2/pl)

Treatment	Shoot dw(g)	Root dw(g)	Root fw(g)	sh/r dw	%dm shoot	%dm root	Dr (mm)	Lrw (m/g)	Arw (m^2/g)	Lrp (m/pl)	Arp (m^2/pl)
Al^{3+} (meq L^{-1})											
0	25.9a	6.40a	126b	4.4a	17.2b	4.6b	0.37c	175a	0.20a	1160a	1.36a
0.33	24.7a	6.57a	155a	3.9b	16.2b	4.5b	0.44b	166a	0.23a	1110a	1.58a
0.56	19.7b	6.64a	126b	3.0c	17.3b	4.9b	0.46b	141a	0.20a	931a	1.35a
1.12	7.6c	3.32b	55c	2.3d	22.0a	6.1a	0.51a	76b	0.12b	253b	0.42b
Species											
M. utilis	21.6a	7.06a	141a	3.0b	18.4b	4.8b	0.48a	145NS	0.21NS	1104a	1.56a
M. deer.	17.4b	4.40b	90b	3.8a	17.9b	5.2a	0.41b	134	0.17	623b	0.80b

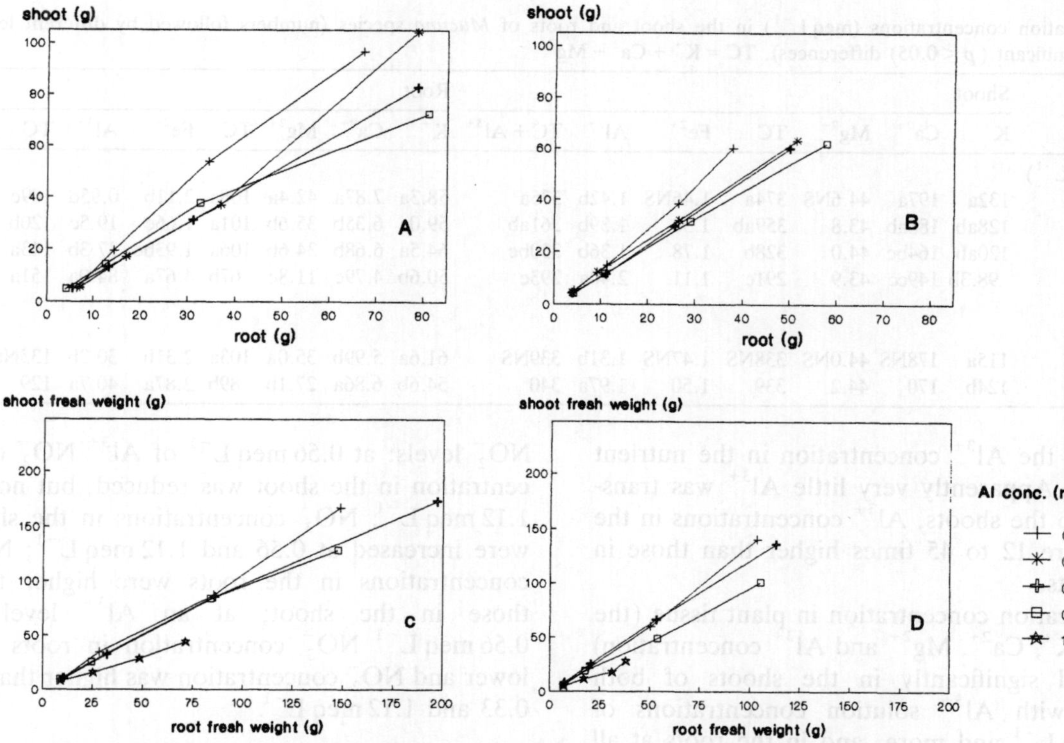

Fig. 2. Al^{3+} effect on the mean root/shoot ratio fresh weight basis of *M. p. utilis* (exp. 2 and exp. 3; **A, C**) and of *M. deeringiana* (**B, D**) at 3, 4, 5 and 6 weeks after planting.

significantly affected by Al^{3+} levels. In the roots total P and 'other forms of P' increased and the concentration of ionic $H_2PO_4^-$ was reduced by Al^{3+} concentrations in the solution of 0.56 and 1.12 meq L^{-1}.

Table 2 shows that Ca concentration in the shoot was significantly reduced at 0.56 meq L^{-1} of Al^{3+}; a significant decrease of K^+, and increase of Al^{3+} concentration in the shoot was only found at the highest Al^{3+} solution concentration of 1.12 meq L^{-1}; for Mg^{2+} no significant effects were found, but the sum of K^+, Ca^{2+}, Mg^{2+} was already reduced by 0.56 meq L^{-1}.

In the roots, Ca^{2+} and Mg^{2+} concentrations were already affected by Al^{3+} at 0.33 meq L^{-1}, the K^+ concentration and the sum of K^+, Ca^{2+} and Mg^{2+} was only reduced at 1.12 meq L^{-1}. Al^{3+} concentration in the root increased propor-

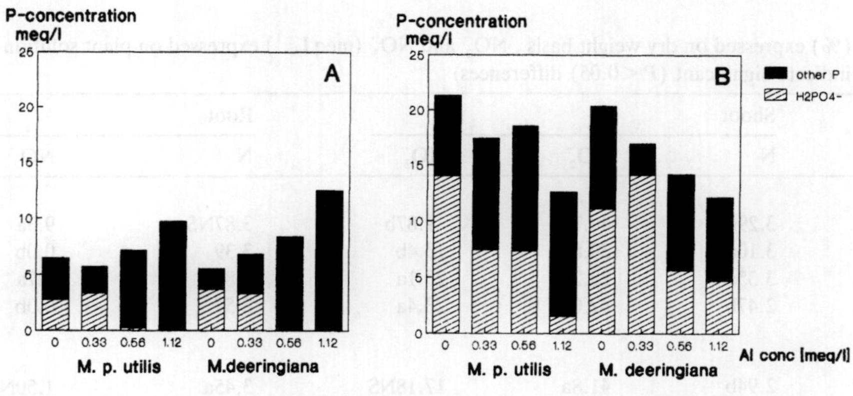

Fig. 3. P-concentration in the root (**A**) and shoot (**B**) of both species; 'others forms of P' = total P − $H_2PO_4^-$.

Table 2. Cation concentrations (meq L^{-1}) in the shoot and roots of *Mucuna* species (numbers followed by different letters indicate significant ($p < 0.05$) differences), TC = $K^+ + Ca^2 + Mg^{2+}$

Treatment	Shoot							Root						
	K^+	Ca^{2+}	Mg^{2+}	TC	Fe^{2+}	Al^{3+}	TC + Al^{3+}	K^+	Ca^{2+}	Mg^{2+}	TC	Fe^{2+}	Al^{3+}	TC + Al^{3+}
Al^{3+} (meq L^{-1})														
0	132a	197a	44.6NS	374a	1.46NS	1.42b	375a	58.3a	7.87a	42.4a	108a	2.11b	0.95d	109c
0.33	128ab	188ab	43.8	359ab	1.58	1.39b	361ab	59.0a	6.35b	35.6b	101a	1.66c	19.5c	120b
0.56	120ab	164bc	44.0	328b	1.78	1.36b	330bc	64.5a	6.68b	34.6b	106a	1.93bc	37.3b	143a
1.12	98.3b	149cc	43.9	291c	1.11	2.40a	293c	50.6b	4.79c	11.8c	67b	4.67a	84.0a	151a
Species														
M. utilis	115a	178NS	44.0NS	338NS	1.47NS	1.31b	339NS	61.6a	5.99b	35.0a	103a	2.31b	30.2b	133NS
M. deer.	124b	170	44.2	339	1.50	1.97a	340	54.6b	6.86a	27.1b	89b	2.87a	40.7a	129

tional to the Al^{3+} concentration in the nutrient solution. Apparently very little Al^{3+} was transported to the shoots, Al^{3+} concentrations in the roots were 12 to 45 times higher than those in the shoots.

Total cation concentration in plant tissue (the sum of K^+, Ca^{2+}, Mg^{2+} and Al^{3+} concentration) increased significantly in the shoots of both species with Al^{3+} solution concentrations of 0.56 meq L^{-1} and more, and in the roots at all Al^{3+} levels.

Fe^{2+} concentration in shoot tissue was not significantly affected, but in the roots it was doubled at the highest Al level in the nutrient solution.

Total N concentration in the shoots was significantly reduced at an Al^{3+} concentration of 1.12 meq L^{-1} for both species when the data were expressed on a plant dry weight basis (Table 3); for the roots no effect was found.

Both species contained remarkably high levels of NO_2^- in the control plants. Al^{3+} concentration of the nutrient solution had a significant effect on NO_2^- levels: at 0.56 meq L^{-1} of Al^{3+} NO_2^- concentration in the shoot was reduced, but not at 1.12 meq L^{-1}. NO_3^- concentrations in the shoot were increased at 0.56 and 1.12 meq L^{-1}; NO_3^- concentrations in the roots were higher than those in the shoot; at an Al^{3+} level of 0.56 meq L^{-1} NO_3^- concentration in roots was lower and NO_2^- concentration was higher than at 0.33 and 1.12 meq L^{-1}.

Discussion

Under the conditions of the present experiments, a solution pH of 4.2 gave the highest shoot and root weight of *M. p. utilis* and *M. deeringiana*. Under field conditions, Kretschmer (1989) found that *M. pruriens* can tolerate pH values of 5.0–6.5. The fact that tolerance of plants to acidity under field conditions was lower than under our controlled conditions, might be due to a higher nutrient availability of our nutrient solution com-

Table 3. N content (%) expressed on dry weight basis, NO_2^- and NO_3^- (meq L^{-1}) expressed on plant solution (numbers followed by different letters indicate significant ($P < 0.05$) differences)

Treatment	Shoot			Root		
	N	NO_2^-	NO_3^-	N	NO_2^-	NO_3^-
Al^{3+} (meq L^{-1})						
0	3.29a	40.7a	9.87b	3.87NS	9.3a	20.5b
0.33	3.10a	39.8a	13.4b	3.39	0.0b	32.9a
0.56	3.35a	28.2b	20.1a	3.86	2.7a	22.0b
1.12	2.47b	42.0a	23.4a	3.57	0.0b	34.7a
Species						
M. utilis	2.94b	41.8a	17.18NS	3.45a	1.50NS	21.2a
M. deer.	3.20a	33.5b	16.16	3.89b	4.50	33.8b

pared with the nutrient conditions in the field and/or to Al-toxicity in the field under acid conditions.

Generally Al^{3+} inhibits root growth (Berggren and Fiskejo, 1987), and plant roots are the tissue immediately affected by high Al^{3+} concentrations. Roots of both *Mucuna* species were influenced by Al^{3+} concentrations of 0.33 meq L^{-1} and higher. They showed black stripes, consisting of damaged cells in the elongation zone by the time the solution was refreshed. The fact that within 1 day after each refreshing normal cell elongation again started, indicated that the plants had adapted to high Al^{3+} concentrations or eliminated the most toxic Al forms from the solutions. Total Al concentration in solution remained almost constant. Horst *et al.* (1983) showed in cowpea (*Vigna unguiculata*) that directly after applying Al^{3+} root elongation was inhibited as a result of inhibited cell division in the root meristem and that after 18 hours cell division started again, although Al still was present. In our experiments we found that this 'primary Al- shock' occurs every time the solution is refreshed.

Despite negative effects on cell elongation in individual roots, root fresh weight of both *Mucuna* species responded positively to Al^{3+} concentrations in solution of 0.33 meq L^{-1}. Posi-

Table 4. Calculation of Al^{3+} uptake from the nutrient solutions at 6 weeks after planting. C_s = solution concentration of Al^{3+}, Water uptake = amount of water added during investigation,

$$\text{uptake concentrations} = C_u = \frac{\text{nutrient uptake/plant}}{\text{water uptake/plant}}.$$

C_s (meq L^{-1})	Water uptake (l/pl)	Al-upt. (meq/pl)	C_u (meq L^{-1})	C_u/C_s
M. p. utilis				
0.33	10.15	3.62	0.36	1.09
0.56	8.12	5.36	0.66	0.85
1.12	3.23	4.99	1.50	0.75
M. deeringiana				
0.33	7.43	2.24	0.30	1.10
0.56	6.54	4.30	0.67	0.85
1.12	2.60	3.69	1.42	0.79

$C_u/C_s > 1$, means relative depletion of the nutrient solution;
$C_u/C_s < 1$, means accumulation in the root environment;
$C_u/C_s = 1$, means no change in the nutrient concentration (De Willigen and Van Noordwijk, 1987).

tive effects of relatively small quantities of Al on root growth in nutrient solutions might be based on one of the following mechanisms.

I) The presence of Al may reduce toxicity of other elements occuring at supra-optimal concentrations in the solution, *e.g.* P or Ca. Asher and Loneragan (1967), found that 24 μM of P in a continuously recirculated nutrient solution gave toxic effects on clover and lupins. It might be that the present concentration of 0.15 mM of P in the solution was supra-optimal for *Mucuna*, and that Al could stimulate plant growth by reducing the $H_2PO_4^-$ concentration of the solution. However, precipitation of Al-PO_4 in the solutions is not likely at 0.15 mM of $H_2PO_4^-$ concentration in solutions with pH 4.2. P concentrations in roots and shoot also did not support this hypothesis.

Al in the solution may have reduced the Ca availability, which might have been supra-optimal. Al-toxicity in the field is associated with acid soil conditions, where the availability of both Ca and Mg is often very low (Sanchez, 1976). *M. p. utilis* did not significantly respond to 1 ton $CaCO_3$/ha on acid soil in Nigeria (Hairiah and van Noordwijk, 1986), which indicates that *M. p. utilis* has a high Ca-affinity. The present concentration of 1.25 mM Ca in the nutrient solution could have been supra-optimal for the plant. Horst (1987), found that increasing Al supply depressed Ca concentrations in all root tissue of *Vigna unguiculata*. We found a similar decrease, but we can only speculate that Ca levels in the control were above optimal, and therefore allow stimulation of root growth at low Al concentrations.

II) The presence of Al may make roots less effective in nutrient or water uptake and the plant then shows a 'functional equilibrium' response (Brouwer, 1963) by increasing its root growth. Two variants of such explanation are:

a. Effects through water uptake. External Al^{3+} concentrations of 0.33 meq L^{-1} and more interfered with cell elongation and resulted in a higher root diameter and lower specific root length (Table 1). To obtain the root length required for water uptake, the plants had to produce a higher root dry weight, when Al was present in the solution. Root surface area (m^2/plant) of both *Mucuna* species with 0.33 and

0.55 meq L^{-1} Al^{3+} concentration was not significantly different with the control. In our experiment water uptake per unit shoot fresh weight was equal for control and 0.33 meq L^{-1} Al^{3+}, and it even increased for higher Al concentration of the solution; no indications of inhibited water uptake were obtained.

b. Effects through N-metabolism. Jarvis and Hatch (1987) found that 0.15 meq L^{-1} of Al^{3+} in the nutrient solution with N supplied as NO_3^-, reduced total N concentration in the shoot of clover from luxury levels >5% in control to <2.5% in Al-plants when expressed on a dry weight basis. In our study total N concentration in the shoot was reduced by Al^{3+} at 1.12 meq L^{-1}, from 3.4% in the control to 2.4% when expressed on a dry weight basis (Table 3), while N concentration in the root was not affected. Reduction of total N concentrations of both *Mucuna* species, indicates that NO_3^- absorption from the solution by the plants has been restricted by high Al^{3+} levels. For the Al concentrations which stimulate root growth, however, no effects on NO_3-uptake were found.

The different hypotheses mentioned above explaining stimulated root growth, will be further tested in future experiments.

Plant species and varieties vary widely in tolerance to excess Al in the growth medium. Species affected by Al concentrations below 0.75 meq L^{-1} are regarded as sensitive, species tolerating concentrations above 9 meq L^{-1} as insensitive (Thornton *et al.*, 1986). The *Mucuna* species tested here belong to an intermediate group according to this classification and show a certain degree of Al tolerance. In general Al-tolerance of plant species can be based on:

a) Al-exclusion by the root system; from calculations we found that at 0.33 meq L^{-1} Al^{3+} in the solution the ratio of 'uptake concentration' (C_u) and concentration in the solution (C_s) of both species was just above 1 and at higher Al^{3+} concentrations it was below 1 (Table 4); this means that only at higher Al concentrations the plants excluded part of the Al carried in the transpirational waterflow.

b) maintaining a high root rhizosphere pH if the plant grows in an acid soil in relation with preferential uptake of N as NO_3^--rather than as NH_4^+ (Taylor and Foy, 1985). NO_3 uptake will increase the pH of the rhizosphere (Marschner and Römheld, 1983), which leads lower levels of soluble of Al. In our study with NO_3 solutions, in the first days after the starting the treatments plants acidified the solution in the presence of Al; a few days later, however, the plants started to increase the solution pH even at the highest Al level; this effect of Al will be discussed in the next article,

c) detoxification of Al^{3+} in the rhizosphere through exudation by the roots of chelating agents *e.g.* organic acids, (Bartlett and Riego, 1972; Clarkson, 1969; Mengel and Kirkby, 1982); the black and white stripe pattern on *Mucuna* roots might be due to such shift in Al-speciation in the solution every time the solution was refreshed.

d) detoxification inside the roots (Foy, 1974; Foy *et al.*, 1978; McCormick and Borden, 1974) *e.g.* as Al-PO_4; cross sections of *Mucuna* roots stained with the aluminon technique showed that Al was accumulated in the root, mainly in the epidermis (results not shown here) as indicated by a pinkish colour, which was not found in the control roots. Only a small part of Al was transported to the shoot. In both *Mucuna* species Al-PO_4 can account for maximally 45, 60 and 40% of the Al in the roots at Al^{3+} concentrations of 0.33, 0.56 and 1.12 meq L^{-1} respectively, as follows from the ratio of non-ionic P (total P minus $H_2PO_4^-$ concentration) and Al concentration, both in meq L^{-1}; this value reflects an overestimate, as not all non-ionic P can be Al-PO_4. By subtracting the non-ionic P concentration of the control plant, from that in the Al-treatments we obtain an underestimate of the possible binding of Al as Al-PO_4: 2, 39 and 30% of P in the roots for the three Al levels respectively. For the two highest Al-levels the over estimate and the under estimate are fairly close together,

e) tolerance to high Al concentrations in the shoot, as for example in tea, Hawaiian grasses, *Atriplex hastata*, pine trees, mangrove (Foy *et al.*, 1978); in *Mucuna* we found shoot growth to be severely affected at 1.12 meq L^{-1}, with only slightly increased Al concentrations in the shoot.

Summarizing, detoxification of Al in the solution by the roots (c) and localisation of Al in relatively non-toxic forms, Al-PO_4 and others, in the root epidermis (d), seem to contribute most to the Al tolerance of *Mucuna* species.

Acknowledgements

This work was made possible by a scholarship from the EEC for the first author. We thank Dr P de Willigen, Dr M van Noordwijk, Drs A J Gijsman, Drs J den Hertog for critical reading of the manuscript, Ir J van der Heide for kindly providing seeds of *M. deeringiana* and Mr W van Hal for assistance in chemical analyses.

References

Asher C J and Loneragan J F 1967 Response of plants to phosphate concentration in solution culture. I. Growth and phosphorus content. Soil Sci. 103, 225–32.

Bailey J L 1962 Techniques in Protein Chemistry. Amsterdam, Elsevier, 299 p.

Bartlett R J and Riego D C 1972 Effect of chelation on the toxicity of Aluminum. Plant and Soil 37, 419–423.

Berggren D and Fiskesjo G 1987 Aluminium toxicity and speciation in soil liquids – experiments with *Allium cepa* L. Environmental Toxicology and Chemistry 6, 771–779.

Brouwer R 1963 Some aspects of the equilibrium between overground and underground plant parts. Jaarb. IBS, Wageningen, 31–39.

Clarkson D T 1969 Metabolic aspects of aluminium toxicity and some possible mechanisms for resistance. *In* Ecological Aspects of the Mineral Nutrition of Plants. Ed. I H Rorison pp 381–397. Blackwell Sc. Publ. Oxford.

De Willigen P and Van Noordwijk M 1987 Roots, Plant Production and Nutrient Use efficiency. Doct. Thesis, Agricultural University of Wageningen, 282 p.

Foy C D 1974 Effects of aluminium on plant growth. *In* The Plant Root and its Environment. Ed. E W Carson. pp 601–642. Charlottesville, University Press Virginia.

Foy C D, Chaney R L and White M C 1978 The physiology of metal toxicity in plants. Annu. Rev. Plant Physiol. 29, 511–566.

Hackett C 1962 Stimulative effects of aluminium on plant growth. Nature 195, 471–472.

Hairiah K and Van Noordwijk M 1986 Root studies on a tropical ultisol in relation to nitrogen management. Institute for Soil Fertility, Rep. 7–86, 116 p.

Hairiah K and Van Noordwijk M 1989 Root distribution of leguminous cover crops in the humid tropics and effects on a subsequent maize crop. *In* Nutrient Management for Food Crop Production in Tropical Farming Systems. Proc. Symp. Malang, 1987. Institute for Soil Fertility, Haren, The Netherlands.

Haridasan M, Paviani T I and Schiavini I 1986 Localization of aluminium in the leaves of some aluminium-accumulating species. Plant and Soil 94, 435–437.

Horst W J, Wagner A and Marschner H 1983 Effect of aluminium on root growth, cell division rate and mineral element contents in root of *Vigna unguiculata* genotypes. Z. Pflanzenphysiol. 109, 95–103.

Horst W J 1987 Aluminium tolerance and calcium efficiency of cowpea genotypes. J. Plant Nutr. 10, 1121–1129.

Jarvis S C and Hatch D J 1987 Differential effects of low concentrations of aluminium on the growth of four genotypes of white clover. Plant and Soil 99, 241–253.

Keltjens W G and van Ulden P S R 1987 Effects of Al on nitrogen (NH_4^+ and NO_3^-) uptake, nitrate reductase activity and proton release in two sorghum cultivars differing in Al tolerance. Plant and Soil 104, 227–234.

Klotz F and Horst W J 1988 Effect of ammonium and nitrate-nitrogen nutrition on aluminium tolerance of soybean (*Glycine max* L.). Plant and Soil 111, 59–65.

Kretschmer Jr A E 1989 Tropical forage legume development, diversity, and methodology for determining persistence. *In* Persistence of Forage Legumes. ASA, CSSA, SSSA, Madison, WI.

Kuiper D and Staal M 1987 The effects of exogenously applied plant growth substances on the physiological plasticity in *Plantago major* ssp *pleiospeima*: Responses of growth, shoot to root ratio and respiration. Physiol. Plant. 69, 651–658.

Maas F M, Hoffmann I, van Harmelen M J and de Kok L J 1986 Refractometric determination of sulphate and other anions in plants separated by High-Performance Liquid Chromatography. Plant and Soil 91, 129–132.

Marschner H and Römheld V 1983 *In vivo* measurement of root-induced pH changes at the soil root interface: Effect of plant species and nitrogen source. Z Pflanzenphysiol 111, 241–251.

Marschner H 1986 Mineral Nutrition of Higher Plants. Academic Press, London, 674 p.

McCormick L H and Borden F J 1974 The occurrence of aluminium-phosphate precipitate in plant roots. Soil Sci. Soc. Am. Proc. 38, 931–934.

Mengel K and Kirkby E A 1982 Principles of Plant Nutrition, 3rd edition. Intern. Potash Inst. Bern, 593 p.

Murphy J and Riley J P 1962 A modified single solution method for the determination of phosphate in natural waters. Anal. Chim. Acta. 27, 31–36.

Payne R W, Lane P W, Ainsley A E, Bicknell K E, Digby P G N, Harding S A, Verrier P J, White R P 1987 Genstat 5 Reference Manual. Clarendon Press, Oxford, 749 p.

Sanchez P A 1976 Properties and Management of Soils in the Tropics. Wiley, New York, 618 p.

Smakman G and Hofstra J J 1982 Energy metabolism of *Plantago lanceolata*, as affected by change in root temperature. Physiol. Plant. 56, 33–37.

Taylor G J and Foy C D 1985 Mechanisms of aluminium tolerance in *Triticum aestivum* (wheat). IV. The role of ammonium and nitrate nutrition. Can. J. Bot. 63, 2181–2186.

Tennant D 1975 A test of modified line intersect method of estimating root length. J. Ecol. 63, 995–1001.

Thornton F C, Schaedle M and Raynal D J 1986 Effect of aluminum on the growth of sugar maple in solution culture. Can. J. For. Res. 16, 892–896.

Van Eyk-Bos C G 1986 Recuperation de tierras invadidas por el *Imperata contracta* (H.B.K.) Hitchc. por medio de la siembra de Leguminosa, especialmente el *Mucuna deeringiana* (Bort.) mall en Uraba, Colombia. Una technologia de bajos insumos. (Reclamation of *Imperata contracta* (H.B.K.) Hitchc. invaded terrains by sowing leguminous species, especially *Mucuna deeringiana* (Bort.) Small in Uraba, Columbia). CONIF report, Bogota, Columbia, 45 p.

Van Hai T, Nga T T and Laudelout H 1989 Effect of aluminium on the mineral nutrition of rice. Plant and Soil 114, 173–185.

Van Noordwijk M 1987 Methods for quantification of root distribution pattern and root dynamics in the field. *In* Methodology in Soil-K Research. pp 263–281. Proc. 20th IPI Colloq. Intern. Potash Inst. Bern.

M. L. van Beusichem (Ed.), *Plant nutrition – physiology and applications*, 375–379.

PLSO IPNC191

Effects of aluminium on tap-root elongation of soybean (*Glycine max*), cowpea (*Vigna unguiculata*) and green gram (*Vigna radiata*) grown in the presence of organic acids

S. SUTHIPRADIT[1], D.G. EDWARDS[2] and C.J. ASHER
Department of Agriculture, University of Queensland, St. Lucia, Queensland 4067, Australia. [1]*Present address: Department of Earth Science, Faculty of Natural Resources, Prince of Songkla University, Haadyai, Thailand.* [2]*Corresponding author*

Key words: aluminium toxicity, cowpea, fulvic acid, *Glycine max* (L.) Merr, green gram, malic acid, monomeric aluminium, oxalic acid, soybean, tap-root elongation, *Vigna radiata* (L.) Wilczek, *Vigna unguiculata* (L.) Walp.

Abstract

The role of fulvic, malic, and oxalic acids in alleviating the toxic effects of aluminium (Al) on tap-root elongation of soybean cv. Fitzroy, cowpea cv. Vita 4, and green gram cv. Berken was studied. Treatments consisted of a factorial combination of four Al concentrations (0, 12.5, 25 and 50 μM as $Al(NO_3)_3 \cdot 9H_2O$) and two concentrations either of malic or oxalic acid (0, 50 μM) or fulvic acid (0, 65 mg L^{-1} of organic carbon). The free monomeric Al in solution was determined using a pyrocatechol violet procedure which distinguishes between monomeric and organically complexed Al. Fulvic acid completely alleviated the toxic effect of Al at all concentrations on soybean and cowpea and at concentrations <25 μM on green gram. The non-toxic Al-fulvate complex remained in solution. Both malic and oxalic acid, at the concentrations tested, failed to alleviate Al toxicity on any species; a much higher proportion of the added Al remained in monomeric form in the presence of these acids.

Introduction

Beneficial effects of both soluble and solid forms of organic matter in decreasing Al toxicity to plants grown in acid soils, sand culture, and solution culture have been reported. Limited available data indicates that Al phytotoxicity is greatly reduced by short chain carboxylic acids at concentrations found in soil solutions (Hue *et al.*, 1986). They reported that tap-root elongation of cotton (*Gossypium hirsutum* L.) was reduced by Al when the molar ratios of citric acid and oxalic acid: Al were <2.7, and when the molar ratio of malic acid: Al was <6.8.

Fulvic acid is a major soil humic compound commonly found in acid soil solutions. It has ligands which are able to form strong complexes with Al (Schnitzer, 1969). Humic and fulvic acids extracted from acid soils have few free carboxyl groups and are present largely as iron and Al complexes (Schnitzer and Skinner, 1963; Hargrove and Thomas, 1981). Limited solution culture evidence suggests that Al complexed by soluble soil organic matter is not phytotoxic to roots of maize seedlings (Bartlett and Riego, 1972). Fulvic acid may be one of the Al detoxifiers in acid soils; however, this role of fulvic acid needs to be confirmed.

The present experiment describes effects of fulvic, oxalic, and malic acids on tap-root elongation of soybean, cowpea, and green gram in the presence of Al. Oxalic acid and malic acid were selected for comparison with fulvic acid because they exist in acid soil solutions; the concentration of each acid was chosen within the mid-range of reported values for acid soil solutions.

Methods

Fulvic acid was prepared from a brownish-coloured natural swamp water of intial pH 4.0. Properties of the concentrated fulvic acid solution after filtration through a 0.45 μm membrane filter to remove microorganisms are shown in Table 1. This solution was stored at 5°C in a sterilized bottle until required.

The experiment was conducted in 4.5 L pots of continuously aerated nutrient solution maintained at 25°C in constant temperature water baths in a glasshouse. Treatments consisted of a factorial combination of four Al concentrations (0, 12.5, 25 and 50 μM as $Al(NO_3)_3 \cdot 9H_2O$), and two concentrations either of oxalic or malic acid (0, 50 μM) or fulvic acid (0, 65 mg L^{-1} of organic carbon). The treatments were replicated four times. Appropriate aliquots of stock solutions of the organic acids were added to the designated pots which contained 3.5 L of basal nutrient solution at pH 5.5. The stock solution of $Al(NO_3)_3$ was then added dropwise with vigorous stirring. The treated solutions in all pots were then made up to 4.5 L and the pH was adjusted to 4.5 ± 0.05 by addition of 0.05 M H_2SO_4. The solutions were then equilibrated under continuous aeration for 7 days prior to planting.

Four seedlings of soybean cv. Fitzroy, five seedlings of cowpea cv. Vita 4, or green gram cv. Berken with roots 25–30 mm long were transplanted into support baskets placed on top of each pot so that the roots were in contact with the solution. Black polyethylene beads were placed around the seedlings in the support baskets. After 4 days, plants were harvested and the length of the tap roots was measured.

The solutions were sampled prior to planting and at intervals throughout the experiment. They were allowed to stand without aeration for 1 h prior to sampling at a depth of 40 mm. The solution samples were filtered through a 0.45 μm membrane filter before analysis. Total Al in the solutions was determined using ICP emission spectroscopy, and the concentration of monomeric Al in those solutions not containing any added organic acid was measured colorimetrically using pyrocatechol violet (Dougan and Wilson, 1974). Monomeric Al in the presence of organic acids was determined using the short-term colorimetric procedure developed by Kerven *et al.* (1989). Solution pH, SO_4-ion concentration, and electrical conductivity were determined also. Molar ionic strength was estimated as 0.013 X electrical conductivity (mS cm^{-1}) (Griffin and Jurinak, 1973). The sum of activities of Al monomers in solution ($\Sigma a_{Al\ mono}$) was calculated from solution parameters as described by Blamey *et al.* (1983). The concentrations of the individual basal nutrients were determined. With the exception of phosphorus, which declined from 1.0 μM to values in the range 0.45 to 0.55 μM over the 4 days, there were no significant changes. The same solutions were reused successively for soybean, cowpea, and green gram. Phosphorus was added to all pots to restore the concentration to 1 μM before new seedlings were planted. Aluminium concentrations in all pots decreased slightly over the period of the experiment, but no further additions of $Al(NO_3)_3$ were made. The pH of the solutions in all pots was readjusted to 4.5 ± 0.05 prior to planting each species. The concentrations of organic acids in the solutions were not determined.

Results

Effect of organic acids on total and monomeric Al in solution

In the absence of organic acids, from 85 to 96% of the total Al in solution was present as mono-

Table 1. Some chemical properties of fulvic acid solution prepared from Beerwah swamp water

pH	Concentration of element (μM)								Organic C (mg L^{-1})	D.M.[a] (mg L^{-1})	Ash %
	Ca	Mg	K	Na	Fe	Mn	Zn	Al			
2.2	15	8	45	13	5	nd[b]	nd[b]	5	4520	6590	2.38

[a] Dry matter
[b] Not detected

Table 2. Concentrations of total and monomeric Al measured and the calculated sum of activities of monomeric Al species in nutrient solutions in the presence and absence of oxalic acid, malic acid or fulvic acid. [Values are means of eight determinations]

Organic acid treatment	Soybean				Cowpea				Green gram			
	Nominal Al concn				Nominal Al concn				Nominal Al concn			
	0	12.5	25	50	0	12.5	25	50	0	12.5	25	50
	Total Al concn (μM)											
Control	0	13	24	47	0	11	22	45	0	10	21	43
Oxalic	0	12	22	44	0	11	19	45	0	11	20	45
Malic	0	12	25	43	0	11	21	44	0	9	22	45
Fulvic	0	17	26	51	0	14	27	50	0	14	25	47
	Monomeric Al concn (μM)											
Control	0	11	23	45	0	10	21	43	0	9	20	41
Oxalic	0	10	21	35	0	9	16	32	0	10	17	30
Malic	0	10	19	39	0	10	17	36	0	9	16	32
Fulvic	0	nd[a]	3.2	5.9	0	nd[a]	3.0	5.2	0	nd[a]	1.0	4.2
	Sum of activities of Al monomers (μM)											
Control	0	9.0	18.3	35.6	0	7.8	16.7	34.7	0	7.0	15.2	32.1
Oxalic	0	8.2	18.2	28.5	0	7.5	13.0	25.7	0	7.9	13.8	23.4
Malic	0	8.0	15.8	31.7	0	8.5	14.2	29.9	0	7.1	13.0	24.9
Fulvic	0	nd[a]	2.5	4.5	0	nd[a]	2.4	4.1	0	nd[a]	0.9	3.3

[a] Not detected.

meric species (Table 2). Only a small proportion of the Al present in solution was complexed by oxalic or malic acid; the measured concentrations of uncomplexed monomeric Al accounted for 73 to 100% of the total Al concentration. In contrast, no monomeric Al was detected in the 12.5 μM nominal Al treatment when fulvic acid was present. At 25 and 50 μM nominal Al concentrations, only 10 to 12% of the added Al remained uncomplexed. Because the ionic strengths of all solutions were low (c. 2.05 mM) and similar to each other, the effects of the treatments on Al activities were closely similar to their effects on concentrations of free monomeric Al remaining in solution (Table 2).

Effects on tap-root elongation

In the absence of any added organic acid, tap-root elongation of soybean, cowpea and green gram decreased markedly with increasing total Al concentration in solution (Fig. 1). Neither oxalic acid nor malic acid at 50 μM had any significant protective effect against Al inhibition of tap-root elongation in any species, even at 12.5 μM Al. In contrast, fulvic acid was highly effective in protecting the roots of all species against Al toxicity. In green gram alone, there was a small but significant depression of root length in the 50 μM Al treatment. The success of fulvic acid in preventing Al toxicity was due to the much reduced sum of activities of Al monomers in its presence (Fig. 1).

Discussion

Fulvic acid at 65 mg L^{-1} of organic carbon was effective in eliminating the toxic effects of 50 μM Al on tap-root elongation totally in soybean and cowpea, and to a very large extent in green gram. The molecular weight of the fulvic acid used in the present experiment was not established; consequently, the molar concentration of fulvic acid responsible for the observed effects on Al speciation and on root growth cannot be determined precisely. However, Kerven (pers. comm.) used gel filtration chromatography to arrive at an approximate mean molecular weight of 2000 for a fulvic acid preparation

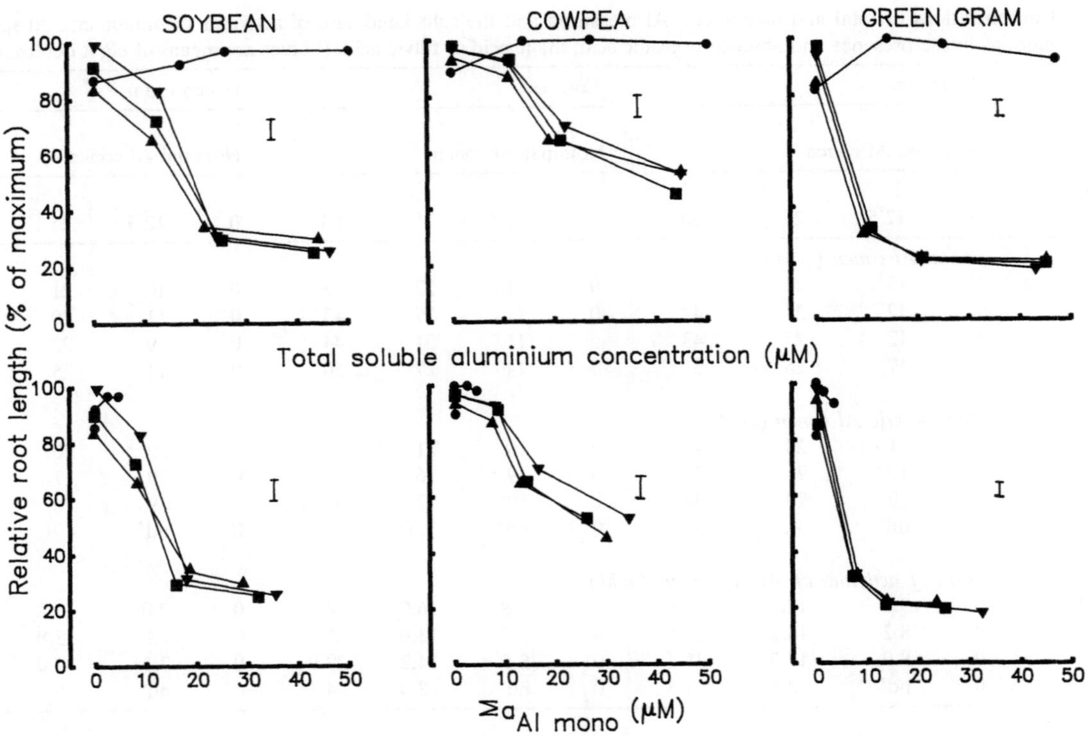

Fig. 1. Relationships between relative root length of soybean cv. Fitzroy, cowpea cv. Vita 4, and green gram cv. Berken and (a) total Al concentration and (b) $\Sigma a_{\text{Al mono}}$ after 4 days' growth in nutrient solutions at pH 4.5. [▼, control; ●, fulvic acid (65 mg organic carbon L^{-1}); ▲, malic acid (50 μM); ■, oxalic acid (50 μM); vertical bars represent LSD at $P = 0.05$].

similar to that used in the present study. Based on this molecular weight, the concentration of fulvic acid used in the present experiment was about 33 μM. This concentration of fulvic acid fully complexed 12.5 μM Al and complexed about 90% of the Al added to give a nominal concentration of 25 μM (Table 2). Even where 50 μM Al was added, the calculated Σa_{Almono} (4.5 μM) in the presence of fulvic acid did not exceed the critical toxicity level for root elongation of soybean (Alva *et al.*, 1986; Bruce *et al.*, 1988).

Unlike fulvic acid, oxalic acid and malic acid failed to show any protective effect against Al toxicity in any species when a nominal 50 μM Al was added (molar ratio 1:1). Tap-root elongation of the three species was strongly depressed even when the molar ratio of oxalic acid or malic acid to Al was 4:1. The present finding for oxalic acid differs from that of Hue *et al.* (1986), who showed that an oxalic acid: Al molar ratio >2.7 totally alleviated Al toxicity. The failure to observe any protective effect of oxalic acid against Al toxicity in the present experiment may be due to the higher calcium concentration (500 μM) used. Hue *et al.* (1986) used 250 μM. The concentration of oxalic acid remaining in solution in both studies was calculated using ionic strength, the activity coefficients of both oxalic acid and calcium, the concentrations of both added into the nutrient solutions, and the solubility product (K_{sp}) of calcium oxalate, viz. 2.34×10^{-9}. These calculations show that only 8 μM of the 50 μM oxalic acid would have remained free in the present experiment to complex with Al, compared with 16 μM of the 50 μM in the study of Hue *et al.* (1986). It thus appears that the concentration of oxalic acid which remained in solution in the present experiment was inadequate to complex most of the Al and alleviate the toxic effects on tap-root elongation. We conclude that complexation of oxalate by calcium is a factor which needs to be considered carefully in any future work on the possible role of oxalate in alleviation of Al toxicity.

The failure of malic acid to alleviate Al inhibition of tap-root elongation of the three species studied at malic acid: Al molar ratios of 1, 2 or 4 was fully consistent with the observations of Hue *et al.* (1986). They reported a protective effect of malic acid against tap-root elongation of cotton only when the malic acid: Al molar ratio was ⩾6.8.

Previous work by Tan and Binger (1986) established that humic acid benefitted growth of corn in sand culture; they suggested that humic acid chelated much of the added Al, rendering it non-phytotoxic. The present experiment has established that fulvic acid is able to alleviate Al toxicity through complexation and a consequent strong reduction in the concentration and activity of phytotoxic monomeric Al species. Although oxalic acid and malic acid failed to alleviate Al toxicity under the conditions of the present experiment, these acids may still play a useful role in alleviating Al toxicity under other circumstances.

References

Alva A K, Edwards D G, Asher C J and Blamey F P C 1986 Effects of phosphorus/aluminium molar ratio and calcium concentration on plant response to aluminum toxicity. Soil Sci. Soc. Am. J. 50, 133–137.

Bartlett R J and Riego D C 1972 Effect of chelation on the toxicity of aluminum. Plant and Soil 37, 419–423.

Blamey F P C, Edwards D G and Asher C J 1983 Effects of aluminum, OH:Al and P:Al molar ratios, and ionic strength on soybean root elongation in solution culture. Soil Sci. 136, 197–207.

Bruce R C, Warrell L A, Edwards D G and Bell L C 1988 Effects of aluminium and calcium in the soil solution of acid soils on root elongation of *Glycine max* cv. Forrest. Aust. J. Agric. Res. 39, 319–338.

Dougan W K and Wilson A L 1974 The absorptiometric determination of aluminium in water: A comparison of some chromogenic reagents and the development of an improved method. Analyst 99, 413–430.

Griffin G P and Jurinak J J 1973 Estimation of activity coefficients from the electrical conductivity of natural aquatic systems and soil extracts. Soil Sci. 116, 26–30.

Hargrove W L and Thomas G W 1981 Effect of organic matter on exchangeable aluminum and plant growth in acid soils. *In* Chemistry in the Soil Environment. Eds. R H Dowdy, J A Ryan, V V Volk and D E Baker. pp 151–166. American Society of Agronomy, Soil Science Society of America, Madison, WI.

Hue N V, Craddock G R and Adams F 1986 Effect of organic acids on aluminum toxicity in subsoils. Soil Sci. Soc. Am. J. 50, 28–34.

Kerven G L, Edwards D G, Asher C J, Hallman P S and Kokot S 1989 Aluminium determination in soil solution. II. Short-term colorimetric procedures for the measurement of inorganic monomeric aluminium in the presence of organic acid ligands. Aust. J. Soil Res. 27, 91–102.

Schnitzer M 1969 Reactions between fulvic acid, a soil humic compound and inorganic soil constituents. Soil Sci. Soc. Am. Proc. 33, 75–81.

Schnitzer M and Skinner S I M 1963 Organo-metallic interactions in soils. I. Reactions between a number of metal ions and the organic matter of a podzol Bh horizon. Soil Sci. 96, 86–93.

Tan K H and Binger A 1986 Effect of humic acid on aluminium toxicity in corn plants. Soil Sci. 141, 20–25.

M. L. van Beusichem (Ed.), *Plant nutrition – physiology and applications*, 381–389.

PLSO IPNC453

Organic acids related to differential aluminium tolerance in wheat (*Triticum aestivum*) cultivars

C.D. FOY[1], E.H. LEE[1], C.A. CORADETTI[1] and G.J. TAYLOR[2]
[1]*Climate Stress Laboratory, USDA-ARS, Beltsville, MD 20705, USA, and* [2]*Department of Botany, University of Alberta, Edmonton, Alberta, Canada T6G 2E9*

Key words: aluminium toxicity, aluminium tolerance, aluminium detoxification, chelation, high performance liquid chromatography, *Triticum aestivum* L.

Abstract

Five cultivars of wheat (*Triticum aestivum* L.), which represented a range of tolerance to Al, were grown in nutrient solutions containing 0 or 2 mg Al L^{-1} at pH 4.5 and analyzed for organic acids by high performance liquid chromatography. Cultivars used were 'Atlas 66' and 'Seneca' (Al-tolerant), 'Centurk' (intermediate) and 'Redcoat' and 'Scout' (Al-sensitive). Aluminium treatment (averaged over five cultivars) decreased concentrations of *c*-aconitic acid and increased those of fumaric and malic acids in plant shoots. Cultivar effects (averaged over two Al levels) were significant for *c*-aconitic, *t*-aconitic, fumaric, succinic and total organic acids in plant shoots. Only two of the organic acids studied, *c*-aconitic and fumaric, were significantly affected by Al × cultivar interactions. Aluminium treatment (averaged over five cultivars) increased concentrations of fumaric, malic, succinic and total organic acids and decreased those of *c*-aconitic acid in plant roots. Cultivars (averaged over two Al treatments) differed significantly in concentrations of fumaric, malic and total organic acids in roots. Concentrations of organic acids in plant roots were not affected by Al × cultivar interactions. Differential Al tolerances among cultivars were not consistently related to differences in concentrations of organic acids in plant shoots or to changes in these concentrations under Al stress.

Introduction

Plant species and cultivars within species differ widely in tolerance to Al toxicity, but the physiological or biochemical mechanisms involved in differential Al tolerance are still not clear (Foy, 1984; 1988; Roy *et al.*, 1988; Taylor, 1988a; 1988b). One hypothesis is that Al-tolerant plants contain and exude organic acids or other ligands that chelate Al in the rhizosphere or within the plant, and thereby reduce its chemical activity and toxicity (Jones, 1961). For example internal detoxification of Al by chelation could explain how Al tolerant wheat (*Triticum aestivum* L). cultivars tolerate intracellular Al concentrations five to seven times those tolerated by sensitive cultivars (Niedziela and Aniol, 1983).

Detoxification of Al by chelation has been demonstrated by many investigators. The Al-detoxification capacities of organic acids correspond with relative positions of OH/COOH groups on the main C chain, that is, those favoring the formation of a stable 5 or 6 bond structure with Al (Hue *et al.*, 1986) in accordance with this principle, citric acid, in particular, forms strong chelates with Al. Hydroxy Al was toxic to maize (*Zea mays* L.) seedlings, but Al citrate was not Lee (1977). Bartlett and Riego (1972) also found that the toxicity of ionic Al^{3+} for maize was prevented by complexing the Al with citrate, EDTA or soil organic matter extract. Citric acid at a molar ratio of 3 citric acid to 1 Al reduced the Al-induced inhibition of pollen tube extension in tea, *Camellia sinensis* L. Ktze, (Konishi *et al.*, 1988). Citric acid reversed the Al inhibition of yeast hexokinase (Wormack and Colowick, 1979), and organic acids prevented Al-induced conformational changes in calmodulin, a Ca protein that regulates ATPase activity in membranes (Haug and Caldwell, 1985; Suhayda and Haug, 1984; 1985; 1986).

Several investigators have correlated organic acids with differential Al tolerances in plants. Klimashevskii and Chernysheva (1980) reported that Al-tolerant cultivars of pea, (*Pisum sativum* L.) maize and barley (*Hordeum vulgare* L.) contained higher concentrations of citric acid than sensitive cultivars of the same species. Aluminium tolerant cells of carrot callus tissue released more citric acid to the culture medium than did cells of the parent line (Ojima *et al.*, 1984). Furthermore, a growth medium conditioned by such cells decreased Al toxicity in other carrot cell cultures (Ojima and Ohira, 1985; 1988). Detoxification of the conditioned medium was duplicated by adding citric acid but not succinic or fumaric acids. Results indicated that the Al tolerance of TA-1 carrot cells may be due to the chelating effects of citric acid. Shoots and roots of plants cultured from Al-tolerant carrot cells were also much more tolerant to 1 m*M* $AlCl_3$ than were those generated from Al-sensitive cells (Ojima and Ohira, 1988). The Al tolerance of carrot, determined by root elongation tests, agreed with that determined by staining of roots with haemotoxylin. However, more recent evidence showed that the supposedly Al tolerant cell line had been selected in the presence of precipitated Al phosphate rather than Al ions (Koyama *et al.*, 1988). The selected cell line was more sensitive to Al ions than the wild type. Furthermore, the excretion of citric acid occurred only in the presence of slowly soluble phosphate (Al or Fe). Hence, it appears that excretion of organic acids was a response to low P availability rather than Al ions. Cambraia *et al.* (1983) observed that Al stress increased concentrations of organic acids, sugars and amino acids in roots of sorghum (*Sorghum bicolor* L. Moench) plants. An Al-tolerant sorghum cultivar accumulated significantly higher concentrations of *t*-aconitic and malic acids than did an Al-sensitive cultivar. These investigators concluded that Al was detoxified by a chelation mechanism.

Roots of an Al-tolerant 'Dade' snapbean (*Phaseolus vulgaris* L.) cultivar contained higher concentrations of total organic acids, either in the presence or absence of Al, than did those of the Al-sensitive 'Romano' cultivar (Lee and Foy, 1986). Aluminium stress reduced concentrations of malic and citric acids more than those of other organic acids studied. Hence, results indicated that Dade had a higher potential for chelating and detoxifying Al than Romano. Aluminium stress induced by 55 and 110 μM Al at pH 4.5 reduced concentrations of citric, succinic and total organic acids in the roots of Al-sensitive 'Kearney' barley but not in those of Al-tolerant 'Dayton' (Foy *et al.*, 1987). Aluminium stress decreased concentrations of levulinic acid more than those of citric acid, and these decreases were greater in Al-sensitive Kearney than in Al-tolerant Dayton. Evidence showed that superior Al tolerance in the Dayton barley (compared with Kearney) was correlated with the ability to maintain normal concentrations of organic acids in the presence of Al. Hence, Dayton either protected its organic acids from destruction more effectively than did Kearney or maintained the production of these acids more effectively under Al stress.

While chelation of Al in the cytoplasm or rhizosphere may be an effective means of reducing the activity of phytotoxic Al species, several authors have suggested that such a tolerance mechanism may involve a high energetic cost to the organism (Taylor, 1987; 1988b). Aluminium tolerance mechanisms involving the production and exudation of chelating agents to complex and detoxify Al would impose a high ecological cost upon species surviving on high Al sites; the consequences are believed to be a reduced growth rate and narrow ecological adaptation limits (Cuenca and Herrara, 1987).

The objective of our study was to determine the effects of Al stress on concentrations of organic acids in shoots and roots of wheat cultivars representing a wide range of tolerance to Al toxicity.

Materials and methods

Growth of plants

Seeds of the five wheat cultivars were soaked in 5% Chlorox for 5 minutes, germinated in deionized water for 24 hours, and transferred to glass trays containing stainless steel screens covered by cheesecloth which dipped into de-ionized water (3 days). Seedlings were elongated for 4 days in an aerated dilute nutrient solution (Taylor and Foy, 1985) before being exposed to the Al treatments. The basal nutrient solution contained the following: 3.3 NO_3^--N, 0.3 NH_4^+-N, 1.0 Ca, 0.8 K, 0.3 Mg, 0.1 SO_4-S, 0.1 HPO_4-P in m*M*, and 64.0 Cl. 60.2 Na, 20.0 Fe, 6.0 B, 2.0 Mn, 0.5 Zn, 0.15 Cu and 0.1 Mo in μM. Iron was supplied as FeHEDTA (ferric hydroxy ethyethylene diamine tetraacetate). Seedlings were supported by cork stoppers and grown in 8 L polyethylene buckets covered by black plexiglass tops containing 4 stoppers each. Each stopper held 3 plants. The experiment was conducted in a growth chamber at 22–25°C; a photosynthetic photon flux density (PPF) of 400 $\mu mol\ s^{-1}\ m^{-2}$, relative humidity of 50–60%; and a light dark cycle of 16/8 h. Light was provided by a combination of incandescent and fluorescent lights. The experimental treatments were 0 and 2 mg Al L^{-1} added as $AlK(SO_4)_2 \cdot 12\,H_2O$. Solutions were adjusted to pH 4.5 daily. The experimental design was a randomized complete block containing 5 cultivars, 2 Al treatments and 3 replications.

Extraction of organic acids

After 15 days of growth in the experimental solutions, plants were harvested and shoot and root fresh weights were determined. Fresh samples of shoots and roots were frozen immediately in dry ice and stored in a freezer at −70°C until organic acids were extracted. Prior to freezing, the roots were washed 3 times in deionized water. During sample preparation, precautions were taken to prevent thawing. During grinding, the mortar and pestle were kept wedged in dry ice, and at intervals, liquid N_2 was poured over the samples. Shoot and root samples were ground to a fine powder, lyophilized, and stored in a desiccator until analyzed. Samples of 0.2 g plant tissues were ground thoroughly (7-mL Ten Broeck tissue grinder) with 3 mL of cold 0.01 *N* H_3PO_4 for 3 min. Samples were transferred to centrifuge tubes and the tissue grinder was rinsed with a minimal volume (not exceeding 10 mL) of 0.01 *N* H_3PO_4. The combined extracts and rinses were centrifuged at 10,000 *g* for 20 min at 4°C. The supernatant was decanted into a 10 mL volumetric flask and brought to volume 0.01 *N* H_3PO_4. The supernatant was filtered through an octadecyl (C_{18}) column with the Baker Extraction System (J. T. Baker Chemical Co., Phillipsburg, NJ)[z]. Samples were then transferred to automatic sample vials through a 0.45 μm filter (Millipore Corp., Bedford, MA)[z].

Organic acid analysis by HPLC

The high performance liquid chromatography (HPLC) apparatus used was a Perkin-Elmer Model 75 equipped with a Model Series 2 solvent delivery system and a 10-μL lock with an automatic sample injector. The HPLC system was equipped with a dual beam, continuously variable wave length detector and a Model LC-420 B auto sampler. The effluent for organic acid analysis was monitored at 210 nm for the detection of carboxyl groups.

Signals from the detector were recorded with a Waters Data Model 730 instrument which integrated the peak area and was programmed (with external standards) to calculate the concentrations of individual organic acids in each extract. Organic acids were separated on a 300 × 7.8 mm HPLC organic analysis column (Bio-Rad Laboratories, Richmond, CA)[z]. The column contained Aminex HPX-87 resin (8% cross-linked resin, 9 μm dia.), a strong cation exchange resin which separates organic acids by ion exclusion and partition chromatography. The HPLC system also included an ion exclusion microguard column which was connected to the analytical column to remove sample contaminants.

Organic acid analyses were performed isocratically at a flow rate of 0.5 mL min^{-1} at 22°C and 100 psi pressure. The mobile phase was 0.01 *N* H_3PO_4 (prepared by diluting HPLC reagent-grade H_3PO_4 with HPLC grade distilled water) and filtering through a 0.45 μm Millipore filter. Analytical grade organic acids (Sigma Chem. Co., St. Louis, MO)[z] were used as stan-

[z] Mention of a trademark or proprietory product does not constitute a guarantee or warranty of this product by the U.S. Department of Agriculture and does not imply its approval to the exclusion of other products that also may be suitable.

dards. Standard mixtures were composed of several representative short chain acids and TCA cycle acids at concentrations of 0.1 to 10 mg mL^{-1} in 0.01 $N\ H_3PO_4$ in HPLC grade distilled water to calibrate the column. Sample loadings (10 μL) on the HPLC column were made by an autosample injector.

Results

Growth

A factorial analysis of variance showed significant Al × cultivar interactions on fresh weights of plant shoots ($P = 0.0112^*$) and on fresh weights of roots ($P = 0.0227^*$). Aluminium treatments reduced both shoot and root fresh weights of Redcoat and Scout cultivars, and the shoot fresh weight of Seneca, but did not affect those of Atlas 66 and Centurk (Tables 1 and 2). Aluminium tolerance rankings for these cultivars are in general agreement with those obtained earlier on acid soils (Foy *et al.*, 1974) and in nutrient solutions (Taylor and Foy, 1985). For example, Atlas 66 was more tolerant to Al in nutrient solution than Redcoat or Scout; however, Seneca was less tolerant and Centurk more tolerant than in a previous experiment (Taylor and Foy, 1985).

pH changes

Final solution pH values at harvest time were lower in treatments with Al than without Al, despite daily adjustments to pH 4.5. In the presence of Al, the Al-sensitive Scout had a lower final solution pH than Al-tolerant Atlas 66 but no lower than that induced by Al-tolerant Centurk. In Al-containing solutions, the final pH values were 5.7, 5.7, 5.1, 4.9 and 5.2 for Atlas 66, Seneca, Centurk, Redcoat and Scout, respectively. In solutions without Al, the corresponding final pH values were 6.0, 6.7, 5.5, 6.1, and 6.4, for the same cultivars, respectively. These patterns of solution pH coincided with those obtained earlier (Taylor and Foy, 1985); however, it is doubtful that differential pH change in the rhizosphere is a primary cause of differential Al tolerance, at least in wheat (Miyasaka *et al.*, 1989; Taylor, 1988a; b; c; d).

Organic acids in plant shoots

Aluminium treatment (averaged over five cultivars) decreased concentrations of *c*-aconitic and increased those of fumaric and malic acids in plant shoots (Table 3). Tendencies for Al to increase concentrations of *t*-aconitic and total organic acids in plant shoots were not significant (data not shown). Cultivars (averaged over treatments with and without Al) differed in concentrations of *c*-aconitic, *t*-aconitic, fumaric, succinic and total organic acids in plant shoots. Considering both treatments with and without Al, Seneca had the highest and Atlas 66 the lowest concentrations of *t*-aconitic acid in plant shoots. The Al-sensitive cultivar Scout had higher *t*-aconitic acid concentrations than Al-tolerant Atlas 66. Seneca had higher succinic acid concentrations than Redcoat, but the other two Al-tolerant cultivars (Atlas 66 and Centurk) were not differ-

Table 1. Effects of Al on shoot and root growth of wheat cultivars in nutrient solution

Al added mg L^{-1}	Atlas 66	Seneca	Centurk	Redcoat	Scout
	Top fresh wt (*g* pot^{-1})[a]				
0	12.75 a	16.89 a	9.43 a	12.97 a	16.24 a
2	14.56 a	13.39 b	9.30 a	8.96 b	9.21 b
	Root fresh wt (*g* pot^{-1})[a]				
0	0.43 a	0.66 a	0.43 a	0.56 a	0.97 a
2	0.43 a	0.50 b	0.44 a	0.29 b	0.40 b

[a] Within a shoot or root column, means having a letter in common are not significantly different at the 5% level by an LSD multiple comparison test.
LSD: Shoot fresh weight- 3.44; Root fresh weight- 0.26.

Table 2. Relative shoot and root fresh weights of wheat cultivars grown in nutrient solution containing 0 Al vs 2 mg Al L^{-1}

Cultivar	Relative top wt[a] Al / no Al	Relative root wt[a] Al / no Al
Atlas 66	114 a	103 a
Seneca	84 abc	93 ab
Centurk	99 ab	104 a
Redcoat	70 bc	53 bc
Scout	58 c	46 c

[a] Within a column, means having a letter in common are not significantly different at the 5% level.
LSD: Shoot fresh weight = 31; Root fresh weight = 44.

ent from Al-sensitive Scout in this regard. Seneca shoots also had higher concentrations of total organic acids than Al-sensitive Redcoat, but Al-tolerant Atlas 66 and Al-sensitive Scout did not differ in this respect. Hence, when cultivar effects were averaged over all treatments, there was no obvious relationship between Al tolerance and total organic acid concentrations in plant shoots. Unexpectedly, citric acid concentrations were not affected by Al treatment, cultivar or Al × cultivar interactions in either shoots or roots (Table 3). Acetic acid was not detected in plant shoots. Only two of the organic acids measured in plant shoots, *c*-aconitic and fumaric, were affected by Al × cultivar interactions (Table 4). With no Al stress, Seneca shoots contained significantly higher concentrations of *c*-aconitic acid than Atlas 66, Centurk, Redcoat and Scout (Table 4). In the absence of Al stress, the Al-tolerant Centurk and Atlas 66 cultivars were not different from Al-sensitive Redcoat and Scout in concentrations of *c*-aconitic acid in their shoots. Under Al stress, Centurk was significantly higher in *c*-aconitic acid concentrations than Atlas 66, Redcoat and Scout and equal to Seneca. Aluminium tolerant and sensitive cultivars were not consistently different in concentrations of *c*-aconitic acid in their shoots when both groups were grown with Al. Aluminium treatment decreased concentrations of *c*-aconitic acid in all five cultivars, and this decrease appeared greatest in Seneca and least in Centurk. However, a separate statistical analysis of relative *c*-aconitic acid concentrations in plant tops (Al/no Al %) showed no significant cultivar differences. For example, the Al-induced decrease in Al-sensitive Scout and Redcoat was no greater than that in Al-tolerant Atlas 66.

The five cultivars did not differ in concentrations of fumaric acid in the shoots of plants grown without Al (Table 4). Aluminium treatment increased concentrations of fumaric acid in the shoots of Al-tolerant Centurk, but not in the other four cultivars, which were either Al-tolerant or sensitive (Table 4). In the presence of Al, the shoots of the Al-tolerant Centurk were higher in concentrations of fumaric acid than those of the other four cultivars which did not differ from each other in this respect. Aluminium tolerant Atlas 66 did not differ from Al-sensitive Redcoat and Scout in concentrations of fumaric acid in the shoots of plants grown with Al.

Table 3. Analysis of variance for organic acid concentrations in tops and roots of wheat cultivars grown in nutrient solutions containing 0 and 2 mg Al L^{-1}

Source of variation	Organic acids ($\mu g\ g^{-1}$ dry wt)						
	c-aconitic	*t*-aconitic	citric	fumaric	malic	succinic	Total
	Tops						
Al treatment	0.0001**	NS	NS	0.0002**	0.026*	NS	NS
Cultivar	0.0005**	0.023*	NS	0.0149*	NS	0.815[+]	0.0699[+]
Al × cultivar	0.0263*	NS	NS	0.0167*	NS	NS	NS
	Roots						
Al treatment	0.0021**	NS	NS	0.0012**	0.0001**	0.0203*	0.0004**
Cultivar	NS	NS	NS	0.0092**	0.0014**	NS	0.0112*
Al × cultivar	NS	NS	NS	NS	NS	NS	NS

* Significant at the 5% level; ** Significant at the 1% level; [+] Significant at the 10% level.

Table 4. Effect of Al on organic acid concentrations in shoots of wheat cultivars in nutrient solution. Mean separation was done only for those values involving significant Al × cultivar interactions (Table 3).

Cultivars	Organic acid concentrations in tops ($\mu g\ g^{-1}$ dry wt*)						
	c-aconitic	*t*-aconitic	citric	fumaric	malic	succinic	Total
	0 Al						
Atlas 66	56.2 bcd	616	92	0.42 c	100	67.4	951
Seneca	90.8 a	855	106	0.72 b	71	80.4	1227
Centurk	67.0 bc	509	85	0.52 c	54	55.1	768
Redcoat	67.8 b	682	101	0.60 c	26	45.4	942
Scout	52.9 cde	740	103	0.72 bc	61	65.5	1053
	2 mg Al L^{-1}						
Atlas 66	31.1 f	607	92	0.98 bc	64	48.7	899
Seneca	39.8 ef	824	110	1.22 b	209	83.2	1269
Centurk	49.0 de	735	114	2.12 a	168	74.4	1209
Redcoat	34.5 f	653	96	0.74 bc	104	45.6	994
Scout	29.8 f	781	94	0.84 bc	116	61.5	1142
Std. error	4.9	66	10	0.19	44	11.8	114

* Within a column, means having a letter in common are not significantly different at the 5% level.

Organic acids in plant roots

Aluminium treatment (averaged over five cultivars) increased concentrations of fumaric, malic, succinic and total organic acids and decreased concentrations of *c*-aconitic acid in plant roots (Table 3). Tendencies for Al to increase root concentrations of *t*-aconitic and citric acids were not significant. Cultivars (averaged over Al treatments) differed significantly in concentrations of fumaric, malic and total organic acids in roots (Table 3). The Al-tolerant Centurk contained lower concentrations of fumaric acid than the other four cultivars which did not differ from each other in this respect. Considering the average all treatments, (with or without Al) the roots of the Al-tolerant cultivars, Atlas 66 and Centurk, contained lower concentrations of malic acid than those of Al-sensitive Scout and Redcoat. Aluminium tolerant Atlas 66 and Centurk contained lower concentrations of total organic acids in their roots than Al-sensitive Redcoat. Organic acid concentrations in plant roots were not affected by Al × cultivar interactions (Table

Table 5. Effects of Al on organic acid concentrations in roots of wheat cultivars in nutrient solution

Cultivar	Organic acid concentration in roots ($\mu g\ g^{-1}$ dry wt)*						
	acetic	*t*-aconitic	citric	fumaric	malic	succinic	Total
	0 Al						
Atlas 66	62.6	4.6	37	1.1	49	28.8	186
Seneca	66.8	5.2	54	1.6	77	31.8	231
Centurk	97.7	4.6	12	0.4	34	22.0	176
Redcoat	51.4	5.7	192	1.2	68	21.6	283
Scout	49.5	5.2	37	1.2	111	27.9	230
	2 mg Al L^{-1}						
Atlas 66	40.8	6.0	60	1.8	110	38.7	278
Seneca	46.8	7.1	78	2.0	157	33.2	319
Centurk	39.4	4.5	54	1.1	101	31.9	233
Redcoat	40.9	5.7	130	1.9	218	37.4	428
Scout	39.3	4.5	63	1.7	251	40.7	405
Std. error	10.6	1.0	54	0.2	3	6.2	41

* Organic acid concentrations in roots were not significantly affected by Al × wheat cultivar interactions.

3), but the data for all treatments are shown in Table 5. *C*-aconitic acid was not detected in plant roots.

Discussion

The potential role of organic acids in Al tolerance has been discussed in several recent publications. With some exceptions (Cambraia *et al.*, 1983), the data indicate that concentrations of organic acids decline under Al stress, with Al tolerant cultivars maintaining higher tissue concentrations that Al sensitive cultivars (Foy *et al.*, 1987; Klimashevskii and Chernysheva, 1980; Lee and Foy, 1986; Suhayda and Haug, 1986). Results of our study were not in agreement with this general trend. When five cultivars of wheat were grown with 0 or 2.0 mg Al L^{-1} to produce distinct differences in Al tolerance, only two of the organic acids measured in plant shoots, *c*-aconitic and fumaric, were affected by Al × cultivar interactions. Although Al treatment decreased concentrations of *c*-aconitic acid in all five cultivars, the decline was no greater in Al sensitive Redcoat and Scout cultivars than in Al tolerant Atlas 66 and less than that in Al tolerant Seneca. Aluminium treatment increased concentrations of fumaric acid in the shoots of Al tolerant Centurk, but not in those of the other four cultivars. The effects of Al on fumaric acid were also not associated with Al tolerance. Organic acid concentrations in the roots of *Triticum aestivum* were not influenced by Al × cultivar interactions; these results are in contrast to those obtained with *Hordeum vulgare*, *Phaseolus vulgaris*, *Pisum sativum*, and *Zea mays* (Foy *et al.*, 1987; Klimashevskii and Chernysheva, 1980; Lee and Foy, 1986; Suhayda and Haug, 1986). Results of our current study indicated that differential Al tolerance in *Triticum aestivum* is not correlated with changes in organic acid concentrations of either shoots or roots.

Failure of our results for wheat to agree with those for other plant species, may reflect true differences in the biochemistry of Al toxicity and tolerance in different species or differences in experimental conditions. These differences have prompted us to question some of the interpretations of experimential data. Do differences in organic acid concentrations between Al tolerant and Al sensitive plant cultivars reflect increased production of potential Al binding ligands in tolerant cultivars as a means of reducing the phytotoxicity of Al, or does the operation of other tolerance mechanisms prevent Al-induced disruption of normal patterns of organic acid synthesis and degradation? In Al sensitive cultivars, inhibition of key enzymes along biosynthesis or degradation pathways could lead to accumulation of reaction substrates and a reduction in concentrations of reaction products. In Al tolerant cultivars, normal concentrations of organic acids might be maintained through immobilization, compartmentalization or detoxification in the cytosol, or a mechanism which limits entry of Al into the cytosol (Taylor, 1988a; 1988b).

The hypothesis that differences in organic acid concentrations among Al-tolerant and Al-sensitive cultivars are results of differential tolerance to Al, rather than its cause, is consistent with the growing body of literature which indicates that organic acid levels in both Al-tolerant and Al-sensitive cultivars decline in response to Al stress, while Al-tolerant cultivars are able to maintain higher tissue concentrations of certain carboxylic acids than sensitive cultivars. This interpretation would also account for reports of both increased and decreased concentrations of specific organic acids under conditions of Al stress. For example, in our study, increased concentrations of *c*-aconitic acid were accompanied by decreased concentrations of fumaric acid. Similarly, Foy *et al.* (1987) reported increased concentrations of malic acid in an Al-sensitive cultivar of *Hordeum vulgare* grown with Al, while concentration of citric, succinic, and levulinic acids decreased.

While we would agree that carboxylic acids in the cytosol could reduce or prevent the toxic effects of Al at the cellular level, it is not clear that this process operates as a primary tolerance mechanism in *Triticum aestivum* or other plant species. An equally plausible interpretation of the data reported in this and other studies is that differences in concentrations of organic acids among Al-tolerant and Al-sensitive cultivars reflects the operation of (an)other tolerance mechanism(s). Furthermore, several important gaps of information about the role of organic acids in Al tolerance have yet to be filled. While the stability of Al^{3+}-organic complexes seems well established, it is possible that the stability of these complexes in the cytosol has been overestimated by failure to recognize that Al^{3+} is not the major species present in neutral solutions. The neutral $Al(OH)_3$ limits the solubility of free $Al^{3+} \cdot 6H_2O$ to less than 10^{-10} *M* at pH 7.0, yet it is the non-hydrolyzed Al^{3+} that has been widely used in calculations of complex stability (Martin, 1986; Taylor, 1988a). Additional information regarding the identity of Al-organic acid complexes in intact plants, as well as information about the toxic effect of Al on the activity of key enzymes will also be required before the potential role of organic acids in conferring tolerance to Al can be adequately tested.

Acknowledgements

We thank Randy Rowland for instruction in the use of HPLC, Stephanie Wilding for help in a preliminary experiment with wheat and Miguel McCloud for other technical assistance.

References

Bartlett R J and Riego D C 1972 Effect of chelation on the toxicity of aluminum. Plant and Soil 37, 419–423.

Cambraia J, Galvani F R, Estevao M M and Santanna R 1983 Effects of aluminum on organic acid, sugar and amino acid composition of the root system of sorghum. J. Plant Nutr. 6, 313–322.

Cuenca G and Herrera R 1987 Ecophysiology of aluminum in terrestrial plants growing in acid, aluminum-rich tropical soils. *In* Ecophysiology of Acid Stress in Aquatic Organisms. Eds. H Witters and O Vanderborght. Ann. Soc. Zool., Belg. 117, 57–74, Supplement 1.

Foy C D 1984 Physiological effects of hydrogen, aluminum and manganese toxicities in acid soils. *In* Soil Acidity and Liming. Agronomy Monograph 12, Second edition. Ed. F Adams, pp 57–97, Am. Soc. Agron., Madison, WI.

Foy C D 1988 Plant adaptation to acid, aluminum-toxic soils. Commun. Soil Sci. Plant Anal. 19, 959–987.

Foy C D, Lafever H N, Schwartz J W and Fleming A L 1974. Aluminum tolerance of wheat cultivars related to region of origin. Agron. J. 66, 751–758.

Foy C D, Lee E H and Wilding S B 1987 Differential aluminum tolerances of two barley cultivars related to organic acids in their roots. J. Plant Nutr. 10, 1089–1101.

Haug A R and Caldwell C R 1985 Aluminum toxicity in plants: The role of the root plasma membrane and calmodulin. *In* Frontiers of Membrane Research in Agriculture. Eds. J B St. John, E Berlin and P C Jackson, pp 359–381. Rowman and Allenheld, Totowa, NJ.

Hue N V, Craddock G R and Adams F 1986 Effect of organic acids on Al toxicity in subsoils. Soil Sci. Soc. Am. J. 50, 28–34.

Jones L H 1961 Aluminum uptake and toxicity in plants. Plant and Soil 13, 292–310.

Klimashevskii E L and Chernysheva N F 1980 Content of organic acids and physiologically active compounds in plants differing in their susceptibility to the toxicity of Al^{3+}. Soviet Agric. Sci. 2, 5–7.

Konishi S, Ferguson I B and Putterill J 1988 Effect of acidic polypeptides on aluminum toxicity in tube growth of pollen from tea (*Camellia sinensis* L.). Plant Sci. 56, 55–59.

Koyama H, Okawara R, Ojima K and Yamaza T 1988 Re-evaluation of characteristics of a carrot cell line previously selected as aluminum tolerant cells. Physiol. Plant. 74, 683–687.

Lee E H and Foy C D 1986 Aluminum tolerances of two snapbean cultivars related to organic acid content evaluated by high performance liquid chromatography. J. Plant Nutr. 9, 1481–1498.

Lee Y S 1977. Aluminum toxicity in corn seedlings. Hanguk Toyang Bilyo Hakhoe Chi. 10, 75–78.

Martin R B 1986 The chemistry of aluminum as related to biology and medicine. Clinical Chem. 32, 1797–1806.

Miyasaka S C, Kochian L V, Schaff J E and Foy C D 1989 Mechanism of aluminum tolerance in wheat: An investigation of genotypic differences in rhizosphere pH, k^+, and H^+ transport and root-cell membrane potentials. Plant Physiol. 91, 1188–1196.

Niedziela G and Aniol A 1983 Subcellular distribution of Al in wheat roots. Acta. Biochim. Pol. 30, 99–104.

Ojima K, Abe H and Ohira K 1984 Release of citric acid into the medium by aluminum tolerant carrot cells. Plant Cell Physiol. 25, 855–858.

Ojima K and Ohira K 1985 Reduction of aluminum toxicity by addition of conditioned medium from aluminum tolerant cells of carrot. Plant Cell Physiol. 26, 281–286.

Ojima K and Ohira K 1988 Aluminum tolerance and citric acid release from a stress-selected cell line of carrot. Commun. Soil Sci. Plant Anal. 19, 1229–1236.

Roy A K, Sharma A and Talukder G 1988 Some aspects of aluminum toxicity in plants. Bot. Rev. 54, 145–178.

Suhayda C G and Haug A 1984 Organic acids prevent aluminum induced conformational changes in calmodulin. Biochim. Biophys. Res. Commun. 119, 376–381.

Suhayda C G and Haug A 1985 Citrate chelation as a potential mechanism against aluminum toxicity in cells: The role of calmodulin. Can. J. Biochem Cell Biol. 63, 1167–1175.

Suhayda C G and Haug A 1986 Organic acids reduce aluminum toxicity in maize (*Zea mays*) root membranes Physiol. Plant. 68, 189–195.

Taylor G J 1987 Exclusion of metals from the symplasm: A possible mechanism of metal tolerance in higher plants. J. Plant Nutr. 10, 1213–1222.

Taylor G J 1988a The physiology of aluminum tolerance. *In*

Metal Ions in Biological Systems, Vol 24: Aluminum and Its Role in Biology. Ed. H. Sigel. pp 165–198. Marcel Dekker, NY.

Taylor G J 1988b The physiology of aluminum tolerance in higher plants. Commun. Soil Sci. Plant Anal. 19, 1179–1194.

Taylor G J 1988c Mechanisms of aluminum tolerance in *Triticum aestivum* (wheat). V. Nitrogen nutrition, plant induced pH and tolerance to aluminum: Correlation without causality? Can. J. Bot. 66, 694–699.

Taylor G J 1988d Aluminum tolerance is independent of rhizosphere pH in *Triticum aestivum* L. Commun. Soil Sci. Plant Anal. 19, 1217–1227.

Taylor G J and Foy C D 1985 Mechanisms of aluminum tolerance in *Triticum aestivum* L. (wheat). I. Differential pH induced by winter cultivars in nutrient solutions. Am. J. Bot. 72, 695–701.

Womack F C and Colowick S P 1979 Proton-dependent inhibition of yeast and bran hexokinase by Al and ATP preparations. Proc. Nat. Acad. Sci., USA, 76, 5080–5084.

M. L. van Beusichem (Ed.), *Plant nutrition – physiology and applications*, 391–396.

PLSO IPNC296A

Mineral element concentrations and grain yield of sorghum (*Sorghum bicolor*) and pearl millet (*Pennisetum glaucum*) grown on acid soil

R.B. CLARK[1], C.I. FLORES[2], L.M. GOURLEY[2] and R.R. DUNCAN[3]
[1]*Department of Agronomy and U.S. Department of Agriculture, Agricultural Research Service, University of Nebraska, Lincoln, NE 68583, USA,* [2]*Department of Agronomy, Mississippi State University, Mississippi State, MS 39762, USA, and* [3]*Department of Agronomy, University of Georgia, Georgia Agricultural Experiment Station, Griffin, GA 30223-1797, USA*

Key words: acid soil tolerance and toxicity, aluminium toxicity, leaf element concentrations, mineral deficiency, *Pennisetum glaucum* (L.) R. Br., *Sorghum bicolor* (L.) Moench, tropical soil, ultisol

Abstract

The adaptation of plants to acid soil conditions may be associated with mineral elements taken into plants. Leaf mineral element concentrations were determined in sorghum (*Sorghum bicolor* (L.) Moench) and pearl millet (*Pennisetum glaucum* (L.) R. Br.) genotypes grown on an acid ultisol (pH 4.0, 60% Al saturation) to better understand mineral element concentrations that might be associated with species and genotypic differences to tolerate acid soil conditions. Sorghum genotypes showed a broad range in grain yield while all pearl millet genotypes had relatively good grain yields when grown on the acid soil. Leaf concentrations of Si, Ca, Mg, S, and Cu were higher; P, K, and Al were lower; and N, Mn, Fe and Zn were similar in pearl millet when relative comparisons were made with sorghum. The relatively high leaf concentrations of Ca, Mg, and especially Si in pearl millet may have contributed to its ability to yield well on the acid soil.

Introduction

Mineral element deficiencies and toxicities occur extensively and are major constraints for production of many crops grown on acid soils (Clark, 1982; Foy, 1984; Foy *et al.*, 1978; Sanchez and Salinas, 1981). Toxicities of Al, Mn, and Fe (especially Al) are common mineral problems for plants grown on acid soils, but deficiencies of P, Ca, Mg, K, S, and Zn often appear and may be as important as the toxic elements to decrease plant growth. Aluminium toxicity is one of the most important factors restricting growth of plants on acid soils. Aluminium interacts with many of the mineral nutrients by interfering with their availability, uptake, and internal function. Aluminium also has specific effects on plant growth processes, especially those concerned with root development and elongation. Information is limited concerning differences among plant species and genotypes for mineral element concentrations and their association with plant ability to tolerate relatively severe acid soil conditions.

The objectives of this study were (1) to determine leaf concentrations of mineral elements in sorghum (*Sorghum bicolor* (L.) Moench) and pearl millet (*Pennisetum glaucum* (L.) R. Br.) genotypes grown on a relatively severe acid soil, and (2) to note mineral element concentrations that might be related with genotypic and species differences for tolerance to acid soil conditions.

Materials and methods

Ten sorghum and twenty pearl millet genotypes were grown on an acid ultisol (clayey, oxidic, isohyperthermic, Typic Palehumult) at Quilichao (80 km southeast of Cali) in Columbia, South America. The sorghum and pearl millet genotypes were from fairly broad backgrounds, and separated into acid soil tolerant (AS-T) and susceptible (AS-S) (sorghum) and into High 5 and Low 5 (pearl millet) groups according to grain yield (Table 1).

The chemical properties of the relatively severe acid soil were: 60% Al saturation; pH 4.0 (1 water : 1 soil); 7.9% organic matter; 4.00 Al, 1.75 Ca, 0.65 Mg, and 0.34 K in +cmol kg^{-1} soil; and 12.6 P, 38.6 Mn, 61.0 Fe, 2.80 Zn, and 1.33 Cu in mg kg^{-1} soil. Solutions used to extract the soil were: 1 *N* KCl for Al, Ca, and Mg and 1 *N* NH_4-acetate for K; Bray II for P; and 0.005 *M* DTPA (diethylenetriaminepentaacetic acid) for Mn, Fe, Zn, and Cu.

Table 1. Names of genotypes, genotypic group, and origin/genetic type of sorghum and pearl millet grown on a relatively severe acid ultisol in Columbia, South America

Plant species	Group[a]	Genotype name	Origin
Sorghum	AS-T	MN 4508	Uganda
		3D X 57/1/1/910	Uganda
		IS 3522	Sudan
		IS 6944	Sudan
		IS 12152	Ethiopia
		Tortillero	Honduras
	AS-S	B-Yellow PI	USA (Mississippi)
		B Wheatland Der.	USA (Mississippi)
		BTX 623	USA (Texas)
		ICA Nataima	Columbia
			Genetic type
Pearl millet	High 5	IVS-P78	Synthetic
		ICMS 7704	Synthetic
		ICMH 440	Hybrid
		ICMH 433	Hybrid
		IVC-A82	Population
		WC-C75	Cultivar
		NELC-P79	Synthetic
		ICMS 8021	Synthetic
		IVS 5454	Synthetic
		ICMS 7857	Synthetic
		ICMV 81111	Cultivar
		ICMV 81237	Cultivar
		ICMH 451	Hybrid
		NELC-H79	Synthetic
		ICMS 7703	Synthetic
	Low 5	ICMH 423	Hybrid
		ICMS 7835	Synthetic
		ICMV 81253	Population
		ICMS 8008	Population
		ICMH 415	Hybrid

[a] AS-T = acid soil tolerant, AS-S = acid soil susceptible, High 5 = Five highest grain yielders, and Low 5 = Five lowest grain yielders.

The soil site had been maintained at 60% Al saturation for two previous years and mineral nutrients other than P and N were considered adequate for plant growth. Phosphorus was added preplant as triple superphosphate at 44 kg P ha^{-1} and N was added as urea 30 days after planting at 100 kg N ha^{-1}.

The genotypes were grown during the dry season (September to January) using a randomized complete block design with three replications. Pearl millet and sorghum were grown in adjoining plots. Each genotype was planted in 4-row plots, 3 m long, and 0.60 m between rows. Plots were overplanted and seedlings thinned to 9 cm between plants (200,000 plants ha^{-1}). Harvested experimental units were four 2-m row segments within each plot.

Weeds were controlled initially with atrazine (2-chloro-4-ethylamino-6-isopropylamine-s-triazine) at 1.0 kg active ingredient ha^{-1} applied after planting with a power sprayer and later by hand. Bird damage was minimized during the grain filling stage and until grain was harvested. At least four leaves (second from the top) were randomly selected per plot and combined from each genotype at flowering (75 days after planting for sorghum and 52 days after planting for pearl millet) to obtain leaf samples for mineral element analysis. Each leaf sample was dried at 60°C for 7 d, ground to pass 0.5-mm screen, and representative amounts taken for mineral element analysis. These samples were sent to the University of Nebraska for mineral element analysis by energy-dispersive x-ray fluorescence (Knudsen *et al.*, 1981). Samples were subjected to Kjeldahl digestion and N determined colorimetrically with Nessler reagent (Jackson, 1958).

Results and discussion

Sorghum genotypes showed a greater range in grain yields than pearl millet genotypes (Table 2). Differences in grain yield among the sorghum genotypes was 11-fold compared to less than

Table 2. Grain yield and leaf concentrations of N, P, S, Ca, Mg, and K of sorghum (S) and pearl millet (PM) genotypes grown on a relatively severe acid ultisol in Colombia, South America

Trait or element	Plant	Range[a]	AS–T[b] High 5	AS–S[b] Low 5	Mean[c]
Grain yield	S	325–3600	3070*	800	2160
($kg\ ha^{-1}$)	PM	1980–3460	3230*	2330	2770
	PM/S		1.1	2.9	1.3
N	S	12.3–20.2	14.2	16.0	14.9
($mg\ g^{-1}$)	PM	14.8–22.4	18.5	19.4	18.7
	PM/S		1.3	1.2	1.3
P	S	0.82–1.42	0.98*	1.17	1.06
($mg\ g^{-1}$)	PM	0.70–1.03	0.85	0.90	0.86
	PM/S		.87	.77	.81
S	S	0.97–1.60	1.04*	1.26	1.13
($mg\ g^{-1}$)	PM	2.09–4.05	2.49*	3.50	3.10
	PM/S		2.4	2.8	2.7
Ca	S	4.7– 8.6	7.4*	6.4	7.0
($mg\ g^{-1}$)	PM	12.3–16.9	13.9	14.7	14.0
	PM/S		1.9	2.3	2.0
Mg	S	1.34–1.96	1.64	1.72	1.67
($mg\ g^{-1}$)	PM	3.49–7.33	4.76	5.19	4.74
	PM/S		2.9	3.0	2.8
K	S	4.5–15.4	12.7	11.9	12.4
($mg\ g^{-1}$)	PM	2.8– 8.6	4.8	3.7	4.0
	PM/S		.38	.31	.32

[a] Range = range for genotypic means.
[b] AS–T and AS–S = acid soil tolerant and susceptible sorghum genotypes for grain yield, respectively; High 5 and Low 5 = highest and lowest five pearl millet genotypes for grain yield, respectively.
[c] Mean for all genotypes.
* = Significance between AS–T and AS–S groups or between High 5 and Low 5 groups at $P < 0.05$ according to t test.

2-fold for the pearl millet genotypes. Higher grain yielding genotypes were considered to be more tolerant to acid soil conditions than lower grain yielders. Using this definition for acid soil tolerance, some of the sorghum genotypes had low tolerance to the acid soil conditions while all of pearl millet genotypes showed relatively good tolerance to the acid soil conditions. The AS-S sorghum group had nearly 4-fold lower grain yields than the AS-T group whereas the Low 5 pearl millet genotypes had 72% the yield of the High 5 group. The AS-T sorghum and High 5 pearl millet had similar grain yields.

Sorghum has generally been considered to be susceptible to acid soil conditions (Sanchez and Salinas, 1981), but sorghum germplasm with good tolerance to acid soil conditions has been identified (Borgonovi *et al.*, 1987; Duncan, 1981; 1982b; 1988; Flores *et al.*, 1988b; Gourley, 1987). Additional studies showed pearl millet genotypes to grow and yield well at 70% Al saturation compared to sorghum (unpublished data). Other studies showed pearl millet to tolerate acid soils better than sorghum (Araujo, 1983; Long *et al.*, 1973; Rao, 1985). Pearl millet grain yields grown at 50–53% were similar to sorghum grown at 40% Al saturation (Flores *et al.*, 1988a; b).

Differences among the sorghum and pearl millet genotypes for grain yield may have been a characteristic of specific genotypes within each species. That is, some genotypes could be considered to be exotic (photoperiod sensitive) compared to others that were improved breeding lines (nonphotoperiod sensitive). The three sorghum genotypes coming from USA breeding programs were relatively susceptible while five of the six AS-T sorghum genotypes originating from Africa (Table 1) had the highest grain yields and were relatively tolerant to the acid soil conditions. The pearl millet genotypes all yielded fairly well and represented synthetics, hybrids, cultivars, and selections from populations (Table 1) mostly of African origin (ICRISAT information).

Table 3. Leaf concentrations of Si, Al, Mn, Fe, Zn, and Cu of sorghum (S) and pearl millet (PM) genotypes grown on a relatively severe acid ultisol in Colombia, South America

Element	Plant	Range[a]	AS–T[b] High 5	AS–S[b] Low 5	Mean[c]
Si	S	8.1–18.8	15.6*	9.5	13.2
($mg\ g^{-1}$)	PM	27.9–43.4	33.5	38.7	37.1
	PM/S		2.2	4.1	2.8
Al	S	429–1855	610*	1064	792
($ug\ g^{-1}$)	PM	160–653	357	436	405
	PM/S		.59	.41	.51
Mn	S	768–1207	980	930	960
($ug\ g^{-1}$)	PM	851–1082	950	968	954
	PM/S		.97	1.04	.99
Fe	S	254–838	312*	491	383
($ug\ g^{-1}$)	PM	299–537	361	338	365
	PM/S		1.16	.69	.95
Zn	S	25.2–36.4	30.1	30.5	30.3
($ug\ g^{-1}$)	PM	20.4–77.0	24.8	38.2	36.2
	PM/S		.82	1.25	1.19
Cu	S	17.4–25.9	22.4	22.1	22.3
($ug\ g^{-1}$)	PM	27.7–36.2	32.9	32.8	32.4
	PM/S		1.47	1.48	1.45

[a] Range = range for genotypic means.
[b] AS-T and AS-S = acid soil tolerant and susceptible sorghum genotypes for grain yield, respectively; High 5 and Low 5 = highest and lowest five pearl millet genotypes for grain yield, respectively.
[c] Mean for all genotypes.
* =Significance between AS-T and AS-S groups or between High 5 and Low 5 groups at $P<0.05$ according to t test.

The sorghum and pearl millet genotypes varied fairly extensively for the mineral element concentrations in their leaves (Tables 2 and 3). The AS-T and AS-S sorghum groups showed greater differences in mineral element concentrations than the High 5 and Low 5 pearl millet groups. It is not known whether mineral element concentrations were related to grain yield in this study, but if they were, greater differences in mineral element concentrations within genotypes might be expected in sorghum than in pearl millet. Wide variability among genotypes for mineral elements has been noted in many sorghum genotypes grown on acid soils (Clark and Gourley, 1987; 1988; Clark *et al.*, 1988; Duncan 1981; 1982a).

Even though some comparisons of leaf mineral element concentrations of sorghum and pearl millet may be questionable because the two crops were grown in adjacent plots and pearl millet matured much earlier than sorghum, relative comparisons can be made and showed some marked differences between the species. These differences showed that the two species varied extensively in uptake and accumulation of mineral elements. These species also differed in tolerance to the acid soil conditions (grain yield).

Sorghum had fairly high leaf concentrations of P, K, and Al compared to pearl millet, pearl millet genotypes had high concentrations of S, Ca, Mg, Si, and Cu compared to sorghum, and both species had relatively similar concentrations of N, Mn, Fe, and Zn (Tables 2 and 3). Differences were noted in some mineral element concentrations among the genotypes considered to be tolerant and susceptible to the acid soil conditions. Among the elements that are normally toxic to plants grown on acid soils (Al, Mn, and Fe), only Al showed any differences between species; the more acid soil susceptible sorghum genotypes had higher Al concentrtions than the acid soil tolerant sorghum and the pearl millet genotypes (Table 3). Differences between the species for P were small compared to differences noted for some of the other mineral elements considered important to plants grown on acid soils, especially Ca and Mg (Table 2). Pearl millet contained 2- to 3-fold higher leaf concentrations of Ca and Mg than sorghum. On the other hand, pearl millet contained only about one-third the concentration of K as sorghum. An inverse relationship between (Ca + Mg) and K is common in many plants (Dibb and Thompson, 1985). Although relatively large differences (3-fold) were noted between the species for S, it is unknown what contribution S may have had on grain yield or plant ability to tolerate acid soil conditions.

The element to show one of the largest differences in leaf concentration between species was Si (Table 3). Pearl millet contained nearly 3-fold higher concentrations of Si than sorghum. The difference between AS-T sorghum and High 5 pearl millet was 2-fold, but the difference between AS-S sorghum and Low 5 pearl millet was 4-fold. High concentrations of Si could possibly have contributed to the relatively good pearl millet growth and grain yields on the acid soil. Silicon can alleviate Al (and Mn) toxicities in sorghum (Galvez *et al.*, 1987; 1989), and higher Si was found in AS-T compared to AS-S sorghum (Clark and Gourley, 1988). Higher Si was also found in leaves of sorghum grown on an acid soil at lower than at higher Al saturations (Clark and Gourley, 1987; 1988).

Acknowledgement

This research was supported in part by the International Sorghum/Millet Collaborative Research Support Program (INTSORMIL) through U.S. Agency for International Development (USAID) grant AID/DAN-1254-G-5065-00 and Project Nos. NE-114A and MS-111. Pearl millet seed for this study was provided by the International Crops Research Institute for Semi-Arid Tropics (ICRISAT), Patancheru, Andhra Pradesh, India. Published as Journal Article No. 8989 from the Nebraska Agricultural Research Division. Present address of C I Flores is Research and Development Department, CERES International, A.P. 484, Los Mochis, Sinaloa, Mexico C.P. 81200.

References

Araujo A G de 1983 Evaluation of cropping systems for the poor soils of the state of Piaui. *In* Proc. 1st Conf. Mix Cropping Systems in Northeast Brazil. Brazilian Inst. Agric. Res. (EMBRAPA), Teresina, Piaui, Brazil. (*in Portuguese*). (Cited by Rao, 1985).

Borgonovi R A, Schaffert R E and Pitta G V E 1987 Breeding aluminum-tolerant sorghums. *In* Sorghum for Acid Soils. Eds. L M Gourley and J G Salinas. pp 271–292. Int. Sorghum/Millet Coop. Res. Sup. Prog. (INTSORMIL)/Int. Crops Res. Inst. Semi-Arid Trop. (ICRISAT)/Int. Center Trop. Agric. (CIAT), Cali, Colombia, SA.

Clark R B 1982 Plant response to mineral element toxicity and deficiency. *In* Breeding Plants for Less Favorable Environments. Eds. M N Christiansen and C F Lewis. pp 71–142. John Wiley & Sons, New York.

Clark R B and Gourley L M 1987 Evaluation of mineral elements in sorghum grown on acid tropical soils. *In* Sorghum for Acid Soils. Eds. L M Gourley and J G Salinas. pp 253–270. Int. Sorghum/Millet Coop. Res. Sup. Prog. (INTSORMIL)/Int. Crops Res. Inst. Semi-Arid Trop. (ICRISAT)/Int. Center Trop. Agric. (CIAT), Cali, Colombia, SA.

Clark R B and Gourley L M 1988 Mineral element concentrations of sorghum genotypes grown on tropical acid soil. Commun. Soil Sci. Plant Anal. 19, 1019–1029.

Clark R B, Flores C I and Gourley L M 1988 Mineral element concentrations in acid soil tolerant and susceptible sorghum genotypes. Commun. Soil Sci. Plant Anal. 19, 1003–1017.

Dibb D W and Thompson W R Jr 1985 Interaction of potassium with other nutrients. *In* Potassium in Agriculture. Ed. R D Munson. pp 515–533. Am. Soc. Agron./Crop Sci. Soc. Am./Soil Sci. Soc. Am., Madison, WI.

Duncan R R 1981 Variability among sorghum genotypes for uptake and elements under acid soil field conditions. J. Plant Nutr. 4, 21–32.

Duncan R R 1982a Concentration of critical nutrients in tolerant and susceptible sorghum lines for use in screening under acid soil field conditions. *In* Genetic Aspects of Plant Nutrition. Eds. M R Saric and B C Loughman. pp 101–104. Martinus Nijhoff Publishers, The Hague, The Netherlands.

Duncan R R 1982b Field screening of genetically variable sorghum genotypes in acid soil stress environments. *In* Plant Nutrition 1982. Ed. A Scaife. Vol. 1, pp 139–144. Proc. 9th Internat. Colloq., Coventry. Commonw. Agric. Bureaux, Slough, UK.

Duncan R R 1988 Sequential development of acid soil tolerant sorghum genotypes under field stress conditions. Commun. Soil Sci. Plant Anal. 19, 1295–1305.

Flores C I, Clark R B and Gourley L M 1988a Agronomic traits of pearl millet grown on infertile acid soil. p. 108. Agron. Abstr., Am. Soc. Agron. Madison, WI.

Flores C I, Clark R B and Gourley L M 1988b Growth and yield traits of sorghum grown on acid soil at varied aluminum saturations. Plant and Soil 106, 49–57.

Foy C D 1984 Physiological effects of hydrogen, aluminum, and manganese toxicities in acid soil. *In* Soil Acidity and Liming. Ed. F Adams. pp 57–97. Am. Soc. Agron./Crop Sci. Soc. Am./Soil Sci. Soc. Am., Madison, WI.

Foy C D., Chaney R L and White M C 1978 The physiology of metal toxicity in plants. Annu. Rev. Plant Physiol. 29, 511–566.

Galvez L, Clark R B, Gourley L M and Maranville J W 1987 Silicon interactions with manganese and aluminum toxicity in sorghum. J. Plant Nutr. 10, 1139–1147.

Galvez L, Clark R B, Gourley L M and Maranville J W 1989 Effects of silicon on mineral composition of sorghum grown with excess manganese. J. Plant Nutr. 12, 547–561.

Gourley L M 1987 Finding and utilizing exotic Al-tolerant sorghum germplasm. *In* Sorghum for Acid Soils. Eds. L M Gourley and J G Salinas. pp 293–309. Int. Sorghum/Millet Coop. Res. Sup. Prog. (INTSORMIL)/Int. Crops Res. Inst. Semi-Arid Trop. (ICRISAT)/Int. Center Trop. Agric. (CIAT), Cali, Colombia, SA.

Jackson M L 1958 Soil Chemical Analysis. Prentice Hall, Englewood Cliffs, NJ.

Knudsen D, Clark R B, Denning J L and Pier P A 1981 Plant analysis of trace elements by x-ray. J. Plant Nutr. 3, 61–75.

Long F L, Langdale G W and Myhre D L 1973 Response of an Al-tolerant and an Al-sensitive genotype to lime, P, and K on three Atlantic coast flatwoods soils. Agron. J. 65, 30–34.

Rao M R 1985 Perspectives of sorghum and pearl millet in the cropping systems of northeast Brazil. *In* Proc. Sorghum in the Production Systems of Latin America. Eds. C L Paul and B deWalt. pp 87–106. Int. Maize Wheat Improv. Center (CIMMYT), Mexico City, Mexico. (in Spanish).

Sanchez P A and Salinas J G 1981 Low-input technology for managing oxisols and ultisols in tropical America. Adv. Agron. 34, 279–406.

M. L. van Beusichem (Ed.), *Plant nutrition – physiology and applications*, 397–401.

PLSO IPNC418A

Effects of aluminium on growth, nutrient uptake, proton efflux and phosphorus assimilation of aluminium-tolerant and -sensitive sorghum (*Sorghum bicolor*) genotypes

K. TAN[1] and W.G. KELTJENS[2,3]
[1]*Department of Soil Science and Agricultural Chemistry, South China Agricultural University, Guangzhou, China, and* [2]*Department of Soil Science and Plant Nutrition, Wageningen Agricultural University, P.O. Box 8005, 6700 EC Wageningen, The Netherlands.* [3]*Corresponding author*

Key words: aluminium distribution, nutrient uptake, phosphorus fractions, proton efflux, root morphology, *Sorghum bicolor* (L.) Moench

Abstract

Biomass and root morphology of the aluminium-tolerant sorghum (*Sorghum bicolor* (L.) Moench) genotype SC0283 were much less affected by Al in nutrient solution than those of the Al-sensitive genotype TAM428. Both genotypes accumulated most Al in the roots. Accumulation of Al increased with increasing Al supply and was greater in TAM than in SC. However, major part (over 60%) of the root Al was removable with 0.05 *M* H_2SO_4. Adsorption/precipitation of phosphorus in/on roots also increased with increased Al levels in the culture solution. Aluminium greatly suppressed Mg uptake by both genotypes and enhanced root proton efflux density. This H-ion efflux density of TAM without Al stress was even greater than that of SC at high Al. In both genotypes the concentrations of major P fractions were either not affected or even increased by Al. Results suggested that the severe depression in growth of TAM under Al stress may be attributed to the profound inhibition of root development due to high accumulation of Al.

Introduction

Plants grown in acid soils often suffer from Al toxicity. An obvious symptom of Al injury is inhibition of root development. The phytotoxicity of Al is related to its interference with cell division, cell elongation and the uptake of nutrients and water. That Al induces phosphorus deficiency through P fixation by Al in both soil and roots has been suggested (Foy *et al.*, 1978). However, limited information has been given on the overall status of major P fractions within plants grown under Al stress.

Plant species and varieties vary greatly in Al tolerance. Mechanism for Al tolerance has been studied in two aspects: the exclusion or detoxification of Al. It has been demonstrated that plant-induced pH changes in rhizosphere, which alters the activity of Al, could be related to Al susceptibility (Foy, 1988; Taylor, 1988). Study on Al distribution in plant's roots would help understanding the mechanism.

Different responses to acid soils among sorghum genotypes have been well documented (Duncan, 1988). Relative seminal root length has been used to measure Al susceptibility (Furlani and Clark, 1981). Difference in proton efflux by roots of Al-tolerant and -sensitive sorghum genotypes under Al stress has been reported in consideration of root surface area (Keltjens, 1987). Little has been known about distribution of Al and P metabolism in the Al-stressed sorghum plants.

This paper aimed at studying (1) Al distribution in the Al-tolerant and -sensitive sorghum genotypes and their reaction to Al toxicity with respect to biomass and root morphology; (2) phosphorus assimilation and the association of nutrient uptake and H-ion efflux under Al stress.

Materials and methods

The experiment was carried out in a growth chamber at a temperature of 25°C, relative air humidity of 75% and with a day length of 16 hours at a light intensity of 155 W m^{-2}.

Seedlings of two sorghum (*Sorghum bicolor* (L.) Moench) genotypes, SC0283 (Al-tolerant) and TAM428 (Al-sensitive), were grown in moist quartz sand for 7 days and then transferred to 50-L plastic containers (30 seedlings of a genotype per container) filled with aerated nutrient solutions (N source: NH_4NO_3) differing in Al levels, being 0, 0.4, 0.8 and 1.2 mg Al L^{-1} added as $AlCl_3$. The solutions were replaced once a week and maintained at pH 4.2 by daily titration with 0.2 *M* NaOH (or H_2SO_4), the amount of which was recorded as net H-ion (or OH-ion) efflux. The culture solutions were sampled three times a week for NH_4, NO_3, P, K and Al analyses and the results were used for adjustment and NH_4/NO_3 uptake calculation.

After 35 days all plants were harvested. Plants of each container were divided into 10 subsamples of 3 plants each. For analysis of the various P fractions (Chapin and Bieleski, 1982) 10-g fresh matter was taken out from shoot and roots of 3 subsamples and treated with liquid N_2. Three other subsamples were used for root length and root volume measurement. Another 3 subsamples of 20-g fresh root each were submerged for 15 minutes in 400-ml 0.05 *M* H_2SO_4. Subsequently, concentrations of Al and P in the desorption solutions were analysed.

The culture solution composition, plant sample treatments and analytic methods have been reported by Keltjens (1987).

Results and discussion

Plant growth and root morphology

Growth of the sorghum genotype TAM428 was severely depressed by Al in solution (Table 1). Compared to control (without Al) its dry matter yield at the lowest Al level (0.4 mg L^{-1}) was almost halved and even less at higher Al levels. The biomass of the genotype SC0283 was not affected at low Al supply and only slightly reduced at higher Al. Thus in this study it was confirmed that the two genotypes, being standard 'checks' in studies on the response of sorghum to acid soil stress (Duncan, 1988), differed greatly in Al-susceptibility.

Root morphology of both genotypes was greatly influenced by Al. Al-stressed plants developed stubby, brittle and brown colored roots with reduced surface area, which was in agreement with earlier results (Grundon *et al.*, 1987; Keltjens, 1987). The effects were more severe with increasing Al supply and usually more profound in TAM than in SC (Table 1).

Table 1. Effects of Al on biomass and root morphology of the two sorghum genotypes SC and TAM grown in solutions for 35 days

Al (mg L^{-1})	Dry matter yield[a] (g $plant^{-1}$)		Root morphological characteristics[b] Specific length (m g^{-1} d.m.)		Average diameter (mm)		Specific surface area (m^2 kg^{-1} d.m.)	
	SC	TAM	SC	TAM	SC	TAM	SC	TAM
0	13.9	12.3	225 a	133 c	0.31 a	0.39 c	216 a	161 cd
0.4	13.7	6.9	179 b	99 d	0.34 ab	0.49 d	193 b	152 d
0.8	11.4	4.2	156 c	72 e	0.36 bc	0.57 e	176 c	128 e
1.2	9.6	3.3	147 c	67 e	0.38 bc	0.56 e	174 c	117 e

[a] Mean of the dry matter yields of 30 plants in a container.
[b] Within item blocks, values followed by the same letter are not significantly different at the 5% level according to the Duncan's new multiple range test (DNMRT).

Aluminium distribution

Concentrations of Al in shoots of both genotypes were very low, being 0.23-0.30 mmol kg^{-1} d.m. for SC and 0.34-0.38 for TAM with Al treatments, which indicated that the upward translocation of Al was rather weak and similar in the two genotypes. On the contrary, Al was located in roots in large amounts (Fig. 1). Most of the Al (over 60%) could easily be removed by 0.05 *M* H_2SO_4, which suggests that Al mainly exists on the outer surface of the roots or in its cortical free space. The accumulation of Al in/on roots

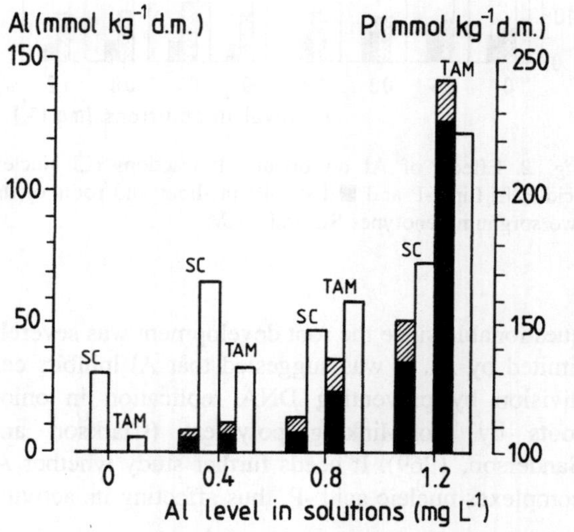

Fig. 1. Acid removable Al (■) and P (□) from roots of SC and TAM genotypes. Total length of the left column represents root total Al.

of both genotypes increased with increasing Al addition and was much greater for TAM than for SC at a same Al level. The Al in roots might cross-link pectins (Foy *et al.*, 1978) or coagulate pectic acid in the wall of young cells (Rorison, 1958), thus inhibiting cell elongation. The greater accumulation of Al in/on TAM's roots might account for its worse development.

In addition, the concentration of extractable phosphate in/on roots increased with increasing Al level and tended to correspond with that of desorbed Al. The molar ratio of extractable Al and Al-increased extractable P reached values of 1.1 and 0.8 at 1.2 mg Al L^{-1} for TAM and SC, respectively. This implicates a possibility of adsorption/precipitation of Al and P in/on the roots as indicated by Clarkson (1966).

Nutrient uptake and proton efflux

In concordance with prior experiments (Keltjens, 1987; Keltjens and van Ulden, 1987), results of plant analyses showed that Al significantly decreased Mg uptake by both genotypes and Ca uptake by SC (Table 2). The Mg concentrations in shoot of TAM grown in the presence of Al and those of SC at high Al levels (0.8 and 1.2 mg L^{-1}) were below the critical level, *i.e.* 0.18% (Chapman, 1966). However, generally there was no obvious effect of Al on N and K absorption by both genotypes. With TAM aluminium significantly increased P uptake. Therefore, with respect to nutrient uptake, Al

Table 2. Plant's nutrient concentrations and root proton efflux

Al (mg L^{-1})	Nutrient concentration (mmol$_c$ kg^{-1} d.m.)[a] N	P	K	Ca	Mg	Ca-Aa[b]	H-ion flux (mmol/plant)[c]
	Al-tolerant sorghum genotype SC283						
0	2410 ab	399 ab	1174 a	248 a	174 a	574(7.97)	8.85(0.74)
0.4	2398 ab	392 abc	1226 a	192 d	134 c	720(9.86)	10.48(0.93)
0.8	2352 b	385 abcd	1245 a	174 e	102 de	628(7.17)	9.02(1.10)
1.2	2356 b	366 d	1205 a	176 e	92 f	732(7.05)	8.21(1.04)
	Al-sensitive sorghum genotype TAM428						
0	2536 a	342 e	1188 a	216 c	150 b	698(8.62)	9.19(1.16)
0.4	2454 ab	373 cd	1238 a	228 bc	106 d	825(5.66)	6.47(1.15)
0.8	2416 ab	405 a	1239 a	240 a	106 d	882(3.67)	4.43(1.35)
1.2	2433 ab	376 bcd	1183 a	222 c	98 ef	862(2.81)	3.74(1.58)

[a] Within columns, values followed by the same letter are not significantly different at the 5% level (DNMRT).
[b] Excess cation (NH_4 + K + Ca + Mg + Na) over anion (NO_3 + H_2PO_4 + SO_4 + Cl) uptake; between brackets: mmol$_c$ $plant^{-1}$.
[c] Between brackets: root H-ion efflux density measured in the last week before harvest (mmol$_c$ m^{-2} surface.d^{-1}).

could only induce Mg deficiency. Under a condition of low Al stress, differences in growth response between the two sorghum lines may partly be explained by Mg deficiency in TAM rather than SC. At higher Al levels this explanation would no longer be valid. Al-reduced Ca uptake seems unable to explain their variation in Al-susceptibility.

In terms of nutrient uptake pattern both genotypes showed excess cation ($NH_4 + K + Ca + Mg + Na$) over anion ($NO_3 + H_2PO_4 + SO_4 + Cl$) absorption, which was enhanced as Al level increased. The plants consequently released protons from their roots to the solutions in correspondence with the excess cation uptake. In agreement with earlier work (Keltjens, 1987) the root H-ion efflux density measured during the last week before harvest was enhanced with increasing Al supply as a result of enhanced excess cation uptake and the root morphological changes as mentioned above. The H-ion flux density of TAM was even greater in the absence of Al than that of SC at the highest Al level (Table 2). This could be an explanation for the Al tolerance of SC in the field (soil) where a rhizosphere can be developed.

Phosphorus assimilation

With sufficient supply phosphorus in the plants mostly remained in inorganic form which made up over 70% of total P. The proportion of P in inorganic phosphate, nucleic acids, phospholipids and phosphate esters decreased successively. In the absence of Al, concentrations of nucleic acid-P and lipid-P in TAM's roots were 37 and 46% higher than those in SC's roots, respectively, while ester-P concentration was 32% lower in TAM's roots. Aluminium did not affect the P conversion greatly. Under Al stress there was only a tendency of increasing ester-P in TAM's shoot and both lipid-P and nucleic acid-P in SC's roots (Fig. 2).

Therefore, under conditions of sufficient P supply, upward translocation and assimilation of P in both genotypes were either not affected or even promoted by Al. The minor effect of Al on nutrient absorption except for Mg may prove normal functioning of phospholipids. However, the functioning of P in nucleic acids remains questionable since the root development was severely limited by Al. It was suggested that Al inhibits cell division by preventing DNA replication in onion roots by cross-linking polymers (Clarkson and Sanderson, 1969). It needs further study whether Al complexes nucleic acid -P, thus affecting its activity.

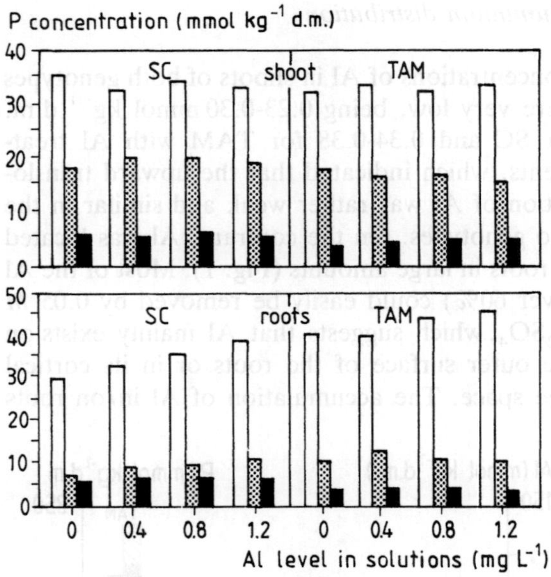

Fig. 2. Effects of Al on organic P fractions (☐ Nucleic acid-P, ⊞ Lipid-P and ■ Ester-P) in shoot and roots of the two sorghum genotypes SC and TAM.

Acknowledgements

The authors would like to thank Mrs. E van Loenen, Mr. E Heij and Mrs. W van Vark for their technical and analytical assistance during the experiment.

References

Chapin F S III and Bieleski R L 1982 Mild phosphorus stress in barley and a related low-phosphate-adapted barleygrass: Phosphorus fractions and phosphate adsorption in relation to growth. Plant Physiol. 54, 309–317.

Chapman H D 1966 Diagnostic Criteria for Plants and Soils. Univ. of California, 698 p.

Clarkson D T 1966 The effect of aluminium on the uptake and metabolism of phosphorus by barley seedling. Plant and Soil 41, 165–172.

Clarkson D T and Sanderson J Jr 1969 The uptake of a

polyvalent cation and its distribution in the root apices of *Allium cepa*. Tracer and autoradiographic studies. Planta 89, 136–154.

Duncan R R 1988 Sequential development of acid soil tolerant sorghum genotypes under field stress conditions. Commun. Soil Sci. Plant Anal. 19, 1295–1305.

Foy C D 1988 Plant adaptation to acid, aluminium toxic soils. Commun. Soil Sci. Plant Anal. 19, 959–987.

Foy C D, Chaney R L and White M C 1978 The physiology of metal toxicity in plants. Annu. Rev. Plant Physiol. 29, 511–566.

Furlani P R and Clark R B 1981 Screening sorghum for aluminium tolerance in nutrient solutions. Agron. J. 73, 587–594.

Grundon N J, Edwards D G, Takkar P N, Asher C J and Clark R B 1987 Nutritional Disorders of Grain Sorghum. ACIAR, Australia. 88 p.

Keltjens W G 1987 Nitrogen source and aluminium toxicity of two sorghum genotypes differing in aluminium susceptibility. J. Plant Nutr. 10, 841–856.

Keltjens W G and van Ulden P S R 1987 Effects of Al on nitrogen (NH_4^+ and NO_3^-) uptake, nitrate reductase activity and proton release in two sorghum cultivars differing in Al tolerance. Plant and Soil 104, 227–234.

Rorison I H 1958 The effect of aluminium on legume nutrition. *In* Nutrition of the Legumes. Ed. E G Hallsworth. pp 43–61. Butterworths, London.

Taylor G J 1988 The physiology of aluminium tolerance in higher plants. Commun. Soil Sci. Plant Anal. 19, 1179–1194.

M. L. van Beusichem (Ed.), *Plant nutrition – physiology and applications*, 403–407.

PLSO IPNC418B

Similarities and differences between aluminium toxicity and phosphorus deficiency in sorghum (*Sorghum bicolor*) plants

W.G. KELTJENS and E. VAN LOENEN
Department of Soil Science and Plant Nutrition, Wageningen Agricultural University, P.O. Box 8005, 6700 EC Wageningen, The Netherlands

Key words: aluminium stress, nutrient uptake, phosphorus metabolism, proton efflux, *Sorghum bicolor* (L.) Moench, specific root length, suboptimal phosphorus supply

Abstract

Since Al toxicity has often closely been associated with disturbed plant P metabolism, plant response to conditions of Al stress and suboptimal P supply was compared. In the first experiment plants were grown in the absence of Al, while phosphorus was supplied at seven levels varying from extremely deficient to excessive. In the second experiment half the plants was grown in the absence, half in the presence of 0.75 mg Al L^{-1}. As expected, both P deficiency and Al stress significantly reduced dry matter yield. However, changes in root morphology were not observed at deficient P supply, thus proving to be Al-specific. In general concentrations of all P fractions declined with decreasing P supply. As aluminium did not affect internal concentrations of ester-P, nucleic acid-P and inorganic-P during the early stage of growth and even increased concentrations of all four fractions later, the conclusion can be drawn that Al-induced reduction in growth can not be the result of a shortage of organic-P in the plant. On the contrary, higher concentrations will just be the result of impaired growth.

Effects of Al stress and P deficiency on nutrient uptake, proton efflux and root respiration were similar. Suppressed net uptake of NO_3, K and Mg might be the result of enhanced leakage of plasma membranes due to (i) shortage of pospholipids as induced by deficient P supply, or (ii) malfunctioning of Al-saturated phospholipids in the presence of Al. Reduced root respiration, if interpreted in terms of deteriorated root energy supply, might also disturb nutrient uptake.

Introduction

Plants grown in culture solutions containing Al often accumulate large amounts of Al in/on the roots in close association with P (Foy *et al.*, 1978). The Al-P interaction appears to be located on the outer root surface and in the walls and cytoplasmic membranes of cortical cells (McCormick and Borden, 1972, 1974; Waisel *et al.*, 1970). By precipitation/adsorption reactions insoluble Al-phosphate will be formed, thereby inhibiting upward transport and assimilation of P (Clarkson, 1966; Cumming *et al.*, 1986). Therefore, roots of Al-injured plants can contain high concentrations of inorganic P, while at the same time upper plant parts show symptoms of P deficiency.

In acid soils the plant P status can be worse, because of a low external P supply due to a similar Al-P interaction in the soil compartment.

The aim of the present work was to investigate whether part of the plants' response to Al stress may be attributed directly to P deficiency. By immobilization of P in the roots, aluminium may inhibit P metabolism and induce deficiencies of essential organic-P fractions in the plant, such as ester-P, lipid-P and nucleic acid-P. If so, symptoms of Al stress should partly resemble those of P deficiency and *vice versa*. Therefore, sorghum plants grown under conditions of Al stress and suboptimal P supply were compared with respect to relevant plant characteristics, such as root morphology, P assimilation, nutrient uptake and root respiration.

To prevent temporal situations of complete P exhaustion of the nutrient solutions, especially at extreme low P levels, plants were continuously supplied with fresh phosphorus. Rates of P supply varied from extremely deficient to excessive.

Materials and methods

One-week-old seedlings of the Al-susceptible sorghum (*Sorghum bicolor* (L.) Moench) genotype TAM428 were transferred to 50-L containers filled with nutrient solutions containing (m*M*): NH_4NO_3 1.0; KCl 1.0; $CaCl_2$ 0.25; $MgSO_4$ 0.25; FeEDTA 0.08 and additional micronutrients. In the one part of the research (Exp. I) seven pots were used to study effects of P supply, while in another four pots part of the plants was exposed to Al (Exp. II). Phosphorus supply was at a daily increase of 10% and continuously provided as a NaH_2PO_4 solution using a peristaltic pump. During the 35-day experimental period of Exp. I total P supply for the various P levels was (mmol P/pot): 0.77 (P1); 1.33 (P2); 1.89 (P3); 3.85 (P4); 7.28 (P5); 13.70 (P6) and 52.86 (P7). In Exp. II phosphorus supply per plant was equal to the P6 level. In the Al experiment for both the control treatment (zero Al) and the Al treatment (0.75 mg Al L^{-1} as $AlCl_3$) two containers were used. Once a week all solutions were completely refreshed, while pH was adjusted to 4.2 one or twice a day with 0.5 *M* NaOH. Twelve and 52 plants were transferred to each pot in Exp. I and II, respectively.

The experiments were carried out in a growth chamber at a constant temperature of 25°C and a relative air humidity of 75%. Plants were illuminated for 16 h each day at a quantum flux density of about 700 μmol $m^{-2}s^{-1}$.

In Exp. I all plants were harvested after 35 days, while in Exp. II plants were harvested 3, 7, 10, 14, 17, 21, 24 and 28 days after transfer. At each harvest plants were divided into two subsamples and subsequently partitioned into shoot and roots. Measurements of dry matter yield, root length, root respiration and chemical composition were carried out in duplicate as described before (Keltjens and van Ulden, 1987). In N_2-treated fresh shoot and root material concentrations of ester-P, nucleic acid-P, lipid-P and inorganic-P were estimated according to Chapin and Bieleski (1982). During both experiments root proton efflux was monitored by titration of the nutrient solutions, while NH_4 and NO_3 absorption were measured by disappearance of these ions from the nutrient solution.

Results and discussion

Plant growth

Plant biomass was severely reduced by limited P supply as well as by Al stress (Table 1). At levels of P supply lower than P6, phosphorus became suboptimal. At the lowest P level shoot and root biomass even decreased to 20–30% of that at optimal supply. Reduction in biomass due to Al stress was about 80–85% at the end of the 28-day period.

Contrary to biomass, effects of P and Al on root morphology were completely different. While Al-stressed plants had short, thick roots with a small specific surface area, plants grown under conditions of suboptimal P supply did not show any of these symptoms (Table 2). This points to the specific role of Al in inhibiting root cell division and elongation.

Phosphorus assimilation

Both in shoot and roots, concentrations of inorganic-P, nucleic acid-P, lipid-P and ester-P were in a decent order, irrespective of P level (Fig. 1). Comparable to results of Hart and Jessop (1982), concentrations of all four P fractions generally increased as P supply increased. Only ester-P in shoot of suboptimal supplied plants (<P6) did not increase. Beyond P6 the increase of most fractions was much stronger than at suboptimal P supply. Obviously, when the need of P for growth was met, increasing P supply still further enhanced uptake and assimilation of P. Extremely high concentrations of inorganic-P at P7, indicating the luxurious P status of these

Table 1. Relative yield and whole plant chemical composition (mmol kg^{-1} d.m.) of sorghum plants after 35 days of growth at various levels of P (Exp. I) or after 28 days of growth under conditions of Al stress (Exp. II)

	Rel. d.m. yield[a]		Chemical composition				
	Shoot	Roots	N-total	P	K	Ca	Mg
Exp. I							
P1	18	31	2229	27	512	65	60
P2	29	55	2215	27	407	55	50
P3	32	47	2106	38	638	63	62
P4	47	70	2306	53	848	54	57
P5	61	104	2525	74	1003	70	66
P6	92	142	2612	96	968	76	81
P7	100	100	2747	395	1158	68	97
Exp. II							
−Al	100	100	2735	73	833	30	77
+Al	19	14	2098	168	396	27	33

[a] Relative to P7 (Exp. I) and −Al (Exp. II), respectively.

plants, probably serve as a storage pool of phosphorus.

That shoot ester-P only started to increase beyond P6, *i.e.* the transition of suboptimal to optimal P supply could indicate this P compound to be the growth limiting factor. Unexpectedly, at extreme low P supply (P1, P2) over 65% of total-P was still present as non-assimilated inorganic-P. Probably, this fraction will be or functional or not yet be assimilated.

Effects of Al on P assimilation were similar in shoot and roots. During the first one to two weeks concentrations of ester-P, nucleic acid-P and inorganic-P were not significantly affected by Al (Fig. 2). Later, concentrations of all four P compounds were significantly higher in Al-stressed plants, proving (i) Al did accumulate large amounts of inorganic-P in the roots not at the cost of organic-P, (ii) Al did not inhibit growth by inducing a shortage of one of the P compounds. On the contrary, higher concentrations of the various organic-P fractions will probably be the result of impaired growth. Since phospholipids play an important role in plant membrane structure (Simon, 1974), lower concentrations of lipid-P in Al-stressed plants, as observed during the first two weeks, might strengthen negative effects of Al by nutritional disorders.

Nutrient uptake and proton efflux

Effects of suboptimal P supply and Al stress on internal nutrient concentrations were very simi-

Table 2. Root characteristics of sorghum plants supplied for 35 days with different levels of P (Exp. I) or grown for 28 days under conditions of Al stress (Exp. II)

	Spec. length (m g^{-1} f.m.)	Average diameter (mm)	Spec. surface area ($m^2 kg^{-1}$ f.m.)
Exp. I			
P4	7.18[a]	0.42[a]	9.49[a]
P5	6.60[a]	0.44[a]	9.11[a]
P6	6.84[a]	0.43[a]	9.26[a]
2×P6	7.24[a]	0.42[a]	9.52[a]
Exp. II			
−Al	4.10[a]	0.56[a]	7.18[a]
+Al	1.95[b]	0.81[b]	4.94[b]

[a,b] Within the column of each experiment values followed by the same letter are not significantly different at the 5% level according to the Student's t-test.

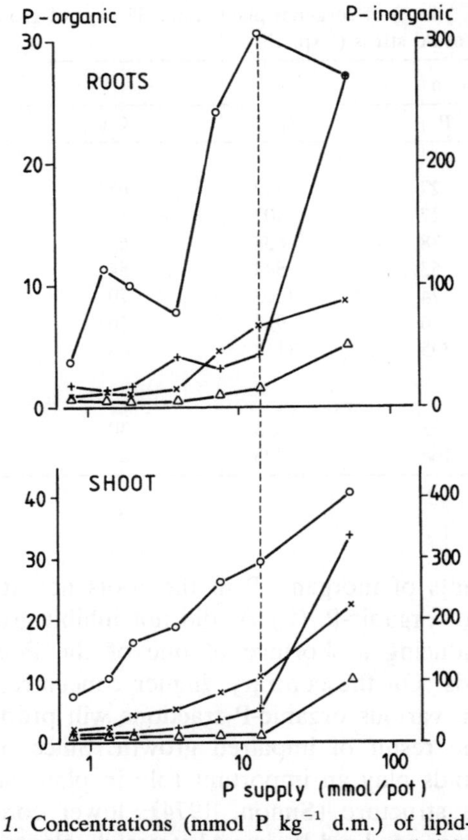

Fig. 1. Concentrations (mmol P kg^{-1} d.m.) of lipid-P (x), nucleic acid-P (0), ester-P (△) and inorganic-P (+) in shoot and roots of 35-day old plants supplied with various levels of P (Exp. I).

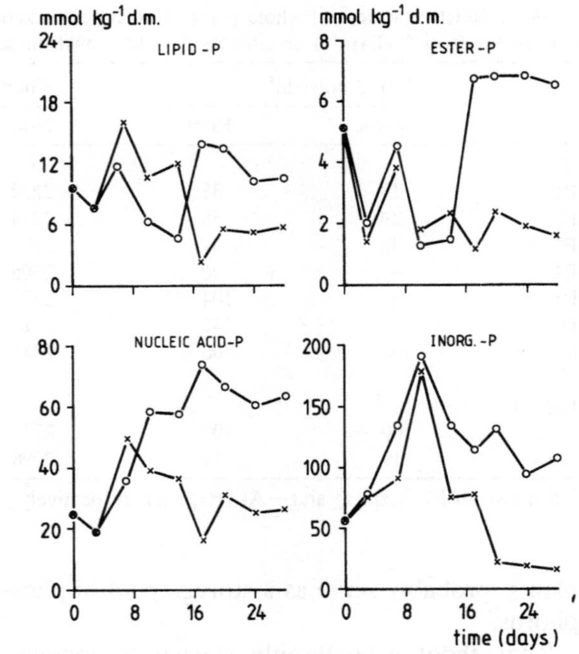

Fig. 2. Concentrations of lipid-P, ester-P, nucleic acid-P and inorganic-P in sorghum plants grown for 28 days in the absence (x) or presence (0) of 0.75 mg Al L^{-1} (Exp. II).

lar, *i.e.* a strong decrease of K, N-total and Mg and no effect on Ca (Table 1). Under both conditions lower N in plants was mainly the result of reduced uptake of NO_3 rather than NH_4. This was also clearly demonstrated by enhanced NH_4/NO_3 uptake (Fig. 3) and suppressed concentrations of free NO_3-N in plants (data not shown). Consistent with previous work (Keltjens and van Ulden, 1987) effects of Al and suboptimal P on NH_4 uptake were absent or positive.

Since nitrogen nutrition, particularly the nitrogen form, plays a dominant role in the plant ionic balance (Troelstra, 1983), increased NH_4/NO_3 uptake resulted in enhanced excess cation uptake and improved root proton efflux (Fig. 3; Table 3). Because of Al-induced reduction of the specific root surface area, the root proton efflux density was increased much stronger by Al than

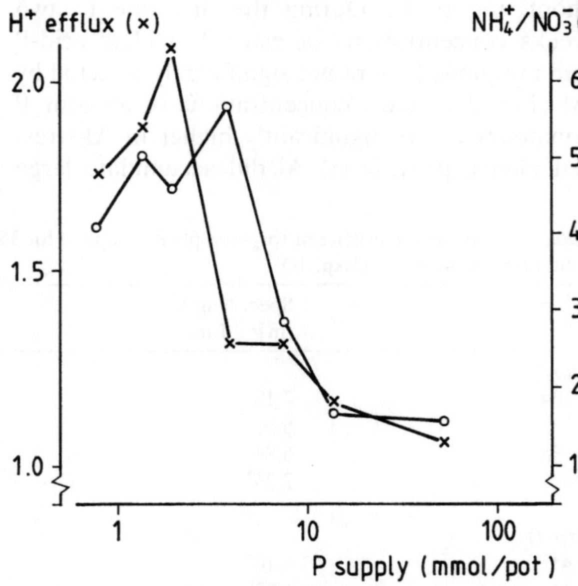

Fig. 3. Proton efflux (mmol g^{-1} d.m.) and NH_4/NO_3 absorption of sorghum plants measured during a 35-day growth at various levels of P supply (Exp. I).

Table 3. Effects of Al on root proton efflux density and NH_4/NO_3 uptake ratio measured during the final four days of the 28-day experimental period (Exp. II)

Treatment	Root H^+ efflux density ($mmol\ m^{-2}\ d^{-1}$)	NH_4/NO_3 uptake (mol/mol)
−Al	5.69	3.24
+Al	14.79	5.11

by suboptimal P supply. Root proton efflux density will be of direct importance for changes in pH and Al-solubility in root rhizosphere.

Explanations of the observed effects on nutrient uptake are still hypothetical. Suppressed net uptake of K, NO_3 and Mg might be explained by increased leakage due to membrane changes. Under conditions of suboptimal P supply a shortage of phospholipids might enhance plasma membrane permeability and make cells leaky (Simon, 1974). Under conditions of Al stress aluminium can bind to phospholipids at the plasma membrane and possibly induce membrane instability and leakage (Haug and Caldwell, 1985). Since NH_4 will normally be converted in the root directly after absorption into large organic nitrogen molecules, leakage of NH_4, just like the rather immobile divalent cation Ca, will be unlikely. Inhibited nutrient uptake due to impaired energy status seems also plausible with this sorghum genotype. Insoluble Al-P compounds in roots might induce a local shortage of inorganic-P for many phosphorylating enzymes (Clarkson, 1966). Both under conditions of Al stress and suboptimal P supply, root respiration was strongly reduced (see also Keltjens, 1988). This, of course, may also have direct negative effects on nutrient uptake.

References

Chapin F S III and Bieleski R L 1982 Mild phosphorus stress in barley and a related low-phosphorus-adopted barley-grass: Phosphorus fractions and phosphate absorption in relation to growth. Physiol. Plant. 54, 309–317.

Clarkson D T 1966 Effect of aluminum on the uptake and metabolism of phosphorus by barley seedlings. Plant Physiol. 41, 165–172.

Cumming J R, Eckert R T and Evans L S 1986 Effect of aluminium on ^{32}P uptake and translocation by red spruce seedlings. Can. J. For. Res. 16, 864–867.

Foy C D, Chaney R L and White M C 1978 The physiology of metal toxicity in plants. Annu. Rev. Plant Physiol. 29, 511–566.

Hart A L and Jessop D 1982 Concentration of total, inorganic, and lipid phosphorus in leaves of white clover and stylosanthes. N.Z. J. Agric. Res. 25, 69–76.

Haug A R and Caldwell C R 1985 Aluminum toxicity in plants: The role of the root plasma membrane and calmodulin. *In* Frontiers of Membrane Research in Agriculture. Eds. J B St. John, E Berlin and P C Jackson. pp 359–381. Rowman and Allanheld, Totowa, NJ.

Keltjens W G 1988 Short-term effects of Al on nutrient uptake, H^+ efflux, root respiration and nitrate reductase activity of two sorghum genotypes differing in Al-susceptibility. Commun. Soil Sci. Plant Anal. 19, 1155–1163.

Keltjens W G and Van Ulden P S R 1987 Effects of Al on nitrogen (NH_4^+ and NO_3^-) uptake, nitrate reductase activity and proton release in two sorghum cultivars differing in Al tolerance. Plant and Soil 104, 227–234.

McCormick L H and Borden F Y 1972 Phosphate fixation by aluminum in plant roots. Soil Sci. Soc. Am. Proc. 36, 799–802.

McCormick L H and Borden F Y 1974 The occurrence of aluminum-phosphate precipitate in plant roots. Soil Sci. Soc. Am. Proc. 38, 931–934.

Simon E W 1974 Phospholipids and plant membrane permeability. New Phytol. 73, 377–420.

Troelstra S R 1983 Growth of *Plantago lanceolata* and *Plantago major* on a NO_3/NH_4 medium and the estimation of the utilization of nitrate and ammonium from ionic-balance aspects. Plant and Soil 70, 183–197.

Waisel Y, Hoffen A and Eshel A 1970 The localization of aluminum in the cortex cells of bean and barley roots by X-ray micro analysis. Physiol. Plant. 23, 75–79.

M. L. van Beusichem (Ed.), *Plant nutrition – physiology and applications*, 409–412.

PLSO IPNC289

Calcium protection against aluminium toxicity

R.F. KORCAK
Fruit Laboratory, Agricultural Research Service, USDA, Beltsville, MD 20705, USA

Key words: acidophiles, aluminium toxicity, blueberry, gypsum, root disorders, *Vaccinium* sp.

Abstract

The occurrence of a slow (three to five year) stunting disorder of blueberry which usually results in plant death or the removal of the plant from the filed due to low productivity has been ascribed to a chronic effect of Al on the root system. This disorder has been observed where blueberries are grown under atypical soil conditions: low organic matter (usually 2% or less) and soil textures heavier than the normal sands and loamy sands. Laboratory tests indicate that the success of gypsum additions, used to alleviate the disorder, will vary depending upon the amount applied and the particular blueberry species under study. Data is presented from field soils, where this disorder is prevalent, that high soil Al reduces root dry mass and top growth and that additional Al aggravates the condition.

Introduction

There is increasing interest in blueberry (*Vaccinium* sp.) production on atypical soil sites, namely soils that are low (2%) in organic matter and/or heavier in texture than the typical sands and loamy sand soils. These sites are either chosen or maintained at low pH values (4.2 to 5.5) for optimal growth of the acid-loving blueberry. A chronic problem has developed on many of these sites whereby plants become stunted and eventually die out within the first three to five years after planting, sometimes without any foliar symptomology. The incidence of this condition may be either sporadically within a plant row or in localized areas within a field.

Evidence has accumulated in this laboratory indicating that aluminium damage to the root systems is the causal factor associated with incidence of stunting and dying off. There has been a high degree of interest in the amelioration of aluminium damage to crops grown on acid soils and subsoils using gypsum (Noble *et al.*, 1988). However, acidophiles, such as blueberry, are known for their adverse reaction to high soil calcium (Korcak, 1988).

This study presents further evidence on the importance of soil aluminium in the incidence of this stunting disorder and the potential of reducing the chronic weakening of blueberry roots by calcium. Root elongation of three blueberry species as affected by Ca pretreatments and Al is reported along with a study of the Al fractions of a soil known to induce the disorder.

Materials and methods

Experiment one

Tissue cultured plants of three blueberry cultivars were used to examine the affect of Al on root elongation under different Ca source/level regimes. The cultivars used were: 'Blueray' (*V. cormybosum* -highbush), 'Top Hat' (*V. angustifolium* -lowbush), and 'Northblue' (a half highbush/half lowbush cross). The 3-week-old plants were pretreated in complete nutrients solutions containing either 0.07 or 1.30 m*M* Ca from either Ca-sulfate or -chloride for 40 days. The complete nutrient solution contained 2.4 m*M* N (as NH_4), 0.1 m*M* P, 4 m*M* K, 0.45 m*M* Mg, 12 μ*M* Fe (as EDDHA), 1.2 μ*M* Mn, 3.25 μ*M* B, 0.3 μ*M* Zn, 0.08 μ*M* Cu and 0.05 μ*M* Mo. Growth chamber conditions were; a 16 hr light period of 6.68 $\mu Es^{-1}m^{-2}$ PAR from a mixture on incandescent and fluorescent bulbs at 25°C day and 23°C night temperatures at 70% relative humidity. The light intensity level is low but prior work has shown this level to be optimal for blueberry. Each treatment combination, per cultivar, was replicated three times. On day 40, the roots were rinsed thoroughly in demineralized water and three root tips per plant were marked with India ink about 0.5 cm behind the root tip. The plants were returned to solution bottles containing the complete nutrient solution without added Ca and one-half of the plants were treated with 133 μM Al (from sulfate). All solutions were adjusted to a pH of 4.7 with the addition of sulfuric acid. After 10 days exposure to the Al treatment root growth measurement were made and recorded as mm of elongation.

Experiment two

Two Florida soils representing affected and non-affected plants a peak sample, used to amend affected areas, were fractionated according to a modified scheme used by Miller *et al.* (1983) to identify aluminium/soil relationships. The procedure was modified by omitting the carbonate-associated metal extraction due to the acidity of the soils under study. These two soils plus a high organic matter, acid sand from a commercial blueberry plantation were also used in controlled, short-term plot (500 g soil per pot) studies with 3-week-old 'Blueray' highbush blueberries. The soils were spiked with the equivalent of 1.0 μg Al/g soil (from sulfate) or left unamended. There were three replicates of each treatment/soil combination. The plants were grown for 70 days at which time total post-treatment growth was measured as well as leaf, stem, and root fresh and dry weights. Roots were gently washed clean of soil and five root tip segments were saved for hematoxylin and localized Fe reduction staining. Roots were stained for 10 min with a mixture of 2 g hematoxylin plus 0.2 g $NaJ0_3$ per liter and destained for 30 min before microscopic examination. Root iron reduction was examined using the ferric reduction localization staining procedure of Bell *et al.* (1988). The air-dried tissues were dry ashed and analyzed for Ca, Mg, Al, and Fe contents by atomic absorption spectroscopy using standard methods.

Results and discussion

Root elongation of 'Blueray' highbush blueberry was differently affected by both Al treatment and Ca level/source pretreatment (Table 1). Under low Ca-sulfate conditions root elongation was significantly decreased by 133 μM Al. The higher Ca-sulfate level or either Ca-chloride level did not exhibit any response to Al treatment. However, root elongation was lower with all treatments compared to low Ca-sulfate treatment. Ballinger (1962) studied the effect of sul-

Table 1. Root elongation after 10 day exposure to 133 μM Al as compared to control (no Al) of Ca-pretreated blueberry plants. Pretreatments were 0.07 (low) or 1.3 (high) mM Ca from either chloride or sulfate

Pretreatment	Al treatment	Cultivar		
		'Blueray'	'Northblue'	'Top Hat'
		mm		
Ca Sulfate – 0.07 mM	–Al	16.6*[a]	4.7	0.0
	+Al	8.3	3.3	0.3
Ca Sulfate – 1.3 mM	–Al	2.0	3.5*	11.7
	+Al	2.0	0.0	6.3
Ca Chloride – 0.07 mM	–Al	2.8	5.7*	7.8*
	+Al	3.3	0.0	0.8
Ca Chloride – 1.3 mM	–Al	3.8	6.7	13.0*
	+Al	5.0	4.3	0.0

[a] Average of 3 roots per plant.
* indicates a significant difference between Al treatment at the 5% level.

fate and chloride levels on highbush blueberry growth in sand cultures. Abnormal growth was observed only with chloride levels of 104 meq L^{-1}, which is more than the chloride levels applied at the high Ca-chloride pretreatment. However, the high chloride levels (Ballinger, 1962) were accompanied by relatively high Na levels which may have confounded his results. Therefore the reduced elongation with high Ca was apparently due to Ca. The cause of the reduction in elongation with low or high Ca-chloride is unclear, whether it was a Ca or chloride or mixed Ca/chloride effect is not discernible.

The lowbush 'Top Hat' blueberry responded differently to both Ca and Al (Table 1). Pretreatment with high Ca independent of source produced the most root elongation which in both cases we reduced by Al. The lowbush blueberry appears to have a greater tolerance to chloride than highbush.

The 'Northblue', a mixed highbush and lowbush cultivar, expressed an intermediate response compared to 'Blueray' and 'Top Hat' (Table 1). Root elongation was always higher under the minus Al treatment independent of Ca source and level.

The reaction of the three cultivars to Ca pretreatments and Al indicate the following: a. there are significant differences between blueberry species to both Ca pretreatments and added Al, b. The use of gypsum in alleviating potential damage to blueberry root systems by Al is questionable due to the adverse effect of elevated Ca on root growth, and c. if gypsum is used, low level additions, perhaps made on a yearly basis, might be better than a single high level addition.

Soil Al fractions from two Florida soils, FLA63 and FLA72, and a commercial blueberry field soil from New Jersey (Berryland) as well as the Al fractions from a peat used to amend the Florida soils are presented in Table 2. The FLA63 soils has a water pH of 4.4 with an organic matter content of 2.1%. Blueberry growth is normal on this soil. The FLA73 soil has a water pH of 4.7 with 1.1% organic matter. Blueberries grown on this soil exhibit a stunting disorder without expression of foliar symptomology which would indicate a nutrient deficiency or toxicity. The Berryland soil produces highly vigorous blueberry growth and has a water pH of 4.3 with 4% organic matter.

All Al fractions were higher from FLA72 compared to either FLA63 or Berryland soil. Of particular importance are the very high soluble and organically bound Al fractions from the FLA72 soil. Attempts have been made in the field to correct the stunting disorder with the addition of peat similar to the sample analyzed in Table 2. These organic amendments have been unsuccessful in reversing the stunting disorder in the field. As indicated in Table 2, the peat contains high levels of exchangeable and organically bound Al as well as high amounts of Fe-oxide and residual Al.

Table 2. Aluminium fractions extracted from two Florida soils (FLA63 and FLA72), a commercial blueberry soil (Berryland) and peat used as a soil amendment. Stunting and die off of blueberries was noted from the FLA72 soils while viborous growth was found from FLA63 sites

Al fraction	Material			
	FLA72	LFA63	Berryland	Peat
	ug Al/g			
Soluble	63 ± 9[a]	2 ± 0	3 ± 0	6 ± 0
Exchangeable	30 ± 6	15 ± 2	16 ± 1	210 ± 9
absorbed			68 ± 2	55 ± 4
6 ± 5	1 ± 1			
Organically bound	414 ± 77	36 ± 12	11 ± 4	488 ± 48
Manganese bound	8 ± 2	2 ± 0	1 ± 0	95 ± 5
Iron-oxide bound	15 ± 2	4 ± 1	2 ± 1	444 ± 102
Precipitated	81 ± 18	8 ± 3	3 ± 2	550 ± 145
Residual	42 ± 8	20 ± 3	12 ± 2	141 ± 21

[a] All values are means of 4 samples ± standard deviation.

Table 3. Growth parameters and leaf and root Al concentrations from the Blueray blueberries grown for 70 days in Al amended and unamended soils

Parameter	Treatment	Soil		
		FLA72	FLA63	Berryland
Top growth (mm)	−Al	64*[a]	197	265
	+Al	27	241	254
Leaf dw (g)	−Al	0.18	0.22	0.36
	+Al	0.11	0.21	0.40
Root dw (g)	−Al	0.07*	0.14	0.08
	+Al	0.04	0.14	0.08
Leaf Al (μg/g)	−Al	179	93	69
	+Al	125	121	47
Root Al (μg/g)	−Al	2601*	621	595
	−Al	3457	694	549

[a] Values are averages of 3 replicates.
* indicates a significant difference between Al treatments.

The distribution of Al among the peat fractions is consistent with the work of Wieder and Lang (1986) and indicates that upon oxidation of the peat the potential for Al release exists. This release would increase the severity of Al damage which is borne out by field experience.

A short-term (70 day) growth study using 'Blueray' highbush blueberries showed the severe growth reductions induced from the FLA72 soil (Table 3) compared to FLA63 and Berryland soils. The addition of an Al spike to these soils further reduced growth on the FLA72 soil but had little effect on the two other soils. Root dry weight and root Al concentrations were both significantly affected by added Al on the FLA72 soil (Table 3).

Tissue concentrations of Ca, Mg, and Fe were not significantly affected by treatments (data not shown) and all values were within known acceptable limits (Korcak, 1988).

Root tip staining with hematoxylin indicated a higher degree of Al associated with roots from the FLA72 soil versus either FLA63 or Berryland soil. Root staining using the localized ferric reduction stain procedure of Bell *et al.* (1988) in order to identify sites of Fe-deficiency were inconclusive (data not shown).

Currently, field trails are in progress on utilizing low level additions of gypsum to ameliorate the stunting disorder. Additional work is also in progress on identifying Ca tolerant blueberry species that could be used in genetic studies on reducing the damage of Al to root systems.

References

Ballinger W E 1962 Studies of sulfate and chloride ion effects upon Wolcott blueberry growth and composition. Proc. Am. Soc. Hort. Sci. 80, 331–339.

Bell P F, Chaney R L and Angle J S 1988 Staining localization of ferric reduction on roots. J. Plant Nutr. 11, 1237–1252.

Korcak R F 1988 Nutrition of blueberry and other calcifuges. Hort. Rev. 10, 183–227.

Miller W P, McFee W W and Kelly J M 1983 Mobility and retention of heavy metals in sandy soil. J. Environ. Qual. 12, 579–584.

Noble A D, Sumner M E and Alva A K 1988 The pH dependency of aluminum phytotoxicity alleviation by calcium sulfate. Soil Sci. Soc. Am. J. 52, 1398–1402.

Wieder R K and Lang G E 1986 Fe, Al, Mn, and S Chemistry of Sphagnum peat in four peatlands with different metal and sulfur input. Water Air Soil Pollut. 29, 309–320.

M. L. van Beusichem (Ed.), *Plant nutrition – physiology and applications*, 413–417.

PLSO IPNC525

Temperature and magnesium effects on aluminium toxicity in annual ryegrass (*Lolium multiflorum*)*

Z. RENGEL[1] and D.L. ROBINSON
Department of Agronomy, Louisiana Agricultural Experiment Station, Louisiana State University Agricultural Center, Baton Rouge, LA 70803, USA. [1]*Present address and address for correspondence: Institute for Agroecology, Faculty of Agricultural Sciences, Simunska 25, YU-41000 Zagreb, Yugoslavia*

Key words: aluminium toxicity, cation uptake, grass tetany, *Lolium multiflorum* Lam., temperature regimes

Abstract

Two ryegrass (*Lolium multiflorum* Lam.) cultivars differing in Al sensitivity were grown under two temperature regimes (10/6 and 22/18°C day/night temperature) in nutrient solution (pH 4.2) with three levels of Al (0, 3.7 and 74 μM) and two levels of Mg (0.1 and 1 mM). Higher Mg concentration greatly alleviated Al toxicity effects on root and shoot dry matter production. Aluminium depressed plant growth more at the higher than at the lower temperature in solution containing 0.1 mM Mg but not in solutions of 1 mM Mg. Added aluminium depressed uptake rates of Mg and Ca but the effect was considerably alleviated with increased Mg concentration in the nutrient solution. Average K uptake remained unaffected by Al at 1 mM Mg in solution. Increasing the nutrient solution Al concentration increased the K/(Ca + Mg) ratio of shoots, thus increasing the grass tetany potential of ryegrass forage. This ratio was larger in cv. Wilo which is more sensitive to Al than cv. Marshall.

Introduction

Low Mg and Ca concentrations and high K/(Ca + Mg) equivalent ratios in forage can lead to hypomagnesemia of grazing ruminant animals (Grunes *et al.*, 1970). Low temperature (Cherney and Robinson, 1985) and exposure to Al (Rengel and Robinson, 1989a) have increased the grass tetany potential of annual ryegrass forage.

Aluminium depresses uptake of Mg and Ca, while the effects on K uptake depend on the growth conditions (for review see Foy, 1988). Increased Mg concentrations in solution alleviated Al impairement of root growth of wheat (Aniol, 1983; Kinraide and Parker, 1987). More efficient accumulation of Mg was noted in Al-tolerant corn inbred lines (Clark, 1977).

The objectives of the present study were to examine the interactions of temperature and nutrient solution Mg and Al concentrations on plant growth and on Mg, Ca and K uptake rates by two annual ryegrass cultivars differing in Al sensitivity.

Materials and methods

Seeds of two ryegrass (*Lolium multiflorum* Lam.) cultivars, Wilo (relatively Al-sensitive) and Marshall (relatively Al-tolerant) (Rengel and Robinson, 1989a; b), were surface sterilized, germinated, and grown in 1/5 Steinberg complete nutrient solution (pH 4.2) in growth chambers as described elsewhere (Rengel and Robinson, 1989a). Nutrient solution was re-

* Approved for publication by the Director of the Louisiana Agricultural Experiment Station as manuscript No. 89-09-3299.

newed on days 10, 13, 15, and 18 after the start of germination, and in 2-day intervals thereafter, preventing pH changes of more than 0.4 in control and 0.1 in Al-containing solutions. On day 15, different nutrient solution concentrations of Mg (0.1 and 1 m*M*) and Al (0, 3.7 and 74 μM) were imposed. Aluminium was added as $Al_2(SO_4)_3$.

For the first 15 days after germination, plants were grown with a 14/10 h light/dark period at $22 \pm 0.5/18 \pm 0.5$°C light/dark temperature in two growth chambers. In one growth chamber temperature was not changed throughout the experiment. At the beginning of the light period on day 15, temperature in the second growth chamber was successively lowered to 10 ± 0.5°C/6 ± 0.5°C (day/night). Fresh nutrient solutions were temperature-equilibrated for 2 h before solution renewal.

At the beginning of cold exposure as well as 13 days later (28-day-old plants), roots and shoots from each pot in each chamber were separated, rinsed briefly three times in distilled water, gently blotted, dried at 60°C for 48 h, weighed, and digested in concentrated HNO_3 for 4 h at 125°C. Chemical analyses of digests were obtained using an inductively coupled plasma emission spectrometer. Average nutrient uptake rates were calculated from the Williams' (1946) growth analysis formula as described elsewhere (Rengel and Robinson, 1989a).

The completely randomized experiment contained a split-plot arrangement of treatments with growth chambers as main plots and temperature as the main plot factor. The subplot factors (cultivars, Mg and Al nutrient solution concentrations) were replicated three times in a complete factorial arrangement. Analysis of variance was used to test for significance of main effects and first order interactions. Multiple comparisons among means were made using Tukey's procedure (Steel and Torrie, 1980).

Results and discussion

Plant Growth Parameters

Lower temperature, increased Al, and decreased nutrient solution Mg concentrations each depressed root and shoot growth (Table 1). The first order interactions (temperature × Mg, temperature × Al, and Mg × Al) were significant ($P < 0.001$). Deleterious Al effects on root and shoot growth were most pronounced at 22°C and 0.1 m*M* Mg concentration (Table 1). Increases in temperature have usually increased Al toxicity in other studies (Aniol, 1983; Konzak *et al.*, 1977).

Table 1. Weight of ryegrass roots and shoots in mg (10 plants)$^{-1}$ as influenced by cultivar, temperature (10/6 and 22/18°C), and nutrient solution concentrations of Mg (m*M*) and Al (μM). Standard errors of the means were omitted but they did not exceed 7% of the corresponding means. Tukey's $HSD_{0.05}$ for the temperature × Mg × Al × cultivar interaction was 56.8 and 282.3 mg (10 plants)$^{-1}$ for roots and shoots, respectively

Treatment		Root dry weight				Shoot dry weight			
Al	Mg	cv. Marshall		cv. Wilo		cv. Marshall		cv. Wilo	
		10/6	22/18	10.6	22/18	10/6	22/18	10/6	22/18
0	0.1	82	320	120	380	370	1800	490	2000
		(100)[a]	(100)	(100)	(100)	(100)	(100)	(100)	(100)
3.7	0.1	75	270	100	250	330	1400	430	1400
		(91)	(84)	(83)	(66)	(89)	(78)	(88)	(70)
74	0.1	65	140	70	160	280	830	320	880
		(79)	(44)	(58)	(42)	(76)	(46)	(65)	(44)
0	1.0	89	320	120	380	390	1600	480	1900
		(100)	(100)	(100)	(100)	(100)	(100)	(100)	(100)
3.7	1.0	80	320	110	350	350	1700	430	1600
		(90)	(100)	(91)	(92)	(90)	(106)	(90)	(84)
74	1.0	76	320	100	340	340	1600	420	1500
		(85)	(100)	(79)	(89)	(87)	(100)	(88)	(79)

[a] Numbers in parentheses indicate percent of the control (zero Al).

Since lower temperature *per se* was a growth limiting factor in the present study, it could be postulated that Al would affect plants with lower growth rates less than those with higher growth rates. This result was observed in an earlier study with ryegrass (Rengel and Robinson, 1989a) and is also consistent with smaller root ($P<0.001$) and shoot ($P<0.021$) weights of more Al-tolerant cv. Marshall than the more Al-sensitive cv. Wilo as observed here (Table 1). For Al-treated plants, the growth improvement caused by increased Mg concentration was greater at 22°C, which might have been due to the greater Mg uptake rates at the higher temperature (Table 2).

Mg, Ca and K Uptake Rates

No significant first order interaction involving cultivar effect was found for Mg, Ca or K uptake rates ($P<0.22$ to $P<0.96$). The interaction of temperature and Mg concentration in solution significantly affected Mg ($P<0.0001$) and Ca uptake rates ($P<0.001$) since increased nutrient solution Mg concentration stimulated Mg and

Table 2. Average net uptake rates in μmol (g root)$^{-1}$ h^{-1} of Mg, Ca, and K as influenced by cultivar, temperature (10/6 and 22/18°C), and nutrient solution concentrations of Mg (mM) and Al (μM). Standard errors of the means did not exceed 10% of the corresponding means. Tukey's $HSD_{0.05}$ for the temperature × Mg × Al × cultivar interaction was 0.53, 0.71, and 9.3 μmol (g root)$^{-1}$ h^{-1} for Mg, Ca, and K, respectively

Treatment		Cation uptake rates					
Al	Mg	Mg		Ca		K	
		10/6	22/18	10/6	22/18	10/6	22/18
cv. Marshall							
0	0.1	1.7	3.3	2.5	7.6	47	86
3.7	0.1	0.9	1.6	2.2	5.8	48	89
74	0.1	0.1	0.5	1.6	3.9	36	67
0	1.0	2.6	5.9	1.8	5.3	46	87
3.7	1.0	2.5	5.8	1.8	4.7	49	84
74	1.0	2.0	4.1	1.7	3.6	47	80
cv. Wilo							
0	0.1	1.5	3.3	2.1	5.8	42	87
3.7	0.1	0.8	1.5	2.1	4.7	46	89
74	0.1	0.1	0.4	1.4	2.8	34	67
0	1.0	2.4	6.2	1.7	4.9	40	86
3.7	1.0	2.2	5.2	1.5	3.3	41	90
74	1.0	1.8	3.8	1.4	2.6	42	88

depressed Ca uptake rates more at higher temperature (Table 2). The Mg–Ca antagonism has also been noted in wheat forage (Ohno and Grunes, 1985).

Plants can absorb Mg^{2+} and Ca^{2+} by passive, transpiration-dependent uptake and by active, metabolically-driven uptake mechanisms (Läuchli, 1976). The speculative limit of the sum of Ca and Mg solution concentrations below which transpiration has little influence on uptake of these cations was estimated at 5 mM (Barber, 1984). Since the Ca and Mg solution concentrations in the present study were 1.28 and 2.27 mM, nutrient uptake was probably influenced more by biochemical processes at the surface of root cells than by transpiration, which was undoubtedly decreased by lower temperature.

Greater Mg ($P<0.03$) and Ca uptake rates ($P<0.0001$) were noted in the more Al-tolerant cv. Marshall (Table 2) which is consistent with results obtained with other species (for the review see Foy, 1988). Rengel and Robinson (1989a) showed that Al in nutrient solution inhibited Ca and Mg uptake by Wilo more than by Marshall ryegrass. It appears that the difference in Al tolerance between Wilo and Marshall ryegrass is more closely related to greater Mg and Ca uptake in the presence of Al stress (Table 2) than to differences in Al uptake (data not shown). Such an observation may be species-dependent since it is consistent with the results with rice (Fageria, 1985) but is at variance with findings of Horst *et al.* (1983) with cowpea.

Uptake of Mg and Ca decreased with Al addition, while K uptake increased slightly and then dropped as Al concentration was raised (Table 2). The temperature × Al interaction significantly affected uptake rates of the three nutrients because deleterious Al effects were more pronounced at higher temperature. This result is probably due to the greater effects of Al stress on nutrient uptake processes if other environmental factors are near optimum for plant growth.

Since the Al effects on Mg, Ca and K uptake rates were greater at the lower nutrient solution Mg concentration, the Mg × Al interaction significantly affected ($P<0.001$) uptake rates of all three nutrients, confirming beneficial Mg effects

Table 3. The equivalent ratio of K/(Ca + Mg) in ryegrass shoots as influenced by cultivar, temperature (10/6 and 22/18°C), and nutrient solution concentrations of Mg (m*M*) and Al (μM). Standard errors of the means did not exceed 9% of the corresponding means. Tukey's $HSD_{0.05}$ of the temperature × Mg × Al × cultivar interaction was 0.52

Treatment Al	cv. Marshall				cv. Wilo			
	0.1 Mg		1.0 Mg		0.1 Mg		1.0 Mg	
	10/6	22/18	10/6	22/18	10/6	22/18	10/6	22/18
0	4.5	3.9	4.1	3.5	4.8	4.3	4.2	3.6
3.7	5.4	5.4	4.4	3.9	5.9	6.4	4.4	4.4
74	5.8	6.5	4.8	5.2	6.5	8.0	5.1	5.5

in alleviating Al toxicity (cf. Aniol, 1983; Kinraide and Parker, 1987). It should also be noted that at the 1 m*M* nutrient solution Mg concentration K uptake rate remained unaffected even at 74 μM Al (Table 2). The Mg effect on depression of K uptake has been reported (Hannaway *et al.*, 1982) but more often no effect at all was noted (Ohno and Grunes, 1985; Ologunde and Sorensen, 1982; and Table 2 of the present study).

K/(Ca + Mg) ratio in ryegrass shoots

Increased nutrient solution Mg concentration decreased K/(Ca + Mg) ratios (Table 3). This effect was more pronounced at 74 μM Al concentration (Mg × Al, $P < 0.01$) because Mg partially alleviated Al depression of Ca and Mg uptake rates. The cv. Wilo had higher K/(Ca + Mg) ratios (Table 3). The significant cultivar effect on shoot cation ratio was noted earlier in wheat forage (Karlen *et al.*, 1978). In the present study, the significant cultivar × Al interaction ($P < 0.02$) refected the higher ratio in Wilo than in Marshall, especially at 74 μM Al and 0.1 m*M* Mg. Hence, higher K/(Ca + Mg) ratios in Wilo resulted from greater Al-sensitivity as indicated by lesser ability to retain reasonably high Mg and Ca uptake rates during Al stress.

It was reported earlier that K/(Ca + Mg) ratios were not increased by lower temperatures (Cherney and Robinson, 1985) but were increased at elevated Al concentrations (Rengel and Robinson, 1989a). In the present study, temperature did not significantly affect the K/(Ca + Mg) ratio ($P < 0.42$) (Table 3). The temperature × Al interaction was significant ($P < 0.0001$) because increased temperature lowered the ratio where no Al was added and elevated the ratio at 74 μM Al. Since the grass tetany potential of forage usually increases following temperature rises after a prolonged cold period (Grunes *et al.*, 1970), the results of the present study suggest that the grass tetany potential of ryegrass forage will increase more under such temperature changes if high levels of active Al are present in soil solution.

References

Aniol A 1983 Aluminum uptake by roots of two winter wheat varieties of different tolerance to aluminum. Biochem. Physiol. Pflanzen 178, 11–20.

Barber S A 1984 Soil Nutrient Bioavailability: A Mechanistic Approach. John Wiley and Sons, New York, 398 p.

Cherney D J R and Robinson D L 1985 Influence of climatic factors and forage age on the chemical components of ryegrass related to grass tetany. Agron. J. 77, 827–830.

Clark R B 1977 Effect of aluminum on growth and mineral elements of Al-tolerant and Al-intolerant corn. Plant and Soil 47, 653–662.

Fageria N K 1985 Influence of aluminium in nutrient solutions on chemical composition in two rice cultivars at different growth stages. Plant and Soil 85, 423–429.

Foy C D 1988 Plant adaptation to acid, aluminum-toxic soils. Commun. Soil Sci. Plant Anal. 19, 959–987.

Grunes D L, Stout P R and Brownell J R 1970 Grass tetany of ruminants. Adv. Agron. 22, 331–374.

Hannaway D B, Bush L P and Leggett J E 1982 Mineral composition of Kenhy tall fescue as affected by nutrient solution concentrations of Mg and K. J. Plant Nutr. 5, 137–151.

Horst W J, Wagner A and Marschner H 1983 Effect of aluminum on root growth, cell division rate and mineral element contents in roots of *Vigna unguiculata* genotypes. Z. Pflanzenphysiol. 109, 95–103.

Karlen D L, Ellis Jr. R, Whitney D A and Grunes D L 1978 Influence of soil moisture and plant cultivar on cation uptake by wheat with respect to grass tetany. Agron. J. 70, 918–921.

Kinraide T B and Parker D R 1987 Cation amelioration of aluminum toxicity in wheat. Plant Physiol. 83, 546–551.

Konzak C F, Polle E and Kittrick J A 1977 Screening several crops for aluminum tolerance. *In* Plant Adaptation to Mineral Stress in Problem Soils. Ed. J M Wright. pp 311–327. Spec. Publ. Cornell Univ. Exp. Sta. Ithaca, NY.

Läuchli A 1976 Symplastic transport and ion release to the xylem. *In* Transport and Transfer Processes in Plants. Eds. I F Wardlaw and J B Passioura. pp 101–112. Academic Press, New York.

Ohno T and Grunes D L 1985 Potassium-magnesium interactions affecting nutrient uptake by wheat forage. Soil Sci. Soc. Am. J. 49, 685–690.

Ologunde O O and Sorensen R C 1982 Influence of K and Mg in nutrient solutions on sorghum. Agron. J. 74, 41–46.

Rengel Z and Robinson D L 1989a Aluminium effects on growth and macronutrient uptake by annual ryegrass. Agron. J. 81, 208–215.

Rengel Z and Robinson D L 1989b Aluminum and plant age effects on adsorption of cations in the Donnan free space of ryegrass roots. Plant and Soil 116, 223–227.

Steel R G D and Torrie J H 1980 Principles and Procedures of Statistics. 2nd ed. McGraw-Hill Book Co., New York.

Williams R F 1946 The physiology of plant growth with special reference to the concept of net assimilation rate. Ann. Bot. N.S. 10, 41–72.

M. L. van Beusichem (Ed.), *Plant nutrition – physiology and applications*, 419–424.

PLSO IPNC421

Bioassay technique to assess acid soil constraints for growth of wheat (*Triticum aestivum*) roots

V.C. BALIGAR, R.J. WRIGHT, K.D. RITCHEY and N.K. FAGERIA
ASWC Research Laboratory, USDA-ARS, P.O. Box 867, Beckley, WV 25802, USA

Key words: aluminium toxicity, cation deficiency, hill land acid soils, root bioassay, *Triticum aestivum* L., wheat

Abstract

A root bioassay technique was adapted to evaluate the relationship between average longest root length (ALRL) of 4 day old wheat (*Triticum aestivum* L.) and soil and soil solution chemical properties of 55 surface and subsurface horizons from fourteen hill land soils of the Appalachian region. The soil pH ranged from 3.6 to 6.7, while exchangeable Al and Ca were to 0.0 to 8.5 and 0.02 to 7.7 cmol kg^{-1} respectively.

Root elongation was inhibited in a number of the acid soil horizons. The ALRL was limited in B horizons relative to A horizons, suggesting that subsoil rooting in these horizons would be restricted. The ALRL was negatively correlated with soil and soil solution Al parameters and positively correlated to Ca parameters.

Stepwise regression equations were successful in predicting ALRL from the soil and soil solution chemical properties of A, E, or B horizons. The root bioassay technique adapted here holds promise of detecting root growth constraints in various types of acid soils.

Introduction

Crops are known to exhibit various degrees of response to chemical constraints often found in acid soils. For successful crop production on acid soils, one needs to evaluate the magnitude of the soil's chemical constraints for growth and the adaptability of chosen species or cultivars to that soil. In order to assess the compatibility of these two variables a short term and reliable plant growth bioassay technique is needed to identify soils (soil horizons) that might have acidity related limiting characteristics. Short term (2–4 days) bioassay techniques using root development of seedlings have been used to identify soils with Al toxicity problems and to differentiate relative Al tolerance in plant species or cultivars within species (Karr *et al.*, 1984; Ritchey *et al.* 1988; Wright *et al.*, 1989). These methods are based on the fact that during the early stages of growth, plants derive their nutrient needs from the seed and only Al and H toxicity and Ca and B deficiencies limit seedling root elongation.

Growth at low pH is not often limited by H^+ ion activity, but toxicity of Al, Mn and/or deficiency of essential nutrient elements such as Ca, Mg, N, P and micronutrients. Aluminum toxicity is probably the most important growth limiting factor for plants in most strongly acid soils. Chemical extraction methods commonly used for soil Al are 1*M* KCl, 1*M* NH_4OAc and 0.01*M* $CaCl_2$ (Hoyt and Nyborg 1971; Khalid and Silva 1979; Bromfield *et al.*, 1983). With the exception of 0.01*M* $CaCl_2$, the other extractants are not very successful in removing amounts of soil Al that are correlated with plant growth. In various types of acid soils the concentrations of Al, Mn, Ca, Mg, and pH of the soil solution have been related to poor plant growth (Evans and Kamprath 1970; Ritchey *et al.*, 1988; Wright *et al.*, 1987a, b; 1988; 1989).

The objective of the present study was to

adapt a root bioassay technique to assess the relationships between wheat seedling root growth and chemical properties in 55 acidic soil horizons from the Appalachian region of the eastern United States.

Materials and methods

In the current study 55 surface and subsurface horizons from 14 major hill land soils of the Appalachian regions were used. Soil samples were collected in early spring, air dried and passed through a 2 mm screen. Selected soil properties of A, B, and E horizons are given in Table 1. Soil pH-H_2O ranged from 3.60 to 6.7, while exchangeable Al and Ca ranges were 0.0 to 8.5 and 0.02 to 7.7 cmol kg^{-1} soil, respectively.

The root bioassay technique consisted of growing acid soil tolerant hard red spring wheat (*Triticum aestivum* L., cv. Yecorra Rojo) seedlings in 200 ml plastic cups containing soil at 33 kPa moisture tension. Wheat seeds were germinated for 24 to 30 h on moist chromatography paper and planted at 5 seeds per plastic cup. There were three replications for each of the soil horizons which were arranged in a randomized complete block design. The cups were placed on trays containing moist paper towels and covered with a plastic dome to provide a humid atmosphere to maintain the desired soil moisture level. Plants were grown for 4 days in a climatically controlled growth chamber set at 70% RH and 21°C with 12 h per day of 115 μmol s^{-1} m^{-2} light illumination. At harvest the longest root of each seedling was measured and for each soil horizon root lengths was averaged over all the seedlings and designated as average longest root length (ALRL).

Soil solutions were obtained at harvest by the centrifugation method of Reynolds (1984). Soil solution pH and electrical conductive (EC) were measured immediately. Total concentrations of Al, and basic cations were determined by ICP emission spectroscopy. Soil samples from different horizons were analyzed for: soil pH (1:1 H_2O and 1:1 0.01*M* $CaCl_2$), exchangeable bases (Thomas 1982), exchangeable acidity and Al (determined from the 1*M* KCl soil extract, Yuan 1959), and organic C using a Leco CHN 600. Extractable Al was determined by extraction with 0.01*M* $CaCl_2$ (soil:solution = 1:2, Hoyt and Nyborg 1971). Statistical Analysis System (SAS) software was used to calculate correlation coefficients and multiple regression equations relating ALRL to soil and soil solution chemical properties.

Table 1. Range of selected soil and soil solution properties of the A, E and B horizons

Soil[a]	Range of values			Soil soln.	Range of values, μ*M*		
	A	E	B		A	E	B
EX–Al	0.00– 7.5	0.97– 4.6	0.00– 8.5	Al	16.7– 1449	27.8– 701	0.7–232
EX–Mn	0.01– 1.2	0.00– 0.6	0.00– 0.3	Mn	16.7– 1003	14.0–1269	0.4–256
EX–H	0.07– 1.5	0.14– 0.6	0.05– 0.5	K	202.8– 2532	168.8– 509	34.3–469
EX–Ca	0.18– 7.3	0.06– 2.3	0.02– 7.7	Ca	116.0–12836	68.9– 808	4.5–422
EX–Mg	0.11– 2.2	0.03– 0.6	0.02– 6.8	Mg	132.0– 4838	60.0– 353	2.1–294
EX–K	0.12– 0.7	0.04– 0.4	0.04– 0.5	pH	3.4– 6.7	4.1– 5.0	4.2–6.8
CEC	2.03– 10.7	1.28– 7.1	0.67–23.8	EC	0.3– 1.0	0.1– 0.3	0.0–0.3
EX–Base Sat	16.90– 98.9	7.96–47.4	3.90–98.7				
Carbon (C)	12.30–159.9	8.70–30.8	11.00–52.3				
pH–H_2O	3.60– 6.1	4.20– 5.0	4.26– 6.7				
pH–$CaCl_2$	3.11– 5.8	3.55– 4.2	3.70– 6.1				

[a] EX = Exchangeable (cmol kg^{-1}); 1*N* NH_4OAC extractant for Ca, K, Mg and Mn and 1*M* KCl extractant for Al and H.
CEC = Σ cmol kg^{-1} of [K + Ca + Mg + Mn + Al + H];
Ex –Base sat, % = [K + Ca + Mg/CEC] × 100;
C = g kg^{-1};
EC = Sdm^{-1}.

Results and discussion

Large differences in ALRL of wheat seedlings were observed across the 55 soil horizons. However, overall mean ALRL among A, E, or B horizons were similar (Table 2). In the current study only three C horizons were used, so they are not included in discussion of the different horizons. Although A horizons had the lowest mean pH, they were also characterized by the highest mean Ca and base saturation and lowest exchangeable Al and Al saturation when compared to the E and B horizons. The slight decline in ALRL in the B horizon could be related to higher levels of Al and lower levels of basic cations.

Correlation coefficients relating ALRL to soil and soil solution properties of A, E and B horizons are shown in Table 3. In various horizons soil properties such as Al, H, pH, Ca and exchangeable base saturation were consistently correlated with ALRL. The ALRL of wheat seedlings in 55 horizons is plotted as a function of the pH ($CaCl_2$) in Figure 1. The ALRL was significantly correlated ($P<0.01$) with soil pH (Table 3). Wright *et al.* (1987b) using similar soils reported a significant ($P<0.01$) negative correlation between 5 week old snapbean root growth and soil Al saturation and total soil solution Al. The ALRL was closely correlated with Al parameters. The Al extracted

Table 2. Range of values for average longest root length (ALRL), of wheat seedlings in all horizons and in A, E or B horizons

Soil horizons[a]	ALRL., cm/plant Mean	Range
All horizons (n = 55)	7.09	3.6–11.1
A horizons (n = 14)	7.83	4.4–10.2
E horizons (n = 8)	7.85	6.4– 8.8
B horizons (n = 30)	6.57	3.6–11.1

[a] A includes A and Ap horizons; E includes E and BE horizons and B includes, BA, Bw, BC and Bt horizons.

Table 3. Correlation coefficient (r) values between average longest root length, ALRL (cm/plant) and soil and soil solution properties of all horizons and A, E or B horizons

Soil[a]	ALRL All	A	E	B
Toxic Constraints				
EX–Al	−0.62**	−0.80**	−0.34	−0.58**
ET–Al	−0.65**	−0.67**	−0.74*	−0.68**
SAT–Al	−0.68**	−0.72**	−0.83*	−0.72**
EX–H	−0.15	−0.68**	0.28	−0.33
SAT–H	0.04	−0.47	0.21	−0.01
EX–Acidity	−0.61**	−0.84**	−0.31	−0.58**
pH–H_2O	0.48**	0.80**	−0.02	0.75**
pH–$CaCl_2$	0.60**	0.80**	0.31	0.80**
Nutrient Constraints				
EX–Ca	0.20	0.55*	0.52	0.01
SAT–Ca	0.54**	0.75**	0.62	0.50**
EX–Base–SAT	0.63**	0.76**	0.60	0.69**
CEC	−0.20	−0.31	0.13	−0.26
Carbon (C)	0.08	−0.67**	0.71*	0.20

Soil solution[b]	ALRL All	A	E	B
Toxic Constraints				
Al	−0.03	−0.72**	0.45	−0.23
pH	0.40**	0.77**	−0.58	0.71**
EC	0.31*	0.08	0.44	0.24
Nutrient Constraints				
Ca	0.25	0.25	0.57	0.29
CAR	0.33*	0.45	0.12	0.28
Ca–Mg	0.07	0.28	0.47	−0.15
CAT–1	0.39**	0.30	0.35	0.30
CAT–2	0.37**	0.33	0.39	0.29

[a] ET = Extractable ($mg\ kg^{-1}$) by 0.01*M* $CaCl_2$;
SAT = (Element/CEC) × 100;
[b] CAR = Ca/[Ca + Mg + K + Mn + Al];
Ca–Mg = Ca/[Ca + Mg];
CAT–1 = Ca/[Ca + Mg + K + Na];
CAT–2 = Ca/[Mg + K + Na];
*,** Significant at 0.05 and 0.01 level of probability, respectively.

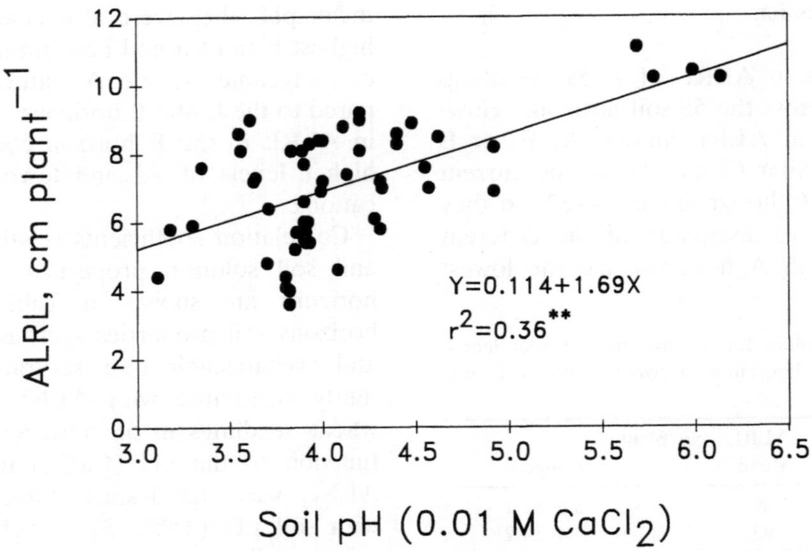

Fig. 1. Average longest root length (ALRL) of wheat seedlings as a function of soil pH (0.01*M* $CaCl_2$).

by 0.01*M* $CaCl_2$ proved to be a better predictor of reduction of ALRL than exchangeable Al determined by 1*M* KCl. The ALRL as a function of 0.01M $CaCl_2$ extractable Al is shown in Figure 2. Extractable Al > 30 mg kg^{-1} has reduced ALRL by about half. Other researchers have also shown significant relationships between plant growth and Al extracted by 0.01*M* $CaCl_2$ (Bromfield *et al.*, 1983; Hoyt and Nyborg 1971). The amount of Al extracted by 0.01*M* $CaCl_2$ represents soluble plus some exchangeable Al and would be more closely related to concentration of Al in the soil solution than the Al extracted by high ionic strength extractants such as 1*M* KCl. High ionic strength extractants are known to extract both exchangeable and non-exchangeable forms of Al. Root growth in short term bioassay techniques should be related to soil solution Al which would be approximated by an extractant such as 0.01*M* $CaCl_2$.

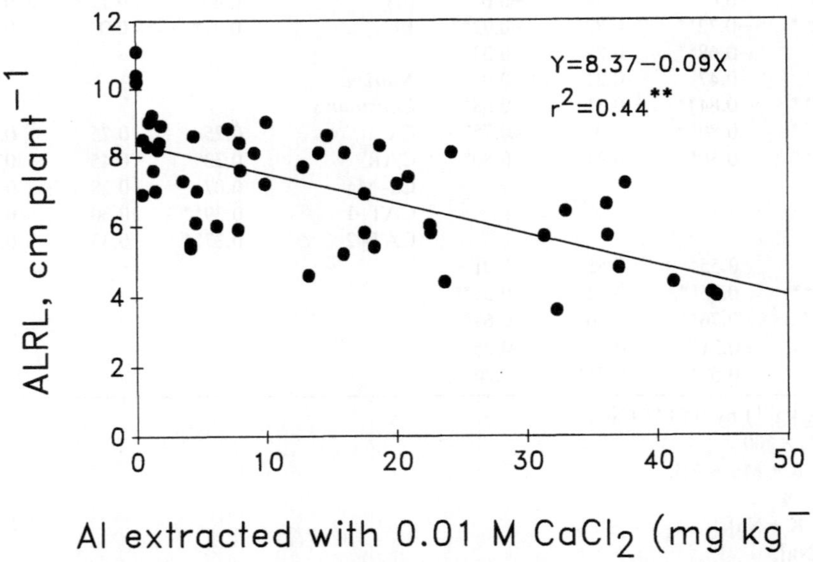

Fig. 2. Average longest root length (ALRL) of wheat seedlings in 55 soil horizons as related to soil Al extracted by 0.01*M* $CaCl_2$.

The ALRL in A and B horizons was significantly correlated with soil Ca saturation. In the current study ALRL was severely affected when Ca saturation was <15% (Fig. 3). Wright *et al.* (1987b) using similar soils reported a positive significant ($P<0.01$) relation between 5 week old snap bean root-growth and soil exchangeable Ca and percent saturation of Ca. Severe depression of snapbean root-growth was noted when soil Ca saturation was <15%. The C content of soil gave a negative significant relationship ($P<0.01$) with ALRL in A horizons and positive relationships in E and B horizons.

In different horizons soil solution Al levels ranged from <1 to 1449 μM, with higher mean values observed in A horizons. In A horizons ALRL gave significant negative correlations with total Al in the soil solutions. The ALRL was positively related to EC values. With the exception of E horizons, the ALRL was positively related ($P<0.01$) to soil solution pH.

The ALRL was positively related to concentrations of cations in soil solution and ratios of Ca and cations. In soil solution, Ca levels ranged from 4.5 to 12836 μM with higher levels being found in the A horizons. The presence of Ca and other cations can ameliorate Al toxicity. The presence of higher levels of Ca might have ameliorated the toxic effects of high levels of Al in the A horizons. Wright *et al.* (1987b) reported a highly significant ($P<0.01$) relationship between 5 week old snapbean root growth and soil solution Ca.

Stepwise regression equations were used to predict the seedling ALRL from several soil and soil solution variables (Table 4). In the A and Ap horizons 87% of the significant ($P<0.01$) variability in ALRL was accounted for by ex-

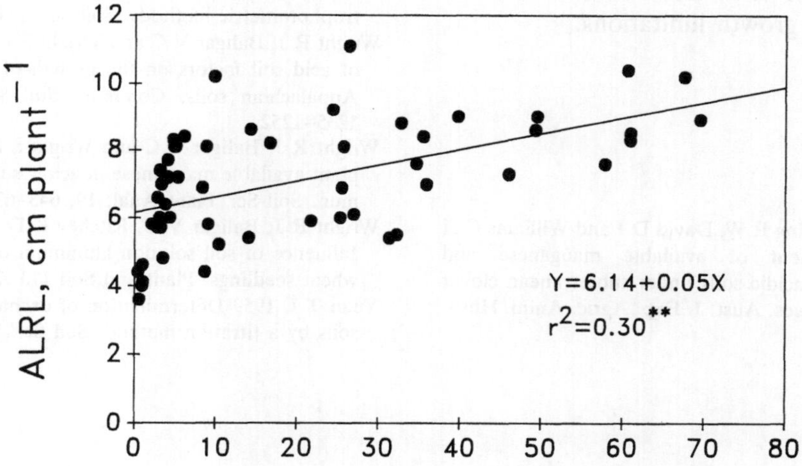

Fig. 3. Average longest root length (ALRL) of wheat seedlings in 55 soil horizons as a function of the percent Ca saturation.

Table 4. Multiple regression equations relating ALRL of wheat (Y) to soil and solution properties of all and A, E or B horizons

Soil horizons	Multiple regression equation[a]	R^2
ALL	Y = 7.65 − 0.43 EX-Acidity − 0.04 ET-Al + 9.37 CAT-1 − 2.61 CAT-2	0.66**
A	Y = 11.80 − 1.64 EX-H − 0.10 ET-Al − 0.002 SS-Al − 3.56 CAR	0.87**
E	Y = 11.81 − 0.54 EX-Al + 1.78 C − 1.07 pH-H_2O − 0.002 SS-Ca	0.98*
B	Y = 0.17 + 0.62 C + 1.83 pH-$CaCl_2$ − 0.04 ET-Al − 2.15 Ca-Mg	0.81**

[a] SS = Soil solution and for the other abbreviations refer to Tables 1 and 3.
*,** Significant at 0.05 and 0.01 level of probability respectively.

changeable H, extractable and soil solution Al and CAR. In subsurface E and BE horizons 98% of the significant ($P < 0.05$) variance in ALRL was related to soil exchangeable Al, C, pH (H_2O) and concentration of soil solution Ca. However, 81% of the significant ($P < 0.01$) variance in ALRL in B horizons (BA, Bw, BC and Bt) was attributable to C, pH($CaCl_2$), extractable Al and soil solution Ca Mg ratios. Wright *et al.* (1987a) found that soil or soil solution Ca and pH values were needed along with Al in regression equations describing root and shoot growth of 5 week old subterranean clover in acidic soil horizons.

In the present experiment the short term root bioassay technique was readily able to identify soil horizons with acidity related limitations to root growth. Root elongation was significantly related to soil and soil solution pH, Ca and Al. These results confirm that acid soil infertility is related to a number of factors and that Al alone cannot explain all growth limitations.

References

Bromfield S M, Cumming R W, David D J and Williams C H 1983 The assessment of available manganese and aluminum status in acidic soils under subterranean clover pastures of various ages. Aust. J. Exp. Agric. Anim. Husb. 23, 192–200.

Evans C E and Kamprath E J 1970 Lime response as related to percent Al saturation, solution Al and organic matter content. Soil Sci. Soc. Am. Proc. 34, 893–896.

Hoyt P B and Nyborg M 1971 Toxic metals in acid soil. 1. Estimation of plant-available aluminum. Soil Sci. Soc. Am. Proc. 35, 236–240.

Karr M C, Coutinho J and Ahlrichs J L 1984 Determination of aluminum toxicity in Indiana soils by petri dish bioassay. Proc. Ind. Acad. Sci. 93, 85–88.

Khalid R A and Silva J A 1979 A study of soil aluminum extraction methods in relation to plant aluminum and yield in tropical soils. Trop. Agric. (Trinidad), 56, 53–63.

Reynolds B 1984 A simple method for the extraction of soil solution by high speed centrifugation. Plant and Soil 78, 437–440.

Ritchey K D, Baligar V C and Wright R J 1988 Wheat seedling response to soil acidity and implications for subsoil rooting. Commun. Soil Sci. Plant. Anal. 19, 1285–1293.

Thomas G W 1982 Exchangeable cations. *In* Methods of Soil Analysis, Part 2, 2nd Ed. Agronomy 9. Eds. A L Page, R H Miller and D R Keeney. pp 159–165. Am. Soc. Agron, Madison, WI.

Wright R J, Baligar V C and Wright S F 1987a. Estimation of phytotoxic aluminum in soil solution using three spectrophotometric methods. Soil Sci. 144, 224–232.

Wright R J, Baligar V C and Wright S F 1987b The influence of acid soil factors on the growth of snapbeans in major Appalachian soils. Commun. Soil Sci. Plant Anal. 18, 1235–1252.

Wright R J, Baligar V C and Wright S F 1988 Estimation of plant available manganese in acidic subsoil horizons. Commun. Soil Sci. Plant Anal. 19, 643–622.

Wright R J, Baligar V C, Ritchey K D and Wright S F 1989 Influence of soil solution aluminum on root elongation of wheat seedlings. Plant and Soil 113, 294–298.

Yuan T L 1959 Determination of exchangeable hydrogen in soils by a titration method. Soil Sci. 88, 164–167.

M. L. van Beusichem (Ed.), *Plant nutrition – physiology and applications*, 425–428.

PLSO IPNC100A

Level of acid soil field stress for sorghum (*Sorghum bicolor*) tolerance development: Comparison among locations

R.R. DUNCAN and L.M. SHUMAN
Department of Agronomy, University of Georgia, Georgia Agricultural Experiment Station, Griffin, GA 30223-1797, USA

Key words: aluminium toxicity, *Sorghum bicolor* L. Moench

Abstract

A dual-phase, sequential breeding program in the Southeastern United States has resulted in development of acid soil tolerant sorghum [*Sorghum bicolor* (L.) Moench] germplasm (GPP4BR, PI 531231) containing source parental lines for hybrid development programs in tropical and subtropical environments. The first phase involved selection for Mn toxicity tolerance at soil pH-H_2O 4.5–4.8 (pH-KCl 4.0–4.3) and 20% Al saturation of the soil CEC. The second phase involved selection for Al toxicity tolerance at soil pH-H_2O 4.1–4.4 (pH-KCl 3.7–4.0) and 50% Al saturation. Four locations and 7 soil types are utilized in the breeding program. The Phase II program in Georgia is comparable to the acid soil stress levels found in Brazil and is apparently more stressful than the Colombia program. An Al saturation of 50% with soil organic matter contents less than 5% is evidently the maximum field stress level which the Sorghum genus can withstand.

Introduction

Quantitating levels of acid soil stress in the field have been difficult for breeders due to variable intra- and inter-species response reactions, genotype-by-environment interactions, and variability among soil types. However, knowledge about specific soil stress characteristics and their influence on plant responses to different levels of stress for a specific soil type is essential for ascertaining mechanisms involved in tolerance and susceptibility reactions and for refining breeding programs to maximize developmental efforts in specific environments.

This paper will not attempt to address the question of mechanisms, but will seek to illucidate soil stress levels and plant response interactions which have led to successful development of improved acid soil tolerant sorghum [*Sorghum bicolor* (L.) Moench] genotypes in various field-stressed environments.

Methods

Soils

Characteristics of the specific acid ultisols and inceptisol in Georgia and the oxisol in Puerto Rico used in the sorghum breeding program are presented in Table 1 (Duncan *et al.*, 1989). The ultisols from the subtropical, humid southeastern United States are characterized by a) low organic matter (<1.5%) b) moderate to strong acidity, c) low base status and high leachability, d) highly weathered with predominately 1 : 1 nonexpanding clays, e) low availability of P, high P fixing capacity, and high Fe- and Al-oxide contents, f) high Mn availability at moderate (water pH 4.5–5.0) acidity levels, g) low water infiltration rates, and h) high Al availability at soil pHs below 4.5 (Duncan, 1988). Soils used for the sorghum breeding program in Georgia were located in the Piedmont region and have the above general

Table 1. Specific acid soil characteristics of the ultisols and inceptisol in Georgia and the oxisol in Puerto Rico used in the sorghum breeding program

Soil type[a]	Extraction								Al[e] Saturation
	Soil pH		Ml[b]		KCl[c]	NH_4OAc[d]			
	H_2O	KCl	Mn	P	Al	Ca	Mg	K	
			mg kg^{-1}		cmol$_c$ kg^{-1}				%
GEORGIA LOCATION									
Dyke clay loam	4.3	3.9	62	6	1.59	0.67	0.12	0.55	54
Congaree silt loam	4.3	4.0	24	7	1.40	0.94	0.20	0.51	45
Cedarbluff silt loam	4.3	3.7	37	26	1.16	0.73	0.17	0.32	48
Cecil sandy clay loam	4.3	3.8	26	20	1.47	0.46	0.09	0.24	63
Pacolet sandy clay	4.4	3.8	10	3	2.02	0.97	0.25	0.34	56
Appling clay loam	4.1	4.0	29	9	0.79	0.76	0.12	0.08	45
PUERTO RICO LOCATION									
Coto clay	4.3	3.6	–	–	1.51	2.03	0.27	0.09	38

[a] Dyke = Typic Rhodudult; Congaree = Fluventic Dystrochrept; Cedarbluff = Fragiaquic Paleudult; Cecil, Pacolet, Appling = Typic Lanhapludult; Coto = Tropeptic Eutrustox (Haplorthox).
[b] Extracted using Mehlich 1.
[c] Extracted using 1 *N* KCl.
[d] Exchangeable cations extracted with 1 *M* NH_4OAc, pH 7.
[e] Percent Al saturation of the CEC = $\frac{\text{Al (cmol}_c\text{ kg}^{-1})}{\text{Al + Ca + Mg + K + Na(cmol}_c\text{ kg}^{-1})} \times 100.$

characteristics, but they were medium-to-fine textured with well-developed structure.

Breeding program

The acid soil tolerance sorghum breeding program has been described in detail elsewhere (Duncan, 1988). A schematic of the acid soil constraints and the dual-phase breeding approach is presented in Figure 1. The maximum level of acid soil stress that sorghum can withstand is 50% aluminium saturation (Borgonovi *et al.*, 1987) when soil organic matter contents remain below 5% (Hargrove, 1986). Higher organic matter contents may reduce the level of stress, and genotypes emerging from such a screening environment may have difficulty surviving more stressful acid soil environments (Gourley, 1987).

Results

Level of soil stress

Soils utilized to impose 45–55% Al saturation of the CEC on segregating genotypes in the breeding program are characterized by water pH levels ranging from 4.1–4.4 (KCl-pH 3.6–4.0), 10–60 μg Mn g^{-1}, 0.46–2.00 cmol$_c$ Ca kg^{-1} and 0.09–0.27 cmol$_c$ Mg kg^{-1}. The pH levels were higher but the Al saturation percentages were comparable with soils used in the sorghum breeding program in Brazil (Table 2). The soils in Colombia had comparable pH levels, but higher Al saturation percentages and substantially higher organic matter contents (Table 3) (<2% in Georgia vs 3–4% in Brazil vs 7–8% in Colombia).

Specific soil constraints

Using two of the same soils employed in the breeding program, Shuman *et al.* (1989) evaluated the effects of soil aluminium on growth and aluminium concentration in sorghum. They determined that plant growth reductions were due to Al toxicity and not to Mn toxicity or Ca deficiency. Relationships between plant growth and soil Al values were significant for topsoils but notsignificant for subsoils. Organic matter in the top soils apparently reduced the harmful effects of Al on sorghum plants and

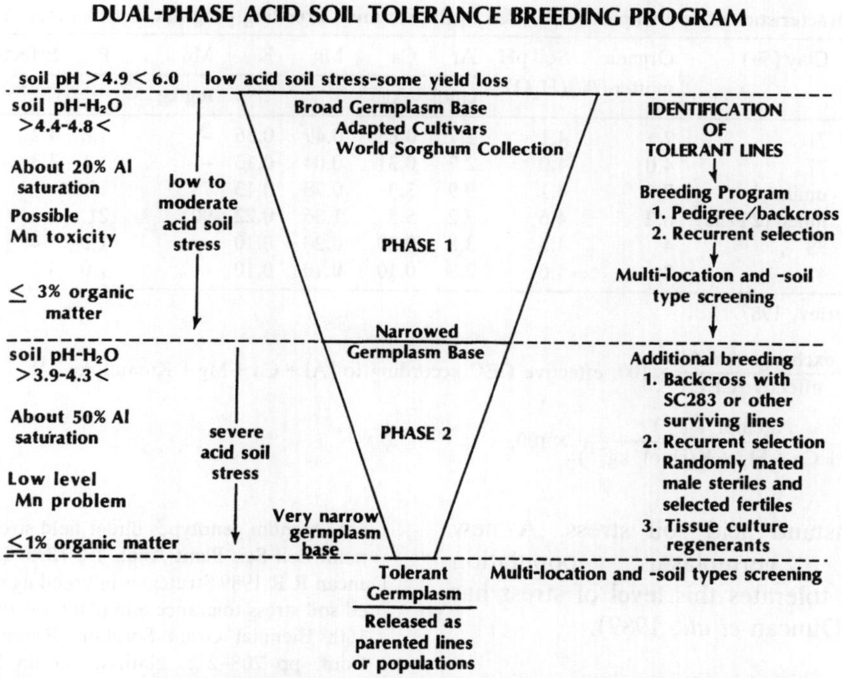

Fig. 1. Schematic of acid soil constraints, narrowing of germplasm base during development, and breeding approaches during the dual-phase program.

plant growth was better than on the subsoils. The best predictors of plant growth and plant Al concentration were soil solution Al^{3+} activity ($r = -0.92^{**}$, 0.83^{**}) and Al saturation of the exchange complex (-0.89^{**}, 0.82^{**}, respectively).

Discussion

Sorghum is extremely sensitive to acid soil stress environmental conditions. Under field conditions, genotypes must tolerate the acid soil stress complex and not just Al toxicity (Duncan, 1989). In addition, the breeding program must incorporate adaptability, desirability, and pest resistance characteristics simultaneously while the acid soil tolerance level is improved. The dual-phase approach allows for these sorghum germplasm improvements. Different soil types at several locations will help to reduce breeding problems with genotype-by-environment interactions. An aluminium saturation of approximately 50% is evidently the peak level at which the *Sorghum*

Table 2. Chemical characteristics of the Latosol used to evaluate sorghum germplasm and develop cultivars at Sete Lagoas, Minas Gerais, Brazil[a]

Lime t/ha	Soil depth cm	Soil pH (H_2O)	Al	Ca	Mg	K	P mg kg^{-1}	Aluminium saturation[b] %
			$cmol_c\ kg^{-1}$					
0	0–20	4.7	2.1	1.1	0.30	0.19	3	62
	20–40	4.6	2.3	0.8	0.25	0.17	2	66
2	0–20	4.8	1.4	1.9	0.27	0.15	3	41
	20–40	4.6	1.9	0.9	0.22	0.12	2	59

[a] Adapted from Borgonovi *et al.*, 1987.

[b] $\%\ \text{Al saturation} = \dfrac{\text{Al}(\text{cmol}_c\ \text{kg}^{-1})}{\text{Al} + \text{Ca} + \text{Mg} + \text{K}(\text{cmol}_c\ \text{kg}^{-1})} \times 100\,.$

Table 3. Typical characteristics of two tropical soils used in the sorghum breeding program in Colombia, South America[a]

Soil order	Depth cm	Clay (%)	Organic matter (%)	Soil pH (H_2O)	Al	Ca	Mg	K	Mn	P	Effective CEC	Al saturation (%)
					$cmol_c\ kg^{-1}$				$mg\ kg^{-1}$			
Ultisol	0–20	71	7.1	4.1	2.7	0.65	0.49	0.36	–	1.8	4.2	64[b]
	20–35	77	4.0	4.0	2.7	0.31	0.04	0.13	–	1.1	3.3	83
Ultisol	0–20	unlimed	7.3	4.1	9.9	3.3	0.78	0.13	78	15	–	60[c]
	0–20	limed (2T/ha)	8.2	4.6	7.2	5.5	1.36	0.22	90	21	–	40
Oxisoil	0–12	38	4	4.5	3.8	0.20	0.20	0.10	–	1.0	4.4	86[b]
	12–32	41	2	4.6	2.8	0.10	0.10	0.10	–	1.0	3.1	89

[a] Adapted from Gourley, 1987.

[b] $\%\ \text{Al saturation} = \dfrac{\text{exchangeable Al}}{\text{effective CEC}} \times 100$; effective CEC according to $[\text{Al} + \text{Ca} + \text{Mg} + \text{K}(\text{cmol}_c\ \text{kg}^{-1})]$.

[c] $\text{Al saturation} = \dfrac{\text{Al}(\text{cmol}_c\ \text{kg}^{-1})}{\text{Al} + \text{Ca} + \text{Mg} + \text{K}(\text{cmol}_c\ \text{kg}^{-1})} \times 100.$

genus can withstand acid soil stress. A new random-mating germplasm population (GPP4BR) that tolerates this level of stress has been released (Duncan *et al.*, 1989).

References

Borgonovi R A, Shaffert R E and Pitta G V E 1987 Breeding aluminium-tolerant sorghums. *In* Sorghum for Acid Soils. Eds. L M Gourley and J G Salinas. pp 271–292. Centro Internacional de Agricultura Tropical (CIAT), Cali, Colombia.

Duncan R R 1981 Registration of GPIR acid soil tolerant sorghum germplasm population (Reg. No. GP 73). Crop Sci. 21, 637.

Duncan R R 1984 Registration of acid soil tolerant sorghum germplasm (Reg. Nos. GP 140–142). Crop Sci. 24, 1006.

Duncan R R 1988 Sequential development of acid soil tolerant sorghum genotypes under field stress conditions. Commun. Soil Sci. Plant Anal. 19, 1295–1305.

Duncan R R 1989 Strategies in breeding sorghum for improved soil stress tolerance and plant nutritional traits. *In* Proc. 16th Biennial Grain Sorghum Research and Utilization Conf. pp 208–216. National Grain Sorghum Producers Assoc., Abernathy, TX.

Duncan R R, Torres-Cordona S, Goenaga R, Sotomayor-Rios A and Shuman L M 1989 Registration of GPP4 BR(H)C5 acid soil tolerant sorghum germplasm population (Reg. No. GP 234). Crop Sci. 29, 1581–1582.

Gourley L M 1987 Finding and utilizing exotic Al-tolerant sorghum germplasm. *In* Sorghum for Acid Soils. Eds. L M Gourley and J G Salinas. pp 293–309. Centro Internacional de Agricultura Tropical (CIAT), Cali, Colombia.

Hargrove W L 1986 The solubility of aluminium-organic matter and its implication in plant uptake of aluminium. Soil Sci. 142, 179–181.

Shuman L M, Ramseur E L and Duncan R R 1989 Effects of soil aluminium on the growth and aluminium concentration in sorghum. Agron J. 82, 313–318.

M. L. van Beusichem (Ed.), *Plant nutrition – physiology and applications*, 429–434.

PLSO IPNC100B

Sequential growth stage adjustment to acid soil stress in sorghum (*Sorghum bicolor*)

R.R. DUNCAN and R.E. WILKINSON
Department of Agronomy, University of Georgia, Georgia Agricultural Experiment Station, Griffin, GA 30223-1797, USA

Key words: acid soil stress, amylase, gibberellic acid, indoleacetic acid, isocitric lyase, *Sorghum bicolor* (L) Moench

Abstract

Shoot growth of sorghum [*Sorghum bicolor* (L.) Moench] cultivars (Funk G522DR, GP140, TAM428, SC283, SC599, and SC574) was shown to respond similarly to increased [H^+] and the response was explicable as a gibberellic acid (GA) sequestering and failure of GA translocation. Root growth inhibition by increased [H^+] was explicable as a failure of indoleacetic acid (IAA) translocation. Mn^{2+} was shown to inhibit GA precursor biosyntheses and possibly, to increase IAA oxidation. Amylase and isocitric lyase activities were sensitive to H^+, Ca^{2+}, and Mn^{2+}. Differences in growth patterns of acid soil tolerant and sensitive sorghum cultivars depended upon the physiology and biochemistry of the individual cultivars.

Introduction

Sorghum [*Sorghum bicolor* (L.) Moench] cultivars vary in susceptibility to acid soil stress (Duncan, 1988). Low pH soils are characterized by high H^{2+}, Mn^{2++}, and Al^{3+} availability plus a deficiency of Ca^{2+} and Mg^{2+} (Foy, 1984). Cultivar responses to these factors may not be uniform. Field observations led to the speculation that there were three physiological phases of sorghum response to acid soils (Duncan, 1988).

The first response (IMPACT) is initiated with water inhibition and may last for as long as one month but is usually considered to be 0–14 days and is basically functioning when the first or second true leaves are formed. The second phase (PRE-FLOWERING) (35–65 days after planting) coincides with continued vegetative development and subsequent panicle initiation. The third phase (POST-FLOWERING) includes panicle exsertion, anthesis, and grain-filling. Axiomatically, phases 2 and 3 are dependent upon successful establishment of the seedling in the impact phase. Any restriction to seedling shoot or root growth would have direct consequences in the latter two phases. Yet, the influences of acid soil parameters on sorghum seedling development have not been documented.

Seed germination and seedling development are critical stages in which plant growth regulators play a critical role. Gibberellic acid (GA) has been shown to be intimately involved in shoot growth (Chrispeels *et al.*, 1967), stored food utilization (Chrispeels *et al.*, 1967; Doig *et al.*, 1985), and in some instances, root elongation (Wilkinson and Duncan, 1989b). Indoleacetic acid (IAA) is known to be intimately associated with root growth (Wilkinson and Duncan, 1989b). Any parameter affecting [GA] or [IAA] synthesis, degradation, and/or translocation from the sites-of-synthesis to the sites-of-utilization could drastically influence seedling development during the impact phase.

Thus, the specific influence of soil acidity on the growth and development of sorghum cultivars with varying field tolerance to acid soil stress was evaluated.

Materials and methods

Six sorghum cultivars were selected for quantitative investigations on the basis of their acid soil stress field reactions (Duncan, 1987; 1988). Cultivar response reactions were: tolerant – SC283, SC574; intermediate – GP140, SC599; susceptible – Funk G522DR, TAM428.

Seeds were planted in quartz sand and watered with 0.01 *M* sodium acetate buffer (pH 6.0, 5.5, 5.0, 4.5, or 4.0) ± calcium ($CaCl_2$) (0, 10, or 100 mg L^{-1} Ca^{2+}) or manganese ($MnCl_2$) (0, 1.4, or 140.0 mg L^{-1} Mn^{2+}). [0.1 *M* sodium acetate inhibits root growth while 0.01 *M* sodium acetate does not influence root growth. (Unpublished data)]. After 7 days, root and shoot growth were measured (Wilkinson and Duncan, 1989 a,b). Amylase and isocitric lyase activities were measured in seed and primary leaf tissues concomitantly (Wilkinson and Duncan, 1989 c,d).

ent-Kaurene synthesis

Cell-free enzyme systems from TAM428 and SC599 etiolated coleoptiles were utilized to evaluate the incorporation of [^{14}C]-isopentenyl pyrophosphate into *ent*-kaurene (Wilkinson, 1982).

Results and discussion

Shoot growth

The sorghum cultivars varied in seedling growth rates (as measured by shoot growth) when grown at pH 6 (Table 1) and increasing [H^+] induced decreased shoot length. Pot eluate ion concentrations were not appreciably altered from those originally added (unreported data). These growth curves resembled bioassay responses to a common factor. Normalization of shoot growth curves at pH 6 as a standard revealed that these growth curves were not statistically different. Although there were different basic growth rates between sorghum cultivars at pH 6, the decrease in growth as [H^+] increased was equal in all tested cultivars. This decrease in shoot growth exactly matched the lipid solubility of GA_3 in water when partitioned between ethyl acetate and water at several pH's (Table 2) and the lines were equivalent statistically.

GA is synthesized in seedling plumules and translocated to the aleurone where it induces the synthesis of amylase (Chrispeels and Varner, 1967). Lipid barriers exist between the plumule cells and the aleurone (Doggett, 1988). Additionally, the spaces between the cells would be exposed to the pH of the external medium while the pH of the cytoplasm would remain relatively uniform (Felle, 1988). Therefore, the sorgum shoot growth data are explicable as a sequestering of GA within the cells where it is synthesized and a Ga deficiency at the site of growth. Proof of this hypothesis would depend upon a reversal of the [H^+] influence by an exogenous supply of GA. In low pH soil susceptible (Funk G522DR) and resistant (SC574 and SC283) cultivars, growth at pH 4.5 was significantly decreased, while exogenous GA_3 (6.7 μM) reversed the shoot growth inhibition at pH 4.5 (Table 3).

Calcium did not influence shoot growth (data not shown). But high [Mn^{2+}] greatly decreased

Table 1. Influence of pH on the growth (7-d) of sorghum cultivars.

	Shoot length (cm) pH		Primary root length[b] (cm) pH	
	6.0	4.0	6.0	4.0
Funk G522DR	9.1a[a]	2.8ab	8.3a	0.8b
GP140	7.9a	3.6a	7.6a	3.0a
TAM428	6.2b	2.7ab	5.8bc	0.9b
SC283	5.2bc	1.5c	6.2b	0.1b
SC599	4.8c	1.8bc	5.6bc	0.5b
SC574	4.0c	1.4c	5.1c	0.1b

[a] Values in a column followed by the same letter are not significantly different at the 5% level.
[b] Secondary roots were not present.

Table 2. Influence of pH on the partitioning of gibberellic acid (GA_3) and indoleacetic acid (IAA) between water and ethyl acetate as compared to Funk G522DR shoot and SC283 root growth

	pH				
	6.0	5.5	5.0	4.5	4.0
	Shoot length (%)				
GA_3 (water)[a]	91	80	55	35	22
Funk G522DR	90	74	58	43	24
	Primary root length (%)				
IAA (water)[b]	100	83	60	38	18
SC283	100	83	55	18	4

[a] $GA_3 = 25\ \mu g/mL$ (Wilkinson and Duncan, 1989a).
[b] $IAA = 25\ \mu g/mL$ (Wilkinson and Duncan, 1989b).

Table 3. Influence of pH and gibberellic acid (GA_3) on the shoot and root growth of Funk G522DR, SC574, and SC283 sorghum cultivars

pH	GA_3 (μM)	Cultivars G522DR	SC574	SC283
		Shoot length (%)		
6.0	–	100a[a]	100a	100a
4.5	–	60c	60b	57c
4.5	6.7	76b	105a	79b
		Primary root length (%)		
6.0	–	100a	100a	100a
4.5	–	68c	21b	13b
4.5	6.7	90b	21b	13b

[a] Values in a column in a box followed by the same letter are not significantly different at the 5% level.

Table 4. Influence of pH and Mn^{2+} on Funk G522DR shoot and SC283 root growth

Mn ($mg\ L^{-1}$)	pH 6.0	5.5	5.0	4.5	4.0
	Funk G522DR shoot (cm)				
0	9.0a[a]	8.7a	7.5a	7.3ab	4.2c
1.4	10.0a	8.6a	9.0a	6.5b	4.5c
140.0	5.2c	4.7c	5.0c	4.5c	3.7c
	SC283 Primary root length (cm)				
0	6.2a	6.0a	4.5b	3.2c	0.7c
1.4	6.6a	5.6a	5.1ab	3.1c	0.1d
140.0	0.9d	0.7d	0.7d	0.6d	0.1d

[a] Values in a box followed by the same letter are not significantly different at the 5% level.

shoot elongation (Table 4). Excess manganese inhibits isoprenoid and GA syntheses (Wilkinson and Ohki, 1988). These data show that excess Mn^{2+} exacerbates the $[H^+]$ sequestering of GA in the cells by preventing GA biosynthesis which thereby decreases the quantity of GA present in the tissue.

Root growth

Root length decreased as $[H^+]$ increased (Table 1); and when normalized, the rates of decreased root length were not significantly different. Exogenous GA did not reverse the $[H^+]$ influence on root length in SC574 and SC283 (Table 3).

Table 5. Root length of SC283 as influenced by pH and indoleacetic acid (IAA)

pH	IAA (nM)	Primary root length (%)
6.0	–	100a[a]
4.5	–	71b
4.5	0.01	74b
4.5	0.1	102a
4.5	1	95a
4.5	10	77b

[a] Values followed by the same letter are not significantly different at the 5% level.

Since GA is known to induce the synthesis of an IAA (auxin) oxidation (auxinase) inhibitor (Hare, 1964), the response to GA in Funk G522DR might be a secondary response. IAA water solubility and SC283 root growth as functions of $[H^+]$ were not significantly different (Table 2). Thus the influence of $[H^+]$ on root growth might be explicable as an IAA translocation problem. Exogenous IAA (10^{-9} and 10^{-10} *M*) reversed the influence of H-ion concentration on SC283 root growth (Table 5). Thus, the influence of H-ion concentration is explicable as either: a) an IAA translocation problem, or b) an IAA oxidation reaction.

Calcium concentration did not influence sorghum root growth (data not shown) but excess manganese induced a greatly decreased SC283 root growth (Table 4). Since Mn^{2+} has been reported to oxidize IAA (Hare, 1964), this response could be explicable as a $[H^+]$-induced sequestering of IAA by excess Mn^{2+}. In both cases, the result would be an IAA deficiency in the cells of the root elongation zone.

Amylase

In cereal grains, amylase synthesis in the aleurone layer is controlled by GA (Chrispeels and Varner, 1967) and release of the amylase into the scutellum is influenced by $[Ca^{2+}]$ (Bush *et al.*, 1986). In 7-d old Funk G522DR seed, total amylase activity: a) decreased as the $[H^+]$ increased, and b) increased as the $[Ca^{2+}]$ increased (Table 6). Additionally, excess Mn^{2+} induced a decrease in total amylase activity (Table 6).

Juvenile seedlings (0–7 d) of sorghum cultivars grow at different rates (Table 1), which might be explicable as differential total amylase activities in seed with high amylose contents. Comparison of total amylase activities showed that Funk G522DR, which grew most vigorously, had about one-half of the total amylase activity of SC283 while SC574 failed, repeatedly, to show any amylase activity (Table 7).

Table 6. Influence of H^+, Ca^{2+}, and Mn^{2+} on total starch hydrolysis by amylase from Funk G522DR seed

	pH				
	6.0	5.5	5.0	4.5	4.0
	(%)				
Starch hydrolyzed per seed/min	100a[a]	100a	92b	86c	44d
		Ca (ppmw)			
		0	10	100	
		%			
Starch hydrolyzed per seed/min		100a	111b	124a	
		Mn (ppmw)			
		0	1.4	140.0	
		%			
Starch hydrolyzed[b] per seed/min		100b	104a	40c	

[a] Values on a line followed by the same letter are not significantly different at the 5% level.
[b] After 5 days germination.

Table 7. Comparison of sorghum cultivar seed total amylase contents and primary leaf isocitric lyase activity contents (%)

	Cultivar		
	Funk G522DR	SC283	SC574
Starch hydrolyzed seed/min	100b[a]	183a	0c
Glyoxylichydrozone product gFW/min	100a	1c	20b

[a] Values on a line followed by the same letter are not significantly different at the 5% level.

Isocitric lyase

In oil bearing seeds, GA induces glyoxysomal utilization of oil for energy and isocitric lyase is a requisite enzyme in this system (Doig *et al*, 1975). Funk G522DR had a relatively high isocitric lyase activity in 4-d old leaves while SC574 leaves had about 20% of the activity found in Funk G522DR; and SC283 isocitric lyase activity was only 1% of that found in Funk G522DR (Table 7). Thus, differences in the growth patterns of sorghum shoots may be strongly influenced by GA synthesis, metabolism, and translocation.

Kaurene synthesis

Comparison of acid soil sensitive (TAM428) and tolerant (SC599) cultivars for gibberellin precursor biosynthesis (*ent*-kaurene) from (^{14}C) isopentenyl pyrophosphate (^{14}C-IPP) showed massive differences in Mn-ion sensitivity between these two cultivars. Incorporation of ^{14}C-IPP into *ent*-kaurene for TAM428 was massively influenced (more active) by [Mn^{2+}] with maximum ^{14}C-IPP at 3 μM, while SC599 incorporation of ^{14}C-IPP into *ent*-kaurene was relatively stable throughout the [Mn^{2+}] range (1–20 μM) even though there was a statistically significant maximum at 1 μM (Table 8). Consequently, TAM428 is apparently more sensitive to excess Mn than SC599.

In summation, these data show that responses of various sorghum cultivars to soil acidity stress are due to physiological and chemical phenomena (*i.e.*-GA and IAA sequestration), and certain biochemical differences (*i.e.*-GA synthesis, IAA degradation, and enzymatic differences in stored energy utilization). [H^+, Ca^{2+}, and Mn^{2+}] are evidently involved in responses among sorghum cultivars to acid soil stress. Their relationships to Al^{3+} toxicity are currently under investigation. Unpublished comparisons of several cultivars in the late vegetative and reproductive phases of the sorghum plant development indicate that other physiological and biochemical factors may also be important. Acid soil susceptibility and tolerance in the latter two growth stages (panicle initiation to boot stage and anthesis to physiological grain maturity) may be influenced by Mn^{2+} sequestering, biochemical compartmentation of isoprenoid syntheses, or differences in enzymatic sensitivities to [Mn^{2+}]. Clearly, the quest for explanation of differences (tolerance mechanisms) among sorghum cultivars to acid soil stress has only just begun. Each major growth stage may involve different acid soil tolerance mechanisms.

Table 8. Influence of [Mn^{2+}] on the incorporation of [^{14}C] isopentenyl pyrophosphate into *ent*-kaurene by cell-free enzyme systems from etiolated coleoptiles of TAM428 and SC599 sorghum cultivars

	Mn(μM)						
	1	3	5	7	10	15	20
	% ^{14}C-IPP incorporation						
TAM428	2.3b[a]	23.5a	0.8b	0.2b	0.3b	0.7b	0.6b
SC599	4.8a	2.0b	1.8b	1.2b	0.9b	0.8b	0.8b

[a] Values on a line followed by the same letter are not significantly different at the 5% level.

Acknowledgements

This work was conducted from funds allocated to State and Hatch Projects 1386, 1408, and 1410 of the University of Georgia College of Agricultural Experiment Stations.

References

Bush D S, Cornejo M-J, Huang C-N and Jones R L 1986 Ca-ions stimulated secretion of α-amylase during development in barley aleurone protoplasts. Plant Physiol. 82, 566–574.

Chrispeels M J and Varner J E 1967 Gibberellic acid-enhanced synthesis and release of α-amylase and ribonuclease by isolated barley aleurone layers. Plant Physiol. 42, 398–406.

Doggett H 1988 Sorghum. Longman Scientific and Technical Journal. John Wiley and Sons, New York, pp 70–122.

Doig R L, Colborne A J, Morris G and Laidman D L 1975 The induction of glyoxysomal enzyme activities in the aleurone cells of germinating wheat. J. Exp. Bot. 26, 387–398.

Duncan R R 1988 Sequential development of acid soil tolerant sorghum genotypes under field stress conditions. Commun. Soil Sci. Plant Anal. 19, 1295–1303.

Duncan R R 1987 Sorghum genotype comparisons under variable acid soil stress. J. Plant Nutr. 10, 1079–1088.

Felle H 1988 Short-term pH regulation in plants. Physiol. Plant Nutr. 12, 1395–1407.

Foy C D 1984 Physiological effects of hydrogen, aluminium, and manganese toxicities in acid soil. *In* Soil Acidity and Liming. Ed. F Adams. pp 57–97. Am. Soc. Agron. Monograph No. 12, 2nd Ed., Am. Soc. Agron., Madison, WI.

Hare R C 1964 Indoleacetic acid oxidase. Bot. Rev. 30, 129–165.

Wilkinson R E 1982 Mefluidide inhibition of sorghum growth and gibberellin precursor biosynthesis. J. Plant Growth Regul. 1, 85–94.

Wilkinson R E and Duncan R R 1989a H-ions, Ca-ions, and Mn-ions influence on sorghum seedling shoot growth. J. Plant Nutr. 12, 1395–1407.

Wilkinson R E and Duncan R R 1989b Sorghum seedling root growth as influenced by H-ions, Ca-ions, and Mn-ions concentrations. J. Plant Nutr. 12, 1379–1394.

Wilkinson R E and Duncan R R 1989c Influence of H-ions, Ca-ions, and Mn-ions on the development of amylase in the endosperm of several sorghum cultivars. J. Plant Nutr. 12, 1483–1501.

Wilkinson R E and Duncan R R 1989d Sorghum cultivar variation in seed triglyceride content and glyoxysomal activity. J. Plant Nutr. 12, 1503–1513.

Wilkinson R E and Ohki K 1988 Influence of manganese deficiency and toxicity on isoprenoid syntheses. Plant Physiol. 87, 841–846.

M. L. van Beusichem (Ed.), *Plant nutrition – physiology and applications*, 435–441.

PLSO IPNC341

Exchangeable aluminium and root growth of wheat (*Triticum aestivum*) as criteria of lime requirement in acid soils of northeast Portugal

J. F. COUTINHO
Departamento Geociências, Universidade Trás-os-Montes e Alto Douro, P-5000 Vila Real, Portugal

Key words: aluminium toxicity, lime requirement, root elongation, soil acidity, *Triticum aestivum* L., wheat

Abstract

The lime requirements of 154 soils were estimated in a biossay method. Different parameters of soil acidity were compared as criteria for liming and for establishing adequate methods for assessing lime requirements, with special reference to pH values and extractable Al. Soil pH ($CaCl_2$) ranged from 3.61 to 7.35, while organic matter (OM) content ranged from 2.8 to 113.1 $g.kg^{-1}$.

Relative root growth (RRG) of wheat cv. Abe seedlings during a 48-h period was established as the bioparameter against which laboratory methods were calibrated. RRG correlated much better with % CEC_e saturated with exchangeable Al than with any other parameter, showing that additional quantities of Al extracted with $LaCl_3$ and $CuCl_2$ may not have a sound biological meaning. It was observed that the toxicity of Al was alleviated by high contents of soil organic matter. The relationship between pH and RRG was found to be unclear.

To assess the lime requirements (LR) of acid soils, titration curves were used to predict LR for a desired pH ($CaCl_2$) of 5.5 (pH_d). With this method LR was found to be overestimated, even for a very Al-sensitive wheat cultivar and considering pH_d values below neutrality. Across all soils, the best estimates of LR were obtained when both exchangeable Al-KCl and OM were taken into consideration.

Introduction

Soil acidity is normally regarded as one of the most important factors impeding crop production (Adams, 1981; Taylor, 1988). However, there continues to be disagreement on the accurate assessment of the lime requirement of a particular soil. This situation may be attributed to the fact that the effect of soil acidity is often not due to any single constraint but rather to a combination of factors involving soil properties (Sumner *et al.*, 1986) and physiological and biochemical characteristics of the plants themselves (Foy, 1988).

Although Al toxicity is considered to be the most important factor limiting crop production on very acid soils (Adams, 1981; Foy, 1988), soil pH is classically the criterion on which most lime requirement methods are based with the objective of raising the pH value to near neutrality. Nevertheless, the significance of soil pH depends on various factors, such as laboratory methodology, organic matter content and type of clay mineral (Coutinho, 1989). Advantages expected from raising the pH to near neutrality are often not obtained (McLean and Brown, 1984) and sometimes yield depressions occur for reasons not yet fully explained (Coutinho and Moreira, 1986; Farina *et al.*, 1980).

In regard to Al, the concepts of soil acidity have been changed considerably after the recognition of the important role of ionic Al-species in the electrochemical properties of the soil (Chernov, 1947; Jackson, 1963). These research workers rediscovered many earlier findings about the presence of Al-ions in soils (Vietch, 1904) and the importance of their phytotoxicity (Miyake, 1915). Al toxicity is presently known to reduce both root and shoot growth, with roots much more affected than shoots (Taylor, 1988). Root elongation and ramification are seriously inhibited, and the roots become thickened, brown and brittle (Foy, 1988). These characteristics of Al-stressed plants are important criteria in plant breeding for selecting tolerant genotypes. Recently the same approach was used by several authors (Karr *et al.*, 1984; Ritchey *et al.*, 1988; Saigusa *et al.*, 1980) for measuring stress intensity due to soil acidity.

The objectives of the present study were to compare values of pH and of Al extracted by different salt solutions and its degree of saturation in CEC_e as a criterion for liming and for establishing lime requirement recommendations for the more characteristic soils of northeast Portugal. As reference parameter, in the study use was made of a biossay measuring root elongation during a 48-h period of seedlings of wheat (*Triticum aestivum* L.) cv. Abe, a plant very sensitive to soil acidity (Campbell and Lafever, 1976).

The choice of initial root growth as reference parameter is based on the correlation between this parameter and final yield for assessing the effect of soil Al (Foy, 1988), taking advantage of the fact that this method is much less expensive in time and equipment than ones involving growth chambers, greenhouses and field experiments.

Table 1. Range of selected properties of the 154 topsoils under study

Parameter	Range of values	
	Minimum	Maximum
$pH(CaCl_2)$	3.61	7.35
exch. Ca[a]	0.08	10.93
exch. Mg[a]	0.04	4.94
exch. K[a]	0.04	0.97
exch. Na[a]	0.01	0.36
exch. bases[a]	0.21	12.93
exch. acidity[a]	<0.01	3.69
exch. Al[a]	<0.01	3.68
CEC_e[a]	0.82	12.97
Al-CEC_e saturation	0.01	0.89
Al-$LaCl_3$[a]	<0.01	9.27
Al-$CuCl_2$[a]	0.26	29.35
OM[b]	2.8	113.1

[a] $cmol(+).kg^{-1}$, [b] $g.kg^{-1}$.

Materials and methods

The study comprised 154 topsoils, mainly from the northeast region of Portugal. Soil samples were air dried and passed through a 2-mm screen. Selected properties of the soils are presented in Table 1. The soils formed a heterogeneous population, and the specific values for the different parameters are given elsewhere (Coutinho, 1989).

The titration curves were obtained with use made of six different levels of 0.02 *M* $Ca(OH)_2$, depending on the characteristics of each soil, added to soil-0.01 *M* $CaCl_2$ suspensions at a 1:10 ratio. There were three replications for each level of base. After three days, with occasionally shaking, the pH values of the supernatant solutions were recorded. Lime requirements for attaining the desired $pH(CaCl_2)$ value of 5.5, LR-$pH_d(CaCl_2)$ 5.5, were then graphically established by plotting the pH values.

The bioassay was performed after 180 days of soil-$CaCO_3$ incubation. Samples of each soil received five different levels of $CaCO_3$ (reprecipitated), depending on the characteristics of the soil. After thorough mixing, the samples were brought to 33 kPa moisture tension and placed in plastic bags. There were three replications for each soil and $CaCO_3$ rate.

The root bioassay technique as described by Karr *et al.* (1984), was modified by using 200-mL plastic cups instead of petri dishes and six seeds were planted in each cup. After 48 h, total root length of each seedling was measured. For each soil the results of the three replications, five $CaCO_3$ levels and six plants were statistically analyzed by a sampling and subsampling hierarchic design procedure. The relative root growth (RRG) was calculated for each level of amendment, considering the highest root growth in each soil as reference values. Values for RRG were plotted against $CaCO_3$ and biological lime requirements (LR_{biol}) were graphically established as the quantity of lime needed to reach a RRG value of 90%.

Original soil samples were analyzed for pH in the supernatant liquid of a 1:2.5 suspension of soil in 0.01 *M* $CaCl_2$, previously shaken for one hour. Exchangeable bases were extracted with 1

M NH_4CH_3OO at pH 7 (Thomas, 1982) and exchangeable acidity and aluminium were determined from a 1 *M* KCl soil extract (McLean, 1982). The effective cation exchange capacity of the soil (CEC_e) was taken as the sum of exchangeable Ca, Mg, K, Na and acidity. Percentage aluminium saturation was calculated on the basis of CEC_e. Aluminium was also obtained by extractions with 0.33 *M* $LaCl_3$ and 0.50 *M* $CuCl_2$ using 1:10 soil/solution ratios (Oates and Kamprath, 1983). Organic matter (OM) was determined in all samples by a modification of the Walkley-Black procedure (Schulte, 1980).

Results and discussion

The root bioassay clearly shows differences among soils and levels of $CaCO_3$ application within soils. In 23 samples of slightly acid soils, liming did not lead to significant ($P<0.05$) differences in root elongation among the five treatments, the average root length being about 26 cm per plant.

For the other soils, the effects of $CaCO_3$, although always significant ($P<0.05$), differed greatly. In some very acid soils, the effect was dramatic. In the most extreme case, average root length of the seedlings, in 48 h, was 2 cm per plant for the non-limed sample, and the amendment was responsible for an increase to a maximum value of 24 cm per plant obtained in the same period.

Treatment effects were not limited to only differences in the root length. Also root morphology was significantly affected by liming. Absence of liming led to a lack of fine branching and the roots were thick, stubby and had brownish tips. These symptoms point to inhibition of cell elongation and lateral ramification, abnormal thickness of the cortex and disruption of the normal meristem functions, typically referrered to as stress injuries in the Al phytotoxicity syndrome (Foy, 1988).

Liming criteria

Regression equations relating RRG to soil parameters are shown in Table 2. The correlation with $pH(CaCl_2)$ values revealed a modest contribution of this parameter to explaining the variance across all soils (r^2). The unfavourable values of r^2 and $s_{y.x}$ of this relationship can be attributed to the scattering of RRG values in the low-pH range, shown in Figure 1. For pH values ranging between 4.0 and 4.5, the RRG parameter showed values varying between 8 and 88%. Nevertheless, the value of $pH(CaCl_2)$ for RRG = 90 is 5.03 (Eq. 1), which coincides with the critical value obtained by application of the graphical method of Cate and Nelson (Nelson and Anderson, 1977).

Meanwhile, the relationship with 1 *M* KCl-Al was not much better (eq. 2) and the r^2 values were considerably reduced when values of Al extracted by 0.33 *M* $LaCl_3$ or 0.5 *M* $CuCl_2$ were considered (Eq. 3 and 4). For these three extractions, the correlation between Al and RRG was inversely related with the capacity of the extractant cation to displace Al bound to organic matter (Coutinho, 1988; Oates and Kamprath, 1983). In other words, the larger the portion of soil Al bound onto organic matter, the lower is the usefulness of an extractant capable of extracting this bound Al for predicting Al-caused phytotoxicity. Such results indicate that Al-humus complexes may exert a suppressing in-

Table 2. Regression equations relating relative root growth (*y*) to several soil parameters

x	Equation	*P*	r^2	$s_{y.x}$	
$pH(CaCl_2)$	$y = 100 - (153849.88 \times 0.147^x)$	<0.001	0.496	21.818	[1]
Al-KCl[a]	$y = 100 - (46.431 \times 0.239^x)$	<0.001	0.609	19.231	[2]
Al-$LaCl_3$[a]	$y = 100 - (25.697 \times 0.173^x)$	<0.001	0.416	23.490	[3]
Al-$CuCl_2$[a]	$y = 100 - (9.647 \times 0.075^x)$	<0.001	0.197	27.503	[4]
Al-CEC_e sat	$y = \frac{2369.24}{1 + 384.97 \times e^{-0.028(1-x)}}$	<0.001	0.824	12.870	[5]

[a] $x = \frac{1}{[Al]}$.

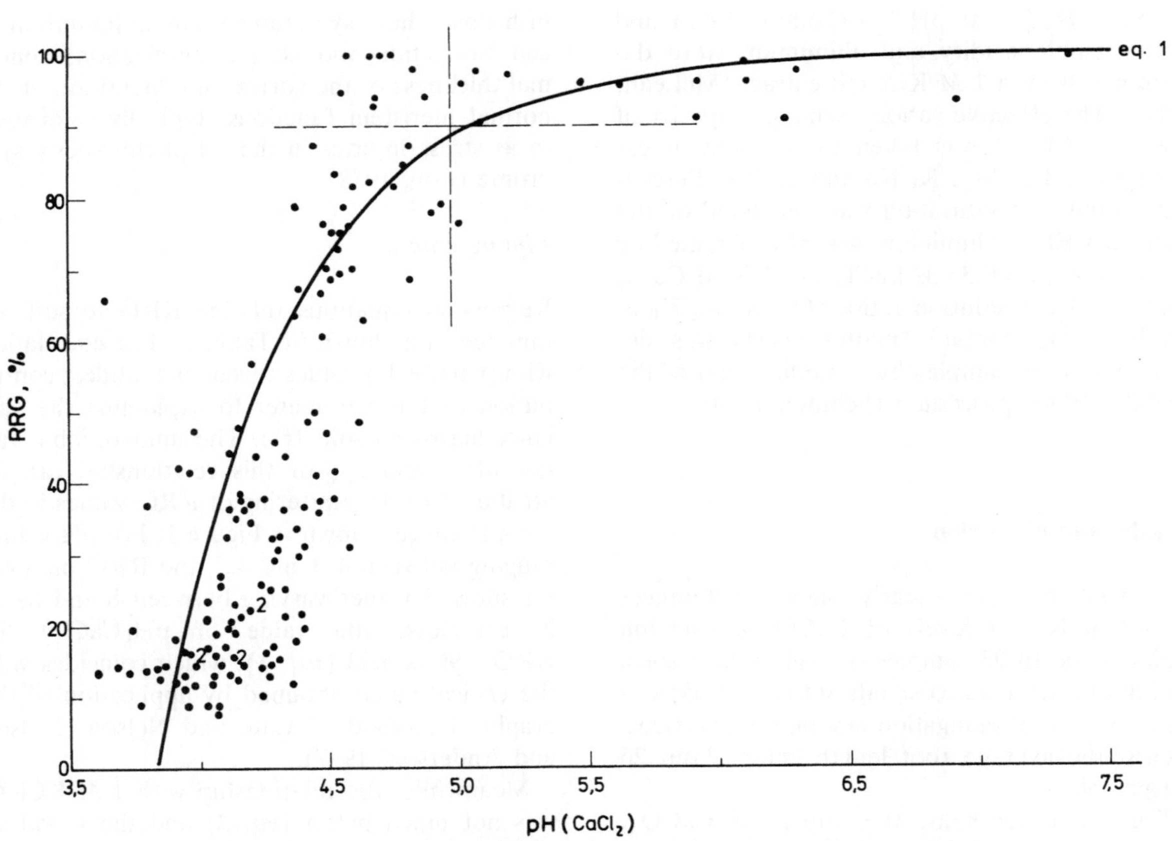

Fig. 1. Relative root growth (RRG) in 154 unlimed soils as a function of soil pH (0.01 *M* $CaCl_2$).

fluence on Al activities in soil solutions (Hue *et al.*, 1986).

The best correlation was achieved with percentage Al saturation values (eq. 5). For the wheat plants, a value of 90% for RRG was obtained with Al saturation values of 5% (Fig. 2), which are lower than those obtained by the graphical method of Cate-Nelson, which were about 10%. These data indicate that wheat cv. Abe is a very Al-sensitive genotype, as pointed out by Campbell and Lafever (1976), reaching its maximum yield capacity only when exchangeable Al is absent.

As percentage Al saturation values are obtained with 1 *M* KCl extraction and Al toxicity is directly related with Al activity in soil solution (Hue *et al.*, 1986), these results tend to agree with the findings of Turner and Clark (1966). These authors pointed out that percentage Al saturation values are better indices of Al activity in soil solutions than are absolute values of exchangeable Al.

Lime requirement

The regression equations relating LR_{biol} with some selected soil parameters are presented in Table 3. For $pH_d(CaCl_2)$ of 5.5, the data show that estimated LR values are generally larger than the requirements observed with the bioassay method to reach a RRG value of 90% on each soil. In eq. 6, the intercept does not differ significantly ($P<0.05$) from 0 and the regression coefficient is lower ($P<0.001$) than 1. As was pointed out earlier, these results were obtained with wheat cv. Abe, a very Al-sensitive plant, and the available evidence, therefore, supports the opinion that a $pH_d(CaCl_2)$ value of 5.5 is high enough to eliminate Al-toxicity. However, this value is lower than the pH_d normally considered for LR methods based on an equilibrium pH of a soil suspension with buffer solutions. The results are shown in Figure 3. As can be observed, the linear model explains fairly well the relationship between the two variables, al-

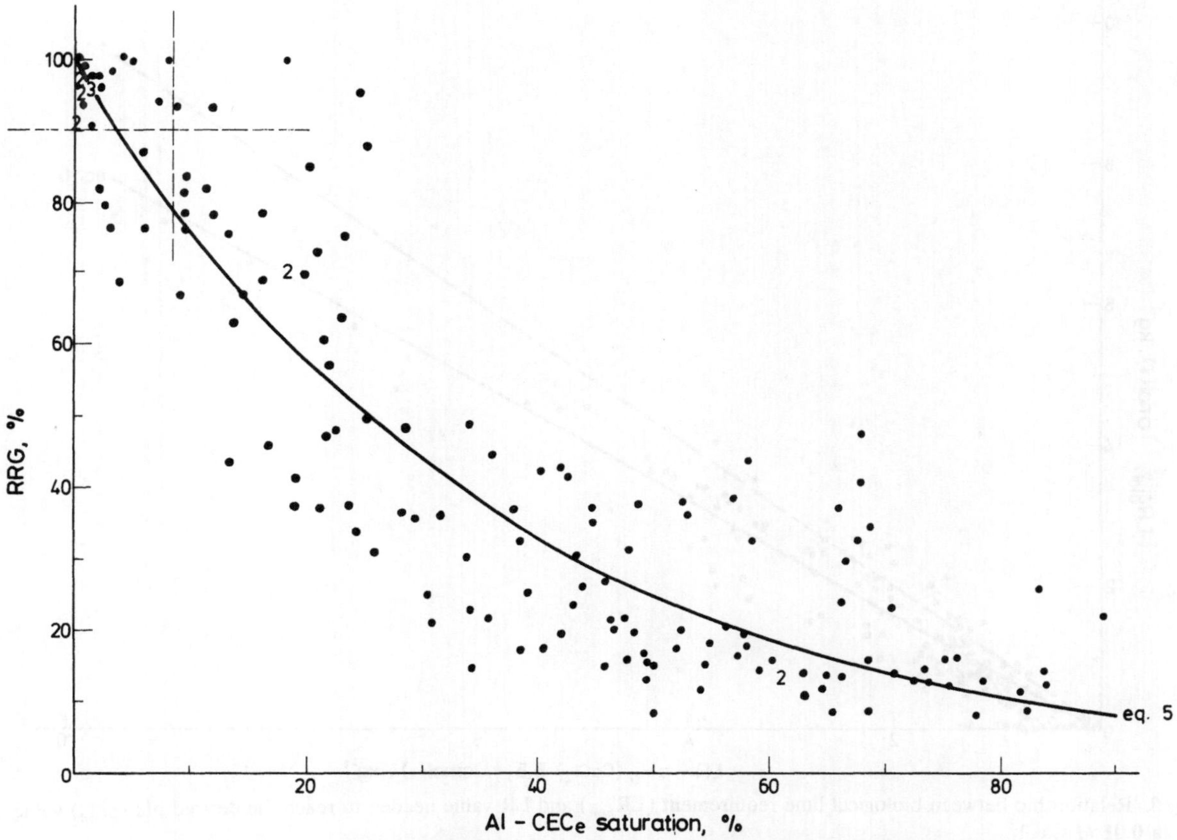

Fig. 2. Relative root growth (RRG) in 154 unlimed soils as a function of Al-CEC_e saturation.

though the scattering tends to increase with increasing LR values.

When considering LR methods suitable for routine purposes, based on extractable Al, the obtained with a 0.50 *M* $CuCl_2$ extraction was found to overestimate the acidity to be neutralized. It was, further, found that the relationship between Al-$CuCl_2$ and LR_{biol} is quite poor (eq. 7). In contrast, Al extracted with 0.33 *M* $LaCl_3$ enables a fairly good estimate to be made of $LR_{biol.}$ In eq. 8, the value of the intercept does not differ significantly ($P < 0.05$) from 0, the value of the regression coefficient is slightly, but significantly ($P < 0.001$), lower than unity, and the percentage of variance explained by the correlation is 79% across all soils. For Al extracted with 1 *M* KCl, the correlation gave a similar r^2 value, although the liming factor (b) is near 2, in agreement with previous results (Coutinho, 1988; Kamprath, 1984).

These results indicate, therefore, that a liming rate, as predicted by 1 *M* KCl, does not neutral-

Table 3. Regression equations relating biological lime requirement (*y*) to several soil parameters

x	Equation	*P*	r^2 or R^2	$s_{y.x}$	
$pH_d(CaCl_2)5.5$	$y = -0.046 + 0.820x$	<0.001	0.901	0.539	[6]
Al-KCl[a]	$y = -0.130 + 1.988x$	<0.001	0.770	0.825	[7]
Al-$LaCl_3$[a]	$y = 0.031 + 0.867x$	<0.001	0.793	0.780	[8]
Al-$CuCl_2$[a]	$y = 0.418 + 0.359x$	<0.001	0.596	1.092	[9]
Al-KCl[a]; OM[b]	$y = -0.506 + 1.381x_1 + 0.029x_2$	<0.001	0.885	0.582	[10]

[a] $cmol(+).kg^{-1}$, [b] $g.kg^{-1}$.

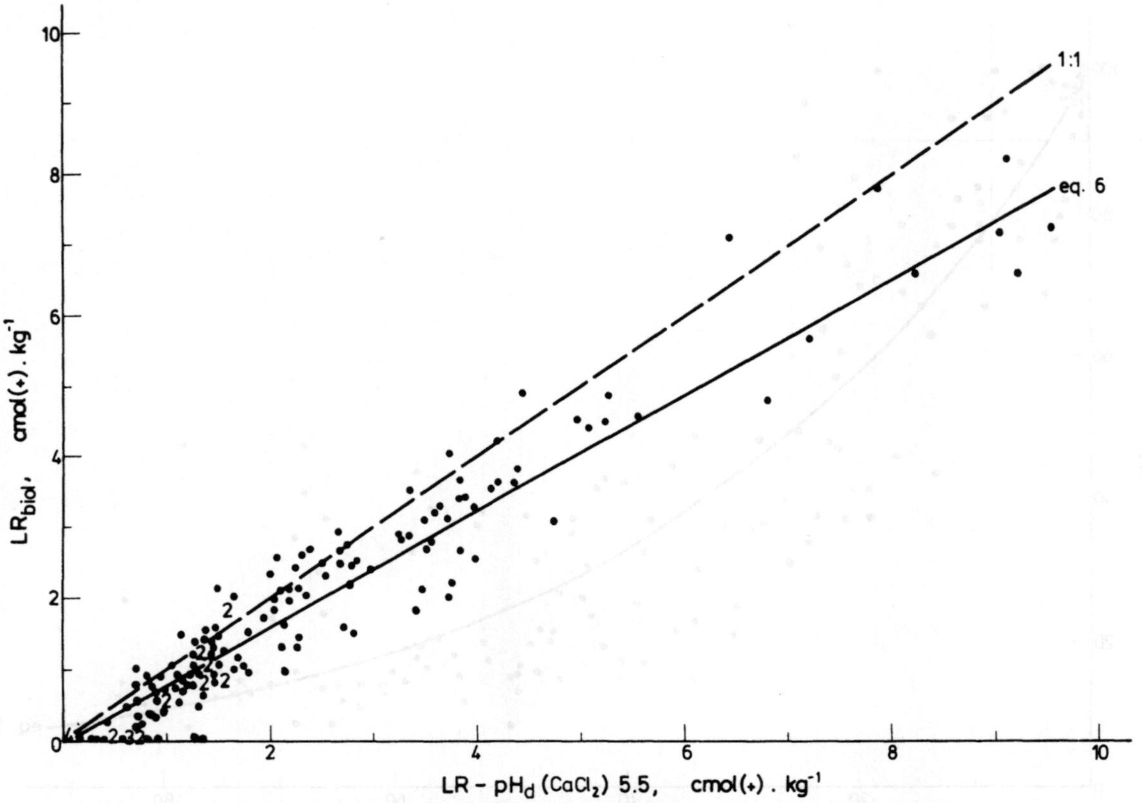

Fig. 3. Relationship between biological lime requirement (LR_{biol}) and LR value needed to reach the desired pH (pH_d) value of 5.5 in 0.01 *M* $CaCl_2$.

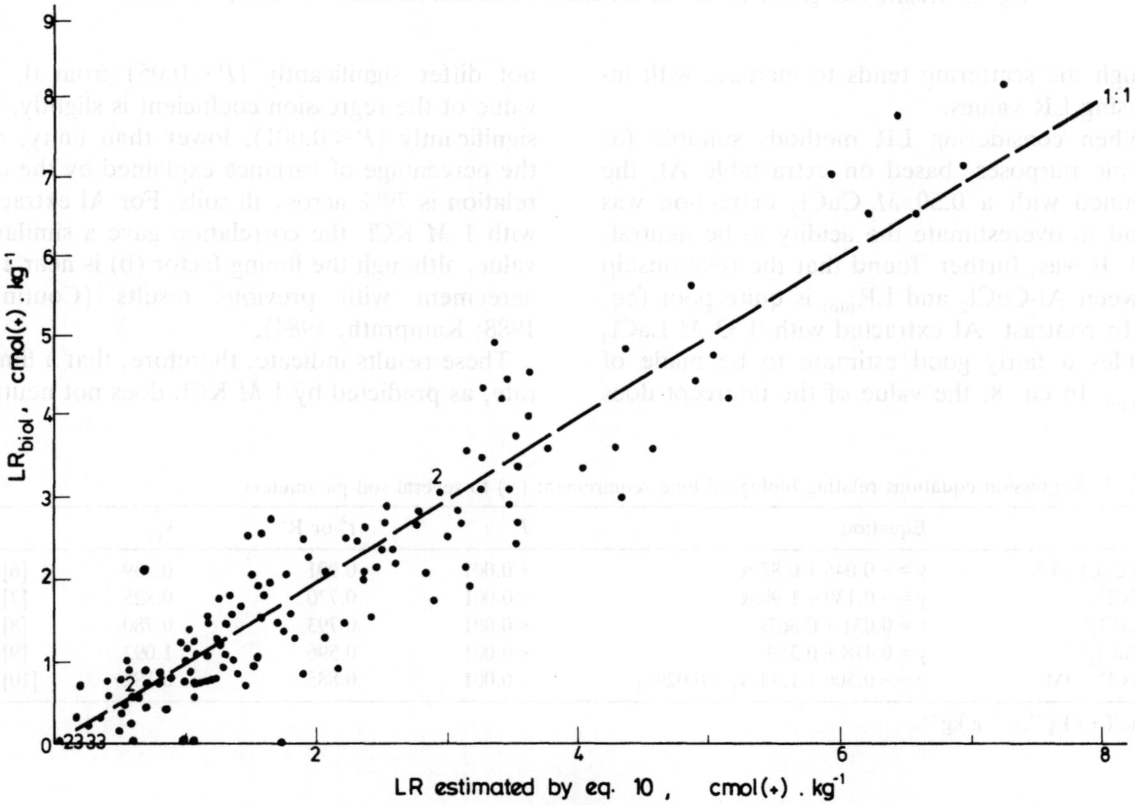

Fig. 4. Biological lime requirement (LR_{biol}) as a function of Al extracted with 1 *M* KCl and organic matter content (eq. 10).

ize all exchangeable aluminium. When lime is added to acid soils at the rate predicted by 1 *M* KCl, it reacts with the readily exchangeable Al, causing a disturbance of the equilibrium between various Al forms, as a result of which non-readily exchangeable Al shifts to more readily exchangeable sites and to the soil solution. If lime is added at a rate predicted by 0.33 *M* $LaCl_3$, it reacts with readily and non-readily exchangeable Al, neutralizing the exchangeable and soil solution Al pool in the soil, particularly the organically bound Al (Oates and Kamprath, 1983).

Meanwhile, a multiple regression equation in which Al-KCl and organic matter (OM) content are considered together (eq. 10) yielded a better prediction of LR_{biol} based on initial root growth. The relationship between these two sets of values is shown in Figure 4. Thus, eq. 10 emphasizes the important role of organic components in acid soils, increasing LR either by binding Al that reacts with lime and is not extracted by 1 *M* KCl, or by buffering the pH value of the soil, thus obstructing the conversion of Al-species to non-phytotoxic forms.

References

Adams F 1981 Alleviating chemical toxicities: Liming acid soils. *In* Modifying the Root Environment to Reduce Crop Stress. Eds. G F Arkin and H M Taylor. pp 269–301. Am. Soc. Agron. Engen., St. Joseph, MI.

Campbell L G and Lefever H N 1976 Correlation of field and nutrient culture techniques of screening wheat for aluminum tolerance. *In* Plant Adaptation to Mineral Stress in Problem Soils. Eds. M J Wright and S A Ferrari. pp 277–286. Cornell Univ. Sp. Publ., Ithaca, NY.

Chernov V A 1947 The Nature of Soil Acidity. English translation. Soil Sci. Soc. Am., Madison, Wisconsin, WI, 170 p.

Coutinho J F 1988 O alumínio extraível pelo KCl e $LaCl_3$ e a necessidade em cal dos solos ácidos. Anais UTAD 1, 167–177.

Coutinho J F 1989 Acidez do Solo e Toxicidade do Alumínio: Calibração de métodos de avaliação da necessidade em cal. PhD Thesis. Univ. Vila Real, Portugal, 708 p.

Coutinho J F and Moreira N T 1986 Soil acidity and the response of vetch to liming. Proc. 11th Meet. European Grassland Feder., pp 243–247.

Farina M P W, Sumner M E, Plank C O and Letzch W S 1980 Exchangeable aluminum and pH as indicators of lime requirement for corn. Soil Sci. Soc. Am. J. 44, 1036–1041.

Foy C D 1988 Plant adaptation to acid, aluminum-toxic soils. Commun. Soil Sci. Plant Anal. 19, 959–987.

Hue N V, Craddock G R and Adams F 1986 Effect of organic acids on aluminum toxicity in subsoils. Soil Sci. Soc. Am. J. 50, 28–34.

Jackson M L 1963 Aluminum bonding in soils: A unifying principle in soil science. Soil Sci. Soc. Am. Proc. 27, 1–10.

Kamprath E J 1984 Crop response to lime in the tropics. *In* Soil Acidity and Liming. Ed. F Adams. pp 349–368. Am. Soc. Agron., Madison, WI.

Karr M C, Coutinho, J F and Ahlrichs J L 1983 Determination of aluminum toxicity in Indiana soils by petri dish bioassay. Indiana Acad. Sci. 93, 405–411.

McLean E O 1982 Soil pH and lime requirement. *In* Methods of Soil Analysis, Part 2: Chemical and Microbiological Properties, 2nd ed. Eds. A L Page, R H Miller and D R Keeney. pp 199–204. Am. Soc. Agron., Madison, WI.

McLean E O and Brown J R 1984 Crop response to lime in the Midwestern United States. *In* Soil Acidity and Liming, 2nd ed. Ed. F Adams. pp 267–303. Am. Soc. Agron., Madison, WI.

Miyake K 1916 The toxic action of soluble Al salts upon the growth of the rice plants. J. Biol. Chem. 25, 23–28.

Nelson L A and Anderson R L 1977 Partitioning of soil test-crop response probability. *In* Soil Testing: Correlating and Interpreting the Analytical Results. Ed. T R Peck, J T Cope and D A Whitney, pp 19–38. Am. Soc. Agron., Madison, WI.

Oates K M and Kamprath E J 1983 Soil acidity and liming. II. Evaluation of using aluminum extracted by various salts for determining lime requirements. Soil Sci. Soc. Am. J. 47, 690–692.

Ritchey K D, Baligar V C and Wright R J 1988 Wheat seedlings responses to soil acidity and implications for subsoil rooting. Commun. Soil Sci. Plant Anal. 19, 1285–1293.

Saigusa M, Shoji S and Takahashi T 1980 Plant root growth in acid andosoils from Northeastern Japan. 2. Exchange acidity Y1 as a realistic measure of aluminum toxicity potential. Soil Sci. 130, 242–250.

Schulte E 1980 Recommended soil organic matter tests. *In* Recommended Chemical Soil Tests Procedures for the North Central Region. Ed. W C Dahnke, pp 28–30. N. Dakota St. Univ. Bull. 499 (revised). NDSU, Fargo, ND.

Sumner M E, Fey M V and Noble A D 1986 Nutrient status and toxicity problems in acid soils. *In* Trans. 13th Congr. Int. Soil Sci. Soc. Hamburg Vol. V, pp 119–125.

Taylor G J 1988 The physiology of aluminum phytotoxicity. *In* Metal Ions in Biological Systems, Vol. 24: Aluminum and Its Role in Biology. Eds. H Siegel and A. Siegel. pp 135–164. Marcel Dekker, New York, NY.

Thomas G W 1982 Exchangeable cations. *In* Methods of Soil Analysis, Part 2: Chemical and Microbiological Properties, 2nd ed. Eds. A L Page, R H Miller and D R Keeney. pp 159–165. Am. Soc. Agron., Madison, WI.

Turner R C and Clark J S 1966 Lime potential in acid clay and soil suspensions. Trans. Soil Chemistry and Fertility. Commun. II and IV Int. Soil Sci. Soc. pp 207–215.

Vietch F P 1904 Comparison of methods for the estimation of soil acidity. J. Am. Chem. Soc. 26, 637–662.

M. L. van Beusichem (Ed.), *Plant nutrition – physiology and applications*, 443–448.

PLSO IPNC368

Effect of low pH of the root medium on proton release, growth, and nutrient uptake of field beans (*Vicia faba*)

S. SCHUBERT, E. SCHUBERT and K. MENGEL
Institute of Plant Nutrition, Justus Liebig University, Südanlage 6, D-6300 Giessen, FRG

Key words: ATPase, ion uptake, proton buffer capacity, proton release, soil pH, solution pH, *Vicia faba* L.

Abstract

The effect of low root medium pH on growth and proton release of field beans (*Vicia faba* L. cv. Kristall) was studied in soil and nutrient solution experiments. Decrease of soil pH due to proton release by roots strongly depended on the proton buffer capacity of 8 different soil types tested in a pot experiment. Whereas in soils of high proton buffer capacity no pH decrease during the growth period was detectable, in soils of low buffer capacity pH in the bulk soil dropped from about pH 7.3 to 6.5, 6.3 or 5.8 during growth until maturity. This decrease in pH was closely correlated with an inhibition of plant dry weight production ($Y = 1.06 \times +3.33$, $r = 0.94^{***}$). Growth reduction was not due to direct inhibition of nitrogen fixation. In short term experiments vegetative growth and proton release were inhibited at pH < 6. At pH 5 or lower proton uptake was observed in 1 m*M* $CaSO_4$. Low pH (4.0 relative to pH 7.0) decreased uptake of all major ions except for Cl the exclusion of which was disturbed. It is concluded that the sensitivity of field beans to low pH is related to a lack of capability to release protons by ATPase activity. This sets limits to nutrient uptake and possibly cytoplasmic pH regulation.

Introduction

Regarding the sensitivity of plants to low root medium pH considerable species differences have been found in solution culture experiments (Moritsugu *et al.*, 1983). Particularly pH demands of various leguminous species have been investigated (Andrew and Johnson, 1976; Munns, 1986). Whereas some tropical legumes tolerate low pH values such as 4.0, *Glycine* and *Medicago* species are rather sensitive to low pH (Andrew, 1976). Field trials in Northern Idaho have shown that yield of important crops decreased at even moderately low pH making an indirect effect of Al toxicity unlikely (Mahler and McDole, 1987). For example, minimal acceptable pH was 5.65 (lens), 5.52 (pea), and 5.23 (winter barley). The authors also reported varietal differences for winter wheat and recommended liming the soils to pH 5.6 to 6.0.

Although inhibition of nutrient uptake, root growth, and nitrogen fixation have been suggested to limit plant growth at low pH (Mahler and McDole, 1987; Moore, 1974; Munns, 1978), the ultimate physiological cause of direct pH sensitivity remains obscure. In this paper, we describe experiments that show strong inhibition of ATPase-driven proton release by field beans at low pH. This limits nutrient uptake and possibly cytoplasmic pH regulation.

Materials and methods

Experiment 1

This pot experiment was designed to test the effect of proton buffer capacity of 8 different soils on proton release, dry matter production, and nitrogen fixation of field beans (*Vicia faba* L.

cv. Kristall). Important soil characteristics (uppermost layer) are listed in Table 1. Before sowing, pH of some soils was adjusted by base addition according to the buffer capacity of the soils (Table 2, 'initial pH'). Buffer capacity of the soils was determined by means of neutralization curves as described by Mengel and Steffens (1982). Soils were put into small Mitscherlich pots and fertilized (after analysis) to 2.13 g K, 1.70 g P, and 0.25 g Mg per pot. Additionally, the following amounts of micronutrients per pot were thoroughly mixed with the soil: 2 mg B, 5 mg Cu, 10 mg Mn, 5 mg Mo, 10 mg Zn, and 0.5 mg Co. Nitrogen was given as starter N (0.1 g N) in liquid form as NH_4NO_3. Each treatment (soil) comprised 15 replicates. After watering the soils to 60–70% of the maximum water-holding capacity seeds were treated with a commercial inoculum of *Rhizobium leguminosarum* ('Radicin') and sown to obtain 6 plants per pot. Plants were grown under open-air conditions except on rainy days when they were kept in the greenhouse. Harvests were taken at the flowering stage (4 pots), three weeks later at the end of flowering (4 pots), two weeks later during seed filling (4 pots), and at full maturity (3 pots). Plants were divided into roots and shoots and dry weight was determined. N concentration was determined with the Kjeldahl method. Total proton release was calculated using the neutralization curves. Whereas results are reported for the last harvest only, a linear regression of dry weight production and proton release was calculated using data of all 4 harvests.

Experiment 2

In contrast to Exp. 1 this experiment was carried out to investigate the effect of soil pH on vegetative growth of field beans supplied with mineral nitrogen. The soil (Dystric Cambisol; pH 4.5; 0% $CaCO_3$; 1.5% organic matter; 8.99 cequ./kg soil CEC; 7.0% clay, 25.3% silt, 67.7% sand) was sterilized by heating at 250°C for 24 hours. Various pH values were adjusted (4.7, 5.4, 6.2, 7.0) as described above. Plants (4 plants per pot) were grown with 3 replicates in a growth chamber at a light intensity of 84 W m^{-2}. The light/dark ratio was 12/12 with temperatures of 18°C and 14°C, respectively. Four weeks after emergence plants had reached a height of about 30 cm and were harvested for shoot dry weight determination.

Experiment 3

Water culture experiments were conducted to separate direct and indirect effects of pH on proton release.

Plant cultivation

Field bean seeds (see above) were soaked in aerated 0.5 m*M* $CaSO_4$ for 2 days and then germinated in quartz sand. After 2 weeks plants (4 plants/pot) were transferred into 5 L nutrient solution (pH 7.0) of the following composition; m*M*: NH_4NO_3, 0.2; NaH_2PO_4, 0.2; K_2SO_4, 0.3; $CaCl_2$, 0.6; $MgSO_4$, 0.3; μM: $MnSO_4$, 0.04, H_3BO_3, 2; $CuSO_4$, 0.04; $ZnSO_4$, 0.04; Na_2MoO_4, 0.6; $CoCl_2$, 0.6; Fe-EDTA, 10. After one week, the nutrient solution was concentrated twofold and, thereafter, it was renewed every week until plants were used for experiments. Plant cultivation and experiments were carried out under controlled conditions in a growth chamber as described.

Table 1. Important characteristics of soils used in Exp. 1

Soil type	pH (0.01 *M* $CaCl_2$)	CEC (cequ./kg soil)	$CaCO_3$ (%)	Org. matter (%)	Clay (%)	Silt (%)	Sand (%)
1. Calcaric Regosol	7.60	17.88	6.90	1.23	18.35	76.86	4.79
2. Calcaric Fluvisol	7.50	46.40	4.29	3.31	49.45	33.40	17.15
3. Fluvial Colluvisol	7.45	18.45	2.10	1.98	17.30	76.81	5.89
4. Calcic Cambisol	7.25	7.11	0.98	1.17	5.30	9.89	84.81
5. Rhodic Acrisol	4.90	20.88	0.00	2.63	45.20	40.97	13.83
6. Eroded Orthic Luvisol	7.00	16.16	1.30	1.79	18.70	76.24	5.06
7. Orthic Luvisol	5.80	17.72	0.47	1.93	19.00	74.52	6.48
8. Dystric Cambisol	6.00	7.99	0.29	1.20	7.10	39.39	53.51

Proton efflux studies
After 3 weeks of cultivation plants had reached a height of about 30 cm and were then transferred to nutrient solutions of pH 4.5 or pH 7.0. Proton release over a perid of 4 days was measured daily by titration to the initial pH with NaOH. Proton net efflux and influx rates in 1 m*M* $CaSO_4$ of different pH (adjusted by addition of either $Ca(OH)_2$ or H_2SO_4) were determined in 1-L pots by measuring the change of proton activity with a pH electrode at 15 minutes intervals. Prior to the experiment, roots of intact plants were washed for 15 minutes in a large volume of 1 m*M* $CaSO_4$.

Experiment 4

Plants were grown for 5 weeks in water culture as described and then transferred to nutrient solution pH 4.0 and pH 7.0. The pH values were kept constant by daily titration with NaOH or H_2SO_4. Each treatment had 4 replicates. After 5 days plants were divided into roots and shoots, dried at 105°C and analysed for cations (atomic absorption) and anions (ion chromatography).

Results

Experiment 1

Soils used in this experiment widely varied in proton buffer capacity, mainly due to differences in $CaCO_3$ content and cation exchange capacity (Tables 1, 2). The former is regarded as the major buffer system for pH range 6.5 to 8.3 (Prenzel, 1985). Whereas in soils no. 1 to 3 (high buffer capacity) no pH change was detectable during the growth period until maturity, pH of soils no. 4 to 8 (low buffer capacity) decreased from about 7.3 to 6.5, 6.3 or 5.8 (Table 2). This pH decrease was related to inhibition of dry weight production with two exceptions, namely plants grown on soils no. 2 and 5. Apparently, for these soils the result was confounded by other factors than pH. Possibly, the high clay content of both soils might have played a role. Correlating proton release and dry weight production of plants for the 4 harvests resulted in a highly significant linear regression ($Y = 1.06 \times +3.33$, $r = 0.94^{***}$), suggesting that both proton release and growth are intimately related. Since N concentrations of plants were optimal and did not differ for plants grown on various soils (Table 2) symbiotic nitrogen fixation can be safely excluded as a limiting factor in this experiment.

Experiment 2

As with nitrogen-fixing plants, dry weight production of field beans supplied with mineral nitrogen was susceptible to low soil pH (Fig. 1). Since these plants formed no nodules it can be concluded that growth of field beans was directly inhibited by a moderately low soil pH of 6.2.

Table 2. Effect of proton buffer capacity ($b = [H^+]c/[H^+]a$, with c = concentration, a = activity) of 8 different soils on the soil pH change, plant dry weight production (shoots + roots), nitrogen concentration of the shoots, and proton release per g plant dry weight. Data are given for the last harvest at maturity (Exp. 1)

Soil type	b	Initial pH (0.01 *M* $CaCl_2$)	Final pH (0.01 *M* $CaCl_2$)	Dry weight (g)	N (%)	mmol H^+ g^{-1} dry weight
1. Calcaric Regosol	15430	7.60	7.56	12.66	2.67	–
2. Calcaric Fluvisol	7644	7.50	7.50	9.44	2.93	–
3. Fluvial Colluvisol	5477	7.45	7.45	11.42	2.74	–
4. Calcic Cambisol	37	7.25	6.52	7.90	2.76	1.09
5. Rhodic Acrisol	38	7.20	6.35	9.48	2.91	1.12
6. Eroded Orhthic Luvisol	23	7.00	6.25	8.56	2.79	1.07
7. Orthic Luvisol	12	7.35	6.25	7.26	2.74	0.92
8. Dystric Cambisol	6	7.30	5.81	8.28	2.61	1.14
LSD *P* = 5%	–	–	–	2.93	0.93	–
P = 1%	–	–	–	3.59	1.10	–
P = 0.1%	–	–	–	4.60	1.41	–

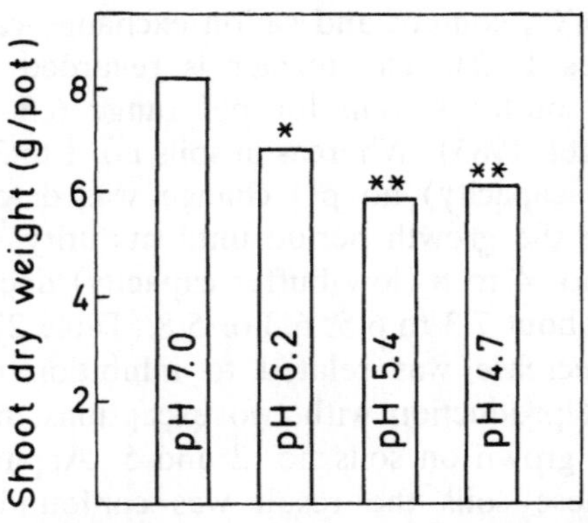

Fig. 1. Effect of soil pH on the shoot dry weight production of young field beans supplied with mineral nitrogen (Exp. 2). Significant differences relative to pH 7.0 at $*p = 5\%$ or $**p = 1\%$ level.

Experiment 3

In order to avoid side effects that are likely to occur in soils, proton release was also studied in nutrient and $CaSO_4$ solutions. Plants grown in nitrogen-containing solution (pH 7), after transfer to nutrient solution, pH 4, instantaneously stopped net proton release and were hardly capable of restoring this process during a 4 day period (Table 3). A similar result was obtained when the pH of 1 m*M* $CaSO_4$ solution was varied from 4 to 8 (Fig. 2). Net proton efflux rates were optimal at pH 7 and 8 but were increasingly inhibited with decreasing pH. At values below pH 6 net proton uptake was observed.

Experiment 4

Transferring plants from nutrient solution pH 7 to pH 4 resulted in decreased uptake of all major ions, except chloride (Table 4). Since during the 5 days experimental period differences in dry matter production were still negligible (not shown), it is justified to compare ion concentrations of plants of the two treatments.

Table 3. Effect of the nutrient solution pH on the daily proton release by roots of young field bean plants. Plants were grown for 3 weeks at pH 7.0 and then transferred to nutrient solution pH of 4.5 or 7.0 (Exp. 3)

pH of nutrient solution	1.	2.	3.	4.	Total
	day of treatment				
	μmol H^+/pot				
7.0	61	36	36	28	170
4.5	0***	3***	4**	3*	10***

Significant difference at $*p = 5\%$, $**p = 1\%$, or $***p = 0.1\%$ level.

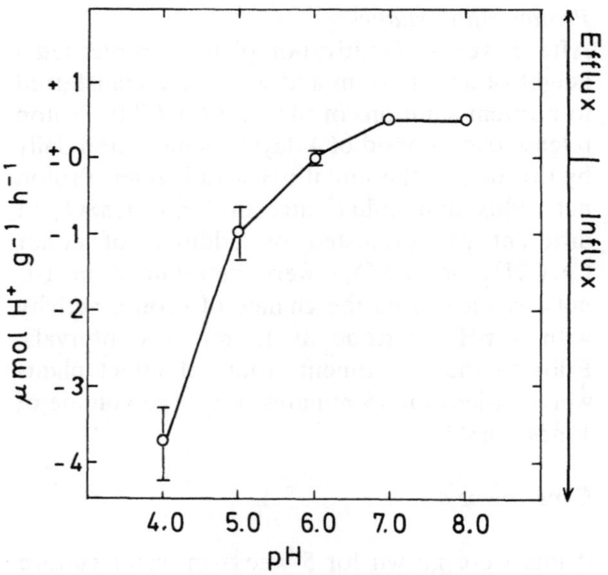

Fig. 2. Effect of pH on net proton efflux and influx rates of roots of young field beans in 1 m*M* $CaSO_4$ (Exp. 3); Vertical bars indicate the standard error.

Table 4. Effect of nutrient solution pH on the concentration (% of dry weight) of cations and anions in roots and shoots of young field beans after 5 days treatment (Exp. 4)

Ion	Roots		Shoots	
	pH 7.0	pH 4.0	pH 7.0	pH 4.0
K	6.80	5.79**	5.27	4.72**
Na	0.57	0.55	0.26	0.22**
Ca	0.68	0.72	0.99	0.90*
Mg	0.28	0.25*	0.39	0.35**
P	1.26	0.63***	0.84	0.70*
Cl	1.56	2.44**	1.68	1.92
NO_3	2.37	1.66**	1.03	1.18
SO_4	1.89	1.44**	0.83	0.75

Significant differences at $*p = 5\%$, $**p = 1\%$ or $***p = 0.1\%$ level.

Discussion

The results have shown that field beans depending on symbiotic nitrogen fixation decrease soil pH as long as the proton buffer capacity is not too high (Table 2). Soil acidification by roots has been demonstrated before for many plant species not only in pot experiments but also in field trials

(Hauter and Steffens, 1985; Hauter and Mengel, 1988; Jarvis and Robson, 1983; Marschner *et al.*, 1986; Schaller and Fischer, 1985; Singh *et al.*, 1987). Acidification of the root medium was related to decreased plant dry matter production (Fig. 1, Table 2). Apparently, symbiotic nitrogen fixation did not limit growth at low pH since N concentrations of plants did not differ for the various treatments (Table 2). Furthermore, vegetative growth of field beans was reduced by low pH when plants were supplied with mineral nitrogen (Fig. 1). On the other hand, net proton release by roots was decreased even at moderately low pH (Fig. 2). It is therefore suggested that at low pH growth of field beans was limited by inhibited net proton release.

It has been demonstrated before that net proton release by plant roots can be primarily explained in terms of ATPase activity (Mengel and Schubert, 1985; Schubert and Läuchli, 1986). Proton extrusion by ATPase is regarded as the energetically driving process for the uptake of nutrient cations (Leonard, 1984; Briskin, 1986) and anions (Lin, 1981; Ullrich-Eberius *et al.*, 1981) as well as for internal pH regulation (Raven and Smith, 1976; Felle, 1988). Consequently, inhibition of proton release at low pH (Fig. 2, Table 3) limited nutrient uptake (Table 4) and plant growth (Fig. 1, Table 2). A similar relationship between net proton extrusion and dry matter production at different pH values was found earlier for pea by Van Beusichem (1982). The author suggested that 'the wide adaptation to soil and climate conditions among legume genotypes is partly based on the intensity of the proton extrusion pump operation.'

Proton release is ascribed to the alkaline nutrient uptake pattern that includes excess uptake of inorganic cations over inorganic anions (Allan and Raven, 1987; Van Beusichem, 1982; Jarvis and Robson, 1983). In a recent article Schubert and Mengel (1989) have reviewed evidence for net proton release that does not depend on excess cation uptake. According to these results proton extrusion by ATPase activity must be considered as an initial step in nutrient uptake. The question arises whether an alkaline nutrient uptake pattern or net proton pumping by ATPase is primarily responsible for pH-dependent proton release and growth. As in the experiments of Van Beusichem (1982) excess cation uptake was more pronounced at pH 7 compared with pH 4 (Table 2). Higher proton release at pH 7 compared with pH 4 (Fig. 2) or pH 4.5 (Table 3) could therefore be explained by an alkaline nutrient uptake pattern. However, the determination of short term net proton release at varying pH in 1 m*M* $CaSO_4$ solution may indicate that low pH directly affected net proton extrusion by ATPase. Net proton release was extremely inhibited at low pH in 1 m*M* $CaSO_4$. In order to increase the pH from 6 to 7 addition of only about 0.5 μM calcium (as $Ca(OH)_2$) to the largely unbuffered $CaSO_4$ solution was required, an amount that is considered unlikely to change the uptake rate of either calcium or sulfate. Also, net proton efflux rates were modified by pH within 15 minutes (Fig. 2). Anion effects on net proton release, however, were measured after a lag phase of at least 2 hours only (Schubert, 1987). On the other hand, the possibility cannot be ruled out that sulfate uptake rates were increased due to the larger proton gradient (Lin, 1981).

Low pH in the root medium changes the electrical membrane potential and acidifies the cytoplasm (Bertl and Felle, 1985). The ability of root cells to restore the original potential and regulate internal pH by proton release may be of eminent significance to adapt to low medium pH. Field beans appear not to be capable of overcoming the restrictions of low medium pH on net proton release (Fig. 2, Table 3). We suggest, therefore, that the sensitivity of field beans to low pH is caused by a lack of capability to release protons by ATPase activity.

It is interesting to note that chloride concentrations in shoots and particularly in roots were increased at low medium pH which is in contrast to all other ions. Since chloride exclusion appears to be essential for legumes (Läuchli, 1984) it is possible that a lack of chloride exclusion may contribute to the sensitivity of field beans to low pH.

Acknowledgement

The authors are grateful for financial support by the DFG (Deutsche Forschungsgemeinschaft).

References

Allan S and Raven J A 1987 Intracellular pH regulation in *Ricinus communis* grown with ammonium or nitrate as N source: The role of long distance transport. J. Exp. Bot. 38, 580–596.

Andrew C S 1976 Effect of calcium, pH and nitrogen on the growth and chemical composition of some tropical and temperate pasture legumes. I. Nodulation and growth. Aust. J. Agric. Res. 27, 611–623.

Andrew C S and Johnson A D 1976 Effect of calcium, pH and nitrogen on the growth and chemical composition of some tropical and temperate pasture legumes. II. Chemical composition (calcium, nitrogen, potassium, magnesium, sodium, and phosphorus). Aust. J. Agric. Res. 27, 625–636.

Bertl A and Felle H 1985 Cytoplasmic pH of root hair cells of *Sinapis alba* recorded by a pH-sensitive micro-electrode. J. Exp. Bot. 36, 1142–1149.

Briskin D P 1986 Plasma membrane H^+-transporting ATPase: Role in potassium ion transport? Physiol. Plant. 68, 159–163.

Felle H 1988 Short-term pH regulation in plants. Physiol. Plant. 74, 583–591.

Hauter R and Mengel K 1988 Measurement of pH at the root surface of red clover (*Trifolium pratense*) grown in soils differing in proton buffer capacity. Biol. Fert. Soils 5, 295–298.

Hauter R and Steffens D 1985 Einfluß einer mineralischen und symbiontischen Stickstoffernährung auf Protonenabgabe der Wurzeln, Phosphat-Aufnahme und Wurzelentwicklung von Rotklee. Z. Pflanzenernaehr. Bodenkd. 148, 633–646.

Jarvis S C and Robson A D 1983 A comparison of the cation/anion balance of ten cultivars of *Trifolium subterraneum* L., and their effects on soil acidity. Plant and Soil 75, 235–243.

Läuchli A 1984 Salt exclusion: An adaptation of legumes for crops and pastures under saline conditions. *In* Salinity Tolerance in Plants: Strategies for Crop Improvement. Eds. R C Staples and G H Toennissen. pp 171–187. John Wiley and Sons, New York.

Leonard R T 1984 Membrane-associated ATPases and nutrient absorption by roots. *In* Advances in Plant Nutrition. Vol. 1. Eds. P B Tinker and A Läuchli. pp 209–240. Praeger, New York.

Lin W 1981 Inhibition of anion transport in corn root protoplasts. Plant Physiol. 68, 435–438.

Mahler R L and McDole R E 1987 Effect of soil pH on crop yield in Northern Idaho. Agron. J. 79, 751–755.

Marschner H, Römheld V, Horst W J and Martin P 1986 Root-induced changes in the rhizosphere: Importance for the mineral nutrition of plants. Z. Pflanzenernaehr. Bodenkd. 149, 441–456.

Mengel K and Schubert S 1985 Active extrusion of protons into deionized water by roots of intact maize plants. Plant Physiol. 79, 344–348.

Mengel K and Steffens S 1982 Beziehung zwischen Kationen/Anionen-Aufnahme von Rotklee und Protonenabscheidung der Wurzeln. Z. Pflanzenernaehr. Bodenkd. 145, 229–236.

Moore D P 1974 Physiological effects of pH on roots. *In* The Plant Root and Its Environment. Ed. E W Carson. pp 135–151. Univeristy Press of Virginia, Charlotteville, VA.

Moritsugu M, Suzuki T, and Kawasaki T 1983 Effect of nitrogen source on growth and mineral uptake in plants under constant pH and conventional culture conditions. Ber. Ohara Inst. Landw. Biol. Okayama Univ. 18, 125–144.

Munns D N 1978 Soil acidity and nodulation. *In* Mineral Nutrition of Legumes in Tropical and Subtropical Soils. Eds. C S Andrew and E J Kamprath. pp 243–263. CSIRO, Melbourne, Australia.

Munns D N 1986 Acid soil tolerance in legumes and rhizobia. *In* Advances in Plant Nutrition. Vol. 2. Eds. B Tinker and A Läuchli. pp 63–91. Praeger, New York.

Prenzel J 1985 Verlauf und Ursachen der Bodenversauerung. Z. Deutsch. Geol. Ges. 136, 293–302.

Raven J A and Smith F A 1976 Nitrogen assimilation and transport in vascular land plants in relation to intracellular pH regulation. New Phytol. 76, 415–431.

Schaller G and Fischer W R 1985 pH-Änderungen in der Rhizosphäre von Mais- und Erdnußwurzeln. Z. Pflanzenernaehr. Bodenkd. 148, 306–320.

Schubert S 1987 Plant nutrition and H^+ extrusion by plant roots. Plant Physiol. (Life Sci. Adv.) 6, 29–33.

Schubert S and Läuchli A 1986 Na^+ exclusion, H^+ release, and growth of two different maize cultivars under NaCl salinity. J. Plant Physiol. 126, 145–154.

Schubert S and Mengel K 1989 Important factors of nutrient availability: Root morphology and physiology. Z. Pflanzenernaehr. Bodenkd. 152, 169–174.

Singh L, Pal U R and Arora Y 1987 Direct and residual effect of liming on yield and nutrient uptake of maize (*Zea mays* L.) in moderately acid soils in the savanna zone of Nigeria. Fert. Res. 12, 11–20.

Ullrich-Eberius C I, Novacky A, Fischer E and Lüttge U 1981 Relationship between energy-dependent phosphate uptake and the electrical membrane potential in *Lemna gibba* G 1. Plant Physiol 67, 797–801.

Van Beusichem M L 1982 Nutrient absorption by pea plants during dinitrogen fixation. 2. Effects of ambient acidity and temperature. Neth. J. Agric. Sci. 30, 85–97.

M. L. van Beusichem (Ed.), *Plant nutrition – physiology and applications*, 449–453.

PLSO IPNC633

Nitrogen fixation of lucerne (*Medicago sativa*) in an acid soil: The use of rhizotrons as a model system to simulate field conditions

J.W.M. PIJNENBORG and T.A. LIE
Department of Microbiology, Wageningen Agricultural University, Hesselink van Suchtelenweg 4, 6703 CT Wageningen, The Netherlands

Key words: acid soil, agar-contact method, *Medicago sativa* L., nitrogen fixation, pH micro-electrode, *Rhizobium meliloti*, rhizotron

Abstract

The effect of pelleting seeds of lucerne with lime was studied in an acid sandy soil. In pot experiments, the fraction of seedlings with crown nodules, *i.e.* nodules on the upper 10 mm of the taproot, increased from 26% to 71%. In rhizotrons, the application of $CaCO_3$ resulted in an even stronger response.

An agar-contact method was used to study pH changes in the rhizosphere during a period up to 12 days. Application of 1.0 μmol of $CaCO_3$, in drops of 12 μL volume, resulted in an initial soil pH of 6.1 and yielded 75% crown nodulation. In the absence of $CaCO_3$, roots induced a pH increase from 5.1 (day 0) to 5.7 (day 12). However, this did not increase nodulation (5%). Obviously, this type of alkalinization does not overcome the acid-sensitive step of the nodulation process.

Introduction

The deficient growth of lucerne on acid soils is partly due to a poor symbiosis with *Rhizobium meliloti* (*e.g.* Mulder *et al.*, 1966). The obvious remedy is to neutralize soil acidity by applying large amounts of lime. Good results were obtained in Australia by pelleting the legume seeds with a layer of $CaCO_3$ (Loneragan *et al.*, 1955). In this way, the amount of lime required could substantially be lowered. More recently, the use of lime-pelleted seeds has been introduced in the Netherlands to cultivate lucerne on acid sandy soils. The dry matter yield in the first growing season due to lime-pelleting (30 kg of $CaCO_3$/ha), was similar to the yield of soil limed with 1000 kg $CaCO_3$/ha (Deinum and Eleveld, 1986). In this paper, nitrogen fixation is studied in more detail using small root boxes (rhizotrons) as a model system.

Material and methods

Soil

The sandy soil has the following characteristics: pH-H_2O, 5.2; organic matter, 2.2%; cation exchange capacity (CEC), 3 meq/100 g. The soil did not contain native *R.meliloti*. Air-dried soil was mixed with a sterile solution of 0.34 g KH_2PO_4 plus 0.25 g $MgSO_4$-$\cdot 7H_2O$ per litre of demineralized water, to obtain a moisture content of 12% (w/w).

Rhizobium

For inoculation, *R.meliloti* strain K-24 was grown during 4 days (30°C) in yeast extract mannitol (YEM) medium (Vincent, 1970).

Plant

Lucerne (*Medicago sativa* L. cv Resis), obtained

from Van der Have (Kapelle, The Netherlands), was grown in a climate room under a 16 h light (200 lux)/8 h dark cycle at 20°C and a relative humidity of 70% in either pots or rhizotrons, as described in more detail by Pijnenborg and Lie (1990).

Pots, 10 cm in diameter and 10 cm in height, were heat-sterilized (120°C) and filled with 600 g of moist soil. Plant holes of 5 mm depth were made and 24 seeds were sown per pot. The plants were harvested 25 days later. Each treatment consisted of 4 pots.

Rhizotrons were made of plastic petri dishes (9 cm ϕ), by removing the top 2 cm. Lucerne seeds were surface-sterilized (1 min in 70% ethanol, followed by 20 min in 6% H_2O_2), and pre-germinated during one day at 30°C on agar-water (1.0%, w/v) to a root length of 3 to 5 mm. Each rhizotron was filled with 50 g of moist soil and then 7 (nodulation experiments) or 4 (measurement of soil pH) seeds were sown. The rhizotrons were placed in the climate room at an angle of 60° to induce the roots to grow towards the lid. The plants were allowed to grow for 12 to 14 days. In nodulation experiments, each treatment was replicated 4 times; one replication consisted of 4 rhizotrons (28 seedlings). Soil pH was determined in 4 independent measurements.

Inoculation and lime-pelleting

The seeds were inoculated with rhizobia alone, or were supplied with rhizobia and a lime pellet. Inoculation was carried out either with a peat-based inoculum (in pots) or in a liquid form (in rhizotrons). A sterile peat-loam mixture (Van Schreven, 1970) was mixed with an fullgrown culture of *R.meliloti* in YEM medium to obtain a moisture content of 60% (w/w). This mixture was incubated during a week at 30°C. An equal amount (weight) of the peat-bacteria mixture was added to an aqueous solution of 2% methyl cellulose (Tylose, Fluka A.G., Buchs, Switzerland). Four parts (per weight) of seeds were mixed with one part of the peat-bacteria-methyl cellulose mixture. Methyl cellulose was used to fix the rhizobial cells to the seed coat. Final pelleting was done by dusting the wet seed in a ratio of 2 to 1 with dry lime, until evenly coated pellets (10 μmol $CaCO_3$ per seed) were obtained.

In rhizotrons, the soil was inoculated directly at the positions where the seeds would be sown, by using a syringe to add 12 μL of a suspension of *R.meliloti* in 70 m*M* NaCl plus 10 m*M* KH_2HPO_4/K_2HPO_4 buffer of pH 5.2. Lime-pelleting was simulated prior to inoculation with 12 μL of a $CaCO_3$ suspension.

Analyses of nitrogen fixation

The nitrogen content of the dried (24 h, 80°C) shoots was determined by the indophenol-blue method (Novozamsky *et al.*, 1974). The acetylene reduction activity (ARA) was determined by analyzing the ethylene content in samples of 100 μL of the gaseous phase, after incubating the seedlings in 250 mL bottles for 60 minutes in air containing 10% (v/v) acetylene.

Determination of soil pH

In rhizotrons, local changes in soil pH were quantified using an agar-contact method, slightly modified according to Marschner and Römheld (1983). The lids of petri dishes were filled with 20 mL of agar solution, containing 7.5 g agar (Difco Bacto agar, Brunschwig Chemie, Amsterdam) 0.06 g bromocresol purple (Merck) and 0.17 g $CaSO_4$ per litre of demineralized water. After solidification, the 3 mm layer was covered with a nylon gauze (53 mesh) and brought into contact with the soil. After three hours of incubation, a micro-electrode (type MI 410, Microelectrodes Inc., Londonderry, U.S.A.) was inserted into the agar. The values obtained with the agar method are in close agreement with the actual soil pH (unpublished results).

Results and discussion

Nodulation

At first, the effect of inoculation and lime-pelleting on the nodulation of lucerne seedlings was studied in pot experiments. The results are given in Table 1. Since the soil is free of *R.meliloti* bacteria, no nodules were formed on the non-

Table 1. Effect of inoculation and lime-pelleting on the number of crown-nodulated seedlings, amount of nitrogen in shoots and acetylene reduction activity (ARA) of the seedlings, 25 days after sowing. Data (mean ± SE) are averaged for 4 pots containing 24 plants

	Inoculation[a]			
	–		+	
	Lime pelleting[b]			
	–	+	–	+
Crown nodulation (%)	0	0	26 ± 18	71 ± 7
Shoot-N (mg pot^{-1})	2.9 ± 0.9	2.8 ± 0.5	7.3 ± 1.0	19.0 ± 0.9
ARA (μmol ethylene pot^{-1} h^{-1})	0	0	1.7 ± 0.3	3.2 ± 0.7

[a] Inoculated with 4.0×10^5 cells of *R.meliloti* per seed.
[b] Pelleted with 10 μmol of $CaCO_3$ per seed.

inoculated control plants. When inoculated, 26%, and with additional lime-pelleting, 71% of the plants became nodulated. Under these conditions, nodule formation was confined to the upper 10 mm of the taproot and on laterals in this section, within 3 mm distance from the taproot (crown nodules). Nitrogen fixation, as estimated by the amount of shoot-N and ARA, was closely correlated to crown nodulation (Table 1).

For more detailed experiments, seedlings were grown in rhizotrons. Through the transparent plastic lid, continuous observations could be made of the early root developments and nodule formation. The effect of increasing amounts of locally applied $CaCO_3$ on the nodulation is given in Table 2. Inoculation resulted hardly in nodulation (5%). Optimal crown nodulation (75%) was already achieved when 1.0 μmol of $CaCO_3$ was supplied.

The nodulation response in rhizotrons was much more sensitive to addition of 10 μmol of lime than in pots: from 5% to 87% (Table 2) *versus* 26% to 71% (Table 1). The fact that rhizotron-grown seedlings nodulated less when only *R.meliloti* bacteria were applied, is probably due to the way the seeds were inoculated. It is generally accepted that peat-based inoculation is superior to liquid inoculation (*e.g.* Burton and Curely, 1965).

Table 2. The effect of local addition of $CaCO_3$ in rhizotrons on the nodulation of 14-day-old lucerne seedlings, inoculated with 4.0×10^7 cells of *R.meliloti*. Data (mean ± SE) are averaged for 4 × 28 seedlings

Amount of $CaCO_3$ (μmol $seed^{-1}$)	Number of crown-nodulated seedlings (%)
0	5 ± 2
1.0	75 ± 7
2.0	80 ± 4
10.0	87 ± 6

Changes in soil pH

The changes in soil pH due to lime were measured around the root crown of lucerne seedlings by making a contact print of the rhizotron-soil in a layer of indicator agar. Qualitative changes in pH could be seen by a change in color. Bromocresol purple is yellow at pH 5.2 and becomes purple at pH 6.8 (Clark, 1928). This is shown in Plate 1, which is made 9 days after sowing seeds that received 10 μmol of $CaCO_3$.

The changes in pH were measured during a period of 12 days after sowing. A comparison was made of seeds, either untreated or treated with 1.0 μmol of $CaCO_3$. The results are presented in Figure 1. Without lime, the rhizosphere-pH remained at 5.1 during the first 3 days. However, an increase was observed at day 6 (pH 5.5) and day 12 (pH 5.7). By adding 1.0 μmol of $CaCO_3$, the treated soil spots were neutralized to pH 6.1. Within one day, the surrounding bulk of acid soil leveled this increase down to pH 5.3 (Fig. 1). At later stages, no significant differences with the untreated seedlings were observed; soil pH was raised to 5.7 (day 6) and 5.8 (day 12).

From the above is becomes clear, that the root of a lucerne seedling is capable of inducing a marked increase in rhizosphere pH. Blanchar and Lipton (1986) found that the soil along the older taproot of 10-day-old lucerne seedlings in a

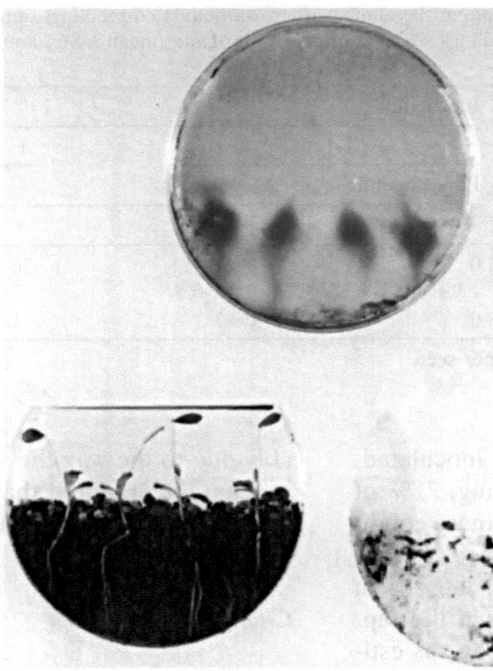

Plate 1. A layer of indicator-agar (top), after 3 h of contact with the rhizotron-soil (bottom, left), 9 days after sowing 4 lucerne seeds, treated with 10 μmol of $CaCO_3$. The soil was separated from the agar by means of the nylon gauze (bottom, right).

soil of pH 5.5 was more basic (pH 6.8). The environment around the younger lateral roots was more acidic (pH 4.2). They suggested that the older parts of the root system absorb more anions than cations. To maintain electroneutrality in the plant, OH-ions have to be excreted (*e.g.* Dijkshoorn, 1962).

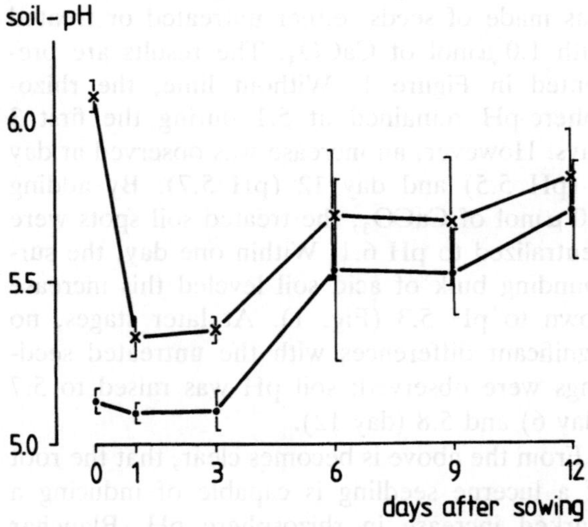

Fig. 1. Dynamics in soil pH in the root crown of seedlings supplied with 0 μmol of $CaCO_3$ (●) or 1.0 μmol of $CaCO_3$ (×). The measurements were carried out in four-fold; bars indicate SE.

When the seeds were supplied with a small amount of lime, significantly higher pH values were measured during the first 3 days (Fig. 1), and an excellent nodulation was found (Table 2). After 6 days, inoculated seedlings that had not received lime, increased soil pH to 5.5. Lucerne can nodulate well at pH 5.5 (Munns, 1965). However, only 5% of the seedlings nodulated (Table 2). This failure cannot be due to restricting numbers of *R.meliloti*, since previous studies showed that the inoculated bacteria multiplied just as well in the presence or absence of a lime-pellet. Apparently, the nodulation process is restricted by low pH only during the initial phase (day 0 to day 3). This phenomenon has been studied previously in nutrient solution with pea (Lie, 1969) and lucerne plants (Munns, 1968). The temporal acid sensitivity was attributed to the 'acid sensitive step' in the nodulation process.

Conclusions

In the acid soil (pH-H_2O 5.2) used in this study, lime-pelleting more than doubled crown nodulation of pot-grown lucerne plants (26% to 71%).

This is similar to the results obtained under field conditions (Pijnenborg and Lie, 1990).

Rhizotron-grown plants are more sensitive to addition of lime: with only 1.0 μmol of $CaCO_3$, crown nodulation increased from 5% to 75%. The rhizotron system allows continuous observation of the development of root nodules. Furthermore, with this sytem *in-situ* soil pH values can locally be quantified. The small amount of lime (1.0 μmol) significantly increased initial soil pH followed by equilibration during the first 3 days. At later stages, unlimed seedlings also started to alkalinize their rhizosphere. However, this did not result in a better nodulation. Obviously, rhizosphere pH during the initial 3 days is decisive for the success of the nodule formation in the soil which is in agreement with earlier experiments conducted in water culture (Lie, 1969; Munns, 1968).

Acknowledgements

We thank the Department of Soil Science and Plant Nutrition, especially Mr J Nelemans, for the opportunity to use the micro electrode. These investments were supported by the EEC (contract no. EEC/AUW 1914).

References

Blanchar R W and Lipton D S 1986 The pe and pH in alfalfa seedling rhizospheres. Agron. J. 78, 216–218.

Burton J C and Curley R L 1965 Comparative efficiency of liquid and peat-base inocualants on field-grown soybeans (*Glycine max*). Agron. J. 57, 379–381.

Clark W M 1928 The Determination of Hydrogen Ions, 3rd Ed. Williams and Wilkins, Baltimore, MD, pp 94–103.

Deinum B and Eleveld J 1986 Effects of liming and seed pelleting on the growth of lucerne (*Medicago sativa* L.) on sandy soils. *In* Proc. 11th Symp. Europ. Grassland Fed. Eds. F M Borba and J M Abreu. pp 270–273. Sousa Ferradeira Artes Graficas, Lisboa, Portugal.

Dijkshoorn 1962 Metabolic regulation of the alkaline effect of nitrate utilization in plants. Nature 194, 165–167.

Lie T A 1969 The effect of low pH on different phases of nodule formation in pea plants. Plant and Soil 21, 391–405.

Loneragan J F, Meyer D, Fawcett R G and Anderson A G 1955 Lime pelleted clover seed for nodulation on acid soils. J. Aust. Inst. Agric. Sci. 21, 264–265.

Marschner H and Römheld V 1983 *In vivo* measurement of root-induced pH changes at the root-soil interface: Effect of plant species and nitrogen source. Z. Planzenphysiol. 111, 241–245.

Mulder E G, Lie T A, Dilz K and Houwers A 1966 Effect of pH on symbiotic nitrogen fixation of some leguminous plants. *In* Proc. of the Ninth International Congress Microb. Moscow, pp 133–151.

Munns D N 1965 Soil acidity and growth of a legume. I. Interaction of lime with nitrogen and phosphate on the growth of *Medicago sativa* L. and *Trifolium subterraneum* L. Aust. J. Agric. Res. 16, 733–741.

Munns D N 1968 Nodulation of *Medicago sativa* in solution culture. I. Acid sensitive steps. Plant and Soil 28, 129–146.

Novozamsky I, Eck R van, Schouwenburg J C van and Walinga I 1974 Total nitrogen determination in plant material by means of the indophenol-blue method. Neth. J. Agric. Sci. 22, 3–5.

Pijnenborg J W M and Lie T A 1990 Effect of lime-pelleting on the nodulation of lucerne (*Medicago sativa* L.) in an acid soil: A comparative study carried out in the field, in pots and rhizotrons. Plant and Soil. 121, 225–234.

Van Schreven D A 1970 Some factors affecting growth and survival of *Rhizobium* spp. in soil-peat cultures. Plant and Soil 32, 113–130.

Vincent J M 1970 A Manual for the Practical Study of Root Nodule Bacteria. IBP Handbook 15, Blackwell Scientific Publications, Oxford, UK, pp. 3–4.

M. L. van Beusichem (Ed.), *Plant nutrition – physiology and applications*, 455–461.

PLSO IPNC426

Role of plant aerenchyma in wet tolerance of and methane emission from plants

T. WAGATSUMA, T. NAKASHIMA, K. TAWARAYA, S. WATANABE, A. KAMIO and A. UEKI
Faculty of Agriculture, Yamagata University, Tsuruoka 1–23, Yamagata 997, Japan

Key words: aerenchyma, air permeability within plant, barnyardgrass, methane emission, reed, rice, tolerance to low oxygen, wet tolerance, upland crops

Abstract

The objective of the present investigation was to clarify the role of plant aerenchyma in wet tolerance of and methane emission from plants. Wet tolerance was investigated both in a soil culture (18% and 44% moisture content) and in a water culture at four treatments, *i.e.*, 1 or 7 mg dissolved oxygen per liter, and 1 or 50 mg Mn per liter. Methane emission from reed was investigated in a soil culture.

Wet tolerance was positively correlated with combined tolerance to Mn and low oxygen concentration.

Low oxygen treatment decreased considerably the K content of excised roots and slightly the P content; the roots of intact plants were not affected. No correlation was observed among wet tolerance, tolerance to low oxygen concentration and the decrease in K content of excised roots. These results suggest that plant tolerance to a low oxygen concentration and high moisture level in soil may not be due to the tolerance of the root cells.

The presence of a continuous aerenchyma throughout the plant was observed in plant species, which tolerate wet conditions, *i.e.*, two *Echinochloa* sp., reed and rice. Air permeability within plants is generally associated with the development of a continuous aerenchyma throughout the plant. Adzuki bean especially has adventitious roots that are able to tolerate a high moisture level in soil.

It was estimated that one gram of reed shoot was able to decrease the methane concentration in the soil atmosphere by 1 vol%.

These results suggest that the presence of a continuous aerenchyma throughout the plant may play a significant role in the tolerance to wet conditions and methane emission in waterlogged soil.

Introduction

In soil with a high moisture level, various changes occur; the concentration of Mn-ions increases with the decrease of Eh to 250 mV, the content of free oxygen and nitrate decreases or becomes zero (Marschner, 1986).

Tolerance to wet soil conditions in its different aspects varies widely among plant species. (Tadano *et al.* 1979). Differences in manganese tolerance which were reported (Tanaka and Hayakawa, 1975; Tanaka *et al.*, 1975), have not been investigated in connection with the oxygen concentration in the media.

In wet soils or marshes, various weeds, especially reed and *Echinochloa* sp. (barnyardgrass), grow luxuriantly. Strong reductive conditions with Eh below −200 mV develop when organic soils become waterlogged. Under such conditions, there is a high concentration of methane in the soil atmosphere. (Takai, 1980) In paddy fields a large proportion of methane is emitted to the atmosphere through the rice plants (Holzapfel-Pschorn *et al.*, 1986; Minami and Yagi, 1988). Reed and several weeds emit less methane than rice (Holzapfel-Pschorn *et al.*, 1986). In unvegetated paddy fields methane was emitted almost exclusively by ebullition (Holzapfel-Pschorn *et al.*, 1986). Nowadays, there is an increasing concern about emission of methane into the atmosphere in connection with the global greenhouse effect and the destruction of the ozone layer.

The objectives of the present investigation are (1) to clarify the mechanism of wet tolerance and (2) to estimate the methane emission ability from several weeds, especially reed.

Materials and methods

Soil culture experiment

Deionized water was added to normally fertilized soil collected in the farm of Yamagata University (Fluvisols). This soil was not Mn toxic (exchangeable Mn: 3.7 mg per kg dry soil, 0.5% tannic acid reducible Mn: 7.0 mg per kg dry soil). Two moisture levels, 18% and 44% (W/W), were maintained for a period of 16 days. Eight plant species were used: two barnyardgrass cultivars (*Echinochloa crus-galli* Beauv., *Echinochloa oryzicola* Vasing), reed [*Phragmites australis* (Cav.) Trin. ex Steudel], rice (*Oryza sativa* L. cv. Sasanishiki), adzuki bean (*Vigna angularis* cv. Takara), head lettuce [*Lactuca sativa* L. (capitata group) cv. Great lake], wheat (*Triticum aestivum* L. cv. Hanagasa) and Japanese radish [*Raphanus sativus* L. (daikon group) cv. Minowase No. 3]. The harvested plant tops were dried and analyzed for the mineral content.

Water culture experiment

Four treatment solutions were prepared: normal oxygen concentration (7 mg dissolved oxygen per liter of solution) and normal Mn concentration (1 mg Mn per liter) [A]; normal oxygen and high Mn concentrations (50 mg Mn per liter) [AM]; low oxygen (1 mg oxygen per liter) and normal Mn concentrations [N]; low oxygen and high Mn concentrations [NM]. Other conditions were standard (Wagatsuma *et al.*, 1988) and the treatments were applied for 17 days. A part of the plant samples cultured under normal conditions was used for investigations of the anatomical structure of the aerenchyma by scanning electron microscopy, light microscopy or stereoscopic microscopy.

Experiment on tolerance to low oxygen concentration

Intact plants or excised roots were placed in tap water (adjusted to pH 5.0) or a concentration of 7 or 1 ppm oxgyen at 30°C, and treated for 24 hr. The contents of P, K, Ca and Mg were analyzed in the harvested roots.

Air permeability within plants

Air permeability within plants was measured under negative pressure; the petiole (Japanese radish) or culm (other plant species) was inserted into a glass column (inside diameter of 1.5 cm and 100 cm long) filled with water, and the volume of the evolved gas was measured.

Methane emission from reed shoot

Marshy and heavy clay soil collected from Hachirogata in Japan (Gleysols) was mixed with fertilizers. Reed seedlings grown in a lysimeter for 8 years were turned upside down and transferred to wagner pots in spring. About 3 months after the waterlogging treatment, the soil atmosphere was periodically collected in a test tube filled with water by agitating the soil with a stick. Concentrations of oxygen, di-nitrogen and methane, were analyzed by a gas chromatographic technique (Hitachi 163 Gas chromatograph, 3 mm × 2.0 m steel column packed with 80–100 mesh Molecular Sieve 5 A).

All the experiments were replicated and the results were expressed as mean values.

Results and discussion

Wet tolerance and tolerance to other treatments

Wet tolerance was in the following order: two *Echinochloa* sp. (268, 255), reed (253) > rice

Table 1. Correlation coefficients between each tolerance and ratio of mineral contents in the corresponding two treatments

Tolerance	Mineral content ratios for tops					
of tops[a]	P	K	Ca	Mg	Fe	Mn
W.T.	0.490	0.936**	−0.794*	−0.912**	−0.664	−0.759*
N/A	0.955**	0.918*	0.617	0.839*	0.585	0.929*
AM/A	−0.515	0.045	0.291	0.762*	0.125	−0.787*
NM/N	−0.324	−0.772*	−0.499	−0.273	−0.679	−0.801*
NM/A	0.652	0.770*	0.262	0.639	−0.513	−0.629
Tolerance	Mineral content ratios for roots					
of roots	K	Ca	Mn			
N/A	0.958**	−0.267	−0.090			
AM/A	−0.241	0.265	0.061			
NM/N	−0.743*	0.371	0.287			
NM/A	0.896**	0.364	0.015			

In the soil experiment, roots were not harvested. Significant levels were 5% (*) and 1% (**). [a]: W.T. = wet tolerance, N/A = tolerance to low oxygen concentration, AM/A = Mn tolerance at normal oxygen concentration, NM/N = Mn tolerance at low oxygen concentration, NM/A = combined tolerance to high Mn and low oxygen concentration.

(215), adzuki bean (183) ≫ head lettuce (87) > wheat (58) > Japanese radish (10). Tolerance to low oxygen concentration (N/A) was strong in rice (90), adzuki bean (87) and *E. oryzicola* (85), and weak in Japanese radish (44). Mn tolerance (AM/A) was strong in the two *Echinochloa* sp. (105, 97) and reed (91), and weak in adzuki bean (43) and wheat (48). Combined tolerance to Mn and low oxygen concentration (NM/A) was strong in reed (98) and *E. oryzicola* (97), and weak in wheat (50), adzuki bean (47) and Japanese radish (37).

The relating other data not indicated in the present report and the detailed procedures and explanations have been published elsewhere (Wagatsuma *et al.*, 1990).

Plants with a tolerance to a low oxygen concentration had a high content of P, K, Mg and Mn in the tops and a high content of K in the roots (Table 1). The tops of the Mn tolerant plants contained a large amount of Mg, and low amount of Mn. Plants with combined tolerance had a high content of K only in both plant parts. Plants with wet tolerance had a high content of K and low content of Mn.

Wet tolerance was positively correlated with the combined tolerance, and weakly correlated with the tolerance to low oxygen concentration (Table 2). Oxygen concentration in the medium did not affect the Mn tolerance (r = 0.722*).

Low oxygen concentration decreased considerably the K content of excised roots in all the plant species (Fig. 1). Decrease in K content was pronounced after exposure for 10 hr to the low oxygen treatment, especially in adzuki bean and wheat. Ca content of excised roots, on the contrary, increased following a 10 hr exposure to the low oxygen treatment especially in Japanese radish, adzuki bean and head lettuce. Virtually no decrease in K content or increase in Ca content was observed in whole roots of intact plants even 24 hr after exposure to the low oxygen treatment (data not shown). P content was decreased, and Mg content increased though less considerably than that of K and Ca. K and P losses from excised roots at low oxygen concentrations may be ascribed to the destruction of the plasma membrane of root cells; Ca and Mg increase on a dry weight basis may result from the loss of

Table 2. Correlation coefficients for tolerance

	W.T.	N/A	AM/A	NM/N	NM/A
W.T.	1				
N/A	0.623	1			
AM/A	0.473	−0.286	1		
NM/N	0.478	−0.186	0.722*	1	
NM/A	0.763*	0.387	0.536	0.827*(*)	1

For symbols, see Table 1. Figure underlined indicates significance near 5% level, and *(*) indicates the significance between 5 and 1% levels.

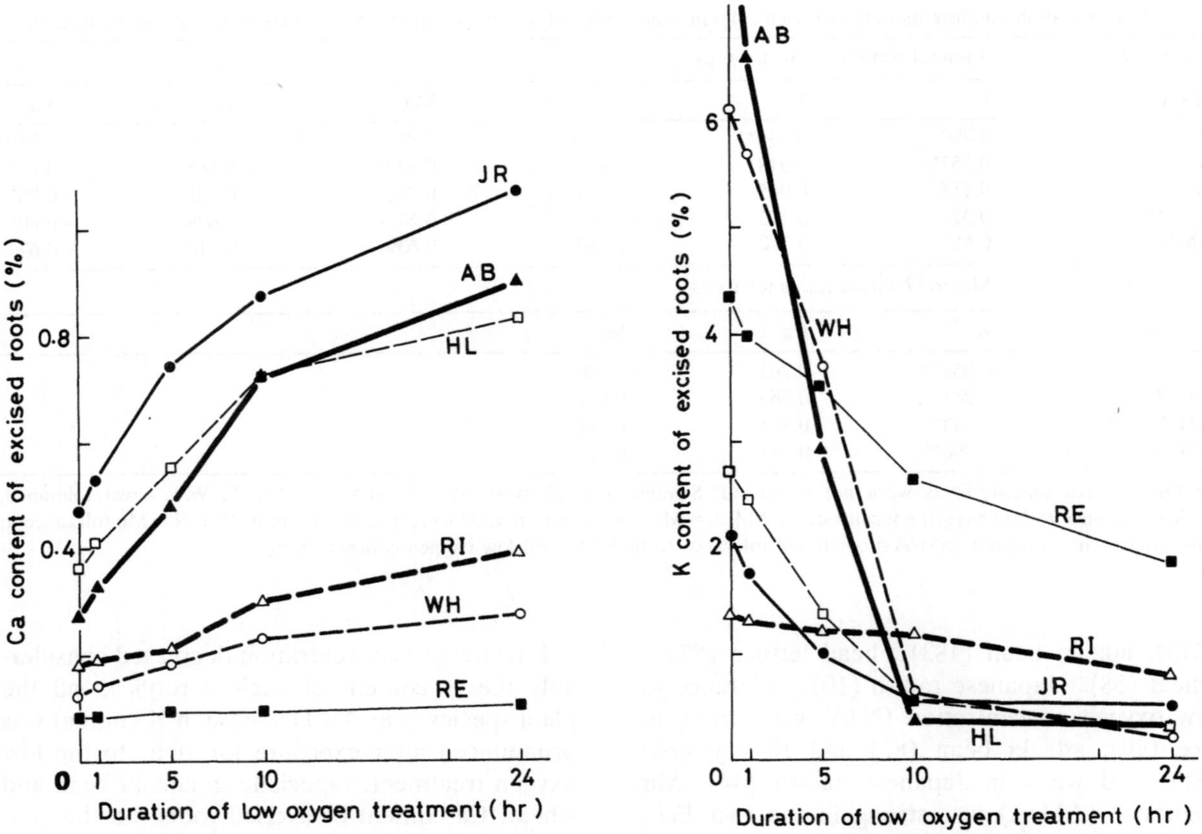

Fig. 1. Effect of low oxygen treatment on mineral contents of excised roots. JR: Japanese radish, AB: adzuki bean, HL: head lettuce, RI: rice, WH: wheat, RE: reed.

soluble compounds from root cells. In whole roots of the intact plants, low oxygen concentration in the medium (at least for 24 hr) may not affect the plasma membrane of root cells due to the existence of air in the intercellular spaces, especially in the mature basal region of roots. No correlation was observed between the extent of decrease in the K content and low oxygen tolerance or wet tolerance. These results suggest the importance of the presence of air spaces in roots, *i.e.*, intercellular spaces, lysigenous intercellular spaces or aerenchyma.

Anatomical structure and air permeability within plant

Reed roots originating from subterranean stem developed an aerenchyma in the cortex except in the tip portion even when grown in water culture at normal oxygen concentration (Fig. 2a). Rice and the two *Echinochloa* sp. also developed an aerenchyma more considerably in the root cortex (data not shown). The development of the lysigenous aerenchyma in wheat was less conspicuous and was seen only in the basal portion of roots. In the roots of adzuki bean, head lettuce and Japanese radish, non-lysigenous intercellular spaces were developed in the cortex. Aerenchyma was developed even in the nodes of the subterranean stem of reed (Fig. 2b), and the stems of rice (Fig. 2c) and reed (Fig. 2d). A special spongy net structure was observed between the epidermis layers of the leaf sheath of the two *Echinochloa* sp. (Fig. 2e). Though less considerably, the aerenchyma was developed in the stem of wheat even under normal oxygen conditions (data not shown). In head lettuce, the aerenchyma was developed especially in the lower parts of leaves. The two types of aerenchyma were prominent in the midrib of rice leaf (Fig. 2f).

In conclusion, the development of a continu-

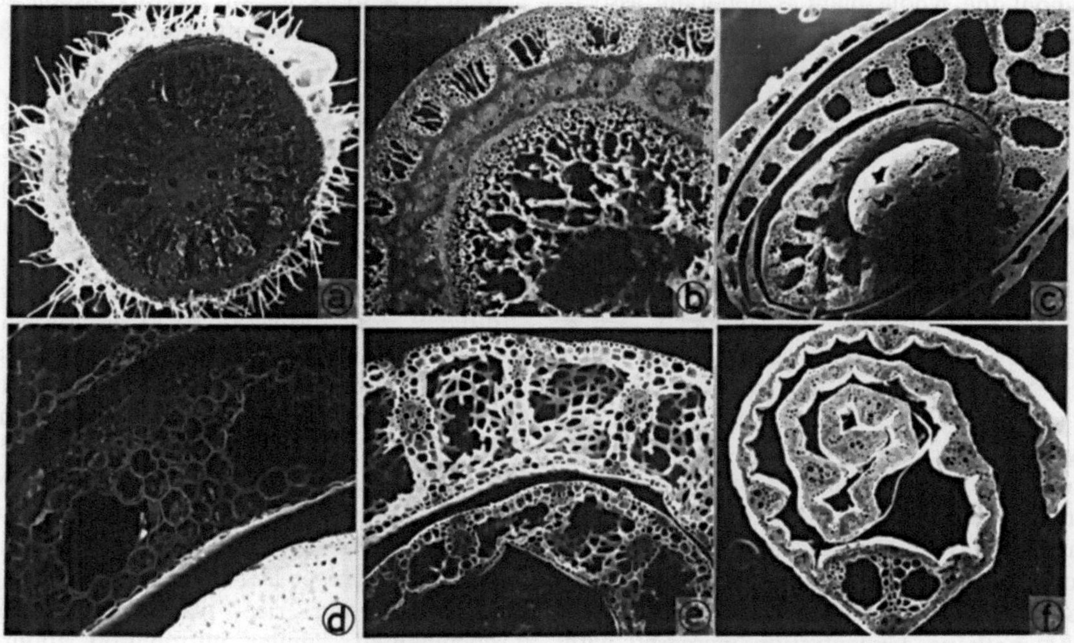

Fig. 2. Anatomical structure of various parts of plants observed by scanning electron microscopy. ⓐ basal root generated from subterranean stem in reed (×100), ⓑ node of subterranean stem in reed (×76) ⓒ rice stem (×74), ⓓ reed upper stem (×400) ⓔ stem of *E. oryzicola* (×300), ⓕ rice leaf, with curling (×140).

ous aerenchyma throughout the whole plant could be seen in rice, reed and the two *Echinochloa* sp., and to a lesser extent in wheat and lettuce. No aerenchyma was recognized in the stem of adzuki bean and in the petiole of Japanese radish.

Air permeability was very high in the two *Echinochloa* sp., medium in reed and rice, and low in wheat; no gas emission was detected in Japanese radish (Table 3). The perennial reed seedlings have thick stems and are able to produce a larger amount of gas (data not shown).

These data suggest that the presence of a continuous aerenchyma throughout the whole plant may play an important role in the transport of atmospheric oxygen to the root system exposed to a low oxygen concentration and plant survival under such conditions. Adventitious roots were developed abundantly from the stem base on and in the upper soil which was rich in oxygen even in wet conditions; this mechanism may contribute to the survival of adzuki bean under such conditions.

The presence of a continuous aerenchyma

Table 3. Air permeability within plant

Plant species	Mean length of stems or petioles cm	Shoot weight g/plant	Gas emission from shoot mL/min/stem or petiole
E. crus-galli	84	5.41	14.2
E. oryzicola	85	4.27	10.4
Reed	71	3.75	4.4
Rice	64	4.20	2.4
Wheat	n.d.	3.77	0.02
J. radish	n.d.	1.56	0

Gas emission from shoot was measured under negative pressure with a glass column filled with water.

throughout the whole plant is generally associated with wet tolerance. Although adzuki bean shows a relatively high wet tolerance, no aerenchyma is present in the shoot. Instead, it has a remarkable ability to generate adventitious roots in the basal stem.

Methane emission from reed

After the thick stems were fully emerged and the redox potential of the soil dropped below −200 mV, all the shoots were cut below the water surface. Nitrogen, oxygen and methane contents in the soil atmosphere were analyzed periodically during ca. 50 days after the shoots of a different dry weight were cut; shoot weights of harvested reed ranged from 4 g to 47 g. The oxygen concentration was very low, and more than 80% of the soil atmosphere consisted of nitrogen and methane. The nitrogen concentration decreased and the methane concentration increased with time after the shoots were cut; methane concentration in the soil atmosphere at harvest (23 Aug.) ranged from 24% to 68%, and those ca. 50 days after the shoots cut (11 Oct.) ranged from 59% to 78%. An almost constant concentration was reached after one week of incubation; at harvest, methane concentration in the soil atmosphere was lower in pot grown luxuriantly. Other detailed data were shown in elsewhere (Wagatsuma *et al.*, 1990). It was estimated that one gram dry weight of reed shoot was able to decrease the methane concentration in the soil atmosphere by 1 vol% (Fig. 3). In fact, methane was detected in the gas collected directly from the shoot (data not shown). Gas emission from the plant shoot in yellow waterlily has been ascribed to the presence of gradients in the temperature (thermal transpiration) and water vapor (hygrometric pressure) between the atmosphere and the lacunae of the youngest leaves (Dacey, 1981). Gas emission from reed was reported only under field conditions (Holzapfel-Pschorn et al., 1986). Although methane emission from rice plants has been well documented, similar information about reed, *Echinochloa* sp. or other hygrophytes is limited. It appears that a large amount of methane was emitted from the latter plants, based on the current experiments on the anatomical structures

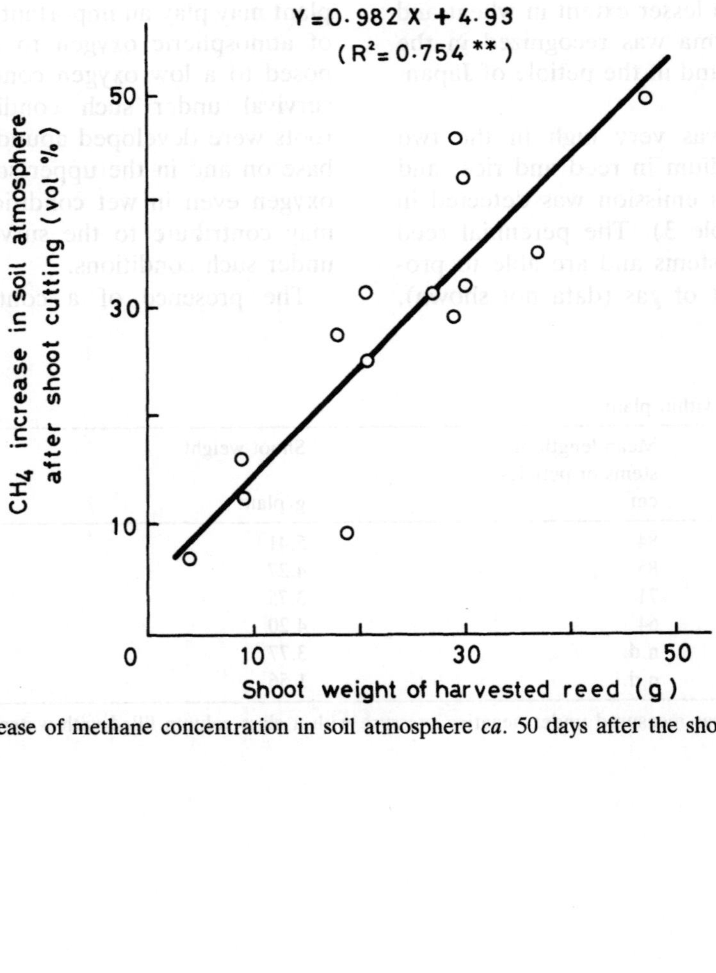

Fig. 3. Increase of methane concentration in soil atmosphere *ca.* 50 days after the shoots were cut.

and air permeability of the organs of these plants. In conclusion, it is suggested that the presence of an aerenchyma within the whole plant plays an important role at low oxygen concentration in both the oxygen supply from the atmosphere to roots and methane emission from soil to the atmosphere.

Acknowledgments

The authors thank Dr Gotoh M, Yamagata University, for supplying the seeds of the two varieties of *Echinochloa* sp. The authors also thank Drs Minami K and Yagi K, National Institute of Agro-Environmental Sciences, Japan, for their valuable suggestions. This work was supported in part by a Grant-in-Aid from the Ministry of Education, Science and Culture of Japan (No. 63560222).

References

Dacey J W H 1981 Pressurized ventilation in the yellow waterlily. Ecology 62, 1137–1147.

Holzapfel-Pschorn A, Conrad R and Seiler W 1986 Effects of vegetation on the emission of methane from submerged paddy soil. Plant and Soil 92, 223–233.

Kobayashi Y 1981 On the extractants of available manganese in grassland soil. Jap. J. Soil. Sci. Plant Nutr. 52, 539–542 (*In Japanese*).

Kobayashi Y 1982 Dissolution of some manganese oxide with tannic acid. Jap. J. Soil Sci. Plant Nutr. 53, 50–52 (*In Japanese*).

Marschner H 1986 Mineral Nutrition of Higher Plants. Academic Press, London, 498 p.

Minami K and Yagi K 1988 Method for measuring methane flux from rice paddies. Jap. J. Soil Sci. Plant Nutr. 59, 458–463 (*In Japanese with English summary*).

Tadano T, Kirimoto K, Aoyama I and Tanaka A 1979 Comparison of tolerance to high moisture conditions of the soil among crop plants: Studies on the comparative plant nutrition. J. Sci. Soil Manure, Jpn. 50, 261–269 (*In Japanese*).

Takai Y 1980 Microbial dynamics of paddy soils. Fertilizer Science 3, 17–56 (*In Japanese*).

Tanaka A and Hayakawa Y 1975 Comparison of tolerance to soil acidity among crop plants (Part 2) Tolerance to high levels of aluminum and manganese: Studies on the comparative plant nutrition. J. Sci. Soil Manure, Jpn. 46, 19–25 (*In Japanese*).

Tanaka A, Tadano T and Fujiyama H 1975 Comparison of adaptability to heavy metals among crop plants (Part 1) Adaptability to manganese: Studies on the comparative plant nutrition. J. Sci. Soil Manure, Jpn. 46, 425–430 (*In Japanese*).

Wagatsuma T, Kawashima T and Tawaraya K 1988 Comparative stainability of plant root cells with basic dye (methylene blue) in association with aluminum tolerance. Commun. Soil Sci. Plant Anal. 19, 1207–1215.

Wagatsuma T, Nakashima T, Tawaraya K, Watanabe S, Kamio A and Ueki A 1990 Relationship between wet tolerance, anatomical structure of aerenchyma and gas exchange ability among several plant species. Bull. Yamagata Univ., Agric. Sci. 11, 121–141.

M. L. van Beusichem (Ed.), *Plant nutrition – physiology and applications*, 463–467.

PLSO IPNC476

Growth and chemical composition of barley (*Hordeum vulgare*) cultivars on saline substrate as compared with a salt tolerant variety of wheat (*Triticum aestivum*)

R. ANSARI
Atomic Energy Agricultural Research Centre, Plant Physiology Division, Tandojam, Pakistan

Key words: barley, *Hordeum vulgare* L., ion relations, salinity, salt tolerance, *Triticum aestivum* L., wheat

Abstract

Growth and chemical composition of one cultivar (Pak-70) of wheat (*Triticum aestivum* L.) and two (Weeah and Proctor) of barley (*Hordeum vulgare* L.) were studied in sand culture where Pak-70 was observed to be the most salt tolerant followed by Weeah and Proctor. The resistance to salinity so observed was accompanied with restricted uptake of Na and Ca and a preferential accumulation of K and P in the straw. The capacity of the plants to maintain in their tissues a balance between these ions was vital for plant growth under conditions of salinity. Although plants at 60 days from sowing are rapidly growing and showed little difference among cultivars, chemical analysis at this stage could be used to predict the pattern of plant growth up to maturity.

Introduction

Growth of many crops may be adversely affected under saline conditions due to disturbance in the electrolyte balance resulting in deficiency of some essential nutrients and excess of certain unwanted salts in the plant tissue. The reduced water potential at high salt concentrations may further aggravate the effects. Salt tolerance under such conditions is generally related to the ability to regulate Na and Cl uptake by plant roots and subsequent translocation to the shoot (Bajwa and Bhumbla, 1978; El-Shaikh and Ulrich, 1971; Erdie and Kuiper, 1979; Lessani and Marschner, 1978). Very meager information is, however, available on how ionic ratios in tissue are affected by substrate salinity in relation to the salt tolerance of a species. The present study reports the growth, ion uptake and ratio between ions in plant tissue in two barley and one wheat cultivar in response to increasing external salt concentration.

Methods

A pot culture experiment was conducted using barley (*Hordeum vulgare* L., cvs. Proctor and Weeah) and wheat (*Triticum aestivum* L., cv. Pak-70) grown in acid washed fine river sand supplemented with 0, 50, 100 or 150 m*M* NaCl. Ammonium sulphate (120 kg N/ha), single super phosphate (60 kg P_2O_5/ha) and potassium sulphate (40 kg/ha) were also added to the sand prior to sowing. The pots were placed in a pot house in a completely randomized design with eight replicates. Pregerminated seeds (48 h after imbibition) were sown and later thinned to eight plants per pot. Irrigation with 1L of full-strength Hoagland solution (Hoagland and Arnon, 1950) was applied every month. Other irrigations, whenever needed, were made with distilled water.

Two plants were harvested from each pot after 30, 60 and 90 days from sowing for recording shoot dry weight while the remaining two plants were left for grain production. The straw from the harvest at 60 days was analysed for P, K, Ca and Na (Jackson, 1962).

Results

Growth

Shoot weights at various harvests and grain weights at maturity, calculated as percent of the respective controls, show (Fig. 1) that, irrespective of species and cultivars, growth and yield of plants progressively decreased with increasing salinity. Barley, cv. Proctor, seemed comparatively more sensitive than its counterparts from the very beginning, but the adverse effect of salt on its growth became more pronounced with time.

At the first harvest at 30 days (Fig. 1A) there was no significant difference between the growth of Weeah barley and Pak-70 wheat under salinity, but Proctor barley had a significantly lower dry weight than the other two varieties at each salt concentration used. At 60 days (Fig. 1B), growth among the barley varieties was not significantly different. Wheat, Pak-70, had a significantly higher dry weight at the highest salinity used.

At the later growth stages (90 days and at maturity Fig. 1C, 1D), Weeah barley and Pak-70 wheat were both better than Proctor barley at 50 m*M* NaCl, but at higher salt levels Pak-70 exhibited significantly better performance than both barley cultivars. At these high salt levels (100, 150 m*M* NaCl), Weeah was more salt tolerant than Proctor. Plants of Proctor did not survive above 100 m*M* NaCl beyond 90 days and produced grain only under 50 m*M* NaCl, while Weeah produced grain at 100 m*M* NaCl also. Pak-70 plants survived to the end of the experiment with grain production at all salt levels used (Fig. 1D).

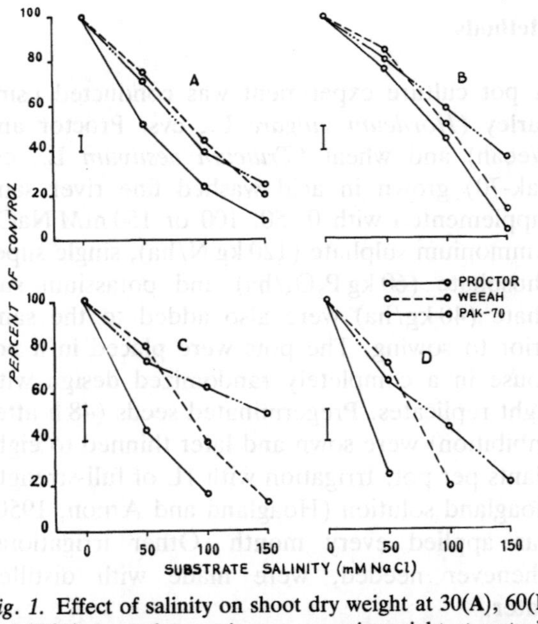

Fig. 1. Effect of salinity on shoot dry weight at 30(A), 60(B) and 90(C) days after sowing or on grain weight at maturity (D), calculated as percent of control. Vertical bars indicate the L.S.D. for significance of effects of salinity levels and cultivars.

Ionic relations

Accumulation

Analysis of straw at 60 days after sowing showed (Fig. 2) that wheat, cv. Pak-70, generally accumulated much less Na than the barley cultivars under control as well as saline conditions. K in wheat was also lower under control conditions and stayed so with increasing NaCl. Phosphorus

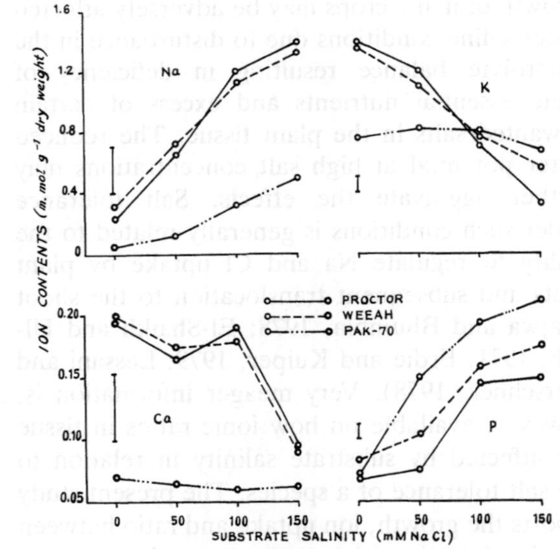

Fig. 2. Ionic concentrations (mmol g^{-1} dry weight) in shoots at 60 days after sowing. Vertical bars indicate the L.S.D. for significance of effects of salinity levels and cultivars.

accumulated more in both species under saline conditions. Differences in ionic concentrations between barley cultivars at each salt level were not significant except for K at 150 m*M* and P at all salt levels, with cv. Weeah always accumulating more than cv. Proctor.

Increase in salinity caused Na and P to increase in both barley and wheat, with concomitant decreases in Ca and K but in only barley cultivars. In wheat, Ca and K remained almost unchanged. These trends of increases or decreases in ion accumulation were of a greater magnitude in barley than in wheat, except for P which under saline conditions increased more in the latter than in the former species.

Ratios

Low ratios between the various analysed ions (Fig. 3) were found to be associated with high salt tolerance in both species and their cultivars. An increase in these ratios was observed with increasing salinity, except for Ca/P which decreased with salinity in both species, and Ca/K which increased in barley but remained almost unchanged in wheat under saline conditions. All these ratios were always lower in wheat than in barley and, among barley cultivars, lower in Weeah than in Proctor. Differences at each salt level between the two barley cultivars were, however, not very pronounced in case of Na/Ca and were evident only at 150 m*M* external salt level for Na/K and Ca/K. The barley cultivars could, however, be clearly separated according to their tolerance to salts on the basis of Na/P and Ca/P for which clear-cut differences at each salt level were observed.

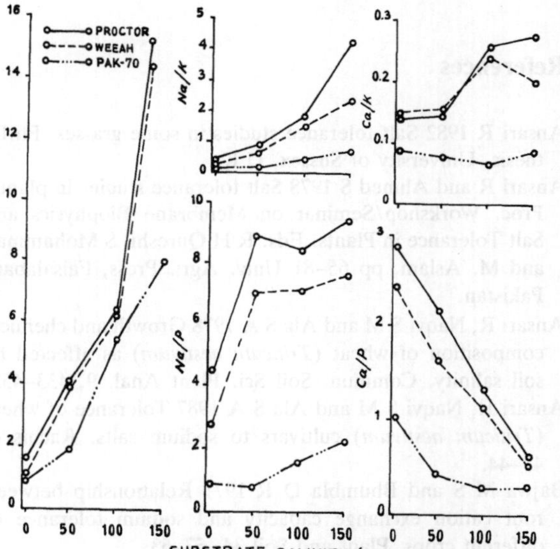

Fig. 3. Ratios of various ions in shoots at 60 days after sowing.

Discussion

Crop plants are known to show a wide range of resistance to salinity, with considerable differences among cultivars. The sensitive varieties of a tolerant species may, hence, be affected more by salinity than a tolerant variety of a comparatively sensitive species. This was probably the case in the present study where Pak-70 wheat was found to be more salt tolerant than the barley cultivars, although among field crops the latter is considered more salt tolerant than the former (Ansari, 1982; Maas and Hoffman, 1977). Resistance in crops is ascribed mainly to restricted translocation of both Na and Cl to the shoot (Greenway and Rogers, 1963). Preferential accumulation of either Na (Rush and Epstein, 1976) or both Na and Cl (Tal, 1971) was also reported to account for salt tolerance in some tomato species.

In the present study resistance was accompanied by restricted uptake of Na and Ca and preferential accumulation of K and P (Fig. 2). Tolerance to salinity is, however, not simply a matter of ion avoidance or accumulation, but is also a function of ion regulation allowing osmotic adjustment to avoid imbalance in the tissue which could cause further disturbances in plant metabolism (Flowers *et al.*, 1977, Greenway and Munns, 1980). Potassium, which is an essential cytoplasmic element (Flowers and Läuchli, 1983), because of its involvement in osmotic regulation and its competitive effect against Na (Rains, 1976), is frequently considered important under saline conditions. Calcium, which like K is also an essential mineral nutrient, helps in maintaining membrane integrity (Jones and Lunt, 1967), is important in senescence processes (Pooviah and Leopold, 1973) and is also known to counter the harmful effects of Na (Lahaye and Epstein, 1971). Reports on a correlation between Ca requirement and Na tolerance (Bajwa and Bhumbla, 1971), on disturbance of Ca nutrition in corn (Maas and Grieve, 1987) and sorghum (Grieve and Maas, 1988) at high external Na, as well as on reduced Ca and K concentrations in plant tissue in the case of salinity are also present (Ansari *et al.*, 1978; Erdie and Kuiper, 1979).

The high K and Ca levels in the salt sensitive barley cultivars in the control treatment of the present study were signficantly reduced due to salts whereas the lower K and Ca levels in tolerant wheat remained almost unchanged under similar conditions (Fig. 2). The decrease in K and Ca in barley cultivars may, therefore, have caused disturbance in many metabolic processes dependent upon these ions, and may have shown their effect on plant growth. Wheat was probably not affected in this manner and, hence, fared better.

Phosphorus, which has a central role in energy metabolism of cells and is involved in a number of anabolic and catabolic pathways (Bieleski, 1973), accumulated under salinity conditions (Fig. 2), as was also reported earlier (Ansari and Ahmed, 1978). The high P content in plant tissue may stimulate growth by meeting the additional energy requirement of plants imposed by salinity stress (Poljakoff-Mayber and Gale, 1975). In the present study the increase in tissue-P under saline conditions was greater in wheat than in barley cultivars (Fig. 2). In barley, P concentration increased under salinity conditions but the increase was probably not high enough to offset the deleterious effects of salts to an extent as observed in wheat.

The importance of balanced nutrient solution for proper plant growth needs no emphasis. It is evident, however, from the present study that for a plant growing under saline conditions not only a balanced external supply of nutrients but also a balance among ions in its tissue is important for a proper performance under stress. The increase in Na and P and decrease in K and Ca in the straw under salinity conditions in the present study (Fig. 2) were directly related with the reduction in the straw and grain weights (Fig. 1). The capacity to maintain a balance between not only Na and K and Ca, as stressed in earlier studies (Epstein, 1962; Flowers *et al.*, 1977; Greenway and Munns, 1980; Pitman, 1965), but also regarding P, as well as a balance between other ions (*e.g.* Ca: P, Ca: K, Fig. 3) might be essential for a proper performance of these cultivars under salinity conditions.

In the present study, chemical plant analysis was carried out 60 days after sowing when the plants were growing rapidly. Although at that time significant differences among cultivars in dry weight were present only at the highest salinity level (Fig. 1B) the chemical compositions were found to be indicative of later straw (Fig. 1C) and grain yields (Fig. 1D). It was thus evident that ionic composition at 60 days, irrespective of how salinity affected plant growth, did set the stage for future performance of the plants up to grain production. This corroborates the findings of another study (Ansari *et al.*, 1987) in which two wheat cultivars were exposed in sand culture to substrate salinity composed of various sodium salts.

Acknowledgements

Financial support from the IAEA, Vienna and the British ODA, London is gratefully acknowledged. I am also thankful to Dr T J Flowers, School of Biological Sciences, University of Sussex, UK for helpful comments on the manuscript.

References

Ansari R 1982 Salt tolerance studies in some grasses. Ph.D. thesis, University of Sussex, U.K.

Ansari R and Ahmed S 1978 Salt tolerance studies in plants. Proc. Workshop/Seminar on Membrane Biophysics and Salt Tolerance in Plants. Eds. R H Qureshi, S Mohammad and M. Aslam, pp 65–81 Univ. Agri. Press, Faisalabad, Pakistan.

Ansari R, Naqvi S M and Ala S A 1978 Growth and chemical composition of wheat (*Triticum aestivum*) as affected by soil salinity. Commun. Soil Sci. Plant Anal. 9, 433–453.

Ansari R, Naqvi S M and Ala S A 1987 Tolerance of wheat (*Triticum aestivum*) cultivars to sodium salts. Rachis 6, 41–44.

Bajwa M S and Bhumbla D R 1971 Relationship between root cation exchange capacity and sodium tolerance of different crops. Plant and Soil 34, 57–63.

Bajwa M S and Bhumbla D R 1974 Growth and cation absorption by plants under different levels of sodium. Ind. J. Agric. Sci. 44, 598–601.

Bieleski R L 1973 Phosphate pools, phosphate transport and phosphate availability. Annu. Rev. Plant Physiol. 24, 252–255.

El-Shaikh A M and Ulrich A 1971 Sodium absorption by intact sugar beet plants. Plant Physiol. 48, 747–751.

Epstein E 1962 Mutual effects of ions in their absorption by plants. Agrochimica 6, 293–322.

Erdie L and Kuiper P J C 1979 The effect of salinity on growth, cation contents, Na uptake and translocation in salt sensitive and salt tolerant *Plantago* species. Physiol. Plant. 47, 95–99.

Flowers T J and Läuchli A 1983 Sodium versus potassium: Substitution and compartmentation. *In* Encyclopedia of Plant Physiology, New Series, Vol. 15B. Eds. A. Läuchli and R.L. Bieleski. pp 651–680. Springer-Verlag, Berlin.

Flowers T J, Troke P F and Yeo A R 1977 The mechanism of salt tolerance in halophytes Annu. Rev. Plant Physiol 28, 89–121.

Greenway H and Rogers A 1963 Growth and ion uptake of *Agropyron elongatum* on saline substrates as compared with salt tolerant variety of *Hordeum vulgare*. Plant and Soil 18, 21–30.

Greenway H and Munns R 1980 Mechanism of salt tolerance in nonhalophytes. Annu. Rev. Plant Physiol. 31, 149–190.

Grieve C M and Maas E V 1988 Differential effect of sodium/calcium ratio on sorghum genotypes. Crop Sci. 28, 659–665.

Hoagland D R and Arnon D I 1950 The water culture method for growing plants without soil. Calif. Agri. Exptl. Stn. Circular No. 347.

Jackson M L 1962 Soil Chemical Analysis. Prentice Hall, NJ.

Jones R G W and Lunt O R 1967 The function of calcium in plants. Bot. Rev. 33, 407–426.

Lahaye P A and Epstein E 1971 Calcium and salt tolerance by bean plants. Plant Physiol. 25, 213–218.

Lessani H and Marschner H 1978 Relation between salt tolerance and long distance transport of sodium and chloride in various crop species. Aust. J. Plant Physiol. 5, 27–37.

Maas E V and Grieve C M 1987 Sodium induced calcium deficiency in salt stressed corn. Plant Cell Env. 10, 559–564.

Maas E V and Hoffman G J 1977 Crop salt tolerance: Evaluation of existing data. *In* Managing Saline Water for Irrigation. Ed. H. E. Dregne. pp 187–198. Texas Technical University, Lubbock, TX.

Pitman M G 1965 Transpiration and the selective uptake of potassium by barley seedlings (*Hordeum vulgare* cv. Bolivia). Aust. J. Biol. Sci. 18, 987–998.

Poljakoff- Mayber A and Gale J 1975 Plants in Saline Environments. Springer-Verlag, Berlin.

Pooviah B W and Leopold A C 1973 Deferral of leaf senescence with calcium. Plant Physiol. 52, 236–239.

Rains D W 1976 Mineral metabolism. *In* Plant Biochemistry. Eds. J. Varner and J. Bonner. pp. 561–597. Academic Press, New York.

Rush D W and Epstein E 1976 Genotypic responses to salinity: Differences between salt sensitive and salt tolerant genotypes of tomato. Plant Physiol. 57, 162–166.

Tal M 1971 Salt tolerance in the wild relative of tomato: Responses of *Lycopersicon esculentum*, *L. peruvianum* and *L. esculentum minor* to sodium chloride salinity Aust. J. Agr. Res. 22, 631–638.

M. L. van Beusichem (Ed.), *Plant nutrition – physiology and applications*, 469–472.

PLSO IPNC365A

Effect of salinity on calcium transport in tomato (*Lycopersicon esculentum*)

P. ADAMS and L.C. HO
AFRC Institute of Horticultural Research, Littlehampton, W. Sussex, BN17 6LP, UK

Key words: calcium, fruit, *Lycopersicon esculentum* L., salinity, tomato

Abstract

Increasing salinity in the growing medium decreased the Ca content (%) of tomato fruit. Salinization of NFT solutions with either NaCl or major nutrients (K, Ca and NO_3-N) had a similar effect on Ca accumulation, suggesting a common osmotic effect on Ca transport. Fruit grown at high salinity (8 mS cm^{-1}) during the day (to reduce water accumulation) and low salinity (3 mS cm^{-1}) at night (to encourage Ca accumulation) had lower Ca contents than those grown at a constant 5.5 mS cm^{-1}. This difference increased with fruit age and plant age. Results obtained with ^{45}Ca uptake studies support the notion that any short-term stimulation of Ca uptake due to a reduction in salinity at night is nullified by the long-term effect of salinity on the development of xylem vessels within the fruit.

Introduction

Increasing salinity has little effect on the accumulation of dry matter but reduces that of both water and calcium (Ehret and Ho, 1986a,b). Hence the improvement in fruit quality due to the increased % dry matter is accompanied by a reduction in % Ca. Most of the water enters the fruit during the day via the phloem whilst a higher proportion of the Ca is acquired at night via the xylem (Ho, 1989). Thus, it was suggested that use of a high salinity during the day should reduce the water content of the fruit and changing to low salinity at night should encourage Ca accumulation, partly compensating for the high day salinity.

The purpose of this study was to assess the relative importance of the short-term and long-term effects of salinity (*i.e.*, on the transport process and xylem structure, respectively) that determine the calcium status of tomato fruit.

Methods

In the glasshouse, 15 plants per plot were grown in separate NFT systems (7 m channels). Constant salinities of 3, 5.5 and 8 mS cm^{-1} were compared with a fluctuating salinity of 8 mS cm^{-1} during the day and 3 mS cm^{-1} at night, referred to as the 8/3 mS treatment. The latter was achieved by draining the system for 1 h before sunrise and for 1 h after sunset before switching to the other salinity. Salinities above 3 mS cm^{-1} were obtained by addition of either major nutrients (K, Ca and NO_3-N) or NaCl (Adams and El-Gizawy, 1986). Three replicates of the treatments were arranged in a randomised block design. The Ca content was measured by atomic absorption spectrophotometry.

Uptake of ^{45}Ca was studied with plants in a growth room using deep pots of flowing nutrient solution (5.5 mS cm^{-1}). The salinity was increased with NaCl. Plants were supplied with 0.8 MBq ^{45}Ca in nutrient solutions of 3, 5.5 and 8 mS cm^{-1} at the start of either the light or dark period; some plants received a different salinity in the dark from that in the preceding light period. All plants were harvested after 12 h feeding with ^{45}Ca. Detached fruit grown at different salinities (3, 5.5, 8/3 and 8 mS cm^{-1}) were arranged with the pedicels dipping into a solution of ^{45}Ca (0.2 MBq at 2 mS cm^{-1}). The distribution of ^{45}Ca in the fruit was assessed after 24 h. ^{45}Ca in the ashed samples was measured by liquid scintillation counting (Ehret and Ho, 1986b).

Results

Comparison of salinity sources

The average rate of dry matter accumulation of proximal fruit (second from main stem on trusses 2, 5 and 8) decreased with fruit age from 0.18 g d^{-1} at 15 d after anthesis to 0.04 g d^{-1} after 55 d; this was not affected by increasing the salinity up to 8 mS with either enhanced major nutrients or added NaCl. In contrast, the rate of Ca accumulation decreased with increasing salinity, the average rates being 0.121, 0.100 and 0.078 mg d^{-1} respectively for fruit grown in the 3, 8/3 and 8 mS treatments. There was, however, little effect of fruit age on these rates. Fruit grown with the salinity fluctuating diurnally accumulated amounts of Ca intermediate between those grown at 3 and at 8 mS cm^{-1}. Similar responses were obtained with both osmotica. The

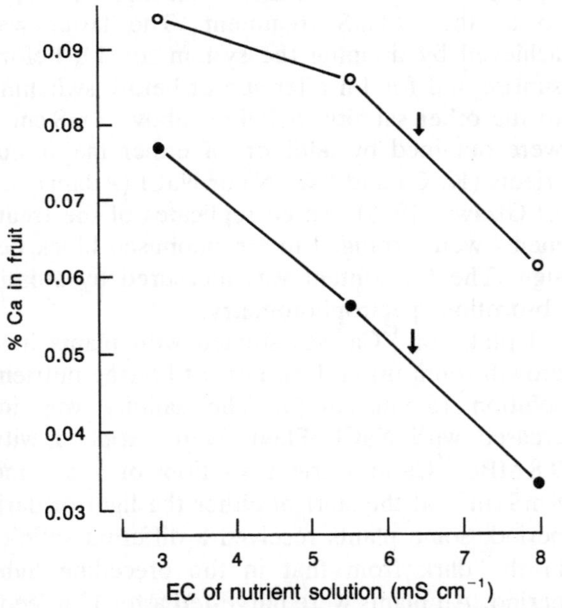

Fig. 1. Effect of salinity treatments on the Ca content of the distal tissues (0) and the rest (●) of tomato fruit 44 d after anthesis. The values for the 8/3 mS treatment are indicated by the arrows. Data are averaged for trusses 2, 5, 8 and 11.

reduction in Ca content of the distal tissue was greater than in the rest of the fruit (Fig. 1).

Comparison of fluctuating and constant salinities

The average daily rate of accumulation of Ca by the fruit was 10% higher at 5.5 mS cm^{-1} than in

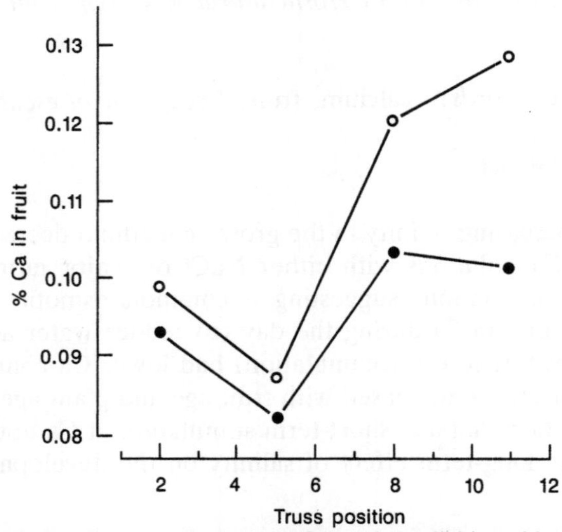

Fig. 2. Effect of constant and fluctuating salinities with the same average value on the Ca content (% dry matter) of whole ripe tomato fruit. 0, constant 5.5 mS cm^{-1}; ●, 8 mS cm^{-1} during the day and 3 mS cm^{-1} at night.

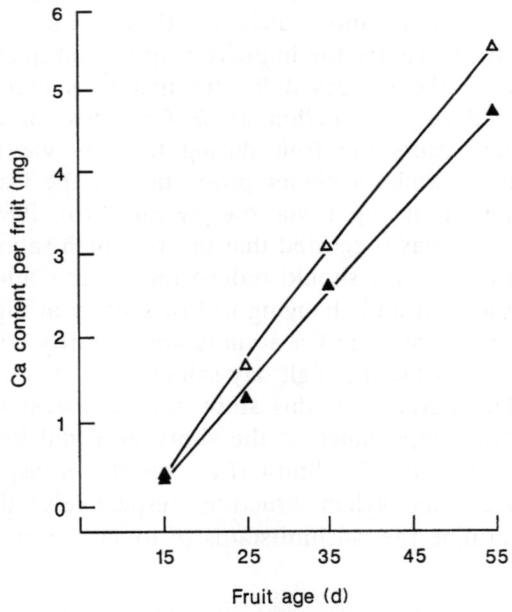

Fig. 3. Ca accumulation (mg) by whole tomato fruit grown at constant 5.5 mS cm^{-1}, △, and at 8 mS cm^{-1} during the day and 3 mS cm^{-1} at night, ▲. The data are averages of trusses 2, 5, 8 and 11.

the 8/3 mS treatment on truss 2 and 37% higher on truss 11 (Fig. 2). Thus, the total Ca content of fruit grown at constant 5.5 mS cm^{-1} was always higher than in those grown with the fluctuating salinity (8/3 mS), although the mean salinity value in the solution was also 5.5 mS cm^{-1}. The average Ca content of ripe fruit from trusses 2, 5, 8 and 11 was 17% higher in fruit grown at 5.5 mS cm^{-1} than at 8/3 mS (Fig. 3) whereas the corresponding difference in the dry weight of the fruit was less than 8% (Fig. 4).

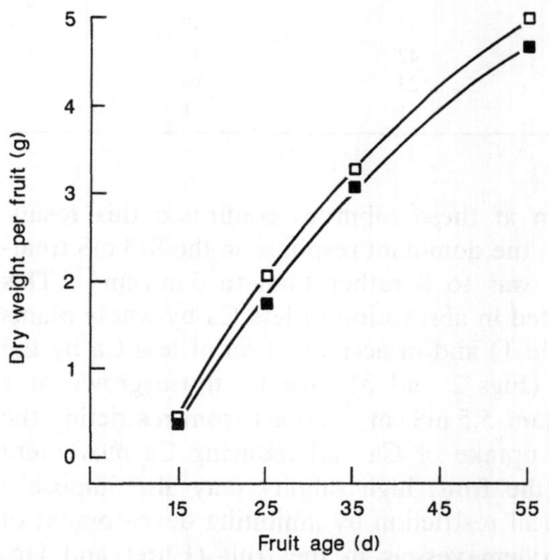

Fig. 4. Dry matter accumulation by tomato fruit grown at constant 5.5 mS cm^{-1}, □, and at 8 mS cm^{-1} during the day and 3 mS cm^{-1} at night, ■. The data are averages of trusses 2, 5, 8 and 11.

Short-term effects of salinity

The accumulation of ^{45}Ca by whole tomato plants transferred to high salinity was reduced by 42% at 8 mS cm^{-1} during the light period as compared with those grown at 3 mS cm^{-1}; the reduction in the fruit was 56% (Table 1). The corresponding reductions in the dark period were 44% and 80%. Plants transferred from 8 mS in the day to 3 mS at night absorbed similar amounts of ^{45}Ca to those at constant 3 mS cm^{-1} during the dark period. Plants treated with the fluctuating salinity (8/3 mS) accumulated less ^{45}Ca in 24 h than those at a constant 5.5 mS cm^{-1} and were similar to those at 8 mS cm^{-1}; the amount of ^{45}Ca that moved into the fruit was

Table 1. Accumulation of ^{45}Ca by whole plants and transport of ^{45}Ca to fruiting trusses of tomatoes grown at different salinities during a light (12 h) and dark (12 h) period. ^{45}Ca was supplied at either the beginning of the light period or of the dark period

Period	Plant part	Salinity (mS cm^{-1})			
		3	5.5	8	
Light	Whole plant (10^5 dpm/g dry weight)	3.1	3.0 (−3)[a]	1.8 (−42)	
	Fruit (10^4 dpm/g dry weight)	1.6	0.9 (−44)	0.7 (−56)	
		3	5.5	3[b]	8
Dark	Whole plant (10^5 dpm/g dry weight)	1.6	0.9 (−44)	1.2 (−25)	0.9 (−44)
	Fruit (10^4 dpm/g dry weight)	4.6	3.5 (−24)	1.1 (−76)	0.9 (−80)
		3/3	5.5/5.5	8/3	8/8
Whole day (24 h)	Whole plant (10^5 dpm/g dry weight)	4.6	4.0 (−13)	3.0 (−35)	2.7 (−41)
	Fruit (10^4 dpm/g dry weight)	6.2	4.4 (−29)	1.8 (−70)	1.6 (−74)

[a] % Difference from 3 mS cm^{-1} given in parenthesis.
[b] Grown at 8 mS cm^{-1} during the preceding light period.

Table 2. Accumulation (10^6 dpm) and distribution (%) of ^{45}Ca in detached tomato fruit grown at different salinities (truss 23)

	Salinity (mS cm^{-1})			
	3	5.5	8/3	8
Accumulation per g dry weight	0.50	0.35	0.41	0.31
% ^{45}Ca in:				
Pedicel	31	32	34	37
Calyx	29	35	42	44
Proximal end of berry	33	31	23	19
Distal end of berry	7	2	1	<1

intermediate between that at 5.5 and 8 mS cm^{-1} (Table 1).

Long-term effects of salinity

Detached fruit from truss 23 of plants grown at 8/3 mS accumulated similar amounts of ^{45}Ca to those grown at 5.5 mS cm^{-1}, which were intermediate between those grown at 3 and 8 mS cm^{-1}. However, the proportion of the ^{45}Ca absorbed by fruit that moved into the berry was reduced more by the fluctuating salinity than by the constant 5.5 mS cm^{-1} treatment, particularly in the distal tissues (Table 2). In this respect, the response of Ca movement to the fluctuating salinity was intermediate between those at 5.5 and 8 mS cm^{-1}.

Discussion

High salinity (8 mS cm^{-1}) always reduced Ca import into the fruit, particularly to the distal end, irrespective of the period of treatment or the osmoticum used. This response was most marked at night, when more of the newly absorbed Ca moved into the fruit than during the day. On the bottom trusses, *i.e.*, on young plants, diurnal fluctuations in salinity between 8 and 3 mS cm^{-1} resulted in Ca contents of the fruit similar to those at 5.5 mS cm^{-1}. The response on the higher trusses, however, was intermediate between that of fruit grown at constant salinities of 5.5 and 8 mS cm^{-1}. Studies of ^{45}Ca absorption by detached fruit from older plants grown at these salinities confirmed this result. Thus, the dominant response in the 8/3 mS treatment was to 8 rather than to 3 mS cm^{-1}. This resulted in absorption of less Ca by whole plants (Table 1) and in accumulation of less Ca by the fruit (Figs 2 and 3) than by those grown at a constant 5.5 mS cm^{-1}. Apart from restricting the total uptake of Ca and reducing Ca movement into the fruit, high salinity may also impose a physical restriction by inhibiting development of the xylem vessels in the fruit (Ehret and Ho, 1986b). Therefore, even though a reduction in salinity at night would favour an increase in Ca uptake, the total daily uptake of Ca at 8/3 mS would still be less than at 5.5 mS cm^{-1}. The additional restriction to xylem transport inside the fruit therefore precludes the possibility of any increase in Ca import by the fruit. Hence, any short-term advantage of low salinity at night after a high salinity during the day is nullified by the long-term effect of salinity on xylem development.

References

Adams P and El-Gizawy A M 1986 Effect of salinity and watering level on the calcium content of tomato fruit. Acta Hortic. 190, 253–259.

Ehret D L and Ho L C 1986a The effects of salinity on dry matter partitioning and fruit growth in tomatoes grown in nutrient film culture. J. Hortic. Sci. 61, 361–367.

Ehret D L and Ho L C 1986b Translocation of calcium in relation to tomato fruit growth. Ann. Bot. 58, 679–688.

Ho L C 1989 Environmental effects on the diurnal accumulation of ^{45}Ca by young fruit and leaves of tomato plants. Ann. Bot. 63, 281–288.

M. L. van Beusichem (Ed.), *Plant nutrition – physiology and applications*, 473–476.

PLSO IPNC365B

Effect of salinity on the distribution of calcium in tomato (*Lycopersicon esculentum*) fruit and leaves

P. ADAMS
AFRC Institute of Horticultural Research, Littlehampton, W. Sussex, BN17 6LP, UK

Key words: calcium, electrical conductivity, fruit, *Lycopersicon esculentum* L., salinity, sodium, tomato

Abstract

Ca accumulation in tomato plants was studied at five constant salinities (3–9 mS cm^{-1}), achieved by addition of NaCl to the NFT solutions. Increasing salinity reduced total Ca accumulation by the fruit but had little effect on their dry weight, resulting in a progressive decline in the Ca content (% in the dry matter), particularly at the distal end. In contrast, salinity had little effect on total Ca accumulation by the leaves but reduced their dry weight, so increasing the % Ca. This response was similar for both the laminae and petioles, but the % Ca was always higher in the laminae.

The Na content of the distal tissue was slightly lower than in the rest of the fruit, but the Na content increased with salinity at a similar rate in both tissues. As salinity reduced the Ca content of the distal tissue more than that of the rest of the fruit, the Na/Ca ratio increased more rapidly in the distal tissue. In the leaves, Na accumulated mainly in the petioles. Thus, the increase in Na/Ca ratio with Na level was greater in the petioles than in the laminae.

Introduction

Previous studies with tomatoes grown at constant salinities in NFT showed that increasing the salinity above 4 mS cm^{-1} by addition of major nutrients improved the quality of the fruit (Ehret and Ho, 1986; Massey *et al.*, 1984) but reduced their Ca content (Adams and El-Gizawy, 1986). Similar results were obtained using NaCl to increase the salinity (Adams, 1989).

Ca distribution is not even in either the fruit or the leaves. For example, Bradfield and Guttridge (1984) found that the Ca content of the proximal part of the fruit was twice that at the distal end. In the leaves, Ward (1964) found that the laminae usually have higher Ca contents than the petioles. These differences may be accentuated by increased salinity.

This paper presents the effects of salinity on Ca accumulation in tomato fruit and leaves, and shows that they differ because dry matter accumulation by the fruit remains virtually constant whilst that of the leaves decreases as the salinity rises.

Methods

Seedlings of cv. Marathon, sown in February, were grown in rockwool cubes and transferred to NFT channels in early April when the buds of the first truss were just visible. Each plot consisted of a separate NFT system containing 150 L of a complete nutrient solution and 15 plants. Four levels of Na (21.8, 37.0, 52.2 and 67.4 m*M*), achieved by adding NaCl to the basic nutrient solution, were compared with a control maintained at 6.5 m*M* Na; the corresponding electrical conductivities (EC) of these treatments were 4.6, 6.7, 7.6, 9.1 and 3.3 mS cm^{-1}. Two replicates of the treatments were provided in randomised blocks. The volume, pH and EC of the recirculating solutions were adjusted daily, the N, P and K contents twice a week, and the Na, Ca and Mg contents once a week. The glasshouse atmosphere was enriched with CO_2 to 1000 vpm until late April.

Samples of the fifth leaf below the head of the plants, *i.e.*, leaves not fully expanded, were divided into laminae and petioles, dried at 80°C and ground to <2 mm. Fruit samples were collected 44 d after anthesis from truss 6, and divided into two portions: (a) a disc 3.0-3.5 cm in diameter cut from the distal end of the fruit and (b) the remainder of the fruit. These were dried at 80°C in a forced draught oven for 48 h and ground to <2 mm. Ca was determined by atomic absorption spectrophotometry and Na by flame photometry.

Results

The Ca content (% of dry matter) of the distal part of the fruit was always lower than in the remainder of the fruit, and was depressed more substantially than in the rest of the fruit as the salinity increased (Fig. 1). Over the EC range tested, the Ca content of the distal tissue decreased by 60% as compared with 20% in the rest of the fruit. The corresponding rates of decrease were 0.0074% and 0.0040% Ca per mS cm^{-1} increase.

In contrast, the Ca content of the leaves increased with the salinity (Fig. 2). The laminae had higher Ca contents than the petioles, but both had similar rates of increases with salinity (0.095% and 0.088% Ca per mS cm^{-1} respectively).

The Na content of the fruit increased with the Na concentration in the nutrient solution, but was always lowest in the distal tissue (68-78% of that in the rest of the fruit). The ratio of Na/Ca in the fruit increased more in the distal part of the fruit (Fig. 3) due to the more marked decline in Ca content. In the leaves, Na accumulated in the petioles resulting in Na concentrations that were more than three times those in the laminae. Thus, the Na/Ca ratio in the petioles increased rapidly with Na level whilst that in the laminae increased slowly (Fig. 4).

Dry matter accumulation by the fruit was not reduced progressively by the increasing salinity as was the dry weight of the leaves (Table 1). In contrast, the total accumulation of Ca by whole fruit decreased progressively with increasing salinity whereas that of the leaves was not affected.

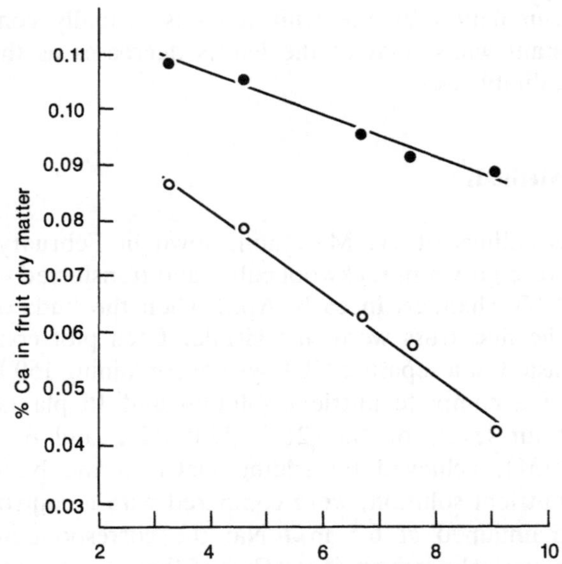

Fig. 1. Effect of salinity on the Ca content of the distal portion (○) and the remainder (●) of mature green tomato fruit (44 d after anthesis).

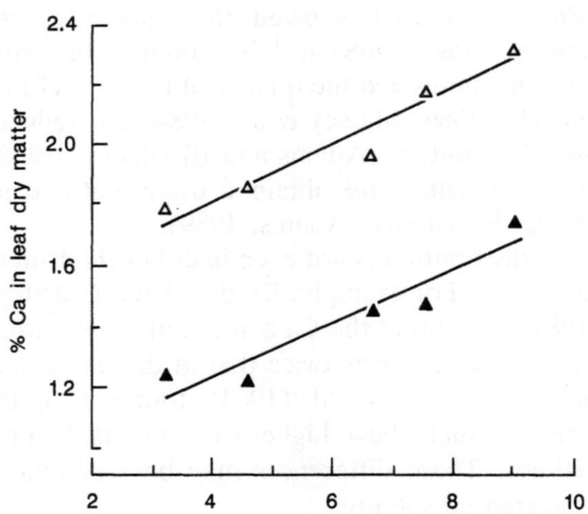

Fig. 2. Effect of salinity on the Ca content of the laminae (△) and petioles (▲) of tomato leaves (fifth below the head).

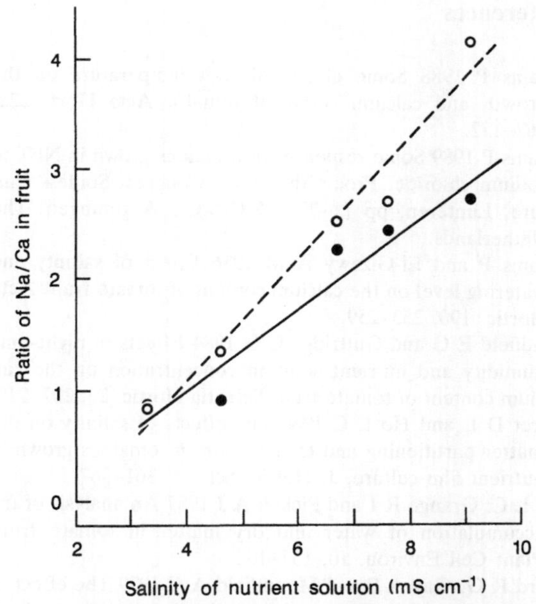

Fig. 3. Effect of salinity on the ratio of Na/Ca in the distal portion (○) and the remainder (●) of mature green tomato fruit (44 d after anthesis).

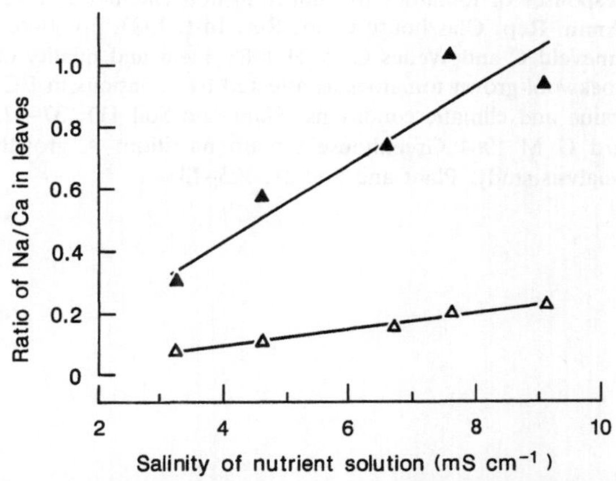

Fig. 4. Effect of salinity on the ratio of Na/Ca in the laminae (△) and petioles (▲) of tomato leaves (fifth below the head).

Discussion

Increasing the salinity progressively reduced Ca accumulation by the fruit but had little effect on dry weight, hence resulting in a decrease in % Ca. In contrast, there was no effect of salinity on total Ca accumulation by the leaves. As the dry weight decreased with increasing salinity, however, the % Ca increased. The difference in response to salinity by the fruit and leaves supports the hypothesis that the sink strength of the fruit trusses is an important factor in maintaining the partitioning of assimilates in favour of the fruit and at the expense of vegetative growth (Hurd *et al.*, 1979).

The movement of Ca in tomato plants appears to be confined almost exclusively to the xylem (Ho *et al.*, 1987). The data presented here suggest that as water absorption by the roots is reduced by increasing salinity, preferential partitioning of the water, and hence Ca, to the leaves progressively reduces the Ca available to the fruit while maintaining the supply to the leaves.

The distribution of Ca in the leaves was similar to that found by Ward (1964). Under poor light conditions, however, there was no difference between the Ca contents of the laminae and petioles (Adams, 1988). Thus Ca distribution within the leaves depends on the rate of transpiration, high rates favouring Ca accumulation in the laminae. Unpublished data (Adams and Holder) showed that increasing the vapour pressure deficit from 0.2 kPa to 0.4 and 0.8 kPa increased the Ca content of the terminal leaflet of young expanded tomato leaves by 35 and 46% respectively. Sonneveld and Welles (1988) reported an increase in the Ca content of young leaves of about 100% when the prevailing average vapour

Table 1. Effect of salinity on the accumulation of (a) dry matter and (b) Ca by whole tomato fruit (44 d after anthesis) and complete leaves (fifth below the top). Relative values (%) are given in parenthesis for comparison with the control (3.3 mS cm^{-1})

Plant part	EC (mS cm^{-1})				
	3.3	4.6	6.7	7.6	9.1
Dry matter (g)					
Fruit	5.11(100)	5.55(109)	5.94(116)	5.74(112)	4.83(95)
Leaves	1.70(100)	1.56(92)	1.52(89)	1.38(81)	1.31(77)
Ca content (mg)					
Fruit	6.0(100)	5.8(97)	5.6(93)	4.8(82)	4.1(68)
Leaves	27.0(100)	25.8(96)	26.8(99)	26.2(97)	26.9(100)

pressure deficit was increased from 0.35 to 0.52 kPa.

Sonneveld and Welles (1988) found little effect of salinity in the range 2.6-5.9 mS cm^{-1} on the Ca content of the fruit whereas Adams and El-Gizawy (1986) found that the Ca content declined progressively as the EC increased from 4 to 10 mS cm^{-1}. Since the effect of increasing the salinity with major nutrients (Adams and El-Gizawy, 1986) was similar to that reported here using NaCl, the response appears to be a general one, *i.e.*, to decreasing osmotic potential rather than to specific cations.

Blossom-end rot was not induced at 9.1 mS cm^{-1}, probably because the EC was maintained at a constant value in the recirculating nutrient solution. Young fruit from this treatment, sampled 22 d after anthesis, contained 0.034% Ca in the distal tissues, but did not develop the disorder. Hence, very low Ca contents need not result in the development of blossom-end rot when the salinity of the substrate, *i.e.*, the water stress on the plant, does not fluctuate.

The data presented show that assessments of the effects of salinity on Ca status that are based entirely on % Ca can be misleading since the accumulation of dry matter and Ca may be affected differently.

References

Adams P 1988 Some effects of root temperature on the growth and calcium status of tomato. Acta Hort. 222, 167–172.

Adams P 1989 Some responses of tomatoes grown in NFT to sodium chloride. Proc. 7th Intern. Congress Soilless Culture, Lunteren, pp 71–79. I.S.O.S.C., Wageningen, The Netherlands.

Adams P and El-Gizawy A M 1986 Effect of salinity and watering level on the calcium content of tomato fruit. Acta Hortic. 190, 253–259.

Bradfield E G and Guttridge C G 1984 Effects of night-time humidity and nutrient solution concentration on the calcium content of tomato fruit. Scientia Hortic. 22, 207–217.

Ehret D L and Ho L C 1986 The effects of salinity on dry matter partitioning and fruit growth in tomatoes grown in nutrient film culture. J. Hortic. Sci. 61, 361–367.

Ho L C, Grange R I and Picken A J 1987 An analysis of the accumulation of water and dry matter in tomato fruit. Plant Cell Environ. 10, 157–162.

Hurd R G, Gay A P and Mountfield A C 1979 The effect of partial flower removal on the relation between root, shoot and fruit growth in indeterminate tomato. Ann. Appl. Biol. 93, 77–89.

Massey D M, Hayward A C and Winsor G W 1984 Some responses of tomatoes to salinity in nutrient film culture. Annu. Rep. Glasshouse Crops Res. Inst. 1983, pp 60–62.

Sonneveld C and Welles G W H 1988 Yield and quality of rockwool-grown tomatoes as affected by variations in EC-value and climatic conditions. Plant and Soil 111, 37–42.

Ward G M 1964 Greenhouse tomato nutrition: A growth analysis study. Plant and Soil 21, 125–133.

M. L. van Beusichem (Ed.), *Plant nutrition – physiology and applications*, 477–480.

PLSO IPNC613

Sodium, chloride and potassium allocation in an annual sweetclover (*Melilotus segetalis*) from the Guadalquivir salt marsh in southwest Spain

T. MARAÑÓN, J.M. ROMERO and J.M. MURILLO
Instituto de Recursos Naturales y Agrobiología, CSIC, Apartado de Correros 1052, E-41080 Sevilla, Spain

Key words: ionic content, *Melilotus segetalis* (Brot.) Ser., salinity, sweetclover

Abstract

Annual sweetclover (*Melilotus segetalis* (Brot.) Ser. plants from the Guadalquivir salt marsh (S.W. Spain) were grown in a glasshouse, on sandy soil with nutrient solution and four salt treatments (0, 10, 50 and 100 m*M* NaCl). Plants (five replicates per treatment and age) were harvested at 50, 90 and 120 days after germination. Sodium, chloride and potassium were determined for each plant in leaves, stems and roots.

Total dry weight reduction with salinity was associated with increasing levels of sodium and chloride in plants and decreasing potassium uptake.

Introduction

Salt-affected soils cover about 15 million ha in the Mediterranean Basin (Le Houerou, 1986). High levels of salinity reduce plant growth and may be lethal due to the combined effects of a high external osmotic pressure, which limits water uptake, a high internal concentration of toxic ions, mainly sodium and chloride, and the imbalance in nutrient uptake (Greenway and Munns, 1980).

This study is part of a broader project aimed at evaluating forage genetic resources for salt-affected areas under Mediterranean climate (Marañón *et al.*, 1989b). We study the response to NaCl salinity of a *Melilotus segetalis* population from the Guadalquivir salt marsh (S.W. Spain). Dry weight and mineral elements (Na, Cl and K) allocation were measured in plants grown in a glasshouse under different salt treatments.

Methods

Seeds of *Melilotus segetalis* (annual legume) were collected in June 1987 from a salt marsh grassland in the Guadalquivir delta (see description of site no. 3 in Marañón *et al.*, 1989a), and stored in paper bags at room temperature.

The experiment was carried out in a glasshouse during January–June 1988. Seeds were sown in pots of 10-L volume filled with a calcareous sandy soil, watered bi-daily with a 20% Hoagland solution no 2 (Benton Jones, 1982) and three salt treatments, adding 10, 50 and 100 m*M* NaCl. Plants were harvested at 50, 90 and 120 days after germination, and separated into roots, stems, leaves and reproductive output (flower buds, flowers and fruits).

Plant materials were overdried at 70°C for at least 48 hours, ground to powder, ashed, and digested with HCl. Sodium and potassium concentrations were measured by flame spectrophotometry (Comité Inter-Instituts, 1969), and chloride by thiocyanate colorimetry (Florence and Farrar, 1971).

Plant growth parameters (absolute and relative growth rates, specific absorption and utilization rates) were calculated according to Hunt (1982), for the control and the three salt treatments.

Results and discussion

Salinity and plant growth

Total dry weight of Melilotus plants at the reproductive stage (120 days after germination) decreased with salinity, 83% and 38% of control for 50 and 100 m*M* NaCl respectively (Table 1). The reduction was lower in shoots than in roots, but no clear trend was found in leaf weight ratio (Table 1).

Changes in dry weight allocation with age and salinity are shown in Figure 1. It is remarkable that the reproductive effort increased with salinity, from 18% (control) to 39% (100 m*M* NaCl). Therefore, salt-affected plants, compared to control, were smaller in size, allocating relatively less to roots and stems but more to reproduction.

The salt treatments caused a reduction in absolute growth rate, from 2.7 to 1.0 g/week, as well as in relative growth rate, from 0.5 to 0.4 g/g week (values for the period between 50 and 120 days after germination). Despite this growth reduction, it is worth mentioning that Melilotus plants grown under relatively high salinity levels (100 m*M* NaCl) attained up to 10.4 g of total dry weight, with a relative growth rate of 0.43, supporting the potential value of this ecotype as a forage genetic resource for salt-affected soils.

Table 1. Growth analysis of *Melilotus segetalis* plants grown in a glasshouse under four NaCl salinity treatments. Derived quantities were calculated between 50 and 120 days (growth rates) and 50–90 days (mineral absorption and utilization rates). Values are means of 5 replicates

Quantity	Salinity treatments (m*M* NaCl)			
	0	10	50	100
Height (cm)	78.20	79.60	68.30	61.80
Total dry weight (g)	27.13	18.80	22.60	10.36
Root-shoot ratio	0.24	0.20	0.19	0.17
Leaf weight ratio	0.09	0.06	0.12	0.10
Absolute growth rate (g/week)	2.70	1.86	2.25	1.02
Relative growth rate (g/g week)	0.53	0.47	0.52	0.43
Sodium absorption rate (mg/g week)	5.95	16.05	42.80	42.04
Potassium absorption rate (mg/g week)	107.35	65.33	73.02	44.19
Sodium utilization rate (mg/mg week)	690.55	64.67	38.02	22.49
Potassium utilization rate (mg/mg week)	21.42	18.05	24.62	25.92

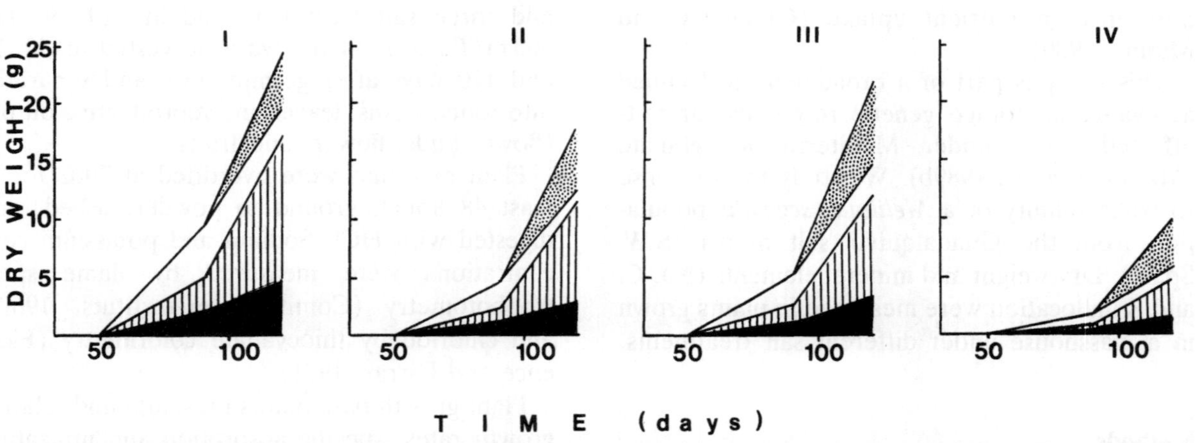

Fig. 1. Absolute allocation of dry weight to roots (black), stems (vertical lines), leaves (open) and reproduction (stippling) in *Melilotus segetalis* grown in a glasshouse with 0 (I), 10 (II), 50 (III) and 100 (IV) m*M* NaCl treatments. Values are means of five replicates.

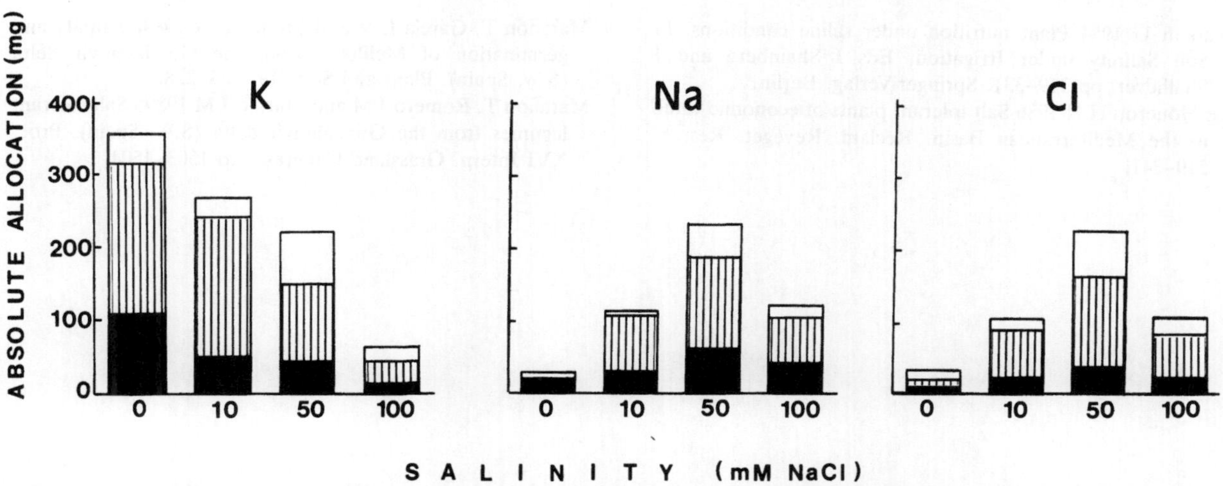

Fig. 2. Absolute allocation of potassium, sodium and chloride to roots (black), stems (vertical lines) and leaves (open), in *Melilotus segetalis* plants (120 days after germination) grown under four NaCl salt treatments. Values are means of five replicates.

Mineral allocation and salt tolerance

The absolute content of potassium per plant decreased with salinity, from 359.4 to 64.2 mg (82% reduction), for control and 100 m*M* NaCl respectively (Fig. 2). Despite this general trend, the proportional leaf allocation of potassium at 50 m*M* NaCl treatment was higher than in the control.

Sodium and chloride absolute allocations showed a similar pattern (Fig. 2), increasing with salinity up to 50 m*M* NaCl but decreasing for the 100 m*M* NaCl treatment. Comparing the relative contents of sodium and chloride between plants grown at 50 and 100 m*M* NaCl treatments, we observe that these increased in roots and stems, but slightly decreased in leaves. However, due to the drastic reduction in total biomass (55.9%), the absolute allocation of both elements decreased (Fig. 2).

The specific absorption rate, that is, the rate of nutrient uptake per unit weight of root (Hunt, 1982), increased for sodium (7 fold) and decreased for potassium (2.4 fold) along the salinity gradient (Table 1).

On the other hand, the rate of dry weight increment per unit of absorbed nutrient, the 'specific utilization rate' (Hunt, 1982) was relatively constant for potassium, 18–26 mg/mg week, and decreased strongly (30.7 times) for sodium (Table 1).

Potassium uptake seems to act as a limiting factor for plant growth in saline environments (Kafkafi, 1984). When the amount of sodium cations increases in the external root medium, they compete and replace potassium cations in the root influx. The ability to withdraw sodium and to retranslocate potassium is crucial for salt tolerance (Jechske, 1984).

References

Benton Jones J Jr 1982 Hydroponics: Its history and use in plant nutrition studies. J. Plant Nutr. 5, 1003–1030.

Comité Inter-Instituts 1969 Méthodes de référence pour la détermination des éléments minéraux dans les végétaux: N, P, K, Na, Ca, Mg. Oléagineux 24, 497–504.

Florence T M and Farrar Y S 1971 Spectrophotometric determination of chloride by the mercury thiocyanate method. Anal. Chim. Acta 54, 373–377.

Greenway H and Munns R 1980 Mechanisms of salt tolerance in nonhalophytes. Annu. Rev. Plant Physiol. 31, 149–190.

Hunt R 1982 Plant Growth Curves. Arnold, London.

Jeschke W D 1984 K^+–Na^+ exchange at cellular membranes, intracellular compartimentation of cations, and salt tolerance. *In* Salinity Tolerance in Plants. Eds. R C Staples and G H Toenniessen, pp 37–66. John Wiley and Sons, New York.

Kafkafi U 1984 Plant nutrition under saline conditions. *In* Soil Salinity under Irrigation. Eds I Shainberg and J Shalhevet. pp 319–331. Springer-Verlag, Berlin.

Le Houerou H N 1986 Salt tolerant plants of economic value in the Mediterranean Basin. Reclam. Reveget. Res. 5, 319–341.

Marañón T, García L V and Troncoso A 1989a Salinity and germination of Melilotus from the Guadalquivir delta (S.W. Spain). Plant and Soil 119, 223–228.

Marañón T, Romero J M and Murillo J M 1989b Salt tolerant legumes from the Guadalquivir delta (S.W. Spain). Proc. XVI Intern. Grassland Congress. pp 1503–1504.

M. L. van Beusichem (Ed.), *Plant nutrition – physiology and applications*, 481–485.

PLSO IPNC731

The effect of magnesium sulphate and calcium sulphate on yield and nutrient composition of flax (*Linum usitatissimum*) grown on chernozemic soils

C.A. GRANT and L.D. BAILEY
Agriculture Canada, Brandon Research Station, P.O. Box 610, Brandon, Manitoba, Canada R7A 5Z7

Key words: calcium, flax, *Linum usitatissimum* L., magnesium, manganese, phosphorus, potassium, salinity

Abstract

$CaSO_4$ and $MgSO_4$ at rates up to 126 mmol kg^{-1} were applied to two Chernozemic soils, in a growth chamber study. Dry matter yield of flax was reduced by application of 126 mmol Mg kg^{-1} which produced slightly saline conductivities of 4.6 and 5.2 dS m^{-1} in the two soils. Applications of $CaSO_4$ or $MgSO_4$ reduced tissue concentration of K. Tissue concentration of K was negatively correlated with soil conductance. Addition of $CaSO_4$ did not influence Ca or Mg concentration of the tissue, while addition of $MgSO_4$ increased tissue content of Mg and reduced tissue content of Ca. Tissue concentration of Ca was more closely related to Mg level in the soil than to soil Ca or conductance. Application of Mg increased the proportion of tissue Mg relative to Ca and K. Extractable Zn and Mn were reduced by application of $CaSO_4$ and $MgSO_4$ but tissue concentrations were not influenced.

Introduction

An estimated 2.2 million ha of land in western Canada is affected by accumulation of soluble salts and the area is expanding by approximately 10% per year (Prairie Farm Rehabilitation Administration, 1983). Many of these soils contain excess levels of $MgSO_4$ (Chang *et al.*, 1983). In addition to the soils severely influenced by salts, there are many which are marginally saline and others which contain high levels of $CaSO_4$. These soils may still be productive, yet the salts present may influence nutrient uptake by the plant.

Flax is moderately tolerant to salinity and is frequently grown in Canada on soils that are slightly saline (2 to 4 dS m^{-1}) (McKeague, 1981). However, there is little information on the effect of $MgSO_4$ and $CaSO_4$, at concentrations producing marginally saline conditions, on nutritional balance and crop growth of flax. This study, therefore, was designed to evaluate the effects of increasing levels of $CaSO_4$ and $MgSO_4$ on soil conductance, flax dry matter production and nutrient availability on two Chernozemic soils.

Materials and methods

Soil was collected in the fall from the upper 15 cm of two Orthic Black Chernozem soils, a Newdale clay loam and a Beresford clay loam. Seven fertilizer treatments were replicated 3 times in the two soils. The treatments were: (1) Control; (2) 42 mmol Mg kg^{-1} as $MgSO_4$; (3) 84 mmol Mg kg^{-1} as $MgSO_4$; (4) 126 mmol Mg kg^{-1} as $MgSO_4$; (5) 42 mmol Ca kg^{-1} as $CaSO_4 \cdot 2H_2O$; (6) 84 mmol Ca kg^{-1} as $CaSO_4 \cdot 2H_2O$; (7) 126 mmol Ca kg^{-1} as $CaSO_4 \cdot 2H_2O$.

Soils were amended with 100 μg K g^{-1} as KCl, 40 μg P g^{-1} as $NH_4H_2PO_4$, 50 μg S g^{-1} as $(NH_4)_2SO_4$, 10 μg Zn g^{-1} as $ZnSO_4 \cdot 2H_2O$ and 139 μg N g^{-1} as NH_4NO_3. The N, P, K, S, Ca and Mg salts were blended thoroughly with the soil. The $ZnSO_4 \cdot 2H_2O$ was dissolved in distilled water and sprayed onto the soil, during blending. Soil samples were removed from the fertilized soils and analyzed for pH, conductivity, NH_4OAc-extractable K, Ca and Mg, and DTPA-extractable Zn and Mn (McKeague, 1981). Soil solution samples were removed by centrifugation from soils at field capacity and analyzed for Ca, Mg and K. Cations were analyzed by atomic absorption spectroscopy. Twenty flax seeds were sown 1.0 cm deep in 4 kg of air-dried soil in $15 \times 15 \times 20$ cm wooden pots lined with plastic. The soil was watered to field capacity with No-Damp (oxine benzoate 2.5% solution), to prevent damping off. Pots were covered with plastic and stored in the dark at 22°C and the soil kept damp until the seedlings emerged. The pots were then transferred to a growth chamber (16 h light/8 h dark, 16°C day/10°C night). The soils were watered to field capacity with distilled water whenever the moisture content fell below 70% of field capacity, by weight. Pots were arranged in a completely randomized design and rerandomized after each watering. Seedlings were thinned to 12 per pot after emergence and to 6 per pot after 4 weeks of growth. An additional 100 $\mu g\,N\,g^{-1}$ as NH_4NO_3 who dissolved in distilled water and watered into the soil surface after 4 weeks of growth.

The above ground portion of the plants were harvested after 8 weeks of growth, at early bloom. Dry weights were measured and the tissue ground, digested and analyzed for Ca, Mg, K, Zn and Mn using nitric acid/perchloric acid digestion and atomic absorption spectrophotometry.

Contrast analysis was conducted using the GLM procedure and correlation coefficients using the CORR procedure of SAS.

Results and discussion

Soil conductivity increased with each increment of $MgSO_4$ applied, to a maximum of 4.6 $dS\,m^{-1}$ in the Newdale soil and 5.2 $dS\,m^{-1}$ in the Beresford soil, with 126 mmol Mg kg–$^{-1}$ (Tables 1 and 2). The high conductivity levels are classified as moderately saline (McKeague, 1981). Lower levels of Mg and all levels of $CaSO_4$ produced conductivities higher than the control, but in the slightly saline classification. Conductivity did not differ between high and low $CaSO_4$ levels. Magnesium salts are more soluble than Ca salts and the solubility of $MgSO_4$ is 260 $g\,l^{-1}$ as compared to 2.41 $g\,l^{-1}$ for $CaSO_4 \cdot 2H_2O$ (Weast, 1976). Therefore, the $MgSO_4$ would have a much greater influence on solution conductivity than would the $CaSO_4$. The pH of both soils were increased by the application of Ca or Mg. In both soils, the pH increased with each increment of Mg applied but was higher with 42 than 126 m kg^{-1} $CaSO_4$.

Dry matter yield of flax on both soils was lowest with the addition of 126 mmol Mg kg^{-1} (Tables 1 and 2). Yield on this treatment did not differ significantly from the yield of the control on the Newdale soil but was lower than that of the lower Mg rates. In the Beresford soil, flax yield on the high Mg soil was lower than that on the control ($p < 0.08$) or on the high Ca treatment. Addition of up to 84 mmol Mg kg^{-1} or 126 mmol Ca kg^{-1} did not reduce dry matter yield of flax. This is marginally saline and would not be expected to severely limit plant growth (McKeague, 1981). Janzen and Chang (1987) found that barley yield decreased when salinity was greater than 6.0 $dS\,m^{-1}$, regardless of the salt type present. Carter *et al.* (1979) found that a solution Mg/Ca ratio of between 1.3 and 2.5 reduced the yield of barley. A solution Mg/Ca ratio of 1.58 observed with the 84 mmol kg^{-1} rate of $MgSO_4$ in the Beresford soil, in the current study did not reduce dry matter yield of flax, while yield was reduced in the Newdale soil at a Mg/Ca ratio of 0.77, which occurred with the 126 mmol kg^{-1} rate of $MgSO_4$ (data not presented). Therefore, Mg/Ca ratio was not a suitable predictor of yield reduction in flax in this study.

Extractable K in the soil was generally higher in the Newdale soil treated with $MgSO_4$ than with $CaSO_4$ (Tables 1 and 2). In the Beresford soil, application of the 126 mmol kg^{-1} rate of $MgSO_4$ or $CaSO_4$ increased extractable K. However, application of Mg reduced the tissue concentration of K, while uptake was reduced at the highest Mg rate. Application of Ca tended to reduce tissue concentration of K in the Newdale soil. Potassium concentration in the tissue was

Table 1. Soil pH, conductance and extractable nutrients, and dry matter yield and tissue nutrient composition of flax as influenced by additions of $MgSO_4$ and $CaSO_4$

Treatment (mmol kg^{-1})	Conductance (dS m^{-1})	pH	Extractable soil nutrient ($\mu g\,g^{-1}$)					Dry matter yield (g pot^{-1})	Tissue nutrient concentration				
			K	Ca	Mg	Zn	Mn		K (%)	Ca (%)	Mg (%)	Zn ($\mu g\,g^{-1}$)	Mn ($\mu g\,g^{-1}$)
NEWDALE													
Control	1.14	7.38	882	7155	708	2.35	28.3	8.7	2.95	2.17	0.44	6.95	132
Magnesium													
42	2.74	7.74	904	7058	1734	2.34	28.9	10.1	2.90	1.77	0.67	8.02	150
84	3.63	7.85	872	7082	2325	2.29	24.7	10.5	2.76	1.38	0.91	7.73	147
126	4.61	8.23	915	7133	3492	2.22	25.7	8.4	2.68	1.12	1.10	8.32	134
Calcium													
42	2.62	8.65	838	8475	709	2.15	24.2	10.1	2.87	2.25	0.43	7.53	130
84	2.98	7.34	899	10610	736	2.17	23.8	10.2	2.74	2.01	0.41	6.38	130
126	2.99	7.77	862	11230	737	2.20	24.6	9.6	2.80	2.11	0.42	7.05	119
MSE	1.47	0.0142	1054	40214	11294	0.0045	0.93	0.997	9.4×10^{-3}	0.030	9.4×10^{-4}	1.86	219
BERESFORD													
Control	0.76	7.72	1045	5002	994	1.79	28.0	10.9	2.89	1.64	0.50	8.19	223
Magnesium													
42	2.22	7.76	1062	5021	1756	1.64	29.8	11.7	2.85	1.28	0.76	7.88	219
84	3.80	7.95	1089	4851	2906	1.63	26.7	11.1	2.76	0.84	0.92	8.45	202
126	5.15	8.32	1080	4924	3811	1.87	27.5	9.4	2.67	0.74	1.12	8.02	237
Calcium													
42	3.05	8.52	1066	6508	1321	1.65	25.0	11.9	2.82	1.53	0.49	7.43	208
84	2.94	7.50	989	7154	1055	1.69	28.4	10.4	2.89	1.50	0.47	6.90	199
126	2.88	7.88	1109	9079	1195	1.81	24.2	12.2	2.81	1.63	0.50	7.72	234
MSE	1.25	0.0108	1564	34001	6794	0.0050	9.63	0.907	8.7×10^{-4}	0.0118	0.0163	0.952	1061

Table 2. Contrast analysis for soil pH, conductance and extractable nutrients, and dry matter yield and tissue nutrient composition of flax as influenced by $CaSO_4$ and $MgSO_4$

Contrast	Conductance	pH	Soil nutrient concentration					Dry matter yield	Tissue nutrient concentration				
			K	Ca	Mg	Zn	Mn		K	Ca	Mg	Zn	Mn
NEWDALE													
Control *vs* Ca	0.0001[a]	0.0001	ns[b]	0.0001	ns	0.0017	0.0001	0.0700	0.0564	ns	ns	ns	ns
Control *vs* Mg	0.0001	0.0001	ns	ns	0.0001	ns	0.0123	ns	0.0205	0.0001	0.0001	ns	ns
High *vs* low Ca	ns	0.0001	ns	0.0001	ns	ns	ns	ns	ns	ns	ns	ns	ns
High *vs* low Mg	0.0001	0.0002	ns	ns	0.0001	0.0573	0.0013	0.0500	0.0147	0.0004	0.0001	ns	ns
Control *vs* high Ca	0.0001	0.0012	ns	0.0001	ns	0.0202	0.0004	ns	ns	ns	ns	ns	ns
Control *vs* high Mg	0.0001	0.0001	ns	ns	0.0001	0.0363	0.0057	ns	0.0046	0.0001	0.0001	ns	ns
High Ca *vs* high Mg	0.0001	0.0003	0.0600	0.0001	0.0001	ns	ns	ns	ns	0.0001	0.0001	ns	ns
Low Ca *vs* low Mg	ns	0.0001	0.0201	0.0001	0.0001	0.0042	0.0001	ns	ns	0.0047	0.0001	ns	ns
BERESFORD													
Control *vs* Ca	0.0001	0.0030	0.0835	0.0001	ns	ns	ns	ns	ns	ns	ns	ns	ns
Control *vs* Mg	0.0001	0.0009	0.0805	ns	0.0001	ns	ns	ns	0.0867	0.0001	0.0001	ns	ns
High *vs* low Ca	0.0324	0.0001	ns	0.0001	ns	0.0206	ns	ns	ns	ns	ns	ns	ns
High *vs* low Mg	0.0001	0.0001	0.0808	ns	0.0001	0.0010	ns	ns	0.0576	0.0004	0.0001	ns	ns
Control *vs* high Ca	0.0007	0.0865	0.0288	0.0001	ns	ns	ns	ns	ns	ns	ns	ns	ns
Control *vs* high Mg	0.0001	0.0001	0.0261	ns	0.0001	ns	ns	0.0802	0.0236	0.0001	0.0001	ns	ns
High Ca *vs* high Mg	0.0004	0.0001	ns	0.0001	0.0001	ns	ns	ns	ns	0.0001	0.0001	ns	ns
Low Ca *vs* low Mg	0.0023	0.0001	ns	0.0001	0.0001	ns	0.0834	0.0031	ns	0.0308	0.0001	ns	ns

[a] p value.
[b] ns indicates non significant at the 0.10 level.

negatively correlated with conductance in the two soils ($r = -0.63 p < 0.0001$). The relationship with conductance was better than that with extractable Mg ($r = -0.53$). Janzen and Chang (1987) observed that K concentration in barley tissue was depressed by soil salinity.

Extractable Ca was increased by the application of $CaSO_4$ but not by $MgSO_4$. Extractable Mg was increased by the application of $MgSO_4$ but not by $CaSO_4$ (Tables 1 and 2). Application of $MgSO_4$ reduced tissue concentration and uptake of Ca, with the reduction increasing with the rate of Mg. Even 'nonsaline' levels of $MgSO_4$ strongly depressed tissue levels of Ca, although the tissue concentrations were still well within the sufficiency range (Manitoba Provincial Soil Testing Laboratory, 1982). Tissue concentration of Ca was much more strongly correlated with extractable Mg in the soil ($r = -0.84$) than with conductance or extractable Ca ($r = -0.61$ and 0.67, respectively). Magnesium appeared to compete directly with Ca for uptake by the plant. This is in contrast to the results of Janzen and Chang (1987) who found that Ca concentration was inversely related to conductivity, but was not differentially affected by the type of salt present. Application of Ca did not influence tissue levels of Ca. Concentration and uptake of Mg increased with Mg application but was unaffected by Ca addition. Tissue concentration of Mg was highly correlated with extractable Mg ($r = 0.98$) and to a lesser extent, with conductance ($r = 0.70$, $p < 0.0001$). Janzen and Chang (1987) found that Mg concentration in barley tissue was proportional to the activity of Mg ions in the soil solution and the relationship between Mg uptake and electrical conductivity was dependent on type of salt present.

The sum of the charge equivalents of Ca, Mg and K in the tissue, calculated as concentration of K + 2(concentration of Mg + Ca), differed by a maximum of 7.2% from that of the control, regardless of the treatment. Total uptake of cations varied little with treatment, but the proportion of the cations was influenced by treatment, with Mg substituting for both Ca and K.

In the Newdale soil, DTPA extractable Zn and Mn were reduced by the application of $CaSO_4$ and by the application of 126 mmol Mg kg^{-1} (Tables 1 and 2). Extractable Mn was not influenced by application of either Ca or Mg in the Beresford soil, while extractable Zn in the Beresford soil was lower in the low than in the high Ca and Mg treatments or the control. Addition of $CaSO_4$ or $MgSO_4$ did not influence Zn or Mn concentration or uptake of flax tissue.

Conclusions

Applications of 126 mmol Mg kg^{-1} as $MgSO_4$ reduced dry matter yield of flax in two Chernozemic soils. Applications of $CaSO_4$ up to 126 mmol Ca kg^{-1} and $MgSO_4$ up to 84 mmol Mg kg^{-1} did not influence flax yield. Reduced yield was associated with slightly saline conductivities of 4.6 and 5.2 dS m^{-1} in the two soils. Applications of $CaSO_4$ or $MgSO_4$ did not influence extractable K, but reduced tissue concentration of K. Tissue concentration of K was negatively correlated with soil conductance. Addition of $CaSO_4$ did not influence Ca or Mg concentration of the tissue. Addition of $MgSO_4$ increased tissue content of Mg and reduced tissue content of Ca. Tissue Ca concentration was more closely related to Mg level in the soil than to soil Ca or conductance. Application of Mg increased the proportion of tissue Mg relative to Ca and K. Extractable Zn and Mn were reduced by application of $CaSO_4$ and $MgSO_4$ but tissue concentrations were not influenced.

References

Carter M R, Webster G R and Cairns R R 1979 Calcium deficiency in some Solonetzic soils of Alberta. J. Soil Sci. 30, 161–174.

Chang C, Sommerfelt T G, Carefoot J M and Schaalje G B 1983 Relationships of electrical conductivity with total dissolved salts and cation concentration of sulfate-dominated soil extracts. Can. J. Soil Sci. 63, 79–86.

Janzen H H and Chang C 1987 Cation nutrition of barley as influenced by soil solution composition in a saline soil. Can. J. Soil Sci. 67, 617–629.

Manitoba Provincial Soil Testing laboratory 1982 Interpretive Guidelines for Plant Tissue Analysis. Winnipeg, Manitoba, 13 p.

McKeague J A 1981 Manual on Soil Sampling and Methods of Analysis, 2nd Edition. Canadian Society of Soil Science, Ottawa, 212 p.

Prairie Farm Rehabilitation Administration 1983 Land Degradation and Soil Conservation Issues on the Canadian Prairies. Agriculture Canada, Regina, Saskatchewan.

Weast R C 1976 CRC Handbook of Chemistry and Physics. CRC Press. Cleveland, OH, 2393 p.

M. L. van Beusichem (Ed.), *Plant nutrition – physiology and applications*, 487–493.

PLSO IPNC640

Gas exchange of *Ficus carica* in response to salinity

S.D. GOLOMBEK and P. LÜDDERS
Institute of Crop Science, Technical University Berlin, Albrecht-Thaer Weg 3, D-1000 Berlin 33

Key words: *Ficus carica* L., non-stomatal conductance to CO_2, photosynthesis, respiration, salinity, sodium chloride, stomatal conductance to CO_2, water use efficiency

Abstract

Gas exchange characteristics were studied in two fig clones (*Ficus carica* L.) grown with NaCl (up to 100 m*M*) in the nutrient solution. Net photosynthesis was negatively correlated with Na- and Cl- content of the leaf. Light intensity and clone influenced the extent of reduction in net photosynthesis. Short-term exposure to salinity resulted in enhanced net CO_2-assimilation caused by increased nonstomatal conductance to CO_2 (G_m). It also reduced transpiration rate resulting from a reduced stomatal conductance (G_s). This short-term response to salinity seemed to be determined primarily by osmotic adaptation to the reduced osmotic potential of the nutrient solution. Long-term exposure to salinity decreased net photosynthesis primarily caused by the decrease of G_m and to some extent of reduced G_S resulting in a slightly decreased intercellular CO_2 concentration. With long-term salt treatment transpiration rate decreased and water use efficiency was enhanced because of the greater reduction of transpiration than net photosynthesis.

Introduction

The growth of the fig (*Ficus carica* L.), a moderately salt tolerant species, is reduced under high salinity. The importance of this problem enhances with increasing significance of the cultivation of the fig under irrigation. Changes in relative growth rate are associated with changes in leaf area and/or gas exchange rate. There are no reports on the gas exchange rate responses of the fig to salinity. Therefore the objective of the present investigations was to examine the way in which gas exchange of leaves of the fig is changed under salinity.

Rates of photosynthesis are usually decreased in NaCl-treated glycophytes (Long and Baker, 1986). There are few studies investigating the relationship between Na- and Cl- content of the leaf and net photosynthesis. Until now an unequivocal distinction between the relationships of Na and Cl on net photosynthesis *in vivo* was only done with species which excluded the transport of one of these ions into the leaf (Downton, 1977).

There is little information about the combined action of radiation intensity and salinity on net photosynthesis. The importance of salinity-induced reduction of net photosynthesis with increasing light intensity seems to be dependent on species (Downton *et al.*, 1985; Gale, 1975).

Reported effects of NaCl on respiration vary and both stimulation and repression have been observed (Shone and Gale, 1983).

Many studies have concluded that the decline of photosynthesis in response to salinity is to some extent the result of an increased stomatal conductance (G_S) and low intercellular CO_2 levels *e.g.* Robinson *et al.* (1983), Ball and Farquhar (1984). However often nonstomatal limitations are responsible for most of the decline in photosynthesis (Long and Baker, 1986, Pezeshki *et al.*, 1987). Salinity often reduces transpiration rate and enhances water use efficiency (Ball and Farquhar, 1984; Lloyd *et al.*, 1987; Richardson and McCree, 1985).

Materials and methods

Plants

Cuttings from clones of *Ficus carica* L. of the variety 'Bardajik' (Turkey) and the origin 'Faro' (Portugal) were struck with bottom heat (25°C) in a box with high relative humidity. After rooting they were selected for uniformity and were transfered to 50 liter pots (eight plants per pot) in a climate chamber with nutrient solution. The climate chamber was maintained under the following conditions: 23°C, RH 60 to 80%, 14 h daylength, light intensity of 380 μmol m^{-2} s^{-1} (PAR) at canopy height provided by HQI-TS 400 W/D lamps (Osram). The nutrient solution contained in m*M*: 2 K_2SO_4, 3 $CaNO_3$, 2,5 KH_2PO_4, 3 $MgSO_4$, and in μM: 50 FeEDTA, 30 H_3BO_3, 2 $MnSO_4$, 1 $CuSO_4$, 1,4 $(NH_4)_6MO_7$, 1 $ZnSO_4$. The pH was 6.5 and the pots were continuously aerated. The solutions were replaced weekly and were topped up daily with deionized H_2O. All plants were trimmed to a single shoot. Salinity treatments were begun when the plants had grown for about three months. In all salt treatments the NaCl-concentration was 50 m*M* on the first day and was stepped up to the final treatment concentration on the third day. Control plants were grown without addition of NaCl.

Experiments

Experiment 1

The plants were grown in NaCl-concentrations of 50, 75 and 100 m*M* for three weeks with six replicates of each treatment. Measurements were made using the first fully expanded leaf from the apex of the shoot at the beginning of the treatment (seven leaves per plant at this time).

Experiment 2

In this experiment the NaCl-treatment was 100 m*M* for six weeks with twelve replicates of each treatment. Plants were used with only one big leaf, which was orientated horizontally to the light.

Gas exchange

Short-term determination

The short-term determinations of the gas exchange rates were made using portable CO_2-H_2O-analysers in an open gas system (Porometer, Walz). This system consisted of a differential infrared gas analyser (Binos, Leybold-Heraeus) and a leaf chamber, which enclosed 2 cm^2 of the leaf. Measurements were made of rates of photosynthesis and transpiration of the lower side of the leaf. Leaf temperature, relative humidity and temperature inside the cuvette were recorded.

Continuous determination

For the continuous determination of net photosynthesis, an open gas exchange system was used with an infrared gas analyser (WA 225 Mk3, Analytical Development Co.). A gas exchange unit (WA 161 Mk3, Analytical Development Co.) enabled the measuring of six leaves in sequence. In the cuvette a tangential fan provided rapid heat exchange and maximum boundary layer conductances.

CO_2- and H_2O-exchange rates of experiments 1 and 2 were determined short-term with the portable gas exchange analysing system. These measurements were performed 3-4, 5 h after the beginning of the photoperiod. In experiment 2, additional continuous determination of net photosynthesis of leaves enclosed in cuvettes was carried out.

Calculations

Net photosynthesis (P_n), transpiration (E), stomatal conductance to CO_2 (G_S) and non-stomatal conductance (G_m) to CO_2 were calculated as in Jarvis (1971), while the intercellular CO_2 concentration (C_i) was calculated as in Farquhar and Sharkey (1982).

$$G_S = \frac{E}{\Delta H_2O \times 1{,}6}$$

$$G_m = \frac{1}{C_a/P_n - 1/G_S}$$

$$C_i = C_a - \frac{P_n}{G_S}$$

ΔH_2O = difference between the water contents of the air in the intercellular and ambient air ($mol\ m^{-3}$)

C_a = CO_2 concentration of the ambient air (μLL^{-1})

Light intensity was varied either by adjusting the distance between the light source and the leaf or by screening.

Ion analyses

Sodium ions were measured with a flame photometer and chloride ions by silver tiration with a chloridometer.

Statistics

A randomized complete block design was used. For comparison of two means, the t-test was used, when twelve replicates were available and the means ± SE are presented when three replicates were available.

When the reactions of the two clones were similar, only the clone 'Bardajik' is presented for sake of clarity.

Results and discussion

Relationship between leaf Na- and Cl-content and net photosynthesis

Net photosynthesis was negatively correlated with the Na- and Cl- content of the leaf (Fig. 1). The reduction of net photosynthesis under salinity was not caused by a decrease in chlorophyll content (data not shown). The linear relationship of the Na- and Cl- contents of the leaf ($r = 0{,}99^{**}$) does not allow a distinction between the effects of Na and Cl on photosynthesis. A relationship between net photosynthesis and the Na- and Cl- content of the leaf has also been shown for *e.g.* 'Valencia' orange (*Citrus sinensis* L.) (Lloyd *et al.*, 1987) and bald cypress (*Taxodium distichum* L.) (Pezeshki *et al.*, 1987). In grapevine (*Vitis vinifera* L.), a relatively Na excluding species, Downton (1977) demonstrated a

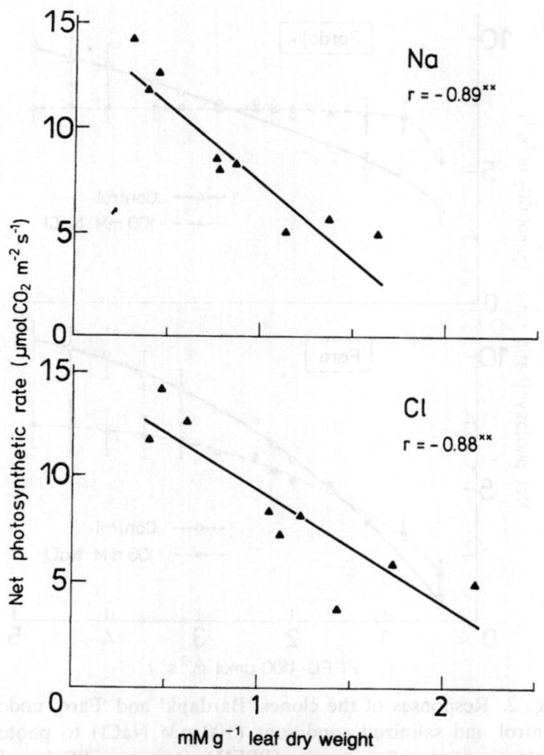

Fig. 1. Changes in net photosynthesis as a function of Na- and Cl- content of leaves of the clone 'Bardajik' after three weeks of salt treatment (saturating light intensity) (Exp. 1).

negative relationship between Cl and net photosynthesis. In comparison with the investigations on several other species the moderately salt tolerant fig showed relatively high Na- and Cl- concentrations of the leaf at a comparable reduction of net photosynthesis. Thus the relatively high salt tolerance of fig in relation to for example, citrus, grapevine and 'Valencia' orange was not caused only by exclusion of Na and Cl from the shoot to a level comparable to these other species.

Effect of light intensity on the influence of salinity on net photosynthesis

With decreasing light intensity the salt-induced reduction of net photsynthesis is mitigated absolute and relative to the control. At lower light intensities the decrease of net photosynthesis with salinity diminished. The two clones showed clear differences in the development of the effect of salinity on net photosynthesis with increasing light intensity (Fig. 2). So light intensity and the

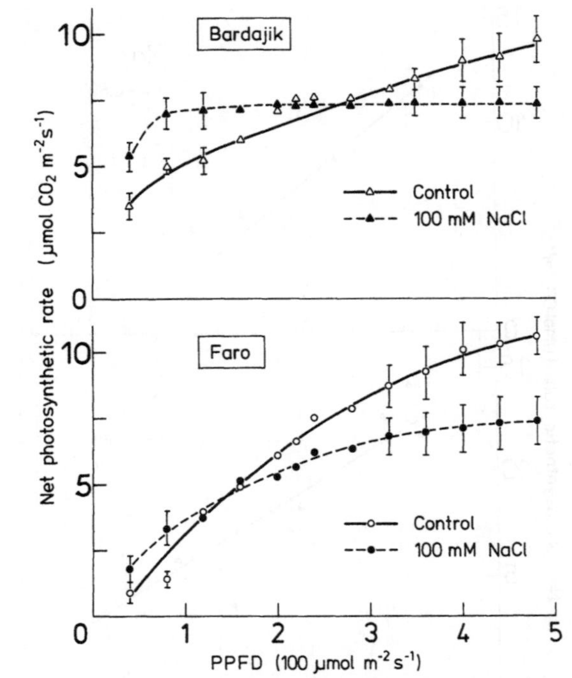

Fig. 2. Responses of the clones 'Bardajik' and 'Faro' under control and salinized conditions (100 m*M* NaCl) to photosynthetic photon flux density (PPFD). Bars are $^+$SE (n = 3) (Exp. 2).

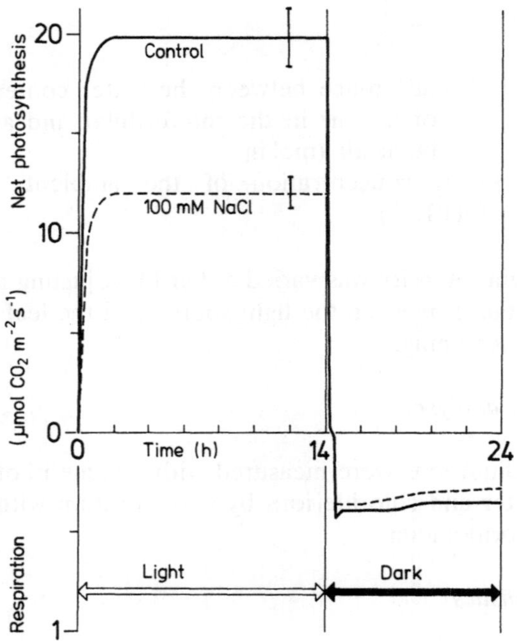

Fig. 3. Daily course of CO_2 exchange of the clone 'Bardajik' under control and salinized (100 m*M* NaCl) conditions (saturating light intensity). Bars are $^+$SE (n = 3) (Exp. 2).

choice of the clone had an essential influence on the extent of the reduction of net photosynthesis under salinity. Downton *et al.* (1985) and Gale (1975) also found for spinach (*Spinacea oleracea* L.) and bean (*Phaseolus vulgaris* L.) that net photosynthesis increased to a greater extent in control than in salt stressed leaves with increasing photon flux density. Gale (1975) however demonstrated for bean a different development of salt-induced reduction in net photosynthesis with increasing light intensity. Further investigations have to be made on the reason of the stimulation of net photosynthesis by NaCl at low light intensity.

Daily course of CO_2 exchange

In the present investigation there was no day rhythm of net photosynthesis found. The dark respiration was reduced tendencially by the NaCl treatment (Fig. 3). Salinity-induced repression of dark respiration (Shone and Gale, 1988) and stimulation (Curtis *et al.*, 1988) of leaves has been reported.

Development of gas exchange parameters with time of exposure to salt

In the first days of salt treatment, net CO_2 uptake was enhanced because of increased non-stomatal conductance (G_m) (Fig. 4, 5). The transpiration was reduced as a result of reduced stomatal conductance (G_S) (Fig. 5) and was probably an adaptation to the reduced osmotic potential of the nutrient solution. The decrease of G_S and increase of G_m resulted in a slowly decreasing intercellular CO_2 concentration (Fig. 5). The cause of the initially enhanced net photosynthesis is probably a higher production of organic solutes to assist osmotic adaptation to the altered water relations. The water use efficiency increased in the first days of salinization because of the increased net photosynthesis and decreased transpiration (Fig. 5). The subsequent short-term decline in water use efficiency of salt-treated plants resulted from the further progressive reduction of net CO_2-assimilation (Fig. 5). The short-term response to salinity seems to be determined primarily by reactions of the plant to the decreased osmotic potential of the nutrient

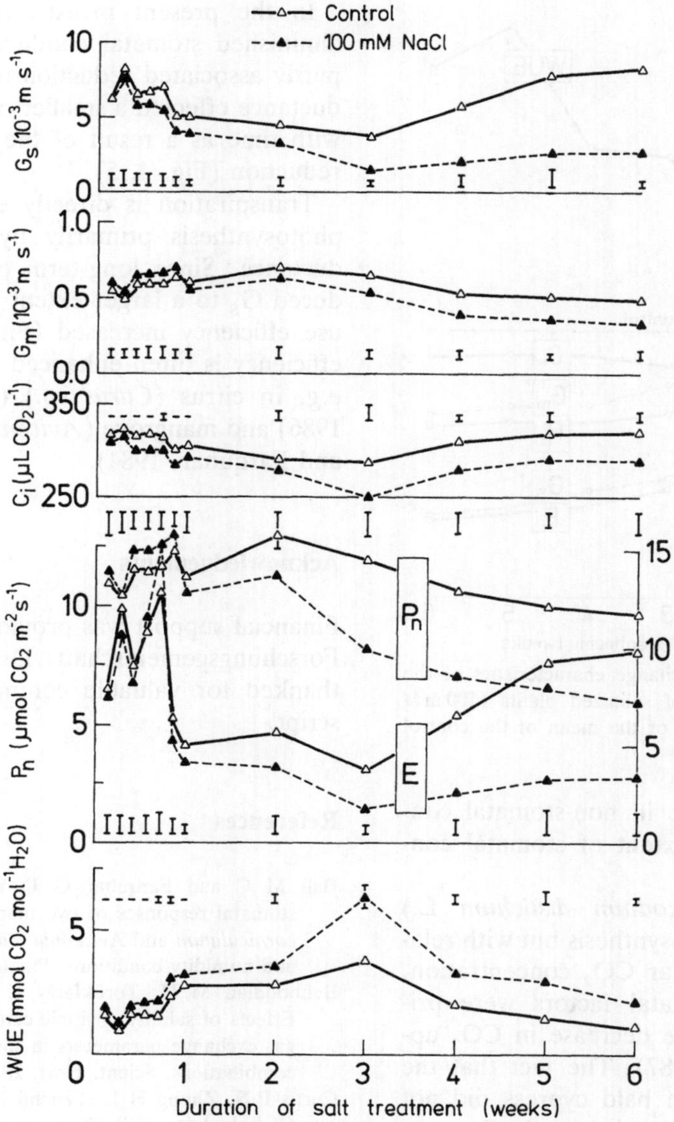

Fig. 4. Time course of gas exchange characteristics of the clone 'Bardajik' under control and salinized (100 m*M* NaCl) conditions (saturating light intensity). Bars indicate the LSD at $P = 0,05$ (n = 12) (Exp. 2).

solution, a suggestion, which is made by Munns and Termaat (1986) as a result of observations on several physiological reactions to salinity.

Long-term exposure (>2 weeks) to salinity decreased G_S and G_m (Fig. 4, 5). Salt treatment limited G_S more than G_m resulting in a 70% reduction of G_S and a 30% reduction of G_m after six weeks of salt treatment (Fig. 5). Because G_m was about ten times smaller than G_S (Fig. 4), changes of G_m were of much greater importance for the CO_2-assimilation despite the greater reduction of G_S with salinity. This is also reflected by the similar changes of net photosynthesis and G_m (Fig. 4). If a decreased G_S in response to salinity results in a reduced intercellular CO_2 level, stomata impose a limitation to assimilation. In this investigations salinity reduced G_S resulting in a slightly lower intercellular CO_2 level and consequently the negative effect of G_S on assimilation was small (Fig. 5). Reduction in net photosynthesis with longer duration of salt treatment was here therefore primarily a con-

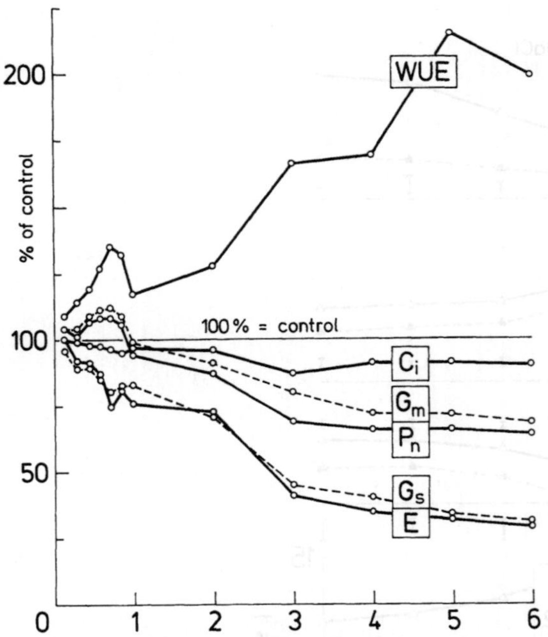

Fig. 5. Time course of gas exchange characteristics of the clone 'Bardajik', the results of salinized plants (100 m*M* NaCl) presented as percentage of the mean of the control (Exp. 2).

sequence of the decrease in non-stomatal conductance and to some extent of stomatal conductance.

In bald cypress (*Taxodium distichum* L.) salinity reduced net photosynthesis but with relatively constant intercellular CO_2 concentration, indicating that non-stomatal factors were primarily responsible for the decrease in CO_2 uptake (Pezeshki *et al.*, 1987). The fact that the intercellular CO_2 level in bald cypress did not decrease indicates that the reduction in G_S compensated for the reduced capacity for CO_2-assimilation to optimize the efficiency of water use. For some other plant species there were described a co-limiting of G_S and G_m together with the assimilation rate during salt treatment *e.g.* in two mangrove species (*Avicennia marina* L., *Aegiceras corniculatum* L.) (Ball and Farquhar, 1984). In salt-treated spinach (*Spinacia oleracea* L.) the more significant limiting factor in total CO_2 fixation was G_S (Robinson *et al.*, 1983). The relative importance of G_S and G_m on the decline in photosynthesis seems to be dependent on plant species, salt concentration and the length of exposure (Long and Baker, 1986).

In the present investigations salt treatment diminished stomatal conductance with time. A partly associated reduction in non-stomatal conductance effected a smaller or no reduction of C_i with time as a result of the decreased stomatal reduction (Fig. 4, 5).

Transpiration is directly effected by G_S and photosynthesis primarily by nonstomatal conductance. Since long-term exposure to salt reduced G_S to a larger extent than G_m, the water use efficiency increased (Fig. 4, 5). Water use efficiency is often enhanced with salt treatment, *e.g.* in citrus (*Citrus* sp.) (Behboudian *et al.*, 1986) and mangrove (*Avicennia marina* L.) (Ball and Farquhar, 1984).

Acknowledgements

Financial support was provided by the Deutsche Forschungsgemeinschaft. Dr D P Hucklesby is thanked for valuable comments on the manuscript.

References

Ball M C and Farquhar G D 1984 Photosynthetic and stomatal responses of two mangrove species, *Aegiceras corniculatum* and *Avicennia marina* to long-term salinity and humidity conditions. Plant Physiol. 73, 238–242.

Behboudian M H, Törökfalvy E and Walker R R 1986 Effects of salinity on ionic content, water relations and gas exchange parameters in some citrus scion rootstock combinations. Scient. Hort. 28, 105–116.

Curtis P S, Zhong H L, Läuchli A and Pearcy R W 1988 Carbohydrate availability, respiration, and the growth of kenaf (*Hibiscus cannabinus*) under moderate salt stress. Am. J. Bot. 75, 1293–1297.

Downton W J S 1977 Photosynthesis in salt-stressed grapevines. Aust. J. Plant Physiol. 4, 183–192.

Downton W J S, Grant W J R and Robinson S P 1985 Photosynthesis and stomatal responses of spinach leaves to salt stress. Plant Physiol. 77, 85–88.

Farquhar G D and Sharkey T D 1982 Stomatal conductance and photosynthesis. Annu. Rev. Plant Physiol. 33, 317–345.

Gale J 1975 The combined effect of environmental factors and salinity on plant growth. *In* Plants in Saline Environments. Eds. A Poljakoff-Mayber and J Gale. pp 186–192. Springer-Verlag, New York.

Jarvis P G 1971 The estimation of resistances to carbon dioxide transfer. *In* Plant Photosynthetic Production, Manual of Methods. Eds. Z Sestak, J Catsky and P G

Jarvis. pp 556–631. Dr. W. Junk, The Hague, The Netherlands

Lloyd J, Kriedemann P E and Syversten J P 1987 Gas exchange, water relations and ion concentration of leaves on salt stressed 'Valencia' orange, *Citrus sinensis* (L.) Osbeck. Aust. J. Plant Physiol. 14, 387–396.

Long S P and Baker N R 1986 Saline terrestrial environments. *In* Photosynthesis in Contrasting Environments. Eds. N R Baker and S P Long. pp 63–102. Elsevier, Amsterdam, New York, Oxford.

Munns R and Termaat A 1986 Whole-plant responses to salinity. Aust. J. Plant Physiol. 13, 143–160.

Pezeshki S R, DeLaune R D and Patrick W H 1987 Effect of salinity on leaf ion content and photosynthesis of *Taxodium distichum* L.. Am. Midl. Natural. 119, 185–192.

Richardson S G and McCree K J 1985 Carbon balance and water relations of sorghum exposed to salt and water stress. Plant Physiol. 79, 1015–1020.

Robinson S P, Downton W J S and Millhouse J A 1983 Photosynthesis and ion content of leaves and isolated chloroplasts of salt-stressed spinach. Plant Physiol. 73, 238–242.

Shone M G T and Gale J 1983 Effect of sodium chloride stress and nitrogen source on respiration, growth and photosynthesis in lucerne (*Medicago sativa* L.). J. Exp. Bot. 34, 1117–1125.

Appendix

Abbreviations

P_n	net photosynthesis rate	(μmol CO_2 m^{-2} s^{-1})
C_i	intercellular CO_2 concentration	(μLL^{-1})
E	transpiration rate	(mmol H_2O m^{-2} s^{-1})
G_S	stomatal conductance to CO_2	(m s^{-1})
G_m	nonstomatal conductance to CO_2	(m s^{-1})
PPFD	photosynthetic photon flux density	(μmol m^{-1} s^{-1})
WUE	water use efficiency	(mmol CO_2 mol^{-1} H_2O)

M. L. van Beusichem (Ed.), *Plant nutrition – physiology and applications*, 495–499.

PLSO IPNC390

Ionic balance and osmotic status in carrot (*Daucus carota*) cell suspensions grown under sodium chloride, osmotic and water stress

E. W. BACHMANN
Department of Plant Sciences, ETH Zürich, Eschikon 33, CH-8315 Lindau, Switzerland

Key words: carrot, cell suspensions, *Daucus carota* L., ionic balance, mannitol, organic acids, polyethylene glycol, salinity, sodium chloride

Abstract

The stress induced by salinity can be separated in chemical, osmotic and water stress components. In this study, these stress components were investigated separately. The osmotic component can be studied by adding mannitol to the growth medium and the water stress by externally applying a solution of high molecular polyethylene glycol (PEG 6800). Our results show that in carrot cell cultures, all three treatments decrease the cell water potential, with concomitant increase of osmotic components in the cell. However the salinity, osmotic and water stresses reduce the intracellular level of the nutrient ions nitrate and sulfate. The sodium chloride and PEG stresses decrease the intracellular potassium level. Under 150 mmol NaCl and 17% PEG stresses, the sucrose levels are increased but the organic acid contents are decreased and the ionic balance shows an increased excess of positive charges.

Introduction

Drought is one of the most serious stresses that limits plant production. Approximately 40 percent of the land surface is considered arid or semiarid and approximately a thousand million hectares of arable land are salt-affected. Many crop plants are grown in areas in which they suffer from stresses because they are not well adapted to the ecological site conditions. With the increasing demand of a growing world population, agriculture will have to expand further into stress-prone environments. Breeding for stress tolerance is therefore becoming more and more important and new cultivars are desired that perform better under stress.

The objective of this research project was to study the physiology of salt tolerance. External salinity induced by sodium chloride can interfere with the plant's physiology in different ways. The sodium chloride may be taken up by the cell where it causes a beneficial osmotic adjustment but it may simultaneously interfere with biochemical and biophysical processes. This form of injury is referred to as chemical stress. Alternatively, the sodium chloride may penetrate the plant but restricted to the apoplast. Plasmolysis occurs once the solute concentration in the cell wall becomes sufficiently high. This type of injury is called an osmotic stress. This situation can be simulated in cell cultures by a solute that penetrates the cell wall space but enters the cell itself at a very low rate. Mannitol may serve as such a solute. Finally, external sodium chloride can be prevented from entering the plant at all. Then, a water stress develops within the plant that cannot be distinguished from regular, non-osmotic water stress. Plasmolysis does not occur in this case. In cell cultures, this situation can be simulated by applying to the medium a solute (PEG 6800) which, because of its size, cannot penetrate cell walls (Oertli, 1987).

It is thus possible to investigate separately the effects of salinity, osmotic and water stresses in cell cultures. In this study we compared the effect of NaCl, mannitol and PEG stresses on the intracellular content of cations, anions, organic acids, carbohydrates and amino acids in carrot cell suspension cultures.

Materials and methods

Modified Murashige and Skoog media (Murashige and Skoog, 1962) were used in this study. Iron was supplied as ferric chloride (0.1 mmol L^{-1}) coupled with the disodium salt of ethylenediaminetetraacetic acid (EDTA; 0.2 mmol L^{-1}). The media contained the following most important ions in mmol L^{-1}: ammonium: 20.61, nitrate: 39.40, phosphate: 1.25, potassium: 20.04, calcium: 2.99, magnesium: 1.50, sulfate: 1.50, chloride: 5.99. The sucrose content was 58.4 mmol L^{-1}. The hormones added were 2,4 dichlorophenoxyacetic acid (2,4 D; 1.81 μmol L^{-1}) and 6-furfurylaminopurine (kinetin; 0.65 μmol L^{-1}). Three modified MS-Media were prepared: the first was supplemented with 150 mmol L^{-1} NaCl (measured total osmolality 450 mOsm/kg water), the second with 271 mmol L^{-1} mannitol (measured total osmolality 429 mOsm/kg water), and the third with 17% polyethylene glycol (PEG) of average MW 6800 (Aldrich Chemical Co. Ltd, Gillingham, Dorset-England), resulting in a measured total osmolality of 582 mOsm/kg water (531 after 5 days of culture). The osmolality was determined with a freezing point depression osmometer (Roebling, D-Berlin). All media were sterilized by filtration.

The carrot cell line under study was previously used by Harms and Oertli (1985). These cells were selected for increased contents of tryptophane and proline. The carrot cells were adapted to the different stress media according to the method described by Harms and Oertli (1985). The cell suspensions were subcultured weekly in their respective media for more than one year before the onset of the experiments. The media of the suspension cultures were renewed two days before and at the beginning of the sampling period. An aliquot of 20 mL of dense cell suspension was added to 200 mL of the corresponding sterile filtered culture medium. Two samples of 10 mL each were taken from each of the cell suspension cultures after 5 days. The cells were then analyzed. The following methods were used for the analyses: atomic absorption spectrophotometry for the cations, ion chromatography for the anions (Column: Waters Anionen IC-PAK™), GC for the organic acids (silyl derivatives; Mollica and Morselli, 1984), HPLC for the sugars (Column Bio Rad HPX-87P) and HPLC for the amino acids (Dabsyl derivatives; Knecht and Chang, 1986). The ammonium and bicarbonate ions were not determined. The ionization state of the molecules were calculated at the prevailing intracellular pH.

Results and discussion

The growth of the carrot cells in reference and stress media are shown in Table 1. It can be noted that the growth rates in the mannitol and PEG media were higher than in the reference medium. It is possible that the normal cells were starved at the moment of the subculture and had a longer lag phase.

Nutrient content of the cell

Under salt stress the intracellular sodium and chloride levels were lower than those of the media (Table 2). Both NaCl and PEG stress reduced the intracellular potassium level. The stronger reduction of potassium under NaCl stress can partly be explained by a competition of sodium with potassium. Nitrate and sulfate were decreased unspecifically by all considered stresses. Phosphate was the only anion increased under stress.

Table 1. Relative growth of carrot cell suspensions cultivated under reference and stress conditions. At day 0, the index is 100% for all media

Days of culture	MS	Relative growth (%)		
		Culture media		
		MS + NaCl	MS + Mannitol	MS + PEG
0	100	100	100	100
5	148	103	219	233
10	177	140	384	297

Table 2. Osmotic balance of the intracellular content in carrot suspension cultivated under reference and stress conditions: 150 mmol L^{-1} NaCl, 271 mmol L^{-1} mannitol, 17% PEG

	Intracellular content in mmol/kg fresh weight			
	Culture media			
	MS	MS + NaCl	MS + Mannitol	MS + PEG
Calcium	8.23	9.46	9.62	8.63
Potassium	96.54	60.06	91.75	87.92
Magnesium	8.48	15.30	12.55	11.92
Sodium	2.08	105.53	4.59	5.46
Total mmol/kg FW	115.33	190.35	118.51	113.93
Phosphate	10.19	15.66	18.73	14.66
Chloride	4.01	83.34	2.25	3.06
Nitrate	28.62	17.38	9.65	11.28
Sulfate	1.42	0.8	0.99	0.93
Total mmol/kg FW	44.24	117.18	31.62	29.93
Succinic acid	0.81	0.66	5.66	1.29
Fumaric acid	0.33	0.59	5.56	0.36
Malic acid	4.10	5.72	16.52	6.80
Citric acid	19.92	5.28	11.15	3.79
Total mmol/kg FW	25.16	12.25	38.89	12.24
Sucrose	6.81	15.46	36.12	22.18
Glucose	12.4	12.14	17.23	15.6
Fructose	6.27	5.42	6.98	13.85
Mannitol			40.72	
Total mmol/kg FW	25.48	33.02	101.05	51.63
Proline	6.74	15.34	9.13	10.51
Total Amino acids mmol/kg FW	30.71	35.65	41.55	49.00
Sum osmotic components mmol/kg FW	240.91	388.45	331.62	256.73
Osmolality of the media mOsm/kg H_2O	158	450	429	531

Sugar and organic acid contents of carrot cells

Stress induced an increase of intracellular sucrose. Carrot cells could adapt to salt, osmotic and water stresses by increasing the uptake of sucrose from the medium. The intracellular levels of glucose and fructose are slightly reduced by NaCl stress. Under NaCl and PEG stresses an elevated level of sugar is associated with a reduced level of organic acids; this could indicate a metabolic block. This hypothesis is supported by the fact that under mannitol stress the sugars and organic acids contents of the cells are both increased: in this case the enzymes of the metabolic pathway were not inhibited by mannitol.

Amino acid content of the cells

Both organic acid and amino acid metabolism are linked by 2-oxoglutaric acid that is aminated to form glutamic acid; this glutamic acid is then the precursor of glutamine, proline and arginine.

Our results (Table 2) show that proline and

Table 3. Balance of the charges of carrot cells cultivated in suspension under reference and stress conditions: 150 mmol L^{-1} NaCl, 271 mmol L^{-1} mannitol, 17% PEG. The ionisation states of the molecules were calculated at the intracellular pH

	Culture media			
	MS	MS + NaCl	MS + Mannitol	MS + PEG
Intracellular pH	5.90	5.76	5.87	5.90
Cations Total meq/kg FW	+132.04	+215.11	+140.68	+134.48
Anions Total meq/kg FW	−46.15	−118.53	−33.45	−31.56
Organic acids Total meq/kg FW	−51.48	−23.14	−74.01	−23.46
Amino acids Total meq/kg FW	+1.35	+1.43	+3.01	−0.97
Balance of the charges meq/kg FW	+35.76	+74.87	+36.23	+78.49

the sum of the amino acids were increased under NaCl, mannitol and PEG stresses. The highest proline level was observed under NaCl stress.

Osmotic balance and charge balance

Under NaCl stress, the cells take up sodium chloride from the medium; these ions account for 48% of the osmotic active components in the cells. Under all stresses, sucrose was increased but it accounted for only 3 to 10% of the total osmotic pressure (Table 2). Under mannitol stress, mannitol penetrates slowly the cells and contributes to the osmotic balance (a level of 40.7 mmol/kg FW was observed in the cells). The osmotic (mannitol) stress increased the pool of organic acids available for synthesis of glutamic acid and this could explain the increase in the total amino acids content.

Calculation of the charge balances shows that NaCl stress increased the excess of intracellular cations; this can be partly explained by the reduction of the organic acids (Table 3). Since the cells were washed with water only, cations bound to the surface of the cells were not removed.

In conclusion, this study shows that the chemcial, osmotic and water depletion aspects of NaCl stress have several common effects on the content of carrot cells. Our results for water stress agree with those of Fallon and Phillips (1989): in both cases accumulation of intracellular solutes compared to controls are observed. As samples were taken after 5 days of culture, lower levels of sugars were observed. Further experiments are necessary to determine the cause of the decreased level of organic acids under salt and water stresses. The synthesis of organic acids could be blocked or the organic acids could be used for the increased synthesis of amino acids.

References

Fallon K M and Phillips R 1989 Responses to water stress in adapted and unadapted carrot cell suspension cultures. J. Exp. Bot. 40, 681–687.

Harms C T and Oertli J J 1985 The use of osmotically adapted cell cultures to study salt tolerance *in vitro*. J. Plant Physiol. 120, 29–38.

Knecht R and Chang J Y 1986 Liquid chromatographic determination of amino acids after gas-phase hydrolysis and derivatization with (dimethylamino)azobenzenesulfonyl chloride. Anal. Chem. 58, 2375–2379.

Mollica J N and Morselli M F 1984 Gas chromatographic determination of non volatile organic acids in sap of sugar maple (*Acer saccharum* Marsh.) J. Assoc. Off. Anal. Chem. 67, 1125–1129.

Murashige T and Skoog F 1962 A revised medium for rapid growth and bioassays with tobacco tissue cultures. Physiol. Plant. 15, 473–497.

Oertli J J 1987 Measurement of the resistance of cell walls to collapse during moisture stress. Proc. of the Intern. Conf. Measurement of Soil and Plant Water Status, Vol. 2, pp 193–198, Utah State University.

D

Fertilizer application in relation to yield formation and quality characteristics

M. L. van Beusichem (Ed.), *Plant nutrition – physiology and applications*, 503–507.

PLSO IPNC332

Response of grass swards to fertilizer nitrogen under cutting and grazing

P.J.A.G. DEENEN
Department of Field Crops and Grassland Science, Wageningen Agricultural University, Haarweg 333, 6709 RZ Wageningen, The Netherlands

Key words: cutting, grassland, grazing, *Lolium perenne* L., nitrogen, nitrogen balance, nitrogen response, perennial ryegrass

Abstract

The dry matter (DM) response of a perennial ryegrass sward to fertilizer nitrogen (N) was studied under cutting and rotational grazing in two long-term experiments on a clay and a sandy soil. For both experiments the results of the third experimental year are presented. Under cutting, a significant DM response to N was obtained up to 550 kg N ha^{-1} yr^{-1} in both experiments. However, under grazing the response was lower and differed between the sites. On the sandy soil no response was obtained above 250 kg N ha^{-1} yr^{-1}, whereas on the clay soil there was a significant response from 250 to 550 kg N ha^{-1} yr^{-1}.

The low recovery of N in the herbage resulting from sward deterioration (especially on the sandy soil) and a low retention of N in the animal products caused an increasing amount of N susceptible to loss to the environment with increasing N applications.

Introduction

In the Netherlands fertilizer nitrogen (N) is extensively used on grassland. Current recommended rates are about 400 kg N ha^{-1} yr^{-1} for grasslands on sand, clay and wet peat soils and about 250 kg N ha^{-1} yr^{-1} on well-drained peat soils (Anon., 1984). These recommendations are derived mainly from cutting trials with an approximate cutting interval of four weeks (Prins, 1983). However, it is generally accepted that the response of a grass sward to fertilizer N differs depending on whether the sward is defoliated by cutting or by grazing animals. Grazing cattle can exert positive and negative effects on pasture production. The return of nutrients in excreta can have a positive effect on pasture production if the fertilizer N application level is low (*e.g.* During and McNaught, 1961; Norman and Green, 1958). However, with increasing N fertilization the negative effects of sward damage through treading, poaching, urine scorching and tiller pulling increase (Edmond, 1966; Keuning, 1981; Wilkins and Garwood 1986; Tallowin *et al.*, 1986).

This paper reports DM response to fertilizer N in grazing experiments in the third experimental year. Comparisons are made with cutting experiments and a N balance of each experiment is presented.

Materials and methods

Experiment 1 was carried out on a sandy soil near Wageningen with four levels of fertilizer N on both the cutting and grazing treatment; 250, 400, 550 and 700 kg N ha^{-1} yr^{-1}. The cutting treatment also included 0 N ha^{-1} yr^{-1}. Experiment 2 was conducted on a young marine clay soil near Swifterbant with two fertilizer N levels (244 and 540 kg N ha^{-1} yr^{-1}). Both experiments were conducted on a permanent sward with perennial ryegrass (*Lolium perenne* L.) being the main species. In both experiments a rotational grazing system was adopted. The grazing animals were steers in experiment 1 and dairy cows in experiment 2.

The annual DM yield under grazing was assessed as the sum of the pre-grazing herbage mass including the herbage accumulation during the grazing period less the sum of the post-grazing herbage mass except that following the last grazing in autumn. The harvests in the cutting trial were made at the same time as the determination of the residual herbage mass at the end of each grazing period. Herbage mass was measured above 4 cm cutting height. All harvested herbage was analyzed for total N content. For both experiments the results of the third experimental year are presented (experiment 1: 1987; experiment 2: 1988).

Results and discussion

The experimental results (1987) on the sandy soil are presented in Figure 1. This diagram relates supply of fertilizer N, N uptake in the herbage and DM yield. Quadrant IV in Figure 1 illustrates the difference in N uptake between the cutting and the grazing treatments. Except for the lowest N application rate the uptake of N under grazing is lower than under cutting, but uptake levels off beyond a N application of 400 kg N ha^{-1} yr^{-1}. Quadrant I in Figure 1 presents the relationship between N uptake and DM yield, it shows that above an uptake of about 400 kg N ha^{-1} yr^{-1} the sward gave a higher yield under cutting than under grazing.

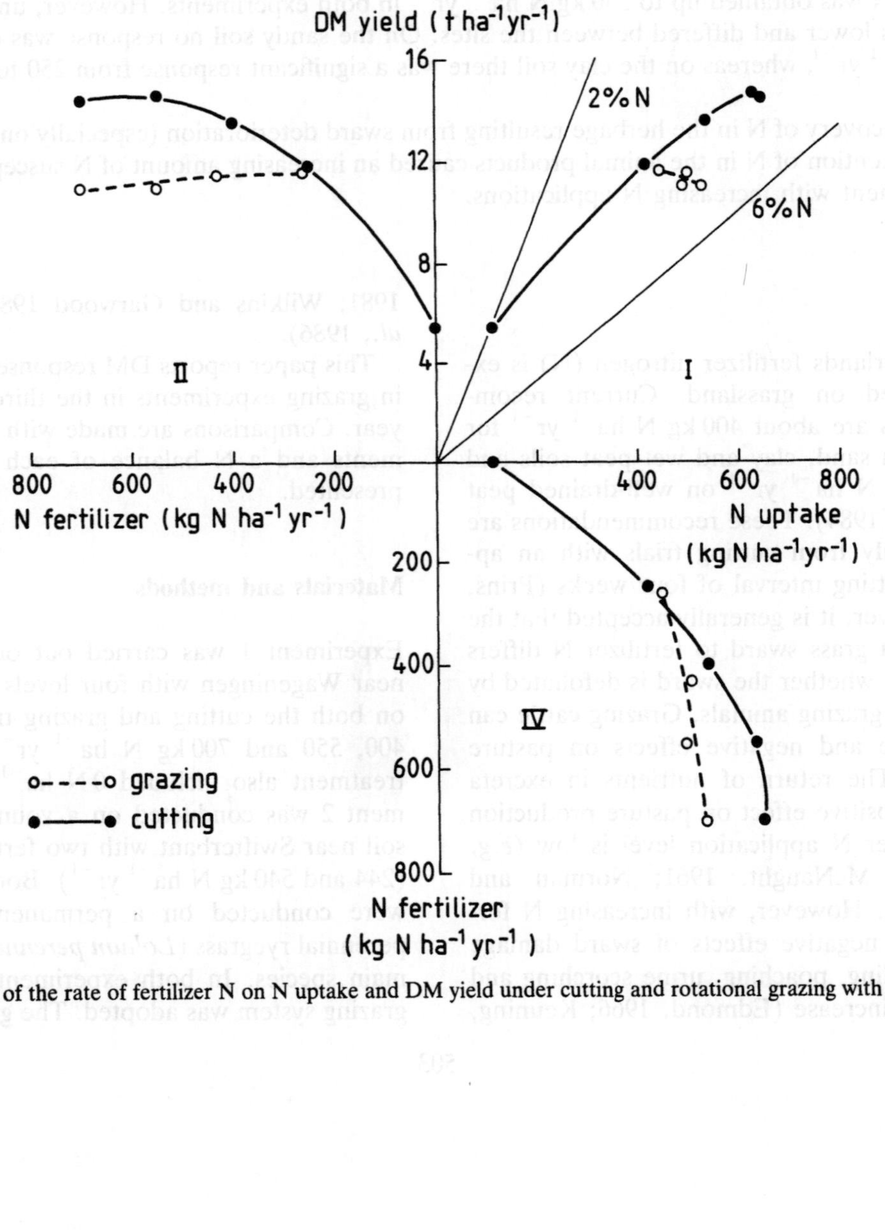

Fig. 1. The effect of the rate of fertilizer N on N uptake and DM yield under cutting and rotational grazing with steers on a sandy soil (1987).

The lower recovery of N and the lower effectiveness of the absorbed N (Quadrant I) in the grazing treatment is probably due to sward damage resulting from urine scorching and poaching (1987 was extremely wet). The DM response to applied fertilizer N under cutting and grazing is different (Fig. 1 Quadrant II). Under cutting there is a positive response up to 550 kg N ha^{-1} yr^{-1} but under grazing there is a slightly negative response of the DM yield to fertilizer N applied. The difference in N fertilization to achieve maximum yields under both management systems is greater than the previously published differences of about 200 kg N ha^{-1} yr^{-1} (Jackson and Williams, 1979). Our differences can be ascribed to increased sward damage at the higher N application rates due to a more vulnerable sward. The magnitude of the difference depends on the degree and time of occurrence of the sward damage.

Table 1 presents the N balance of this experiment. Since the marginal recovery of N decreases strongly (Fig. 1 Quadrant IV) there is not a large increase in the amount of N voided in the excrements (viz. 396 kg N ha^{-1} yr^{-1} at the low N and 486 kg N ha^{-1} yr^{-1} at the high N application). The efficiency of N utilization by the grazing steers declines from 9.7% to 2.8% of the total N input.

The results of experiment 2 are presented in Figure 2. At the low N level the difference in DM yield between the grazing and cutting treatments can be associated with a combined positive effect of excretal N (mainly urine N) and a better regrowth facilitated by a relative large mass of residual herbage at the end of each grazing period. At the high N level the postive effect of the residual herbage on regrowth might have been counteracted by the negative effects of sward damage.

Quadrant IV in Figure 2 shows that the recovery of N was relatively high in both treatments and exceeded even 100% with cutting. This reflects a possibly larger allocation of DM and hence N to the herbage at the high N application.

Table 2 presents the N balance of grazing experiment 2. When total N input increased by 306 kg N ha^{-1} yr^{-1}, output in animal products only increased by 42 kg N ha^{-1} yr^{-1}. The ef-

Table 1. Effect of fertilizer N on the N balance under grazing with steers on a sandy soil (1987) in kg N ha^{-1} yr^{-1}

	Level of fertilizer N			
	N1	N2	N3	N4
A: N INPUT				
fertilizer	259	426	547	701
precipitation	40	40	40	40
	299	466	587	741
B: N INTAKE				
animals	425	487	472	506
C: N RETENTION				
animal product	29	26	26	20
D: N EXCRETION				
dung	88	90	83	83
urine	308	371	363	403
	396	461	446	486
E: N EFFICIENCY				
(% of input retained)	9.7	5.6	4.4	2.8

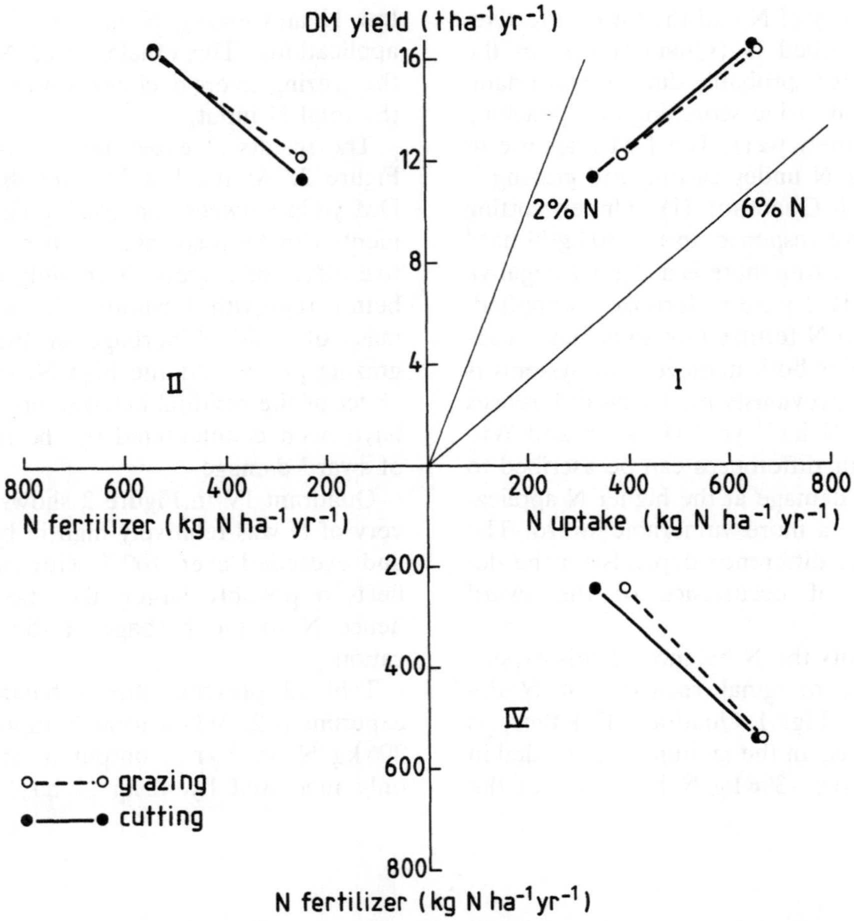

Fig. 2. The effect of the rate of fertilizer N on N uptake and DM yield under cutting and rotational grazing with dairy cows on a clay soil (1988).

ficiency of the N utilization decreased from 24% to 19% of the total N input. This implies a strong increase in excretal N (236 kg N ha^{-1} yr^{-1}). However, since the recovery of excretal N is low, the potential amount of N susceptible to loss from the system through ammonia volatilization, denitrification and leaching will be doubled.

It may be concluded that the optimum N application under grazing on the sandy soil was not greater than 250 kg N ha^{-1} yr^{-1}. A lack of response to higher N doses was due to sward deterioration, especially urine scorch and poaching. On a soil less susceptible to sward damage the economical optimum N application might be higher. The low recovery of N in the herbage (especially on the sandy soil) resulting from sward deterioration and low retention of N in the animal products caused an increasing amount of N to be susceptible to loss to the environment with increasing N applications.

The results of this study suggest that for an optimum N application under grazing a distinction should be made between soil types in relation to their susceptibility to sward deterioration to ensure a high recovery of N by the herbage. However, if reduction of N losses from the system is emphasized this can only be attained by reducing fertilizer N input.

Table 2. Effect of fertilizer N on the N balance under grazing with dairy cows on a clay soil (1988) in kg N ha^{-1} yr^{-1}

	Level of fertilizer N N1	N2
A: N INPUT		
fertilizer	244	540
concentrates	19	29
precipitation	40	40
	303	609
B: N INTAKE		
animals	361	639
C: N RETENTION		
animal growth	5	8
milk	69	108
	74	116
D: N EXCRETION		
dung	98	130
urine	189	393
	287	523
E: N EFFICIENCY		
(% of input retained)	24	19

References

Anonymous 1984 Handboek voor de Rundveehouderij. Proefstation voor de Rundveehouderij, Schapenhouderij en Paardenhouderij. Lelystad, The Netherlands.

During C and McNaught K J 1961 Effects of cow urine on growth of pasture and uptake of nutrients. N.Z. J. Agric. Res. 4, 591–605.

Edmond D B 1966 The influence of animal treading on pasture growth. Proc. 10th Int. Grassl. Congr., Helsinki, pp 453–458.

Keuning J A 1981 Urinebrandplekken in grasland. 1. Bedrijfsontwikkeling 12, 453–458.

Norman M J T and Green J O 1958 The local influence of cattle dung and urine upon the yield and botanical composition of permanent pasture. J. Brit. Grassl. Soc. 13, 39–45.

Prins W H 1983 Limits to nitrogen fertilizer on grassland. Doctoral thesis, 132 p. Wageningen Agricultural University, Wageningen.

Tallowin J R B, Kirkham F W and Brookman S K E 1986 Sward damage by sod-pulling -the effect of nitrogen. *In* Grazing. Ed. J Frame. pp 44–48. Occasional Symposium No. 19, British Grassland Society.

Wilkins R J and Garwood E A 1986 Effects of treading, poaching and fouling on grassland production and utilization. *In* Grazing. Ed. J Frame. pp 19–31. Occasional Symposium No. 19, British Grassland Society.

M. L. van Beusichem (Ed.), *Plant nutrition – physiology and applications*, 509–514.

PLSO IPNC301

Response of tomatoes (*Lycopersicon esculentum*) to an unequal distribution of nutrients in the root environment

C. SONNEVELD and W. VOOGT
Glasshouse Crops Research Station, P.O. Box 8, 2670 AA Naaldwijk, The Netherlands

Key words: electrical conductivity, fruit quality, *Lycopersicon esculentum* Mill., nutrient distribution, split root, substrate sampling, tomato yield, water absorption

Abstract

Tomato (*Lycopersicon esculentum* Mill.) plants were grown in a split root system. The plants were rooted in two separate cubes of rockwool, which were subsequently irrigated with nutrient solution of equal (control) or different EC values. Besides optimal values, too low and too high values for maximal production were included.
The yield was determined by the EC value considered optimal for plant nutrition if present in one of both rockwool cubes. The quality of the fruits was primarily determined by standard EC values available in part of the root environment. Water was preferably taken up from the low EC compartment, nutrients from the high EC compartment. Samples of leaves and fruits were analyzed to get information about uptake and translocation of nutrients in the plant.

Introduction

In the Dutch greenhouse industry many crops are grown in rockwool slabs. As the nutrient solution is supplied by means of a trickle irrigation system, great differences occur in the nutrient levels from spot to spot. Generally, the concentrations of nutrients at spots between emitters are higher than at spots under emitters, just like in soils (Bernstein and Francois, 1975; Hoffman, 1986; Oster *et al.*, 1984).

When a sample of solution is gathered for analysis, the question arises how the effect of this heterogeneity in nutrient concentration in the root environment must be interpreted in terms of yield and fruit quality. In two experiments, effects of an unequal distribution of nutrients in rockwool slabs were studied. The test crop was tomato. Besides, the effects on yield, effects on fruit quality and on uptake of water and nutrients were investigated.

Methods and materials

Experiments

In the experiments each tomato plant was grown in two rockwool cubes with length, width and height of 0.15, 0.10 and 0.10 m, respectively. The two cubes were separated by placement in different gullies as shown in Figure 1. Each cube was provided with a dripper so that it was possible to realize different nutrient levels in the cubes. The nutrient solutions were recirculated. The quantity of nutrient solution supplied was 20 L per m^2 greenhouse area per day. In the first experiment it was found that transport of nutrient solution from one to the other growing cube was possible, through the propagating cube. So, in the second experiment the propagating cubes were placed on small polystyrene strips, to prevent such transport.

The tomato crops were grown from January

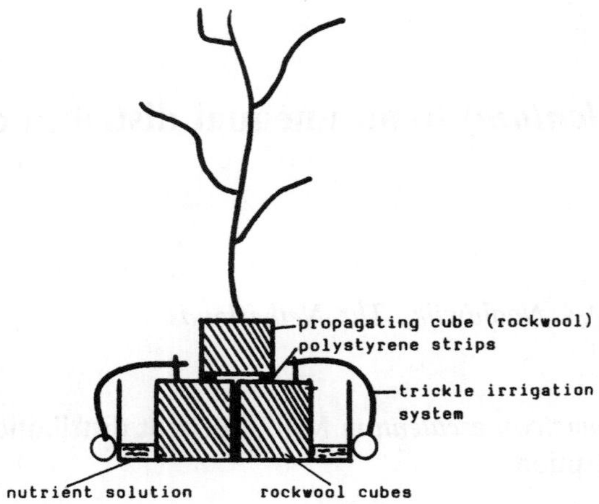

Fig. 1. The growing system used

till August. For some time after planting out, all cubes were supplied with the same nutrient solution having an EC value of 2.5–3.0 $dS.m^{-1}$ (25°C). Thus each plant was enabled to develop equal root parts in the two cubes. The treatments were set up in the beginning of March.

In both experiments five treatments were laid out in four parallels in a youden scheme. The EC values maintained in the pairs of rockwool cubes were as follows.

Experiment 1: 0.75/0.75, 2.5/2.5, 5.0/5.0, 0.75/2.5 and 2.5/5.0 $dS.m^{-1}$ (25°C).

Experiment 2: 0.75/3.0, 3.0/3.0, 5.0/3.0, 7.5/3.0 and 10.0/3.0 $dS.m^{-1}$ (25°C).

In experiment 1, the round Dutch tomato cultivar Counter was grown and in experiment 2 the beefsteak type cultivar Dombito. The experiments were carried out in a greenhouse in which as setpoint for heating 15°C was maintained during night and 19°C during day. The ventilation temperature was 24°C.

Nutrient solution

At an EC value of 3.0 in the recirculation basin, the nutrient solution contained following ions in $mmol. L^{-1}$: NO_3 17, H_2PO_4 1, SO_4 5, K 7, Ca 7, Mg 3.5, which values are operative for tomato growing in rockwool (Sonneveld and De Kreij, 1987). The different EC values were realized through proportionally higher or lower concentrations of nutrient elements. Measurements of EC values in the rockwool cubes showed no differences with the values in the recirculation basins, which is understandable in view of the ample supply of nutrient solution.

Micro elements were added proportionally to the macro elements. At an EC value of 3.0 following concentrations in $\mu mol. L^{-1}$ were pursued: Fe 15, Mn 7, Zn 7, B 50, Cu 0.7, Mo 0.5.

The pH was roughly controlled by additions of NH_4NO_3 and adjusted by additions of HNO_3 or a mixture of KOH and $Ca(OH)_2$ (mol ratio 2:1). On average NH_4 was added in a ratio to NO_3 of 1:7.

The water used in the experiment was rainwater or demineralized water, with an EC value of about 0.3 $dS.m^{-1}$ and a concentration of sodium chloride of 0.5 $mmol\ L^{-1}$.

Absorption of water and nutrients

In the second experiment the uptake of water and nutrients was determined over the period March-July. The composition of the recirculating nutrient solution was checked frequently and kept constant by addition of fertilizers and water. The absorption of water and nutrients was calculated from the quantities of water and nutrients needed to keep volume and ionic composition constant. The nutrients present in the raw water were taken into account.

Tissue analysis

The nutrient status of the tomato plants was checked by analysing samples of laminae and fruits. The samples were gathered in the second part of the growing period, thus at a moment that the plants had grown for a rather long period under treatment conditions. The samples were rinsed in a detergent solution, dried, ground and analysed. The analytical methods used are described by De Bes (1986).

Crop observations

At harvest, the number and weight of fruits were determined. Fruit colouring was expressed in terms of number of days elapsing between picking and reaching colouring stage 100% orange. Shelf life was expressed as the number of days between 100% orange and fruit softening. In the fruit sap, EC value, acid content and refraction were measured and expressed as dS.m^{-1} (25°C), mmol. L^{-1} and percentages Brix, respectively. Russetting, gold specks and irregular colouring were judged visually. The index used for russetting and gold specks ranged from 0, unaffected, till 3, heavily affected fruits. Irregularly coloured fruits, being the fruits with insufficiently regular colouring at picking stage for first class quality, were expressed as percentages.

Results

Fruit yields of both experiments are listed in Table 1. In experiment 1, yield was highest for plants grown, completely or partly, at an EC value of 2.5 in the root environment. At a value of 0.75, yield was not significantly reduced by the low nutrient status in the root environment, but at a value of 5.0 yield was substantially reduced by the low osmotic potential. In experiment 2, no significant differences between yields were found. Fruit weights tended to be highest where EC values in the root environment were completely or partly low.

The data of fruit quality characteristics are summarized in Table 2. For treatments in which an equal EC value was maintained in the root environment, quality was improved by increasing EC values, for such an increase shortened the colouring period, extended shelf life, increased

Table 1. Yield and fruit weight of tomato in both experiments

Experiment 1			Experiment 2		
EC value	Yield (kg.m^{-2})	Fruit weight (g)	EC value	Yield (kg.m^{-2})	Fruit weight (g)
0.75/0.75	22.7	82	0.75/3.0	23.8	188
2.5 /2.5	24.0	77	3.0 /3.0	24.0	180
5.0 /5.0	21.1	71	5.0 /3.0	25.1	177
0.75/2.5	24.2	83	7.5 /3.0	24.6	178
5.0 /2.5	23.7	80	10.0 /3.0	23.6	173
LSD 0.05	2.4	ns	LSD 0.05	ns	ns

Table 2. The effect of variation in EC values in the root environment on quality of tomato fruits

Characteristics	EC values					LSD
	0.75/0.75	2.5/2.5	5.0/5.0	0.75/2.5	2.5/5.0	0.05
Experiment 1						
Colouring in days	4.3	3.9	3.4	4.0	3.6	0.3
Shelf life in days	6.2	6.6	9.1	7.0	7.4	1.4
EC fruit sap, dS.m^{-1}	4.5	5.1	5.5	5.0	5.1	0.2
Acids in fruit sap, mmol.L^{-1}	5.9	6.6	7.6	6.4	6.8	0.3
Refraction fruit sap, %Brix	4.1	4.1	4.6	4.1	4.2	0.1
Russetting index	0.44	0.43	0.28	0.43	0.42	ns
Gold specks index	1.82	2.27	1.17	2.36	2.34	0.63
Irregular colouring, %	21	17	2	10	11	14
	EC values					LSD
	0.75/3.0	3.0/3.0	5.0/3.0	7.5/3.0	10.0/3.0	0.05
Experiment 2						
Colouring in days	3.3	2.9	2.9	2.8	2.9	0.4
Shelf life in days	4.1	5.1	5.5	5.5	5.4	1.3
Russetting index	1.30	1.21	1.20	1.28	1.25	ns
Gold specks index	1.24	0.84	0.96	0.95	0.85	0.29
Irregular colouring, %	42	12	9	12	14	ns

EC values, acid content and refraction of the fruit sap and decreased the indexes for russetting (not significant) gold specks and the percentage irregularly coloured fruits. With different EC values in the root environment, in experiment 1 the quality characteristics tended to adjust to the standard EC value of 2.5. In experiment 2, the EC value of 0.75, supplied in part of the root environment, tended to affect fruit quality negatively.

Analytical data of tissue samples gathered in experiment 1, are listed in Tables 3 and 4. With equal EC values in the root environment, the Na, Ca, Mg and Cl contents of leaves were higher at EC = 0.75 than at EC = 2.5. The P and K contents on the contrary were lower. At EC = 5.0, especially Ca content was low. As to fruits, a high Na content and a low P content were found at the low EC value and a low Ca content at the high EC value.

In the treatments in which a low or a high EC value was maintained in part of the root environment, element contents in laminae and fruits tended to adjust to those at the standard EC value. In experiment 2, high P contents were found in plant tissues of the treatments with high EC values (5–10) in part of the root environment. In comparison with the other two treatments, the tissue contents were on average 27% higher. As to the other elements, plant tissue contents did not show real differences between treatments.

The quantitative uptake of water and nutrients in the treatments of experiment 2 is shown in Table 5. Water absorption was strongly reduced by EC values above the standard value of 3.0. The uptake of nutrients was mostly highest in root halves with EC values above the standard value. At the low EC value (0.75), nutrient absorption appeared to be very low. For some

Table 3. Analytical data of laminae of young leaves of tomato grown in experiment 1. Element contents are expressed as $mmol.kg^{-1}$ dry matter and dry matter content as % of fresh material

Elements	EC values				
	0.75/0.75	2.5/2.5	5.0./5.0	2.5/0.75	2.5/5.0
Na	193	58	39	73	55
K	658	953	1080	888	972
Ca	858	794	587	698	748
Mg	274	161	160	184	156
Cl	66	32	57	47	31
N	3340	3476	3738	3561	3545
P	137	192	210	190	191
S	483	473	423	442	440
Dry matter	11.0	10.4	10.8	10.8	11.0

Table 4. Analytical data of tomato fruits grown in experiment 1. Element contents expressed as $mmol.\ kg^{-1}$ dry matter and dry matter content as % of fresh material

Elements	EC values				
	0.75/0.75	2.5/2.5	5.0/5.0	2.5/0.75	2.5/5.0
Na	59	20	18	28	21
K	940	1116	1086	1107	1123
Ca	34	36	26	34	32
Mg	65	64	60	66	68
Cl	86	60	70	62	60
N	1300	1298	1302	1368	1443
P	128	169	160	170	175
S	56	56	51	54	57
Dry matter	4.4	4.6	5.2	4.6	4.9

Table 5. Absorption of water and nutrients by the different root parts in the treatments of experiment 2. The quantity of water is expressed as L per m^2 greenhouse area per day and the quantity nutrients in mmol per m^2 greenhouse area per day. Mean values over 150 days

Elements	EC values				
	0.75/3.0	3.0/3.0	5.0/3.0	7.5/3.0	10.0/3.0
Water	1.1/ 1.5	1.3/ 1.3	0.5/ 2.1	0.4/2.2	0.2/2.4
NO_3	2 /20	11 /11	8 /14	16 /6	14 /8
P	0.5/ 2.8	1.6/ 1.6	2.4/ 1.8	3.5/0.7	3.7/0.5
K	2 /15	8 /8	7 /10	13 /4	13 /4
Ca	−0.4/ 5.4	2.5/ 2.5	1.6/ 3.4	4.5/0.5	4.1/0.9
Mg	0.0/ 1.4	0.7/ 0.7	0.2/ 1.2	1.1/0.3	0.6/0.8

elements, not any absorption at all was found in that root half. As for calcium, even a negative value was measured.

In both experiments, the root development was judged visually at the end of the growing period. No differences were visible between treatments and between root halves within treatments.

Discussion

Results of both experiments described in this paper showed that tomatoes grown under an unequal distribution of nutrient concentrations in the root environment primarily responded to standard nutrient levels available in part of the root environment. Sometimes effects of the low EC value dominated. This seemed to be the case with the fruit weight. Although no significant differences were found, the fruit weight was highest in the three treatments in which an EC value of 0.75 was present in the root environment (Table 1). In experiment 2, this was accompanied by some negative effects on fruit quality. Effects of high EC values did not occur at all where a root half had the disposal of a standard EC value. However, in experiment 2 phosphate absorption was aggravated by a local high EC value. In experiment 1, such an effect on phosphate uptake was not found. This difference might be due to the cultivars grown (Howell and Bernard, 1961; Zijlstra *et al.*, 1987).

Nutrient and water absorption differed strongly between root halves of different EC value. Water was preferably absorbed from spots of low concentration, which is in agreement with results of salinity experiments (Bingham and Garber, 1970; Kirkham *et al.*, 1969; Lunin and Gallatin, 1965). For nutrient uptake the reverse was the case. These findings lead to the conclusion that plants absorb water and nutrients independently, which is in agreement with results of Allerton (1954), who stated that tomatoes make a double root system in his special "container-gravel" culture.

In treatments in which the roots had the disposal of both standard and low concentrations, only a restricted absorption of nitrate, phosphate and potassium and no absorption of calcium and magnesium were realized from the low concentration root part. The negative value found for calcium could indicate transport of calcium from one root to the other. Nutrient transport from one root to the other is possible indeed (De Jager, 1984), but not likely for calcium. So, the negative value for this ionic species is considered as a measuring error. For root halves with high EC values, the quantity of nutrients absorbed per unit water taken up appeared to be very high. So, for the root half grown at an EC value of 10, it was 70, 18, 65, 20 and 3 mmol per litre for nitrate, phosphate, potassium, calcium, and magnesium, respectively.

The tomato yield did not respond to high EC values in 50% of the root environment. This is in agreement with the work of Klapwijk and Wubben (1989), who removed a large part of the roots of a full grown tomato crop on rockwool and could not observe any crop reaction. This points to an overcapacity of root activity, making the plant less sensitive to stress situations in part of the root system.

Returning to the question posed in the introduction, on which spot a sample of solution must be gathered, the information presented suggests random sampling. This means that a sample should be composed of solution gathered from sufficient different spots selected at random, for a sample gathered in this way best reflects the quantity of nutrients available in the rockwool slabs. Avoidance of high concentrated spots between emitters seems to be incorrect, for such spots probably play an important role in nutrient absorption. With respect to fruit quality low concentrations have to be avoided in the nutrient solution supplied via the emitters, thus preventing spots of low EC values, which may adversely affect fruit quality.

References

Allerton F W 1954 *In* Tomato Growing. pp 87–96. Faber and Faber, London.

Bernstein L and Francois L E 1975 Effects of frequency of sprinkling with saline waters compared with daily drip irrigation. Agron. J. 67, 185–190.

Bes S S de 1986 A summary of methods for analysing glasshouse crops. Glasshouse Crops Research Station, Naaldwijk, The Netherlands, 5 p.

Bingham F T and Garber M J 1970 Zonal salinization of the root system with NaCl and boron in relation to growth and water uptake of corn plants. Soil Sci. Soc. Am. Proc. 34, 122–126.

Hoffman G J 1986 Salinity. *In* Trickle Irrigation for Crop Production. Eds. F S Nukayama and D A Bucks. pp 345–362. Elsevier, Amsterdam.

Howell R W and Bernard R L 1961 Phosphorus response of soybean varieties. Crop Sci. 1, 311–313.

Jager A de 1984 Effects of localized supply of H_2PO_4, NO_3, Ca and K on the concentration of that nutrient in the plant and the rate of uptake by roots in young maize plants in solution culture. Neth. J. Agric. Sci. 32, 43–56.

Kirkham M B, Gardner W R and Gerloff G C 1969 Leaf water potential of differentially salinized plants. Plant Physiol. 44, 1378–1382.

Klapwijk D and Wubben C F M 1989 The effect of root pruning and additional root growth on production of tomato. Glasshouse Crops Research Station, Naaldwijk, The Netherlands, Annual Report 1988, p 19.

Lunin J and Gallatin M H 1965 Zonal salinization of the root system in relation to plant growth. Soil Sci. Soc. Am. Proc. 29, 608–612.

Oster J D, Hoffman G J and Robinson F E 1984 Dealing with salinity: Management alternatives: Crops, water, and soil. California Agriculture 38, No 10, 29–32.

Sonneveld C and Kreij C de 1987 Nutrient solutions for vegetables and flowers grown in water or substrates, sixth edition. Glasshouse Crops Research Station, Naaldwijk, The Netherlands. Series Voedingsoplossingen glastuinbouw, no 8.

Zijlstra S, Nijs A P M den, Sonneveld C and Vos G 1987 Een nieuwe aanpak van het necroseprobleem bij meeldauwresistente komkommers. Prophyta 41, 138–140.

M. L. van Beusichem (Ed.), *Plant nutrition – physiology and applications*, 515–518.

PLSO IPNC206

Effect of increasing nutrient concentration on growth and nitrogen uptake of container-grown peach and olive

M. TATTINI
Institute for Woody Plants Propagation, National Research Council, Via Donizetti 6, I-50144 Florence, Italy

Key words: biomass partitioning, nitrogen uptake, nutrient concentration, olive, peach, soilless culture

Abstract

Five nutrient solutions of increasing ionic strength were supplied to container-grown self-rooted peach and olive, during 16 weeks of culture. Dry matter production of peach was closely related to nutrient supply, whereas olive plants did not correlate with the ionic strength of the nutritive solution. Peach and olive plants exhibited large differences in biomass partitioning: peach always showed a higher root/shoot ratio. As the nitrate supply was increased an enhanced nitrogen absorption was observed in peach. On the other hand, olive appeared to be unable to take up and assimilate such supplies.

Introduction

The pot production of fruit trees, using artificial substrates, has not received the attention that has been devoted to several other species.

In the Mediterrean area the peach and olive culture in high density planting, as well as the use of relatively new propagation techniques (*in vitro*), implies that the container-growth is an essential step in the nursery practices. However there is little knowledge about the nutrient requirements supporting optimum olive growth (Tattini *et al.*, 1986). Regarding peach plants, hydroponic culture experiments have been carried out, investigating basic aspects of mineral nutrition (Edwards, 1986) but large differences exist regarding duration and type of supply.

This experiment was focused on the growth responses of container-grown peach and olive plants, supplied with nutrient solutions of different concentrations.

Materials and methods

Two months old self-rooted olive (cv 'Maurino') and peach (cv 'Armking') plants, headed back to 8–10 cm, were grown in a drip fertigation system, in 3.5 L pots containing peat and sand (50/50). The plants were supplied during a 16 week period with nutrient solutions of increasing ionic strength: 0-8-16-24 and 32 meq/pot/week. The 8 meq solution supplied the following macronutrients: $NH_4^+ = 0.75$ $NO_3^- = 3.75$ $K^+ = 1.75$ $Ca^{++} = 0.80$ $Mg^{++} = 0.30$ $H_2PO_4^- = 0.35$ $SO_4^{--} = 0.30$.

The control treatment was supplied with deionized water.

Sixteen plants for each treatment were harvested, after 8 and 16 weeks, for growth and nutrient uptake measures.

Dry weights of the shoot, thin (<1 mm) and thick (> mm) roots were determined after a 3 day period at 65°C. Root fresh weight was also measured for nutrient uptake calculations.

Relative growth rate ($\bar{R}$) was estimated using the equation (Causton and Venus, 1981): $\bar{R} = \ln W_2 - \ln W_1/t_2 - t_1$ where W is the whole plant dry weight and t is the time. Subscripts 1 and 2 denote initial and final harvest.

Nitrogen uptake, per gram of fresh root weight, over the experimental period, was estimated using the equation (Edwards, 1986):

$$I = N_2 - N_1 \times \ln(WR_2/WR_1)/WR_2 - WR_1$$

N is total nitrogen content (stem + leaves + roots); WR is root fresh weight.

Total nitrogen content was carried out using Kjeldahl digestion and steam distillation of ammonia. Free nitrate in the tissues was evaluated according to Cataldo *et al.* (1975).

Results

Data in Table 1 and Figure 1 refer to peach and olive growth responses to the increasing nutrient supply. At the end of the culture period an increasing ionic strength in the fertilizing solution corresponds to an increased dry matter production of peach plants. On the other hand olive plants, supplied with the most concentrated solution (32 meq) showed a reduced growth compared to the 24 meq treatment (Fig. 1). Dry matter production vs time, referred as 'relative growth rate' ($\bar{R}$) was again different for peach and olive (Table 1).

Peach plants showed increased values of $\bar{R}$ in the period 0–8 weeks related to increasing nutrient supplies, whereas in the period 8–16 weeks a reduced dry matter production was induced by the highest nutrient solutions.

Regarding olive plants no growth differences were observed among 16, 24 and 32 meq/pot/

Table 1. Effect of increasing nutrition (meq/pot/week) on growth of peach and olive. 'Relative growth rate' ($\bar{R}$) = mg/g/w. All the growth parameters are on dry weight basis

	$\bar{R}$ (0–8 week)		$\bar{R}$ (8–16 week)		Root/shoot		Thin/thick roots	
Treatment	Olive	Peach	Olive	Peach	Olive	Peach	Olive	Peach
0	22.8 c	86.6 d	63.8	41.2	0.31 a	0.90 a	0.70 a	1.50 a
8	137.7 b	177.3 c	103.7	151.9	0.15 b	0.41 b	0.50 b	1.30 b
16	174.8 a	253.7 b	104.9	133.0	0.13 b	0.34 c	0.40 c	1.10 c
24	173.3 a	299.8 a	136.1	125.0	0.10 c	0.29 d	0.40 c	1.00 c
32	181.4 a	318.9 a	106.8	123.0	0.08 c	0.29 d	0.30 d	1.00 c

Means in a column not accompanied by the same letter, are signficantly different at the 5% level, using LSD test.

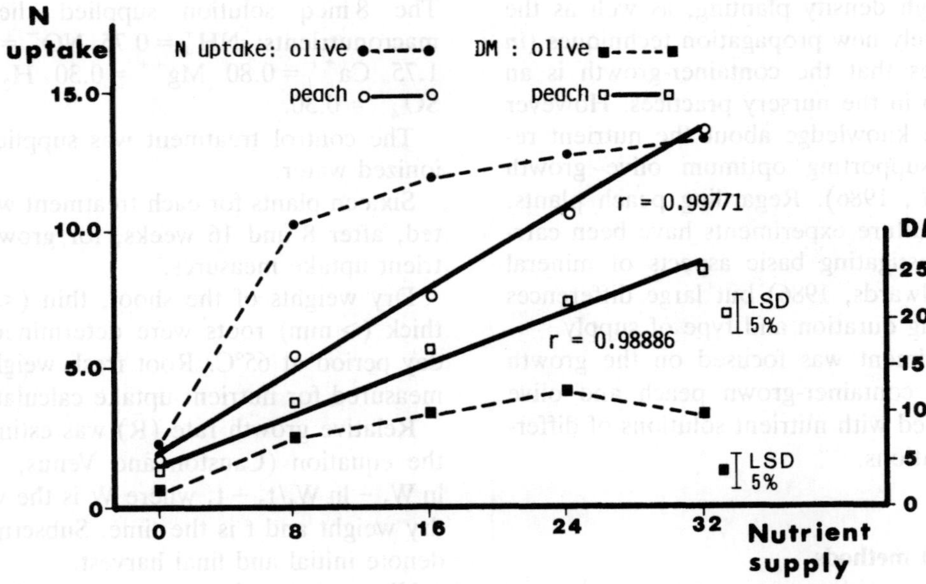

Fig. 1. Effect of increasing nutrient supply (meq/pot/week) on nitrogen uptake (m moles/g fr root wt) and on dry matter production (g) at final harvest.

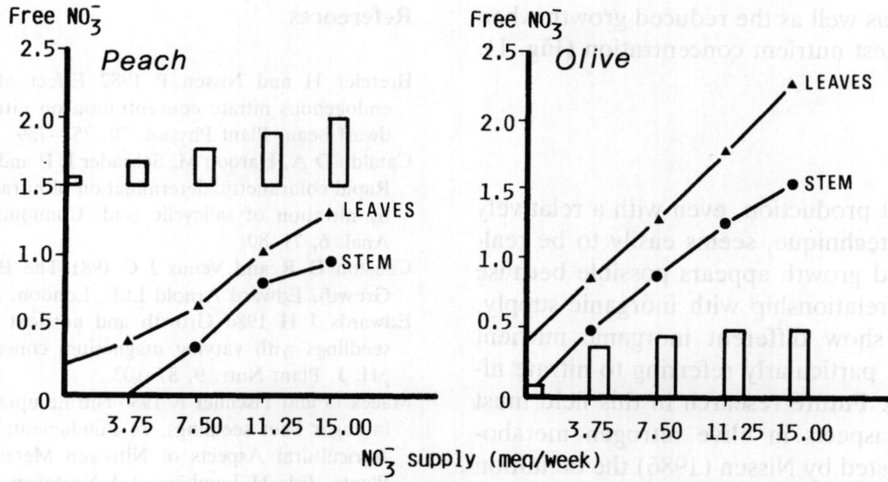

Fig. 2. Effect of increasing NO_3^- supply on free nitrate content (mg/g DM) of peach and olive tissues. The bars refer to nitrogen uptake (per cent scale).

week treatments, during the first 8 weeks of culture.

Peach plants showed a stronger reduction of the relative growth rate in time than olive plants.

The two species also showed large differences in biomass partitioning: independent of the relevant effect of the strength of the nutrient solution, peach plants always exhibited a higher root/shoot ratio than olive (Table 1). In a similar manner, the thin/thick roots ratio was strongly reduced in olive plants and the resulting root apparatus had a large proportion of old, lignified roots.

Regarding nitrogen uptake, peach plants showed a linear relationship with nitrate supply, while olive plants did not absorb in relation to the increasing supply (Fig. 1).

Free nitrate in stem and leaves of both species, increased with the nitrate supply (Fig. 2). However, in olive plants, free nitrate content increased even when nitrogen uptake was constant (NO_3^- supplies from 7.5 to 15.0 meq).

Discussion

Stimulation of growth of peach plants appear to be possible in a wide range of nutrient concentrations, which could exceed those tested in this experiment.

On the other hand container-growth of olive plants presents several problems, principally related to a naturally reduced development of the root apparatus, especially production of thin, metabolically active, roots. This appears to be responsible for the reduced and slow olive growth compared to peach.

Peach plants showed significant influence of nitrate supply on nitrogen uptake and dry matter production: thus the nitrate concentrations tested in this experiment appear to be in a suitable range where the nitrate uptake is not in the 'concentration independent uptake' phase (Breteler and Nissen, 1982; Nissen, 1986). On the other hand, olive plants exhibited a relatively high nitrogen uptake when using the lowest nitrate supply (Figs. 1, 2) but the nitrate absorption was saturated when the nitrate nitrogen supply exceeded 7.5 meq/pot/week.

Maeck and Tischner (1986) showed that when the 'concentration independent uptake' phase was reached, the nitrate-reductase activity sharply decreased and the uptake largely exceeded the assimilation (Nissen, 1986). This could explain the very high nitrate content in olive stem and leaves even when the nitrogen uptake was constant (Fig. 2) as well as the reduced growth when using the highest nutrient concentration (Fig. 1).

Conclusions

The peach pot production, even with a relatively new growing technique, seems easily to be realized; increased growth appears possible because of a positive relationship with inorganic supply. Olive plants show different inorganic nutrient requirements, particularly referring to nitrate nitrogen supply. Future research in this field must clarify basic aspects in olive nitrogen metabolism; as suggested by Nissen (1986) the definition of nitrate 'independent concentration uptake' phase is of primary importance.

Acknowledgements

The technical assistance of Mr R Tafani and Miss L Traversi is gratefully acknowledged.

References

Breteler H and Nissen P 1982 Effect of exogenous and endogenous nitrate concentration on nitrate utilization in dwarf bean. Plant Physiol. 70, 754–759.

Cataldo D A, Haroon M, Schrader L E and Young V L 1975 Rapid colorimetric determination of nitrate in plant tissues by nitration of salicyclic acid. Commun. Soil Sci. Plant Anal. 6, 71–80.

Causton D R and Venus J C 1981 The Biometry of Plant Growth. Edward Arnold Ltd., London, 307 p.

Edwards J H 1986 Growth and nutrient uptake of peach seedlings with varying magnesium concentrations at low pH. J. Plant Nutr. 9, 87–102.

Maeck G and Tischner R 1986 Nitrate uptake and reduction in sugar beet seedlings. *In* Fundamental, Ecological and Agricultural Aspects of Nitrogen Metabolism in Higher Plants. Eds H Lambers, J J Neeteson and I Stulen. pp 33–36. Kluwer Academic Publishers, Dordrecht, The Netherlands.

Nissen P 1986 Nutrient uptake by plants: Effect of external ion concentrations. Acta Hortic. 178, 221–228.

Tattini M, Mariotti P L and Fiorino P 1986 Fertigation growth and foliar analysis of self-rooted olive (cv "Frangivento") grown in container. Riv. Ortoflorofrutt. It., 70 439–445.

M. L. van Beusichem (Ed.), *Plant nutrition – physiology and applications*, 519–523.

PLSO IPNC060

Nutrient uptake, production and quality of *Rosa hybrida* in rockwool as affected by electrical conductivity of the nutrient solution

C. DE KREIJ[1,3] and TH.J.M. VAN DEN BERG[2]
[1]*Institute for Soil Fertility Research, P.O. Box 30003, 9750 RA Haren, The Netherlands, and* [2]*Research Station for Floriculture, Linnaeuslaan 2A, 1431 JV Aalsmeer, The Netherlands.* [3]*Present address: Glasshouse Crops Research Station, P.O. Box 8, 2670 AA Naaldwijk, The Netherlands*

Key words: electrical conductivity, flower quality, glasshouse, rockwool, *Rosa hybrida* L., rose

Abstract

For fruit-vegetable crops it has been found that a raised electrical conductivity (EC) of the nutrient solution improves fruit quality. To test this effect for flower crops, an experiment was conducted with rose (*Rosa hybrida*) grown in rockwool in six treatments: EC-values ranging from 1.0 to 5.0 $dS.m^{-1}$ in the drip-irrigation water and from 1.1 to 7.1 $dS.m^{-1}$ in the drainage water; the leaching fraction was 30–42%. Stem thickness, length and firmness, and vase life were negatively affected by a high EC. Optimum flower production was obtained at an EC of the irrigation water of 1.4 $dS.m^{-1}$, corresponding to 2.4 $dS.m^{-1}$ in the drainage water. Low EC gave a stronger reduction in summer than in winter, and an EC higher than the optimum gave the severest reduction in winter. Nutrient uptake, based on the mineral composition of the plant, relative to water uptake was, in m*M*: N 5.8, P 0.4, K 2.0, Mg 0.3 and Ca 0.7. With the leachate a considerable amount of the nutrients was lost.

Introduction

In Dutch floriculture, glasshouse rose takes the leading position, occupying 811 ha, or 23% of the total area in cut flowers in 1988. Until 1985 roses were mostly grown in soil. However, rockwool culture offers several advantages, for instance control of the composition of the nutrient solution to obtain optimum flower growth and quality. Consequently, cultivation of roses in rockwool expanded rapidly and the crop occupied 57 ha in 1987. However, most research on nutrient requirements of roses was done in soil or substrates, which, in contrast to rockwool, have an adsorption capacity (Armitage and Tsujita, 1979; Arnold Bik, 1970, 1972; Borelli, 1981; Brown and Ormrod, 1980; Fernandez Falcon *et al.*, 1986; Gabriëls and Meneve, 1973; Yaron *et al.*, 1969, Young *et al.*, 1973).

As cultivation in rockwool was already common in vegetable production, its concepts were also introduced in flower production. In tomato production a high electrical conductivity (EC) of the nutrient solution results in a better fruit quality (Sonneveld and Welles, 1988). In flower production a considerable portion of the vegetative part is harvested as well. It would be desirable if a high EC would also improve the vegetative part of the plant, especially in winter, when flowers have weak stems.

When crops have only a small rooting volume at their disposal, the nutrient solution has to be well balanced. Information about nutrient uptake can help to determine the most suitable solution composition. For this reason, and to get more information about the amount of nutrients lost with the leachate, a nutrient balance sheet was determined.

Materials and methods

The trial was conducted from March 1986 until April 1988. Rooted cuttings (9 plants per m^2) of

Rosa hybrida L. cultivar 'Sonia', were placed on rockwool slabs (20.4 L rockwool per m^2 glasshouse). The slabs, wrapped in plastic foil, were laid on a slightly sloping surface. On the low front side drainage holes in the foil permitted free drainage of excess nutrient solution. Trickle irrigation was used with one emitter per plant. Watering frequency was the same in every treatment: 1–5 times per day, adjusted to the transpiration rate of the crop, so that each day a quantity of leachate was produced of approximately 30% of the amount of solution supplied. There were six treatments, with 5 replicates, based on the EC of the leachate; for treatments 1–4 the target values were 1, 2, 3 and 5 $dS.m^{-1}$ (25°C), respectively. In treatments 5 and 6 the target value in summer was 2 $dS.m^{-1}$, which was raised in winter to 5 and 8 $dS.m^{-1}$, respectively. To calculate the annual effect of EC on production, winter EC values were counted for 30%, because winter production was 30% of the annual production. Rainwater was used to make up the nutrient solution. At EC = 1.5 $dS.m^{-1}$ the composition was, in m*M*: NO_3 11.0, H_2PO_4 1.25, SO_4 1.25, NH_4 0.75, K 5.0, Ca 3.75, Mg 0.75, and in u*M*: Fe 25, Mn 5, Zn 3.5, B 20, Cu 0.75, Mo 0.5.

For different EC-values the concentrations of the major elements were proportionally changed; the trace element concentrations remained constant. Production (two grades, determined on a visual basis), stem length, weight and vase life (in a controlled climate) of flowers were recorded. Stem strength was determined according to the degree of curvature of a flower with 60 cm stem length in a vase.

For the balance sheet of N, P, K, Ca and Mg the supply was calculated by multiplying the amount of irrigation water by the respective concentrations. The same was done for the leachate. Plant uptake was calculated from the harvested weight of different plant parts (stem, leaf, bud and wood, but not roots) and their respective mineral contents. Uptake of nutrients was divided by uptake of water to calculate the influx concentration.

Results

EC values, water uptake and leaching fractions are given in Table 1. The highest water uptake was found in treatment 2. EC of the leachate (y) was significantly correlated with EC of the nutrient solution in the root environment (x), collected with a syringe: $y = 1.05x - 0.067$; $r = 0.985$; d.f. = 400.

In winter, at high EC, in particular in treatments 5 and 6, many leaves had severe white spots and were shed; the symptoms resembled those of Mg-deficiency. These "sick" leaves from treatment 5 and 6 had lower Mg-content than healthy leaves from treatment 1 and 2 of the same age, but P and K-contents were higher (Table 2). At the end of the trial it was found that at high EC there were more dead and fewer living bottom breaks (Table 3).

Highest production was found in treatment 2.

Table 1. EC of irrigation water and leachate, water uptake by the crop and leachate fraction, July 1986 – April 1988. Leachate fraction is amount of drainage water relative to amount of solution supplied

Treatment		EC realized		Water-uptake	Leachate fraction
No.	EC target value in leachate	irrigation water	leachate		
		summer/winter, $dS\ m^{-1}$		$l\ m^{-2}\ day^{-1}$	$m^3\ m^{-3}$
1	1	1.0	1.1	1.03	0.33
2	2	1.4	2.4	1.12	0.30
3	3	1.8	3.3	1.05	0.33
4	5	2.9	4.9	0.86	0.42
5	2/5	1.5/3.1	2.6/4.9	1.03	0.31
6	2/8	1.5/5.0	2.4/7.1	0.95	0.37

Table 2. Composition of healthy and sick leaves (mmol/kg DM)

	N	P	K	Ca	Mg
Target value	2200	100	800	400	120
Winter 1986/1987					
healthy	2950	120	960	400	100
sick	2480	350	1550	370	30
Winter 1987/1988					
healthy	3400	230	925	170	130
sick	3840	260	1614	260	90

Table 3. Bottom breaks (number per square metre) at the conclusion of the trial

Treatment	Bottom breaks Dead	Living
1	4.8	24.1
2	5.5	22.8
3	7.9	18.1
4	9.8	16.2
5	10.4	15.8
6	15.6	12.1
sign.	* * *	* * *
LSD ($p = 0.05$)	3.1	3.6

A significant linear decrease in total mass production and first grade flowers was found when EC of the leachate exceeded 2.4 $dS.m^{-1}$ (Fig. 1). This corresponded to EC = 1.4 $dS.m^{-1}$ in the drip-irrigation water (Table 1). Above EC = 1.4 $dS.m^{-1}$ in the drip irrigation water the correlation between EC in irrigation water (x) and total mass production (y1) in $kg.m^{-2}$ was: $y1 = 11.6 - 1.04\,x$ ($r = 0.958$, $p < 0.01$), and number of first grade flowers (y2) in $number.m^{-2}$: $y2 = 328 - 26.1\,x$ ($r = 0.954$, $p < 0.01$), for July 1986 – April 1988. There was an effect of season on production. Relative to treatment 2, the lowest EC (treatment 1) reduced production by 7% in winter, and by 13% in summer. The high EC

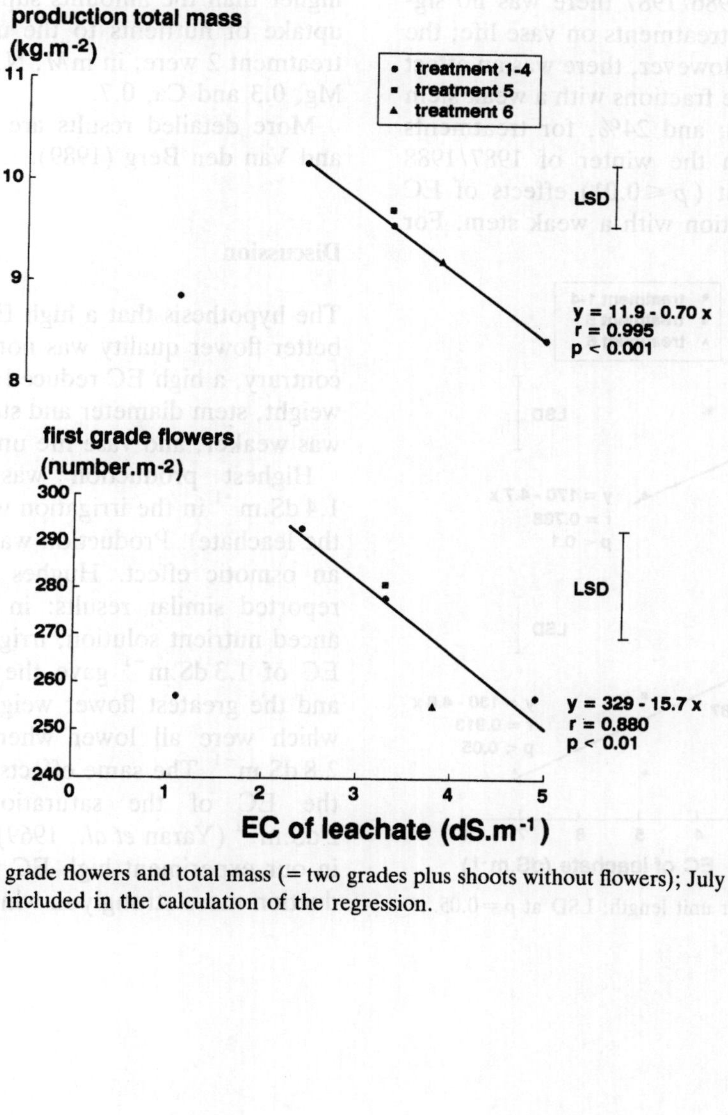

Fig. 1. Production of first grade flowers and total mass (= two grades plus shoots without flowers); July 1986 – April 1988; LSD at $p = 0.05$; treatment 1 not included in the calculation of the regression.

values in treatments 3 and 4 gave a reduction of 15–22% in winter, and of 2–12% in summer, relative to treatment 2. So the low EC resulted in the greatest reduction in summer, the high EC in winter.

There were significant ($p \leq 0.05$) negative effects of increasing EC on flower weight and stem length. For treatments 1–4 the mean flower weights were 25.4; 25.3; 24.4 and 22.8 g, respectively (LSD ($p = 0.05$) = 1.1), and the mean stem lengths were 59.2; 59.6, 58.6 and 56.5 cm (LSD ($p = 0.05$) = 1.7).

Quality characteristics of the stem are given in Figure 2 separately for summer and winter. At high EC, stem weight per unit length was low. Similar effects were found for stem length, stem diameter and stem weight. In the summer of 1987, treatments 5 and 6, grown at a high EC in the preceding winter, still had a lower stem weight per unit length than treatment 2, although the differences were not significant.

In the winter of 1986/1987 there was no significant effect of the treatments on vase life; the mean was 8.7 days. However, there was an effect on stem strength. The fractions with a weak stem were 4; 0; 8; 12; 28; and 24%, for treatments 1–6, respectively. In the winter of 1987/1988 there were significant ($p \leq 0.01$) effects of EC on vase life and fraction with a weak stem. For treatments 1–6 the vase life was 10.9; 10.0; 8.1; 10.4; 7.6 and 6.0 days, respectively (LSD ($p = 0.05$) = 2.4); the fractions with weak stems were 24; 32; 52; 28; 60, and 68% (LSD ($p = 0.05$) = 24). High EC, in particular treatments 5 and 6, resulted in a shorter vase life and a higher fraction with weak stems than low EC, except for treatment 4.

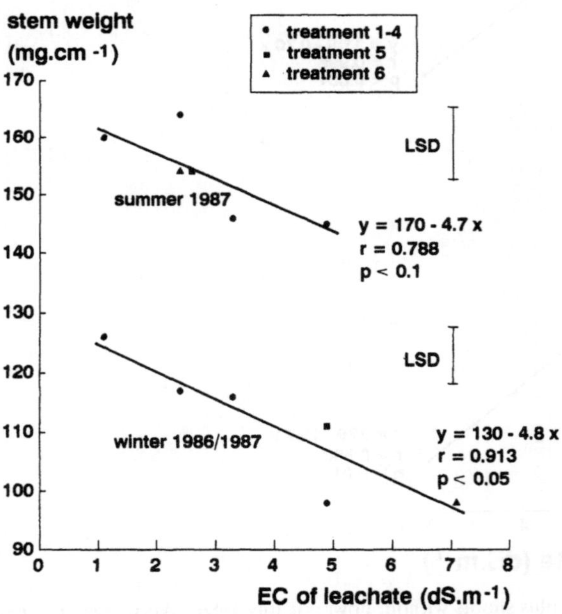

Fig. 2. Stem weight per unit length; LSD at $p = 0.05$.

Table 4. Nutrient balance sheet for treatment 2 (kg per ha per year)

	N	P	K	Mg	Ca
Supply	690	161	854	77	506
Leachate	332	26	438	53	317
Plant uptake	331	49	326	29	110
Leachate + uptake	654	75	764	82	427

The nutrient inputs and outputs did not balance (Table 4). The amounts of N, P, K and Ca taken up by the plants plus the amounts leached were 5; 53; 11 and 16% lower, and of Mg 5% higher than the amounts supplied. The ratios of uptake of nutrients to the uptake of water for treatment 2 were, in m*M*: N, 5.8; P, 0.4; K, 2.0; Mg, 0.3 and Ca, 0.7.

More detailed results are given by De Kreij and Van den Berg (1989).

Discussion

The hypothesis that a high EC would result in a better flower quality was not confirmed. On the contrary, a high EC reduced flower weight, stem weight, stem diameter and stem length; the stem was weaker, and vase life unaffected or shorter.

Highest production was found at EC = 1.4 $dS.m^{-1}$ in the irrigation water (2.4 $dS.m^{-1}$ in the leachate). Production was lower at high EC, an osmotic effect. Hughes and Hanan (1978) reported similar results: in gravel with a balanced nutrient solution, irrigation water with an EC of 1.3 $dS.m^{-1}$ gave the highest production and the greatest flower weight and stem length, which were all lower when EC was 1.9 and 2.8 $dS.m^{-1}$. The same effects occur in soils when the EC of the saturation extract exceeds 2 $dS.m^{-1}$ (Yaran *et al.*, 1969). It is not clear why in our experiment high EC values reduced production more strongly in winter than in summer.

Perhaps it is due to an induced Mg deficiency caused by a high root temperature relative to the low light intensity. It is known that the first flush in spring, grown under lower light intensities than in summer, has a higher P-content than the third or fourth flush, grown in mid-summer (Johansson, 1979a, b). Bakly (1974) found a severe toxic effect when P fertilizer was applied at three times the optimum rate. In tomato, an increase in P concentration of the nutrient solution resulted in a decrease in Mg content of the leaves (Hipp and Gerard, 1969). At low EC, in particular in summer, production was reduced. This was caused by a lack of nutrients, probably N.

With the leachate a considerable amount of the nutrients supplied was lost. For P there was a great difference between the amount supplied and the amount taken up by the plant and contained in the leachate; a reasonable explanation is that P precipitated in the substrate. If this occurs it is impossible to calculate plant uptake from the difference between P-supply and P-leachate, in the way Heisel (1981) did. Perhaps this is the reason for the higher P uptake (0.27 Mol P per Mol N) which Heisel (1981) found, as compared with our results (0.07 Mol P per Mol N).

References

Armitage A M and Tsujita 1979 Supplemental lighting and nitrogen nutrition effects on yield and quality of Forever Yours roses. Can. J. Plant Sci. 59, 343–350.

Arnold Bik R 1970 Fertilization trial with 'Baccara'. (*In Dutch*). Vakblad Bloemisterij 25, 13, 478–479.

Arnold Bik R 1972 Effects of nitrogen and potassium nutrition on flower yield and quality of the glasshouse rose Carol. Colloq. Proc. no. 2 pp. 89–92. Intern. Potash Inst., Berne.

Bakly S A 1974 Effects of fertilization treatments on the yield of Chryslar Imperial rose plants. Agric. Res. Rev. 52, 9, 95–99.

Borelli A 1981 The influence of moisture regime and nitrogen fertilization on glasshouse rose production. (*In Italian: with English summary*) Riv. Ortoflorofrutt. 65, 109–117.

Brown W W and Ormrod D P 1980 Soil temperature effects on greenhouse roses in relation to air temperature and nutrition. J. Am. Soc. Hortic. Sci. 105, 57–59.

De Kreij C and Van den Berg Th 1989 EC, production, quality and balance sheet of minerals for rose grown in rockwool. (*In Dutch*). Report no. 80, Research Station for Floriculture, Aalsmeer, The Netherlands.

Fernández Falcón M, Alvarez González C E, García V and Báez J 1986 The effect of chloride and bicarbonate levels in irrigation water on nutrition content, production and quality of cut roses 'Mercedes'. Scientia Hortic. 29, 373–385.

Gabriëls R and Meneve I 1973 Nutritional requirements of roses grown in peat. Scientia Hortic. 1, 341–349.

Hipp B W and Gerard C J 1969 Magnesium – phosphorus interrelationships in tomatoes. Agron. J. 61, 403–405.

Heisel C 1981 Ein Beitrag zur Ermittlung der Nährstoffaufnahme bei Hausrosen. Arch. Gartenbau 29, 189–200.

Hughes H E and Hanan J J 1978 Effect of salinity in water supplies on greenhouse rose production. J. Am. Soc. Hortic. Sci. 103, 694–699.

Johansson J 1979a Leaf composition of flowering shoots from different greenhouse rose cultivars as influenced by rootstock and season. Acta Agric. Scand. 29, 85–92.

Johansson J 1979b Main effects and interactions of N, P, and K applied to greenhouse roses. Acta Agric. Scand. 29, 191–208.

Sonneveld C and Welles G W H 1988 Yield and quality of rockwool-grown tomatoes as affected by variations in EC values and climatic conditions. Plant and Soil 111, 37–42.

Yaron B, Zieslin N and Halevy A H 1969 Response of Baccara roses to saline irrigation. J. Am. Soc. Hortic. Sci. 94, 481–484.

Young T W, Snyder G H, Martin F G and Hayslip N C 1973 Effects of nitrogen, phosphorus and potassium fertilization of roses on Oldsmar fine sand. J. Am. Soc. Hortic. Sci. 98, 109–12.

M. L. van Beusichem (Ed.), *Plant nutrition – physiology and applications*, 525–531.

PLSO IPNC350

Synchronisation of supply and demand is necessary to increase efficiency of nutrient use in soilless horticulture

M. VAN NOORDWIJK
Institute for Soil Fertility Research, P.O. Box 30003, 9750 RA Haren, The Netherlands

Key words: cucumber, nutrient balance, rockwool, root volume, soilless horticulture, synchronisation, tomato, water uptake

Abstract

In modern horticulture on artificial substrates effective systems for plant nutrition have been developed for many crops. The efficiency of nutrient use, however, is often low: on average 40 to 80% of all nutrients applied to tomato and cucumber grown on rockwool slabs is leached from the root environment. In soilless systems only the volume of nutrient solution provides buffering for nutrients when supply and uptake of nutrients are not equal. From a simple model of water and nutrient balance for a well-mixed nutrient solution system, relations between nutrient and water leaching were derived, explaining the low nutrient use efficiencies obtained in practice on soilless media. Improvement of nutrient use efficiency largely depends on improving synchronisation of nutrient supply with nutrient demand. The ratio of "uptake concentration" (current nutrient uptake rate divided by current water uptake rate) and the nutrient concentration in the system is a key parameter. Understanding of fluctuations in this ratio, as determined by daily rhythms, weather conditions and growth stages, may lead to intensive plant nutrition systems that are not only effective, but also efficient.

Introduction

The introduction of techniques for soilless culture in glasshouse horticulture has led to a large reduction in the buffering capacity of the root environment for nutrients and water, both through a reduction in size of the root environment and through a low buffering per unit volume. The small buffering capacity offers possibilities for manipulating and rapidly changing the root environment, but it also imposes a need for frequent replenishment and for regulating the nutrient concentration of the solution. Analysis of the minimum required root volume has shown that further reductions are possible. Nutrient uptake is no problem even when root size is restricted, provided that the concentration is in the usual range and the supply of nutrients to the roots is continuously maintained.

As a plant rarely takes up water and the various nutrients in proportion to the external supply, it is continuously changing the composition of the nutrient solution. The smaller the rooted volume, the greater these disturbances are. Problems of maintaining an "ideal" root environment in poorly buffered systems pose a major obstacle to obtaining maximum plant production as well as a high nutrient use efficiency, as will be discussed in this article. As systems with a fast continuous recirculation of nutrient solution are not yet feasible in commercial practice as long as cheap and reliable sterilisation techniques for large amounts of solution are not available, rockwool slabs with drainage-to-waste are the most common system in practice (at least in the Netherlands). Along with frequent trickle irrigations during the day, excess water and nutrients drain from the rockwool slabs continuously. Recently, however, a trend has developed to collect the drainage water for re-use after sterilization (Runia *et al.*, 1988).

Van Noordwijk and Raats (1982) determined

the nutrient balance for a cucumber crop grown on rockwool under semi-practical conditions. Only about 30% of the N, P, Mg and K and about 10% of the Ca and Mg applied during the growing season was actually taken up by the crop (harvested fruits plus crop residue at harvest time). Figure 1 shows the results of a survey of water and nutrient use among tomato and cucumber growers (Van der Burg and Hamaker, 1984). Total nutrient (fertilizer) use is directly correlated with the amount of water given in excess of plant transpiration and follows the recommended recipe for the nutrient solutions, which are different for tomato and cucumber. From the estimated uptake of nutrients it can be seen that only 20–60% of the nutrients applied is taken up by the crop, the remainder (40–80%) being lost to soil, surface water or groundwater. The amount of nutrients leached per hectare (up to 1000 kg of nitrogen per growing season) is a matter of serious concern.

In this article I hope to clarify the apparent logic for this low nutrient use efficiency. Improvement of nutrient use efficiency primarily depends on a better synchronisation of nutrient supply with nutrient demand. The point of view expressed here was developed on the basis of detailed observation of root and nutrient distribution in rockwool slabs, analysis of flow patterns of nutrient solution in a rockwool slab between trickler and drains, estimation of nutrient balances over a whole growing season, and a theory on leaching requirements (Van Noordwijk and Raats, 1980; 1982; Van Noordwijk, 1983).

Minimum root volume

Analysis of the minimum root volume for soilless culture systems (De Willigen and Van Noordwijk, 1987) suggests that the required size of the root system for maximum growth is determined by water uptake, according to equation 1:

$$A_r = \frac{E_p}{L_p[\Delta H_p - 2\pi_0\sigma_r^2/(1-\sigma_r)]} \quad (1)$$

where:

A_r = required root surface area per plant [m^2],
E_p = maximum transpiration rate per plant [L/hour],
L_p = hydraulic conductance per unit root surface area [$Lm^{-2}s^{-1}Pa^{-1}$],
ΔH_p = maximum acceptable matric suction in the leaves [Pa],
π_0 = osmotic potential of the nutrient solution [Pa],
σ_r = reflection coefficient for solutes ($0 < \sigma_r < 1$).

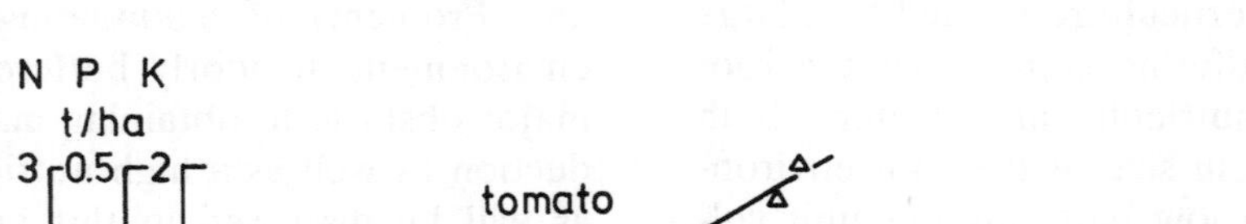

Fig. 1. Fertilizer use in 15 glasshouses in one growing season in relation to the amount of water leached (after Van der Burg and Hamaker, 1984); estimated nutrient uptake for both cucumber and tomato is indicated.

In a number of glasshouse experiments with tomato and cucumber, the following parameters were found to describe the effects of physical limitations to root extension on shoot growth. Reduced shoot growth and/or fruit production was found for root sizes less than A_r:

	L_p 1/(m^2 s Pa)	ΔH_p MPa	σ_r	E_p L/hr	π_0 MPa	A_r m^2
Tomato	$1.2\,10^{-10}$	0.5	0.3	0.2	0.04	0.94
Cucumber	$0.9\,10^{-10}$	0.4	0.3	0.3	0.04	2.3

Cucumber requires 2.5 times more root surface area than tomato. For a uniform root diameter of 0.3 mm, root surface areas of 0.9 and 2.3 m^2 are equivalent to root volumes of 0.14 and 0.35 dm^3, respectively. The external volume required to obtain this minimum amount of roots is also larger for cucumber than for tomato, but both are much smaller than the 10 dm^3 of rockwool currently used.

Reasons for leaching

Losses of water and nutrients are partly due to uneven delivery of solution by the tricklers and uneven growth and uptake by the plants, which cause leaching of excess nutrient solution when the water supply is adjusted to the most-demanding plant with the slowest trickler. Even in completely homogeneous systems, however, an apparent need for leaching of nutrient solution stems from the salt accumulation which would otherwise occur. If irrigation water is used which contains NaCl, accumulation of salts can only be avoided by leaching or by including specific NaCl absorbers (*e.g.* saltmarsh plants) in a recirculation system (P.J.C. Kuipers, pers. comm.). When water of good quality is used, salts accumulating are mainly nutrients supplied in excess of crop demand. This problem consists of two parts: accumulation of all nutrients contained in the solution if the total salt concentration C_s exceeds the "uptake concentration" C_u (nutrient uptake/water uptake), and shifts in the relative concentration of individual ions as the ratio in which they are supplied may not be equal to the ratio in which they are taken up in different growth stages (Van Goor *et al.*, 1989). Analysis of salt distribution patterns in rockwool slabs (Van Noordwijk and Raats, 1980; 1982) showed that small-scale differences in total salt content, as indicated by electrical conductivity (EC) occurred: over a 5 cm distance EC varied by a factor of 2. Relatively high EC values were found in the lowest zone of the slab, between two plants and between tricklers. Ions accumulating in the dead corners are mainly Cl^-, SO_4^{2-}, Ca^{2+} and Mg^{2+}. In a nutrient solution of the recommended composition Ca and Mg are supplied in higher concentrations relative to plant uptake than K, so as to maintain suitable K/Ca and K/Mg ratios for adequate uptake. The necessity to maintain a K/Ca ratio in the root environment which differs from the uptake ratio may be due to the fact that probably only young parts of the root system are involved in Ca uptake, while the whole root length will be active in K uptake. For the young roots the K/Ca uptake ratio may be equal to the ratio supplied, while around the older roots Ca accumulates, unless the solution is thoroughly mixed.

As we may assume that plant nutrient uptake is independent of the external concentration over a considerable range of concentrations, at least for N, P and K (Clarkson, 1985; De Willigen and Van Noordwijk, 1987), the plant has a destabilizing effect on the nutrient solution. As soon as $C_u < C_s$, C_s will tend to rise even more, and as soon as $C_u > C_s$, C_s will decrease further. The ratio of C_u and C_s is therefore an important characteristic of the system. The "ideal" situation of $C_u/C_s = 1$ throughout the growing season is difficult to realise as C_u is continuously changing, mainly due to variation in water uptake during the day and between days of different insolation. Figure 2 shows weekly recordings of C_u/C_s for tomato and cucumber experiments, on a recirculating nutrient solution. When C_s is considered to be constant throughout the growing season, C_u apparently exceeds C_s in the initial part of the growing season, approximately equals C_s during cloudy periods and is considerably lower than C_s during sunny periods. During sunny periods we may therefore expect nutrient accumulation to occur in the root environment

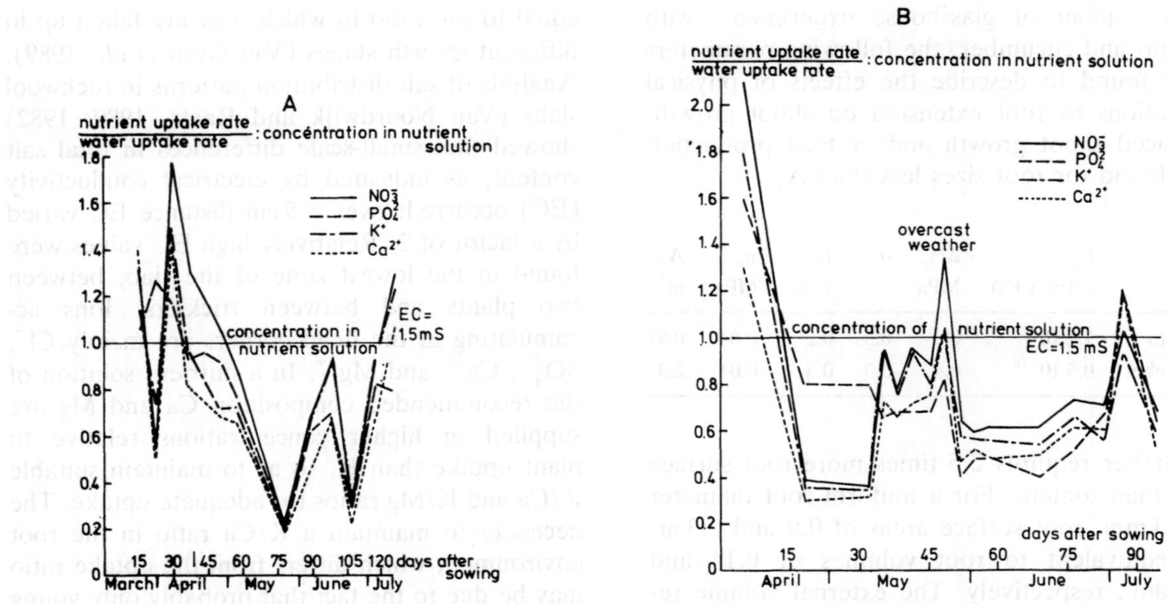

Fig. 2. Uptake concentration C_u (nutrient uptake divided by water uptake) divided by external concentration C_s in tomato (A) and cucumber (B) experiments on a recirculating nutrient solution.

and the apparent need for washing out these excess salts with nutrient solution to arise.

The concentration of all nutrients in the nutrient solution needed to obtain maximum yields (and/or quality) has been established for many plant species in experiments in which many concentration levels, maintained throughout the growing season, were tested (Sonneveld and Van der Wees, 1980). If in such an experiment a concentration C_s would be tested which equals the average C_u over the whole growing period, the plants would receive insufficient nutrients during some parts of the growing season and would respond by sub-maximum growth and/or quality. From the fluctuations in the C_u/C_s ratio in Figure 2 it is understandable that in the current advice C_s exceeds the average C_u by 30 to 40%. For such a value of C_s nutrient shortages can be avoided. As a consequence, however, during large parts of the growing season nutrients will accumulate in root environment. The relation between C_u/C_s, the degree of salt accumulation that can be tolerated in the root environment, and the nutrient use efficiency obtained can be quantified in a simple model of nutrient and water balance for a well-mixed system.

Nutrient and water balance

The relationship between leaching of water and nutrients and fractional uptake (C_u/C_s) can be formulated simply for a perfectly mixed system such as a rapidly recirculating nutrient solution. For imperfectly mixed systems of low buffering capacity, such as rockwool slabs, this algebraic description may still be a reference. For the water and nutrient balance of a perfectly mixed system (Raats, 1980) we may write:

$$N_a = N_u + N_l\,, \qquad (2)$$

$$W_a = W_u + W_l\,, \qquad (3)$$

where:

N_a and W_a = input of nutrients and water, respectively, N_u and W_u = uptake of nutrients and water, respectively, N_l and W_l = leaching of nutrients and water, respectively.

We then define "leaching fractions" l_n and l_w for nutrients and water as:

$$l_n = N_l/N_a = 1 - N_u/N_a\,, \qquad (4)$$

$$l_w = W_l/W_a = 1 - W_u/W_a\ , \qquad (5)$$

the required relation between l_n and l_w as a function of relative utilisation of the nutrients in the solution c_u is given by:

$$l_n = 1 - c_u(1 - l_w),\ \text{for}\ l_w > 0\ , \qquad (6)$$

where:

$c_u = C_u/C_n$,
$C_s = N_l/W_l$, the system concentration,
$C_n = N_a/W_a$, the concentration added,
$C_u = N_u/W_u$, the uptake concentration.

From (6) we see that for $c_u = 1$, when roots of the plant effectively act as a sponge absorbing the solution as it comes, $l_n = l_w$; for $c_u > 1$ the nutrient solution is depleted and $l_n < l_w$, and for $c_u < 1$ salts are accumulating in the solution and $l_n > l_w$. Figure 3 shows the relation between l_w and l_n according to (6).

The data of Van der Burg and Hamaker (1984) on the nutrient and water balance of tomatoes and cucumber grown on rockwool, presented in Figure 1, fit in well with this scheme for a value of c_u of about 0.6 (compare Table 1). For horticultural practice we thus obtain:

$$l_n = 0.4 + 0.6\, l_w,\ \text{for}\ l_w > 0\ . \qquad (7)$$

From (6) we see that c_u affects l_n in two ways: c_u determines the nutrient excess for $_w$ approaching 0 and it determines the extra nutrient loss when nutrient solution is washed through

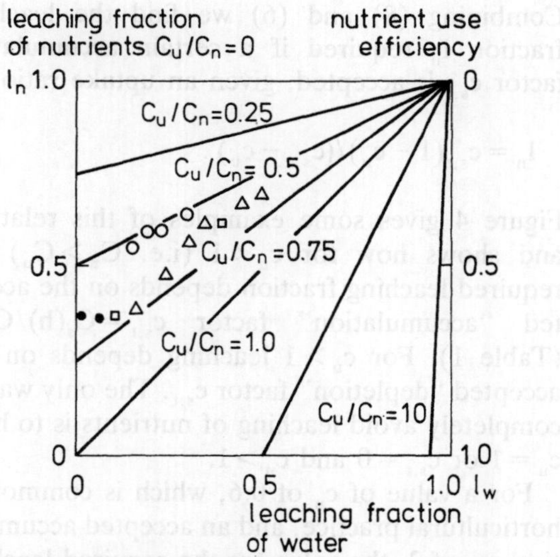

Fig. 3. Relation between leaching fraction of nutrients l_n and leaching fraction of water l_w for a well mixed system, as influenced by $c_u = C_u/C_n$, (equation 6); points refer to the data of Figure 1.

the system to the drains. In fact, a third effect exists, as the required leaching of water is at least partly determined by the accumulation of nutrients in the root environment. This last relation can be formulated as follows.

Salt accumulation or depletion in the system is limited to a tolerance factor $c_{s,t} = C_s(h)/C_s(l)$ (Table 1). Such tolerances apparently vary from 1.75 for K and Ca to 5.0 for SO_4. A relation exists between l_n, l_w and $c_{s,t}$:

$$c_{s,t} = C_s/C_n = l_n/l_w\ . \qquad (8)$$

Table 1. Estimated daily uptake concentration, C_u, of nutrients by a tomato crop compared with the recommended composition of the nutrient solution in the rockwool slab (Sonneveld and Van der Wees, 1980). Three system concentrations are shown for each nutrient: lowest $C_s(l)$, desired $C_s(d)$ and highest $C_s(h)$. The range of concentrations tolerated is defined as $C_s(h)/C_s(l)$. Uptake concentration C_u calculated for a daily transpiration of 2.5 l/plant is compared with the desired system concentration $C_s(d)$ in the last column

	Daily uptake per plant (mg)	Uptake conc. C_u	System concentration C_s $C_s(l)$	$C_s(d)$	$C_s(h)$	Tolerance $C_s(h)/C_s(l)$	Utilisation $C_u/C_s(d)$
			($mg\,L^{-1}$)				
N	225	90	84	130	210	2.5	0.69
P	45	18	15	31	47	3.0	0.58
S	50	20	32	64	160	5.0	0.31
K	450	180	160	200	270	1.75	0.90
Ca	100	40	160	200	280	1.75	0.20
Mg	30	12	24	48	72	3.0	0.25

Combining (8) and (6) we find the leaching fraction l_n required if a certain accumulation factor $c_{s,t}$ is accepted, given an uptake ratio c_u:

$$l_n = c_{s,t}(1 - c_u)/(c_{s,t} - c_u)\,. \tag{9}$$

Figure 4 gives some examples of this relation, and shows how for $c_u < 1$ (i.e. $C_n > C_u$) the required leaching fraction depends on the accepted "accumulation" factor $c_{s,t} = C_s(h)/C_s(l)$ (Table 1). For $c_u > 1$ leaching depends on the accepted "depletion" factor $c_{s,t}$. The only way to completely avoid leaching of nutrients is to have $c_u = 1$ or $c_{s,t} = 0$ and $c_u > 1$.

For a value of c_u of 0.6, which is common in horticultural practice, and an accepted accumulation $c_{s,t}$ of 2, the value for the required leaching fraction of nutrients, l_n, is 57%. If N-uptake is 700 kg/ha, nitrogen losses to the environment are 930 kg/ha in this case. If c_u could be increased to 0.8, 0.9 or 0.95, leaching could be reduced to 33, 18 and 9.5% of the amount applied and losses of nitrogen to the environment to 350, 156 and 73 kg/ha, respectively. For $c_{s,t} = 5$ the values would be 24, 12 and 6.2%, respectively, and N-losses would amount to 220, 97 and 46 kg/ha.

In systems which are imperfectly mixed, heterogeneity of salt distribution can be built up; the plant, however, has a tendency to reduce the existing heterogeneity by preferentially taking up water in the area with the lowest salt concentrations (Cerda and Roorda van Eysinga, 1981).

On further analysis the uptake concentration C_u can be split up into two components:

$$C_u = N_u/W_u = \text{NUTE}/\text{WUTE} \tag{10}$$

where:

NUTE = (nutrient concentration in the plant)$^{-1}$ = dry weight per unit nutrient uptake = Nutrient Utilisation Efficiency,

WUTE = transpiration ratio = Water Utilisation Efficiency = dry weight produced per unit water transpired.

Both NUTE and WUTE are quantities about which much information is available. WUTE values in the range 3–7 g L^{-1} are common (De Wit, 1958), relatively high values being typical of C4-types of photosynthesis, relatively low values for plants having a C3 photosynthesis. A value of WUTE of 4 g L^{-1} in combination with an N-content of 2.5% gives a $C_u = 4 \times 0.025 = 0.1$ g L^{-1}, which agrees with the value assumed in Table 1. Variation of C_u around this average value can be accounted for by short-term fluctuations in transpiration ratio and internal concentration.

In equation (1) we used the concept of a reflection coefficient σ_r, which is defined as the fraction of solutes arriving at the root membrane but not passing it. In fact, σ_r is equal to $1 - C_u/C_s$. When $C_u = C_s$, no reflection occurs and when $C_u = 0$, there is a complete reflection. Using this definition and substituting $\pi_0 = RTC_s$

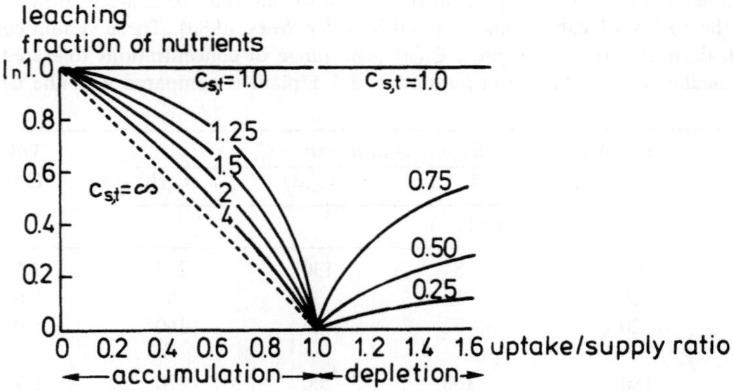

Fig. 4. Leaching fraction of nutrients l_n required in a well-mixed system as a function of uptake/supply ratio c_u and accumulation or depletion factor $c_{s,t}$, according to equation (9).

(where R is the gas constant and T the absolute temperature), we obtain (for $C_s > C_u$):

$$A_r = \frac{E_p}{L_p[\Delta H_p - 2\,R\,T(C_s - C_u)^2/C_u]} \;. \qquad (11)$$

This equation shows that values of C_s considerably exceeding C_u increase the required root surface area, or impede water uptake if the size of the root system remains the same. Although restrictions on water uptake in certain parts of the cropping cycle may help to induce flowering or fruit set, the present use of high EC values is probably far from ideal and other ways of regulating plant growth should be investigated as they may lead to higher nutrient use efficiencies.

References

Cerda A and Roorda van Eysinga J P N L 1981 Tomato plant growth as affected by horizontally unequal osmotic concentrations in rockwool. Neth. J. Agric. Sci. 29, 189–197.

Clarkson D T 1985 Factors affecting mineral nutrient acquisition by plants. Annu. Rev. Plant Phys. 36, 77–115.

De Willigen P and Van Noordwijk M 1987 Roots, plant production and nutrient use efficiency. Doct. thesis, Wageningen Agricultural University, 282 p.

De Wit C T 1958 Transpiration and crop yields. Versl. Landbouwk. Onderz. 64.6, 88 p.

Raats P A C 1980 The supply of water and nutrients in soilless culture. Proc. Fifth Intern. Congr. on Soilless Culture, Wageningen, 53–62.

Runia W Th, Van Os E A and Bollen G J 1988 Disinfection of drainwater from soilless cultures by heat treatment. Neth. J. Agric. Sci. 36, 231–238.

Sonneveld C and Van der Wees A 1980 Voedingsoplossingen voor de teelt van tomaten in steenwol, 5th ed. Proefst. Tuinbouw onder Glas, Naaldwijk, 42 p.

Van der Burg A M M and Hamaker Ph 1984 Water- en mineraalhuishouding bij teelten op substraat in de praktijk. Inst. Cultuurtechniek en Waterhuishouding, Nota 1520, 44 p.

Van Goor B J, De Jager A and Voogt W 1989 Nutrient uptake by some horticultural crops during the growing period. Proc. Seventh Intern. Congr. on Soilless Culture, Flevohof, pp 163–176. ISOSC, Wageningen.

Van Noordwijk M 1983 Meststofgebruik in de substraatteelt: Benutting of verspilling? Groente en Fruit 39.23, 42–45.

Van Noordwijk M and Raats P A C 1980 Drip and drainage systems for rockwool cultures in relation to accumulation and leaching of salts. Proc. Fifth Intern. Congr. on Soilless Culture, Wageningen, pp. 279–287. ISOSC, Wageningen.

Van Noordwijk M and Raats P A C 1982 Zoutophoping en -uitspoeling in samenhang met het druppelsysteem bij de teelt op steenwool (The influence of the drip system upon the accumulation and leaching of salts in rockwool cultures). Inst. Bodemvruchtbaarheid, Rapp. 9–81, 36 p.

M. L. van Beusichem (Ed.), *Plant nutrition – physiology and applications*, 533–537.

PLSO IPNC671

Cultivation of cut flowers with ammonium as nitrogen source

P. KASTEN and K. SOMMER
Institute of Agricultural Chemistry, University of Bonn, Meckenheimer Allee 176, D-5300 Bonn 1, FRG

Key words: ammonium nutrition, ammonium-depot, brompton stocks, chrysanthemums, nitrogen fertilization

Abstract

Ammonium-depot-fertilization of chrysanthemums and brompton stocks was compared with a conventional fertilization system in greenhouse experiments. Plants which were exclusively grown on the basis of ammonium showed no signs of incompatibility. No significant differences in yield level between the two fertilization systems were observed. Chrysanthemums which were fertilized with ammonium showed an improved quality. The durability was extended by 4 days and the percentage of first rate quality plants was increased. A much lower nitrate level was found in the substrate following ammonium fertilization. Therefore ammonium-depot-fertilization turned out to be at least equivalent to common fertilization systems and besides it comprises economical and ecological advantages.

Introduction

In general cut flower fertilization in greenhouses is performed via splitted and equally distributed treatments. Frequently an uncontrolled application of fertilizers can be observed, significantly exceeding plant requirements (Scharpf, 1989). As a result excessive nutrient levels remain in the substrate and consequently leaching, especially of nitrate, may occur. Nitrate leaching represents much less a financial loss rather than a serious problem for ground water quality.

High levels of nitrate nutrition may cause problems with regard to cultivation and marketing. After excessive nitrate nutrition plant tissues have generally bigger cells with thinner walls. Such plants show reduced stability and are more susceptible to parasitic diseases (Saur and Schönbeck, 1976). These problems can be overcome by nitrogen nutrition based solely on ammonium. This encouraged us to compare the ammonium-depot-fertilization, which previously had delivered promising results for vegetable and crop production (Sommer, 1986; Titz and Sommer, 1989), with a conventional fertilization system.

In general, cut flowers which are exclusively treated with ammonium fertilizer as a nitrogen source, are presumed to demonstrate incompatibility combined with a decrease in quality and yield (Bowe, 1965; Fleming *et al.*, 1987; Röber and Schaller, 1985). However, this applies only in the case of homogeneous distribution of the ammonium in the root zone. In the ammonium-depot-fertilization system only a small part of the roots (<10%) is exposed to the local supply of ammonium (Sommer, 1986).

Therefore the goal of our study was to investigate the effects of the ammonium-depot-fertilization on yield and quality of cut flowers under greenhouse conditions as well as on the nitrate content in the substrate.

Materials and methods

The experiments were carried out in the greenhouse of the experimental horticultural station Auweiler-Friesdorf. Regoltime chrysanthemum species (*Chrysanthemum indicum* hybridum) and Centum brompton stocks (*Matthiola incana*) were used as experimental plants and cultivated as bedding plants. The substrate used was composed of 30% peat and 70% soil.

Chrysanthemums

The chrysanthemums were cultivated from September 1987 to January 1988. The conventional fertilization (=control) with 6 fluid and equally distributed nitrogen applications (60% nitrate-N, 40% ammonium-N) was compared with three different systems of ammonium-depot-fertilization:

1. ammonium-depot fertilization
2. ammonium-depot-fertilization with previously added 6 g straw manure L^{-1} *without* balancing nitrogen fertilization
3. application of an urea/ammonium sulphate strip to the soil surface into every second strip of soil between the rows at the time of planting (content of urea-N: 66%).

In system 1 and 2 the ammonium fertilizer (in aqueous form and stabilized with a nitrification inhibitor) was manually injected into every second strip of soil between the rows at a depth of 10 cm measured from the soil surface at the time of planting. The localized application caused a spatially limited absorption. The fact that ammonium is hardly mobile in soils implies continuous presence of fertilized ammonium in the root zone of the plants. With it the plants had a durable nitrogen source.

Fertilization in the above described way gives ammonium the character of a depot fertilizer. Figure 1 illustrates how roots enclose such an *ammonium-depot*.

In this system the depot-effect is based on an interaction between root and soil in contrast to conventional depot-fertilizers which are characterized as slow releasing nutrient sources. The ammonium uptake occurs according to the need *i.e.* the assimilatory potential of the plants, because plants need carbon skeletons for the ammonium neutralization. A detailed description of the underlying mechanism has been provided by Sommer (1986).

Throughout all treatments the amount of N-fertilization was 40 g N/m^2. The nitrogen from straw manure was not regarded in system 2. The other macro- and micronutrients were previously added to the substrate so that no further fertilization was necessary.

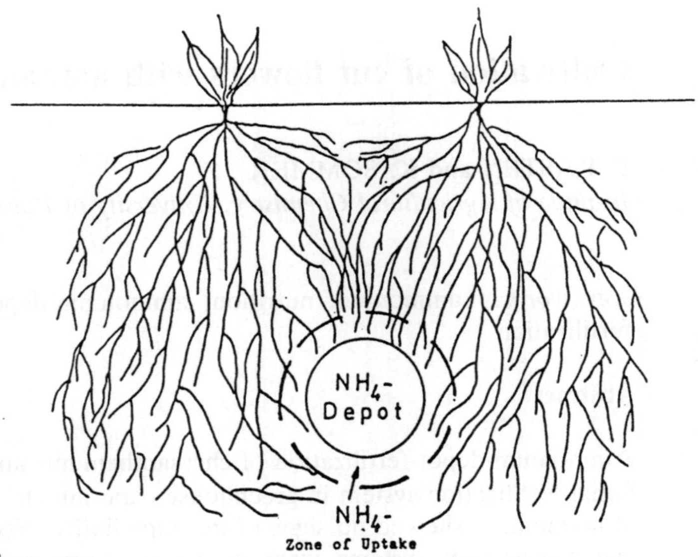

Fig. 1. The ammonium-depot which is enclosed by roots.

Brompton stocks

The brompton stocks were cultivated from February 1988 to May 1988. The N-fertilization level in this experiment was 30 g N/m^2. The empolyed fertilization systems were similar to those used in the chrysanthemum experiment.

As commonly found the entire amount of N in the conventional fertilization was supplied within the first half of the culturing period.

Results and discussion

Chrysanthemums as well as brompton stocks showed no incompatibility caused by ammonium nutrition. The plants developed normally and reached very good market quality. The conveyed experiments reveal that cut flowers can indeed by produced in bed or soil culture with ammonium as the only N-source. Prevailing prejudices concerning this specific N-form must be repudiated with regard to the ammonium-depot-fertilization.

Chrysanthemums

The ammonium-fed plants reached the stage of harvest half a week earlier than the control plants. This fact has been reported by Röber (1971), but only for nutrient solution cultures.

Table 1. Fresh- and dry matter yields of chrysanthemums in dependency on the N-fertilization system

	Treatment				
	Convent. system	NH_4-dep.– fertiliz.	NH_4-dep.+ straw m.	$(NH_4)SO_4$/ urea-strip	LSD 5%
Fresh weight/pl.(g)	48.99	45.92	49.69	47.24	
Dry weight/plant(g)	4.76	4.60	4.75	4.56	
Rel. dry matter contents (%)					
Stem	19.69	21.52	20.50	19.89	0.94
Leaf	7.00	7.29	6.96	6.93	0.24
Blossom	10.68	11.05	10.66	10.63	
Total plant	9.71	10.02	9.56	9.69	

The finding is reflected by the number of open blossoms per stalk at harvest time (Table 2). This can be explained by high remaining amounts of nitrogen in the substrate, which are well known to inhibit the process of ripening (Schwemmer, 1986). This effect cannot be observed after ammonium-depot-fertilization with self-regulatory N-uptake.

Fresh- and dry matter yields did not differ significantly (Table 1). However, differences were found in the dry matter content. Thus a tendentious increase of dry matter content was measured following ammonium nutrition in blossom and total plant, and a significant increase in leaf and stem. After treatment with urea/ammonium sulphate (strip application) similar dry matter contents were obtained as in the conventional treatment.

Plain differences were obtained between the fertilization systems in the plant quality with the exception of flowers diameter. Ammonium-depot-fertilization achieved the highest share of first rate quality plants, whereas the combination with straw manure proved to be the most successful (Fig. 2). The number of blossoms and buds per stalk was the most important factor for selling quality while the slightly increased stem length following conventional nutrition (Table 2) had no influence.

The very good plant quality after ammonium-depot-fertilization connected with previously added straw manure could be the result of a better carbon dioxide supply. It is well known that plants are able to absorb carbon dioxide, when it is produced during straw rotting (Führ, 1962). Especially during winter time the carbon

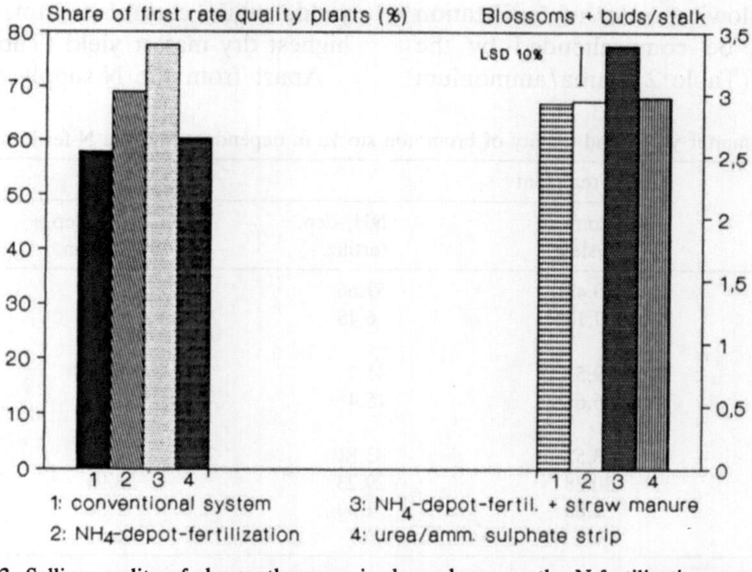

Fig. 2. Selling quality of chrysanthemums in dependency on the N-fertilization system.

Table 2. Quality and N-content of chrysanthemums in dependency on the N-fertilization system

	Treatment				
	Convent. system	NH_4-dep.– fertiliz.	NH_4-dep.+ straw m.	$(NH_4)_2SO_4$/ urea-strip	LSD 5%
Blossoms/stalk	2.09	2.51	2.48	2.48	
Buds/stalk	0.84	0.43	0.91	0.50	
Flow. diameter (cm)	5.8	5.7	5.8	5.7	
Stem length (cm)	71.8	67.7	70.2	68.5	
Durability (days)	27.1	30.9	25.8	25.5	3.27
%N in dry matter					
Total plant	3.04	3.17	3.13	3.38	
Stem	1.71	1.81	1.64	2.19	0.22
Leaf	4.58	4.77	4.59	4.98	0.17
Blossom	2.59	2.42	2.59	2.49	

dioxide supply can easily become a growth limiting factor in greenhouses with a closed ventilation system. Hence straw manure could result in a higher energy level of the plants. Durability tests showed that plants fed with ammonium without supplementary straw manure were significantly superior (Table 2). In general the durability of the used species was very good. However, main reasons for the end of durability were bended blossoms and withered leaves and therefore a loss of stability. The dry matter content which can be regarded as an indicator for plant stability was significantly increased due to the ammonium-depot-fertilization system (Table 1).

Also the green colour of the leaves is an important factor for the quality of cut flowers. It decreased in the following order of fertilization systems, which can be comprehended by the Kjeldahl-N-content (Table 2): urea/ammonium sulphate strip application > ammonium-depot-fertilization > ammonium-depot-fertilization + straw manure and conventional fertilization.

Brompton stocks

The brompton stocks were cultivated as usual in the period between February and May. In contrast to the chrysanthemums, brompton stocks reached the stage of harvest faster after conventional fertilization, whereat leaves of these plants of the lower tertial were found to be extremely chlorotic unlike the results of the other treatments. Both could be connected with the N-metabolism before and during flower period.

Ammonium-depot-fertilization combined with straw manure caused an increase in fresh matter yield, while control treatment resulted in the highest dry matter yield (Table 3).

Apart from the N-supply all fertilization sys-

Table 3. Fresh- and dry matter yields and quality of brompton stocks in dependency on the N-fertilization system

	Treatment			
	Convent. system	NH_4-dep.– fertiliz.	NH_4-dep.+ straw m.	$(NH_4)_2SO_4$/ urea-strip
Fresh weight/pl. (g)	60.43	59.66	70.94	62.92
Dry weight/plant (g)	7.16	6.48	6.69	6.70
Stem length (cm)	49.5	45.7	47.6	47.6
Raceme length (cm)	16.6	15.4	15.4	16.2
Rel. share of rac. length on total length (%)	33.52	33.81	32.27	34.09
Blo. +buds/stalk	30.28	29.23	28.94	29.15
Raceme density	1.83	1.90	1.88	1.80
Durability (days)	6.7	6.7	6.6	5.8

Table 4. N-content of brompton stocks and nitrate content of the substrate in dependency on the N-fertilization system

	Treatment				
	Convent. system	NH_4-dep. − fertiliz.	NH_4-dep. + straw m.	$(NH_4)_2SO_4$/ urea-strip	LSD 5%
%N in dry matter					
Total plant	3.18	3.66	3.46	3.61	0.15
Stem	1.57	1.73	1.83	1.77	0.13
Leaf	3.91	4.20	4.36	4.14	0.19
Blossom	4.88	5.38	5.21	5.49	
Substrate					
NO_3-N (g/m²)	34.8	20.1	27.5	26.3	

tems produced a similar plant quality. Decisive quality properties (stem length, proportion of raceme of total length, number of blossoms and buds in relation to raceme length = "raceme density", durability) were not influenced by the permanent N-supply following ammonium-depot-fertilization (Table 3). The Kjeldahl-N-content after conventional treatment was significantly lower, as well in the total plant as in single plant parts, than after the other treatments (Table 4). The high N-content of the brompton stock blossoms was striking.

The N-content of the plant material was inversely proportional to the nitrate-N-content of the substrate at the end of cultivation period. The highest remaining nitrate levels were found after conventional fertilization (Table 4).

Nitrate contents of the substrate indicate the possibility to decrease the extreme nitrate leaching with the help of ammonium-depot-fertilization. Moreover this system is favourable in terms of labour since only one treatment at the time of planting is necessary.

These results confirm that cut flower cultivation based on ammonium-depot-fertilization are at least equivalent to common fertilization systems in yield and quality with substantial economical and ecological advantages.

Acknowledgements

The investigations were kindly supported by the "Chamber of Agriculture of the Rhineland". The authors thank the employees of the experimental station Auweiler-Friesdorf for assistance during the realization of the experiments.

References

Bowe R 1965 Beitrag zum Einfluß verschiedener Stickstoffarten auf die Entwicklung und den Blumenertrag von *Gerbera jamesonii*-Sämlinge in Kieskultur, Dissertation Berlin, 195 p.

Fleming A L, Krizek D T and Mirecki R W 1987 Influence of Ammonium Nutrition on the Growth and Mineral Composition of two Chrysanthemum Cultivars Differing in Drought Tolerance. J. Plant Nutr. 10, 1869–1881.

Führ F 1962 Untersuchungen zur Aufnahme von Kohlendioxid und Strohabbauprodukten durch die Pflanzenwurzel, Dissertation Bonn, 156 p.

Röber R 1971 Der Einfluß der N-Ernährung auf die Blumenbildung und -entwicklung von Chrysanthemen. Gartenbauwiss. 36, 189–200.

Röber R and Schaller K 1985 Pflanzenernährung im Gartenbau. Eugen Ulmer GmbH and Co. Stuttgart, 352 p.

Saur R and Schönbeck F 1976 Untersuchungen über den Einfluß von Stickstoff und Ethirimol auf den Befall von Gerste durch *Helminthosporium sativum*. Z. Pflanzenkrankheiten Pflanzenschutz 83, 519–528.

Scharpf H C 1989 Pflanzenernährung im Zierpflanzenbau – wohin geht die Entwicklung? *In* Düngen im Zierpflanzenbau. Lehr- und Versuchsanstalt für Gartenbau Hannover Ahlem. 7–16. Taspo-Praxis 16. Verlag Bernhard Thalakker, Braunschweig.

Schwemmer E 1986 Stickstoff-Hauptmotor des Pflanzenwachstums. Dt. Gartenbau 9, 407–408.

Sommer K 1986 Ammonium – Eine überzeugende Alternative in der Stickstoffdüngung. Vortragsveranstaltung zum "Dies Academicus" WS 1986/87 der Universität Bonn.

Titz R and Sommer K 1989 Ertragsstruktur sowie Nitratgehalte in Pflanzen und Böden bei Freilandsalaten auf Ammonium-Basis gegenüber konventioneller Düngung. VDLUFA-Schriftenreihe, Kongressband 1988.

M. L. van Beusichem (Ed.), *Plant nutrition – physiology and applications*, 539–544.

PLSO IPNC413

Conceptual evaluation of intensive production systems for tomatoes

C. M. GERALDSON
IFAS, University of Florida, Gulf Coast Research and Education Center, 5007 – 60th Street East, Bradenton, FL 34203, USA

Key words: fertigation, gradient concept, nutritional stability

Abstract

A constant source of nutrients (gradient or fertigation) in conjunction with a constant source of water (water table or microirrigation) has been used to provide nutritional stability in the development of intensive production systems. The maximum buffer capacity of the gradient concept compared to the less than maximum buffer capacity of fertigation procedure provides the differential in nutritional stability that is significant for advances in productivity. The average statewide tomato yield in Florida has doubled with the adoption of the mulched gradient system (reevaluation of the entire production system including cultivars and cultural procedures as well as advances in nutritional stability). The development of a containerized gradient system which is currently being evaluated may have the potential to further advance productivity by reducing the unit cost of production and, at the same time, increasing environmental and conservational compatibility.

Introduction

The more extensive crop production systems depend on the soil as a primary source of nutrients and water, whereas the more intensive system decreases or eliminates that dependency. In the transition to intensive production systems, nutritional stability becomes increasingly more dependent on the controlled integration of water and nutrients into the system. In an approach to provide nutritional stability, intensive production systems were developed using a constant source of water in conjunction with a constant source of nutrients as basic components. The designed nutritional stability of the intensive production systems was protected with a full-bed plastic mulch; the concept is now being extended to a containerized culture. The purpose of this paper is to compare intensive production systems to grow tomatoes.

Materials and methods

Providing a constant source of water and a constant source of nutrients has been used as a basic procedure in developing intensive production systems. A constant water table or microirrigation procedures in conjunction with nutrients supplied by gradient or fertigation has been evaluated (Geraldson, 1970; 1972; 1973). The electrical conductivity and the ionic concentrations in the saturation extract has been used to nutritionally evaluate the distribution of ions in the root environment (Geraldson, 1977).

The basic components used in the production systems are described as follows:

1. *Mulched gradient*: A reservoir of soluble nutrients (primarily N and K) at the soil bed surface in conjunction with a constant water table provided the gradient concept. (Fig. 1). The pH of the soil was maintained at 6.5 to 7.0.

Fifteen hundred kg/ha of an 18-0-25 was banded at the soil bed surface; 500 kg of an 0-20-0 and 20 kg of a minor element mix were mixed in the soil bed and the prepared root environment was covered with a full bed mulch. This concept was initiated and developed in the 1960's, projected to the growers in the 1970's (Geraldson, 1970), and was associated with a doubling of the average tomato production statewide in the 1970's and 80's (D. Conner, 1987). The data presented in Table 1 are from experimental plots in 1987 and are typical of the ionic composition from a mulched gradient system. Samples were obtained at 3 depths in the plant row and in the fertilizer band twice during the growing season.

2. *Mulched micro fertigation*: Providing nutrients with or in conjunction with a micro irrigation system (Fig. 2) was evaluated in the 1970's (Geraldson, 1972; 1973) and the fertigation procedure was advanced in the 1980's (Pitts *et al.*, 1988). The data presented in Table 2 were from experimental plots in 1987 with an ionic composition in the rhizosphere typical of a fertigation procedure. Water requirement was based on tensiometer measurements. Soluble nutrients (primarily N and K) were supplied with the micro irrigation and increased chronologically according to a projected requirement to provide a total of 300 kg of N/ha/season. Samples were obtained from portions of the soil bed similar to those obtained for the gradient culture.

3. *Containerized gradient*: All of the soluble fertilizer 450 gms of (10-0-20) was placed in 1 end of a plastic bag (15.0 × 17.5 × 35 cm) containing 15 liters of a peat based medium (Fig. 3 – Spring 1989). Phosphorus and minor elements were mixed in the medium which was limed to a pH of 6.5 to 7.0. A micro irrigation tube supplied water to each container delivering 500 cc/application chronologically increasing as required from 1 to 8 applications/day. The data are presented in Table 3a. Samples were obtained from lateral portions of the containerized medium moving from the plant end to the fertilizer end.

4. *Containerized fertigation*: Micro irrigation was applied equivalent to the schedule indicated

Table 1. Electrical conductivity and concentrations (mg L^{-1}) of NO_3, K, Ca and Mg in the saturation extract from fractionated portions of the soil bed profile utilizing the mulched gradient system

Depth cm[a]		ECe dSm^{-1}	NO_3	K	Ca	Mg
Feb. 20, 1987						
Fert	0–5	29.0	3572	4000	760	488
	5–10	16.4	2200	1700	488	192
	10–20	4.3	494	313	129	45
Row	0–5	0.9	59	63	56	21
	5–10	0.7	29	50	48	20
	10–20	0.5	14	36	29	11
June 24, 1987						
Fert	0–5	6.6	400	1410	434	187
	5–10	5.3	407	752	436	247
	10–20	2.6	203	293	219	114
Row	0–5	3.8	192	39	566	328
	5–10	2.4	53	22	321	178
	10–20	1.4	15	20	116	87

[a] Sampled at 3 depths in the fertilizer band and the plant row (Fig. 1).

Table 2. Electrical conductivity and concentrations (mg L^{-1}) of NO_3, K, Ca and Mg in the saturation extract from fractionated portions of the soil bed profile utilizing a Micro-fertigation System in a commercial field

Depth cm[a]		ECe dSm^{-1}	NO_3	K	Ca	Mg
March 25, 1987						
Edge	0–5	0.8	30	36	80	22
	5–10	0.6	18	26	73	19
	10–20	0.9	19	40	105	30
Row	0–5	1.9	56	103	93	27
	5–10	1.8	81	236	132	40
	10–20	2.0	60	160	82	25
May 27, 1987						
Edge	0–5	0.7	4	18	99	27
	5–10	1.5	11	57	206	56
	10–20	1.8	8	169	183	64
Row	0–5	1.2	16	78	131	48
	5–10	1.0	10	42	100	44
	10–20	0.9	9	79	70	29

[a] Sampled at 3 depths at the edge of the bed and in the plant row (see Fig. 2).

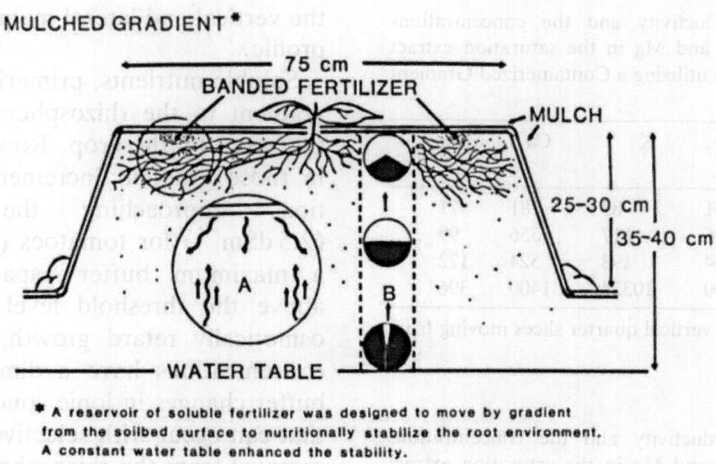

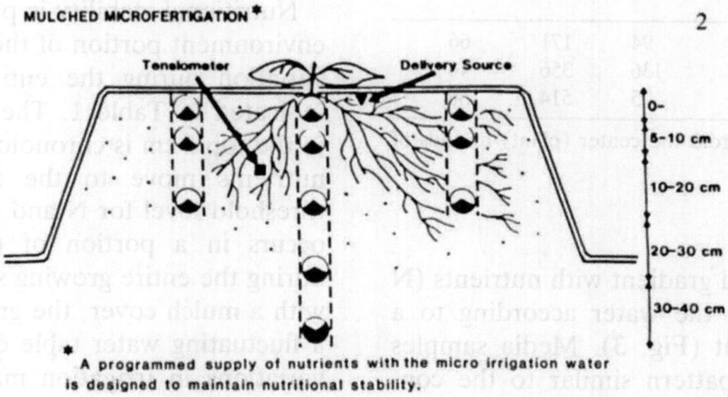

2

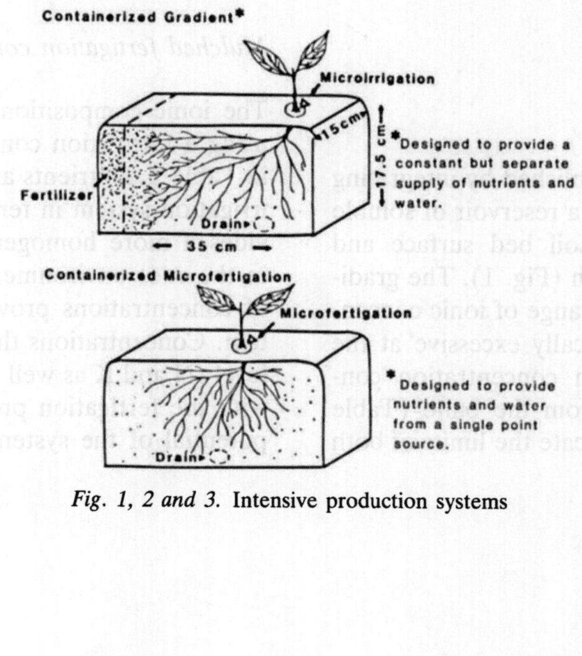

3

Fig. 1, 2 and 3. Intensive production systems

Table 3a. Electrical conductivity and the concentrations (mg L^{-1}) of NO_3, K, Ca and Mg in the saturation extract from portions of the media utilizing a Containerized Gradient Concept

Sample location[a]	ECe dSm^{-1}	NO_3	K	Ca	Mg
Plant 1	1.6	161	83	181	71
2	2.2	36	117	356	99
3	15.0	24	198	524	122
Fert 4	38.0	4600	10324	1400	396

As in Figure 3 sampled as vertical quarter slices moving from the plant to the fertilizer.

Table 3b. Electrical conductivity and the concentrations (mg L^{-1}) of NO_3, K, Ca and Mg in the saturation extract from portions of the medium utilizing a Containerized Fertigation System[a]

Sample location[a]	ECe dSm^{-1}	NO_3	K	Ca	Mg
End 1	1.7	12	94	171	66
Plant	2.9	14	136	356	46
End 2	3.4	0	25	514	186

[a] As in Figure 3 moving from the center (plant) to ends of the container.

with the containerized gradient with nutrients (N and K) applied with the water according to a projected requirement (Fig. 3). Media samples were obtained in a pattern similar to the containerized gradient and the data are presented in Table 3b.

Results and discussion

Mulched gradient concept

A gradient system was established by integrating a constant water table with a reservoir of soluble fertilizer banded on the soil bed surface and covered with a plastic mulch (Fig. 1). The gradient provides a predictable range of ionic concentrations and ratios, osmotically excessive at the surface and diminishing in concentration concentrically with distance from the band (Table 1). The data presented indicate the limits of both the vertical and lateral maintained in the soil bed profile.

Soluble nutrients, primarily N and K, move by gradient to the rhizosphere solution to repalce that used by the crop. Roots tend to proliferate in those gradient increments where concentrations approaching the threshold level (2.5 dSm^{-1}) for tomatoes (Maas, 1986) provide a maximum buffer capacity. Concentrations above the threshold level for any given crop osmotically retard growth, whereas decreasing concentrations have a diminishing potential to buffer changes in ionic concentrations and ratios that can occur with selective and increasing crop removal from the rhizosphere solution. The purpose of the gradient is to extend the buffer capacity of any soil or media beyond its inherent potential.

Nutritional stability is provided when the root environment portion of the gradient has minimal variation during the entire growing season as indicated in Table 1. The reservoir of nutrients in the top 5 cm is chronologically depleted as the nutrients move to the root environment. A threshold level for N and K as well as total salts occurs in a portion of the root environment during the entire growing season. However, even with a mulch cover, the gradient is vulnerable to a fluctuating water table due to periodic rain or variations in irrigation management. Maximum commercial tomato yields of 20 kg/plant (Geraldson, 1980) and a doubling of the average yield statewide (Conner, 1987) has been associated with the full bed mulch culture.

Mulched fertigation concept

The ionic composition of the root environment using a fertigation concept (Fig. 2) is presented in Table 2. Nutrients are supplied with the micro irrigation system in fertigation. This system provides a more homogeneous ionic concentration in the root environment compared to the range of concentrations provided by the gradient system. Concentrations that approach the threshold level (N and K as well as total salts) do not occur with the fertigation procedure, thus limiting the potential of the system to buffer variations due to differential and increasing plant removal. The threshold portions of the gradient provide a N or K concentration that always exceeds the maximum provided by fertigation. The ECe in the gradient ranges from 29.0 to less than 1.0 dsm^{-1} and fluctuates from 2.0 to less than 1.0 with the fertigation procedure. It is suggested that the fertigation concept compared to the gradient is nutritionally more vulnerable to stress when the buffer potential ceases to be adequate. Yields from fertigation procedures have been less than (Geraldson, 1973) and only infrequently equivalent (Pitts *et al.*, 1988) to the gradient system.

Containerized gradient concept

A micro-source of water used to provide a functionally constant moisture integrated with an in-container concentrated source of soluble nutrients provides a containerized gradient system for advancing productivity (Fig. 3). The ionic composition of root environments from this system is presented in Table 3a and is nutritionally similar to the gradient presented in Table 1.

With this enclosed system, water requirement is equivalent to that transpired which is about 6 to 8 acre inches compared to 60 acre inches (Geraldson, 1970) utilized by the conventional seepage system (constant water table) and about 20 acre inches (Pitts, 1988) utilized by the micro-fertigation procedure. Further potential advantages of a containerized system include: plastic mulch and fumigation can be eliminated; cultivation can be minimized; and the use of marginal or non-productive land for location and a semi-permanent plant support system add to the economic feasibility of the system but only if yields are sufficient to maintain or reduce unit costs.

With one season of evaluating the concept, yields of 8 to 10 kg/plant (average commercial yields) have been encouraging and, by commercial standards, has potential feasibility. The inherent buffer capacity of the containerized culture, even though enhanced by the use of synthetic media, is basically about 1/3 of the soil culture just because of comparative volume of the root environment solution (15 liters compared to perhaps 45 on a per plant basis). Thus, nutritional stability in the containerized culture is that much more dependent on the potential of the gradient to maintain the buffer capacity. Conversely, a variance in stability and the associated plant response has that much more of a potential for exposure.

Containerized fertigation concept

With programmed fertigation procedures, the ionic concentration in the root environment was significantly less than threshold levels (nutrients supplied were equivalent to that supplied by gradient) (Table 3b, Fig. 3). It is not unexpected that the containerized fertigation procedure lacks the potential to buffer inefficiencies. With one season of evaluating the concept, it has a zero commercial potential.

Research advances for intensive production

The transition from an extensive soil-based production system to an intensive gradient-based concept shifts the emphasis from a multitude of interdependent variables to the ionic composition of the rhizosphere as a common denominator of productivity. It is suggested that nutritional control of the production system is centered in those gradient increments where the range of concentrations approach the threshold level.

In the past, soil has provided a physical buffer against inefficiency. In the transition to intensive production, the gradient provides a buffer that functionally enhances nutritional efficiency. By eliminating soil as a contributing variable, nutritional control of the rhizosphere enables plants to be grown to exacting and predetermined specifications. Providing nutrients and water is only the first step in developing an intensive production system. Concepts of nutrition and nutritional research must be altered accordingly if such advances are to continue.

References

Conner D 1987 Florida Agricultural Statistics Vegetable Summary 1986–87.

Geraldson C M 1970 Precision nutrient gradients: A component for optimal production. Soil Sci. Plant Anal. 1, 317–331.

Geraldson C M 1972 Changing concepts in nutrition associated with the transition from extensive to intensive crop production. Soil Crop Sci. Soc. of Fla. Proc. 32, 84–86.

Geraldson C M 1973 Nutritional studies utilizing a constant micro sources of moisture. Proc. Trop. Region Am. Soc. Hortic. Sci. 17, 359–362.

Geraldson C M 1977 Nutrient intensity and balance soil testing – correlating and interpreting the analytical results. Am. Soc. Agron. (Special publication).

Geraldson C M 1980 Importance of water control for tomato production using the gradient mulch system. Proc. Fla. State Hortic. Soc. 93, 278–279.

Maas E V 1986 Salt tolerance of plants. Appl. Agric. Res. 1, 12–26.

Pitts J, G A Clark, J Alverez, P H Everett and J M Grimm 1988 A comparison of micro to sub-surface irrigation of tomatoes. Proc. Fla. State Hortic. Soc. 101, 393–399.

M. L. van Beusichem (Ed.), *Plant nutrition – physiology and applications*, 545–549.

PLSO IPNC276

The effects of continuity of early nitrogen nutrition on growth and development of *Lactuca sativa*

I. G. BURNS
AFRC Institute of Horticultural Research, Wellesbourne, Warwick CV35 9EF, UK

Key words: *Lactuca sativa* L., lettuce transplants, maturity date, nitrogen, seedling nutrition, temporary deficiency

Abstract

The effects of a temporary interruption in N supply during the seedling stage on subsequent growth and development was examined in two experiments with transplanted lettuce. Early growth was strongly affected by withholding N for a short period, and once differences in plant size had been established they were maintained until maturity, long after the N supply had been restored. The N-stress treatment had little effect on the rate of crop development; maturity date was only slightly affected by the temporary N stress and, although the plants continued to grow as maturity approached, they were unable to recover the earlier losses in growth. Most of the yield differences were concentrated in the trimmed heads. The pattern of response was not affected by the age of the seedlings at the time that the N treatment was imposed.

Introduction

Recent studies have shown that young seedlings can often suffer from temporary deficiencies of N, even when quite heavy dressings of granular N fertilizer are broadcast and incorporated in the soil shortly before drilling (Greenwood *et al.*, 1989). Young plants are particularly vulnerable to shortages of N if the N supply is limited, and evidence is accumulating which suggests that this can affect their subsequent performance. Experiments with lettuce and celery have shown that restricting the early N supply not only produced smaller seedlings, but also reduced yields (Kratky and Mishima, 1981; Tremblay *et al.*, 1987). However, similar effects were not observed with cauliflower, where reductions in seedling growth were largely recovered by maturity (Wurr *et al.*, 1986). Preliminary studies with lettuce seedlings have also shown that even a short period (of a few days) without N can seriously affect growth, and that the resulting weight differences are maintained throughout the lifetime of the crop, long after normal N supplies have been restored (Burns, 1988). However, it was not clear whether the maturity date of the previously N-stressed plants was delayed, which might have allowed them to recover their earlier losses of growth.

The object of this paper is to outline results of two further experiments with lettuce designed to examine in greater detail the effects of temporary interruptions in the N supply early in growth on the rate of dry matter production and plant development. Particular attention was given to the effects on the average maturity date and on the marketable yield of the crop in order to determine the extent to which it can recover from adverse effects of early N stress. Consideration was also given to the influence on the response of the size of the seedlings when the N supply was interrupted. The first experiment was designed to measure the changes in growth rate throughout the lifetime of the crop and consisted of harvesting batches of plants on each of a series of dates from transplanting until after maturity (singlepick experiment). In the second experiment, individual plants were harvested only when they became mature and total yields determined by accumulating the results from each cut (multipick experiment). Both experiments were carried out using plants raised in the glasshouse (where the N treatments were imposed) and transplanted into the field.

Methods

Lettuce (*Lactuca sativa* L. cv Pennlake) seeds were sown under glass in multi-cell modules containing a compost specially prepared from a general horticultural peat in which P, K, trace elements and lime had been mixed. There were a series of 8 sowing dates from which 4 were selected (designated sowings 4 to 7, each at 2 or 3 day intervals from 2 to 9 May) for use in these experiments. The module trays were initially irrigated daily with 3.0 mM $Ca(NO_3)_2$ solution and then switched to a 6.0 mM $Ca(NO_3)_2$ solution after 7 days. The N-stress treatment was imposed on half of the plants in each sowing by replacing the N feed with 6.0 mM $CaCl_2$ solution between 28 May and 6 June. The remaining plants were irrigated with the $Ca(NO_3)_2$ solution throughout (control treatment). All plants were thoroughly watered with $Ca(NO_3)_2$ solution on 6 June and were transplanted into the field the following day.

The field site was divided into two separate areas (for the singlepick and multipick experiments respectively) on a sandy loam soil which had been uniformly fertilized with an adequate level of NPK fertilizer for unrestricted growth. The singlepick experiment was arranged in a split-plot design with 2 replicates by 6 harvest dates on the main-plots and 4 sowing dates by 2 N-stress treatments on the sub-plots. The multipick experiment had a factorial combination of 4 sowing dates and 2 N-stress treatments and was arranged in a randomized block design with 3 replicates. The plants were laid out on a 30 cm square spacing with 25 plants per plot or sub-plot.

In the singlepick experiment, the plants were harvested at 7 to 13 day intervals starting 12 days after transplanting and finishing after most heads had split. On each date all plants were cut off at ground level and weighed fresh. In the multipick experiment, the plants were harvested individually only when each was mature (based on heart firmness). Total and trimmed fresh weights were determined in each case and yields were calculated by accumulating the individual plant weights over all cuts.

Results

Transplant size

Average shoot fresh weights of the lettuce plants on the day before transplanting ranged from c. 2 to 6 g, see Table 1. Analysis of variance on the log-transformed data showed that the different sowing dates had a significant effect on their fresh weights ($P < 0.001$), creating a 2.5-fold ratio in plant size between the earliest and latest sowings for both treatments. The differences between the two N treatments were smaller, with the temporary N stress causing average reductions in fresh weights of 26% (significant at $P < 0.001$). There was no interaction between sowing date and N treatment, showing that the relative weights in the two treatments remained approximately constant over all sowing dates.

Singlepick experiment

The weights of the plants in all treatments increased until penultimate harvest 5; thereafter the stressed plants continued to grow, whereas the weights of the control plants declined. This loss of weight was caused by an outbreak of botrytis, which infected the control plants almost exclusively after this date. However, since measurements of heart firmness and stem elongation showed that the majority of the plants had reached maturity by harvest 5, treatment effects upto and including maturity were unlikely to have been affected by the infection.

Table 1. Mean shoot fresh weights of seedlings on the day before transplanting

	Fresh weight (g)			
Treatment	Sowing 4	Sowing 5	Sowing 6	Sowing 7
Control	6.0	4.5	3.9	2.5
N stress	4.8	3.3	2.5	1.9

Analysis of variance on the log-transformed data from all harvests showed that the fresh shoot weights were significantly affected by size differences created both by N treatment ($P < 0.001$) and by sowing date ($P < 0.01$); there were no second or third order interactions between the effects of harvest date, N treatment or sowing date. At each harvest, the N-stressed plants were consistently smaller than control plants sown on the same date (upto harvest 5), with average reductions of 17% in fresh weight, largely reflecting the effects that were present at transplanting. In contrast, however, most of the weight differences resulting from the various sowing dates at transplanting were substantially modified soon after the plants were transferred to the field, with the largest plants at each harvest normally originating from sowing 6 and the smallest from sowing 7. Furthermore, the average differences in fresh weight between plants from these sowings was only 9.8% compared with an average of 30% at transplanting (cf. Table 1). This pattern of response was clearly evident at harvest 1 (12 days after transplanting) suggesting that the older plants were either more susceptible to transplanting shock or were slower to establish, and that the resulting differential effects were maintained throughout subsequent growth. Thus the size differences in the field that were originally attributed to sowing date were, in reality, caused by a combination of sowing date and post-transplanting stress.

The effects of the N treatments were examined in greater detail by averaging the fresh yield data over all sowing dates in Figure 1. This shows that the temporary N stress caused a depression in the absolute growth rate of the plants even though the N supply had been restored, so their yields gradually diverged from those of the control, at least until harvest 5 (Fig. 1a). However, when this data was replotted in a semi-logarithmic form (Fig. 1b), the growth curves for the two treatments became virtually parallel, showing that there was little difference between their relative growth rates. These results confirm those of a previous experiment (Burns, 1988) , that one growth has been checked by a short period of N stress the effects are maintained throughout the subsequent lifetime of the crop. In addition, equivalent graphs for plants from the different

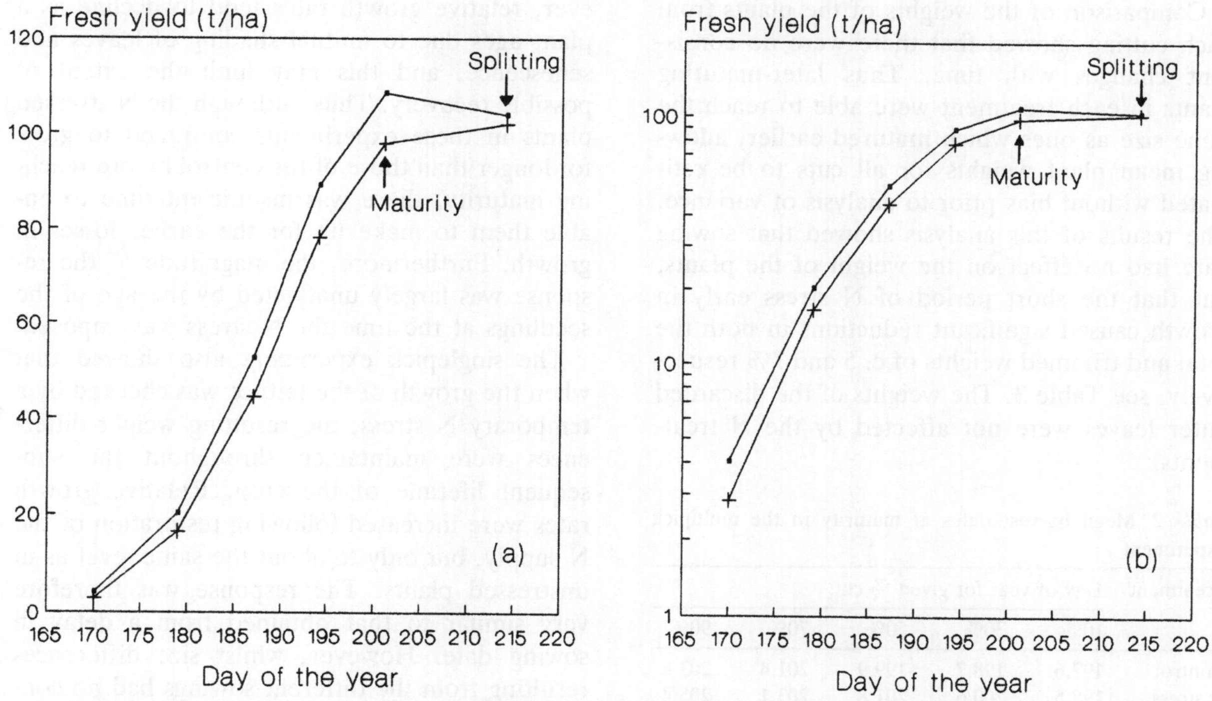

Fig. 1. Mean growth curves for shoot fresh yield in the singlepick experiment: (a) on a linear basis; (b) on a semi-logarithmic basis. Key: ● unstressed control treatment; +N-stressed treatment.

sowing dates (data not presented) also showed a similar pattern of response, with relative growth rates for the various sowing dates virtually identical. Thus, on the basis of these results, it was not possible to distinguish between size effects caused by different sowing dates from those caused by temporary N stress.

Multipick experiment

Analysis of the frequency distribution of cutting showed that the mean date on which a given proportion of the crop had been cut was significantly affected by N treatment (normally $P < 0.001$), but not by sowing date or by any other source of variation in the experiment. The cutting dates were therefore averaged over data for all sowing dates to obtain the best estimates of N treatment effects. The results, presented in Table 2, show that a given proportion of the N-stressed crop always reached maturity between 1 and 2 days later than in the control treatments, with the biggest delays occurring when 30 to 90% had matured. The duration of cutting (*i.e.* the time between the first and last cut) was unaffected by the N treatment.

Comparison of the weights of the plants from each cutting showed that there were no consistent changes with time. Thus later-maturing plants in each treatment were able to reach the same size as ones which matured earlier, allowing mean plant weights for all cuts to be estimated without bias prior to analysis of variance. The results of this analysis showed that sowing date had no effect on the weight of the plants, but that the short period of N stress early in growth caused significant reductions in both the total and trimmed weights of c. 5 and 7% respectively, see Table 3. The weights of the discarded outer leaves were not affected by the N treatments.

Table 2. Mean harvest dates at maturity in the multipick experiment

Treatment	Day of year for given % cut:				
	10%	30%	50%	70%	90%
Control	197.6	198.7	199.9	201.4	203.8
N stress	198.5	200.6	201.8	203.1	205.4
SED[a]	0.18	0.42	0.44	0.35	0.47

[a] Standard errors of differences of means (for 14 degrees of freedom).

Table 3. Time-averaged shoot and trimmed head weights in the multipick experiment

Treatment	Mean shoot weight (g)	Mean head weight (g)
Control	932	613
N stress	885	571
SED[a]	20.1	13.2

[a] Standard errors of differences of means (for 14 degrees of freedom).

Discussion

Wurr *et al.* (1987) have shown that the duration of lettuce growth from transplanting to maturity is mainly dependent on the combined effects of accumulated temperature and radiation. However, the factors controlling maturity are complex, and their data also indicated that environmental conditions during plant raising could often be important. The results of the current experiments show that the early N-nutrition of the crop can also have an effect, causing maturity date to be slightly delayed by temporary N stress and giving the plant at least some opportunity to recoup earlier losses of growth. However, relative growth rates tend to decline as a plant ages due to mutual shading of leaves and senescence, and this may limit the extent of possible recovery. Thus, although the N-stressed plants in these experiments continued to grow for longer than those of the control before reaching maturity, there was insufficient time to enable them to make up for the earlier losses in growth. Furthermore, the magnitude of the response was largely unaffected by the age of the seedlings at the time the N stress was imposed.

The singlepick experiment also showed that when the growth of the lettuce was checked by a temporary N stress, the resulting weight differences were maintained throughout the subsequent lifetime of the crop. Relative growth rates were increased following restoration of the N supply, but only to about the same level as in unstressed plants. The response was therefore very similar to that obtained from a delay in sowing date. However, whilst size differences resulting from the different sowings had no consistent effect on maturity date, a short period of N stress caused a small but significant delay to maturity. This suggests that the response to the temporary N stress may not have been due simply to the effect of plant size *per se*, but also to a specific effect of the N treatment. Similar conclusions may be drawn from the data for plant weight at maturity, which show that both total and marketable yields were significantly affected by the N treatement, but not by sowing date.

References

Burns I G 1988 Effects of interruptions in N, P or K supply on the growth and development of lettuce. J. Plant Nutr. 11, 1627–1634.

Greenwood D J, Stone D A, Burns I G and Rowse H R 1989 Improving the efficiency of N-fertilizer use for vegetable crops. J. Sci. Food Agric. 48, 125–127.

Kratky B A and Mishima H Y 1981 Lettuce seedling and yield response to preplant and foliar fertilization during transplant production. J. Am. Soc. Hortic. Sci. 106, 3–7.

Tremblay N, Yelle S and Gosselin A 1987 Effects of CO_2 enrichment, nitrogen and phosphorus fertilization on growth and yield of celery transplants. HortScience 22, 875–876.

Wurr D C E, Cox E F and Fellows J R 1986 The influence of transplant age and nutrient feeding regime on cauliflower growth and maturity. J. Hortic. Sci. 61, 503–508.

Wurr D C E, Fellows J R and Pittam A J 1987 The influence of plant raising conditions and transplant age on the growth and development of crisp lettuce. J. Agric. Sci., Camb. 109, 573–581.

M. L. van Beusichem (Ed.), *Plant nutrition – physiology and applications*, 551–556.

PLSO IPNC069

Influence of nitrogen nutrition on powdery mildew (*Sphaerotheca fuliginea*) infection and metabolism of cucumber (*Cucumis sativus*)

M.K. SCHENK and A. BETTIN
Institute of Plant Nutrition, University of Hannover, Herrenhäuser Str. 2, D-3000 Hannover 21, FRG

Key words: amino acid, carbohydrate, cucumber, *Cucumis sativus* L., nitrogen nutrition, phenol metabolism, *Sphaerotheca fuliginea*

Abstract

The relationship between N nutrition, plant metabolism and powdery mildew infection, an important disease in glasshouse cucumber production, was studied on plants grown in nutrient solution. Strong N deficiency did not restrain infection but instead restrained growth of the fungus after successful infection. This was presumably due to a shortage of some amino acids. NO_3 excess diminished hyphae development. This was not caused by factors which are generally discussed in relation to powdery mildew development like carbohydrate content and activity of enzymes of phenol metabolism. Powdery mildew infection could not be reduced by controlled N nutrition without yield reduction.

Introduction

Powdery mildew in glasshouse cucumber production is generally controlled by fungicides. However, for powdery mildew on cereals (*Erysiphe graminis*) it is known, that the disease can be controlled by avoiding N excess nutrition (Springer and Heitefuss, 1985). N nutrition influences powdery mildew on cereals via plant metabolism (Bainbridge, 1974). The reason might be that infection barriers are weakened by higher N supply because the phenol content decreases (Graham, 1983). Phenols are essential for the lignification and thus for the stability of the cell wall (Vance *et al.*, 1980). Furthermore it is discussed that chinones, which are also synthesized from phenols, might be toxic to powdery mildew (Fric, 1976). The development of fungus could also be influenced by nutrient supply. Organic nitrogen compounds as well as carbohydrates, especially sucrose, are considered important (Bushnell and Gay, 1978; Gay and Manners, 1981). Both fractions are influenced by N supply.

The objective of this study was to investigate the influence of N supply of cucumber on yield and powdery mildew development in relation to plant metabolism.

Material and methods

Young cucumber plants were grown in a growth chamber using a day length of 16 h with a quantum flux density of 100 μmol m^{-2} s^{-2} (PAR). Day/night temperature was 25/20°C and relative air humidity 60–80%. Experiments with fruiting plants (Table 2) and with varied light intensity (Fig. 1) were done in a glasshouse during summer. Seeds of the variety 'Corona' were germinated in sand and transplanted into pots with 5 L nutrient solution (fruiting plants 8 L pots). The basic nutrient solution contained: 2 m*M* K H_2PO_4, 2 m*M* K_2HPO_4, 1.5 m m*M* $MgSO_4$, 1.5 m*M* $CaCl_2$, 18 μM Fe-chelate, 4 μM $MnCl_2$, 20μM H_3BO_3, 0.4 μM $ZnSO_4$, 0.4 μM $CuSO_4$, 0.3 μM MoO_3, 5 mg L^{-1} dicyandiamide, pH 6–6.5. The aerated solution was renewed 1–4 times per week.

The 4th leaf from the top was inoculated with

fresh spores. To evaluate growth of fungus, leaf disks were cut 9 days after inoculation and hyphae were dyed according to Wolf and Fric (1981).

Total nitrogen was determined after Peterson and Chesters (1964) including NO_3. Glucose and fructose were measured enzymatically in a water extract (Bergmeyer, 1974). Amino acids were determined fluorimetrically after separation by chromatogaphy (Morvan, 1987). Phenyl-alanine-ammonia-lyase was determined according to Zucker (1965), water soluble peroxidase after Förster (1980) and cell wall-bound peroxidase according to Wissemeier (1988). Leaf samples to measure activity of enzymes of phenol metabolism were taken 4 days after inoculation, because initial development of fungus was not affected (Bettin, 1989).

Results

Yield and powdery mildew development

Cucumber plants were grown in different N concentrations supplied as either calcium nitrate or ammonium sulfate (Table 1). Maximum shoot yield was obtained with 4 m*M* N for both N forms. The N form did not affect yield significantly. N content in shoot matter increased with N supply. About 80% of the powdery mildew conidia infected the cucumber plants successfully, that means they penetrated the cell wall and developed secondary hyphae. Thus, N supply did not affect initial development. To assess the growth after infection, the number of conidiophores was counted per colony developed from one conidium. Table 1 shows that the conidiophores were already significantly reduced at high nitrate supply 5 days after inoculation and at strong N deficiency after 9 days. The other treatments did not differ. In contrast to nitrate, the highest N supply in form of ammonium did not influence the number of conidiophores.

Table 2 confirms that growth of conidiophores was reduced by excess nitrate supply whereas mild N deficiency had no effect on powdery mildew. In contrast, fruit yield was decreased at both excess and deficient N supply. These results suggest that it is not possible to control powdery mildew on cucumber without yield reduction. Similar results were obtained with other experiments in the glasshouse, which are not reported in this paper.

Table 2. Influence of NO_3 supply on N content, fruit yield and number of conidiophores of *S. fuliginea* of cucumber plants

m*M* NO_3	N in dm (%)	Conidiophores per colony	Fruit yield per plant (kg)
1	3,1	187	1,8
4	4,7	146	3,7
20	5,4	81	2,9
LSD 0,05	0,4	63	0,3

Table 1. Influence of N form and N concentration on cucumber growth, infection and development of *S. fuliginea*

m*M* N	Shoot yield (g fm/plant)	N in dm (%)	Growing conidia at day 1 (%)	Conidiophores per colony 5d[c]	7d	9d
0	27,4	1,5	79,1	8,7	63	253
1 NO_3[a]	94,0	2,3	81,2	7.9	76	471
4	136,5	3,8	77,6	7,5	74	439
20	134,6	5,4	77,9	4,2	45	185
1 NH_4[b]	98,8	2,1	76,7	8,1	76	473
4	128,1	4,6	79,2	8,5	77	436
20	141,5	5,6	78,2	7,8	82	463
LSD 0,05	28,3	0,7	n.s.	2,4	26	80

[a] supplied as $Ca(NO_3)_2$; [b]supplied as $(NH_4)_2SO_4$; [c]days after inoculation.

Powdery mildew at N deficiency

Powdery mildew growth at strong N deficiency was probably reduced because of a lack of some amino acids, as can be seen from Table 3.

Another reason for restricted growth of powdery mildew at N deficiency could be a change of phenol metabolism. Table 4 shows that the phenylalanine-ammonia-lyase activity was increased with strong N deficiency. Thus more phenols could have been synthesized which are precursors of the barrier lignine or compounds like chinone which are thought to be toxic to funghi. The activity of the corresponding enzymes, the cell wall-bound peroxidase and the water soluble peroxidase, respectively, was similarily increased (Table 4). It might be that phenol metabolism was related to change of powdery mildew growth. The activity increase of all three enzymes after infection, however, was independent of N supply.

Powdery mildew at NO_3 excess

Reduced powdery mildew growth at excess NO_3 nutrition was not related to phenol metabolism (Table 4). Thus, infection barriers based on phenols seemed not to be involved in the observed growth reduction. High nitrate content of leaf tissue was also not the cause, as was shown elsewhere (Bettin, 1989).

Another reason for decrease of fungus growth could be that Ca content increased from 2.1% up to 5% in dry matter because nitrate was supplied as Ca $(NO_3)_2$. Other nutrients were not significantly influenced. A high Ca content of leaf tissue could impair uptake of nourishment and thus growth of fungus, because membrane permeability is diminished. In order to investigate the influence of nitrate excess separately from Ca supply, ammonium nitrate was used without nitrification inhibitor. Table 5 shows that fungus growth was reduced with extreme N deficiency as

Table 3. Influence of N concentration and N form on water soluble amino acid content (μ moles/g fm) of cucumber leaves

m*M* N	Alanine	Phenylalanine	Lysine	γ-amino-butyric acid	Aspartic acid
0	0,160	0,026	0,569	0,127	0,099
1 NO_3	0,276	0,110	1,691	0,234	0,146
1 NH_4	0,318	0,084	1,365	0,214	0,148
LSD 0,05	0,111	0,038	0,733	0,076	0,038

not significantly affected:

– glycine	– histidine	– serine
– valine	– methionine	– ornithine
– leucine	– tryptophane	– glutamic acid (glutamate)
– isoleucine	– tyrosine	– glutamine
– proline	– threonine	– asparagine.

Table 4. Activity of enzymes of phenol metabolism as affected by nitrate supply

m*M* NO_3	Conidiophores per colony	Phenylalanine-ammonia-lyase nmoles cinnamic acid/g fm h		Cell wall-bound peroxidase $\Delta E \cdot 10^3$/cm mL min		Water soluble peroxidase $\Delta E \cdot 10^3$/g fm min	
0	70	38[a]	(0)[b]	135[a]	(167)[b]	232[a]	(469)[b]
1	190	24	(2)	55	(210)	63	(479)
4	207	14	(2)	59	(169)	134	(1015)
20	68	14	(1)	61	(204)	124	(1067)
LSD 0.05	49	12	n.s.	77	n.s.	103	n.s.

[a] Enzyme activity on leaf half without infection.
[b] Activity increase on leaf half after infection.

Table 5. Influence of N concentration on number of conidiophores of *S. fuliginea* as well as N and Ca content in the shoot of cucumber

m*M* N as NH_4NO_3	Conidiophores per colony	Shoot dry matter	
		N (%)	Ca (%)
0	229	1,5	1,7
1	468	2,8	1,5
4	496	4,8	1,5
20	201	6,0	1,3
LSD 0,05	123	0,3	0,2

well as N excess. However, Ca content decreased with N supply. Thus, growth inhibition of powdery mildew at NO_3 excess was not due to calcium. Therefore, the relationship between NO_3 supply, carbohydrate content and powdery mildew growth was examined on plants grown at 100, 50, 25 and 12% global radiation in a glasshouse in order to vary carbohydrates in plants. Glucose content of leaves was reduced at NO_3 excess compared to optimum NO_3 nutrition (Fig. 1). Powdery mildew growth increased at both NO_3 concentrations with glucose content of leaves, however, the number of conidiophores was on a lower level for NO_3 excess. Thus, powdery mildew growth was related to glucose content, but, the reduced development of *S. fuliginea* at NO_3 excess could not be explained by less glucose. The same was true for fructose.

Discussion

Powdery mildew at N deficiency

Decreasing N supply inhibited initial development of powdery mildew on cereals (Bainbridge, 1974). In contrast, the infection of *S. fuliginea* on cucumber was not affected either with N deficiency or with NO_3 excess. The condition of the epidermis had no influence on the first development stages of cucumber powdery mildew. The cell wall seems to be more important as a barrier for cereals. The further development of *S. fuliginea*, however, was inhibited by strong N deficiency of cucumber plants as was also found with powdery mildew on cereals (Bainbridge, 1974).

Reduced fungus growth at N deficiency is discussed in relation to phenol metabolism of plants (Kiraly, 1964). Low N supply increases phenol content and phenol metabolism (Goodman *et al.*, 1986; Graham, 1983; Kiraly, 1964). Graham (1983) assumes, that it is due to an increased

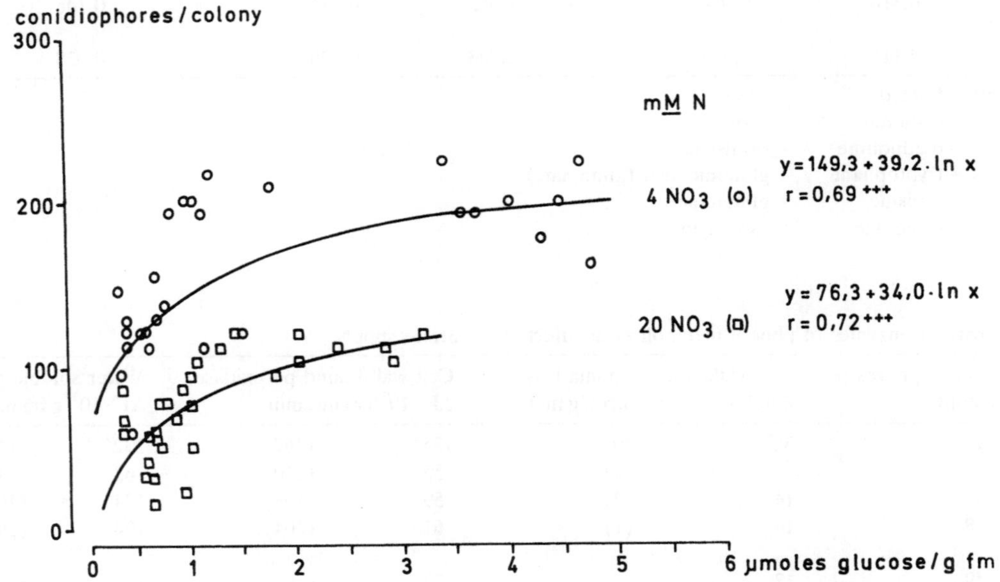

Fig. 1. Relationship between glucose content of cucumber leaves and number of conidiophores of *S. fuliginea* at varied radiation and N supply.

flow of C skeletons into N metabolism that less secondary compounds can be synthesized. However, Kutsch (1986) emphasized that a lot of details, upon which the hypothesis is based, are not settled.

The intensity of phenol metabolism is often characterized by the activity of the key enzyme phenylalanine-ammonia-lyase (PAL) (Green *et al.*, 1975; Mishagi, 1982). The activity of this enzyme was increased at N deficiency. Thus, more phenols could have been built during strong N deficiency. However, according to Margna (1977) and Da Cunha (1987), phenol metabolism is not limited by enzyme activity but availability of the substrate phenylalanine.

The activity of the cell wall-bound peroxidase (POD), which catalyzes the polymerisation of phenol derivates to lignine (Gross, 1980) was similarily increased with N deficiency. The same was true for water soluble peroxidase which can oxidize phenols to chinones (Fric, 1976; Mishagi, 1982). However, initial development of powdery mildew was not impaired. This indicates that the considered processes of phenol metabolism were not significant for infection. But this could cause inhibition of the further development of the fungus.

Growth inhibition of powdery mildew at strong N deficiency was probably due to a lack of some amino acids wherease total amino acid content was not significantly affected (Bettin, 1989). This assumption is supported by the observation that uredospores of *Puccinia graminis*, another obligate biotropic fungus, are not able to synthesize the amino acids arginine and lysine (Jäger and Reisener, 1969).

Powdery mildew at NO_3 excess

NO_3 excess inhibited growth of *S. fuliginea* wherease NH_4 excess had no effect. In contrast, NO_3 excess supply to wheat increased powdery mildew (Last, 1954). Similarily, Sturm (1959) observed on clover an increased development of *E. polygoni* with supply of NH_4NO_3. However, Wijngaarden and Ellen (1969) found with peas an inhibition of fungus growth by an increase of N supply.

Initial development of fungus was not influenced by NO_3 excess. This indicates that not infection barriers were fortified but nourishment conditions were impaired. In agreement with this conclusion no significant changes of phenol metabolism due to NO_3 excess were found. High nitrate content of leaf tissue was also not the reason (Bettin, 1989). This was also not probable though high nitrate supply increases mainly the NO_3 concentration of vacuole sap whereas the fungus is only in contact with cytoplasm (Gay and Manners, 1981). Changed contents of other nutrients were also not the cause.

Reduced fungus growth at NO_3 excess was also not related to carbohydrate metabolism. Total carbohydrate content as well as glucose and fructose content decreased with NO_3 excess. Furthermore, fungus growth was promoted by glucose and fructose content varied by radiation, confirming, that the fungus needs low molecular carbohydrates (Gay and Manners, 1981). But, the same glucose/fructose contents resulted in less growth at NO_3 excess than at optimum NO_3 supply indicating, that other reasons were responsible for the growth reduction. It is speculated that microscopical examinations could give further information.

Control of powdery mildew

Deficient N supply reduced cucumber fruit yield more severely than powdery mildew. A control of *S. fuliginea* without yield losses was not possible in other experiments (Bettin, 1989). Powdery mildew on winter wheat, however, could be reduced without yield reduction (Springer and Heitefuss, 1985). Reason for the differing results could be, that N supply influences efficiency of resistance mechanisms or nutrient supply to powdery mildew in relation to plant variety.

References

Bainbridge A 1974 Effect of nitrogen nutrition of the host on barley powdery mildew. Plant Pathol. 23, 160–161.

Bergmeyer H U 1974 Methoden der enzymatischen Analyse. Vol. 2, pp 1221–1224. Verlag Chemie, Weinheim.

Bettin A 1989 Stickstoffernährung von Gurken – Einfluß auf Stoffwechsel und Ertrag sowie Entwicklung von Echtem Mehltau. Diss. Univ. Hannover.

Bushnell W R and Gay J 1978 Accumulation of solutes in

relation to the structure and function of haustoria in powdery mildews. *In* The Powdery Mildews. Ed. D M Spencer. pp 183–235. Academic Press, London.

Da Cunha A 1987 The estimation of L-phenylalanine ammonia-lyase shows phenylpropanoid biosynthesis to be regulated by L-phenylalanine supply and availability. Phytochemistry 26, 2723–2727.

Förster H, Buchenauer H and Grossmann F 1980 Nebenwirkungen der systemischen Fungizide Triadimefon und Triadimenol auf Gerstenpflanzen. III. Weitere Beeinflussungen des Stoffwechsels. Z. Pfl. krankh. Pfl. schutz 87, 717–730.

Fric F 1976 Oxidative enzymes. *In* Physiological Plant Pathology. Eds. R. Heitefuss and P H Williams. pp 617–631. Springer-Verlag, Berlin.

Gay J L and Manners J M 1981 Transport of host assimilates to the pathogen. *In* Effects of Disease on the Physiology of the Growing Plant. Ed. P G Ayres. pp 85–100. Society for Experimental Biology, Seminar Series 11, University Press, Cambridge.

Goodman R N, Kiraly Z and Wood K R 1986 The biochemistry and physiology of plant disease. pp 211–244. University of Missouri Press, Columbia, MO.

Graham R D 1983 Effects of nutrient stress on susceptibility of plants to disease with particular reference to the trace elements. Adv. Bot. Res. 10, 221–276.

Green N E, Hadwiger L A and Graham S O 1975 Phenylalanine ammonia-lyase, tyrosine-ammonialyase and lignin in wheat inoculated with *Erysiphe graminis* f. sp. *tritici*. Phytopathology 65, 1071–1074.

Gross G G 1980 The biochemistry of lignification. Adv. Bot. Res. 8, 25–63.

Jäger K and Reisener H J 1969 Untersuchungen über Stoffwechselbeziehungen zwischen Parasit und Wirt am Beispiel von *Puccinia graminis* var. *tritici* auf Weizen. I. Aufnahme von Aminosäuren aus dem Wirtsgewebe. Planta 85, 57–72.

Kiraly Z 1964 Effect of nitrogen fertilization on phenol metabolism and stem rust susceptibility of wheat. Phytopathol. Z. 51, 252–261.

Kutsch H 1986 Möglicher regulatorischer Zusammenhang zwischen assimilatorischer Nitratreduktion und Produktion sekundärer Pflanzenstoffe. Landw. Forsch. 39, 97–102.

Last F T 1954 The effect of time of application of nitrogenous fertilizer on powdery mildew of winter wheat. Ann. Appl. Biol. 41, 381–392.

Margna U 1977 Control at the level of substrates supply: An alternative in regulation of phenylpropanoid accumulation in plant cells. Phytochemistry 16, 419–426.

Mishagi I J 1982 Physiology and Biochemistry of Plant Pathogen Interactions. pp 103–108. Plenum Press, New York.

Morvan Y 1987 Untersuchungen zur Resistenz von *Vicia faba* L. Sorten gegenüber *Aphis fabae* (Scop.). Diss. Univ. Hannover.

Peterson L A and Chesters G 1964 A reliable total nitrogen determination on plant tissue accumulating nitrate nitrogen. Agron. J. 56, 89–90.

Springer B and Heitefuss R 1985 Optimierung von Stickstoffdüngung und Pflanzenschutz im Winterweizen. DLG-Mitt. 100, 252–256.

Sturm H 1959 Untersuchungen über das Auftreten von Echtem Mehltau (*Erysiphe polygoni* D.C.) an Kleearten bei verschiedenen Umweltverhältnissen. Z. Acker- und Pflanzenbau 107, 203–239.

Vance G 1975 Effect of powdery mildew on the level of endogenous cytokinins in barley with regard to resistance. Phytopathol. Z. 84, 105–114.

Wijngaarden T and Ellen J 1969 An observation on the influence of nitrogen fertilization on the attack of peas by powdery mildew (*Erysiphe polygoni* D.C.) Plant and Soil 30, 143–144.

Wissemeier A 1988 Beziehung zwischen Mangantoleranz und Oxidation von Mangan in Blättern von Cowpea-Genotypen (*Vigna unguiculata* (L.) Walp.). Diss. Univ. Hohenheim.

Zucker M 1965 Induction of phenylalanine deaminase by light and its relation of chlorogenic acid synthesis in potato tuber tissue. Plant Physiol. 40, 779–784.

M. L. van Beusichem (Ed.), *Plant nutrition – physiology and applications*, 557–559.

PLSO IPNC691

Fertilizer nitrogen budgets of ^{15}N-labelled sugarbeet (*Beta vulgaris*) tops and $Na^{15}NO_3$ dressings split-applied to winter wheat (*Triticum aestivum*) in microplots on a loam soil

J.P. DESTAIN, E. FRANÇOIS and J. GUIOT
Centre de Recherches Agronomiques de l'État, B-5800 Gembloux, Belgium

Key words: *Beta vulgaris* L., nitrogen recovery, split application, sugarbeet tops, *Triticum aestivum* L.

Abstract

The fate of N from sugarbeet (*Beta vulgaris* L.) tops returned to the soil (50 T ha^{-1}) in autumn 1986 before sowing winter wheat (*Triticum aestivum* L.), and from $NaNO_3$ split-applied in 3 equal dressings (at tillering, stem elongation and flag leaf stages) was studied using isotopically labelled ^{15}N in open stainless-steel cylinders pressed into the soil.

At harvest, the percentage utilization (PU) of N from sugarbeet was very low (6.66%) and negatively influenced by fertilizer N (5.59%), while that of fertilizer N was rather high (69.64%) and unchanged by addition of tops. Residual N in soil represented 25.9% of the amount applied in tops and ranged from 33% for the tillering application to 21% for the flag leaf application. N losses (mainly denitrification) from sugar beet tops amounted to 67% and were very low for mineral fertilizer (less than 5%).

Introduction

In the loam region of Belgium, sugarbeet (*Beta vulgaris* L.) is the main preceding crop for winter wheat. Very often, sugarbeet tops, containing a considerable amount of N (more than 200 kg N ha^{-1}), are returned to the soil before sowing the cereal crop. The aim of the present work is to assess the contribution of N from soil-incorporated sugarbeet tops, to quantify its immobilization and losses, and study its interaction with fertilizer N.

Materials and methods

Experimental design

Tops from sugarbeets having received 80 kg N ha^{-1} either as $Na^{15}NO_3$ containing 20 atom% ^{15}N or 80 kg N ha^{-1} unlabelled $NaNO_3$ were incorporated into the soil simulating plough action (50 T ha^{-1}) in microplots (open stainless-steel cylinders; length, 55 cm; internal diameter 29 cm) previously pressed into the soil in autumn 1986 before sowing winter wheat (var. Odéon).

In spring 1987, a total N dose of 135 kg N ha^{-1} as $Na^{15}NO_3$ (5 atom% ^{15}N) or as unlabelled $NaNO_3$ was split-applied in three equal dressings at tillering (T; Feekes growth stage 3), stem elongation (SE; Feekes stage 5), and flag leaf (FL; Feekes stage 10) and compared with a control. The experimental treatments (Table 1) were replicated six times and distributed according to the randomized block layout.

Soil and climate

The soil of the experimental site was a well-drained silty loam (clay 12.0%, silt 85.7%, sand 2.3%). Climatic conditions during the 1986–87 experimental period were characterized by a rainy autumn and winter (450 mm rainfall vs 384 mm normally) and a very rainy growing

Table 1. Experimental treatments

Treatment	Sugarbeet tops ($50\ T\ ha^{-1} = 363\ kg\ N\ ha^{-1}$)	N dressing ($135\ kg\ ha^{-1}$)
1. Control	–	–
2. Tops	Unlabelled	–
3. Tops	Labelled	–
4. Tops + fertilizer N	Labelled	Unlabelled
5. Fertilizer N	–	Labelled[a]
6. Fertilizer N + tops	Unlabelled	Labelled[a]

[a] N doses($kg\ ha^{-1}$) applied at following Feekes stage:

T	SE	FL
45 (^{15}N)	45	45
45	45 (^{15}N)	45
45	45	45 (^{15}N)

period (473 mm vs 308 mm). Temperature was nearly normal for autumn and winter (6.7°C) but very low during the growing period (9.8°C vs 13.9°C normally).

Sample preparation and analytical methods

At maturity (August 17, 1987) plants were harvested quantitatively, cut into pieces, dried, weighed, and finely ground. The soil was removed in 10-cm layers, weighed, forced through a 4-mm sieve, and carefully mixed. One sample was taken for determination of humidity, another was freeze dried. Both soil and plant samples were analyzed for total N and their isotopic composition determined by mass spectrometry (Sira 12-VG Isogas, Manchester, UK).

Results and discussion

Percent recovery of tops and fertilizer N by winter wheat

Percent recovery of N from tops by winter wheat reached only 6.66% (Table 2) and was negatively influenced by fertilizer N (5.59%). These figures are markedly lower than the 27% found by Abshahi *et al.* (1985); similarly, Nordmeyer and Richter (1985) reported that more than 30% of organic N in leaves was available in the subsequent vegetation period.

Percent recovery of fertilizer N was rather high and not influenced by the presence of sugarbeet tops. In agreement with previous findings (Riga *et al.*, 1988; Destain *et al.*, 1989), recovery

Table 2. Percent recovery of tops- nd fertilizer-N by winter wheat

Treatment	Stage	Grain	Straw	Total (mean ± SD)
3. Labelled tops		5.43	1.23	6.66 ± 0.74
4. Labelled tops + fertilizer N		4.46	1.13	5.59 ± 0.33
5. Labelled fertilizer N	T	49.58	12.80	61.38[aa] ± 1.75
	SE	59.55	14.62	74.17[b] ± 3.00
	FL	60.01	13.37	73.38[b] ± 4.73
	Total:			69.64
6. Labelled fertilizer N + tops	T	46.05	13.58	59.63[a] ± 5.82
	SE	57.61	14.92	72.53[b] ± 9.86
	FL	59.02	13.61	72.63[b] ± 7.96
	Total:			68.26

[a] Values within each column followed by the same letters are not significantly different at the 0.05 probability level.
b ???

Table 3. Dry matter yield and N uptake by winter wheat (kg ha^{-1})

Treatment	Yield (Mean ± SD)	Total N	Tops- + Fertilizer-N	Soil N
1. Control	11,515 ± 2,600	117.8		117.8[aa]
2. Unlabelled tops	15,085 ± 2,190	148.9	31.1[b]	117.8
3. Labelled tops	14,355 ± 2,200	142.0	24.4	117.6[a]
4. Labelled tops + fertilizer N	23,482 ± 1,892	213.9	112.4[c]	101.5[b]
5. Fertilizer N	22,550 ± 2,182	208.3	94.0	114.3[a]
6. Labelled fertilizer N + tops	23,571 ± 2,057	214.3	112.4[c]	101.9[b]

[a] Values within each column followed by the same letter are not significantly different at the 0.05 probability level.
[b] Calculated by difference between 2 and 1.
[c] This amount represent ^{15}N derived from labelled material (fertilizer and tops) and takes into account percent recovery of N for treatment 4 and 6 in Table 2.

Table 4. Residual N in soil and N-losses

Treatment	Stage	N remaining in soil (% of amount applied)	N losses (% of amount applied)
3. Labelled tops		25.90[a]	67.38[aa]
4. Labelled tops + fertilizer N		25.56[a]	68.85[a]
5. Labelled fertilizer N	T	33.42	5.20
	SE	24.98	0.85
	FL	21.14	5.49
	Total:	26.51[a]	3.85[a]
6. Labelled fertilizer N + tops	T	34.33	6.15
	SE	23.84	3.62
	FL	22.13	5.23
	Total:	26.77[a]	5.00[a]

[a] Values within each column followed by the same letter are not significantly different at the 0.05 probability level.

was less for the T dressing than for the SE and FL applications.

Soil- and fertilizer-derived N

Soil-derived N was neither influenced by the presence of sugarbeet tops nor by mineral N, averaging approximately 120 kg N ha^{-1} (Table 3). Nevertheless a slight reduction in uptake of soil mineral N seemed to occur when sugarbeet leaves and N fertilizer were applied together.

Residual N in soil from the tops averaged 26% and was not influenced by fertilizer N (Table 4). Residual fertilizer N ranged from 33% for the T dressing to 21% for the FL dressing without any detectable influence of tops.

The percentage of sugarbeet N unaccounted for was very high (67%) probably due to denitrification in autumn after application, and perhaps in spring when high rainfall occurred. During winter, leaching losses of mineral N originating from the greens could have occurred, but were not measured. Total losses were twice as high as those reported by Abshahi *et al.* (1984) (34%), but which were measured in California under drier climatic conditions.

As in earlier studies (Destain *et al.*, 1989; Riga *et al.*, 1988), losses from mineral fertilizer were minimal (<5%).

References

Abshahi A, Hills F J, and Broadbent F E L 1984 Nitrogen utilization by wheat from residual sugarbeet fertilizer and soil incorporated sugarbeet tops. Agron. J. 76, 954–958.

Destain J P, Guiot J, François E and Riga A 1989 Fertilizer nitrogen budgets of two doses of $Na^{15}NO_3$ dressings split-applied to winter wheat in microplots on a loam soil. Plant and Soil 117, 177–183.

Nordmeyer H and Richter J 1985 Der Einfluss einer Rübenblattdüngung auf den Stickstoffhaushalt von Lössböden. VDLUFA – Schriftenreihe 16, 121–127.

Riga A, François E, Destain J P, Guiot J and Oger R 1988 Fertilizer nitrogen budget of $Na^{15}NO_3$ and $(^{15}NH_4)_2SO_4$ split-applied to winter wheat in microplots on a loam soil. Plant and Soil 106, 201–208.

M. L. van Beusichem (Ed.), *Plant nutrition – physiology and applications*, 561–564.

PLSO IPNC218A

Nitrate reductase activity, grain yield and grain protein in wheat (*Triticum aestivum*) as affected by nitrogen fertilization under semi-arid conditions

R. GONZÁLEZ PONCE, M.L. SALAS and A. LAMELA
Instituto de Edafología y Biología Vegetal, CSIC, Serrano 115-dpdo., E-28006 Madrid, Spain

Key words: nitrogen, nitrate reductase, protein, *Triticum aestivum* L., wheat

Abstract

In a field experiment under semi-arid conditions, doses from 0 to 280 kg N ha^{-1} were applied to wheat (*Triticum aestivum* L.) cultivar 'Pané 247', a tall-stem variety. The results obtained were the following: a) an increase of N fertilization increased the dry weight (DW), the NO_3^- content and the '*in vivo*' nitrate reductase activity (NRA) of the wheat plant in growth stage 1 (3 leaves), b) the NRA in this stage, depended on the NO_3^- content in the plant and both were related to the dry weight, c) this NRA, may provide yield of the wheat and, to a lesser extent, the protein content, d) up to 170 kg N ha^{-1}, both, the grain yield and protein content grew with the N dose, at higher fertilization the yield decreased, while the grain protein content continued to grow, with the optimum protein yield being reached at 211 kg N ha^{-1}, e) an increase of N fertilization increased the total accumulation of N in the plant, but due to lodging, N translocation to the grain was reduced.

Introduction

The capacity of the plants to use the nitrogen available in the soil, especially nitrates, is the main factor limiting protein synthesis. Thus, the possibility was examined to use the assay of nitrate reductase activity (NRA) as an index of the availability of N reduced by the plant for protein synthesis (Dalling and Lyon, 1977).

The use of the NRA as a N nutritional index of the plant is also based on the fact that its level is regulated by the NO_3^- concentration in the plant tissue (Croy and Hageman, 1970), and because the enzyme is substrate-inducible and catalyses the rate-limiting step in the nitrate assimilation (Schrader *et al.*, 1968). On the other hand, the NRA assayed in the seedling stage was correlated positive and significantly with the dry matter weight in this state (Reilly, 1976), with the protein content in the grain (Eilrich, 1968) and with the protein yield of the grain (Dalling and Lyon, 1977; Eilrich, 1968), but there was no correlation with the grain yield (Kubanek and Cerny, 1978).

Growing doses of N fertilizer increased the NO_3^- content and NRA in wheat seedlings, although the grain yield was either correlated with the NRA during the anthesis or the total NRA of the cycle, while the protein content was correlated with the NRA assayed at a late, *i.e.*., the wax-ripe stage (Rossi, 1978).

On the other hand, in semi-arid conditions, up to a high N does an increase of N fertilization increased the yield and protein content in the grain, but at higher N doses the yields decreased while the protein content continued to increase (Johnson *et al.*, 1973).

The aim of the present paper is an initial study of how N fertilization affects the yield, protein content and protein yield of the wheat cultivated in semi-arid conditions, where there is a water deficit in spring. It is also studied how such fertilization affects the NO_3^- and NRA levels in an early growth stage of the wheat plant and whether determining the NRA at this stage may be used as a predictor of the yield, protein content and protein yield of the wheat grain.

Materials and methods

In the 1986–87 period, an experiment was carried out in a 15 m^2 plot as part of a larger plot, leaving a separating passage of 0.5 m. The plot belonged to a farm in the province of Toledo (Spain) and was chosen because its soil was very homogeneous and had a very low N level in the soil (0.07%).

For the present initial study, the plot was subdivided into 8 microplots of 1 m^2 each, which were separated by passages. There were as many micro-plots as there were N fertilizer treatments, since the latter were not repeated in view of the reduced area of homogeneous soil available. These treatments were distributed at random.

On December 1st, 1986, wheat (*Triticum aestivum* L.) cv. Pané 247 was sown, a tall-stem variety and commonly used in this area. Eight treatments were given, a control (0) and seven doses of N fertilizer: 40, 80, 120, 160, 200, 240 and 280 kg N ha^{-1}. 60% of the fertilizer was added before sowing, in the form of ammonium sulphate and a 40% in the form of calcium ammonium nitrate at tillering, *i.e.* growth stage 3 according to the Feekes scale (Large, 1954). Before sowing 39 kg P ha^{-1} were added as calcium phosphate and 41 kg ha^{-1} as potassium chloride.

At growth stage 1 (3 leaves), 3rd February, 10 plants of wheat per micro-plot were taken, and after drying at 70°C, the dry weight (DW) of the above-ground part was measured. The NO_3^- content was also determined in an autoanalyser after extraction by the method of Bremmer (1965). The nitrate reductase activity (NRA) was assayed by the '*in vivo*' method on the upper fully expanded leaf blade, according to the technique described by Guerrero (1985). Following this technique, 200 mg of leaf blade were introduced in tubes with 10 mL of an incubation medium containing 0.1 *M* KNO_3, 0.1 *M* potassium phosphate buffer (pH = 7.7) and 1% (v/v) 1-propanol. The tubes were stoppered and vacuum-infiltrated twice for 5 min and then flushed with argon for 2 min. Finally after incubation at 30°C in darkness for 90 min, the nitrite released in the incubation medium was determined by the method of Snell and Snell (1949). The nitrate reductase activity was expressed as nmoles of NO_2^- formed (g. fresh wt^{-1} hr^{-1}).

At growth stage 11.4 (ripening), 18th June, the straw and grain yields were measured and the N content determined in both by using the Kjeldahl method. The protein content (N × 5.7), and the protein yield were calculated.

The N translocation efficiency was calculated as the ratio of total N accumulated in the grain to total N accumulated in the above-ground part of the plant, and expressed as a percentage.

Results

Figure 1 shows that there were parabolic curvilinear responses of the DW, NO_3^- content and

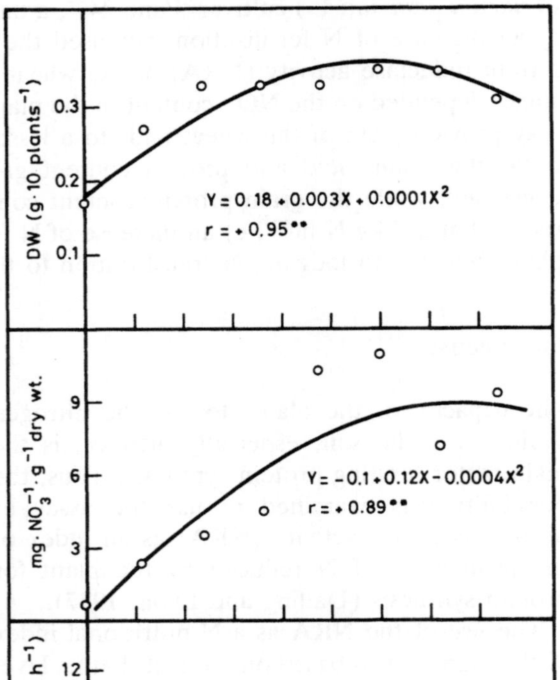

Fig. 1. Effects of N fertilization on nitrate reductase activity (NRA), NO_3^- content and dry weight (DW) of 10 wheat plants cv. Pané 247 at growth stage 1.

NRA of the wheat plant in growth stage 1 to the N fertilization applied before sowing, *i.e.*, only to 60% of the total N dose. These responses were lower as the N doses increased, with optimum values being reached at 121, 150 and 150 kg N ha^{-1} respectively. In view of the similarity of the responses, positive and very significative linear correlations were found between NO_3^- and NRA (r = +0.98**), NO_3^- and DW (r = +0.92**) and NRA and DW (r = +0.90**)

Figure 2 shows that both grain yield and protein content grew up to 170 kg N ha^{-1}, while at higher N doses the protein content continued to grow, but the grain yield decreased. The maximum protein yield was reached at 211 kg N ha^{-1}.

Positive and significant linear correlations of N fertilization were also found with the total N accumulation in the wheat plant at ripening stage (r = +0.78*), but there was a negative correlation with the efficiency of N translocation from the plant to the grain (r = −0.84**). This efficiency was positively correlated with the weight of 1000 grains (r = +0.87**).

Also, positive and significant linear correlations, were found between the NRA assayed '*in vivo*' at growth stage 1 and the grain yield (r = +0.81**), the total protein yield (r = +0.92**) and, somewhat less, the protein content in the grain (r = +0.69*).

Discussion

The positive responses found in the present study of the NO_3^- content and NRA in wheat seedlings to N fertilization were described earlier (Rossi, 1978). In addition, a positive and significant correlation has been found between the NO_3^- content in the plant and the NRA as has been found previously at different growth stages of the wheat (Eilrich and Hageman, 1973). This

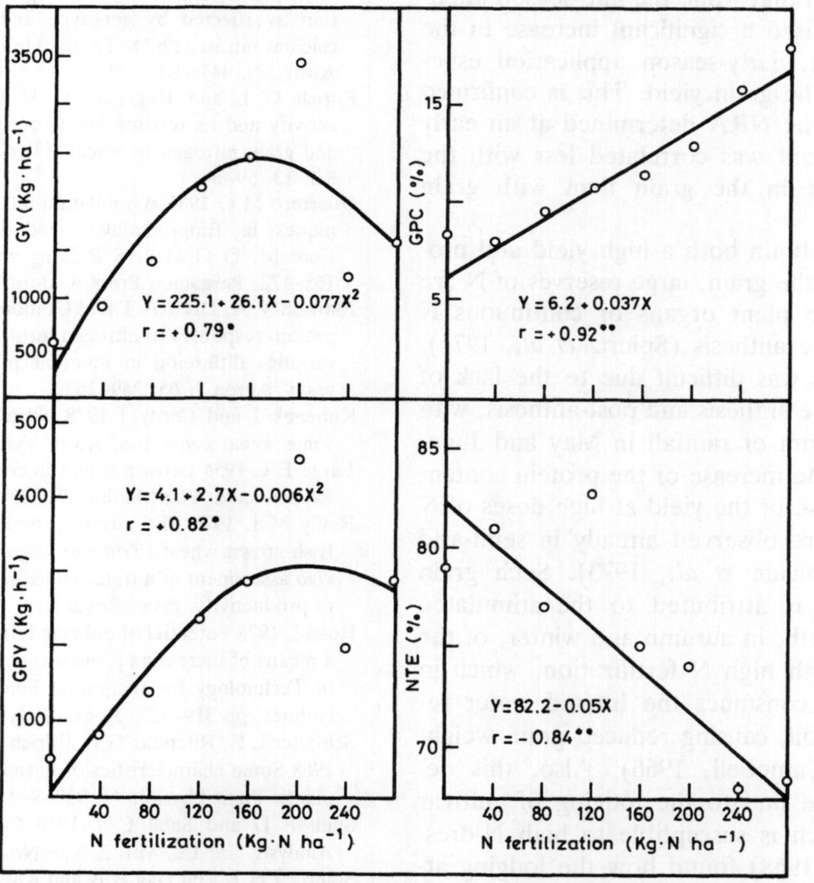

Fig. 2. Effects of N fertilization on the grain yield (GY), grain protein content (GPC), grain protein yield (GPY) and N translocation efficiency (NTE) in the wheat cv. Pané 247 at ripening.

demonstrates the dependence of this enzyme activity on the availability of NO_3^- in the tissue of the plant.

Here, a positive correlation exists between the NRA and the dry weight of wheat plants at growth stage 1, the same was observed in wheat plants growing in a nutritive solution (Reilly, 1976). In addition, a good correlation was obtained between such early NRA and the straw weight at ripening ($r = +0.97^{**}$). There were also a positive correlations between the NRA assayed at growth stage 1 and the grain yield, protein content and protein yield. All this, indicates that the '*in vivo*' method of the NRA at an early growth stage of the plant may serve as a predictor of these parameters. However, other findings indicate that the heading stage is the best moment to assay the NRA and to obtain a good correlation with the grain yield and grain protein yield (Brunetti *et al.*, 1976).

On the other hand, Eilrich and Hageman (1973) indicated that while the late-season application of N caused a significant increase in the protein content, early-season application especially affected the grain yield. This is confirmed here, because the NRA determined at an early stage of the plant was correlated less with the protein content in the grain than with grain yield.

In order to obtain both a high yield and protein content in the grain, large reserves of N are required in the plant organs or continuous N uptake after the anthesis (Spiertz *et al.*, 1971). Such N uptake was difficult due to the lack of water during the anthesis and post-anthesis, with 14.3 and 23.3 mm of rainfall in May and June, respectively. The increase of the protein content and the decrease of the yield at high doses of N fertilization were observed already in semi-arid conditions (Johnson *et al.*, 1973). Such grain yield decrease is attributed to the stimulated vegetative growth, in autumn and winter, of the wheat plant with high N fertilization, which in spring quickly consumes the limited water reserves in the soil, causing reduced grain weight (Brown and Campbell, 1966). Also, this decrease could be due to the lodging of cultivar Pané 247, which is susceptible to high N-dressings. Eilrich (1968) found how the lodging affected the N translocation efficiency. Here, with increasing N-dressings the total N accumulation increased in the plant, but due to lodging, the N translocation efficiency to the grain decreased and the 1000 grains weight was reduced.

References

Bremmer J M 1965 Methods of soil analysis. Am. Soc. Agron. 21, 1215–1219.

Brown P L and Campbell R E 1966 Fertilizing dryland spring and winter wheat in the Brown Soil zone. Agron. J. 58, 348–351.

Brunetti N, Ferrari L, Bozzoni A and Mosconi C 1976 Effect of nitrogen fertilization on nitrate reductase activity on grain, straw and protein yields in durum wheat. Rivista di Agron. 10, 171–177.

Croy L I and Hageman R H 1970 Relationship of nitrate reductase activity to grain protein production in wheat. Crop Sci. 10, 280–285.

Dalling M J and Lyon R H 1977 Level of activity of nitrate reductase at the seedling stage as a predictor of grain nitrogen yield in wheat (*Triticum aestivum* L.) Aust. J. Agric. Sci. 28, 1–4.

Eilrich G L 1968 Nitrate reductase activity in wheat (*Triticum aestivum* L.) and its relationship to grain protein production as affected by genotype and spring application of calcium nitrate. Ph.D. Thesis. University of Illinois (Diss. Abstr. 29, 4478-B).

Eilrich G L and Hageman G H 1973 Nitrate reductase activity and its relationship to accumulation of vegetative and grain nitrogen in wheat. (*Triticum aestivum* L.) Crop Sci. 13, 59–66.

Guerrero M G 1985 Assimilation nitrate reduction. *In* Techniques in Bioproductivity and Photosynthesis. Eds J Coombs, D O Hall, S P Long and J M D Scurlok. pp 165–172. Pergamon Press, Oxford, UK.

Johnson V A, Dreir A T and Grabouski P H 1973 Yield and protein responses to nitrogen fertilizer of two winter wheat varieties differeing in inherent protein content of their grain. Agron J. 65, 259–263.

Kubanek J and Cerny J 1978 Nitrate reductase activity of some wheat genotypes. Rostl. Vyrob. 24, 1031–1038.

Large E C 1954 Growth stages in cereals: Illustration of the Feekes scale. Plant Pathol. 3, 128–129.

Reilly M L 1976 The nitrate assimilation capacity of some Irish-grown wheat (*Triticum vulgare*) varieties. III. An *in vivo* assessment of nitrate reductase activity and its relation to productivity. Proc. Royal Irish Acad. 76 (34), 567–576.

Rossi L 1978 Potential of utilizing nitrate reductase activity as a means of increasing protein content and yield of cereals. *In* Technology for Increasing Food Production. Ed. J C Holmes. pp 319–323. Rome, Italy.

Schrader L E, Ritenour G L, Eilrich G L and Hageman R H 1968 Some characteristics of nitrate reductase from higher plants. Plant Physiol. 43, 930–940.

Snell F D and Snell C T 1949 Colorimetric Methods of Analysis. 3rd Ed. Vol. 2. Van Nostrand, NJ, 804 p.

Spiertz J H J, Ten Hag B A and Kupers L J P 1971 Relation between green area duration and grain yield in some varieties of spring wheat. Neth. J. Agric. Sci. 19, 211–222.

M. L. van Beusichem (Ed.), *Plant nutrition – physiology and applications*, 565–568.

PLSO IPNC218B

Effects of nitrogen and herbicide on grain yield and protein content of a tall and a semi-dwarf wheat (*Triticum aestivum*) cultivar in semi-arid conditions

R. GONZÁLEZ PONCE, A. LAMELA and M.L. SALAS
Instituto de Edafología y Biología Vegetal, CSIC, Serrano 115-dpdo., E-28006 Madrid, Spain

Key words: nitrogen, protein, *Triticum aestivum* L., wheat, varieties

Abstract

Nitrogen fertilization produced increases in grain yield only when tri-allate was used for the control of a heavy infestation of wild oats in the wheat crop. Grain protein content and grain protein yield increased with N fertilization, using or not the herbicide.

The use of herbicide resulted in higher grain yield and protein yield; these increases were greater for the semi-dwarf variety than for the tall variety.

The semi-dwarf variety was more sensitive to wild oat competition and the lack of N fertilization than the tall variety and due to the lack of rainfall during milky-ripening of the grain its greater productive potentiality did not show.

Introduction

In semi-arid regions with water deficit in the growing season, the response of wheat to nitrogen fertilizer is low.

The semi-dwarf varieties of the wheat developed for their higher response to nitrogen and greater productivity levels if compared with the standard varieties (Russell *et al.*, 1970). However, it was found that semi-dwarf varieties responded better that tall varieties to nitrogen in the arid conditions of South Dakota (Ward *et al.*, 1974), but other researchers have no found such clear difference (McNeal *et al.*, 1971).

In general, there is an inverse relationship between the grain yield of cereal crops and its protein content. This seems to be due to an effect of N dilution in the grain as starch production increases (Evans and Wardlaw, 1975). Some researchers have found that nitrogen fertilizer increases the grain yield and its protein content (Bauer, 1970; Bishop and McEachern, 1971) which occurs with intermediate water availability in the soil (Terman *et al.*, 1969). But if water conditions are favourable, nitrogen fertilizer increases the grain yield and decreases the protein content (Kramer, 1979).

On the other hand, cereal fields are often infested by wild oat, a weed which competes with wheat for the N available in the soil (González Ponce *et al.*, 1984; González Ponce and Lamela, 1987) as well as for the N fertilizers (Chesalin and Timofeeva, 1974). The competition affects the grain yield of wheat (González Ponce *et al.*, 1984; González Ponce, 1987; González Ponce and Lamela, 1987) and even the protein content of the grain (Friesen *et al.*, 1960), although other authors did not find that the protein content of the grain was affected (Bell and Nalewaja, 1968; González Ponce, 1987; González Ponce and Lamela, 1987).

The aim of the present study is to find out whether in semi-arid conditions a control of wild oat improves the effect of N fertilization on wheat and whether the use of this nutrient and herbicide increase the grain yield, the protein content and grain protein yields of wheat.

Materials and methods

In a semi-arid area of the Province of Toledo (Spain), an experiment was carried out on a haploxeralf soil, low in N (0.06%). In the previous year the field was chosen for an even and heavy infestation of wild oat (*Avena sterilis* L.). Consequently, in the year of the experiment it was heavily reinfested. The field was divided into 64 micro-plots of 2 × 4 m, where the treatments were applied.

In October, the following wheat varieties were used. Pané 247 (standard and of tall-stem) and Anza (Mexican origin, semi-dwarf, high response to N and great productivity potential). These varieties were sown at 125 and 100 kg grain ha^{-1} respectively, in view of the different size of the seed and in order to obtain about 300 wheat plants per m^2.

Before the emergence of wild oat and wheat, the plots were treated with or without herbicide. The herbicide used was Avadex BW-40 (triallate), which is specially used to control the weed in the wheat crop. The dose applied was $3L.ha^{-1}$.

In addition, each variety treated or not with herbicide received the following treatments of nitrogen fertilizers 0, 42, 84 and 126 kg N ha^{-1}. 60% of the doses were applied as ammonium sulphate before sowing, while 40% were as calcium ammonium nitrate applied in the growth stage 2 of the wheat, according to Feekes scale (Large, 1954). Also before sowing 43.7 kg P ha^{-1} in the form of calcium superphosphate and 41.5 kg K ha^{-1} in the form of potassium chloride were applied to all micro-plots. The treatments were distributed in radomized blocks with 4 replicates per treatment.

At ripening of the wheat, in July, the grain yield per treatment was measured and average samples were taken from every treatment to determine the crude protein content in the grain and the protein yield. The crude protein content was calculated after drying the grain at 70°C and the N content was determined by Kjeldahl method, which was multiplied by a factor 5.7.

Results

At ripening the control plots contained 208 wild oat plants per m^2. The autumn rains helped to distribute the herbicide in the soil, so that the weeds in the treated plots were reduced to 2% of the control.

During stem extension and heading in April and May the plants developed well because of the rainfall. But in June, during milky-ripening, rainfall was low. This lack of rainfall together with high temperatures and reduced rooting depth caused an important water deficit in the plants during this period (Table 1).

Table 2 shows the level of significance of herbicide, doses of N and varieties treatments. The interaction between treatments was not significant.

Table 3 shows that nitrogen fertilizer increased grain yield of both varieties only when the herbicide was used and especially for 0 and 42 kg N ha^{-1}. The semi-dwarf variety yielded less than the tall variety when it received no N fertilizer. But when it was fertilized with any dressing of N, it produced similar yields. On the other hand, the use of herbicide increased the grain yields of both varieties, the effect being less noticeable at 0 dose of nitrogen and more in the semi-dwarf variety.

The protein content in the grain shown in Table 4 increased in both varieties after applying N fertilizer, with or without application of herbicide. However, while the use of herbicide did not affect the protein content in the tall variety, it caused a lowering of the protein content in the semi-dwarf variety as a result of N dilution in the grain. This was due to the fact that the semi-dwarf variety produced a greater increase of

Table 1. Mean temperatures and rainfall in the October 1984–July 1985 period

	Oct	Nov	Dec	Jan	Feb	Mar	Apr	May	Jun	Jul
Temp (°C)	14.3	9.8	7.0	4.0	9.2	8.5	12.6	14.7	21.8	26.4
Rainfall (mm)	41.5	128.6	11.4	105.5	54.8	2.7	59.8	43.1	2.7	0

Table 2. Analysis of variance of grain yield for herbicide application, nitrogen application, and variety

Source	Degrees of freedom	F values
Herbicide (H)	1	47.7***
N doses (N)	3	7.14***
Varieties (V)	1	4.95*
H×N	3	2.04^{ns}
H×V	1	1.58^{ns}
N×V	3	0.06^{ns}
H×N×V	3	0.09^{ns}
Replicates	3	1.74^{ns}
Error	45	
Total	63	

* Significant at $P \leq 0.05$; *** Significant at $P \leq 0.01$; ^{ns} no significant.

grain yield with the use of herbicide than the tall variety.

Finally, Table 5 shows how the protein yield increased in both varieties with increasing N fertilizer doses, with or without herbicide, a feature which is more noticeable in the semi-dwarf variety, especially with the use of herbicide. The use of herbicide increased the protein yield in both varieties. This increase was more noticeable in the semi-dwarf variety.

Discussion

When there was a water deficit during the milky-ripening stage of wheat, which is a typical phenomenon in spanish semi-arid areas, grain yields of wheat increased with N fertilization with the use of herbicide. However, this increase was not higher for the semi-dwarf variety as might have been expected in view of the water deficit in question, something that had been found by other researchers (McNeal *et al.*, 1971). The fact that the yields of the semi-dwarf variety were lower than that of the tall variety, when both were not fertilized with nitrogen, is due to the greater needs of the semi-dwarf variety as regards this element. In these climatic

Table 3. Mean wheat grain yield of two varieties (kg/ha ± s.e.) in relation to use of herbicide and application of increasing amounts of N fertilizer

Kg N ha^{-1}	Pané 247.			Anza.		
	Without herbicide	With herbicide	%Δ	Without herbicide	With herbicide	%Δ
0	1009 ± 44	1272 ± 141	26	608 ± 129	1025 ± 84	69
42	1132 ± 93	1857 ± 113	64	626 ± 240	1845 ± 324	195
84	1184 ± 146	2215 ± 111	87	730 ± 228	1997 ± 153	174
126	1369 ± 292	2292 ± 157	67	1076 ± 363	2284 ± 350	112

Table 4. Mean protein contents (%) of two wheat varieties in relation to use of herbicide and application of increasing amounts of N fertilizer

Kg N ha^{-1}	Pané 247		Anza	
	Herbicide application			
	no	yes	no	yes
0	9.1	8.6	8.6	8.0
42	9.1	9.1	12.0	10.3
84	10.8	10.8	13.1	10.3
126	10.8	10.3	13.1	12.0

Table 5. Mean grain protein yield of two varieties (kg/ha) in relation to use of herbicide and application of increasing amounts of N fertilizer

Kg N ha^{-1}	Pané 247			Anza		
	Without herbicide	With herbicide	% Δ	Without herbicide	With herbicide	% Δ
0	91	108	18.7	51	80	56.9
42	103	171	66.0	74	188	154.1
84	131	239	82.4	97	205	111.3
126	148	239	61.5	143	274	91.6

conditions, higher doses of N that those indicated here are not expected to produce a yield response, and less so in the semi-dwarf variety.

Such good response of wheat to N with application of herbicide is logical since the competition of wild oat with wheat for N affects the grain yield of the wheat (González Ponce *et al.*, 1984; González Ponce and Lamela, 1987). Also the use of herbicide produced an increase of grain yield.

The fact that in the absence of N fertilizer the use of herbicide has produced a lesser increase of the grain yield than in the presence of N fertilizer is due to the great need of the weed as regards N (Thurston, 1959) which turns out not to compete very much for this nutrient with the wheat when it is not fertilized. On the other hand, the effect of the herbicide was more obvious on the yield of the semi-dwarf variety than on the tall variety, which shows that the semi-dwarf variety is more sensitive to wild oat competition than the tall variety.

The increase of the grain yields and protein contents here obtained as a result of N fertilization coincides with results obtained in conditions of intermediate water supply (Terman *et al.*, 1969). These conditions happened in our experiment, since if the rainfall was enough in autumn and winter it was low at milky-ripening stage of the plant in late spring. The N fertilization increased the grain protein yield per area.

The fact that the use of the herbicide has not increased the protein content of the grain coincides with earilier results for the same tall variety and with similar environmental conditions (González Ponce and Lamela, 1987). Previously, it was observed that wild oat competition does not affect the protein content of the grain (Bell and Nalewaja, 1968).

The use of herbicide increased the protein yield of the wheat grain due, above all, to the increase of the grain yield. The increase of the protein yield was higher in the semi-dwarf variety than in the tall variety due to the greater increase of the grain yield obtained in the semi-dwarf variety than in the tall variety.

References

Bauer A 1970 Effect of fertilizer nitrogen rate on yield of six spring wheats. North Dak. Farm. Res. 27, 3–9.

Bell AR and Nalewaja JD 1968 Competition of wild oat in wheat and barley. Weed Sci. 16, 505–509.

Bishop RF and McEachern CR 1971 Response of spring wheat and barley to nitrogen, phosphorus and potassium. Can. J. Soil. Sci. 51, 1–11.

Chesalin GA and Timofeeva AA 1974 The utilization of fertilizer nutrients by crop and weed plants. Khim. Sel'sk Koz. 12, 298–299.

Evans LT and Wardlaw IF 1975 Aspects of the comparative physiology of grain yield in cereals. Adv. Agron. 28, 301–359.

Friesen G, Shebeski LH and Robinson AD 1960 Economic losses caused by weed competition in Manitoba grain fields. II. Effect of weed competition on the protein content of cereal crops. Can J. Plant Sci. 40, 652–658.

González Ponce R, De Andrés M and Lamela A 1984 Nitrogen nutrition of winter wheat subjected to the mixed action of fertilizer and herbicide. Proc. 6th Int. Coll Optim. Plant Nutr., Montpellier 3, 927–934.

González Ponce R 1987 Competition for N and P between wheat and wild oat (*Avena sterilis* L.) according to the proximity of their time of emergence. Plant and Soil 102, 133–136.

González Ponce R and Lamela A 1987 Improving the efficiency of nitrogen fertilizer in wheat treated with herbicide for the control of wild oats. J. Plant Nutr. 10, 1771–1778.

Kramer Th 1979 Environmental and genetic variation for protein content (*Triticum aestivum* L.). Euphytica 28, 209–218.

Large EC 1954 Growth stages in cereals: Illustrations of the Feekes' scale. Plant Pathol. 3, 128–129.

McNeal FH, Berg MA, Brown PC and McGuire CF (1971) Productivity and quality response of five spring wheat genotypes (*Triticum aestivum* L.) to nitrogen fertilizer. Agron. J. 63, 908–910.

Russell DA, Henshaw DM, Schauble CE and Diamond RB 1970 High yields cereals and fertilizer demand. Tenn. Valley Auth. USA.

Terman GL, Raming RE, Dreier AF and Olson RA 1969 Yield protein relationship in winter grains, as affected by nitrogen and water. Agron. J. 61, 755–759.

Thurston JM 1959 A comparative study of the growth of wild oats (*A. fatua* and *A. ludoviciana* Dur.) and of cultivated cereals with varied nitrogen supply. Ann. Appl. Biol. 47, 716–739.

Ward RC, Carson PL, Pylman RW and Hoeft RG 1974 When fertilizer supply runs short match spring wheat variety to available nitrogen. South Dak. Farm Home Res. 25, 23–26.

M. L. van Beusichem (Ed.), *Plant nutrition – physiology and applications*, 569–575.

PLSO IPNC169

Seed characters in oilseed rape (*Brassica napus*) in relation to nitrogen nutrition

A. KULLMANN[1], V.B. OGUNLELA[2] and G. GEISLER[1]
[1]*Institute of Crop Science and Plant Breeding, University of Kiel, Christian-Albrechts-University Olshausen Str. 40/60, D-2300 Kiel 1, FRG, and* [2]*Department of Agronomy, Ahmadu Bello University, PMB 1044, Zaria, Nigeria*

Key words: assimilate, *Brassica napus* L., greenhouse, hydroponics, nitrogen supply, seed number, seed weight

Abstract

A greenhouse experiment was conducted to study the influence of N supply on seed weight and number of seeds per pod in oilseed rape (*Brassica napus* L.) in relation to position of raceme on the plant. Three levels of N supply (2, 7 and 12 m*M* N as NH_4NO_3) were applied as treatments. Weight of 1000 seeds and number of seeds per pod were estimated from the terminal raceme and five axillary racemes at 27, 44 and 61 days after bloom (DAB). Seed weight was increased significantly by N supply up to 7 m*M* N irrespective of raceme position on the plant. Number of seeds per pod was increased by N only on the lower racemes at 27 and 61 DAB but was increased on every raceme at 44 DAB. Seed numbers for the lower portions of racemes were generally fewer than those for the upper portions. Nitrogen influence on seed weight was greater than it was on seed number, and the lower the raceme position was on the plant, the fewer the number of seeds per pod was likely to be.

Introduction

Oilseed rape (*Brassica napus* L.) is an important oil crop in the temperate regions of the world mainly because of the many economic uses to which its oil and by-products are put. However, if a high yield production level in this crop is to be attained or maintained, regular applications of adequate amounts of fertilizers would be necessary.

Experimental evidence continued to support the view that for maximum seed yield, oilseed rape requires heavy fertilizer dressings (Scott *et al.*, 1973). Nitrogen nutrition of this crop is important to its productivity and optimum requirements of this major nutrient is relatively high and has been shown to vary from soil to soil. A dressing of 210 kg N per ha produced maximum seed yield in spring-sown rapeseed crop (Allen and Morgan, 1972) while Kumar and Gangwar (1985) fairly recently reported increased seed yield in Indian rape with increasing N supply but only up to 90 kg N per ha. However, in only very few instances was seed yield response to nitrogen application explained in terms of variations in the important yield components.

Allen and Morgan (1972), after analysing the physiological effects of nitrogenous fertilizer on yield in oilseed rape, concluded that the higher yields of seed in this crop achieved through increasing levels of N application were due to a greater production of seeds by a larger number of pods which carried more seeds. At present there is a dearth of information on the above aspect of seed and pod development in oilseed rape. The position of the sink (pods) relative to that of the source (leaves) is believed to be of importance in the yield physiology of the rape plant. The supply of assimilate to developing seeds or pods is therefore an important factor detemining yield development in oilseed rape (Tayo, 1974). Although there is some reported work on the significance of position within the plant on number of seeds per pod and seed weight (Clarke, 1979; Mendham and Scott, 1975), how this is likely to be modified by the nitrogen nutrition of the plant is yet to be examined. In order to interpret more precisely how nitrogen increases yield in oilseed rape there is need to investigate the relative importance of competition for assimilates.

The present investigation was therefore conducted to examine the influence of nitrogen nutrition of the oilseed rape plant on seed weight and seed number per pod at different positions of the terminal and axillary racemes.

Materials and methods

A greenhouse experiment was conducted using an hydroponic system. Young seedlings of summer oilseed rape cv. 'Callypso' were raised and later transplanted onto 5-plastic pots containing nutrient solution (Hoagland and Arnon, 1950). Details of the composition of the modified Hoagland solution used in this study have been previously given by Ogunlela *et al.* (1989). The greenhouse conditions were 20°/15°C day/night temperture, 12-h daylength and 80 Wm^{-2} (400–700 nm) illumination.

Nutrient solution was aerated continuously and replaced weekly and its pH maintained at about neutral by adding $CaCO_3$. The standard solution contained no nitrogen. The treatments applied were three levels of nitrogen supply, namely, 2, 7 and 12 m*M* N as NH_4NO_3 and these were added to the standard solution at weekly intervals. Treatments were replicated three times in a randomised complete block design. Each N level (treatment) was assigned 30 pots, each of which contained one plant.

Rape plants were allowed to grow normally until bloom stage, when the number of axillary racemes (branches) on each was artificially regulated to five and the numbers of flowers on the main and axillary racemes maintained at 60 and 40 respectively. At three growth stages, namely, at 27, 44 and 61 days after bloom (DAB), rape plants (6 plants per N level) were harvested separately. The terminal and the five axillary racemes for each plant were separated and then sub-divided into upper and lower portions (Fig. 1). The pods on each plant part were harvested separately and from these were estimated number of seeds per pod and weight of 1000 seeds.

Data collected were subjected to analysis of variance to test significance of treatment effects and treatment means compared using the test of least significant difference (LSD) at 5% probability level.

Results

Seed weight

Nitrogen supply influenced seed weight in oilseed rape significantly as determined at different growth stages (Table 1). Regardless of whether the seeds came from pods on the terminal raceme or from any of the five axillary racemes, weight of 1000 seeds at 27 DAB increased as nitrogen supply was increased from 2 to 12 m*M* N but increment in seed weight was significant only up to the first N increment *i.e.* 7 m*M* N. Seeds from the terminal raceme were generally heavier than those from the axillary racemes at 27 and 44 DAB.

At 44 and 61 DAB, with the exception of those seeds which came from the terminal raceme at the earlier growth state, seed weight increased significantly when N supply was increased from 2 to 7 m*M* N but as N nutrition was increased to 12 m*M* N, seed weight tended to decrease. It was only under conditions of N deficiency (*i.e.* 2 m*M* N) that the seeds from the terminal raceme were significantly larger than those from the axillary racemes at 61 DAB.

Seed number

Responses to nitrogen supply by seed number per pod at three growth stages of oilseed rape is shown in Table 2. At 27 DAB, number of seeds per pod on the terminal raceme and on branches 1 and 3 did not respond to N supply. However, in the case of branches 2, 4 and 5, seed number increased significantly as N nutrition increased up to 7 m*M* N; but increasing N supply to 12 m*M* N gave only marginal further increase in seed number. Pods located on branch 5 (*i.e.* 5th axillary raceme) gave the greatest response to N

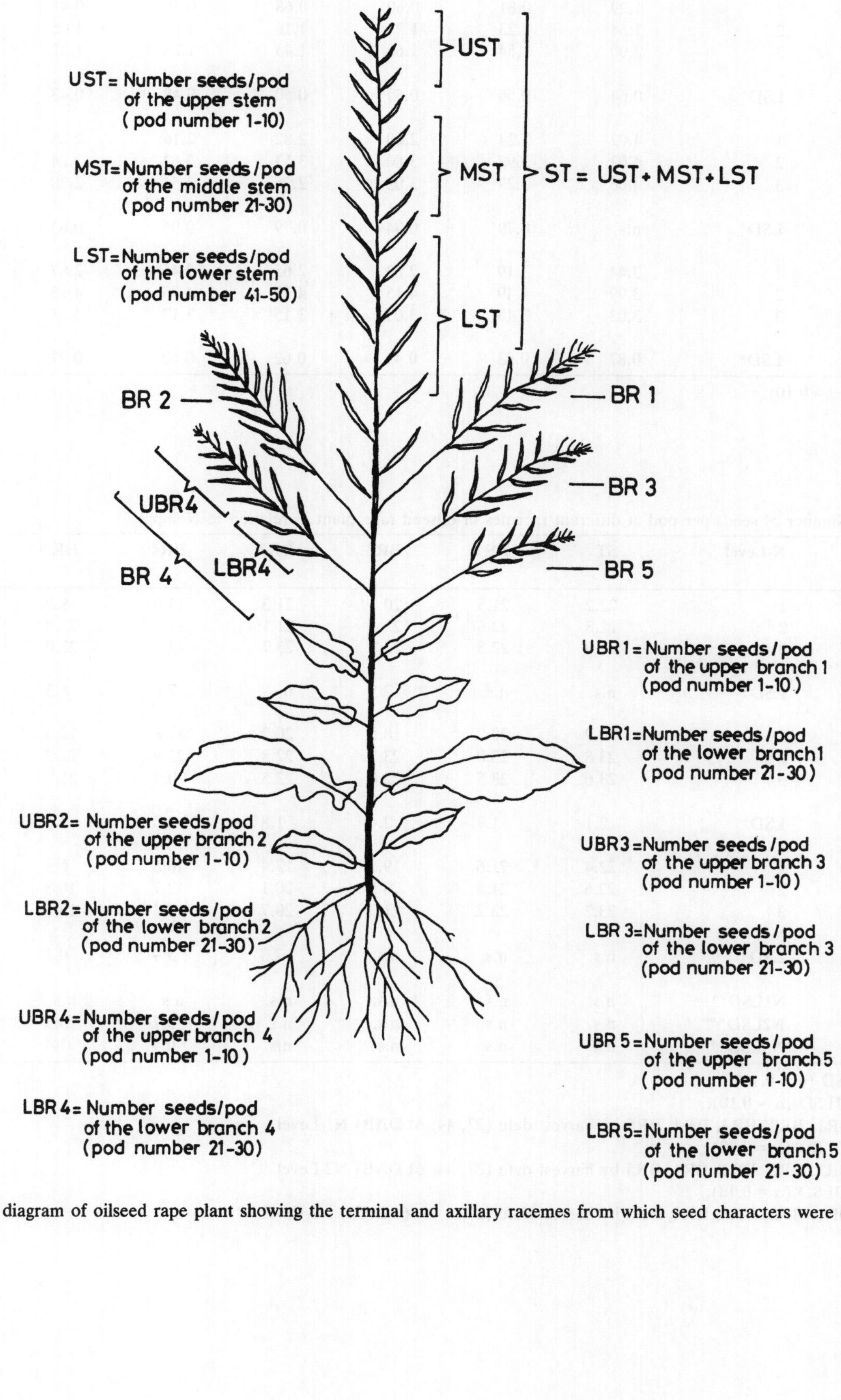

Fig. 1. A diagram of oilseed rape plant showing the terminal and axillary racemes from which seed characters were collected.

Table 1. Weight of thousand seeds (g) at different racemes of oilseed rape plant in relation to N supply

Harvest date	N-Level	ST	BR1	BR2	BR3	BR4	BR5	LSD*
	1	1.29	0.81	0.60	0.68	0.50	0.41	0.25
27 DAB	2	1.54	1.23	1.34	1.16	1.07	1.06	0.28
	3	2.00	1.54	1.66	1.43	1.26	1.21	n.s.
	LSD*	0.64	0.56	0.59	0.50	0.48	0.43	
	1	4.02	3.24	2.60	2.82	2.16	2.26	1.34
44 DAB	2	4.02	3.98	3.80	3.53	3.62	3.24	0.53
	3	4.06	3.27	3.02	2.76	2.54	2.28	1.18
	LSD*	n.s.	0.79	0.94	0.79	0.96	0.83	
	1	3.44	3.19	2.83	2.65	2.45	2.87	0.74
61 DAB	2	3.99	4.49	4.15	4.22	4.05	4.38	n.s.
	3	3.03	3.13	3.03	3.15	3.17	3.15	n.s.
	LSD*	0.87	0.83	0.49	0.62	0.60	0.77	

* =LSD ($\alpha = 0.10$).

Table 2. Number of seeds per pod at different racemes of oilseed rape plant in relation to N supply

Harvest date	N-Level	ST	BR1	BR2	BR3	BR4	BR5	LSD*
	1	22.2	21.5	20.1	21.3	13.0	8.9	9.2
27 DAB	2	22.8	22.6	22.4	21.1	21.9	22.8	n.s.
	3	21.9	22.5	22.8	23.7	23.3	23.0	n.s.
	LSD*	n.s.	n.s.	2.5	n.s.	7.6	6.3	
	1	20.0	20.2	18.5	20.2	17.8	12.6	5.3
44 DAB	2	23.8	23.0	23.2	22.8	22.9	21.3	n.s.
	3	23.6	23.5	22.0	23.5	20.9	21.7	1.3
	LSD*	2.1	1.7	2.4	1.8	3.0	4.6	
	1	22.4	21.6	19.1	17.4	10.5	5.9	11.3
61 DAB	2	22.6	21.5	20.5	20.1	17.7	16.8	n.s.
	3	23.2	23.2	21.3	20.7	19.4	19.6	n.s.
	LSD*	n.s.	n.s.	n.s.	7.3	5.9	0.2	
	N1LSD**	n.s.	n.s.	n.s.	n.s.	n.s.	n.s.	
	N2LSD***	n.s.	n.s.	n.s.	n.s.	3.7	3.3	
	N3LSD****	n.s.	n.s.	n.s.	n.s.	0.4	1.4	

* = LSD ($\alpha = 0.10$).
** = N1LSD ($\alpha = 0.10$).
VA: ST, BR1, BR2, BR3, BR4, BR5 by harvest date (27, 44, 61 DAB) N1-Level.
*** = N2LSD ($\alpha = 0.10$).
VA: ST, BR1, BR2, BR3, BR4, BR5 by harvest date (27, 44, 61 DAB) N2-Level.
**** = N3LSD ($\alpha = 0.10$).
VA: ST, BR1, BR2, BR3, BR4, BR5 by harvest date (27,m 44, 61 DAB) N3-Level.

in terms of seed number. Under very limited N supply, number of seeds per pod for branch 4 or branch 5 was rather low compared to that for any of the other branches (axillary racemes) or the terminal raceme. At 44 DAB, response was significant in respect of pods on all the five axillary racemes and the terminal raceme. Number of seeds per pod increased as N supply increased from 2 to 7 m*M* N.

At 61 DAB, seed number in respect of pods on the terminal raceme and on branches 1, 2 and 3 failed to respond to N supply but in the case of pods on branches 4 and 5, seed number per pod increased significantly as N nutrition of the rape plant increased. Pods on the lowest two axillary racemes contained seeds which were generally fewer in number than those on the upper racemes (*i.e.* branches 1, 2 and 3) and on the terminal raceme.

Number of seeds per pod at the sub-sections of the terminal raceme and the axillary raceme of oilseed rape plant in response to nitrogen supply is shown in Table 3. Generally, at 27 DAB, N effect on number of seeds per pod and the upper or lower portion of the terminal raceme and of axillary racemes 1, 2 and 3 was not significant. However, in the case of axillary racemes 4 and 5, the number of seeds per pod in the upper or lower portion was increased significantly by N supply. Response to N was significantly only up to 7 m*M* N. At 44 DAB, seed number at the upper or lower portion of the terminal and axillary racemes was increased significantly by N supply up to 7 m*M* N. As the plant grew older (*i.e.* at 61 DAB), however, the effect of N supply was not significant, except at the lower portion of branch 4 and both portions of branch 5. Seed number was generally less on the lower portion of branch 5 under conditions of N deficiency (*i.e.* 2 m*M* N).

Table 3. Number of seed per pod at sections of different racemes of oilseed rape plant in relation to N supply (61 DAB)

Date DAB	N Lev.	LST	MST	UST	LBR1	UBR1	LBR2	UBR2	LBR3	UBr3	LBR4	UBR4	UBR5	LBR5
	1	22.8	21.6	22.3	22.2	20.7	20.7	19.6	23.5	19.1	14.4	11.7	11.6	6.3
27	2	22.8	23.7	21.9	22.4	22.9	21.7	23.1	22.4	19.7	22.8	21.0	23.2	22.5
	3	21.2	22.0	22.6	22.7	22.3	22.5	23.1	23.7	23.7	23.1	23.4	22.8	23.2
	LSD*	n.s.	n.s.	n.s.	n.s.	n.s.	3.4	n.s.	n.s.	7.9	7.7	7.8	5.2	
	1	22.2	19.4	18.2	20.1	20.4	19.7	17.4	21.2	19.3	18.1	17.5	15.0	10.2
44	2	23.9	24.8	22.7	22.9	23.1	23.4	23.0	22.3	23.4	23.3	22.6	22.5	20.1
	3	25.2	23.9	21.8	23.7	23.2	22.4	21.5	24.2	22.8	21.4	20.5	21.2	22.1
	LSD*	2.8	2.1	2.8	1.7	2.4	2.5	2.6	2.2	2.5	4.0	2.3	6.2	6.0
	1	23.1	21.7	22.5	22.3	20.0	21.7	16.4	19.2	15.7	15.6	5.4	7.8	4.1
61	2	23.0	22.8	22.0	21.2	21.8	23.0	18.0	20.1	20.1	18.4	17.0	17.6	16.1
	3	24.7	23.6	21.2	22.3	24.1	20.9	21.7	20.3	21.0	20.1	18.7	19.4	19.7
	LSD*	n.s.	n.s.	n.s.	n.s.	n.s.	n.s.	n.s.	n.s.	n.s.	6.7	7.7	6.9	

* = LSD ($\alpha = 0.10$).
LST = Lower Stem; MST = Middle Stem; UST = Upper STEM.
LBR* = Lower Branch*; UBR* = Upper Branch*.
See Fig. 1.

Discussion

Oilseed rape plants require adequate nitrogen supply to enable them develop yield components at levels which can translate into high seed yields. The detrimental effects of nitrogen deficiency on seed weight can be very large and more obvious during the earlier part of seed development then during the latter part. This was so regardless of which raceme the seeds came from. Tayo and Morgan (1975) had concluded that the 2–3 week period from anthesis is particularly critical in determining the yielding capacity of an oilseed rape plant.

At 27 DAB, seeds from the lower racemes tended to weigh less than those from the upper or terminal racemes, suggesting that the upper racemes compete for assimilate better than do the lower racemes. It had been suggested that intra-plant differences in numbers and weight of pods and seeds may be caused by differences in assimilate availability within the plant (Clarke, 1979). However, assimilate availability in the plant varied with time and with proximity of sink to source (Major *et al.*, 1978). Pods from the main branch are believed to produce the largest amount of seed dry weight (Diepenbrock and Geisler, 1979).

There were significant increases in seed weight when nitrogen supply was increased to 7 m*M* N in the present study, regardless of the raceme the pods came from; however the magnitude of increase was a function of developmental stage and the position of the pods on the plant. Efficiency of drymatter accumulation in seeds diminished when N supply was increased to 12 m*M* N. This observation is contrary to an earlier observation by Allen and Morgan (1972) that individual rape seed were slightly heavier in the early stages where no nitrogen was applied but there were no differences in average seed weight at maturity. Krogman and Hobbs (1975) also found that the main effect of fertilizer on seed weight was not significant.

The effect of nitrogen supply on seed weight was larger than that on number of seeds per pod, especially at 27 and 61 DAB. This can be explained by the fact that seed and pod numbers are determined during the period three weeks following full flowering (Mendham *et al.*, 1981). Ovules in pods eventually develop to form seeds and as such any factor which prevents the full development of ovules into seed are likely to influence seed number per pod. Nitrogen deficiency in rape plants is therefore likely to influence seed number per pod. Nitrogen deficiency in rape plants is therfore likely to account for the fewer seeds per pod observed at 2 m*M* N. This was more apparent in pods from branches 4 and 5. Analysis of physiological effects of nitrogenous fertilizer on oilseed rape yield by Allen and Morgan (1972) has highlighted the importance of seeds per pod as a yield-determining character. They concluded that the higher yield of seed through increasing level of applied N was due to a greater production of seeds by a larger number of pods which carried more seeds. Under such conditions more embryos in each pod develop into seeds, and the seeds grow larger.

Irrespective of the portion of the terminal raceme or axillary raceme, nitrogen effects on seed number were virtually the same. However the upper portions of the racemes had pods with larger numbers of seeds than did their lower portions. The growth of seed pod hulls and their strong competition with developing seeds for assimilate can under certain circumstances lead to significant seed abortion (Wright *et al.*, 1988). Pod development might be somewhat impeded in the lower portions of racemes or the extent of ovule abortion in such lower portions may be greater. Clarke (1979) also observed that seed number per pod in oilseed rape was highest on the main raceme as was seed weight. There was a general decline in number of seeds per pod and an increase in number of aborted seeds at the lower branch portions (Clarke, 1979). In winter *B. napus* both number of seeds per pod and seed weight were highest in the upper portions and lowest in the lower portions of the main raceme and subtending branches (Mendham and Scott, 1975).

Nitrogen availability in rape plant merely modified whatever influences pod position on the raceme or the plant had on seed number per pod or ovule development into seeds. Positive response to N supply was significant only up to 7 mM N, possible because of the diminished nitrogen use efficiency when N availability is in excess that which is needed for proper yield development. Luxury consumption of N in rape plant may have adversely affected the partitioning of dry matter to the seed.

The present study has increased our knowledge of assimilate distribution within oilseed rape plant as influenced by N nutrition. However, there is need for more in-depth study of the numbers and positions of ovules and seeds which develop or abort as these determine the number of seeds a pod is likely to contain.

Acknowledgements

The work reported here was done when Prof Dr V B Ogunlela was a research fellow of the Alexander von Humboldt-Foundation and a visiting scientist at the Institute of Crop Science and Plant Breeding, University of Kiel. The authors wish to thank Ms Martina Bach for technical assistance, Ms Gisel Scheel for figure drawing, and Ms Elke Büll for manuscript typing.

References

Allen E J and Morgan D G 1972 A quantitative analysis of the effects of nitrogen on the growth, development and yield of oilseed rape. J. Agric. Sci. Camb. 78, 315–324.

Clarke J M 1979 Intra-plant variation in number of seed per pod and seed weight in *Brassica napus* 'Tower'. Can. J. Plant Sci. 59, 959–962.

Diepenbrock W and Geisler G 1979 Compositional changes in developing pods and seeds of oilseed rape (*Brassica napus* L.) as affected by pod position on the plant. Can. J. Plant Sci. 59, 819–830.

Krogman K K and Hobbs E H 1975 Yield and morphological response of rape (*Brassica campestris* L. cv. Span) to irrigation and fertilizer treatments. Can. J. Plant Sci. 55, 903–909.

Kumar A and Gangwar K S 1985 Analysis of growth, development and yield of Indian rape seed (*Brassica campestris* var. 'Toria') in relation to nitrogen and plant density. Indian J. Agron. 30, 358–363.

Major D J, Bole J B and Charnetski W A 1978 Distribution of photosynthates after $^{14}CO_2$ assimilates by stems, leaves and pods of rape plants. Can. J. Plant Sci. 58, 783–787.

Mendham N J and Scott R K 1975 The limiting effect of plant size at inflorescence initiation on subsequent growth and yield of oilseed rape (*Brassica napus*). J. Agric. Sci. Camb. 84, 487–502.

Mendham N J, Shipway P A and Scott R K 1981 The effects of seed size, autumn nitrogen and plant population density on the response to delayed sowing in winter oilseed rape (*Brassica napus*). J. Agric. Sci. Camb. 96, 417–428.

Ogunlela V B, Kullmann A and Geisler G 1989 Leaf growth and chlorophyll content of oilseed rape (*Brassica napus* L.) as influenced by nitrogen supply. J. Agron. Crop Sci. 163, 73–89.

Scott R K, Ogunremi E A, Ivins J D and Mendham N J 1973 The effects of fertilizers and harvest date on growth and yield of oilseed rape sown in autumn and spring. J. Agric. Sci. Camb. 81, 287–293.

Tayo T 1974 The analysis of the physiological basis of yield in oilseed rape (*Brassica napus* L.). Ph.D. Thesis, University of Cambridge.

Tayo T and Morgan D G 1975 Quantitative analysis of growth, development and distribution of flowers and pods in oilseed rape (*Brassica napus* L.). J. Agric. Sci. Camb. 85, 103–110.

Wright G C, Smith J C and Woodroof 1988 The effect of irrigation and nitrogen fertilizer on rapeseed (*Brassica napus*) production in south-eastern Austrialia. I. Growth and seed yield. Irrig. Sci. 9, 1–13.

M. L. van Beusichem (Ed.), *Plant nutrition – physiology and applications*, 577–579.

PLSO IPNC726

Effect of nitrogen on thiocyanate content of *Brassica oleracea* var. *acephala* leaves

J.A. CHWEYA
Department of Crop Science, University of Nairobi, P.O. Box 29053, Nairobi, Kenya

Key words: *Brassica oleracea* L., lamina, petiole, thiocyanate

Abstract

An experiment was carried out to study the effect of nitrogen fertilizer application levels (0, 47, 94 and 188 kg N/ha) on thiocyanate contents of petioles and lamina of two *Brassica oleracea* L. var. *acephala* varieties ('Thousand-headed' kale and 'Georgia' collards).

The results showed that N application significantly decreased thiocyanate contents of both laminae and petioles. Applying 47, 94, 188 kg N/ha reduced thiocyanate of laminae and petioles by 26, 41 and 52% and 2, 9 and 39%, respectively. The results also showed that thiocyanate of leaves from 'Thousand-headed' kale was significantly higher than that of 'Georgia' collards. Lamina and petiole thiocyanates of 'Georgia' collards were 79% and 83% of those of 'Thousand-headed' kale, respectively.

Introduction

Leaves of kale and collards are popular vegetables in Kenya. However, the leaves yield thiocyanates after maceration (Chweya, 1984; 1988; Johnston and Jones, 1966). Thiocyanates, which result from hydrolysis of glucosinolates by the enzyme myrosinase, can be goitrogenic (Van Etten and Tookey, 1983). There is a relationship between the consumption of cruciferous plants, or their products, and incidence of goitre in humans (Benns *et al.*, 1978; Michajlovskij, 1986).

Factors regulating the level of glucosinolates in kale leaves, and hence levels of thiocyanates upon the hydrolysis of the glucosinolates, have not been elucidated yet. Environmental conditions (Johnston and Jones, 1966) and agronomic factors such as plant density and nitrogen fertilization (Chweya, 1984) may play a role in determining glucosinolates levels in kale leaves. In the production of leafy vegetables, application of nitrogen is highly recommended as it increases the vegetativeness of the plants and hence increased leaf yields. Some amino acids such as glutamate, alanine and serine are utilized in the synthesis of glucosinolate (Fowden, 1967). It is therefore possible that nitrogen nutrition could be one of the factors that may affect the levels of glucosinolates in kale leaves. The objective of this study was to show the effect of nitrogen fertilizer application on thiocyanate contents of kale and collard leaves.

Materials and methods

The experiment was conducted at the Kabete Field Station of the Faculty of Agriculture, College of Agriculture and Veterinary Sciences, University of Nairobi, between November, 1985 and February, 1986. The Station lies at altitude of 1940 m above sea level, latitude 1° 15′S and longitude 36° 44′E. The soil is deep, well drained and friable clay derived from tertiary trachytic lava. The soil pH–H_2O was 6.6 and had CEC of 26.6 meq per 100 g soil. Available phosphorus, extracted by a mixture of 0.1 *N* HCl and 0.03 *N* H_2SO_4, was 17 mg P per 100 g soil. Total nitrogen was 2.8 mg N per g dry soil.

Treatments included two *Brassica oleracea* var. *acephala* varieties ('Thousand-headed' kale and 'Georgia' collards) and four nitrogen (N) levels (0, 47, 94 and 188 kg N/ha). The experiment was arranged in a split-plot design with three blocks. The N levels formed the main plots and the varieties the sub-plots. Seedlings of the varieties were raised in a nursery and transplanted to the various plots thirty days after emergence. A spacing of 60 and 30 cm between and within rows, respectively, was used. At transplanting there was a basal application of triple phosphate (46% P_2O_5) at a rate of 200 kg/ha to the various plots. At 18 and 30 days after transplanting, the plants were topdressed with half of each of the various N levels in the form of calcium ammonium nitrate (21% N). During the experimental period care was taken to protect the plants from weed, pest and water stress damages through hand-weeding, frequent spraying with Ripcord and supplemental irrigation, respectively.

First and second leaf harvesting were done at 29 and 42 days after transplanting. During the leaf harvesting, all leaves except the top 4–6 leaves from designated plants were harvested. This is a common practice with kale farmers. After harvesting, leaf samples were separated into petioles and laminae and thiocyanate determined using a modified (Chweya, 1983) method of Bible *et al.* (1980).

Result and discussion

There was a significant effect of N levels on lamina and petiole thiocyanate contents. Table 1 and Figure 1 show that by applying N fertilizer to the plants, the thiocyanate of both petioles and laminae decreased. The effect was, however, more pronounced in the laminae than it was with petioles. For petioles, plants that were top-dressed with the highest level of N contained significantly less thiocyanate than those that were either not top-dressed or top-dressed with low N levels. The picture for the effect of N levels for lamina thiocyanate was different; top-dressing plants with 47 kg N/ha significantly reduced the lamina thiocyanate. Overall, the thiocyanate of leaves decreased with increasing levels of N.

Table 1. Effect of nitrogen application on thiocyanate contents of combined kale and collard leaves

Nitrogen (kg/ha)	Thiocyanate content (μg/g DW)	
	Lamina	Petiole
0	2879c*	1138b
4.7	2121b	1118b
9.4	1708ab	1031b
18.8	1377a	696a

* Values followed by same letter(s) down the column are not significantly different at 5% probability level.

The results also showed that the thiocyanate contents of the two varieties were significantly different. The leaves of 'Thousand-headed' kale had significantly higher thiocyanate than those of the 'Georgia' collards. Lamina and petiole thiocyanate of 'Georgia' collards was 79 and 83% of that of 'Thousand-headed' kale, respectively. The results further indicated that petioles had significantly lower thiocyanate than laminae. Petiole thiocyanate was, on the average, 51% less than that of lamina.

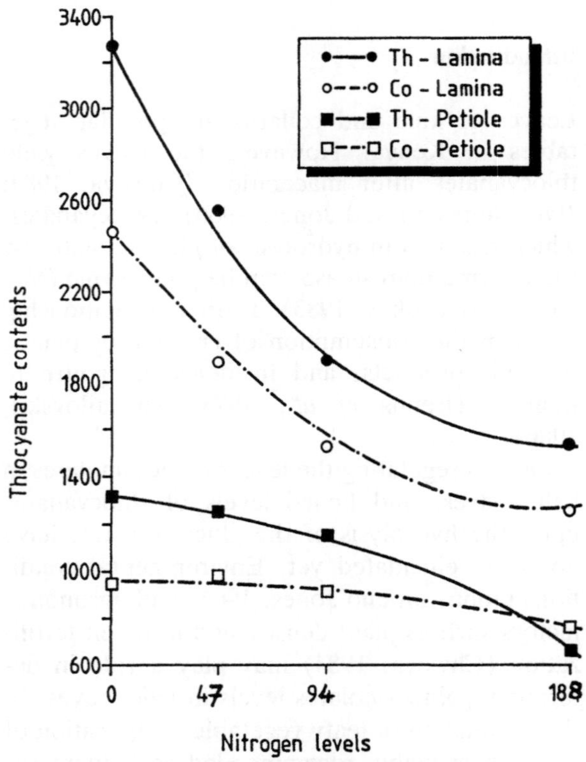

Fig. 1. Effect of N application (kg/ha) on lamina and petiole thiocyanate contents (μg/g) of kale (Th) and collard (Co) leaves.

There were no significant interactions between variety and N levels for either petiole or lamina thiocyanate. There were also no significant correlation between lamina and petiole thiocyanate.

The study shows that increasing N application decreased thiocyanate contents of kale and collard leaves. This implies that N application decreased glucosinolates concentration in the plants. Any factor that inhibits good growth of the plant may lead to increased glucosinolates concentration in the plant and, therefore, the amount of thiocyanates yielded after their hydrolysis. For example, poor soils, unfavourable temperatures and water stress (Bible and Chong, 1975; Bible *et al.*, 1980) and high plant density (Chweya, 1984) have been found to increase the amount of thiocyanate contents of roots of radish and leaves of kale, respectively. Nitrogen deficiency leads to poor vegetative growth and, therefore, may lead to accumulation of glucosinolates in the plant. This then may explain the decrease of thiocyanate contents in the leaves with N application. Nitrogen increased vegetative growth of the plants.

The study also shows that there were significant differences in thiocyanate contents in the leaves of the two varieties studied. Differences between kale varieties in their thiocyanate contents have been reported (Johnston and Jones, 1966). Similar results have also been reported for cabbage varieties (Bible *et al.*, 1980). The results imply that glucosinolates synthesis in plants is genetically determined and it should, therefore, be possible to breed for varieties that synthesize less glucosinolates.

The study further indicates that lamina had significantly higher thiocyanate than petiole did. Similar results have been reported before by Johnston and Jones (1966) and Paxman and Hill (1974). This is an important observation in relation to consumption of the leaves. It is common that petioles are discarded during the preparation of the leaves for cooking. The results above indicate that this may not be advisable as one may increase the concentration, in a given quantity of leaf, of thiocyanate ingested by discarding the petioles.

Acknowledgement

The author is very grateful to International Foundation for Science for financially supporting this project.

References

Benns G B, Hall J W and Beare-Rogers J L 1978 Intake of brassicaceous vegetables in Canada. Can J. Pub. Health 69, 64–66.

Bible B and Chong C 1975 Correlation of temperatures and rainfall with thiocyanate ion content in roots of radishes grown on two soil types. Hortic. Sci. 10, 484–485.

Bible B, Hak-Yoon J and Chong C 1980 Influence of cultivar, season, irrigation and date of planting on thiocyanate ion content in cabbages. J. Am. Soc. Hortic. Sci. 105, 88–91.

Chweya J A 1984 Yield and quality of kale as affected by nitrogen side-dressing, spacing and supplementary irrigation. Acta Hortic. 163, 295–301.

Chweya J A 1988 Nitrate-N and thiocyanate contents in kale (*Brassica oleracea* var. *acephala*, DC) leaves from some kale growing areas in Kenya. Acta Hortic. 218, 181–190.

Fowden L 1967 Aspects of amino acid metabolism in plants. Annu. Rev. Plant Physiol. 18, 85–106.

Johnston T D and Jones D I H 1966 Variations in the thiocyanate content of kale varieties. J. Sci. Food Agric. 17, 70–71.

Michajlovskij N 1986 Naturally occurring goitrogens in foodstuffs and their role in the etiology of endemic goitre. Proceedings of an interdisciplinary conference on natural toxicants in food. Euro Food Tox. II. Zurich, Switzerland. pp 25–40.

Paxman P J and Hill R 1974 Thiocyanate content of kale. J. Sci. Food Agric. 25, 323–328.

Van Etten C H and Tookey H L 1983 Glucosinolates. *In* Handbook of Naturally Occurring Food Toxicants Ed. M. Rechcigl Jr. pp 15–30. CRC Press, Boca Raton, FL.

M. L. van Beusichem (Ed.), *Plant nutrition – physiology and applications*, 581–583.

PLSO IPNC511A

Influence of nitrogen fertilization on sugarbeet (*Beta vulgaris*) quality in an area of southern Spain

A. TRONCOSO and M. CANTOS
Instituto de Recursos Naturales y Agrobiología de Sevilla, C.S.I.C., B.P.O. 1052, E-41080 Sevilla, Spain

Key words: *Beta vulgaris* L., nitrogen, sucrose quality, sugarbeet

Abstract

Influence of soil nitrogen availability on quality components (sucrose, alpha-amino nitrogen, K^+ and Na^+) of sugarbeet was studied. This study was carried out on a homogeneous soil of southern Spain. We have tested 0, 100, 150, 200 and 250 kg N/ha. The use of increasing quantities of nitrogen fertilizer leads to a large decrease in sucrose concentration in the root, but also to a large increase in root production. This inverse relation does not exist at high levels of fertilization. Sugar yield per ha is most increased between 100 and 150 kg N/ha. On the other hand, the increase of nitrogen fertilization leads to a significant increase of alpha-amino nitrogen. High levels of nitrogen fertilization have a decisive negative effect on the juice purity calculated according to the Wieninger and Kubadinov equation, and therefore on the recovery of sucrose from the root tissue.

Introduction

The nitrogen ratio released every year from organic matter is not sufficient to cover the metabolic needs of sugarbeet for an adequate production. For this reason, addition of this element as fertilizer is necessary. However, an excessive increase in nitrogen application results in enhanced levels of α-NH_2-N in the root (Vielemeyer, 1986). This nitrogen form makes the subsequent sucrose extraction difficult. Thus, a better nitrogen nutrition for optimal sucrose recovery is necessary.

A number of experiments were carried out to study the effect of different fertilization levels on root yield and sucrose production of sugarbeet, and on the parameters used to estimate beet root quality, according to Wieninger and Kubadinov (1971).

Materials and methods

The nitrogen fertilization was carried out in the Torre de la Marisma farm in Sevilla (Spain) during three growing seasons. A site with a very homogeneous soil (Pelloxererts) of 25500 m^2 surface was chosen, divided in seven equal blocks (3600 m^2 each). We have tested nitrogen application rates of 0, 100, 150, 200 and 250 kg N/ha, with occasional irrigation.

A Venema autoanalyzer was used to analyze the sugarbeet roots (British Sugar Company method).

Results and discussion

There is an inverse relation between nitrogen fertilization and root sucrose content ($r = -0.920$; $P \leq 0.01$; Fig. 1).

On the contrary, increase of nitrogen fertilization results in higher yields, expressed as weight of roots per ha ($r = 0.787$; $P \leq 0.05$; Fig. 2). However, this positive relation disappears with higher levels of fertilization and, consequently, an excess of nitrogen fertilizers was not profitable from this point of view.

When the nitrogen supply was compared with the sucrose yield (kg per ha), a clear relation was not found (r = 0.439; Fig. 3). In any case, the sugar yield per ha is increased between 100 and 150 kg N/ha, being lower in the control and also especially when the quantity of nitrogen is excessive although with poor statistical significance.

In Figure 4, it was observed that very high levels of nitrogen application, higher than 200 kg/ha, lead to a significant increase of alpha-amino nitrogen (r = 0.908; $P < 0.05$); therefore there is a decrease of juice purity as nitrogen fertilizer application is increased (Fig. 5).

The recoverable sucrose was calculated from

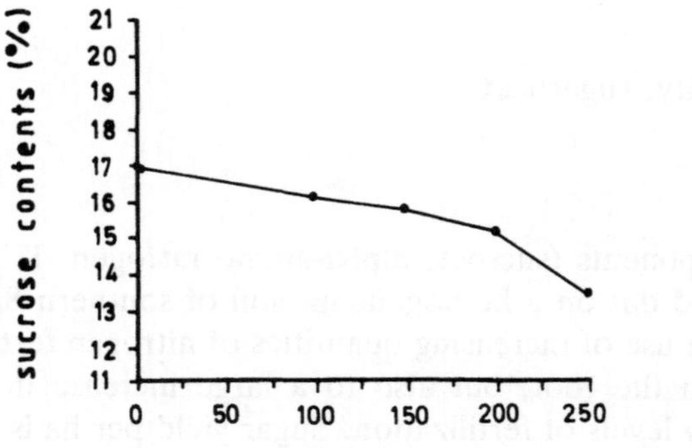

Fig. 1. Relation between nitrogen fertilization and sucrose in root.

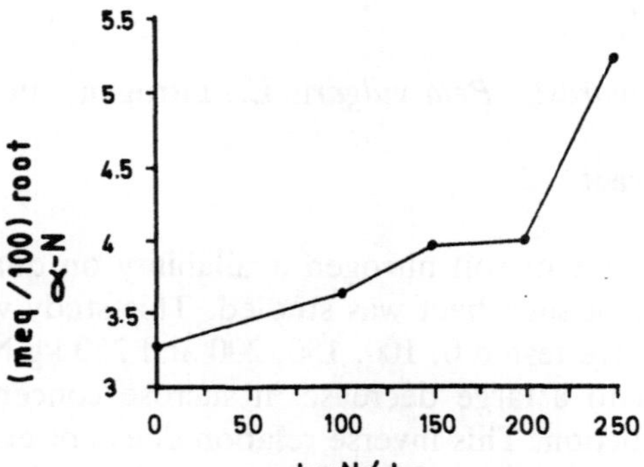

Fig. 4. Relation between nitrogen fertilization and αN concentration in root.

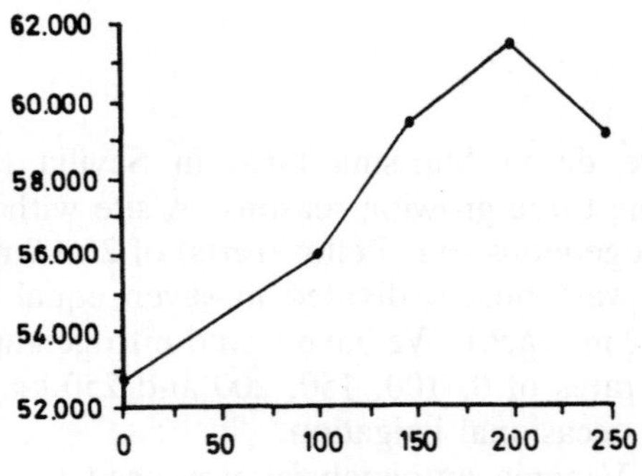

Fig. 2. Relation between nitrogen fertilization and production of roots.

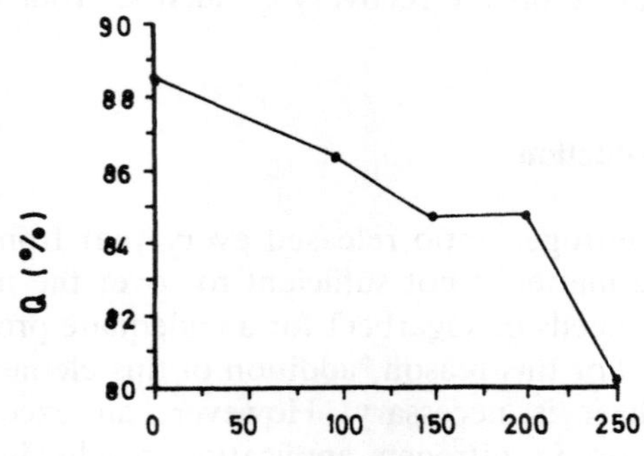

Fig. 5. Relation between nitrogen fertilization and juice purity.

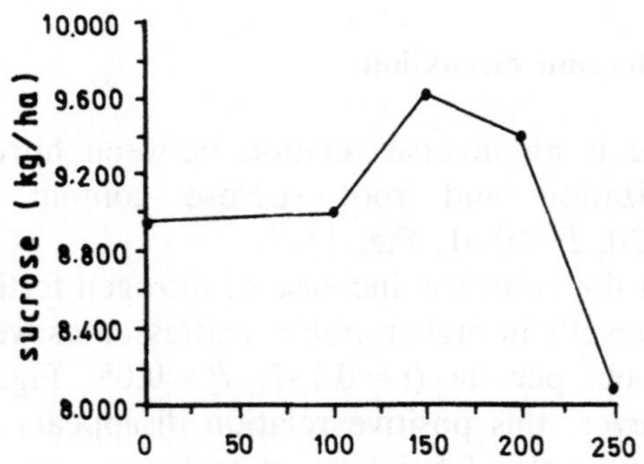

Fig. 3. Relation between nitrogen fertilization and sucrose production.

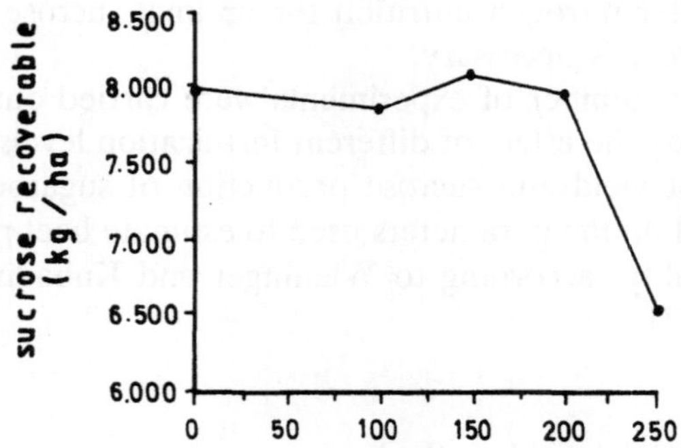

Fig. 6. Relation between nitrogen fertilization and recoverable sucrose.

the juice purity and the sucrose yield per ha (Fig. 6). Doses above 150 kg N/ha lead to a decisive decrease of the sucrose recovery.

These nitrogen fertilization experiments demonstrate the negative effects of nitrogen fertilization levels above 200 kg N/ha on the sugar production of sugarbeet in this area of Spain, since these levels induce decreases in the root yield per ha, juice purity and recoverable sucrose.

References

Vielemeyer H P 1986 Influence de l'époque d'apport de N sur le processus de formation du rendement dans la betterave sucrière. Arch. Acker-Pflanzenbau Bodenk. 30, 131–137.

Wieninger L and Kubadinow N 1971 Beziehungen zwischen Rübenanalysen und technischer Bewertung von Zuckerrüben. Comunication presenteé a la 14émé Assemblée Generale de la CITS.

M. L. van Beusichem (Ed.), *Plant nutrition – physiology and applications*, 585–590.

PLSO IPNC550A

Effect of preceding crops and nitrogen application on safflower (*Carthamus tinctorius*) yield and nutrient uptake, and on soil nitrogen

J.S. BOHRA and K. SINGH
Department of Agronomy, Institute of Agricultural Sciences, Banaras Hindu University, Varanasi-221005, India

Key words: *Carthamus tinctorius* L., fallow, green gram, nitrogen balance, nitrogen uptake, *Phaseolus aureus*, rice, safflower

Abstract

Field investigations under dryland conditions were conducted during 1985–86 and 1986–87 on three safflower (*Carthamus tinctorius*)-based cropping sequences, *viz*. rice (*Oryza sativa*)-safflower, green gram (*Phaseolus aureus*)-safflower, and fallow-safflower. Rice and green gram received recommended fertilizer doses whereas to the succeeding safflower nitrogen was applied in quantities of 0, 20 and 40 kg N ha^{-1} along with uniform applications of P and K. The results revealed that safflower yield, N P K uptake, benefit-cost ratio (BCR) as well as soil N were appreciably increased with green gram as preceding crop. The fallow-safflower sequence proved most uneconomical. An N application of 40 kg N ha^{-1} to safflower significantly enhanced yield and NPK uptake, but failed to maintain soil N.

Introduction

Safflower (*Carthamus tinctorius* L.) has been gaining much popularity in India in recent years as an edible vegetable oil. Owing to its deep root system and drought tolerance the crop is particularly suitable for dryland areas and in most parts of the country it can be successfully grown in dry seasons after early maturing cereals and pulses (Rao, 1989). However, many farmers in India do not like to make safflower a part of their cropping sequences as they believe that it has deleterious effects on the succeeding crop.

Considerable research work has been done to assess the nitrogen requirement of safflower (Abel, 1976; Singh and Singh, 1980). However, the results of recent experiments on fertilizer use in cropping systems revealed that fertilizer recommendations should be based on whole cropping systems rather than on individual crops. The present study was therefore undertaken to investigate the nitrogen requirement of safflower in different cropping sequences, *viz*. green gram-safflower, fallow-safflower and rice-safflower under dryland conditions of Varanasi, India.

Materials and methods

The experiment was conducted during 1985–86 and 1986–87 on the Research Farm, Institute of Agricultural Sciences, Banaras Hindu University, Varanasi, under dryland conditions on a deep sandy loam soil low in organic carbon (0.32%) and with a pH of 7.2. The soil was medium in available nitrogen (175 kg N ha^{-1}), available phosphorus (13 kg P ha^{-1}) and available potassium (180 kg K ha^{-1}). The soil in the 0–15 cm layer was analysed for available N (Subbiah and Asija, 1956); P (Olsen *et al.*, 1954) and K (Jackson, 1973).

The experiment was laid out in a split-plot design with three replications, with green gram (cv. Pusa Baisakhi), fallow and upland rice (cv. Cauvery) as main plots during the rainy season. Each main plot was subdivided into three sub-

Table 1. Monthly rainfall (mm) during the experimental period

Month	Rainfall (mm)	
	1985–86	1986–87
July	476	351
August	285	197
September	95	94
October	34	10
November	4	0
December	0	7
January	6	12
February	44	2
March	1	0
April	2	0
Total	947	673

plots in the dry season. Sowing of green gram and rice was done in the first fortnight of July and that of safflower in the first fortnight of October in both years. Recommended doses of fertilizer, *i.e.* 80 kg N + 17.5 kg P + 33.2 kg K ha^{-1}) to rice and 20 kg N + 17.5 kg P + 16.6 kg K ha^{-1} to green gram were applied as basal applications. To the succeeding safflower nitrogen was applied as urea (0, 20 or 40 kg N ha^{-1}) in each sub-plot, and 13.1 kg P + 15.6 kg K ha^{-1} were applied uniformly to all plots at the time of sowing.

The normal rainfall in Varanasi is 1100 mm of which 87% falls between June and September. The rainfall during 1986–87 was much below normal, and the year was considered a drought year (Table 1).

Results and discussion

In general the performances of safflower and the preceding crops were poor during the second year 1986–87. Rice suffered from drought at the flowering and seed-filling stages during this year (Table 1), resulting in very poor yields. However, the drought had no negative effects on green gram owing to its earliness. The safflower crop also received very little rain during the second year.

Effects of preceding crops

Preceding crops grown in rainy seasons caused considerable yield differences in safflower (Table 2). Yield-attributing characteristics of safflower improved when grown after green gram as compared to those after fallow and rice. However, the differences were significant only in the case of capitula per plant in both years. Similarly, seed yield of safflower was maximal in the green gram-safflower sequence, but yield differences became significant only in the second year. This indicates that green gram took two years to establish a significant legume effect. The beneficial effect of green gram on the succeeding safflower may be due to decomposition of sloughed-off and dead nodules in the soil (Meelu and Rana, 1981; Walker *et al.*, 1954). Safflower after rice performed better than after rainy-season fallow, probably due to the residual fertility left by the adequately fertilized rice crop.

Table 2. Effects of preceding crops and nitrogen application on yield attributing characteristics and seed yield of safflower

Treatment	Seed yield (q ha^{-1})		Capitula $plant^{-1}$		Seeds $capitulum^{-1}$		100-seed wt. (g)	
	1985–86	1986–87	1985–86	1986–87	1985–86	1986–87	1985–86	1986–87
Preceding crop								
Green gram	8.21	6.50	7.71	5.66	59.5	40.1	3.88	3.77
Fallow	7.38	4.78	5.87	4.04	50.8	31.4	3.74	3.68
Rice	7.65	4.85	6.36	4.53	57.9	33.5	3.89	3.58
L.S.D. (0.05)	NS	1.23	0.82	0.93	NS	NS	NS	NS
Nitrogen level (kg N ha^{-1})								
0	7.06	3.99	5.87	4.04	59.0	32.0	3.66	3.58
20	7.87	5.48	6.20	4.64	57.7	35.6	3.81	3.67
40	8.30	6.66	7.87	5.54	56.4	37.3	4.03	3.81
L.S.D. (0.05)	0.90	0.36	0.29	0.50	NS	NS	NS	NS

NS: Not significant.

Table 3. Effects of preceding crops and nitrogen application on oil content, oil recovery and benefit cost ratio of safflower

Treatment	Seed oil content (%)		Oil recovery (kg ha^{-1})		Benefit-cost ratio*	
	1985–86	1986–87	1985–86	1986–87	1985–86	1986–87
Preceding crop						
Green gram	30.3	30.7	249	199	1.2	1.1
Fallow	30.6	30.8	225	147	1.1	0.8
Rice	30.3	30.7	232	149	1.2	0.9
L.S.D. (0.05)	NS	NS	NS	42.7	–	–
Nitrogen level (kg N ha^{-1})						
0	30.6	30.8	216	123	1.0	0.7
20	30.4	30.7	239	168	1.2	1.0
40	30.2	30.6	251	204	1.2	1.2
L.S.D. (0.05)	NS	NS	21.2	16.5	–	–

* $BCR = \frac{\text{Gross return}}{\text{Cost of cultivation}}$.

It is clear from the data presented in Table 3 that preceding rainy season crops as well as fallow failed to affect the seed oil content of safflower. However, oil recovery and BCR were clearly high in both years, when safflower was grown after green gram. Safflower after fallow, because of its poor yield, produced the lowest oil-recovery and BCR values.

Safflower removed significantly more N, P, and K from soil when grown after green gram, as compared to rice and fallow (Table 4). Safflower after rice also tended to produce higher nutrient uptake values than after fallow, but the differences were not significant.

Effect of nitrogen

Seed yield of safflower increased significantly with increasing levels of nitrogen application up to 40 kg N ha^{-1}, but during the first year a significant difference was observed only between 0 and 40 kg N ha^{-1} (Table 2). The increase in seed yield may be attributed to a significant increase in capitula production. However, the 100-seed weight values and the number of seeds per capitulum remained unaffected. These findings are in accord with those of other workers (Arunachalam and Morachan, 1976; Hoag *et al.*, 1968).

Table 4. Effects of preceding crops and nitrogen application on N, P and K uptake by safflower

Treatment	Total nitrogen uptake (kg N ha^{-1})		Total phosphorus uptake (kg P ha^{-1})		Total potassium uptake (kg K ha^{-1})	
	1985–86	1986–87	1985–86	1986–87	1985–86	1986–87
Preceding crop						
Green gram	27.9	22.8	16.4	14.30	23.0	20.8
Fallow	22.9	16.3	13.9	9.6	19.2	14.3
Rice	23.6	17.0	14.1	9.7	20.0	14.1
L.S.D. (0.05)	4.2	3.7	1.4	2.3	3.2	4.3
Nitrogen level (kg N ha^{-1})						
0	21.2	12.8	12.7	7.6	17.0	10.8
20	25.2	18.6	15.0	11.5	20.5	17.0
40	29.0	25.3	17.4	14.7	24.1	22.0
L.S.D. (0.05)	1.6	1.3	0.8	0.9	1.2	1.6

Because of higher seed yields, oil recovery and CBR improved considerably with nitrogen application. Nitrogen applied at 40 kg N ha^{-1} yielded significantly more oil than 0 and 20 kg N ha^{-1} during 1986–87, but in 1985–86, 20 and 40 kg N ha^{-1} remained statistically at par. Singh *et al.* (1983) reported similar effects of nitrogen on oil content of safflower.

The uptake values for N, P and K increased significantly with increasing levels of N applied (Table 4). This was mainly due to the influence of applied nitrogen on biomass production and to the higher contents of these nutrients in seed and stalk. The results are in close agreement with those of Singh and Singh (1980).

Performance of safflower in different sequences

The data presented in Table 5 reveal that the green gram-safflower sequence was most profitable, as it yielded maximum BCR values in both years, followed by the rice-safflower sequence, whereas the fallow-safflower sequence gave the lowest values. Thus, it becomes apparent that leaving the land fallow before safflower is not a good practice. In 1986–87, the BCR values were low in the rice-safflower sequence which can be attributed to the severe drought in that year particularly at flowering and maturity of the rice crop.

It is also clear from Table 5 that at 0 kg N ha^{-1} the yield of safflower was higher in the green gram-safflower sequence than in the other sequences. This may be due to the beneficial legume effect of green gram on the succeeding safflower. In the green gram-safflower sequence, the BCR values remained almost the same at 20 and 40 kg N ha^{-1}, whereas during the second year N at 20 kg ha^{-1} yielded a higher BCR than at 40 kg N ha^{-1}, thus suggesting that 20 kg N ha^{-1} can be saved by growing green gram before safflower.

Nitrogen balance

Data on the available-nitrogen balance (Table 6) indicate that there were considerable losses of nitrogen in the fallow-safflower and rice-safflower sequences, whereas a gain was observed in the green gram-safflower sequence. In the rice-safflower sequence, the N loss was more severe in the second year because of the very low rice yield, so that the major part of the applied nitrogen has remained unutilized by the crops and may have been lost. The uptake of nitrogen by safflower was clearly higher when grown after

Table 5. Yield and cost-benefit ratio of different cropping sequences as influenced by treatments

Cropping sequence	Nitrogen level (kg N ha^{-1})	Seed/grain yield (q ha^{-1}) 1985–86	1986–87	Benefit-cost ratio 1985–86	1986–87
Green gram-safflower	0 R[a]	5.00	5.25		
	D[b]	7.48	5.19	1.32	1.20
	20 R	5.69	5.96		
	D	8.23	7.01	1.43	1.46
	40 R	5.26	5.76		
	D	8.83	7.26	1.43	1.44
Fallow-safflower	0	6.90	3.52	1.09	0.64
	20	7.49	4.48	1.13	0.79
	40	7.84	6.35	1.11	1.09
Rice-safflower	0 R	19.12	4.32		
	D	7.00	3.28	1.11	0.49
	20 R	20.51	4.66		
	D	7.88	4.90	1.20	0.63
	40 R	21.18	5.09		
	D	8.04	6.37	1.21	0.77

[a] Rainy season, [b] Dry season.

Table 6. Soil nitrogen balance as affected by preceding crops and nitrogen application

Cropping sequence		Soil N pool ($kg\ N\ ha^{-1}$)			N removal by crops ($kg\ ha^{-1}$)			Expected balance ($kg\ ha^{-1}$)	Actual balance ($kg\ ha^{-1}$)	Net gain or loss ($kg\ ha^{-1}$)
		Initially available N	N added	Total	Preceding crop	Safflower	Total			
1985–86										
Green gram-safflower	N_0	175	20+0	195	30.2	25.6	55.8	139	160	+21
	N_{20}	175	20+20	215	35.0	28.3	63.3	152	165	+13
	N_{40}	175	20+40	235	32.7	29.9	62.6	172	179	+ 7
Fallow-safflower	N_0	175	0+0	175	–	21.0	21.0	154	145	– 9
	N_{20}	175	0+20	195	–	23.5	23.5	172	157	–15
	N_{40}	175	0+40	215	–	24.2	24.2	191	169	–22
Rice-safflower	N_0	175	80+0	255	37.2	21.9	59.1	196	153	–43
	N_{20}	175	80+20	275	38.3	23.8	62.1	213	162	–51
	N_{40}	175	80+40	295	39.0	25.2	64.2	231	173	–58
1986–87										
Green gram-safflower	N_0	160	20+0	180	34.7	18.7	53.4	127	148	+21
	N_{20}	165	20+20	205	40.0	24.3	64.3	141	158	+17
	N_{40}	179	20+40	234	37.3	25.3	62.6	171	174	+ 3
Fallow-safflower	N_0	145	0+0	145	–	11.8	11.8	133	132	– 1
	N_{20}	157	0+20	177	–	15.7	15.7	161	147	–14
	N_{40}	169	0+40	209	–	21.5	21.5	188	161	–27
Rice-safflower	N_0	153	80+0	233	9.8	11.5	21.3	212	152	–60
	N_{20}	162	80+20	262	11.4	17.7	29.1	233	157	–76
	N_{40}	173	80+40	293	12.7	21.8	34.5	259	168	–91

green gram than after rice and fallow. This could be attributed to the legume effect exerted by green gram. Terman and Brown (1968) and Sadanandan and Mahapatra (1973) reported similar results.

Increasing levels of N applied failed to maintain the soil nitrogen status and resulted in net losses of available nitrogen. This means that safflower failed to utilize the applied N efficiently, particularly at 40 kg N ha^{-1}.

References

Abel G H 1976 Effects of irrigation regimes, planting dates, nitrogen levels and row spacing on safflower cultivation. Agron. J. 68, 448–451.

Arunachalam L and Morachan Y B 1976 Effect of NPK fertilizer on safflower under rainfed conditions. Madras Agric. J. 66, 327–329.

Hoag B K, Zubriski J C and Geiszyler G N 1968 Effect of fertilizer treatment and row spacing on yield, quality and physiological response of safflower. Agron. J. 60, 198–200.

Jackson M L 1973 Potassium determination for soils. *In* Soil Chemical Analysis, pp 111–133. Prentice-Hall of India Pvt. Ltd., New Delhi, India.

Meelu O P and Rana D S 1981 Legumes in crop rotations improve yield and soil productivity. Indian Fmg., 31, 30.

Olsen S R, Cole C V, Watanabe F S and Dean L A 1954 Estimation of available P in soils by extraction with sodium bicarbonate. U.S. Dept. Agric. Circ 939.

Rao V R 1989 Cropping systems researches in India and the new frontiers open for boosting up area and production of safflower. *In* Proceedings Second Intern. Safflower Conference, Hyderabad, India.

Sadanandan N and Mahapatra I C 1973 Studies on multiple cropping: Balance sheet of nitrogen in various cropping pattern. Indian J. Agron. 18, 323–327.

Singh P P, Dubey K M and Kaushal P K 1983 Effect of nitrogen and phosphorus on the seed yield of safflower. JNKVV Res. J. 17, 1–3.

Singh U B and Singh R M 1980 Effect of graded levels of moisture regimes, N and P fertilization on seed yield, oil content and NPK uptake by safflower. Indian J. Agron. 25, 9–17.

Subbiah B V and Asija G L 1956 A rapid procedure for determination of available nitrogen in soils. Curr. Sci. 25, 259–260.

Terman G D and Brown M A 1968 Crop recovery by applied fertilizer nitrogen. Plant and Soil 29, 48–65.

Walker T W, Orchiston H D and Adam A F R 1954 The nitrogen economy of grass legume associations. J. Brit. Grassl. Soc. 9, 249–279.

M. L. van Beusichem (Ed.), *Plant nutrition – physiology and applications*, 591–594.
 PLSO IPNC204

Nitrogen, phosphorus and potassium uptake of two sorghum (*Sorghum bicolor*) cultivars in an acid soil of Venezuela

E. CASANOVA and P. R. SOLORZANO
Postgrado en Ciencia del Suelo, Facultad de Agronomía, Universidad Central de Venezuela, Maracay, Apartado postal 4579, Venezuela.

Key words: nitrogen, phosphorus, potassium, *Sorghum bicolor* (L.) Moench, acid soil, Venezuela

Abstract

Two grain sorghum hybrids were compared in their nutritional requirements and response to fertilization: 1) Chaguaramas III which is a local genotype developed for tropical areas, tolerant to soil acidity with a high yield capacity and good growth and 2) Savanna v, the most imported sorghum seed used in Venezuela. A typic paleustalf, sandy loam, acid soil was used for the experiment with 3 rates of Nitrogen (N) (0, 100, 200 kg/ha), 2 rates of phosphorus (P) (100, 200 kg P_2O_5/ha) and 2 rates of potassium (K) (75, 150 kg K_2O/ha). Plant samples were taken at 25, 38, 46–52 and 89–95 days after planting to measure dry matter, N, P and K uptake over the growing season, nutrient accumulation and grain yield. The results showed in both hybrids that, compared to K, a higher percentage of the total N and P needs was taken up during grain development. This late demand was explained by the large amounts of N and P in the panicles which have to be available and transported during grain development. Maximum yields were obtained with 200, 200, 150 kg/ha of N, P_2O_5 and K_2O, respectively in Chaguaramas and with 75 kg/ha of K_2O and the same rates of N and P_2O_5 in Savanna v.

Introduction

Little is known about the nutrient uptake of sorghum during the growing season in tropical countries where cereals are planted in soils of poor fertility. In Venezuela 60% of sorghum requirements is imported and 40% (450.000 MT) is produced on acid soils where N-P-K fertilizers are needed for profitable production. Information is lacking to advise farmers about the best way to fertilize grain sorghum in the area selected to do this research.

This paper presents the results concerning maximum grain sorghum yields using N-P-K fertilizer in Guarico State and makes comparisons between the nutritional requirements and response to fertilization of two grain sorghum hybrids: Chaguaramas III, which is a local genotype developed for tropical areas, tolerant to soil acidity with a high yield capacity and good growth, and Savanna v, developed in USA which was the most commonly used sorghum seed imported in Venezuela at the time this research was conducted.

Methods

A Typic Paleustalf, sandy loam, acid soil pH 5.5, no lime applied, 1.29% organic carbon, 12 mg/kg available P, 60 mg/kg available K, 29 meq/100 g cation exchange capacity, was used to establish the experiment with 3 replications, 3 rates of N (0, 100, 200 kg/ha), 2 rates of P (100, 200 kg P_2O_5/ha) and 2 rates of K (75, 150 kg K_2O/ha). Plant samples were taken at 25, 38, 46–52 and 89–95 days after planting to measure dry matter, N, P, and K uptake over the growing season, nutrient accumulation and grain yield. The site is classified as a semiarid region with 800-900 mm of annual rainfall.

Results and discussion

Dry matter production

Figure 1 shows the pattern of growth (dry weight) for the various plant parts. There were differences between the two hybrids: Chaguaramas produced more total dry matter because more stalk and leaves but less panicles (grain and other heads parts) than Savanna v. Vertical lines that designate the 50% flowering stage show that Savanna v. flowered a week before Chaguaramas III and both hybrids flowered before 60 days after emergence as described by Vanderlip (1979) for grain sorghum in the United States. There were similarities in these two hybrids: after about 50 days the head increased in weight rapidly and represented more than 50% of the total dry weight at physiological maturity.

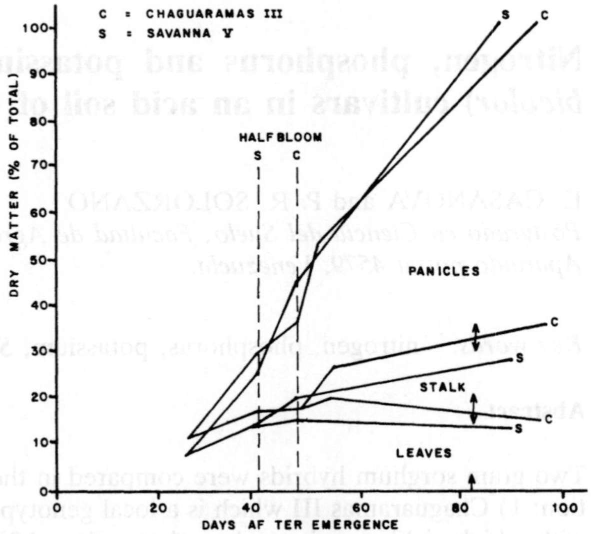

Fig. 1. Total dry matter production and dry weight of the various plant parts in sorghum over the growing season at Chaguaramas, Guárico State, Venezuela (averaged over treatments).

Nutrient accumulation in sorghum plant parts at maturity

Figure 2 shows N-P-K- accumulation at physiological maturity. For both hybrids a large portion of the N and P, but only small portion of K, was moved in the grain. This late-season demand in the panicles has to become available by uptake or redistribution to the grain sink. Clark and Gourley (1987), Anon. (1987), Vanderlip (1979), Solorzano (1986) and Duncan (1981)

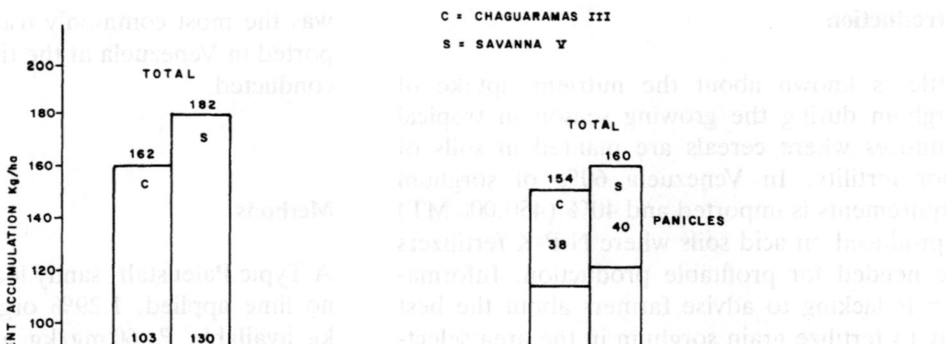

Fig. 2. Nutrient accumulation in sorghum plant parts at maturity at Chaguaramas, Guárico State, Venezuela (averaged over treatments).

Table 1. Average yield response to N-P-K treatments of two hybrids at Chaguaramas, Guárico State, Venezuela

Sorghum[a] hybrid	Levels of nutrient (kg/ha)			Average yield (kg/ha)	Duncan Multiple range test
	N	P_2O_5	K_2O		
C	0	100	75	5258	lm
C	0	100	150	5517	Jklm
C	0	200	75	5575	Jklm
C	0	200	150	5375	Jklm
C	100	100	75	6500	defghi
C	100	100	150	5817	jk
C	100	200	75	6508	defgh
C	100	200	150	6683	def
C	200	100	75	5667	Jkl
C	200	100	150	5917	ij
C	200	200	75	6533	c
C	200	200	150	7300	ab
S	0	100	75	5033	m
S	0	100	150	5450	Jklm
S	0	200	75	6525	cdefgh
S	0	200	150	6542	cdefg
S	100	100	75	6792	bcde
S	100	100	150	6800	bcd
S	100	200	75	6825	bc
S	100	200	150	6600	c
S	200	100	75	6742	b
S	200	100	150	6700	c
S	200	200	75	7600	a
S	200	200	150	7308	ab

[a] C = Chaguaramas III
S = Savanna v
[b] Numbers with the same letters are not significantly different a 5% level.

have reported that for 600 kg/ha of grain sorghum the plant uptake should be 178, 63 and 180 kg/ha of N, P, K, respectively. The amounts taken up in this experiment were similar for N but almost half for P and about 10% lower for K. When leaves were analyzed 64–52 days after emergency the levels of N, P, K (%) were 1.52, 0.20 and 1.66 for Chaguaramas III and 1.71, 0.22 and 1.29 for Savanna v. These levels were lower than the sufficiency ranges reported for this stage of development by Solorzano (1986) and Clark and Gourley (1987) on soils with high P levels and high management.

Yield response to N-P-K treatments

Table 1 shows the yield response to N-P-K fertilization treatment for both hybrids averaged over replications. N fertilization is very important in increasing yield in this area. The application of 100 kg/ha of N at a fixed level of P and K (100 and 75 kg/ha of P_2O_5 and K_2O, respectively) increased grain yield with 1242 kg/ha for Chaguaramas III and 1759 kg/ha for Savanna v. When the rates of N, P, K, were increased to 200, 200, 150 the increase in yields was 2042 and 2275 kg/ha for Chaguaramas III and Savanna v, respectively.

Acknowledgements

The authors appreciate the financial support from Dr. Albert Ludwick, Latin American Coordinator of Potash and Phosphate Institute. We are also grateful for the cooperation of Protinal for letting us use the experimental field at Agricola Chaguaramas, Guarico State, Venezuela.

References

Anon. 1987 Plant food uptake at various yield levels, per acre. Better Crops. Fall 15 p.

Clark R B and Gourley L M 1987 Evaluation of mineral elements in sorghum grown on acid tropical soils. *In* Sorghum for Acid Soils. Eds. L M Gourley and J G Salinas. pp 251–270. Int. Sorghum/Millet Coop. Res. Sup. Prog. (INTSORMIL)/Int. Crops Res. Inst. Semi-Arid Trop. (ICRISAT)/Int. Center Trop. Agric. (CIAT), Cali, Colombia, SA.

Duncan R R 1981 Variability among sorghum genotypes for uptake of elements under acid soil field conditions. J. Plant. Nutr. 4, 21–32

Sere C and Estrada R D 1987 Potential role of grain sorghum in the agricultural systems of regions with acid soils in Tropical Latin America. *In* Sorghum for Acid Soils. Eds. L M Gourley and J G Salinas. pp 145-169. Int. Sorghum/Millet Coop. Res. Sup. Prog. (INTSORMIL)/Int. Crops Res. Inst. Semi-Arid Trop. (ICRISAT)/Int. Center Trop. Agric. (CIAT), Cali, Colombia, SA.

Solorzano P R 1986 El Sorgo Granífero, su producción en Venezuela. Protinal, 140 p.

Vanderlip R L 1979 How a sorghum plant develops. Cooperative Extension Service, Kansas State University, KS, 19 p.

M. L. van Beusichem (Ed.), *Plant nutrition – physiology and applications*, 595–601.

PLSO IPNC643

Genotypic diversity in pearl millet (*Pennisetum glaucum*) for nitrogen, phosphorus and potassium use efficiencies*

S.P. WANI, M.A. ZAMBRE and K.K. LEE
International Crops Research Institute for the Semi-Arid Tropics (ICRISAT), Patancheru P.O. 502 324, A.P., India

Key words: genetic diversity, grain yield, harvest index, nutrient uptake, pearl millet, *Pennisetum glaucum* (L.) R.Br., translocation index

Abstract

Twelve genotypes of pearl millet (*Pennisetum glaucum* (L.) R.Br. comprising of hybrids, composites, varieties, and landraces were evaluated for N, P, and K uptake, efficiency of grain production per unit of N, P, and K absorbed and their efficiency of transfer from vegetative parts to the grain. The genotypes were grown on Alfisols in two fields at the ICRISAT Centre with two N and P levels in a rainy season (20 kg N + 9 kg P ha^{-1} and 80 kg N + 18 kg P ha^{-1}).

Genotypes varied significantly for total dry matter, grain yield, and for uptake, use efficiency and translocation indices of N, P, and K. Genotype × fertility interactions were not observed for all parameters studied, except for grain yield, indicating that genotypes could be evaluated for N, P, and K use efficiency and translocation indices at different soil fertility levels. Hybrid MBH 110 showed the highest use efficiency for N, P, and K with a maximum harvest index, and it was followed by two other hybrids amongst the genotypes tested. Genotypes (composites, and a landrace) which showed lower use efficiency and translocation indices for N, P, and K also had lower grain yields than the hybrids tested. Positive relationships were found between harvest index and phosphorus use efficiency, and N, and P translocation indices. Thus, a future challenge lies in selecting lines with high N, P, and K use efficiency and translocation indices.

Introduction

Pearl millet is an important rainfed cereal crop grown on marginal soils in the semi-arid tropics. Most of it is grown with little or without fertilizer applied. Nitrogen is usually the nutrient limiting crop production and its poor recovery by crops, when applied as fertilizer, is of worldwide concern. With steadily increasing prices of fertilizers, it becomes important to produce maximum pearl millet grain yields per unit of fertilizer applied. In recent years, crop improvement research specifically directed towards increasing the efficiency of mineral nutrition of plants has received increased attention (Devine, 1982; Gabelman and Loughman, 1987; Sarić and Loughman, 1983). The differential response of pearl millet genotypes to applied N (ICRISAT, 1988; Kanwar *et al.*, 1973; Murty, 1967) suggests that differences in nutrient uptake and translocation to the grains exist in pearl millet genotypes. Nutritional differences in genotypes have rarely been related to final economic yield. Increased and efficient production of pearl millet may be charcterised as functions of increased uptake and accumulation of nutrients by the plant, increased production of dry matter per unit of nutrient assimilated, and increased translocation of nutrients from vegetative parts to the grain.

* Submitted as CP # 522 by the International Crops Research Institute for the Semi-Arid Tropics (ICRISAT).

Genotypes differences with respect to N, P, K and other nutrients for the abovementioned traits have been reported in sorghum (Seetharama *et al.*, 1987). This paper deals with differences among a set of 12 pearl millet genotypes grown at two levels of applied N and P fertility. The genotypes were compared for the extent of variation in the abovementioned traits and for the relationships between grain yield and some of these traits.

Materials and methods

Experimental details

Twelve pearl millet genotypes selected for this study consisted of three hybrids (BJ 104, MBH 110, and ICMH 451) released for commercial cultivation in India, five composites (MC-C8, D2 C6, ERC-Co, EC-C6 and IVC-C7) made at ICRISAT, one composite (RCB 2) released for cultivation in Rajasthan, India, a synthetic (Gam 73) from Senegal, and Local landraces from India (Rajasthan Locals 1 and 2). The experiment was conducted in two fields (locations) (Table 1) during the rainy season on Alfisols at ICRISAT Center, Patancheru, India (17°36′N, 78°16′E, 545 m altitude).

Both experiments were conducted in a split-plot design with two N fertility levels as main plots and pearl millet genotypes as sub-plots.

Table 1. Details of Alfisols at ICRISAT Center, 1987 rainy season, and of field experiments

Properties	Location I	Location II
Soil pH	6.60	7.00
EC (M mhos cm^{-1})	0.19	0.24
Organic carbon ($g\ kg^{-1}$)	3.40	3.30
Available P ($mg\ kg^{-1}$)	14.7	14.3
NO_3-N+NH_4-N ($mg\ kg^{-1}$)	19	20.0
Total N ($mg\ kg^{-1}$)	410	410
Gross plot area (m^2)	40.5	24
Harvest area (m^2)	36	13.5
Date of sowing	17 June '87	16 June '87
Date of irrigation	–	22 July and 4 August '87

Each treatment was replicated four times. For the 20 kg N ha^{-1} treatment 20 kg N ha^{-1} as urea and 9 kg P ha^{-1} as single super-phosphate were applied as basal dressings. For the 80 kg N ha^{-1} treatment a basal dose of 40 kg N ha^{-1} and 18 kg P ha^{-1} was applied. The remaining N was applied. The remaining N was applied as urea after thinning (20 DAS). The crop was machine sown on ridges spaced 0.75 m apart, and plant-to-plant spacing of 0.1 m was maintained by thinning the plants 12 DAS. Weeding and inter-row cultivations were carried out as and when required.

At harvest, the above-ground plant parts were harvested. The panicles were separated and threshed. Fresh stover yield was recorded and 10-kg subsamples of stover were collected and chopped. The subsampled stover biomass and the grains were dried at 70°C for 72 h and their dry mass recorded. N and P in ground grains and in stover were determined with the use of a Technicon Autoanalyser, and K was determined using atomic absorption spectrophotometry (Jackson, 1973).

Characterstics used to compare genotypes

The traits used to characterise the genotypes were i) total dry matter production (grain plus stover), ii) harvest index [HI: (grain mass/total dry matter mass × 100), iii) N, P and K contents of grain and stover (dry mass multiplied by concentrations in the respective parts), iv) total N, P, and K in plants (contents in grains + contents in stover), v) N, P, and K use efficiency (NUE, PUE and KUE: grain mass produced per unit of N, P, and K in total dry matter), vi) N, P, and K translocation indices (NTI, PTI and KTI: N, P, and K contents in grains divided by N, P, and K contents in total dry matter × 100).

Results

Grain and total dry matter yield, and total N, P, and K uptake

Mean grain and total dry matter yields of pearl millet across the genotypes and locations increased significantly to 3.10 t ha^{-1} and 7.90 t ha^{-1}

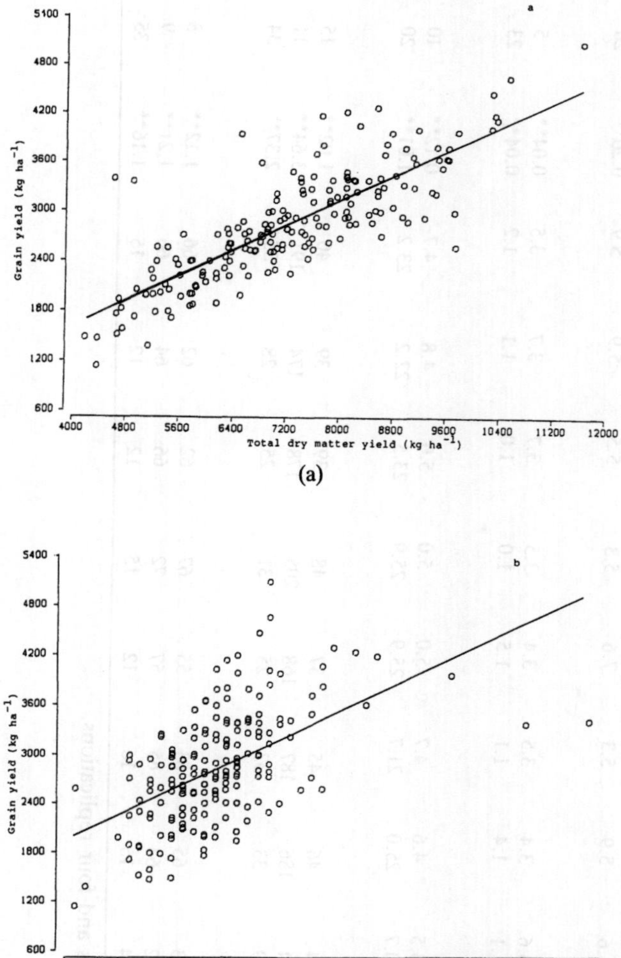

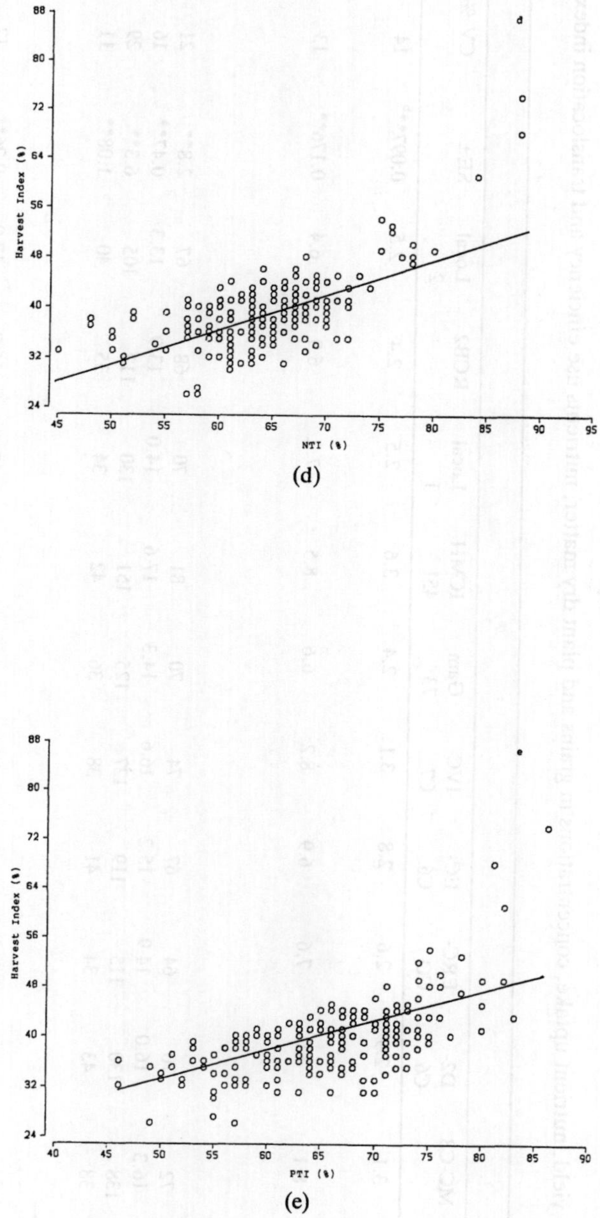

Fig. 1. **a**. Relationship between total dry matter yield (kg ha^{-1}) and grain yield (kg ha^{-1}) of pearl millet genotypes. $y = 84 + 0.380 \times$ ($R^2 = 0.67^{**}$, df = 190). **b**. Relationship between grain yield (kg ha^{-1}) and harvest index (HI %) of pearl millet genotypes. $y = 472 + 60.10 \times$ ($R^2 = 0.28^{**}$, df = 190). **c**. Relationship between phosphorus use efficiency (PUE) and harvest index (HI) of pearl milet genotypes. $y = 8.64 + 0.162 \times$ ($R^2 = 0.48^{**}$, df = 190). **d**. Relationship between harvest index (HI) and nitrogen translocation index (NTI) of pearl millet genotypes. $y = 1.41 + 0.589 \times$ ($R^2 = 0.44^{*}$, df = 190). **e**. Relationship between harvest index (HI) and phosphorus translocation index (PTI) of pearl millet genotypes. $y = 7.86 + 0.478 \times$ ($R^2 = 0.38^{**}$, df = 190).

Table 2. Grain and total dry matter yield, nutrient uptake, concentrations in grains and plant dry matter, nutrients use efficiency and translocation index of pearl millet genotypes[a]

	Genotypes													
	BJ 104	MBH 110	MC-C8	D2-C6	ERC-CO	EC-C6	IVC-C7	Gam 73	ICMH 451	Local 1	RCB2	Local 2	SE±	CV %
Grain yield ($t\ ha^{-1}$)	2.8	3.3	3.1	2.9	2.6	2.8	3.1	2.4	3.6	2.5	2.4	2.5	0.097**[b]	14
Total dry matter yield ($t\ ha^{-1}$)	6.6	6.8	8.1	7.0	7.6	6.9	8.2	6.6	8.5	7.2	6.9	6.4	0.176**	13
Nutrient uptake ($kg\ ha^{-1}$)														
Nitrogen	69	74	72	70	64	67	74	70	81	70	68	67	2.8**	21
Phosphorus	13.5	15.0	16.3	16.0	14.9	15.2	16.6	14.3	17.6	14.0	13.9	13.3	0.47**	16
Potash	124	107	138	130	115	119	127	125	151	130	112	105	6.3**	29
Harvest index	43	48	38	43	34	41	38	36	42	34	35	40	1.08**	11
Nitrogen concentration ($mg\ N\ g^{-1}$)														
Grain	15.8	16.2	14.1	14.3	15.2	14.7	14.9	15.4	14.7	17.0	16.8	17.0	0.26**	12
Stover	6.1	5.3	5.3	6.4	4.6	5.9	5.3	7.6	5.3	5.5	5.9	5.9	0.20**	20
Phosphorus concentration ($mg\ P\ g^{-1}$)														
Grain	3.3	3.4	3.4	3.4	3.6	3.4	3.5	3.4	3.5	3.7	3.7	3.5	0.04**	5
Stover	1.1	1.1	1.1	1.5	1.1	1.4	1.1	1.5	1.0	1.0	1.1	1.2	0.04**	23
Potash concentration ($mg\ K\ g^{-1}$)														
Grain	4.9	4.7	4.6	5.1	4.5	4.6	4.7	5.0	5.0	5.0	4.8	4.7	0.12**	10
Stover	28.7	25.6	24.4	27	20.7	25.0	21.7	25.9	25.9	23.5	22.2	23.2	1.01**	20
Nutrient use efficiency (kg grain produced kg^{-1} nutrient uptake)														
Nitrogen	44	48	47	47	44	46	45	37	48	39	39	40	1.02**	15
Phosphorus	209	219	191	185	174	186	187	168	205	178	174	191	3.64**	11
Potash	32	40	29	30	29	33	32	25	31	25	28	35	2.37*[c]	34
Nutrient translocation index (%)														
Nitrogen	66	74	63	63	63	65	63	55	67	62	62	66	1.12**	8
Phosphorus	69	74	65	62	62	64	65	57	72	66	64	66	1.21**	9
Potash	16	19	14	17	14	15	16	12	15	12	12	15	1.16**	35

[a] Each value is the average of two locations, two N levels and four replications.
[b] ** $P = <0.01$.
[c] * $P = <0.05$.

respectively, with 80 kg N ha^{-1} applied as compared to 2.55 t ha^{-1} and 6.50 t ha¯1 respectively, with 20 kg N ha^{-1} applied. In all genotypes, except D2C6 and IVC-C7, application of 80 kg N ha^{-1} increased the grain yields significantly over those obtained with 20 kg N ha^{-1} treatment. However, for total dry matter, the genotype × fertility level interactions were not significant. Even though grain and dry matter yields varied between locations, there were no genotype × location interaction effects for these parameters. Mean grain and total dry matter yields of pearl milet genotypes across the fertility levels and locations varied significantly. The landraces from India (Rajasthan Locals 1 and 2) produced dry matter yields similar to those of improved varieties and hybrids (Table 2). A maximum mean grain yield of 3.6 t ha^{-1} was observed for hybrid ICMH 451, followed by another hybrid MBH 110. Total N, P, and K uptake varied significantly (Table 2). However, total N, P, and K uptake by landraces and some of the improved genotypes were similar. There was a positive relationship (df = 190) between grain mass and total dry matter ($R^2 = 0.67$**, Fig. 1a), grain mass and HI ($R^2 = 0.28$**, Fig. 1b), grain mass and total N ($R^2 = -0.53$**), P ($R^2 = 0.71$**) and K ($R^2 = 0.31$**), uptake respectively.

Harvest index and nutrient translocation indices

Application of 80 kg N ha^{-1} or field location (or fertility treatment) had no effects on HI of pearl millet genotypes. The HI varied significantly amongst genotypes. The improved hybrids and the composites, except ERC-Co (an ergot-resistant composite), had higher HI than the local landraces (Table 2). Harvest index was positively correlated with PUE ($R^2 = 0.48$**, df = 190, Fig. 1c).

NTI of pearl millet genotypes did not change with 80 kg N ha^{-1} applied in comparison with 20 kg N ha^{-1}. However, NTI varied significantly among locations, but without genotype × location interactions. Mean PTI of pearl millet across the locations increased significantly to 68.2% when N was increased from 20 to 80 kg ha^{-1}, whereas in that situation mean KTI decreased from 16.5% to 13%. Nutrient translocation indices varied significantly amongst genotypes. In general, the genotypes with higher HI showed higher mutrients translocation indices (Table 2). Higher proportions (>60%) of plant N and P were translocated to the grains, whereas <80% of total plant K remained in the stover. More variability for NTI and PTI than for KTI was observed (Table 2). There was a positive relationship (df = 190) between HI and NTI ($R^2 = 0.44$**, Fig. 1d) and also with PTI ($R^2 = 0.38$**, Fig. 1e).

Nutrient (N, P, K) concentrations in grain and stover

Mean N concentrations in grain and stover across the genotypes and locations increased significantly from 14 to 17 and from 5.1 to 6.4 mg g^{-1}, respectively, with 80 kg N ha^{-1} instead of 20 kg N ha^{-1} applied. Phosphorus concentration in grains was not changed, whereas P concentration declined significantly from 1.3 mg g^{-1} to 1.07 mg g^{-1} dry matter when N application was increased from 20 to 80 kg ha^{-1}. Potassium concentration in grains decreased from 4.93 to 4.66 mg g^{-1} when N applied was increased from 20 to 80 kg ha^{-1}, whereas K concentration in stover was not changed. Locations also had a significant effect on nutrient concentrations in grains and stover.

Mean N, P, and K concentrations in grain and stover varied significantly with genotypes. The landraces had higher concentrations in the grains than the improved genotypes. However, N, P, and K concentrations in stover of local landraces were similar to those in stover of improved genotypes, except for BJ 104 and D2C6 (Table 2). There was no genotype × fertility interaction for N concentration in grain and stover, but a significant interaction was observed between genotypes and fertility levels for P and K concentrations in grains and stover. In a few genotypes P and K concentrations in grain and stover increased when N applied was increased from 20 to 80 kg ha^{-1}, whereas in others it decreased. Grain N concentration was positively correlated (df = 190) with a total N uptake ($R^2 = 0.71$**), and also with total K uptake ($R^2 = 0.50$**) while it was negatively correlated with NUE ($R^2 = 0.83$**), and KUE ($R^2 = 0.34$**).

Nutrient use efficiencies

Mean NUE of pearl millet genotypes decreased significantly from 48 to 39 kg grain kg^{-1} plant N uptake when N applied was increased from 20 to 80 kg ha^{-1}. Phosphorus and potassium use efficiencies were not affected by increase in N applied. Nutrient (N, P, and K) use efficiencies varied significantly among genotypes and locations, but there were no genotype × location interactions for N, P, and K use efficiencies. Improved genotypes, *e.g.* ICMH 451, MBH 110, BJ 104, IVC-C7, and MC-C8 with higher grain yields and HI showed higher N, P, and K use efficiencies than the Local landraces. There was no interaction between genotypes and fertility levels for N, P, and K use efficiencies.

Discussion

The pearl millet genotypes tested represented a broad genetic background. They varied significantly in grain and total dry matter yields, and in nutrient use efficiencies and translocation indices. Most of the improved genotypes (hybrids ICMH 451, MBH 110, and BJ 104 and composites MC C8, D2C6, EC-C6, and IVC-C7) showed higher grain yield than the Local landraces (Rajasthan Locals 1 and 2), composites (ERC-Co, RCB 2) and the synthetic variety Gam 73. Differential responses of pearl millet genotypes to N, in terms of grain yield, were been observed earlier (Kanwar *et al.*, 1973; Murty, 1967). Total plant nutrient uptake (N, P, and K) of local landraces was similar to that of some of the improved genotypes. A large diversity amongst pearl millet genotypes for N, P, and K use efficiency and N, P, and K translocation indices suggest that it is possible to identify pearl millet lines with high nutrient use efficiencies and high nutrient translocation indices. The improved genotypes with increased grain yields have shown higher NUE, PUE, KUE, NTI, and PTI (Table 2) than the landraces. However, improvements in some of these traits have come inadvertently along with selections made for improving the HI. This is supported by the strong positive relationships between HI and PUE, NTI, PTI, N, and P (Figs. 1c, d and e). However, no such selection has taken place for N and K use efficiencies even though wide diversity for these traits was known to exist. It means that a future challenge exists to incorporate high N and K use efficiencies in improved genotypes along with improved dry matter production, HI and N, P, and K translocation indices. Further, there are no genotype × fertility interactions indicating that genotyes can be evaluated under diverse fertility conditions. However, for improving N and K use efficiencies, specific selections need to be made different from those for improving PUE or N and P translocation indices. With such directed efforts it should be possible to improve pearl millet grain yields. These results suggest that it is essential to breed genotypes with increased dry matter production, improved nutrient use efficiencies and nutrient translocation indices. A positive relationship between grain yield and total dry matter production and grain yield and HI suggests that for further improvement of grain yield of pearl millet, it is essential to improve HI and total dry matter production. Similarly, Phul *et al.* (1974) observed that in a set of 50 pearl millet genotypes grain yield was positively correlated with tiller numbers, HI and flag leaf area. In the past, improved grain yields of pearl millet genotypes were obtained mainly through improved HI rather than improved total dry matter production and total plant nutrient uptake. The recently improved genotypes, like ICMH 451, IVC-C7, and MC-C8 produced higher grain yields than the local landraces. Some of the improved genotypes have normal HI, and the increased grain yields in these genotypes came through increased total dry matter production. This suggests a potential for improving total dry matter production as well as HI, which would also improve grain production. Wide diversity exists amongst genotypes for N, P, and K concentrations in grain and stover. The local landraces showed higher N concentrations in grains than the N concentrations in high-yielding improved genotypes, which was due to a dilution effect in the grain of improved genotypes. Where pearl millet is grown as fodder or in cases where straw is incorporated into the soil, the priorities will be entirely different. As PUE, NTI and PTI are positively correlated with HI, which in turn determines grain yield, for selection of parents and for testing genotypes with possibly improved other traits, it is important to monitor them for nutrient use efficiency and translocation indices to ensure that such traits are not lost in exchange for others.

These studies were conducted in two locations which were only 5 km apart. Further studies need to be conducted to confirm that no genotype × location interactions exist for the abovementioned traits.

Acknowledgement

We acknowledge Dr K L Sahrawat's help in chemical analysis of the samples.

References

Devine T E 1982 Genetic fitting of crops to problem soils. *In* Breeding Plants for Less Favorable Environments. Eds. M B Christiansen and C F Lewis. pp 143–173. John Wiley and Sons, New York.

Gabelman W H and Loughman B C (Eds.) 1987 Genetic Aspects of Plant Mineral Nutrition. Martinus Nijhoff/Dr. W. Junk Publishers, Dordrecht, The Netherlands. 629 p.

ICRISAT 1989 Annual Report 1987. pp 94–96 Patancheru, A.P. 502 324, India.

Jackson M L 1973 Soil Chemical Analysis. Prentice-Hall of India Pvt. Ltd., New Delhi, 498 p.

Kanwar J S, Das M N, Sardana M G and Bapat S R 1973 Are fertilizer applications to jowar, maize and bajra economical? Fert. News 18, 19–28.

Murty B R 1967 Response of hybrids of *Sorghum* (jowar) and *Pennisetum typhoides* (bajra) to nitrogen. J. Postgrad. Sch. IARI 5, 149–157.

Phul P S, Gupta S K and Gill K S 1974 Association analysis of some morphological and physiological traits in pearl millet. Indian J. Genet. Plant Breed. 34, 346–352.

Sarić M R and Loughman B C (Eds.) 1983 Genetic Aspects of Plant Nutrition. Martinus Nijhoff/Dr. W. Junk Publishers, Dordrecht, The Netherlands. 495 p.

Seetharama N, Clark R B and Marranville J W 1987 Sorghum genotype differences in uptake and use efficiency of mineral elements. *In* Genetic Aspects of Plant Mineral Nutrition. Eds. W H Gabelman and B C Loughman. pp 437–443. Martinus Nijhoff/Dr. W. Junk Publishers, Dordrecht, The Netherlands.

M. L. van Beusichem (Ed.), *Plant nutrition – physiology and applications*, 603–606.

PLSO IPNC635

Nitrogen and potassium dynamics, rooting intensity, and infection with VA mycorrhiza in a haplic podzol at intensive and organic farming

H. GÖMPEL, A. POMIKALKO, S. BEYER, B. SATTELMACHER, M. PETERS and H.-P. BLUME
Institute of Plant Nutrition and Soil Science, University of Kiel, Olshausen Str. 40, D-2300 Kiel 1, FRG

Key words: haplic podzol, nitrogen and potassium dynamics, nitrogen fixation, organic farming, root length density, VA mycorrhizas

Abstract

Two sandy haplic podzols under organic and intensive farming were investigated. In the first experimental year 1987 no differences could be observed for N and K inputs nor for outputs by leaching. Due to a high amount of fertilizer in 1988 the NO_3-N and K^+ concentrations in the soil solution of the intensively farmed soil increased considerably.

Root length densities of rye were approximately equal at both sites (1987). Root infection with VA-mycorrhiza however was much lower in the intensively farmed site.

Introduction

Conventional intensive farming is often blamed for a high contribution to ground water pollution due to excessive use of fertilizer. However, organic fertilizers or cultivation of legumes, as practised in "organic" farming, may result in considerable NO_3-N leaching as well (Bramm, 1981).

The experiments reported here were carried out in order to obtain detailed information on nutrient inputs by fertilization, dinitrogen fixation and precipitation as well as on nutrient outputs by harvest organs and leaching. For better understanding of nutrient depletion from soil horizons, rooting intensity and infection of the roots with VA-mycorrhiza were determined. For this purpose two sites of approximately equal pediological condition under intensive and organic farming were chosen.

Materials and methods

Experimental sites

Two sandy haplic podzols in Schleswig-Holstein (Northern Germany) were chosen.

The organically used site is under "biological dynamic" farming for 40 years. The agricultural acreage amounts to 115 ha, with about 40 ha arable land (Live stock: 0.8 cattle units). A 8-year-crop rotation is used: root crops, summer rye, rye grass, winter rye, root crop, Seradella, vetch-summer rye-grass mixture, winter rye.

The conventionally farmed site has 54 ha agricultural acreage, with about 27 ha pasture-land (Live stock: 1.3 cattle units). A 2-year-crop rotation (maize, winter rye) is used. Both sites have an average precipitation of 750 mm/a and a mean temperature of 8.4°C. Comparative mea-

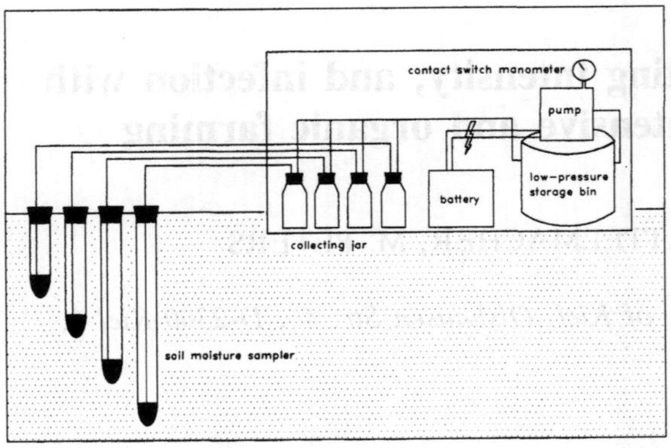

Fig. 1. Schematic representation of the soil moisture sampling.

surements were performed in rye with vetch-rye-grass or maize as the respective precrops.

Sampling and analytical procedures

In bi-weekly intervals precipitation and soil solution were collected in 20, 50, 100, and 200 cm depth, with ceramic cups exposed to a continuously working low pressure system (see Fig. 1). NO_3^-, NH_4^+ and K^+ concentrations were determined with an autoanalyser system (Technicon) and flame photometer.

Nitrogenase activity was evaluated *in situ* by C_2H_2-reduction; ethylene content was measured with a Perkin Elmer Sigma 2B gaschromatograph (Alumina column, oven 160°C, inj. 110°C, det. 200°C).

Root density was determined by core sampling (Eijkelkamp root auger, 750 cm^3) with 4 replicates. Each replicate consisted of 4 samples taken within the rows (2 samples) and between the rows (2 samples). After separating roots from soil the root length was estimated according to the line intersection method (Böhm, 1979; Tennant, 1975). Root length density (RLD) was recorded at shooting (1), flowering (2) and harvest (3) time.

Mycorrhizal infection of roots was evaluated after staining with fuchsin according to Kormanik *et al.* (1980).

Results and discussion

In the first experimental year (1987), N and K fertilizer inputs were almost equal in both cultivation systems as the fertilization at the conventional site was at a comparatively low level (85 kg N ha^{-1}, 104 kg K ha^{-1}). In both sites use of fertilizers increased in 1988. The increase can not be explained only by crop rotation. Due to the cultivation of a vetch-rye-grass mixture in the "organically farmed" site, approximately 60 kg nitrogen were incorporated into the system by biological N_2- fixation (25% of total nitrogen input) (Table 1).

We have determined the water balance for both sides and found little differences (Peters *et al.* unpubl.). Therefore it is suggested that a

Table 1. Nutrient inputs (kg ha^{-1}) by fertilization, deposition respectively biological N_2-fixation, yield (harvested dry matter) as well as accumulated amount of nutrients for an intensively and organically farmed podzol

Nutrient inputs by	Intensive				Organic			
	1987 Rye		1988 Maize		1987 Rye		1988 Vetch-Rye	
	N[c]	K[c]	N	K	N	K	N	K
Fertilizer								
min.	65	99	90					
org.			409	485	59	34	163	174
Deposition[a]	20	5	20	5	21	5	21	5
N_2-fixation							60	
Σ	85	104	520	490	80	39	244	179
Yield (DM)[b]	91dt[b]		190dt		82dt[b]		44dt	
Nutrient yield	89[b]	65[b]	208	285	83[b]	99[b]	77	63

[a] mean values of two years.
[b] seed and straw.
[c] kg ha^{-1}.

comparison of nutrient dynamics can be made on the basis of the concentration in the soil solution.

Nutrients below *100 cm* were considered as being leached. In 1987 no differences in NO_3-N concentration were observed, while 1988 a strong increase due to the application of slurry (168 kg N/ha) is evident in the conventionally farmed site (Fig. 2).

The dynamics of K^+ in soil solution are similar to that of NO_3-N (Fig. 3). While no differences were notable in 1987, a steep increase after slurry application was apparent on the conventionally farmed site in 1988. These observations, however, refer only to the upper layer of soil.

Root length density (RLD) at the first two sampling dates (shooting and flowering) was higher at the organically used site especially in the lower soil horizons (Table 2). At the final harvest the opposite is true. The steep increase in RLD from flowering to maturity is surprising. Comparable results, however, have been reported before by Böhm (1978).

Taking the relatively low $H_2PO_4^-$ concentration (11.2 mg/100 g CAL) of the "organically farmed" soil into account (intensively farmed: 32.8 mg/100 g CAL), data on infection with VA-mycorrhiza appeared of special interest. As can be seen from Table 2 infected root length

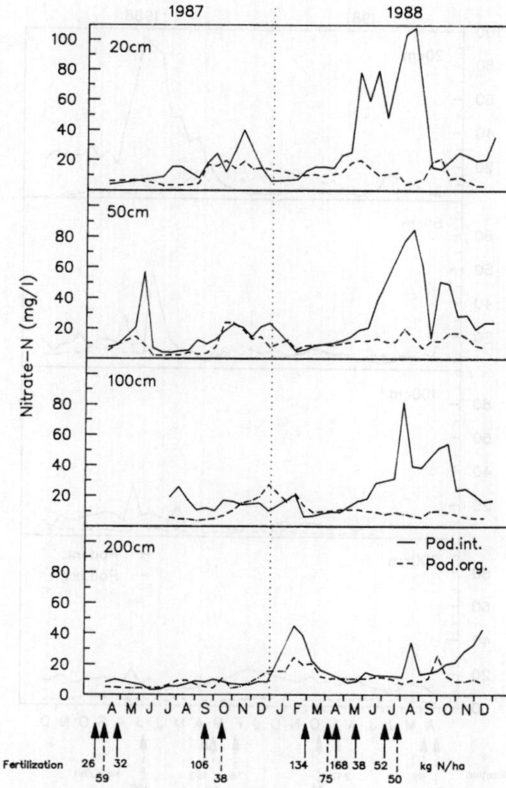

Fig. 2. Comparison of the NO_3-N concentrations of an intensively (Pod. int.) and organically (Pod. org.) farmed haplic podzol. (Arrows give date and amount of nitrogen input).

Table 2. Development of Root Length Density (RLD, cm/cm^3 soil) and Infected Root Length Density (IRL, cm/cm^3 soil) of rye (*Secale cereale*) on two podzols in 1987 for four soil depths and three sampling dates (1 = shooting, 2 = flowering, 3 = harvest)

Farm system	Soil depth (cm)	Sampling date					
		1		2		3	
		RLD	IRL	RLD	IRL	RLD	IRL
Organic	0–15	4.72	0.48	4.41	3.36*	6.86	4.68*
Intensive		2.76	0.08	3.64	0.34	8.13	0.00
	$LSD_{5\%}$	4.06	1.26	2.12	1.80	5.18	2.43
Organic	15–30	2.28	0.43	3.54	2.39*	3.23*	1.39*
Intensive		1.57	0.12	3.02	0.26	6.17	0.26
	$LSD_{5\%}$	1.22	0.83	1.20	0.41	2.17	0.53
Organic	30–45	1.13	0.05	1.29*	0.60*	1.82	0.43
Intensive		0.74	0.06	0.43	0.01	1.28	0.14
	$LSD_{5\%}$	0.87	0.32	0.61	0.40	1.84	0.36
Organic	45–60	2.01*	0.14	0.81*	0.27	1.29*	0.26
Intensive		0.49	0.02	0.33	0.00	0.84	0.15
	$LSD_{5\%}$	1.07	0.33	0.28	2.05	0.60	0.19

* = significant differences.

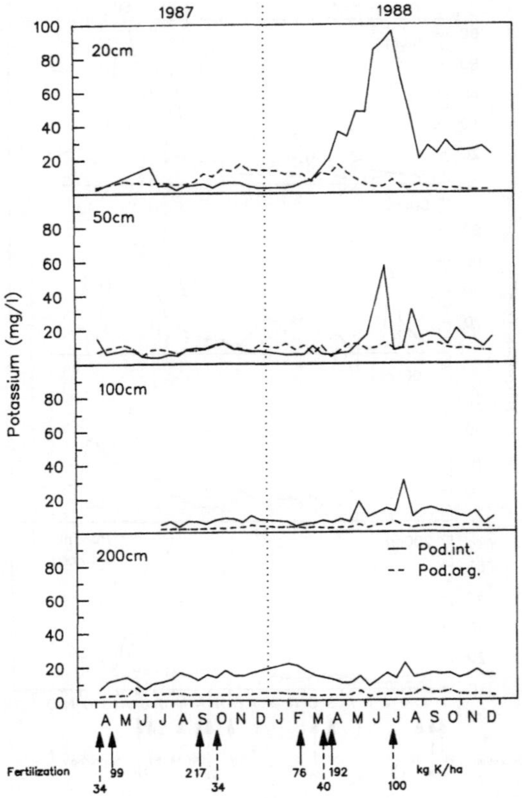

Fig. 3. Comparison of the K concentrations of an intensively (Pod. int.) and organically (Pod. org.) farmed haplic podzol. (Arrows give date and amount of potassium input).

density (IRL) in the organically used site is significantly higher in all soil depths and at all sampling dates. The use of pesticides, host plant diversity and fallow periods within the crop rotation may have affected plant mycorrhization besides of soil $H_2PO_4^-$ content (Baltruschat and Dehne, 1988; Gnekow, 1988; Menge *et al.*, 1979; Thompson, 1987). Our data are not adequate to discriminate between these possible causes.

References

Baltruschat H and Dehne H W 1988 The occurrence of vesicular-arbuscular mycorrhiza in agro-ecosystems. I. Influence of nitrogen fertilization and green manure in continuous monoculture and in crop rotation on the inoculum potential of winter wheat. Plant and Soil 107, 279–284.

Bramm A 1981 Einfluß der Landbewirtschaftung auf die Gewässerqualität. *In* Beachtung ökologischer Grenzen bei der Landbewirtschaftung. pp 162–186. Paul Parey Verlag.

Böhm W 1978 Untersuchungen zur Wurzelentwicklung bei Winterweizen. Z. Acker- und Pflanzenbau 147, 264–269.

Böhm W 1979 Methods of studying root systems. Springer-Verlag, Berlin-Heidelberg-New York, 167 p.

Gnekow M A 1988 Die Rolle von VA-Mycorrhiza bei der Phosphaternährung von Kulturpflanzen in Abhängigkeit von Phosphatdüngung und Bewirtschaftungsweise. Thesis, University Hohenheim, Inst. Plant Nutrition.

Kormanik P C, Bryan W C and Schultz R C 1980 Procedures and equipment for staining large numbers of plant roots for endomycorrhizal assay. Can. J. Microbiol. 26, 536–538.

Menge J A, Johnson E L V and Minassian V 1979 Effect of heat treatment and three pesticides upon the growth and reproduction of the mycorrhizal fungus *Glomus fasciculatus*. New Phytol. 82, 473–480.

Tennant D 1975 A test of a modified line intersection method of estimating root length. J. Ecol. 63, 995–1001.

Thompson J P 1987 Decline of vesicular-arbuscular mycorrhizae in long fallow disorder of field crops and its expression in phosphorus deficiency of sunflower. Aust. J. Agric. Res. 38, 847–867.

M. L. van Beusichem (Ed.), *Plant nutrition – physiology and applications*, 607–610.

PLSO IPNC219

Nutritional stress of maize (*Zea mays*) growing on a gleysol of east Croatia

V. KOVAČEVIĆ, V. KATUŠIĆ and V. VUKADINOVIĆ
Agricultural Institute, Faculty of Agriculture, YU-54000 Osijek, Yugoslavia

Key words: ear-leaf composition, grain yield, magnesium excess, maize, phosphorus fertilization, potassium deficiency, potassium fertilization, *Zea mays* L.

Abstract

A field trial with five rates of K or P fertilization was conducted on a gleysol known to be deficient in K for maize plants. By application of 825 kg K/ha, grain yield of maize (hybrid OsSK 247) was increased by 238%, while response to P (406 kg P/ha) was lower: a 90% increase. Depending on rate of K fertilization, ear-leaf K and Mg levels were as follows: 0.62% K and 1.66% Mg and 1.27% K and 1.07% Mg, for the usual and the highest rate of K (2-year means), respectively.

Introduction

Disorders in maize growth and chlorosis symptoms were obtained on some soils of East Croatia. They were manifested in the form of either P, K or Zn deficiency symptoms (Kovačević *et al.*, 1986; 1988). By adequate fertilization these disorders can be overcome (Kovačević *et al.*, 1987; Bertić *et al.*, 1989). In our study, a response of maize to increased K fertilization was shown (two-year experience). An imbalanced K and Mg supply and K fixation are the factors limiting maize growth under these conditions.

Material and methods

In the early spring of 1986 a field trial with varying rates of K and P fertilizer applied was conducted on a gleysol in Posavina province. Preliminary soil testing had shown that the levels of P and K availability were low (less than 10 mg P and K/100 g of soil, according to the ammonium lactate method). Also, typical symptoms of K deficiency were found in the early growth stages of soybean and maize.

Five rates of K or P were applied (150, 262, 450, 638 and 825 kg K/ha; 52, 111, 210, 308 and 406 kg P/ha). In addition, 240 kg N was applied per ha. In the second year, all plots were fertilized uniformly (240 kg N, 52 kg P, 150 kg K/ha). The field trial was conducted with four replicates. Each experimental plot measured 300 m^2. The maize hybrid OsSK 247 (developed by the Agricultural Institute Osijek) was sown on May 7th 1986 and April 26th 1987 and harvested in October (17th and 24th, for 1986 and 1987, respectively). Grain yields were calculated on a 14% moisture basis.

At the beginning of the silking stage (July 16th 1986 and July 23th, for 1986 and 1987, respectively), ear leaves were collected (25 per plot) for chemical analysis. Soil samples (25 borings per sample) were taken to a 30-cm depth (November 18th 1986 and October 5th 1987).

Plant samples were digested with sulphuric acid (Holz, 1973). Soil samples were extracted with ammonium lactate for P and with 1 *N* ammonium acetate (Whiteside, 1979) for exchangeable K, Ca and Mg. Phosphorus was determined colorimetrically with the molybdate-vanadate method, K flamephotometrically, and Ca and Mg by atomic absorption spectrophotometry.

Results and discussion

Application of K fertilizer resulted in maize grain yield increases up to 238% (two-year mean). Already with the lowest K rate, the grain yield was increased by 52%. The response of maize to P fertilization was less pronounced (Table 1).

Mean yields differed for the two years (4.71 and 3.95 t/ha for 1986 and 1987, respectively). Drought affected the growth of maize during its most intensive growth stage in 1987 (mean temperature 20.3°C and 23.1°C and mean rainfall values 73 and 22 mm during July 1986 and 1987, respectively). As a result, grain yields were lower in 1987 than in 1986. In addition, mean air temperature and rainfall during the growing season of maize (from May to September) were as follows: 18.6°C and 19.5°C, and 275 mm and 292 mm, for 1986 and 1987, respectively. Excessive rainfall was registered in May 1987 (151 mm).

Maize plants on plots with standard quantities of K applied showed signs of K deficiency. This phenomenom was associated with very low ear-leaf K levels. More favourable ear-leaf K levels and avoidance of K deficiency symptoms were encountered in maize plants grown under conditions of increased K supply. However, an ear-leaf K level of 1.32% indicates a latent K deficiency (Mengel and Kirkby, 1987). Excess ear-leaf Mg was associated with K deficiency. However, due to increased K fertilization, ear-leaf K- and Mg relations improved, but the Mg level remained excessively high. Ear-leaf Ca and P levels were less affected by variations in fertilizer supply (Table 2).

Alkaline soil reaction, low levels of plant-available K and P and high levels of Mg and Ca are the main soil characteristics. Imbalance between exchangeable K and Mg is a factor promoting K deficiency in maize (Table 3).

Exchangeable K in the soil was somewhat increased by K applications, but in the second

Table 1. Influences of potassium and phosphorus applications on the grain yield of maize (hybrid OsSK 247)

Treatment in spring 1986 (kg/ha)			Grain yield of maize (t/ha)		Two-year mean	
N	P	K	1986	1987	t/ha	Index
0	0	0	2.65	1.77	2.21	97
240	52	150	2.48	2.10	2.29	100
240	52	262	3.75	3.18	3.47	152
240	52	450	6.08	4.83	5.46	238
240	52	638	7.35	5.39	6.37	278
240	52	825	7.62	7.85	7.74	338
240	111	150	2.36	3.27	3.82	123
240	210	150	4.52	3.75	4.14	181
240	308	150	5.37	3.62	4.50	197
240	406	150	4.91	3.77	4.34	190
	LSD	0.05	0.77	0.86		
		0.01	1.03	1.15		

Table 2. Concentrations of P, K, Ca and Mg in the ear leaf of maize at the start of the silking stage, as affected by P and K fertilizer application

Treatment in spring 1986 (kg/ha)			Percentages on a dry-matter basis							
			1986				1987			
N	P	K	P	K	Ca	Mg	P	K	Ca	Mg
0	0	0	0.36	0.67	1.28	1.32	0.28	0.64	1.09	1.44
240	52	150	0.41	0.54	1.01	1.83	0.33	0.69	0.95	1.48
240	52	262	0.38	0.66	1.17	1.58	0.35	0.66	1.02	1.54
240	52	450	0.34	1.02	1.15	1.27	0.30	0.83	0.86	1.32
240	52	638	0.32	1.07	1.08	1.00	0.31	0.80	0.94	1.33
240	52	825	0.31	1.32	1.10	1.00	0.29	1.21	1.04	1.15
240	111	150	0.45	0.43	0.95	1.82	0.34	0.67	1.19	1.53
240	210	150	0.38	0.74	1.08	1.57	0.34	0.76	1.20	1.56
240	308	150	0.42	0.94	0.99	1.51	0.32	0.89	0.98	1.34
240	406	150	0.50	0.97	0.92	1.42	0.39	0.83	0.97	1.46
	LSD	0.05	0.09	0.11	0.13	0.28	0.02	0.10	0.12	0.18
		0.01	0.13	0.15	0.20	0.37	0.03	0.13	0.16	0.25

Table 3. Chemical characteristics of the soil used in the experiment

Treatment in spring 1986 (kg/ha)				pH		mg/100 g of soil				
N	P	K		H_2O	KCl	P	K	Ca	Mg	K:Mg
Soil status in November 1986										
0	0	0		7.72	6.90	2.9	14.3	659	116	0.12
240	52	150		7.54	6.90	2.9	12.8	828	127	0.10
240	52	262		7.58	6.93	3.0	13.2	759	123	0.11
240	52	450		7.84	6.96	3.1	13.9	751	117	0.12
240	52	638		7.88	7.05	2.8	15.0	747	108	0.14
240	52	825		7.76	7.01	2.3	16.7	719	114	0.15
240	111	150		7.52	7.30	3.2	11.3	764	119	0.10
240	210	150		7.81	7.04	6.1	13.1	865	117	0.11
240	308	150		7.45	6.91	7.9	13.7	764	126	0.11
240	406	150		7.31	6.88	8.5	13.0	705	121	0.11
		LSD	0.05	0.23	0.28	0.2	1.3	154	12	
			0.01	0.30	0.38	0.3	1.7	208	16	
Soil status in October 1987										
0	0	0		7.66	6.76	3.1	10.2	740	146	0.07
240	52	150		7.75	6.77	2.7	10.4	756	126	0.08
240	52	262		7.86	7.03	2.9	9.7	803	143	0.07
240	52	450		7.82	7.18	2.8	9.7	860	147	0.07
240	52	638		7.81	7.00	2.6	10.0	777	138	0.07
240	52	825		7.73	6.85	2.9	10.6	886	142	0.07
240	111	150		7.91	7.11	3.2	8.6	863	137	0.06
240	210	150		7.87	6.99	4.6	10.0	704	139	0.07
240	308	150		8.00	6.98	6.1	10.2	687	137	0.07
240	406	150		7.73	6.91	8.0	10.1	716	137	0.07
		LSD	0.05	0.22	0.30	0.3	0.8	213	19	
			0.01	0.29	0.41	0.5	1.1	287	26	

year the levels had fallen again to the initial value. This phenomenom can be explained by strong K-fixing properties of the soil. For example, Vukadinović *et al.* (1988) found with the wet-fixation method that of 100 mg K/100 g soil added, 76% K was fixed. P-fertilizer application gave a more lasting improvement of the available-P status of the soil.

According to Johansson and Hahlin (1977) an exchangeable K:Mg ratio of less than 1, results in crop responses to K fertilization, whereas with values larger than 3, K application often causes a yield depression. An extremely low K:Mg ratio (Table 3) can be used to explain the very pronounced response of maize to K fertilization.

Similar problems of K nutrition exist in some other soils of East Croatia. An application 920 kg K/ha increased grain yield of maize by 39% (three-year mean). The response of maize was highest in the first year of testing. The ear-leaf K and Mg relations were closely correlated with grain yields of maize. For example, due to a K fertilizer application of 920 kg K/ha, the ear-leaf K:Mg ratio increased from 0.34 to 1.20 in the first year of testing. At the same time, grain yield of maize increased from 4.98 to 8.96 t/ha. A similar comparison in the third year of testing yielded the following values: 0.42 and 0.51 for the fear-leaf K:Mg ratio and 5.30 and 6.71 t/ha for grain yield (Bertić *et al.*, 1989).

For practical purposes we recommend a modification of the usual fertilization practice. About 400 kg K/ha should be added every third year as ameliorative fertilization. In addition, a part of the usual quantity of K applied should be band-placed at the time of sowing.

Conclusions

The maize hybrid OsSK 247 was grown on a gleysol low in P and K. Five rates of K and P were applied in the spring of 1986. The highest rates of K and P (825 kg K and 406 kg P/ha) raised grain yields by 238% and 90%, respectively (two-year means). Excess ear-leaf Mg (1.66%) and a very low ear-leaf K level (0.62%) were associated with K deficiency symptoms in maize grown with normal quantities of fertilizer applied. The highest rate of K application improved the ear-leaf K and Mg status: 1.27% K and 1.07% Mg (two-year means). An imbalanced K and Mg supply (K and Mg antagonism) promotes K deficiency.

References

Bertić B, Vukadinović V and Kovačević V 1989 Excess of magnesium uptake in maize (*Zea mays* L.) plants as promoting factor of potassium deficiency. Magnesium Bulletin 11, 25–28.

Holz F 1973 Die automatische Bestimmung des Stickstoffs als Indophenolgrün in Böden und Pflanzen. Landw. Forsch. Sonderheft 26, pp. 177–192.

Johansson O A H and Hahlin J M 1977 Potassium/magnesium balance in soil for maximum yield. *In* Proceedings Seminar on Soil Environment and Fertilization Management in Intensive Agriculture (SEFMIA), Tokyo. pp 487–499. Society of the Science of Soil and Manure, Tokyo.

Kovačević V, Jurić I and Žugec I 1987 Soybean response to lime and phosphorus under the growing conditions of Eastern Croatia. Eurosoya 6, 58–61.

Kovačević V, Vukadinović V and Bertić B 1988 Excessive iron and aluminium uptake and nutritional stress in corn (*Zea mays* L.) plants. J. Plant Nutr. 11, 1263–1272.

Kovačević V, Žugec I and Bertić B 1986 Poremećaji mineralne ishrane biljaka na tlima Slavonije. Savremena poljoprivreda 34, 133–150.

Mengel K and Kirkby E A 1987 Principles of Plant Nutrition. International Potash Institute, Bern, 687 p.

Vukadinović V, Bertić B and Kovačević V 1988 Kalium- und Phosphorverfügbarkeit auf den Böden im Gebiet von Posavina. Tag. Ber., Akad. Landwirtsch. Wiss. DDR, Berlin 267, 73–80.

Whiteside P J 1979 An introduction to atomic absorption spectrophotometry. Pye Unicam Ltd, Cambridge, 63 p.

M. L. van Beusichem (Ed.), *Plant nutrition – physiology and applications*, 611–617.

PLSO IPNC380

Studies on potassium–magnesium interaction in coconut (*Cocos nucifera*)

M. JEGANATHAN
Soils and Plant Nutrition Division, Coconut Research Institute, Lunuwila, Sri Lanka

Key words: coconut, *Cocos nucifera* L., copra, leaf analysis, nut water analysis, ultisol

Abstract

The widespread deficiency of magnesium in coconut acts as a limiting factor to increased production. Two ongoing experiments on potassium–magnesium (K–Mg) interaction in coconut conducted in lateritic gravels (ultisols), in the wet and intermediate agroclimatic zones of Sri Lanka showed significant yield responses ($P = 0.01$) to differential K treatments, in the wet zone, but not in the intermediate zone. Differential Mg treatments, however, did not give rise to yield responses. Leaf and nut water analysis showed significant changes in the concentrations of Na, K and Cl ($P = 0.001$), with a distinct inverse relationship between Na and K when K was applied. Differential Mg applications showed a significant effect only for leaf Mg ($P = 0.001$), in the fourth year of the experiment, in the wet zone. Results indicate the usefulness of nut water analysis as an additional diagnostic tool, for Na, K and Cl.

Introduction

In the plantation agriculture in Sri Lanka, coconut (*Cocos nucifera* L.) occupies the highest area of nearly 420,000 ha. The estate sector (8 ha or more) comprises 104,000 ha (24.7%) of the area while the balance, 316,000 ha (75.3%) is categorized as small holdings.

Coconut is a very important commodity in Sri Lanka. It is used traditionally as a component in food, and about 70% of the production is consumed locally. About 9% of the total expenditure on food of an average household is spent on coconut, the principal source of edible oil and fat. Coconut provides about 22% of the total intake of calories of an average consumer in Sri Lanka second to rice. These data amply demonstrate the importance of coconut.

In order to meet the high domestic consumption and also maintain the export market, increased production can be achieved with regular and increased fertilization.

Potassium is the most important nutrient for the production of nuts and copra, the latter used in oil extraction. Although the importance of K nutrition has been experimentally shown (Loganathan and Balakrishnamurti, 1979; Salgado, 1950) widespread occurrence of Mg deficiency in the lateritic and the sandy soils has been a limiting factor in coconut production. It is necessary therefore to study the limitations in the use of high levels of K obviously due to K–Mg interactions and their antagonistic effects. This paper describes results of two experiments conducted to elucidate this aspect in coconut nutrition with a view to determine levels of Mg in the soil and leaf for more effective K usage.

Materials and methods

Two experiments were conducted on adult tall coconut (variety typical) of age about 55 years, in 7 and 14 hectare blocks, at Sirikandura Estate, Dodanduwa (Southern Province) and at Heemmeliyagara Estate, Hiruwalpola (North Western Province), from October 1984.

Sirikandura Estate (SE) falls in the wet zone receiving an annual rainfall of 1875 to 2500 mm, while Heemmeliyagara Estate (HE) is in the intermediate zone receiving an annual rainfall of 1000 to 1875 mm. The rainfall is bimodal, with peaks in April/June and October/November. The soils in both experimental sites are Red Yellow Podzolics with soft and hard laterite (Ultisols).

The chemical characteristics of the soils are given in Table 1.

Design and treatments

The experimental design consisted of a 4×4 factorial arrangement for the two nutrients K and Mg in three blocks, each consisting of 16 plots of eight effective palms separated by a single row of guard trees. The planting density is 158 palms/ha.

Fertilizer applications were made annually, starting in 1984, as follows:

Muriate of potash ($60\%\ K_2O$) kg/palm/yr		Kieserite (24% MgO) kg/palm/yr	
K_1	0.0	Mg_1	0.0
K_2	1.2	Mg_2	0.6
K_3	2.4	Mg_3	1.2
K_4	3.6	Mg_4	1.8

Basal applications of urea and rock phosphate, each at the rate of 0.7 kg/palm/yr, equivalent to 0.32 kg N and 0.19 kg P_2O_5, were made in all treatments.

At SE, only the first three levels of treatment were imposed. The fertilizer was applied on the surface of a circular area with a radius of 1.75 m, around the base of the palm, and incorporated into the soil using a fork.

Soil and leaf sampling

Four palms were chosen randomly from each plot. In these palms, the 14th leaf from the top, the 1st being the fully opened leaf with its leaflets separated, was selected and six leaflets were taken from the mid region and composited to form a sample, oven-dried at 85°C and ground for analysis.

Soil samples were collected from the manure circle. Two borings were made at 1 m from the base on either side of the palms and samples were collected from two depths, 0–25 cm, and 25–50 cm and composited separately.

Nut water analysis

Nut water or 'coconut water' is the colourless liquid found within the hard white kernel (the solid endosperm), both of which are enclosed by

Table 1. Soil analysis – October 1984

Parameters	Sirikandura Estate Depth		Heemmeliyagara Estate Depth	
	0–25 cm	25–50 cm	0–25 cm	25–50 cm
Chemical parameters of pretreatment samples in the fertilized circle				
pH	4.2	3.8	5.5	5.1
E.C. (μmhos/cm)	51.39	55.25	46.88	52.26
Exch. Na (meq%)	0.08	0.06	0.08	0.10
Exch. K (meq%)	0.26	0.16	0.30	0.44
Exch. Ca (meq%)	1.27	0.59	1.56	1.04
Exch. Mg (Meq%)	0.20	0.16	0.57	0.38
Avail. Cl (mg/kg)	27.62	27.30	24.01	30.08
Outside the fertilized circle, in the centre of square formed by four palms				
pH	4.0	3.9	5.0	5.0
E.C. (μmhos/cm)	33.39	21.75	36.08	16.34
Exch. Na (meq%)	0.03	0.03	0.07	0.08
Exch. K (meq%)	0.08	0.04	0.13	0.10
Exch. Ca (meq%)	1.06	0.31	0.73	0.65
Exch. Mg (meq%)	0.20	0.16	0.42	0.40
Avail. Cl (mg/kg)	18.94	20.40	16.06	14.33

the hardshell. From each plot, eight nuts representing each of the palms, selected at random, were sampled and the nut water from them was mixed well, filtered and aliquots were taken for analysis.

Chemical analysis

For soil analysis a 1 *M* CH_3COONH_4 extractant was used for the extraction of exchangeable Na, K, Ca and Mg and determined by atomic absorption, Na and K in the emission and Ca and Mg in the absorption modes. Cl was read off the Chlor-O-Counter after extraction in 1:1 soil water ratio.

For leaves, N was determined in 0.1-g samples digested in a Se/H_2SO_4 mixture, and P, Na, K, Ca and Mg, in 0.5-g samples digested in a $HNO_3/HClO_4$ mixture. N and P were estimated colorimetrically on a Technicon Auto Analyzer and Na, K, Ca and Mg, as for soils, by atomic absorption. Cl was determined in 2.0-g samples, ashed in a muffle furnace at 450°C, extracted with water and read off the Chlor-O-Counter.

Nut water was analysed by diluting 20 times for Na and K, and 100 times for Ca and Mg and read off the Atomic Absorption Spectrophotometer (Somasiri *et al.*, 1986). Cl was estimated on the Chlor-O-Counter, using 0.1 mL of nut water (Periathamby, personal communication).

Yield

Commencing January 1985, ripe nuts were harvested from the two mature bunches bimonthly and the total number of nuts recorded. Weights of the husked nuts were obtained from samples of 32 nuts representing each treatment. The yield of copra was determined using the 'copra index factor' (Mathes, 1985, personal communication).

Results and discussion

Soil and leaf analysis – Pretreatment data

The pretreatment soil analytical data in Table 1 show the exchangeable cations in the manure circle higher than in the centres of squares. This is due to the residual fertility in the manure circle, while the centre of squares reflects the inherent fertility.

The mean exchangeable K level in the manure circles of the two sites at the first depth was 0.28 meq%. Earlier experiments in lateritic soils in the wet zone have shown that in spite of such high K levels in the soil, the palms responded to muriate of potash (Loganathan and Balakrishnamurti, 1979). In the sub-soil (25–50 cm) the exchangeable K was much lower at SE than at HE, 0.16 and 0.44 meq%, respectively, which is probably due to leaching losses resulting from higher rainfall in the wet zone.

Information available on exchangeable Mg is scanty. Ollagnier *et. al.* (1983) considered an exchangeable Mg level of 0.46 meq% as high. The critical values for exchangeable K, and Mg in the Phillipines are 0.45, and 2.9 meq%, respectively (Santiago, 1978). Margate *et al.* (1979) did not observe a K–Mg antagonism in a long-term KCl fertilizer study on a clay loam soil in the Philipinnes in spite of high application of KCl (8 kg/palm/year), with soil levels of 0.45 meq% K and 5.3 meq% Mg. On this basis, the soils in SE are not rich.

The data in Table 2 show the average concentrations of nutrients in the 14th leaf. With the

Table 2. Pretreatment leaf nutrient concentrations (14th leaf) on the Sirikandura and Heemeliyagara Estates

	Nutrients (% in dry weight)					
	N	P	K	Ca	Mg	Cl
Sirikandura	2.02	0.15	1.09	0.38	0.16	0.30
Heemmeliyagara	1.98	0.15	1.10	0.50	0.30	0.34
Sufficiency ranges	1.9–2.1	0.11–0.13	1.2–1.5	0.35–0.55	0.25–0.30	0.30–0.40

exception of K in both locations and Mg at SE, the levels of the other nutrients fall within the sufficiency ranges as described by Loganathan and Atputharaja (1986).

Sirikandura Estate Analysis of leaf samples collected in May 1988 (Table 3) showed that the concentrations of both N and P had remained unaltered by the differential treatments.

Applications of K showed significant quadratic responses in leaf K and Mg, a significant linear increase in Cl and decrease in Na and Ca.

Increasing rates of Mg applied showed a highly significant linear increase for Mg only in 1988, the fourth year of the experiment.

At SE leaf K reached the value of 1.49% from the premanurial 1.09% at the highest level, K3 (2.4 kg/palm/year), a significant improvement in K nutrition. Increasing applications of Mg caused significant increases in leaf Mg, from the premanurial 0.16%, to 0.22% at the Mg3 level (1.2 kg/palm/year) still below the critical concentration of 0.25%.

Heemmeliyagara Estate As at SE, both N and P remained unaffected by the differential K–Mg treatments (Table 4).

Increasing rates of K applied showed significant linear increase in leaf K and decreases in leaf Na and Mg. Cl showed a linear response. From, the pretreatment 1.10% level, leaf K increased to only 1.12% at K4, still below the critical level of 1.2%. Unlike at SE, increasing rates of Mg caused no changes in any nutrient levels.

Nut water analysis

Sirikandura Estate Analysis of the nut water collected in March 1988 (Table 5) showed responses to increasing rates of K and Mg. K application caused highly significant positive responses, both linear and quadratic, in the uptake of K and Cl, and a negative response for Na. Mg application only caused a significant linear increase in Cl, which is unusual and difficult to explain.

Heemmeliyagara Estate As at SE, negative linear and quadratic responses of Na to increasing K applications were observed (Table 6). K and Mg showed significant responses, both linear and quadratic, to increasing rates of K. Increased application of Mg caused a linear increase in K, which was unexpected, in view of the fact that usually these two nutrients show antagonistic effects.

In Sri Lanka, chemical analysis of nut water has been successfully used in the study of P and K nutrition of coconut (Salgado, 1955; 1966). In

Table 3. Leaf nutrient concentrations after four years of differential K and Mg applications, at Sirikandura Estate – May 1988

	Concentration of nutrients (% in 14th leaf)						
Treatment	N	P	Na	K	Ca	Mg	Cl
K_1	2.27	0.156	0.30	0.63	0.50	0.27	0.37
K_2	2.38	0.158	0.25	1.18	0.41	0.15	0.49
K_3	2.36	0.156	0.21	1.49	0.40	0.15	0.56
Mg_1	2.32	0.157	0.26	1.14	0.43	0.15	0.49
Mg_2	2.34	0.158	0.25	1.09	0.45	0.20	0.45
Mg_3	2.35	0.157	0.24	1.07	0.43	0.22	0.47
Sign. level							
K l	–	–	***	***	**	***	***
K q	–	–	–	**	–	**	–
Mg l	–	–	–	–	–	***	–
CV%	3.55	2.81	11.89	7.76	9.24	9.41	9.20

* $P=0.05$; ** $P=0.01$; *** $P=0.001$.
l = linear response, q = quadratic response.

Table 4. Leaf nutrient concentrations due to differential applications of K and Mg at Heemmeliyagara Estate – January 1988

		Concentration of nutrients (% in 14th leaf)						
Treatment		N	P	Na	K	Ca	Mg	Cl
K_1		2.14	0.128	0.15	0.83	0.56	0.35	0.35
K_2		2.13	0.129	0.12	0.99	0.51	0.32	0.46
K_3		2.13	0.130	0.11	1.04	0.45	0.29	0.49
K_4		2.11	0.130	0.10	0.12	0.50	0.27	0.52
Mg_1		2.13	0.131	0.12	0.98	0.52	0.30	0.46
Mg_2		2.11	0.129	0.13	1.00	0.52	0.31	0.46
Mg_3		2.16	0.129	0.11	1.00	0.50	0.32	0.43
Mg_4		2.11	0.128	0.12	1.00	0.50	0.32	0.48
Sign. level								
K	*l*	–	–	***	***	*	***	***
K	q	–	–	–	–	*	–	–
CV%		6.21	21.77	12.89	14.95	14.95	11.75	22.70

* $P = 0.05$; ** $P = 0.01$; *** $P = 0.001$.
l = linear response, q = quadratic response.

Table 5. Nut water nutrient concentrations after four years of differential application of K and Mg at Sirikandura Estate – March 1988

		Nutrient (mg L^{-1})				
Treatment		Na	K	Ca	Mg	Cl
K_1		270	1634	186	101	1870
K_2		139	2375	171	105	2095
K_3		91	2627	180	99	2102
Mg_1		179	2195	185	100	2073
Mg_2		167	2238	179	101	2038
Mg_3		154	2202	174	105	1956
Sign. level						
K	l	***	***	–	–	***
K	q	***	***	–	–	**
Mg	l	–	–	–	–	**
CV%		15.73	4.49	13.73	10.72	3.61

* $P = 0.05$; ** $P = 0.01$; *** $P = 0.01$.
l = linear response, q = quadratic response.

Table 6. Nut water nutrient concentrations after four years of differential application of K and Mg at Heemmeliyagara Estate – March 1988

		Nutrient (mg L^{-1})				
Treatment		Na	K	Ca	Mg	Cl
K_1		140	1918	283	156	1959
K_2		80	2128	278	138	1997
K_3		71	2201	282	130	2012
K_4		64	2196	301	142	2062
Mg_1		96	1985	292	143	1983
Mg_2		85	2120	303	143	2037
Mg_3		90	2141	276	140	2027
Mg_4		84	2197	272	141	1984
Sign. level						
K	l	***	***	–	**	–
K	q	*	*	–	***	–
Mg	l	–	**	–	–	–
CV%		41.44	7.55	14.92	9.74	5.48

* $P = 0.05$; ** $P = 0.01$; *** $P = 0.001$.
l = linear response, q = quadratic response.

other countries this approach has also been used with some measure of success (Lockhard *et al.*, 1969; Southern, 1956).

Present indications are that nut water analysis can be used as an additional tool in the interpretation of field experiment data for Na, K, Mg and Cl. Both Ca and Mg have functionally limited roles to perform, both in the liquid and in the solid endosperm, and therefore their concentration will be low and so too the changes.

Yield

Sirikandura Estate Analysis of the second year's data (1986) showed a significant response in terms of copra production at ($P = 0.01$) to increasing K applications. For the years 1987 and 1988, these responses were highly significant at $P = 0.001$ (Table 7). Nut yields also significantly increased during these two years as a result of K application (Table 8).

Table 7. Weight of copra in different years following the start of differential annual applications of K and Mg at Sirikandura Estate, with 158 palms per ha

Treatment	Copra (kg/ha)	%	Difference (kg/ha)
	Yield 1986		
K_1	1668*	100	–
K_2	1882	113	214
K_3	1985	119	317
	Yield 1987		
K_1	1442**	100	–
K_2	1798	125	356
K_3	1843	129	401
	Yield 1988		
K_1	1127***	100	–
K_2	1559	138	432
K_3	1696	150	569

* $P = 0.05$; ** $P = 0.01$; *** $P = 0.001$.

Table 8. Nut yields in different years following the start of differential annual applications of K and Mg at Sirikandura Estate, with 158 palms er ha

Treatment	Nuts/ha	%	Difference
	Yield 1986		
K_1	8091 NS	100	–
K_2	8447	104	356
K_3	8815	109	724
	Yield 1987		
K_1	7433**	100	–
K_2	8545	115	1112
K_3	8888	120	1455
	Yield 1988		
K_1	5728***	100	–
K_2	7290	127	1562
K_3	7773	136	2045

N.S. = Not Significant; * $P = 0.05$; ** $P = 0.01$; *** $P = 0.001$.

There has been an overall decrease in yield of copra, despite the increased application of K. The drop in the 'no fertilizer' plots has been drastic, 32.4% between 1986 and 1988. During the same period, at the K1 and K2 levels, the decreases have been 17.2% and 14.6%, respectively. Fertilizer application has to some extent arrested the decline. The fall in production can be attributed to adverse weather conditions, to the insufficient state of nutrition of the palms and to the low soil nutrient status. Table 9 gives the rainfall and yield data for the 'no fertilizer' plots at both sites. At SE, rainfall was high, except for 1986. The lower rainfall in 1986 was compensated for by a better distribution and therefore yields were not adversely affected in 1987.

From the nutritional aspects, it was observed that with all K application levels the K demand of the trees has been satisfied, and it is yet to be seen whether better yields will be obtained when the Mg concentrations reach the sufficiency level. It is only after K deficiency has been corrected that Mg manuring was found to have a positive effect on production (Brunin, 1970; Coomans, 1977).

Soil analysis data for the years 1987 and 1988 at the Mg3 application level yielded 0.73 and 0.63 meq% Mg, respectively. These fairly high exchangeable Mg levels in the soil can be expected to raise the Mg concentrations in the leaves enough to have a bearing on future production.

Table 9. Yield and rainfall data and corresponding nut and copra yield in the 'no fertilizer' treatments at the two estates

Year	Rainfall	No. of wet days	Nuts/ha	Copra (kg/ha)
		Heemmeliyagara Estate		
1984	1815	87	–	–
1985	1262	55	11179	2074
1986	885	44	9263	1501
1987	1134	59	5649	751
1988	807	42	5267	898
		Sirikandura Estate		
1984	2240	133	–	–
1985	2724	164	9381	1936
1986	1990	174	8907	1783
1987	2676	149	8117	1541
1988	2505	137	6590	1249

Heemmeliyagara Estate Statistical analysis of the yield data for the years 1986, 1987 and 1988, in terms of nuts and copra, showed neither a significant response to the main treatments K and Mg nor to any interaction between them.

A combination of factors contributed to the lack of yield response, such as 'no fertilizer' plots performing much better than the others at the commencement of the studies (Table 9), and delayed application of fertilizers due to prolonged drought periods, compounded by the drought effect itself.

Changes in the concentrations of Na and Cl were observed in both leaf and nut water, resulting from differential fertilizer treatments. Both elements have some functions in the nutrition of coconut, particularly Cl, but discussion of such functions is beyond the scope of this paper.

Conclusions

Significant yield responses were obtained for the experiment at SE, albeit with an overall decrease in yield due to weather and nutritional factors. The experiment at HE failed to display any yield responses as a result of drought.

With respect to leaf and nut water analysis, in both experiments responses to the main treatments K and Mg, were noticeable, with varying trends. Changes in the leaf composition was more pronounced at SE than at HE.

Nut water analysis shows promise as an additional tool for interpretation of field experimental data.

The presentation covers a four-year period. Statistical studies on when to conclude long-term fertilizer trials on coconut yield consider an eight- to ten-year period as sufficient to understand the full response (Mathes, 1980). Both experiments are still in progress.

Acknowledgements

The author is thankful to Dr R Mahindapala, Director, Coconut Research Institute for helpful criticism in the preparation of the paper, and to Mr D T Mathes, Biometrician, Coconut Research Institute, for statistical advice and useful discussions.

Thanks are due to Miss S Pariathamby, Mrs D M D I Wijebandara and Mr A A Fernando, for the maintenance of the laboratory and field records, and to the staff of the Soils and Plant Nutrition Division, Coconut Research Institute, for handling the analyses. Thanks are also due to Mrs H M W S Athauda for typing the manuscript.

References

Brunin C 1970 La nutrition magnesienne des cocoteraies en Côte d'Ivoire. Oléagineux 25, 269–274.

Coomans P 1977 Premiers resultats experimentaux sur la fertilisation des cocotiers hybrides en Côte d'Ivoire. Oléagineux 32, 155–166.

Loganathan P and Balakrishnamurti T S 1979 Effect of NPK fertilizers on the yield and leaf nutrient concentration of adult coconut on a lateritic gravelly soil in Sri Lanka. Ceylon Cocon. Q. 30, 81–90.

Loganathan P and Atputharajah P P 1986 Effects of fertilizers on yield and leaf nutrient concentrations in coconut. Trop. Agric. Trinidad 63, 143–148.

Lockard R G, Ballaux J C and Azucena B 1969 The results of pretreatment leaf and nut water analysis from ten coconut fertilizer experiments. Phil. Agric. 53, 276–288.

Margate R Z, Magat S S, Alforja L M and Habana J A 1979 A long-term KCl fertilization study of bearing coconut in an inland upland area of Davao (Philippines). Oléagineux 34, 235–242.

Mathes D T 1980 A study on when to conclude a long-term fertilizer trial on coconut yield. Ceylon Cocon. Q. 31, 127–133.

Ollagnier M, Ochs R, Pomier M and de Taffin G 1983 Effect of chlorine on the hybrid coconut PB 121 in the Ivory Coast and Indonesia. Oléagineux 38, 309–321.

Santiago R M 1978 Growth of coconut seedlings as influenced by different fertility levels in three soil types. Phil. J. Cocon. Stud. 3, 15–27.

Salgado M L M 1950 A. Rep. Coconut Research Scheme for 1948, Sessional paper XXII, Colombo Govt. Publications Bureau.

Salgado M L M 1955 The nutrient content of nut water in relation to available soil nutrients as a guide to the manuring of the coconut palms; a new diagnostic method. *In* Plant Analysis and Fertilizer Problems. Ed. P Prevot. pp 217–238. IRHO, Paris.

Salgado M L M and Abeywardene V 1964 Nutritional and physiological studies on coconut water. Ceylon Cocon. Q. 15, 95–108.

Somasiri L L W, Warnasiri W H, George G D and Jeganathan M 1986 Report of the Soils and Plant Nutrition Division, Coconut Research Institute. Report for 1986, 87–88.

Southern P J 1956 The flame spectrophotometric determination of potassium, sodium, calcium and magnesium in nut water. Papua New Guinea Agric. J. 11, 69–76.

M. L. van Beusichem (Ed.), *Plant nutrition – physiology and applications*, 619–623.

PLSO IPNC347

Fruit calcium, quality and disorders of apples (*Malus domestica*) and pears (*Pyrus communis*) influenced by fertilizers

J.T. RAESE and D.C. STAIFF
Tree Fruit Research Laboratory, USDA-ARS, 1104 N. Western Avenue, Wenatchee, WA 98801, USA

Key words: apple, bitter pit, cork spot, *Malus domestica* L., mineral analyses, macronutrient fertilizers, pear, *Pyrus communis* L., tree vigor

Abstract

Apple and pear trees fertilized with soil-applied calcium nitrate were associated with higher concentrations of fruit Ca, lower fruit N:Ca or N + P + K + Mg:Ca ratios, more red fruit-color of 'Delicious' apples, better finish, and a lower incidence of fruit disorders (alfalfa greening, bitter pit and cork spot) than with the other fertilizers. High rates of nitrogen fertilizers resulted in greater tree vigor, higher yield of 'Golden Delicious' apples and 'Anjou' pears, but a higher incidence of fruit disorders, and higher fruit N:Ca ratios than the low rate of N fertilizers.

Introduction

Fruit quality and physiological disorders of apples, *Malus domestica* and pears, *Pyrus communis* are related to the mineral composition of the fruit, especially Ca (Bramlage *et al.*, 1980; Fallahi *et al.*, 1988; Mason, 1970; Perring, 1986; Raese, 1982; 1988). While Ca sprays increase fruit Ca and quality and reduce fruit disorders (Raese, 1988; Stahly, 1986), a dearth of information exists showing that soil-applied $Ca(NO_3)_2$ fertilizers increase fruit Ca and reduce fruit disorders of apples and pears (Raese and Staiff, 1983; Shear, 1972). The objectives of this report are to show the effect of different rates and formulations of fertilizers on tree performance, fruit quality, disorders (bitter pit of apples, alfalfa greening and cork spot of pears) and mineral composition of apples and pears.

Materials and methods

Apples

Young, bearing 'Delicious' and 'Golden Delicious' apple trees were fertilized each spring (1985–87) with low, medium and high rates of 4 nitrogenous fertilizers. The two orchards were located near Cashmere, Washington in Cashmont sandy loam soils. The experiments consisted of a randomized split-block design containing 3 and 4 replications, respectively, of up to 10 trees each. Twenty leaves per sample (fruiting-spur or mid-shoot) were collected each June and August, washed, weighed and analyzed for N, P, K, Ca, Mg, Al, B, Cu, Fe, Mn, Na and Zn. Ten fruits per sample were collected just prior to grower-harvest, washed, weighed and, after cold storage at 0°C, analyzed for the above mineral elements in the peel and cortex (2 opposite longitudinal wedges). Fruit quality and mineral analyses procedures were described earlier (Raese and Staiff, 1983; Raese, 1988).

Pears

Mature, bearing 'Anjou' pear trees were fertilized each autumn (1979–85) with low, medium and high rates of 2 nitrogenous fertilizers. The pear orchard was located at Chelan, Washington and planted in a Supplee very fine sandy loam.

The experiment consisted of a randomized strip block containing 10 replications. Leaf and fruit samples were collected and prepared as described for apples. Fruit disorders for apples and pears were observed in the orchard just prior to harvest and twice again after 3 to 8 months in cold storage at 0°C. Apple and pear disorders were rated for severity and percentage of fruit affected. Alfalfa greening of 'Anjou' pears is a disorder showing symptoms of green specks or longitudinal green streaks with or without superficial pits on the stem- or calyx-end or over the entire surface of the fruit (Raese *et al.*, 1979).

Results

Apple

'Delicious'

After 3 years of fertilization, the trees treated with calcium nitrate resulted in slightly higher tree vigor and yield than with the other fertilizers (Table 1). Calcium nitrate treatments also resulted in the greatest red color development on the fruit surface. This is important because red color development was a problem on 'Delicious' apples in 1987 due to above-normal, late-

Table 1. Effect of rate and source of soil-applied fertilizers on tree vigor, fruit yield, size, bitter pit, red-skin color, firmness, cortex Ca, P and N:Ca ratios of 'Delicious' apples, Cashmere, WA – 1987

Fertilizer treatments[z] (1985–87)	Tree vigor (1–5)	Yield /tree (kg)	Fruit size (g)	Bitter pit (%) Orchard	Bitter pit (%) Storage	Red area (%)	Firmness (N)	Fruit cortex (dry wt.) Ca (ppm)	Fruit cortex (dry wt.) P (%)	Fruit cortex (dry wt.) N:Ca (ratio)
Whole plots										
NH_4NO_3	3.4 ab	36.1	203	1.5 c	49	61 ab	57 a	194	0.048 b	12.9 ab
$Ca(NO_3)_2$	3.7 a	38.0	205	1.8 bc	51	73 a	53 ab	215	0.039 c	11.9 b
MAP	3.2 b	36.1	203	7.6 a	58	50 b	54 ab	202	0.065 a	14.9 a
16–16–16	3.5 ab	32.3	208	5.1 ab	49	53 b	52 b	191	0.042 bc	15.5 a
Sub plots										
Low (223g N)	3.1 b	34.2	198	2.9 b	38	57	55	218 a	0.050	12.1 b
Med (446g N)	3.6 a	38.0	207	6.6 a	55	61	54	185 b	0.050	15.1 a
High (669g N)	3.7 a	36.1	209	7.6 a	62	60	52	199 ab	0.046	14.2 ab

[z] Mono-ammonium phosphate (MAP); 16–16–16 = 16% N, 16% P_2O_5, 16% K_2O. N rate = g/tree. Treatment means separated by Waller-Duncan K ratio, t test, 5%.

Table 2. Effect of rate and source of soil-applied fertilizers on tree vigor, yield, fruit size, bitter pit (orchard and storage observation), finish, firmness, titratable acids and internal color of 'Golden Delicious' apples, Cashmere, WA – 1987

Fertilizer treatments[z] (1985–87)	Tree vigor (1–5)	Yield /tree (kg)	Fruit size (g)	Bitter pit (%) Orchard	Bitter pit (%) Storage	Fruit finish (1–8)	Firmness (N)	Titr. acids (% mal.)	Agtron color early storage blue	Agtron color early storage green
Whole plots										
NH_4NO_3	4.0	24 b	171	2.5 ab	15.8 b	5.2 ab	60 a	0.216 a	29	43 ab
$Ca(NO_3)_2$	3.7	34 ab	172	0.7 b	9.6 b	6.0 a	56 b	0.211 a	32	46 a
MAP	3.7	34 ab	175	5.0 a	37.5 a	4.5 b	56 b	0.197 b	29	42 b
16–16–16	3.8	41 a	180	3.3 ab	25.4 ab	5.1 b	54 b	0.218 a	29	43 ab
Sub plots										
Low (113g N)	3.5 b	25 b	174	1.5 b	16.6	5.3	57	0.216	32 a	45 a
Med (345g N)	3.8 ab	35 a	176	4.1 ab	22.2	5.2	55	0.207	31 a	44 a
High (685g N)	4.1 a	40 a	173	5.5 a	27.5	5.1	57	0.208	27 b	41 b

[z] Mono-ammonium phosphate (MAP); 16–16–16 = 16% N, 16% P_2O_5, 16% K_2O. N rate = g/tree. Treatment means separated by Waller-Duncan K ratio, t test, 5%.

summer temperatures in the Pacific Northwest. Fruit from trees treated with ammonium nitrate or calcium nitrate had the lowest N:Ca ratios and the lowest incidence of bitter pit at harvest. However, all samples developed a severe amount of bitter pit after 4 months in cold storage. The higher rates of N fertilizers resulted in the highest tree vigor, bitter pit and N:Ca ratios and the lowest concentration of fruit Ca (Table 1).

'Golden Delicious'

Calcium nitrate-treated fruit had the highest fruit finish rating, whitish flesh color and the lowest incidence of bitter pit at harvest or after 3 months in cold storage (Table 2). Ammonium nitrate treatments resulted in the greatest firmness for both 'Delicious' and 'Golden Delicious' apples (Tables 1 and 2). Tree vigor, yield, bitter pit and green color of apple flesh were increased with the high rate of N fertilizers (Table 2). Fruit from the calcium nitrate treatments had the lowest concentration of N, P, K, Mg, Mn and Zn and the lowest N:Ca or N+P+K+Mg:Ca ratios (Table 3). MAP treatments had the highest fruit N, P, Mn and Zn concentrations and N:Ca ratio while 16–16–16 had the highest fruit K level. Fruit N, Mg and Mn levels increased

Table 3. Effect of rate and source of soil-applied fertilizers on mineral composition and ratios of fruit peel of 'Golden Delicious' apples, Cashmere, WA – 1987

Fertilizer treatments[z] (1985–87)	Mineral composition of fruit peel (dry wt.)									
	Ca (ppm)	N (%)	P (%)	K (%)	Mg (ppm)	B (ppm)	Mn (ppm)	Zn (ppm)	N:Ca ratio	NPKMg :Ca
Whole plots										
NH_4NO_3	383	0.42 bc	0.052 b	0.65 bc	581 bc	30	5.5 b	3.0 ab	11.3 ab	31.5 ab
$Ca(NO_3)_2$	444	0.41 c	0.046 c	0.59 c	552 c	30	4.6 c	2.8 b	9.4 b	25.8 b
MAP	403	0.51 a	0.064 a	0.68 ab	616 ab	29	7.3 a	3.3 a	13.3 a	34.0 a
16–16–16	425	0.47 ab	0.056 b	0.73 a	641 a	32	7.2 a	2.8 b	11.7 ab	33.1 a
Sub plots										
Low (113g N)	390	0.42 b	0.056	0.67	569 b	34 a	5.2 c	2.9	11.0	31.8
Med (345g N)	440	0.45 ab	0.054	0.66	618 a	29 b	6.1 b	3.0	10.8	29.4
High (685g N)	412	0.49 a	0.053	0.66	606 ab	28 b	7.2 a	3.0	12.4	32.2

[z] Mono-ammonium phosphate (MAP); 16–16–16 = 16% N, 16% P_2O_5, 16% K_2O. N rate = g/tree. Treatment means separated by Waller-Duncan K ratio, t test, 5%.

Table 4. Effect of rate and source of soil-applied fertilizers on mineral composition of fruiting-spur leaves of 'Golden Delicious' apples sampled 13 August 1987, Cashmere, WA

Fertilizer treatments[z] (1985–87)	Mineral composition of leaves (dry wt.)									
	Ca (%)	N (%)	P (%)	K (%)	Mg (%)	B (ppm)	Cu (ppm)	Mn (ppm)	N:Ca ratio	NPKMg :Ca
Whole plots										
NH_4NO_3	1.71 ab	1.74	0.25 b	1.90	0.57	44	5 a	63 b	1.04 bc	2.67 bc
$Ca(NO_3)_2$	1.78 a	1.73	0.24 b	1.80	0.55	44	5 a	40 b	0.99 c	2.46 c
MAP	1.62 ab	1.80	0.37 a	1.90	0.54	46	4 b	101 a	1.14 ab	2.90 ab
16–16–16	1.54 b	1.84	0.28 b	2.01	0.53	45	5 a	115 a	1.21 a	3.07 a
Sub plots										
Low (113g N)	1.71	1.67 b	0.28	1.94	0.55 ab	47 a	4.9	51 b	1.00 b	2.67 b
Med (345g N)	1.72	1.79 ab	0.28	1.84	0.58 a	44 ab	4.6	73 b	1.06 b	2.65 b
High (685g N)	1.56	1.86 a	0.29	1.93	0.51 b	43 b	4.4	116 a	1.22 a	3.02 a

[z] Mono-ammonium phosphate (MAP); 16–16–16 = 16% N, 16% P_2O_5, 16% K_2O. N rate = g/tree. Treatment means separated by Waller-Duncan K ratio, t test, 5%.

Table 5. Effect of rate and source of soil-applied fertilizers on tree vigor, yield, fruit size, fruit disorders (cork spot or alfalfa greening), fruit finish, fruit cortex Ca and N:Ca ratios, and Ca in fruiting-spur leaves of 'Anjou' pears, Chelan, WA – 1986

Fertilizer treatments[z] (1979–86)	Tree vigor (1–5)	Yield /tree (kg)	Fruit size (g)	Fruit disorders: cork spot (no./tree)	Alf. greening (1–5)	Alf. greening (1–10)	Fruit finish (1–8)	Fruit cortex Ca (ppm)	Fruit cortex N:Ca (ratio)	Leaf Ca (%)
Low (230g N)										
NH_4NO_3	3.2 bc	171	132 b	0.6 b	2.7 a	0.8 cd	4.6 b	827 a	6.5 c	1.74 b
$Ca(NO_3)_2$	2.7 c	114	132 b	0.3 b	1.4 b	0.4 d	5.5 a	849 a	5.7 c	1.67 b
Med (460g N)										
NH_4NO_3	4.1 a	173	135 b	2.9 a	3.0 a	1.6 b	4.3 b	704 b	8.1 ab	1.80 b
$Ca(NO_3)_2$	3.3 b	143	135 b	1.4 ab	1.2 b	0.9 cd	5.4 a	839 a	6.9 bc	1.67 b
High (690g N)										
NH_4NO_3	4.2 a	230	133 b	2.4 a	3.3 a	2.2 a	4.7 b	731 b	8.8 a	2.15 a
$Ca(NO_3)_2$	3.4 b	190	150 a	0.7 b	1.3 b	1.0 c	5.7 a	764 ab	7.1 bc	1.78 b

[z] N rate = g/tree. Treatment means separated by Duncan's multiple range test, $P = 0.05$. Alfalfa greening (A.G.) orchard rating (1–5) where 5 = very severe; storage rating (1–10) where 10 = greening covering entire surface area. Fruit finish rating of 8 = excellent.

whereas fruit B decreased with the high rates of N fertilizers (Table 3). Concentrations of the above elements in fruiting-spur leaves followed a similar trend for N, P, K, B, Mn, N:Ca and N + P + K + Mg:Ca (Table 4).

Pears

'Anjou'

At the Chelan orchard, tree vigor, yield, fruit disorders and N:Ca ratios increased with the higher rates of N fertilizers (Table 5). However, fruit from trees fertilized with calcium nitrate had lower tree vigor, less fruit disorders, and lower N:Ca ratios and better fruit finish than the ammonium nitrate treatments (Table 5).

Discussion

Apples and pears

In both apples and pears, calcium nitrate fertilization lowered the N:Ca and NPKMg:Ca ratios in fruit and resulted in a lower incidence of fruit disorders than trees receiving the other fertilizer materials (Table 1–5). Calcium nitrate fertilization is apparently associated with not only an increase in fruit Ca, though not always significant, but also a lowering of fruit N, P, K and Mg. Fruit disorders of apples and pears are almost always related to low fruit Ca but also frequently associated with high concentrations of N, P, K and Mg (Raese, *et al.* 1979; Raese 1988). Therefore, low N:Ca or N + P + K + Mg:Ca ratios in fruit are important considerations for control of fruit disorders, and this may explain the benefit of calcium nitrate fertilization.

In every case, the analyses of fruit and leaves show an increase of Mn concentration with the increased rate of N fertilizer while no increase in Mn concentration was evident for trees treated with the less acid-forming fertilizer, calcium nitrate. While these 3 studies show evidence that calcium nitrate-fertilized apple and pear trees produce fruit with lower N:Ca ratios and a low incidence of fruit disorders, other orchards with severe fruit disorder problems may not respond to soil-applied fertilizers. In these situations, Ca sprays may be more affective for control of disorders (Raese, 1988).

Acknowledgements

The authors appreciate the assistance of Agricultural Technicians, P Fletcher, D Frederick, S Ivanov, P Terry and A Yazdaniha.

References

Bramlage W J, Drake M and Lord W J 1980 The influence of mineral nutrition on the quality and storage performance of pome fruits grown in North America. *In* Mineral Nutrition of Fruit Trees. Eds. D Atkinson, J E Jackson, R O Sharples and W M Waller. pp 29–40. Butterworths, London–Boston.

Fallahi E, Righetti T L and Raese J T 1988 Ranking tissue mineral analyses to identify mineral limitations on quality in fruit. J. Am. Soc. Hortic. Sci. 113, 382–389.

Mason J L and Welsh M F 1970 Cork spot (pit) of 'Anjou' pears related to calcium concentration in fruit. HortScience 5, 447.

Perring M A 1986 Incidence of bitter pit in relation to the calcium content of apples. J. Sci. Food Agric. 37, 591–606.

Raese J T 1982 Disorders of Anjou pears related to mineral content. *In* Plant Nutrition 1982. Ed. A. Scaife, pp 510–514. Proc IXth Intern. Colloq. (Coventry). Commonwealth Agricultural Bureaux, Slough, UK.

Raese J T 1988 Calcium: Effects on apple and pear disorders and fruit quality. Proc. Wash. State Hortic. Assoc. 84, 247–257.

Raese J T, Pierson C F and Richardson D G 1979 Alfalfa greening of 'Anjour' pear. HortScience 14, 232–234.

Raese J T and Staiff D C 1983 Effect of rate and source of nitrogen fertilizers on mineral composition of d'Anjou pears. J. Plant Nutr. 6, 769–779.

Shear C B 1972 Incidence of cork spot as related to calcium in the leaves and fruit of 'York Imperial' apples. J. Am. Soc. Hort. Sci. 97, 61–64.

Stahly E A 1986 Time of application of calcium sprays to increase fruit calcium and reduce fruit pitting of apples sprayed with TIBA. HortScience 21, 95–96.

M. L. van Beusichem (Ed.), *Plant nutrition – physiology and applications*, 625–631.

PLSO IPNC597

Energy flow in an apple plant-aphid (*Aphis pomi* De Geer) (Homoptera: Aphididae) ecosystem, with respect to nitrogen fertilization

I. Life table analyses

C.H. RUTZ[1], U. HUGENTOBLER[1], H. CHI[2], J.U. BAUMGÄRTNER[3] and J.J. OERTLI[1]
[1]*Department of Plant Sciences, ETH Zürich, Eschikon 33, CH-8315 Lindau, Switzerland,*
[2]*Department of Entomology, National Chung-Hsing University, Taichung, Taiwan, ROC, and*
[3]*Department of Plant Sciences, ETH Zürich, CH-8092 Zürich, Switzerland*

Key words: aphid population, aphids, apple, nitrogen fertilization

Abstract

The effect of nitrogen fertilization on the life table parameters of green apple aphids, feeding on apple saplings of different nutritional quality, was investigated. The experiments were carried out with cloned apple plants (cv. Golden Delicious), all originating from one seed, growing in a climate chamber under constant conditions. The apple saplings were irrigated with nutrient solutions containing different nitrogen levels (0.2, 0.5, 1, and 3 *N*), and infested with *Aphis pomi* De Geer. The 1 *N* treatment corresponded to a 15 m*M* nitrogen concentration, containing NO_3^- and NH_4^+ in a 14:1 ratio. The levels of nitrogen fertilization studied here influenced the life table parameters of *A. pomi* in the following way: the mean generation time of *A. pomi*, growing on the 0.5 *N* treatment, was lower than the corresponding values found on the 0.2 and 1 *N* treatments. The highest net reproduction rate was produced on low (0.2 *N*) nitrogen nutrition. The intrinsic rate of natural increase was highest on the 0.5 *N* treatment.

Introduction

As phloem feeding insects, aphids are affected by the nutritional status of their host plants. Several authors have suggested that nitrogen could be the limiting growth nutrient for phloem feeding insects and, therefore, account for the excess sap intake necessary to acquire sufficient nitrogen (*e.g.*, Dixon, 1975; Lindemann, 1948; Mittler, 1958). This dependence can be studied via energy budgets.

Research on energy distribution within an insect population has been quite common. The qualitative aspects of food requirements by insects were reviewed by Waldbauer (1968), whereas Wiegert and Petersen (1983) included insect energy budgets and their use in population modelling. In order to quantify the impact that resulted from the feeding of *Oryzaephilus surinamensis* (L.) on rolled oats, White and Sinha (1981) determined an energy budget for this insect species. Llewellyn and Leckstein (1978) showed for *Aphis fabae* Scop. that the aphids reared on synthetic diets used energy less efficiently than those feeding on broad beans. Llewellyn and Hargreaves (1984) presented differences in the biology and energetics of *Macrosiphum euphorbiae* (Thomas) feeding on different sites of the host plant. An interspecific comparison by Llewellyn (1982) revealed a higher production efficiency for *Aphis fabae* Scop. than for leaf chewing insects. This increased efficiency was ascribed to the higher quality of the food consumed by fluid feeders. Randolph *et al.* (1975) found a high production efficiency for *Acyrthosiphon pisum* (Harris), which they attributed to the high nutritive quality of the host plant.

In an ongoing project, we are studying the effect of nitrogen fertilization on the energy flow in green apple aphid colonies (*Aphis pomi* De Geer). This paper presents the preliminary results with respect to energy budgets by collecting and analysing basic life table data.

Materials and methods

Plants

Apple plants (cv. Golden Delicious), all originating from one seed, were propagated by cell culture techniques under sterile conditions. The propagation and rooting media were composed according to Broome and Zimmerman (1984) and Wermelinger (1985). The rooting media were composed according to Wermelinger (1985). For adaptation, the plants were kept in aerated hydroculture consisting of 1 *N* nutrient solution diluted with demineralized water in a 1:7 ratio. The macronutrients of the 1 *N* nutrient solution were those of a Hoagland solution 2 (Hoagland and Arnon, 1938). Micronutrients were applied as suggested by Hewitt and Smith (1975). The plants were then potted in 15-cm diameter plastic pots with a 2-liter volume filled with quartz sand. The apple plants were irrigated twice daily with 50 mL of nutrient solution containing different nitrogen levels (0.2 $N = 3$ m*M*; 0.5 $N = 7.5$ m*M*; 1 $N = 15$ m*M*; and 3 $N = 45$ m*M* nitrogen). These solutions were diluted with demineralized water in a 1:1 ratio before irrigation. Demineralized water was used for additional irrigations. The experiments were carried out during 60 days.

Aphids

A group of parthenogenetically reproducing, viviparous *A. pomi* was raised on cloned apple plants (cv. Golden Delicious) growing in earth-filled pots.

The experiments were carried out in a controlled environment at 20°C/16°C (sinuous day/night temperature curve with a minimum at 6 a.m. and a maximum at 4 p.m.), a L:D = 16:8 regime, and 70% RH.

Life tables and analyses

The plants were irrigated with the four respective treatments for 11 days. Thereafter at each nitrogen fertilization level 15 plants were infested with *A. pomi*. The life table data, *i.e.* individual nymphal development time, individual number of nymphs born per virginopara, and individual mortality of two aphids per plant, were collected. The plants were inspected every two to four days for live adults and newly born nymphs. Between the sampling dates the survivorship was interpolated, and the number of nymphs born per virginopara was assigned equally to the days of the interval. At each sampling date the newly born nymphs were removed. This procedure was continued until the virginoparae died. Between the beginning of the experiments and the first sampling date some aphids were lost, and some others were damaged by manipulations. These aphids were not considered in the further evaluations.

The survivorship and the cumulative number of nymphs born per virginopara were plotted against aphid age in days.

In the first analysis, an interactive program, originally developed by A.P. Gutierrez and coworkers (University of California, Berkeley), was used to evaluate the mean generation time (T), the net reproduction rate (R_0), and the intrinsic rate of natural increase (r_m) of the *A. pomi* colonies.

In our study, the mean generation time is given in days from the birth of a virginopara until 50% of her nymphs were born. The cumulation of the products of survivorship and number of nymphs born per virginopara to each day equals R_0. The division of $\log_{10} R_0$ by T gives an approximate value of r_m. The algorithm used for calculating the life table parameters iteratively approached the exact value of r_m. For a detailed description of these parameters and their use in ecological studies, see Birch (1948), Messenger (1964a), and Southwood (1978).

In the second analysis, the variability within the life table data was considered in the computations of T, R_0, and r_m, as suggested by Chi (1988). The theoretical bases of that method were described by Chi and Liu (1985). In those studies the life table parameters are calculated with respect to both sexes, incorporating variable developmental rates among individuals. In the present study, only virginoparae and no males were considered.

Using the jackknife technique (Efron, 1982; Meyer *et al.*, 1986), the standard deviations of each life table parameter were estimated. The differences between the life table parameters among different nitrogen treatments were tested for significance by using Duncan's new multiple range test (Duncan, 1955).

Results and discussion

Growth of apple plants

A qualitative evaluation of the apple plants' performance revealed that the 0.2 *N* fertilized apple plants had a reduced growth rate and showed slightly chlorotic leaves. The apex growth of the 0.5, 1, and 3 *N* fertilized plants was nearly identical. The 1 *N* and 3 *N* fertilized plants tended to produce lateral shoots. The upper leaves of aphid infested plants were typically curled (see photograph 1).

Life tables

The losses of *A. pomi* between the beginning of the experiments and the first sampling date were 4 and 10% on the 0.5 and 3 *N* treatments, respectively, and zero on the 0.2 and 1 *N* treatments. Accidental losses due to manipulations varied between 11 and 27% on the four treatments. As stated above, these losses were not considered in the further evaluations.

The survivorship and the cumulative number of nymphs born per virginopara to each day are shown in Figure 1. Between the first sampling date and the beginning of the reproductive period, preadult mortality was observed on the lowest (0.2 *N*) and the highest (3 *N*) nitrogen fertilization level (4 and 9%, respectively). No preadult mortality was found on the intermediate treatments (0.5 and 1 *N*). The survivorship of *A. pomi* at adult stage was highest on low (0.2 *N*) nitrogen nutrition. The aphids growing on the 0.2, 0.5, and 1 *N* treatments tended to have a higher number of nymphs born per virginopara than those growing on the 3 *N* treatment.

In contrast to the results reported by Graf *et al.* (1985), survivorship started to decrease before the reproductive period of *A. pomi* was

Photograph 1. Cloned apple plants (cv. Golden Delicious), all originating from one seed, after 23 days of irrigation with nutrient solutions containing four different nitrogen levels (0.2, 0.5, 1, and 3 *N*). The plants marked "A" had been infested with *A. pomi* for 10 days, whereas the plants marked "−A" were uninfested.

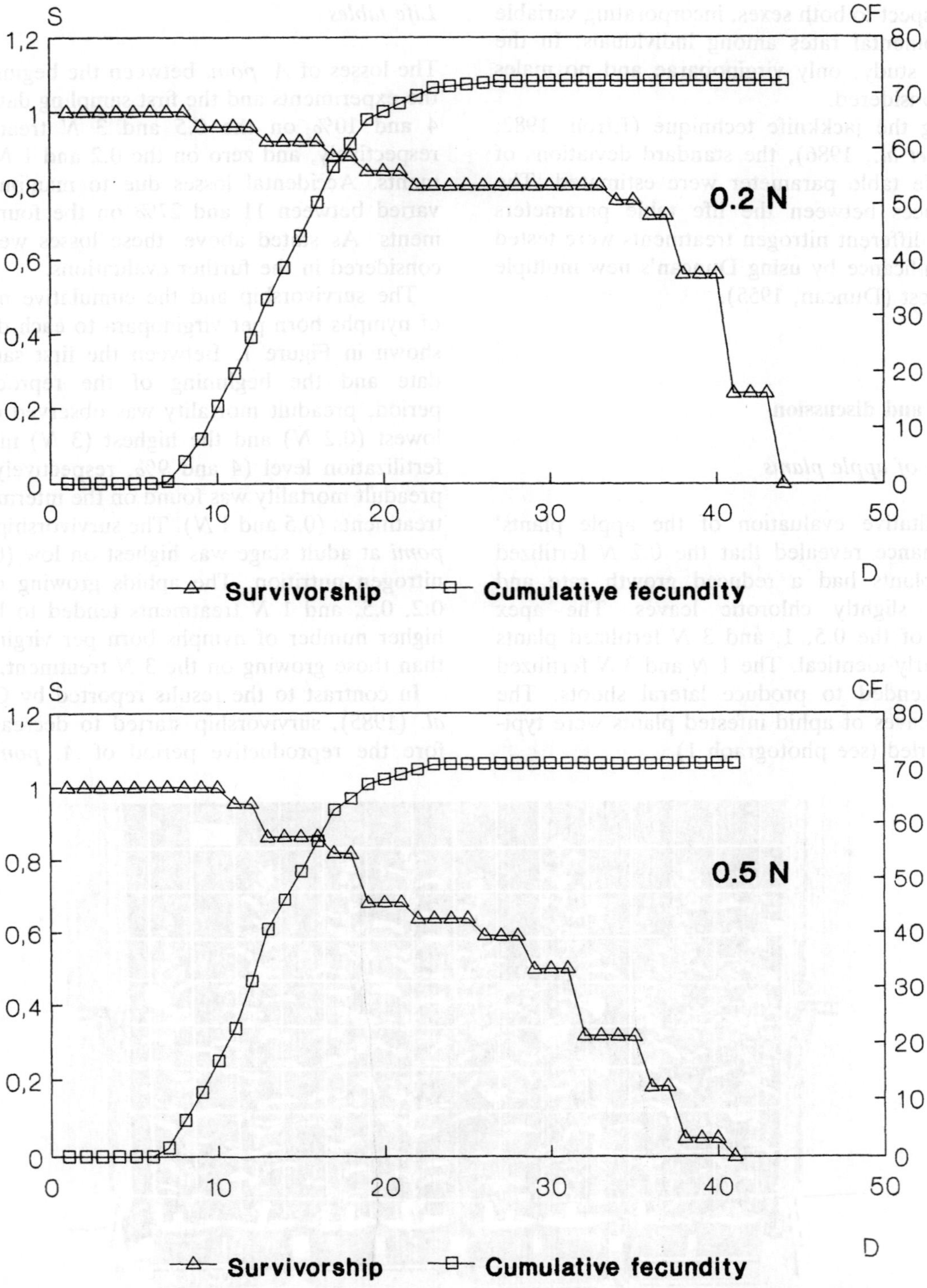

Fig. 1. Survivorship (S) and cumulative number of nymphs born per virginopara (cumulative fecundity) (CF) of *A. pomi*, growing on apple plants fertilized with nutrient solutions containing four different nitrogen levels (0.2, 0.5, 1, and 3 *N*), against aphid age (D) in days. (D = days at 20°C/16°C with a sinuous day/night temperature curve).

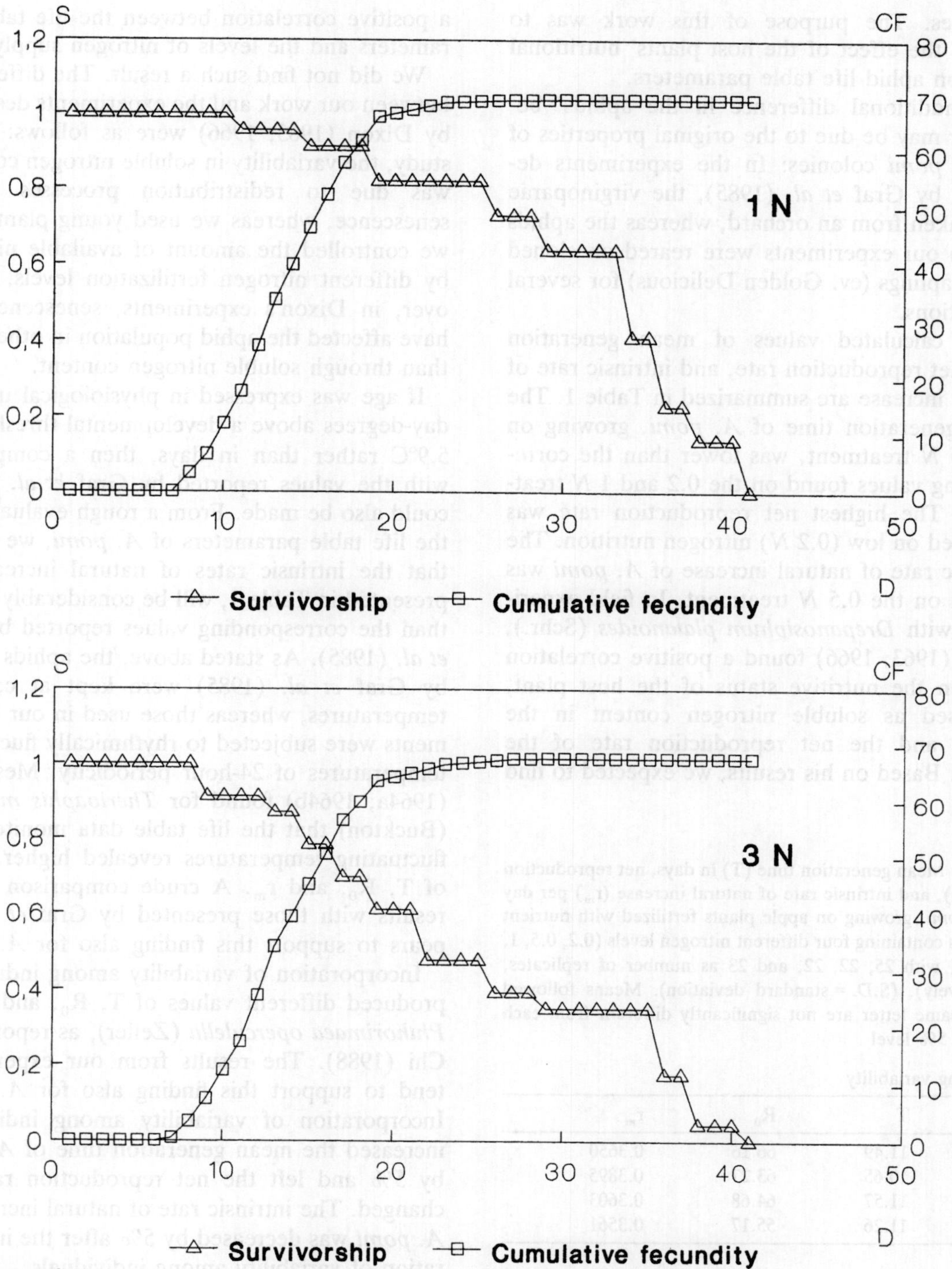

Fig. 1. continued

terminated (see Fig. 1). However, in the experiments reported by Graf *et al*. (1985), the plants were replaced weekly in order to maintain a constant food quality, and the influence of different constant temperatures was investigated. In our study, the aphids remained on the same plants throughout the experiments, and they were exposed to rhythmically fluctuating tem-

peratures. The purpose of this work was to analyse the effect of the host plants' nutritional status on aphid life table parameters.

An additional difference in the aphids' behaviour may be due to the original properties of the *A. pomi* colonies: In the experiments described by Graf *et al.* (1985), the virginoparae were taken from an orchard, whereas the aphids used in our experiments were reared on cloned apple saplings (cv. Golden Delicious) for several generations.

The calculated values of mean generation time, net reproduction rate, and intrinsic rate of natural increase are summarized in Table 1. The mean generation time of *A. pomi*, growing on the 0.5 *N* treatment, was lower than the corresponding values found on the 0.2 and 1 *N* treatments. The highest net reproduction rate was produced on low (0.2 *N*) nitrogen nutrition. The intrinsic rate of natural increase of *A. pomi* was highest on the 0.5 *N* treatment. In field experiments with *Drepanosiphum platanoides* (Schr.), Dixon (1963; 1966) found a positive correlation between the nutritive status of the host plant, expressed as soluble nitrogen content in the leaves, and the net reproduction rate of the aphids. Based on his results, we expected to find a positive correlation between the life table parameters and the levels of nitrogen supply.

Table 1. Mean generation time (T) in days, net reproduction rate (R_0), and intrinsic rate of natural increase (r_m) per day of *A. pomi*, growing on apple plants fertilized with nutrient solutions containing four different nitrogen levels (0.2, 0.5, 1, and 3*N*, with 25, 22, 22, and 23 as number of replicates, respectively). (S.D. = standard deviation). Means followed by the same letter are not significantly different from each other at 5% level

Excluding variability

N	T	R_0	r_m
0.2	11.49	66.16	0.3650
0.5	10.65	63.27	0.3895
1	11.57	64.68	0.3603
3	11.26	55.17	0.3561

Including variability

N	T ± S.D.	R_0 ± S.D.	r_m ± S.D.
0.2	12.06 ± 1.04a	66.16 ± 17.94	0.3476 ± 0.0294a
0.5	11.24 ± 0.83b	63.77 ± 16.50	0.3698 ± 0.0307b
1	12.14 ± 0.84a	64.68 ± 13.33	0.3437 ± 0.0230a
3	11.72 ± 1.25ab	55.17 ± 22.99	0.3424 ± 0.0496a

We did not find such a result. The differences between our work and the experiments described by Dixon (1963; 1966) were as follows: In his study, the variability in soluble nitrogen contents was due to redistribution processes during senescence, whereas we used young plants, and we controlled the amount of available nitrogen by different nitrogen fertilization levels. Moreover, in Dixon's experiments, senescence may have affected the aphid population in other ways than through soluble nitrogen content.

If age was expressed in physiological units of day-degrees above a developmental threshold of 5.9°C rather than in days, then a comparison with the values reported by Graf *et al.* (1985) could also be made. From a rough evaluation of the life table parameters of *A. pomi*, we expect that the intrinsic rates of natural increase, as presented in Table 1, will be considerably higher than the corresponding values reported by Graf *et al.* (1985). As stated above, the aphids reared by Graf *et al.* (1985) were kept at constant temperatures, whereas those used in our experiments were subjected to rhythmically fluctuating temperatures of 24-hour periodicity. Messenger (1964a; 1964b) found for *Therioaphis maculata* (Buckton) that the life table data monitored at fluctuating temperatures revealed higher values of T, R_0, and r_m. A crude comparison of our results with those presented by Graf *et al.* appears to support this finding also for *A. pomi*.

Incorporation of variability among individuals produced different values of T, R_0, and r_m for *Phthorimaea operculella* (Zeller), as reported by Chi (1988). The results from our experiments tend to support this finding also for *A. pomi*. Incorporation of variability among individuals increased the mean generation time of *A. pomi* by 5% and left the net reproduction rate unchanged. The intrinsic rate of natural increase of *A. pomi* was decreased by 5% after the incorporation of variability among individuals.

A statistical analysis of the values of T and r_m revealed the following significant differences: The mean generation time of *A. pomi*, growing on the 0.5 *N* treatment, was significantly lower ($P < 0.05$) than the corresponding values found on the 0.2 and 1 *N* treatments (see Table 1). The intrinsic rate of natural increase was highest ($P<0.05$) on the 0.5 *N* treatment. The differences between the other values of T and r_m were not significant at 5% level.

In conclusion, the life table analyses of *A. pomi*, growing on apple plants fertilized with nutrient solutions containing different nitrogen levels, produced different values of T, R_0, and r_m. The mean generation time of *A. pomi*, growing on the 0.5 *N* treatment, was lower than the corresponding values found on the 0.2 and 1 *N* treatments. The highest net reproduction rate was produced on low (0.2 *N*) nitrogen nutrition. Within the range of nitrogen fertilization levels investigated here, the intrinsic rate of natural increase reached a maximum on the 0.5 *N* treatment. Therefore we consider the range of nitrogen treatments evaluated here appropriate for further experimentation.

Life table analyses alone cannot reveal the causes underlying the differences between the values of T, R_0, and r_m. A more complete investigation of the energy flow in the apple plant-aphid ecosystem, as affected by nitrogen fertilization, is necessary.

References

Birch L C 1948 The intrinsic rate of natural increase of an insect population. J. Anim. Ecol. 17, 15–26.

Broome O C and Zimmerman R H 1984 Culture of shoot meristems: Fruit plants. *In* Cell Culture and Somatic Cell Genetics in Plants, Vol. 1. Ed. I K Vasil. pp 118–119. Academic Press, London.

Chi H 1988 Life-table analysis incorporating both sexes and variable development rates among individuals. Environ. Entomol. 17, 26–34.

Chi H and Liu H 1985 Two new methods for the study of insect population ecology. Acad. Sin., Bull. Inst. Zool. 24, 225–240.

Dixon A F G 1963 Reproductive activity of the sycamore aphid, *Drepanosiphum platanoides* (Schr.) (Hemiptera, Aphididae). J. Anim. Ecol. 32, 33–48.

Dixon A F G 1966 The effect of population density and nutritive status of the host on the summer reproductive activity of the sycamore aphid, *Drepanosiphum platanoides* (Schr.). J. Anim. Ecol. 35, 105–112.

Dixon A F G 1975 Aphids and translocation. *In* Encyclopedia of Plant Physiology, Vol. 1. Eds. M H Zimmermann and J A Milburn. pp 154–170. Springer-Verlag, Berlin.

Duncan D B 1955 Multiple range and multiple *F* tests. Biometrics 11, 1–42.

Efron B 1982 The Jackknife, the Bootstrap and Other Resampling Plans. CBMS-NSF Regional Conference Series in Applied Mathematics. Society for Industrial and Applied Mathematics, Philadelphia, PA.

Graf B, Baumgärtner J and Delucchi V 1985 Life table statistics of three apple aphids, *Dysaphis plantaginea*, *Rhopalosiphum insertum*, and *Aphis pomi* (Homoptera, Aphididae), at constant temperatures. Z. ang. Ent. 99, 285–294.

Hewitt E J and Smith T A 1975 Plant Mineral Nutrition. The English Universities Press, London, 298 p.

Hoagland D R and Arnon D I 1938 The water-culture method for growing plants without soil. Agric. Exp. Sta. Calif. Circ. 347.

Lindemann C 1948 Beitrag zur Ernährungsphysiologie der Blattläuse. Z. Vergl. Physiol. 31, 112–133.

Llewellyn M 1982 The energy economy of fluid-feeding herbivorous insects. *In* Proc. 5th Int. Symp. Insect-Plant Relationships, Wageningen, 1982. Eds. J H Visser and A K Minks. pp 243–251. Pudoc, Wageningen.

Llewellyn M and Hargreaves C E M 1984 The biology and energetics of the potato aphid *Macrosiphum euphorbiae*, living in galls of the apple aphids *Dysaphis devecta* and *Aphis pomi*. Entomol. Exp. Appl. 35, 147–158.

Llewellyn M and Leckstein P M 1978 A comparison of energy budgets and growth efficiency for *Aphis fabae* Scop. reared on synthetic diets with aphids reared on broad beans. Entomol. Exp. Appl. 23, 66–71.

Messenger P S 1964a Use of life tables in a bioclimatic study of an experimental aphid-braconid wasp host-parasite system. Ecology 45, 119–131.

Messenger P S 1964b The influence of rhythmically fluctuating temperatures on the development and reproduction of the spotted alfalfa aphid, *Therioaphis maculata*. J. Econ. Entomol. 57, 71–76.

Meyer J S, Ingersoll C G, McDonald L L and Boyce M S 1986 Estimating uncertainty in population growth rates: Jackknife vs. bootstrap techniques. Ecology 67, 1156–1166.

Mittler T E 1958 Studies on the feeding and nutrition of *Tuberolachnus salignus* (Gmelin) (Homoptera, Aphididae). III. The nitrogen economy. J. Exp. Biol. 35, 626–638.

Randolph P A, Randolph J C and Barlow C A 1975 Age-specific energetics of the pea aphid, *Acyrthosiphon pisum*. Ecology 56, 359–369.

Southwood T R E 1987 Ecological Methods. Chapman and Hall, London, 548 p.

Waldbauer G P 1968 The consumption and utilization of food by insects. Adv. Insect Physiol. 5, 229–288.

Wermelinger B 1985 Einfluss des Ernährungszustandes der Wirtspflanze auf den Befall durch die Gemeine Spinnmilbe (*Tetranychus urticae* Koch). Diss. ETH Nr. 7821, Zürich, 99 p.

White N D G and Sinha R N 1981 Energy budget for *Oryzaephilus surinamensis* (Coleoptera: Cucujidae) feeding on rolled oats. Environ. Entomol. 10, 320–326.

Wiegert R G and Petersen C E 1983 Energy transfer in insects. Annu. Rev. Entomol. 28, 455–486.

M. L. van Beusichem (Ed.), *Plant nutrition – physiology and applications*, 633–637.

PLSO IPNC226

Influence of potassium nutrition on concentrations of water soluble carbohydrates, potassium, calcium, and magnesium and the osmotic potential in sap extracted from wheat (*Triticum aestivum*) ears during preanthesis development

M. KRUMM, V. MOAZAMI and P. MARTIN[1]
Institute of Plant Nutrition, University of Hohenheim, P.O. Box 70 05 62, D-7000 Stuttgart 70, FRG.
[1]*Corresponding author*

Key words: calcium, ear development, magnesium, potassium, tissue press sap, *Triticum aestivum* L., wheat

Abstract

For floret development and final grain number ear elongation in wheat (10–17d before anthesis) is the most sensitive period to adverse growth conditions. Sugars are supposedly the main factor determining floret numbers and grain set, and play next to potassium a significant role in establishing turgor pressure in young tissues. In view of this osmotic function, the influence of K on the concentration of water soluble carbohydrates (WSC, total of sucrose and reducing sugars), Ca and Mg was investigated in pot experiments. Further, the osmotic potential of sap extracted from young ears was determined and compared to sap extracted from the leaf blade. Plants supplied with low amounts of K (moderate K-deficiency) had a considerably lower K concentration in the press sap of the flag leaf and the ear than plants well supplied with K. Concentrations of WSC, Ca and Mg were higher in press sap of the flag leaf in K deficient plants than in plants adequately supplied with K. This indicates a substitution of K in its osmotic role. In press sap from ears, however, WSC, Ca and Mg were not influenced by the K application. Therefore, substances other than those measured must have been responsible for the osmoregulation in the young ear. WSC and the osmotic potential increased (more negative) independently on K supply during ear elongation, while K, Ca and Mg concentrations decreased. Whereas grain number was not influenced by the treatment, single grain weight at maturity was reduced by low K availability in the soil.

Introduction

Grain number and yield of wheat is greatly influenced by environmental conditions during preanthesis ear development. There is evidence that between 10 to 17 days before anthesis ears react more sensitively to adverse environmental conditions, like shading, water stress or high temperature, than in other growth stages (Fischer and Stockman, 1980; Saini and Aspinall, 1982). This most sensitive stage includes the phase of rapid ear elongation, when the ear is still enclosed by leaf sheaths.

K is a mineral nutrient of specific importance in developing tissue, where it is involved in cell wall acidification, cell wall extension and osmotic regulation (for review see Lindhauer, 1989). In mature tissue the osmotic function of K can alternatively be performed by low molecular weight sugars or by other ions, particularly Ca and Mg (Leigh *et al.*, 1986; Lindhauer, 1987; 1989). This, however, only occurs if K supply is suboptimal. It is not known whether young expanding ears react in a similar way on insufficient K supply. Many experiments, including those in which photosynthesis was reduced (Fischer and Stockman, 1989), point to sugars as the main factor determining floret numbers and final grain set. In the present study amounts of K, Ca, Mg and water soluble carbohydrates (WSC) were quantified in ears during the phase of rapid ear elongation under conditions of moderate K deficiency. The data are compared with those of the flag leaf. In addition, the osmotic potential was measured. When different growth stages of a rapidly growing plant organ, or developing and developed plant organs are compared, data on concentration related to dry matter are of limited value. Therefore, tissue sap was extracted and data are given as concentrations in press sap.

Materials and methods

Spring wheat (*Triticum aestivum* L., cv. Schirokko) was cultivated in pots containing a mixture of 2 kg clayey loam soil (11 mg exchangeable K_2O (CAL) 100 g^{-1} soil), 1 kg quartz sand and 0.5 kg peat. The completely randomized design of the experiment involved two treatments, K_0 and K_1. Each pot received a basal dressing of 0.4 g N, 1.2 g Ca, 1.6 g P and 0.5 g Mg as NH_4NO_3, $CaHPO_4$ and $MgHPO_4 \cdot 3H_2O$. Two additional applications of N (0.4 g each) and K (0.4 g each as K_2SO_4, only for the K_1-treatment) were given at the end of tillering and during inflorescence emergence. Pots were watered to weight daily (70% of max. water holding capacity). Plants were grown under constant environmental conditions (16 h day, 18°C, irradiance at the level of the flag leaf ligule 35 W m^{-2} by cool white fluorescent tubes). Under the given temperature regime light conditions proved to be adequate in regard to concentrations and absolute amounts of WSC when compared with plants grown outdoors at somewhat higher temperatures. Tillers were constantly removed, to leave the main culm. Ear samples were taken at four growth stages and for characterization the EUCARPIA decimal code was used: EC 39 (flag leaf ligule just visible), EC 43 (boots just visibly swollen), EC 49 (first awns visible), EC 92 (caryopsis hard). Each harvest included four replicate pots with ten plants. For the determination of grain number and single grain weight 40 ears per treatment were used. The data were statistically analysed using analyses of variance. For the measurement of solute concentrations and osmotic potential, cell sap was extracted by means of a hydraulic press after freezing and thawing. The osmotic potential of the squeezed cell sap was determined by a cryoscopic technique. K, Ca and Mg were measured by atomic absorption spectroscopy. For the determination of WSC, cell sap was incubated with yeast invertase (Sigma) and the resulting sugars determined by reaction with p-hydroxybenzoic acid hydrazide (Blakeney and Mutton, 1980). By using this method sucrose and reducing sugars are measured.

Results

Ear length and grain number

In the very first stages of development ear growth is very slow. Rapid elongation starts at stage EC 39 (flag leaf ligule just visible) and continues for about 6 days with a daily increase of about 2 cm. Figure 1 shows that different K-supply did not influence ear length but in K_0 plants (moderate K deficiency) the onset of rapid elongation was delayed. However, dry weights of the ears did not differ at the corresponding development stages. Grain number per ear was not influenced by the different K supply (Fig. 2). Single grain weight, however, was reduced.

Sap extracted from flag leaves

In leaves of K_0-plants the concentration of K in extracted sap was less than one half of that of K_1-plants (Fig. 3). However, concentrations of Ca and Mg increased considerably in these plants. Concentrations of WSC were also signifi-

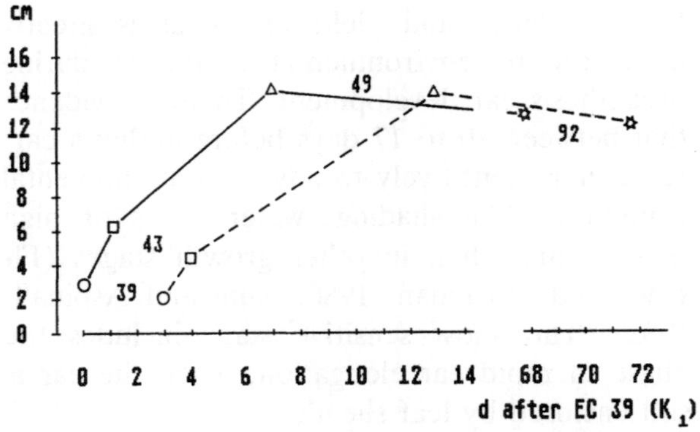

Fig. 1. Length of ears at development stages EC 39, 43, 49 and 92. K_0, interrupted line; K_1, solid line.

cantly increased in K_0 compared to K_1 plants (Fig. 4) (see also Lindhauer, 1987; Pitman *et al.*, 1971). The osmotic potential in K_0 plants was slightly increased (more negative), possibly due to an altered water status of K_0 plants (Fig. 4).

Sap extracted from ears

Sap extracted from ears of K_0 plants contained considerably less K than sap of the corresponding K_1 plants (Fig. 5). However, other than in

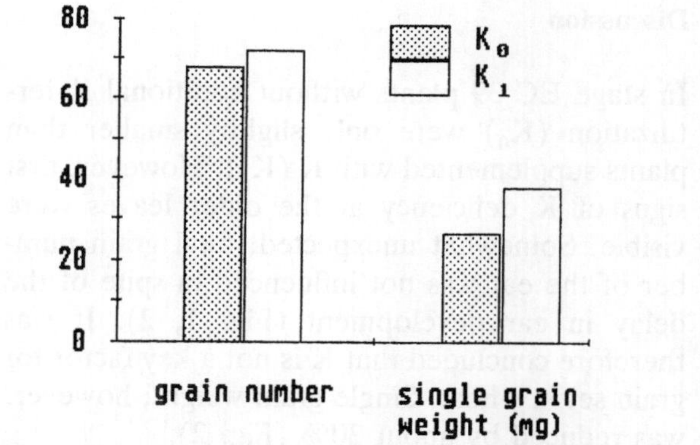

Fig. 2. Grain number and single grain weight. Differences in single grain weight between K_0 and K_1 plants are significant at $P<0{,}001$.

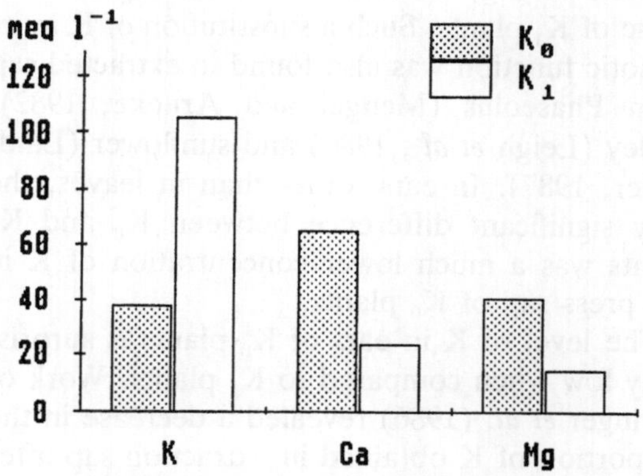

Fig. 3. Concentrations of K, Ca and Mg in sap extracted from the flag leaf blade (EC 49). All differences between K_0 and K_1 plants are significant at $P<0{,}001$.

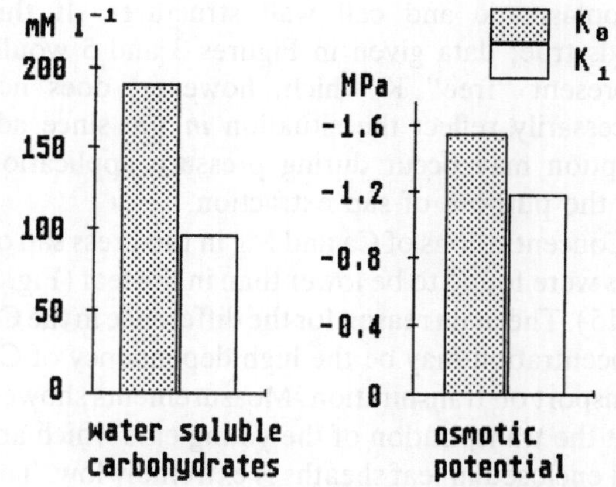

Fig. 4. Concentration of WSC and the osmotic potential in extracted sap from the flag leaf blade (EC 49). Differences between K_0 and K_1 plants are significant at $P<0{,}001$.

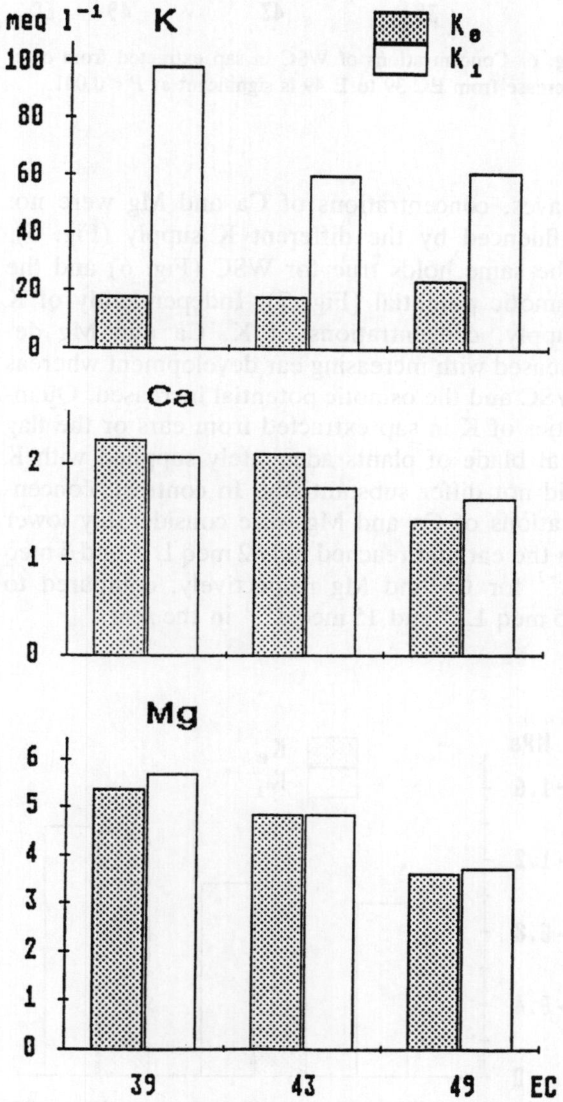

Fig. 5. Concentration of K, Ca and Mg in extracted sap from ears. Differences in concentrations of K between K_0 and K_1 plants and decreases in concentrations of $K(K_1)$, Ca and Mg from EC 39 to 49 are significant at $P<0{,}001$.

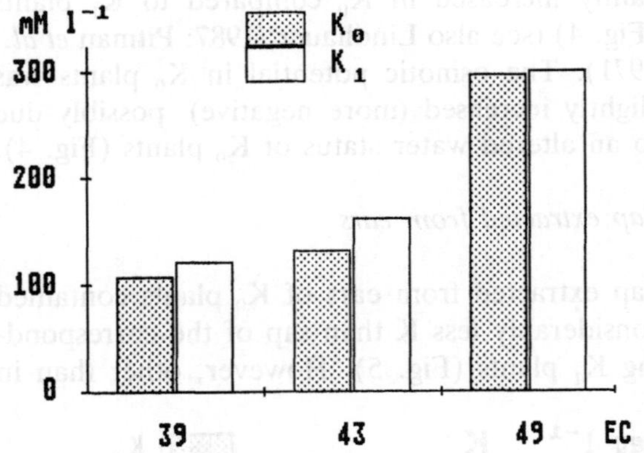

Fig. 6. Concentrations of WSC in sap extracted from ears. Increase from EC 39 to E 49 is significant at $P < 0{,}001$.

leaves, concentrations of Ca and Mg were not influenced by the different K supply (Fig. 5). The same holds true for WSC (Fig. 6) and the osmotic potential (Fig. 7). Independently of K supply, concentrations of K, Ca and Mg decreased with increasing ear development whereas WSC and the osmotic potential increased. Quantities of K in sap extracted from ears or the flag leaf blade of plants adequately supplied with K did not differ substantially. In contrast, concentrations of Ca and Mg were considerably lower in the ear and reached only 2 meq L^{-1} and 5 meq L^{-1} for Ca and Mg respectively, compared to 25 meq L^{-1} and 15 meq L^{-1} in the leaf.

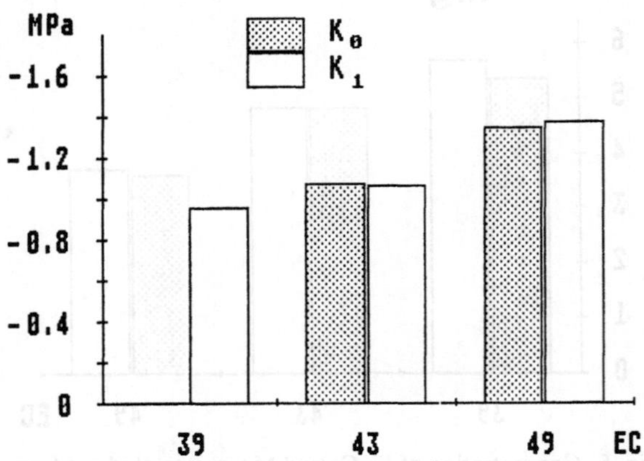

Fig. 7. Osmotic potential of sap extracted from ears. Increase from EC 39 to EC 49 is significant at $P < 0{,}001$.

Discussion

In stage EC 39 plants without additional K fertilization (K_0) were only slightly smaller than plants supplemented with K (K_1). However, first signs of K deficiency at the older leaves were visible. Somewhat unexpected, final grain number of the ear was not influenced in spite of the delay in ear development (Fig. 1, 2). It was therefore concluded that K is not a key factor for grain set in wheat. Single grain weight, however, was reduced by about 30% (Fig. 2).

Analyses of extracted sap from the flag leaf revealed distinct differences in K concentrations between K_0 and K_1 plants. Whereas the K-concentration in K_0 plants was lower, concentrations of Mg, Ca and WSC were higher than those of K_1 plants. Such a substitution of K in its osmotic function was also found in extracted sap from Phaseolus (Mengel and Arneke, 1982), barley (Leigh *et al.*, 1986) and sunflower (Lindhauer, 1987). In ears, other than in leaves, the only significant difference between K_0 and K_1 plants was a much lower concentration of K in the press sap of K_0 plants.

The level of K in ears of K_0 plants is surprisingly low when compared to K_1 plants. Work of Beringer *et al.* (1986) revealed a decrease in the proportion of K obtained in extraction sap after freezing and thawing relative to the calculated K in tissue water with decreasing K application. Such an effect is understandable if it is assumed that a certain amount of K is withheld in the tissue by adsorption at cation exchange sites at cytoplasmatic and cell wall structures. If this holds true, data given in Figures 3 and 5 would represent "free" K which, however, does not necessarily reflect the situation *in situ* since adsorption may occur during pressure application for the purpose of sap extraction.

Concentrations of Ca and Mg in the press sap of ears were found to be lower than in the leaf (Fig. 3 and 5). The main reason for the difference in the Ca concentration may be the high dependency of Ca transport on transpiration. Measurements showed that the transpiration of the young ears which are still enclosed in leaf sheaths is extremely low. The reason for the (smaller) difference in Mg remains to be elucidated.

Water soluble sugars in cells are components of carbohydrate metabolisms but also contribute to the osmotic pressure in the cell. In press sap of the flag leaf of K_0 plants the concentration of WSC was about twice as much as in K_1 plants. In the young ears, however, there was no significant difference between K_0 and K_1 plants. In view of the possible importance of sugars for turgor pressure in growing tissue it is a remarkable result that the concentrations of WSC in press sap of ears increased during the experimental time from about 100 (stage EC 39) to about 300 m*M* (EC 49) (Fig. 6). At the same time the osmotic pressure increased significantly (Fig. 7). However, concentrations of K, Ca and Mg decreased and seem to lag behind the rapid growth of the ear.

The results indicate that osmoregulation as a response to suboptimal K concentrations in plant tissue occurs in different ways in the flag leaf blade and the young expanding ear. Whereas the known pattern of K replacement was found in leaves (*i.e.* an increase in WSC, Ca and Mg), none of the investigated components of tissue sap changed in the young ear. Nevertheless, the osmotic potential was maintained at the same level as in plants well supplied with K. In spite of the low K concentration in sap extracted from ears of plants suffering from K deficiency, conditions were maintained in the ears of such plants which permitted the same number of grains to develop as in ears of plants well supplied with K (Fig. 2). Possibly fructans recently found in our laboratory in relatively high concentrations in young ears play a regulatory role for the control of osmotic potential during extension growth. At stage EC 39 the proportion of fructans relative to total soluble sugars was as high as 73% (weight basis), but subsequently decreased to 33% at stage EC 49 and an even lower level in later stages.

The results presented here highlight the complex situation in rapidly expanding tissue in which high amounts of sugars are needed for cell structures and, in interaction with K, also for the establishment of turgor pressure. Further research is needed to understand the role of sugars in the regulation of floret development in young ears with regard to later grain set and yield.

References

Beringer H, Koch K and Lindhauer M G 1986 Sucrose accumulation and osmotic potentials in sugar beet at increasing levels of potassium nutrition. J. Sci. Food Agric. 37, 211–218.

Blakeney A B and Mutton L L 1980 A simple colorimetric method for the determination of sugars in fruit and vegetables. J. Sci. Food Agric. 31, 889–897.

Fischer R A and Stockman Y M 1980 Kernel number per spike in wheat (*Triticum aestivum* L.): Response to preanthesis shading. Aust. J. Plant Physiol. 7, 169–180.

Leigh R A, Chater M, Storey R and Johnston A E 1986 Accumulation and subcellular distribution of cations in relation to the growth of potassium deficient barley. Plant Cell Environ. 9, 595–604.

Lindhauer M G 1987 Solute concentrations in well-watered and water-stressed sunflower plants differing in K nutrition. J. Plant Nutr. 10, 1965–1973.

Lindhauer M G 1989 The role of K^+ in cell extension, growth and storage of assimilates. *In* Methods of K-Research in Plants. Ed. by International Potash Institute, Bern, Switzerland, 21st Colloquium of the Int. Potash Inst. Louvain-la-Neuve, Belgium 1989, pp 161–187.

Mengel K and Arneke W W 1982 Effect of potassium on the water potential, the pressure potential, the osmotic potential and cell elongation in leaves of *Phaseolus vulgaris*. Physiol. Plant. 54, 402–408.

Pitman M G, Mowat J and Nair H 1971 Interactions of processes for accumulation of salt and sugar in barley plants. Aust. J. Biol. Sci. 24, 619–631.

Saini H S and Aspinall D 1982 Sterility in wheat (*Triticum aestivum* L.) induced by water deficit or high temperature: Possible mediation by abscisic acid. Aust. J. Plant Physiol. 9, 529–537.

M. L. van Beusichem (Ed.), *Plant nutrition – physiology and applications*, 639–642.

PLSO IPNC221

Source: sink relationships in potato (*Solanum tuberosum*) as influenced by potassium chloride or potassium sulphate nutrition

H. BERINGER, K. KOCH and M.G. LINDHAUER
Agricultural Research Station Büntehof, Bünteweg 8, D-3000 Hannover 71, FRG

Key words: cell sap, chloride, fertilizer, potassium, potato, *Solanum tuberosum* L., solute potential, source: sink, sulphate, tuber

Abstract

In pot experiments with *Solanum tuberosum* L. (cv Saturna) the application of KCl as compared to K_2SO_4 delayed tuber development. The solute composition of leaves of the KCl treated plants was significantly lower in K^+ and NO_3^-, but higher in Mg^{2+}, Ca^{2+} and Cl^-. Since the solute potential in the KCl treated plants was more negative and associated with a higher water content, a higher turgor pressure can be assumed. This could explain the enhanced shoot growth observed with KCl. Application of K_2SO_4, on the other hand, accelerated the development of tubers. This might result from a less competitive shoot sink in K_2SO_4 treated plants and a stimulated phloem loading and translocation of assimilates by higher concentrations of leaf-K.

Introduction

Yield and quality of potato tubers are affected by many factors. A moderate K supply, preferentially as K_2SO_4, is recommended if tubers with high starch and high dry matter percentages are to be produced. Generous K applications should be applied if the risk of blue spot disease, which results partly as a consequence of mechanical stress of tubers during harvesting and handling, is to be minimized (Anonymous, 1984). High water content of tubers on the one hand reduces the starch concentration in the tuber by dilution, but – on the other hand – probably increases turgor pressure, so that the tuber is better able to resist mechanical stress.

Many field experiments have compared the effects of KCl versus K_2SO_4 on tuber quality. The oldest ones were published by Maercker (1892) who as early as then reported that tubers of plants treated with KCl contain more moisture and are at a more juvenile physiological stage at harvest than tubers from plants receiving K_2SO_4. Apart from enzymatic differences osmotic events may therefore be involved too in the regulation of tuber metabolism. Recent *in vitro* experiments by Oparka and Wright (1988a, b) demonstrated an osmotic optimum for starch synthesis. The decline of starch accumulation at K concentrations >2–2.5% K in tuber dry weight might be explicable by such findings (Marschner and Krauss, 1980).

The favourable effect on tuber starch content with K_2SO_4 as compared to KCl has also been explained by higher translocation of assimilates into the tubers (Haeder, 1976). This, however, is probably less the cause rather than the consequence of a higher tuber : shoot ratio in the K_2SO_4 treatments. In order to study the contribution of the type of K fertilizer to source-sink relationships in more detail pot experiments were carried out during two seasons with harvests taken at weekly intervals during tuber growth.

Materials and methods

Solanum tuberosum (cv. Saturna) was planted in 12 kg of a sandy soil supplied uniformly with adequate macro- and micronutrients. Only the 2.5 g K/pot (1987) and 5.0 g K/pot (1988) were varied as KCl and K_2SO_4, respectively.

After flowering 4 replicate pots/treatment were harvested at weekly intervals and plants were separated into tubers, stolons and shoots. After fresh weight determinations had been made the tubers were chopped and aliquots either frozen at −18°C or dried at +60°C. Shoots were also dried at +60°C. At some harvest dates leaves were also frozen at −18°C for determination of solute potential and composition.

Cations in plant dry matter or in cell sap were analyzed by atomic absorption spectrometry. Inorganic anions in cell sap were determined by ion chromatography and organic anions as well as sugars by HPLC (Beringer *et al.*, 1986).

Results

In both seasons application of potassium sulfate resulted in significantly less shoot growth and increased tuber yields as compared to KCl (Fig. 1). After the 8th harvest, however, this advantage in tuber development was matched by KCl. Nevertheless the harvest index expressed by the tuber : shoot ratio and, in general, also the percent tuber dry matter were higher in the K_2SO_4 treated plants until final harvest (Fig. 2).

The stimulated shoot growth in the KCl treatments is very likely a consequence of Cl^- uptake and of preferential Cl^- compartmentation in the shoot. Leaves taken at three dates show significantly lower osmotic potentials and higher water contents (Table 1) suggesting the slightly higher turgor potential to be the reason for the higher shoot growth. In the leaf cell sap of the KCl treatment the anion fraction is mainly inorganic, in the K_2SO_4 plants organic anions dominate, whereas for both treatments the sums of inorganic and organic anions are closely balanced by cations (Table 2). The data in Table 3, which were confined at the other two harvests, are most interesting in relation to source/sink regulation. Application of K_2SO_4 resulted in a significantly higher concentration of K^+, malate and hexoses in leaf cell sap.

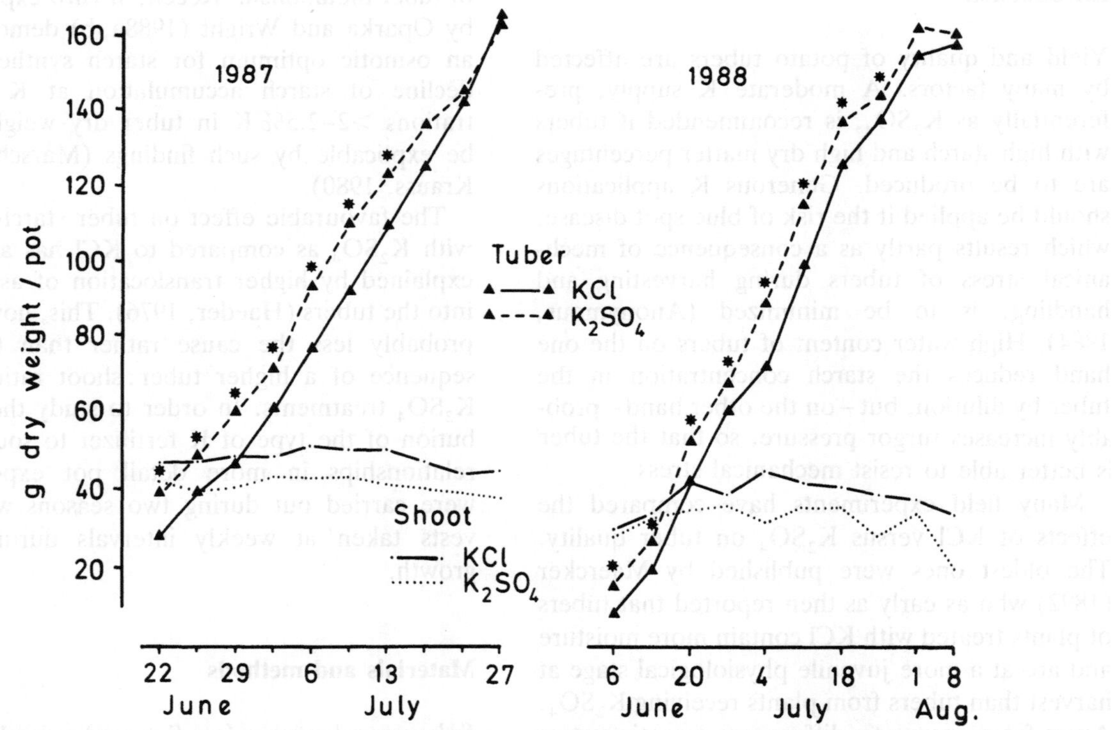

Fig. 1. Shoot and tuber development of potato fertilized with KCl and K_2SO_4, respectively (1987: 2.5 g K/pot; 1988: 5.0 g K/pot) * significant at LSD 5%.

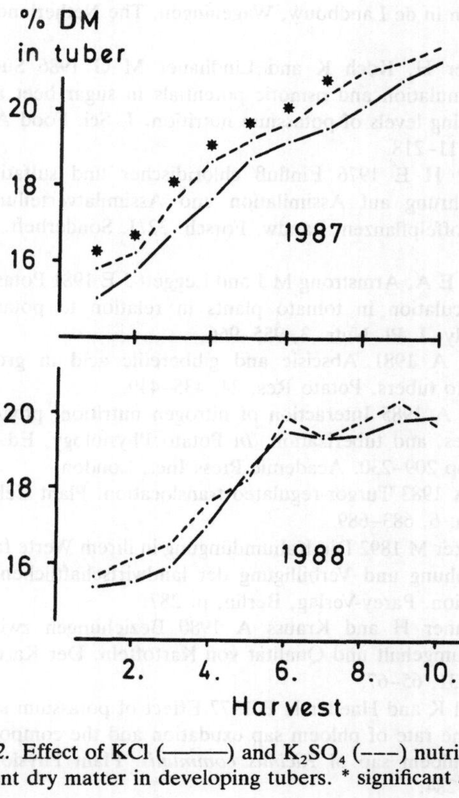

Fig. 2. Effect of KCl (———) and K_2SO_4 (----) nutrition on percent dry matter in developing tubers. * significant at LSD 5%.

Discussion

Less shoot growth, earlier tuber development and a slightly higher tuber dry matter in plants supplied with K_2SO_4 suggest an ontogenetic effect of the form of K fertilizer mediated by the water status of the crop. Cell sap in leaves of K_2SO_4 treated plants had a significantly higher osmotic potential and contained less water. This suggests a lower turgor potential, less cell extension in shoot tissue and consequently less competition for assimilates between shoot- and tuber-sinks. Whether osmotic or turgor potential are metabolic messengers for the earlier start of tuber growth in K_2SO_4 plants must still remain an open question, because less shoot growth could also be associated with a lower gibberellic/abscisic acid ratio known to favour tuber initiation (Krauss, 1981; 1985).

Interesting from the viewpoint of source-sink regulation is also the much higher concentration of K^+ in cell sap of the leaves of K_2SO_4 as compared to KCl plants. Whereas Cl^- is obviously compartmented in the vacuoles of leaf cells and neutralized there by Ca^{2+} and Mg^{2+} K^+ is most likely balanced by malate. Due to the high

Table 1. Osmotic potentials of cell sap and water content of leaves in relation to plant development and K nutrition (means of 4 replicates)

	Fertil.	04/07/88	25/07/88	01/08/88
Ψ_s (MPa)	KCl	−0.96	−1.12	−1.24
	K_2SO_4	−0.84[a]	−0.84[b]	−0.95[b]
g H_2O/g dry wt.	KCl	8.9	9.5	7.8
	K_2SO_4	8.6	7.6	6.2[a]

[a,b] Significant at LSD 5% and 1%, respectively.

Table 2. Effect of K nutrition and developmental stage on anions and cations in leaf cell sap (me/L)

	Fertil.	04/07/88	25/07/88	01/08/88
Inorganic	KCl	227	289	316
anions	K_2SO_4	149[b]	182[b]	219[b]
Organic	KCl	75	120	149
anions	K_2SO_4	203[b]	245[b]	304[b]
Cations	KCl	300	424	459
	K_2SO_4	327[a]	417	498

[a,b] Significant at LSD 5% and 1%, respectively.

Table 3. Concentration of selected cations and anions (me/L) in leaf cell sap (harvest date 04/07/88)

	KCl	K_2SO_4
K	94.8	164.6[b]
Mg	76.3	67.4
Ca	128.2	94.0[a]
Cl	202.2	18.9[b]
SO_4	18.3	115.8[b]
NO_3	4.6	14.0
Malate	37.1	169.7[b]
Citrate	37.9	32.5
Fructose	10.1	17.9[a]
Glucose	7.5	13.9[a]
Sucrose	2.5	0.5[a]

[a,b] Significant at LSD 5% and 1%, respectively.

mobility of K malate in plants (Kirkby *et al.*, 1981; Touraine *et al.*, 1988), it seems that plants supplied with K_2SO_4 contain a larger pool of recirculating K^+. This may favour phloem loading and phloem transport (at all three sampling dates sucrose concentration in leaf sap was significantly less in K_2SO_4 treatment) (Lang, 1983; Mengel and Haeder, 1977). Further studies on the unloading and accumulation of assimilates in the tubers are in progress. They reveal differences in Cl and SO_4 contents as well as in solute potential (Ricke, pers. commun.).

References

Anonymous 1984 Adviesbasis voor bemesting van landbouwgronden. Consultentschap voor Bodemaangelegenheden in de Landbouw, Wageningen, The Netherlands, 41 p.

Beringer H, Koch K and Lindhauer M G 1986 Sucrose accumulation and osmotic potentials in sugar beet at increasing levels of potassium nutrition. J. Sci. Food Agric. 37, 211–218.

Haeder H E 1976 Einfluß chloridischer und sulfatischer Ernährung auf Assimilation und Assimilatverteilung in Kartoffelpflanzen. Landw. Forsch. 32/I. Sonderheft, 122–131.

Kirkby E A, Armstrong M J and Leggett J E 1981 Potassium recirculation in tomato plants in relation to potassium supply. J. Pl. Nutr. 3, 955–966.

Krauss A 1981 Abscisic and gibberellic acid in growing potato tubers. Potato Res. 24, 435–439.

Krauss A 1985 Interaction of nitrogen nutrition, phytohormones, and tuberization. *In* Potato Physiology, Ed. P H Li. pp 209–230. Academic Press Inc., London.

Lang A 1983 Turgor-regulated translocation. Plant Cell Environ. 6, 683–689.

Maercker M 1892 Die Kaliumdüngung in ihrem Werte für die Erhöhung und Verbilligung der landwirtschaftlichen Produktion. Parey-Verlag, Berlin, p. 287.

Marschner H and Krauss A 1980 Beziehungen zwischen Kaliumgehalt und Qualität von Kartoffeln. Der Kartoffelbau 31, 65–67.

Mengel K and Haeder H E 1977 Effect of potassium supply on the rate of phloem sap exudation and the composition of phloem sap of *Ricinus communis*. Plant Physiol. 59, 282–284.

Oparka K J and Wright K M 1988a Osmotic regulation of starch synthesis in potato tubers? Planta 174, 123–126.

Oparka K J and Wright K M 1988b Influence of cell turgor on sucrose partitioning in potato tuber storage tissue. Planta 175, 520–526.

Touraine B, Grignon N and Grignon C 1988 Charge balance in NO_3^--fed soybean. Estimation of K^+ and carboxylate recirculation. Plant Physiol. 88, 605–612.

M. L. van Beusichem (Ed.), *Plant nutrition – physiology and applications*, 643–647.

PLSO IPNC424

Starch synthesis in potato (*Solanum tuberosum*) tubers: Activity of selected enzymes in dependence of potassium content in storage tissue

M.G. LINDHAUER and M.A.R. DE FEKETE
Agricultural Research Station Büntehof, Bünteweg 8, D-3000 Hannover 71, FRG, and Botany Department, T.H. Darmstadt, Schnittspahnstr. 3–5, D-6100 Darmstadt, FRG

Key words: enzyme activity, potassium nutrition, potato, *Solanum tuberosum* L., starch synthesis

Abstract

Starch synthesis in potato tubers grown at varied K nutrition (0.1 (K_1), 0.25 (K_2) and 1.0 mmol K L^{-1} nutrient solution (K_3) was investigated with particular regard to the activity of selected enzymes (sucrose synthase, UDP-D-glucose pyrophosphatase, starch phosphorylase, amylases) in dependence on tuber K content. Allocation of K to the tubers was nearly the same in all treatments. The activity of enzymes related to tuber K content did not differ significantly. Starch and K content of tubers increased with progressing age, whereas a decrease was observed in growth rate, starch synthesis per day and K uptake per day. Positive correlations between the rates of K uptake, starch production and growth indicate that the dynamic phase of K supply to the tubers is of greater importance for starch synthesizing processes than the influence of total K content.

Introduction

Starch content of the tubers is one of the main selection criteria in potato breeding for industrial starch production. As far as K nutrition of potato plants is concerned, its significance for starch synthesis and starch accumulation within the tubers is liable to controversion. According to Nitsos and Evans (1969) and Hawker *et al.* (1979) starch synthesizing enzymes have a specific requirement for K ions. Experience shows that for high starch potato tubers 1.8% K in tuber dry matter seems to be optimal. K contents exceeding this level were still able to increase tuber yield per plant (Forster and Beringer, 1983), but starch content per tuber dry weight was slightly decreased. The physiological reason for this depression in starch content is unknown. The present experiment was therefore intended to investigate starch synthesis in potato tubers at varied K nutrition, with particular regard to the activity of selected enzymes in dependence on tuber K content.

Material and methods

Plant growth

Potato plants (*Solanum tuberosum* L. cv Saturna) were grown in the greenhouse in nutrient solution of the following composition: 2 m*M* $MgSO_4$, 2 m*M* $Ca(NO_3)_2$, 1 m*M* $CaCl_2$, 1 m*M* NaH_2PO_4, and micronutrients. Only K, supplied as K_2SO_4, differed among the treatments: 0.1 m*M* (K_1), 0.25 m*M* (K_2), and 1.0 m*M* K (K_3), respectively. The nutrient solution was replaced once a week. The design of the vessels containing the nutrient solution allowed to separate the stolon-bearing section of the stem from the roots, so that the tubers developed on a sieve above the nutrient solution, while the roots grew through the sieve into the solution. The tubers on the sieve were in slight contact with the nutrient solution to ensure sufficient Ca supply. Free access to the tuber compartment permitted daily observation to assess the exact date of tuber initiation and the growth rate of individual tubers.

Sample preparation

Individual tubers of different age (K_1: N = 21; K_2 and K_3: N = 18) were harvested and cut longitudinally into halves. One half was used for the determination of water and K content. A slice of 1 g representing all tissues of the tuber (cortex, medulla) was taken from the square cut in the central part of the other half.

Analytical methods

For the preparation of enzymes 1 g of potato tuber tissue was homogenized in a Potter-Elvehjem glass homogenizer with 0.1 g Polyclar AT (Serva, Heidelberg) in 5 mL of 0.03 *M* glycylglycine buffer (pH 7.6) containing 1.2 m*M* EDTA, 0.02 *M* mercaptoethanol and 40% by vol. glycerol. The whole homogenate was frozen until the activity of enzymes was measured. Sucrose synthase activity was measured by the method of Cardini *et al.* (1955). The assays of starch phosphorylase, amylases and UDP-D-glucose pyrophosphatase activity were performed as described by Büttner (1983). Furthermore, the content of inorganic phosphate (P_i) and the presence of glucose-1-phosphate and glucose-6-phosphate were determined on aliquots of the total homogenate as described by Büttner (1983). The measurement of starch, reducing sugars and sucrose was performed according to Büttner (1983) on the methanol extract of the tuber tissue. Potassium was analysed by means of atomic absorption spectroscopy (AAS) after dry ashing of aliquots of tuber dry matter.

Results

Irrespective of significant differences in K availability in the nutrient solution the K content in tubers of comparable age was nearly the same (Table 1). In all treatments the developing tubers had obviously been preferentially supplied with K in spite of the seriously inhibited aboveground fresh matter production in K_1 and, to a lesser degree, in K_2 plants due to K deficiency (data not shown). Similar to potassium no statistically significant treatment-dependent differences could be found neither between tuber

Table 1. Age and components of tubers grown in nutrient solution at varied K concentrations (K_1 = 0.1 m*M* K^+; K_2 = 0.25 m*M* K^+; K_3 = 1.0 m*M* K^+)

	K_1		K_2		K_3	
	Means	Range	Means	Range	Means	Range
Age (days)		6 - 31		4 - 25		6 - 24
K-content (mg/g fr.wt.)	3.9	2.9– 4.6	4.2	3.4– 4.8	4.2	3.7– 5.3
H_2O-content (g/100 g fr.wt.)	81.1	80.2– 85.5	83.7	80.4– 87.0	83.7	80.2– 86.1
Starch (mg/g fr.wt.)	98.1	67.2–136.6	88.4	68.4–129.5	83.4	51.5–109.1
Sucrose (μmol/g fr.wt.)	25.6	8.5– 39.4	29.6	16.2– 51.3	30.5	11.0– 54.1
Red. sugars (μmol/g fr.wt)	19.9	2.5– 62.9	34.4	7.9– 77.4	42.6	11.1–108.9
P_i (μmol/g fr.wt.)	12.7	10.3– 17.3	11.3	8.1– 14.0	12.4	9.7– 15.1
Glu-1-P + Glu-6-P (μmol/g fr.wt.)	4.4	2.3– 7.3	3.5	1.9– 5.8	3.0	1.7– 5.2

contents of water and starch nor between tuber concentrations of sucrose, reducing sugars, inorganic phosphate or glucose-1-phosphate and glucose-6-phosphate, although a tendency to higher starch and glucose-1-phosphate plus glucose-6-phosphate concentrations could be observed in K_1 plants, the concentrations of these components being lowest in K_3 plants (Table 1). In contrast, the concentrations of sucrose, reducing sugars and K^+ rose from K_1 to K_3.

Table 2 shows that the average activity of starch phosphorylase was the same in all K treatments. In K_1 tubers the activities of UDP glucose pyrophosphatase, amylase and sucrose synthase were tendentially, but not significantly, lower than the respective enzyme activities in K_2 or K_3 tubers.

Starch and K contents of tubers increased with progressing age until a maximum was attained (data not shown), whereas a decrease was observed in growth rate, starch synthesis per day and K uptake per day (Table 3). The tuber growth rate dependend on the daily rate of K uptake. The shape of amylase activity curves as a function of tuber age is similar to that of growth rate, starch synthesis per day or K uptake per day as a function of age.

Discussion

The enzymes tested in our investigation were chosen for the following reasons (see Fig. 1): Sucrose imported from the leaves is the basic

Table 2. Age and enzyme activitives of tubers grown in nutrient solution at varied K concentrations ($K_1 = 0.1\,mM\,K^+$; $K_2 = 0.25\,mM\,K^+$; $K_3 = 1.0\,mM\,K^+$)

	K_1		K_2		K_3	
	Means	Range	Means	Range	Means	Range
Age (days)		6 - 31		4 - 25		6 - 24
Amylase*	23.3	16.5– 32.0	31.3	20.4– 52.2	31.0	16.3– 46.6
Sucrose-synthase	96.4	18.6–191.0	134.3	53.9–258.3	127.3	46.0–207.0
Starch-phosphorylase	218.2	174.0–305.4	216.3	197.9–273.4	227.6	130.9–267.2
UDPG-pyrophosphatase	23.7	9.4– 39.4	28.0	14.5– 51.6	27.8	9.2– 42.3

* Enzyme activities in $\mu mol\,g^{-1}\,fr.wt\,h^{-1}$.

Table 3. Linear regression equations and correlation coefficients (r) for different characteristics of potato tubers grown at varied K supply. For further details see Table 2

Correlation between	K treatment	Regression equation	Correlation coefficient
Tuber age (d):	K_1	y = − 0.502 × + 31.5	−0.746*
amylase activity	K_2	y = − 0.385 × + 35.9	−0.275
(μmol/g fr.wt.h)	K_3	y = − 1.59 × + 52.4	−0.836*
Tuber age (d):	K_1	y = − 0.122 × + 3.6	−0.809*
growth rate (g/day)	K_2	y = − 0.138 × + 3.65	−0.816*
	K_3	y = − 0.112 × + 3.54	−0.646*
Tuber age (d):	K_1	y = −10.98 × +332.9	−0.813*
starch production (mg/tuber.day)	K_2	y = − 9.27 × +279.7	−0.744*
	K_3	y = − 8.15 × +276.0	−0.704*
Tuber age (d):	K_1	y = − 0.014 × + 0.52	−0.860*
K uptake (mg/g fr.wt.day)	K_2	y = − 0.018 × + 0.6	−0.779*
	K_3	y = − 0.021+ 0.64	−0.912*
K uptake (mg/g fr.wt.day):	K_1	y = 611.5 × − 23.1	0.743*
starch production (mg/tuber.day)	K_2	y = 299.4 × + 55.7	0.549*
	K_3	y = 379.7 × + 27.9	0.742*
K uptake (mg/g fr.wt.day):	K_1	y = 7.05 × − 0.4	0.769*
growth rate (g day)	K_2	y = 3.68 × − 0.6	0.496*
	K_3	y = 5.87 × − 0.09	0.768*

* Correlations significant at P = 0.05.

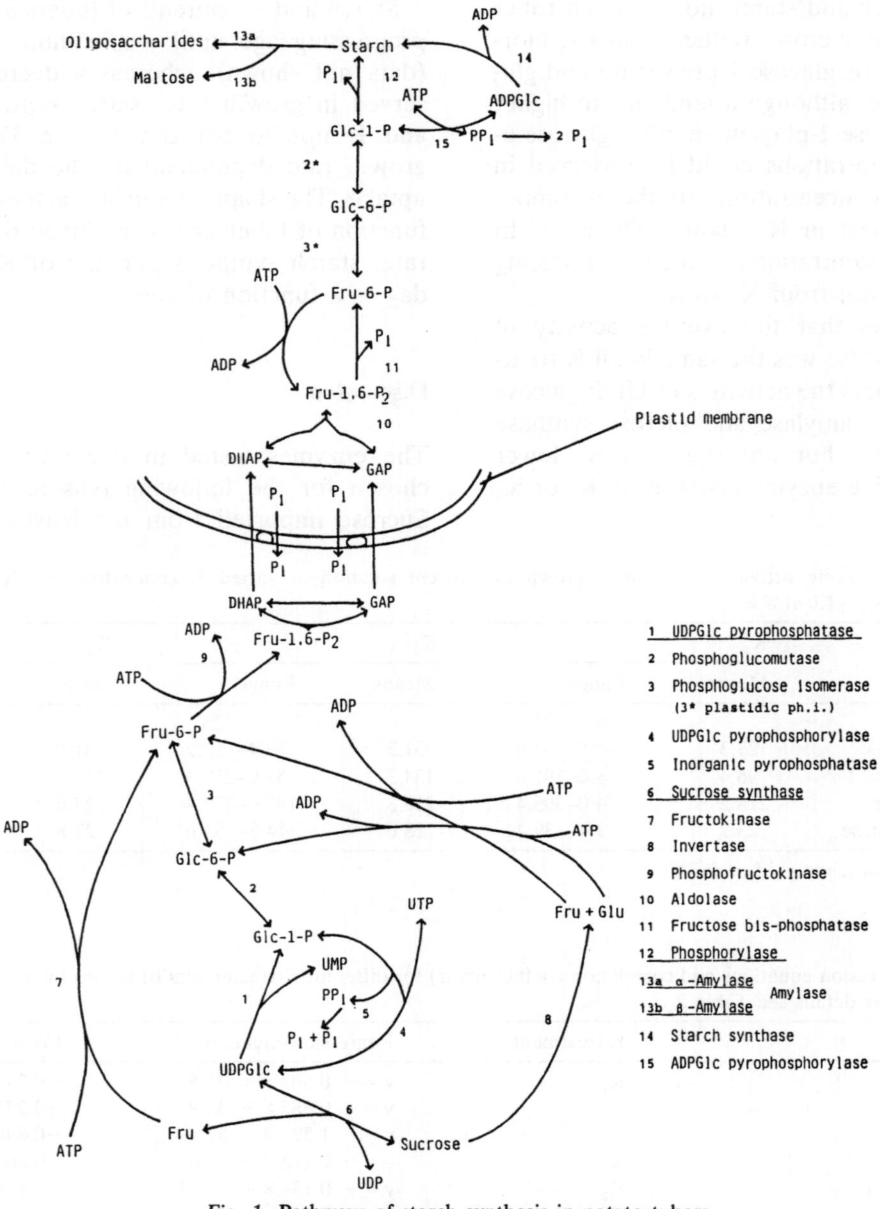

Fig. 1. Pathways of starch synthesis in potato tubers

substrate for starch synthesis in potato tubers. As invertase is inhibited in young potato tubers (Anderson *et al.*, 1980) sucrose is hydrolysed by sucrose synthase. The step from the resulting UDPglucose to glucose-1-phosphate is most probably catalyzed by UDPglucose pyrophosphatase, since the high activity of inorganic phyrophosphatase causes a very low cytoplasmic pyrophosphate (PP_i) concentration within potatoes, an adequately high concentration of PP_i being the prerequisite for the activity of the UDPglucose phryophosphorylase. Due to the facts that starch is produced in the amyloplast and that only triose phosphates can be transported through the membrane in exchange with inorganic phosphate (P_i), the glucose-1-phosphate produced in the cytoplasmic compartment must be transformed into triose-phosphates and these must be converted again into glucose-1-phosphate in the plastid (Fig. 1). Starch synthesis may proceed then either via ADPglucose pyrophosphorylase and starch synthase or through the starch phosphorylase catalyzed reaction. The concentration of P_i was found to be sufficiently low and that of glucose-1-phosphate plus glucose-6-phosphate to be high enough (Table 1) to enable either of the two pathways to proceed.

The surprisingly high amylase activity at the time of intense starch synthesis can be explained with the need of primers at the beginning of starch accumulation (De Fekete and Vieweg, 1974; Kumar and Singh, 1980; McGeachin, 1980), whereas later on there are enough chains of α-(1→4)-linked glucose residues.

No fundamental difference was found in enzyme activity determined in dependence on K content in tubers originating from different K treatments (Table 2). This is not surprising, as the tubers in all treatments had accumulated the same amounts of K ions. Even in K_1 the tuber K content was higher than 1.8% in dry matter considered as optimum for maximum starch accumulation (Forster and Beringer, 1983). Consequently, no correlation between tuber K and starch content could be measured. Tuber starch content did also not differ significantly between the K treatments (Table 1), though there was a slight decreasing tendency of starch content with increasing K supply. Sucrose concentrations showed a slightly increasing tendency with increasing K treatments. Nevertheless, sucrose concentrations were not limiting for starch synthesis.

From linear regression calculations (Table 3) it can be derived that the influence of tuber age on starch synthesizing processes is much more pronounced than the influence *e.g.* of K content on these processes. K uptake into tubers decelerates with increasing age. Postive correlations between K uptake, starch production and growth rate, respectively, indicate that the dynamic phase of K supply to the tubers is of greater importance for these processes than the static status of total K content.

References

Anderson R S, Ewing E E and Senesac A H 1980 Inhibition of potato tuber invertase by an endogenous inhibitor: Effect of salts, pH, temperature, and sugars on binding. Plant Physiol. 66, 451–456.

Büttner G 1983 Stoffwechselwege von der Saccharose zur Stärke in wachsenden Gersten- und Maiskörnern. Dissertation, Techn. Universität, Darmstadt.

Cardini C E, Leloir L F and Chiriboga J 1955 The biosynthesis of sucrose. J. Biol. Chem. 214, 149–155.

De Fekete M A R and Vieweg G H 1974 Starch metabolism: Synthesis versus degradation pathways. *In* Plant Carbohydrate Biochemistry. Ed. J B Pridham. pp 127–144. Academic Press, London, New York, San Francisco.

Forster H and Beringer H 1983 Stärkegehalt von Kartoffelknollen in Abhängigkeit von K-Ernährung und Entwicklungsstadium. Z. Pflanzenernähr. Bodenkd. 146, 572–582. 582.

Hawker J S, Marschner H and Krauss A 1979 Starch synthesis in developing potato tubers. Physiol. Plant. 46, 25–30.

Kumar R and Singh R 1980 The relationship of starch metabolism to grain size in wheat. Phytochem. 19, 2299–2303.

McGeachin R L 1980 A possible role for liver α-amylase in glycogenesis. *In* Mechamism of Saccharide Polymerization and Depolymerization. Ed J J Marshall. pp 209–214. Academic Press, New York.

Nitsos R E and Evans H J 1969 Effects of univalent cations on the activity of particulate starch synthetase. Plant Physiol. 44, 1260–1266.

M. L. van Beusichem (Ed.), *Plant nutrition – physiology and applications*, 649–652.
 PLSO IPNC591

The nutrient requirements of Ornithogalum and Lachenalia, two indigenous South African flowering bulbs

A.S. CLAASSENS
Department of Soil Science and Plant Nutrition, University of Pretoria, Pretoria 0002, Republic of South Africa

Key words: fertilization, Lachenalia, nutrient requirements, Ornithogalum

Abstract

The nutrient requirements of two new flowering bulbous plants, Ornithogalum and Lachenalia, were studied using a sand culture technique. For the short-season Lachenalia, sufficient nutrients must be supplied at an early growth stage. The long-season Ornithogalum requires a good nutrient supply over a longer period, for both vegetative and bulb growth. The P and K requirements for both genera are not very high but adequate K levels are important at an early growing stage due to the low K content of the bulbs. The same is true for N in the case of Lachenalia. However, the N content of Ornithogalum bulbs is high and a low N supply at the early growing stages had little effect on yield.

Introduction

South Africa has a very extensive flora from which flowering plants with a commercial potential are frequently selected. Two bulbous plants that have been selected for this purpose belong to the genera Ornithogalum and Lachenalia.

Because they are newly cultivated plants there is no information in the literature on their nutrient requirements and growers tend to fertilize them in the same way as the more familiar bulbous plants such as tulips or lilies. Preliminarily investigations have shown that the nutrient requirements of these newly cultivated plants are lower than the more familiar bulbous plants. In their natural environment the growing conditions are usually harsh and the supply of nutrients from the soil is very low. They are, however, adapted to these conditions and a high nutrient supply could possibly be detrimental. The object of this research was to obtain some information on their nutrient requirements.

Materials and methods

The experiments were conducted using sand cultures in a glasshouse. Sixteen kg washed quartz sand was used in Mitscherlich pots. Three bulbs were planted in each pot and nine dm^3 of nutrient solution supplied two weeks after planting. The excess nutrient solution was collected and regularly returned to the pots. Water losses due to evapo-transpiration were replenished using deionised water. The nutrient solutions were replaced every four weeks.

The experiments were repeated over four seasons. The sources of the bulbs used for each season differed for practical reasons, while Lachenalia bulbs were not available every season to include in the experiments.

A modified Hoagland no 2 nutrient solution (Bonner & Galston, 1952), was used as a control and was further modified to obtain solutions for different treatments. The main treatments consisted of (i) different levels of N, P and K, (ii)

withholding all nutrients for certain growth periods and (iii) applying low levels of N and K (5 and 0.25 as opposed to the 15 and 6 mmol dm^{-3} N and K of the control) for different periods at the beginning of the growing season, as indicated in Figures 1 to 6.

Flowers were harvested at the marketing stage, dried at 65°C and weighed. At the end of the growing-season the leaves were separated from the bulbs, and after washing with deionised water they were dried, weighed and milled, then chemically analysed. The bulbs were treated in the same way. The yields were expressed as the total dry weight of leaves and flowers.

Nitrogen and phosphorus were determined on a Technicon auto analyzer following wet digestion in H_2SO_4. K was determined by atomic absorption spectroscopy after wet digestion in a $HClO_4$–HNO_3 mixture. When too little sample was available for analysis the replicates were pooled, otherwise replicates were analysed individually.

Where possible all results were statistically analysed as a randomised blockless design. Duncan's multiple range test was used to test for significant differences between treatements (SAS Computer Programme, SAS User's Guide, 1982).

Results and discussion

P supply

Ornithogalum The yield data for various experiments in different seasons are given in Figure 1. On studying the results of any one season one gets the impression that varying the P-supply has no consistent effect on yield. On considering the data for all seasons, it is clear that except for the 1986 season, the lowest P-supply level of 0,25 mmol $H_2PO_4^-$ dm^{-3} always gave the best yields. At the next higher levels 0,5 of 1,0 mmol $H_2PO_4^-$ dm^{-3}), which were used as controls, lower yields (which were in some cases significant) were obtained. A further increase in the P-supply caused insignificant increases and decreases in yields. It can therefore be concluded that the P requirements are low and although the highest P levels did not necessarily reduce the yield significantly, lower P levels could reduce the possibility of inconsistent yields.

The fact that these conclusions could only be reached when the data of more than one growing season were available, means that all the controls as well as all the other nutrient treatments, which were based on the concentrations of the controls, actually were supplied with P at levels (0, 5 and 1 mmol H_2PO_4 dm^{-3}) which were too high and possibly detrimental to the plants.

The yield reduction at relatively low P supply levels may be due to an imbalance in nutrient element supply. This could, however, not be confirmed from the chemical analyses of the leaves.

Lachenalia According to Figure 1 there is no specific trent in the Lachenalia yield due to P supply and it appears that Lachenalia could grow well with a relative high P supply.

Effect of the nutrient content of bulbs

Ornithogalum Although it is widely believed that bulbs contain a fair amount of reserve nutrients (Hartman and Kester, 1983), it does not seem to be the case for all nutrient elements in Ornithogalum bulbs (Table 1). For K, Ca and Mg the inorganic nutrient content of bulbs was lower than that of the leaves, indicating a lack of appreciable reserves. When the bulbs accummulate higher concentrations of elements than in the leaves, as in the case of N and P, it can be assumed that there are reserve nutrients.

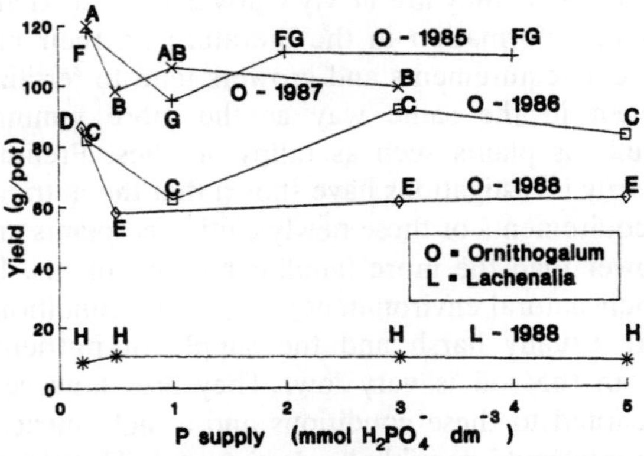

Fig 1. The effect of P supply on top growth of Ornithogalum and Lachenalia. (Points on each graph with the same letter do not differ significantly.)

Table 1. Nutrient elements composition (%) of bulbs and leaves

Nutrient	Ornithogalum		Lachenalia	
	Leaf	Bulb	Leaf	Bulb
N	2.4	4.15	3.86	3.22
P	0.2	0.54	0.56	0.51
K	4.2	0.16	5.8	0.11
Ca	1.5	0.9	1.31	0.13
Mg	0.4	0.28	0.57	0.12

Withholding all the nutrients from bulbs at the beginning of the growing season, soon resulted in poor growth and chlorosis. The yield reduction increased with increasing length of time of withholding nutrients (Fig. 2). When nutrients were reapplied, the plants recovered from their chlorosis symptoms and growth improved, but yields were generally lower. The relative long growing season of Onithogalum probably gave it a chance to recover and reduce yield differences. The 1986 data in Figure 2 showed a different pattern in that the yields increased when the nutrients were withheld for four weeks after planting before applying them again, and then decreased to the same level as the control (which received nutrients two weeks after planting) when withholding the nutrients for longer periods. The reason for this apparent contradiction is probably that for this particular Ornithogalum selection the P supply was too high. The P supply used in this case coincides with the P level (1 mmol $H_2PO_4^-$ dm^{-3}) which gave the lowest yields in 1986 (Fig. 1). It can be reasoned that withholding nutrient supply results in 'no P' versus 'high P' treatments during the early growing stages. The 'no P', associated with withholding the nutrients, was beneficial compared with the 'high P' of the control for a short period. When withholding nutrients for longer periods, the 'no nutrients' effect became dominant and resulted in a reduction in yield.

Lachenalia The nutrient element content of Lachenalia bulbs were lower than those of the leaves (Table 1) indicating that the bulbs contain no appreciable reserves. Withholding nutrients from Lachenalia at the early growth stage caused similar growth and chlorosis tendencies as observed with Ornithogalum (Fig. 2). When reapplying nutrients the growth rate increased and chlorosis symptoms disappeared. Due to a relative short growing period, the plants had less time to recover than is the case with Ornithogalum.

Influence of low N and K supply at an early growth stage

Ornithogalum According to Figure 3 low K supply (0.25 against 6 mmol K dm^{-3} of the control) at an early growth stage tends to decrease the yield. This decrease could be due to the low K reserves of the bulbs (Table 1 and Fig. 5) which could not compensate for the low K supply level in the nutrient solution. On the other hand, low N levels (5 against 15 mmol N dm^{-3} of

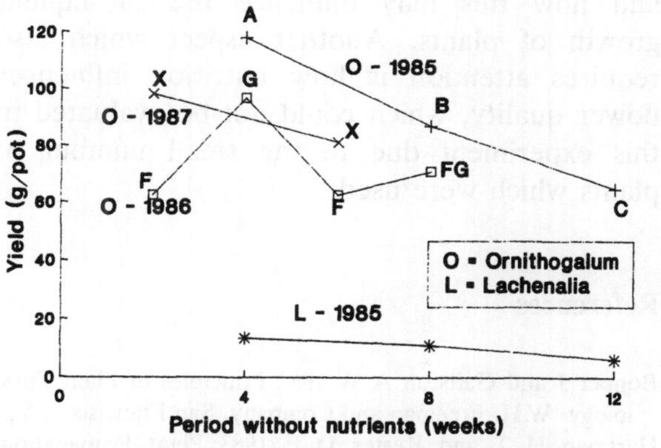

Fig. 2. The effect of withholding nutrients at the early growing stages on top growth of Ornithogalum and Lachenalia. (Points on each graph with the same letter do not differ significantly.)

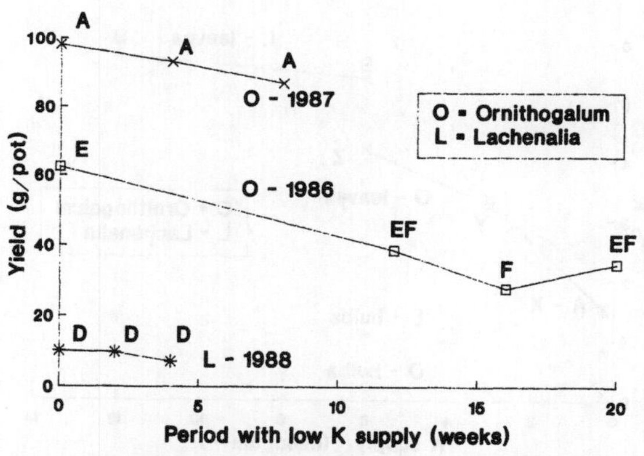

Fig. 3. The effect of low K supply at the early growing stages on top growth of Ornithogalum and Lachenalia. (Points on each graph with the same letter do not differ significantly.)

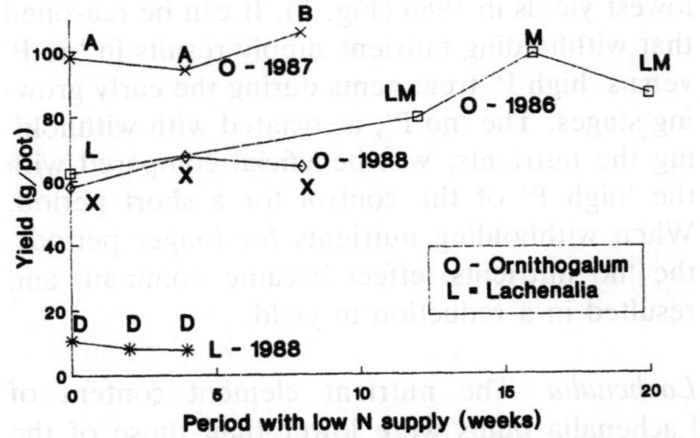

Fig. 4. The effect of low N supply at early growing stages on top growth of Ornithogalum and Lachenalia. (Points on each graph with the same letter do not differ significantly.)

the control) in the early growth stages did not result in a decrease in yields. (Fig. 4). This trend could be due to the relative high N content of bulbs (Table 1 and Fig. 6) which could act as reserves. From Fig. 5 it is evident that with increasing K supply bulbs could not accumulate any K reserves.

Lachenalia As opposed to Ornithogalum, Lachenalia yields were slightly reduced by both low N and K supply levels at the early growth stages (Figs. 3 and 4). This was probably due to the low content of both N and K in the bulbs (Figs. 5 and 6). According to Figures 5 and 6, with increasing K and N supply the bulbs were not able to accumulate any appreciable K and N reserves.

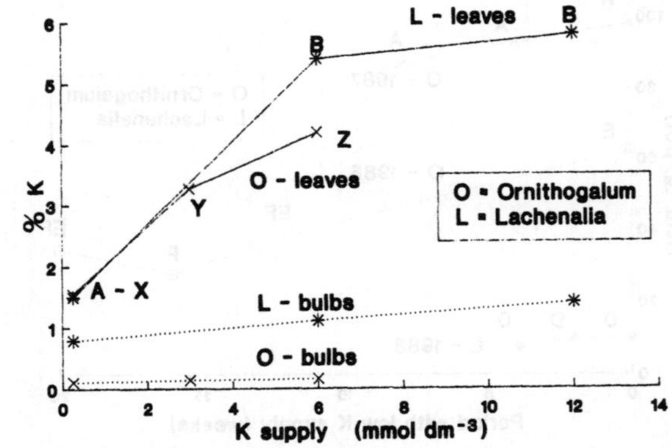

Fig. 5. The effect of K supply on the K content of leaves and bulbs of Ornithogalum and Lachenalia. (Points on each graph with the same letter do not differ significantly.)

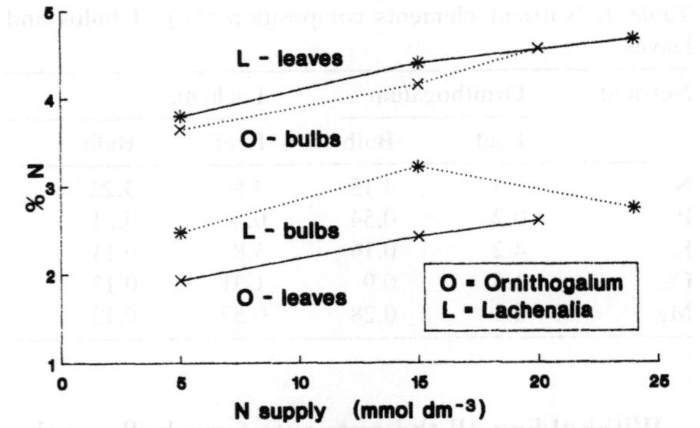

Fig. 6. The effect on N supply on the N content of leaves and bulbs of Ornithogalum and Lachenalia.

Conclusions

Results showed that although high P levels were not directly detrimental to the plants, low P levels were adequate for both plants and very low levels actually gave the best yields with Ornithogalum.

The nutrient content of the bulbs is lower than that of the leaves for most nutrients, indicating that the bulbs do not store appreciable amounts of inorganic nutrient element reserves. The nutrient content of the bulbs could have an important effect on plant growth. When the content of any one nutrient in the bulbs is low, the supply during the early growth stages is important, especially in the case of the short growing season Lachanalia.

Future research should be concentrated on factors influencing the nutrient content of bulbs and how this may influence the subsequent growth of plants. Another aspect which also requires attention is how nutrition influences flower quality, which could not be evaluated in this experiment due to the small number of plants which were used.

References

Bonner J and Galtston A W 1952 Principles of Plant Physiology. W.H. Freeman and Company, San Francisco, 55 p.

Hartman H T and Kester D E 1983 Plant Propagation: Principles and Practices, 4th. Ed. Prentice Hall International, Inc., 491 p.

S.A.S. USER'S GUIDE 1982 Statistics. S.A.S. Institute Inc. Cary, NC, pp 140–199.

M. L. van Beusichem (Ed.), *Plant nutrition – physiology and applications*, 653–654.
 PLSO IPNC511B

Growth and mineral composition of grape-vine rootstock cultured *in vitro* with different levels of ammonium nitrate

A. TRONCOSO, A. VILLEGAS[1], C. MAZUELOS and M. CANTOS
Instituto de Recursos Naturales y Agrobiología de Sevilla, CSIC, B.P.O. 1052, E-41080 Sevilla, Spain. [1]*Colegio Postgraduados, MEX-56230 Chapingo, México*

Key words: grape-vine, *in vitro* culture, nitrogen nutrition

Abstract

Growth and mineral composition of grape-vine explants (13.3 EVEX rootstock) cultured on a basal medium with 10.4 m*M* of NO_3^-, and respective additions of 5, 10, 15, 20 and 25 m*M* of NH_4NO_3 were analysed.

Low N-availability (10.4 m*M*) induced low shoot formation. Addition of NH_4NO_3 up to 15 m*M* increased number, quality and N levels of shoots. Further increase of N in the medium induced very high contents of N and water in tissues, and led to a bad shoot quality (large and abnormally shaped leaves and dark colour-fragile tissues, difficult to handle in further propagations). The content of K in explants was negatively affected by the concentration of NH_4NO_3 in the substrate and by the level of N in tissues.

Introduction

In a previous work (Troncoso *et al.*, 1988) the influence of N on growth and mineral composition of grape-vine rootstocks 41 B and 161–49, were studied. Although the response of the rootstock 161–49 was much more active than that of 41 B, increasing N in the medium increased number, size and N contents of shoots and decreased formation of callus in both rootstocks, but excess of N produced non-desired shoots.

The aim of the present work was to study the response to N of a new grape-vine rootstock (13.3 EVEX) from the "Estación de Viticultura y Enologa de Jerez (Spain)", and to study more in detail the influence of N availability on the quality and mineral composition of shoots.

Materials and methods

Homogeneous, 10 mm long explants of grape-vine rootstock 13.3 EVEX were cultured at a temperature of 25°C, light intensity of 2500 lux and 16 h photoperiod.

The treatments were:
—Control, with basal substrate (Table 1);
—Five NH_4NO_3 concentrations added to the

Table 1. Composition of the basal nutritive medium

Chemical compound	m*M*	Chemical compound	μM	Chemical compound	m*M*
KNO_3	7.91	$MnSO_4.4H_2O$	5.0	M-inositol	27.75
$Ca(NO_3)_2.4H_2O$	1.27	H_3BO_3	100.0	Thiamine	2.96
KH_2PO_4	1.25	$ZnSO_4.7H_2O$	30.0	6-BAP	4.43
$MgSO_4.7H_2O$	1.50	$Na_2MoO_4.2H_2O$	1.0	IBA	0.48
$FeSO_4.7H_2O$	0.09	$CuSO_4.5H_2O$	0.1	Sucrose	30 $g.L^{-1}$
Na_2EDTA	0.10	$CoCl_2.6H_2O$	0.1	Agar	6 $g.L^{-1}$

Table 2. Influence of NH_4NO_3 on number, growth, hydration and mineral composition of shoots

NH_4NO_3 added (m*M*)	Shoots per explant (average)	% of shoots >10 mm	Fresh weight (mg/plant)	% hydration	Mineral composition		
					N (%)	K (%)	Fe (mg/kg)
0	2.60a	25.93a	125	81.25	1.32	2.95	85
5	4.01b	43.09b	220	82.50	2.54	2.52	161
10	5.76b	43.84b	345	82.92	3.39	2.20	234
15	5.00b	45.84b	539	83.23	4.03	1.77	336
20	5.37b	51.19b	541	85.50	4.79	1.49	431
25	5.49b	49.74b	499	89.75	5.50	1.48	530

Identical letters indicate non-significant difference ($P<0.05$) by the Tukey test.

control medium: 5, 10, 15, 20 and 25 m*M* of the salt, respectively.

Each treatment involved 24 explants (8×3 replicates).

After 30 days, the number and size (per cent under and over 10 mm) of shoots, fresh and dry weight, tissue hydration (fresh weight minus dry weight/fresh weight × 100) and mineral composition (Pinta *et al.*, 1969; 1973) of explants were analysed. To determine the quality of shoots, size and shape of leaves, colour and fragility of tissues, and facility to handle in further propagation processes were also considered.

Results and discussion

As shown in Table 2, low concentration of N in the medium (control) produced low number and size of shoots. Additions of NH_4NO_3 up to 10 m*M* increased both number and length of shoots, and higher N additions did not cause further increase of these characteristics. The rootstock 13.3 EVEX showed a lower response to N than 161–49 and 41 B (Troncoso *et al.*, 1988).

Addition of NH_4NO_3 to the substrate up to 15 m*M*, resulted in a linear increase of shoot fresh weight (r = 0.99). Further increase in N addition had a negative effect on the weight of shoots. Addition of NH_4NO_3 resulted in increased hydration percentage (r = 0.93), N content (r = 0.97) and Fe content (r = 0.99) of explant tissues. In contrast, the levels of K in tissues were negatively affected by NH_4NO_3 additions (r = −0.96) and by N in tissues (r = −0.99). The other nutrients considered were not clearly influenced by N, their average levels being 3.3 g/kg for P, 5.8 g/kg for Ca, 3.0 g/kg for Mg, 43 mg/kg for Cu, 124 mg/kg for Mn, and 241 mg/kg for Zn, all on a dry weight basis.

N-applications clearly affected the quality of shoots. Low N availability produced few and short shoots. Addition of 10 or 15 m*M* of NH_4NO_3 gave more, larger, well-formed shoots, with normally sized-shaped leaves, intensive green colour and flexible tissues, very easy to handle and with good response in further propagations. Larger N-additions, 20 m*M* and more, produced non-desired shoots with very dark colour, abnormally large shaped leaves and fragile tissues, difficult to handle in further propagation processes.

References

Pinta M et Membres du Comité Inter-Instituts d'Etude des Techniques Analytiques du Diagnostic Foliare 1969 Méthodes de référence pour la détermination des éléments minéraux dans les végetaux: N, P, K, Ca, Mg. Oléagineaux 24, 497–504.

Pinta M et Membres du Comité Inter-Instituts d'Etude des Techniques Analytiques du Diagnostic Foliare 1973 Méthodes de référence pour la détermination des éléments minéraux dans les végétaux. Determination des éléments Ca, Mg, Fe, Mn, Zn et Cu par absortion atomique. Oléagineaux 28, 87–92.

Troncoso A, Villegas A, Mazuelos C and Cantos M 1988 Influenza del livello di azoto (NH_4NO_3) del mezzo sullo sviluppo e composizione minerale di meristemi di vite "*in vitro*". III Convegno sui portainnesti della vite. Potenza (Italia). NOTAE Accademia Italiana della Vite e del Vino 5, 144.

M. L. van Beusichem (Ed.), *Plant nutrition – physiology and applications*, 655–658.

PLSO IPNC464B

Phosphorus uptake characteristics of a world collection of white clover (*Trifolium repens*) cultivars

A.D. MACKAY[1], J.R. CARADUS[1], A.L. HART[1,3], G.S. WEWALA[2], J. DUNLOP[1], M.G. LAMBERT[1], J. VAN DEN BOSCH[1] and M.C.H. MOUAT[1]
[1]*Grasslands Division and* [2]*Applied Mathematics Division, Department of Scientific and Industrial Research, Private Bag, Palmerston North, New Zealand.* [3]*Corresponding author*

Key words: cultivars, intraspecific variation, nitrogen fixation, phosphate fertilizer, *Trifolium repens* L.

Abstract

As part of a selection and breeding programme to improve the phosphorus (P) nutrition of white clover (*Trifolium repens* L.), 119 white clover cultivars and lines from 25 countries were screened for their dry matter response to added P. A large amount of variation for shoot dry weight response to added P was found. The 119 cultivars were clustered into eight groups on the basis of fitted quadratic P response curves. The objective of this study was to measure shoot P uptake and examine the relationship between P and nitrogen (N) of white clover cultivars from within each of these clusters. Twenty six of the 119 cultivars were selected to represent the eight clusters.

There were significant differences in P uptake, N accumulation and dry matter production when expressed as a function of added P among the 25 cultivars selected to represent the 8 clusters. This variation was found across a range of added P levels and provides evidence for intraspecific variation for the absorption and utilisation of P. From the most to the least P responsive clusters, P uptake and N accumulation decreased, as did the amount of P absorbed, N accumulated and dry matter when expressed as a function of added P. Cultivars in the two most responsive clusters (A and B) also utilised P more effectively for growth than less responsive cultivars. There appears sufficient scope to improve P use by breeding in white clover.

Introduction

White clover (*Trifolium repens* L.), an important component of temperate pastures, utilises fertiliser and soil phosphorus (P) inefficiently, especially when grown in a mixed sward with grasses (Jackman and Mouat, 1972). Maintenance P fertiliser requirements of a white clover-based sward must be set at a level that ensures its survival and growth. With most New Zealand soils being P deficient (During, 1972), P fertiliser costs represent a major item of on-farm expenditure. An improvement in the P nutrition of white clover, whether through increased P uptake or more effective use of plant P, would have an immediate and positive impact on the profitability of pastoral farmers.

As part of a selection and breeding programme to improve the P nutrition of white clover (*Trifolium repens* L.), 119 white clover cultivars and lines from 25 countries were screened for their dry matter response to added P (Mackay *et al.*, 1990). A large amount of variation in shoot dry weight response to added P was found. The 119 cultivars were divided into eight groups on the basis of fitted quadratic P response curves. In this paper shoot P uptake and the relationship between shoot P and N in representative white clover cultivars from within these clusters are reported.

Materials and method

Twenty-six cultivars were selected to represent the eight clusters. These and the clusters they represent are listed in Table 1. A description of each cultivar and its characteristics are given by Caradus (1986).

The study was conducted during April and May (autumn) 1987 and temperatuares in the glasshouse ranged from 15 to 30°C. The soil used was a Wainui silt loam (Typic Dystrochrept), with a pH of 5.0 and Olsen P of 6 mg kg^{-1}. It was collected from the field to a depth of 100 mm and passed moist through a 4-mm sieve.

Rates of 0, 20, 40, 80, 120, 200, 300, 400, 500, 600 and 750 mg P kg^{-1} were used, the 0, 40, 120 and 200 rates being replicated 3 times, the 400, 500 and 600 rates twice, and 20, 80, 300 and 500 only once. Sulphur was added as $CaSO_4$ and potassium as KCl. Soil was incubated at a gravimetric moisture content of 48% for 30 d, before packing into 120 mm diameter pots (753 g oven dried) to a bulk density of 0.8 Mg m^{-3}.

Table 1. List of cultivars and lines, country of origin and cluster grouping

Cultivar/breeding line	Country of origin	Cluster[b]
Atoliaj	USSR	F
Barbian	Netherlands	C
Cultura	Netherlands	C
Dusi[a]	South Africa	F
G23[a]	New Zealand	G
G26[a]	New Zealand	E
Gandalf	Denmark	B
Gigant	Italy	D
Gomelskij	USSR	B
Gwenda	United Kingdom	A
Huia	New Zealand	C
Isolation V[a]	New Zealand	D
Kersey	United Kingdom	D
Kopu	New Zealand	F
Luclair	France	H
Menna	United Kingdom	C
Pitua	New Zealand	E
Podkowa	Poland	D
Regal	USA	D
Simone	Italy	F
Spreading Selection[a]	New Zealand	C
Tahora	New Zealand	E
Tillman	USA	D
Trifo	Denmark	A
Viglasska	Czechoslovakia	B
Whatawhata	New Zealand	E

[a] Breeding lines.
[b] Refer to Mackay *et al.*, 1990 for description of clusters.

Seeds of each cultivar were germinated at 25°C in petri-dishes, transplanted after 48 hours to sand trays and grown for 10 days before planting into pots. Three seedlings were transplanted into each pot and pots were arranged in a randomised block design in a glasshouse. A mixture of *Rhizobium trifolii* strains NZP 540, 541, 542, 547, 548 and 560 was added to each pot one week after planting. Three times a week pots were watered to a gravimetric moisture content of 55%. At each watering, pots were re-randomised within blocks, and blocks were moved within the glasshouse. A more complete description of experimental techniques is given by Mackay *et al.* (1990).

A destructive harvaest was made 56 d after planting. Total P and N concentration in shoots were determined by an autoanalysis method (Twine and Williams, 1971) following Kjeldahl digestion. Phosphate uptake, nitrogen accumulation, and the N:P ratio were calculated, as were P uptake, N accumulation and dry matter production as a function of added P. Analysis of variance was used to examine differences between cultivar and the groups representing the clusters for the various attributes.

Results and discussion

Cultivar differences

Cultivar differences in % N, % P, P uptake and N accumulation were generally non-significant when compared across all P levels. By restricting the comparison to the added P level (400 mg kg^{-1} of soil) at which all cultivars had reached at least 90% of maximum growth, there were large differences in these characteristics amongst the 26 cultivars (Table 2), indicating that intraspecific variation for these two trials exists.

When P uptake and N accumulation of each cultivar was expressed as a function of added P there were also large significant differences for these two characteristics amongst the cultivars (Table 2). More than a 3.5-fold variation in P

Table 2. Attributes of the 26 cultivars at the added P level giving 90% of maximum growth

Cultivar	P (%)	N (%)	N:P	P uptake/ added P ($\mu g\ mg^{-1}$)	N accumulation/ added P ($\mu g\ mg^{-1}$)
Atoliaj	0.240	2.45	10.2	1.34	13.7
Barbian	0.300	3.78	12.5	1.24	15.4
Cultura	0.280	2.25	8.0	2.10	16.9
Dusi	0.310	2.58	8.3	1.67	13.9
G23	0.280	3.55	12.7	0.91	11.6
G26	0.330	3.35	10.2	1.18	12.0
Gandalf	0.295	2.40	8.1	1.80	14.6
Gigant	0.285	2.35	8.3	1.80	14.9
Gometskij	0.320	2.58	8.0	1.97	15.6
Gwenda	0.355	2.68	7.6	1.93	14.6
Huia	0.305	2.75	9.0	1.47	13.2
Isolation V	0.280	2.85	10.2	1.74	17.7
Kersey	0.265	1.98	7.5	1.85	13.9
Kopu	0.280	2.78	10.0	1.67	16.4
Luclair	0.320	3.80	11.9	0.67	7.9
Menna	0.275	2.90	10.5	1.15	12.3
Pitau	0.260	2.45	9.4	1.39	13.1
Podkowa	0.290	2.10	7.2	2.00	14.5
Regal	0.260	1.95	7.5	1.03	7.7
Simone	0.240	1.80	7.5	1.87	14.1
Spreading Selection	0.300	3.50	11.7	1.06	12.3
Tahora	0.325	2.63	8.2	1.68	13.4
Tillman	0.220	1.70	7.7	1.91	14.7
Trifo	0.250	2.08	8.4	2.13	17.1
Viglasska	0.305	2.28	7.6	2.34	17.5
Whatawhata	0.310	4.10	13.2	0.99	13.1
SED	0.032	0.43	1.6	0.34	1.4
P	*	***	**	**	***

*, **, ***, significant at the 5%, 1%, and 0.1% level, respectively.

uptake and a 2.3-fold variation in N accumulation was measured among cultivars when expressed as a function of added P. This variation was not restricted to one P level but was found across a wide range of added P levels and provides further evidence for intraspecific variation both for the absorption and utilisation of P.

Cluster differences

The 119 cultivars were originally clustered into the eight groups on the basis of principal component analysis of dry matter response to added P (Mackay *et al.*, 1990). For clusters A to G response to P at very low P levels ranged from very high to very low, and change in response to P with increasing level of added P ranged from a rapid reduction in the slope of the response (cluster A) through only a slight reduction (cluster F) to a slight increase in the slope of the response curve (cluster G) (Fig. 1). Cluster H was unusual in that it had both a low response to P at very low P levels and a moderate reduction in change in response as P level increased.

Some distinct differences in the P and N characteristics of cultivars were associated with the differences noted in the shoot dry weight response of cultivars across the eight clusters. While the N concentration did not differ across the 8 clusters, the P concentration and uptake and N accumulation all decreased from the most (cluster A) to the least (cluster H) P responsive clusters (Fig. 1). The more responsive cultivars (cluster A and B) absorbed more P and accumu-

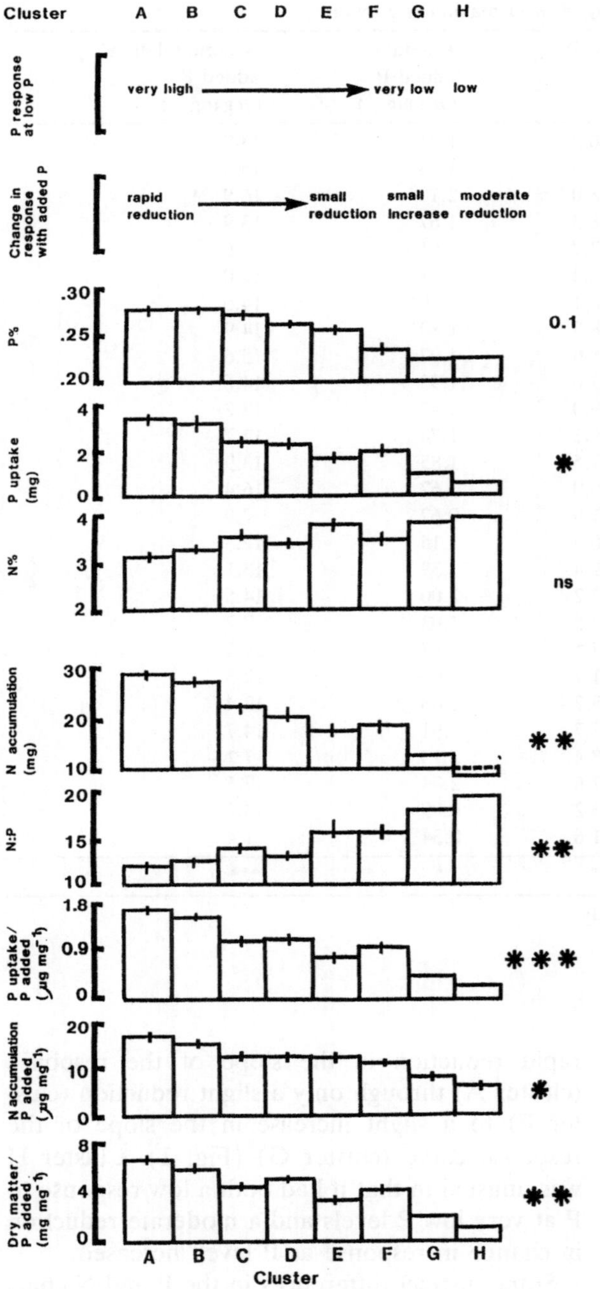

Fig. 1. Shoot dry matter response to added P of 119 cultivars and the P and N characteristics of 26 cultivars. Values are means of cultivars within each cluster, over all P levels. Bars indicate standard errors. Clusters G and H were represented by one cultivar. ns, not significant; 0.1, *, **, ***, significant at the 10%, 5%, 1% and 0.1% level, respectively.

lated more N. These trends were found across the whole range of added P. Surprisingly the N:P ratio increased across clusters A to H, despite the fact that P uptake, N accumulation and dry matter production when expressed as a function of added P increased in the opposite direction (Fig. 1). In addition to absorbing more P and accumulating more N, the most responsive cultivars (clusters A and B) utilised P and N more effectively for growth.

Tests of heritability will determine the usefulness of differences in the P and N characteristics noted in this study for use in any ongoing breeding programme to improve P use in white clover.

Acknowledgements

To Craig Anderson, Ian Black, Philip Budding, Roger Claydon, Elli Cornege, Des Costall, Miranda Filemoni, Margaret Knighton, Allison MacKay, Venessa Pokaia, Metiria Turei, Brett Watson for technical assistance, and to Cherry Liddane, Biotechnology Division, DSIR, for supplying Rhizobium inoculum.

References

Caradus J R 1986 World checklist of white clover varieties. N.Z. J. Agric. Res. 14, 119–164.

During C 1972 Fertilisers and Soils in New Zealand Farming. Government Printer, Wellington, N.Z.

Jackman R H and Mouat M C H 1972 Competition between grass and clover for phosphate. I. Effect of browntop (*Agrostis tenuis* Sibth) on white clover (*Trifolium repens* L.) growth and nitrogen fixation. N.Z. J. Agric. Res. 15, 653–666.

Mackay A D, Caradus J R, Dunlop J, Wewala G S, Mouat M C H, Lambert M G, Hart A L and van den Bosch J 1990 Response to phosphorus of a world collection of white clover cultivars. *In* Genetic Aspects of Plant Mineral Nutrition. Eds. N El Bassam, M Dambroth and B C Loughman. pp 553–558. Kluwer Academic Publishers, Dordrecht, The Netherlands.

Twine J R and Williams C H 1971 The determination of phosphorus in Kjeldahl digests of plant material by automatic analysis. Comm. Soil Sci. Plant Anal. 2, 485–489.

M. L. van Beusichem (Ed.), *Plant nutrition – physiology and applications*, 659–665.

PLSO IPNC550B

The effects of phosphorus and bio-fertilizers on physiological parameters, yield and quality of cowpea (*Vigna sinensis*)

J.S. BOHRA, K. SINGH and J.S. JAMWAL
Department of Agronomy, Institute of Agricultural Sciences, Banaras Hindu University, Varanasi-221005, India

Key words: cowpea, leghaemoglobin, nitrogen fixation, nodulation, phosphorus, physiological parameters, protein recovery, *Vigna sinensis* Savi

Abstract

Field investigations undertaken on cowpea (*Vigna sinensis* Savi) revealed that phosphorus applied at 13.1 and 26.2 kg P ha^{-1} had positive effects on physiological growth parameters, such as LAI, NAR, RGR and LAD as well as on nodulation, leghaemoglobin content of root nodules, protein recovery, protein translocation efficiency and grain yield. Among the bio-fertilizers, seed treatment with Rhizobium significantly enhanced nodulation, protein recovery and grain yield. High positive correlations were observed between LAI, NAR and yield. Thus, seed inoculation with Rhizobium and application of phosphorus may improve yield and quality of cowpea on alluvial soils.

Introduction

Cowpea (*Vigna sinensis* Savi), an important legume crop in India, is grown both for its grain and its green pods and provides a good source of protein to the predominantly vegetarian population. In the last few years, major changes have occurred in the varietal composition and agronomic practices of cowpea. However, only few studies are available regarding the effect of phosphorus (Ahlawat *et al.*, 1979; Mbowe, 1975) and bio-fertilizers (Gowda *et al.*, 1979) on the yield of cowpea. Moreover, in none of these studies physiological growth parameters, nodulation and leghaemoglobin content of root nodules in response to phosphorus and bio-fertilizer application were investigated. The aim of the present study was to get more insight into the above effects of phosphorus and bio-fertilizers.

Materials and methods

The field experiment was conducted under dryland condition to study the effect of phosphorus and bio-fertilizers on physiological parameters, yield and quality of cowpea (*Vigna sinensis* Savi cv. Pusa Dophasli) during the rainy seasons (June–September) of 1984 and 1985 at the Research Farm, Institute of Agricultural Sciences, Banaras Hindu University, Varanasi, India. Total rainfall received during the period of crop growth was 667 mm in 1984 and 856 mm in 1985. The soil of the experimental field was an Entisol low in available nitrogen (162 kg ha^{-1}) and medium in available phosphorus (10 kg P ha^{-1}) and potassium (194 kg K ha^{-1}).

The experiment was laid out in a randomized block design with three levels of phosphorus (0, 13.1 and 26.2 kg P ha^{-1}) and three bacterial cultures (no bacteria, Rhizobium and Phosphobacterin). The treatments were replicated thrice. The gross plot size was 16.5 m^2. Phosphorus according to treatment and a uniform dose of 24.9 kg K ha^{-1} were applied as basal dressings. Seeds were sown in rows 45 cm apart using 30 kg seed ha^{-1}.

For recording biometric observations at different stages of crop growth, five plants in the net plot area were randomly selected and tagged.

For determinations of dry matter production and nodule weight per plant, four plants were randomly selected from sampling rows at different stages of crop growth. Furthermore, measurements were made on physiological growth parameters, *viz*. LAI (Rhodes and Bloodworth, 1964), NAR (Briggs *et al*., 1920), RGR (Blackman, 1979), CGR (Radford, 1967), and LAD (Radford, 1967). Data on nodule weight, leghaemoglobin content of fresh nodules (Proctor, 1963) and protein translocation efficiency (Gowda *et al*., 1979) were also obtained. Coefficients of correlation between ancillary and growth parameters and grain yield were computed (Panse and Sukhatme, 1978).

Results

Nodule dry weight

Increasing levels of phosphorus application improved nodule dry weight at 25 and 40 days after sowing in both years (Table 1). Significant increases in nodule dry weight were also recorded due to Rhizobium inoculation of seeds as compared to phosphobacterin and no culture treatments.

Leghaemoglobin content of root nodules.

Phosphorus application failed to affect the leghaemoglobin content of nodule tissue in both years (Table 1). However, seed inoculation with Rhizobium markedly improved the leghaemoglobin content or root nodules over that of other treatments.

Growth parameters

Dry matter accumulation
Significantly higher dry matter accumulation was recorded with increasing levels of phosphorus application throughout the crop growth period in both years (Table 2). Among the bio-fertilizers, phosphobacterin failed to affect the dry matter accumulation, but Rhizobium significantly promoted it.

Leaf area index (LAI)
Phosphorus application tended to improve the LAI throughout the growth period, but the effect was significant only at 40 and 55 days after sowing in both years (Table 3). The difference between 26.2 and 13.1 kg P ha^{-1} was also significant at these stages, except at 55 days after sowing during the second year. LAI was also

Table 1. Nodule dry weight $plant^{-1}$ and leghaemoglobin content of cowpea as affected by phosphorus and bio-fertilizers

Treatments	Nodule dry weight (mg)				*Leghaemoglobin content ($mg.g^{-1}$ fresh weight)			
	Days after sowing				Days after sowing			
	25		40		25		40	
	1984	1985	1984	1985	1984	1985	1984	1985
Phosphorus level (P kg ha^{-1})								
0	64.3	64.3	123.9	120.1	2.9	2.8	3.0	2.9
13.1	64.8	66.0	133.9	130.5	3.1	3.0	3.3	3.1
26.2	65.8	66.5	136.7	134.0	3.4	3.1	3.5	3.3
LSD (0.05)	0.45	1.44	5.73	5.76	–	–	–	–
Bio-fertilizers								
No culture	63.9	64.0	121.1	121.1	2.5	2.5	2.6	2.6
Rhizobium culture	66.2	68.4	150.0	142.2	3.6	3.5	3.7	3.6
Phosphobacterin culture	64.9	64.4	123.3	122.2	3.2	2.9	3.3	3.1
LSD (0.05)	0.45	1.44	5.73	5.76	–	–	–	–

* Average of three replications.

Table 2. Dry matter accumulation (g $plant^{-1}$) of cowpea as affected by phosphorus and bio-fertilizers

Treatments	Days after sowing							
	25		40		55		70	
	1984	1985	1984	1985	1984	1985	1984	1985
Phosphorus level (P kg ha^{-1})								
0	0.72	0.72	4.93	5.10	10.15	10.12	14.72	14.82
13.1	0.77	0.76	5.05	5.62	10.31	10.41	14.92	15.18
26.2	0.86	0.78	5.21	5.81	10.44	10.73	15.08	15.25
LSD (0.05)	0.055	0.031	0.118	0.385	0.233	0.216	0.134	0.298
Bio-fertilizers								
No culture	0.74	0.74	4.91	5.40	10.11	10.24	14.74	14.81
Rhizobium culture	0.88	0.77	5.35	5.92	10.62	10.75	15.22	15.53
Phosphobacterin culture	0.78	0.75	4.93	5.44	10.17	10.32	14.77	14.92
LSD (0.05)	0.055	0.031	0.118	0.385	0.233	0.216	0.134	0.298

significantly influenced by Rhizobium inoculation at all growth stages except at 25 days after sowing in both years and at 70 days after sowing in the second year.

Net assimilation rate (NAR)
NAR values were improved due to increasing levels of phosphorus application only during the 25–39 days and 40–54 days growth periods of the first year (Table 4). Rhizobium inoculation improved the NAR in all growth periods of the first year.

Relative growth rate (RGR)
Phosphorus application significantly improved RGR values only during the 40–54 days growth period of the first year and the 25–39 days growth period of the second year; phosphorus at 26.2 kg P ha^{-1} was significantly superior to 13.1 kg P ha^{-1} during these stages (Table 4). Among bio-fertilizers, Rhizobium significantly improved the RGR value only during the 40–55 days growth period of 1984.

Crop growth rate (CGR)
CGR values improved significantly with increasing levels of phosphorus during the 40–54 and 55–69 days growth periods of both years (Table 5). Rhizobium also had a significant effect on CGR at these growth periods in both years.

Table 3. Leaf area index (LAI) of cowpea as affected by phosphorus and bio-fertilizers

Treatments	Days after sowing							
	25		40		55		70	
	1984	1985	1984	1985	1984	1985	1984	1985
Phosphorus level (P kg ha^{-1})								
0	0.29	0.31	1.38	1.40	3.20	2.95	1.53	1.42
13.1	0.29	0.32	2.49	2.62	3.40	3.54	1.53	1.44
26.2	0.30	0.32	3.14	3.21	3.52	3.68	1.68	1.48
LSD (0.05)	N.S.	N.S.	0.282	0.263	0.083	0.330	N.S.	N.S.
Bio-fertilizers								
No culture	0.29	0.32	2.17	2.22	3.18	3.22	1.47	1.42
Rhizobium culture	0.30	0.32	2.62	2.64	3.64	3.61	1.68	1.48
Phosphobacterin culture	0.29	0.31	2.22	2.31	3.28	3.20	1.48	1.44
LSD (0.05)	N.S.	N.S.	0.282	0.263	0.083	0.330	0.045	N.S.

Table 4. Net assimilation rate (NAR) and relative growth rate (RGR) of cowpea as affected by phosphorus and bio-fertilizers

	Net assimilation rate (NAR) g dm^{-2} day^{-1}						Relative growth rate (RGR) $g.g^{-1}$ day^{-1}					
	Days after sowing						Days after sowing					
	25–39		40–54		55–69		25–39		40–54		55–69	
	1984	1985	1984	1985	1984	1985	1984	1985	1984	1985	1984	1985
Phosphorus level (P kg ha^{-1})												
0	0.062	0.058	0.073	0.070	0.032	0.030	0.053	0.051	0.035	0.034	0.005	0.004
13.1	0.073	0.074	0.080	0.078	0.034	0.033	0.062	0.063	0.039	0.037	0.005	0.005
26.2	0.078	0.078	0.086	0.085	0.036	0.034	0.068	0.068	0.041	0.040	0.005	0.005
LSD (0.05)	0.0035	N.S.	0.0045	N.S.	N.S.	N.S.	N.S.	0.0013	0.0007	N.S.	N.S.	N.S.
Bio-fertilizers												
No culture	0.066	0.062	0.079	0.076	0.033	0.031	0.056	0.051	0.038	0.035	0.005	0.005
Rhizobium culture	0.079	0.077	0.081	0.079	0.036	0.034	0.059	0.056	0.039	0.038	0.005	0.005
Phosphobacterin culture	0.069	0.066	0.079	0.077	0.034	0.034	0.057	0.056	0.038	0.036	0.005	0.005
LSD (0.05)	0.0035	N.S.	0.0045	N.S.	N.S.	N.S.	N.S.	N.S.	0.0007	N.S.	N.S.	N.S.

Leaf area duration (LAD)

Phosphorus application significantly improved LAD values only during the 40–54 days growth periods in both years, but the two phosphorus levels remained statistically at par (Table 5). As regards the bio-fertilizers, only Rhizobium significantly increased the LAD value during 40–54 days growth period of the second year.

Grain and straw yields

Marked effects of phosphorus application were observed on grain and straw yields of cowpea (Table 6). Increasing levels of phosphorus application significantly enhanced both the yeilds of grain and straw in both years. Among the bio-fertilizers, seed inoculation with Rhizobium im-

Table 5. Grop growth rate (CGR) and leaf area duration (LAD) of cowpea as affected by phosphorus and bio-fertilizers

Treatments	Crop growth rate (CGR) gm^{-2} day^{-1}						Leaf area duration (LAD) dm^2 day^{-1} $plant^{-1}$					
	Days after sowing						Days after sowing					
	25–39		40–54		55–69		25–39		40–54		55–69	
	1984	1985	1984	1985	1984	1985	1984	1985	1984	1985	1984	1985
Phosphorus level (P kg ha^{-1})												
0	7.24	7.19	14.61	14.58	4.28	4.25	42.32	42.28	75.54	75.49	55.12	55.08
13.1	7.25	7.21	15.10	15.00	4.38	4.33	42.33	42.29	79.95	79.92	55.14	55.10
26.2	7.25	7.24	15.33	15.28	4.45	4.42	42.31	42.27	79.80	79.70	55.13	55.09
LSD (0.05)	N.S.	N.S.	0.095	0.124	0.063	0.065	N.S.	N.S.	0.318	0.311	N.S.	N.S.
Bio-fertilizers												
No culture	7.24	7.20	14.92	14.90	4.31	4.21	42.32	42.32	77.53	77.42	55.12	55.10
Rhizobium culture	7.25	7.22	15.20	15.15	4.44	4.37	42.35	42.34	77.65	77.76	55.19	55.16
Phosphobacterin culture	7.24	7.21	14.95	14.92	4.34	4.33	42.31	42.30	77.61	77.62	55.15	55.12
LSD (0.05)	N.S.	N.S.	0.095	0.124	0.063	0.065	N.S.	N.S.	N.S.	0.311	N.S.	N.S.

Table 6. Grain and straw yield, protein recovery and protein translocation efficiency of cowpea as affected by phosphorus and bio-fertilizers

Treatments	Yield (q ha^{-1})				Protein recovery (kg ha^{-1})						Protein translocation efficiency (%)	
	Grain		Straw		Grain		Straw		Total			
	1984	1985	1984	1985	1984	1985	1984	1985	1984	1985	1984	1985
Phosphorus level (P kg ha^{-1})												
0	5.3	4.2	14.1	11.0	124.1	101.8	99.9	84.3	223.9	186.1	55.4	54.7
13.1	6.3	5.2	16.3	13.7	150.9	130.8	118.5	110.2	269.4	241.0	56.0	54.5
26.2	6.9	5.7	17.4	14.7	165.7	141.6	126.8	117.6	292.6	259.2	56.6	54.6
LSD (0.05)	0.29	0.46	0.79	0.99	9.22	4.70	6.77	8.33	–	–	–	–
Bio-fertilizers												
No culture	5.6	4.7	14.7	12.4	134.2	111.1	104.6	99.1	238.9	210.2	56.2	52.8
Rhizobium culture	7.0	5.5	17.8	14.4	159.2	145.4	121.2	113.9	280.4	259.2	56.8	56.0
Phosphobacterin	5.9	4.8	15.2	12.6	136.1	113.9	105.4	100.0	241.5	213.9	56.3	50.0
LSD (0.05)	0.29	0.46	0.78	0.99	9.22	4.70	6.77	8.33	–	–	–	–

proved the grain and straw yields significantly, contrary to phosphobacterin.

Protein recovery and protein translocation efficiency

Protein recovery, in both grain and straw, improved significantly with increasing levels of phosphorus application in both years (Table 6). However, phosphorus application only slightly improved protein translocation efficiency in the first year.

As regards the bacterial cultures, Rhizobium inoculation of seeds led to significantly higher protein recovery. Protein translocation efficiency was also improved due to Rhizobium inoculation and the effect was more pronounced during the second year.

Table 7. Correlation coefficient ('r' values) for the relationship between yield and growth and ancillary characters

Character	'r' values	
	1984	1985
1. Leghaemoglobin at flowering	0.994**	0.319
2. LAI at flowering	0.707**	0.648*
3. NAR at flowering	0.632*	0.339
4. RGR at flowering	0.412	0.316
5. CGR at flowering	0.316	0.382
6. LAD at flowering	0.400	0.408
7. Dry matter accumulation at flowering	0.903**	0.614*

** Significant at 1 per cent level of probability.
* Significant at 5 per cent level of probability.

Correlation studies

Leghaemoglobin contents at flowering were positively correlated with grain yield of cowpea only during the first year (Table 7). Of the growth parameters, only dry matter accumulation, LAI and NAR were significantly correlated with grain yield of cowpea in both years.

Discussion

Phosphorus fertilization favourably affected the nodulation of cowpea. This may be attributed to the fact that phosphorus, being an important constituent of energy-rich compounds, is required for energy transfer during certain metabolic processes of plant growth. Rhizobium bacteria in the nodules require energy for fixing atmospheric nitrogen and an improvement in nodulation in the presence of applied phosphorus is thus to be expected. Moreover, phosphorus is required to maintain the mobility and flagellation of bacterial cells thereby promoting their migration and favouring an early infection of legume roots (Whyte and Trumble, 1953). However, phosphorus application failed to affect the leghaemoglobin content of the fresh nodules. This is in conformity with the findings of Sidhu *et al.*, 1967.

The favourable effect of phosphorus on certain growth parameters, particularly dry matter accumulation, LAI and CGR, might be attributed to the complex phenomena of phosphorus utilization in plant metabolism. Moreover, phosphorus helps the plant in developing meristematic tissues, promotes cell division and leaf area development, and is also required for an efficient absorption and utilization of other plant nutrients (Meyer and Anderson, 1952).

Significant improvements in grain- and straw yields were observed with phosphorus application levels increasing from 0 to 26.2 kg P ha^{-1}. The response to phosphorus was in agreement with the low available phosphorus status of the experimental soil as mentioned earlier. The favourable effects of phosphorus on the grain yield of cowpea might be mainly due to stimulatory effects of phosphorus on various physiological growth parameters and on yield-promoting characteristics, such as number of pods per plant and 1000-grain weight (data not reported). Phosphorus application improved protein recovery in both grain and straw, obviously due to higher biomass production and higher protein contents of grain and straw. Protein translocation efficiency was also improved with phosphorus application. It seems that phosphorus might have helped in the efficient translocation of nitrogen from vegetative parts to developing seeds as well as the synthesis of protein from the reduced nitrogen compounds within the seeds (Huffaker and Rains, 1978).

The results of the present experiment revealed that inoculation of cowpea seeds with Rhizobium significantly increased dry matter production, nodulation and physiological growth parameters probably due to initial help rendered by Rhizobium in the establishment of cowpea (Moolani and Jana, 1965; Okan *et al.*, 1972). Grain and straw yields were also increased significantly through Rhizobium inoculation of seeds. The favourable effect of Rhizobium on yield may be attributed to the higher leghaemoglobin content of Rhizobium in treated plots which might have resulted in higher nitrogen fixation and a more continuous supply of nitrogen to the plants (Bhatnagar, 1971; Saxena *et al.*, 1975; Srivastava, 1981). Protein translocation efficiency was improved through Rhizobium inoculation, probably due to its favourable effect on continuity of the supply of nitrogen from vegetative parts to the reproductive 'sinks'.

The phosphobacterin culture, however, did not influence growth and yield of cowpea, probably because the existing soil and environmental conditions were not conducive to normal functioning of these bacteria. Multiplication of this organism probably requires higher levels of soil organic carbon (Goswami, 1976) than encountered in the present investigation. Lack of response may also have been due to antagonistic effects of the native soil flora.

A high positive correlation was observed between LAI and yield (Table 7). The correlation between NAR and yield was also significant in the first year. All these growth parameters are interrelated and treatment differences observed either in total dry matter production or economic yield are resulting from interactions between all these factors (Kona, 1979).

References

Ahlawat I P S, Singh A and Saraf C S 1981 Effect of winter legumes on the nitrogen economy and productivity of succeeding cereals. Exp. Agric. 17, 57–62.

Bhatnagar R S 1971 Rhizobium inoculation effects on selected pulse crops. Ph.D. Thesis, Post-Graduate School, I.A.R.I., New Delhi, India, 142 p.

Briggs, G E, Kidd F and Wert C 1920 A quantitative analysis of plant growth. Ann. Appl. Biol. 7, 202–223.

Blackman V H 1919 The compound interest law and plant growth. Ann. Bot. 33, 353–360.

Goswami K P 1976 Worth of Azotobacter as a bacterial fertilizer. Fert. News 21, 32–34.

Gowda S T, Hegde S V and Bagyraj D J 1979 Rhizobium inoculation and seed pelleting in relation to nodulation, growth and yield of cowpea. Curr. Res. 8, 42–43.

Huffaker R C and Rains D W 1978 Factors affecting nitrate acquisition by plants: Assimilation and fate of reduced nitrogen. *In* Nitrogen in the Environment: Soil-Plant-Nitrogen Relationships, Vol. II. Eds. R N Donald and J G MacDonald. pp 1–43. Academic Press, New York.

Kona G S S R 1979 Physiological variation in moong (*Vigna radiata* L.) Ph.D. Thesis, Dept. Plant Physiol., B.H.U., Varanasi, India, 134 p.

Meyer B S and Anderson B D 1952 Plant Physiology. Van Nostrand Reinhold Co, New York, 312 p.

Mbowe F F A 1975 Grain legume research in Tanzania. Tropical Grain Legume Bull. 2, 2–3.

Moolani M K and Jana M K 1965 A note on response of green gram to fertilizers in laterite soil. Indian J. Agron. 10, 43–44.

Okon Y, Eshel Y and Heins Y 1972 Cultural and symbiotic properties of Rhizobium strains isolated from nodules of *Cicer arietinum*. Soil Biol. Biochem. 4, 165–.

Panse V G and Sukhatme P V 1978 Statistical Methods for Agricultural Workers. I.C.A.R., New Delhi, India, 104 p.

Proctor M H 1963 A note on the haemoglobin estimation. N.Z. J. Sci. 6, 60–63.

Radford P J 1967 Growth analysis formulae, their uses and abuses. Crop Sci. 7, 171–175.

Rhoads F M and Bloodworth M E 1964 Area measurement of cotton leaves by dry weight method. Agron. J. 56, 520.

Saxena M C, Tilak K V B R and Yadav D S 1975 Response of pigeon pea to inoculation and pelleting. Indian J. Agron. 21, 321–324.

Sidhu G S, Nirmaljit Singh and Rangil Singh 1967 Symbiotic nitrogen fixation by some summer legumes of Punjab. J. Res. Punjab Agric. Univ. 4, 244–248.

Srivastava S N L 1981 Effect of phosphate fertilization on growth, yield and quality of pea (Var. T. 163) in relation to nitrogen, molybdenum and Rhizobium inoculation. Ph.D. Thesis, Dept. Agron. B.H.U., Varanasi, India. 123 p.

Whyte R O and Trumble H C 1953 Legumes in Agriculture F.A.O. Agric. Studies, No. 21, F.A.O., Rome.

M. L. van Beusichem (Ed.), *Plant nutrition – physiology and applications*, 667–674.

PLSO IPNC609

Effect of successive cutting on nodulation and nitrogen fixation of *Leucaena leucocephala* using ^{15}N dilution and the difference methods

N. SANGINGA[1], F. ZAPATA[1], S.K.A. DANSO[2] and G.D. BOWEN[2,3]
[1]*Soil Science Unit ,and* [2]*Soil Fertility, Irrigation and Crop Production Section, International Atomic Energy Agency, P.O. Box 100, A-1400 Vienna, Austria.*[3] *Corresponding author*

Key words: cuttings, isotope dilution, *Leucaena leucocephala* isolines, ^{15}N, nodulation, nitrogen fixation, Rhizobium strains

Abstract

Despite the widespread use of *Leucaena leucocephala* in alley cropping systems, little information is available on the effect of different magagement practices on its nitrogen fixation. We studied the effects of three successive cuttings at 12 week intervals on nodulation and nitrogen fixation of *L. leucocephala* isolines K636 and K28 inoculated with three Rhizobium strains. The ^{15}N isotope dilution and the total N difference methods were used to estimate the proportions and amounts of N_2 fixed. Nodule senescence and decay occurred within 3 weeks after each cutting, new ones being formed to continue N_2 fixation during regrowth. At the first and the second cuttings, only K636 responded to inoculation, producing an average of five times more biomass than K28. A response of K28 to inoculation occurred only at the third cutting at 36 weeks after planting. Marked differences also occurred between Rhizobium strains in both N_2 fixed and the time course of fixation. With *L. leucocephala* K636 cut at 24 weeks, plants inoculated with Rhizobium strain IRc 1045 derived 57% of their N from the atmosphere which decreased to 36% at 36 weeks. Opposite results were obtained with Rhizobium strain USDA 3409 on the same isoline during the two periods. No significant differences in % N_2 fixed occurred when the three Rhizobium strains were inoculated on K28. Although the average proportion of N derived from the atmospheric N_2 fixation remained constant, the total N_2 fixed increased significantly with successive cuttings. In K28, there was a good agreement between both methods of estimating N_2 fixed, but with K636, the difference method gave a higher estimate of N_2 fixed and was less precise.

Introduction

There is an increased interest in the use of leguminous woody species as sources of green manure, animal fodder and firewood in agroforestry systems such as in alley cropping (Kang *et al.*, 1981; 1985; Wilson *et al.*, 1986). Alley cropping is the growing of crops, usually food crops, in alleys formed by trees or woody shrubs that are cut back at crop planting and maintained as hedge-rows by frequent trimming during cropping (Wilson and Kang, 1981). The leaves and twigs from the cut trees are added to the soil as green manure or mulch.

Leucaena leucocephala has been one of the most commonly used trees in alley cropping, its high symbiosis with N_2 fixing Rhizobium being one highly desirable attribute. A well nodulated *L. leucocephala* tree can fix about 40 to 60% of its N from the atmosphere (Sanginga *et al.*, 1986) and its prunings can yield about 500 kg N/ha in a year (Guevarra, 1976). Increases in maize grain yield by up to 30% over a control have been reported (Kang *et al.*, 1981; 1985; Mulongoy and Van der Meersch, 1988; Sanginga *et al.*, 1986).

Although *L. leucocephala* can withstand repeated and drastic prunings, its regrowth and the amounts of green manure produced are affected by the frequency and height of cutting (Duguma *et al.*, 1988; Guevarra *et al.*, 1978; Pathak and

Patel, 1982). There is however little information on the corresponding effect of successive cutting on nodulation and nitrogen fixation of *L. leucocephala*. We therefore investigated the effect of successive cuttings on nodulation and N_2 fixation by *L. leucocephala*, and Rhizobium strains, plant genotypes interactions on these processes at each cutting. Values of N_2 fixed obtained using the ^{15}N isotope dilution and the total N difference methods were compared.

Materials and methods

The experiment was conducted in a glasshouse at the International Atomic Energy Agency (IAEA), Seibersdorf Laboratory, Austria. Mean day and night temperatures were 28°C and 20°C; the light intensity was approximately 10000 lux for a 12 h photoperiod and the relative humidity varied between 60 and 70% (day-night amplitude).

i) Soil characteristics and preparation

The soil, from Seibersdorf, Lower Austria, classified as a Typic Eutrocrept (calcareous, clay loam with abundant gravel in the top layer, pH 8.3; total N, 0.3%; extractable P, 55.8 ppm; humus, 6.7%) was air-dried, sieved (2 mm), mixed with sand (1/1 ratio) and then transferred to plastic pots (5 kg of a mixture soil: sand). It was kept moist by watering with deionized water to approximately field capacity. A basal fertilizer consisting of 50 mg K/kg soil as muriate of potash and 1 mL of a combination of a micronutrient solution (B 0.05%; Mg 0.05%; Zn 0.005%; Mo 0.005% and Cu 0.002%) per kg of soil was applied to all pots before planting.

ii) Seed sources and scarification

Seeds of *L. leucocephala* K28 and K636, were scarified and sterilized in concentrated H_2SO_4 for 30 min, thoroughly rinsed with distilled water and planted two per pot.

iii) Rhizobium strains, inoculation

Strains of Rhizobium used were TAL 1145 isolated from *L. leucocephala* in Colombia, IRc 1045 isolated from *L. leucocephala* in Nigeria and USDA 3409 also isolated from *L. leucocephala* in the Philippines. These strains had been shown to be effective on *L. leucocephala*. They were cultured in yeast extract mannitol broth (Vincent, 1970) at 28°C for 5 days. Inoculation was performed at sowing by adding near to the seed 1 ml (approximately 10^9 cells) of a suspension of the above Rhizobium strains.

iv) Experimental design and treatments

Treatments:

i) Plant isolines: *L. leucocephala* isolines K28 and K636 with different nitrogen fixing capacities (Sanginga *et al.*, unpublished data).

ii) Inoculation: seedlings were either uninoculated or inoculated with each of the 3 different Rhizobium strains described above.

iii) Cuttings: plants were cut at 5 cm above the ground three times at 12 weeks intervals.

Five replicate pots per treatment for each of the two isolines uninoculated or inoculated with each of the three Rhizobium strains were used. They were arranged randomly within blocks. A single application of fifty milligrams of N per kg of soil containing 10 atom % ^{15}N excess KNO_3 was performed as a solution to all pots soon after the first cutting. Uninoculated and non-nodulated *L. leucocephala* isolines were used as reference plants.

v) Harvest

At each cutting the above ground plant material was harvested. Leaves, branches and stems were separated, chopped into 10 to 20 mm pieces, then oven-dried at 70°C and weighed. The dried samples were ground to pass through a 0.5 mm sieve. At the final cutting, the roots were separated from the soil, and the soil particles adhering on the roots were carefully removed. Nodules were also collected and counted. Roots and nodules were also oven-dried and weighed.

Separate pots were used to study the effect of cutting on nodule senescence or decay at each harvest period. At each cutting two pots were harvested and nodules were separated into live (active, pink with leghemoglobin) and dead (senescing, decaying or empty) and then counted and oven-dried. The remaining duplicate pots were allowed to regrow for 3 weeks after which the same nodule observations were made.

Analytical methods

Total N and N isotope ratio determinations of the different plant parts were done on an Automatic Nitrogen Analyzer 1500 Carlo Erba coupled to a SIRA mass spectrometer (Fiedler and Proksch, 1975). The isotope dilution equation of Fried and Middelboe (1977) was used to calculate the proportions of N derived from atmospheric N_2 (% Ndfa) and the actual amount of N_2 fixed (mg N/plant).

Statistical analyses

Data were analyzed statistically using the PC statistical package (Mohan and Blane, 1985). Two way analyses of variance were performed for each isoline/Rhizobium strain combination to determine treatment effects. Separate one-way analyses of variance were performed for each genotype to test the effect of cutting treatments. When a significant ($P < 0.05$) treatment effect was found, an LSD was calculated to compare treatment means. Student's test was used within treatments to examine differences between the ^{15}N-dilution and N-difference methods of estimating symbiotic N_2 fixation for both second and third cuttings.

Results

Plant growth and total N

Shoot dry weight and total N of both *L. leucocephala* isolines increased between the three successive cuttings over the 36 weeks after planting (WAP). Mean shoot dry matter at successive cuttings was 4, 8, 30 g/pot and 0.5, 4, 31 g/pot for inoculated K636 and K28, respectively. There was a significant correlation between shoot dry weight and total nitrogen (r = 0.90; $P < 0.05$).

Both plant isoline and inoculation with Rhizobium significantly affected the growth of *L. leucocephala* at all three successive cuttings. For the first two harvests, K636 had higher (5 fold greater) shoot dry weight and total N than K28. At the third harvest both isolines produced similar shoot biomass and total N yield (Figs. 1 and 2).

At each measurement time, growth of uninoculated plants was similar for both isolines and lower than that of inoculated plants, except for K28 at the first two cuttings. K28 showed no inoculation response until the third cutting. At the first two cuttings, inoculation with Rhizobium more than doubled shoot and total N yield in K636. At the third cutting growth and total N of K28 were similar to that of inoculated K636 (Figs. 1 and 2).

Marked differences occurred between Rhizobium strains in plant growth response with time. At the first cutting, Rhizobium strains IRc 1045 and TAL 1145 produced similar growth responses on K636, increasing shoot dry weight and total N by 6 and 11 fold respectively above that of the control. Rhizobium strain USDA 3409 was even more effective in stimulating K636 growth, resulting in shoot dry weight and total N of 12 and 23 fold that of the uninoculated controls and up to double that of strain TAL 1145. It was not until the third harvest that the symbiotic effectiveness of TAL 1145 was equal to the other two strains. With *L. leucocephala* K28, no significant differences occurred between rhizobial strains and their effect on both shoot growth and total N at each harvest.

Nodulation and nitrogen fixation

Nodules occurred on all inoculated treatments 3 weeks after planting, while up to the last harvest, none of the uninoculated plants was nodulated. No significant differences in nodulation (number and weight) occurred between the two isolines and between the different Rhizobium strains at the first cutting. Within 3 weeks after cutting there was a marked senescence of the preharvest nodules and newly formed nodules replaced the senescent ones (Table 1).

The proportion and amounts of N_2 fixed by K636 and K28 at the second and third cuttings are given in Table 2. Data from the isotope

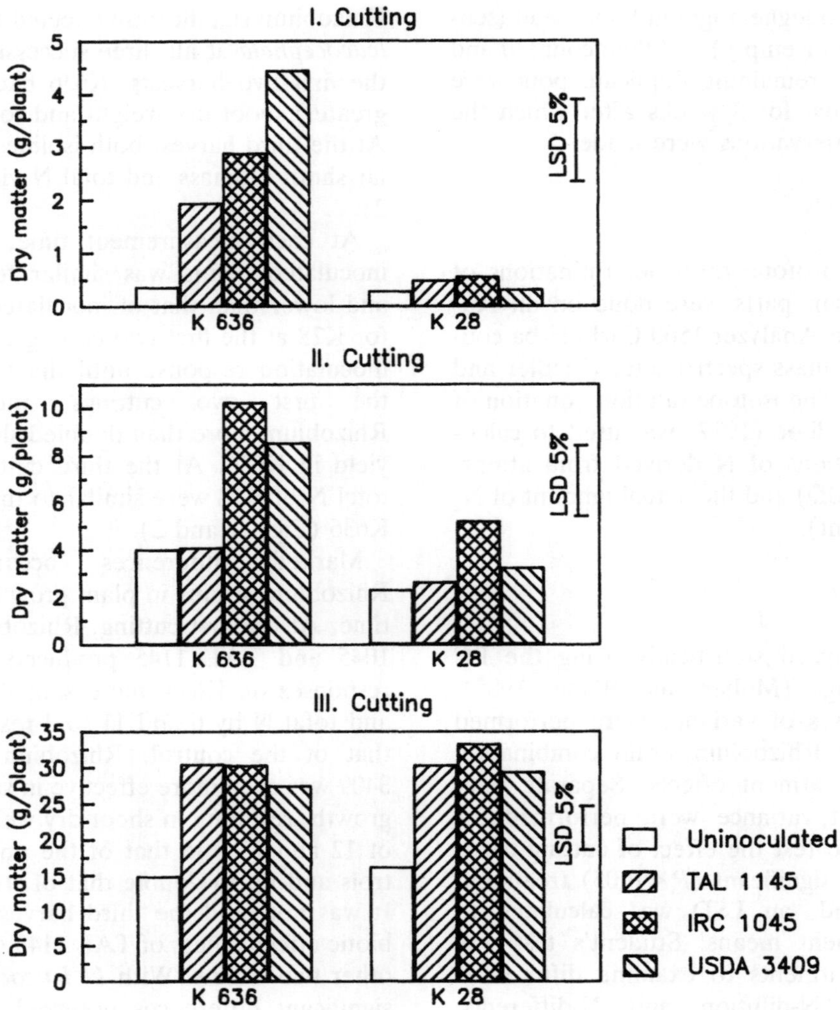

Fig. 1. Dry matter (g/plant) of *Leucaena leucocephala* genotypes K636 and K28 inoculated with different Rhizobium strains during successive cuttings.

dilution method show that while no significant differences in % Ndfa by K28 and K636 inoculated with TAL 1145 and USDA 3405 occurred between the 2 cuttings, a lower fixation estimate (17% decrease) was found with IRc 1045. Total N_2 fixed increased significantly by 83% for K636 and 277% for K28 with the successive cuttings.

Significant interactions occurred between plant isolines and Rhizobium strains on these two N_2 fixation parameters. At the second cutting *L. leucocephala* K636 had fixed an average of 105 mg/plant compared with 65 mg/plant for K28. However, the % Ndfa values were not significantly different. At the third cutting, K28 inoculated with TAL 1145 and IRc 1045 had higher % Ndfa and average amounts of N derived from fixation than K636. With K636, IRc 1045 produced the highest % Ndfa (57%) at the second cutting whilst at the last cutting its % Ndfa was low (36%) compared to that of USDA 3409 (53%). No significant differences in % Ndfa occured between the different Rhizobium strains when inoculated on K28. A significant correlation was found between the amounts of N_2 fixed and the shoot dry weight or root dry weight produced by these Rhizobium strains ($r = 0.95$; $P < 0.05$).

There was a close agreement between both methods of estimating N_2 fixation in K28 except for the 24 WAP harvest of plants inoculated with

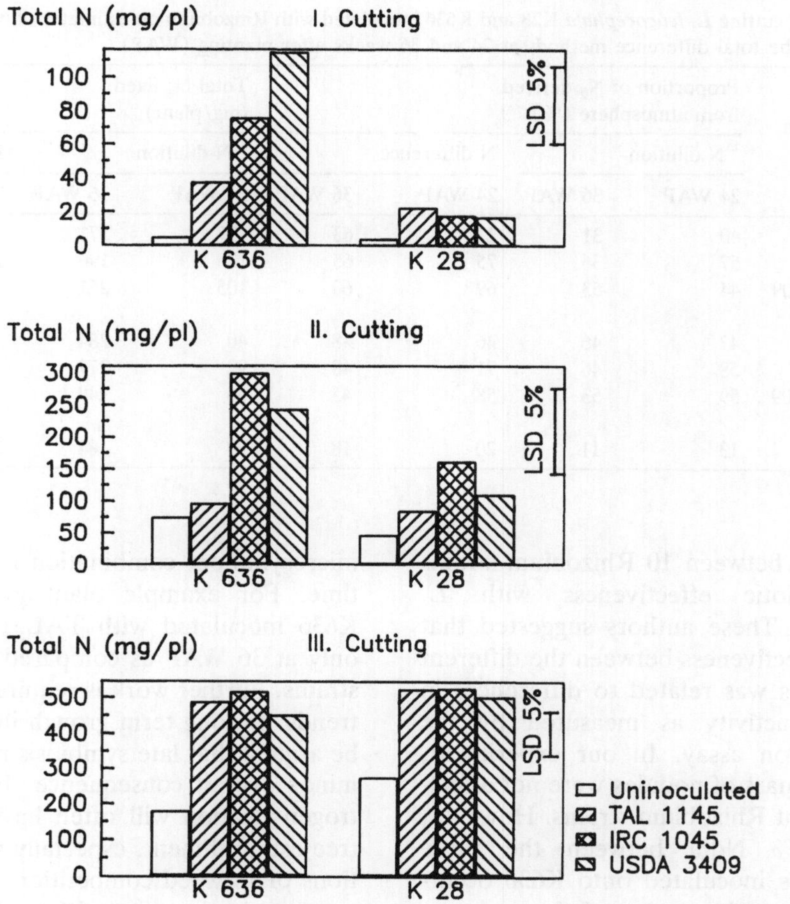

Fig. 2. Total nitrogen (mg/plant) of *Leucaena leucocephala* genotypes K63 and K28 inoculated with different Rhizobium strains during successive cuttings.

IRc 1045 where there was a 20% difference. With K636, however the difference method resulted in higher estimates of N_2 fixation with two of the three Rhizobium strains. The difference method however was less precise as shown by its higher standard error (Data not shown) and a large coefficient of variation ($CV > 80\%$) compared to 20% given by the isotope dilution.

Table 1. Nodulation of *L. leucocephala* K28 and K636 as affected by cutting treatment at 3 weeks after regrowth

Plant genotypes	Nodule (No/plant)		Nodule dry weight (mg/plant)	
	Dead	Live	Dead	Live
Cut				
K28	30	31	95	153
K636	20	33	80	160
Uncut				
K28	2	54	ND*	107
K636	6	48	ND*	114

* not determined

Discussion

Our study clearly demonstrates the importance of tree genotypes or lines, Rhizobium strain and management practices such as cutting or prunings on the responsiveness of *L. leucocephala* to inoculation.

The effect of different Rhizobium strains on the effectiveness of N_2 fixation in the same host species, as were observed for K636 are consistent with previous cross-inoculation studies performed under greenhouse and field conditions in Nigeria (Sanginga *et al.*, 1986; 1989a) demon-

Table 2. The effect of cutting *L. leucocephala* K28 and K636 inoculated with Rhizobium on estimates of N_2 fixation calculated by the ^{15}N-dilution and the total difference methods at 24 and 36 weeks after planting (WAP)

Plant genotype	Rhizobium strains	Proportion of N_2 derived from atmosphere (%)				Total N_2 fixed (mg/plant)			
		^{15}N dilution		N difference		^{15}N-dilution		N difference	
		24 WAP	36 WAP	24 WAP	36 WAP	24 WAP	36 WAP	24 WAP	36 WAP
K636	TAL 1145	40	31	22	63	37	171	21	308
	IRc 1045	57	36	75	65	170	196	224	336
	USDA 3409	44	53	69	60	105	251	168	271
K28	TAL 1145	47	46	46	48	40	284	39	251
	IRc 1045	58	46	71	48	92	277	115	253
	USDA 3409	59	53	58	45	66	240	64	226
LSD 5%		13	11	20	18	99	84	179	58

strating variation between 10 Rhizobium strains in their symbiotic effectiveness with *L. leucocephala* K8. These authors suggested that differences in effectiveness between the different Rhizobium strains was related to differences in nodule specific activity as measured by the acetylene reduction assay. In our experiment, the number and mass of nodules were not affected by the different Rhizobium strains. However, differences in % Ndfa between the three Rhizobium strains inoculated onto K636 occurred and this may explain some of the variation found in plant growth. In contrast to K636, no differences in effectiveness of fixed N_2 was obtained between the three Rhizobium strains on K28, which responded to inoculation only at the third cutting. We are uncertain as to the cause of the genotypic differences in response to inoculation that occurred in the present study. Nodulation time and nodule growth were similar and this and other data (N. Sanginga, G.D. Bowen and S.K.A. Danso, unpublished) indicate that differences in plant growth between K28 and K636 are more related to nitrogen fixation abilities than to the speed and degree of nodulation. This emphasizes the importance of the host plant in influencing nitrogen fixation and its consequence on plant growth. The basis of differential time responses of *L. leucocephala* isolines to inoculation with Rhizobium requires more detailed study.

Our experiment illustrates well that the assessment of the effectiveness of a particular host/microsymbiont combination is also a function of time. For example plant growth response of K636 inoculated with TAL 1145 was observed only at 36 WAP as compared to other rhizobial strains. Further work is required to confirm these trends. In long term growth in the field, it could be argued that late symbiosis response may be of minor overall consequence. However, early nitrogen fixation will often have major effects in tree establishment, especially under stress conditions or in weed competition. This point will be essential in the successful establishment of nitrogen fixing trees in field plantations. Late nodulation and nitrogen fixation could be undesirable especially under stress conditions. Early fixation will increase chances of the plant to survive under these constraints. It is highly appropriate that these studies be transferred to the field.

Our study confirmed the ability of *L. leucocephala* to withstand repeated cuttings, as reported by other investigators (Guevarra *et al.*, 1978; Kang *et al.*, 1985). *Leucaena leucocephala* biomass and total N yield increases after successive cuttings confirm studies of Duguma *et al.* (1988). An interesting finding of our work is that successive cutting treatments did not significantly affect the proportion of N_2 derived from the atmosphere while the amounts of N_2 fixed were significantly enhanced at each regrowth following cutting. Nodule senescence and decay occurred within 3 weeks following each cutting, with new ones being formed to sustain N_2 fixation during regrowth.

Table 3. The effect of excluding roots and nodules on estimates N_2 fixation calculated by the ^{15}N method on *L. leucocephala* at the 36 WAP

Plant genotype	Rhizobium strains	Proportion of N_2 derived from atmosphere (%)			Total N_2 fixed (mg/plant)		
		Roots	Above ground	Whole plant	Roots	Above ground	Whole plant
K636	TAL 1145	23	34	31	171	171	342
	IRc 1045	34	37	36	194	196	390
	USDA 3409	45	56	53	310	251	561
K28	TAL 1145	29	51	46	186	284	470
	IRc 1045	29	51	46	177	276	453
	USDA	28	48	53	146	240	386
LSD 5%		13	14	11	103	81	137

One might expect the % Ndfa to increase as soil N is depleted. It must be remembered however, that cutting also returns N to the soil as sloughed nodules and probably sloughed roots (Bowen, 1959). In some legumes, normal soil N probably is supplemented with organic N from roots and nodules that degenerate after clipping. In other legumes, root nodules remain intact after defoliation and symbiotic N_2 fixation continues at depressed levels until new foliage develops although much must depend on the severity of cutting (Phillips *et al*, 1983). In our study it is likely that the % Ndfa was maintained constant during the successive cuttings because N from sloughed nodules and roots after cutting replenished soil N. Losses and decomposition of roots and nodules and the fate of N in the soil and its translocation to plant need to be monitored. This may help to understand N_2 fixation and the whole N cycling processes as affected by various management practices in alley cropping. In the other case, the constant % Ndfa during the successive cuttings could be explained by the newly formed nodules following each cutting. During the early stage (3 weeks after cutting) symbiotic N_2 fixation could be partially reduced until it reaches its maximum when new foliage develops.

Quantification of N_2 fixed which ignores the contribution of roots and nodules turnover would result in serious errors (Sanginga *et al.*, 1989b). In most studies of N_2 fixed retained N in roots have been ignored because of difficulties in harvesting and sampling. However, where total N balance or the impact of crop growth on soil fertility is at issue such as alley cropping, the errors would have greater impact and as shown by the data in Table 3, the amount of N from roots, including decomposing fine roots and nodules has to be measured.

There was a good agreement between both methods of estimating N fixation in K28. Previous reports also noted the similarity between the ^{15}N dilution and the difference estimates of N_2 fixation in plant shoots of few NFTs (Cornet *et al.*, 1985; Gauthier *et al.*, 1985; Sanginga *et al.*, 1989b). However, with K636 such agreement did not occur. The difference method gave a higher estimate of N_2 fixation but was less precise. The one instance when total N difference and isotope dilution yielded identical % Ndfa and N_2 fixed occured when the fixing and non fixing *L. leucocephala* had the identical fertilizer use efficiency (Data not presented). The causes of these differences, and the different conclusions with the two isolines needs to be resolved. The great CV of the difference method suggests to us that the ^{15}N dilution method is the one to be preferred. However, this may well pose logistic difficulties in field studies. This is at present being investigated. Where quantification of total N_2 fixed is required such as in N balance studies in ecosystems or farming systems such as agroforestry then the isotope dilution is the method of choice.

Our study had emphasized that in programmes aimed at enhancing symbiotic N_2 fixation in nitrogen fixing trees, there is need not only to select the host and the Rhizobium combination for their N_2 fixing abilities with time, but also to critically assess the potential and actual impacts of management practices such as cuttings or prunings on N_2 fixation processes.

Acknowledgements

The authors gratefully acknowledge the assistance of Helga Axmann and the staff of Soil Science Unit, IAEA Seibersdorf Laboratory, for their technical assistance, Mehrnaz Tadjbakhsh for typing the manuscript and for secretarial services and Dr. G. Hardarson for helpful suggestions. Dr. K. Mulongoy and B.T. Kang from IITA kindly provided the Rhizobium isolates and seeds used in this experiment.

References

Bowen G D 1959 Field studies on nodulation and growth of *Centrosema pubescens*. Q. J. Agric. Sci. 16, 253–265.

Cornet F, Otto C, Rinaudo G, Diem H G and Dommergues Y 1985 Nitrogen fixation by *Acacia holosericea* grown in field simulated conditions. Acta Oecol. 6, 211–218.

Duguma B, Kang BT and Okali D U U, 1988 Effect of pruning intensities of three woody leguminous species grown in alley cropping with maize and cowpea on an alfisol. Agrofor. Syst. 6, 19–35.

Fiedler R and Proksch G 1975 The determination of ^{15}N by emission and mass spectrometry in biochemical analysis: A review. Anal. Chim. Acta 78, 1–62.

Fried M and Middelboe V 1977 Measurement of amount of nitrogen fixed by a legume crop. Plant and Soil 47, 713–715.

Gauthier D L, Diem H G, Dommergues Y R and Ganry F 1985 Assessment of N_2 fixation by *Casuarina equisetifolia* inoculated with *Frankia* ORS021001 using ^{15}N methods. Soil Biol. Biochem. 17, 375–379.

Guevarra A B 1976 Management of *Leucaena leucocephala* (Lam.) de Wit for maximum yield and nitrogen contribution to intercropped corn. Honolulu, HI. Ph.D. thesis, University of Hawaii, 126 p.

Guevarra A B, Whitney A S and Thompson J R 1978 Influence of intra-row spacing and cutting regimes on growth and yield of leucaena. Agron. J. 70, 1033–1037.

Kang B T, Wilson G F and Sipkens L 1981 Alley cropping maize (*Zea mays* L.) and leucaena (*Leucaena leucocephala* (Lam.) de Wit in Southern Nigeria. Plant and Soil 63, 165–179.

Kang B T, Grimme H and Lawson T T 1985 Alley cropping sequentially cropped maize and cowpea with leucaena on sandy soil in Southern Nigeria. Plant and Soil 85, 267–277.

Mulongoy K and Van der Meersch M K 1988 Nitrogen contribution by leucaena (*Leucaena leucocephala*) prunings to maize in an alley cropping system. Biol. Fert. Soils 6, 282–285.

Mohan R and Blane K 1985 Statistical programmes for micro-computers. University of Georgia, GA.

Ndoye I and Dreyfus B 1988 N_2 fixation by *Sesbania rostrata* and *Sesbania sesban* estimated using ^{15}N and total N difference methods. Soil Biol. Biochem. 20, 209–213.

Pathak P S and Patel B D 1982 Leucaena research at the Indian Grassland Fodder Research Institute. *In* Leucaena Research in the Asian-Pacific region, pp 83–88. IDRC Ottawa, Canada.

Phillips D A, Center D M and Jones M B 1983 Nitrogen turnover and assimilation during regrowth in *Trifolium subterraneum* L. and *Bromus mollis* L. Plant Physiol., 71 472–476.

Sanginga N, Mulongoy R and Ayanaba A 1986 Inoculation of *Leucaena leucocephala* (Lam.) de Wit with *Rhizobium* and its nitrogen contribution to a subsequent maize crop. Biol. Agric. Hort. 3, 347–352.

Sanginga N, Mulongoy K. and Ayanaba A. 1989a Effectivity of indigenous rhizobia for nodulation and early nitrogen fixation with *Leucaena leucocephala* growth in Nigerian soils. Soil Biol Biochem. 21, 231–237.

Sanginga N, Zapata F, Danso S K A and Bowen G D 1989b Effect of successive cuttings on uptake and partitioning of ^{15}N among plant parts of *Leucaena leucocephala*. Biol. Fert. Soils. (*In press*)

Wilson G F and Kang B T 1981 Developing stable and productive biological cropping systems for the humid tropics. *In* Biological Husbandry: A Scientific Approach to Organic Farming. Ed. B. Stonehouse. pp 193–203. Butterworth, London.

Wilson G F, Kang B T and Mulongoy K 1986 Alley cropping: Trees as sources of green-manure and mulch in the tropics. Biol. Agric. Hortic. 3, 251–267.

Vincent J M 1970 A Manual for the Practical Study of Root Nodule Bacteria. IBP Handbook No. 15. Blackwell, Oxford, 164 p.

M. L. van Beusichem (Ed.), *Plant nutrition – physiology and applications*, 675–682.

PLSO IPNC614

Inorganic N × *Bradyrhizobium* strain interaction on cowpea (*Vigna unguiculata*) varietal performance in the field, using the ^{15}N isotope dilution technique

K.O. AWONAIKE[1], K.S. KUMARASINGHE[2] and S.K.A. DANSO[2]
[1]*Institute of Agricultural Research and Training, Obafemi Awolowo University, PMB 5029, Ibadan, Nigeria, and* [2]*Joint FAO/IAEA Division, International Atomic Energy Agency, P.O. Box 100, A-1400 Vienna, Austria*

Key words: *Bradyrhizobium*, cowpea, inorganic nitrogen, ^{15}N, isotope dilution, nitrogen fixation, *Vigna unguiculata* (L.) Walp

Abstract

Genotypic differences among legume species and varieties can be exploited for increased nitrogen (N_2) fixation and yields under normal and stress conditions. The symbiotic performance and yield of inoculated and uninoculated cowpea varieties IT84E-124, TVx 3236 and AFB 1757 were examined under low (20 kg N/ha) and high (100 kg N/ha) levels of inorganic N fertilization. Large genotypic variations were observed in dry matter and total N yields, N_2 fixation and response to cowpea *Bradyrhizobium* inoculation. Inoculation resulted in greater N_2 fixed in var. IT84E-124 than in uninoculated control at the two applied N levels. For var. AFB-1757 inoculation enhanced N_2 fixed only at the high N level, while N_2 fixation in TVx 3236 was not affected by bradyrhizobial inoculation. On the average, N_2 fixation in these cowpea/*Bradyrhizobium* associations was not inhibited by the higher N rate (54.5 kg N fixed/ha as against 65.4 kg N/ha at the 20 and 100 kg N/ha rates, respectively).

There were, however, varietal differences (sometimes influenced by inoculation) in the susceptibility or tolerance of N_2 fixation to high inorganic N. For uninoculated AFB 1757, N_2 fixation was enhanced at the higher N rate by 80%, compared to more than 200% when inoculated. In contrast, the N_2 fixed in TVx 3236 was about 27% lower (irrespective of the inoculation treatment) at the higher than at the lower N rate, while that of IT84E-124 increased by 100% when uninoculated and decreased by 29% when inoculated. Our results indicate that in soils containing indigenous rhizobia, some cowpea genotypes may still respond to inoculation and that some cowpea varieties have undiminished abilities to fix N_2 in soils containing high levels of inorganic N.

Introduction

Cowpea plays a significant role in the farming systems in many developing countries in the tropics. Its importance lies mainly in its ability to provide protein-rich diets, and much of its N requirement for growth, through symbiotic N_2 fixation. Rhizobial inoculation is usually essential for high N_2 fixation in many legumes. However, several reports have shown that because cowpea *Bradyrhizobium* strains abound in most soils of the major cowpea growing areas (Doku, 1969, Sellschop, 1962), cowpeas rarely respond to bradyrhizobial inoculation (Awonaike and Onyemaobi, 1982; Doku, 1969; Ezedinma, 1963; Minchin *et al.*, 1978; Rhodes and Nangju, 1979).

Several soil factors influence nodulation and nitrogen fixation in legumes. While low levels of N fertilization have been reported to enhance N_2

fixation in some legumes (Awonaike *et al.*, 1980; Dart and Wildon, 1970; Pate and Dart, 1961), high levels generally inhibit nodulation and N_2 fixation in most legumes (Awonaike *et al.*, 1980; Summerfield *et al.*, 1976). Large genotypic variations have however been observed in many legume species for tolerance to various levels of inorganic N (Danso *et al.*, 1987; Hardarson *et al.*, 1984; Senaratne *et al.*, 1987). This is important in farming systems of the tropics where cowpea is commonly grown in mixtures with cereals which require high amounts of inorganic N for optimum growth. To derive maximum benefits from cowpeas used in intercrops, it is therefore essential to select cowpea-*Bradyrhizobium* strain symbiotic associations that are tolerant of high inorganic N.

We report here the results of a field study to examine differences in the response of three cowpea varieties to inoculation and their symbiotic effectiveness under low and high inorganic N condition.

Materials and methods

The study was conducted in 1987 at the Ballah substation of the Institute of Agricultural Research and Training in Ilorin, Nigeria (8°30′N, 5°35′E). The soil was a well drained, leached ferrogenous loamy sand (D'Hoore, 1964). Some of its physical and chemical properties are shown in Table 1. Single superphosphate (300 kg/ha) and muriate of potash (100 kg/ha) were applied as basal fertilizers at planting.

Table 1. Some physical and chemical properties of the soil

(i) Physical properties	
Sand (%)	81
Silt (%)	8
Clay (%)	11
(ii) Chemical properties	
pH	6.5
Organic C (%)	0.24
N (%)	0.02
Available P-Bray 1(mg/kg)	2.1
Exchangeable cations (meq/100 g)	
Ca	2.5
K	0.19
Na	0.09

Three erect cowpea varieties (IT84E-124, TVx 3236 and AFB 1757) which differed in days to maturity (60, 75 and 75, respectively) were used. Maize (*Zea mays* L.) var. Kewesoke was used as the reference non-N_2 fixing crop.

A mixture of *Bradyrhizobium* strains, designated as BR_E, consisting of peat-based strains IRc 252 (isolated from Onne in South-eastern Nigeria) and IRc 430^A (isolated from Maradi in Niger Republic) was used to inoculate surface sterilized cowpea seeds before planting. These strains were known to perform well in areas with no previous history of cowpea production (Asanuma and Ayanaba, 1985). The uninoculated control seeds (BR_0) were surface-sterilized and planted directly into the soil without inoculation.

Two fertilizer rates, 20 kg N/ha and 100 kg N/ha (designated as low N or N_L, and high N or N_H, respectively) in the form of ammonium sulphate with 5 and 1% ^{15}N atom excess, respectively, were uniformly applied in solution form to the plots immediately after sowing. A split-plot randomized complete block design was used, with inoculation/N fertilizer rates as main plots and cowpea varieties as subplots; these were replicated 5 times. Each subplot measured $2m^2$ with interrow spacing of 40 cm and 20 cm within the row. Seedlings were thinned from two to one plant per hill one week after planting. Due to drought caused by the early cessation of rainfall in the late planting season of 1987, the above ground portions of all plants were harvested 54 days after planting. Samples were oven dried at 70°C. Percent N was determined in a Kjeldahl digest (Eastin, 1978) and the N isotopes ratio analyses were performed on a VG-isogas mass spectrometer (Fiedler and Proksch, 1975). The isotope dilution equation of McAuliffe *et al.*, (1958) was used to calculate the %N derived from the atmosphere (%Ndfa) and to obtain estimates of N uptake from soil and fertilizer.

Data collected were analysed statistically using the computer statistical programme package of Mohan and Blane (1985). Two-way analyses of variance were performed for nitrogen rate/*Bradyrhizobium* strain combination effects on the varieties of cowpea (results are not shown). Separate one-way analyses of variance were then performed for *Bradyrhizobium* strain × varietal interactions at each N level, where significant ($P<0.05$) interaction effects were found in the two way analyses.

Results

(a) *Dry matter and nitrogen yield*

Uninoculated (BR_0) *treatments*
Data on dry matter yields are shown in Table 2. Due to the drought, plant stands were not uniform and consequently dry matter yields were characterized by high variability. Thus, although uninoculated (BR_0) TVx 3236 outyielded the other two varieties by about 25% at the low fertilizer N rate, the differences were not significant ($P<0.05$). At the high N rate also, no varietal difference was observed although AFB 1757 outyielded IT84E-124 by more than 50%. The dry matter yields of uninoculated IT84E-124 and TVx 3236 were not influenced by N rates, whereas that of AFB 1757 was enhanced by more than 30% at the high N rate. Overall, N rate had little effect on the yield of uninoculated cowpeas (average 3.0 and 3.1 t/ha at the low and high N rates respectively).

Inoculated (BR_E) *treatments*
Inoculation responses were variable and were influenced by both plant genotype and N rate. The highest yield increases over the uninoculated controls were obtained with IT84E-124 (54 and 63% at the N_L and N_H rates, respectively). Yields of TVx 3236 on the other hand remained virtually unaffected by inoculation at both N levels, while a significant reduction ($P<0.05$) of over 50% was recorded for AFB 1757 at the low N level, in contrast to a non-significant increase of 35% at the high N rate. On the average, there was only a slight increase in the yield of all plants (<10%) due to inoculation at the low N level, compared to a marked increase (32%) at the higher N rate.

The data on N yields for the uninoculated and inoculated treatments are shown in Table 3. They were strongly influenced by dry matter yield ($R = 0.96$). Thus, the trends and magnitudes of N yield differences among all treatments were virtually identical to those reported above for dry matter yields (Table 2).

(b) *Nitrogen fixation*

Data on total N_2 fixed and the proportion of N derived from fixation are shown in Table 4. Except in one case (AFB 1757 at the N_L level) these two parameters were closely linked. The overall correlation coefficient between total N and %Ndfa was 0.78 which was highly significant ($P<0.003$).

Uninoculated (BR_0) *treatments*
At the N_L level, average N_2 fixed was 42% and TVx 3236 was the only variety with more than 50% of its N derived from atmospheric N_2 fixation (Table 4). This was almost twice and about 5 times more N_2 fixed (total) than in uninocu-

Table 2. Total dry matter yield of IT84E-124, TVx 3236 and AFB 1757 varieties of cowpea, uninoculated (BR_0) and inoculated with a mixture of *Bradyrhizobium* strains (BR_E), at two levels of inorganic N application

Variety	Inoculation	Dry matter yields (ton/ha) at two rates of N applied (kg N/ha)	
		20	100
IT84E-124	BR_O	2.8^{bc}*	2.4^{b}
TVx 3236		3.5^{ab}	3.3^{ab}
AFB 1757		2.8^{b}	3.7^{ab}
IT84E-124	BR_E	4.3^{a}	3.9^{ab}
TVx 3236		3.8^{ab}	3.5^{ab}
AFB 1757		1.8^{c}	5.0^{a}
CV (%)		28.1	32.4

* Mean values in a vertical column followed by the same letter(s) do not differ significantly according to Duncan's Multiple Range Test ($P<0.05$).

Table 3. Total nitrogen yield for IT84E-124, TVx 3236 and AFB 1757 varieties of cowpea, uninoculated (BR_0) and inoculated by a mixture of *Bradyrhizobium* strains (BR_E), at two levels of inorganic N application

Variety	Inoculation	Total yield of N (kg M/ha) at two rates of N applied (kg N/ha)	
		20	100
IT84E-124	BR_I	75.1^{cd}*	67.8^{b}
TVx 3236		127.0^{ab}	103.7^{ab}
AFB 1757		93.2^{bc}	113.6^{ab}
IT84E-124	BR_E	142.1^{a}	112.3^{ab}
TVx 3236		124.0^{ab}	107.5^{ab}
AFB 1757		53.4^{d}	152.5^{b}
CV (%)		25.7	30.2

* Mean values in a vertical column followed by the same letter(s) do not differ significantly according to Duncan's Multiple Range Test ($P < 0.05$).

lated AFB 1757 and IT84E-124, respectively. Although the %Ndfa in AFB 1757 at the low N rate was significantly ($P < 0.05$) higher than in the uninoculated IT84E-124, the corresponding values of total N_2 fixed were not significantly different ($P < 0.05$), apparently due to high variability in plant growth.

At the high N level, average %Ndfa was 54 and the only uninoculated variety with less than half of its N derived from fixation was IT84E-124, although this still represented an increase of more than twice the proportion of N derived from N_2 fixation in the corresponding IT84E-124 treatment at the low N level. The trend for uninoculated AFB 1757 and TVx 3236 at the high N level was the reverse of that found at the low N level, with AFB 1757 now deriving a higher proportion of its N from fixation than TVx 3236. The differences were however not significant ($P < 0.05$). The trend for total N_2 fixed in the uninoculated plants at the high N level mirrored that for %Ndfa ($R = 0.91$).

The highest N_2 was fixed in AFB 1757, and was almost 3 times that fixed in the poorest, IT84E-124. Also, total N_2 fixed by uninoculated AFB 1757 at the high N level was almost 80% more than that fixed at the low N level. For uninoculated TVx 3236, however, there was a

Table 4. N fixed in IT84-124, TVx 3236 and AFB 1757 varieties of cowpea, uninoculated (BR_0) and inoculated by a mixture of *Bradyrhizobium* strains (BR_E), at two levels of inorganic N application

Variety	Inoculation	Total N fixed (kg N/ha) and fixed N expressed as % of total N in plants at two rates of N applied (kg N/ha)			
		20		100	
		Total	%	Total	%
IT84E-124	BR_0	14.3^{c}	19^{b}*	29.2^{c}	43^{cd}
TVx 3236		81.3^{a}	64^{a}	53.9^{b}	52^{bcd}
AFB 1757		40.1^{bc}	43^{a}	72.7^{ab}	64^{ab}
IT84E-124	BR_E	85.3^{a}	60^{a}	52.8^{b}	47^{cd}
TVx 3236		69.4^{ab}	56^{a}	62.4^{bc}	58^{abc}
AFB 1757		34.2^{c}	64^{a}	109.8^{a}	72^{a}
CV (%)		33.1	38.5	17.8	28.3

* Mean values in a vertical column followed by the same letter(s) do not differ significantly according to Duncan's Multiple Range Test ($P < 0.05$).

27% reduction in total N_2 fixed with increased N rate, and total N_2 fixed was also about 17% lower than that fixed in a similarly treated AFB 1757 at the high N level.

Inoculate (BR_E) treatments
With Bradyrhizobial inoculation, an average of 60% of total N was fixed at the low N level and none of the varieties derived less than 50% of total N from atmospheric N_2 fixation (Table 4). The greatest increase in %Ndfa (over 3-fold) occurred in IT84E-124 which, when nodulated by native bradyrhizobial strains, derived little N from fixation. Inoculated AFB 1757 achieved a modest increase in %Ndfa over the uninoculated treatment, compared to a slight drop in inoculated TVx 3236. Similar to %Ndfa, total N_2 fixed increased substantially (more than five-fold) when IT84E-124 was inoculated, compared to the uninoculated treatment. For TVx 3236, inoculation did not make much difference in total N_2 fixed, while a drop of 20% in total N_2 fixed resulted from the inoculation of AFB 1757. Mean total N_2 fixed was 64.5 kg N/ha.

High inorganic N application resulted in variable responses in N_2 fixation in the inoculated treatments. Compared to the lower N level, %Ndfa increased from 64 to 72 in AFB 1757, was virtually unchanged in TVx 3236 and dropped from 60 to 47% in IT84E-124. Average N_2 fixed was 76.7 kg N/ha for inoculated varieties at the high N level. Differences in total N_2 fixed in the inoculated plants at the low and high N levels in general appeared to be more pronounced than for %Ndfa. Variety AFB 1757 gave more than a three-fold increase in total N_2 fixed at the high compared to the low N level. In contrast, total N_2 fixed in TVx 3236 and IT84E-124 decreased at the higher N level by about 48 and 29%, respectively.

(c) *Nitrogen uptake from soil sources (fertilizer and native soil N)*

Uninoculated (BR_0) treatments
Data on total N accumulated from soil sources are shown in Table 5. Substantially more fertilizer N (almost ten-fold on the average) was absorbed at the high than at the low N level. There were, however, no varietal differences ($P < 0.05$) in the fertilizer N uptake at each of the two levels of applied N. In contrast, more soil N was absorbed at the low than at the high N level. Although there was no varietal difference in native soil N uptake at the low N level (with an average of 44.6 kg N/ha), TVx 3236 significantly ($P < 0.05$) absorbed more N from the soil than either of the other two varieties at the high N level of application.

Inoculated (BR_E) treatments
With the exception of AFB 1757, in which inoculation significantly ($P < 0.05$) reduced N fertilizer uptake at the low N level, neither the level of N applied nor inoculation had any significant influence on the quantity of fertilizer N absorbed by the different cowpea varieties (Table 5).

Table 5. N uptake from soil sources by IT84E-124, TVx 3236 and AFB 1757 varieties of cowpea, uninoculated (BR_0) and inoculated by a mixture of *Bradyrhizobium* strains (BR_E), at two levels of inorganic N application

Variety	Inoculation	N uptake (kg/ha) at two rates of N applied (kg/ha)			
		20		100	
		Fert. N	Native soil N	Fert. N	Native soil N
IT84E-124	BR_0	4.4[a]*	56.4[a]	25.4[a]	13.2[cd]
TVx 3236		3.9[a]	41.8[a]	31.1[a]	18.7[b]
AFB 1757		3.6[a]	49.5[a]	30.4[a]	10.5[d]
IT84E-124	BR_E	4.8[a]	52.0[a]	32.0[a]	27.7[a]
TVx 3236		3.7[a]	50.9[a]	34.5[a]	10.6[d]
AFB 1757		1.5[b]	17.7[b]	27.1[a]	15.6[bc]
C (%)		20.5	16.8	15.8	17.4

* Mean values in a vertical column followed by the same letter(s) do not differ significantly according to Duncan's Multiple Range Test ($P < 0.05$).

Again, the only variety for which inoculation resulted in a significantly ($P<0.05$) reduced native soil N uptake was AFB 1757. At the high N level however, inoculation increased native soil N uptake in IT84E-124 by over 100% and by more than 50% in AFB 1757. In TVx 3236 on the contrary, a reduction of over 40% was observed.

Discussion

Most reports indicate that in areas where cowpea is normally cultivated, and where rhizobial strains occur naturally in the soil, inoculation with bradyrhizobial strains usually does not result in increased nodulation and dry matter yield (Asanuma and Ayanaba, 1985; Awonaike and Onyemaobi, 1982; Ezedinma, 1963). The fact that we observed many nodules on the uninoculated plants (data not presented), which were of different ages and N_2 fixing activity (as determined by size and interior pigmentation) suggests that the soil contained sufficient numbers of indigenous cowpea rhizobia. This was not unexpected as cowpea is a traditional crop in this area. However, our results do not fully support the several reports that postulate no response to inoculation in such a soil. Rather, our data indicate that response to inoculation in the presence of native rhizobia can occur with some genotypes. This finding will be discussed in the light of the three different responses displayed by the three varieties at the low N level (which is closer to the situation in many farmers fields than the high N level).

These responses ranged from the frequently reported lack of increase in dry weight for TVx 3236 through a greater than 50% increase for IT84E-124 to a more than 50% reduction in dry matter yield of AFB 1757. The lack of an increase or the presence of only a slight increase in dry matter yield with increasing N level in the uninoculated plants also suggests that native bradyrhizobial strains were effective in N_2 fixation. Our data and those of Eaglesham *et al.* (1982) which are among the few available quantitative field estimates available for N_2 fixed by cowpea indicate that like soybean (Rennie, 1984), cowpea on the average obtained about half of its N through symbiotic N_2 fixation. However for soybean, because of its normally higher dry matter yield, the 50% Ndfa was equivalent to 100 kg N/ha, compared to the 60 kg N/ha for all the cowpea treatments in the present study.

Few studies have dealt with plant genotypic differences in N_2 fixation among cowpeas and how these could be utilized for optimum N_2 fixation under different situations. Even though a limited gene pool was examined, the results we obtained were interesting. Plant genotypic variation had a more pronounced effect on N_2 fixation in the uninoculated than in the inoculated plants grown at the low N level. From this study, we deduce that, although IT84E-124 would be a poor choice in farming without bradyrhizobial inoculation, it would be a profitable cultivar in case a suitable rhizobial strain is introduced. The opposite is observed for AFB 1757, which performed better without than with inoculation. A suitable compromise is, however, TVx 3236, which at the low N level performed almost equally well with and without inoculation.

High levels of available N have been reported to commonly limit the potential of the legume-*Rhizobium* symbiosis to fix N_2. However, some species and varieties support greater N_2 fixation than others when soil inorganic N is high (Copeland and Pate, 1969; Hardarson *et al.*, 1984; Senaratne *et al.*, 1987). Proper use made of this known genetic variability among existing species and cultivars should therefore result in high N_2 fixation and high crop yield, when soil N is high or when fertilizer N needs to be applied to an associated non-N_2 fixing crop. In our studies cowpea generally displayed a remarkably undiminished ability to support N_2 fixation as the fertilizer N rate was increased from 20 to 100 kg N/ha, a level that was reported to severely decrease N_2 fixation in common bean (Ruschel *et al.*, 1979) and soybean (Hardarson *et al.*, 1984; Wagner and Zapata, 1982). In our studies, nitrogen fixation (both % and total) was generally slightly increased when N fertilization was increased from 20 to 100 kg N/ha. These results, if confirmed in subsequent studies and for other varieties of cowpea, would indicate that cowpea is a suitable species for intercropping, also in situations of high soil inorganic N levels.

The interaction between plant genotypes and

bradyrhizobial strains (inoculated versus uninoculated) yielded some large differences in N_2 fixation. Variety AFB 1757 was the most outstanding in its tolerance of high inorganic N and, irrespective of being inoculated or not, gave substantial (80 and >200%) increases in total N_2 fixed at the higher than at the lower N rate. However, with the other two varieties, except in the uninoculated IT84E-124 where high N resulted in more than 100% increase in total N_2 fixed compared to low N, the high N rate slightly depressed N_2 fixation. Our results further indicate the great potential of screening within a given plant species for variability in N_2 fixation under stress conditions.

Sometimes the impression that N_2 fixation in a variety or species is tolerant of high inorganic N may be false. This could be the case when leaching of added N fertilizer is so excessive that differences in N rates added become narrow. In our study, however, this was not the case as the average N uptake (for both inoculated and uninoculated treatments) at the low N rate was 3.7 kg/ha or 18.5% fertilizer use efficiency (FUE) compared to 30.1 kg/ha at the high N rate or 30.1% FUE (*i.e.* an almost 10-fold difference in N uptake for a 5-fold difference in N applied). These % FUE values are either similar to or higher than many values reported for grain legumes which in addition to the uptake of N from soil sources, fix substantial amounts of atmospheric N_2 (Fried and Broeshart, 1975; Rennie *et al.*, 1982; Wagner and Zapata, 1982).

Acknowledgements

The technical assistance of the IAEA, Vienna through the Africa Core BNF Programme of the Agency is gratefully acknowledged. Dr K Mulongoy of IITA kindly supplied the strains of *Bradyrhizobium*. Ms H Axmann of the IAEA Laboratory, Seibersdorf, Austria analysed the samples for N content and N isotopic ratios. The staff of the Institute of Agricultural Research and Training, Ballah-Ilorin substation provided field support. We thank Drs G Hardarson and N Sanginga for their useful comments and Ms M Tadjbakhsh for typing the script.

References

Ahmad M H, Eaglesham A R J, Hassouna S and Seaman B 1981 Examining the potential for inoculant use with cowpeas in West African soils. Trop. Agric. Trinidad 58, 325–335.

Asanuma S and Ayanaba A 1985 Evaluation of selected *Rhizobium* strains for symbiotic effectiveness on cowpea under tropical conditions. *In* BNF in Africa. Proc. 1st AABNF. Eds. H Ssali and S Keya. pp. 504–511

Awonaike K O, Lea P J, Day J M, Roughley R J and Miflin B J 1980 Effects of combined nitrogen on nodulation and growth of *Phaseolus vulgaris*. Exp. Agric. 16, 303–311.

Awonaike K O and Onyemaobi N 1982 Effect of nitrogen fertilizer application and *Rhizobium* strains on nodulation and yield of Ife Brown var. of cowpea (*Vigna unguiculata*) Nig. J. Sci. 16, 1–9.

Copeland R and Pate J S 1969 Nitrogen metabolism of nodulated white clover in presence and absence of nitrate nitrogen. *In* White Clover Research. Ed. J Lowe. pp 71–77. Occasional Symposium. No. 6, British Grassland Society.

Danso S K A, Hera C and Doucka C 1987 Nitrogen fixation in soybean as influenced by cultivar and *Rhizobium* strain. Plant and Soil 99, 163–174.

Danso S K A, Hera C and Doucka C 1987 Nitrogen fixation in soybean as influenced by cultivar and *Rhizobium* strain. Plant and Soil 99, 163–174.

Dart P J and Wildon D C 1970 Nodulation and nitrogen fixation by *Vigna sinensis* and *Vicia purpurea*: The influence of concentration, form and site of application of combined nitrogen. Aust. J. Agric. Res. 21, 45–46.

D'Hoore J L 1964 Soil map of Africa. Commission for Technical Cooperation in Africa Joint Project No. 11.

Doku E V 1969 Host specificity among five species in the cowpea cross-inoculation group. Plant and Soil 30, 126–128.

Eaglesham A R J, Ayanaba A, Ranga Rao V and Eskew D L 1982 Mineral N effects on cowpea and soybean crops in a Nigerian soil. II. Amounts of N fixed and accrual to the soil. Plant and Soil 68, 183–192.

Eastin E F 1978 Total nitrogen determination for plant material. Anal. Biochem. 85, 591–594.

Ezedinma F O C 1963 Notes on the distribution and effectiveness of cowpea rhizobia in Nigerian soils. Plant and Soil 21, 134–136.

Fiedler R and Proksch G 1975 The determination of ^{15}N by emission and mass spectrometry in biochemcial analysis. A review. Anal. Chim. Acta 78, 1–62.

Fried M and Broeshart H 1975 An independent measurement of the amount of nitrogen fixed by a legume crop. Plant and Soil 43, 707–711.

Hardarson G, Zapata F and Danso S K A 1984 Effect of plant genotype and nitrogen fertilizer on symbiotic nitrogen fixation by soybean cultivars. Plant and Soil 82, 397–405.

McAuliffe C, Chamblee D S, Uribe-Arango H and Woodhouse W W 1958 Influence of inorganic N on nitrogen fixation by legumes as revealed by ^{15}N. Agron. J. 50, 334.

Minchin F R, Summerfield R J and Eaglesham A R J 1978 Plant genotype and *Rhizobium* strain interactions in cowpea (*Vigna unguiculata* (L.) Walp.) Trop. Agric. Trinidad 55, 107–115.

Mohan R and Blane K 1985 Statistical programmes for microcomputers. University of Georgia, Athens, GA.

Pate J S and Dart P J 1961 Nodulation studies in legumes. IV. The influence of inoculum strain and time of application of ammonium sulphate on symbiotic response. Plant and Soil 15, 329–346.

Rennie R J 1984 Comparsion of N balance and ^{15}N isotope dilution to quantify N_2-fixation on field-grown legumes. Agron. J. 76, 785–790.

Rennie R J, Dubetz S, Bole J B and Muendel H H 1982 Dinitrogen fixation measured by ^{15}N isotope dilution in two Canadian soybean cultivars. Agron. J. 74, 725–730.

Rhodes E R and Nangju D 1979 effects of pelleting cowpea and soybean seeds with fertilizer dusts. Exp. Agric. 15, 27–32.

Ruschel A P, Salati E and Vose P B 1979 Nitrogen enrichment of soil and plant by *Rhizobium phaseoli-Phaseolus vulgaris* symbiosis. Plant and Soil 51, 425–429.

Sellschop J P F 1962 Cowpea, *Vigna unguiculata* (L.) Walp. Field Crops Abstract 15, 259–266.

Senaratne R, Amornpimol C and Hardarson G 1987 Effect of combined nitrogen on nitrogen fixation of soybean (*Glycine max* L. Merrill) as affected by cultivar and rhizobial strain. Plant and Soil 103, 45–50.

Summerfield R J, Huxley P A, Dart P J and Hughes A P 1976 Some effects of environmental stress on seed yield of cowpea (*Vigna unguiculata* (L.) Walp.) cv. Prima. Plant and Soil 44, 527–546.

Wagner G H and Zapata F 1982 Field evalaution of reference crops in the study of nitrogen fixation by legumes using isotope techniques. Agron. J. 74, 607–612.

E

Methods and techniques aiming at optimization of crop nutrition and growth

M. L. van Beusichem (Ed.), *Plant nutrition – physiology and applications*, 685–692.
 PLSO IPNC202

Prospects for improvement of nitrogen fertilizer recommendations for cereals: A simulation study

J.J.R. GROOT[1] and H. VAN KEULEN[2]
[1]*Institute for Soil Fertility Research, P.O. Box 30003, 9750 RA Haren, The Netherlands, and* [2]*Centre for Agrobiological Research, P.O. Box 14, 6700 AA Wageningen, The Netherlands*

Key words: immobilization, leaching, mineralization, soil mineral nitrogen, split nitrogen application

Abstract

A simulation model for cereal growth, crop nitrogen dynamics and soil nitrogen supply was used to compare the effects of alternative fertilizer application strategies on nitrogen use efficiency, crop yield and grain quality. Based on the simulations it is concluded that alternative strategies in terms of rate and timing do not improve the N fertilization recommendations currently used in agricultural practice. A better understanding of organic matter dynamics in the soil and nitrogen transport in the rhizosphere are necessary for further improvement of fertilizer recommendations and nitrogen use efficiency.

Introduction

In the Netherlands, average grain yields exceed 8 t ha^{-1}. In addition to breeding efforts and better pest and disease control, the increase in cereal yields over the last decades is the result of high nitrogen inputs. The nitrogen requirement of a crop yielding 10.000 kg grain dry matter per ha is approximately 270 kg ha^{-1}. The recommended rate of fertilizer application in the Netherlands includes a mineral N supply (soil + fertilizer N) of 140 kg ha^{-1} in early spring, followed by topdressings up to 100 kg ha^{-1} (Prins *et al.*, 1988).

In cereals, two phases with respect to the nitrogen economy can be distinguished. Before anthesis, increased nitrogen uptake of the crop results mainly in increased photosynthetic surface, larger numbers of tillers and larger numbers of grains per ear, while photosynthetic capacity per unit leaf area is hardly affected (Morgan, 1988). After anthesis, nitrogen uptake normally ceases and most of the nitrogen for grain growth is translocatead from vegetative plant parts, resulting in a decline in photosynthesis (Groot and Spiertz, 1990).

The effect of nitrogen application not only depends on its rate, but also on the time of application. Fertilizer experiments have shown that splitting the total N dressing improves N use efficiency, but as long as the total amount applied is close to the optimum for crop performance, the effects on yield between different methods of splitting the total amount applied are small (Sylvester-Bradley *et al.*, 1987).

At present, both grain-quality and environmental considerations have resulted in reanalysis of existing results of fertilizer experiments or even in new series of field experiments. We believe that, due to the multiple interactions between various physiological processes affecting yield formation and the fluctuations in environmental and nutritional limitations, the extent to which fertilizer affects yield and quality cannot be easily predicted from experimental results only. It has been illustrated that simulation of nitrogen dynamics in soil and crop nitrogen uptake facilitates evaluation of the effects of fertilizer applications within a given set of constraints (Groot and Spiertz, 1990; Neeteson *et al.*, 1989).

In the present study a simulation model.is used to examine the prospects for improved nitrogen fertilizer recommendations, both in terms of yield and nitrogen uptake.

Simulation model

Crop growth

Simulation of crop growth is based on Spitters *et al*. (1989). Gross canopy photosynthesis is calculated as a function of leaf area index, radiation distribution within the canopy, and the photosynthesis-light response curve of individual leaves. Maintenance requirements for various plant organs are first subtracted and the partitioning of the remaining assimilates to leaves, stems, roots and grains is varied with the stage of development of the crop according to fixed empirical functions. The rate of crop development is a function of ambient air temperature, but is modified to account for the effects of vernalization and photoperiod. Assimilates allocated to various plant organs are converted into structural plant material, taking into account the energy required for conversion (growth respiration) as a function of the protein content of the growing material. In the model, both photosynthesis and maintenance respiration increase with increasing nitrogen content of vegetative plant parts (Van Keulen and Seligman, 1987).

The rate of root extension is related to soil moisture content and temperature of the soil compartment in which root extension occurs. It is assumed that root length density decreases exponentially with depth in the profile.

Crop nitrogen demand is based on the concept of nitrogen deficiency. If the nitrogen content of a given plant part is below its maximum value corresponding to the current stage of development, a sink for nitrogen exists. Maximum values of nitrogen content were assessed in a series of field experiments in which nitrogen supply was considered not to be limiting (Groot, 1987). Actual nitrogen uptake proceeds at a maximum rate until crop demand is satisfied if nitrogen transport through the soil is not limiting, otherwise uptake is determined by nitrogen availability.

Soil moisture

The soil is treated as a multilayer system. Changes in soil water status are derived from the combined effect of infiltration, extraction due to soil surface evaporation, extraction by the roots and downward movement through the compartments if their moisture storage capacity is exceeded. Potential soil surface evaporation and potential crop transpiration are calculated according to a modified Penman approach. When actual transpiration, which depends on soil moisture status, is lower than potential transpiration, gross canopy assimilation is reduced proportionally.

Soil mineral nitrogen

The amount of available mineral nitrogen depends on the balance of fertilizer application, decomposition of old organic matter (humus) and fresh organic matter (crop residues), crop nitrogen uptake, and transport of nitrate through the profile. Denitrification is ignored. Decomposition of both fresh and old organic matter is treated as a first-order kinetics process, where the specific rate of decomposition is modified to account for the effects of soil temperature and soil moisture content. Decomposition results in either mineralization or immobilization of nitrogen, depending on the C/N ratio of the substrate.

According to De Willigen and Van Noordwijk (1987), nitrogen uptake proceeds according to crop demand if transport through the soil is not limiting. Alternatively, the roots act as a "zero-sink", *i.e.* uptake has a maximum under the prevailing conditions. Diffusion of nitrate to the root surface is considered to be the major limiting process in nitrogen uptake, because soil water content strongly affects the diffusion coefficient (Barraclough and Tinker, 1981).

Simulations

Simulations were based on a series of field experiments with winter wheat where the soil properties of the experimental sites were well-known, and for which weather data were available (Anon., 1982–1985). The total amount of nitrogen applied ranged from 0 to 200 kg ha^{-1}. During the growing season, dry matter production and nitrogen uptake were measured at three-weekly intervals in each of the treatments. In Fig. 1 simulated and measured values of dry matter production and nitrogen uptake in 1982 and 1984 are compared. For these two years, 82% of the simulated values for dry matter production (Fig. 1a) and 74% of those for nitrogen uptake (Fig. 1b) were within 20% of the measured values.

In 1983 and 1984, another series of detailed

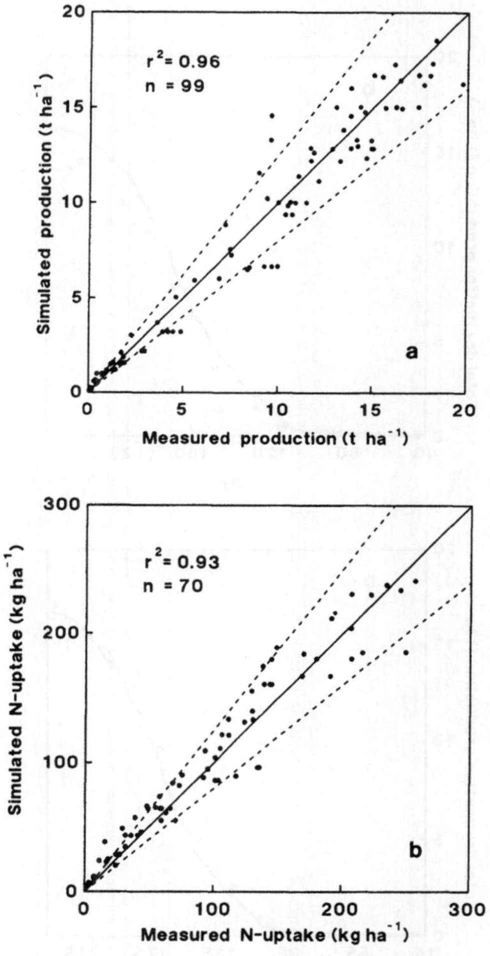

Fig. 1. Comparison of measured and simulated total dry matter yield **(a)** and measured and simulated nitrogen uptake **(b)** for a series of fertilizer experiments with winter wheat with nitrogen applications ranging from 0 to 200 kg per ha. Dotted lines indicate a 20% deviation from the 1:1 line.

field experiments was carried out, mainly to obtain crop-specific parameters for the simulation model (Groot, 1987). In these experiments, crop production, nitrogen uptake and soil mineral nitrogen were measured at intervals of two to three weeks. Although measurements and simulations are not completely independent, these data sets have been used to validate simulations of the time course of soil mineral nitrogen over the growing season. For situations without nitrogen application, amounts of soil mineral nitrogen and nitrogen uptake are simulated accurately (Fig. 2a). The same holds for nitrogen applied early in the season (February): Figure 2b. However, when fertilizer is applied later in the season, the simulated sharp increase in total soil mineral nitrogen is not reflected in the measurements: Figure 2c. Despite these discrepancies, simulated and measured crop nitrogen are in reasonably good agreement (Fig. 2c). Dry matter production is simulated satisfactorily for each of the situations studied (Fig. 3).

In the Netherlands, the recommended rate of application is based on a mineral N supply (soil + fertilizer N) of 140 kg ha^{-1} early in spring, followed by applications of 60 kg ha^{-1} at the start of stem elongation and 40 kg ha^{-1} at heading. The simulation model was used to compare, for a sandy soil and a clay soil, the effect of nitrogen either applied as a single application of 180 kg ha^{-1} on February 15th, or as split dressings on February 15th (80 kg ha^{-1}), May 10th (60 kg ha^{-1}) and June 15th (40 kg ha^{-1}). The amount of mineral nitrogen at the start of the simulations was 60 kg ha^{-1}.

Results of the simulations are given in Table 1. For both the clay soil and the sandy soil, the simulated differences in yield between single and split applications are negligible. For the clay soil simulated nitrogen uptake is similar for the two treatments, but for the sandy soil, split application increases nitrogen uptake by 10%. On both soil types and in both treatments mineral nitrogen had accumulated in the soil at harvest. The simulated results for the clay soil show that both for the single and the split application, most of the mineral nitrogen at harvest is in the top 30 cm of the soil, while the layer 30–60 cm is almost depleted (Fig. 4). In the sandy soil, min-

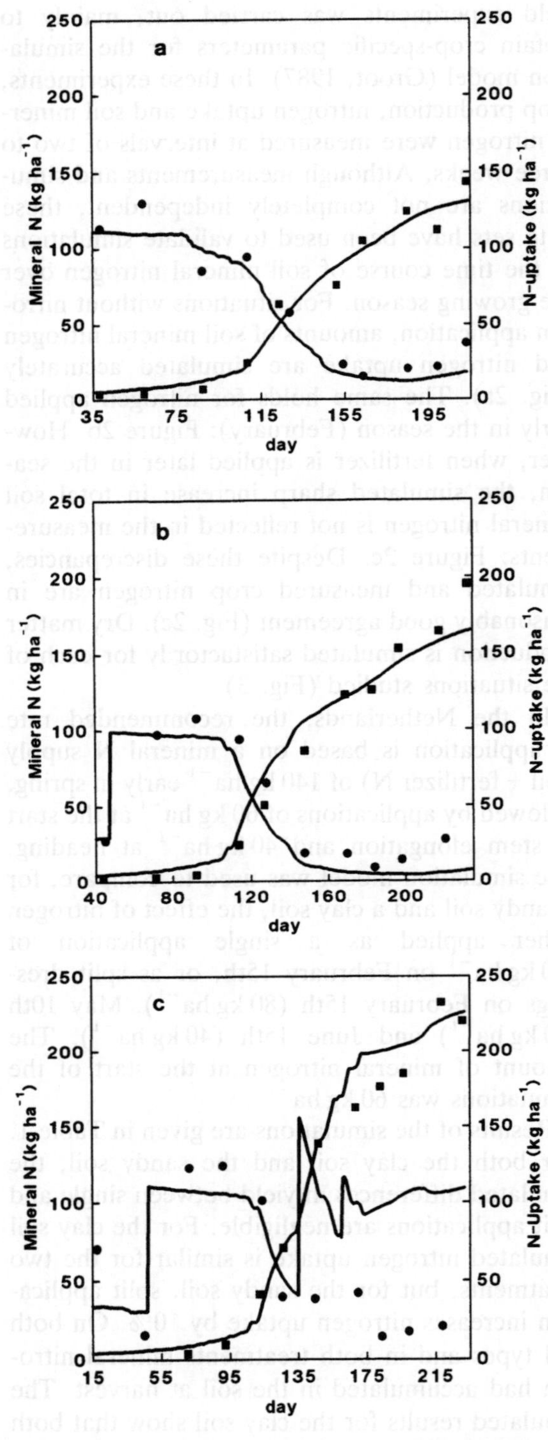

Fig. 2. Time course of soil mineral nitrogen in the 0–100 cm layer and crop nitrogen uptake for field experiments in which no nitrogen was applied (a), nitrogen was applied early in the season (b) or was applied in three split applications (c). ●: measurements of soil mineral nitrogen, ■: measurements of crop nitrogen uptake, ——: simulations.

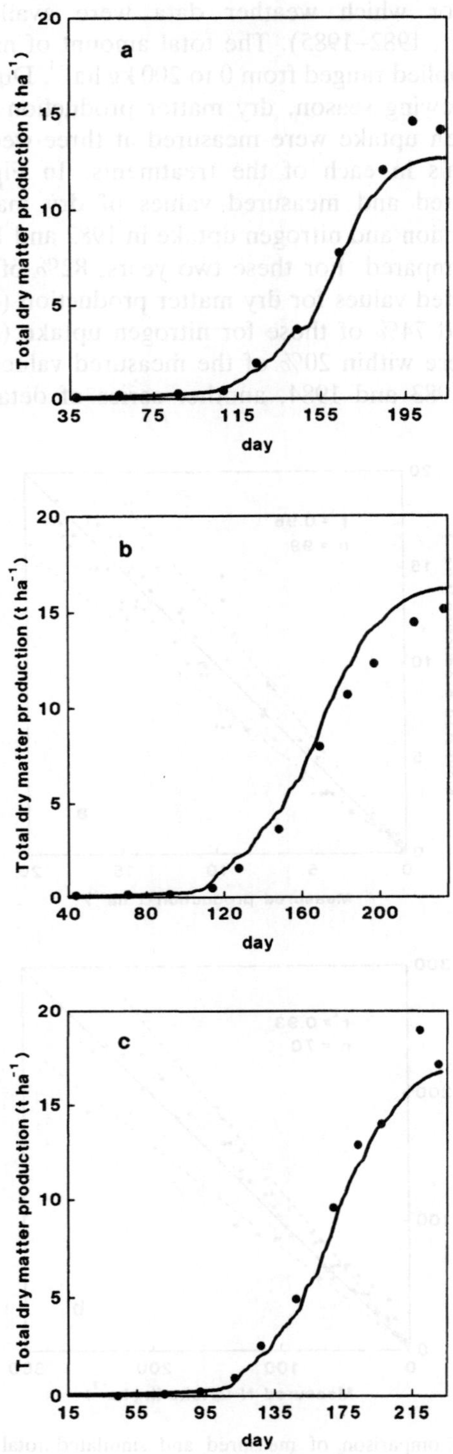

Fig. 3. Time course of total dry matter production for a field experiment in which no nitrogen was applied (a), an experiment with a single application early in the season (b) and an experiment with three nitrogen applications (c). ●: measurements of total dry matter production, ——: simulations.

Table 1. Simulated values (kg ha^{-1}) of grain yield, crop nitrogen uptake (N_{upt}), amount of mineral nitrogen in the soil at harvest in the layers 0–30 cm (N_{0-30}), 30–60 cm (N_{30-60}) and 60–100 cm (N_{60-100}) and amount of nitrogen leached between sowing and harvest for two fertilizer regimes on clay and sand

	Yield	N_{upt}	N_{0-30}	N_{30-60}	N_{60-100}	Leaching
Clay, 80 + 60 + 40	7807	242	49.1	0.2	15.8	21.4
Clay, 180 + 0 + 0	7806	242	40.1	5.9	17.6	21.4
Sand, 80 + 60 + 40	5372	191	20.7	29.7	14.3	79.5
Sand, 180 + 0 + 0	5330	174	13.9	48.1	20.3	79.5

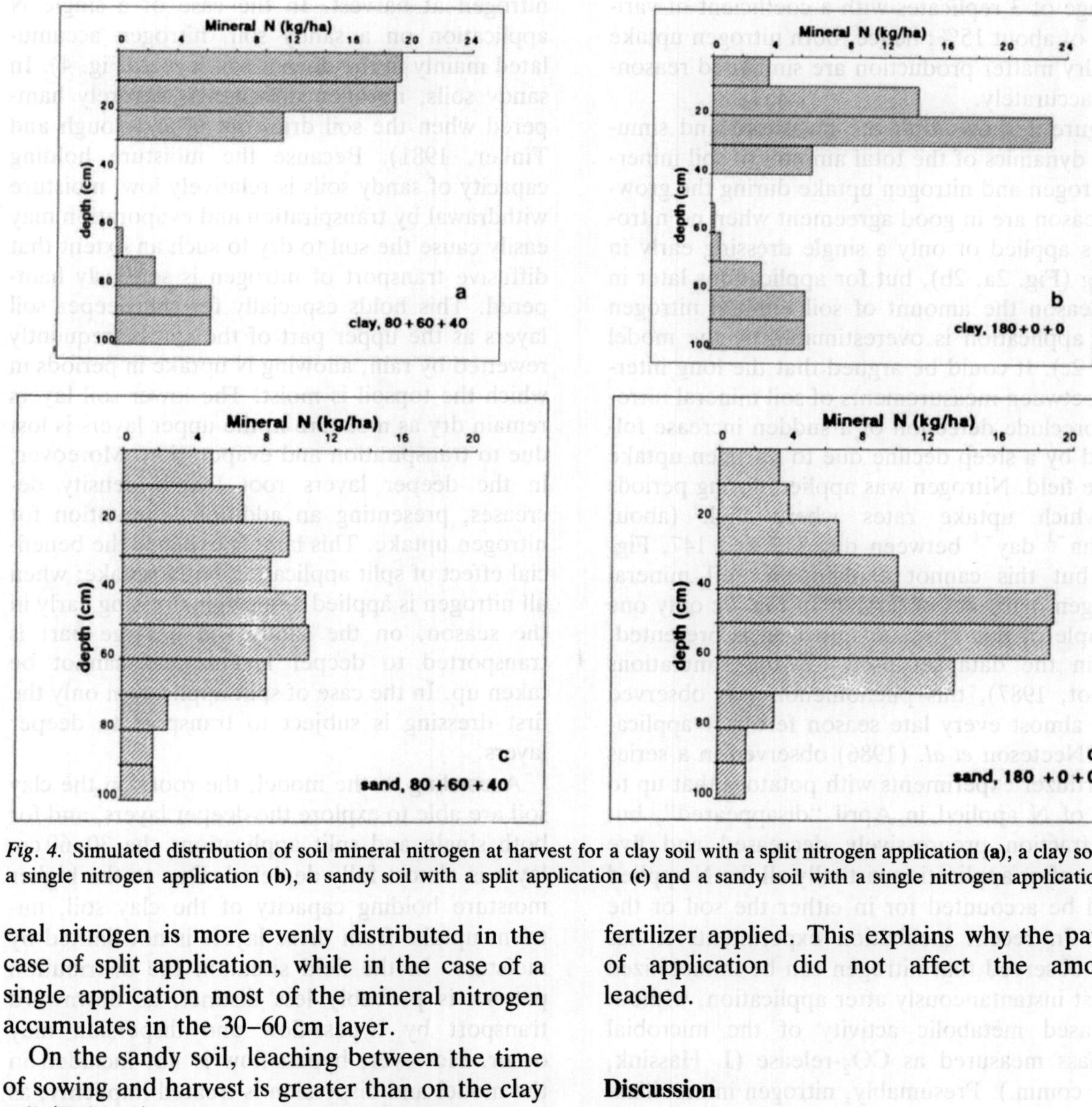

Fig. 4. Simulated distribution of soil mineral nitrogen at harvest for a clay soil with a split nitrogen application (a), a clay soil with a single nitrogen application (b), a sandy soil with a split application (c) and a sandy soil with a single nitrogen application (d).

eral nitrogen is more evenly distributed in the case of split application, while in the case of a single application most of the mineral nitrogen accumulates in the 30–60 cm layer.

On the sandy soil, leaching between the time of sowing and harvest is greater than on the clay soil (Table 1). In the model, nitrogen lost due to leaching (*i.e.* nitrogen moved to depths greater than 100 cm) originated from the store present in the soil at the time of sowing, and not from fertilizer applied. This explains why the pattern of application did not affect the amounts leached.

Discussion

When nitrogen fertilizer experiments are carried out, the criteria used to evaluate the effects of fertilizer applications are generally total dry matter production, grain yield, nitrogen uptake and soil mineral nitrogen dynamics. When a simulation model is used to evaluate the effects of various fertilizer application strategies, it should be able to reproduce these measurable outputs realistically. Figure 1 shows that for the situations studied, measurements and simulated results were strongly correlated. Each of the measured data points in Figure 1 is based on an average of 3 replicates with a coefficient of variation of about 15%; hence, both nitrogen uptake and dry matter production are simulated reasonably accurately.

Figure 2 shows that the measured and simulated dynamics of the total amount of soil mineral nitrogen and nitrogen uptake during the growing season are in good agreement when no nitrogen is applied or only a single dressing early in spring (Fig. 2a, 2b), but for applications later in the season the amount of soil mineral nitrogen after application is overestimated by the model (Fig. 2c). It could be argued that the long intervals between measurements of soil mineral nitrogen preclude detection of a sudden increase followed by a steep decline due to nitrogen uptake in the field. Nitrogen was applied during periods in which uptake rates where high (about 3 kg ha^{-1} day^{-1} between day 113 and 147, Fig. 2c), but this cannot explain the soil mineral nitrogen dynamics observed. In Fig. 2c only one example of this "loss" of nitrogen is presented, but in the data set used for the simulations (Groot, 1987), this phenomenon was observed after almost every late season fertilizer application. Neeteson *et al.* (1986) observed in a series of fertilizer experiments with potatoes that up to 80% of N applied in April "disappeared", but this fraction progressively decreased and five weeks after application virtually all the N applied could be accounted for in either the soil or the crop. In recent incubation experiments it has been observed that nitrogen can be immobilized almost instantaneously after application, without increased metabolic activity of the microbial biomass measured as CO_2-release (J. Hassink, pers. comm.). Presumably, nitrogen immobilization by microbial action in the upper centimeters of the soil profile is the cause of the "disappearance". As crop response in terms of N uptake and dry matter production is simulated accurately it could even be argued that simulated soil mineral nitrogen is a more reliable indicator of N availability than measured soil N.

A simulated comparison between single and split N application showed that on a clay soil simulated dry matter production and nitrogen uptake were similar. On a sandy soil, however, split application increased nitrogen uptake by 10%, resulting in lower amounts of soil mineral nitrogen at harvest. In the case of a single N application on a sandy soil, nitrogen accumulated mainly in the deeper soil layers (Fig. 4). In sandy soils, nitrogen diffusion is severely hampered when the soil dries out (Barraclough and Tinker, 1981). Because the moisture holding capacity of sandy soils is relatively low, moisture withdrawal by transpiration and evaporation may easily cause the soil to dry to such an extent that diffusive transport of nitrogen is seriously hampered. This holds especially for the deeper soil layers as the upper part of the soil is frequently rewetted by rain, allowing N uptake in periods in which the topsoil is moist. The lower soil layers remain dry as moisture in the upper layers is lost due to transpiration and evaporation. Moreover, in the deeper layers root length density decreases, presenting an additional limitation for nitrogen uptake. This largely explains the beneficial effect of split application on N uptake: when all nitrogen is applied as a single dressing early in the season, on the sandy soil a large part is transported to deeper layers, and cannot be taken up. In the case of split application only the first dressing is subject to transport to deeper layers.

According to the model, the roots on the clay soil are able to explore the deeper layers, and for both single and split applications the 30–60 cm layer is almost fully depleted. Due to the higher moisture holding capacity of the clay soil, nutrient uptake from these layers is not limited by moisture. In the field situation the distribution pattern is probably less distinct, as downward transport by bypass-flow and dispersion may occur. However, bypass-flow is not included in the model and dispersion is treated implicitly, as a finite layer thickness is used. As in the Netherlands cereals are grown mostly on clays and loams with relatively high moisture holding capacities, it is not expected that split applications will increase nitrogen uptake dramatically. This confirms the findings of Sylvester-Bradley *et al.* (1987) who concluded, on the basis of the results of 104 fertilizer experiments in the UK, that the effect of different methods of splitting the total amount on yield was small.

In field situations, high N applications in spring might increase the number of tillers to such an extent that not all of them are able to form heads. In the model, tillering is not treated explicitly, and therefore the model might not include some of the advantages of split applications.

The current N fertilizer recommendation seems to be successful with respect to yield and nitrogen uptake. Simulations of grain yield and nitrogen uptake on a clay soil as a function of the amount of N applied show that applications exceeding the amount of N recommended do not increase yield and nitrogen uptake (Fig. 5). Due to the high mineralization rate for the situation studied the application could even be lowered by about 50 kg ha^{-1} without negative effects on yield and N uptake, while at the same time the risk of nitrate leaching in winter is reduced.

Simple methods to predict mineralization accurately would make it possible to correct recommendations for mineralization during the season, thus increasing nitrogen use efficiency in many situations. According to the simulations,

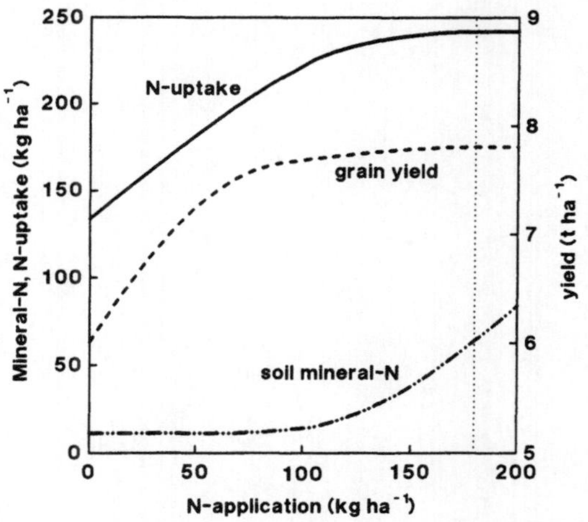

Fig. 5. Simulated nitrogen uptake, soil mineral nitrogen and grain yield as a function of the amount of nitrogen applied on a clay soil. The dotted line represents the current recommendation.

the amount of nitrogen required is lower for maximum grain yield than for maximum nitrogen uptake (Fig. 5). This illustrates that it is possible to produce high-protein grain with dressings higher than the rate required for optimum grain yield. Aiming at high grain quality, however, easily leads to accumulation of nitrogen in the soil, with subsequent leaching during winter.

Based on the simulations presented we conclude that future nitrogen fertilizer experiments in the field, in which crop production and nitrogen uptake are measured as a function of the amount of nitrogen applied, will not contribute substantially to improvement of fertilizer recommendations. In our opinion, a better understanding of processes such as nitrogen transport in the rhizosphere, mineralization and immobilization is required to improve nitrogen fertilizer recommendations and nitrogen use efficiency.

References

Anon. 1982–1985 Experimental field Hélécine, Measurements 1982, 1983, 1984, 1985. K.U. Leuven, Europese Zaadmaatschappij N.V., Bodemkundige Dienst België and Belgisch Instituut tot Verbetering van de Biet. (*In Dutch*).

Barraclough P B and Tinker P B 1981 The determination of ionic diffusion coefficients in field soils. I. Diffusion coefficients in sieved soils in relation to water content and bulk density. J. Soil Sci. 32, 225–236.

De Willigen P and Van Noordwijk M 1987 Roots, Plant Production and Nutrient Use Efficiency. Doct. Thesis, Agricultural University, Wageningen, The Netherlands, 282 p.

Groot J J R 1987 Simulation of nitrogen balance in a system of winter wheat and soil. Simulation Reports CABO-TT no. 13, Centre for Agrobiological Research and Department of Theoretical Production Ecology, Wageningen, 69 p.

Groot J J R and Spiertz J H J 1990 The role of nitrogen in yield formation and achievement of quality standards in cereals. *In* Plant Growth: Interactions with Nutrition and Environment. Eds. J R Porter and D W Lawlor. Cambridge University Press (*In press*).

Morgan J A 1988 Growth and canopy carbon dioxide exchange rate of spring wheat as affected by nitrogen status. Crop Sci. 28, 95–100.

Neeteson J J, Greenwood D J and Habets E J M H 1986 Dependence of soil mineral N on N-fertilizer application. Plant and Soil 91, 417–420.

Neeteson J J, Greenwood D J and Draycott A 1989 Model calculations of nitrate leaching during the growth period of potatoes. Neth. J. Agric. Sci. 37, 237–256.

Prins W H, Dilz K and Neeteson J J 1988 Current recommendations for nitrogen fertilisation within the EEC in

relation to nitrate leaching. Proceedings no. 276, The Fertiliser Society, London, 27 p.
Spitters C J T, Keulen H van and Kraalingen D W G van 1989 A simple and universal crop growth simulator: SUCROS87. *In* Simulation and Systems Management in Crop Protection. Eds. R Rabbinge, S A Ward and H H Van Laar. pp 147–181. Simulation Monographs. PUDOC, Wageningen.
Sylvester-Bradley R, Addiscott T M, Vaidyanathan L V, Murray A W A and Whitmore A P 1987 Nitrogen advice for cereals: present realities and future possibilities. Proceedings no. 263, The Fertiliser Society, London, 27 p.
Van Keulen H and Seligman N G 1987 Simulation of Water Use, Nitrogen Nutrition and Growth of a Spring Wheat Crop. Simulation Monographs. PUDOC, Wageningen, 310 p.

M. L. van Beusichem (Ed.), *Plant nutrition – physiology and applications*, 693–697.

PLSO IPNC035

Weather, nitrogen-supply and growth rate of field vegetables

D.J. GREENWOOD, D.A. STONE and A. DRAYCOTT
AFRC Institute of Horticultural Research, Wellesbourne, Warwick CV35 9EF, UK

Key words: field vegetables, growth rate, models, plant nitrogen-content, radiation, soil moisture deficit, temperature

Abstract

A new procedure was developed for interpreting data from multi-harvest N-fertilizer experiments on 5 different vegetable crops. Per cent N in the plant dry matter of each N-deficient crop was, throughout growth, almost proportional to relative growth rate (standardized to the average weather). After correcting for the effects of plant mass, growth rate of each N-sufficient crop varied considerably during the growth period and approximately in proportion to the growth rate when N-fertilizer was withheld. Some of the variation was associated with small changes in soil water which also greatly influenced C-partition between foliage and storage roots of at least one crop. Some of the variation was related to temperature and radiation. The significance of these findings to modeling the effects of N-nutrition and environment on growth is discussed.

Introduction

N-fertilizer and other agronomic practices must be adjusted for differences in conditions so as to maximize productivity and minimize environmental pollution. Models are needed to calculate the best way of meeting these objectives in any given situation. Central to any model of N-response is the treatment of the dependence of growth rate on plant-N content and the weather. One approach to providing the necessary information is to carry out well-monitored, multi-harvest fertilizer-N experiments and attempt to establish relationships between the changes in plant growth and composition and the changes in the soil and aerial environment that take place during the growing period. A method of carrying out this data analysis is suggested by a set of simple equations that have defined the dependence of growth rate of some arable crops on their dry weights and N contents in West Europe (Greenwood *et al.*, 1986). The purpose of the paper is to describe the outcome of this approach on 5 different crops.

Experiments

Cabbage (summer), carrots, leeks, onions and red beet were grown in a field experiment with zero (F0), the optimum (F1) and 1.5 × optimum (F2) rates of fertilizer-N and harvested at intervals during the growing season. Optima, taken as the minimum rates of N-fertilizer at which maximum yields were expected, were calculated using a simulation model (Greenwood and Draycott, 1989). With the exception of cabbage, all crops were grown at each of two plant densities. The soil was a sandy loam, it had a mineralization rate of about 0.7 kg N $ha^{-1}d^{-1}$ at 15.9°C and contained 19 cm of water m^{-1} at field capacity. There were no barriers to rooting. At the final harvest there was little difference between the plant dry weights on the F1 and F2 treatments; dry weights on the F0 treatments were between 26% (cabbage) and 71% (carrot) of those on the F1 treatments.

Theory

The rate of increase of plant dry weight (dW/dt) of many crops can often be defined by

$$\frac{dW}{dt} = \frac{K1 \times W}{K + W} \tag{1}$$

where K1 and K are coefficients that, in practice, are highly correlated. Nevertheless good fits to data from field crops are generally obtained with K set equal to $1\ t\ ha^{-1}$ (*e.g.* Greenwood *et al.* 1986). Rewriting Eq. (1) gives

$$\text{Relative growth rate} = \frac{1}{W}\frac{dW}{dt} = \frac{K1}{1 + W} \tag{2}$$

Under approximately constant environmental conditions, K1 has been found to be constant for long periods and thus appears to be a measure of growth rate corrected for the effects of plant mass. Integrating Eq. (2) gives

$$K1 \times T = (\ln W + W) + \text{a constant} \tag{3}$$

Results

For each plant density and N level (lnW + W) of equation (3) was regressed against time. The poorest correlation coefficient for any treatment for cabbage was 0.980 (df = 8), for onion 0.974 (df = 7), for carrot 0.989 (df = 6) for red beet 0.958 (df = 7), and for leek 0.744 (df = 7). Harvest dates were within the period 147–281 days from January 1 except for leeks when they extended from day 215–386. Fits were thus very good for all crops harvested during the main summer growing period. Correlation coefficients were poorer for leeks, almost certainly because, in contrast to other crops, harvests extended over a wide range of ambient temperatures. The values of the regression coefficients (Kl of Eq. (3)) for cabbage, carrot and red beet grown on the F1 and F2 plots were between 0.17 and $0.20\ t\ ha^{-1} d^{-1}$ which are similar to those usually found for crops that gave high yields; the intercepts (*i.e.* constant of Eq. (3)) gave a measure of starting weights.

Substitution in Eq. (2) of these average values of K1 during the growing period and the corresponding crop weights at each harvest date provides estimates of the relative growth rates (standardized to the average weather) at each harvest date. Such calculations were carried out for each crop and treatment combination. When no fertilizer was applied relative growth rate for every crop was proportional to %N in the dry matter (*e.g.* Fig. 1a). For all crops, this simple model with a separate regression coefficient for each crop and spacing treatment removed 86.9% (df = 69) of the variance in %N. The corresponding linear model removed 88.8% (df = 60) and the quadratic model 91.0% (df = 51). These more complex models gave only just significantly better fits ($P < 0.05$) than the simple proportional model. When either the F1 or F2 rates of N-fertilizer were applied, results similar to those given in Fig. 1b were obtained; relative growth rate initially declined without appreciable decline in %N for a period that varied from experiment to experiment; thereafer %N was linearly related to relative growth rate with an intercept that increased with rate of fertilizer-N.

In addition, instead of calculating a mean value of K1 over the entire period, we also calculated the corresponding values of K1, designated as KT, over each of the intervals between successive harvests. KT was calculated from the finite difference form of Eq. (3), as $\Delta(\ln W + W)/\Delta T$, where ΔT is the interval between successive harvests. Even though for each treatment, KT varied substantially during the season, the values of KT on the F0 treatment were, as a close approximation, always proportional to the values of KT on the F1 and F2 treatments (Table 1). There was no significant improvement by adding intercepts or terms to bring about curvature in the relationships. Subsequent presentation is, unless otherwise stated, confined to values of KT averaged over the F1 and F2 treatments. Attempts were made to relate these between harvest values of KT for each crop to mean soil temperature, mean incoming radiation, and effective days (Scaife *et al.*, 1987) calculated from both daily air temperature and

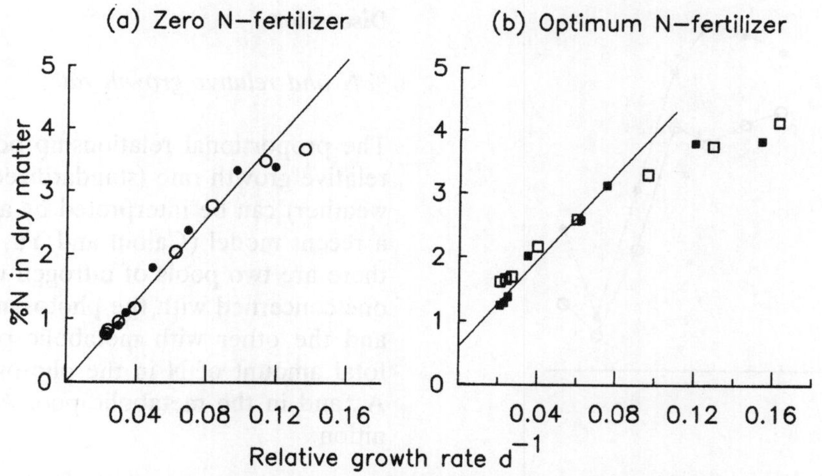

Fig. 1. Relationships between %N and relative growth rate (standardized to the average weather) of carrots grown with zero-N fertilizer (F0) or with the optimum level (F1). Closed symbols refer to low density and open symbols to high density crop spacing.

Table 1. Relationships between KT with ample N fertilizer (Y) and KT with zero N fertilizer (X) measured at intervals throughout the growing period

Model	% Variance accounted for	Mean square due to additional parameters to preceeding model	Residual d.f. appropriate to model	Residual mean square of model
$Y = c$	–	–	36	0.0270
$Y = m_iX$	81.4	0.193	32	0.0011
$Y = m_iX + c_i$	84.4	0.002	27	0.0009
$Y = m_iX + d_iX^2 + c_i$	85.7	0.001	22	0.0009

m, c, d are coefficients; subscript i indicates they have a separate value for each of the experiments. Y (KT) has been averaged over the F1 and F2 treatments and X (KT) has been averaged over the F0 treatments. Both Y and X were averaged over both plant spacings.

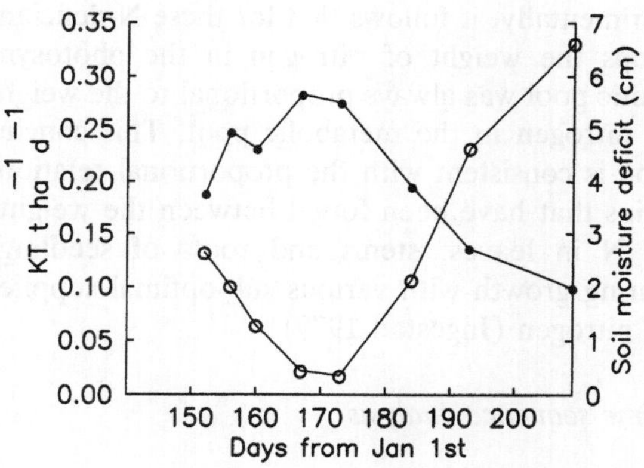

Fig. 2. Relationships between K1 for cabbage (closed circles) and between soil moisture deficit (open circles) and time. Soil moisture deficit was calculated from the model described in Greenwood and Draycott (1989).

daily incoming radiation. About half the % variance in KT of carrots and leeks was removed by each of these weather parameters. Much less was removed for other crops. The poor correlation for onions could be attributed to a sudden reduction in growth rate, possibly due to the onset of senescence, mid-way during the growing season. Unexpectedly, values of KT for cabbage almost precisely mirrored the soil moisture deficit (Fig. 2) and regression of KT against soil moisture deficit removed 91.8% of the variance in KT. A more complicated relationship between KT and soil moisture existed for red beet. Measurements of soil moisture content in the top 30 cm were made during growth on the F0 plots and these together with the corresponding values of KT on

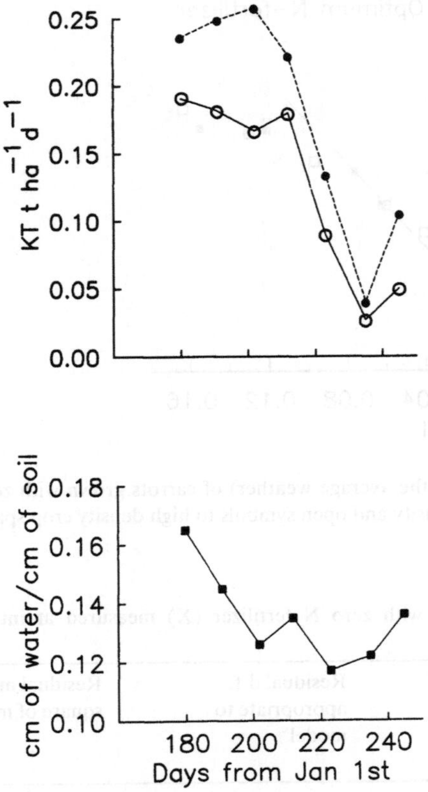

Fig. 3. Mean values of K1; open circles (F0 treatment), closed circles (average of F1 and F2 treatments) and soil moisture content in the top 30 for the F0 treatment (closed squares) over successive intervals during the growth of red beet. Maximum soil moisture deficit, calculated as in Figure 2 was 9 cm.

the F0 and F1 and F2 treatments are given in Figure 3. Variation in KT on the F0 treatment paralleled that on the F1 and F2 treatments. Initially, soil moisture content fell and KT remained approximately constant. Then when the soil moisture content fell below a critical value of about 0.13 cm of water/cm of soil, KT fell very sharply. Increase in plant dry weight almost ceased but N-uptake was reduced to a smaller extent so that the %N of the whole plant and especially of the leaves increased. There was a massive net translocation of C but not N from leaf to root so that the rate of root dry matter production continued almost unchecked. Later on there was an increase in soil moisture content and KT increased sharply.

Discussion

%N and relative growth rate

The proportional relationship between %N and relative growth rate (standardized to the average weather) can be interpreted by a modification of a recent model (Caloin and Yu, 1984). Suppose there are two pools of nitrogen within the plant; one concerned with the photosynthetic processes and the other with metabolic processes. If the total amount of N in the photosynthetic pool is A_P and in the metabolic pool A_M then by definition

$$A_P = \frac{1}{\gamma} \times \frac{dW}{dt} \tag{4}$$

where γ is a coefficient that remains constant throughout growth. Also if A_M is linearly related to A_P then

$$A_M = \propto A_P + C \tag{5}$$

where $\propto$ and C are constants

$$\%N = (A_P + A_M) \times \frac{100}{W}$$

$$\%N = \frac{1}{\gamma} \times \frac{1}{W} \times \frac{dW}{dt}(1 + \propto) \times 100 + \frac{C}{W} \times 100 \tag{6}$$

If $C = 0$, %N is proportional to relative growth rate. As this was found to be the case experimentally, it follows that for these N-deficient crops the weight of nitrogen in the photosynthetic pool was always proportional to the weight of nitrogen in the metabolic pool. This conclusion is consistent with the proportional relationships that have been found between the weights of N in leaves, stems and roots of seedlings during growth with various sub-optimal supplies of nitrogen (Ingestad 1979).

Time sequence analysis

Growth rate depends on plant mass *per se*. Thus any time sequence analysis aimed at elucidating how growth is influenced by environmental changes, must be carried out with a function of growth rate that is independent of plant mass. This requirement is met by relative growth rate when plants are small but not when they are large, and conversely by absolute growth rate when plants are large, but not when they are small. It appears to be met throughout by K1 as is indicated by the good fits to Eq. (3). The reason is that K1 changes from being approximately equal to relative growth rate when plants are small, to being approximately equal to absolute growth rate when they are large. It is a consequence of the basic assumption in the definition of K1 that growth rate increases asymptotically with increase in plant masss per unit area.

The correlations between KT and soil water suggest that the incidence of water stress, sometimes transient, obscured any good correlations that might have been expected between KT and measures of the aerial environment such as incoming radiation or ratio of total evaporation to saturation deficit (Monteith, 1986).

Law of the minimum type relationships usually define the effects of different nutrients on growth (*e.g.* Wood *et al.* 1972). Our work suggests that by contrast, the net effects of lowering radiation, temperature and water availability are multiplicative with those of %N in the plant on growth. On this basis, a 50% reduction, say of radiation, would cause the same fractional reduction in growth rate of an N-deficient plant as of an N-sufficient plant of the same size; a conclusion that is supported by entirely independent work (Agren, 1985). If these multiplicative relationships are found to be of general applicability, they could be useful for modeling the interactive effects of weather and N-supply on growth rate.

Acknowledgements

We are grateful to Mr J Hunt for chemical analyses and to Mr K B Niendorf and Miss Mary Turner for help with the experiments and collating the data.

References

Agren G I 1985 Theory for growth of plants derived from the nitrogen productivity concept. Physiol. Plant 64, 17–28.

Caloin M and Yu O 1984 Analysis of the time course change in nitrogen content in *Dactylisglomera L.* using a model of plant growth. Ann. Bot. 54, 69–76.

Greenwood D J, Neeteson J J and Draycott A 1986 Quantitative relationships for the dependence of growth rate of arable crops on their nitrogen content, dry weight and aerial environment. Plant and Soil 91, 281–301.

Greenwood D J and Draycott A. 1989 Experimental validation of an N-response model for widely different crops. Fert. Res. 18, 153–174.

Ingestad T 1979 Nitrogen stress in birch seedlings. 2. N, K, P, Ca and Mg nutrition. Physiol. Plant. 45, 149–157.

Monteith J L 1986 How do crops manipulate water supply and demand. Phil. Trans. R. Soc. Lond., A316, 245–259.

Scaife A, Cox E F and Morris G E L 1987 The relationship between shoot weight, plant density and time during the propogation of four vegetable species. Ann. Bot. 59, 325–334.

Wood J T, Greenwood D J and Cleaver T J 1972 Interactions between the beneficial effects of nitrogen, phosphate and potassium plant growth. J. Agric. Sci. Camb. 78, 389–391.

M. L. van Beusichem (Ed.), *Plant nutrition – physiology and applications*, 699–703.

PLSO IPNC254A

Modelling the response of crops to fertilizers

B.H. JANSSEN and F.C.T. GUIKING
Department of Soil Science and Plant Nutrition, Wageningen Agricultural University, P.O. Box 8005, 6700 EC Wageningen, The Netherlands

Key words: actual nutrient uptake, economically optimum fertilizer rates, maize, nutrient accumulation, nutrient dilution, potential nutrient supply, *Zea mays* L.

Abstract

Relations between nutrient uptake and yield and the balance between nutrients form the basis of the presented model. The model calculates the yield of a crop as a function of the uptake of N, P, and K. For each nutrient a range of possible yields is calculated, and the yield ranges are combined to one yield estimate.

The actual uptake of a nutrient is calculated as a function of the potential supplies of all three nutrients. Several methods for the assessment of the potential supplies of nutrients are discussed. Once the supplies are known the effects of fertilizer application can be calculated. Fertilizers increase the potential supply but not necessarily the uptake of a nutrient. Setting prices for fertilizers, the most economic combinations can be found.

Introduction

In the course of some projects on soils and crops in Kenya and Suriname, a system has been developed for quantitative evaluation of the fertility of tropical soils (Janssen *et al.*, 1990). The principles of this so-called QUEFTS system can be applied also to calculate the response to fertilizers (Janssen *et al.*, 1989). The pivot of the system is formed by the relations between nutrient uptake and yield of maize. In principle, other nutrients than N, P and K can be included, provided the relationships between uptake of those nutrients and yield have been established. The methodology of the model can be used for each crop of which the relationships between nutrient uptake and the yield of the economic product are known.

Main concepts used in the system

Some of the concepts used in the model need an explanation: actual uptake and potential supply, maximum recovery fraction, maximum accumulation and maximum dilution of a nutrient in the crop.

In the soil, nutrients are made available to crops by processes like mineralization of organic matter and weathering of minerals. Often the supplies of the different nutrients are not in balance with the needs of a crop. When the supply of a particular nutrient is small compared to those of other nutrients, the whole supply of that nutrient will be taken up by the crop. When the supply of a particular nutrient is large compared to those of other nutrients, crop growth is limited by the low availability of the other nutrients and the crop cannot make use of the whole supply of the particular nutrient. Then the *actual uptake* is less than the *potential supply*.

The potential supply is enlarged by application of fertilizers. A part of the applied nutrient is made unavailable to crops by processes like microbial immobilization and chemical precipitation, and a part is lost by *e.g.* leaching and volatilization. The fraction that is recovered by the crop varies from zero to about 0.8. Under otherwise comparable conditions the recovery of a fertilizer nutrient by the crop is the higher, the

Table 1. Equations for calculating grain yields of maize (kg/ha) corresponding with maximum accumulation (YNA, YPA, YKA) and maximum dilution (YND, YPD, YKD) of N, P and K in the crop. UN, UP and UK stand for actual uptake of N, P and K in kg/ha

YNA = 30 × (UN-5)	Eq. 1	YND = 70 × (UN-5)	Eq. 2
YPA = 200 × (UP-0.4)	Eq. 3	YPD = 600 × (UP-0.4)	Eq. 4
YKA = 30 × (UK-2)	Eq. 5	YKD = 120 × (UK-2)	Eq. 6

lower the supply of that nutrient by the soil is. The *maximum recovery fraction* for a fertilizer nutrient is a function of soil, weather and crop properties.

When a nutrient is poorly available compared to the other nutrients and growth factors, it is diluted in the plant and its content goes down to a minimum value. On the contrary, when a nutrient is abundantly available, it accumulates in the plant till its content reaches a maximum value. Given a certain uptake of N, P or K, the possible yields range between the yields that correspond with *maximum accumulation* and *maximum dilution*, respectively, which are denoted by YNA, YND, YPA, YPD, YKA and YKD (Table 1).

Scheme of calculations

First the *potential supplies* of N, P and K have to be assessed. This can be done along various ways which are discussed in the next section.

The *actual uptake* of a nutrient is calculated as a function of the potential supply of that nutrient in relation to the potential supplies of the other nutrients. The curve of actual uptake against potential supply of a nutrient consists of three parts. At low values of the potential supply (situation A) actual uptake equals supply, at medium values (situation B) the curve is parabolic, and at high values (situation C) the actual uptake has reached a maximum value that is determined by the potential supply of the other nutrients (Table 2). The actual uptake of each nutrient is calculated twice, taking into account the potential supplies of the second and the third nutrient, respectively. The lower of the two values is considered the more realistic one, in conformity with the Law of the Minimum.

With the data for actual uptake the yields that correspond with *maximum accumulation* and *maximum dilution* are calculated (Table 1). To produce any grain, maize plants must have a minimum size, for which the required uptakes per ha are 5 kg N, 0.4 kg P and 2 kg K. Therefore, these values are subtracted from UN, UP and UK in the equations of Table 1.

Next the *yield ranges* found for the three nutrients are *combined* to one yield estimate. Principles for this combining are that the estimated yield lies in the overlap if yield ranges overlap, and that it can never exceed the upper limit of the lowest yield range if there is no

Table 2. Equations for the calculation of the actual uptake of Nutrient 1 as a function of the potential supplies of Nutrients 1 and 2. S1 and S2 are the potential supplies of Nutrients 1 and 2. The symbols a, d and r refer to the constants in the equations of Table 1; their values are also indicated. Situations A, B and C are explained in the text.

Situation	Condition		
A	$S1 < r1 + (S2 - r2)(a2/d1)$		
C	$S1 > r1 + (S2 - r2)(2 \times d2/a1 - a2/d1)$		
B	S1 in between		
	Equation for U1(2)		
A	$U1(2) = S1$		Eq. 7
C	$U1(2) = r1 + (S2 - r2)(d2/a1)$		Eq. 8
B	$U1(2) = S1 - \dfrac{0.25(S1 - r1 - (S2 - r2)(a2/d1))^2}{(s2 - r2)(d2/a1 - a2/d1)}$		Eq. 9
Nutrient	Values of constants		
	a	d	r
N	30	70	5
P	200	600	0.4
K	30	120	2

overlap. For the calculation of the yield within an overlap of the yield ranges of two nutrients, a parabolic equation has been derived (Janssen *et al.*, 1990). After calculation of yields for nutrient pairs, (N and P, N and K, P and K), the yields for paired nutrients are averaged. This average is the final yield estimate.

Assessment of potential nutrient supplies and maximum fertilizer recovery fractions

Various procedures can be followed to estimate the potential supplies of nutrients by the unfertilized soil and the maximum recovery fractions of applied fertilizer nutrients.

Nutrient uptake in fertilizer trials

This method is the most thorough but also the most expensive one. The experiments should be designed in such a way that each of the nutrients is in turn the most limiting growth factor. The concept is illustrated with the examples of Table 3, presenting the N uptakes and yields obtained in a fictitious factorial trial with four N, three P and three K rates. The actual uptake of N is expected to equal or closely approach the potential supply of N for the units receiving no N and high P and K applications. If no N is applied, the uptakes of N are 38, 65, and 66 kg/ha at P rates of 0, 100 and 200 kg/ha, respectively. The N uptake strongly depends on P, while an effect of K shows up only at the highest P level. In view of the very small difference in N uptake between P100 and P200, it is not likely that considerably higher N uptakes will be found at still higher rates of P and K. Hence, the potential supply of N by the soil alone can be estimated at 66 kg.

In a similar way it can be argued that the potential N supply is about 91, 115 and 160 kg/ha, when fertilizer N is applied at rates of 50, 100 and 200 kg/ha, respectively. The corresponding apparent recovery fractions of fertilizer N are 0.50, 0.49 and 0.47, respectively. They are calculated as the difference in N uptake between the units with and without fertilizer N, divided by the applied amount of N. A higher recovery fraction might be found at N rates less than 50 kg/ha, but from the available data it is concluded that 0.5 is the highest possible recovery fraction. The increase in potential N supply by fertilizer N is thus calculated as 0.5 times the quantity of fertilizer N, resulting in supplies of 116 and 166 kg/ha at N rates of 100 and 200 kg. The actual uptakes increase less than the potential supplies. If no P is applied, application of N does not increase uptake of N.

Yields in fertilizer trials

Usually the uptake of nutrients is not determined in fertilizer trials or for only a limited number of the experimental units. Then one must do with yield data. Applying the equations of Table 1 some approximate estimates of nutrient supplies can be made. The yield data of Table 3 serve to illustrate the reasoning.

It is obvious that P is extremely limiting if no P is applied. Hence, it can be assumed that all available P has been taken up and that P is maximally diluted in the crop. Thus the uptake

Table 3. **Nitrogen uptake and maize yield in a factorial experiment with fertilizer rates of 0, 50, 100 and 200 N, and 0, 100 and 200 P and K. All data in kg per ha**

Rate of		N uptake at N rate of				Yield at N rate of			
P	K	0	50	100	200	0	50	100	200
0	0	38	38	38	38	990	990	990	990
0	100	38	38	38	38	990	990	990	990
0	200	38	38	38	38	990	990	990	990
100	0	65	88	109	146	3354	4035	4476	4967
100	100	65	88	109	146	3666	4480	5001	5553
100	200	65	88	109	146	3843	4792	5365	5891
200	0	65	88	109	147	3743	4648	5284	5931
200	100	66	91	114	159	4043	5193	6090	7433
200	200	66	91	115	160	4158	5446	6455	7951

and the supply of P by the soil is equal to: 0.4 + 990/600 = 2.05 kg/ha (Equation 4 in Table 1).

More difficulties are met when it is tried to derive the N supply from the yield data, because the yields of the N0 treatments do not reach a plateau. It can be concluded that the N uptake at N0 P200 K200 must be at least: 5 + 4158/70 = 64.4 (Equation 2). The potential supply of N is certainly higher.

It is still more difficult to estimate the potential supply of K. In view of the yield increases obtained when the application of N or P is raised from 100 to 200 kg/ha, it is likely that considerable higher yields than 5931 kg (Table 3) are possible for K0 treatments, So, K is certainly not maximally diluted in the plant at N200 P200 K0. Therefore the uptake of K must be noticeably higher than 2 + 5931/120 = 51.4 kg/ha (Equation 6) and the supply of K still higher.

As a first approximation values of 65 kg N and 60 kg K for the potential supplies of N (SN) and K (SK) are substituted in the equations of Table 2 and yields calculated. This procedure is continued with higher values for SN and SK until the calculated yields agree with the measured yields.

Chemical soil analysis

In the QUEFTS system, empirical relationships between the potential supplies of N, P and K and certain combinations of chemical soil properties are applied. The used soil (0-20 cm) properties are: pH(H_2O), organic carbon, P-Olsen and exchangeable K. Additional information can be obtained from organic N and total P (Janssen *et al.*, 1990).

Nutrient uptake from an unfertilized soil

If nutrient contents in the crop have been determined, Equations 7, 8 and 9 (Table 2) can be used in a reverse way to calculate the potential supplies from the actual uptakes of the nutrients. Details of the procedure will be published later.

Maximum recovery fraction

The maximum recovery fraction of fertilizer nutrients can only exactly be found by fertilizer experiments, including chemical analysis of crops. This was shown above for N. If no estimates of the maximum recovery fractions are available, a fraction of 0.5 is used as standard for N and K, and of 0.1 for P.

Economics

Having assessed the potential supplies of N, P and K by the unfertilized soil and the maximum recovery fractions of fertilizer N, P and K, yields can be calculated for any combination of fertilizer N, P and K. Of practical interest is to find the most economic combination. Table 4 shows the results of the calculation of the most profitable way of allocating available money to N, P or K. Prices per kg maize, N, P and K were set at 1, 4, 8 and 2 money units. Table 4 refers to the

Table 4. Calculation of the most profitable combinations of fertilizer N, P and K, marginal yield increase (MYI), and net returns on fertilizer costs (NR). Prices per kg maize, N, P and K are 1, 4, 8 and 2 units

Fertilizer costs, unit	Rates (kg/ha) of			Yield, kg/ha	MYI, kg/ha	NR, unit
	N	P	K			
0	0	0	0	990		
100	0	12.5	0	1659	669	559
200	0	25	0	2161	502	971
300	0	37.5	0	2532	371	1242
400	0	50	0	2819	287	1429
500	9	58	0	3044	225	1554
600	18	63	12	3265	221	1675
700	27	68	24	3485	220	1795
800	36	73	36	3705	220	1915

same example as Table 3. The supplies of N, P and K by the soil were 66.5, 2.05 and 67.0 kg/ha, respectively. The above mentioned standard recovery fractions were used. Since P is the most limiting nutrient, maximum benefit is obtained by applying only P till the yield reaches a level of about 3 t/ha where the second nutrient (N) becomes equally limiting. By applying these two nutrients in a balanced ratio the yield increases a little and then also K is limiting. For still higher yields a balanced application of all three nutrients is required: per 100 units of money it is 9 kg N, 5 kg P and 12 kg K, resulting in a yield increase of 220 kg. The calculated K rate is relatively high due to the low price assumed for fertilizer K.

The data of Table 4 suggest a linear yield increase if nutrients are applied in balanced ratios. With increasing rates other growth factors like water availability or solar radiation might become limiting. Such yield limits can be introduced in the model (in the example of Table 4, it was set at 10000 kg/ha). If the calculated yields approach that limit, diminishing returns are found.

References

Janssen B H, Guiking F C T, Van der Eijk D, Smaling E M A, Wolf J and Van Reuler H 1989 A system for quantitative evaluation of soil fertility and the response to fertilizers. *In* Land Qualities in Space and Time. Eds. J Bouma and A K Bregt. pp 185–188. Pudoc, Wageningen, The Netherlands.

Janssen B H, Guiking F C T, Van der Eijk D, Smaling E M A, Wolf J and Van Reuler H 1990 A system for quantitative evaluation of the fertility of tropical soils (QUEFTS). Geoderma. (*In press.*)

M. L. van Beusichem (Ed.), *Plant nutrition – physiology and applications*, 705–709.

PLSO IPNC337

Modelling potassium uptake by wheat (*Triticum aestivum*) crops

P. SEWARD[1,2], P.B. BARRACLOUGH [1] and P.J. GREGORY[2]
[1]*Biochemistry and Physiology Department, AFRC Institute of Arable Crops Research, Rothamsted Experimental Station, Harpenden, Herts., AL5 2JQ, UK, and* [2]*Department of Soil Science, University of Reading, London Road, Reading, UK*

Key words: models, potassium uptake, *Triticum aestivum* L., wheat

Abstract

Spring wheat was grown in the field under deficient and sufficient levels of soil K and with high and low supplies of fertiliser nitrogen. Measurements were made of K uptake, soil nutrient supply parameters, root growth and, in solution culture, root influx parameters. Mechanistic models predicted uptake reasonably well under K-deficient conditions, but over-predicted uptake, by as much as 4 times, under K-sufficient conditions. The over-prediction was apparently due to poor characterisation of plant demand.

Introduction

Nutrient uptake from soil depends both on plant demand and soil supply and several mathematical models simulating the dynamic-interaction between these processes have been developed (Baldwin *et al.*, 1973; Claassen and Barber, 1976; Barber and Cushman, 1981; Claassen *et al.*, 1986). These models are based on the same principles and differ only in detail. In the models, ions are transported to roots by mass flow and diffusion and are absorbed at rates which depend on their concentration at the root surface as expressed in a Michaelis-Menten type relationship.

Potassium concentration gradients near roots growing in blocks of soil have been successfully predicted with the model (Claassen *et al.*, 1986). In pot experiments the models have predicted K uptake reasonably well for young maize and soyabean plants, providing that allowance was made for root competition (Barber, 1984). The Barber-Cushman model has successfully predicted the uptake of K by soyabeans in the field (Silberbush and Barber, 1984). There are considerable benefits in being able to apply such models to field situations, but their ability to predict K uptake under wide-ranging conditions of soil supply and plant demand, and over extended growing periods, has hardly been tested. The objective of the present study was to compare measured uptake of K by spring wheat crops with that predicted by the models. The crops were grown either on K-deficient or K-sufficient soil with high and low supplies of fertiliser nitrogen.

Materials and methods

Field experiment

The field experiment was conducted on a sandy loam soil (10% clay) at Woburn Experimental farm, Bedfordshire, UK, and utilized long-established plots in which K had been allowed to deplete or accrue. The plots contained either a low level of K (LK, 45 mg kg^{-1} exchangeable K) or a high level (HK, 160 mg kg^{-1} exchangeable K) and were considered to be deficient or sufficient in K respectively (K availability index, MAFF 1986). Spring wheat (var. Alexandria) was sown on 9 March 1988 in rows 12 cm apart at a rate of 310 seeds m^{-2}; each plot was 6 m^2. 50

kg N/ha (LN) was applied to the seedbed of all plots and a further 100 kg N/ha (HN) applied on 6 May to half of the plots to give 4 treatments (LKLN, LKHN, HKLN, HKHN). The treatments were replicated 4 times in a randomised design, and the plots kept near field capacity with frequent irrigation. The crops were first sampled on 11 May, just before stem extension, and thereafter, at two-weekly intervals until anthesis.

Models and model parameters

The models of Barber and Cushman (1981) and Claassen *et al.* (1986) were used to predict the uptake of K by spring wheat from stem extension to anthesis (40–80 days after emergence). These models are available to run on a personal computer. The former model predicts uptake directly whilst the latter predicts influx (uptake rate per unit root area) and requires a separate calculation of uptake from the predicted influx and the measured root growth. The models require inputs of upto 11 soil and plant parameters and some values used for the present predictions are given in Table 1.

The initial concentration of K in the soil solution (C_{li}) was determined by ICP in solution extracted by drainage centrifugation. The buffer power of the soil (b) was derived from a Q/I relationship and varied with concentration. Diffusion coefficients (D) were calculated from the equation of Nye (1968) which requires values of volumetric soil water content, buffer power and the impedance factor. The latter was calculated from a relationship derived by Barraclough and Tinker (1981) for this soil.

The root influx parameters (I_{max}, K_m, C_{min}) were determined with 18-day old plants in solution culture using the depletion technique of Claassen and Barber (1974). Growing conditions were 20°C, 16 hour daylength and a light intensity of 350 μE m^{-2} s^{-1}. The plants were starved of K for 12 hours immediately prior to measurement. Water influx was calculated from the potential evapotranspiration and the measured root surface area.

Roots in the field were measured 4 times during the experiment. Procedures for coring, washing and counting of roots have been given by Barraclough and Leigh (1984). Mean root radius (r_0) and mean half-distance between roots (r_1) were derived from formulae given by Claassen *et al.* (1986).

Uptake was predicted from the measured parameters for the 0–20 and 20–40 cm soil layers,

Table 1. Model parameters and their values for the period 40–80 days in the top 20 cm of soil

	LKLN	LKHN	HKLN	HKHN
Soil supply parameters				
C_{li}, initial soil solution concentration (μmol cm^{-3})	0.040	0.109	0.466	0.827
D, diffusion coefficient (cm^2 s^{-1} × 10^{-8})	4.71	4.11	13.03	11.80
b, buffer power	16.4	16.4	5.4	5.4
Root growth parameters				
L_0, initial root length (cm × 10^5)	4.10	4.73	5.62	5.03
r_0, mean root radius (cm)	0.016	0.016	0.016	0.016
k, rate of root growth (cm s^{-1})	0.221	0.211	0.198	0.161
r_1, mean half-distance between roots (cm)	0.282	0.274	0.264	0.268
Root influx parameters				
I_{max} maximum rate of influx (μmol cm^{-2} s^{-1} × 10^{-5})	1.53	1.53	1.53	1.53
K_m, Michaelis constant (μmol cm^{-3} × 10^{-2})	5.35	5.35	5.35	5.35
C_{min}, conc. when influx = 0 (μmol cm^{-3} × 10^{-4}	4.00	4.00	4.00	4.00
v_0, water influx (cm^3 cm^{-2} s^{-1} × 10^{-7})	3.30	3.30	3.30	3.30

respectively, and added to give total uptake. Uptake was predicted for 3 periods, 40–53, 40–67 and 40–80 days from emergence. For each period and each soil layer, initial values of C_{li}, b and L_o were used and time-averaged values of D and r_i. The depth of 40 cm was considered to be the extent of the K-enriched zone. The soil had little or no structure and it seems reasonable to assume that the roots in each 20 cm layer were uniformly distributed.

Results and discussion

The treatments had a large effect on potassium uptake which ranged from 8 g m^{-2} for LKLN to 19 g m^{-2} for HKHN. High nitrogen applications increased potassium uptake only at the high K level. Initial root length was 50% greater in the high nitrogen treatments, but subsequent root growth rates were greater in the low nitrogen treatments so that root length in the top 40 cm of soil at anthesis was similar in all treatments, about 16 km m^{-2}. Differences in root growth was not a major factor therefore in causing differences in K uptake. Observed and predicted uptakes for the three periods are compared in Figure 1. There was good agreement in the case of the LKLN treatment, but uptake was increasingly over-predicted as soil nitrogen and potassium levels increased; the over-prediction was as much as four times for the HKHN treatment. Predicted uptake for the HN treatments was higher than that for the LN treatments because of higher values of C_{li} for K (Table 1). The increased K in solution was due to the application of ammonium nitrate fertiliser which caused K^+ to be displaced from soil exchange sites by NH_4^+. The models predicted uptake reasonably well where uptake was expected to be determined by supply, that is the LK treatments, but where plant demand was expected to be controlling uptake, that is in the HK treatments, the models did not predict uptake very well.

Over-prediction would occur if the measured root length was not all active in uptake, or if older roots were less efficient, or if I_{max} were too large. The first two possibilities are unlikely. The root systems were healthy and growing in moist soil and Clarkson (1981) has shown that roots of

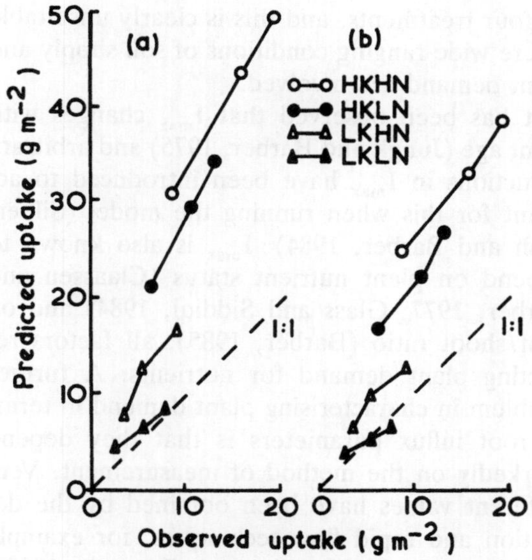

Fig. 1. A comparison of observed potassium uptakes on three occasions with those predicted by the models of (a) Barber and Cushman (1981), and (b) Claassen *et al.* (1986) for four soil treatments.

different ages have similar uptake efficiencies. The problem appears to lie in the value of I_{max}. The measured field influxes were an order of magnitude less than the value of I_{max} used in the models. When an I_{max} value 10 times smaller was used, the models gave slight under-predictions of uptake for all treatments.

The environmental conditions and ages of the plants on which the root influx parameters (I_{max}, K_m and C_{min}) were measured were very different from those in the field. In the growth room, the temperature was 20°C, the plants were young and growing rapidly with a relative growth rate (RGR) of 0.2 d^{-1}. Moreover, the plants had been starved of K for 12 hours before measurement, further accentuating their demand. By contrast, in the field the crops were growing in soil deficient or sufficient in N and K, with average soil temperatures of 14°C, and were 40–80 days of age. The average RGR from the most favourable treatment was 0.05 d^{-1}. It appears that the conditions under which I_{max} was measured in the growth room gave an inordinately high value that was inappropriate for the prediction of K uptake by the field crops. In addition, one set of values for the root influx parameters was used to make predictions across all four treatments, and this is clearly unsuitable where wide-ranging conditions of soil supply and plant demand are involved.

It has been observed that I_{max} changes with plant age (Jungk and Barber, 1975) and arbitrary reductions in I_{max} have been introduced to account for this when running the model (Silberbush and Barber, 1984). I_{max} is also known to depend on plant nutrient status (Claassen and Barber, 1977; Glass and Siddiqi, 1984) and on root/shoot ratio (Barber, 1985), all factors reflecting plant demand for nutrients. A further problem in characterising plant demand in terms of root influx parameters is that they depend markedly on the method of measurement. Very different values have been obtained by the depletion and rapid flow techniques, for example (Wild *et al.*, 1979; Mullins and Edwards, 1989). The current way of expressing demand in terms of root influx parameters requires that the parameters are measured at all stages of plant development and under very similar growing conditions to the test plants if uptake is to be satisfactorily simulated. In the field, growing conditions can change markedly during plant development and the root influx parameters would have to be measured many times.

An alternative approach to characterising plant demand is to specify the relationship between nutrient concentration in the dry matter and growth rate (Van Noordwijk and De Willigen, 1979). Essentially, uptake proceeds according to demand whilst the potential supply rate is adequate and when it is inadequate, uptake proceeds at the permitted supply rate. This simple, practical approach requires a knowledge of potential growth rates and of critical nutrient concentrations in plants, which are known to vary with the stage of development and the plant part (Tisdale *et al.*, 1985). The characterisation of demand for N and P is complicated by their assimilation and storage in organic forms, but the situation may be simpler for K which is largely present as the inorganic cation with an osmotic function. Evidence for the plant regulation of K uptake was reported by Leigh (1989) who found that some crop plant species maintain a constant K concentration in their fresh tissues when grown under a very wide-range of soil K levels. This might form the basis of a new and simpler physiological approach to characterising plant demand.

Current models have proved invaluable for validating the principles of nutrient uptake from soil, but if mechanistic models are to be routinely and successfully applied to crops under wide-ranging conditions of supply and demand, a simpler way of characterising the regulation of nutrient uptake by plants is necessary.

Acknowledgement

This work was partly funded by the award of a research grant from the Agricultural and Food Research Council to P Seward.

References

Baldwin J P, Nye P H and Tinker P B 1973 Uptake of solutes by multiple root systems from soil. III. A model for calculating the solute uptake by a randomly dispersed root system developing in a finite volume of soil. Plant and Soil 38, 621–635.

Barber S A 1984 Soil Nutrient Bioavailability: A Mechanistic Approach. John Wiley and Sons, New York, 398 p.

Barber S A 1985 Potassium availability at the soil-root interface and factors influencing potassium uptake. *In* Potassium in Agriculture. Ed. R D Munson. pp 309–324. ASA-CSSA-SSSA, Madison, WI.

Barber S A and Cushman J H 1981 Nitrogen uptake model for agronomic crops. *In* Modelling Waste Water Renovation-Land Treatment. Ed. J K Iskandar. pp 382–409. John Wiley and Sons, New York.

Barraclough P B and Tinker P B 1981 The determination of ionic diffusion coefficients in sieved soils in relation to water content and bulk density. J. Soil Sci. 32, 225–236.

Barraclough P B and Leigh R A 1984 The growth and activity of winter wheat roots in the field: The effect of sowing date and soil type on root growth of high-yielding crops. J. Agric. Sci., Cambr. 103, 59–74.

Claassen N and Barber S A 1974 A method for characterising the relation between nutrient concentration and flux into roots of intact plants. Plant Physiol. 54, 564–568.

Claassen N and Barber S A 1976 Simulation model for nutrient uptake from soil by a growing plant root system. Agron. J. 68, 961–964.

Claassen N and Barber S A 1977 Potassium influx characteristics of corn roots and interaction with N, P, Ca and Mg influx. Agron. J. 69, 860–864.

Claassen N, Syring K M and Jungk A 1986 Verification of a mathematical model by simulating potassium uptake from soil. Plant and Soil 95, 209–220.

Clarkson D T 1981 Nutrient interception and transport by root systems. *In* Physiological Processes Limiting Plant Productivity. Ed. C B Johnson. pp 307-330. Butterworths, London.

Glass A D M and Siddiqi Y 1984 The control of nutrient uptake rates in relation to the inorganic composition of plants. *In* Advances in Plant Nutrition, Vol. 1. Eds. P B Tinker and A Läuchli. pp 103–147. Praeger, New York.

Jungk A and Barber S A 1975 Plant age and the phosphorus uptake characteristics of trimmed and untrimmed corn root systems. Plant and Soil 42, 227–239.

Leigh R A 1989 Potassium concentrations in whole plants and cells in relation to growth. *In* Methods of K Research in Plants. Proceedings of the 21st Colloquium of the International Potash Institute, pp 117–126. IPI, Berne.

Ministry of Agriculture Fisheries and Food 1986 Fertilizer Recommendations 1985–86. MAFF Reference Book 209. HMSO, London.

Mullins G L and Edwards J H 1989 A comparison of two methods for measuring potassium influx kinetics by intact corn seedlings. J. Plant Nutr. 12, 485–496.

Nye P H 1968 The use of exchange isotherms to determine diffusion coefficients in soil. Transactions of the IXth International Congress Soil Science 1, 117–126.

Silberbush M and Barber S A 1984 Phosphorus and potassium uptake of field-grown soybean cultivars predicted by a simulation model. Soil Sci. Soc. Am. J. 48, 592–596.

Tisdale S L, Nelson W L and Beaton J D 1985 Soil Fertility and Fertilizers. 4th edition. Macmillan, New York.

van Noordwijk M and de Willigen P 1979 Calculation of the root density required for growth in soils of different P-status. *In* The Soil-Root Interface. Eds. J L Harley and R Scott Russell. pp 381–390. Academic Press, London.

Wild A, Woodhouse P J and Hopper M J 1979 A comparison between uptake of potassium by plants from solutions of constant potassium concentration and during depletion. J. Exp. Bot. 30, 697–704.

M. L. van Beusichem (Ed.), *Plant nutrition – physiology and applications*, 711–715.

PLSO IPNC601

Simulation of fertilizer influence on yield and quality of potato (*Solanum tuberosum*) tubers by nonlinear optimization method

H. KOLBE, WEI-LI ZHANG and L. BALLÜER
Institute of Agricultural Chemistry, Georg-August University, Von-Siebold-Str. 6, D-3400 Göttingen, FRG

Key words: chemical compounds, field and pot trials, mathematical optimization model, mineral nutrients, potato, *Solanum tuberosum* L., threshold values, tuber yield

Abstract

Data of yield characteristics and the main chemical compounds of potato tubers, received from special NPK-fertilized pot and field trials, were individually evaluated by multiple regression analysis. The best suited mathematical functions, including linear, quadratic and interactive terms of N-, P- and K-concentrations in tuber dry matter, were integrated as independent variables into the models and the SUMT-method was applied for the nonlinear programming.

Particular nutrient requirements could be derived from different NPK-ratios which were obtained by optimization. These ratios led to maximum tuber yield (=objective function) while at the same time observing special threshold values (*e.g.* nitrate) or limiting indices (=constraints) of tuber quality. By means of simulation it was possible to demonstrate characteristic differences between the effect of N-fertilization and varied N-concentrations on tuber yield and composition.

Calculation of dependent variables by fixed NPK-ratios of the tubers resulted in excellent correspondence of the values for several parameters with the predictions made by both models, which were based on pot or field experiments.

Introduction

Increasing NPK-fertilization results in different reactions of tuber yield and composition. Therefore, it seems to be much more complicated to optimize yield subject to limits of special indices of tuber quality, threshold values or ecological requirements than to optimize tuber yield only.

In order to solve such problems by modeling, the effects of nutrients on important tuber parameters had first to be quantified exactly. The resulting mathematical functions obtained were then used to set up a computer model. A selection of initial calculated results of optimization and simulation with respect to fertilizer influences on tuber yield and quality are presented here.

Materials and methods

Outdoor pot (Mitscherlich) and field trials with potato (*Solanum tuberosum* L.) were carried out in 1981 and 1982 by special widely graded and combined NPK-fertilizer plots. After harvesting (stage of dry petiols) tuber yield characteristics, chemical compounds as well as discolouration were investigated as shown in Table 1. Exact experimental layout have already been reported by Kolbe and Müller (1985), Müller and Hippe (1987) and Zhang (1989).

Multiple nonlinear regression analyses were carried out to get mathematical functions with high multiple r^2, such as equation (1), which describes the relations between the concentrations of nitrogen (N), phosphorus (P) and potas-

Table 1. Calculation of maximum yield subject to different constraints for dependent variables, effects of N-fertilization compared with related N-concentrations on NPK-ratios and compounds examined and accuracy of simulated results obtained using the established optimization models based on pot and field experiments for potato tubers

		Pot trials				Field trials				Model comparison			
		Nitrate threshold values[a]				Quality index[b]	N-fertilization/ N-content variation[c]			8		9	
	Lower bounds	1	2	3	Upper bounds	4	5	6	7	Pot	Field	Pot	Field
Objective function:													
Total tuber yield (g/pot; dt/ha FM)		1180.21	1077.25	886.50		450.61	471.83	442.93	423.36	778.90	491.31	721.33	460.25
Independent variables:													
N (% DM)	1.20 ≤	2.40	2.04	1.85 ≤	2.60	1.73	1.80	1.65	1.65	1.90	1.90	1.75	1.75
P (% DM)	0.18 ≤	0.48	0.48	0.48 ≤	0.48	0.32	0.30	0.35	0.30	0.29	0.29	0.27	0.27
K (% DM)	1.20 ≤	3.01	3.08	1.47 ≤	3.10	1.62	1.78	2.09	1.78	1.50	1.50	2.03	2.03
Dependent variables (constraints):													
Tuber yield ≥ 80 g (g/pot; dt/ha FM)	250.00 ≤	847.45	666.85	493.54 ≤	850.00	206.01	264.98	237.89	229.00	462.06	251.39	415.28	313.20
Tuber yield 80–30 g (g/pot; dt/ha FM)	150.00 ≤	307.27	397.44	317.65 ≤	400.00	177.45	168.32	168.46	171.62	253.23	174.33	245.33	158.95
Tuber yield ≤ 30 g (g/pot; dt/ha FM)	30.00 ≤	56.64	68.77	91.16 ≤	100.00	14.00	12.31	13.44	13.55	60.37	12.49	56.61	11.21
Tuber number (per pot; m^2)	12.00 ≤	19.83	20.35	20.70 ≤	25.00	48.51	49.92	52.17	47.55	15.24	48.95	14.79	50.19
Tuber weight (g FM)	35.00 ≤	72.12	56.81	39.38 ≤	80.00	84.93	87.13	85.36	84.29	45.36	87.74	46.55	87.59
Dry matter (%)	15.50 ≤	16.07	15.50	19.62 ≤	22.00	22.00	21.41	21.15	21.88	19.91	21.90	18.85	20.95
Starch (% DM)	63.00 ≤	64.80	64.70	68.69 ≤	75.00	69.82	69.79	70.86	70.15	68.07	70.23	67.06	69.17
Crude protein (% DM)	7.00 ≤	14.71	12.66	11.58 ≤	16.50	10.75	11.13	10.23	10.25	11.80	11.76	10.86	10.84
Pure protein (% DM)	3.50 ≤	5.63	5.44	5.11 ≤	8.00	4.68	4.98	4.73	4.75	5.36	4.98	5.21	5.16
NPN-compounds (% DM)	3.00 ≤	9.48	7.39	6.55 ≤	10.00	5.77	6.16	5.33	5.48	6.40	6.30	5.68	5.88
Ratio crude/pure protein (%)	35.00 ≤	39.27	42.30	42.98 ≤	65.00	45.00	44.93	48.26	47.19	45.27	42.00	47.78	47.01
Ascorbic acid (mg/100 g DM)	60.00 ≤	117.60	104.74	64.33 ≤	120.00	74.03	76.73	80.02	77.15	79.22	71.60	81.68	80.91
Vitamin C (mg/100 g DM)	70.00 ≤	123.49	116.71	90.67 ≤	150.00	93.91	97.88	100.05	97.99	97.88	92.67	103.65	103.65
Citric acid (% DM)	1.00 ≤	5.40	4.82	1.00 ≤	5.40	1.56	1.69	1.87	1.76	1.48	1.44	2.48	1.97
Glucose (% DM)	0.50 ≤	0.85	1.19	1.66 ≤	2.30	0.76	0.58	0.93	1.04	1.39	0.74	1.39	0.44
Fructose (% DM)	0.30 ≤	0.89	0.78	1.04 ≤	1.80	0.49	0.36	0.61	0.86	0.98	0.33	1.04	0.46
Ratio glucose/fructose	0.70 ≤	1.33	1.17	1.89 ≤	2.00	1.94	1.32	1.82	1.46	1.86	1.53	1.66	0.69
Reducing sugars (% DM)	0.90 ≤	1.50	1.64	2.59 ≤	4.00	1.08	0.86	1.33	1.73	2.42	0.50	2.47	1.60
Saccharose (% DM)	0.50 ≤	2.63	2.03	3.23 ≤	5.00	1.70	1.40	1.30	1.75	3.29	1.46	2.92	1.13
Total sugars (% DM)	2.40 ≤	6.17	6.01	9.56 ≤	10.00	2.81	2.87	2.75	3.26	5.38	2.90	5.79	3.35
Nitrate (mg/1000 g DM)	100.00 ≤	2700.00	1350.00	270.00 ≤	2700.00	385.56	494.39	399.89	382.80	524.77	511.72	535.43	536.86
Magnesium (mg/1000 g DM)	700.00 ≤	948.99	1264.82	905.64 ≤	1800.00	1077.64	1085.70	1201.60	1079.61	1146.19	1016.89	1327.10	1106.34
Manganese (mg/1000 g DM)	7.00 ≤	9.49	10.15	8.20 ≤	15.00	7.50	7.52	7.91	7.54	9.27	7.26	10.08	7.62
Cell wall compounds (% DM)	4.00 ≤	10.99	12.54	5.82 ≤	15.00	11.02	10.24	10.81	10.67	10.91	10.39	10.76	9.62
Enzymatic browning (index)	1.00 ≤	1.91	1.70	6.08 ≤	7.50	5.15	4.82	4.46	4.63	5.48	5.40	3.97	4.20
Additional variables:													
N-li							−0.25	−0.25					
K-li							−0.13	−0.13					

[a] Nitrate threshold values: No. 1 ≤ 2700 mg/1000 g DM = 100%, No. 2 ≤ 1350 mg/1000 g DM = 50%, No. 3: ≤ 270 mg/1000 g DM = 10%.
[b] Quality index: Yield of small tubers ≤ 14.0 dt/ha FM, Dry matter ≥ 22.0%, Ratio crude/pure protein ≥ 45.0%, Citric acid ≥ 1.5% DM, Reducing sugars ≤ 1.1% DM.
[c] N-fertilization: No. 5–6, N-content variation: No. 5–7.

sium (K) in dry matter (DM) as independent variables ($x_1, \ldots, x_n$; regression coefficients b_i, c_{ij}, d_j) and further tuber parameters as dependent variables (y):

$$y = a + b_1 x_1 + , \ldots , + b_n x_n + c_{11} x_1^2 + c_{12} x_1 x_2 + , \ldots , + c_{n-1,\cdot n} x_{n-1} x_n + c_{nn} x_n^2 \quad (1)$$

The general model of mathematical optimization (Operations Research) applied here can be formulated as follows (2, 3, 4):

$$z = f(x_1, \ldots, x_n) \rightarrow \text{maximum/minimum!} \quad (2)$$

subject to the constraints

$$A_{i,min} \leq g_m(x_1, \ldots, x_n) \leq A_{i,max} (i = 1, \ldots, m) \quad (3)$$

$$x_j = d_j(x_1, \ldots, x_n)(j = 1, \ldots, n) \quad (4)$$

Optimal values of n independent variables $x_1, \ldots, x_n$ (=NPK concentrations in DM) will be determined that maximize the objective function Z (=total tuber yield of fresh weight, FM) subject to m constraints g_i $(x_1, \ldots, x_n)$ (=tuber yield characteristics except total yield, contents of chemical compounds in DM, visual estimated discolouration of raw potato pulp). The bounds A_i for $i = 1, \ldots, m$ can be set with the lower (min) and upper (max) observed values in the experiments (3), and with given proportions determined by functions which were defined between independent variables (4).

Because of using mixed quadratic terms like equation (1) for numerical solution of (2) subject to (3, 4), the SUMT-algorithm (Sequential Unconstrained Minimization Techniques) was applied by an IBM-compatible personal computer implementation in the nonlinear programming (Fiacco and McCormick, 1968; Zimmermann, 1987).

Results and discussion

After completion of the models by integration of the best suited mathematical objective function and constraints including the experimental determined bounds, optimization led to absolute maximum yield of the trials. As shown in Table 1 (example 1) a maximum tuber yield of 1180 g/pot is obtained through high NPK-concentrations. Under these conditions the values of parameters such as P, citric acid, as well as a nitrate content of 2700 mg/1000 g DM in the solutions of the calculation reach their upper bounds.

If high nitrate contents in potato tubers cannot be accepted (Kolbe, 1987), a further calculation, for example, can be carried out by adjusting the upper bound of this component to 1350 mg/1000 g DM (Table 1, example 2). Comparing the solution from yield maximization subject to this nitrate threshold value with that from absolute yield maximization (example 1), a reduction of the nitrate content of 50% is realized by a decrease of the N-content by about 15%. Moreover, a slight decrease of tuber yield of 10% is registered. This result is in agreement with those obtained from other experiments with potatoes, Chinese cabbage and spinach (Pieters *et al.*, 1984; Zhang *et al.*, 1988; Zhang, 1989).

But if the average nitrate content in potato tubers is not higher than 10% of the first calculation (Table 1, examples 1–3), a marked reduction in N-content of about 23% and a decrease of the K-content is observed. In this case a decrease in tuber yield of about 25%, compared with the absolute yield maximum, must be accepted.

The different reactions of dependent variables to the given conditions are caused by special types of relationships between NPK-concentrations and yield, nitrate or other components of the plants, that have some general validity not only for potato tubers, but also for other crops (Fig. 1).

With respect to N- and K-concentration increases (excluding P and corresponding interactions), the response curves of some dependent parameters such as tuber yield, dry matter content (and even starch, sugars and cell wall compounds) react mainly in parabolic functions as a diminishing return in yield, a phenomenon first reported by Mitscherlich (1909). The reactions of other chemical ingredients such as nitrate (but even NPN-compounds) appear as curves of exponential or logarithmic types. For parameters such as ascorbic acid (and some yield characteris-

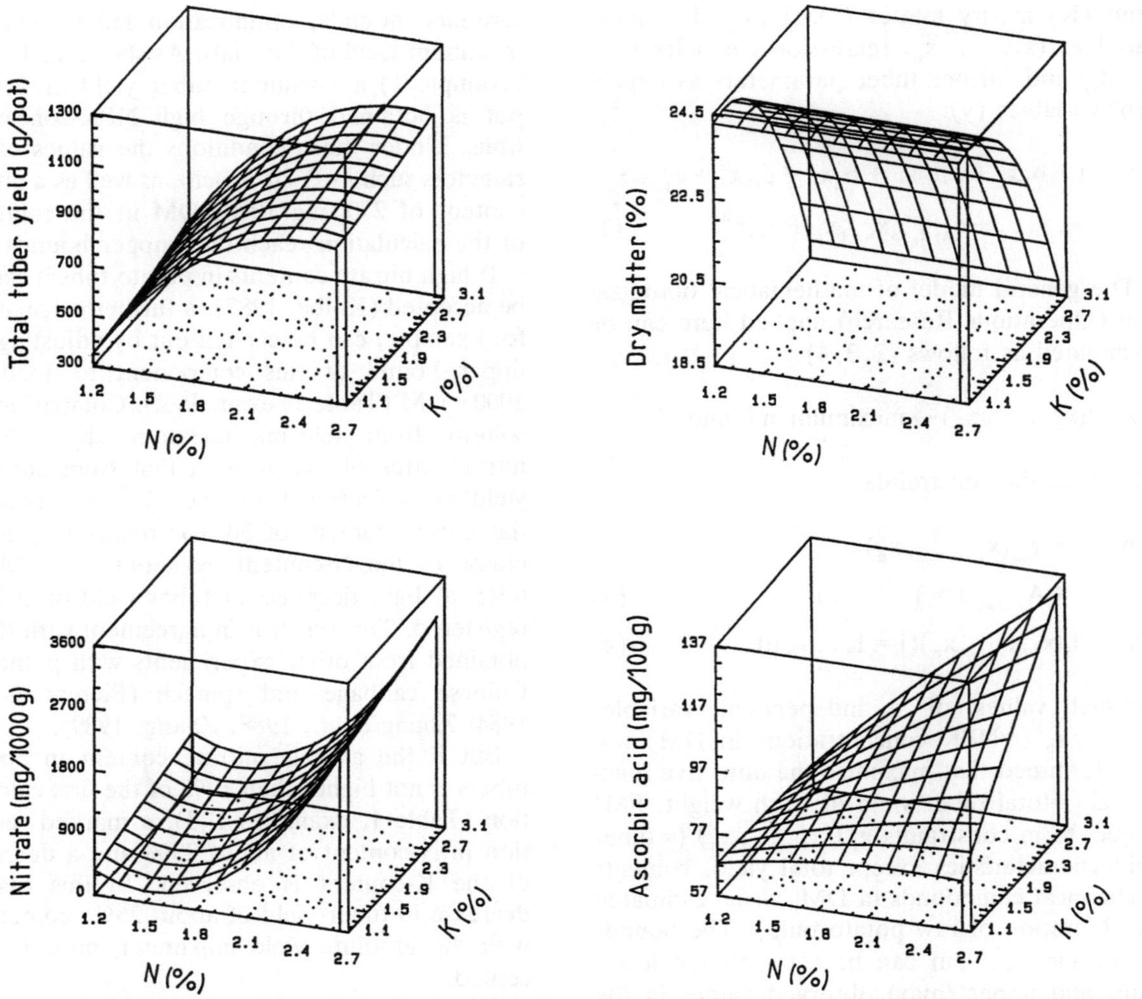

Fig. 1. Relationships between effects of increasing concentrations of nitrogen and potassium on yield, dry matter, nitrate and ascorbic acid content of potato tubers harvested from pot experiments and evaluated by multiple regression analysis.

tics, sugars, citric acid and discolouration) the interaction terms of nutrient supply are of great importance.

By means of optimization models it is also possible to calculate the nutrient requirements by fixing several lower or uppers bound of constraints in order to maximize tuber yield and satisfy special quality demands, *e.g.* for industrial processing of potato tubers (Table 1, example 4).

Further calculations demonstrate differences between results obtained from simulations of "N-fertilization" and from corresponding N-concentrations (Table 1). By fixing the determined relations between NPK-contents, which is described in equation (4), the simulation of plant reactions to different amounts of N-fertilization will be conducted (examples 5–6). Simulation of reactions through N-fertilization, which can be calculated by fixing the N-concentration first to 1.80% DM and then to 1.65% DM, resulted in a typical variation of P- and K-content in tuber DM, as was also reported in the results of previous field experiments (Simpson, 1962; Vertregt, 1968; Zänker *et al.*, 1975).

Variations of the N-compounds, tuber yield and graded yields as well as sugars and cell wall components are influenced mainly (through high values of regression coefficients) by N-contents and can be calculated with a relatively high degree of accuracy using different N-concentrations only (examples 5–7). Other parameters, such as dry matter, starch, vitamin C, citric acid, reducing sugars, magnesium, manganese and enzymatic browning, or tuber number, yield of small tubers, ratio of glucose/fructose and even cell wall components reach more sensitively to the variation of K- or P-contents. Altogether, the effects of N-fertilization are presented through an integrated expression of the single reactions mentioned above (examples 5–6): increasing values of tuber yield and weight, of all N-compounds and discolouration; decreasing values of small tubers and tuber number, content of starch, vitamin C, citric acid, reducing sugars, minerals (including P- and K-contents) and cell wall compounds.

A first test of the reliability of the optimization models established can be made by comparing calculated results between models based on pot and field trials with fixed NPK-ratios (Table 1, examples 8, 9). The plants of pot and field experiments grew up in almost the same weather conditions and most of the parameters investigated corresponded to the values obtained by using the two models, with the exception of graded yields and sugar components.

It should be mentioned that in spite of the extremely wide ranges of nitrate content from 100 to 2700 mg/1000 g DM and ascorbic acid from 60 to 120 mg/100 g DM which were determined in the experiments, there are only insignificant differences between values obtained by the two models.

It can be concluded that potato tuber parameters are directly or indirectly dependent on different concentrations of the main nutrient elements in the tuber. By means of mathematical optimization techniques it seems to be possible to calculate NPK-ratios (instead of fertilizer requirements) that optimize tuber yield subject to a special tuber quality. Furthermore, it is possible to simulate complex fertilization reactions of tuber yield and composition and also to separate fertilizer influences into single causal relations as well as to test several hypothesis of agricultural research.

References

Fiacco A V and McCormick G P 1968 Nonlinear Programming: Sequential Unconstrained Minimization Techniques. John Wiley and Sons, New York, 210 p.

Kolbe H 1987 Untersuchungen zur Bedeutung des Nitratgehaltes in Kartoffelknollen. Kartoffelbau 38, 105–109.

Kolbe H and Müller K 1985 Einfluß differenzierter Nährstoffgaben auf einige wertgebende Inhaltsstoffe in Speisekartoffeln. Kartoffel-Tagung 6, 12–21.

Mitscherlich E A 1909 Das Gesetz des Minimums und das Gesetz des abnehmenden Bodenertrages. Landwirtsch. Jahrb. 38, 537–552.

Müller K and Hippe J 1987 Influence of differences in nutrition on important quality characteristics of some agricultural crops. Plant and Soil 100, 35–45.

Pieters J H, Van der Boon J, Slangen J H G and Titulaer H H H 1984 Adviezen voor de stikstofbemesting van groentegewassen in de volle grond, ter voorkoming van te hoge nitraatgehalten. Bedrijfsontwikkeling 15, 245–247.

Simpson K 1962 Effects of soil-moisture tension and fertilizers on the yield, growth and phosphorus uptake of potatoes. J. Sci. Food Agric. 13, 236–248.

Vertregt N 1968 After-cooking discolouration of potatoes. Eur. Potato J. 11, 226–234.

Zänker J, Gall H, Mirswa W, Ebert K and Töpfer S 1975 Einfluß hoher mineralischer Stickstoffdüngung und Beregnung auf Ertrag und Qualität der Kartoffel. 3. Mitteilung: Blaufleckigkeit, Rohverfärbung, Speisewert, Gehalt an verschiedenen Inhaltsstoffen sowie Schlußfolgerungen für die optimale Stickstoffversorgung bei Speisekartoffeln. Arch. Acker- u. Plfanzenbau u. Bodenkd. 19, 811–825.

Zhang W-L 1989 Einfluß unterschiedlicher Nährstoffgaben (Stickstoff, Phosphat und Kalium) und deren Wechselwirkungen auf den Gehalt an einigen qualitätsbestimmenden Inhaltsstoffen von Chinakohl, Kartoffeln und Hirse sowie Einsatz eines mathemathischen Optimierungsmodelles zur Förderung der Produktion qualitativ hochwertiger landwirtschaftlicher Erzeugnisse. Ph.D. thesis, Göttingen, 282 p.

Zhang W-L, Kolbe H, Ballüer L and Severin H 1988 Produktion landwirtschaftlicher Erzeugnisse mit Hilfe eines mathematischen Optimierungsmodells (vorgestellt am Beispiel von Chinakohl). VDLUFA-Schriftenreihe 27, 44–45.

Zimmermann H-J 1987 Methoden und Modelle des Operations Research. Friedr. Vieweg und Sohn, Braunschweig, 364 p.

M. L. van Beusichem (Ed.), *Plant nutrition – physiology and applications*, 717–721.

PLSO IPNC294

Determination of fruiting capacity of apple trees (*Malus domestica*) by DRIS

E. SZÜCS and T. KÁLLAY[1]
Research Institute for Fruit Growing and Ornamentals, Park u.2., H-1223 Budapest, Hungary.
[1]*Corresponding author*

Key words: DRIS, fruit load, apple trees, *Malus domestica* Borkh., nutritional imbalance index

Abstract

With use made of the Diagnosis and Recommendation Integrated System (DRIS) contradictory relationship between nutritional status and fruit load of Jonathan apple trees were encountered. Data were collected from commercial orchards, and results were verified in an experimental orchard. Correlations between leaf NP^{-1}, NK^{-1}, KP^{-1} and fruit load ($kg.tree^{-1}$) could be described by quadratic equations. Maximum values of the equations ($NP^{-1} = 14$, $NK^{-1} = 3$, $KP^{-1} = 5$) were used as norms for DRIS index calculations. Nutritional Imbalance Indices (NII) were computed to evaluate actual fruit load and nutrition.

Evaluation of annual orchard data led to the conclusion that our norms must be considered as limits for the Highest Permissible Load (HPL). A HPL value is characteristic of a given orchard and is a parameter determined by agro-ecological conditions.

Introduction

The fruiting capacity of apple trees is reflected in the highest yield that the grower may harvest without the risks of overloading.

The correlation between nutrition and yield was best demonstrated by Lamb *et al.* (1959) with the example of biennially bearing trees. For regularly bearing trees the same phenomenon was proved (Szücs and Kállay, 1987).

Evaluating the nutritional status of apple trees is a complex problem, taking into account that both the current fruit load and preceding yields may influence the nutritional balance of a given orchard in a given year.

The Concept of Balance (Beaufils, 1973) seems to be interpretable for apple trees, provided nutrition and yield are both considered. In a related paper dealing with this topic (Szücs *et al.*, 1989) DRIS indices, as obtained with two methods, *viz*. the traditional method (Sumner, 1977) and the quadratic regression method were compared. In the present paper, attention is given to determine the fruiting capacity of apple trees by means of DRIS calculations.

Materials and methods

Data for basic DRIS calculations were obtained from 18 commercial orchards in three consecutive years. The results of DRIS analyses were evaluated with data obtained in a long-term fertilization trial (Szücs and Kállay, 1978; Terts, 1970). The commercial orchards represented various soil and cultivation types, but were all fertilized similarly. In the experimental orchard, various levels of fertilization were practiced, as follows:

0 = Not fertilized
N = 130 kg N/year
NPK = 130 kg N + 26 kg P + 166 kg K/year
2/NPK/ = 260 kg N + 52 kg P + 332 kg K/year

The nutritient levels in soils showed significant differences (Kállay and Szücs, 1982).

In each orchard Jonathan trees were sampled. Mid-shoot leaves were collected in the middle of summer for foliar analyses. The yields were estimated from quantities sold in the commercial orchards, and by weighing in the experimental plots. All yield data were expressed in $kg.tree^{-1}$ (295 trees/ha).

DRIS calculations according to Sumner (1977) were made for nutrient ratios corresponding to yield (NP^{-1}, NK^{-1}, KP^{-1}), while nutrient ratios independent of yield were omitted. The correlation between nutrient ratios and yield could be expressed by use of quadratic equations. The equations allowed estimates to be made of maximum yields and of correlating nutrient ratios used as norms for DRIS index calculations (Szücs *et al.*, 1989). The norms obtained with the traditonal method and the quadratic regression method were:

	NP^{-1}	(CV)	NK^{-1}	(CV)	KP^{-1}	(CV)
Traditonal method	15.0	(15.4)	2.20	(26.4)	7.1	(20.4)
Quadratic regression method	14.0	(-)	3.00	(-)	5.0	(-)

DRIS index calculations based on norms obtained with quadratic regression followed the general rule (Sumner, 1977) with the omission of coefficient of variance (CV) values.

The Nutrition Imbalance Index (NII), the sum of DRIS indices, regardless of sign, was calculated in each method. Though the values of NII obtained with the different methods differed greatly, the meaning of NII remained the same.

Results

Application of DRIS to apple trees under different nutritional conditions

In the experimental orchard, large differences were observed in both soil and leaf nutrient contents, but fruit yield was little affected. These results indicated that yields were influenced by other factors than nutrition. DRIS indices, obtained with the traditional method are presented in Table 1, together with the corresponding Nutritional Imbalance Indices and yields.

During the first four years yields showed little variation, so that the influences of fertilization on the nutritional status of the trees were clearly evident. In the 0 and N treatments, the relative deficiency of potassium is obvious, whereas in the NPK and 2(NPK) treatments the relative oversupply of that nutrient is clear. The significant yield increases in 1980 and 1981 accentuated the potassium deficiency in the 0 and N treatments, while in the NPK and 2(NPK) treatments, the oversupply of potassium was somewhat alleviated. The influence of the second high yield was even stronger than that of the first one.

The NII values were smaller in the 0 and N treatments, signifying that yields of 60 kg were near fruiting capacity. As yields increased in the years 1980 and 1981, potassium deficiency increased considerably and caused appreciable increases in NII as well. This phenomenon is understandable, although not in agreement with the Concept of Balance (Sumner, 1977). DRIS index calculations of the norms based on quadratic regressions, seem to avoid the above mentioned contradiction (Table 2).

The general tendencies of nutritional balance are similar to those appearing from traditional DRIS calculations, but the discrepancy caused by the higher yields was not encountered.

Determination of fruiting capacity of apple trees

The NII values—according to the Concept of Balance—tend to decrease when yields increase. As the norms obtained with the quadratic regression method signify the highest possible fruit yield of the population observed, NII values near zero will correspond with high yields only. A yield of NII = 0 may be considered as the highest yield that a given population of apple trees can

Table 1. DRIS indices, NII values and yields for the experimental orchard in the period 1976–1981, as obtained with the traditional method

Year	Treatment	N_i	P_i	K_i	NII	Yield, kg. tree^{-1}
1976	0	−14.5	9.1	5.4	29.0	61
	N	−12.5	6.8	5.7	25.0	58
	NPK	−14.9	−0.6	15.5	31.0	61
	2(NPK)	−16.2	−8.6	24.8	49.6	70
1977	0	4.0	0.4	−4.4	8.8	65
	N	6.0	0.6	−6.6	13.2	62
	NPK	−0.6	−21.5	22.1	44.2	70
	2(NPK)	−6.5	−18.1	24.6	49.2	73
1978	0	−0.4	6.3	−5.9	12.6	66
	N	0.2	1.5	−1.7	3.4	60
	NPK	−11.5	−12.0	23.5	47.0	74
	2(NPK)	−10.6	−13.9	24.5	49.0	67
1979	0	0.5	5.0	−5.5	11.0	58
	N	7.6	5.0	−12.6	25.2	58
	NPK	−7.3	−10.9	18.2	36.4	67
	2(NPK)	−5.3	−27.4	32.7	65.4	69
1980	0	4.1	13.6	−17.7	35.4	86
	N	6.1	7.8	−13.9	27.8	74
	NPK	−7.9	−11.5	19.4	38.8	106
	2(NPK)	−11.4	−19.3	30.7	61.4	96
1981	0	7.4	5.7	−13.1	26.2	92
	N	8.5	4.6	−13.1	26.2	78
	NPK	−3.7	−4.8	8.5	17.0	106
	2(NPK)	−4.9	−10.1	15.0	30.0	96

produce, if its nutritional status corresponds with the "ideal" ratios as obtained by quadratic regression analysis. This level can be evaluated when nutrient data together with yield data are assessed in a given orchard for consecutive years.

The fruiting capacity of the orchard may be estimated by simple linear regression analysis, when NII is plotted against yield. In Table 3 estimates of fruiting capacity of apple trees in the experimental orchard are given, based on six years' averages of NII and yield.

Fruiting capacity of 0 and N trees seems to be limited by poor nutritional status, but the fruiting capacity, as determined by the current nutritional status was utilized almost completely. On the other hand, NPK and 2(NPK) trees could not fully utilize their possibilities, as created by massive nutrition.

Orchard 1 is an old orchard where the actual yields amount to approx 81% of the capacity. Orchard 2 has the same age, with fertilizer applied regardless of demands. Orchard 3 is a young plantation with favorable conditions. Twice in six years the fruiting capacity was already met.

Discussion

Following the traditional DRIS method for apple trees, results indicated that the Nutritional Imbalance Index decreased when yield tended to increase, but that above a certain yield level NII started to increase again. This phenomenon is against the Concept of Balance. NII calculated with the quadratic regression method permitted determination of the fruiting capacity of apple trees at NII = 0.

Table 2. DRIS indices, NII values and yields for the experimental orchard in the period 1976–1981, as obtained with the quadratic regression method

Year	Treatment	N_i	P_i	K_i	NII	Yield, kg. tree^{-1}
1976	0	−0.48	−0.11	0.59	1.08	61
	N	−0.45	−0.15	0.60	1.20	58
	NPK	−0.57	−0.34	0.91	1.82	61
	2(NPK)	−0.67	−0.54	1.21	2.42	70
1977	0	−0.04	−0.21	0.25	0.50	65
	N	−0.015	−0.19	0.205	0.41	62
	NPK	−0.30	−0.79	1.09	2.18	70
	2(NPK)	−0.45	−0.73	1.18	2.36	73
1978	0	−0.13	−0.11	0.24	0.48	66
	N	−0.14	−0.21	0.35	0.70	60
	NPK	−0.55	−0.60	1.15	2.30	74
	2(NPK)	−0.53	−0.65	1.18	2.36	67
1979	0	−0.11	−0.13	0.24	0.48	58
	N	0.04	−0.12	0.08	0.24	58
	NPK	−0.43	−0.55	0.98	1.96	67
	2(NPK)	−0.462	−0.959	1.421	2.84	69
1980	0	0.017	0.027	−0.044	0.09	86
	N	0.018	−0.073	0.054	0.15	74
	NPK	−0.45	−0.57	1.02	2.04	106
	2(NPK)	−0.44	−0.79	1.23	2.46	96
1981	0	0.039	−0.109	0.07	0.22	92
	N	0.059	−0.129	0.07	0.26	78
	NPK	−0.286	−0.379	0.665	1.33	102
	2(NPK)	−0.355	−0.518	0.873	1.75	96

Table 3. Determination of fruiting capacity of apple trees in the experimental orchard

Treatment	Equation kg. tree^{-1}	Fruiting capacity, t.ha^{-1}	Average yields, t.ha^{-1}
0	$Y' = 85.6 - 24NII$	25.2	21.1
N	$Y' = 71.3 - 12NII$	21.0	19.2
NPK	$Y' = 124.3 - 23NII$	36.6	23.6
2(NPK)	$Y' = 132.2 - 23NII$	39.0	23.2

Table 4. Determination of fruiting capacity of apple trees in commercial orchards

Orchard	Equation kg. tree^{-1}	Fruiting capacity, t.ha^{-1}	Average yields, t.ha^{-1}
Orchard 1	$Y' = 139 - 24NII$	41.0	33.4
Orchard 2	$Y' = 189 - 44NII$	55.7	28.6
Orchard 3	$Y' = 197 - 97NII$	58.1	39.4

Nutrition-fruit yield interactions provide a tool for fruit growers to influence the nutritional status of their trees either by fertilization or by crop regulation. The authors suggest that fruiting capacity at NII = 0 be designated as Highest Permissible Load, above which fruit thinning is needed to avoid overloading.

References

Beaufils E R 1973 Diagnosis and Recommendation Integrated System (DRIS): A general scheme for experimentation and calibration based on principles developed from research in plant nutrition. Soil Sci. Bull. No. 1. Univ. of Natal.

Kállay T and Szücs E 1982 Mineral composition of Jonathan apples related to storage quality in a fertilization experiment. *In* Plant Nutrition 1982. Ed. A Scaife. Vol. 1, pp 256–261. Proc. 9th Int. Plant Nutr. Coll., Coventry. Commonw. Agric. Bureaux, Slough, UK.

Lamb G D, Golden J D and Power M 1959 Chemical composition of the shoot leaves of apple Laxton's Superb as affected by biennial bearing. J. Hortic. Sci. 4, 193–199.

Sumner M E 1977 Use of DRIS system in foliar diagnosis of crops at high yield levels. Commun. Soil Sci. Plant Anal. 8, 251–268.

Szücs E and Kállay T 1978 Néhány tápelem hatásának vizsgálata a Jonathán terméshozamára és tárolhatóságára (The influence of some nutrients on the yield and keeping quality of Jonathan). Kertgazdaság 10, 43–52.

Kállay T and Szücs E 1987 Almafák terhelési viszonyai és a gyümölcsminöség néhány összefüggése (Some relations of crop load and fruit quality on apple trees.) "Lippay János" Tud. Ülés. Kertészeti Egyetem, Budapest, I. 625–633.

Szücs E, Kállay T and Szenci Gy 1989 Determination of DRIS indices for apple. Int. Symp. Diag. Nutr. St. Dec. Fruit Orch. Warsaw. (*In press.*)

Terts I 1970 Beitrag zur Beurteilung der Nährstoffversorgung obstbaulich genutzter Böden. Arch. Gartenb. 18, 463–476.

M. L. van Beusichem (Ed.), *Plant nutrition – physiology and applications*, 723–727.

PLSO IPNC197

Criteria for fertilization of vining peas in Portugal

M. E. SARAIVA FERREIRA and M. A. C. FRAGOSO
INIA, Horticultura e Floricultura, Quinta do Marquês, P-2780 Oeiras, Portugal

Key words: cultivar behaviour, fertilization, macronutrient content, nutrient removal, vining peas

Abstract

It is imperative that production costs of vining peas be reduced in Portugal. Because data on nutrient requirements for different cultivars are scarce, correct fertilization is difficult and fertilizers are applied wrongly, both as regards type and quantity. Over a 5-year period, fertilizer trials have been carried out with three pea cultivars—Dark Skin Perfection, Jof and Vitalis—to assess plant behaviour, yield and product quality for the freezing industry. Results presented here are from those sites where there was no response to fertilizer applications, and consist of measurements of macronutrient uptake throughout the growing period. From such selected plots, located at various sites, plant samples were obtained and analysed for dry matter and macronutrient content.

The nutritional behaviour of those varieties was thus characterized. It is concluded that such data will improve fertilizer recommendations and also assist in the diagnosis of nutrient unbalances.

Introduction

In the late seventies, peas became the leading crop for the country's freezing industry where they represent 45% of the frozen products. Industries continue to demand larger supplies of high-quality produce at lower prices.

As potatoes and winter cereals were the traditional crops in the pea growing areas, farmers proved reluctant to rely entirely on symbiotic N_2 fixation for the N nutrition of peas, despite an evidence of good nodulation. Excessive use of N fertilizers affected pea quality and caused other disturbances, sometimes aggravated by limited water resources, and both nitrate and salinity problems. Besides, heavy fertilizer applications, from 50 to 100 kg of N per hectare, and low plant populations, around 50 plants per m^2, resulted in high production costs.

Fertilizer trials were initiated, mostly with N and P. Results from sites with responses to fertilizers are to be reported elsewhere. Trials in which no response to fertilizer applications was obtained, provided the present data. This study aims at defining the nutrient uptake patterns for vining peas in Portugal, hitherto unknown, and at comparing cultivars regarding dry matter production, nutrient concentration and removal values, nutrient ratios, and dilution effects.

Materials and methods

In the western part of the country, the main pea growing area, fertilizer trials have been conducted over a five-year period under similar conditions of soil, climate, and production technology. Different pea cultivars were chosen with a view to comparing responses to fertilization. They were sown in February-March, with 70 plants per m^2, and grown, as non-irrigated crops, on loams and sandy soils with low organic matter and pH varying from 4.0 to 7.5. In most soils, frequently calcareous, cation exchange capacity was usually below 20 meq. per 100 g soil.

As already mentioned, only those sites which failed to show response to fertilizers are used in this work. A total of 277 plots were selected where crop productivity levels of 4 to 5 t. ha^{-1} were considered satisfactory for establishing uptake patterns for vining peas. Plots per cultivar were 152 for DSP, 69 for J, and 56 for V.

The cultivars studied were Dark Skin Perfection (DSP), a commonly grown cultivar and chosen as reference, Vitalis (V), initially used for comparison and later substituted by Jof (J), a recently more popular cultivar.

In selected plots, 5 to 9 samples were collected at regular intervals throughout the growth period. Samplers walked through the plots pulling plants at random (around 1 plant per 10 m^2), a usual practice in pea sampling (Gritton and Chi, 1972). Samples were recorded as days from sowing and node counting was practiced, besides other biometric measurements.

The evaluation of stages of development for each pea cultivar followed codes established by Biddle and Knott (1988). Only reference codes 004, 203, 206, and 208, representing emergence, first open flower, pod swell, pod fill, and green wrinkled pod, respectively, were used in the present study. In the reproductive stage, codes referred to the first podding truss. The last code, representing maturity of produce fit for industrial processing, coincided with crop harvesting dates when tenderometer readings (T.R.) for peas were from 120 to 130. At this stage yields were recorded.

The samples of plant tops were dried at 70–80°C in a forced-air oven for dry matter (DM) evaluation, and then ground in a Wiley mill (40-mesh) for macronutrient determinations expressed as percentage in DM. Analytical procedures were: the micro-Kjeldahl technique for N; nitroperchloric acid digestion (0.2 g/50 mL) for elemental analyses; emission spectrophotometry (EEL) for K; atomic absorption spectrophotometry (Perkin-Elmer 403) for Ca and Mg, with Sr as releasing agent; the yellow phosphovanadomolybdate method, using a Unicam Pye spectrophotometer, for P.

For each cultivar, regression equations were calculated for DM (g per 10 plants) and macronutrient contents (% in DM), as average values, referring to sampling dates throughout the growth period. Effects of soil, climate and plant population were confounded.

Furthermore, to assess the nutritional status of the crops, sums and ratios of macronutrient concentrations were examined. Nutrient removal was calculated and expressed in kg. ha^{-1}.

Results and discussion

Table 1 shows the different behaviour of the three cultivars. Mean values of days from sowing are presented as rounded off figures. The interval in days found between codes 207 and 208 (pod fill and green wrinkled pod, respectively) were 8 for DSP, 19 for J, and 11 for V. A longer reproductive period for cultivar Jof makes the harvesting operation less critical, with better chances to improve pea quality.

Regarding dry matter and macronutrient values per sampling date, fitted equations were constructed for each cultivar, as presented in Figures 1 and 2.

Dry matter contents

The dry matter accumulation (Fig. 1) in the reproductive stage, from first open flower (code 203) to green wrinkled pod (code 208), was 80% of total, as also observed by McMahon and Price (1982). Among cultivars, the highest values were obtained for Vitalis. Increases in dry matter percentages between codes 207 and 208 were 28.2, 50.0, and 82.3 for DSP, J, and V cultivars, respectively.

Macronutrient contents

In Figures 1 and 2, Vitalis showed the highest values for N, P and K percentages in the vegetative stage, and the lowest for Mg. This finding supports earlier results obtained by Ferreira

Table 1. Days needed to reach selected stages of development for the three cultivars

Cultivar	Codes for selected stages of development				
	004	203	206	207	208
DSP	16	65	72	82	90
J	17	56	66	78	97
V	17	65	71	81	92

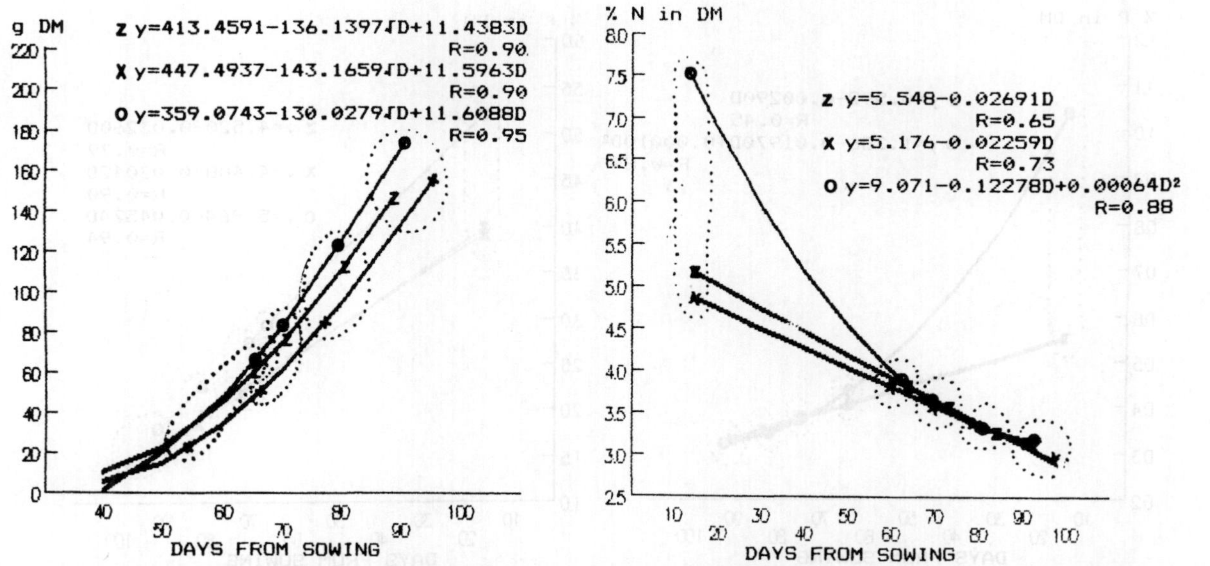

Fig. 1. Fitted equations representing five-year average values for dry matter production and N in plant tops as functions of time at four and five development stages, respectively. In the case of dry matter, the data for code 004 were omitted. The data pertaining to one code are each time encircled. Z = Dark Skin Perfection; X = Jof; O = Vitalis.

(1988) for both Ca and Mg when comparing this cultivar with DSP.

Nitrogen concentrations were within the usual 3–4% range. Since there was no response to N applications and pea plants usually showed remarkable nodulation, it is evident that biological fixation was the main N source. Average N values at harvest (code 208) were 2.83% for Jof and slightly higher—3.00%—for the other two cultivars.

Phosphorus concentrations in Jof—below 0.40%—were the lowest throughout the growth period and did not allow a statistically significant curve adjustment. In the reproductive stage, P values for the DSP and V cultivars varied from 0.32% to 0.51%, in agreement with ranges presented by McMahon and Price (1982). In the three cultivars, K levels were below 3.00% in the reproductive stage, in agreement with findings of some other authors (Ferreira, 1988; Geraldson *et al.*, 1973) but exceeding the values found by McMahon and Price (1982).

Regarding Mg concentrations, Jof had higher values than the DSP and V cultivars, the three becoming closer in the reproductive stage and falling within the range 0.28–0.50%. In all cultivars, values were lower than those stated by McMahon and Price (1982), who reported a range of 0.73–0.87%. Such findings for Mg, together with the examination of K-Ca-Mg triangular diagrams, as resulting from selected data, suggested that low Mg availablility together with Ca excess might occur in the main pea growing area. Information on soil cations availability has not proved conclusive. Neutral to alkaline soils were frequently related to lower values for the K/(Ca + Mg) ratio and higher values for the (N + P + K) and (K + Ca + Mg) sums. In all cultivars, the ranges of nutrient concentration (as % in DM) in the reproductive period agreed with those of Geraldson *et al.* (1973): 3.10–3.60 for N; 0.30–0.35 for P; 2.20–2.80 for K; 1.20–1.50 for Ca, and 0.27–0.35 for Mg.

Values of nutrient sums, such as (N + P + K + Ca + Mg), (N + P + K), and (K + Ca + Mg), as well as several ratios, proved independent of cultivar and varied within each one.

Regarding nutrient removal, as mean values in kg. ha^{-1}, Table 2 shows that the (N + P + K + Ca + Mg) sums were moderate and rather similar in DSP and J cultivars, viz. 336 and 344, respectively, and remarkably higher in the V cultivar with a value of 437, due to a greater DM accumulation (Fig. 1).

Biometric measurements per plant gave 18–20 nodes and 7 pods for DSP, while V and J cultivars were equal, with 22–23 nodes, both with 9 pods. This emphasizes the advantage of J cultivar as being more productive and less demanding in nutrients.

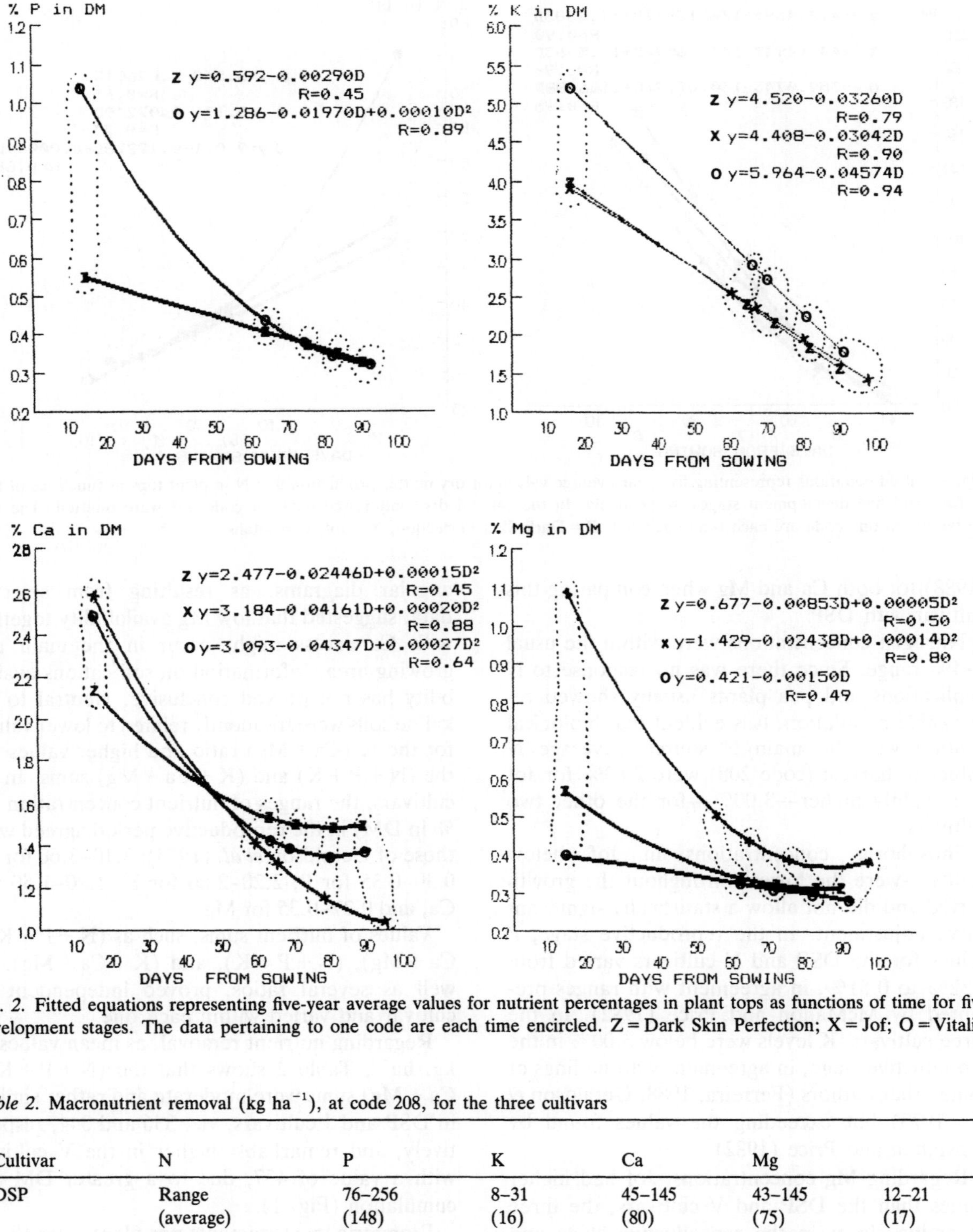

Fig. 2. Fitted equations representing five-year average values for nutrient percentages in plant tops as functions of time for five development stages. The data pertaining to one code are each time encircled. Z = Dark Skin Perfection; X = Jof; O = Vitalis.

Table 2. Macronutrient removal (kg ha^{-1}), at code 208, for the three cultivars

Cultivar		N	P	K	Ca	Mg
DSP	Range	76–256	8–31	45–145	43–145	12–21
	(average)	(148)	(16)	(80)	(75)	(17)
J	Range	103–186	9–30	53–164	35–82	10–25
	(average)	(148)	(20)	(107)	(53)	(16)
V	Range	134–236	16–26	84–130	51–119	14–26
	(average)	(198)	(22)	(112)	(86)	(19)

Mean $N:P_2O_5:K_2O$ ratio values were 4.34:1:2.73 for DSP; 4.03:1:2.98 for J; and 4.02:1:2.75 for V. With different cultivars, McMahon and Price (1982) found 4.06:1:1.36, while values reported by Rankov and Uzunova (1988) were 3.68:1:2.03 and 3.58:1:1.77.

Conclusions

It is concluded that:

For assessing nutrient uptake patterns in vining peas, it is fundamental that samples are collected at various development stages.

For similar levels of productivity, the DSP and J cultivars required less nutrients than the V cultivar.

Besides, with its longer interval between pod fill and green wrinkled pod, the J cultivar makes harvesting less critical, with better chances for improving pea quality. Thus, Jof is a good choice for further studies.

Nutrient removal values are expected to be characteristic of new cultivars. The $N:P_2O_5:K_2O$ removal ratio of 4:1:3 for Jof seems acceptable.

In the main pea growing area, the soil cations K, Ca, and Mg, should receive further attention. The values for K concentrations in plant tops were found to be higher than in other studies.

The nutrition of vining peas in Portugal was evaluated with the use of regression equations. Further trials with newer cultivars are expected to provide more reliable nutritional indices. Because of biological N_2 fixation the use of N fertilizers can be omitted, but also P and K applications are hardly justified. Fertilizers are thus to be used very judiciously.

Acknowledgements

The authors wish to thank Dr. A Scaife for criticism, their colleague M. L. Fernandes for statistical aid, and Mrs. Fernanda Vargues for assistance with chemical analyses and data compilation.

References

Biddle A J and Knott C M 1988 Pea Growing Handbook. PGRO, Peterborough, 264 p.

Ferreira M E 1988 Variation in macronutrient content in peas for processing. Acta Hortic. 220, 267–274.

Geraldson C M, Klacan G R and Lorenz O A 1973 Plant analysis as an aid in fertilizing vegetable crops. *In* Soil Testing and Plant Analysis. Eds. L M Walsh and J D Beaton. pp 365–379. Soil Science Society of America, Madison, WI.

Gritton E T and Chi P 1972 Sampling procedures and optimum sample size for estimating yield components in peas (*Pisum sativum* L.). J. Am. Soc. Hortic. Sci. 97, 451–453.

McMahon C R and Price G H 1982 Nutrient Uptake Studies in Some Processing Vegetable Crops. Consolidated Fertilizers Limited, Australia, 39 p.

Rankov V and Uzunova E 1988 On biological removal of nutrients with green pea yield. Acta Hortic. 220, 275–280.

M. L. van Beusichem (Ed.), *Plant nutrition – physiology and applications*, 729–733.

PLSO IPNC398

Problems of potassium nutrition in citrus (*Citrus sinensis*) orchards in Egypt*

A.F.A. FAWZI, M.M. EL-FOULY and F.K. EL-BAZ
Botany Department, National Research Centre, Dokki-Cairo, Egypt

Key words: *Citrus sinensis* L., clay, magnesium, nutritive status, orange, potassium fertilization

Abstract

The content of exchangeable pottasium (K) in soil is a widespread international parameter of the K-status of a soil. The use of this parameter in Egypt suggests that alluvial soils are extremely rich in K, and that sandy and calcareous soils are low in this nutrient. However, the content of total K in citrus leaves was found to be unsatisfactory in orchards on alluvial soils. The reason is that K uptake by citrus is hindered by K fixation in clay and is antagonized by other cations. Annual applications of fertilizer K, combined with a balanced nutrition regarding other nutrients was found to increase fruit yield in citrus orchards. It was concluded that for higher and better citrus orchards, not only those on poor sandy and calcareous soils, but also those on Nile-alluvial soils are in need of annual K-fertilizer applications. It could further be concluded that the use of exchangeable soil K as index of K availability should be reconsidered under Egyptian conditions.

Introduction

In spite of increasing cropping intensification in Egypt, the belief that crops in the South-Mediterranean region do not need K fertilization is still common (Saurat, 1987). The use of soil NH_4-acetate-extractable K as sole criterion for K availability led to the generalized concept of K sufficiency in the alluvial soils of Egypt (Balba, 1979). Little information was published on determining the actual K needs of crops in the South-Mediterranean region (Ait Housa 1986; Daoud and Mokhtar, 1985; El-Fouly *et al.*, 1987; Fawzi and El-Fouly, 1979; Shehata *et al.*, 1985). Positive responses of some field crops to K application were reported in Egypt (Abdel Hadi *et al.*, 1987; Fawzi *et al.*, 1983).

The present work was conducted to investigate K status of soils and yield responses to K-fertilizer application in orange orchards in different areas of Egypt.

* NRC/GTZ Project micronutrients and other plant nutrition problems in Egypt.

Materials and methods

Sampling and yield determinations

Soil and leaves were sampled in the 1980–1987 period from orange orchards (*citrus sinensis* L.) on variably textured soils. Total numbers of 1000 and 700 soil samples and same numbers of leaves were taken from orchards on Nile alluvial soil and on sandy or calcareous desert soils, respectively. Sampels were collected in Upper Egypt (Souhag, Assiut), Middle Egypt (Minia, Beni-Sweif, Fayoum) and Lower Egypt (Delta: Qaliobia, Monofia, Dakahlia, Gharbia, Kafr El-Sheikh) as well as from the Egyptian Eastern Desert (Sharkia, Ismailia) and Western Desert (Behaira). Surface (0–30 cm depth) soil and 4–6 month-old leaves from spring growth cycles were sampled to represent entire orchards or 2-ha sections of them. Soil and leaf samples were prepared for analysis according to Chapman and Pratt (1978). Yields were expressed in tons of commercially ripe fruits per ha.

Chemical analysis

Soil pH and E.C. (mmhos/cm) in 1:2.5 soil/water, $CaCO_3$, organic matter (O.M); NH_4 acetate-extractable K and −Mg, according to Jackson (1973).

Leaf Total-K, using a wet digestion procedure with an 8:1:1 ratio of nitric, perchloric and sulphuric acid (Chapman and Pratt, 1978).

Field trials

Two on-farm field trials were conducted to investigate the effect of K fertilizer on fruit yield. The first one (Experiment I) was carried out in the 1983–1988 period in a private 4-ha Navel-orange orchard (loam soil) in Qualiobia, Delta. Trees were 16 years old in 1983. The following treatments were included (quantities applied annually to 400 trees per ha):

(1) Control (farmer's practice): 25 m^3 farmyard manure + 75 kg P_2O_5 (as superphosphate), in December + 300 kg N as ammonium sulphate, split into 3 equal parts in March, May and July; all applied to soil.

(2) Control + K: 187 kg K_2O (as K-sulphate) split into 2 parts in May and July as soil application. Trees were irrigated by flooding with Nile water.

(3) Treatment (2) + 12.5 kg Fertilon Sahara (BASFCo.), containing 625 g Zn + 437 g Mn + 214 Fe (pure element) EDTA-chelated. The formulation of the compound was based upon previous soil-and leaf analyses conducted in the project "Micronutrients and Plant Nutrition Problems in Egypt". The total "Fetrilon Sahara" dose was split into 3 equal parts, applied in March, May and Middle of May, as foliar spray.

The second experiment (Experiment II) was carried out in 2 consecutive seasons (1984 and 1985) in 3 public-sector orchards, aged 10 years in 1984. The orchards were located in Ahmed-Orabi village, South Tahrir, on very poor sandy soils. Balady, Shamouti and Valencia oranges were growing in the 1st, 2nd and 3rd orchard, in areas of 80, 120 and 200 ha, respectively. The trees were irrigated by flooding with underground water.

The following fertilizers were yearly applied/ha (300 trees):

(1) Control (farmer's practice): 75 kg P_2O_5 + 250 kg N;

(2) Control + K (120 kg K_2O) + 10 kg Fetrilon Sahara. All fertilizer were applied as in Experiment 1.

In the Experiments 1 and 2, the treatments were randomized with 5 replicates/treatment.

Statistical analysis

Yield data were statistically analyzed using a simple F test for Experiment I for every season (Snedecor and Cochran, 1976) and the New LSD method for combined analysis (Waller and Duncan, 1969) for results of the 2 seasons in Experiment (II).

Results

Soil characteristics

The data in Table 1 show the prevalence of high pH and low organic matter values in the pertinent soils, but without serious salinity problems. $CaCO_3$ contents were above the moderate level in some desert areas. Exchangeable K was very high (>41 mg/100 g) in the alluvial soils of the Nile Valley and the Delta, and below the adequate level (<16 mg/100 g) in newly-reclaimed desert areas.

Soil – and leaf K and soil clay content

The concentration of exchangeable soil K increases and that of total leaf K decreases with increasing clay content in soils of the different locations (Fig. 1). Deficiency of K in orange leaves prevails on alluvial soils of the Nile Valley and the Delta (Table 2).

Exchangeable soil K/Mg ratio and total leaf K

Total-K concentration in leaves increases with increasing exchangeable K/Mg ratio in soil, and vice versa (Table 3).

Table 1. Soil characteristics in orange orchards in Egypt

Location	Clay (%)	Silt (%)	pH	E.C. (mmhos/cm)	$CaCO_3$ (%)	O.M. (%)	Exchangeable K (mg/100 g)
Old alluvial areas							
Upper Egypt	33–56	30–34	8–8.9	1–1.6	2–3	1–2	56–78
Middle Egypt	29–46	21–38	8–8.9	1–1.6	2–3	1–2	91–116
Lower Egypt	29–46	21–38	8–8.9	0.2–1.0	2–3	1–2	41–80
Newly reclaimed desert areas							
Ismailia	5–10	1–10	8–8.8	0.1–0.6	0.5–1.6	0.4–0.8	5–15
Sharkia	5–10	1–10	8–8.6	0.1–0.6	0.5–1.6	0.4–0.8	5–15
Behaira	4–10	2–5	8–8.5	0.1–0.6	0.5–1.6	0.4–0.8	5–15
North Tahrir	10–20	10–18	8–9.0	1.0–1.5	25–30	0.4–1.0	9–15

Response of yield to K application

On the loamy soils (Experiment I), the trees with +K and +K + micronutrients treatments produced more yield than those of the control (−K) treatment. The differences were significant all-over 1983–1987 (Fig. 2). The highest yield occurred in the trees treated with K + micronutrients, being highly significant more than that of the control treatment. It could be also observed in figure 2 that in 1985–87, K application prevented yields from the decline already occured in the control trees. Adding K alone or with micronutrients, also resulted in highly significant yield increase on the poor sandy soils in Experiment 2, where the yield was almost doubled (Table 4).

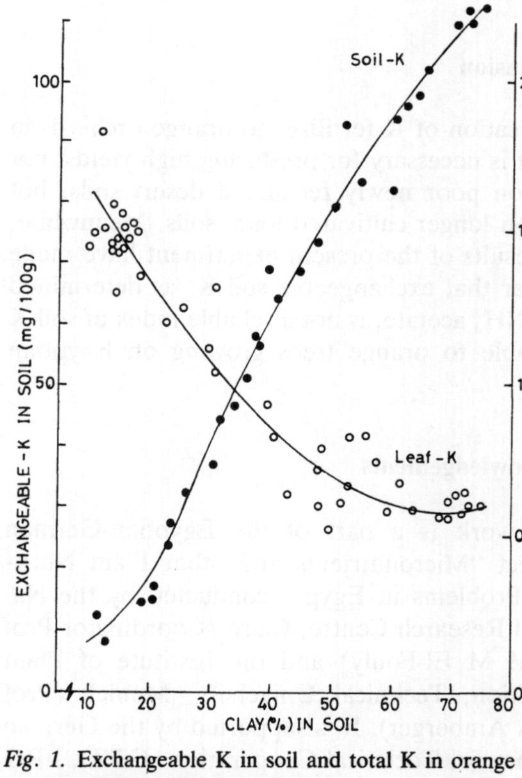

Fig. 1. Exchangeable K in soil and total K in orange leaves for soils differing in clay contents.

Table 2. Percent of K–deficient samples in orange orchards of Egypt

Sample	Nile Valley and Delta (1000 samples)	Newly reclaimed desert areas (700 samples)
(% of total samples)		
Soil	5	82
Leaves	70	24

Satisfactory K levels (Chapman, 1973):
Soil: 21–30 mg/100 g.
Leaves: 1.0–1.7%.

Table 3. Exchangeable soil K/Mg ratio and total leaf K in orange orchards on loamy soils of Egypt (36–43% clay)

Exchangeable quantities (mag/100 g)			Leaf (%)	
K	Mg	K/Mg	K	
0.9463	9.868	0.10	0.70	L
2.1739	12.335	0.18	0.78	L
1.0230	3.2896	0.31	0.84	L
0.6650	2.13816	0.31	0.92	L
1.2020	2.13816	0.56	1.09	L/S
1.7391	2.2204	0.78	1.38	S

L: Below satisfactory level, S: Satisfactory level.

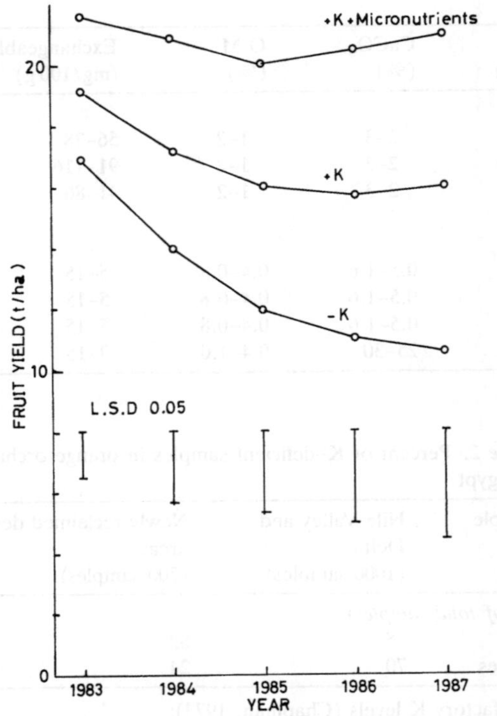

Fig. 2. Response of navel orange to potassium and micro nutrients application (Experiment 1: loamy soil).

Discussion

Soil characteristics in Egyptian orange orchards are similar to those for comparable orchards in arid and semi-arid regions; where $CaCO_3$ and pH are high and organic matter is low (El-Fouly *et al.*, 1988) particularly in some desert areas (Amberger *et al.*, 1988). The results suggest that K is strongly fixed in soils of the Nile-alluvial area of Egypt. Several authors (*e.g.* Abdel-Salam *et al.*, 1979) reported the abundance of K-fixing minerals in Egyptian soils. Such fixation is aggravated by wetting and drying cycles (Gaultier, 1983), a condition which is locally met in irrigated orchards. Therefore, K deficiency prevails in orange leaves, even though the soil is rich in exchangeable K. Low K- supplying power is not only a function of K fixation by clay, but also of competition among soil cations. For instance, Mg can play a considerable role in hindering K uptake by trees (Mengel and Kirkby, 1987).

Consequently, K in orange leaves decreased with decreasing exchangeable-K/Mg ratio in soil. These results were confirmed by the positive responses of orange to application of K-sulphate, especially within a balanced nutirition regime in which the negative effects of other nutritional disorders, *e.g.* micronutrient deficits, on yield were eliminated. Such yield increments prove the need of K-fertilizer application in orange orchards at least for avoiding yield losses due to K deficiency. Similarly positive responses of orange to balanced nutrition, including K, on different soil types in Egypt were reported elsewhere (El-Fouly *et al.*, 1988; El-Sherif *et al.*, 1983; Firgany *et al.*, 1983; Hassanein *et al.*, 1980).

Table 4. Response of orange trees to K application with additional micronutrients. (Experiment 2: Sandy soil)

C.V.	Fruit yield (t/ha)		
	Control	K + micronutrients	L.S.D. 0.05
Balady	2.15	5.43	1.41
Shamouti	6.75	11.0	2.85
Valencia	2.5	6.0	1.92

Conclusion

Application of K fertilizer to orange orchards in Egypt is necessary for producing high yields, not only on poor newly reclaimed desert soils, but also on longer cultivated loam soils. Meanwhile, the results of the present experiment have made it clear that exchangeable soil K, as determined with NH_4 acetate, is not a reliable index of soil K available to orange trees growing on Egyptian soils.

Acknowledgements

This work is a part of the Egyptian-German Project "Micronutrients and other Plant Nutrition Problems in Egypt" conducted by the National Research Centre, Cairo (Coordinator Prof Dr M M El-Fouly) and the Institute of Plant Nutrition, Technical University, Munich (Prof Dr A Amberger). It is supported by the German Agency for Technical Cooperation (GTZ), FRG and the Egyptian Academy for Scientific Research and Technology (ASRT), A.R.E.

References

Abdel Hadi A H, Asy K G and Mohamed Y M 1987 Effect of fertilizer on the production of *Trifolium alexandrinum* (berseem) in Egyptian soils. Med. Pot. News 3, 2–4.

Abdel-Salam M A, El-Demerdash S E and El-Kady M 1979 Potassium status in extension area soils. Proc. Int. Workshop: Role of K in Crop Production. Ed. A Saurat and M M El-Fouly. Cairo, pp 139–151.

Ait Houssa A 1986 Contribution a L'etude du potassium dans les sols marocains. These-Nancy, France.

Amberger A, El-Fouly M M and Fawzi A F A 1988. Diagnostic and remedial measures of microelement problems in grapes grown on calcareous soils in Egypt. Agrochimica, 32, 41–53.

Balba A M 1979 Potassium status in soils and water and sufficiency for crop nutrition in Egypt. Proc. Int. Workshop: Role of K in Crop Production. Eds. A Saurat and M M El-Fouly. Cairo, pp 133–137.

Chapman H D 1973 Diagnostic Criteria for Plant and Soils. Dept. Soils and Plant Nutrition, University of California/ Citrus Research Centre, Agric. Exp. Sta., Riverside pp 610–621.

Chapman H D and Pratt P F 1978 Methods of Analysis for Soils, Plants and Water. University of California, Dept. of Agric. Sci., USA. 309 p.

Daoud Y and Mokhtar A 1985 Le regime due potassium dans quelques types de sol du Hodna (Algerie). Potash Rev. 6, 1–7.

El-Fouly M M, Amberger A and Fawzi A F A 1988 Response of Balady orange to macro and microelement fertilization in Egypt. Agrochimica 32, 27–40.

El-Fouly M, Fawzi A F A and Firgani A H 1987 Removal of nutrients by *Trifolium alexandrinum* and effect of NPK fertilization on yield. Med. Pot. News 3, 8–10.

El-Sherif A F, Fawzi A F A and Firgany A H 1983 Effect of micronutrients rational supply on Balady oranges. Egypt. J. Bot. 26, 53–61.

Fawzi A F A and El-Fouly M M 1979 Soil and leaf analysis of potassium in different areas of Egypt. Proc. Int. Workshop: Role of K in Crop Production. Eds. A Saurat and M M El-Fouly. Cairo, pp 73–80.

Fawzi A F A, Firgany A H, Rezk A I, Kishk M H and Shaaban M M 1983 Response of *Vicia faba* bean to K and micronutrient fertilizers. Egypt. J. Bot. 26, 113–121.

Firgany A H, El-Sherif A F, Fawzi A F A and El-Baz F K 1983 Nutritive requirements of "Balady" oranges. Egypt. J. Bot. 26, 37–51.

Gaultier J P 1983 study of a mechanism of potassium fixation in soils: The structural reorganization of montmorillonite application to bionic K-Ca montmorillonite. 6th IPI-Competition for Young Research Workers, p 25. Intern. Potash Institute, Bern.

Hassanein H G, El-Garably G S and Ghoneim M F 1980 Response of citrus trees in Assiut to soil and foliar fertilization with NPK and certain micronutrients. Proc. 2nd Workshop: Micronutrients and Plant Nutrition. Ed. M M El-Fouly. pp 225–232.

Jackson M L 1973 Soil Chemical Analysis. Prentice Hall Inc. N.J., California.

Mengel K and Kirkby E A 1987 Nutrient uptake and assimilation. *In* Principles of Plant Nutrition. pp 136–138. Intern. Potash Institute, Bern.

Saurat A 1987 Potash fertilization in Arab Countries: Results of agricultural research. Afr. J. Agric. Sci. 14, 1–21.

Shehata H M, Abdel Salam M A, Abed F 1985 Detailed study on K dynamics from some soils of Egypt. II. Q/I relations and EUF-K of soils as meassures of potassium availability. Desert Inst. Bull., 35, No. 1. pp 189–204.

Snedecor G W and Cochran W G 1976 Statistical Methods, Iowa State University Press, Ames, IA, 593 p.

Waller R A and Duncan D B 1969 A bays rule for the symmetric multiple comparison problem. Am. Stat. Assoc. J. 64, 1485–1503.

M. L. van Beusichem (Ed.), *Plant nutrition – physiology and applications*, 735–739.

PLSO IPNC129

Some criteria for the assessment of genetic specificity for nitrogen concentration in wheat (*Triticum aestivum*)

V. MOMČILOVIĆ[1], M.R. SARIĆ[2,3] and A. STOJANOVIĆ[1]
[1] *Institute of Field and Vegetable Crops, Faculty of Agriculture, University of Novi Sad, YU-21000 Novi Sad, Yugoslavia, and* [2] *Institute of Biology, Faculty of Sciences, University of Novi Sad, YU-21000 Novi Sad, Yugoslavia.* [3]*Corresponding author*

Key words: genetic specificity, nitrogen concentration, *Triticum aestivum* L., wheat

Abstract

Among nutrients, nitrogen is the one characterized by the smallest variations among cultivars. Three wheat cultivars were grown in a greenhouse in 10 combinations of nutrients. Dry matter mass, nitrogen concentration, nitrogen indices and harvest index were analysed at the stage of milk-wax maturity and full maturity. Results show that variations in cultivar, organ, age and nutrition all exert their influences on the parameters examined.

Introduction

It is well known that nitrogen concentration in cultivars varies less than concentrations of mineral elements. Although cultivars differ in nitrogen requirements, Greenwood (1986) reported that 22 widely different plant species which had received optimum nitrogen doses, showed similar nitrogen percentages in dry matter at the time of harvest. Cultivars differed in nitrogen concentration at the level of individual plant organs. These differences varied according to plant age and organs examined and their ages. Furthermore, the differences also depended on nitrogen content of the medium in which the tested plants were grown (Day *et al.*, 1985; Needham, 1982; Nielsen, 1982; Silvester-Bradley *et al.*, 1984).

Our earlier studies on wheat (Sarić *et al.*, 1986; 1987) indicated the complexity of problems of genetic specificity in mineral nutrition, irrespective of whether we dealt with either newly-bred cultivars specific in respect to the conditions of soil and mineral nutrition, or with their economic production.

Materials and methods

Three cultivars (Novosadska rana 2, Jugoslavija, and Žitnica) of wheat were grown in a greenhouse with 10 different combinations of mineral nutrients added to a chernosem-type soil which contained considerable amounts of nitrogen, phosphorus, and potassium. The plants grew in Mitcherlich pots. Results of five combinations will be presented (control, NP, NK, PK, and NPK_2). The results of the remaining five combinations (N, P, K, NPK_1, and NPK_3) are not presented in the present paper.

Individual and dual elements were added: 1.5 g of each element per pot (index 1), 3 g/pot (index 2), and 4.5 g/pot (index 3). The following salts were used for the different mineral nutrition combinations: NH_4NO_3, K_2CO_3, and KH_2PO_4. In each pot 5 kg of soil and 1 kg of sand were mixed. The mentioned salts were added as one application at the tillering stage. Each combination of mineral nutrients was replicated five times. The test plants were analyzed at milk-wax maturity and full maturity. The following parameters were assessed: nitrogen concentration by

Kjeldahl's method, dry matter mass, nitrogen indeces, and harvest index. The results presented are two-year average values.

Results

Dry matter mass

Dry matter mass of leaves (Fig. 1) was larger at the time of the first assessment than at the time of the second assessment. Export of photosynthates from the leaves varied with nutritional regime and cultivar. It was lower in the unfavourable nutritional regimes. The largest difference in the dry matter mass of leaves between the first and the second harvest date was recorded in the NP treatment. This means that this treatment had the largest translocation of photosynthates from old to younger leaves. A similar but inverse pattern was obtained for the differences in the dry matter mass of grains. It was characteristic for Novosadskana rana 2 that it had the highest grain mass in unfavourable nutritional treatments, while Jugoslavija had the highest grain dry matter mass in the NK and NPK treatments. The smallest difference in dry matter of grains between the first and the second harvest date was recorded in the PK treatment, and the largest in the NP, NK, and NPK treatments.

Dry matter mass was markedly lower in the unfertilized control and PK treatment than in the other treatments. The application of nitrogen, regardless of treatment, significantly increased dry matter mass. It should be mentioned that there were no significant differences in dry mat-

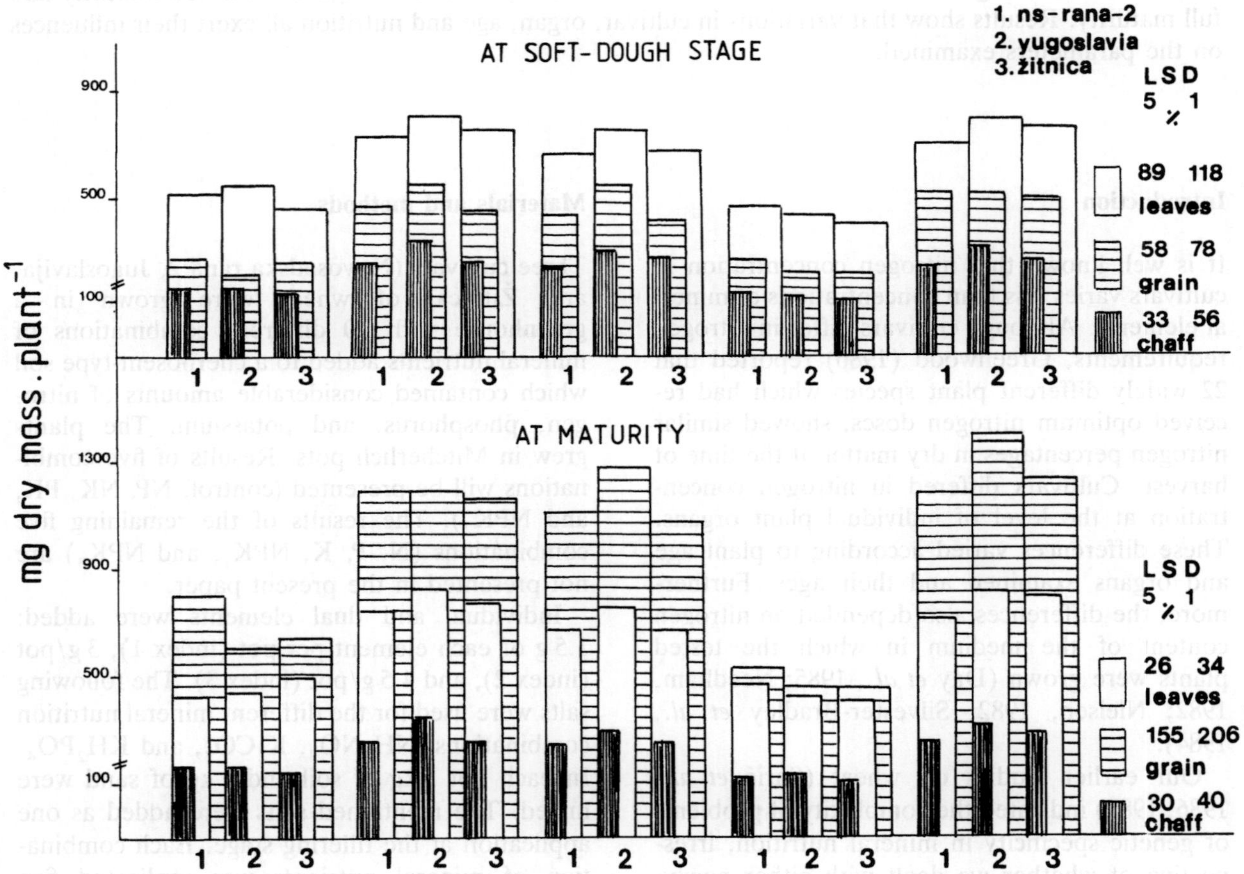

Fig. 1. Dry matter mass as a function of variation in nutrients added and in cultivar used.

ter mass among the NP, NK, and NPK treatments at the second harvest date. However, differences occurred for the cultivar Jugoslavija in favour of the NK treatment in comparison with the NP treatment, and in favour of the NPK treatment in comparison with the NK treatment.

Nitrogen concentration

On both dates of analyses, the concentration of nitrogen (Fig. 2) was lowest in the PK treatment and the control. The differences between the other three treatments were negligible. The concentration of nitrogen was highest in the grain at both harvest dates. Differences among cultivars were larger at the first harvest date, regardless of plant organ, especially in the treatments with inferior nutrition.

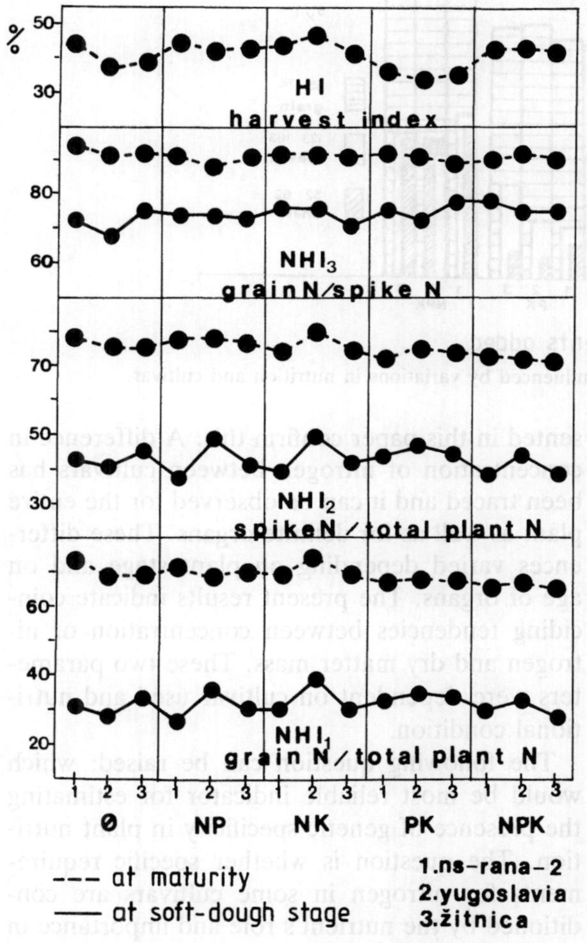

Fig. 2. Nitrogen concentration as a function of variation in nutrients added and in cultivar used.

At the first harvest, the concentration of nitrogen in the grain was highest for the cultivar Jugoslavija in all treatments, except the control in which the cultivar Novosadska rana 2 was superior. At the second harvest however, Jugoslavija had significantly lower nitrogen concentrations than the other cultivars in the NP and NK treatments. There were no differences among cultivars in the other nutritional regimes.

Conversely, at the first harvest the concentration of nitrogen in the leaves was significantly lower in Jugoslavija than in the other cultivars in all three nitrogen treatments. At the second harvest, Novosadska rana 2 had the highest nitrogen concentrations in the nitrogen treatments and Žitnica was in the first place in the PK treatment, while no differences occurred among the cultivars in the control treatment. It may be concluded that under conditions of inferior nutrition the tested cultivars were fairly uniform in nitrogen concentration regardless of the date of assessment and plant organ. On the other hand, in the treatments with nitrogen, the differences among cultivars were highly significant in all organs at the time of the first assessment while the level of significance and rank of the cultivars had changed at the time of the second assessment.

Nitrogen indices

Nitrogen indices (NHI) (Fig. 3) were much higher at the second than at the first harvest which is logical because the grain has reached its maximum mass and nitrogen concentration at the time of the second assessment. There were differences among the nutritional regimes regardless of cultivar. In particular NHI_1 (grain N/total plant N) was lowest when the test plants were optimally treated with NPK. This is to be expected since the nitrogen contents in the vegetative plant organs were high under conditions of optimum nitrogen supply. Similar differences were obtained with the indices NHI_2 (spike N/total plant N) and NHI_3 (grain N/spike N), but absolute values were higher because the concentration and content of nitrogen decreased in the vegetative organs and increased in the grain.

Harvest indices were higher with favourable nutrition than in the control and PK treatment.

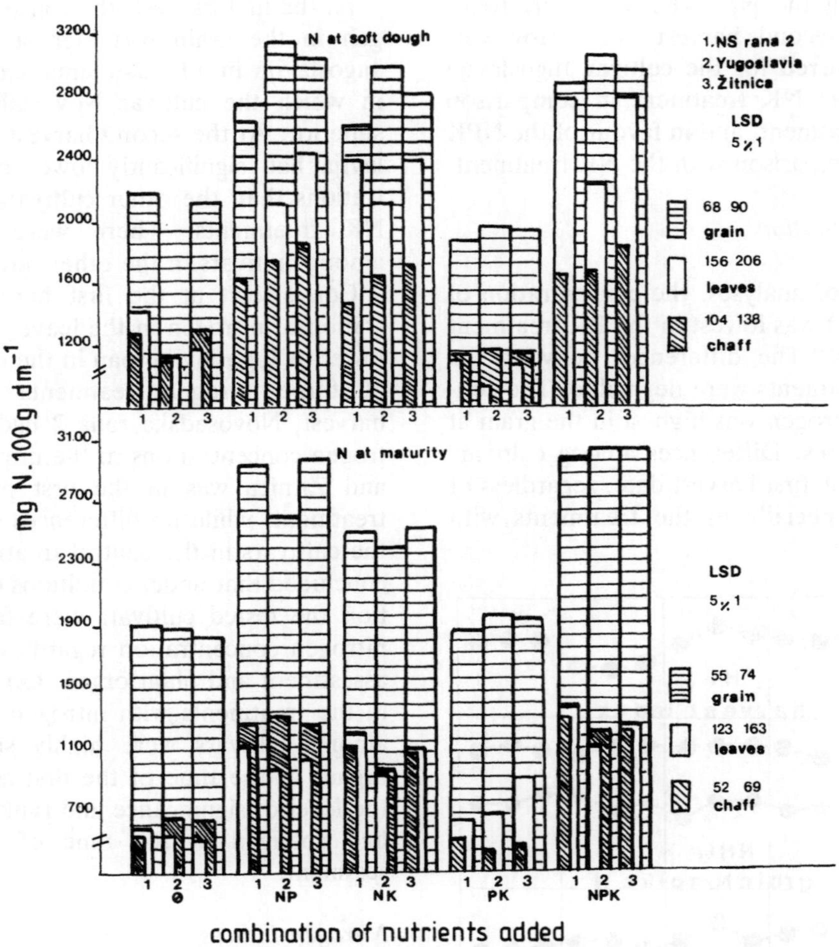

Fig. 3. Nitrogen harvest indices and harvest index, as influenced by variations in nutrition and cultivar.

Under favourable conditions of nutrition differences in harvest index between cultivars were lower than under unfavourable conditions.

Discussion

A maximal production of primary organic matter with a minimum input of nutrients, especially nitrogen, in various cultivars today presents a problem of exceptional significance. Therefore, various criteria used to determine genotype specificity with respect to nutrition are being examined. Nutrient concentrations have up to now been used most frequently as parameters of the nutrient requirements of various cultivars. However, as was emphasized before (Sarić, 1983), this is a complex problem. Results presented in this paper confirm this. A difference in concentration of nitrogen between cultivars has been traced and it can be observed for the entire plant as well as for definite organs. These differences varied depending on plant stage and on age of organs. The present results indicate coinciding tendencies between concentration of nitrogen and dry matter mass. These two parameters were dependent on cultivar used and nutritional condition.

The following question can be raised: which would be most reliable indicator for estimating the presence of genetic specificity in plant nutrition. The question is whether specific requirements for nitrogen in some cultivars are conditioned by the nutrient's role and importance in the physiological and biochemcial processes reflected in structure, metabolism, synthesis of organic matter, and finally in yield. In other words, it is important to know whether nitrogen promotes an increased synthesis of organic matter, or whether a cultivar with a reduced concentration of nitrogen utilizes this element more efficiently due to favourable translocation, retranslocation, or reutilization, leading to a more harmonic relationship between source and sink. In such a situation, a satisfactory dry matter production could be achieved with a low nitrogen concentration.

Acknowledgement

These investments are an integral part of the Project P-620, supported by the J.B. Yugoslavia – U.S.A.

References

Day E G, Paulsen G M and Sears R 1985 Nitrogen relations in winter wheat cultivars differing in grain protein percentage and stature. J. Plant Nutr. 8, 555–566.

Greenwood D J 1986 Prediction of nitrogen fertilizer needs of arable crops. *In* Advances in Plant Nutrition, Vol. 2. Eds. B Tinker and A Läuchli. pp 1–61. Praeger, New York.

Needham P 1982 The role of nitrogen in wheat production: Response interaction and prediction of nitrogen requirements in the United Kingdom. Proc. Fert. Soc. (London) 211, 125–147.

Nielsen Y M 1982 Evaluation and control of nutritional status of cereals. Plant and Soil 64, 403–423.

Sarić M R 1983 Theoretical and practical approaches to genetic specificity of mineral nutrition of plants. Plant and Soil 72, 137–150.

Sarić M R, Mišić T, Vulić B. Momčilović V 1986 Genetski aspekti mineralne ishrane pšenice. II. Koncentracija N, P, K, Ca i Mg u lišću biljaka gajenih u poljskim uslovima. Savremena Poljoprivreda 34, 289–384.

Sarić M R, Krstić B, Stanković Ž 1987 Genetic aspects of mineral nutrition of wheat. I. Concentrations of N, P, K, Ca and Mg in leaves. J. Plant Nutr. 10, 1539–1545.

Silvester-Bradley R, Dampney P M R and Murray A W A 1984 The response of winter wheat to nitrogen. *In* The Nitrogen Requirements of Cereals. MAFF/ADAS Reference Book 385, 233–238. Her Majesty's Stationery Office, London.

M. L. van Beusichem (Ed.), *Plant nutrition – physiology and applications*, 741–746.

PLSO IPNC438

Yield and nitrogen concentrations in wheat (*Triticum aestivum*) as affected by split nitrogen application and growth stage*

W.E. SABBE[1] and J.T. BATCHELOR[2]
[1]*University of Arkansas, Fayetteville, AR 72701, USA, and* [2]*Marketing and Research, Southern Farmers Association, North Little Rock, AR 72119, USA*

Key words: Feekes scale, monitoring, plant analysis, *Triticum aestivum* L.

Abstract

A monitoring program was instituted that provided a N fertilizer recommendation for wheat (*Triticum aestivum* L.) based on the N concentration at a specific growth stage. The study was conducted by growing wheat under nitrogen (N) fertilizer regimes varying both as to total N fertilizer amounts and application at different growth stages. Nitrogen concentration in whole plant samples revealed that N fertilizer uptake occurred when applied at all growth stages. Grain yields were increased with an increase in N fertilizer rate but decreased as application was delayed until the latter reproductive growth stages.

Introduction

The winter growing season of the approximate 3.3 million hectares of soft winter wheat (*Triticum aestivum* L.) in the southern and southeastern areas of the United States is characterized by temperatures mostly above freezing and unpredictable rainfall patterns during the wheat crop season. Therefore, the wheat growth stage at normal N fertilization application in mid- to late-winter can vary dependent upon the preceding weather. Also, the rapidity of growth, utilization of fertilizer N, and availability of fertilizer N after application are subject to weather conditions.

Numerous authors have reported on the effects of improper N fertilization practices on winter wheat. Among the disadvantages are an increase in lodging with excessive N rates (Prutshova and Ukanova, 1976), reduction in tillering (Power and Alessi, 1978), N leaching and runoff (Hucklesby, 1971), denitrification, and a reduction in grain yields (Ayoub, 1974). In both the hard winter wheat and the hard spring wheat areas where leaching is minimal, suitable predictive indices involving soil nitrates are being utilized (Keeney and Nelson, 1982). However, soil inorganic N levels have not been a good predicator of N availability during the growing season (Fox and Piekielek, 1978). Rainfall amounts and temperature fluctuations dramatically alter the soil nitrates levels in Southern U.S. soils as opposed to the more constant level of soil nitrates found in soils of the Great Plains. Plant tissue tests have been used to evaluate response to N fertilizer rates and the N status of wheat, but only recently have these analyses been used to predict the fertilizer rate during the current growing season (Baethgen and Alley, 1989). The objective of this research was to determine the yield and N content response of wheat to N fertilizer rates applied at various growth stages. The goal would be utilized this data into a predictive model where the N fertilizer rate could be recommended if the plant N concentration at a specific growth stage were known.

Methods

Wheat (cv. Rosen) was planted on October 10, 1981, at a 100-kg/ha rate on a Sharkey clay (very

* Contribution for the Agronomy Department and published with the approval of the Director of the Agricultural Experiment Station, University of Arkansas, Fayetteville.

fine, montmorillic, thermic, Vertic Haplaquepts) at the University of Arkansas Northeast Research and Extension Center, Keiser, Arkansas. The primary study involved an initial N fertilizer application (NR1) on February 5, 1982, at a rate of 45, 67, or 90 kg N/ha. An additional second N rate (NR2) was applied at a Feekes growth stage (Large, 1954) of S-4, S-6, S-8, or S-10.51 at 22, 45, or 67 kg N/ha. A companion N rate study utilized a single N application on February 5, 1982, using rates of 0, 45, 67, 90, 112, 134, and 157 kg N/ha. Urea was the N source for all applications and all applications were evenly broadcast on the soil surface. The experimental design was a randomized complete block with a factorial arrangement. Four replications were utilized for both studies. Plant samples were taken prior to the initial N application and at the S-2, S-4, S-8, S-10.51, and S-11.1 growth stages prior to a second application of N fertilizer. Plant samples consisted of above-ground shoots cut at ground level that were then dried and analyzed for total N concentration (Lindner, 1944). Grain yield data was collected at harvest with a plot combine.

Results and discussion

Temperatures through the S-4 growth stage were slightly lower than the 30-year normal with rainfall amounts being adequate (Arkansas Agricultural Statistics, 1983). However, the remaining growth stages through harvest experienced higher temperatures than normal with rainfall being adequate. Grain yields in the N rates study increased with N rates up to 134 kg N/ha (Fig. 1). The 157-kg-N/ha rate produced a non-significantly lower grain yield than the 134-kg-N/ha rate. The plant N concentrations (except for the 0-kg-N/ha rate) generally increased between application and the S-2 or S-4 growth stage and decreased from the S-4 to the S-11.1 growth stages. Between the S-2 or S-4 growth stages, plant N increased or remained constant for the three highest N fertilizer rates. Except for the 45-kg-N/ha rate, the lowest three rates (other than 0-kg-N/ha) produced plant N concentrations that decreased starting with the S-2 growth stage. The plant N concentration of the 0-kg-N/

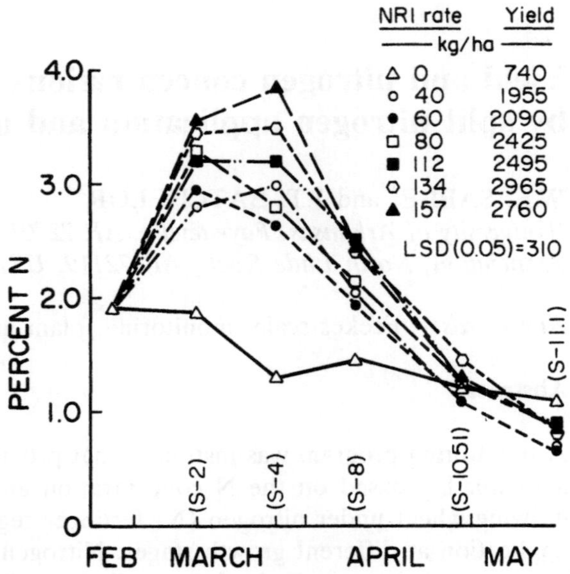

Fig. 1. Grain yields and plant nitrogen concentrations at various growth stages as affected by nitrogen fertilizer rates.

ha rate decreased from about 1.8% N on February 5, 1982, to 1.1% N at the S-11.1 growth stage – a duration of approximately 90 days. The plant N concentrations where N fertilizer was applied were greater than the no-N treatment until the S-10.51 growth stage when all treatments demonstrated essentially the same plant N concentration.

In the multiple application study the initial N rate (NR1) of 90 kg N/ha produced grain yields superior to the two lower NR1 rates (Table 1). Within each secondary N rate (NR2), the application at the S-10.51 growth stage resulted in lower grain yields than secondary applications made at earlier growth stages. Secondary application at the S-2, S-4 and S-8 growth stages increased grain yield for all NR1 rates. The S-4 and S-8 secondary applications were superior to S-2, but no yield differences were observed between the S-4 and S-8 for any NR1 rates. Generally, the higher the NR1 rate, the greater the increase due to the NR2 rates; and also with the higher NR1 rate the later growth stages responded to the NR2 application. Highest grain yields generally occurred within the highest NR1 rates coupled with the highest NR2 rates. Efficiency of application is demonstrated with the split of 112 kg N/ha. The 45–67 split was superior to the 67–45 and 90–22 splits when the

Table 1. Effect of initial N rate (NR1), secondary N rate (NR2), and feekes growth stage (GS) on wheat grain yields (kg/ha)

	NR1 (kg N/ha)														
	45					67					90				
	NR2 (kg N/ha)														
GS	0	22	45	67	Mean	0	22	45	67	Mean	0	22	45	67	Mean
S-2	1955	2426	2898	2696	2494	2090	2763	2561	3033	2612	2426	2628	2628	3303	2679
S-4	1955	2561	3168	3168	2713	2090	2426	2898	3437	2712	2426	2830	3235	3437	2982
S-8	1955	2494	2830	3168	2612	2090	2898	2830	3033	2713	2426	2830	2966	3640	2966
S-10.51	1955	1887	1685	1685	1803	2090	2359	2224	2359	2258	2426	2426	2359	2561	2443
Mean					2405					2574					2784

LSD (.05) = 145

NR2 was applied at the S-4 and S-8 growth stage.

Generally the pattern of plant N concentration with time regardless of treatment was: 1) to increase from the date of NR1 through S-2; 2) to increase or very slightly decrease from S-2 to S-4; and 3) to decrease from S-4 to S-11.1 stage (Fig. 2). When N concentrations declined between S-4 and S-11.1, the overall treatments average was decreased about 0.04% N per day. The N concentration within the plant reflected the fertilizer N additions. At the NR1 rate and with the S-2 growth stage applications, the plant N concentration increased whereas fertilizer application at the later growth stages (S-8 and S-10.51) resulted in the plant nitrogen concentration decreasing less than when no N was applied. The fertilizer N application at the S-4 growth stage resulted in the 67 kg N/ha rate increasing the plant N concentration and the other two NR2 rates decreasing the N concentrations (but at a loss less than the 0 kg N/ha rate).

Figure 3 illustrates the effect of a constant NR2 rate (67 kg N/ha) applied at growth stages S-4, S-6, S-8, and S-10.51. Each NR2 application affected the N concentration of the plant when applied at each growth stage. At growth stages S-4 and S-6, the N concentration of the plant increased whereas at the S-8 and S-10.51 growth stages the response – while not an increase – was a reduction in the rate of plant N concentration decline. The respective grain yields for the 67-kg-N/ha NR2 application at growth stages S-2, S-4, S-8, and S-10.51 were 3033, 3437, 3033, and 2359 kg/ha (Table 1).

Monitoring program

The results of the aforementioned studies contributed to a statewide wheat monitoring program where whole plants were sampled. Additionally, previous work (Ayoub, 1974) had indicated that N fertilizer at tillering or jointing resulted in a greater plant N uptake than fertilizing at seeding or head emergence. Power and Alessi (1978) reported that tillers on the same plant had about the same N concentration regardless of the N fertilizer rate (*i.e* 0, 56, 302 kg N/ha) and that the final yield was closely correlated with the N content of a given tiller at the tillering stage. The monitoring program allows growers to submit whole plant samples from a wheat field along with information regarding the growth stage at each sampling and previous N fertilizer applications. The plant samples were analyzed for the N concentration, and that data plus the growth stage resulted in a N fertilizer recommendation (Table 2). The program has three N concentration categories within each growth stage, except for growth stage S-9. This allows for a possibility of a low or moderate N fertilizer application in addition to a recommendation of no N fertilizer. The information furnished by growers regarding the previous N fertilizer application allows a recommendation as to when to collect another plant sample. For example, if the plant N concentration of a sample would normally lead to a fertilizer recommendation, but N fertilizer had been applied less than 2 weeks prior to sample collection, then the recommendation was to collect another plant sample immediately. This recommendation is based on

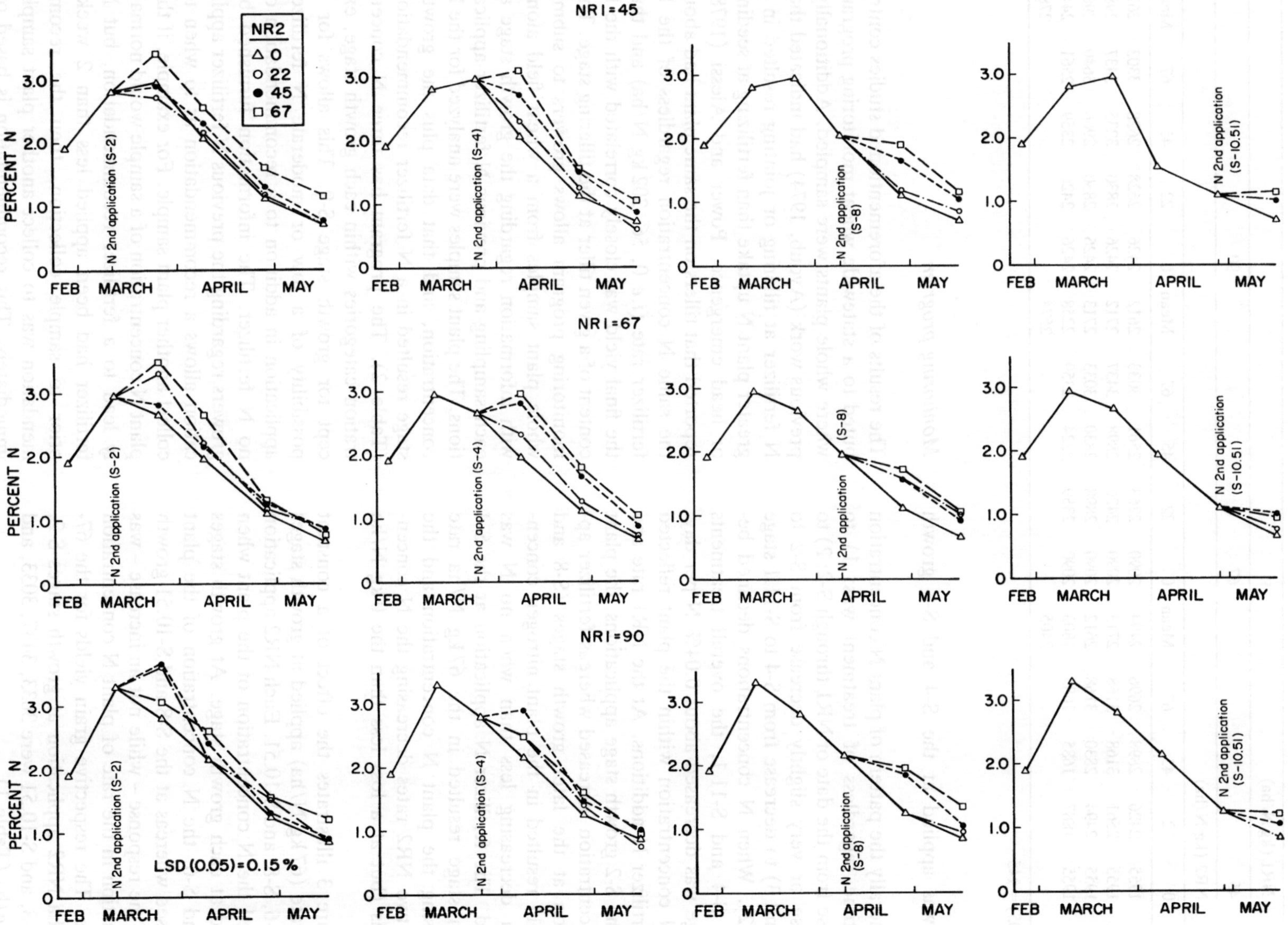

Fig. 2. Plant nitrogen concentration of various growth stages as affected by initial nitrogen rates (NR1) and various secondary nitrogen rates (NR2) applied at selected growth stages. LSD (0.05) = 0.15%

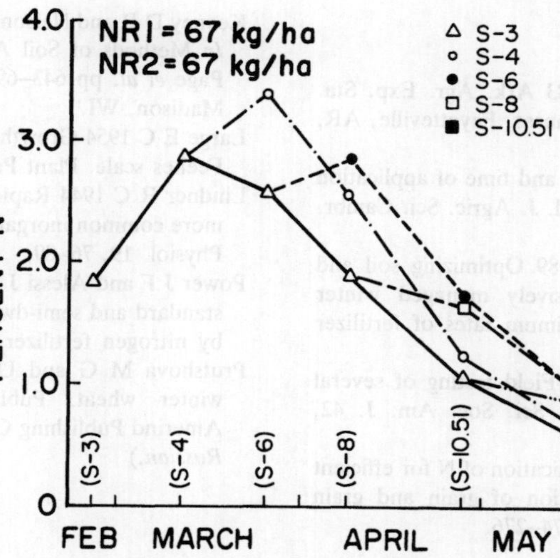

Fig. 3. Plant nitrogen concentration at various growth stages as affected by a constant secondary nitrogen rate.

Table 2. Nitrogen fertilizer (kg N/ha) recommendations for wheat grown in Arkansas based on plant nitrogen concentration and Feekes growth stage

		Tillering		Shoot elongation			Early boot
%N	GS	2 or 3	4	5	6	7	9
3.0+		0	0				
2.25–3.0		50	50				
1.5–2.25		90	100				
2.75+				0			
2.0–2.75				50			
1.5–2.0				67			
2.25+					0		
1.75–2.25					50		
1.25–1.75					67		
2.0+						0	
1.5–2.0						34	
1.0–1.5						50	
1.25+							0
0.75–1.25							22
Total maximum N		90	100	134	134	145	155

the possibility that the previous N fertilizer application had not been utilized by the wheat crop. Another feature is to limit the amount of N fertilizer applied by a specific growth stage (*i.e.* total maximum N). To use this table, simply locate the wheat growth stage at top of Table 2, identify the N concentrations on the left hand side that bracket the plant test results, and use the recommended N rate. This monitoring program developed by the Experiment Station was adapted by the University of Arkansas Extension Service and has been utilized by Arkansas wheat growers for the last seven growing seasons.

References

Arkansas Agricultural Statistics, 1983 Ark. Agr. Exp. Sta. Rep. Ser. 299. University of Arkansas, Fayetteville, AR, 56 p.

Ayoub A T 1974 Effect of N source and time of application on wheat N uptake and grain yield. J. Agric. Sci. Cambr. 82, 567–569.

Baethgen W E and Alley M M 1989 Optimizing soil and fertilizer nitrogen use by intensively managed winter wheat. II. Critical levels and optimum rates of fertilizer nitrogen. Agron. J. 81, 120–125.

Fox R H and Piekielek W P 1978 Field testing of several nitrogen availability indices. Soil Sci. Soc. Am. J. 42, 747–750.

Hucklesby P P 1971 Late spring application of N for efficient utilization and enhanced production of grain and grain protein of wheat. Agron. J. 63, 274–276.

Keeney D R and Nelson D W 1982 Nitrogen inorganic forms. *In* Methods of Soil Analysis. Part 2. 2nd Ed. Eds. A L Page *et al.* pp 643–698. American Society of Agronomy, Madison, WI.

Large E C 1954 Growth stages in cereals – Illustration of the Feekes scale. Plant Pathol. 3, 128–129.

Lindner R C 1944 Rapid analytical methods for some of the more common inorganic constituents of plant tissues. Plant Physiol. 19, 76–89.

Power J F and Alessi J 1978 Tiller development and yield of standard and semi-dwarf spring wheat varieties as affected by nitrogen fertilizer. J. Agric. Sci. Cambr. 90, 97–108.

Prutshova M G and Ukanova O I 1976 New varieties of winter wheat. Published for USDA-ARS, NSF by Amerind Publishing Co. Springfield, VA. (*Translated from Russian.*)

M. L. van Beusichem (Ed.), *Plant nutrition – physiology and applications*, 747–751.

PLSO IPNC230

The influence of available nitrate nitrogen in the soil profile on the nitrate contents of blanching celery

P. DEMYTTENAERE[1], G. HOFMAN[1,2] and G. VULSTEKE[3]
[1]*Laboratory of Agricultural Soil Science, Faculty of Agriculture, State University of Ghent, Coupure 653, B-9000 Ghent, Belgium, and* [3]*Centre for Agricultural and Horticultural Research, Ieperweg 87, B-8810 Roeselare, Belgium.* [2]*Corresponding author*

Key words: nitrogen advice, nitrate content, plant available nitrogen, vegetables

Abstract

The ingestion of nitrate via drinking water or via consumption of vegetables should for health reasons be as low as possible. For field grown vegetables that tend to accumulate nitrates, it is necessary to understand the N-economy in soil and plant. High NO_3-N residues in the soil profile at the start of the growing season, the application of large quantities of pig slurry, as well as an important mineral N-fertilization and a high rate of N-mineralisation, add up to the mean theoretical available NO_3-N for blanching celery in the studied region, to about 600 kg NO_3-N/ha. These amounts greatly exceed the N-demand of the crop, resulting in high NO_3-N residues in the soil profile at harvest. Correspondingly, the NO_3-amounts in the harvested crop ranged between 2500 and 11500 mg NO_3/kg fresh material. To diminish the nitrate contents in the crop and to avoid the pollution of ground- and surfacewater by drainage during winter, a scientifically justified nitrogen advice based on a mineral-N balance is proposed.

Introduction

High NO_3-contents in human food should be avoided for health reasons (Aldershoff, 1982; Suzuki and Mitsuoka, 1981). Nitrates are mainly taken up via vegetables and drinking water (Maynard *et al.*, 1976). Therefore, more and more countries take measures to limit the nitrate contents of vegetables. These limits depend on the relevant crop and differ between countries. In some Western European countries, the tolerated amounts for leafy vegetables vary between 2000 and 4000 mg NO_3/kg fresh material (Viaene, 1987). Recently, NO_3-limits for some leafy vegetables (such as lettuce, spinach, endive and celery) have been introduced in Belgium (Anonymous, 1989). These amounts vary between crops and harvest times from 2500 to 5000 mg NO_3/kg fresh weight. The maximum tolerated NO_3-concentration of blanching celery, used as a test crop, amounts to 4000 mg/kg in Belgium (Anonymous, 1989). The aim of this investigation was to link the nitrate content in the crop studied with the amount of available nitrogen in the soil and to propose criteria minimizing the harmful NO_3-concentrations without significant yield reductions.

Materials and methods

The field trials were all situated in mid West-Flanders on loamy sand and light sandy loam soils. Some average characteristics of the arable layer of the studied fields are given in Table 1. The %C values are relatively high, due to regular applications of large amounts of fresh organic matter.

Soil mineral nitrogen analyses were carried out on bulk soil samples, taken in layers of 0.30 m until a depth of 0.60 m. After thorough mixing, the NO_3-N contents were measured in a 1%

Table 1. Some physico-chemical characteristics of the arable layer of the experimental fields and the mean NO_3-N residues in the soil profile before planting

Year	No.	Layer (m)	pH-KCl	C(%)	C/N	NO_3-N kg ha^{-1}
1983	12	0.0–0.3	5.86 ± 0.27	1.27 ± 0.35	10.49 ± 0.33	93 ± 46
	12	0.3–0.6				91 ± 31
	12	0.0–0.6				184 ± 64
1984	9	0.0–0.3	6.01 ± 0.38	1.47 ± 0.32	10.56 ± 0.34	93 ± 70
	9	0.3–0.6				96 ± 57
	9	0.0–0.6				189 ± 111

$KAl(SO_4)_2$ extraction, using a potentiometric method (Cottenie and Velghe, 1973). Measurements of NH_4-N were omitted as the NH_4-N contents in the soil profile were known to be very low. Only after a recent application of an ammoniacal fertilization, NH_4-N was measured.

Nitrate contents in the crop were determined by the Kjeldahl method after extraction of the dry plant material with distilled water.

Results and discussion

The total available nitrogen in the rooting zone (limited to about 0.60 m for blanching celery) influences the nitrate content in the harvested crop. This mineral nitrogen depends on a number of factors which will be discussed separately.

NO_3-N residues in the soil profile

Table 1 gives the NO_3-N residues in the rooting zone of blanching celery before any application of fertilizers. Considerable differences between the fields are found, depending on crop rotation and variations in nitrogen application rates.

Nitrogen supply from mineralization of soil humus

The yearly application of large quantities of pig slurry and the incorporation of considerable amounts of crop trashes result in the formation of more "young" humus with a high mineralization rate (Greenwood *et al.*, 1985; Hofman, 1988; Stevenson, 1982; Van Dijk, 1982).

Mean mineralization coefficients of 3.44 ± 0.48% (Demyttenaere *et al.*, 1988) were found in situ in this region, corresponding with a N-supply of about 200 kg N/ha. year on soils with a %C of about 1.35. Regular determination of the NO_3-N evolution in the soil profile on uncropped, unfertilized plots showed that about half of this amount is mineralized during the growing season of blanching celery.

Nitrogen release from pig slurry

Pig slurry is frequently applied just before planting time. Knowledge of the N-supply from this organic material is necessary for a reliable estimation of the total available nitrogen. Therefore, during the growing season the NO_3-N amounts in the soil profile of three uncropped, unfertilized plots are compared with the NO_3-N quantities found on plots with an application of 50 ton ha^{-1} pig slurry.

The total nitrogen in the applied slurry corresponds with 296 ± 20 kg N ha^{-1}. During the growing season of blanching celery, a supplementary augmentation of 102 kg N ha^{-1} was measured on the treated plots. The nitrogen restitution, given as percentage of the amended nitrogen via the slurry was 34.5 ± 2.6. This N-restitution out of slurry corresponds quite well with the results mentioned by Van Faassen and Van Dijk (1985). Especially high temperatures in the period of the slurry applications are responsible for a high ammonia volatilization and for a low NO_3-N recovery in the treated soil profile (Van den Abbeel *et al.*, 1989).

Total available nitrogen

Assuming no N-losses to occur from mineral nitrogen fertilization, the mean theoretically

Table 2. Mean theoretically available nitrogen (kg N ha^{-1}) in the soil profile during the growing season of celery

Year	NO_3-N residues[a]	N-fertilisation	N-mineralisation	N-release pig slurry[b]	Theoretical available N (total)
1983	184 ± 64	241 ± 55	93 ± 24	85 ± 41	603 ± 85
1984	189 ± 111	227 ± 46	108 ± 22	77 ± 45	601 ± 154

[a] NO_3-N residue up to 0.60 m before fertilizing
[b] Taking into account a N release out of 50 ton pig slurry of 102 kg N

available amount of nitrogen could be calculated for both years (Table 2).

The total amounts of theoretically available mineral-N exceed, by about 250 kg N/ha, the uptake requirements of the crop (Demyttenaere *et al.*, 1988). Considerable differences exist in amount of available nitrogen between fields.

Nitrate levels in blanching celery

As only stalks (0–22 cm length) are economically valuable for industrial conservation, only the NO_3-contents of that plant part are mentioned. In spite of the approximately equal theoretically available amounts of nitrogen in the soil profile in both years (Table 2), the nitrate contents were significantly lower in 1984 compared to 1983 (Table 3). High rainfall in September and October 1984 (274 mm), compared to 135 mm in 1983, caused a substantial leaching of NO_3-N below the rooting zone. This was confirmed by the lack of correspondence between the theoretical (supposing no N-losses) and measured NO_3-N amounts in the rooting zone in 1984. Furthermore, a considerable increase of the NO_3-N concentrations in the layers between 0.60 and 1.20 m was found in the wet growing season 1984 compared to quite constant amounts of NO_3-N in the subsoil between planting and harvest time in 1983.

These N-losses by drainage and possibly by denitrification may be responsible for the lower degree of significance and for a different slope of the linear relationship between the theoretical available NO_3-N in the rooting zone and the nitrate contents of the stalks in 1984 (Fig. 1). However, highly significant linear relationships were found between the NO_3-N residues in the soil profile (0–0.6 m) at harvest time and the NO_3-level in the stalks for both years (Fig. 2). In both figures, results for two unfertilized objects in 1983 and for one in 1984 are included. This confirms the close relationship between the soil available nitrogen and the NO_3-concentrations in the crop. However, this relationship is influenced by many other factors, such as insoltion, soil type, yield level and the growing stage of the crop. It is known that the nitrate levels in the crop decreases close to maturity and that this drop occurs earlier at lower amounts of available nitrogen in the soil profile (Breimer, 1982; Vulsteke, 1982).

In practice, high yields with low nitrate concentration in the crop and in the soil profile are aimed at harvest. These objectives can be obtained by the application of a mineral-N balance (Demyttenaere *et al.*, 1988). A production of 100 ton/ha fresh stalks can be considered as an economically acceptable yield. To obtain this proposed production, the N-need of the crop is of the order of 350 to 375 kg N/ha and can only be realised with sufficient amounts of mineral nitrogen in the rooting zone. This includes a quite important latent mineral-N residue in the

Table 3. NO_3-N residues in the soil profile and NO_3-concentrations in celery stalks

Year	Fresh stalk production (ton/ha)	Total N uptake (kg N/ha)	NO^3-N residue at harvest (kg/ha up to 0.6 m) calculated	measured	NO_3-conc. in stalk (mg/kg f.m.)
1983	85 ± 21	330 ± 70	273 ± 103	299 ± 158	4958 ± 2297
1984	71 ± 14	269 ± 53	332 ± 145	169 ± 94	2784 ± 617

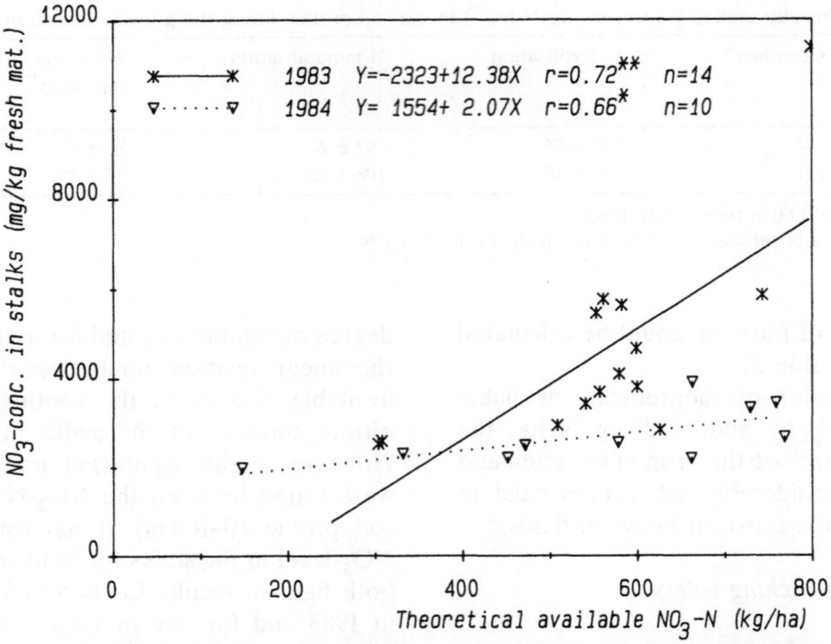

Fig. 1. NO_3-concentrations in stalks in function of the total theoretically available nitrogen.

soil profile (Hofman *et al.*, 1984). However, optimum yields were already found on soils with about 100 kg NO_3-N/ha in the rooting zone at harvest, which is far below the mean NO_3-N residues mentioned in Table 3. Furthermore, restricted soil mineral-N residues result in acceptable nitrate levels in the crop (Fig. 2).

In the given circumstances with mean pig slurry amendments of 50 ton/ha, farmers can only interfere in the mineral-N fertilization. A reduc-

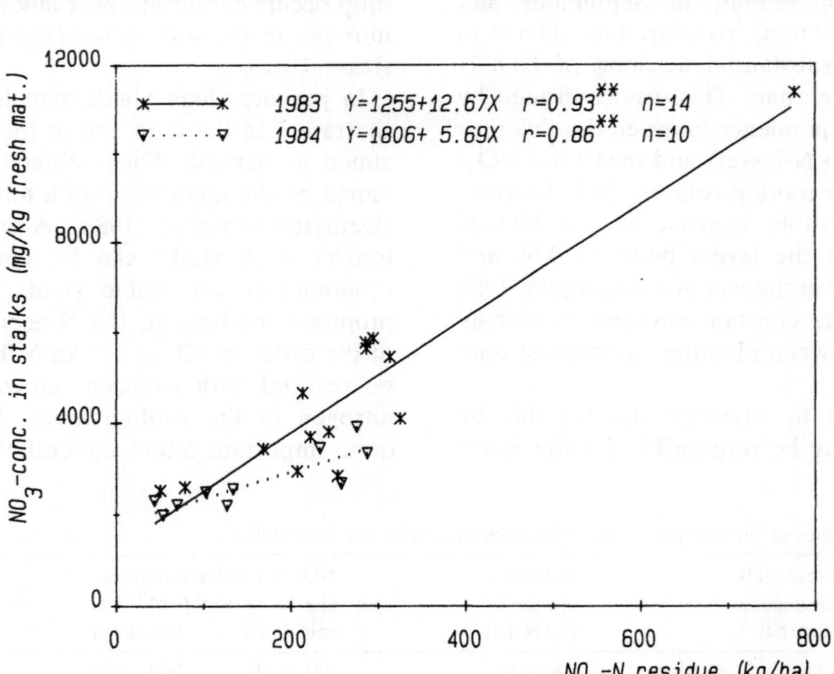

Fig. 2. NO_3-concentrations in stalks in function of the NO_3-N residues in the rooting zone at harvest.

tion of the N-fertilizer amount to about 100 kg N/ha is realistic. An adjustment of this advice is only necessary in wet weather conditions with substantial N-losses during the growing season.

Acknowledgement

Financial support by I.W.O.N.L. (Institute for Encouraging Scientific Research in Industry and Agriculture, Brussels) is gratefully acknowledged.

References

Aldershoff W 1982 Nitraat in groenten en onze gezondheid. Bedrijfsontwikkeling 13, 273–280.

Anonymous 1989 Koninklijk Besluit tot vaststelling van het maximale gehalte aan nitraten in sommige groenten. Belgisch Staatsblad 25.04.1989, 6885–6886.

Breimer T 1982 Environmental factors and cultural measures affecting the nitrate content in spinach. Fert. Res. 3, 193–292.

Cottenie A and Velghe G 1973 Het gebruik van de specifieke nitraat elektrode voor de bepaling van nitraten in gronden en planten. Mededelingen Faculteit Landbouwwetenschappen Gent 38, 560–568.

Demyttenaere P, Hofman G, Vulsteke G and Ossemerct C 1988 A nitrogen fertilisation advice for blanching celery. Mededelingen Faculteit Landbouwwetenschappen Gent 53, 113–121.

Greenwood D, Neeteson J and Draycott 1985 Response of potatoes to N fertilizer: Quantitative relations for components of growth. Plant and Soil 85, 163–183.

Hofman G 1988 Nitrogen supply from mineralization of organic matter. Biological Wastes 26, 315–324.

Hofman G, Ossemerct C, Ide G. and Van Ruymbeke M 1984 Significance of the latent mineral N-residue in the soil profile on nitrogen fertilization advices. *In* Proceedings of the 9th World Fertilizer Congress. Vol. 2, pp 225–229. C.I.E.C., Budapest.

Maynard D, Barker A, Minotti P and Peck N 1976 Nitrate accumulation in vegetables. Adv. Agron. 28, 71–118.

Stevenson F 1982 Organic matter and nutrient availability. *In* Transactions of the 12th International Congress of Soil Science. pp 137–151. Symposia papers 1, New Delhi.

Suzuki K and Mitsuoka T 1981 Increase of faecal nitrosamines in Japanese individuals given a Western diet. Nature 294, 453-456.

Van den Abbeel R, Paulus D, De Ruysscher C and Vlassak K 1989 Stikstofverliezen uit mengmest. *In* Mengmest-problematiek, K.VIV., Antwerpen, Belgium, pp 5.1–5.14.

Van Dijk H 1982 Survey of Dutch soil organic matter research with regard to humification and degradation rates in arable land. *In* Proceedings ECC, Land Use and Rural Resources Management Committee, Soil Degradation Group, Seminar on Soil Degradation. pp 1–12. Wageningen.

Van Faassen H and Van Dijk H 1985 Mineralisation of organic nitrogen (and phosphate) from animal manures in soil. I.B. rapport 7(85). pp 90. Institute for Soil Fertility, Haren, The Netherlands.

Viaene M 1987 Het standpunt van het departement van volksgezondheid ten aanzien van het nitraatprobleem. *In* Nitraten in groenten. Ed. I.W.O.N.L. pp 153–164. Brussel.

Vulsteke G 1982 Nitraten in vollegrondsgroenten. *In* Nitraten in bodem, plant en dier. Ed. K.VIV. pp 7.1–7.21, Genootschap Plantenproduktie en Ekosfeer, Antwerpen, België.

M. L. van Beusichem (Ed.), *Plant nutrition – physiology and applications*, 753–758.

PLSO IPNC623

Control of nitrogen supply of cucumber (*Cucumis sativus*) grown in soilless culture

H. SCHACHT and M.K. SCHENK
Institute of Plant Nutrition, University of Hannover, Herrenhäuser Str. 2, D-3000 Hannover 21, FRG

Key words: cucumber, *Cucumis sativus* L., fertigation model, nitrate sap test, nitrogen supply, soilless culture

Abstract

The aim of this research was the development of a model to control nutrient supply of cucumber grown in soilless culture systems. Nitrogen uptake of cucumber during fruiting was linearly correlated to radiation in the investigated range of 900 to 4000 Wh m^{-2} × day. The relation between radiation and N uptake differed during periods of stem fruit and sucker fruit harvest.

Nitrogen uptake at the same radiation level was less during sucker fruit period. This may be caused by retranslocation of nitrogen from older leaves to growing parts of the plant. During vegetative growth nutrient uptake depended on both radiation and plant age. During summer other climatic factors like temperature and humidity had no effect on nitrogen uptake.

The relation between radiation and nitrogen uptake in a certain growth stage was used for development of a regression model to control N supply. Daily nitrogen supply calculated using this model can be related to nitrate sap test. Optimum nitrate content in petiole sap of young leaves of cucumber was about 4500–5500 mg L^{-1}.

Introduction

Growing vegetable crops in soilless culture systems requires complementation of nutrients at short intervals. Addition of nutrients to a system with recirculating nutrient solution should be done according to nutrient requirement of the plant to avoid accumulation or depletion of nutrients in the solution.

The nutrient requirement of a plant can be related to growth (Asher and Blamey, 1987), which can be described for cucumber as a function of climatic conditions and plant age (Liebig, 1984). These relationships can be used to calculate nutrient uptake and control nutrient supply by a model. N status of plants can be evaluated by nitrate content in petiole sap (Scaife and Stevens, 1983). This gives the opportunity to prove N supply controlled by a model.

The objectives of this study were the development of a model to control N supply of cucumber, and investigation of optimum nitrate content in petiole sap.

Materials and methods

Cucumbers (*Cucumis sativus* L.) cv. 'Camirex' were grown using modified nutrient film technique (NFT) with expanded clay ('Lecaton') as substrate. Substrate was necessary, because root growth of cucumber is injured in NFT (Graves, 1983). Supply of nutrient solution was done by a trickle irrigation system with a water outlet at every plant. Excess solution was gathered in a catchment tank and recirculated. Experiments were carried out in greenhouses at different N supply and varied greenhouse climatisation and were repeated during a period of two years. Cucumbers were planted 5 weeks after sowing with plant densities of 1.7 and 1.9 plants per m^2, 56 plants in each treatment.

Nutrient content of the solution was controlled by a computer measuring EC value and adding stock solution if the EC value fell below the setpoint. Nutrient composition of the solution was analysed twice a week. To reach the desired nitrate concentrations in petiole press sap, N content in the solution was varied from 5 to 100 $mg\ L^{-1}$. Nitrogen was supplied at about 85% as nitrate and 15% as ammonium to keep the pH in a range of 5.5 to 6.5. The concentration ranges of the other nutrients were 30–50 $mg\ L^{-1}$ P, 200–250 $mg\ L^{-1}$ K, 80–100 $mg\ L^{-1}$ Ca, 20–30 $mg\ L^{-1}$ Mg, 6 $mg\ L^{-1}$ Fe, 0.4 $mg\ L^{-1}$ Mn, 0.2 $mg\ L^{-1}$ B, 0.15 $mg\ L^{-1}$ Zn and Cu and 0.01 $mg\ L^{-1}$ Mo.

Temperature was varied by different ventilation setpoints (24° and 28°C) and humidity by using a fogging system and heating at opened ventilation.

Vegetative yield was determined on 10 plants including plant parts harvested during growth period. Fruits were harvested twice a week at a fruit weight of 400 to 600 g. Nutrient uptake was also calculated twice a week by measuring consumption of stock solution and change of concentration in the nutrient solution. Nutrient uptake calculated by this method reflected total nutrient amount in the plant at final harvest with an error of 10%.

Nutrient composition of the first fully developed young leaf was measured weekly as well as nitrate concentration in petiole sap in a sample of 8 leaves. Total N content was determined by a modified Kjeldahl method including nitrate.

Temperature, radiation and humidity were daily monitored and related to the average nutrient uptake of 3 or 4 days by multiple regression analysis and response surface method (Liebig, 1984).

Results and discussion

N uptake and total N content in leaf dry matter increased with N supply as reflected in petiole nitrate concentration, whereas vegetative growth and fruit yield followed an optimum shaped curve as shown in Figure 1. Maximum fruit yield was obtained at 4500–5500 $mg\ L^{-1}$ nitrate in petiole sap. Yield was in the usual range for this fruit size and picking period. Total N content at optimum was about 5.9% N in leaf dry matter. This agrees with data of Wetzold (1972), who found nitrogen deficiency of cucumber below 5% N in leaf dry matter. Fruit yield reduction at high N supply reflected by nitrate concentration of

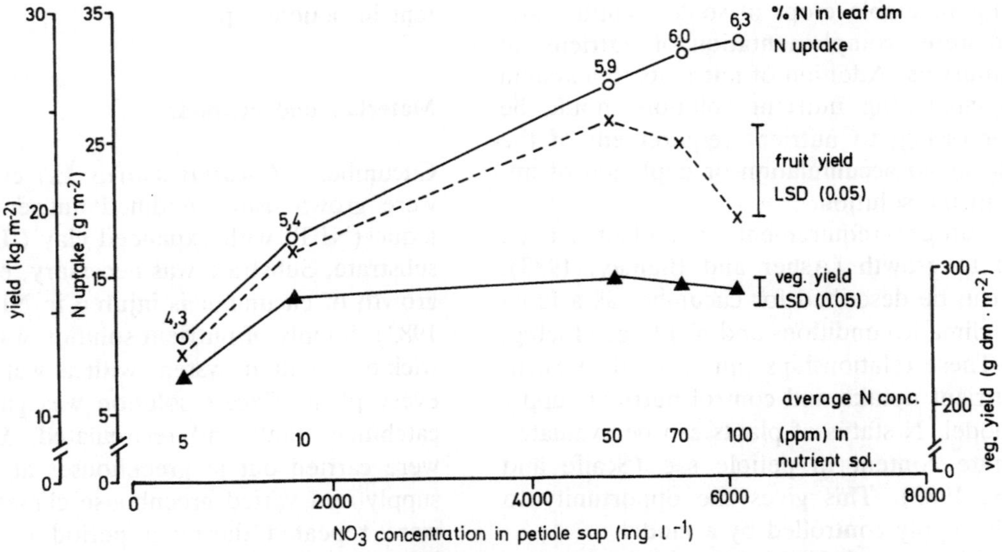

Fig. 1. N uptake, growth and yield of cucumber as affected by nitrate content in petiole sap.

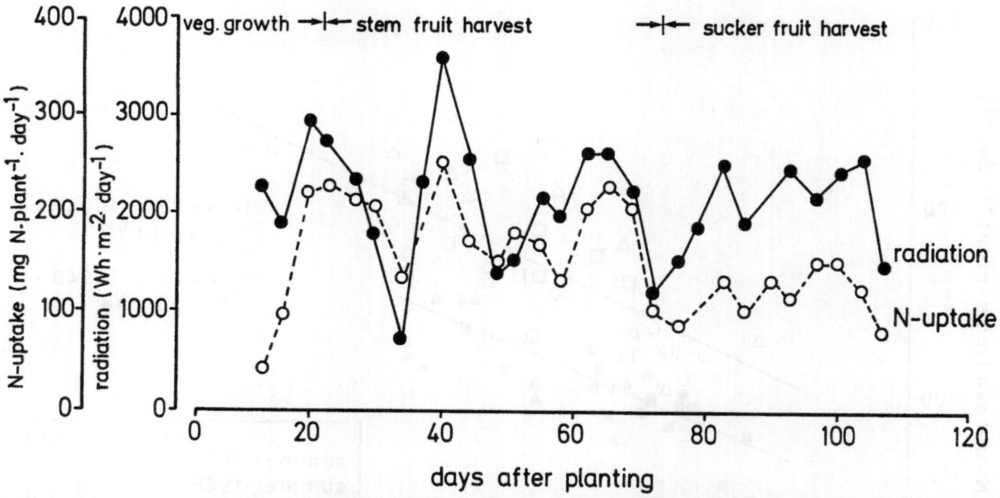

Fig. 2. Daily radiation and nitrogen uptake of cucumber at optimum N supply (4500–5500 mg L^{-1} nitrate in petiole sap).

about 5500–6000 mg L^{-1} in petiole sap could be caused by induced Ca deficiency. This was suggested by low Ca content of young leaves (1.1%), which was half of that at optimum N supply. Furthermore, the developing young leaves had chlorotic and necrotic leaf edges so that leaf expansion was inhibited. These symptoms were similar to those as described by Bakker (1984). Winsor and Massey (1978) also reported Ca deficiency at the top of tomatoes grown at high N supply in NFT.

Optimum N concentration of nutrient solution averaged about 50 mg L^{-1} (range 30–70) in this system with circulating nutrient solution, whereas in conventional hydroponic systems 150–300 mg N L^{-1} were required (Winsor and Massey, 1978). In systems with circulating nutrient solution low nutrient concentrations are sufficient, because nutrients are moved to the root surface.

Daily nutrient uptake and radiation at optimum N supply is given in Figure 2. It is obvious that N uptake depended on radiation. However, the relation between radiation and N uptake differed between stages of development. At vegetative growth stage until day 22 after planting N uptake was determined by both plant age and radiation. During fruiting N uptake was determined by radiation only. A period of stem fruit harvest and a period of sucker fruit harvest can be distinguished. This is also supported by Figure 3.

Daily nitrogen uptake was linearly correlated in both harvest periods to radiation in the investigated range of 900–4000 Wh $m^{-2} \times$ day with the same regression coefficient and squared correlation coefficient. However, during sucker fruit harvest the intercept was half of that during stem fruit harvest. Thus at the same radiation level nitrogen uptake of cucumber was about 40 mg N per plant and day less during sucker fruit period. This may be caused by retranslocation of nitrogen from older leaves to growing parts of the plant. Adams (1980) reported similar N uptake per plant at 3 radiation levels. N uptake was similarily correlated to radiation like fruit yield of cucumber (Liebig, pers. com. based on data published in Liebig, 1984). The residuals were not caused by plant age or fluctuating climatic conditions during these experiments like temperture and humidity. This is supported by Figure 4, which gives results of an experiment, where cucumbers were grown at different temperatures and humidities.

The average temperatures, obtained at the ventilation setpoints reasonable for cucumber production in summer, were in a closed range. The humidity was extremely different. The relationship between N uptake and radiation in this experiment was similar to that shown in Figure 3. Furthermore, the different planting density, carbon dioxide concentration and type of greenhouse had no influence on N uptake.

Table 1 shows, that fruit yield, vegetative

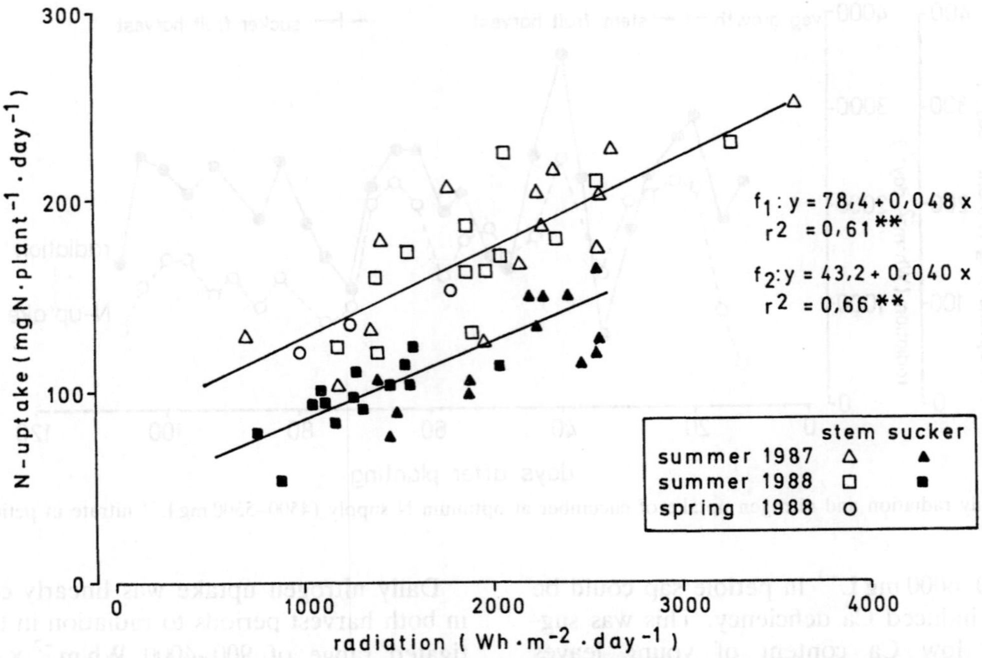

Fig. 3. **Relation between radiation and nitrogen uptake of cucumber during period of stem fruit and sucker fruit harvest.**

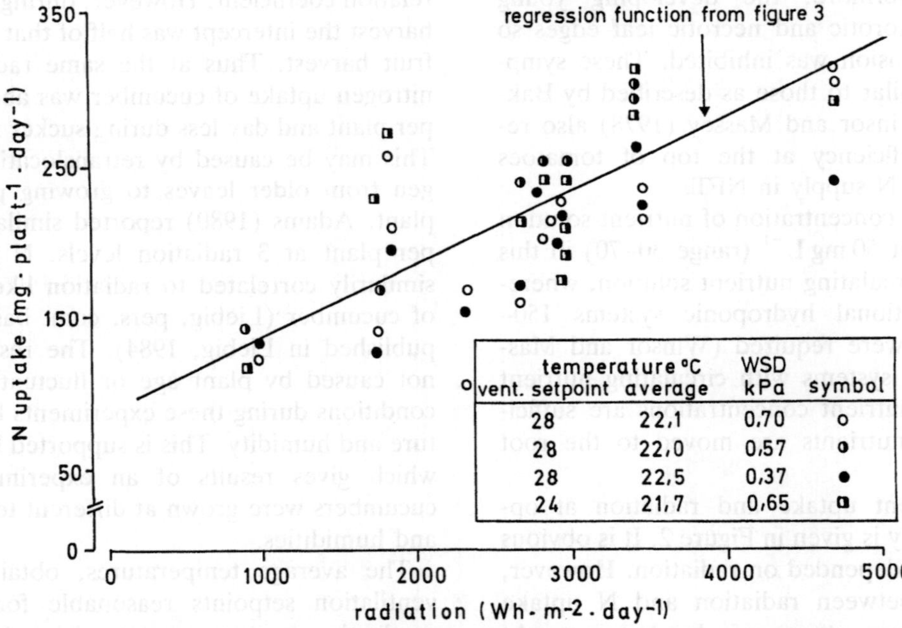

Fig. 4. **Relation between radiation and nitrogen uptake during stem fruit period as affected by different climatisation.**

Table 1. **N uptake, growth and yield of cucumber as affected by different greenhouse climatisation**

Ventilation setpoint (°C)	24	28	28	28
Average temperature (°C)	21.7	22.0	22.1	22.5
Water saturation deficit (kPa)	0.65	0.57	0.70	0.37
Fruit yield (kg fm m^{-2})	22.5	20.5	19.5	19.0
Vegetative growth (g dm m^{-2})	129.3	133.6	124.3	108.5
Total N uptake (g N m^{-2})	30.0	32.1	27.8	27.1

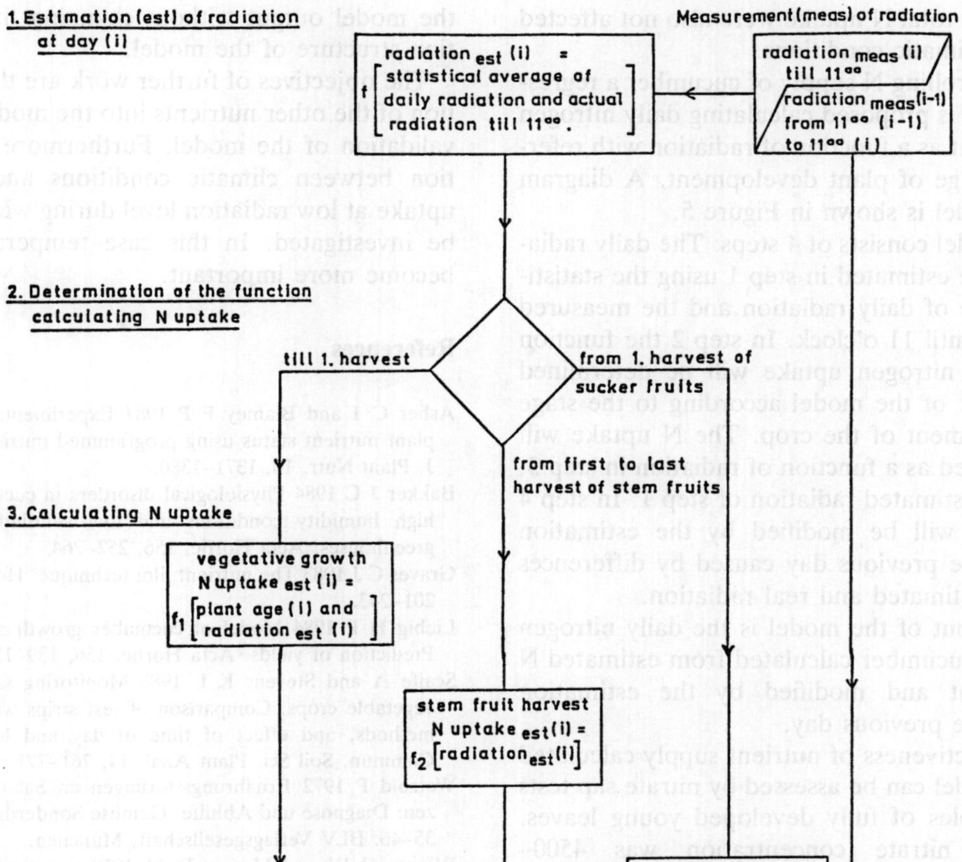

Fig. 5. Diagram of a model to calculate daily nitrogen supply of cucumber.

growth and total N uptake were also not affected by these climatic conditions.

For controlling N supply of cucumber a regression model is proposed calculating daily nitrogen requirement as a function of radiation with reference to stage of plant development. A diagram of this model is shown in Figure 5.

This model consists of 4 steps: The daily radiation will be estimated in step 1 using the statistical average of daily radiation and the measured radiation until 11 o'clock. In step 2 the function calculating nitrogen uptake will be determined by the user of the model according to the stage of development of the crop. The N uptake will be calculated as a function of radiation in step 3, using the estimated radiation of step 1. In step 4 this result will be modified by the estimation error of the previous day caused by differences between estimated and real radiation.

The output of the model is the daily nitrogen supply of cucumber calculated from estimated N requirement and modified by the estimation error of the previous day.

The effectiveness of nutrient supply calculated by this model can be assessed by nitrate sap tests using petioles of fully developed young leaves. Optimum nitrate concentration was 4500–5500 mg L^{-1} as shown in Figure 1. Deviations from this optimum range require a correction of the model output without changing the calculation structure of the model.

The objectives of further work are the integration of the other nutrients into the model and the validation of the model. Furthermore, the relation between climatic conditions and nutrient uptake at low radiation level during winter has to be investigated. In this case temperature may become more important.

References

Asher C J and Blamey F P 1987 Experimental control of plant nutrient status using programmed nutrient addition. J. Plant Nutr. 10, 1371–1380.

Bakker J C 1984 Physiological disorders in cucumber under high humidity conditions and low ventilation rates in greenhouses. Acta Hortic. 156, 252–264.

Graves C J 1983 The nutrient film technique. Hortic. Rev. 5, 201–243.

Liebig H P 1984 Model of cucumber growth and yield. 2. Prediction of yields. Acta Hortic. 156, 139–154.

Scaife A and Stevens K L 1983 Monitoring sap nitrate in vegetable crops: Comparison of test strips with electrode methods, and effect of time of day and leaf position. Commun. Soil Sci. Plant Anal. 14, 761–771.

Wetzold P 1972 Ernährungsstörungen an Salatgurkenpflanzen: Diagnose und Abhilfe. Gemüse Sonderdruck, Heft 2, 35–46. BLV Verlagsgesellschaft, München.

Winsor G W and Massey D M 1978 Some aspects of the nutrition of tomatoes grown in recirculating solution. Acta Hortic. 82, 121–132.

M. L. van Beusichem (Ed.), *Plant nutrition – physiology and applications*, 759–763.

PLSO IPNC254B

A double-pot technique as a tool in plant nutrition studies

B.H. JANSSEN
Department of Soil Science and Plant Nutrition, Wageningen Agricultural University, P.O. Box 8005, 6700 EC Wageningen, The Netherlands

Key words: Gramineae, macronutrients, micronutrients, nutrient deficiency symptoms, perennials, relative growth rate, soil testing

Abstract

The double-pot technique was introduced as a simple method for soil testing. The principle is that seedlings take up nutrients simultaneously from the soil to be investigated and from nutrient solutions of different compositions. The difference in growth between plants on a complete solution and on a solution missing one nutrient is a measure of the availability of that nutrient in the soil.

The method has been applied also for other purposes, using various plant species. Maize and millet were most practical as test crops. It is possible to rank soils by nutrient availability for most nutrients. For P, K and Mg, the results can be interpreted in a more quantitative way, provided that the experiment is carried out according to standard procedures.

Introduction

In the early seventies a double-pot technique was introduced as a practical method of soil testing (Janssen, 1970; 1973b; 1974), enabling the identification of nutrients in short supply without the use of chemical analysis.

It was thought to be useful especially in situations where no laboratory facilities exist or where it is not yet possible to interpret chemical soil data for the purpose of fertilizer recommendations. The method was developed as a sequel to the technique used by Bouma (1965) for the assessment of nutritional stress in plants. Both methods use so called minus-one nutrient solutions. The principle of the double-pot technique, referred to also as the Bouma-Janssen method (Muller, 1976; 1979), is that plants take up nutrients simultaneously from the soil to be investigated and from a nutrient solution. A schematic section of the equipment is shown in Figure 1. The pot, which has a gauze bottom, is filled with soil. Seeds or seedlings are planted in the soil. Their roots pass through the gauze and reach the nutrient solution in the container below. When a nutrient is omitted from the solution, plants can take it up from the soil only. The difference in growth between plants on a deficient and on a complete nutrient solution is a measure of the availability of the relevant nutrient in the soil.

The method has been used for soil testing in a number of tropical countries (Table 1), but also for other purposes. Because it provides an expedient way to create nutrient deficiency symptoms in seedlings, the technique can serve as a tool in the diagnosis of symptoms observed in the field . This proved useful especially for perennial crops (Janssen, 1973a; Muller, 1979). The technique has been applied also in plant physiological and ecological studies (Hoffland *et al.*, 1989; Pegtel, 1983), and for demonstration and teaching purposes.

Technical details

Pot sizes

In the original version the content of the upper

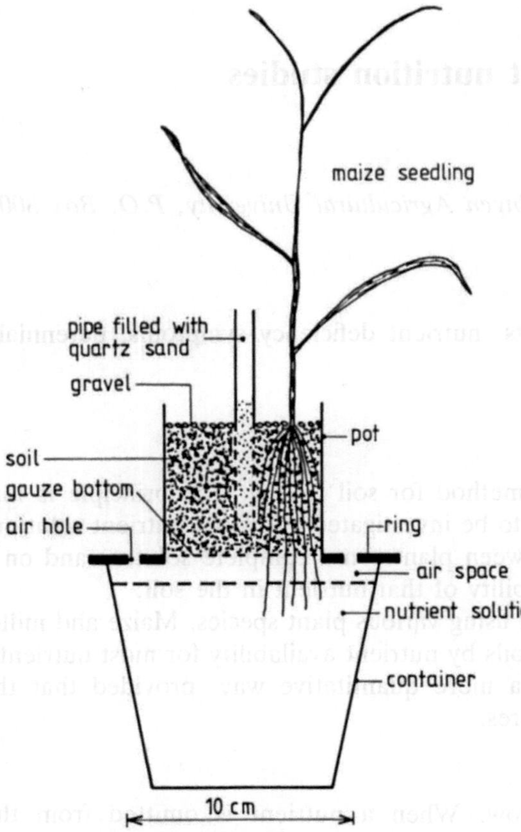

Fig. 1. Schematic section of the equipment.

pot is about 200 cm^3 and that of the lower pots about 700 cm^3 (Fig. 1). Because of the small dimensions of the upper pots quite a number of pots can be filled with a restricted quantity of soil. This is a practical advantage when the soils must be transported over long distances between sampling site and greenhouse.

Depending on the plant species and the goals of the experiments, other pot sizes were used: *e.g.*, 0.75 and 1.2 (Muller, 1974), 3 and 6 litres (Pegtel, 1983) for the upper and lower pots, respectively.

Growth measurement

Plant growth is determined in a non-destructive way by measuring the increase in leaf area of leaf length. In the case of Gramineae, it has been found convenient to measure the lengths of the leaves from the base (soil) to the apex, that is blade and sheath. The sum of the lengths of the individual leaves is termed plant size, S. The relative increase in plant size per unit of time, t, is used as the plant growth parameter, R_S. Therefore

$$R_S = \frac{1}{S}\frac{dS}{dt} \tag{1}$$

Table 1. Survey of plant species and nutrients studied and the origin of soils tested in experiments according to the double-pot technique

Soils from	Plant species	Nutrients	Reference
Cameroun	millet, ryegrass, sorghum	P, K	Aelterman, 1984
Colombia	maize	N, P, S, K, Ca, Mg	Muller and Giama, 1977
Ecuador	cocoa	N, P, S, K, Ca, Mg, Fe, Mn, Zn, Cu, Mo, B	Muller, 1974, 1979
Indonesia	maize	N, P, S, K, Ca, Mg	Brunt, 1982
Indonesia	maize	N, P, S, K, Ca, Mg, Fe, Mn, Zn, Cu, B	Van Staveren and Suwono, 1987
Nepal	cotton	N, P, S, K, Ca, Mg, Fe, Mn, Zn, Cu, Mo	Muller *et al.*, 1979
Netherlands	rape, sunflower	P	Hoffland *et al.*, 1989
Netherlands	semi-natural grasses	N, P, S, K, Ca, Mg	Pegtel, 1983
Suriname	*Pinus caribea*	N, P, K, Ca, Mg, Fe, Cu, B	Janssen, 1973a
Suriname	maize	N, P, K, Ca, Mg	Janssen, 1973b, c
Turkey	wheat	N, P	Janssen, 1970
Zaire	cinchona	N, P, S, K, Ca, Mg, Fe, Mn, Zn, Cu, Mo, B	Muller, 1986

The parameter R_S is comparable with the relative growth rate (RGR) and the same growth functions as developed for dry matter increase are applicable.

Integration of Equation (1) between t_1 and t_2 gives:

$$R_S = (\ln S_2 - \ln S_1)/(t_2 - t_1) \quad (2)$$

Equation (2) gives the mean value of R_S in the period of t_1 to t_2, which is identical with the actual value of R_S at any instant only under conditions of exponential growth.

Sufficiency quotient

As index of the difference in growth between plants on a deficient and on a complete solution, and hence of the availability of the nutrient in the soil, the so-called sufficiency quotient (SQ) was introduced, *i.e.*,

$$SQ_E = \frac{(R_S)_{-E}}{(R_S)_C} \quad (3)$$

where SQ_E = sufficiency quotient for element E, $(R_S)_C$ and $(R_S)_{-E}$ are relative increase in plant size per unit of time of plants on a complete solution, and on a solution without E, respectively.

Because SQ depends on time, it is necessary to standardize the time interval in which SQ is determined. For maize the interval between the four-leaf and the eight-leaf stages was most appropriate in case N and P were examined; for other nutrients the time interval should be longer. The first measurement should not take place until the plant has exhausted the nutrient reserve of the seed. Deficiences of N are the first that show up, followed by those of P. It takes somewhat more time to obtain growth reduction by K or Mg deficiency, and far more time to discover shortage of micronutrients. The duration of double-pot experiments depends also on plant species. The experiments of Table 1 lasted about 3 to 4 weeks for maize, wheat, millet and sorghum, 4 weeks for rape and sunflower, 7 weeks for cotton, 8 weeks for cinchona, 12 to 20 weeks for cocoa. The experiment with *Pinus caribea* was continued for about 9 months, but deficiency symptoms were visible within 3 months. In the experiment with semi-natural grasses the turfs were clipped at intervals of 60 days.

Plant species and growth indices

Ideally, the test plant used in double-pot experiments should have a regular germination, an 'easy' parameter like leaf length for the assessment of plant growth and small seeds with a low nutrient reserve. In practice a test crop was chosen because of its agricultural importance in the area.

Maize is a favourite test crop, although it has the drawback of large seeds, because it does not tiller and is easy to handle. Plant height data of maize resulted in about the same values of the sufficiency quotients as data on plant size. Because the determination of plant size is very time-consuming, the measurement of plant height is preferred.

Aelterman (1984) concluded that millet was a better test crop than sorghum or ryegrass.

In the experiments with cotton and perennial crops (Table 1), seeds were sown in separate seed boxes. When they were large enough to handle, equally sized seedlings were transferred to the upper pots. For the growth measurement of cocoa and cinchona, Muller (1974, 1976) introduced an 'index of leaf surface', being the product of the maximum length and width of each leaf. The increase in the sum of the indices of leaf surface of all leaves per plant was taken as growth index. For cotton the increase of the index of leaf surface as well as the increase in plant height proved suited as growth indices, but the latter was preferred because of its simplicity (Muller *et al.*, 1979).

The only practical characteristic to measure the growth of *Pinus caribea* seedlings was plant height. It was not satisfactory because the omission of nutrients affected also the morphology of the seedling. A qualitative description of the plant's appearance was the best that could be done, and the determination of the dry weight at the end of the experiment (Janssen, 1973a).

In various other studies dry weights of the plants at the end of an experiment were taken as characteristic for plant growth. Usually dry weights show a larger coefficient of variation than relative rates of increase in plant size (R_S), which is mainly due to the fact that an irregular germination has a strong effect on the final dry weight and only a slight effect on R_S (Janssen, 1973b).

Interpretation of sufficiency quotients

The aim of the double-pot technique is to make at least a qualitative distinction between soils that differ in the availability of the examined nutrient. In other words the ranking of soils by sufficiency quotients must result in the same order (but in opposite direction) as their ranking by fertilizer needs.

In some of the studies mentioned in Table 1, it was tried to relate sufficiency quotients to chemical soil properties, in others to results obtained in fertilizer trials in the field. The following discussion refers only to experiments with Gramineae.

Nitrogen

In some experiments a relation between SQ_N and soil organic N and between SQ_N and NO_3 was found (Janssen 1970, 1973b), in other experiments between SQ_N and NH_4 (Brunt, 1982). It is likely that in such short-term experiments nitrogen availability depends more on the quantity of inorganic N present at the start of the experiment than on mineralizable organic N. Therefore SQ_N is only of limited value for the estimation of nitrogen availability to field crops.

Phosphorus

Aelterman (1984) concluded that the value of SQ_P was usually less than 0.80 in P-deficient soils. In such soil, P-Olsen was less than 11, and P-48 (desorption in 48 hours) less than 24 mg/kg. Janssen (1970) compared SQ_P and P-AL, soil P extractable with ammonium lactate and acetic acid. Although the SQ_P values obtained depended on weather conditions during the experiments, in all experiments a relationship between SQ_P and P-AL was found. It was concluded that standard soils are needed as references in all experiments. No relations were found between SQ_P and P-Bray 1 in soils from Suriname (Janssen, 1973b).

Potassium

For most soils with less than 2 mmol/kg exchangeable K, SQ_K was less than 0.85 (Aelterman, 1984). Brunt (1982) found that SQ_K was related to K extractable with 2% citric acid, but not to exchangeable K and K in Morgan Venema extracts. In Suriname SQ_K was below 0.75 for soils with less than 1 mmol exchangeable K; maize responded clearly to K application in field trials on these soils (Janssen, 1973b; c).

Calcium

Because plant roots hardly grow into minus-Ca solutions, the double-pot technique is not suited to investigate Ca availability.

Magnesium

No relationships between SQ_{Mg} and soil Mg could be ascertained by Brunt (1982), which might at least partly be contributed to the fact the range in soil Mg contents was rather narrow.

Micronutrients

Experiments with micronutrients so far were performed with the purpose of ranking the soils under study or of describing deficiency symptoms. No attempts were made to relate the obtained sufficiency quotients to soil contents of micronutrients.

Discussion

The double-pot technique is a helpful method that can be used in plant nutrient research and teaching. When using it for soil testing, standard procedures should be applied: 200 g of soil in the upper pot, and maize, sorghum or millet as test crop. The optimum points of time for the measurement of plant size can best be indicated by the leaf stages of the plants on the complete solution. The first measurement is when they have 4 leaves, and the last measurement when they have 8, 9 or 10 leaves, depending on whether P, K or Mg availability is to be assessed.

If these standard procedures are followed, a SQ value of 0.8 can be considered as critical for these three nutrients. This value has been derived not only from the studies cited in this paper, but also from about ten MSc theses by students of the Wageningen Agricultural University.

As far as it concerns micronutrients, the double-pot technique enables a ranking of soils, but the present data are insufficient to indicate critical SQ values and standard periods for the measurement of the SQ's for these nutrients.

References

Aelterman G 1984 Potproeftechnieken voor het bepalen van beschikbaar K en P in sterk verweerde gronden van de vochtige tropen. Ph.D. thesis, Gent, Belgium, 195 p.

Bouma D 1965 Growth changes of plants following the removal of nutritional stresses. Doct. thesis, Agricultural University, Wageningen, The Netherlands, 98 p.

Brunt J 1982 Principles and application of the 'double-pot' technique for rapid soil testing. AGOF/INS/78/006 Technical note no 14, Centre for Soil Research, Bogor, Indonesia, 48 p.

Hoffland E, Findenegg G R and Nelemans J A 1989 Solubilization of rock phosphate by rape. I. Evaluation of the role of the nutrient uptake pattern. Plant and Soil 113, 155–160.

Janssen B H 1970 Soil fertility in the Great Konya Basin, Turkey. Agric. Res. Rep. 750. Pudoc, Wageningen, The Netherlands, 115 p.

Janssen B H 1973a A study on the nutritional status of *Pinus caribea* Morelet by means of the technique used for the assessment of sufficiency quotients. Centre for Agricultural Research in Suriname, Paramaribo, Celos Bull. 18, 49.

Janssen B H 1973b Een oriënterend onderzoek naar de geschiktheid van mais voor de bepaling van het sufficiency quotient (SQ). Interne Meded. Lab. Landb. Scheik. No 11. Wageningen, The Netherlands, 35 p.

Janssen B H 1973c Onderzoek naar de vruchtbaarheid van enkele terrasgronden langs de Suriname rivier. Interne Meded. Lab. Landb. Scheik. No 13. Wageningen, The Netherlands, 88 p.

Janssen B H 1974 A double-pot technique for rapid soil testing. Trop. Agric. Trinidad 51, 161–166.

Muller A 1974 Nutrient deficiencies in a volcanic ash soil from Ecuador. *In* Plant Analysis and Fertilizer Problems. Ed. J Wehrmann. pp 327–341. Proc. 7th Intern. Colloq., Hannover. Publisher German Society of Plant Nutrition, Hannover, FRG.

Muller A 1976 Testing cinchona seedlings from Zaire on their suitability for use in Bouma-Janssen pot experiments. Internal Report BO 76-6. Royal Tropical Institute, Amsterdam, The Netherlands, 17 p.

Muller A 1979 Deficiency symptoms in cocoa seedlings observed in pot experiments with the Bouma-Janssen method. Neth. J. Agric. Sci. 27, 211–220.

Muller A and Giama S S 1977 Nutrient deficiencies in a soil from the Llanos Orientales, Colombia, tested with the Bouma-Janssen method. Internal Report AO/BO-9. Royal Tropical Institute, Amsterdam, The Netherlands, 15 p.

Muller A, Giama S S and Schelhaas R M 1979 Evaluation of macro and micro nutrient status of soils from the Cotton Development Project, Nepal. Internal Report BO 1979-1. Royal Tropical Institute, Amsterdam, The Netherlands, 19 p.

Pegtel D 1983 Ecological aspects of a nutrient-deficient wet grassland. Verhandl. Gesells. Ökologie, Band X, pp 217–228 Mainz 1981.

Van Staveren J Ph and Suwono 1987 Maize on young volcanic upland soils: Nutrient availability and response to fertilization. Technical Paper Series MARIF No 8, Res. Inst. Food Crops, Malang, Indonesia, 18 p.

M. L. van Beusichem (Ed.), *Plant nutrition – physiology and applications*, 765–768.

PLSO IPNC185

In situ nitrate reductase activity in winter wheat (*Triticum aestivum*) as an indicator of nitrogen availability

M.O. VOUILLOT, J.M. MACHET[1] and B. MARY
INRA de Laon, Station d'Agronomie, B.P. 101, F-02004 Laon cédex, France. [1]Corresponding author

Key words: fertilizer, nitrate, nitrate reductase, nitrogen uptake, *Triticum aestivum* L., wheat

Abstract

The nitrogen nutrition status of a winter wheat crop (cv Arminda) was investigated by using an *in situ* nitrate reductase (NR) assay, performed with or without exogenous nitrate during the incubation. In this study, the establishment of a nitrate assimilation potential curve was attempted for the whole development cycle of the crop, under field conditions. We also tested the capacity of the NR assay to be an early indicator of a temporary nitrogen deficiency of a crop and a physiological criterion for studying the relationships between nitrate absorption and assimilation.

Introduction

Nitrogen is central to the growth of wheat crops particularly at some critical periods where it is essential in the formation of yield components (Croy and Hageman, 1970; Dalling *et al.*, 1975; Eilrich and Hageman, 1973; Harper *et al.*, 1987; Masle, 1980; Meynard, 1985).

Nitrate is the major source of nitrogen for higher plants. Its reduction to nitrite, catalysed by the enzyme nitrate reductase (EC 1.6. 6.1) is considered to be the rate-limiting step in N assimilation in plants (Beevers and Hageman, 1980). Moreover, nitrate uptake by roots could be one of the factors controlling nitrogen assimilation in plants (Rodgers and Barneix, 1988) since the flux coming from vacuole or xylem regulates the nitrate reduction (Robin *et al.*, 1983; Shaner and Boyer, 1976).

The present work had two aims:

1. to establish a potential nitrate assimilation curve for the whole development cycle of a winter wheat crop, under field conditions.
2. to test the capacity of an *in situ* NR assay to be an early indicator of a temporary nitrogen deficiency of a crop and a physiological criterion for studying the relationships between nitrate absorption and assimilation.

Materials and methods

Experimental conditions

Winter wheat (*Triticum aestivum* L. cv Arminda) was sown on October 16, 1987 on a sandy loamy soil in a randomized block design consisting in four nitrogen fertilizer levels. Each N treatment was replicated four times. Each plot was 5 m wide and 6 m long. The fertilizer (ammonium nitrate in solution) was applied as indicated at Table 1.

Analytical methods

The *in situ* nitrate reductase assay was conducted all along the development cycle of the crop, using the method proposed by Robin *et al.*, (1983). The whole shoots (sampled until end of tillering) or the separated organs (leaves, stem, ear, sampled later on), were weighed and incubated for 60 min. under dark, anaerobic conditions at field temperature. At the end of the incubation period, the nitrite accumulated in the tissues was determined colorimetrically after extraction with boiling water. Two treatments were made: during the incubation, the base of organs was plunged either into distilled water or into

Table 1. Dates and rates (kg N ha^{-1}) of N fertilizer applications

Treatments	Tillering 28/02/88	Ear at 1 cm 28/03 1038 °C d	2 cm 18/04 1282 °C d	10 cm 27/04 1327 °C d	20 cm 9/05 1494 °C d	60 cm 24/05 1704 °C d	earing 6/06 1887 °C d	Total rate
N1	60							60
N2	60	50						110
N3	60	80		40				180
N4	60	80	30	30	30	30	40	300

0.1 *M* KNO_3 solution, yielding two values:

NRA_i^- (*in situ* nitrate reductase activity without exogenous nitrate), which represents a minimum assimilation rate sustained with the accumulated NO_3;

NRA_i^+ (*in situ* nitrate reductase activity with exogenous nitrate) which measures the potential assimilation rate when the NO_3 supply to the enzyme is not limiting.

The Δ value was defined as follows:

$$\Delta = 100 \times \frac{NRA_i^+ - NRA_i^-}{NRA_i^+}$$

At each harvest, the dry weight of the whole plant and organs was measured. Total nitrogen content was determined by a salicylate Kjeldahl procedure.

Results

N uptake

The N uptake curve (more exactly the accumulated N in shoots), shows a rapid variation of the absorption rates in all treatments (Fig. 1). This variation was found even in the N4 treatment where N was applied regularly until anthesis to avoid any limitation in the soil N supply. In this treatment, an important absorption took place from the beginning of stem elongation (1070°C d) up to early earing (1640°C d):230 kg N ha^{-1} were absorbed in 48 days. During the next period, until anthesis (1640–2020°C d), no net accumulation of N occurred. This indicated that the absorption was very slow, although 70 kg N ha^{-1} was applied onto the soil surface during this period. The absorption started again after anthesis, leading to a maximum net uptake of 400 kg N ha^{-1}.

N uptake in the N1 treatment was much smaller but its variations followed a similar pattern.

Potential nitrate assimilation rate

The evolution of NRA_i^+ is given at Fig. 2. In the treatment N4, which received seven N applications at approximately 2 weeks interval, two peaks of activity were found: one at the stage 'ear at 5 cm', the other at anthesis. The first peak

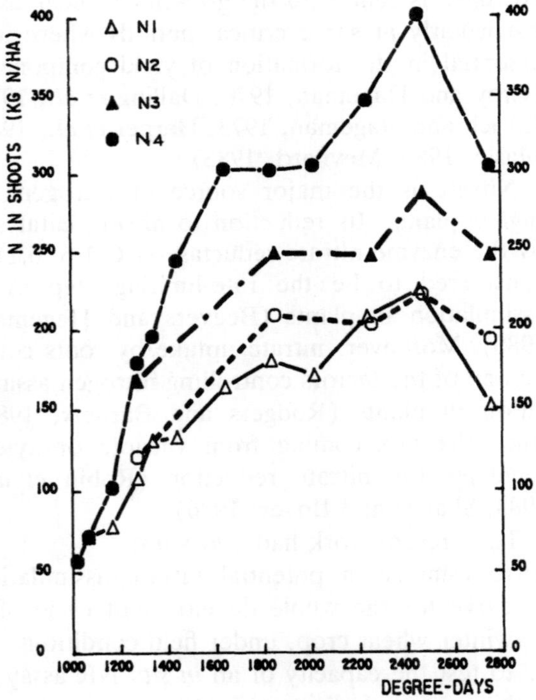

Fig. 1. Evolution of total accumulated nitrogen in shoots all along the development cycle, in the 4 treatments. The abscissa is in cumulated degree-days from the sowing time. Each point is the mean of 8 replicates.

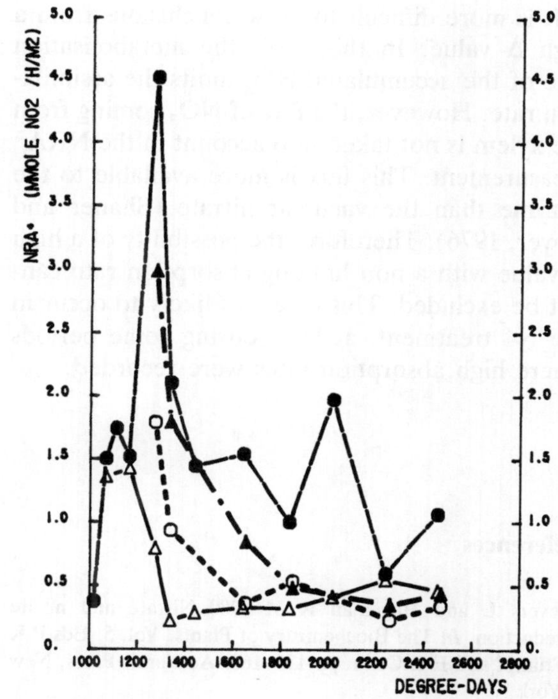

Fig. 2. Evolution of the *in situ* NR activity (with exogenous nitrate). Same symbols as in Figure 3.

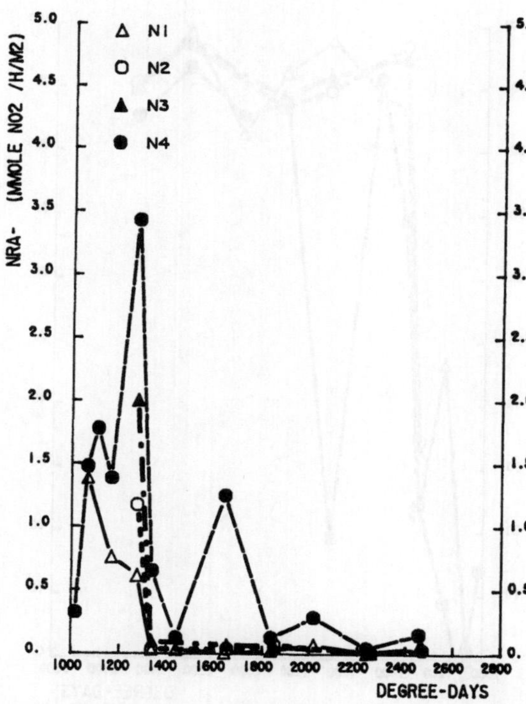

Fig. 3. Evolution of the *in situ* NR activity (without exogenous nitrate). Each point represents the mean of 8 replicates.

had an intensity of 4.5 mmole NO_2 h^{-1} m^{-2}, corresponding to an assimilation rate of 15 kg N ha^{-1} day^{-1}. At anthesis, the peak was smaller, equal to 2.0 mmole NO_2 h^{-1} m^{-2} equivalent to 6.6 kg N ha^{-1} day^{-1}.

In the other treatments, the NRA_i^+ declined steadily during stem elongation; the level was related to the N rates. The activity remained more or less constant after earing, at a value close to 0.5 mmole NO_2 h^{-1} m^{-2}.

In the N4 treatment, there was no correlation between the NRA_i^+ values and the date of N application. Therefore, the activities were not strongly dependant of the fertilizer applications.

Minimum nitrate assimilation rate and Δ value

The *in situ* NR activity without exogenous NO_3^- (NRA_i^-) in the N4 treatment also exhibited 2 peaks (Fig. 3). The first and highest peak was found at the same time ('ear at 5 cm' stage), corresponding to the greatest absorption and assimilation rates, respectively 9.6 and 8.8 kg N ha^{-1} day^{-1}. The second peak occurred earlier, almost one month before anthesis, at 1650°C d. It was due to an enhanced activity of the lower leaves (upper leaf excluded). This high activity was found at the end of the wettest period (70 mm in 13 days) during which the mean absorption rate was still high: 4.5 kg N ha^{-1} day^{-1}.

The evolution of the Δ value for the whole plant is shown in Fig. 4. It was identical for the different organs. Surprisingly, it was also almost independent of the N treatment (except the date 1650°C d already mentioned). The Δ value was lower than 10 until the stage 'ear at 2 cm' (1200°C d). Then, it increased quickly up to about 90, reached at 1400°C d ('ear at 10 cm').

Discussion

Can the NRA_i^+ measurements (Fig. 2) represent the potential assimilation rate, depending only on the phenological stage?

The occurrence of two peaks of NRA_i^+ was not specific to the experiment described here. Simi-

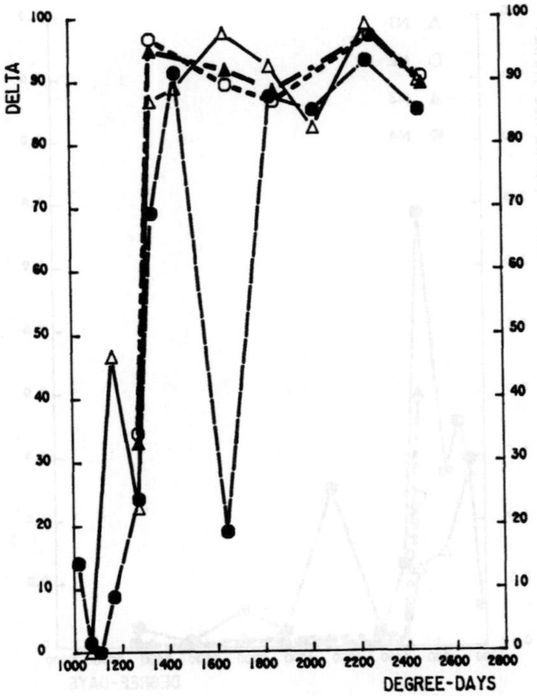

Fig. 4. Evolution of the Δ value. Same symbols as in Figure 2.

lar peaks were also found at the same phenological stages in other experiments in which the sowing date and cultivar were varied (data not shown).

This potential rate can be obtained only if N availability in soil is not limiting. This condition may have been unsatisfied at least in one period where total N in shoots remained constant (Fig. 1). The NRA_i^+ value found in the middle of this period was indeed smaller than the adjacent values, suggesting that the potential activity should have been higher.

Additional experiments in which the soil N availability would be more controlled are necessary to clarify this point.

The Δ value seems to be a useful physiological indicator of the relationships between the assimilation and the absorption rates. When the Δ value is close to zero, the available accumulated NO_3 in the tissues enables a maximal assimilation rate. *A fortiori*, the absorption rate cannot be a limiting factor of the assimilation process.

It is more difficult to draw conclusions from a high Δ value. In this case, the metabolisation rate of the accumulated NO_3 limits the assimilation rate. However, the flux of NO_3 coming from the xylem is not taken into account in the NRA_i^- measurement. This flux is more available to the enzyme than the vacuolar nitrate (Shaner and Boyer, 1976). Therefore, the possibility of a high Δ value with a non limiting absorption rate cannot be excluded. This case was likely to occur in the N4 treatment, at least during some periods where high absorption rates were recorded.

References

Beevers L and Hageman R M 1980 Nitrate and nitrite reduction. *In* The Biochemistry of Plants. Vol. 5. Eds P K Stumpf and E E Conn. pp 115–168. Academic Press, New York.

Croy L I and Hageman R H 1970 Relationship of nitrate reductase activity to grain protein production in wheat. Crop. Sci. 10, 280–285.

Dalling M J, Halloran G M and Wilson J H 1975 The relation between nitrate reductase activity and grain nitrogen productivity in wheat. Aust. J. Agric. Res. 26, 1–10.

Eilrich G L and Hageman R H 1973 Nitrate reductase activity and its relationship to accumulation of vegetative and grain nitrogen in wheat (*Triticum aestivum* L.). Crop Sci. 13, 59–66.

Harper L A, Sharpe R R, Langdale G W and Giddens J F 1987 Nitrogen cycling in a wheat crop: Soil, plant and aerial nitorgen transport. Agron. J. 79, 965–973.

Masle J 1980 L'élaboration du nombre d'épis chez le blé d'hiver. Influence de différentes caractéristiques de la structure du peuplement sur l'utilisation de l'azote et de la lumière. Thèse de Docteur-Ingénieur, INA-PG, Paris, 274 p.

Meynard J M 1985 Construction d'itinéraires techniques pour la conduite du blé d'hiver. Thèse de Docteur-Ingénieur, INA-PG, Paris, 297 p.

Robin P, Conejero G, Passama L and Salsac L 1983 Evaluation de la fraction métabolisable de nitrate par la mesure *in situ* de sa réduction. Physiol. Vég. 21, 115–122.

Rodgers C O and Barneix A J 1988 Cultivar differences in the rate of nitrate uptake by intact wheat plants as related to growth rate. Physiol. Plant. 72, 121–126.

Shaner D L and Boyers J S 1976 Nitrate reductase activity in maize (*Zea mays* L.) leaves. I. Regulation by nitrate flux. Plant Physiol. 58, 499–504.

M. L. van Beusichem (Ed.), *Plant nutrition – physiology and applications*, 769–772.

PLSO IPNC429B

Tissue test of rice plant (*Oryza sativa*) nitrogen using near infrared reflectance

G. D. BATTEN, A. B. BLAKENEY and A. C. McCAFFERY
NSW Agriculture and Fisheries, Yanco Agricultural Institute, Yanco, NSW 2703, Australia

Key words: near infrared reflectance, nitrogen analysis, *Oryza sativa* L., rice

Abstract

To maximize the yield of rice, nitrogen must be applied at the optimum rate and at the appropriate stage of plant development. Traditional nitrogen tissue tests are not suited to crop diagnosis on a large commercial scale. This paper reports a tissue testing system based on rapid drying of samples, using domestic microwave ovens, and rapid, simple and cheap analysis, using near infrared reflectance spectroscopy.

Introduction

In southern Australia rice is grown in the irrigation areas associated with the Murrumbidgee and Murray river systems of New South Wales. The crop is grown exclusively under flood irrigation and operations are entirely mechanised. Planting is carried out in September and October either by combine or sod-seeder into a dry seedbed, or by aerial sowing of germinated seed into shallow water. Flood water is maintained on the crop for approximately four months. Before harvest the water is drained and the soil allowed to dry; the rice is then mechanically header harvested. The crop is essentially disease-free and is troubled by few insects. Management practices exist for controling the major weed species, dirty dora (*Cyprus difformis*) and barnyard grass (*Echinochloa* sp.) (McDonald, 1979).

The physiology of rice growth in southern New South Wales has been described by Boerema (1973, 1974) and is illustrated in Figure 1. Temperatures at both ends of the growing season are low. Seedling emergence is slow but further vegetative development takes place during a period of increasing temperature and radiation. Panicle differentiation and growth occur during the period of highest temperature and radiation. Panicle initiation coincides with the development of roots on the soil surface. It occurs about 65 to 70 days after seedling emergence through either the soil or water surface depending on the method of sowing.

The recommended times for nitrogen fertilizer application to rice are just before permanent flood (see A in Fig. 1) and just after panicle initiation (see B in Fig. 1). If farmers are confident of the appropriate fertilizer rate, all or most is applied before permanent flood. Most farmers are unsure of the appropriate fertilizer rates necessary and fear over-fertilization or low-temperature sterility, and so usually apply one-half to two-thirds of the total nitrogen at permanent flood and the remainder at panicle initiation (Bacon, 1985). Nitrogen needs to be applied within 7 to 10 days after panicle initiation (see C in Fig. 1) to ensure that it is used efficiently by the plant to increase yield. Later application of nitrogen preferentially increases grain protein (Matsushima, 1976). Therefore the amount of nitrogen to apply to maximise yield is a critical management decision at panicle initiation. This paper describes the operation of a tissue testing system that has been developed to assist farmers estimate the nitrogen top-dressing requirements of crops.

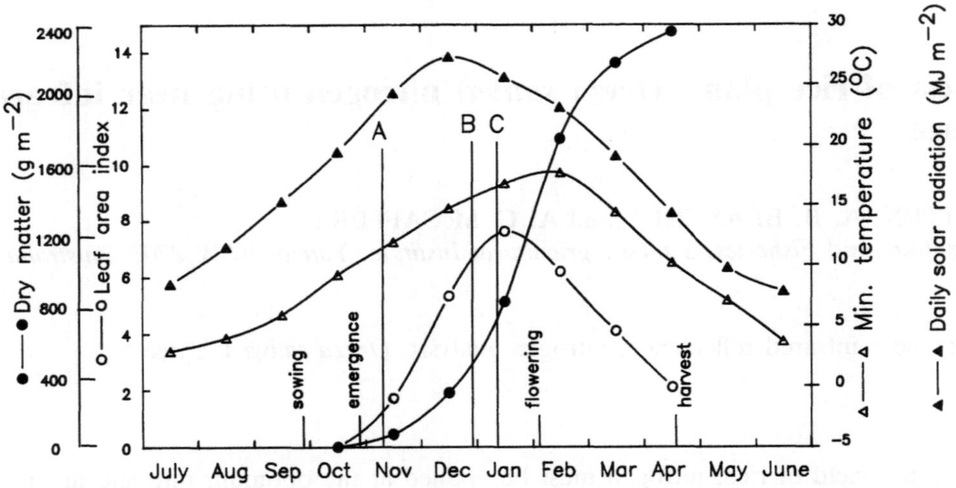

Fig. 1. Diagrammatic representation of the growth stages of rice in southern Australia showing the times of nitrogen fertilizer application and tissue testing. (after Boerema 1973). A. Permanent water, first nitrogen fertilizer application. B. Panicle initiation, tissue testing for nitrogen. B-C. Best period for second nitrogen fertilizer appication.

The tissue testing system

Collection of the sample

Farmers are required to take samples from their crops at the panicle initiation stage of development (Zadoks *et al.*, 1974). Instructions are given on the recognition of panicle initiation at field days and an illustrated leaflet is provided. It is important that panicle initiation is determined accurately because the concentration of nitrogen in shoots changes with the age of the plant. Fertilizer recommendations are based on the nitrogen in shoots at panicle initiation as it is a well defined stage of growth related to plant development rather than age.

Rice growers choose the area of the crop which they propose to fertilize and then take shoot samples from several 0.1 m square quadrat areas chosen at random. Details on the crop and the number of shoots in the sample are recorded on a form supplied by the laboratory.

Drying the sample

A sample drying procedure that quickly kills the tissue was chosen to minimise errors associated with many different people drying their own samples. After some experimentation, drying in domestic microwave ovens has been found to be a reliable technique. A sub-sample of 100 g fresh weight from each quadrat area is cut into short lengths, placed on a shallow dish and dried. Using an oven with a power rating of 650 watts on full power the 100 g sample can be dried to constant weight in 6 to 8 minutes. However as microwave ovens vary widely in power output and internal dimensions the correct drying time should be determined for each oven type. This drying procedure rapidly denatures enzymes and stabilizes the tissue. Our recommendations are designed to be conservative as over-drying results in charring of the sample and loss of nitrogen. Hence, as a precaution against incompletely dried samples, farmers are asked to place the samples in large paper bags for transport. These containers allow escape of moisture during transport and can be dried in the laboratory without repacking.

Transport of the sample to the laboratory

Some farmers deliver samples to the central analytical laboratory in person and wait for the analysis but samples are collected up to 250 km from the central laboratory. Therefore most samples are transported by a courier service operated by the Ricegrowers Cooperative Limited.

Laboratory operations

The samples arrive at the central laboratory with details on the crop and the sample at the panicle initiation stage. Each sample is assigned a number, then checked for state of dryness. Final drying is done, if necessary, in an airing cabinet.

The samples are ground to pass a 0.5 mm screen in a cyclone mill (Cyclotec, Tecator, Sweden). The total nitrogen concentration in the sample is determined by near infrared reflectance spectroscopy using an InfraAlyzer 400 (Technicon Instruments, USA) calibrated to estimate nitrogen in rice shoots by the model developed by Blakeney *et al.*, (1988) and Batten *et al.*, (1989). This calibration has been correlated against the Kjeldahl method ($r^2 = 0.97$) and has a standard error of prediciton of 0.13% nitrogen. The calibration obtained has been verified over three seasons and all five rice varieties currently grown in New South Wales.

Fertilizer recommendation

The nitrogen analysis, together with data on the crop supplied with the sample, is used to arrive at the amount of nitrogen required to achieve the optimum crop yield. The recommended amount of nitrogen is then discussed with the farmer in person, or over the telephone. Most rice growers were advised of the recommended rates of top-dressing within 48 hours of sampling.

The recommended nitrogen fertilizer rates are based on data from time and rate of nitrogen fertilizer studies conducted throughout the rice growing area. Currently, recommendations are based on data from three seasons (Bacon, Batten and Blakeney, pers. com.). Provided the crop has an adequate population density and good weed control, a total plant nitrogen content of more than 1.8% indicates sufficient nitrogen is available to achieve maximum potential yield.

Discussion

The use of microwave drying and analysis by near infrared reflectance spectroscopy has enabled us to overcome the problems inherent in commercial tissue testing for total nitrogen in plants. While traditional wet chemical methods are skilled-labour intensive, use potentially dangerous chemicals and are expensive, the NIR technique is rapid, cheap, clean and can be operated by unskilled staff. During the 1989–1990 rice season, over 3000 samples were tested.

It is imperative that the recommended nitrogen rate is determined after due consideration has been made of the variety, plant density, weed control status and general management of the crop. Sparsely planted crops can often give anomalous results. However the readings are proving to be a valuable aid in crop management. In a survey of rice growers who used the test, 93% reported that they found the test useful and 42% said that they changed the amount of fertilizer they applied. Half those who changed the rate applied less fertilizer as a result of the test.

Farmers who aim to achieve the potential yield of modern nitrogen-responsive rice varieties need to carefully manage crop fertilizer requirements. The approach reported here for rice nitrogen is applicable to improved management of nitrogen and other nutrients in horticultural and field crops. The over-use of nitrogen and other fertilizers results not only in lower yields and a waste of resources, but also contributes to the accumulation of minerals in ground waters. This contribution to environmental pollution could be reduced by monitoring crop fertilizer requirements and applying only the amount of nutrient required by the crop.

Acknowledgements

We thank the Rice Research Committee for financial support of the project and Lynsey Welsh for assistance in preparation of this manuscript.

References

Bacon P E 1985 The effect of nitrogen application time on Calrose rice growth and yield in south-eastern Australia. Aust. J. Exp. Agric. 25, 183–90.

Bacon P E and Heenan D P 1984 Response of Inga rice to application of nitrogen fertilizer at varying growth stages. Aust. J. Exp. Agric. Anim. Husb. 24, 250–254.

Batten G D and Blakeney A B 1989 Prediction of crop fertilizer requirements based on NIR analysis. Proc. 2nd Int. Near Infrared Spectroscopy Conf. Tsukuba, Japan (*In press.*)

Blakeney A B, Batten G D, Bacon P E and Glennie Holmes M R 1988 Tissue test of rice plant nitrogen. Int. Rice Res. Newsl. 13, 26–7.

Boerema E B 1973 Rice cultivation in Australia. Il Riso 22, 131–150.

Boerema E B 1974 Climatic effects on growth and yield of rice in the Murrumbidgee Valley of New South Wales. Il Riso 23, 385–397.

Heenan D P and Lewin L G 1982 Response of Inga rice to nitrogen fertilizer rate and timing in New South Wales. Aust. J. Exp. Agric. Anim. Husb. 22, 62–66.

McDonald D J 1979 Rice. *In* Australian Field Crops. Vol 2. Eds. J V Lovett and A Lazenby. pp 70–94. Angus and Robertson Publishers, Sydney, Australia.

Matsushima S 1976 High-Yielding Rice Cultivation. Japan Scientific Societies Press, Japan, 377 p.

Zadoks J C, Chang T T and Konzak C F 1974 A decimal code for the growth of cereals. Weeds Res. 74, 415–421.

M. L. van Beusichem (Ed.), *Plant nutrition – physiology and applications*, 773–778.

PLSO IPNC128

Pattern of nitrate assimilation and grain nitrogen yield in field-grown wheat (*Triticum aestivum*)

Y.P. ABROL
Division of Plant Physiology, Indian Agricultural Research Institute, New Delhi-110012, India

Key words: grain nitrogen, nitrate, nitrate assimilation, nitrate reductase, *Triticum aestivum* L., wheat

Abstract

To evolve management technology aimed at optimizing fertilizer N use, field-based physiological investigations were conducted. It was observed that following prevalent management practices, NO_3^- content and *in vivo* nitrate reductase activity were high in the first formed leaf blades and declined in the subsequently formed ones. The pattern was by paralleled by soil NO_3^- concentration and its total content. Incubation of excised leaf blades in a nutrient solution containing 15 m*M* NO_3^- showed that while there was slight increase in the NR activity of the lower leaf blades, the activity of the upper ones was enhanced manifold. The level of enhancement was higher in 'high NR' cultivar than in 'low NR' one. Improving the availability of soil NO_3^- at later growth stages by increasing the frequency of splitting the total quantity of fertilizer N applied resulted in more N than in the whole plant and in the grain.

Introduction

Millions of tonnes of nitrogenous fertilizers are produced annually to help boost the yield of various crop plants. Their manufacturing involves high-cost technology which requires a whole range of feed stock-lignite, coal, coke oven gas, electrolysis of water, natural gas, naphta and fuel oil (Anonymous, 1980). The major feed stock is naphta, a petroleum product. Its import results in drainage of foreign exchange reserves of a country, like India. Coupled with it are the problems of storage, transport and distribution (Ramanathan, 1980).

The utilization efficiency of nitrogenous fertilizer applied to a farmer's field is low. The values vary with the crop, type of fertilizers, edaphic and climatic factors, and management practices. The remainder is lost from the soil by volatilization, denitrification, leaching and conversion to unavailable forms. Recently, significant losses of nitrogen, mainly in reduced form, from canopies of various crop and weed species have been reported (Abrol *et al.*, 1984). In actual terms, this amounts to losses of not only millions of rupees spent on the manufacturing of fertilizers, but also of an element essential to crop plants. Further, there may be problems of groundwater pollution. Even if the efficiency remains at the present level, the losses will increase enormously in the near future, as the consumption of fertilizer N is expected to double by 2000 AD.

To improve fertilizer N use efficiency, particularly in rice soils, where the efficiency is very low, agronomists and soil scientists are engaged in developing new management practices and new technologies, viz. use of slow-release and coated ureas, and of nitrification inhibitors. The objective of all such investigations is to evolve methodology whereby the supply of fertilizer N matches its demand/requirement by the crop at various stages of growth and development.

To provide a scientific basis to new management technologies aimed at maximizing fertilizer N use, we have been conducting a series of field-based physiological and biochemical investigations to get answers to some of the questions which agricultural scientists have been asking from time to time.

These are: a) Is the nitrogen being applied at the right time and in sufficient quantity so as to meet the requirements of the growing crop? b) What are the constraints in fertilizer N utilization? c) Can we on the basis of physiological and biochemical investigations explain crop demands more precisely and evolve fertilizer N utilization technologies applicable to different field situations? d) What are the potentials of N utilization and total/grain N harvest of presently available commercial cultivars? How far can we achieve these targets?

Our studies have been confined primarily to wheat (*Triticum aestivum* L.) and barley (*Hordeum vulgare* L.). In this communication, the work done on wheat is reported. Nitrate-N and some exchangeable ammonium-N are the prime sources of nitrogen available to the plant. However, these quantities represent a very small fraction of total N in soil. The rest of it is in unavailable (organic) form. For maximizing crop production, nitrogenous fertilizers, mostly in the form of urea or ammonium sulphate, are applied. Irrespective of the form in which it is applied, in neutral and alkaline soils, it is converted to NO_3^- by nitrifying bacteria. Its assimilation is mediated by a series of enzymes. The first step involves the cytosolic enzyme nitrate reductase (NR), which plays a key role in view of the fact that it represents a port of entry for inorganic N, initiates the reaction sequence, is an unstable enzyme with a short half life, is substrate inducible, has a high Km value (there are reports of accumulation of NO_3^- when heavy doses of N fertilizer are supplied) and is present in all parts of the plant (Abrol *et al.*, 1976). The next step, *viz.*, conversion of nitrite to ammonium is mediated by the enzyme nitrate reductase which is situated in the chloroplasts and

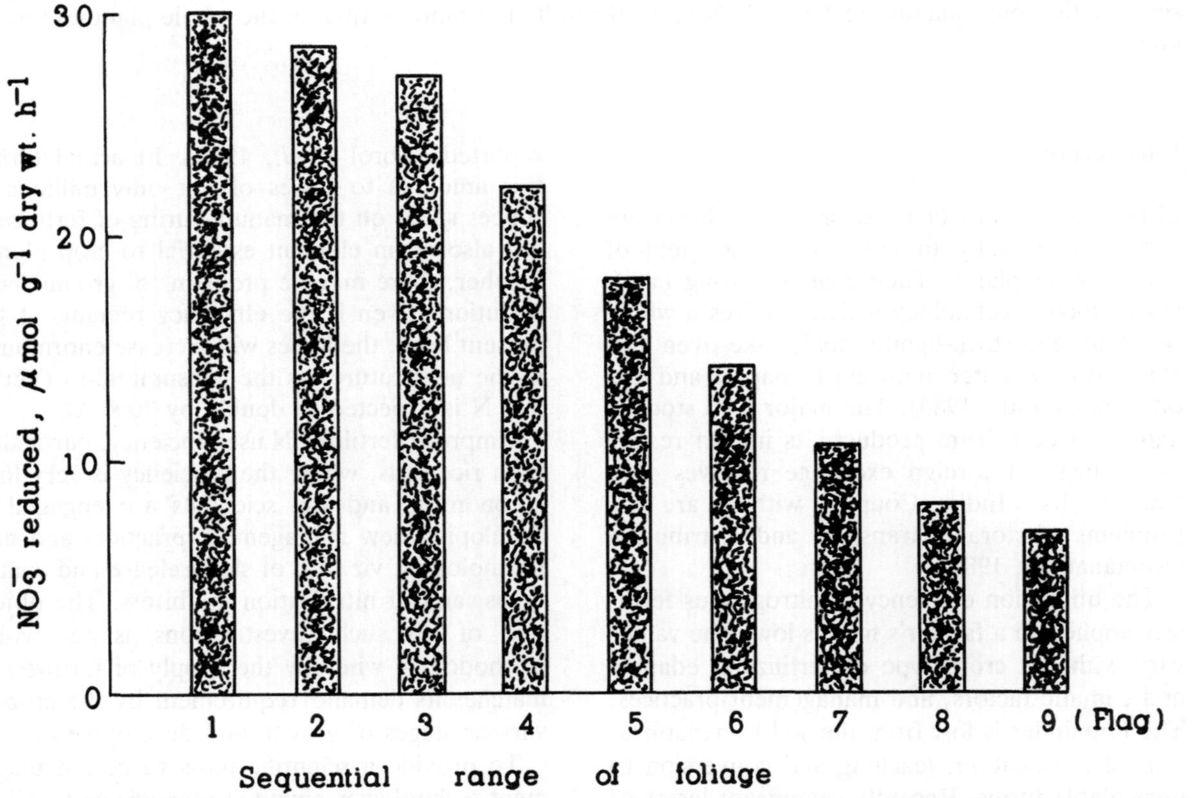

Fig. 1. In vivo nitrate reductase activity (μ mol g^{-1} dry with wt h^{-1}) in the sequential range of foliage of field-grown wheat (*Triticum aestivum* L.) cv HD 2204. Each value is a mean of three replicates. The activity was determined in the fully expanded leaf blades.

makes use of photosynthetically reduced ferredoxin as its electron donor.

For reasons described above, these studies aimed at optimizing fertilizer N utilization, were confined to the above enzymes only. The *in vivo* method for nitrate reduction was followed because it is known to closely approximate the physiological levels of reduction of nitrate (Kaim *et al.*, 1990). The investigations were conducted on field-grown wheat, with nitrogen, phosphorus and potassium supplied as recommended for attaining maximum yield. Fertilizers and irrigation water were supplied following standard practices.

Results and discussion

Pattern of NO_3^- assimilation

Analysis of the above-ground plant parts of field-grown wheat showed that NO_3^- is assimilated mainly by the leaf blades. The leaf sheaths, internodes and reproductive parts, all combined, reduced 25 per cent while the roots contributed approximately 10 per cent. A look at the pattern of nitrate content and *in vivo* NR activity in the successively formed leaf blades revealed that the content/activity is high in the first formed ones and declines in the subsequently formed leaf blades (Fig. 1). This pattern was paralleled by changes in soil nitrate concentration and total nitrate content (See Pokhriyal *et al.*, 1978).

Potential for NO_3^- assimilation

The results of the above analysis suggested the likelihood of NO_3^- shortage being a causative factor in declining assimilatory activity of the upper leaf blades at progressing age. To test this assumption, each of the leaf blades, at and around its fully expanded stage, was excised from the wheat plants and placed in Hoagland's

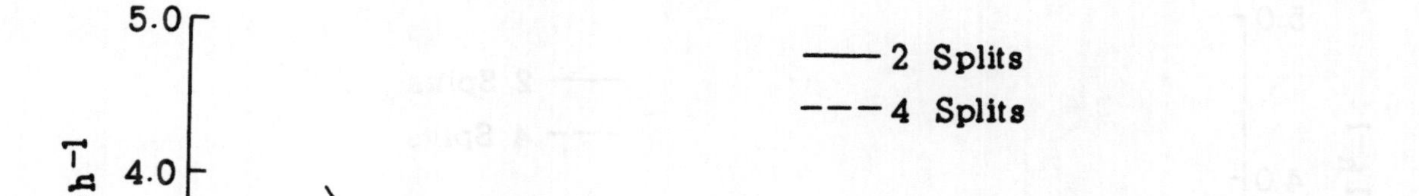

Fig. 2. In vivo nitrate reductase activity (μ mol g^{-1} fr. wt. h^{-1}) in the fully expanded leaf blades (pooled) at different intervals during the ontogeny of field-grown wheat (*Triticum aestivum* L.) cv. HD 2204. Nitrogen was applied in two and four splits. Arrows indicate the time of application (See text for details).

nutrient solution containing 15 m*M* NO_3^-. While there were slight increases in the assimilatory activities of the lower leaf blades, the activities of the upper leaf blades enhanced manifold. An observation of interest was that the level of enhancement was higher in 'high NR' cultivars than in 'low NR cultivars' (For details see Kumar *et al.*, 1981).

If one takes dry weight and the duration for which the leaf blades remain green into consideration, since these are known to be the factors contributing to total NO_3^- reduced, one observes that the upper leaf blades have a vast un-realized potential for nitrate assimilation which can be realized if the substrate *viz*. NO_3^- is available (data not given).

Split application aimed at maximizing NO_3^- assimilation

Application of 120 kg nitrogen per ha in four splits (30 kg N each at 0, 28, 54 and 76 days after sowing) as compared to two splits (60 kg N each at 0 and 28 days after sowing) resulted in enhancement of the NR activity of the upper leaf blades. The magnitude of enhancement was higher in 'high NR' cultivar than in 'low NR' one (Figs. 2 and 3).

Total and grain N yields

Total and grain N yield was higher in 'high NR' cultivars, HD 2177 and HD 2204 as compared to 'low NR' cultivars, Pusa lerma and UP 301 (Table 1). Splitting the N fertilizer dose into four resulted in enhancement in total and grain N yield in both the sets of cultivars. There was little or no enhancement in 'harvest nitrogen index' except in cv HD 2204.

Percentage additional N in grain due to split application varied from 73.8 to 91.3 (Table 2). There was increase in the grain protein percen-

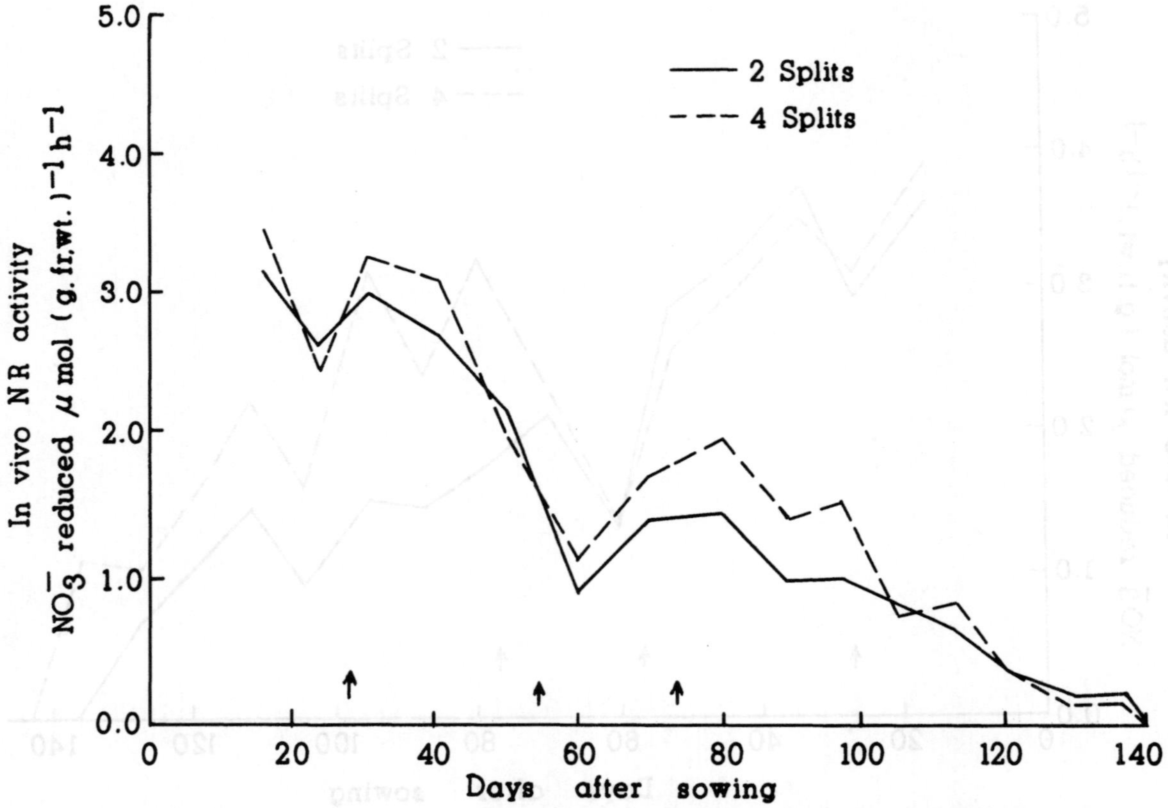

Fig. 3. In vivo nitrate reductase activity (μ mol g^{-1} fr. wt h^{-1}) in the fully expanded leaf blades (pooled) at different intervals during the ontogeny of field-grown wheat (*Triticum aestivum* L.) cv. UP 301. Nitrogen was applied in two and four splits. Arrows indicate the time of application. (See text for details).

Table 1. Effect of splitting a nitrogen fertilizer dose on nitrogen accumulation in whole plant and grains of wheat cultivars differing in NR activity

Cultivar	Frequency of splitting	Nitrogen accumulated in (mg N/plant) as affected by variation in splitting frequency of 120 mg N applied per ha.		
		Whole plant	Grains	Harvest nitrogen index
Low NR				
Pusa lerma	S_2	98.5	78.4	81.4
	S_4	131.2	108.1	80.8
UP 301	S_2	90.0	72.2	80.2
	S_4	107.8	87.3	81.0
High NR				
HD 2177	S_2	122.5	99.1	80.8
	S_4	179.9	141.5	80.9
HD 2204	S_2	122.4	101.0	82.5
	S_4	166.2	141.0	84.5

Table 2. Additional N accumulated as a result of increasing the splitting frequency of the N dose

Cultivar	mg additional N in		% additional N in grain
	whole plant	grains	
Low NR			
Pusa lerma	32.7	29.7	90.8
UP 301	17.8	15.1	84.8
High NR			
HD 2177	57.4	42.4	73.8
HD 2204	43.8	40.0	91.3

Table 3. Effect of splitting frequency of the N fertilizer dose on protein percentage in the grain

Cultivar	Splitting frequency	Grain protein %
Low NR		
Pusa lerma	S_2	10.2
	S_4	10.8
UP 301	S_2	10.1
	S_4	11.3
High NR		
HD 2177	S_2	9.4
	S_4	11.2
HD 2204	S_2	9.6
	S_4	11.3

Table 4. Effect of variation in frequency of splitting the N fertilizer dose on various crop characteristics

Cultivar	Splitting frequency	Grain wt g plant^{-1}	Ears per plant	Grain wt g ear^{-1}	Grains per ear	Spikelets per ear total	Spikelets per ear fertile
Low NR							
Pusa lerma	S_2	4.38	3.36	1.304	38.6	19.7	16.0
	S_4	5.59	3.59	1.557	42.9	20.6	17.9
UP 301	S_2	4.06	3.17	1.281	35.0	21.0	16.8
	S_4	4.42	3.08	1.435	41.4	21.3	18.2
High NR							
HD 2177	S_2	6.01	4.17	1.441	35.6	18.6	14.8
	S_4	7.22	4.55	1.587	38.5	18.7	15.2
HD 2204	S_2	6.04	3.43	1.761	44.7	22.9	17.5
	S_4	7.14	3.48	2.052	54.3	23.4	19.4
Cultivar		0.631	0.284	0.120	2.98	0.51	0.74
N		0.398	NS	0.076	1.88	NS	0.47
Cultivar XN		NS	NS	NS	NS	NS	NS

tage (Table 3). Data on the efect of variation in frequency of N splitting on various crop characteristics are presented in Table 4. Four split applications of N increased the grain weight per plant by virtue of increase in grain weight per ear mediated through increase in grain number per ear. The latter was brought about predominantly by more fertile spikelets at the higher splitting frequency.

Conclusions

Field-based physiological investigations aimed at evolving management technology for optimum fertilizer N use suggest that availability of substrate (NO_3^-) at later stages of growth results in enhancement of nitrate-assimilating activity and higher grain nitrogen yield. The wheat cultivars differ in their potential for NO_3^- reduction and in the capacity to realise the potential under field conditions.

References

Abrol Y P, Kaim M S and Nair T V R 1976 Nitrogen assimilation, its mobilization and accumulation in wheat (*Triticum aestivum* L.) grains. Cereal Res. Comm. 4, 431–440.

Abrol Y P, Kumar P A and Nair T V R 1984 Nitrate uptake, assimilation and grain nitrogen accumulation. Adv. Cereal Sci. and Techn. 6, 1–48.

Anonymous 1980 Fertilizer Production in India. Fertilizer Association of India Publication, New Delhi, 80 p.

Kaim M S, Nair T V R and Abrol Y P 1990 Nitrate uptake, its assimilation and redistribution of reduced nitrogen in uniculm 'Gigas' (*Triticum aestivum* L.) wheat. Proc. Int. Congress Plant Physiology, New Delhi. (*In press*).

Kumar P A, Grover H L and Abrol Y P 1981 Potential for NO_3^- reduction in wheat (*Triticum aestivum* L.). J. Plant Nutr. 3, 843–852.

Pokhriyal T C, Sachdev M S, Grover H L, Arora R P and Y P 1980 Nitrate assimilation in leaf blades of different age in wheat. Physiol. Plant. 48, 477–481.

Ramanathan K V 1980 Fertilizer demand and supply situation in India. Fertilizer News 25, 25–29.

M. L. van Beusichem (Ed.), *Plant nutrition – physiology and applications*, 779–784.

PLSO IPNC588

Inorganic leaf phosphorus as an indicator of phosphorus nutrition in cereals

I. SAARELA
Agricultural Research Centre, SF–31600 Jokioinen, Finland

Key words: *Avena sativa* L., barley, *Hordeum vulgare* L., oats, phosphorus indicators, plant phosphorus test, turnip rape, wheat

Abstract

A plant extraction test which measures the inorganic fraction of phosphorus in fresh leaf samples (Pi) was studied with use made of spring cereals as test crops. Leaf Pi values correlated most closely with relative grain yields of pot-grown barley in mineral soils ($R^2 = 0.89$). For all soils in pot experiments, the R^2 value was 0.59 for barley and 0.58 for oats. Grain yields in field experiments were also significantly correlated with Pi values ($R^2 = 0.28$). Critical Pi values at 95% of maximum yield were 128 mg P per kg fresh weight for pot-grown barley, 95 mg/kg for field-grown barley and 83 mg/kg for field-grown oats. This plant test was found to be more reliable than the acid ammonium acetate soil test for estimating soil phosphorus availability.

Introduction

In Finland as well as in many other countries, the requirement for fertilizer application is usually estimated on the basis of soil tests. The Finnish method which acid ammonium acetate as the extractant (Vuorinen and Mäkitie, 1955) and other tests have appeared to be rather reliable for P (Sippola and Saarela, 1986), but a more accurate diagnostic indicator would improve the economics of fertilization as well as decrease the detrimental nutrient load of surface waters. Attempts have been made to use chemical plant analysis in order to obtain a more accurate diagnosis of P nutrition.

Attempts to assess soil P availability by means of total P concentration in plants have not been very successful, because total P in plants decreases steeply with plant age and is therefore not easy to interpret. A leaf extraction test presented by Bouma and Dowling (1982), Bouma (1983) and Irwing and Bouma (1984) is, according to the authors, independent of plant age. In this method, the inorganic fraction of phosphorus (Pi) in fresh leaf samples is extracted with sulphuric acid and determined colorimetrically. The leaf Pi test was used in spring cereals and compared to the acid ammonium acetate soil test.

Materials and methods

Determination of inorganic leaf P

The analytical method to obtain the Pi values was essentially the same as described by Bouma (1983), but all reagents were added by pipetting. The samples were taken from about 30 cm-high barley, wheat and oat plants at stage 6–7 in Feekes' scale and consisted of the youngest and second youngest fully expanded leaves. Five 2–3 mm long sections of the middle 10–15 cm of ten leaf blades were cut to form the analyzed sample of 400 mg fresh material. Oilseed turnip rape (*Brassica campestris* L.) that occurred in rotation at four sites was sampled at the beginning of flowering. Twenty leaves were randomly collected from the middle section of plants and one half of each was taken to form a composite sample.

The fresh 400-mg leaf samples were each soaked with 0.2 mL of 5 *M* sulphuric acid and macerated with a glass rod for 1–2 min. Then each sample was extracted with 5 mL of distilled water for one minute and filtered, diluted to 50 mL, of which 5 mL was coloured according to the molybdenum blue method using ascorbic acid as the reductant. Citric acid was added 10 min after colouring and the absorbances were measured at 660 nm 30 min later.

The fresh samples were analyzed within five hours after collecting, or stored at +5 °C until the next day. Deep-freezing was found to increase the test values by 30–35%, but the results correlated closely with the values of fresh samples ($r = 0.98$). Drying caused many times larger values, poorly correlated with the ones obtained on fresh samples.

Phosphorus fertilization experiments

Pot experiments

Samples of 21 different soil types were collected from different parts of Finland in 1983. Six of the soils were clays (more than 30% particles smaller than 0.002 mm), 12 coarse-textured mineral soils (silts, loams, sands with less than 30% particles smaller than 0.002 mm) and three organic soils (more than 12% organic C). Each soil was placed in five-liter plastic pots and treated with two amounts of $CaCO_3$ (0 and 4.8 g L^{-1}) and $(NH_4)_2HPO_4$ (0 and 120 mg P L^{-1}) in three replicates and barley was grown on the soils. In 1984 the P application was repeated (0 and 80 mg L^{-1}) and oats was grown.

In the third year (1985) with barley (var. Pomo) grown, a large amount of P (160 mg L^{-1}) was added to the previous P_0 control pots of 17 soils so that P availability did not restrict growth and the residual effects of the applied P (160 + 80 mg L^{-1}) could be tested by means of relative yields. The earlier treatments (0 and 80 mg L^{-1}) were continued in four high-P soils. In 1986, oats were grown using 0 and 80 mg P L^{-1}. Sufficient amounts of N (300 mg L^{-1}), K, Mg, Na, S, Cl, B, Mo, Fe, Mn, Zn and Cu were applied annually. Soils were tested for pH and P and leaves for Pi in 1985 and partially in 1986. Soil characteristics and data on barley are presented in Table 1.

Field experiments

A series of long-term experiments with P rates of 0, 15, 30, 45 and 60 kg/ha placed annually in rows as single superphosphate was conducted with four replicates in 1977–88. Potassium fertilizer was placed separately and a moderate rate of nitrogen (average 80 kg N/ha) was combidrilled with seeds in accordance with the common Finnish practice. The soils were tested for P and pH. Leaf Pi was determined at 18 sites in two years during 1985–86 or 1986–87. The material included 32 yields of spring cereals and four yields of oilseed turnip rape. The relative grain yields for the P_0 control and for the three lowest P rates were calculated by expressing the yield as a percentage of that obtained with the highest P amount (Figs. 3 and 4). Soil and crop data of the field experiments are presented in Tables 2 and 3.

Results and discussion

Pot experiments

The relative grain yields of barley (y) obtained without P application (Fig. 1) were fitted to the logarithmic curve $y = 381.7 \times \log Pi - 1.07 \times$

Table 1. Mean data on soil characteristics and on barley grown in the 1985 pot experiment

Soil group	n	Soil clay (%)	Org. C (%)	Soil pH-H_2O	Soil P (mg L^{-1})	Leaf Pi (mg/kg) Without P	Leaf Pi (mg/kg) With P	Grain yield (g/pot) Without P	Grain yield (g/pot) With P
Unlimed clay soils	6	43.3	3.7	6.0	11.2	137	252	44.1	61.3
Limed clay soils	6	43.3	3.7	7.3	17.8	165	275	55.8	61.8
Unlimed coarse soils	12	21.1	4.7	5.9	11.3	134	278	42.7	55.8
Limed coarse soils	12	21.1	4.7	7.1	17.6	145	253	48.9	56.7
Unlimed organic soils	3	—	23.5	5.4	9.5	143	242	46.6	50.1
Limed organic soils	3	—	23.5	6.1	12.6	117	209	45.3	57.2

Table 2. **Soil pH and acid ammonium acetate extractable P values in field experiments with different P fertilization rates**

Crop and soil	n	Soil pH-H_2O	Extractable P (mg L^{-1}) for different quantities of P applied (kg/ha)				
			0	15	30	45	60
Barley in mineral soils	15	5.8	7.6	9.5	11.4	12.3	14.4
Barley in organic soils	6	4.9	6.7	9.1	10.7	12.7	15.1
Oats in mineral soils	6	5.6	5.7	6.9	7.8	9.4	10.7
Oats in organic soils	2	5.1	5.4	7.1	8.1	9.8	12.0
Wheat in mineral soils	3	6.2	19.2	23.2	24.0	27.3	26.3
Turnip rape in min. soils	4	6.3	15.0	18.4	18.7	22.3	20.4

Table 3. **Inorganic leaf phosphorus and grain yields in the field experiments with different P fertilization rates**

Crop and soil	n	P fertilization rates (kg/ha)										
		0	15	30	45	60	0	15	30	45	60	60
		Leaf Pi (mg/kg)					Relative grain yield					Grain yield (kg/ha)
Barley in mineral soils	15	84	100	121	132	138	80	92	96	100	100	3160
Barley in organic soils	6	99	139	160	178	198	86	95	98	102	100	3280
Oats in mineral soils	6	79	90	96	111	120	90	97	99	100	100	3770
Oats in organic soils	2	103	144	159	189	227	86	100	101	102	100	3370
Wheat in mineral soils	3	72	80	90	95	95	97	98	102	104	100	3350
Turnip rape in min. soils	4	117	176	208	236	240	81	95	98	99	100	2000

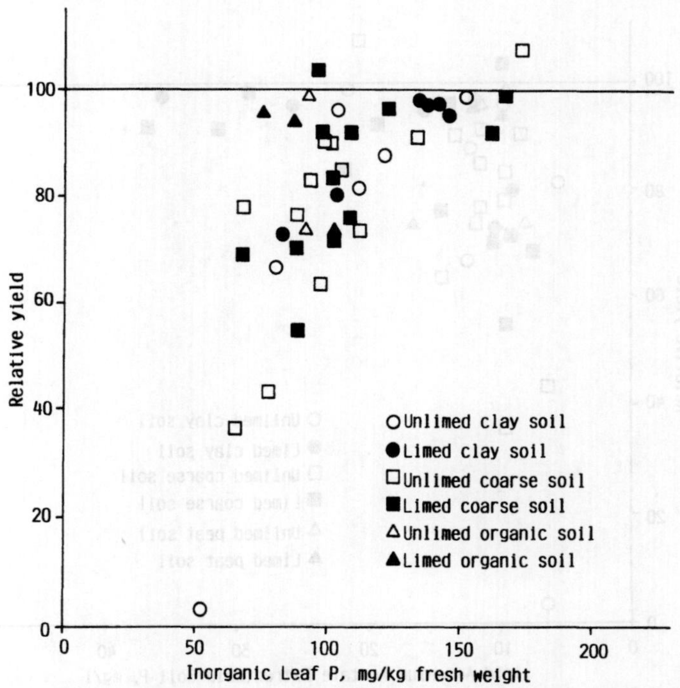

Fig. 1. Relative grain yield of pot-grown barley as a function of leaf Pi values.

Pi − 571.7 (n = 42, R^2 = 0.59). The relationship was also affected by soil organic matter. When the organic C% was inserted in the regression equation, the R^2 value increased to 0.67. In mineral soils with less than 6% organic C the relationship could be expressed as $y = 450.2 \times \log Pi - 1.14 \times Pi - 708.5$ with the multiple determination of $R^2 = 0.89$ (n = 26).

Relative yields were higher in organic soils and in humous mineral soils than in other mineral soils, when Pi values were equal. The influence of organic matter was probably indirect. As a result of its large capacity to hold water and of daily addition of liberal amounts of water to the freely drained pots, as required for soils with smaller water holding capacity, organic matter possibly caused poor aeration and hampered P uptake temporarily prior to sampling, but did not significantly retard growth. The Pi values were relatively low even in those organic soils to which large amounts of P had been applied (Table 1).

The critical Pi value for barley at 95% of maximum yield was 128 mg/kg fresh weight. In pot-grown oats the equation $y = 78.9 \times \log Pi - 53.2$ (n = 18, $R^2 = 0.58$) gave the relative yield of 95% when Pi was 76 mg/kg. Grain yield was more closely correlated with plant Pi than with acid ammonium acetate extractable soil P, as shown in Fig. 2. The R^2 value for soil P was 0.24 and the critical content at 95% was 23.2 mg L^{-1}.

Field experiments

The coefficients of determination (R^2 value) for the relationship between relative grain yields of field-grown crops and their leaf Pi values (Fig. 3 and 4) were 0.28 for all field crops, 0.29 for barley (n = 84), 0.31 for oats (n = 32) and 0.71 for oilseed turnip rape (n = 16). Critical Pi values at 95% of maximum yield were 95 mg/kg fresh weight for barley, 83 for oats and 128 for turnip rape. Wheat did not respond to P application and is therefore not mentioned. Its Pi values ranged from 66 to 100, which means that much lower test values than those earlier obtained in Australia gave maximum yields (Bouma, 1983).

In agreement with earlier studies the results showed that the leaf Pi test is very sensitive to changes in P nutrition (Bouma and Dowling, 1982; Bouma, 1983). Pi is considered to be present in two distinct pools, metabolic Pi in the

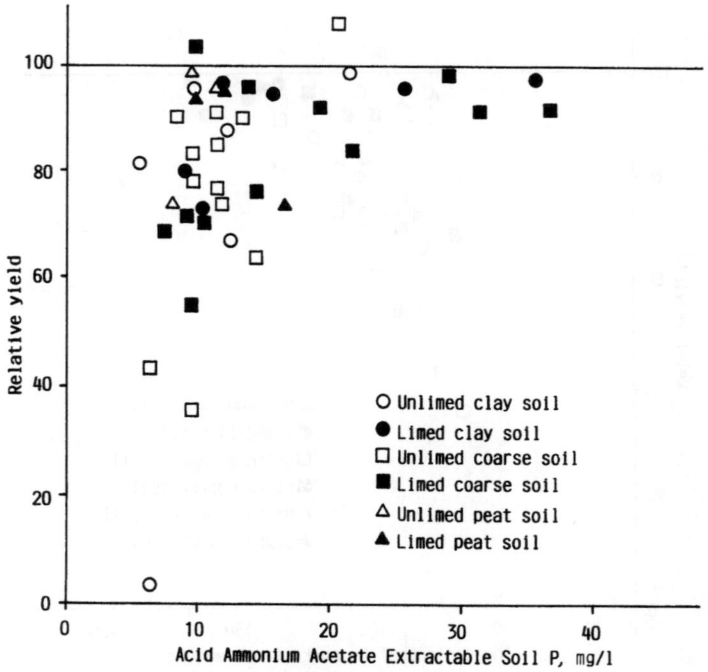

Fig. 2. Relative grain yield of pot-grown barley as a function of soil P test values.

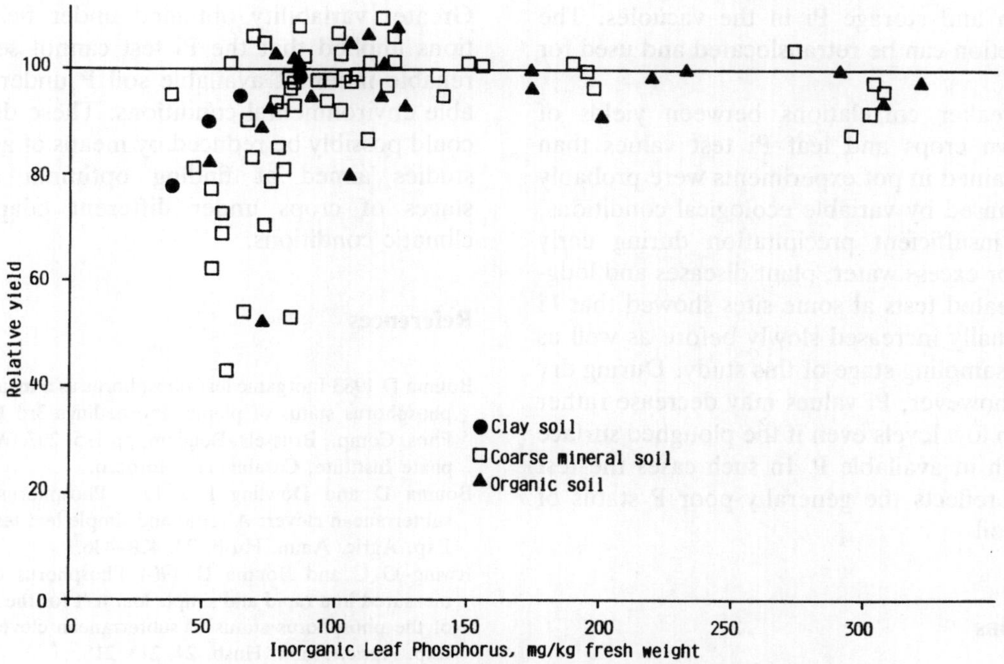

Fig. 3. Relative grain yield of field-grown barley as a function of leaf Pi values.

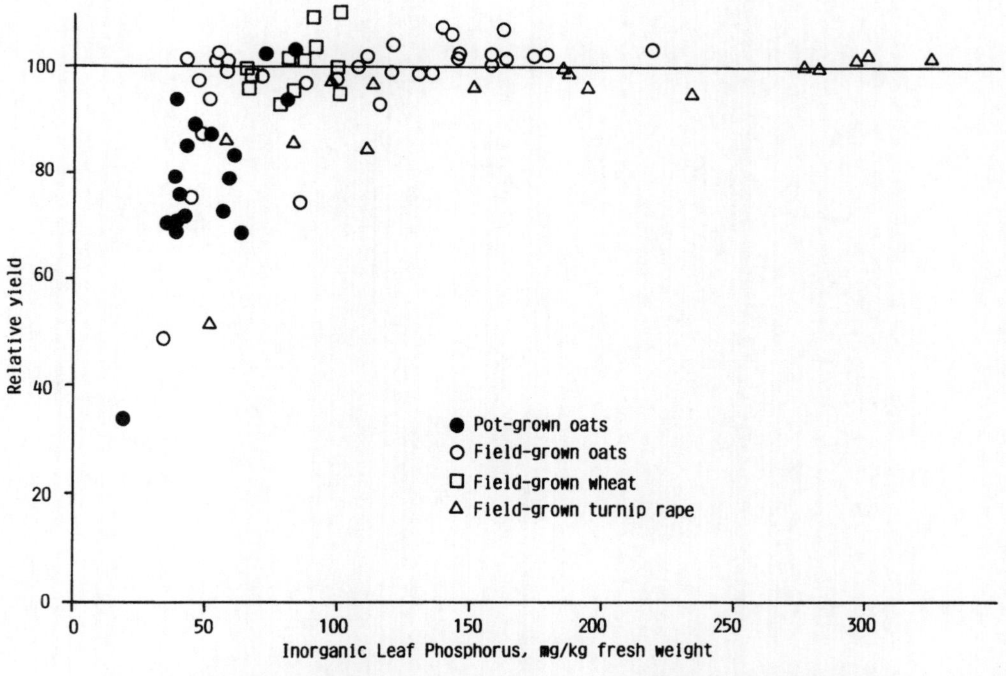

Fig. 4. Relative grain yield of field-grown crops as a function of leaf Pi values.

cytoplasm and storage Pi in the vacuoles. The latter fraction can be retranslocated and used for growth.

The weaker correlations between yields of field-grown crops and leaf Pi test values than those obtained in pot experiments were probably in part caused by variable ecological conditions, such as insufficient precipitation during early summer or excess water, plant diseases and lodging. Repeated tests at some sites showed that Pi values usually increased slowly before as well as after the sampling stage of this study. During dry periods, however, Pi values may decrease rather quickly to low levels even if the ploughed surface soil is rich in available P. In such cases the test probably reflects the generally poor P status of the sub soil.

Conclusions

Close correlations between grain yields and leaf Pi test values in pot-grown barley and oats showed that this method is a sensitive and reliable indicator of the phosphorus status of plants. Greater variability obtained under field conditions implied that the Pi test cannot serve as a reliable index of available soil P under changeable environmental conditions. These difficulties could possibly be reduced by means of additional studies aimed at finding optimum sampling stages of crops under different edaphic and climatic conditions.

References

Bouma D 1983 Inorganic leaf phosphorus as a measure of the phosphorus status of plants. Proceedings 3rd Int. Congr. Phos. Comp., Brussels, Belgium, pp 195–207. World Phosphate Institute, Casablanca, Morocco.

Bouma D and Dowling F J 1982 Phosphorus status of subterranean clover: A rapid and simple leaf test. Aust. J. Exp. Agric. Anim. Husb. 22, 428–436.

Irwing G C and Bouma D 1984 Phosphorus compounds measured in a rapid and simple leaf test for the assessment of the phosphorus status of subterranean clover. Aust. J. Exp. Agric. Anim. Husb. 24, 213–218.

Sippola J and Saarela I 1986 Some extraction methods as indicators of need for phosphorus fertilization. Ann. Agr. Fenn. 25, 265–271.

Vuorinen J and Mäkitie O 1955 The method of soil testing in use in Finland. Agrogeol. Publ. 63, 1–44.

M. L. van Beusichem (Ed.), *Plant nutrition – physiology and applications*, 785–789.

PLSO IPNC184

Potassium deficiency in sunflower (*Helianthus annuus*) is associated with changes in protein profiles: An approach to diagnosis of mineral status

Y. M. HEIMER, A. GOLAN-GOLDHIRSH and S. H. LIPS
The Jacob Blaustein Institute for Desert Research, Ben-Gurion University of the Negev, Sede Boqer Campus, Israel 84990

Key words: *Helianthus annuus* L., potassium deficiency, protein markers, SDS-PAGE, sunflower

Abstract

Limited availability of K^+ by hydroponically-grown sunflower plants was associated with increased abundance of two polypeptides of molecular mass of 60 and 62 KDa detected mainly in young subapical region of the stem. The increased abundance could be detected at least as early as three days after the transfer of plants from complete growth medium (5 m*M* K^+) to media containing suboptimal concentrations (1 m*M* K^+ or less). It is proposed that these polypeptides could be the basis for early diagnosis of K^+ deficiency.

Introduction

K^+ is the most abundant cation in plant cells. Its concentration in plant cells is frequently maintained at 100–200 moles · m^{-3} (Clarkson and Hanson, 1980; Evans and Sorger, 1966). Although K^+ is abundant in most soils, the K^+-containing compounds available for plants, are often deficient in soils (Larson, *et al*., 1985). Consequently, there is increasing use of K^+ fertilizers in agriculture (Pretty and Stungel, 1985). The K^+ requirement for normal plant performance and maximum yield of crops stems from its involvement in several aspects of plant metabolism. K^+ is required in catalytic amounts for activation and stabilization of enzymes and membranes (Clarkson and Hanson, 1980; Evans and Sorger, 1966; Marschner, 1983; Robson and Pitman, 1983; Suelter, 1985; Wyn Jones and Pollard, 1983), protein and starch synthesis (Blevins, 1985; Hsiao, 1976; Wyn Jones and Pollard, 1983) and membrane transport systems (Cheeseman and Hanson, 1980; Hodges, 1976). It is required in larger concentrations for charge balance (Ben-Zioni *et al*., 1971; Clarkson and Hanson, 1980), osmoregulation, turgor generation and volume maintenance (Raschke, 1979), and for the operation of the K^+-shuttle system which mediates transport of nutrients and photosynthates between roots and shoots (Ben-Zioni, Vaadia and Lips, 1971; John M. Cheeseman, personal communication; Lips *et al*., 1987; Touraine *et al*., 1988).

K^+ deficiency in plants brings about changes in metabolic and physiological activities which are manifested in morphological characteristics known as deficiency symptoms such as extensive yellowing of leaves and development of necrotic regions (Clarkson and Hanson, 1980; Evans and Sorger, 1966; Heimer *et al*., unpublished data). The outcome of K^+ deficiency is reduction in vegetative growth and yield and under severe deprivation the plant degenerates (Clarkson and Hanson, 1980; Evans and Sorger, 1966). When the visual symptoms are apparent the damage to the plant is either irreversible or a valuable period of vegetative growth, usually reflected in the final yield, is lost. Thus, a continuous and accurate assessment of K^+ status is required and currently achieved by the following approaches:

1) determination of K^+-compounds available to plants in the soil; 2) chemical analysis of plants material; 3) assay of physiological–biochemical parameters known to correlate with K^+ status of the plant (Bouma, 1983). These procedures are not without drawbacks as outlined in the review by Bouma (1983).

In consideration of the effects of nutrient imbalance, we hypothesized that changes in protein profiles occur upon K^+ deficiency which must ultimately be the basis for the observed changes in growth and metabolism. These changes if detected early and reliably, could provide the basis for molecular diagnosis and allow rational fertilizer management. This is significant with respect to increasing yield of crops, cutting down production cost and reducing the harmful effect on the environment caused by excessive use of fertilizers. Analysis of changes in protein profiles brought about by K^+ deficiency is expected to yield information relevant to other areas: a) the role of K^+ in plant metabolism; and b) the response of plants to nutritional stress analogous to other environmental stresses (Sachs and Ho, 1986) with the ultimate goal of improving deficiency tolerance. We report here our initial observations on the changes in protein profiles triggered by K^+ deficiency in sunflower.

Materials and methods

Sunflower (*Helianthus annuus* L. cv. Saffola) seedlings were grown in Long Ashton nutrient solution (Hewitt, 1952) with various concentrations of K^+ as indicated, under 12 h light (250□E $\cdot$ $m^{-2} \cdot s^{-1}$) at 23°C and 12 h darkness at 18°C with continuous aeration. Organs were quickly removed, weighed, placed in liquid nitrogen and stored at −70°C. Samples were transferred to liquid nitrogen and ground to a fine powder with mortar and pestle. The powder was washed once with cold (−20°C) 80% acetone and 4 times with cold (−20°C) 100% acetone and air dried at room temperature. Proteins were extracted from the dried powder with 1X Laemmli sample buffer (Laemmli, 1970) containing 2 moles $\cdot$ m^{-3} EDTA, 5 moles $\cdot$ m^{-3} ascorbate, 5 moles $\cdot$ m^{-3} $NaHSO_3$ and 10 mmoles $\cdot$ m^{-3} leupeptin. SDS-PAGE was carried-out according to Laemmli (Laemmli, 1970) on 10% polyacrylamide gel. Protein was determined by the Folin phenol reagent (Lowry *et al.*, 1951).

Results and discussion

The protein profiles of extracts taken from plants grown on 5 moles $\cdot$ m^{-3} K^+ and from plants transferred to 0 K^+ and harvested at the time of appearance of the visual symptoms of deficiency (14 days after the transfer) (Evans and Sorger, 1966; Clarkson and Hanson, 1980; Heimer *et al.*, unpublished data) are shown in Figure 1. Changes in the protein profiles could be observed. The most pronounced changes in the aerial parts were the increase in the intensity of two polypeptides with molecular mass of 60 and 62 KDa, as well as decrease in the intensity of two polypeptides with molecular mass of 52 and 54 KDa in plants transferred to 0 K^+. These changes could be clearly observed in young sub-

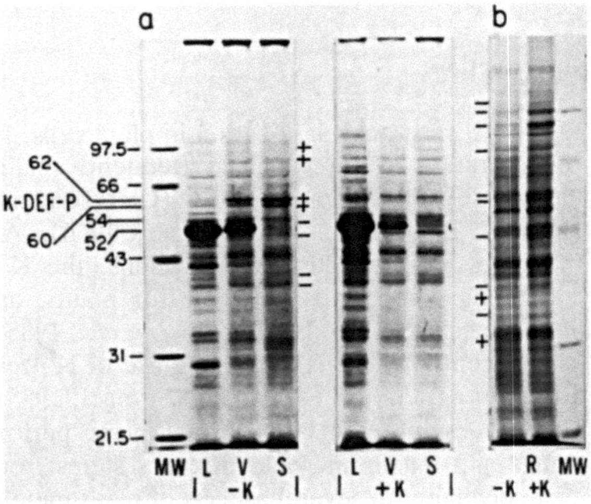

Fig. 1. Sodium dodecylsulfate polyacrylamide gel electrophoresis (SDS-PAGE) of proteins extracted from various organs of sunflower plants grown on K^+ deficient (0 moles $\cdot$ m^{-3}) or K^+ sufficient (5 moles $\cdot$ m^{-3}) media. The nutrient solutions were replaced every 2 days. Plants were harvested when the visual symptoms (necrosis) of K^+ deficiency were observed. The same amounts of proteins were applied to each lane. (+): polypeptides which increased in abundance in organs of deficient plants; (−): polypeptides which decreased in abundance. K-DEF-P designates the putative K^+ deficiency polypeptides. +K = with K^+; −K = without K^+, MW = molecular weight markers in kilodaltons. a) L = leaf, V = veins + petiol, S = stem; b) R = root.

apical regions of stems. The increased abundance of the 60 and 62 KDa polypeptides which we putatively named K-DEF-P for K^+ deficiency proteins, could also be observed in blades and petiols + main veins of young leaves (Fig. 1a). Changes in the protein profiles were also noticed in root extracts where the most pronounced one was the decrease in abundance of the 50 KDa polypeptide in K^+ deficient plants (Fig. 1b). The 60 and 62 KDa polypeptides increased also in hypocotyls of K^+ deficient seedlings grown in the dark (Fig. 2).

The potential use of the K-DEF-P as markers for K^+ deficiency requires their early detectability, at a time when corrective measures can be employed effectively. We have concentrated our initial studies on the K-DEF-P of the stem. The increase in abundance of these two polypeptides as dependent on the concentration of K^+ in the nutrient solution and on the length of time of growth in K^+ deficient media, is summarized in Figure 3. The increase of the K-DEF-P was apparent at least as early as 3 days after the transfer of plants to K^+ deficient nutrient solutions, long before the appearance of the known morphological symptoms (Evans and Sorger, 1966; Clarkson and Hanson, 1980; Heimer *et al.*, unpublished data), usually on the 14th day or more after the transfer. The magnitude of the increase of the abundance of the K-DEF-P, as judged by visual examination of the profiles after staining the SDS gels with Coomassie Brilliant Blue, was similar at all suboptimal concentrations of K^+ used and at all time points examined. It seems as if the response of the treated plants to conditions of reduced K^+ availability was 'all or none'.

The biological significance of the K-DEF-P is not at all clear at this stage. They may reflect impairment of normal development of various

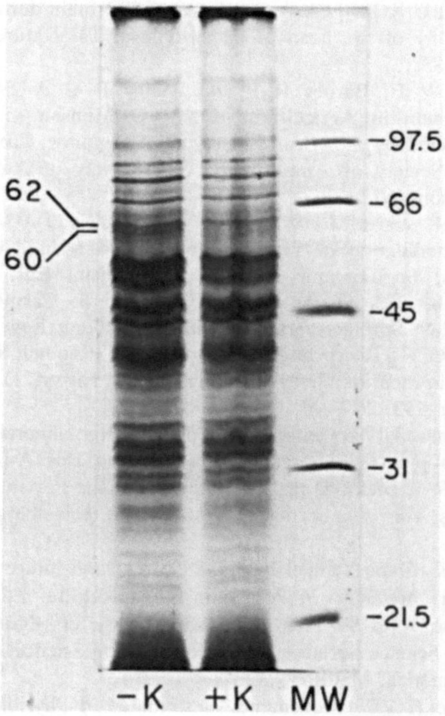

Fig. 2. The appearance of the putative K-DEF-P in hypocotyls of dark grown, K^+ deficient seedlings. Sunflower seedlings were maintained under absolute darkness during germination on vermiculite and subsequent growth on K^+ sufficient or K^+ deficient nutrient solutions. Seedling were harvested under low green light and hypocotyls were removed and placed in liquid nitrogen. Subsequent steps were carried out as described in the legend to Figure 1.

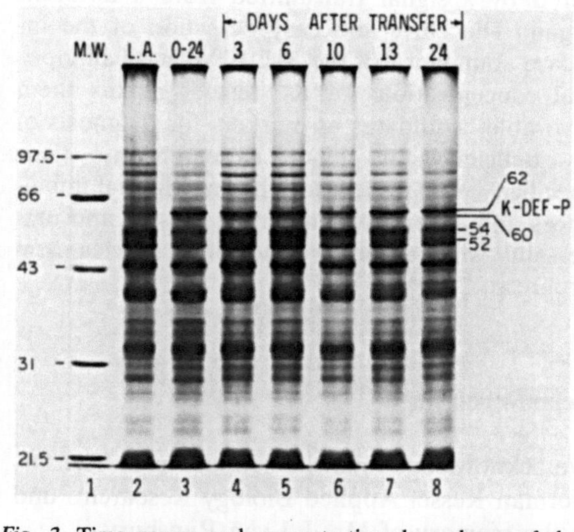

Fig. 3. Time course and concentration dependence of the changes in the abundance of the 60 and 62 KDa polypeptides. Plants were grown for 18 days on K^+ sufficient (5.0 moles · m^{-3}) nutrient solution and then were transferred to nutrient solutions containing 0; 0.01; 0.1 and 1.0 moles · m^{-3} K^+. At time intervals (days) thereafter, plants were harvested. The results shown are from the very young sections of the stem of plants transferred to 1.0 moles · m^{-3} K^+ and are very similar to all other suboptimal concentrations tested. L.A. = Long Ashton nutrient solution (5.0 moles · m^{-3} K^+); 0–24 designates protein profile of extract taken from young sections of the stem 24 days after transfer of sunflower plants to 0 moles · m^{-3} K^+.

organs, mainly stem, brought about by various factors including K^+ deficiency. On the other hand, they may be related more directly to K^+-mediated or K^+-dependent metabolism and acquisition. The increased abundance may indicate a response of the cells to scavenge any K^+ ion available in the ambient, or maintain the highest endogenous concentration possible under conditions of low availability of the cation. In this context the K-DEF-P may be related to K^+ transport or binding. It has been shown recently, that withdrawal of K^+ caused rapid modification in the phosphorylation of plasma membrane associated 100 KDa polypeptide and changes in the profile of soluble proteins (A.D.M. Glass, personal commun. 22, – Plant Physiol. 86: S-477, S-491, 1988). K^+ is known to be translocated between various organs and tissues of the plant (Clarkson and Hanson, 1980; Evans and Sorger, 1966; Touraine *et al.*, 1988). Thus, the K-DEF-P could be increased in response to K^+ status within the responding organ directly, or triggered by a signal transmitted from a different organ. The early and easy detection of the increase abundance of the K-DEF-P at all suboptimal concentrations of K^+ tested, makes them favorable candidates as markers for diagnosis of K^+ deficiency in plants. Research into their specificity with respect to other nutritional imbalances and other environmental stresses and universality with respect to other plant species, is in progress.

Acknowledgement

We acknowledge financial support from The Dr. Herman Kessel Applied Biology Research Fund in the memory of Mr. J J van Ransbury.

References

Ben-Zioni A, Vaadia Y and Lips H S 1971 Nitrate uptake by roots as regulated by nitrate reduction products of the shoot. Physiol. Plant. 24, 288–290.

Blevins D G 1985 Role of potassium in protein metabolism in plants. *In* Potassium in Agriculture. Ed. R D Munson. pp 413–424. American Society of Agronomy, Crop Science Society of America and Soil Science Society of America, Madison, WI.

Bouma D 1983 Diagnosis of mineral deficiencies using plant test. *In* Inorganic Plant Nutrition. Eds. A Läuchli and R L Bieleski. pp 120–146. Encycl. Plant Physiol., New Series, Vol. 15A. Springer-Verlag, Berlin, Heidelberg, New York.

Cheeseman J M and Hanson J B 1980 Does active K^+ influx to root occur? Plant Sci. Lett. 18, 84–87.

Clarkson D T and Hanson J B 1980 The mineral nutrition of higher plants. Annu. Rev. Plant Physiol. 31, 239–298.

Evans H J and Sorger G J 1966 Role of mineral elements with emphasis on the univalent cations. Annu. Rev. Plant Physiol. 17. 47–76.

Hewitt E J 1952 Sand and water culture methods used in the study of plant nutrition. Commonw. Bur. Hortic. Plant Crops (GB), Tech. Commun., 22 East Malling, Kent, UK.

Hodges T K 1976 ATPase associated with membranes of plant cells. *In* Transport in Plants. Eds. U Lüttge and M G Pitman. pp 260–283. Encycl. Plant Physiol., New Series, Vol. 2A, Springer-Verlag, Berlin, Heidelberg, New York.

Hsiao T C 1976 Stomate ion transport. *In* Transport in Plants. Eds. U Lüttge and M G Pitman. pp 195–221. Encycl. Plant Physiol., New Series, Vol. 2B, Springer-Verlag, Berlin, Heidelberg, New York.

Laemmli U K 1970 Cleavage of structural protein during the assembly of the head of bacteriophage T4. Nature 227, 680–685.

Larson W E, Barnes R F and Runge E C A 1985 *In* Potassium in Agriculture. Ed. R D Munson. p. xvii. Published by American Society of Agronomy, Crop Science Society of America and Soil Society of America, Madison, WI.

Lips S H, Soares M I M, Kaiser J J and Lewis O A M 1987 K^+ modulation of nitrogen uptake and assimilation in plants. *In* Inorganic Nitrogen Metabolism. Eds. W R Ullrich, P J Aparicio, P J Syrett and F. Castillo. pp 233–239. Springer-Verlag, Berlin, Heidelberg, New York.

Lowry O H, Rosebrough N, Farr A and Randall R 1951 Protein measurement with Folin phenol reagent. J. Biol. Chem. 193, 265–275.

Marschner H 1983 General introduction to the mineral nutrition of plants. *In* Inorganic Plant Nutrition. Eds. A Läuchli and R L Bieleski. pp 5–60. Encycl. Plant Physiol. New Series, Vol. 15A. Springer-Verlag, Berlin, Heidelberg, New York.

Pretty K M and Stangel P J 1985 Current and future use of world potassium. *In* Potassium in Agriculture. Ed. R D Munson. pp 99–128. American Society of Agronomy, Crop Science Society of America and Soil Science Society of America, Madison, WI.

Raschke K 1979 Movements of stomate. *In* Physiology of Movements. Eds. W Haupt and M E Feinleib. pp 383–441. Encyc. Plant Physiol. New Series, Vol. 7 Springer-Verlag, Berlin, Heidelberg, New York.

Robson A D and Pitman M G 1983 Interactions between nutrients in higher plants. *In* Inorganic Plant Nutrition. Eds. A Läuchli and R L Bieleski. pp 147–180. Encyc. Plant Physiol. New Series, Vol. 15A. Springer-Verlag, Berlin, Heidelberg, New York.

Sachs M M and Ho T-H D 1986 Alteration of gene expression during environmental stress in plants. Annu. Rev. Plant Physiol. 37, 363–376.

Suelter C H 1985 Role of potassium in enzyme catalysis. *In* Potassium in Agriculture. Ed. R D Munson. pp 337–349. American Society of Agronomy, Crop Science Society of America and Soil Science Society of America, Madison, WI.

Touraine B N, Grignon N and Grignon C 1988 Charge balance in NO_3-fed soybean. Estimation of K^+ and carboxylate recirculation. Plant Physiol. 88, 605–612.

Wyn Jones R G and Pollard A 1983 Proteins, enzymes and inorganic ions. *In* Inorganic Plant Nutrition. Eds. A Läuchli and R L Bieleski. pp 528–562. Encycl. Plant Physiol., New Series, Vol. 15B. Springer-Verlag, Berlin, Heidelberg, New York.

M. L. van Beusichem (Ed.), *Plant nutrition – physiology and applications*, 791–795.
 PLSO IPNC086A

Diagnosis of sulfur deficiency in peanut (*Arachis hypogaea*) by plant analysis

N. SUPAKAMNERD, B. DELL[1] and R.W. BELL
School of Biological and Environmental Sciences, Murdoch, University, Murdoch, WA 6150, Australia.
[1]*Corresponding author*

Key words: *Arachis hypogaea* L., deficiency diagnosis, peanut, plant analysis, sulfur

Abstract

Critical values of sulfur (S) concentration in young leaves were established for the diagnosis of S deficiency in peanut by examining the relationship of S concentration in leaves to shoot dry matter (DM) at pegging and pod formation. Plants were grown in pots of a S deficient sand at seven levels of S supply (0, 2, 4, 8, 16, 32 and 64 mg S/3 kg soil) in a glasshouse. Sulfur deficiency symptoms first appeared in young leaves and spread to mature leaves when plants become severely deficient. Sulfur concentrations in leaves, except the young shoot apex, increased with increasing S supply and declined marginally with plant age. Sulfur concentrations in the youngest fully expanded leaf (YFEL) when related to shoot DM gave less variable estimates of the critical S concentration than other leaf categories. The critical S concentration in the YFEL, obtained from hand-fitted curves, was 1.8 and 2.0 mg S/g DM for plants at pegging (R2) and pod formation (R4) growth stages, respectively. The YFEL is recommended as a suitable leaf for deficiency diagnosis in peanut.

Introduction

Peanut (*Arachis hypogaea* L.) is an important economic plant ranking second to soybean as a source of vegetable oil (Bunting *et al.*, 1985). Coarse-textured soils have been favoured for good pod development (Chapin *et al.*, 1975) but some of these leached soils have low fertility. S deficiency in crops is widespread globally (Tisdale *et al.*, 1986) and has been reported in field crops of peanut (Bockelee-Morvan and Martin, 1966; Hago and Salama, 1987). Studies on S nutrition of peanut, however, have been largely concerned with fertilizer rates and little is known about its internal S requirements for growth. The present experiment was undertaken to investigate the relationship within peanut between S supply, shoot growth and S distribution in order to determine the most suitable plant part for the diagnosis of S deficiency and to estimate its critical value.

Materials and methods

A glasshouse pot trial was conducted using a S deficient sand (pH – H_2O 4.5; sulfate extractable in 10 mM $Ca(H_2PO_4)_2$, 8.3 mg S/kg) collected from Badgingarra, Western Australia (Gilbert and Robson, 1984). A randomized block design was used with seven rates of S application (0, 2, 4, 8, 16, 32 and 64 mg S/3 kg soil; referred to in the text as SO, S2, *etc.*), each replicated eight times.

The air-dried soil was sieved (5 mm screen), thoroughly mixed and 3 kg portions placed in polythene-lined, 16.5 cm diameter pots. Basal nutrients were applied in solution to the pots at the following rates (mg/pot): $KHCO_3$ (500), $MgCl_2\,6H_2O$ (107), $MnCl_2\,4H_2O$ (21), $ZnCl_2$ (15), $CuCl_2\,H_2O$ (11), H_3BO_3 (1.3), $Na_2MoO_4\,2H_2O$ (1.07) and $CoCl_2\,7H_2O$ (1.07). After drying, $Ca(H_2PO_4)_2\,2H_2O$ (250 mg/pot) and $CaSO_4\,2H_2O$ were added as powders and the soil was mixed by shaking. Plants were watered daily with deionized water to maintain the soil near field capacity.

Twelve uniform, germinated seedlings of peanut cv. White Spanish were sown (DO) in each pot at a depth of 2 cm and inoculated with *Bradyrhizobium* sp. (*Arachis*) strain NC92. The pots were inserted into water baths which maintained soil temperatures in the range 25–30°. Air temperatures ranged from 20–35°C. Plants were thinned to 4 plants/pot 5 days after emergence, and 30 mg N as NH_4NO_3 was applied in solution to each pot in order to meet the requirement for N before the commencement of N fixation (15 days after sowing).

Harvests, each consisting of four replicates, were taken at pegging (R2: after Boote 1982) and pod formation (R4) growth stages on D43 and D55, respectively. Plants were separated into the shoot apex including the youngest folded leaf (SA), the young open leaf (YOL), the youngest fully expanded leaf (YFEL), the leaf immediately older than the YFEL (YFEL + 1), and the two most basal leaves (L1 + 2). Plant parts were dried overnight at 70°C and ground by hand with a mortar and pestle. Oven dried plant samples were ashed by combustion in an O_2 flask to convert S to the sulfate form using H_2O_2 as the absorption solution (Macdonald, 1961). Sulfate levels in the extracts were determined by standard turbimetric procedures.

Critical concentrations in plant parts at D43 and D55 were obtained by plotting S concentrations in plant parts against shoot dry matter (DM) expressed in g DM/plant. Critical S concentrations were estimated from hand-fitted curves at 90% maximum yield, and from two statistical procedures (Cate and Nelson, 1971; Smith and Dolby, 1977).

Results and discussion

Effects of S supply on symptom development and growth in peanuts

The new growth (YFEL and younger leaves) on SO plants turned pale green at early flowering (D29) as is typical for S deficiency in peanut (Reid and York, 1958) and in other species supplied with inadequate S but an adequate or luxury supply of nitrogen (Gilbert and Robson, 1984). By D43, symptoms were also present on young leaves of S2, S4 and S8 plants, and these had extended to include S16 and S32 plants by D55. As plants became more severely S-deficient, the pale green new growth turned yellow and leaves older than the YFEL became pale green. Hence, by D43 all leaves of SO plants had S deficiency symptoms. Associated symptoms in S-deficient plants included decreased leaf size and shortened internodes on the main stem.

Deficiency symptoms occurred well in advance of a suppression in dry matter (DM). At D43, shoot DM was depressed in plants grown with SO, S2 and S4. By D55, DM was also depressed at S8 (Table 1). Unlike shoot DM, root DM was not depressed in S-deficient plants (data not shown). Nodules were well-developed across all treatments and total N concentrations in the YFEL were in the range regarded by Small and Ohlrogge (1973) as adequate (3.5–4.5% N) in upper stems and leaves of peanut plants at early pegging.

The distribution of S among plant parts

For all leaf categories, S concentrations increased with S supply (Table 2). Increases in S concentration with increasing S supply were greatest in the mature leaves. The YFEL as a typical example averaged 2.4, 1.8 and 1.6 mg S/g DM (D43) in S adequate, marginally deficient and deficient plants, respectively. At each level of S supply, S concentrations in defined leaf

Table 1. Effect of S supply on shoot dry matter (g/plant) at 43 and 55 days after sowing peanut cv. White Spanish on a S deficient soil. Values are means of 4 replicates

S treatment	Days after sowing	
mg S/3 kg soil	43	55
0	1.87	2.18
2	2.05	2.41
4	2.15	2.70
8	2.41	2.91
16	2.60	3.23
32	2.50	3.25
64	2.58	3.22
	0.18	0.29

LSD ($P = 0.05$).

Table 2. Effect of S supply on S concentration (mg S/g DM) within plant parts of peanut cv. White Spanish at 43 and 55 days after sowing. Values are means of 4 replicates

S treatment mg S/3 kg soil	Plant part[a]				
	SA	YOL	YFEL	YFEL + 1	L1 + 2
	43 days after sowing				
0	2.1	1.7	1.7	1.8	1.8
2	2.1	1.7	1.6	2.0	2.1
4	1.9	1.7	1.8	2.1	2.2
8	2.0	1.7	1.8	2.3	2.5
16	2.1	2.0	2.3	2.8	2.6
32	2.5	2.4	2.7	3.1	3.1
64	2.6	2.8	3.0	3.2	3.3
LSD ($P = 0.05$)	0.2	0.4	0.2	0.2	0.2
	55 days after sowing				
0	1.7	1.3	1.8	1.5	1.7
2	2.2	1.3	2.0	1.6	1.7
4	1.6	1.3	1.9	1.5	1.7
8	2.0	1.5	2.0	2.0	2.0
16	2.0	1.6	2.1	2.1	2.3
32	2.0	1.8	2.4	2.6	2.5
64	2.3	2.2	2.6	2.7	2.9
LSD ($P = 0.05$)	0.3	0.2	0.2	0.3	0.3

[a] See Materials and methods for detailed descriptions of plant parts.

categories fell slightly as plants aged. At D43 and D55, S concentrations in the SA (Table 2) were insensitive to increases in the external S supply (S0-S16) and in S-deficient plants, exceeded those in all other leaf categories.

Relationship between S concentrations in leaves and plant growth

At D43 and D55, S concentrations in leaf categories older than the SA were closely related to shoot DM (Fig. 1). The L1 + 2, though similar in its response, was not considered further because it is difficult to identify in the field, can easily be contaminated from the soil and is liable to senescence due to shading in closed canopies.

Critical concentrations varied slightly with the method of estimation (Table 3). Critical concentrations derived from the statistical Cate and Nelson were the same (YFEL + 1) or marginally greater (YOL, YFEL) than those derived by the hand-fitted method. The mathematical models did not fit the data particularly well, with R^2 values <50% on 4 occasions. The two-phase

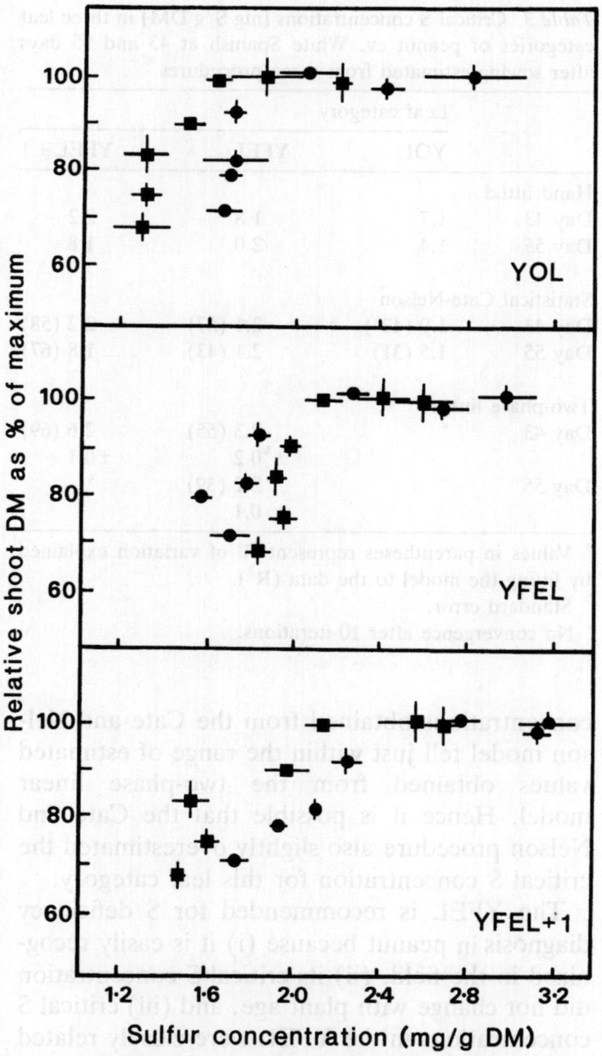

Fig. 1. Relationship between shoot DM and S concentration (mg S/g DM) in three leaf categories of peanut cv. White Spanish harvested 43 (●) and 55 (■) days after sowing. Values are means of 4 replicates with standard error bars.

linear model (Smith and Dolby, 1977) did not fit all the data sets as convergence was not always achieved after iterating the data for 10 cycles. Although deriving the critical concentration by hand-fitted curves involved subjective assessment this procedure appeared to be appropriate for the data set. In contrast, all of the critical S values obtained from the two-phase linear model were those corresponding to S concentrations in plants with adequate S supply (see Fig. 1), and are considered to overestimate the internal S requirement for peanut. For the YFEL, critical

Table 3. Critical S concentrations (mg S/g DM) in three leaf categories of peanut cv. White Spanish at 43 and 55 days after sowing estimated from three procedures

	Leaf category		
	YOL	YFEL	YFEL + 1
Hand fitted			
Day 43	1.7	1.8	2.2
Day 55	1.4	2.0	1.8
Statistical Cate-Nelson			
Day 43	1.9 (42[a])	2.1 (57)	2.2 (58)
Day 55	1.5 (31)	2.1 (43)	1.8 (67)
Two-phase linear			
Day 43	[c]	2.3 (55) ±[b]0.2	2.6 (69) ±0.1
Day 55	[c]	2.2 (39) ± 0.1	[c]

[a] Values in parentheses represent % of variation explained by fitting the model to the data (R^2).
[b] Standard error.
[c] No convergence after 10 iterations.

concentrations obtained from the Cate and Nelson model fell just within the range of estimated values obtained from the two-phase linear model. Hence it is possible that the Cate and Nelson procedure also slightly overestimated the critical S concentration for this leaf category.

The YFEL is recommended for S deficiency diagnosis in peanut because (i) it is easily recognized in the field, (ii) its critical S concentration did not change with plant age, and (iii) critical S concentrations in the YFEL were closely related to S deficiency symptoms. For example, the YFEL of S4 plants became pale green at D43 and its S concentration was the same as the derived critical value (1.8 mg S/g DM). Critical S concentrations in the YFEL + 1 were similar to those in the YFEL, whereas those in the YOL were lower. Thus care should be taken to avoid sampling leaves younger than the YFEL, and if doubt exists about the stage of development of the YFEL, to err in favour of older leaves. The YOL was unsuitable for S deficiency diagnosis in peanut because its S concentrations varied more between replicate pots than those in the YFEL at the pegging harvest (Table 2).

There are few comparable critical values in the recently matured leaves for the diagnosis of S deficiency in legumes. Andrew (1977) established that critical S concentrations in forage legumes ranged from 1.4 to 2.0 mg S/g DM but these values were for whole shoots sampled at early flowering. In subterranean clover during early vegetative growth, critical S concentrations in the YOL were lower (1.4 mg S/g DM: Gilbert and Robson, 1984) than those in the YOL in peanut in the present study.

The critical S concentration in the YFEL established in the present study for the diagnosis of S deficiency was similar to the lower end of the range of S concentrations regarded as adequate for peanut seed yield (2.0–3.0 mg S/kg: Small and Ohlrogge, 1973). The two standards are surprisingly close considering that those proposed by Small and Ohlrogge (1973) were for the prognosis of final seed yield limitations rather than diagnosis, and were for samples collected at early pegging comprising stems and leaves from the upper stem.

Critical values in whole leaves and leaf blades for diagnosis of S deficiency are also reported to decrease with advancing growth stages of plants. For example, critical values in young leaves of subclover decreased with plant age from 1.3 mg S/g DM at D28 to 1.0 mg S/g DM at D42 (Gilbert and Robson, 1984). Though peanuts were not grown through to maturity in the present experiment, this trend, though present in the YOL and YFEL + 1, was not evident in the YFEL. Further work is required to determine critical S concentrations in the YFEL of peanut at later growth stages and to evaluate in the field the critical range found for peanut in the present glasshouse study.

Acknowledgements

The senior author is grateful to the Australian International Development Assistance Bureau for the award of a Postgraduate Studentship.

References

Andrew C S 1977 The effect of sulphur on the growth, sulphur and nitrogen concentrations and critical sulphur concentrations of some tropical and temperate pasture legumes. Aust. J. Agric. Res. 28, 807–820.

Bockelee-Morvan A and Martin G 1966 Sulphur requirements of peanuts. Sulphur Agric. 12, 2–8.

Boote K J 1982 Growth stages of peanut (*Arachis hypogaea* L.). Peanut Sci. 9, 35–40.

Bunting A H, Gibbons R W and Wynne J C 1985 Groundnut (*Arachis hypogaea* L.). *In* Grain Legume Crops. Eds. R J Summerfield and E H Roberts. Collins, London.

Cate Jr R B and Nelson L A 1971 A simple statistical procedure for partitioning soil-test correlation data into two classes. Soil Sci. Soc. Am. Proc. 35, 658–660.

Chapin J C, Gray C and Anderson W B 1975 Nutritional requirements. *In* Plant Production in Texas. pp 22–25. Texas Agric. Exp. Stat. R.M. 3.

Gilbert M A and Robson A D 1984 Sulphur nutrition of temperate pasture species. II. A comparison of subterranean clover cultivars, medics and grasses. Aust. J. Agric. Res. 35, 389–398.

Hago T M and Salama M A 1987 The effects of elemental sulphur on flowering and pod maturation in groundnut (*Arachis hypogaea* L.) under irrigation. Fert. Res. 13, 71–76.

Macdonald A M G 1961 The oxygen flask method: A review. The Analyst 86, 3–12.

Small Jr H G and Ohlrogge A J 1973 Plant analysis as an aid in fertilizing soybeans and peanuts. *In* Soil Testing and Plant Analysis. Eds L M Walsh and J D Beaton. pp 315–327. Soil Sci. Soc. Am., Inc., Madison, WI.

Smith F W and Dolby G R 1977 Derivation of diagnostic indices for assessing the sulphur status of *Panicum maximum* var. Trichoglume. Commun. Soil Sci. Plant Anal. 8, 221–240.

Tisdale S L, Reneau Jr R B and Platou J S 1986 Atlas of sulfur deficiencies. *In* Sulfur in Agriculture. Ed. M A Tabatabai. pp 295–322. Am. Soc. Agron., Madison, WI.

M. L. van Beusichem (Ed.), *Plant nutrition – physiology and applications*, 797–803.

PLSO IPNC356

The diagnosis of manganese deficiency in barley (*Hordeum vulgare*)

N.E. LONGNECKER[1] and R.D. GRAHAM
Department of Agronomy, Waite Agricultural Research Institute, Glen Osmond, S.A., Australia.
[1] *Present address: Department of Soil Science and Plant Nutrition, The University of Western Australia, Nedlands, W.A., Australia*

Key words: barley, chemical analysis, diagnosis, fluorescence, *Hordeum vulgare* L., manganese deficiency

Abstract

Manganese (Mn) deficiency can be diagnosed by analysis of leaf tissue, changes in chlorophyll 'a' fluorescence and visual symptoms. The youngest expanded leaf blade (YEB) is the tissue of choice for chemical analysis of Mn and for fluorescence measurements. Some practical concerns of using YEBs were investigated.

YEBs were collected from main stems and from tillers and analyzed for Mn concentration. The Mn concentration in the YEB from the main stem was not different from the YEB from tillers in Mn-deficient or adequate plants. Thus, YEBs can be collected from any stem for Mn analysis.

The first leaf should not be used for diagnosis of adequacy of the growth medium. It consistently had a higher Mn concentration than YEBs harvested during mid and late tillering in both −Mn and +Mn-treated plants. The first YEB is probably influenced by seed Mn content either directly or through an effect on root growth. The first YEB from Mn-deficient barley grown from seed with higher Mn content had a higher Mn concentration than the first YEB of barley from low Mn seed.

Visual symptoms of Mn deficiency begin on the youngest leaves but necrosis can take time to develop and thus be worse on the next older leaf. We tested whether or not change in chlorophyll 'a' fluorescence (Fo/Fv) due to Mn deficiency was greater on the YEB or YEB + 1. There was no appreciable difference in the fluorescence measurements from the YEB or YEB + 1. Increased fluorescence worked well to detect Mn deficiency over a range of barley cultivars with varying severity of Mn deficiency as measured by growth depression and degree of visual symptoms.

Introduction

In screening for Mn efficiency in barley (Longnecker *et al.*, 1990), we required a quantitative, non-destructive measurement to distinguish between different levels of Mn deficiency stress. The youngest expanded leaf blade with ligule showing (YEB) is the tissue of choice for diagnosis of Mn deficiency in cereals by plant analysis (Hannam and Ohki, 1988). In the routine use of YEBs in our laboratory for diagnosis using chemical analysis of plant tissue and chlorophyll 'a' fluorescence, a number of practical concerns arose which we investigated. The results are reported here.

In sampling field experiments, we frequently bring vegetative samples in from the field and remove YEBs in the laboratory for plant analysis. In this case, one cannot distinguish between the YEBs from the main stem or from tillers. We conducted an experiment to test the hypothesis that the Mn concentration in YEBs from the main stem does not differ from that in tillers.

The first leaf to emerge is taken as a YEB when experiments are sampled soon after sowing. There was some concern that this leaf might not be representative of the Mn availability in the growth medium because of mobilization of the seed reserves to that leaf. For example, Tiffin and Chaney (1973) have shown that iron reserves in soybean cotyledons were readily mobilized to the first leaf. Therefore, we compared Mn concentration in the first expanded leaf with YEBs collected at early and mid-tillering from plants grown in soil to which Mn was only added before sowing.

Fluorescence of YEBs can be used for diagnosis of Mn deficiency since Mn deficiency causes disruption of photosynthesis (Terry and Ulrich, 1974) and a characteristic increase in chlorophyll 'a' fluorescence (Kreidemann *et al.*, 1985; Kriedemann *et al.*, 1989). Visual symptoms of Mn deficiency occur on the younger leaves and YEBs are usually selected for measurements of fluorescence (Hannam *et al.*, 1987). If the plant continues to grow while Mn deficiency persists, necrosis can develop and symptoms can appear worse on the next older leaf (YEB + 1). We thus hypothesized that the increase of chlorophyll 'a' fluorescence caused by Mn deficiency would be greater on the YEB + 1 than on the YEB and could be useful for distinguishing between plants with different levels of Mn deficiency stress.

Materials and methods

Soil was collected from Wangary, South Australia which has a history of Mn deficiency. The soil is approximately 80% $CaCO_3$ with a pH = 8.5 (1:5 H_2O) and was air-dried, sieved and stored for later use in growth cabinet experiments. Plants were grown at 15/10°C, with 10 hour days. Soil was incubated before sowing as described in Uren *et al.* (1988). Time is given as days (D) from sowing in all experiments, with date of sowing designated D0. Data were analyzed statistically using standard analysis of variance methods.

Experiment A

Ten replicate pots were grown with either no added Mn (−Mn) or 100 mg Mn kg^{-1} soil added as $MnSO_4$ (+Mn). Seeds of two barley (*Hordeum vulgare* L.) cultivars, Weeah and Clipper, and an advanced breeder's line (obtained from Dr. D.H.B. Sparrow), WI 2585, were obtained from Wangary, South Australia and from Urrbrae. Seed Mn contents are given in Table 1. Seed from Wangary had lower Mn content than seed from Urrbrae and the two are designated low and high Mn seed, respectively. Pots contained 160 g of incubated soil and were sown with two pregerminated seeds. Basal nutrients were added as in Longnecker *et al.* (1989) except that 102 mg kg^{-1} soil of KH_2PO_4 was added in this experiment. Nitrogen was added as KNO_3 (30 mg pot^{-1}) on both D14 and D24.

We have previously shown that Mn treatment did not affect the rate of leaf emergence. By D11, the first leaves of all treatments had fully emerged and were harvested. At early tillering (D28), five replicate pots were harvested and YEBs from main stems and from tillers were collected. At mid-tillering (D41), YEBs from main stems and from tillers of the other five replicates were collected. YEBs were dried in a forced air oven at 70°C for at least 24 hours, were weighed and digested in concentrated HNO_3. They were analyzed for Mn concentration using an inductively-coupled plasma emission spectrometer.

Plants were scored for Mn deficiency symptoms on D11, D28 and D41 with 1 = healthy, dark green plants, 2 = chlorotic plants, 3 = chlorotic with some necrotic spotting and/or in-

Table 1. Manganese content and concentration in seed of two barley varieties and one breeders' line obtained from two sites

Genotype	Mn concentration ($\mu g\ g^{-1}$)		Mn content (μg $seed^{-1}$)	
	Wangary	Urrbrae	Wangary	Urrbrae
WI 2585	4.0	14.1	0.19	0.65
Clipper	6.1	19.8	0.27	0.80
Weeah	5.4	21.2	0.26	0.97

terveinal chlorosis, 4 = severe chlorosis and necrosis and 5 = apparently dead growing points (youngest leaves dry and shrivelled).

Experiment B

Five replicate pots of ten barley cultivars and breeders' lines were grown with no added Mn (−Mn) or 100 mg Mn kg^{-1} soil (+Mn) as described in Longnecker *et al.* (1990). Plants were scored for Mn deficiency symptoms as described above on D29. Chlorophyll 'a' fluorescence was measured as in Kriedemann *et al.* (1985) at the base of the YEB and the YEB + 1. The YEBs and YEB + 1 were harvested, dried and analyzed for Mn concentration as above.

Results

There was a good correlation between the Mn concentration of YEBs from the main stem and from the tillers ($r^2 = 0.732$; Fig. 1). Analysis of variance showed that there was no significant effect of which stem the YEB is collected from on the Mn concentration of that YEB.

The Mn concentration of the YEBs harvested on D11, D28 and D41 from plants grown from seed with low Mn content are given in Table 2. The Mn concentration in the first emerged leaf (YEB on D11) was higher than the Mn concentration of YEBs harvested later in the experiment, for both soil Mn treatments. The Mn concentration of all of the YEBs from −Mn plants were below the critical Mn concentration given for barley (11 mg kg^{-1} DM, Hannam *et al.*, 1987; and 12 mg kg^{-1} DM, Reuter and Robinson, 1986). There were no significant differences in Mn concentration among WI 2585, Clipper and Weeah in the later harvests. However, WI 2585 developed more severe visual symptoms of Mn deficiency than Clipper and the visual symptoms of Weeah and were the least severe.

In the −Mn treatment, the Mn concentration of the first leaf of plants grown from high Mn seed was higher than that of plants grown from low Mn seed (compare Tables 2 and 3). Additionally, the Mn concentrations of the first expanded leaves of −Mn plants were all above the critical levels for barley. By 28 D, these Mn concentrations had dropped to values similar to those from plants grown from seed with low Mn content. Once again, the Mn concentration of the first leaf was higher in all treatments than the Mn concentration of the YEBs collected later.

Because all of the barley varieties and breed-

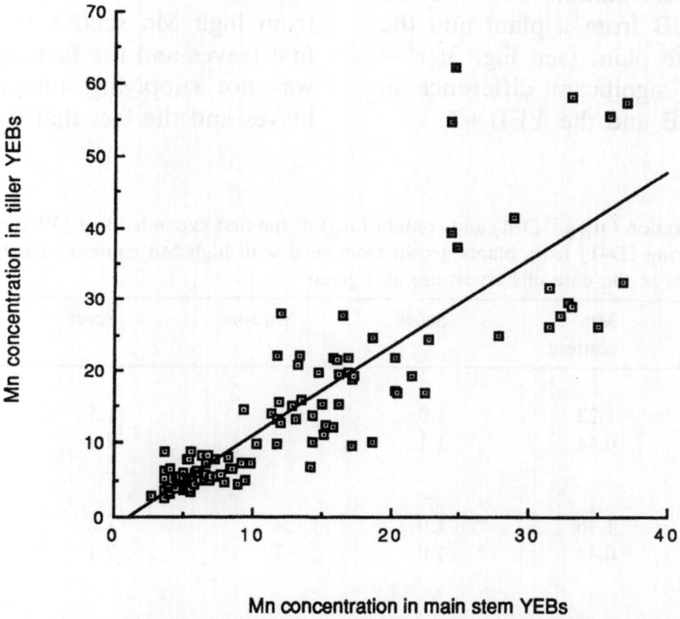

Fig. 1. The relationship between the Mn concentration (μg g^{-1} DM) of YEBs from the main stem and the Mn concentration of YEBs from tillers of the same plants ($r^2 = 0.732$).

Table 2. Manganese concentration ($\mu g\,g^{-1}$ DM) and content (μg) of the first expanded leaf (YEB at D11) and YEBs harvested at early (D28) and mid-tillering (D41) from plants grown from seed with low Mn content receiving two Mn treatments. Mn deficiency scores of the plants at the date of harvest are also given

Treatment	D11			D28		D41	
	Mn conc.	Mn content	Score	Mn conc	Score	Mn conc	Score
WI 2585							
+Mn	34	1.08	1.1	17	1.2	16	1.5
−Mn	7	0.17	1.8	5	2.8	4	4.2
Clipper							
+Mn	34	1.11	1.0	33	1.1	14	1.4
−Mn	10	0.23	1.1	6	2.4	6	3.3
Weeah							
+Mn	47	1.72	1.0	29	1.5	12	1.7
−Mn	10	0.35	1.0	6	2.3	5	2.8

ers' lines used in Experiment B had the same relationship between fluorescence and leaf Mn concentration, they were bulked for presentation. There was a good correlation between increased fluorescence and low leaf Mn concentration. Almost all of the leaves with Mn concentration lower than 10 mg kg^{-1} DM had a higher chlorophyll 'a' fluorescence (Fo/Fv) than those with a Mn concentration of 10 or more (see Fig. 2).

There was a good correlation between the fluorescence of the YEB from a plant and the YEB + 1 from the same plant (see Fig. 3; $r^2 = 0.903$). There was no significant difference in fluorescence of the YEB and the YEB + 1.

Discussion

Plant analysis

These data show that YEBs from any stem are suitable for chemical analysis of tissue Mn concentration. The Mn concentration of YEBs from main stems and tillers did not differ.

If the first leaf is the tissue sampled, the use of plant analysis for detection of Mn-deficient growing conditions can be misleading. Plants grown from high Mn seed had adequate Mn in their first leaves and the fact that the growth medium was not supplying adequate Mn in their first leaves and the fact that the growth medium was

Table 3. Manganese concentration ($\mu g\,g^{-1}$ DM) and content (μg) of the first expanded leaf (YEB at D11) and YEBs harvested at early (D28) and mid-tillering (D41) from plants grown from seed with high Mn content receiving two Mn treatments. Mn deficiency scores of the plants at the date of harvest are also given

Treatment	Mn conc.	Mn content	Score	Mn conc.	Score	Mn conc.	Score
WI 2585							
+Mn	29	1.23	1.0	18	1.5	15	1.8
−Mn	12	0.44	1.1	5	2.8	5	4.5
Clipper							
+Mn	51	1.96	1.0	24	1.2	14	1.4
−Mn	14	0.44	1.0	7	2.1	7	2.5
Weeah							
+Mn	46	2.26	1.0	29	1.7	18	1.8
−Mn	16	0.65	1.0	8	1.8	7	2.3

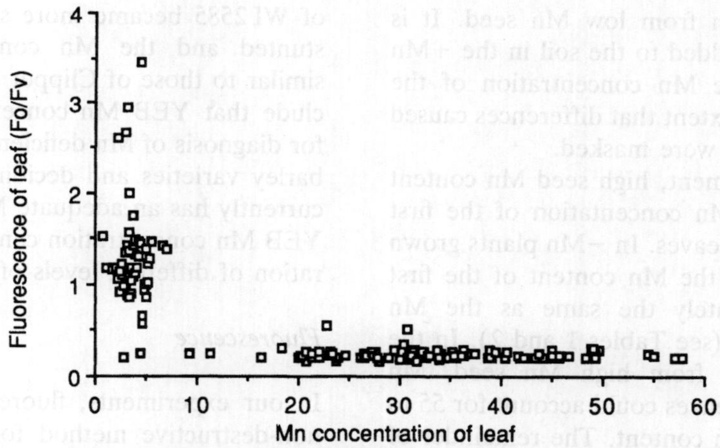

Fig. 2. The relationship between the chlorophyll 'a' fluorescence (Fo/Fv) and the Mn concentration ($\mu g\, g^{-1}$ DM) of the same leaf.

not supplying adequate Mn to prevent deficiency could not be assessed based on the plant tissue test. A correct assessment would have been made for plants grown from low Mn seed based on chemical analysis of the first leaf.

The Mn concentration of the first leaf was higher than that of YEBs collected at subsequent harvests in all treatments. This is a function both of the seed reserves and of the availability of soil Mn. Reuter and Alston (1975) have shown that Mn is readily oxidized in calcareous soils. It is likely that Mn^{2+} added to the soil in the +Mn treatments was being oxidized and rendered unavailable for plant growth since the soil used in these experiments was highly calcareous. The higher availability of soil Mn early in the experiment is illustrated by the comparable Mn concentration in the first leaves of plants grown from low or high Mn seed which received the +Mn treatment. This contrasts with the −Mn treatments in which the first leaf of plants grown from higher Mn seed had a higher concentration than

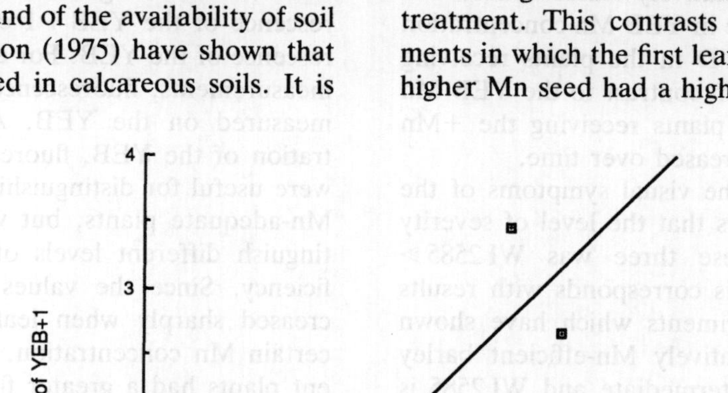

Fig. 3. The relationship between the chlorophyll 'a' fluorescence (Fo/Fv) of YEBs and that of YEB + 1 from the same plants ($r^2 = 0.903$).

that of plants grown from low Mn seed. It is likely that the Mn added to the soil in the +Mn treatment raised the Mn concentration of the first leaf to such an extent that differences caused by seed Mn content were masked.

In the −Mn treatment, high seed Mn content resulted in higher Mn concentation of the first leaf but not of later leaves. In −Mn plants grown from low Mn seed, the Mn content of the first leaf was approximately the same as the Mn content of the seed (see Tables 1 and 2). In the −Mn plants grown from high Mn seed, Mn content of the first leaves could account for 55 to 68% of the seed Mn content. The remainder of the Mn from the seed is likely to have been in roots while some could possibly remain in the seed coat. The similar Mn content of the seed and of the first leaf of the −Mn, low Mn seed plants is surprising since there was likely to also be Mn present in the root systems.

In order for any of the seed Mn to be used for root growth, some of the Mn in the first leaf must have been supplied by root uptake from the soil. The Mn availability of the soil in the −Mn treatments appears relatively stable. There was no significant decrease in YEB Mn concentration between 22 D and 41 D in the plants receiving the −Mn treatments in contrast to the YEB Mn concentration of the plants receiving the +Mn treatments which decreased over time.

A comparison of the visual symptoms of the three genotypes shows that the level of severity of deficiency of these three was WI 2585 ≫ Clipper > Weeah. This corresponds with results of many other experiments which have shown that Weeah is a relatively Mn-efficient barley variety, Clipper is intermediate and WI 2585 is highly susceptible to Mn deficiency (Graham *et al.*, 1983; Longnecker *et al.*, 1990; Sparrow *et al.*, 1983). That one could visually detect differences between the deficiency symptoms of the plants while the Mn concentrations were not significantly different corresponds to results from other experiments (Longnecker *et al.*, 1990). The Mn concentration of YEBs can be used to separate Mn-deficient plants from Mn-adequate plants but cannot necessarily separate Mn-deficient plants with different levels of severity of the stress (as measured by symptoms and growth depression). A likely explanation is that as the Mn deficiency of WI 2585 became more severe its growth was stunted and the Mn concentration remained similar to those of Clipper and Weeah. We conclude that YEB Mn concentration can be used for diagnosis of Mn deficiency in a wide range of barley varieties and deciding whether the plant currently has an adequate Mn supply. However, YEB Mn concentration cannot be used for separation of different levels of severity of stress.

Fluorescence

In our experiments, fluorescence was a useful, non-destructive method for diagnosing Mn deficiency. We obtained a similar relationship between chlorophyll 'a' fluorescence and leaf Mn concentration as did Hannam *et al.* (1987). Those workers defined a critical level of 14 mg kg^{-1} DM for the Mn concentration below which fluorescence would increase. We did not have enough data points in this experiment at the area of the curve needed to accurately define the critical level, but our data are consistent with that value.

There was no significant difference in the fluorescence of the YEB + 1 compared to the fluorescence of the YEB. For consistency with other measurements, fluorescence should probably be measured on the YEB. As with Mn concentration of the YEB, fluorescence measurements were useful for distinguishing Mn-deficient from Mn-adequate plants, but were not able to distinguish different levels of severity of Mn deficiency. Since the values for fluorescence increased sharply when leaf tissue fell below a certain Mn concentration, almost all Mn-deficient plants had a greater fluorescence and there was not a continued increase for plants that had more severe symptoms or greater growth depression.

The fluorescence measurement is useful in diagnosing Mn deficiency because it measures a perturbation in the electron transport of photosystem II (Anderson and Pyliotis, 1969; Spencer and Possingham, 1960). Apparently, photosynthesis is impaired at a higher tissue Mn concentration than is growth (Hannam *et al.*, 1987; Nable *et al.*, 1984; Ohki, 1985). The effect of Mn deficiency on photosynthesis is reversible upon resupply of Mn to the deficient leaves (Campbell and Nable, 1988; Simpson and Robinson, 1984). It is likely that there would be different levels of recovery of severely Mn-deficient tissue and moderately-deficient tissue upon resupply of Mn. It is possible that fluorescence after resupply of Mn may be useful for distinguishing different levels of severity of stress. This remains to be tested.

References

Anderson J M and Pyliotis N A 1969 Studies with manganese deficient spinach chloroplasts. Biochim. Biophys. Acta 189, 280–293.

Campbell L and Nable R O 1988 Physiological functions of manganese in plants. *In* Manganese in Soils and Plants. Eds. R D Graham, R J Hannam and N C Uren. pp 139–154. Kluwer Academic Publishers, Dordrecht, The Netherlands.

Graham R D 1988 Genotypic differences in tolerance to manganese deficiency. *In* Manganese in Soils and Plants. Eds. R D Graham, R J Hannam and N C Uren. pp 261–276. Kluwer Academic Publishers, Dordrecht, The Netherlands.

Graham R D, Davies W J, Sparrow D H B and Ascher J S 1983 Tolerance of barley and other cereals to manganese-deficient calcareous soils of South Australia. *In* Genetic Aspects of Plant Nutrition. Eds. M R Saric and B C Loughman. pp 339–345. Martinus Nijhoff/Dr W Junk, The Hague.

Hannam R J, Riggs J L and Graham R D 1988 The critical concentration of manganese in barley. J. Plant Nutr. 10, 2039–2048.

Hannam R J and Ohki K 1988 Detection of manganese deficiency and toxicity in plants. *In* Manganese in Soils and Plants. Eds. R D Graham, R J Hannam and N C Uren. pp 243–260. Kluwer Academic Publishers, Dordrecht, The Netherlands.

Kriedemann P E, Graham R D and Wiskich J T 1985 Photosynthetic dysfunction and *in vivo* changes in chlorophyll 'a' fluorescence from manganese-deficient wheat leaves. Aust. J. Agric. Res. 36, 157–169.

Longnecker N E, Graham R D, McCarthy K W, Sparrow D H B and Egan J P 1990 Screening for manganese efficiency in barley (*Hordeum vulgare* L.). *In* Genetic Aspects of Plant Mineral Nutrition. Eds N El Bassam, M Dambroth and B C Loughman. pp 273–280. Kluwer Academic Publishers, Dordrecht, The Netherlands.

Nable R O, Bar-Akiva A and Loneragan J F 1984 Functional manganese requirement and its use as a critical value for diagnosis of manganese deficiency in subterranean clover (*Trifolium subterraneum* L., c.v. Seaton Park). Ann. Bot. 54, 39–49.

Ohki K 1985 Manganese deficiency and toxicity effects on photosynthesis, chlorophyll, and transpiration in wheat. Crop Sci. 25, 187–191.

Reuter D J and Robinson J B 1986 Plant Analysis: An Interpretation Manual. Inkata Press, Melbourne, 218 p.

Reuter D J and Alston A M 1975 Immobilization of divalent manganese in calcareous soil. J. Aust. Inst. Agric. Sci. 41, 61–62.

Simpson D J and Robinson S P 1984 Freeze-fracture ultrastructure of the thylakoid membranes in chloroplasts from manganese-deficient plants. Plant Physiol. 74, 735–741.

Sparrow D H B, Graham R D, Davies W J and Ascher J S 1983 Genetics of tolerance of barleys to manganese deficiency. pp 66–70. Proc. Aust. Plant Breed. Conf. Adelaide.

Spencer D and Possingham J V 1960 The effect of nutrient deficiencies on the Hill reaction of isolated chloroplasts from tomato. Aust. J. Biol. Sci. 13, 441–455.

Terry N and Ulrich A 1974 Photosynthetic and respiratory CO_2 exchange of sugar beet leaves as influenced by manganese deficiency. Crop Sci. 14, 502–504.

Tiffin L O and Chaney R L 1973 Translocation of iron from soybean cotyledons. Plant Physiol. 52, 393–396.

Uren N C, Asher C J and Longnecker N E 1988 Techniques for research of manganese in soil plant systems. *In* Manganese in Soils and Plants. Eds. R D Graham, R J Hannam and N C Uren. pp 309–333. Kluwer Academic Publishers, Dordrecht, The Netherlands.

M. L. van Beusichem (Ed.), *Plant nutrition – physiology and applications*, 805–808.

PLSO IPNC526

Iron(II) determination in leaves of strawberry (*Fragaria vesca*)

M. MANZANARES, J.J. LUCENA and A. GÁRATE[1]
Departamento de Química Agrícola. Universidad Autónoma de Madrid, E-28049 Madrid, Spain.
[1]*Corresponding author.*

Key words: *Fragaria vesca*, iron indexes, iron(II)

Abstract

Iron total content in leaves may not reflect the Fe nutritional status of a plant. In our work Fe(II) concentration in strawberry leaves is determined as an index of the iron nutrition of the plant. Two methods are tested using o-phenantroline (o-Ph) and HCl 1 *M* as extractants. Fe assessment in the extracts is achieved by both colorimetry (510 nm) and by Atomic Absorption spectrophotometry. Results obtained in the HCl solutions were abnormally high. Previous washing of samples with chloroform partially solved colour interference of organic compounds. No Fe was detected at the organic washing. An acetate-acetic buffer at pH 3 must be included in the o-Ph extracting solution. Strawberry leaves with different degree of iron chlorosis were submitted to the o-Ph extraction. Measured Fe(II) concentration does not seem to be well correlated with chlorotic symptoms and with other indexes as Fe/Mn ratio in leaves of strawberry plants.

Introduction

The concentration of an element in plant material is normally a reliable index of its nutritional status. However total concentration of Fe in leaves may not reflect the iron nutritional situation of the plant. Green leaves of plants grown in hydroponics under experimental conditions normally present higher Fe concentration than chlorotic leaves (Lucena *et al.*, 1989; Pierson and Clark, 1984). Nevertheless, no significant differences in iron concentration between chlorotic and non chlorotic leaves are observed at field conditions where iron chlorosis is mainly produced by large bicarbonate concentrations in the medium (Abadía *et al.*, 1984). Therefore other indexes have been proposed to study and diagnose iron nutrition (Bar-Akiva, 1969; DeKock, 1981; Hernando and Casado, 1973). Several authors have suggested the use of a fraction of the total iron also called "active Fe" as a better parameter than total iron concentration to diagnose iron chlorosis. This fraction has been closely related to Fe^{2+}. "Active iron" is defined as the portion of the leaf Fe that is available for metabolic reactions or for its incorporation into molecular structures (Pierson and Clark, 1984). Extractants as o-phenanthroline, HCl 1 *M*, HCL 6 *M*, 2,2'-bipyridil, EDTA, batophenanthroline, PDTS, etc, have been proposed to obtain this iron fraction (Rao *et al.*, 1987; Takkar and Kaur, 1984).

In this paper Fe^{2+} concentration in strawberry leaf is determined using two extractants: o-phenantroline and HCl 1 *M*. Katyal and Sharma (1980) employed o-phenantroline (o-Ph) as extractant due its high specificity for Fe^{2+}. Fe(II) assessed with this method has been well related with Fe-chlorosis in different plant species (Abadía *et al*, 1984; Katyal and Sharma, 1984). HCl 1 *M* is also used as extractant. Takkar and Kaur (1984) found that this extractant gave satisfactory results, and is cheaper than o-phenantroline. The application of both extraction methods to strawberry plants grown under different experimental conditions is tested to determine their applicability to diagnose the different iron status of the plants.

Materials and methods

Fe(II) extraction and analysis

Homogenized leaf samples of a strawberry crop (*Fragaria vesca*, cv. Douglas) developed under controlled conditions were collected and then freezed to −18°C. In a first trial, washed, chopped leaf tissue was extracted with 1.5% o-phenantroline adjusted to pH 3.0 with HCl (Katyal and Sharma, 1980) or extracted with HCl 1 *M* (Takkar and Kaur, 1984). The sample:extractant ratio was 1:10 (w:v). Iron(II) assessment was achieved by direct colorimetry at 510 nm in the o-phenantroline extracts using a Beckmann ACTA CIII spectrophotometer and after adding 1 mL o-phenantroline solution when HCL 1 *M* was the extractant. Total Fe in the extracts was also determined in a Perkin Elmer 4000 atomic absorption spectrophotometer, using air-acetylene flame. Five replicates were carried out.

Although pH was initially adjusted in the o-phenantroline extractants, the final pH of the extracts was always larger than 3 (6–7). In a second trial, leaf tissues were extracted with distilled water, HCL 1 *M* and different sodium acetate-acetic acid buffer solutions at pH 3.6, 4.0, 4.6, 5.0 and 5.6, to test the effect of the pH of the Fe(II) extracts. Absorbencies before and after adding o-phenantroline were measured at 510 nm.

In the last trial chopped fresh leaves were initially washed with organic solvents for 30 minutes with occasional shaking before the Fe^{2+} extraction. Chloroform or ether were tested as organic solvents. The Fe^{2+} extraction was carried out on the plant material with o-phenantroline according to the method of Katyal and Sharma (1980). Colorimetric and A.A. determinations were made as mentioned before.

In the second and third trials three repetitions were carried out, and analysis of variance was used to determine the main statistical differences.

Plant material

Strawberry plants were grown under different Fe treatments in two types of conditions. In the first, plants were grown in sand culture and in a controlled growth cabinet. Other plants were grown on an aerated nutrient solution in an experimental greenhouse. Different Fe-chelates and dosages were assayed in order to obtain several degrees of chlorosis. Plant material was sampled at different growth stages and total micronutrient concentration was determined by A.A. spectrophotometry. Also Fe(II) was extracted and measured by the o-Ph method as described in this paper.

To assess chlorotic levels of the samples Fe/Mn ratio was used. This ratio has been shown as a good index in determinate iron chlorosis for this type of biological assays (Lucena *et al.*, 1989).

Results and discussion

Fe(II) extraction and analysis

Iron(II) concentration extracted by o-phenantroline and HCl 1 *M* from 5 replicates are shown in Table 1. The higher values colorimetrically measured were obtained after extraction with HCl 1 *M*. However, Fe concentrations assessed by A.A. in the same HCl 1 *M* samples were about 10 times lower than the colorimetric ones, suggesting that in HCl extracts some colored compounds are present, which would increase values estimated by colorimetry. The same trend was observed in the o-phenantroline extracts, although differences are smaller than with the HCl 1 *M* method.

Plant samples were extracted in distilled water and HCl 1 *M* and absorbencies at 510 nm assessed without adding o-phenantroline to de-

Table 1. Fe concentration (μg/g) in strawberry leaf extracts obtained with o-phenantroline and 1 *M* HCl measured both colorimetrically and by A.A.

Replications	Fe-colorimetry		Fe-AA	
	o-ph	HCl(1 *M*)	o-ph	HCl(1 *M*)
1	3.9	259	3.1	20
2	3.8	295	2.7	20
3	4.7	271	3.0	23
4	3.9	222	3.4	19
5	4.1	263	3.0	20
Avg.	4.1	263	3.0	20
C.V.(%)	9	10	8	7

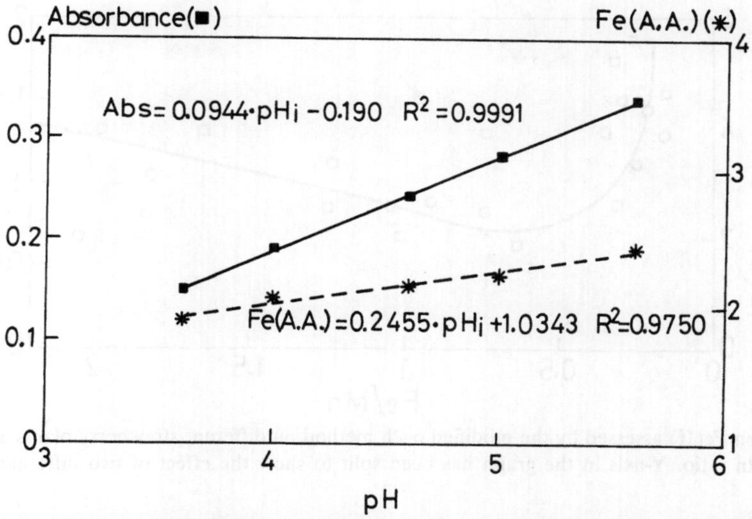

Fig. 1. Effect of pH on the Fe(II) extraction with o-Ph and determination by colorimetry and A.A. Samples are average of five replicates.

termine the amount of colour from sources different from the Fe(II)-o-Ph complex. Results indicated that interferences were higher in water (0.343 absorbance units) than in the acid solution (0.054 a.u.).

Thus the colored interferences in strawberry leaf extracts seem to be pH dependent. Measurements at 510 nm in extracts carried out at pH 3.6, 4.0, 4.6, 5.0 and 5.6 showed higher absorbencies at higher pH (Fig. 1). Iron measured by A.A. in the extracts also increases as the pH increases, but to a lower rate (Fig. 1). Results suggest that the pH increase would cause either a greater extraction of colored compounds or a higher color development of the Fe(II) complex. So the pH control in the Fe^{2+} extraction from our plant tissues seems therefore to be more critical than in grass plants as studied by Katyal and Sharma (1980). The pH control by a buffer solution could improve these methods.

Also, attempts to avoid interferences of organic compounds both naturally present or formed during the extraction procedure were accomplished. Leaf tissue was washed with chloroform or ether before the F^{2+} extraction. Samples previously washed with ether did not prevent interferences of organic compounds (data not shown). Concentration of the metal in the chloroform extracts was negligible. Fe(II) extracted with o-Ph from the washed samples and colorimetrically measured was lower than the Fe(II) estimated from tissues without a previous $CHCl_3$ washing (Table 2).

Table 2. Fe^{2+} (μg/g) in o-phenantroline extracts with and without previous $CHCl_3$ washing and colorimetrically measured

Replications	With washing	Without washing
1	3.52	5.06
2	3.23	7.07
3	3.23	6.11
Avg.	3.37	6.08
C.V.(%)	5	16

Thus, the following modifications to the o-Ph method of Katyal and Sharma (1980) are suggested for Fe(II) determination in strawberry leaf:

- Washing of leaf tissue for 30 minutes with chloroform before the extraction with o-Ph.
- Preparation of the extracting solution at pH = 3.0 in buffer solution, i.e. acetic-acetate.
- Fe assessment in the o-Ph extracts by atomic absorption.

Relation between Fe(II) and chlorosis in strawberry plants

Figure 2 shows a plot of Fe(II) determined by the modified o-Ph method versus Fe/Mn ratio obtained from strawberry plants. The two different experimental conditions in which plants were grown are presented at the same figure. Fe extracted by the o-Ph method has been positively

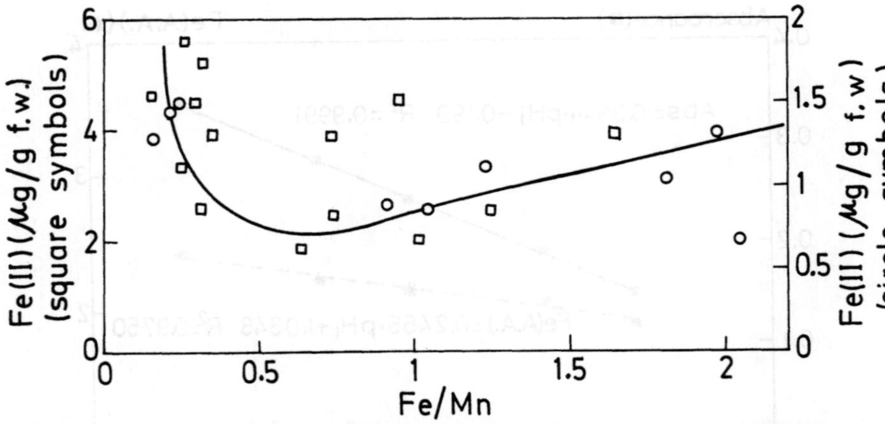

Fig. 2. Relation between Fe(II) assessed by the modified o-Ph method on different strawberry plants and the degree of chlorosis expressed by the Fe/Mn ratio. Y-axis in the graph has been split to show the effect of two different experimental conditions.

correlated with the degree of chlorosis in other plant species (Abadía *et al.*, 1984). In our case this occurs only when the Fe/Mn ratio is greater than approximately 0.5 and leaves were green or dark green. Then the use of the Fe(II) method in strawberry plants as an index of the iron nutrition would be limited to non-chlorotic leaves. Consequently, in our opinion this method is not suitable to screen iron chlorosis in strawberry plants.

The line shows that in chlorotic plants (yellow leaves) Fe(II) extracted with o-Ph decreases as Fe/Mn ratio increases (degree of chlorosis decreases). One possible explanation for the raise of Fe(II) as Fe stress increases is that when strawberry plants are affected by chlorosis, a greater proportion of the total iron is put in active form as an attempt to solve the deficiency effects. But other explanations can also suit in this case, since in severe deficiency plants generally are intensely altered.

We can conclude that improvements have been made to the o-Ph method that need to be tested in other cultures. From our results, Fe(II) determination in strawberry leaves does not offer a satisfactory information of the degree of chlorosis.

References

Abadía J, Monge E, Montañés L and Heras L 1984 Extraction of iron from plant leaves by Fe(II) chelators. J. Plant Nutr. 7, 777–784.

Bar-Akiva A 1969 The use of the activity of metalloenzyme systems for the appraisal of nutritional requirements of citrus trees. Proc. I Inter. Citrus Symposium. Vol. III. Ed. H.D. Chapman. University of California. Riverside pp 1551–1557.

DeKock P C 1981 Iron nutrition under conditions of stress. J. Plant Nutr. 3, 513–521.

Hernando V and Casado M 1974 Clorosis férrica inducida por Mn en plantas de fresón. Anal. Edaf. Agrob. 32, 13–24.

Katyal J C and Sharma B D 1980 Some modification in the analysis to resolve iron chlorosis. Plant and Soil 55, 104–119.

Katyal J C and Sharma B D 1984 Some modification in the assay of Fe^{2+} in 1–10, o-phenantroline extracts of fresh plant tissues. Plant and Soil 79, 449–450.

Lucena J J, Gárate A, Ramón A M and Manzanares M 1989 Iron nutrition of a hydroponic strawberry culture supplied with different Fe chelates. Plant and Soil 123, 9–15.

Pierson E E and Clark R B 1984. Ferrous iron determination in plant tissue. J. Plant Nutr. 7, 107–116.

Rao R C, Sahrawat K L and Burford J R 1987 Diagnosis of iron deficiency in groundnut. Plant and Soil 97, 353–359.

Takkar P N and Kaur N P 1984 HCl method for Fe^{2+} estimation to resolve iron chlorosis in plant. J. Plant Nutr. 3, 561–578.

M. L. van Beusichem (Ed.), *Plant nutrition – physiology and applications*, 809–812.

PLSO IPNC122

An expert system for diagnosing Brussels sprout (*Brassica oleracea*) disorders in the UK

C. PARKER and A. SCAIFE[1]
AFRC Institute of Horticultural Research, Wellesbourne, Warwick CV35 9EF, UK. [1]*Corresponding author*

Key words: *Brassica oleracea* L., computer, disease, deficiency, expert system, pest, pollution

Abstract

An expert system is a computerised knowledge base encapsulating expertise elicited from human experts in a fairly small and clearly-defined field. It is intended to bring expertise in that field quickly, cheaply, and to a wider public than is possible with human experts. 'Sprout-Doctor' is a program, written in the expert system shell 'Crystal', which contains information supplied by about ten pathologists/entomologists/mineral nutrition experts at Wellesbourne. It runs on an IBM or compatible PC, and aims to diagnose all disorders, pests, and diseases of Brussels sprouts which occur in the U.K. In field tests it has aroused considerable interest: however it seems advisable to subject it to further testing and modification if it is to produce a satisfactory success rate.

Introduction

The diagnosis of crop disorders (by which we shall mean diseases, pests, mineral deficiencies and toxicities, physiological disorders, and phytotoxic effects of herbicides and pollutants, etc) by farmers can sometimes be slow and require the intervention of advisers who in turn may need to call for the help of specialists. If the specialist (*e.g.* a pathologist) fails to arrive at a diagnosis it may be necessary to back-track and call in a different specialist (such as a mineral nutrition expert). Since most disorders tend to spread rapidly, these delays can have serious consequences for the crop. Thus it is desirable that the farmer himself should be able to identify and treat as many as possible of the disorders likely to afflict his crops.

The aim is to bring expertise to its site of use, namely the farm, cheaply and effectively. A new approach to this problem involves computers and 'artificial intelligence': specifically, an 'expert system' designed to hold all the necessary knowledge to diagnose any of the disorders likely to occur.

The objectives of the current research were twofold:
(1) Scientific: i.e. is it feasible to create an expert system capable of diagnosing most disorders, including the rare ones which cause most difficulties? (2) Commercial: can such an exercise repay the investment?

The reader may wonder why a system concerned to a large extent with pests and diseases should appear in the proceedings of a plant nutrition colloquium. The answer is that the whole point of any expert system is to approach the problem from the point of view of a layman, and unless there is some obvious way of distinguishing mineral disorders from all other disorders (which is not the case) then it is necessary to put all possible disorders into the system.

France has probably invested more in this approach to crop diagnosis than any other country so far: expert systems have been written for some 15 crops, (Andro *et al.*, 1987) and these have been made available to growers via the 'Minitel' system of micro-computer terminals which are widely distributed in that country.

By contrast, very little such work has been

done in the U.K. In fact, to the best of our knowledge this is the only 'completed' British expert system for crop disorder diagnosis at the time of writing. It has taken the senior author (a psychologist/knowledge engineer) about one year to produce, drawing on advice from several specialists at IHR Wellesbourne.

Methods

An expert system 'shell' known as Crystal, running on an IBM compatible PC, was used to develop the system. This shell allows the formulation of rules, such as 'if spots are target-like, then disease is ring-spot'. These rules become the permanent 'knowledge-base' which is to be confronted with the 'facts' about each disorder as it is encountered. These facts are established by the presentation of 'multiple-choice menus' from which the user selects one or more symptoms which are apparent on the crop. On arriving at a diagnosis, (of which there may be several possibilities) the user can move to descriptions of the diseases etc diagnosed, to see which is the most likely. 71 disorders are covered by the program, which apart from diagnosing the problem, also contains official recommendations for treatment. A print-out is obtainable from any 'session' showing (a) the symptoms entered by the user; (b) those used to diagnose the disorder; and (c) the various possible diagnoses, with comments such as 'common' or 'rare' to assist the user to decide between them.

The rules were formulated after interviewing about eight specialists at Wellesbourne who have experience of Brussels sprout disorders. During the interviews, techniques borrowed from psychology, such as the 'card sort' and repertory grid technique were employed. In the card sort, the expert is shown three cards from a pack of cards each bearing the name of a disorder with which that expert is familiar. He is then asked to say how two of the disorders resemble each other and differ from the third. The aim is to gain an understanding of the variables the expert used to distinguish between the various disorders. The repertory grid is one in which the disorders are listed on one axis and the variables used to distinguish them on the other: ticks indicate whether a disorder exhibits a particular variable (*i.e.* symptom) and thus draw attention to cases where two or more disorders show the same symptom and therefore require additional features to distinguish between them. The variables, *e.g.* part of plant affected, colour of spot *etc.*, were used to build the rules and also to organize them in the most efficient way. For example, diseases and disorders are gathered, among other things, according to the plant parts which they affect and the age of the plants affected. The computer has only to search for a diagnosis under the appropriate headings thus cutting search time considerably.

It became apparent that whilst the presence of a single symptom might be enough to suggest a diagnosis for some disorders, that symptom could not be relied upon to appear without fail. To overcome this problem, the majority of disorders in Sprout Doctor are represented by more than one rule, the simplest of which allows the diagnosis to be made swiftly on the basis of obvious symptoms, whereas the complex rules are used where such symptoms are not present.

The knowledge acquisition process continued throughout the development of the program. Prior to coding, the experts were shown paper representations of the rules elicited by the interviews and further information and corrections were added. Later on, additions to the rule base were made after comment from experts running the program and sessions using field examples. A portable computer was then purchased (Amstrad PPC640DD) and used to test the system in two widely-separated areas of Brussels sprout production – Bedfordshire in East/Central England and Preston in Northwest England. As a result of these visits further modifications were made to the program.

For the Preston test, which was done very late in the growing season (13 Feb 1989), the ADAS national Brussels sprout specialist collected and laid out plants exhibiting six disorders for which he knew the diagnosis, and Sprout-Doctor was used by a complete novice to such problems to arrive at a diagnosis.

Results of the Preston visit

These are shown in Table 1. Roughly speaking, one could say that of the six problems, two (Nos

Table 1. Results of the Preston test of 'Sprout-Doctor'

Disorder as diagnosed by the human expert	Diagnosis according to 'Sprout-Doctor'
1. Light leaf spot (*Pyrenopeziza brassicae*)	i. Dark leaf spot i. (*Alternaria brassicicola*) ii. Light leaf spot iii. Air pollution
2. Slug damage on buttons	i. Large White caterpillar ii. Cabbage moth
3. Rat damage to stem	i. Common canker (*Leptosphaerella maculans*) ii. Boron deficiency
4. Ring Spot (*Mycosphaerella brassicicola*)	i. Dark leaf spot
5. White Blister (*Albugo candida*)	i. White Blister ii. Grey mould (*Botrytis cinerea*)
5. Pigeon damage	i. Pigeon damage

5 and 6) were correctly diagnosed, three (2, 3, 4) incorrectly, and one (1) semi-correctly (in that the correct answer was Sprout-Doctor's second choice). Problem 3 (rat damage) was unknown to us at that time and has now been included. Problem 2 (slug grazing) was missed because no slime trails were visible, as then required by Sprout-Doctor for a 'slug' diagnosis. (It appears that these may not always be left, or else they may disappear).

Discussion

The essential question is whether a computer-based expert system such as this is preferable to other ways of presenting the information, notably books. The computer has the advantage of being able to trace through a very elaborate set of rules very quickly. A book, on the other hand is (1) usable by more people than a computer; (2) more portable-especially to the field, where it should be used; (3) able to present clear colour pictures at a price which is not yet possible with computers. (See Andro (1987) for a discussion of the use of video-disks). Pictures are very desirable for diagnostic purposes. A very valuable aspect of expert systems is that they remind the user of the existence of disorders that he/she may have overlooked: but that is also true of books, provided that they are sufficiently comprehensive. The computerised system is also more readily updated than a book, especially if it is held on a single central computer with a network of users.

The lessons learned from this exercise are:

(1) If the objective is to create a commercially viable expert system, it is important to identify a suitable subject area. One requirement is a large potential market of computer literate people who are themselves inexpert in the field concerned but who are willing to pay to make use of the expertise. In our case it may prove that there are too few such potential customers, because most sprout growers normally have access to human experts, are themselves very knowledgeable about Brussels sprouts but are often unfamiliar with computers.

(2) Correct diagnosis by an expert system is heavily dependent on subtleties of wording which should be perfected by repeated testing. When consulting a human expert, numerous possible misunderstandings are clarified by a constant two-way flow of information which cannot possibly be anticipated or built into a computerised system. In particular, one should note that whereas a particular symptom may indicate a certain disorder, the lack of that symptom may not indicate the absence of the disorder. For example, slime trials indicate presence of slugs; but absence of slime trails does not imply absence of slugs. This same stricture would presumably apply to any complex diagnostic key, even if printed in a conventional book.

(3) The system should be designed to detect and diagnose disorders at the earliest possible stage in order to prevent their spread. (In fact some users would have liked a system based on weather etc to predict the likelihood of an outbreak of a disorder). However it should also recognize the advanced stages or aftermath of problems (such as the leaf distortion left by aphids) so that such problems may be anticipated in future seasons.

(4) Particular attention must be paid to finding ways of distinguishing between disorders having very similar symptoms.

(5) It may happen that the human experts from whose advice the knowledge base is con-

structed are themselves uncertain about the diagnosis: in such cases one cannot expect the computer to succeed. It may indicate the need for more research on the problem concerned, or perhaps merely to find a better expert!

(6) It is desirable to be able to take the expert system to the field in order to refer to the crop whilst answering the questions: furthermore one really needs a hard disk computer with 640 K of memory to cope with most expert systems. Therefore a fairly expensive portable machine is required with the capability of running for several hours away from a mains power source. At present, these considerations clearly favour a conventional book. However, as more people become computer-literate and as computers become smaller, more portable, and better able to display colour images, this situation could change.

It is proposed to continue testing and refining the system by giving it to selected users (growers and advisers) before attempting to sell it.

Acknowledgements

We are extremely grateful to the following (and others), who contributed generously to the knowledge base or helped us in other ways: W Bond, D Clark, R A Cole, P R Ellis, J A Hardman, F M Humpherson-Jones, R B Maude, J D Taylor, F Tyler, C Senior, D G A Walkey and J G White.

Reference

Andro T *et al* 1987 Un ensemble de Systemes Experts pour l'aide au diagnostic des Maladies des Plantes cultivées en France. Proc. 7th International Workshop 'Expert Systems and their Applications', Avignon, France, pp. 173–186. INRA, Paris.

Author index

Subject index

This index lists the key words given in the headings of the papers

Proceedings of the previous Plant Nutrition Colloquia

1. PARIS, 1954
Prevot, P. (Ed.) 1955 Analyse des Plantes et Problèmes des Engrais Minéraux. Institut de Recherches pour les Huiles et Oléagineux (IRHO), Paris, 263 p.

2. PARIS, 1956
Prevot, P. (Ed.) 1957 Plant Analysis and Fertilizer Problems. Institut de Recherches pour les Huiles et Oléagineux (IRHO), Paris, 410 p.

3. MONTREAL, 1959
Reuther, W. (Ed.) 1961 Plant Analysis and Fertilizer Problems. American Institute of Biological Sciences, Washington, Publication No. 8, 454 p.

4. BRUSSELS, 1962
Bould, C., Prevot, P. and Magness, J.R. (Eds.) 1964 Plant Analysis and Fertilizer Problems IV. W.F. Humphrey Press Inc., New York, 430 p.

5. MARYLAND, 1966
Abstracts published as appendix to Proceedings of the VIth Colloquium, vol. 2, pp. 693–704.

6. TEL AVIV, 1970
Samish, RM. (Ed.) 1971 Recent Advances in Plant Nutrition. Gordon and Breach Science Publishers Inc., New York, 2 vols., 736 p.

7. HANNOVER, 1974
Wehrmann, J. (Ed.) 1974 Plant Analysis and Fertilizer Problems. German Society of Plant Nutrition, Hannover, 3 vols., 598 + 40 p.

8. AUCKLAND, 1978
Ferguson, A.R., Bieleski, R.L. and Ferguson, I.B. (Eds.) 1978 Plant Nutrition 1978. New Zealand Department of Scientific and Industrial Research, Information Series No. 134, 2 vols., 629 p.

9. COVENTRY, 1982
Scaife, A. (Ed.) 1982 Plant Nutrition 1982. Commonwealth Agricultural Bureaux, Slough, United Kingdom, 2 vols., 750 p.

10. ROCKVILLE, 1986
Korcak, R.F. (Ed.) 1987 Proceedings Tenth International Plant Nutrition Colloquium. Journal of Plant Nutrition 10 (9–16), 963–2168. Marcel Dekker Inc., New York.

Developments in Plant and Soil Sciences

1. J. Monteith and C. Webb (eds.): *Soil Water and Nitrogen in Mediterranean-type Environments.* 1981 ISBN 90-247-2406-6
2. J. C. Brogan (ed.): *Nitrogen losses and Surface Run-off from Landspreading of Manures.* 1981 ISBN 90-247-2471-6
3. J. D. Bewley (ed.): *Nitrogen and Carbon Metabolism.* 1981 ISBN 90-247-2472-4
4. R. Brouwer, I. Gašparíková, J. Kolek and B. C. Loughman (eds.): *Structure and Function of Plant Roots.* 1981 ISBN 90-247-2510-0
5. Y. R. Dommergues and H. G. Diem (eds.): *Microbiology of Tropical Soils and Plant Productivity.* 1982 ISBN 90-247-2624-7
6. G. P. Robertson, R. Herrara and T. Rosswall (eds.): *Nitrogen Cycling in Ecosystems of Latin America and the Caribbean.* 1982 ISBN 90-247-2719-7
7. D. Atkinson, K. K. S. Bhat, M. P. Coutts, P. A. Mason and D. J. Read (eds.): *Tree Root Systems and Their Mycorrhizas.* 1983 ISBN 90-247-2821-5
8. M. R. Sarić and B. C. Loughman (eds.): *Genetic Aspects of Plant Nutrition.* 1983 ISBN 90-247-2822-3
9. J. R. Freney and J. R. Simpson (eds.): *Gaseous Loss of Nitrogen From Plant-Soil Systems.* 1983 ISBN 90-247-2820-7
10. United Nations Economic Commission for Europe (ed.): *Efficient Use of Fertilizers in Agriculture.* 1983 ISBN 90-247-2866-5
11. J. Tinsley and J. F. Darbyshire (eds.): *Biological Processes and Soil Fertility.* 1984 ISBN 90-247-2902-5
12. A. D. L. Akkermans, D. Baker, K. Huss-Danell and J. D. Tjepkema (eds.): Frankia *Symbioses.* 1984 ISBN 90-247-2967-X
13. W. S. Silver and E. C. Schröder (eds.): *Practical Application of* Azolla *for Rice Production.* 1984 ISBN 90-247-3068-6
14. P. G. L. Vlek (ed.): *Micronutrients in Tropical Food Crop Production.* 1985 ISBN 90-247-3085-6
15. T. P. Hignett (ed.): *Fertilizer Manual.* 1985 ISBN 90-247-3122-4
16. D. Vaughan and R. E. Malcolm (eds.): *Soil Organic Matter and Biological Activity.* 1985 ISBN 90-247-3154-2
17. D. Pasternak and A. San Pietro (eds.): *Biosalinity in Action.* Bioproduction with Saline Water. 1985 ISBN 90-247-3159-3
18. M. Lalonde, C. Camiré and J. O. Dawson (eds.): Frankia *and Actinorhizal Plants.* 1985 ISBN 90-247-3214-X
19. H. Lambers, J. J. Neeteson and I. Stulen (eds.): *Fundamental, Ecological and Agricultural Aspects of Nitrogen Metabolism in Higher Plants.* 1986 ISBN 90-247-3258-1
20. M. B. Jackson (ed.): *New Root Formation in Plants and Cuttings.* 1986 ISBN 90-247-3260-3
21. F. A. Skinner and P. Uomala (eds.): *Nitrogen Fixation with Non-Legumes* (Proceedings of the 3rd Symposium, Helsinki, 1984). 1986 ISBN 90-247-3283-2
22. A. Alexander (ed.): *Foliar Fertilization.* 1986 ISBN 90-247-3288-3
23. H. G. v.d. Meer, J. C. Ryden and G. C. Ennik (eds.): *Nitrogen Fluxes in Intensive Grassland Systems.* 1986 ISBN 90-247-3309-X
24. A. U. Mokwunye and P. L. G. Vlek (eds.): *Management of Nitrogen and Phosphorus Fertilizers in Sub-Saharan Africa.* 1986 ISBN 90-247-3312-X
25. Y. Chen and Y. Avnimelech (eds.): *The Role of Organic Matter in Modern Agriculture.* 1986 ISBN 90-247-3360-X
26. S. K. De Datta and W. H. Patrick Jr. (eds.): *Nitrogen Economy of Flooded Rice Soils.* 1986 ISBN 90-247-3361-8
27. W. H. Gabelman and B. C. Loughman (eds.): *Genetic Aspects of Plant Mineral Nutrition.* 1987 ISBN 90-247-3494-0
28. A. van Diest (ed.): *Plant and Soil: Interfaces and Interactions.* 1987 ISBN 90-247-3535-1

Developments in Plant and Soil Sciences

29. United Nations Economic Commission for Europe and FAO (eds.): *The Utilization of Secondary and Trace Elements in Agriculture*. 1987 ISBN 90-247-3546-7
30. H. G. v.d. Meer, R. J. Unwin, T. A. van Dijk and G. C. Ennik (eds.): *Animal Manure on Grassland and Fodder Crops*. Fertilizer or Waste? 1987 ISBN 90-247-3568-8
31. N. J. Barrow: *Reactions with Variable-Charge Soils*. 1987 ISBN 90-247-3589-0
32. D. P. Beck and L. A. Materon (eds.): *Nitrogen Fixation by Legumes in Mediterranean Agriculture*. 1988 ISBN 90-247-3624-2
33. R. D. Graham, R. J. Hannam and N. C. Uren (eds.): *Manganese in Soils and Plants*. 1988 ISBN 90-247-3758-3
34. J. G. Torrey and J. L. Winship (eds.): *Applications of Continuous and Steady-State Methods to Root Biology*. 1989 ISBN 0-7923-0024-6
35. F. A. Skinner, R. M. Boddey and I. Fendrik (eds.): *Nitrogen Fixation with Non-Legumes* (Proceedings of the 4th Symposium, Rio de Janeiro, 1987). 1989 ISBN 0-7923-0059-9
36. B. C. Loughman, O. Gašparíková and J. Kolek (eds.): *Structural and Functional Aspects of Transport in Roots*. 1989 ISBN 0-7923-0060-2; Pb 0-7923-0061-0
37. P. Planquaert and R. Haggar (eds.): *Legumes in Farming Systems*. 1990 ISBN 0-7923-0134-X
38. A. E. Osman, M. M. Ibrahim and M. A. Jones (eds.): *The Role of Legumes in the Farming Systems of the Mediterranean Areas*. 1990 ISBN 0-7923-0419-5
39. M. Clarholm and L. Bergström (eds.): *Ecology of Arable Land – Perspectives and Challenges*. 1989 ISBN 0-7923-0424-1
40. J. Vos, C. D. van Loon and G. J. Bollen (eds.): *Effects of Crop Rotation on Potato Production in the Temperate Zones*. 1989 ISBN 0-7923-0495-0
41. M. L. van Beusichem (ed.): *Plant Nutrition – Physiology and Applications*. 1990 ISBN 0-7923-0740-2

Kluwer Academic Publishers – Dordrecht / Boston / London

MIX
Papier aus verantwortungsvollen Quellen
Paper from responsible sources
FSC® C105338

If you have any concerns about our products,
you can contact us on
ProductSafety@springernature.com

In case Publisher is established outside the EU,
the EU authorized representative is:
Springer Nature Customer Service Center GmbH
Europaplatz 3, 69115 Heidelberg, Germany

Printed by Libri Plureos GmbH
in Hamburg, Germany